PLASMIDS IN BACTERIA

BASIC LIFE SCIENCES

Alexander Hollaender, General Editor

Council for Research Planning in Biological Sciences, Inc., Washington, D.C.

Recent volumes in the series:

Volume 18 TRENDS IN THE BIOLOGY OF FERMENTATIONS FOR FUELS AND CHEMICALS
Edited by Alexander Hollaender, Robert Rabson, Palmer Rogers, Anthony San Pietro, Raymond Valentine, and Ralph Wolfe

Volume 19 GENETIC ENGINEERING OF MICROORGANISMS FOR CHEMICALS
Edited by Alexander Hollaender, Ralph D. DeMoss, Samuel Kaplan, Jordan Konisky, Dwayne Savage, and Ralph S. Wolfe

Volume 20 MOLECULAR AND CELLULAR MECHANISMS OF MUTAGENESIS
Edited by J. F. Lemontt and W. M. Generoso

Volume 21 GENETIC TOXICOLOGY: An Agricultural Perspective
Edited by Raymond A. Fleck and Alexander Hollaender

Volume 22 PLANT CELL CULTURE IN CROP IMPROVEMENT
Edited by S. K. Sen and Kenneth L. Giles

Volume 23 INDUCED MUTAGENESIS: Molecular Mechanisms and Their Implications for Environmental Protection
Edited by Christopher W. Lawrence

Volume 24 ORGAN AND SPECIES SPECIFICITY IN CHEMICAL CARCINOGENESIS
Edited by Robert Langenbach, Stephen Nesnow, and Jerry M. Rice

Volume 25 BASIC BIOLOGY OF NEW DEVELOPMENTS IN BIOTECHNOLOGY
Edited by Alexander Hollaender, Allen I. Laskin, and Palmer Rogers

Volume 26 GENETIC ENGINEERING OF PLANTS: An Agricultural Perspective
Edited by Tsune Kosuge, Carole P. Meredith, and Alexander Hollaender

Volume 27 NEUTRONS IN BIOLOGY
Edited by Benno P. Schoenborn

Volume 28 GENETIC CONTROL OF ENVIRONMENTAL POLLUTANTS
Edited by Gilbert S. Omenn and Alexander Hollaender

Volume 29 SISTER CHROMATID EXCHANGES: 25 Years of Experimental Research
Part A: The Nature of SCEs
Part B: Genetic Toxicology and Human Studies
Edited by Raymond R. Tice and Alexander Hollaender

Volume 30 PLASMIDS IN BACTERIA
Edited by Donald R. Helinski, Stanley N. Cohen, Don B. Clewell, David A. Jackson, and Alexander Hollaender

Volume 31 GENETIC CONSEQUENCES OF NUCLEOTIDE POOL IMBALANCE
Edited by Frederick J. de Serres

A Continuation Order Plan is available for this series. A continuation order will bring delivery of each new volume immediately upon publication. Volumes are billed only upon actual shipment. For further information please contact the publisher.

PLASMIDS IN BACTERIA

Edited by

Donald R. Helinski
University of California, San Diego
La Jolla, California

Stanley N. Cohen
Stanford University
Stanford, California

Don B. Clewell
University of Michigan
Ann Arbor, Michigan

David A. Jackson
Genex Corporation
Gaithersburg, Maryland

Alexander Hollaender
Council for Research Planning
in Biological Sciences, Inc.
Washington, D.C.

with
Lowell Hager, Samuel Kaplan,
and Jordan Konisky

Claire M. Wilson, *Technical Editor*
Council for Research Planning
in Biological Sciences, Inc.
Washington, D.C.

PLENUM PRESS • NEW YORK AND LONDON

Library of Congress Cataloging in Publication Data

Main entry under title:

Plasmids in bacteria.

(Basic life sciences; v. 30)

"Proceedings of a conference on plasmids in bacteria, held May 13–18, 1984, at the University of Illinois, Urbana, Illinois"—T.p. verso.

Includes bibliographical references and index.

1. Plasmids—Congresses. 2. Bacteria—Congresses. I. Helinski, Donald R. II. Series.

QR76.6.P57 1985 589.9′087322 84-26397

ISBN-13: 978-1-4612-9487-0 e-ISBN-13: 978-1-4613-2447-8

DOI: 10.1007/978-1-4613-2447-8

Proceedings of a conference on Plasmids in Bacteria, held May 13–18, 1984, at the University of Illinois, Urbana, Illinois

Softcover reprint of the hardcover 1st edition 1985
A Division of Plenum Publishing Corporation
233 Spring Street, New York, N.Y. 10013

DEDICATION

The Editorial Committee of the conference on "Plasmids in Bacteria" dedicates this volume to Dr. I. C. Gunsalus in recognition and appreciation of his distinguished contributions to the fields of biochemistry and bacteriology. Gunny's independent scientific career spans some four decades and includes a number of innovative and fundamental contributions to our understanding of basic biochemical processes and bacterial metabolism. His early biochemical studies led to the isolation of the coenzymes pyridoxal phosphate and the discovery of lipoic acid. His later classical studies on the *Pseudomonads* provided the foundation for both fundamental and applied studies on the genetics and biochemistry of alkane and terpene catabolic systems in bacteria. This pioneering research included the characterization of plasmids as carriers of the genes specifying the degradative enzymes responsible for these catabolic pathways. His illustrious career as a biochemist and bacteriologist has earned him a number of awards and honors. We add this dedication and extend our wishes for his continued success in the application of his extraordinary research ability to those research projects that currently command his interest.

ACKNOWLEDGMENTS

We gratefully acknowledge the generous support of the National Science Foundation (Grant NSF PCM 84-06530), the National Cancer Institute (Grant 1 R13 CA 38747), the Argonne Universities Association Trust Fund, and the University of Illinois Research Board.

This material was prepared with the support of the U.S. Department of Energy (Grant DOE DEF G 2-84ER13201). However, any opinions, findings, conclusions, or recommendations expressed herein are those of the author(s) and do not necessarily reflect the views of DOE.

This work was also supported by the Department of Navy Grant N00014-84-G-0031 issued by the Office of Naval Research. The United States Government has a royalty-free license throughout the world in all copyrightable material contained herein.

Financial support was also derived from a number of industrial contributors as cited in alphabetical sequence: Abbott Laboratories; AMGen; Beckman; Biogen Corporation; CIBA-Geigy; Genex Corporation; International Biotechnologies, Inc.; Koppers Company, Incorporated; Eli Lilly; Merck, Sharp and Dohme Research Laboratories; Monsanto; New England Nuclear; Pfizer; Pioneer Hi-Bred; and UpJohn.

CONTENTS

SPECIALIZED FUNCTIONS, STRUCTURE, AND EVOLUTION

POSTERS
(Poster abstracts have been reproduced as submitted)

ROSTERS

INTRODUCTION

The study of bacterial plasmids has not always been as popular as it is today. For many years, the molecular biology of procaryotes was focused heavily on bacteriophage and plasmid investigations which were carried out in only a few laboratories. Whatever interest existed in plasmids concerned the role of these extrachromosomal elements in bacterial conjugation, genetic exchanges, and antibiotic resistance, as well as in the structure of plasmids themselves. Gradually, however, it became increasingly evident that many of the special characteristics displayed by bacteria of medical, agricultural, industrial, and environmental importance are determined by genes carried by plasmids, and this interest in plasmid-encoded functions, such as bacterial virulence properties (exotoxin production, serum resistance, adhesiveness), metabolism of organic compounds, plant tumor formation, and biological nitrogen fixation, led to increasing study of the plasmids that carry these genes. Investigations of other plasmid-related properties such as replication and recombination have yielded much information about fundamental biological processes; information having implications that extend far beyond the particular plasmids under study. Concurrently, plasmids were playing a key role in the discovery of bacterial transposable elements and were proving to be increasingly useful in the elucidation of mechanisms responsible for a variety of chromosomal rearrangement events in bacteria and plants. Their status as "minichromosomes" that could be isolated easily from bacterial cells and then reintroduced into other cells by transformation is of fundamental importance in this regard.

During the same time period, the major role of plasmids in the discovery and development of DNA cloning methodology and the importance of plasmids as vectors for the propagation and expression of a wide variety of foreign genes in bacteria was established and exploited with great success by basic and industrial scientists working in a number of biological and medical disciplines. As a consequence of these developments, interest in the biology of bacterial plasmids has broadened to the point where the results of research on plasmids are followed also by individuals who work primarily on the biology of eucaryotes, and who view bacterial plasmids entirely as research tools. For plasmid biologists, DNA cloning techniques have provided a more detailed molecular understanding of the plasmids themselves, including their structure, replication, conjugal transfer, and inheritance. Thus, in a variety of ways, bacterial plasmids have played and continue to play a central role in the current revolution in biological research.

Significant recent progress in understanding the fundamental properties of plasmids, an explosive rise in the experimental use of plasmids as important tools in basic and applied studies, and the expanding list of specialized bacterial functions determined by plasmids provided the impetus for organizing this conference on "Plasmids in Bacteria." This is the first such international conference in the United States since the meeting held in Jackson, Wyoming in 1977. An international meeting of plasmids was held in Santa Domingo, Dominican Republic, in the early part of 1981.

It was the intention of the Organizing Committee to gather together outstanding workers in plasmids biology from throughout the world and to provide a forum to enable these scientists to present their work and interact with post-doctoral fellows and graduate students, as well as with other senior scientists. Whatever success has been achieved in facilitating the exchange of information and ideas has been due largely to the efforts of the Council for Research Planning in Biological Sciences, the faculty of the University of Illinois, and the staff of the Division of Conferences and Institutes of the University. Special gratitude is extended to Professors Sam Kaplan and Jordan Konisky of the University of Illinois, who have devoted an enormous amount of time and energy to the organization of this meeting. The enthusiasm and extensive scientific interactions that have occurred among the conference participants, as well as the high quality of the contributions to this volume, attest to the highly successful outcome of the efforts by the above individuals, and provide a sense of genuine satisfaction to all those who have played a role in the organization and conduct of this meeting.

Stanley N. Cohen
Donald R. Helinski

PLASMIDS AS ORGANISMS

Naomi Datta

Department of Bacteriology
Royal Postgraduate Medical School
Hammersmith Hospital
Du Cane Road
London W12 OHS, England

It is a quarter century since the publications of evidence for the transmissibility of drug resistance among Enterobacteria (1,48). Although it was not the beginning of plasmid research, it was that discovery that led to the great expansion of work on plasmids, their natural history, and the uses to which they can be put. By their practical importance, resistance or "R" plasmids forced themselves on the attention of microbiologists. The biochemical mechanisms of plasmid-determined resistance were investigated for their biological interest and because of the need to counteract their effects. Because drug-resistant clones can so easily be selected from large populations of bacteria, resistance genes are used to label DNA molecules, providing a practical basis for the genetic analysis of many prokaryotic and eukaryotic systems.

In 1959, no plasmid structure had been seen; the nature of plasmids was inferred from their genetic effects though, soon after that date, plasmid DNA was visualized in gradients (42) and conjugative pili were seen by electron microscopy (16). Close parallels between plasmids and viruses were recognized from the first, and "R factors" were classified, together with F and λ, as episomes, or genetic elements that could exist either autonomously or integrated into the chromosome of their host bacterium (34). Later the term episome, to denote one class of genetic element, was found to be more confusing than useful (30); plasmids do not necessarily integrate into

---

The title is adapted from F. M. Burnet, Virus as Organism (1945) Harvard University Press.

host chromosomes and their mode of integration differs from that of λ. Parallels between conjugative plasmids and bacteriophages are evident [Brinton (15) called the former, "epiviruses" or "sex viruses"]. However, plasmids without transfer genes cannot be excluded from a unifying concept. They, too, are extra genetic elements, transmissible by natural means and with characteristic host ranges among bacteria. The term plasmid as defined by Lederberg (38) includes nearly all of the organisms to be discussed in this volume, although possibly not conjugative transposons. I consider the latter, as well as orthodox plasmids, to be organisms according to the definition of Luria and co-authors (40), i.e., a unit of a continuous lineage with an individual evolutionary history.

Plasmids, such as R plasmids, that determine a specialized bacterial character, were at first considered to be analogous to F-primes, having acquired one or more bacterial genes by a process of insertion into and imprecise excision from the chromosome. We now know much more about the means by which plasmids acquire new genes; the discovery of the moveable genetic elements of prokaryotes, insertion sequences and transposons, that provide opportunities for recombination, both legitimate and illegitimate, and for other genetic rearrangements have, more than anything since 1959, changed our view of the natural history of plasmids.

PLASMID EPIDEMIOLOGY

Many studies of resistance plasmids have traced their spread through different bacteria either in hospitals or in the community (18,47,52,63). Where a plasmid remains unaltered during its spread, its identity can be established fairly easily by the pattern of resistance conferred and by its molecular weight. Quantitative or semiquantitative measurement of resistances can provide an accurate index of the particular resistance genes carried, especially if a number of related antibiotics are used in the tests. For example, the relative levels of resistance to various penicillins and cephalosporins determined by plasmids indicate which β-lactamase is present. Similarly, differential levels of resistance to a variety of aminoglycosides indicate which drug-modifying enzymes are present (56). The method used can be a a simple measurement of zones of inhibition around drug-impregnated discs, provided that test conditions are properly controlled (47). Interpretation is complicated if the plasmid determines gene products with overlapping effects, but the resistance patterns of 2 plasmids can still be compared and their identity established. Confirmation of identity is by showing that examples from different sources yield the same-sized fragments when digested with restriction enzymes. Often, however, variations from the original occur, usually DNA deletions or insertions. Some plasmids undergo such regular deletions, both in nature and in laboratory strains, that this behavior itself helps to identify them for epidemiological purposes. For example, a plasmid (pHH1457) of about 110 megadaltons (Md), encoding multiple drug resistances, was found in isolates of Klebsiella aerogenes serotype K21 in an outbreak of hospital infection. It regularly underwent a deletion, losing about 25 Md of DNA, conjugative ability, and most of its resistance genes, though retaining clinically important genetamicin resistance. This deletion occurred in wild strains and in Rec+ and recA strains of Escherichia coli K12, at frequencies of about 0.25% per bacterial generation. The resistance pattern and this consistent behavior afforded recognition of the plasmid, whose Inc group was unknown. Its spread was traced to other hospital bacteria and, in the K21 Klebsiella, to several hospitals (12). Other examples of such consistent variations are known (18). Usually, however, plasmids undergo variation unpredictably and recognition of relationships between nonidentical plasmids requires some kind of classification.

Plasmids that inhabit a particular bacterial host can be divided into groups (Inc groups) by their incompatibility relationships. If each plasmid had a single DNA region encoding its determinants for replication and incompatibility, that region could be used to identify other plasmids of the same Inc group, either as a probe or as a miniplasmid or basic replicon with an inserted distinguishing marker in compatibility tests. But where replication and incompatibility functions have been analyzed they have been found to be controlled by complex genetic systems. We should not, then, expect that every naturally occurring plasmid will be classifiable into a discrete Inc group. However, many plasmids are so classifiable and DNA reassociation studies indicate that the Inc groups are respectable biological taxa.

Moreover, it is generally true that conjugative plasmids within an Inc group have identical or closely related transfer genes and their conjugative pili are morphologically and serologically similar, whereas plasmids of different groups have different conjugative genes and unrelated pili (13). There are exceptions. Take plasmids of groups IncFI and IncFII; they have closely related tra genes and pili, yet are compatible. This relationship could have resulted from recombination between 2 ancestral plasmids, let us say F itself (IncFI) and another (having the inc genes of IncFII, but without the F-like tra operon). Analysis shows that their lineage was less simple. More than one basic replicon is included in plasmids of IncFI and IncFII. Those of F (IncFI) and R100 (IncFII) are determined by sequences located at different sites (relative to their tra operons) (37,50). Yet recently, sequences have been found in F and in another IncFI plasmid, P307, homologous with a basic replicon of R100 and at the expected plasmid locus. Although they are there, the IncFII-like rep sequence of F seems to be nonfunctional (perhaps as the result of a $\gamma\delta$ insertion) but it is functional in P307 (50,53). The way replication is controlled by F differs from that of the

IncFII R plasmids (43). These plasmids may well be recombinational assemblages of more than one replicon (24), but the evolutionary pathways cannot be traced on present data. Plasmids determining I pili, not only IncI but also IncB, IncK, and IncZ, belong to an analogous complex of groups. These plasmids all determine 2 kinds of pilus, thin and thick, serologically distinguishable and with different functions in conjugation (14). Relationships between the *rep* (vegetative replication) and *inc* genes of plasmids in these Inc groups have not been analyzed. Small serological differences among F-like pili or I-like pili are not related to the Inc groups within the F and I complexes.

The IncF and IncI complexes of conjugative plasmids are exceptional in that closely similar pili are determined by members of different Inc groups. Exceptions of the opposite kind--unrelated pili determined by plasmids of the same Inc group--have rarely been reported in nature. A plasmid of IncM, pDT201, specified F-like pili, quite different from typical M pili (60) and there are examples of naturally occurring cointegrations, giving plasmids with dual incompatibilities and expressing one or both sets of *tra* genes (46,60). Incompatibility grouping, although only an approximate criterion for relatedness of plasmids (53), has been successfully used in numerous studies of the epidemiology and natural history of plasmids (3,20,26).

"PREANTIBIOTIC" PLASMIDS OF ENTEROBACTERIACEAE

A collection of bacterial strains was made by E. D. G. Murray over the period 1917-1954, and when they were opened in 1980, over 400 separate isolates were viable. A high proportion (24%) of these possessed transfer genes, as demonstrated by their ability to mobilize small nonconjugative R plasmids in *E. coli* K12 (Tab. 1). Although the dates of collection were from 1917, some strains had been isolated earlier and were acquired by Murray from other collections; for example, one *E. coli* strain was isolated by Escherich about

Tab. 1. Murray Collection 1917-1954.

MURRAY COLLECTION 1917 - 1954

Genus	No. of strains	Colicin-ogenic	Tra^+ (mobilized tra^- R plasmid)
Salmonella	210	15	47
Shigella	140	31	35
Escherichia	32	10	8
Klebsiella	30	3	9
Proteus	16	0	3
Unidentified	5	0	2
Total	433	59	104 (24%)

Tab. 2. Plasmids from Murray Collection established in *E. coli* K12.

PLASMIDS FROM MURRAY COLLECTION ESTABLISHED IN *E. coli* K12	
ColI	35
ColB	1
Tellurite-resistant	1
Cryptic	47
Total	84

1885. We do not know the dates and places of isolation of all the strains but we know they originated from many parts of the world including Europe, the Middle East, Northern Russia, India, and North and South America. From these strains, 84 conjugative plasmids were established in *E. coli* K12. None carried antibiotic-resistance, mercury-resistance, or haemolysin, 1 determined tellurite-resistance, and 36, colicinogeny (Tab. 2) (31). We classified the plasmids by incompatibility-testing with 25 R plasmids (21 belonging to known Inc groups and 4 odd, unclassified ones). We used the natural markers, col^+ or tellurite-resistance where possible, and labeled the cryptic plasmids with resistance transposons, converting them into R plasmids. The 35 ColI plasmids all belonged to IncI1. This result revealed likely epidemiological relationships. Clusters of $ColIa^+$ *Shigella sonnei* from 1937 and $ColIb^+$ *Salmonella typhimurium* from 1935 suggest that Murray collected multiple isolates from outbreaks in these years (Tab. 3). (Colicin-typing had not been introduced then, so he may not have been aware of any relationship.) If we count only 1 isolate from each of these clusters, the total number of plasmids in the study is reduced to 59 of which we could recognize or label 53. Of these, 40 (> 75%) belonged to Inc groups defined for R plasmids since 1970 (Tab. 4). Thus, these plasmids were common in enterobacteria and their acquisition of resistance genes came with the intense selective pressure exerted by the increasing use of antibiotics. It may be inferred that the resistance genes were acquired in a manner analogous to that in which we labeled the plasmids in the laboratory, as transposons (19).

Tab. 3. ColI plasmids from the Murray strains, all IncI1.

COlI PLASMIDS FROM THE MURRAY STRAINS, ALL IncI1

From:	Year	ColIa	ColIb
Shigella sonnei	1937	17	0
	1939	1	0
	1954	1	0
	?	1	0
Shigella flexneri	?	1	0
Salmonella typhimurium	1935	1	10
	1941	0	1
Salmonella spp.	1938	1	1

Tab. 4. Murray plasmids incompatibility results.

MURRAY PLASMIDS INCOMPATIBILITY RESULTS	
Known groups	40
New groups	7
Unclassified	6
Total	53

The distribution of the plasmids classified was not exactly as expected for R plasmids today (Tab. 5). Plasmids of the F and I complexes and of IncN, which are common today, were represented, but plasmids of IncX were the most common. R plasmids of this group have been identified relatively infrequently but from diverse enterobacterial genera, from widely separate parts of the world and with different resistance genes (26,32,33). There was no epidemiological clustering of strains that yielded IncX plasmids. They were from 13 Salmonella strains of various serotypes, including S. typhi, collected at dates from 1917 to 1939, and 2 Klebsiella spp. dated 1932 and 1940.

Among the Murray plasmids there were none of Inc groups C, H, M, P, or W, all of which are common today as R plasmids. Their absence may be a simple sampling effect. We certainly do not assume that plasmids of these groups did not exist before the era of R plasmids. Some may have been common in bacterial genera other than those collected by Murray. Or they may have been relatively infrequent in enterobacteria and have spread widely only after acquiring resistance genes.

The 13 Murray plasmids that did not belong to any known group were tested for compatibility with one another, each against all.

Tab. 5. Classification of the Murray plasmids.

Classification of the Murray plasmids

Inc group	No. of plasmids	Mol. wt (Md)
F II	3	35-50
F IV	4	35-50
I1	13	45-60
I2	1	40
B	2	50
N	2	25
X	15	30
Murray I	5	25-35
Murray II	2	?
Unclassified	6	30-110
Total	53	25-110

Two new Inc groups were revealed, containing 5 and 2 plasmids, apparently unrelated epidemiologially. Six, including the tellurite-R plasmid, were left unclassified.

No DNA reassociation study has been made to compare the Murray plasmids with present day R plasmids of the same Inc groups, but their close relationship seems probable. They had molecular weights of the order expected for their group (Tab. 5) and their conjugative pili were also as expected, recognized by acting as receptors for known pilus-specific phages and by immune electron microscopy (D. E. Bradley, pers. comm.). It was not only the conjugative pili of the Murray IncI1 and IncB plasmids that was as expected. Like known plasmids of the IncI complex, they determined, as part of their conjugative mechanism, the DNA primase that can substitute for host _dnaG_ product (Ref. 14,64; D. E. Bradley, pers. comm.).

Our survey certainly underestimated the number of conjugative plasmids present in the Murray strains and made no study of nonconjugative ones. Plasmid profiles were not examined for all the original strains but DNA from 20 _E. coli_ strains, negative in the mobilization test, was extracted (11) and run in gel electrophoresis. Plasmid bands were seen in 16 of these extracts, and in 8, migration distances indicated plasmids of over 30 Md, large enough to be conjugative. Moreover, electron microscopy (D.E. Bradley, pers. comm.) revealed the unexpected presence of constitutively produced F-like pili by a strain that had yielded an IncB plasmid. The latter, in _E. coli_ K12, determined the pili, both thick and thin varieties, typical of IncB plasmids. The donor strain presumably harbored 2 unrelated conjugative plasmids, discovered accidentally since our protocol was not designed to isolate more than 1 transconjugant from each donor strain.

The conjugative plasmids in the Murray strains are not the only ones identified in "preantibiotic" cultures. Some were found in old cultures of _Neisseria gonorrhoeae_ (51) and the prototype, F, was discovered in _E. coli_ K12, isolated in 1922 (4).

STABILITY AND MUTABILITY

The plasmids in the Murray collection compared with present day R plasmids show that the functions designated _inc_ and _tra_ are conserved through many generations, vertical and horizontal. Specialized functions of plasmids are sometimes carried very stably through many generations, sometimes rapidly acquired or lost. Colicinogeny is an example of stable carriage. The ColI plasmids among the Murray strains closely resembled other early examples (59) as well as ColI plasmids (sometimes with additional drug resistance genes) found today. The association of one group of colicins (group II) with conjugative plasmids (colicins I with those of IncI, colicins V or B with the IncF complex) and another group (group I) with small nonconjugative ones is well established (29). But the association is not immutable. Colicin I is determined by the IncFI, ColV+ plasmid ColVI-K94 and genes for colicins of group II have also been found on small nonconjugative plasmids (39). Antibiotic-resistance genes are those most evidently acquired and lost by plasmids within our working experience. Selection for drug resistance, especially in hospitals and on farms, has been intense, but it still seems almost incredible that resistance genes to so many unrelated antibiotics should have been accumulated in so short a time. For example, in London hospitals, now and for several years past, one can find bacteria harboring a plasmid, pTH1, of about 65 Md, of group IncM, with at least 9 resistance genes. Labigne-Roussel et al. (36) have demonstrated 4 steps by which these plasmids probably evolved from an ancestor, pIP135, identified in Paris in 1972. The current form has 4 more resistances and is about 20 Md larger than pIP135.

PLASMID LINEAGES

Plasmids can be considered as individual organisms and also as members of a large biological family divisible into smaller taxa. Results of research on the structure and function of some individual plasmids, such as ColE1, NR1 (or R100), R1, R6, RP4 (or RK2) throws light on the nature of plasmids generally and also demonstrates important differences between them.

It seems that some individual plasmids reside permanently in their host organisms. The history of bacteriology is short but that of bacterial disease is much longer. Should we believe that bubonic plague as described in the Bible (41) was caused by Yersinia pestis and, if so, was its pathogenicity plasmid-determined (8)? Was the dysentery that decimated armies dependent upon virulence plasmids in Shigella (35,54)? What is the lineage of the prophage that determines toxin-production in Corynebacterium diphtheriae (7)? There are no answers to such questions.

Probably the first subdivision of plasmids into taxa should be on the basis of host range. The plasmids of gram-positive bacteria have not been found in natural isolates of gram-negative ones, nor vice versa. Plasmids of gram-positive bacteria will replicate and be maintained in a variety of gram-positive hosts (23) and plasmids of gram-negative bacteria have various host ranges among gram-negative genera. [However, the replicative functions of Staphylococcal plasmids may be capable of expression in E. coli (5,27), and laboratory-constructed cointegrates of E. coli plasmids and Streptomyces plasmids will replicate in both hosts (55,61).]

It is often true that the G + C content of plasmid DNA resembles that of the host chromosome, indicating long-term residence (24,33). However, promiscuous plasmids such as RP4 or R300B can be maintained in gram-negative genera with chromosomal DNA of very different G + C ratios. The base composition of plasmids could indicate their origins within bacterial genera or, equally, a tendency of what was originally foreign DNA to mutate to conform with that of the host cell.

The plasmid gene products that determine drug resistance also differ in gram-positive and gram-negative bacteria. This is somewhat surprising since gram-positive resistance genes, cloned in appropriate vectors, are expressed in *E. coli* (28,52). There must be opportunities in mixed populations of bacteria for miscegenation of genetic information, by means of transformation and rescue by transposition, yet the plasmid-determined genes such as encode penicillinases or chloramphenicol transacetylases are not found to cross the gram-positive/negative barrier (2,21,57). Miscegenation of this kind must be rare. Penicillinase was common in strains of *Staphylococcus aureus* by 1950 (6), although it was not until 1963 that its plasmid nature was recognized (44). In contrast, gonorrhea was treated with penicillin for 30 years before penicillinase plasmids appeared in *Neisseria gonorrhoeae* (49). This long delay illustrates lack of genetic exchange between gram-positive and gram-negative bacteria; the gonococcus had to wait for the world-wide dissemination of Tn*3* in gram-negative bacteria before it had the opportunity of acquiring the TEM penicillinase. I do not mean to imply that the resistance genes evolved separately in gram-positive and gram-negative bacteria, only that in recent history interchange has been lacking. Common ancestors for the resistance genes seem almost certain given the similarities in amino acid sequences of penicillinases (2), chloramphenicol transacetylases (57) and aminoglycoside phosphotransferases (21,61) from gram-positive and gram-negative sources. The differences currently seen can hardly have developed in the last 40 years. One is led to conclude that clinically important resistance genes were established in gram-positive and gram-negative bacteria much longer ago than the human use of antibiotics, perhaps on plasmids, though not necessarily in the bacterial species known to man as pathogens.

I am not inclined to speculate on the ultimate origins of plasmids or related organisms. Small plasmids such as p15 must be among the simplest organisms in the world but whether they have contracted, discarding unnecessary complexities or alternatively, are in the first stages of evolution, I do not know.

CONCLUSION

Plasmids, like viruses, are organisms and so are some creatures that do not fit Lederberg's 1952 definition of plasmid; the conjugative transposons of gram-positive bacteria (25) and, in *Haemophilus influenzae*, the amplifiable sequences that can be identified as circularized copies (58). These have so many of the characteristics of plasmids that they must be classed with them, yet do not replicate autonomously. In the enterobacteria, also, "plasmids" have been found without autonomous replication but they differ from conjugative transposons in behaving, at conjugation, like Hfrs (45), whereas the latter leave the host chromosome behind when they transfer themselves to a new host cell. Phages, plasmids and all transposing elements down to the smallest IS can logically be considered as one, unnamed, family of parasitic organisms with 2 modes of reproduction, one that results in vertical inheritance by the growing and dividing host and the other, an infective process that disseminates the organism horizontally, either intra- or intercellularly. I reject the hypothesis that transposition is nonreplicative (9), not on direct evidence but because I find it incredible. The unity of this unnamed family, then, is defined as an act of faith. Within it there is, obviously, great variation. Virulent phages, though not vertically inherited by their known hosts, cannot be classified separately. The degree of autonomy (independence, self-government) varies with the complexity of the organism but all have autonomy in their dissemination. I once considered R plasmids, and perhaps all plasmids, as bacterial symbionts, but although they may confer advantages on their host cells, I now see no need to assume that they must do so in order to survive. The advantages and disadvantages to a host cell of carrying extraneous DNA has been much debated (17,22)--an extra metabolic burden is disadvantageous, and special functions such as antibiotic-resistance may be advantageous. It has been shown that some transposons confer an unexplained survival advantage on their bacterial hosts in some circumstances (10). But if, as I believe, they all have the capacity for replicative dissemination, they are infective and will survive even if they make no direct return for the hospitality of their host cells.

Whether or not they confer advantages on individual host cells, they certainly favor bacterial populations in providing at each level: intergeneric, intercellular, and intracellular, the genetic flexibility necessary for adaptation to changing environments. Plasmids, as I have shown, can be stable organisms. Plasmids can also recombine, reassociate, rearrange, and be drastically altered. The same is probably true of their bacterial hosts of which we really know very little.

REFERENCES

1. Akiba, T. (1959) Mechanisms of resistance in Shigella. Proc. 15th General Meeting of the Japan Medical Association 5:229-305 [In Japanese, quoted by T. Watanabe (1963) Bact. Rev. 27:87-115].
2. Ambler, R.P. (1980) The structure of β-lactamases. Philos. Trans. R. Soc. Lond. (Biol.) 289-321-331.
3. Anderson, E.S., E.J. Threlfall, J.M. Carr, M.M. McConnell, and H.R. Smith (1977) Clonal distribution of resistance plasmid-

carrying Salmonella typhimurium, mainly in Middle East. J. Hyg., (Camb.) 79:425-448.
4. Bachmann, B.J. (1972) Pedigrees of some mutant strains of Escherichia coli K12. Bact. Rev. 36:525-557.
5. Barany, F., J.D. Boeke, and A. Tomasz (1982) Staphylococcal plasmids that replicate and express erythromycin resistance in both Streptococcus pneumoniae and Escherichia coli. Proc. Natl. Acad. Sci., USA 79:2991-2995.
6. Barber, M., and J.E.M. Whitehead (1949) Bacteriophage types in penicillin-resistant staphylococcal infection. Brit. Med. J. 2:565.
7. Barksdale, L. (1970) Corynebacterium diphtheriae and its relatives. Bact. Rev. 34:378-422.
8. Ben-Gurion, R., and A. Shafferman (1981) Essential virulence determinants of different Yersinia species are carried on a common plasmid. Plasmid 5:183-187.
9. Berg, D.E., and C.M. Berg (1983) The prokaryotic transposable element Tn5. Biotechnology 1:417-435.
10. Biel, S.W., and D.L. Hartl (1983) Evolution of transposons: Natural selection for Tn5 in Escherichia coli K12. Genetics 103:581-592.
11. Birnboim, H.C., and J. Doly (1979) A rapid alkaline extraction procedure for screening recombinant plasmid DNA. Nucl. Acids Res. 7:1513-1523.
12. Blenkharn, J.I., and V.M. Hughes (1982) Suction apparatus and hospital infection due to multiply-resistant Klebsiella aerogenes. J. Hosp. Infect. 3:173-178.
13. Bradley, D.E. (1980) Morphological and serological relationships of conjugative pili. Plasmid 4:155-169.
14. Bradley, D.E. (1984) Characteristics and function of thick and thin conjugative pili determined by transfer-repressed plasmids of incompatibility groups I1, I2, I5, B, K and Z. J. Gen. - Microb. 130:1489-1502
15. Brinton, C.C. (1971) The properties of sex pili the viral nature of "conjugal" genetic transfer systems, and some possible approaches to the control of bacterial drug resistance. Crit. Rev. Microbiol. 1:105-160.
16. Brinton, C.C., P. Gemski, and J. Carnanan (1964) A new type of bacterial pilus gentically controlled by the heredity factor of E. coli K12 and its role in chromosomal transfer. Proc. Natl. Acad. Sci., USA 52:776-783.
17. Campbell, A. (1981) Evolutionary significance of accessory DNA elements in bacteria. Ann. Rev. Microbiol. 35:55-83.
18. Chau, P.Y., J. Ling, E.J. Threlfall, and S.W.K. Im (1982) Genetic unstability of R plasmids in relation to the shift of drug resistance patterns in Salmonella johannesburg. J. Gen. Microbiol. 128:239-245.
19. Datta, N., and V.M. Hughes (1983) Plasmids of the same Inc groups in enterobacteria before and after the medical use of antibiotics. Nature 306:616-617.
20. Datta, N., V.M. Hughes, M.E. Nugent, and H. Richards (1979b)

Plasmids and transposons and their stability and mutability in bacteria isolated during an outbreak of hospital infection. Plasmid 2:182-196.

21. Davies, J., and G. Gray (1984) Evolutionary relationships among genes for antibiotic resistance. In Origins and Development of Adaptation, D. Evered and G.M. Collins, eds., pp. 219-228, CIBA Foundation Symposium 102, Pitman, London.
22. Doolittle, W.F., and C. Sapienza (1980) Selfish genes, the phenotype paradigm and genome evolution. Nature 284:601-603.
23. Engel, H.W.B., N. Soedirman, J.A. Rost, W.J. Leeuven, and J.D.A. van Embden (1980) Transferability of macrolide, lincomycin and streptogrammin resistance between group A, B and D streptococci, Streptococcus pneumoniae and Staphylococcus aureus. J. Bact. 142:407-413.
24. Falkow, S. (1975) Infectious Multiple Drug Resistance, Pion Ltd., London.
25. Franke, A.E., and D.B. Clewell (1981) Evidence for a chromosome-borne resistance transposon (Tn916) in Streptococcus faecalis that is capable of "conjugal" transfer in the absence of a conjugative plasmid. J. Bact. 145:494-502.
26. Frost, J.A., B. Rowe, J. Vandepitte, and E.J. Threlfall (1981) Plasmid characterization in the investigation of an epidemic caused by multiply resistant Shigella dysenteriae type 1 in Central Africa. Lancet 2:1074-1076.
27. Goze, A., and S.D. Ehrlich (1980) Replication of plasmids from Staphylococcus aureus in Escherichia coli. Proc. Natl. Acad. Sci., USA 77:7333-7337.
28. Hardy, K., and C. Haefeli (1982) Expression in Escherichia coli of a staphylococcal gene for resistance to MLS antibiotics. J. Bact. 152:524-526.
29. Hardy, K.G., G.G. Meynell, and J.E. Dowman (1973) Two major groups of colicine factors: Their evolutionary significance. Mol. Gen. Genet. 125:217-250.
30. Hayes, W. (1969) The Genetic of Bacteria and their Viruses 2nd edition, Blackwell, Oxford, pp.747-748.
31. Hughes, V.M., and N. Datta (1983) Conjugative plasmids in bacteria of the "pre-antibiotic" era. Nature 302:725-726.
32. Hughes, V.M., N. Datta, and M.C. Faiers (1981) Interhospital spread of a multiply resistant klebsiella. Brit. Med. J. 282:696.
33. Jacob, A.E., J.A. Shapiro, L. Yamamoto, D.I. Smith, S.N. Cohen, and D. Berg (1977) Plasmids studied in Escherichia coli and other enteric bacteria. In DNA Insertion Elements, Plasmids and Episomes, A.I. Bukhari, J.A. Shapiro, and S.L. Adhya, eds. Cold Spring Harbor Laboratory, USA, pp. 607-638.
34. Jacob, F., and E.L. Wollman (1958) Les épisomes, éléments génétiques ajoutés. C.R. Acad. Sci., Paris 247:154-156.
35. Kopecko, D.J., J. Holcombe, and S. Formal (1979) Molecular characterization of plasmics from virulent and spontaneously occurring avirulent colonial variants of Shigella flexneri. Infect. Immun. 24:580-582.

36. Labigne-Roussel, A., J. Witchitz, and P. Courvalin (1982) Evolution of disseminated Inc7-M plasmids encoding gentamicin resistance. Plasmid 8:215-231.
37. Lane, D., and R.C. Gardner (1979) Second EcoRI fragment of F capable of self-replication. J. Bact. 139:141-151.
38. Lederberg, J. (1952) Cell genetics and hereditory symbiosis. Physiol. Rev. 32:403-430.
39. Lehrbach, P.R., and P. Broda (1984) Molecular comparisons of plasmids isolated from colicinogenic strains of Escherichia coli. J. Gen. Microbiol. 130:401-410.
40. Luria, S.E., J.E. Darnell, D. Baltimore, and A. Campbell (1978) Gen. Virol., John Wiley and Sons, New York, p.4.
41. MacArthur, W.P. (1952) The plague of the Philistines (1 Sam. 5,6). Transactions of the Royal Society for Tropical Medicine and Hygiene, England, 46:209-212.
42. Marmur, J., R. Rownd, S. Falkow, L.S. Baron, C. Schildkraut, and P. Doty (1961) The nature of intergenetic episomal infection. Proc. Natl. Acad. Sci., USA 47:972-979.
43. Nordström, K. (1983) Review: Control of plasmid replication. Plasmid 9:1-7.
44. Novick, R.P. (1963) Analysis by transduction of mutations affecting penicillinase formation in Staphylococcus aureus. J. Gen. Microbiol. 33:121-136
45. Nugent, M.E. 1981) A conjugative 'plasmid' lacking autonomous replication. J. Gen. Microbiol. 126:305-310.
46. Nugent, M.E., K. Ellis, N. Datta, and D.E. Bradley (1982) pHH502, a plasmid with IncP and IncIα characters, loses the latter by a specific recA-independent deletion event. J. Gen. Microbiol. 128:2781-2790.
47. O'Brien, T.F., D.G. Ross, M.A. Guzman, A.A. Medeiros, R.W. Hedges, and D. Botstein (1980) Dissemination of an antibiotic resistance plasmid in hospital patient flora. Antimicrob. Ag. Chemother. 17:537-543.
48. Ochiai, K. (1959) Distribution and clinic of desentery caused by antibiotic-resistant strains. Proceedings 15th General Meeting of the Japan Medical Association 306-316 [In Japanese, quoted by T. Watanabe (1963) Bact. Rev. 27:87-115].
49. Phillips, I. (1976) β-lactamase-producing penicillin-resistant gonococcus. Lancet 2:656-657.
50. Picken, R.N., A.J. Naziatis, S. Saadi, and W.K. Maas (1984) Characterization of the basic replicons of the chimeric R/Ent plasmid pCG86 and the related Ent plasmid, P307. Plasmid.
51. Roberts, M., L. Elwell, and S. Falkow (1978) Introduction to the mechanisms of genetic exchange in the gonococcus: Plasmids and conjugation in Neisseria gonorrhoea. In Immunobiology of Neisseria gonorrhoea, G.F. Brooks, E.C. Gotschlich, K.K. Holmes, W.D. Sawyer, and F.E. Young, eds. A.S.M., Washington, D.C.
52. Roberts, M., P. Piot, and S. Falkow (1979) The ecology of gonococcal plasmids. J. Gen. Microbiol. 114:491-494.
53. Saadi, S., W.K. Maas, and P.L. Bergquist (1984) RepFIC, a basic

replicon of IncFI plasmids that has homology with a basic replicon of IncFII plasmids. Plasmid.
54. Sansonetti, P.J., D.J. Kopecko, and S.B. Formal (1982) Involvement of a plasmid in the invasive ability of Schigella flexneri. Infect. Immun. 35:852-860.
55. Schottel, J.L., M.J. Bibb, and S.N. Cohen (1981) Cloning and expression in Streptomyces lividans of antibiotic resistance genes derived from Escherichia coli. J. Bact. 146:360-368.
56. Shannon, K., and I. Phillips (1982) Mechanisms of resistance to aminoglycosides in clinical isolates. J. Antimicrob. Chemother. 9:91-102.
57. Shaw, W.V. (1984) Bacterial resistance to chloramphenicol. Brit. Med. Bull. 40:36-41.
58. Spies, T. and R. Laufs (1983) Circularized copies of amplifiable resistance genes from Haemophilus influenzae plasmids. J. Bact. 156:1263-1267.
59. Stocker, B.A.D. (1966) Heterogeneity of I colicines and I colicine factors. Heredity 21:166-168.
60. Taylor, D.E., J.G. Levine, and D.E. Bradley (1981) In vivo formation of a plasmid cointegrate expressing two incompatibility phenotypes. Plasmid 5:233-244.
61. Thompson, C.J. and G.S. Gray (1983) Nucleotide sequence of a streptomycete aminoglycoside phosphotransferase gene and its relationship to phosphotransferases encoded by resistance plasmids. Proc. Natl. Acad. Sci, USA 80:5190-5194.
62. Timmis, K., F. Cabello, and S.N. Cohen (1975) Cloning, isolation and characterisation of replication regions of complex plasmid genomes. Proc. Natl. Acad. Sci., USA 72:2242-2246.
63. Thompkins, L.S., J.J. Plorde, and S. Falkow (1980) Molecular analysis of R-factors from multiresistant nosocomial isolates. J. Infect. Dis. 141:625-636.
64. Willetts, N., and B. Wilkins (1984) Processing of plasmid DNA during bacterial conjugation. Microbiol. Rev. 48:24-41.

REPORT ON A WORKSHOP: STRUCTURE AND FUNCTION

Rüdiger Schmitt

Lehrstuhl für Genetik
Universität Regensburg
D-8400 Regensburg, Germany

Mobile genetic elements responsible for various genome rearrangements have been identified in prokaryotes and eukaryotes. They have been implicated in genome evolution, speciation, virus-induced cancers, and in the control of gene expression and thus were the focus of intense studies in recent years. These studies have reshaped our concepts of DNA. The storage of invariant genetic information suddenly turned into a dynamic structure, flexibly responding to adoptive needs and selective pressures. Due to their relative simplicity, bacterial transposons and IS elements have been used as paradigms for molecular analyses that revealed distinct features common to these elements: (i) their termini carry inverted repeats of 10-40 bp thought to serve as recognition sites for the transposition enzymes (transposases); (ii) the inverted repeats flank a central region containing genes or open reading frames that encode transposition functions; (iii) insertion into target replicons is accompanied by a short duplication of target DNA flanking the transposed element as direct repeats.

The central feature of mobile elements is their ability to integrate into foreign DNA sequences. This "illegitimate" recombination involving no apparent DNA homology is clearly distinguishable from the "legitimate" homologous recombination. The underlying mechanisms are still a matter of speculation, and it appears likely that different mechanisms apply to different elements and may even be operative in the same elements under changing conditions. At present, 2 alternatives are discussed, namely, <u>conservative</u> and <u>replicative</u> transposition. These are diagrammed in Fig. 1. Conservative transposition has been claimed for Tn5 (Berg) and Tn10 (Kleckner) to occur, whereas replication is involved in the transposition of Tn3 and its relatives. In the latter, cointegrates are

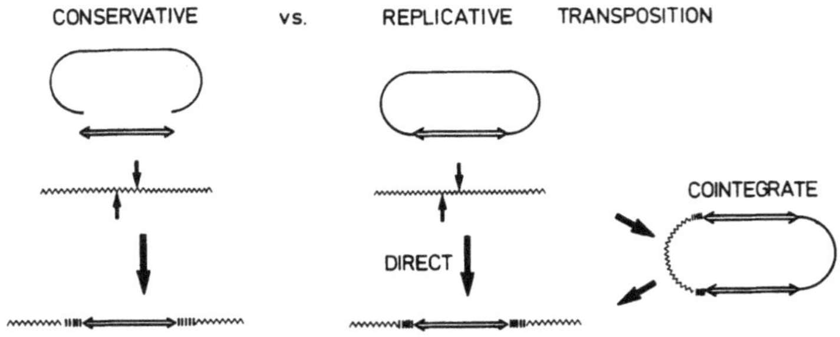

Fig. 1. Diagram of alternative mechanisms of transposition. Symbols: transposable element (⟺), donor (——), target (∿∿∿), direct repeats (▪▪), single-strand cuts (↑↓).

the preferred intermediates; they are resolved by a specific recombination event. Sometimes a bypass, termed "direct transposition," may be used instead. The duplication of target DNA flanking the transposed element is thought to be generated by the introduction of staggered nicks and subsequent DNA synthesis (Grindley & Sherratt). Replicative mechanisms are further subdivided into symmetric and asymmetric models depending on whether both ends of the element are nicked and ligated to the target DNA simultaneously (Shapiro; Arthur & Sherratt) or whether initiation occurs by nick-ligation at one end (Harshey & Bukhari; Galas & Chandler). There is supporting evidence for the latter to occur in phage Mu and in Tn*1721*, a relative of Tn*3* (Schmitt).

Although transposable elements insert into various loci of a target replicon, most IS elements exhibit a preference for certain sites (Arber). Individual insertions into such "hot spots" do not necessarily involve the same target site, but rather center about distinct regions thought to contain a "recognition sequence" for transposase. This selection of target sites is reminiscent of mechanisms operative among class I restriction enzymes. In addition, AT-rich sequences (bearing resemblance to the termini of the element) are frequently found to flank the insertion.

IS elements and transposons are components of many plasmids. IS-mediated fusions, duplications, deletions, inversions, and the acquisition of new genes via transposons are still contributing to the rapid evolution of plasmids, that confer selective advantage to their hosts. On the other hand, little is known about the evolution of transposable elements themselves. An approach to this question relies on a comparison of members of a transposon family that have related sequences and interchangeable functions. Thus, Tn*3*- like elements may have evolved in discrete steps involving the inverted ends, the transposition functions, and finally, different resistance determinants (Schmitt et al.). These elements have succeeded in crossing the barrier between gram negatives and gram positives as document by Tn917 that bears certain similarities tc Tn3 (Shaw & Clewell).

Questions relating to host-element interactions and to the control of transposition frequencies are still unanswered. To the geneticist these elements thus open an exciting filed, with many challenges to the investigator, promising unprecedented insights into the structure and evolution of the genetic material.

EVOLUTIONARY RELEVANCE OF GENETIC REARRANGEMENTS INVOLVING PLASMIDS

Werner Arber, Christian Sengstag, Patrick Caspers,
and Brian Dalrymple

Department of Microbiology
Biozentrum of the University of Basel
Klingelbergstrasse 70, CH-4056
Basel, Switzerland

INTRODUCTION

Early molecular genetic studies with drug-resistance plasmids had indicated that transposable genetic elements, together with natural gene vectors, might play important roles in the biological evolution of microorganisms. More recently, these views have been widely confirmed. IS (insertion sequence) elements and transposons (1) have the characteristic ability to transpose to new locations on DNA, and often they also bring about additional genetic rearrangements, such as deletion, inversion, replicon fusion, and gene amplification (reviewed in Ref. 2). Alterations of gene functions, ranging from complete loss of an activity to improved gene expression, as well as gene fusion, can be the direct or indirect consequences of such genomic rearrangements. These, therefore, contribute to the vertical evolution of bacterial strains in which transposable elements reside. Via intermediary natural gene vectors, such as conjugative plasmids and viral genomes, and probably by still other means of horizontal gene flux (such as transformation and cell fusion), transposable elements can also make essential contributions to horizontal biological evolution. In this process, entire genes or functional domains of genes are acquired from other organisms and can provide a bacterium with new functional capacities (3).

Target selection criteria used by transposable elements are of great relevance for these evolutionary processes. Relatively little precise knowledge is available on these criteria, but it seems that they vary from one transposable element to another. Other relevant questions concern the frequencies with which transpositional events and, subsequently, other genomic rearrangements occur, as well as their dependence on physiological and other environmental conditions. Here again, suggestive evidence is available that not all transposable elements follow the same rules.

The knowledge available on existence, structure, and functions of transposable elements comes largely from investigations with well-defined targets, mostly particular genes, operons, and small plasmids) on the one hand, and with particular, often selectable, transposable elements on the other. But, for obvious reasons, this approach could not answer to full satisfaction the questions raised above. Ideally, one would like to know of the existence of all transposable elements residing in a particular bacterial strain, to determine the frequency with which each of them mediates genetic rearrangements under various physiological conditions, and to understand the criteria applied in target selection. The importance of genomic rearrangements due to transposable elements could then be compared with that of other mechanisms of spontaneous mutagenesis.

For a number of technical reasons, it is practically impossible to carry out such investigations with the entire bacterial genome as target. Instead, large plasmids can offer a welcome opportunity to get at least partial answers to the questions raised. We would like to report here on the progress of relevant work carried out in recent years in our laboratory with the bacteriophage P1 genome, which is carried as a plasmid in the lysogenic condition in its host, *Escherichia coli* (*E. coli*) K12.

EXPERIMENTAL SYSTEM

Strategy for the Study of Spontaneous Prophage Mutants

The genome of the bacteriophage P1 and its relatives has a size of about 90 kb. A considerable part of the genetic information contained on the genome appears essential for vegetative phage reproduction, but not for maintenance of the lysogenic condition, for which relatively few genes are needed. In addition, the P1 DNA carries information for some functions unimportant for either vegetative reproduction or lysogeny, at least in *E. coli* K12 (4).

In the lysogenic condition, spontaneous mutations in most P1 genes can accumulate without affecting the prophage replication and maintenance. Many of these mutations may affect vegetative phage growth after induction of the lytic cycle. Others may alter the phenotype due to the expression of inessential functions, e.g., P1-specific restriction and modification. Still other alterations in the genetic information may remain silent under any experimental conditions applied.

Since many P1 genes are essential for the vegetative phage reproduction, an important fraction of the first class of mutations listed can, in principle, be considered as lethal. These mutations are easy to screen for by induction of lysogenic bacteria replicated or spotted onto a plate seeded with P1-indicator bacteria. This approach reveals the overall frequency of lethal spontaneous mutations of the P1 genome rather than of a particular gene. We consider this to be an interesting parameter, providing information on mutational activities affecting the phage as an individual organism.

Some nonlethal mutations can also be screened for, if they alter particular phenotypic traits. Most of such mutations, however, will fall into one or a few particular genes determining the phenotype considered. Such mutations will not be discussed further in this paper.

Frequency of Spontaneous Loss of the Ability of Phage P1 to Reproduce Vegetatively

In most of our experiments carried out so far, P1 lysogenic bacteria were kept for prolonged periods in rich growth medium either at room temperature or at 30°C. Every few days the cultures were diluted into fresh medium, each time allowing them to grow for 7 to 10 generations before again reaching the stationary phase. Screening for spontaneous P1 mutants after about 10 weeks, or 100 generations, revealed in the order of 10^{-3} to 10^{-2} mutant prophages per subclone tested.

Early experiments showed that roughly two-thirds of all P1 mutants isolated in this way were due to deletions. All studied deletions started at the single IS1 element resident in the wild-type P1 genome, and they extended to various sites on one or the other side of IS1 (5). We concluded that this IS1 element must be very deletogenic to the P1 genome, as had also been shown for IS1 elements at other locations (6). In later experiments we, therefore, used a close relative of P1, which is called P1-15 and does not carry any IS element in its genome (7). As expected, the overall frequency of spontaneous mutations in P1-15 is about 3 times lower than in P1, and the class of IS1-mediated deletion mutants is indeed missing.

One experiment was done with a P1-15 lysogenic strain, which was grown without interruption at 30°C in a turbidostat. Care was thereby taken to maintain the cell density constant, at 5 x 10^7 per ml. After a total of 200 generations spontaneous mutations affected about 10^{-4} prophages per subclone tested. This frequency is significantly lower than that found with the same strain grown with interruptions as described above in the same medium for 100 generations. This difference suggests a dependence of spontaneous mutagenesis on the physiology of the cell.

Most Spontaneous Prophage Mutants Are Due to Transposition of IS Elements

Several different mechanisms are expected to contribute to the spontaneous mutagenesis of the prophage (3). Since our main interest was directed towards transpositional events, the P1 plasmids of mutant prophages were extracted and analyzed for alterations in their size by restriction cleavage analysis. Care was taken that each individual mutant studied came from a different culture insuring its independent origin. Not taking into account the already described IS1-mediated deletion mutants of P1, this analysis showed that at least 90% of spontaneous prophage mutations carry an insertion or sequence duplication. Only a few mutants had no alteration in their genome size. Control isolates of nonmutated prophages usually had the wild-type genome size.

Further structural analysis and Southern hybridization experiments indicated that the increased genome size of the mutants is usually due to the transposition of a chromosomal IS element into the prophage. The most frequently found element is IS2 (28 of 44 studied independent mutants, Ref. 8). IS30 was found 3 times (9). Among the other insertions, IS1, IS5, and $\gamma\delta$ were found, but not all insertions have yet been identified.

Study of the Target Sites of IS2 and IS30 Insertions

The locations of independently isolated insertions on the phage genome are strikingly nonrandom (Fig. 1). This applies particularly to IS2 and to IS30, which insert with high preference into the same distinct segment of the P1 genome. To investigate the target sites used by IS2 and by IS30, a 1756-bp segment covering the hot region for insertion of these elements was sequenced, as well as the junctions between some of the IS elements and their target (10). This study revealed that the target selection criteria used by IS2 and by IS30 must differ fundamentally.

Insertion sites of IS2. The insertion sites of 9 independent IS2-mediated mutations in the sequenced DNA segment appear to be more or less randomly distributed, in contrast to their distribution with regard to the entire phage genome (Fig. 1). The sequences at these 9 insertion sites show no common homology (10). This situation points to rather complex rules in target selection by IS2. Sengstag et al. (10) discussed 2 possible elements which they thought could contribute to attract IS2 preferentially into the hot region for insertion. These are, on the one hand, a segment of considerable, although imperfect, homology with an 85-bp long inside segment of IS2, and, on the other hand, the abundant occurrence of a pentanucleotide, GGTAT, in relative, although variable, proximity (less than 100 bp distant) to all studied IS2 insertions.

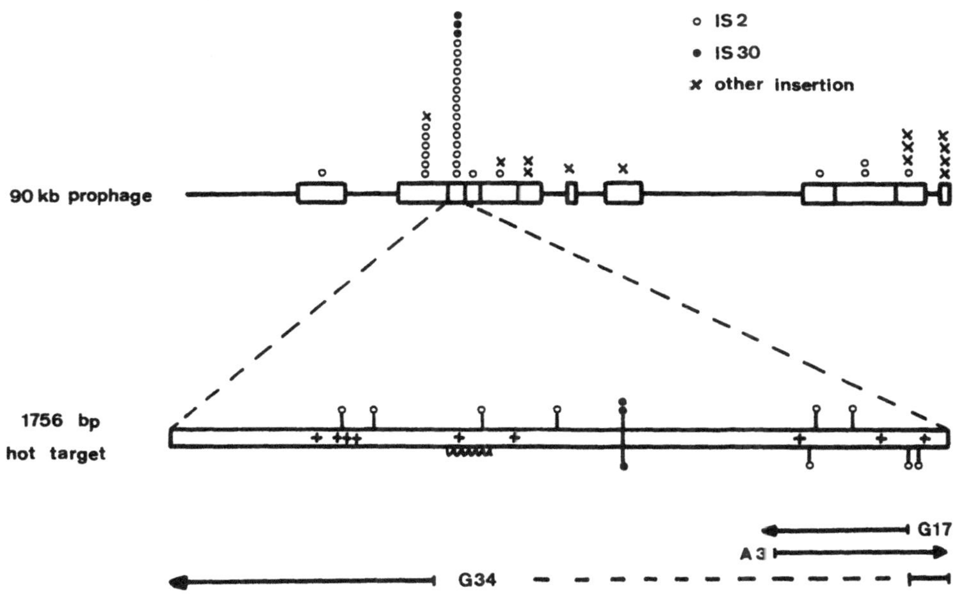

Fig. 1. Location of 44 independently isolated spontaneous insertion mutations on the phage P1 (or its close relative, P1-15) genome. In the upper map boxes represent the precision as obtained by restriction cleavage analysis. The locations of the 9 IS2 and the 3 IS30 insertions shown in the lower map were determined by nucleotide sequencing of the hot region for insertion. All 3 independent IS30-mediated mutations occurred at exactly the same target site. Symbols drawn above the map represent one orientation of the IS insertion, those below the map the other orientation. Data are taken from Ref. 8, which can be consulted for the reading frames in the 1756-bp segment. The wavy line indicates the location of a partial homology with an internal IS2 sequence. Crosses indicate the location of GGTAT sequences. The lines with arrow drawn below the map identify the cloned subfragments used in the experiments described in Fig. 2. The arrows point in the direction in which RNA synthesis proceeds in the constructions shown in Fig. 2.

We have started to investigate the possible importance of the homology to IS2, the pentanucleotide, and still other potential factors which could help to render the considered genome region a hot target for insertion. For this purpose, subfragments were cloned into a small plasmid which permits the selection of spontaneous mutations in the cloned segment or adjacent to it. The strategy used is outlined in Fig. 2 and it is based on the selection for galactose-resistant mutants among cells that are killed by the ex-

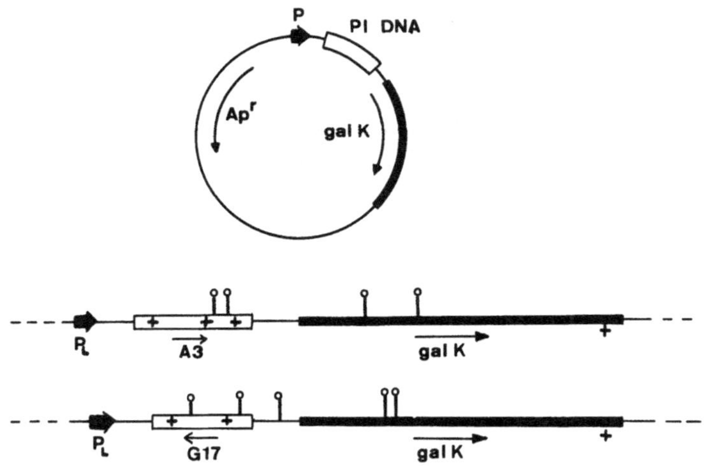

Fig. 2. Analysis of the influence of a DNA segment cloned into the leader sequence of galK on spontaneous, IS2-mediated mutagenesis. A selected DNA segment is cloned as schematically represented above into a small plasmid. The resulting plasmid is tested to see if expression of galK is still normal. A bacterial host deleted for the gal operon and grown in the presence of galactose is killed under conditions which induce the expression of galK in this plasmid. Rare galactose-resistant mutants are then selected and the proportion of mutations mediated by transpositional events is determined by restriction cleavage analysis of the plasmid. In the analysis shown in the lower part, the galK gene expression was under the control of the P_L promoter of phage λ. Among 5 galactose-resistant mutants of the plasmid carrying insert A3 (see Fig. 1), 4 had an increased genome size, in all cases due to the insertion of IS2. Similarly, 5 IS2 insertions and 4 mutations not affecting the plasmid size were analyzed in the plasmid carrying segment G17, which is a part of segment A3, but carried in the opposite orientation. The map positions of independent IS2 insertions are indicated by vertical lines with circles. Crosses indicate the location of GGTAT sequences. Another experiment with segment G34 (a fusion of the left part of the hot region for insertion with its extreme right part), cloned between the lacUV5 promoter and galK, gave similar results, while many fewer IS2 insertions were obtained with the plasmid not carrying any inserted Pl DNA. (Data are taken from C. Sengstag, Ph.D. thesis, University of Basel, 1984.)

pression of the *galK*$^+$ gene carried on the plasmid (11). Galactose-resistant mutants are expected to have lost the *galK* activity.

The available data indicate that both of 2 fragments of the hot spot region so far studied stimulate the appearance of galactose-resistant mutants, and that an important part of these mutations is caused by IS2 insertion. In contrast, galactose-resistant subclones studied in a control experiment were only rarely mediated by IS2 transposition. Interestingly, only about half of the IS2 insertions responsible for galactose-resistance map within the cloned subfragments. The other IS2 elements have either inserted into the *galK* structural gene or into its leader sequence. Since the nucleotide sequence of *galK* and its leader is known, we know that the GGTAT pentanucleotide occurs only once, at the 3' end of the gene. The IS2 insertions occurred quite far from this site (Fig. 2). This suggests that the pentanucleotide is not involved in target selection by IS2, or is at least less important than was previously thought.

The same statement may apply to the partial homology to IS2, since the subfragments cloned into the *galK* operon do not contain this part of the hot region for insertion. However, we have realized that the vector used to clone the subfragments also carried a 31-bp homology to IS2. As a matter of fact, this is a short internal sequence of IS2 (M. Rosenberg, pers. comm.), and it is split into a 19-bp and a 12-bp stretch upon in vitro insertion of the subfragments into the plasmid. Therefore, the potential role of homologies to IS2 remains uncertain. However, they cannot be the sole reason for the hot target property.

We expect to discover in further investigations, whether other particular structural elements in the cloned DNA segments may render the *galK* region a good target for IS2 insertion. If so, an influence at a distance might have to be assumed to explain the distribution of IS2 insertion sites as shown in Fig. 2. As a working hypothesis we assume that IS2 insertion may depend on the activation of the transposition process by a specific macromolecular interaction, e.g., of the transposase being bound to the ends of an IS2 element with a particular "activation" site on a potential target DNA, or vice versa. After activation, however, the actual site of insertion might then be selected in a more random process, at variable distances from the activation site. This hypothesis is based on the actual mechanism with which restriction endonuclease *EcoK* and its relatives function. They require an activation at a specific recognition site before they will cleave the same DNA molecule at a site chosen in each reaction at a variable distance from the activation site, often several thousand base pairs away (12).

Insertion site of IS30. In contrast to the situation with IS2, all 3 independently isolated mutations mediated by IS30 carry IS30 at exactly the same site of the P1 genome, and both orientations of IS30 were found. This site is in the hot region for IS2 insertion, but this may be a coincidence (Fig. 1).

Some years ago we had also found an IS30 by chance in an NR1 plasmid (5). This allowed us not only to delimit IS30, but also to determine that IS30 duplicates 2 bp of its insertion target (9). IS30 is 1221 bp long and has the characteristics of a typical IS element (13). Several E. coli K12 strains analyzed carry 2 to 4 copies of IS30 in the chromosome (5,9).

In Fig. 3, we compare the insertion site of IS30 in the P1 genome with the site at which IS30 is carried in the variant of NR1. One sees some striking homologies. However, while the integration site in P1 DNA is obviously a preferred insertion target, nothing is known about the frequency with which the target in NR1 is used. The latter could perhaps be a secondary target site. As do other IS elements, IS30 appears also to mediate deletion formation. A deletion which occurred in the NR1 variant was analyzed and its new junction with IS30 is also shown in Fig. 3. It reveals additional homology to the preferred target site in P1 DNA (9).

A few other targets for IS30 transposition were recently identified on the P1-15 genome. They served for IS30-mediated cointegration of a small plasmid carrying IS30 with the P1-15 genome, without abolishing the ability of P1-15 to reproduce vegetatively (9). We expect that sequencing of these sites might help to further elucidate the target specificity of IS30.

DISCUSSION

The prophage P1 plasmid represents the entire genome of a large bacteriophage with a relatively complex genetic setup. Most of

....tCAtAAgTGCCAt ←—1221 bp IS30 —→ AtcTCctTTctta.... (in P1 DNA)

....gCgAAaTGCCAgc ←—1221 bp IS30 —→ gcgTCgaaTtgcg.... (in NR1)

deletion in NR1: 1221 bp IS30 —→ AgtaCCTTTaTTA....

Fig. 3. Insertion sites of IS30 in P1 DNA (see Fig. 1) and in NR1. To facilitate comparison, homologies between the target sites in P1 DNA (first line) and NR1 (second line) are written in larger letters. The 2 bp duplicated upon transposition are underlined. The third line gives the sequence adjacent to an IS30-mediated deletion in the NR1 variant. Here, larger letters identify those nucleotides homologous with the corresponding positions in the IS30 target of P1 DNA. (Data extracted from Ref. 9.)

these phage functions are essential for vegetative phage reproduction, but not for prophage propagation. The P1 prophage thus offers an ideal system for the study of spontaneous mutations affecting phage functions. This strategy is well known to bacteriophage geneticists (14,15). Although such mutations may in principle often be lethal, the lethality mainly affects phage reproduction. Therefore, mutations can be accumulated over the course of time in the lysogenic condition, and phenotypic manifestation can be studied at any chosen time by the induction of phage production. Subsequent analysis of the mutational alteration on the phage genome can reveal the mechanism underlying mutagenesis.

In the experiments reported here, transposition of IS elements obviously played the major role in spontaneous mutagenesis. Although IS2 was the most active in the production of spontaneous mutations, several other IS elements residing in the host chromosome also contributed to the production of mutations. The sum of these transpositional activities must be considered as more important for spontaneous mutagenesis of the phage genome as a whole than any other mechanism of mutagenesis. The study indeed allows one to directly compare frequencies of the different events affecting the functioning of the entire genome.

The analysis of transpositional mutations mediated by IS2 and by IS30 clearly revealed that these types of mutations do not occur at random on the P1 genome. Therefore, some genes will undergo spontaneous mutation much more frequently than others. One may therefore expect that some particular genes may represent bad targets for transposition and that these genes will not only show a lower rate of spontaneous mutagenesis, but that their spontaneous mutations may rather be caused by mechanisms other than transposition.

It has been known for some time that target selection criteria vary from one IS element to another (see Ref. 2). Some IS elements appear to select their target by specific recognition of a particular nucleotide sequence. IS30 may belong to this class. Other IS elements appear to proceed by a more complex mechanism, and IS2 may represent a prototype of this class. Although the distribution of their transposition targets in a short DNA segment may appear random, this does not necessarily hold for long DNA molecules, such as entire genomes. A regional specificity for transposition, together with the possibility of choosing a large number of particular sites for insertion, may be of great evolutionary relevance. Indeed, genetic rearrangements other than simple transposition may often also depend on the same criteria of target selection. And at least some of these events play important roles in gene fusion and operon fusion, which can give rise to new biological functions and to altered rules for the expression of existing genes. These mechanisms are of relevance in vertical as well as in horizontal evolution.

It is generally admitted that plasmids are excellent breeding grounds for new biological functions, particularly since their genetic content is often inessential for the host bacterium. This last point may be of relevance in attempts to generalize the conclusions made in this paper. One may expect that chromosomal genes, including the essential ones, may be used as transposition targets according to the same rules as discussed for phage genes. Such mutations might often also be lethal and thus not be propagated. However, there is good experimental evidence that transposition often occurs also at functionally inessential sites, both in bacterial chromosomes as well as in bacteriophage genomes. We, therefore, think that the transpositional mechanisms discussed here can be considered to make an important general contribution to the development of altered and new functions in the wide gene pool of microorganisms.

ACKNOWLEDGMENT

The continued support of our research by the Swiss National Science Foundation is acknowledged.

REFERENCES

1. Campbell, A., D. Berg, D. Bottstein, E. Lederberg, R. Novick, P. Starlinger, and W. Szybalski (1977) Nomenclature of transposable elements in prokaryotes. In DNA Insertion Elements, Plasmids, and Episomes, A.I. Bukhari, J.A. Shapiro, and S.L. Adhya, eds. Cold Spring Harbor Laboratory, pp. 15-22.
2. Iida, S., J. Meyer, and W. Arber (1983) Prokaryotic IS elements. In Mobile Genetic Elements, J.A. Shapiro, ed. Academic Press, New York, pp. 159-221.
3. Arber, W. (1984) Natural mechanisms of microbial evolution. In Genetic Manipulation: Impact on Man and Society, W. Arber, K. Illmensee, W.J. Peacock, and P. Starlinger, eds. The ICSU Press and Cambridge University Press, Cambridge, pp. 1-14.
4. Yarmolinsky, M. (1982) Bacteriophage P1. In Genetic Maps, Vol. 2. S.J. O'Brien ed., Natl. Cancer Inst., Frederick, Maryland, pp. 34-43.
5. Arber, W., S. Iida, H. Jütte, P. Caspers, J. Meyer, and C. Hänni (1978) Rearrangements of genetic material in *Escherichia coli* as observed on the bacteriophage P1 plasmid. Cold Spring Harbor Symp. Quant. Biol. 43:1197-1208.
6. Reif, H.J., and H. Saedler (1975) IS1 is involved in deletion formation in the *gal* region of *E. coli* K12. Mol. Gen. Genet. 137:17-28.
7. Meyer, J., S. Iida, and W. Arber (1983) Physical analysis of the genomes of hybrid phages between phage P1 and plasmid p15B. J. Mol. Biol. 165:191-195.

8. Sengstag, C., and W. Arber (1983) IS2 insertion is a major cause of spontaneous mutagenesis of the bacteriophage P1: Non-random distribution of target sites. EMBO 2:67-71.
9. Caspers, P., B. Dalrymple, S. Iida, and W. Arber (1984) IS30, a new insertion sequence of Escherichia coli K12. Mol. Gen. Genet. 196:68-73.
10. Sengstag, C., J.C.W. Shepherd, and W. Arber (1983) The sequence of the bacteriophage P1 genome region serving as hot target for IS2 insertion. EMBO 2:1777-1781.
11. Miller, J.H. (1972) Experiments in Molecular Genetics. Cold Spring Harbor Laboratory, Cold Spring Harbor, New York.
12. Yuan, R. (1981) Structure and mechanism of multifunctional restriction endonucleases. Ann. Rev. Biochem. 50:285-315.
13. Dalrymple, B., P. Caspers, and W. Arber (1984) Nucleotide sequence of the prokaryotic mobile genetic element IS30. EMBO (in press).
14. Jacob, F., and E.L. Wollman (1956) Recherches sur les bactéries lysogènes défectives. I. Déterminisme génétique de la morphogenèse chez un bactériophage tempéré. Ann. Inst. Pasteur 90:282-302.
15. Arber, W., and G. Kellenberger (1958) Study of the property of seven defective-lysogenic strains derived from Escherichia coli K12. Virology 5:458-475.

MECHANISMS OF TRANSPOSITION IN BACTERIA

Douglas E. Berg

Departments of Microbiology and Immunology and of Genetics
Washington University Medical School
St. Louis, Missouri 63110

ABSTRACT

The transposable elements of bacteria are diverse in size, functional arrangement, DNA sequence, and in their modes of transposition. We review here data suggesting that the kanamycin-resistance transposon Tn5 moves without replicating (conservative transposition), but the ampicillin-resistance transposon Tn3 is duplicated when it transposes, and that both the chloramphenicol-resistance transposon Tn9 and bacteriophage Mu are replicated in some events but not in others. A model is presented in which conservative and replicative transposition are alternative branches of a single pathway.

INTRODUCTION

Transposable elements are DNA segments which move to new genomic sites by pathways distinct from those leading to homologous recombination. These elements were discovered by McClintock in the 1940s as the agents responsible for the intriguing variability of a class of unstable alleles in maize, and have subsequently been found in all groups of organisms. Transposable elements cause spontaneous mutations and genome rearrangements, and alter or modulate the expression of nearby immobile genes. Studies in bacteria have shown that transposition is independent of the recombination (rec) genes and of DNA sequence homology, and is mediated by element-specific proteins termed transposases which probably bind short DNA sequences repeated at each end of the element. Transposition in bacteria facilitates the flow of genes between distantly related species and transposable elements have become valuable laboratory tools, especially for work with bacterial species which had not been extensively analyzed using classical genetic approaches (for reviews, see Ref. 5,21).

There has been a great surge of interest in transposable elements during the last decade, much of it focused on the mechanism and regulation of transposition. Here we review studies of several groups of bacterial elements, and models of how they may transpose. The elements discussed are: (i) the kanamycin-resistance transposon Tn5 and its component IS50 elements, (ii) the chloramphenicol-resistance transposon Tn9 and its component IS1 elements, (iii) the ampicillin-resistance transposon Tn3, and (iv) bacteriophage Mu. Each of these elements encodes its own transposase and duplicates a short segment of target DNA when it inserts. The evidence suggests that IS50 and Tn5 transpose by a conservative (nonreplicative) mechanism, that Tn3 transposes by a replicative mechanism, and that Mu, IS1, and Tn9 transpose by both mechanisms.

MODELS OF TRANSPOSITION

Three general classes of transposition models have been advanced. The first and simplest, conservative transposition (Fig. 1), was developed to account for Tn5 transposition (3,4). It involves double-strand breaks at each end of the element, accompanied by staggered cuts (9 bp apart for Tn5) in the complementary strands of target DNA. The protruding target strands are joined to the element, and transposition is completed by repair synthesis which fills in the gaps and generates the 9-bp target sequence duplication which brackets Tn5. The ends of the vector DNA are not rejoined, and may serve as sites of exonucleolytic attack leading to loss of the vector. Because of the sensitive way in which replicon copy numbers are maintained, this loss would generally be compensated by the premature replication of a sibling of the vector molecule, one which had not participated directly in transposition and thus still retained an intact copy of the mobile element. This model accounts for the transposition of Tn5 and of its component IS element, IS50.

The second model of transposition, the bidirectional replicative model (Fig. 2) was developed to account for results obtained with Tn3 and with prophage Mu (2,20). In this model one DNA strand at each end of the element is cut and joined to staggered ends in the target DNA, forming a pair of primers for the replication of the element. The product, called a cointegrate, consists of vector and target replicons joined by directly repeated copies of the element. These cointegrates are unstable because they can be broken down (resolved) at a low rate by homologous recombination. In addition, Tn3 encodes a protein termed resolvase which breaks down cointegrates rapidly by a site-specific recombination reaction (12).

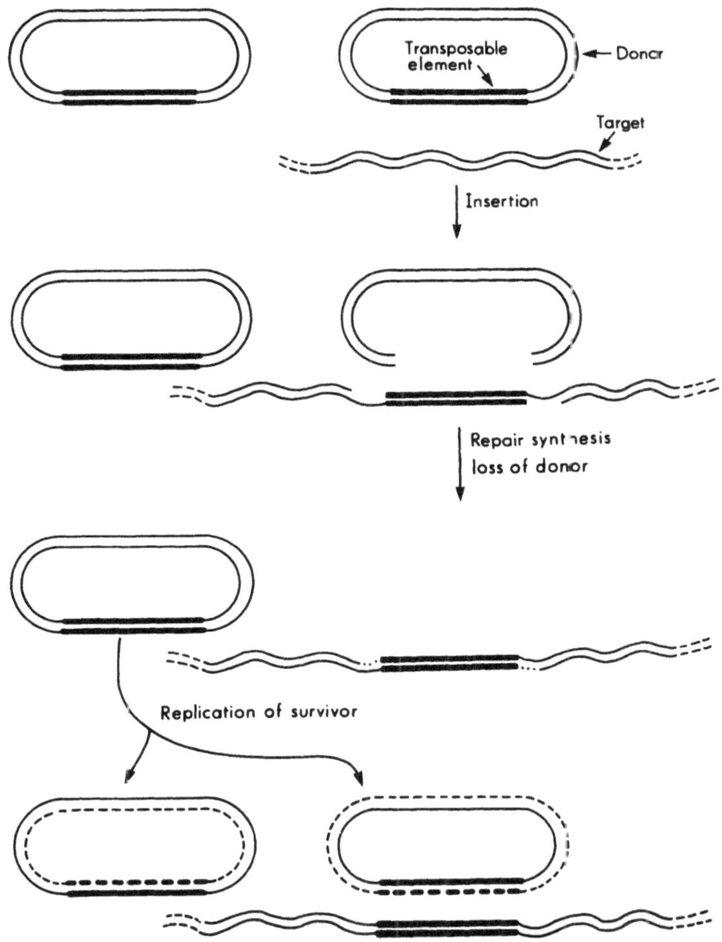

Fig. 1. Model of conservative transposition (adapted from Ref. 3-5).

The third model of transposition, the rolling circle model (Fig. 3), is based on the premise that all transposition is replicative, and was developed to account for results obtained with IS1 and with phage Mu (9,11). Transposition is initiated by the joining of one strand at just one end of the element to the target. Replication from the resulting fork proceeds unidirectionally and spins off a DNA tail consisting of alternating element and vector sequences. Termination occurs at a transposase recognition site which is opposite in orientation to that used to initiate transposition. No specific resolvase is required: either separate donor and target molecules (Fig. 3, left) or cointegrates (Fig. 3, right) result, depending on the way in which DNA strands are joined during termination.

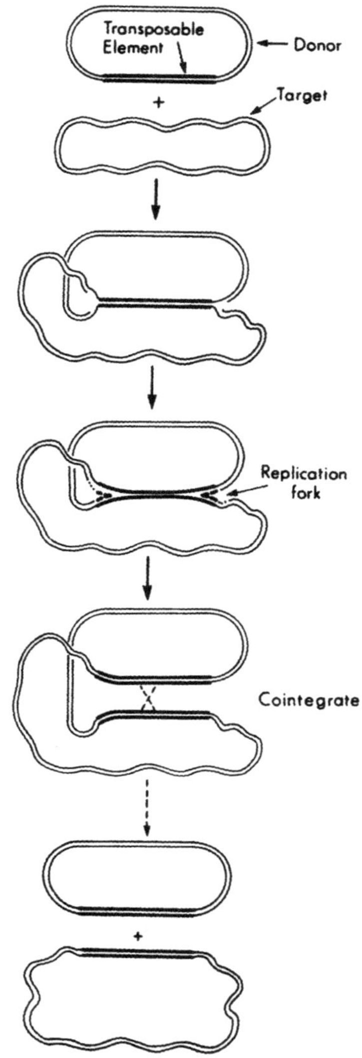

Fig. 2. Model of transposition associated with bidirectional replication of the mobile element (adapted from Ref. 2, 20). The final reaction, resolution, is indicated by a dashed arrow, since this reaction is separate from transposition, and for some elements occurs only by homologous recombination.

RESULTS BEARING ON TRANSPOSITION MODELS

Tn5: Conservative Transposition

The transposition of Tn5 in *recA*$^-$ *E. coli* generates simple insertions of fragments of the donor molecule, not cointegrates con-

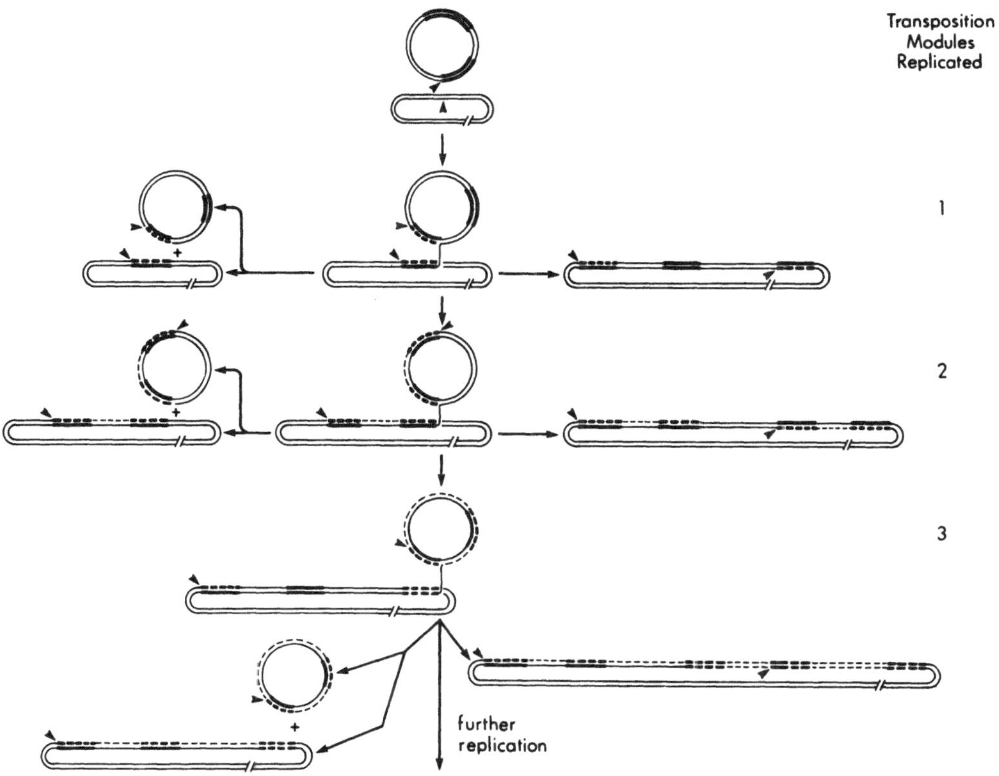

Fig. 3. Model of transposition associated with unidirectional (rolling circle) replication of the element (adapted from Ref. 9, 11).

taining the entire vector molecule, the target molecule, and duplicate copies of the element. For example, transposition of the kan determinant of Tn5 from monomeric pBR322::Tn5 plasmids is not associated with transposition of the vector amp or tet markers. In contrast, transposition of Tn5 from pBR322::Tn5 plasmids in recombination-proficient ($recA^+$) cells is often associated with the co-transposition of pBR322 vector sequences to give cointegrate-like structures (Fig. 4, B-D) (Ref. 3,4,15). To explain this, Hirschel et al., (15) proposed that Tn5 transposes by a rolling circle replicative mechanism (Fig. 3), and that at the termination of replication the recA+ protein regulates how DNA strands are to be broken and then joined to give simple insertions or cointegrate products. In formulating their model they unfortunately did not realize that dimeric and higher oligomeric forms of pBR322::Tn5 plasmids arise by homologous recombination and generally comprise at least one-fourth of any plasmid population in $recA^+$ cells. In reconstruction experiments using dimeric pBR322::Tn5 plasmids in $recA^-$ cells as transposition donors we found that about half of the transposition products contained vector as well as Tn5 sequences (4). However, the

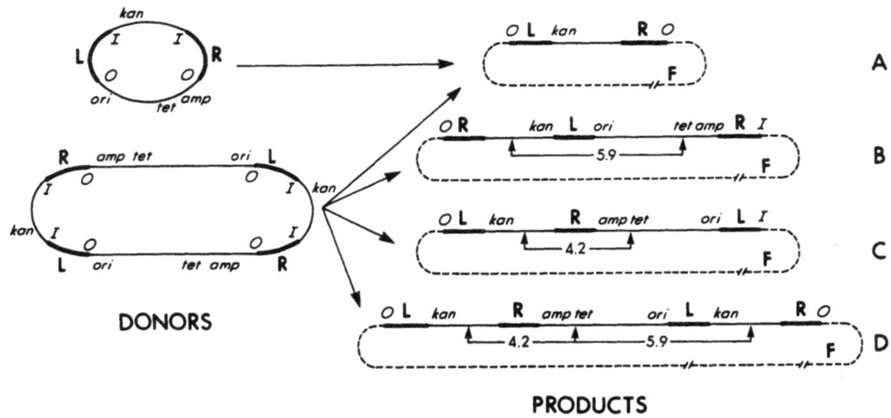

Fig. 4. Test of conservative model for IS50 and Tn5 transposition using monomeric and dimeric pBR322::Tn5 plasmids in recombination-deficient bacteria. Thickened lines designated L and R correspond to IS50L and IS50R, respectively (adapted from Ref. 4).

complete dimeric plasmid was never transposed (see Fig. 4). Thus, the cointegrate-like transposition products obtained in $recA^+$ cells probably result from: (i) simple insertion of segments of dimeric and larger plasmids, and (ii) from homologous recombination following transposition. There is no convincing evidence that Tn5 ever transposes by a replicative process (4).

The Tn3 element which is unrelated to Tn5, and which transposes replicatively and generates cointegrates, encodes a protein called resolvase which catalyzes site-specific recombination between direct repeats of Tn3. Because of this protein cointegrates generated by Tn3 transposition are unstable in $recA^-$ as well as in $recA^+$ cells, unless the gene for resolvase or its site of action is mutant (2,10).

In contrast to the instability of plasmids containing direct repeats of Tn3, dimeric pBR322::Tn5 plasmids and more complex structures containing direct repeats of IS50 or Tn5 (Fig. 4, 5) were easily generated in $recA^+$ cells and stably maintained in $recA^-$ cells (4,13,14,19). We concluded, therefore, that Tn5 does not encode a resolvase (4,13,14). But, a recent report had concluded that the transposase of Tn5 does in fact possess a resolvase activity (22). This new interpretation was based on the studies of a particular plasmid with direct repeats of IS50 elements, and on the rapid accumulation of bacterial subclones harboring recombinant derivatives of this plasmid. The accumulation of recombinants was said to be blocked by mutations in the transposase gene, and thus it was argued that the transposase protein encoded by Tn5 could also function as a resolvase (22).

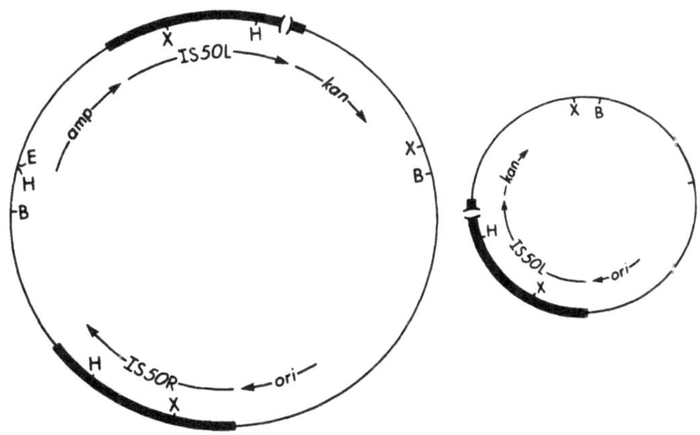

Fig. 5. The structure of pTn5-DR1 (left) and a plasmid (right) derived from it by recombination between directly repeated IS50 elements (thickened lines) in vivo. The letters B, E, H, and X indicate the sites cleaved by the restriction endonucleases BamHI, EcoRI, HindIII, and XhoI, respectively (adapted from Ref. 13). Only IS50R encodes transposase in the strains used in our studies. IS50L contains an ochre mutant allele of the transposase gene as a result of a single base-pair substitution 112 bp from the inside end of the 1534-bp element indicated by ( ) (adapted from Ref. 18).

Because this interpretation contradicts the body of evidence suggesting that Tn5 does not encode a resolution function, we reevaluated the role of transposase in the breakdown of plasmids containing directly repeated copies of IS50. Our analysis indicates that the interpretation of Zupancic et al. (22) is incorrect; there is no evidence that Tn5 encodes a resolution function. We used pTn5-DR1 (Fig. 5; Ref. 13), a plasmid containing direct repeats of IS50 generated from pBR322::Tn5-wild-type by the in vitro rearrangement of BamHI fragments. Occasional breakdown of pTn5-DR1 in the absence of recA function might be expected because homologous recombination does occur in RecA$^-$ cells, although at a very low rate (8). We observed that cells carrying pTn5-DR1 form extremely small colonies [and Zupancic et al. (22) noted a similar slowing in growth rate caused by their plasmid]. We have found, in addition, that: (i) growth rate differences between cells carrying pTn5-DR1 and its recombinant are strongest on kanamycin-containing agar, (ii) recombinants are only rarely obtained after growth on kanamycin-free agar, (iii) poor growth with kanamycin is not corrected by insertions of Tn5-wild-type elsewhere in the genome, and (iv) simple point mutations in the transposase gene of pTn5-DR1 (at the XhoI and HindIII sites marked X and H, respectively, in IS50R in Fig. 5) do not relieve the growth defect (17). [The "transposase" mutations, which Zupancic et al. (22) believed stabilized their plasmid, were complex and incompletely characterized.] Thus, the apparent instability of pTn5-DR1-like plasmids is caused by a still mysterious selection against cells carrying the complete plasmid, superimposed on a low level of recA-independent homologous recombination which in other systems is independent of IS elements. Cells carrying the recombinant plasmid simply outgrow their progenitors, and there is no evidence to suggest that Tn5 encodes a resolution function (17).

Tn3 and Mu: Replicative Transposition

The final products of transposition of mutants of Tn3 defective in either the synthesis or the site of action of resolvase are nearly always cointegrates (2,10). The final products of transposition Tn3-wild-type are usually donor and target molecules which each contain one copy of Tn3 because cointegrates formed as the initial products of transposition are broken down by resolvase. Similarly the induction of lytic growth of Mu prophages leads to transposition concomitant with Mu replication, and cointegrates equivalent to those mediated by Tn3 are formed whenever Mu transposes between replicons (7). These findings led Arthur and Sherratt (2) and Shapiro (20) to propose the bidirectional replicative transposition model (Fig. 2).

IS1 and Mu: Rolling-Circle-Replicative Transposition

In contrast to the cointegrates found after induction of Mu prophages, only single copies of Mu are found inserted after phage Mu infection (11). In addition, in the case of IS1 which is unrelated to Mu, only a fraction of transpositions result in cointegrates (9). Since neither Mu nor IS1 seem to encode an active resolution function, the findings of simple insertions appear to be inconsistent with the bidirectional replicative transposition model. This inconsistency led Harshey and Bukhari (11) and Galas and Chandler (9) to propose the rolling circle model (Fig. 3). The rolling circle model assumes that all transposition is replicative, and provides an explanation for simple (noncointegrate) transposition products as well as cointegrates as primary products of transposition.

There has always been reason to question the assumption that all transposition is replicative which underlies the rolling circle model. Mu insertion after infection has recently been shown to be conservative not replicative (1,16). In addition, we have examined the products of transposition from monomeric and dimeric forms of pBR322::Tn9 (IS1) plasmids (6), and our results indicate that the noncointegrate products of IS1 transposition probably do not arise by a rolling circle mechanism. Rather, with IS1, as with IS50, the simple insertions appear to arise by events in which the element failed to duplicate. Thus, recent results with Mu and with IS1 call for the consideration of new models, ones in which transposition is sometimes conservative and sometimes replicative.

We propose that a single intermediate can lead to cointegrate or noncointegrate transposition products as depicted in Fig. 6. In this model the element and target DNAs are joined by single strands, generating incipient replication forks which are susceptible to nucleolytic attack. Cleavage of vector-transposable element junctions at any time before replication is complete (left pathway) destroys the replication fork, blocks cointegrate formation, and results in transposition without duplication of the element. Successful completion of replication without nucleolytic cleavage leads to cointegrates (right pathway). The conservative insertion of Mu after infection may be a reflection of the linearity of the injected Mu DNA,

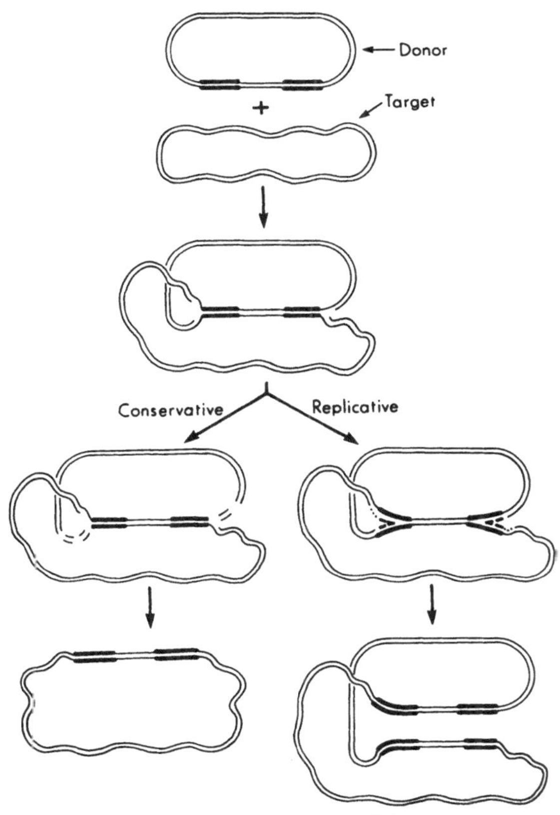

Fig. 6. A model of transposition in which replication (right fork) and nucleolytic cleavage of the donor-element junction (left fork) compete. Shown on the left is the case in which cleavage prior to the onset of replication results in a simple (noncointegrate) structure. However, breakage any time before replication is completed would have the same formal consequences (adapted from Ref. 6).

or of a different balance of the Mu insertion proteins. It is not necessary to invoke a rolling circle model to explain simple insertions as well as cointegrates.

It is possible that even IS50 and Tn5 transpose via the mechanism outlined in Fig. 6, with the caveat that the structure of the transposase leads invariably to destruction of the vector-element junctions prior to completion of replication, and hence to apparently conservative transposition. However, since IS50 appears to be unrelated to the other elements discussed here, it need not transpose by exactly the same mechanism. We still favor our original idea, that double-strand breaks at the vector-element junctions at the start of transposition irrevocably commits IS50-related elements to insertion without replication.

SUMMARY AND CONCLUSIONS

The diversity of bacterial transposable elements indicates that a number of different mechanisms of transposition may have evolved. The available evidence suggests that: (i) Tn3 and related elements move by a replicative process which invariably generates cointegrate DNAs consisting of donor and target replicons joined by direct duplications of the mobile element. (ii) Bacteriophage Mu, and the chloramphenicol-resistance transposon Tn9 and its component IS1 elements, appear to move in 2 ways. One leads to cointegrates, and the other does not. (iii) Tn5 and its component IS50 elements do not generate cointegrates when they move, and hence transpose by a nonreplicative (conservative) process.

It is possible, although not necessary, to consider all types of transposition studied to date as branches of a single pathway. One branchpoint occurs at the onset of transposition, when the structure of the transposase complex determines whether one or both strands at the junctions between each end of the element and its vector are cleaved. The cleavage of both strands leads to transposition which is not associated with replication (Fig. 1). The cleavage of just one strand at each end of the element causes the formation of replication forks (Fig. 2, 6). Whether the element is then duplicated or not depends on the speed of DNA synthesis relative to the chance of double-strand breakage before transposition is completed by the meeting of the replication forks.

ACKNOWLEDGEMENTS

I am grateful to C.M. Berg for critical readings of this manuscript. This work was supported by U.S. Public Health Research Grants 5 Rol AI 14267 and 1 Rol AI 18980.

REFERENCES

1. Akroyd, J.E., and N. Symonds (1983) Evidence for a conservative pattern of transposition of bacteriophage Mu. Nature 303:84-86.
2. Arthur, A., and D.J. Sherratt (1979) Dissection of the transposition process: A transposon-encoded site-specific recombination system. Mol. Gen. Genet. 175:267-274.
3. Berg, D.E. (1977) Insertion and excision of the transposable kanamycin resistance determinant Tn5. In DNA Insertion Elements, Plasmids and Episomes, A.I. Bukhari, J.A. Shapiro, and S.L. Adhya, eds. Cold Spring Harbor Press, New York, pp. 205-212.
4. Berg, D.E. (1983) Structural requirement for IS50-mediated gene transposition. Proc. Natl. Acad. Sci., USA 80:792-796.
5. Berg, D.E., and C.M. Berg (1983) The prokaryotic transposable element Tn5. Biotechnology 1:417-435.
6. Biel, S.W., and D.E. Berg (1984) Mechanism of IS1 transposition in E. coli: Choice between simple insertion and cointegration Genetics (in press).
7. Chaconas, G., R.M. Harshey, J. Sravetnick, and A.I. Bukhari (1981) Predominant end products of prophage Mu DNA transposition during the lytic cycle are replicon fusions. J. Mol. Biol. 150:341-359.
8. Doherty, M.J., P.T. Morrison, and R. Kolodner (1983) Genetic recombination of bacterial plasmid DNA: Physical and genetic analysis of the products of plasmid recombination in Escherichia coli. J. Mol. Biol. 167:539-560.
9. Galas, D.J., and M. Chandler (1981) On the molecular mechanism of transposition. Proc. Natl. Acad. Sci., USA 78:4858-4862.
10. Gill, R., F. Heffron, G. Dougan, and S. Falkow (1978) Analysis of sequences transposed by complementation of two classes of transposition deficient mutants of Tn3. J. Bact. 136:742-756.
11. Harshey, R.M., and A.I. Bukhari (1981) A mechanism of DNA transposition. Proc. Natl. Acad. Sci., USA 78:1090-1094.
12. Heffron, F. (1983) Tn3 and its relatives. In Mobile Genetic Elements, J.A. Shapiro, ed. Academic Press, New York.
13. Hirschel, B.J., and D.E. Berg (1982) A derivative of Tn5 with direct terminal repeats can transpose. J. Mol. Biol. 155:105-120.
14. Hirschel, B.J., D.J. Galas, D.E. Berg, and M. Chandler (1982) Structure and stability of transposon 5-mediated cointegrates. J. Mol. Biol. 159:557-580.
15. Hirschel, B.J., D.J. Galas, and M. Chandler (1982) Cointegrate formation by Tn5, but not transposition, is dependent on recA. Proc. Natl. Acad. Sci., USA 79:4530-4534.
16. Liebhart, J., P. Ghelardini, and P. Paolozzi (1983) Conservative integration of bacteriophage Mu DNA into pBR322 plasmid. Proc. Natl. Acad. Sci., USA 79:4361-4366.

17. Phadnis, S.H., and D.E. Berg (1984) RecA-independent recombination between direct repeats of IS50 elements is not mediated by the IS50-encoded transposase (submitted for publication).
18. Rothstein, S.J., and W.S. Reznikoff (1981) The functional differences in the inverted repeats of Tn5 are caused by a single base pair nonhomology. Cell 23:191-199.
19. Sasakawa, C., J.B. Lowe, L. McDivitt, and D.E. Berg (1982) Control of transposon Tn5 transposition in *Escherichia coli*. Proc. Natl. Acad. Sci., USA 79:7450-7454.
20. Shapiro, J.A. (1979) Molecular model for the transposition and replication of bacteriophage Mu and other elements. Proc. Natl. Acad. Sci., USA 76:1933-1937.
21. Shapiro, J.A., ed. (1983) *Mobile Genetic Elements*. Academic Press, New York.
22. Zupancic, T.J., S.L. Marvo, J.H. Chung, E.G. Peralta, and S.R. Jaskunas (1983) RecA-independent recombination between the direct repeats of IS50. Cell 3:619-637.

INSERTION OF TRANSCRIPTIONAL ELEMENTS OUTSIDE THE REPLICATION REGION CAN INTERFERE WITH REPLICATION, MAINTENANCE, AND STABILITY OF ColE1-DERIVED PLASMIDS

Hermann Bujard, Dietrich Stueber, Reiner Gentz, Ulrich Deuschle, and Ursula Peschke

Molekulare Genetik
Universität Heidelberg
INF 230, 69 Heidelberg
Federal Republic of Germany

INTRODUCTION

The primary goal of our work is to contribute to a better understanding of structure-function relationships of nucleotide sequences defining promoters in the E. coli system. Since strong, unregulated promoters can be assumed to exhibit pertinent structural features most clearly, we investigated, in vitro, more than 50 promoters for their ability (a) to compete for the binding of RNA polymerase (an indication of their "signal strength") and (b) to produce RNA ("overall efficiency" of a promoter) under conditions of limiting amounts of RNA polymerase. Promoters fulfilling these requirements best were found primarily in the genome of coliphage T5 (1). Some of these promoters show extremely high rates of complex formation with RNA polymerase (1,2) and at the same time were identified as being very efficient initiators of RNA synthesis in vitro and in vivo (1,3).

For a better understanding of such transcriptional signals, their sequences had to be cloned to (a) obtain amounts sufficient for biochemical and physical studies in vitro and (b) investigate their properties in vivo under genetically comparable conditions. However, none of the commonly available cloning vehicles permitted

---

Present Address: F. Hoffmann-La Roche and Co. AG, 4002 Basel, Switzerland.

the stable integration of these promoter classes (3), and it became obvious that intense transcriptional activities can interfere in various ways with plasmid maintenance and structural stability. We therefore constructed 2 families of plasmid vectors which not only allow us to stably integrate promoters independently of their strength, but also enabled us to identify some of the problems which heavily transcribed regions can cause to ColE1-type vector systems (3,4).

The principle of these plasmid systems is shown in Fig. 1 and their main properties are described in the corresponding legend.

Copy Number and Plasmid Maintenance

The pLBU3 vector (Fig. 1) permitted the cloning of various phage T5 promoters (3). However, those found to be most efficient in vitro appeared to be either not clonable or they caused a drastic reduction in the copy number, often accompanied with the loss of the entire plasmid upon prolonged propagation of the host cells. Since the reason for this phenomenon could have been an overproduction of tet gene products as well as a possible transcriptional readthrough into the replication region, we developed the pDS1 plasmid family (4). As has been shown (4), integration of a Xba I linker into the Pvu II site of pBR322 [which destroys the coding sequence of the ROP-protein (5)] increased the plasmid copy number 2- to 3-fold, indicating that a possible overproduction of this regulator protein by readthrough from upstream promoters can reduce plasmid copy number and subsequent maintenance. This is demonstrated best by the fact that T5 promoter P_{207} is stably maintained in pDS1, albeit at a reduced copy number (Fig. 2). Insertion of an efficient terminator at site 1 or 2 of pDS1 (Fig. 1), however, increases the copy number again (Fig. 2), showing that transcriptional readthrough into the replication region by itself has a profound effect on plasmid replication. We have not analyzed whether this is due to an overproduction of RNA I (6), a possible "mechanical" interference with the transcriptional unit for RNA II (7), or both.

Strong transcriptional readthrough into the replication region towards the primer promoter (producing RNA II, Ref. 7) also interferes with plasmid replication. Whereas promoters up to 4 times the efficiency of the β-lactamase promoter (P_{bla} or P_3, in Ref. 8) can be tolerated, sequences which exceed this promoter's efficiency 20-fold or more cannot stably be integrated in place of P_{bla} under conditions where translation of β-lactamase m-RNA would have been impossible (9,10).

In summary, for the construction of highly efficient transcriptional units within plasmid vectors carrying ColE1-type replicons, it has to be taken into account that promoter efficiency has to be "balanced" for the unit to be stably maintained. This can be achieved best by inserting compatible transcriptional signals, preferably in addition to repressible promoter elements.

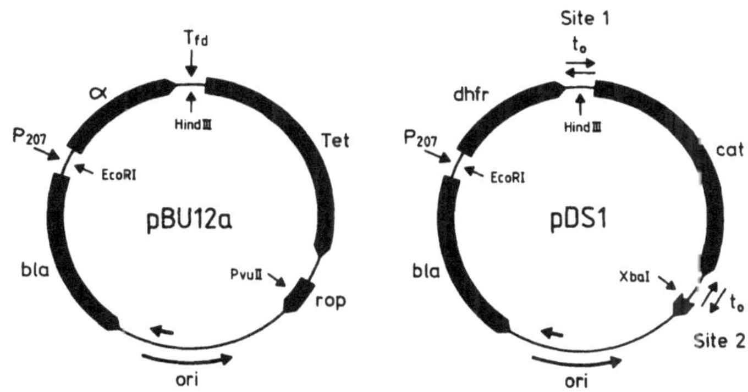

Fig. 1. Genetic and physical map of pBU12a and pDS1. Both plasmids contain 2 easily assayable functions which can be brought under the control of a single promoter (P) but which can also be separated by the insertion of a terminator (T). These functions are for the pBU series: β-galactosidase activity (via α-complementation α) and tetracycline resistance (tet); for the pDS series: trimethoprime (via dihydrofolate reductase, dhfr) and chloramphenicol (chloramphenicol acetyl transferase, cat) resistance. Both plasmids confer ampicillin resistance (via β-lactamase, bla) and carry the ColEl replication region of pBR322 (ori). A major difference of the 2 plasmid systems is the presence of the intact rop gene in pBU12a and a second site for the integration of a terminator (site 2) in pDS1. The direction of transcription of the coding sequences is indicated. Positions and directions of transcription for RNA II (primer RNA) and RNA I are represented by arrows. The physical maps of both plasmids are not drawn to scale. Integration of coliphage fd terminator into pBU12a results in pLBU3 which was used for the cloning of coliphage T5 promoters (3). Insertion of terminators like t_o of coliphage λ (4) at site 1, 2 or at both sites makes this plasmid an even better cloning vehicle for promoter cloning. Promoters and other signals can be placed in front of the dhfr gene into various restriction sites (only the EcoRI site is shown here). Whereas the cat gene carries its own translational initiation site, there is no translation of dhfr-specified message unless a ribosomal binding site has previously been fused to the dhfr gene.

The Influence of Terminator Position on Plasmid Dimerization in pDS1

As shown above, placing the terminator t_o in position 1 or 2 of pDS1 carrying the phage T5 promoter P_{207} re-establishes the copy number to that of the promoter-free plasmid (Fig. 2). However, in analyzing the ratio of monomeric to dimeric forms of the plasmid, it can be clearly recognized that t_o in position 1 increases the number

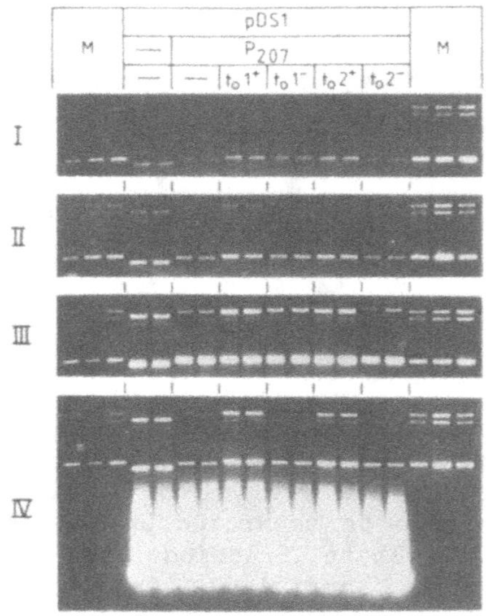

Fig. 2. Electrophoretic analysis of pDS1 derivatives at various cell stages (for details, see Ref. 4). The roman numerals designate the nucleic acid preparations from cells of stage I to IV, M are concentration standards. The duplicate probes can be identified by the headings whereby the following abbreviations are used: P_{207}, coliphage T5 promoter P_{207} inserted into the EcoRI site of pDS1 (Fig. 1); t_o1^+ and t_o1^- describes the insertion of the coliphage λ terminator t_o at site 1 in functional (1^+) and nonfunctional (1^-) orientation, respectively; the same holds for t_o2^+ and t_o2^-. The first band on the top of each gel represents the dimeric, the lowest the monomeric form of the plasmid. In comparing the dimeric forms in stage II and stage IV, the influence of the positioning of t_o can be seen most clearly.

of dimeric structures. This is confirmed by the experiment shown in Fig. 3. Since it has been shown that multimerization decreases the maintenance of ColE1-derived plasmids (11), these observations show that even such seemingly trivial topographic rearrangements of otherwise identical signals can affect the stable propagation of a vector. This result is puzzling in a special way: proposing that heavily transcribed regions may have some impact on a plasmid's replication, a size reduction of this region by moving t_o from position 2 to 1 would be expected to have the opposite effect.

Increase of Recombinational Events in Heavily Transcribed Regions

We have constructed a pDS1 derivative containing 3 identical T5 promoter sequences tandemly fused around 100 bp apart (12). Since a

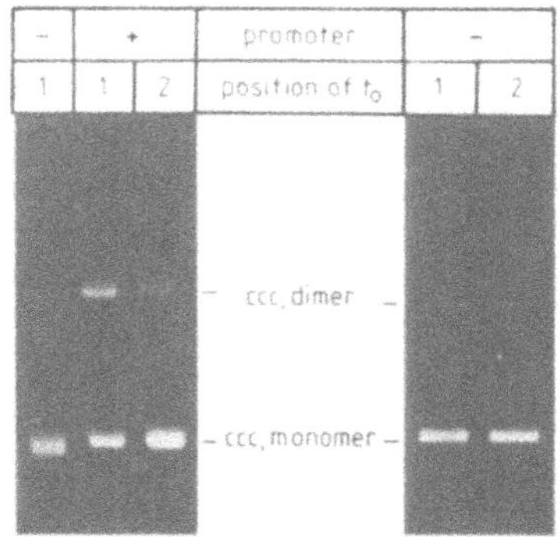

Fig. 3. Influence of the positioning of terminators on the dimerization of pDS1. The different probes of the electrophoretic analysis in agarose gels can be identified by the heading of the figure. The promoter used is P_{207} from coliophage T5. The increase in plasmid dimerization in presence of P_{207} and terminator t_o in position 1 (Fig. 1) is seen in lane 2 (from the left).

battery of 3 strong promoters was assumed to possibly create problems, the promoters were previously fused to a lac operator sequence. In addition, terminator t_o was placed in position 1 and no ribosomal binding site was present to allow the production of DHFR. This construct is perfectly stable if grown in the presence of sufficient amounts of lac repressor (Fig. 4). Upon induction with IPTG, colonies appear at high rates wherein the plasmid has lost one or 2 promoters by homologous recombination (evidenced by fine mapping of restriction sites and sequence analysis). Transferring the plasmid into rec A or rec F strains markedly increased the stability of the construct (Fig. 4). There are 2 interpretations to those results:

- Increased intensity of transcription has an impact on the frequency of recombinational events whereby it is not clear whether the tandemly repeated promoter regions represent a special situation, or whether similar results can be obtained with tandemly repeated sequences further downstream of the transcriptional initiation sites.

- There is a normal low frequency of elimination of promoters but, under condition of transcription, plasmids with less intense transcription (i.e., lower promoter number) have a replicative advantage.

In favor of the first interpretation, we have never observed differences in copy numbers between plasmids in which promoters differ severalfold in their efficiency--as long as translation of the overproduced m-RNA and readthrough into the replication region is prevented.

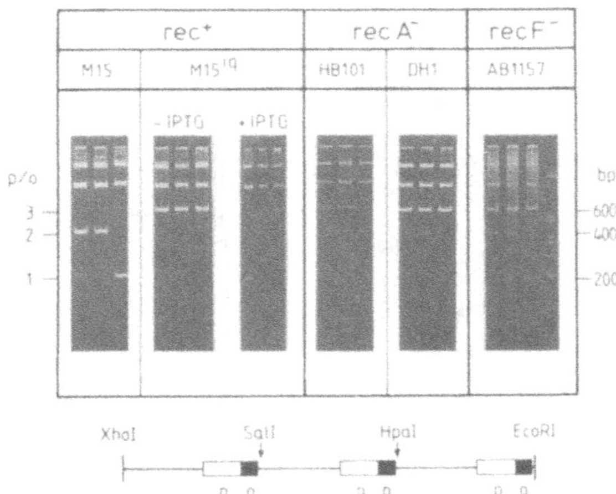

Fig. 4. Stability of a tandemly repeated promoter/operator construct. The upper part of the figure shows the electrophoretic analysis (6% polyacrylamide) of pDS1 containing t_o in position 1 (Fig. 1) and the triple promoter/operator construct shown in the lower part of the figure. The distance between the promoter/operator regions is 100 bp. The host system and the conditions of growth (± IPTG) of the different plasmid isolates can be taken from the headings. At the left side the numbers of promoter/operator regions within the different isolates are indicated, which are 600, 400, or 200 bp in length, respectively, as pointed out at the left side of the figure. The cultures were grown to $OD_{600} \triangleq 2.0$ in LB medium.

Plasmid Carried Translational Initiation Signals and the Possible Limits of the Translational Capacity of the Host System

The pDS1 system carrying a proper promoter insertion permits quantification of different translational initiation signals by inserting the unknown signal in front of the dhfr gene. The resulting polycistronic m-RNA (terminator in position 2) codes for 2 proteins: DHFR and CAT. Since the amount of CAT depends on its original translation initiation signal, the ratio of CAT to DHFR can be used to compare translation initiation signals of unknown activity (13). The products of both indicator genes (dhfr and cat) are tolerated by the cell in concentrations of up to 50% of the total soluble protein if only one of the genes is translated or up to 30% each if both gene products are allowed to be synthesized. Using this system, we have identified some unusually efficient translational initiation signals. Some of these, however, can only be cloned in presence of a strong promoter if an efficient transcriptional terminator is placed in position 2 of the plasmid (Fig. 1). Attempts to integrate such signals into plasmids carrying the identical terminator in position 1 failed. Our interpretation of these experiments is that the amount of DHFR produced under these conditions may become toxic to the cell. However, having stoichiometric amounts of CAT-ribosomal binding sites (terminator in position 2) competing with the signal in front of the dhfr gene appears to allow the survival of the cell. Since under these conditions around 90% of the newly synthesized proteins consist of CAT and DHFR, these experiments seem to indicate that, by use of the proper expression signals in multicopy plasmids, at least the translational machinery of the host can be exhausted at certain stages of growth. Consequently, the stable propagation of a plasmid can also depend on physiological parameters of this type.

CONCLUDING REMARKS

The results described here were obtained as "side products" in a program in which plasmids are merely being used as tools to produce and to study signals of gene expression in vitro and in vivo. Even in such simple constructs as discussed here, where we know the DNA sequences and the topography of the main functions, it appears that we often cannot predict the outcome of seemingly homologous experiments, and puzzling questions accumulate. They indicate that even such straightforward artificial plasmid constructs cannot be looked at in a static way. Instead, they appear as dynamic entities in which, for example, the extent of transcription in various regions may represent only one such variable to be taken into account should such a system become predictable.

REFERENCES

1. Gabain, A.V., and H. Bujard (1979) Interaction of E. coli RNA polymerase with promoters: The relative rates of complex formation and decay with promoters of several phage and plasmid DNA's. Proc. Natl. Acad. Sci., USA 76:189-193.
2. Brunner, M. (1984) Untersuchungen zur Wechselwirkung von E. coli RNA-Polymerase mit Promotoren. Diplomarbeit, Universität Heidelberg, 69 Heidelberg, FRG.
3. Gentz, R., A. Langner, A.C.Y. Chang, S.N. Cohen, and H. Bujard (1981) Cloning and analysis of strong promoters is made possible by the downstream placement of an RNA termination signal. Proc. Natl. Acad. Sci., USA 78:4936-4941.
4. Stueber, D., and H. Bujard (1982) Transcription from efficient promoters can interfere with plasmid replication and diminish expression of plasmid specified genes. EMBO 1:1399-1404.
5. Cesarini, G., M.A. Muesing, and B. Polisky (1982) Control of ColE1 DNA replication: The Rop gene product negatively affects transcription from the replication primer promoter. Proc. Natl. Acad. Sci., USA 79:6313-6317.

6. Muesing, M., J. Tamm, H.M. Shepard, and B. Polisky (1981) A single base-pair alternation is responsible for the DNA over-production phenotype of a plasmid copy-number mutant. Cell 24:235-242.
7. Tomizawa, J., and T. Itoh (1981) Plasmid ColE1 incompatibility determined by interaction of RNA I with primer transcript. Proc. Natl. Acad. Sci., USA 78:6096-6100.
8. Stueber, D., and H. Bujard (1981) Organization of transcriptional signals in plasmids pBR322 and pACYC184. Proc. Natl. Acad. Sci., USA 78:167-171.
9. Deuschle, U. (1984) Entwicklung einer Methode zur quantitativen Bestimmung der in vitro- als auch in vivo-Effizienzen von Transkriptionssignalen aus dem E. coli-system. Diplomarbeit, Universität Heidelberg, 69 Heidelberg, FRG.
10. Deuschle, U., R. Gentz, and H. Bujard (unpublished results).
11. Summers, D.K., and D.J. Sherrat (1984) Multimerization of high copy number plasmids causes instability: ColE1 encodes a determinant essential for plasmid monomerization and stability. Cell 36:1097-1103.
12. Peschke, U. (1984) Konstruction von tandemartig angeordneten Promoter/Operator sequenzen zur analyse der wechselwirkung diser sequenzen mit RNA-polymerase bzw. Repressoren. Diplomarbeit, Universität Heidelberg, 69 Heidelberg, FRG.
13. Bujard, H., C. Baldari, M. Brunner, U. Deuschle, R. Gentz, J. Hughes, W. Kammerer, and D. Stueber (1983) Integration of efficient promoters of the E. coli system into plasmid vectors. In Gene Amplification and Analysis, T.S. Papas, M. Rosenberg, and J.C. Cirikjian, eds. Elsevier/North Holland, New York, Vol. 3, pp. 65-88.

STUDIES ON THE TRANSPOSITION OF IS1

Michael Chandler[1] and David J. Galas[2]

[1]Centre de Recherche de Biochimie et
Génétique Cellulaire du C.N.R.S.
118, Route de Narbonne
31062 Toulouse Cedex, France

INTRODUCTION

The insertion sequence, IS1, is among the smallest insertion elements known. It is a member of a class of elements which includes IS10, IS50, and IS903 which are frequently found in pairs flanking a nontransposable DNA segment. In concert, the 2 flanking elements are able to promote transposition of the entire structure. This type of transposable element has been called a compound or composite transposon. The best characterized IS1-flanked compound transposon is perhaps Tn9 in which 2 directly repeated copies of IS1 flank a 1.1-kb segment of DNA which encodes chloramphenicol acetyltransferase (CAT) (1). Unlike most other elements of this class, however, IS1 is present in the chromosomes of a variety of enterobacteria, in some cases in many copies (2).

IS1 also forms part of several clinically important multiple antibiotic-resistance plasmids where it flanks, in direct repeat, the majority of the resistance genes and delimits a structure having a molecular length of the order of 20 kb (the resistance determinant or r-det) (3-5). The r-det is itself composed of a series of intercalated transposable elements. The flanking IS1 elements give the resistance genes a structural autonomy enabling the r-det to behave as an integral unit. Indeed, derivatives of the resistance plasmid R100.1 have been isolated in *Escherichia coli* from which supernumerary circular copies of the r-det are generated (9). This process

[2] Molecular Biology University of Southern California, Los Angeles, California 90089-1481.

requires the host recA and recBC gene products and can be repressed in trans by the presence of a closely related plasmid in the same cell. The r-det can also undergo tandem duplication (amplification) (6) and transmission to other replicons, by a mechanism which depends largely on homologous recombination between IS1 elements, but may also involve IS1-mediated transposition at a very low frequency (7,8). The insertion sequence IS1 therefore provides a large degree of fluidity for the antibiotic-resistance genes carried by these plasmids, a feature which has undoubtedly contributed to their rapid dissemination. The presence of multiple copies of this insertion sequence in the host chromosome might also be expected to provide similar flexibility within the host genome.

In spite of, or perhaps because of, its presence in the chromosome of most laboratory strains of E. coli, the analysis of the behavior and structure of IS1, and of the organization and expression of IS1-encoded genes is not as advanced as is that of some other elements of this class such as IS50 and IS10. Transposition of IS1 has many aspects in common with that of other insertion elements: it generates a duplication in the target DNA at the point of insertion (10-12) and it induces replicon fusions (cointegrate formation) (13), deletion, and inversion of DNA segments (14,15). For certain transposons, for example those of the Tn3 family, cointegrate formation is an integral part of the major transposition pathway. It is thought to be a replicative mechanism in which the element is copied into its target site while a copy is retained at the original site (see Ref. 16 for review). Where donor and recipient replicons are circular plasmids, the resulting cointegrate structures are composed of fused copies of the donor and recipient separated at each junction by a copy of the element, in a directly repeated configuration. Subsequent recombination between the 2 elements (resolution) completes the transposition, resulting in the regeneration of the donor replicon and of the recipient which then carries a single copy of the element. On this scheme deletions and inversions are viewed as the result of intramolecular transposition.

While IS1 mediates cointegrate formation, several lines of evidence suggest that this is not the sole, nor the most efficient, pathway for transposition of this element (17,18). Thus, in contrast to Tn3-mediated cointegrates, Tn9 (IS1)-mediated cointegrates are relatively stable and, under some circumstances, their frequency of formation is lower than that of Tn9 transposition (17). Cointegrates mediated by other members of this class of transposon are also relatively stable. This suggests that cointegrates are not intermediates in the transposition of these elements, and has led to the proposal that they are alternative end products to direct transposition of the element in the transposition process (18). Although several models have been proposed to explain the relationship between cointegrate formation and direct transposition (for reviews, see Ref. 16 and 19), insufficient experimental evidence is available at present to allow definite conclusions to be drawn.

IS1-flanked compound transposons exhibit at least one other interesting property. Comparison of the transposition frequency of a series of IS1-flanked transposons from a lambda prophage to 2 different target plasmids revealed an approximately exponential decrease in transposition frequency with increasing size (20). Similar results have been obtained with other transposons, in particular Tn10 (21). The simplest interpretation of this length dependence is that some aspect of the transposition process occurs in a processive manner. This could occur during replication of the element as we initially suggested (18) or at any stage in the process which involves one-dimensional tracking of the transposition complex across the entire compound transposon.

In the following, we present a summary of the results of an ongoing study of some of the factors which affect the transposition activity of IS1 and the relative frequencies of cointegrate formation and transposition. We also present an analysis of the structure and coding capacity of the element. The implications of these and other results on the evolution and maintenance of IS1-flanked compound transposons are discussed.

THE EXPERIMENTAL SYSTEM

The experimental system employed throughout these studies uses 2 plasmids: a conjugative plasmid, pOX38, derived from the plasmid F which has been deleted for all known insertion sequences carried by the parent plasmid (22); and the nontransmissible multicopy plasmid, pBR322 or its derivatives. The plasmid pOX38 itself carries no selectable marker and in certain experiments we have therefore used a derivative of this plasmid, pOX38Km, which carries a HindIII fragment of Tn5 specifying resistance to kanamycin cloned into the unique HindIII site of pOX38. The transposon under study can be placed on either of the 2 plasmids by transposition or by in vitro techniques. Using this system, transposition from pBR322 derivative plasmids can be simply monitored by pOX38-mediated transfer of the transposon-associated resistance gene. Cointegrate formation can likewise be monitored by pOX38-mediated conjugal transfer of pBR322 to a suitable recipient. Such transfer can only occur after prior fusion of the 2 plasmids mediated by a transposable element carried by one or the other (17,23). In all experiments, the donor strain harboring a pOX38 and pBR322 derivative plasmid carried a recA mutation. In experiments in which the structures of various transposition products were to be analyzed, the independence of each event was ensured by the use of a single exconjugant colony from each of a series of independent matings using separate isolated colonies of the donor strain.

MODULATION OF COINTEGRATE-FORMING ACTIVITY BY TRANSCRIPTION

In view of the fact that the flanking insertion elements associated with compound transposons retain some of their functional independence, it is interesting to consider the factors which determine the relative frequencies of transposition events involving either the entire compound transposon or the flanking sequences. Since IS1 has no selectable phenotypic marker, it is difficult to measure its frequency of transposition directly. We have therefore made use of the fact that Tn9 induces cointegrates in which the element involved in the fusion (either of the individual IS1 elements or the entire Tn9) is duplicated and appears as a directly repeated copy at the junctions between the 2 replicons (Fig. 1). The relative frequency of appearance of cointegrates in which Tn9 is repeated at both junctions or in which Tn9 occurs at one and IS1 at the other can be used as a measure of the relative activities of the entire Tn9 and its component IS1 elements in cointegrate formation. This in turn presumably reflects the relative transposition frequency of the components.

In the following, the IS1 elements flanking Tn9 have been called IS1A and IS1B, respectively (17). The relative position of the elements with respect to transcription of the CAT gene is shown

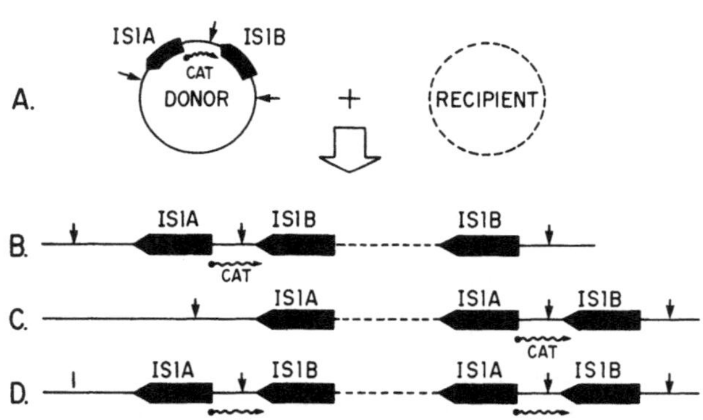

Fig. 1. Cointegrate formation by Tn9. This figure shows schematically the way in which various cointegrate structures are formed: ▬►, IS1; ~~→, CAT gene transcript; ↓, restriction site used for structure determinations. These structures were analyzed by digestion with a restriction enzyme which cleaves within the CAT gene of Tn9 (EcoRI) and within the donor plasmid. The use of a particular IS1-element is characterized by which of the specific IS1-carrying fragments of the donor replicon (A) is no longer present in the cointegrate. A fuller explanation of this analysis will be found in Ref. 17 and 23. (B) IS1B-mediated; (C) IS1A-mediated; (D) Tn9-mediated.

in Fig. 1. Cointegrates generated by Tn9 or by one or the other flanking IS1 element can be simply distinguished by digestion with suitable restriction enzymes (17,23). The analysis of 40 cointegrates generated between pOX38::Tn9 and pBR322 indicated that the overall frequency of utilization of the IS1's was about 10 times that of Tn9 (17). Surprisingly, however, one of the flanking IS1 elements, IS1A (Figs. 1,2), was almost 10-fold more active than the other. Thus the activities of the flanking IS1 elements are either inherently different, as is the case with the flanking elements of Tn5 and Tn10, or the difference in activity is due to differences in the nucleotide sequence environment of the 2 elements.

In order to distinguish between these 2 possibilities, we cloned Tn9 (isolated as a BglII-XhoI fragment together with about 150 bp of surrounding DNA of the pOX38 donor plasmid) from pOX38::Tn9 into various positions and orientations within the plasmid pBR322. We subsequently analyzed the structure of cointegrates generated between these pBR322-Tn9 derivative plasmids and pOX38 (23). A summary of these results is shown in Fig. 2. It can be seen that the relative frequency with which IS1 elements are used in the formation of cointegrates varies enormously from plasmid to plasmid. There is, however, a strong correlation between the relative activity of the IS1 elements and the orientation of Tn9 with respect to the known transcription units of pBR322. The plasmids pXB, pR2.4, and pR2.5 carry Tn9 cloned between the BamHI and SalI sites (pXB), or into the EcoRI site in either orientation (pR2.4 and

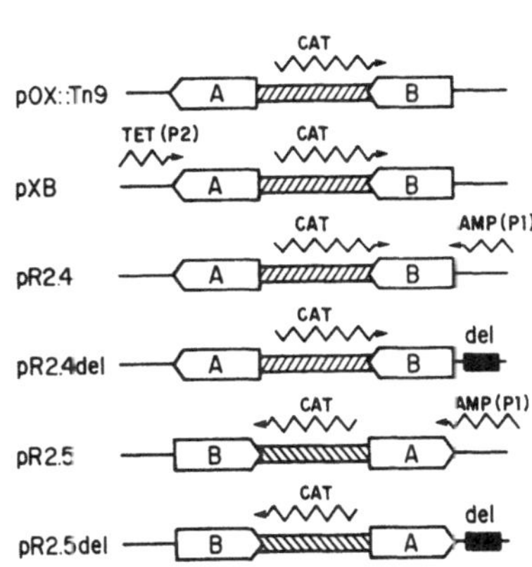

	IS1A	IS1B	Tn9
pOX::Tn9	33	4	3
pXB	4	27	3
pR2.4	25	5	1
pR2.4del	17	4	3
pR2.5	9	19	4
pR2.5del	13	4	2

Fig. 2. Relative frequency of IS1A, IS1B and Tn9-mediated cointegrates. This figure shows the relative frequency of 3 types of cointegrate generated by pOX38::Tn9 and by various pBR322:Tn9 derivative plasmids. ⁄\/\/\→, transcripts; ■, deletion of promoter P1; pXB; Tn9 cloned between the BamHI and SalI sites of pBR322 (see Fig. 3); pR2.4 and pR2.5, Tn9 cloned into the EcoRI site of pBR322. The construction of these plasmids is described in Ref. 23.

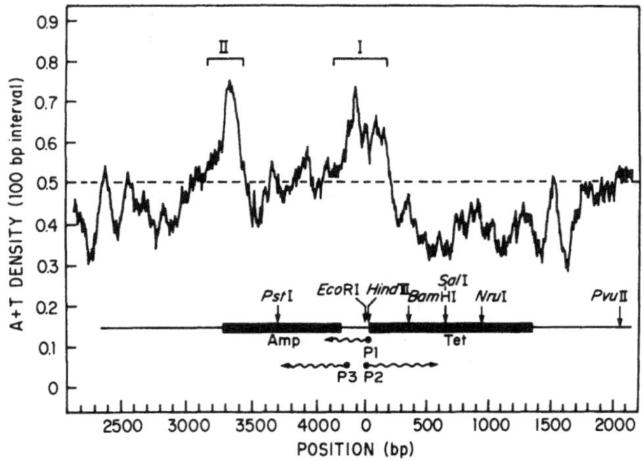

Fig. 3. Physical and genetic map of plasmid pBR322. This figure shows: the position of the promoters P1, P2, and P3; the direction of transcription from these promoters, ~~~; location of the Ampicillin and Tetracycline resistance genes, ▬; and the position of the sites of cleavage of the various restriction enzymes used in the generation of the derivative plasmid described here. Also included is the relative proportion of A+T base pairs in the molecule using a window of 100 bp.

pR2.5). In all 3 plasmids, the IS1 proximal to an oncoming transcript (from promoter P2 for pXB and from promoter P1 for pR2.4 and pR.2.5) is the least active (24) (see Fig. 3 for a physical and genetic map of pBR322). Removal of this transcription, for example, by deletion of the promoter (pR2.5del), results in an increase in the activity of the proximal IS1. Similar results have been obtained by insertion of the strong transcription terminator of phage T4 gene 32 into these and other pBR322 derivative plasmids (Ref. 23; data not shown). It should be noted that in the case of pR2.4 in which transcription enters IS1B, removal of this transcription does not significantly change the activity of the element (23). This is presumably due to the fact that transcription of the CAT gene of Tn9 continues into IS1B (25) reducing its activity with respect to IS1A even in the absence of external transcription.

Thus the difference in activity of the IS1 elements is determined largely by external transcription and is not inherent to the element. The effect of transcriptional inhibition can also be seen in the frequency of cointegrate formation. The frequencies of IS1-mediated cointegrate formation between pOX38Km and in vitro deletion derivatives of pR2.4 and pR2.5 retaining a single IS1 are shown in Fig. 4. Both pR2.4IS1 and pR2.5IS1 show an approximately 10-fold

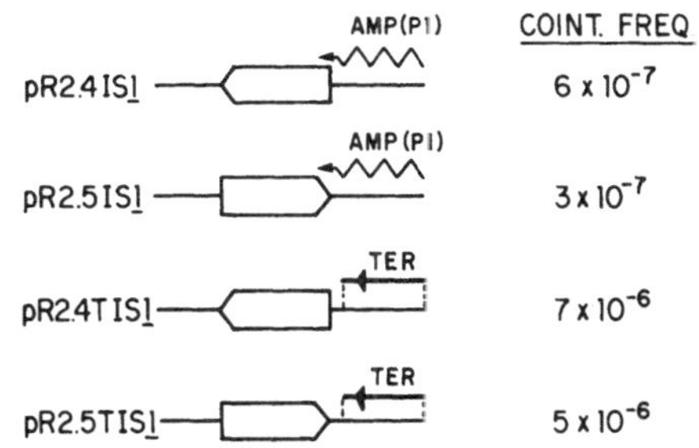

Fig. 4. Cointegrate formation frequency of pBR322::IS1 derivative plasmids. Plasmids carrying a single copy of IS1 were derived from the respective Tn9-carrying plasmids by in vitro deletion using the unique BstEII site in the IS1 elements (23).

The strong transcription terminator of phage T4 gene 32 was introduced into these plasmids on a HindIII/BamHI fragment of pBH857 (57) by substitution between the pBR322 HindIII and BamHI sites (Fig. 3) (23).

Cointegrate formation frequencies were measured using pOX38Km as a target molecule and the results are the mean values of 10 independent experiments (23). Transcription from promoter P1 (Fig. 3) is indicated, ⁓⁓→; and the terminator from T4 gene 32 is indicated, ←.

lower frequency of cointegrate formation than do derivatives in which transcription from the promoter P1 is blocked by substitution of the T4 gene 32 terminator for P1. Similar results have recently been obtained by Machida et al. (25). In the case of the Tn5, Sasakawa et al. (26) have demonstrated an analogous effect on transposition.

EFFECTS OF TRANSCRIPTION ON THE SPECTRUM OF TRANSPOSITION PRODUCTS

The data presented above demonstrate an inhibitory effect of external transcription on IS1-mediated cointegrate formation. In order to determine whether similar inhibitory effects are also exerted on transposition, we have determined the relative frequencies of cointegrate formation and transposition using the various pBR322-Tn9 derivative plasmids as donors and pOX38Km as a target molecule. Transposition from the circular donor plasmids can in principle occur in 2 senses: transposition of Tn9 itself or transposition of the pBR322 segment of the plasmid (inverse transposition) (27-29). Experimentally, all 3 types of event can be simply monitored by determining the relative frequency of transfer of various combinations of pBR322-Tn9-associated markers: Cm^r together with Ap^r

and/or Tc^r (cointegrates), Cm^r (Tn9 transposition), and Ap^r and/or Tc^r (inverse transposition of pBR322), relative to that of pOX38Km alone. Representative results are presented in Fig. 5. Several features of these data are worthy of note: 1) the relative frequencies of transposition, inverse transposition, and cointegrate formation are distinct and characteristic of each of the 3 plasmids; 2) deletion of the promoter (pXBdel) or substitution of the promoter with the T4 gene 32 terminator (pR2.4T, pR2.5T) results in convergence of these spectra to give indistinguishable patterns; 3) The plasmids pXB and pR2.5 in which transcription enters IS1A exhibit quite different patterns. These results demonstrate that transcription influences all 3 types of transposition events.

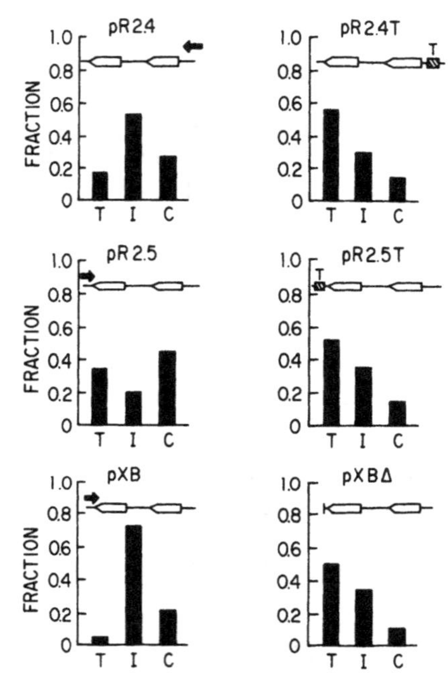

Fig. 5. Spectra of Tn9-mediated recombination products. Tn9-mediated transposition (T), inverse transposition (I) and cointegrate formation (C) were measured as described in the text, using pOX38Km as a target molecule. The relative frequencies are represented here as fractions of the total observed recombination products. The donor plasmids are indicated above each histogram and the environment of the Tn9 represented: ➡, transcription; ▨, transcription terminator from T4 gene 32 (57); the vertical line in the pXBΔ panel indicates the deletion endpoint near the left end of Tn9. The construction of these plasmids is described in Ref. 23.

The principal features of the absolute frequencies (Tab. 1) of each of the transposition events are as follows: 1) the overall frequency of events increases by about an order of magnitude following suppression of incoming transcription; 2) In all 3 cases, there is a marked increase in the absolute frequency of transposition; 3) In both cases where external transcription enters IS1A (pR2.5 and pXB), its removal also has a marked effect on the frequency of inverse transposition; 4) Only a small effect is observed in the case of pR2.4 in which transcription enters IS1B. 5) In all 3 plasmids only a small effect on cointegrate formation is observed.

A coherent explanation for the above results is that transcription into either end of IS1 reduces the probability that it is recognized (or bound) by a transposition complex. The magnitude of this effect must be comparable at either end. Thus the cointegrate formation frequencies of pR2.4IS1 and pR2.5IS1 are similar and increase to a similar extent when transcription inhibition is relieved. The effect seems to be transmitted through the element to some extent since both transposition (using the "outer" ends of the IS1

Tab. 1. Frequencies of transposition, cointegrate formation, and inverse transposition.

	TRANSP.	COINT.	INV.	TOTAL
pR2.4	2.3×10^{-7}	3.5×10^{-7}	6.9×10^{-7}	1.3×10^{-6}
pR2.4T	4.2×10^{-6}	1.0×10^{-6}	2.2×10^{-6}	7.5×10^{-6}
pR2.5	3.2×10^{-7}	4.0×10^{-7}	1.9×10^{-7}	9.0×10^{-7}
pR2.5T	3.8×10^{-6}	1.1×10^{-6}	2.6×10^{-6}	7.5×10^{-6}
pXB	3.9×10^{-8}	1.7×10^{-7}	5.7×10^{-7}	7.8×10^{-7}
pXBdel	4.2×10^{-6}	9.7×10^{-7}	2.9×10^{-6}	8.1×10^{-6}
pBRIS1.11 (monomer)	--	2.0×10^{-7}	--	--
pBRIS1.11 (dimer)	--	3.1×10^{-7}	--	--

The plasmids and methods used in measuring these frequencies have been described (Ref. 23 and text). The plasmid pBRIS1.11 carries an IS1 insertion at pBR322 coordinate 3270 and was isolated from a cointegrate between pOX38::Tn9 and pBR322 as described. Following the introduction of the pBR322 derivative plasmids by transformation in the construction of the donor strains, the presence of monomeric and dimeric plasmids was ascertained by agarose gel electrophoresis of small preparations of plasmid DNA from the transformants. The numbers are the means of ten independent measurements.

elements flanking Tn9) and inverse transposition (using the "inside" ends) are affected by the same transcript (see pXB and pR2.5). In this light it is perhaps significant that no sequences characteristic of strong transcription terminators are obvious in the sequence of IS1 (30,31), neither are all IS1 insertions polar (32,33). This raises the possibility that some transcription could continue through the element to affect the distal end.

The effect of release of transcription inhibition in the case of pR2.4 is interesting. The relatively large increase in transposition accompanied by a small increase in inverse transposition suggests that the "outside" end of IS1B is affected to a greater extent by incoming transcription than is the "inside" end. Presumably, transcription of the CAT gene into the inside end of IS1B is the major factor in determining the activity of this end (25).

The consistently smaller effect of transcription on cointegrate formation in all 3 plasmids is not unexpected. Transposition is presumably determined by the activity of both flanking IS1 elements whereas cointegrate formation can occur using one or the other. Transcription entering one IS1 element would thus inhibit transposition of the entire compound transposon while it would reduce cointegrate formation only by the contribution of the inhibited element.

Finally, an interesting possibility suggested by these results is that translation of the incoming transcript also inhibits transposition activity. An increase of over 2 orders of magnitude in transposition frequency is observed following relief of transcription inhibition by deletion of the Tet gene promoter in pXB (pXBdel) whereas only a 10-fold increase is observed for pR2.5 (pR2.5T). Since pXB carries Tn9 cloned within the Tet gene of pBR322, the oncoming Tet transcript is presumably also translated. For pR2.5 in which Tn9 is cloned into the EcoRI site, no translation of the incoming transcript would be expected. The difference in behavior of pXB (pXBdel) and pR2.5 (pR2.5T) could reflect a supplementary inhibition in pXB due to translation. In contrast to the transcription effect, the hypothetical translation effect would seem not to be transmitted through the element since the change in inverse transposition in pXB is similar to that found with pR2.5. Translation inhibition of events requiring the "inside" ends would thus be lower than that of events requiring the "outside" ends. Here it should be noted that nonsense codons are found at both ends of IS1 in all 3 reading frames, preventing translation into the element. We are investigating the effect of removing translation of the truncated Tet gene in the behaviour of pXB. It indicates that the complex changes in the spectrum of products can be explained by simple changes in the activities of the ends.

The above descriptive treatment is of necessity oversimplied and is presented only to provide a framework. A detailed quantitative interpretation of these results has been performed but presentation of this analysis would be too lengthy to include here (Galas and Chandler, in preparation).

REPLICATIVE AND NONREPLICATIVE EVENTS

One of the strongest pieces of evidence that at least some transpositional events are replicative is the appearance of cointegrates, since the cointegrate of donor and target replicon carries 2 copies of the transposon. There is good evidence of replication in the case of Tn3 (34). Analysis of adjacent deletions mediated by IS903 provides direct evidence that these events also occur by a replicative mechanism (35). On the other hand, the observation of the strong dependence of cointegrate formation on recA function in the case of Tn5 (36) may indicate that simple transposition from dimers of the donor replicon are involved in cointegrate formation, rather than replicative fusion from a monomeric donor molecule. Cointegrate formation from dimeric pBR322::Tn5 plasmid molecules has been reported to be significantly greater than from monomeric molecules (37). While there is some question of the above interpretation of the Tn5 observations, the results do suggest that Tn5 and IS1 may function differently (Chandler and Galas, in preparation). We have measured the frequency of IS1- mediated cointegrate formation from monomeric and dimeric donor molecules and found them to be indistinguishable (Tab. 1). This indicates that cointegrate formation is primarily replicative for IS1 since the element must be replicated in the case of the monomeric donor.

INSERTION SPECIFICITY

The influence of the donor molecule on the outcome of transposition described above prompted us to investigate more closely features of the target molecule which might also have an influence.

It has been noted that IS1 exhibits a strong preference for regions of DNA which are rich in A+T (38,39); that insertion in the lacZ gene of *E. coli* occurs close to a region of homology with the extremity of the element; and that there is a strong preference for cleavage of the target DNA next to a G C base pair (38). It is probable, therefore, that several factors act in concert to determine the final choice of target site, perhaps on different scales of sequence length.

In order to investigate in a more systematic manner the various factors that influence the choice of target site, we have analyzed the distribution of IS1 insertions in the fully sequenced plasmid pBR322 (40,41) and certain derivatives. Insertions of IS1 were isolated from cointegrate plasmids generated by fusion between pBR322 and pOX38::Tn9 following mating to a suitable recombination- proficient recipient strain. Selection was for the transfer of the pBR322-associated tetracycline resistance marker in most cases, and thus precluded the isolation of insertions within this gene. Since such cointegrates are relatively stable, we have adopted the following strategy for isolating pBR322:IS1 derivative plasmids from them. Cointegrates are isolated in a recombination-proficient (recA$^+$) host and plasmid DNA prepared from such strains (42) following copy number amplification. This treatment results in an enrichment for pBR322 derivative plasmids in the population, i.e., for those derivatives which have excised from the cointegrate in general by homologous recombination between the directly repeated copies of the flanking transposable element. The recombination can give rise to both pBR322::Tn9 and pBR322::IS1 derivatives (see Fig. 1). The plasmid DNA was then used to transform *E. coli* C600 to tetracycline resistance. Crystal violet was included in the plates to screen for chloramphenicol-sensitive transformants presumably carrying a plasmid with a single IS1 (43). Plasmid DNA isolated from chloramphenicol-sensitive transformants by the rapid lysis method (42) was screened for size by agarose gel electrophoresis and plasmids exhibiting the size expected for pBR322:IS1 were retained for further analysis. Digestion with either HinfI and HinfI+PstI or PstI and PstI+EcoRI was performed to determine the orientation and position of IS1 in each of the pBR322::IS1 plasmids. Typical results are shown in Fig. 6 and a summary of these results is presented in Fig. 7.

The plasmid pBR322 carries 2 regions of high A+T content (Fig. 3). One, region I, is situated (approximately) between coordinates 4150 and 150 and includes the promoter P3 of the β-lactamase gene. The other, region II, is located between coordinates 3200 and 3400 and covers part of the 3' end of the β-lactamase gene. It can be seen from the results presented in Fig. 7a that just over half (15/25) of the IS1 insertions occur in region I while 6 of the insertions fall in region II and the remaining 4 fall in the relatively A+T-rich β-lactamase gene. Since the region(s) of pBR322 which exhibit strong homology with the ends of IS1 (7/7 base matches) are located at coordinates 406, 415, 1871, and 2084, some distance from the preferred insertion site, this result suggests that A+T richness is more important in determining the sites of insertion than are regions exhibiting homology with the end of the element. More interestingly, of the 15 insertions within region I, 13 occur in one orientation. The insertions shown in this figure were isolated following selection for the transfer of pBR322-associated tetracycline resistance. We have also analyzed 9 plasmid derivatives which were isolated in the same experiment, but following selection for transfer of pBR322-associated ampicillin (carbenicillin) resistance so that the tetracylcine resistance marker was no longer excluded as a target. All 9 isolates proved to carry IS1 inserted in region I and 8 of these were in the prevalent orientation.

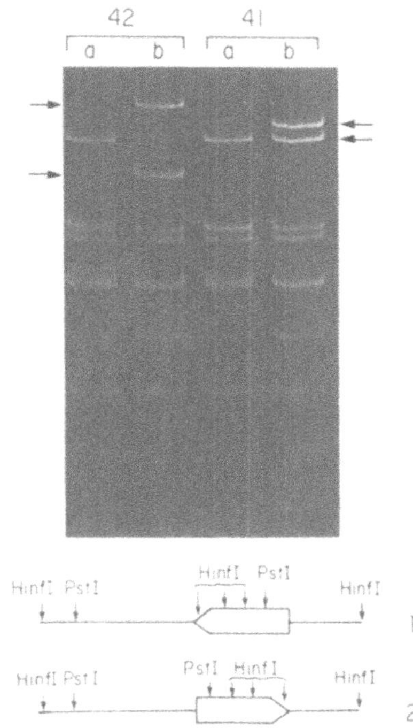

Fig. 6. Analysis of insertion specificity into pBR322 and pKH47. This figure illustrates the analysis of IS1 insertions into the plasmid pKH47, using HinfI + PstI (a) and HinfI alone (b). The plasmids, 41 and 42, carry IS1 insertions into the major "hot spot" region I (Fig. 3) in opposite orientations. For insertions in one orientation, both junction fragments (→) are cleaved by PstI (plasmid 1) while in the other, only one of these fragments is cleaved (plasmid 2).

The position of the relevant restriction sites for IS1s in the region I "hotspot" are shown schematically below.

Thus, of 24 insertions isolated in region I, 21 proved to occur in one orientation. It is also clear from this result that there is no "hotspot" in the Tc^r gene.

The preferential orientation of insertion of IS1 in region I of pBR322 derivative plasmids is further illustrated by the results presented in Fig. 7b. In this series of experiments, 31 independent pOX38::Tn9-mediated IS1 insertions were isolated in the plasmid pKH47. This plasmid is a derivative of pBR322 which carries a re-

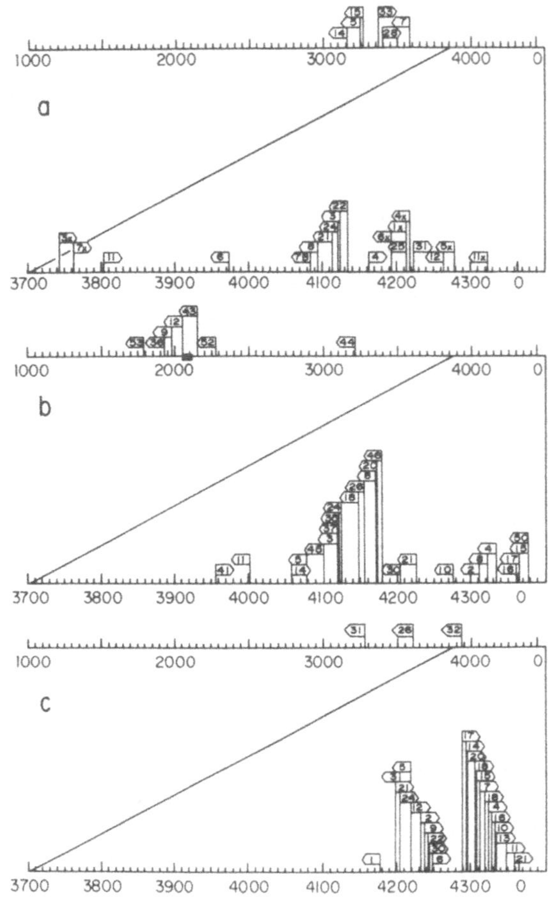

Fig. 7. Distribution of IS1 Insertions into pBR322 and pKH47. a) Insertions of IS1 (Tn9) in pBR322. b) Insertions of IS1 (Tn9) in pKH47. c) Insertions of IS1 (Tn2671del72) in pBR322. ■, 60 bp poly (dA) ·poly(dT) insert.

gion of poly dA:dT residues introduced by polynucleotide kinase tailing (44). The plasmid used in our experiments was found to carry an insertion of approximately 60 bp within the unique PvuII site of pBR322. Note that, of 31 insertions mapped, only 1 occurs in region II of the plasmid while 6 occur in the neighborhood of the poly A:T segment. Thus, as expected, the synthetic poly dA:dT region attracts insertions of IS1 (45). Region I of the plasmid remains, however, the most favored region. Twenty-two of the 31 insertions are found in this region. More significantly, perhaps, almost all (21 of the 22 insertions) are found in the preferred orientation observed in the pBR322 target. Thus, of a total of 46 insertions occurring in this region in the 2 experiments, 42 are observed in one orientation. The probability that such a distribution occurs by chance is low.

We have also analyzed the specificity of insertion using Tn2671.72, another IS1-flanked transposon closely related to Tn9 (46). Using pOX38::Tn2671.72 in place of pOX38::Tn9 in the mating system, we found that the distribution of insertions into pBR322 was the same as that obtained with pOX38::Tn9. The results of an analysis of 27 IS1 insertions into pBR322 are shown in Fig. 7c. Twenty-four insertions fall in region I. They also exhibit a strong orientation preference (21 occur in one orientation). This orientation is opposite, however, to that observed in the case of pOX38::Tn9 donor. It is clear, then, that the donor plasmid/transposon can determine the orientation of insertion. A possible explanation for the orientation reversal is that it reflects a difference in the activities of the extremities of the IS1 elements in the 2 transposons due to the effect of transcription inhibition described previously. For example, the "activity" of an extremity could represent the probability that the extremity is recognized and nicked prior to insertion into its target site. The orientation of insertions in pBR322 cannot depend only on the donor molecule, however. An observed orientation preference implies that a "direction" must be distinguished both in the donor and the target molecule. This could be determined by an oriented recognition sequence in the target, or a directional process such as transcription or replication. We are at present investigating these possibilities.

While it is reasonable to expect that the strong inhibitory effect of transcription could provide one of the required preferences in the donor, it is not at all clear what factors could provide the strand preference in the target. To test one possibility, that the transcription through region I from promoter P1 was responsible, we inserted the T4 gene 32 transcription terminator in the EcoRI site of pBR322 to block this transcription and isolated a set of IS1 inserts into this plasmid using both donor plasmids. This collection exhibited the same orientation preference as before (data not shown), indicating that transcription is not the orientation-determining factor in the target.

STRUCTURE AND ORGANIZATION OF IS1

Analysis of the nucleotide sequence of IS1 has shown that the extremities are composed of 2 imperfectly homologous segments of about 35 bp inverted with respect to each other (30,31). To begin to determine the role of these extremities, we have constructed a group of pBR322-derived plasmids. These carry an IS1 element or a 76 bp fragment of one end of the element cloned into the PvuII site of pBR322 and a second IS1 cloned into the EcoRI site in one or the other orientation. These constructions should yield transposons carrying the pBR322 Tet gene. In order to improve selection for the transposon at low copy number, part of the intervening Tet gene (between the HindIII and NruI sites; Fig. 3) of these 4 plasmids was replaced by a 1.8 kb Hind III/PvuII fragment of the plasmid pKp6 (47) specifying resistance to streptomycin and spectinomycin (P. Gamas, unpub. data).

The frequency of transposition (Sp^r, Sm^r) onto, and of cointegrate formation (Sp^r, Sm^r, Ap^r) with pOX38Km as determined for each of these plasmids is presented in Fig. 8. It can been seen that the plasmids pOC17 and pOC18 which each carry 2 complete copies of IS1 exhibit similar transposition and cointegrate formation frequencies. This indicates that there is no dramatic difference in the activity of the 2 extremities of IS1. Since we do not at present know the transcription pattern of the inserted Sp^rSm^r fragment, we are unable to ascertain the effects of transcription on the activity of the IS1 elements. It is therefore difficult to interpret the differences between the cointegrate formation frequency and that of transposition observed for both plasmids.

While we do not yet fully understand the reason for the differences in activity of pOC15 and pOC16, it is clear from these results that 76 bp at one extremity of IS1 are sufficient for transposition of the synthetic compound transposon. It should be noted here that

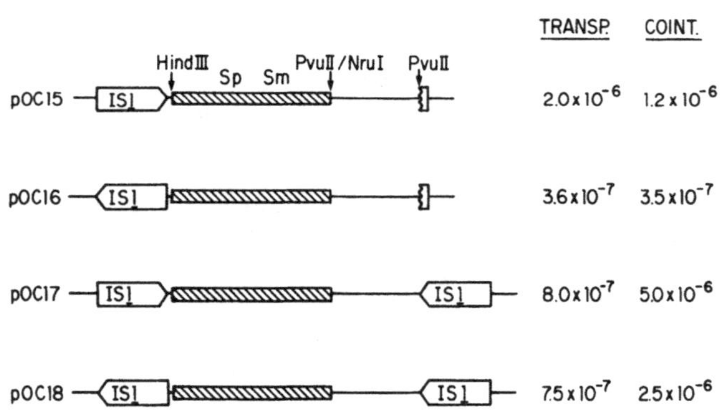

Fig. 8. Transpostion of Sm^r/Sp^r transposons. Transposition and cointegrate formation frequencies of the 4 transposons which form part of a pBR322 based replicon was measured using pOX38Km as a target plasmid. The construction of these plasmids is briefly described in the text and will be described in detail elsewhere. {], 76 bp extremity of IS1 delimited by the IS1 PvuII site; ▨, spectinomycin, streptomycin resistance fragment. The frequencies are the means of 10 independent experiments.

inverse transposition of Ap^r from these plasmids occurred at a frequency of $<10^{-9}$ (the lower limit of detection in these experiments) while it occurred at about 10^{-7} in the case of pOC17 and pOC18. This demonstrates that the extremities of IS1 used in a transposition event must be in the correct relative orientation for successful transposition. Experiments to define the minimum size of the extremity for successful transposition are in progress.

Sequence analysis has also shown that IS1 cannot code for any large polypeptides but indicates that it could code several small polypeptides (Fig. 9). As pointed out by Rak et al. (48) and others (49), many insertion elements have coding frames arranged in a characteristic manner: a large open reading frame coding in one direction and a significantly shorter overlapping open frame coding in the opposite direction with an identical frame configuration (codons in phase). IS1 is peculiar in this respect in not exhibiting the long reading frame. It is like the others, however, in having 2 opposite-strand, aligned reading frames (A and D) in the left half of the molecule (Fig. 9). Because of their conservation in 2 iso-

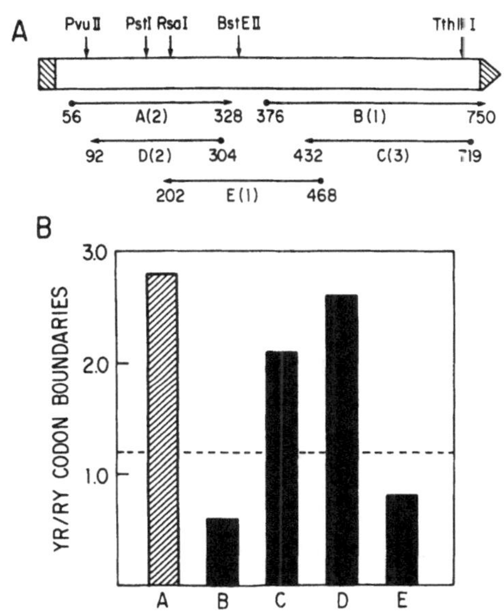

Fig. 9. Organization of IS1. A. Map of IS1 showing relevant restriction sites, and several major open reading frames (ORF) together with their reading phases. The inverted homologous extremities are also indicated. B. Histogram of the number of codon boundaries found with the pyrimidine-purine (YR) character to those with the purine-pyrimidine (RY) character. The expected ratio if codon usage were completely random is 1.2 (52).

insertion elements of IS1, it has been suggested that frames A and B are good candidates for encoding functional IS1 polypeptides (50). However, theoretical considerations of the codon boundary preference exhibited by each potential open reading frames ha led to the suggestion that, while frames A, C and D show a strong positive correlation with the preferred pyrimidine/purine (YR) codon boundary often noted in coding regions (51), B does not. The degree of preference for YR codon boundaries is shown for the open reading frames in IS1 (greater than 50 amino acid equivalents in length) in Fig. 9b. This had led to the suggestion that B may not code for a functional polypeptide (19,52).

Experimental definition of all the functional reading frames and the roles of their products is not yet available. By using S1 digestion at unique restriction cleavage sites in IS1, Ohtsubo and collaborators (53) have demonstrated that deletions in both the A (at PstI) and B (at Tth111I) reading frames render the element inactive. Results obtained in our laboratories in which we have used the Sm^r fragment of pKP6 to insert 14 bp of DNA [see Prentki and Krisch (47)] into several unique restrictions sites in IS1 (PvuII, 76 bp; RsaI, 209 bp; and BstEII, 328 bp; see Fig. 9a) are similar. We were unable to detect cointegrates mediated by these IS1 insertion mutants in the pBR322::IS1/pOXc38Km assay, to a lower limit of 10^{-9} (J. Robert and H. Krisch, unpub. results). Whatever the organization of functional reading frames it is clear that the integrity of much of the sequence of IS1 is required for the activity of the element.

In order to characterize IS1-encoded gene products we have therefore attempted to clone each of the major reading frames, A, B, C, and D, in the plasmid expression vector, pPLC236 (54), under the control of the strong P_L promoter of bacteriophage λ. A detailed description of the construction of these clones will be presented elsewhere. Briefly, the A and D reading frames were separated from B and C by using the unique BstEII site of IS1 in the cloning. The 2 halves defined by the BstEII cleavage site were cloned separately in both orientations downstream from the P_L promoter. Initial results obtained using the minicell system have indicated that one of the clones (that which carries the A reading frame in the correct orientation for expression) specifies a protein species of approximately 10 kd as judged by SDS-PAGE (Fig. 10). The protein InsA has an expected molecular weight of 10.6 kd. Further characterization of this and other potential IS1 gene products is in progress. Another open reading frame (designated F in Ref. 52) is correctly positioned to have produced a protein from the above construct. This hypothetical protein has not been observed.

We wish to point out that the InsA protein carries a region which exhibits a significant amino acid homology with the region of

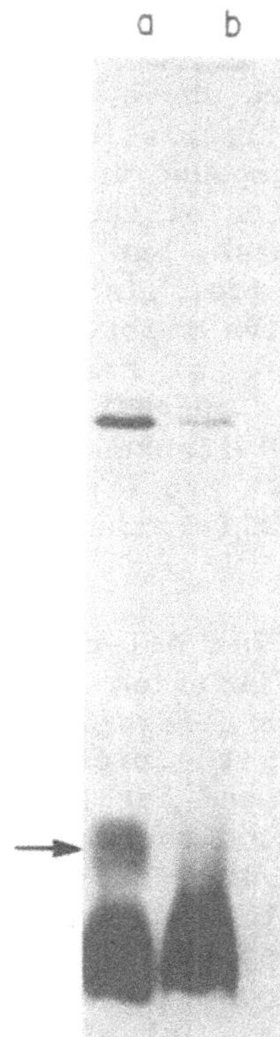

Fig. 10. Autoradiogram of proteins synthesized from a clone carrying reading frame A in the minicell system. The isolation and labeling of minicells is described in Ref. 58. Electrophoresis was on 18% polyacrylamide SDS gels (58) and the proteins were visualized by fluorography.

The construction of the plasmid pOC 7 is described briefly in the text and will be described in detail elsewhere. a) pOC; 7b) vector pPLC236. The small abundant protein observed in both channels in presumably rop (59), the synthesis of which is probably driven by the cloned P_L promoter.

proteins such as cI and cro of bacteriophage λ, and CAP which are thought to be involved in the recognition of specific DNA sequences (55,56). InsA is also highly basic, a characteristic of many DNA binding proteins.

CONCLUSIONS

In the results presented above, we have provided evidence that transcription, and perhaps translation, into either end of IS1 inhibits its activity. For compound transposons such as Tn9, this effect not only determines which of the flanking IS1 elements is more likely to generate cointegrates, but may determine how insertion into a target region occurs, i.e., in one orientation or the other, and it determines the relative importance of fusions and transpositions proper as products of the transposition process. The fact that this effect is so marked suggests that it may play an important role in the evolution of IS1-flanked compound transposons and, together with some of the other properties of IS1 discussed above, it has interesting implications for the formation and maintenance of such transposons.

The preference of IS1 for A+T-rich regions would tend to increase the probability of insertion of the element in promoter or terminator regions, which are generally both A+T-rich (38,39). Such a tendency is illustrated by the preferential insertion of IS1 upstream from the β-lactamase and tet promoters of pBR322. This behavior increases the likelihood that the element will insert between transcription units. The inhibitory effect of transcription into IS1 means that elements which have inserted into a transcription unit will subsequently transpose at a reduced frequency and would tend to favor further transposition of elements which had inserted within nontranscribed regions. Inhibition by translation would further increase this selective effect. Finally, the decrease in the frequency of transposition with increasing length would be expected to favor relatively small compound transposons. Thus, in concert, these properties would tend to favor the formation and dissemination of compound transposons in which the flanking elements "isolate" relatively small regions of DNA carrying integral transcription units. Our studies have also demonstrated a rather dramatic preference in the orientation of IS1 insertions into the major hot spot of pBR322, a phenomenon which has not previously been noted (e.g., Ref. 38). While the sense of the orientation can be determined by the donor molecule, it is evident that the influence of the donor alone is not sufficient. The target region must also influence this asymmetry. One way in which this could occur is by donor modulation of the activities of the extremities of IS1 coupled with a preference for initial cleavage of one strand in the target region. The identical orientation preferences we obtained using a pBR322 derivative carrying terminator that stops transcription from P1 indicates that transcription through the hot spot is not a determining factor (data not shown). Alternative possibilities are that the direction of replication in the target, sequence-dependent DNA conformation, or oriented recognition sequence determine the orientation.

We are currently studying the structural and coding organization of IS1. Thus far, we have shown that the terminal 76 act as a functional end in transposition. From other experiments we have evidence for the synthesis of at least one IS1-coded protein in mini-cells. The biological activity of this protein remains unknown. The coding organization of IS1 is unusual among known insertion sequences in being composed of 5 major, overlapping reading frames (Fig. 9a). Among these major, open reading frames are 3 that exhibit a strong statistical preference for pyrimidine/purine (Y/R) codon boundaries, and 2 that do not (Fig. 9b). One of those showing this Y/R preference is insA, coding for the protein that is apparently expressed by one of our recombinant plasmids. Though hardly conclusive, a comparison of the amino acid sequence of InsA protein with that of several known DNA-binding proteins does suggest that there may be a structural motif in InsA similar to that of the sequence-specific binding regions of these known proteins. In particular, model building analysis has revealed that the carboxyterminal end of the protein may form a characteristic α-helix-turn-α-helix in which one face of the carboxyterminal helix is predominantly hydrophobic while the other is composed of predominantly polar residues. It remains for future studies to define the nature and functions of the IS1-coded proteins that are involved in the transposition activities of this element.

ACKNOWLEDGMENTS

We would like to thank A. Epstein, D. Rifat, B. Marty, and S. Wilkins for expert technical assistance, and P. Gamas, D. Zerbib, L. Kozar, and S. Bass for work on the various aspects of the work reported here. We would also like to thank P. Prentki, O. Fayet, G. Churchward, and R. Alazard for constructive discussions. The work reported here was supported by NIH Grant GM19036 to D.G., by grant 3.169.0.81 to L. Caro and M.C. from the Swiss National Science Foundation and by an "Action Thématique Programée" from the C.N.R.S. (France) to M.C.

REFERENCES

1. MacHattie, L.A., and J.B. Jakowski (1977) Physical Structure and deletion properties of the chloramphenicol resistance transposon Tn9 in phage lambda. In DNA Insertion Elements, Plasmids and Episomes, A.I. Bukhari, J.A. Shapiro, and S.L. Adhya, eds. Cold Spring Harbor Laboratory, New York, pp. 219-228.
2. Nyman, K., K. Nakamura, H. Ohtsubo, and E. Ohtsubo (1981) Distribution of the insertion sequence IS1 in gram-negative bacteria. Nature 289:609-612.
3. Ptashne, K., and S.N. Cohen (1975) Occurrence of insertion sequence (IS) regions on plasmid deoxyribonucleic acid as direct and inverted nucleotide duplications. J. Bact. 122:776.

4. Kopecko, D.J., J. Brevet, and S.N. Cohen (1976) Involvement of multiple translocating DNA segments and recombinational hot-spots in the structural evolution of bacterial plasmids. J. Mol. Biol. 108:333-360.
5. Sharp, P.A., S.N. Cohen, and N. Davidson (1973) Electron microscope heteroduplex studies of sequence relations among plasmids of E. coli: Structure of drug resistance (R) factors and F factors. J. Mol. Biol. 75:235-255.
6. Rownd, R.H., and S. Mickel (1971) Dissociation and reassociation of RTF and r-determinants of the R factor NR1 in Proteus mirabilis. Nature 234:40-43.
7. Chandler, M., E. Boy de la Tour, D. Willems, and L. Caro (1979) Some properties of the chloramphenicol resistance transposon Tn9. Mol. Gen. Genet. 176:221-231.
8. Iida, S., C. Hanni, C. Echarti, and W. Arber (1981) Is the IS1-flanked r-determinant of the plasmid NR1 a transposon? J. Gen. Microb. 126:413-425.
9. Chandler, M., J. Sechaud, and L. Caro (1982) A mutant of the plasmid R100.1 capable of producing autonomous circular forms of its resistance determinant. Plasmid 7:251-262.
10. Grindley, N.D.F. (1978) IS1 Insertion generates a duplication of a nine base pair sequence at its target site. Cell 13:419-426.
11. Calos, M.P., L. Johnsrud and J.H. Miller (1978) DNA Sequence at the Integration Site of the Insertion Element IS1. Cell 13:411-418.
12. Iida, S., R. Marcoli, and T.A. Bickle (1981) Variant insertion element IS1 generates 8-base pair duplications of the target sequence. Nature 294:374-376.
13. Ohtsubo, E., M. Zenilman, and H. Ohtsubo (1980) Plasmids containing insertion sequences are potential transposons. Proc. Natl. Acad. Sci., USA 77:750-754.
14. Reif, H.J., and H. Saedler (1975) IS1 is involved in deletion formation in the gal region of E. coli K12. Mol. Gen. Genet. 137:17-28.
15. Cornelis, G., and H. Saedler (1980) Deletions and inversion induced by a resident IS1 of the lactose transposon Tn951. Mol. Gen. Genet. 178:376-374.
16. Kleckner, N. (1981) Transposable elements in procaryotes. Ann. Rev. Genet. 15:341-404.
17. Galas, D.J., and M. Chandler (1982) Structure and stability of Tn9-mediated cointegrates: Evidence for two pathways of transposition. J. Mol. Biol. 154:245-272.
18. Galas, D.J., and M. Chandler (1981) On the molecular mechanism of transposition. Proc. Natl. Acad. Sci., USA 78:4858.
19. Iida, S., J. Meyer, and W. Arber (1983) Procaryotic IS elements. In Movable Genetic Elements, J. Shapiro, ed. Academic Press, New York, pp. 159-221.
20. Chandler, M., M. Clerget, and D.J. Galas (1982) The transposition frequency of IS1-flanked transposons is a function of their size. J. Mol. Biol. 154:229-243.

21. Morisato, D., J.C. Way, H.-J. Kim, and N. Kleckner (1983) Tn10 transposase acts preferentially on nearby transposon ends in vivo. Cell 32:799-807.
22. Guyer, M.S., R.R. Reed, J.A. Steitz, and K.B. Low (1980) Identification of a Sex Factor Affinity Site in E. coli as γδ. Cold Spring Harbor Symp. Quant. Biol. 45:135-140.
23. Chandler, M., and D.J. Galas (1983) Cointegrate formation mediated by Tn9 II: Activity of IS1 is modulated by external DNA sequences. J. Mol. Biol. 170:61-91.
24. Stüber, D., and H. Bujard (1981) Organization of transcriptional signals in plasmids pBR322 and pACYC184. Proc. Natl. Acad. Sci., USA 78:167-171.
25. Machida, C., Y. Machida, H.-C. Wang, K. Ishizaki, and E. Ohtsubo (1983) Repression of cointegration ability of insertion element IS1 by transcriptional readthrough from flanking regions. Cell 34:135-142.
26. Sasakawa, C., J.B. Loew, L. McDivitt, and D.E. Berg (1982) Control of transposon Tn5 transposition in Escherichia coli. Proc. Natl. Acad. Sci., USA 79:7450-7454.
27. Chandler, M., E. Roulet, L. Silver, and L. Caro (1979) Tn10-mediated integration of the plasmid R100.1 in the bacterial chromosome: Inverse transposition. Mol. Gen. Genet. 173:23-30.
28. Reif, H.J. (1980) Genetic evidence for absence of transposition functions from the internal part of Tn981 a relative of Tn9. Mol. Gen. Genet. 177:667-674.
29. Rosner, J.L., and M.S. Guyer (1980) Transposition of IS1- bio-IS1 from a bacteriophage derivative carrying the IS1-cat-IS1 transposon (Tn9). Mol. Gen. Genet. 178:111-120.
30. Ohtsubo, E., and H. Ohtsubo (1978) Nucleotide sequence of an insertion sequence, IS1. Proc. Natl. Acad. Sci., USA 75:615-619.
31. Johnsrud, L. (1978) DNA sequence of IS1, a transposable element. Mol. Gen. Genet. 169:213-218.
32. Jaskunas, S.R., and M. Nomura (1977) Mapping of the ribosomal protein genes in E. coli by means of insertion mutations. In DNA Insertion Elements, Plasmids and Episomes, A.I. Bukhari, J.A. Shapiro, and S.L. Adhya, eds., Cold Spring Harbor Laboratory, New York, pp. 487-495.
33. Brewster, J.M., and E.M. Morgan (1981) Tn9 and IS1 inserts in a ribosomal ribonucleic acid operon of E. coli are incompletely polar. J. Bact. 148:897-903.
34. Grindley, N.F.D. (1983) Transposition of Tn3 and related transposons. Cell 32:3-5.
35. Wienert, T.A., N.A. Schaus, and N.D.F. Grindley (1983) Insertion sequence duplication in transpositional recombination. Science 222:755-765.
36. Hirschel, B.J., D.J. Galas, and M. Chandler (182) Cointegrate formation by Tn5, but not transposition, is dependent on recA. Proc. Natl. Acad. Sci., USA 79:4530-4532.

37. Berg, D.E. (1983) Structural requirements for IS30-mediated gene transposition. Proc. Natl. Acad. Sci., USA 80:792-796.
38. Galas, D.J., M.P. Calos, and J.H. Miller (1980) Sequence Analysis of Tn9 insertions in the lacZ gene. J. Mol. Biol. 144:19-41.
39. Meyer, J., S. Iida, and W. Arber (1980) Does the insertion element IS1 transpose preferentially into A+T rich DNA segments? Mol. Gen. Genet. 178:471-473.
40. Sutcliffe, J.G. (1978) Complete nucleotide sequence of the Escherichia coli plasmid pBR322. Cold Spring Harbor Symp. Quant. Biol. 43:77-90.
41. Peden, K.W.C. (1983) Revised sequence of the tetracycline resistance gene of pBR322. Gene 22:275-280.
42. Kline, R.D., E. Selsing, and R.D. Wells (1980) A rapid microscale technique for isolation of recombinant plasmid DNA suitable for framing. Plasmid 3:88-91.
43. Proctor, G.N., and R.H. Rownd (1982) Rosanilins: indicator dyes for cloramphenicol resistant enterobacteria containing chloramphenicol acetyltransferase. J. Bacter. 150:1375-1382.
44. Hyashi, K. (1980) A cloning vehicle suitable for strand separation. Gene 11:109-155.
45. Devos, R., R. Contreras, J. van Emmelo, and W. Fiers (1979) Identification of the translocatable element IS1 in a molecular chimera constructed with plasmid pBR322 DNA in which a bacteriophage MS2 DNA copy was inserted by the poly(dA)•Poly(dT) linker method. J. Mol. Biol. 128:621-632.
46. Arber, W., S. Iida, H. Jutte, P. Caspers, J. Meyer, and C. Hanni (1979) Rearrangements of genetic material in escherichia coli as observed in the bacteriophage P1 plasmid. Cold Spring Harbor Symp. Quant. Biol. 43:1197-1208
47. Prentki, P., and H.M. Kirsch (1982) A modified pBR322 vector with improved properties for the cloning, recovery, and sequencing of blunt-ended DNA fragments. Gene 17:189-196.
48. Rak, B., M. Lusky, and M. Hable (1982) Expression of two proteins from overlapping and oppositely oriented genes on transposable DNA insertion element IS5. Nature 297:124-128.
49. Klaer, R., S. Kuhn, H.-J. Fritz, E. Tillmann, I. Saint-Girions, P. Haberman, D. Pfeifer, and P. Starlinger (1980) The sequence of IS4. Cold Spring Harbor Symp. Quant. Biol. 45:215-224.
50. Ohtsubo, H., K. Nyman, W. Doroskiewicz, and E. Ohtsbuo (1981) Multiple copies of isoinsertion sequence IS1 in gram-negative bacteria. Nature 292:640-643.
51. Shepherd, J.C.W. (1981) Method to determine the reading frame of a protein from the purine pyrimidine genome sequence and its possible evolutionary justification. Proc. Natl. Acad. Sci., USA 78:1596-1600.
52. Galas, D.J., and T.F. Smith (1984) The relationship between codon boundaries and multiple reading-frame preferences: Coding organization of bacterial insertion sequences. Mol. Biol. Evol. 1:260-268.

53. Machida, Y., C. Machida, H. Ohtsubo, and E. Ohtsubo (1982) Factors determining frequency of plasmid cointegration mediated by insertion sequence IS1. Proc. Natl. Acad. Sci., USA 79:277-281.
54. Remaut, E., P. Stanssens, and W. Fiers (1981) Plasmid vectors for high efficiency expression controlled by the P1 promoter of coliphage lambda. Gene 15:81-93.
55. Anderson, W.F., Y. Takeda, D.H. Olendorf, and B.W. Matthews (1982) Proposed α-helical super-secondary structure associated with protein DNA recognition. J. Mol. Biol. 159:745-751.
56. Shaw, D.J., D.W. Rice, and J.R. Guest (1983) Homology between CAP and Fnr, a regulator of anaerobic respiration in *Escherichia coli*. J. Mol. Biol. 116:241-47.
57. Prentki, P., and H.M. Krisch (1984) *In vitro* insertional mutagenesis with a selectable DNA fragment. Gene (in press).
58. Roulet, E., B. Allet, and M. Chandler (1984) Cloning and expression of the Mu a gene. Gene (28:65-72).
59. Cesareni, G., M.A. Muesing, and B. Polisky (1982) Control of ColE1 DNA replication: The *rop* gene product negatively affects transcription from the replication primer promoter. Proc. Natl. Acad. Sci., USA 79:6313-6317.

ON THE TRANSPOSITION AND EVOLUTION OF Tn1721 AND ITS RELATIVES

Rüdiger Schmitt,[1] Sabine Mötsch,[1] Peter Rogowsky,[1]
Fernando de la Cruz,[2] and John Grinsted[3]

[1]Lehrstuhl für Genetik
Universität Regensburg
D-8400 Regensburg, Germany

INTRODUCTION

Transposition of Tn3-like elements involves cointegrate formation catalyzed by transposase and resolution of cointegrates catalyzed by resolvase (1).

The tetracycline-resistance transposon (tet) Tn1721 belongs to a group of Tn3-like elements also comprising Tn21 and Tn501 that encode interchangeable transposition functions but specify different resistances (2-4). We report here:

(i) on the arrangement of Tn1721-borne transposition genes (structure);

(ii) on transposase-mediated replicon fusion involving "single-ended" elements (mechanism); and

(iii) on the pedigree of Tn1721 and its relatives (evolution).

STRUCTURE OF Tn1721

Tn1721 consists of a minor transposon (Tn1722) encoding transposition functions and an amplifiable tet region encoding resistance

[2]Departamento de Bioquímica, Universidad de Santander, Santander, Spain; [3]Department of Microbiology, University of Bristol, Bristol BS8 1TD, United Kingdom.

(Ref. 2,5,6; Fig. 1). There are 3 38-base pair (bp) inverted repeats (IR) that allow transposition of either the whole element or the minor transposon alone. Transposition of Tn1721 generates 5-bp direct repeats at the target site (7). Amplification of the tet genes is strictly recA-dependent. Up to 9 tandem duplications of tet have been isolated and a positive gene dosage effect on tetracycline resistance has been shown (9).

Transposition genes are arranged in the order:

res (resolution site) - tnpR (resolvase) - tnpA (transposase)

with the latter genes having identical transcriptional polarity (from left to right; Ref. 10). Resolvase catalyzes the resolution of cointegrates by site-specific recombination at res. By binding to res that overlaps the tnpR promoter it represses its own transcription, but not that of the tnpA gene suggesting that they have independent promoters. When using transcriptional fusion to galactokinase, the tnpR promoter showed 40% of lac promoter efficiency, whereas the weak tnpA promoter was only detectable in the transposase-mediated transposition reaction. Transposase itself is temperature-sensitive; its activity being optimum at 30°C, but only one hundredth at 37°C and not detectable at 40°C (R. Schmitt, unpub. data).

Resolution of cointegrates depends on the presence of 2 res sites in 1 replicon having the same orientation. We have sequenced

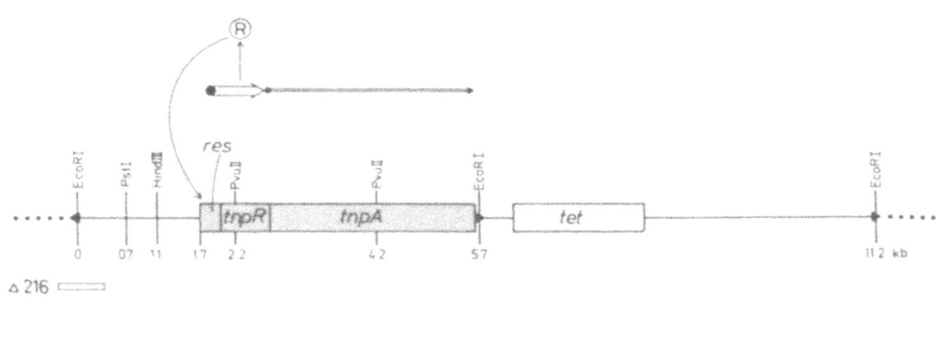

Fig. 1. Diagram of Tn1721 showing central and terminal 38-bp IR (◄►) flanking the minor transposon (Tn1722) and the repetitious tet region (tet: tetracycline resistance determinant). The homology with Tn501 (shaded) and the location of transposition genes (res: resolution site; tnpR: resolvase; tnpA: transposase), the extent of deletion Δ216 used to construct Tn1722Δ1K (13), relevant restriction sites and their coordinates are indicated. R denotes TnpR protein acting at res (9).

and mapped the crossover site (site I) within a stretch of 11 base pairs, 161-172 bp upstream of the tnpR gene (10). This crossover sequence bears almost no resemblance to the crossover site of Tn3. Footprinting experiments using purified resolvase suggest that there are 2 additional binding sites for resolvase located between the 3' end of tnpR and the crossover point that overlaps the -35 (site II) and -10 boxes (site III) of the tnpR promoter.

"SINGLE-ENDED" TRANSPOSITION

Transposons are characterized by inverted repeats at their ends shown to be essential for transposase-catalyzed transposition (11,12). We have recently reported that truncated derivatives of Tn1721 and Tn21 with a deletion of 1 IR ("single-ended") are capable of mediating transposition of resistance markers to target replicons, if an active tnpA gene (transposase) is present in cis or in trans (13,14). This process has a number of features in common with conventional transposition:

(i) it requires transposase;

(ii) it requires 1 intact IR;

(iii) insertions of donor into target DNA occur at random sites with a strong bias towards AT-rich sequences;

(iv) insertions generate 5-bp direct repeats of target DNA.

Restriction mapping and sequence analyses suggested to us that the resulting fusions between donor and target replicons may be generated by an asymmetric mode of replicative transposition. Accordingly, transposition:

(i) initiates at the intact repeat;

(ii) progresses by replication of the element and the donor vector; and

(iii) terminates at or near the duplicated single IR.

We have analyzed 15 random insertions of Tn1722Δ1K, a "single-ended" derivative of Tn1721 into R plasmid RSF1010 (Fig. 2A). It was found that the element mediates fusions of the donor (pJOE100::-Tn1722Δ1K) and target (RSF1010) replicons; the fused portions are nonpermuted (13). In each case, the 2 replicons were joined at the single IR. The second junction was located at or somewhat beyond the duplicated IR in 11 products (class A); 4 fusions ended at various sites within the donor plasmid (class B). Possibly, the latter arose by secondary deletions of class A fusions. The insertions are flanked by a 6-bp direct repeat in 1 and by 5-bp direct repeats of target DNA in 2 independent class A recombinants that have been se-

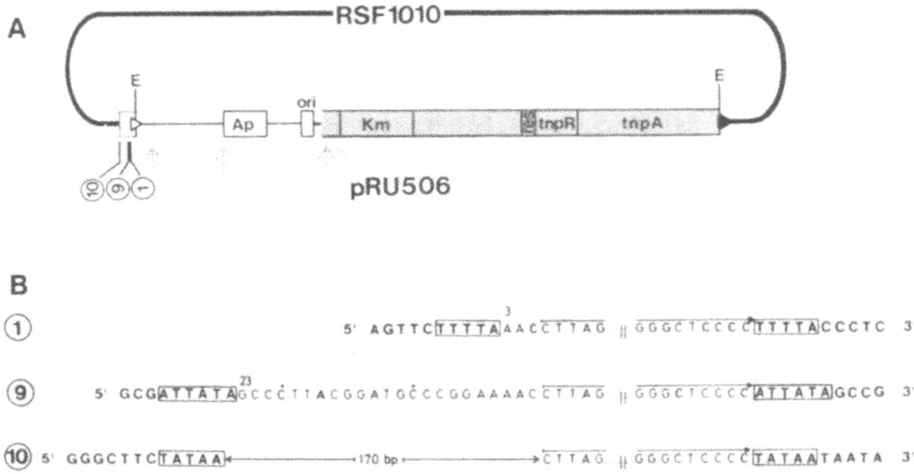

Fig. 2. A Replicon fusion of pRU506 (pJOE100::Tn1722Δ1K; Ref. 7) and RSF1010 mediated by the "single-ended" element Tn1722Δ1K (13). The latter is indicated by a shaded box; the intact IR (▶) and the duplicated IR (▷) plus their EcoRI sites (E), and relevant gene loci are shown (Ap: ampicillin resistance: Km: kanamycin resistance; ori: origin of replication). All fused replicons are joined at the intact IR; the locations of the second junction in 7 independent fusions are marked by dotted arrows and numbered lines, respectively.

B Sequencing from the EcoRI sites across the right-hand and left-hand junctions of 3 fusion products, marked by circled numbers in Fig. 2A. Normal and boldface lettering distinguish between donor and target DNA, respectively; 5-bp (and 6-bp) direct repeats are boxed and lines above sequences indicate the respective portions of intact IR (arrowhead) and duplicated IR (arrowtail; see Fig. 2A). The number of nucleotides of donor DNA extending beyond the duplicated IR is indicated in each sequence.

quenced so far. The target sites are highly AT-rich, a feature also observed in transposition of the wild-type element Tn1721 (R. Schmitt, unpub. data).

It is therefore reasonable to assume that replicon fusion mediated by transposase and a single IR reflects a process operative in normal transposition. The results are compatible with an asymmetric mode of transposition as proposed by Harshey and Bukhari (15) and Galas and Chandler (16). A model involving rolling circle-type replication is shown in Fig. 3. It assumes:

(i) initiation by nick-ligation at the intact repeat;

(ii) progression of the replication fork along the donor transposon; and

(iii) termination at a signal, ideally an intact IR, which, however, may be replaced by secondary signals like a direct repetition of the original IR sequence.

The latter sequences are always less efficient; they are thought to trigger termination by slowing down the progression of the enzyme complex mediating replicative transposition until it stops (imprecisely) at or beyond the duplicated IR. Replicon fusion is then completed by a second nick (involving the same strand that was initially cut) and subsequent ligation to the target strand. This process formally corresponds to direct transposition as opposed to cointegrate formation that requires cutting of the other strand (16). We suggest that cutting of the same strand is a consequence of the fact that the initiating and terminating IR sequences have the same orientation in "single-ended" transposition.

EVOLUTION

Tn21, Tn501, and Tn1721 belong to a subgroup of transposons more closely related to each other than to Tn3 (3). They are closely related by their transposition functions encoded by a 4-kb region of homology and by the sequence of their 38-bp terminal repeats, but they also specify different resistance determinants. Tn21 is 20 kb in size and specifies resistance to mercuric ions (Hg), streptomycin/spectinomycin (Sm/Sp), and sulfonamides (Su) (17). Tn501 is 8.2 kb and specifies Hg (18), and Tn1721 is 11.2 kb and specifies Tc (5). The structures of these elements are compared in Fig. 4.

The close structural similarities in this group of elements suggests that their transposition functions have descended from a common ancestral transposable element. The following discussion of transposon evolution will consider 2 aspects:

(i) the possible origin of such an ancestral element and

(ii) modes, by which other genes are acquired to generate the transposable elements that are seen today.

Origin of Transposable Elements

An essential feature of a transposable system is a protein (transposase) that recognizes a specific site (IR) and produces a specific nick at the site acting as an origin for replication. The occurrence of a second copy of the recognition sequence (IR) by dup-

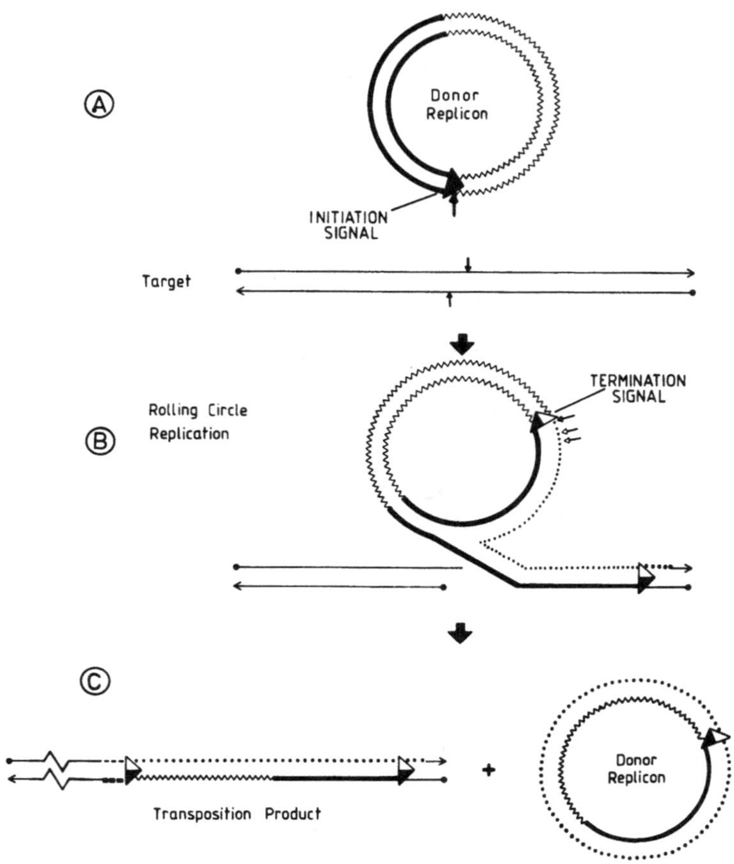

lication would generate a "basic transposon" composed of a *tnpA* gene flanked by copies of the IR.

The specific recombination system involved in transposition of Tn*3*-like elements is specified by *tnpR*/*res*. The *tnpR* genes that have been examined show similarities to genes that code for proteins involved in inversion of segments of DNA (4,9) such as the *hin* gene contained by an invertible element that controls phase variation in *Salmonella* (19). Such an element may have been incorporated into the "basic transposon." Initially, the segment of DNA retained its invertibility, i.e., it had a recognition site on each side. Deleting 1 of these flanking sites would lock the inversion gene into one orientation or the other, with respect to the *tnpA* gene. This ex-

Fig. 3. Model of asymmetric replicative transposition (15,16) adapted to "single-ended" replicon fusion (13,14).

A Transposition is initiated by nick-ligation at the single intact repeat (▶) and by 5-bp staggered nicks at the target site.

B Nick-ligation of single donor and target DNA strands and rolling circle type replication proceed along the transposable element (▬▬) and the donor vector (〰〰), until a termination signal [duplicated IR (▶) is encountered].

C Cutting of the same single donor DNA strand (arrows) and ligation to free end of target DNA. Nick-ligation involving the same strand bypasses cointegrate formation, unlike in the transposition of "2-ended" elements, where different DNA strands are nicked at the initiation and termination steps, respectively (16). We suggest that the orientation of the terminating IR sequence, whether direct or inverted, is responsible for directing the endonucleolytic activity (of transposase).

planation is in accord with the fact that the orientation of tnpR and res with respect to tnpA in Tn3 is opposite to that in Tn21, Tn501, and Tn1721 (Ref. 4,9; Fig. 4). The ancestral transposable element thus generated consists of (38-bp) inverted repeats flanking a central region that contains tnpA, tnpR, and res. Present day paradigms of such elements are γδ (Tn3 branch) and Tn1722 (Tn21-Tn501-Tn1721 branch).

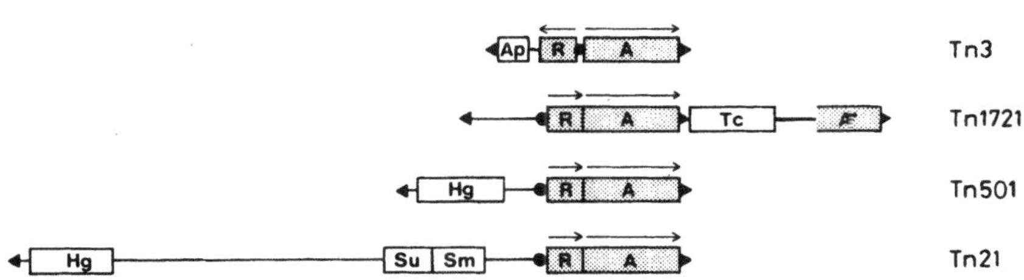

Fig. 4. Structure of Tn3-like transposons superimposed relative to their transposition genes (shaded). All members have related 38-bp ends (◀ ▶), resolvase (R), and transposase (A; A' = truncated). Their transcriptional polarities are indicated by arrows; res is marked by a dot (●). Approximate locations of resistance determinants are shown (resistances to ampicillin, Ap; mercuric ion, Hg; streptomycin, Sm; sulfonamide, Su; tetracycline, Tc).

Acquisition of Other Genes by Transposable Elements

The ancestral transposable element discussed above has acquired various other genes, such as antibiotic- and heavy metal-resistance determinants, during evolution. It is this acquisition that is the most immediately important, and which can be discussed with reasonable confidence. We identify 3 major ways in which the genes can be picked up. In general, subsequent deletions of various parts of the new elements are necessary to generate present day transposons.

(i) Acquisition by concerted action of 2 elements. The structure of Tn1721 originally suggested a general way in which transposons might pick up the DNA sequences (20). Thus, Tn1721 consists of a complete copy of a transposable element (Tn1722) and also a 1.8-kb fragment of the same element with the Tc determinant interspersed (see above). This suggests that, originally, 2 complete copies of a Tn1722-like progenitor inserted on each side of the Tc determinant (still present in RP4; Ref. 21) and that subsequent partial deletion of 1 element resulted in the structure of Tn1721. It is not known yet if deletion formation in this case involved specific sites.

(ii) Acquisition by transposition. Insertion of a pre-existing transposable element into another frequently results in the acquisition of new resistance genes by the latter. A good example is Tn4 (22,23). Here, Tn3 (Ap) is inserted into the Hg^r determinant of an element almost identical with Tn21. Thus, Tn4 is simply Tn21 with an intact Tn3 element in it. This type of reaction has also been implicated in the generation of Tn501 (24). This element contains 1 entire end of Tn2613 identical with this end of Tn21 (25). Thus, the resistance determinant of Tn501 originates from Tn2613. The latter or an element very like it is inserted into a progenitor of the Tn1722-type 80 bp from the end; a subsequent resolvase-catalyzed deletion generated Tn501 (Fig. 5B).

(iii) Acquisition by other mechanisms. Elements closely related to Tn21 such as Tn4, Tn2410, Tn2424, Tn2613, Tn4000, and BP201 (25,26) are clinically important, especially in hospitals where aminoglycosides are used. These elements are very flexible and appear to be able to acquire antibiotic resistance genes very easily. In addition to *tnp* genes, Tn21-like elements usually contain Hg and Su/Sm determinants (Fig. 4). The variation resulting in the different elements frequently occurs near the latter; it is as though there is a "hot spot" into which other genes can be incorporated (26). Compared with Tn21, various elements differ by insertion of segments of DNA into 2 discrete regions on each side of the *aadA* gene (Sm). The latter may once have inserted into a single "hot spot," such that it is now divided into 2. The mode of integration into these sites may be by homologous recombination (F. Schmidt, pers. comm.).

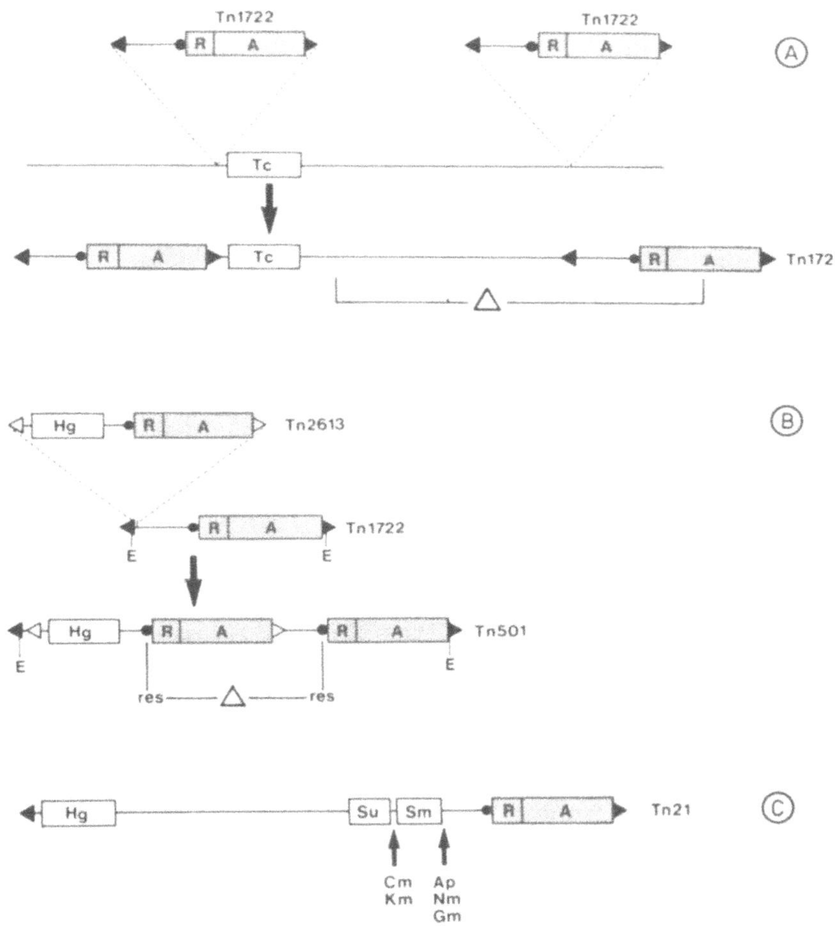

Fig. 5. Modes of acquisition of resistance determinants by transposable elements.

A Concerted action of 2 elements (Tn1722) and subsequent deletion (Δ) generates Tn1721 (6).

B Insertion of the pre-existing Hg transposon Tn2613 (25) into Tn1722; resolvase-catalyzed interaction of the 2 res loci (•) deletes intervening DNA and results in Tn501 (10,24).

C Insertion of various resistance determinants (chloramphenicol, Cm; neomycin, Nm; gentamycin, Gm) into "hot spots" flanking the Sm determinant of Tn21 (26).

Tab. 1. Homologies between transposition genes with reference to Tn501.

Transposon	TnpR[1]	TnpA[1]	IR_R[2]
Tn501	100% (4)	100%[3]	38/38 (100%)
Tn1721	100%[4]	≤100%[5]	36/38 (95%) (7)
Tn21	79% (4)	n.d.	27/38 (71%) (3)
Tn3	32%	30%	19/38 (50%)

[1]Deduced aminoacid sequences.
[2]Identical base pairs of right-hand IRs (Fig. 4).
[3]N. Brown, personal communication.
[4]P. Rogowsky and R. Schmitt, unpublished data.
[5]Inferred from hybridization data (2,20).

Relationships Between Tn3-like Elements

The genes (tnpA, tnpR) and sites (IR, res) involved in transposition are the essential features of a transposable element and were thus used as appropriate indicators for defining evolutionary relationships. In this context the antibiotic resistance genes are less important and can be exchanged relatively easily, as discussed above. Our knowledge of relationships between the Tn3-like elements

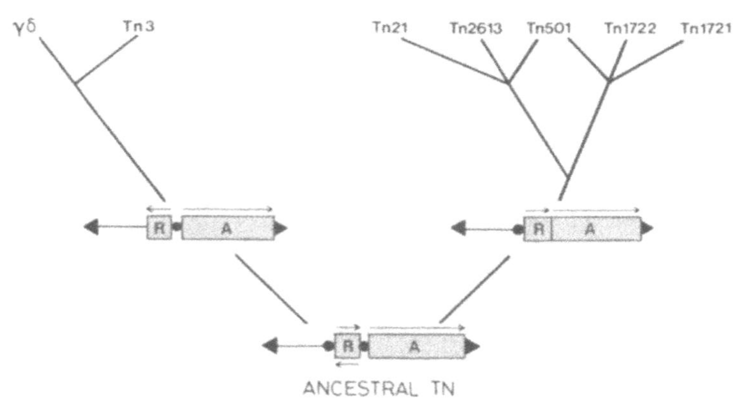

Fig. 6. Pedigree of the Tn3 family derived from restriction, sequence, and complementation data. The 2 main branches diverged early by locking the tnpR-res segment in one or the other orientation relative to tnpA. In each line, diversification occurred by the acquisition of various resistance genes as illustrated in Fig. 5.

depends primarily on the known physical maps and sequences of various parts of the elements (Tab. 1). In some instances, too, complementation of tnpA and tnpR functions can be used.

There are 2 main groups of such elements that are well characterized: there is Tn3 itself and γδ on the one hand, and Tn21, Tn501, Tn1721, Tn1722, and Tn2613 on the other. These 2 branches are defined by sequence relationships and by complementation of tnp functions. Within the Tn21/Tn1721 branch the transposition functions of Tn501 and Tn1721 are interchangeable, while those of Tn21 are related but different (3). Thus, the line Tn3/γδ diverged from the other a long time ago; Tn21 branched from Tn501/Tn1721 before these 2 diverged from each other. Tn501 represents a special case, since it evolved by the convergence of 2 lines (Tn1722, Tn2613). These relationships are shown in Fig. 6.

ACKNOWLEDGMENTS

We thank Mrs. Helene Beier for excellent illustrations. This work was supported by grants from the Deutsche Forschungsgemeinschaft and the Fonds der Chemie to R.S., from the Fundación March to F.C., and from the Medical Research Council to J.G.

REFERENCES

1. Heffron, F. (1983) Tn3 and its relatives. In Mobile Genetic Elements, J.A. Shapiro, ed. Academic Press, New York. p. 223.
2. Schmitt, R., J. Altenbuchner, and J. Grinsted (1981) Complementation of transposition functions encoded by transposons Tn501(Hg^R) and Tn1721(Tc^R). In Molecular Biology, Pathogenicity, and Ecology of Bacterial Plasmids, S.B. Levy, R.C. Clowes, and E.L. Koenig, eds. Plenum, New York, pp. 359-370.
3. Grinsted, J., F. de la Cruz, J. Altenbuchner, and R. Schmitt (1982) Complementation of transposition of tnpA mutants of Tn3, Tn21, Tn501 and Tn1721. Plasmid 8:276-286.
4. Diver, W.P., J. Grinsted, D.C. Fritzinger, N.L. Brown, J. Altenbuchner, P. Rogowsky, and R. Schmitt (1983) DNA sequences of and complementation by the tnpR genes of Tn21, Tn501 and Tn1721. Mol. Gen. Genet. 181:189-193.
5. Schmitt, R., E. Bernhard, and R. Mattes (1979) Characterisation of Tn1721, a new transposon containing tetracycline genes capable of amplification. Mol. Gen Genet. 172:53-65.
6. Schmitt, R., J. Altenbuchner, K. Wiebauer, W. Arnold, A. Pühler, and F. Schöffl (1981) Basis of transposition and gene amplification by Tn1721 and related tetracycline-resistance transposons. Cold Spring Harbor Symp. Quant. Biol. 45:59-65.
7. Schöffl, F., W. Arnold, A. Pühler, J. Altenbuchner, and R. Schmitt (1981) The tetracycline resistance transposons Tn1721 and Tn1771 have three 38-base pair repeats and generate five-base pair direct repeats. Mol. Gen. Genet. 181:87-94.

8. Wiebauer, K., S. Schraml, S.W. Shales, and R. Schmitt (1981) Tetracylcine resistance transposon Tn1721: recA-dependent gene amplification and expression of tetracycline resistance. J. Bact. 147:851-859.
9. Altenbuchner, J., and R. Schmitt (1983) Transposon Tn1721: Site-specific recombination generates deletions and inversions. Mol. Gen. Genet. 190:300-308.
10. Rogowsky, P., and R. Schmitt (1984) Resolution of a hybrid cointegrate between transposons Tn501 and Tn1721 defines the recombination site. Mol. Gen. Genet. 193:162-166.
11. Heffron, F., B.J. McCarthy, H. Ohtsubo, and E. Ohtsubo (1979) DNA sequence analysis of the transposon Tn3: Three genes and three sites involved in transposition of Tn3. Cell 18:1153-1164.
12. Foster, T.J., M.A. Davis, D.E. Roberts, K. Takeshita, and N. Kleckner (1981) Genetic organization of transposon Tn10. Cell 23:201-213.
13. Mötsch, S., and R. Schmitt (1984) Replicon fusion mediated by a single-ended derivative of transposon Tn1721. Mol. Gen. Genet. (in press).
14. Avila, P., F. de la Cruz, E. Ward, and J. Grinsted (1984) Plasmids containing one inverted repeat of Tn21 can fuse with other plasmids in the presence of Tn21 transposase. Mol. Gen. Genet. (in press).
15. Harshey, R.M., and A.J. Bukhari (1981) A mechanism of DNA transposition. Proc. Natl. Acad. Sci., USA 78:1090-1094.
16. Galas, D.J., and M. Chandler (1981) On the molecular mechanism of transposition. Proc. Natl. Acad. Sci., USA 78:4858-4862.
17. de la Cruz, F., and J. Grinsted (1982) Genetic and molecular characterisation of Tn21, a multiple resistance transposon from R100.1. J. Bact. 151:222-228.
18. Bennett, P.M., J. Grinsted, C.-L. Choi, and M.H. Richmond (1978) Characterization of Tn501, a transposon determining resistance to mercuric ions. Mol. Gen. Genet. 159:101-106.
19. Zieg, J., and M. Simon (1980) Analysis of the nucleotide sequence of an invertible controlling element. Proc. Natl. Acad. Sci., USA 77:4196-4200.
20. Altenbuchner, J., C.-L. Choi, J. Grinsted, R. Schmitt, and M.H. Richmond (1981) The transposons Tn501 (Hg) and Tn1721 (Tc) are related. Genet. Res., Camb. 37:285-289.
21. Waters, S., P. Rogowsky, J. Grinsted, J. Altenbuchner, and R. Schmitt (1983) The tetracycline resistance determinants of RP1 and Tn1721: Nucleotide sequence analysis. Nucl. Acids Res. 11:6089-6105.
22. Kopecko, D.J., J. Bravet, and S.N. Cohen (1976) Involvement of multiple translocating DNA sequences and recombinational hot spots in the structural evolution of bacterial plasmids. J. Mol. Biol. 108:333-360.
23. Foster, T.J., H. Nakahara, A.A. Weiss, and S. Silver (1979) Transposon A-generated mutations in the mercuric resistance genes of plasmid R100.1. J. Bact. 140:176-181.

24. Grinsted, J., and N.L. Brown (1984) Tn501 contains a terminus of Tn21: Consequences for evolution and for complementation. Mol. Gen. Genet. (submitted for publication).
25. Tanaka, M., T. Yamamoto, and T. Sawai (1983) Evclution of complex resistance transposons from an ancestral mercury transposon. J. Bact. 153:1432-1438.
26. Schmidt, F. (1984) The role of insertions, deletions, and substitutions in the evolution of R6 related plasmids encoding aminoglycosides transferase ANT-(2''). Mol. Gen. Genet. (in press).

REPEATED DNA SEQUENCES RECOMBINE 1,000 TIMES MORE FREQUENTLY IN A PLASMID THAN IN THE CHROMOSOME OF *BACILLUS SUBTILIS*

L. Jannière, B. Niaudet,[1] and S.D. Ehrlich

Institut Jacques Monod, C.N.R.S.
Université Paris VII
2, Place Jussieu
75251 Paris Cedex 05

INTRODUCTION

Escherichia coli chromosome and small ColE1-like plasmids recombine via different pathways. The chromosome follows mainly the RecBC pathway, engages in the RecF pathway only in the absence of exonuclease I and requires the RecA protein to follow the RecE pathway (1,2). The plasmids, on the contrary, recombine predominantly by the RecF and not the RecBC pathway, and engage very efficiently in the RecE pathway (opened by the *sbcA* mutation) in the absence of the RecA protein (3-6).

Plasmids and chromosomes appear to follow different recombination pathways in *Bacillus subtilis* as well, as indicated by the fact that (a) 3 out of 12 tested *rec* genes had different effects on plasmid and chromosome recombination (7); (b) *recB* and *recD* mutations inhibited recombination of transforming DNA with the plasmid more than with the chromosome (8); (c) *recE* mutation had the opposite effect (8-10).

In the present study, we compared recombination frequency between 2 homologous sequences 3.7-3.8 kb long, carried either on a small plasmid or on the chromosome of *B. subtilis*. We found that the duplication in the plasmid recombined 10^3 times more often than the duplication in the chromosome. The difference was not due to the higher plasmid copy number (about 35 in our study), since the chromosomal sequences tandemly repeated 30 times still recombined

---

[1] Present address: Institut Pasteur, 28, rue du Dr. Roux, 75724 Paris Cedex 15.

much less often than the plasmid sequences. These results support the hypothesis that *B. subtilis* plasmids and the chromosome recombine in a different way.

APPROACH

To measure recombination frequencies we decided to construct on a *B. subtilis* genome (plasmid, chromosome) a structure composed of a kanamycin resistance (Km^R) gene flanked by 2 copies of pBR322-derived sequences having the same orientation. Recombination between the repeated sequences should give rise to a circular pBR322-derived molecule carrying the Km^R gene, unable to replicate in *B. subtilis* and a replicative genome devoid of the Km^R gene. Recombination frequency could then be determined either by following the rate of appearance of the kanamycin-sensitive cells, or by comparing the amounts of the nonreplicative recombinant and the replicative parental Km^R-encoding molecules. The structure needed to measure recombination frequencies was created by a Campbell-like recombination between a transforming circular pBR322 molecule carrying the Km^R gene and a genome (plasmid, chromosome) resident in *B. subtilis* recipient cells, carrying sequences derived from pBR322. The appropriate resident genomes were constructed and recombined with the transforming plasmid as described below.

CONSTRUCTIONS

The first step involved constructing a plasmid composed of pBR322 and the Km^R gene of pUB110, unable to replicate in *B. subtilis*. This plasmid was named pHV429 (Fig. 1). The second step consisted of obtaining 2 very similar plasmids, one able to replicate in *E. coli* and *B. subtilis*, the other in *E. coli* only. This was achieved by cleaving the plasmid pHV33 (a hybrid between pBR322 and pC194, Ref. 11) by *Bam*HI, and treating it then with BAL31 and T4 ligase (Fig. 1). A plasmid named pHV33Δ1 able to replicate in *E. coli* and *B. subtilis* and a plasmid named pHV33Δ81, able to replicate in *E. coli* only were thus obtained (another plasmid pHV33Δ11, replicating in both hosts was also obtained). pHV33Δ1, carries 400 bp more of pC194 sequences than pHV33Δ81, part of which is necessary for its replication in *B. subtilis* (12). The latter plasmid carries about 100 bp more of pBR322 sequences than the former.

To create a structure which could be used to measure recombination frequency in a plasmid we transformed *B. subtilis* competent cells harboring pHV33Δ1 (or pHV33Δ11) to kanamycin resistance with pHV429. The latter plasmid, which cannot replicate in *B. subtilis*, must recombine with the homologous sequences (pBR322) of the resident plasmid in order to transform. This recombination gives rise to a plasmid pHV467 (or pHV468, Fig. 1) in which the Km^R gene is flanked by directly repeated pBR322-derived sequences.

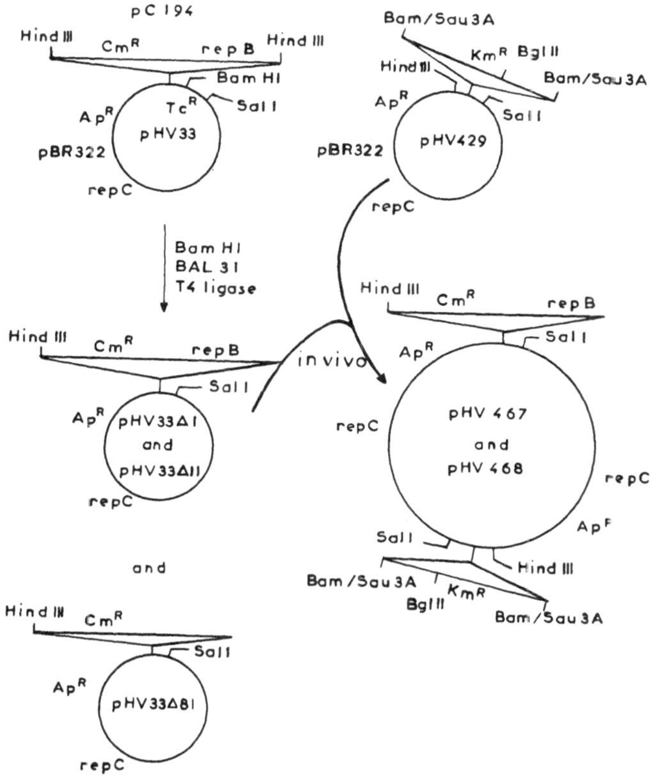

Fig. 1. Construction of plasmids carrying repeated pBR322 sequences.

To create a similar structure in B. subtilis chromosome we followed the procedure outlined in Fig. 2, which consists of 3 steps. First, 2 segments of the B. subtilis chromosome, one carrying the thyB gene, the other of unknown genetic content (labeled X, Fig. 2) were excised from pHV438 (13) and inserted in pHV33Δ81. A plasmid named pHV452 was thus obtained. Second, pHV452 cleaved with BglII was used to transform B. subtilis competent cells to chloramphenicol resistance. This plasmid, unable to replicate in B. subtilis, must recombine with the homologous chromosomal sequences in order to transform. The recombination event replaces wild-type chromosomal sequences between thyB and X with pHV33Δ81 sequences (13) and creates the chromosome named α (Fig. 2). Third, competent cells harboring the α chromosome were transformed to kanamycin resistance with pHV429. Recombination of this plasmid with homologous sequences of the α chromosome (pBR322) gives rise to a chromosome labeled β (Fig. 2) in which the Km^R gene is flanked by directly repeated pBR322 sequences.

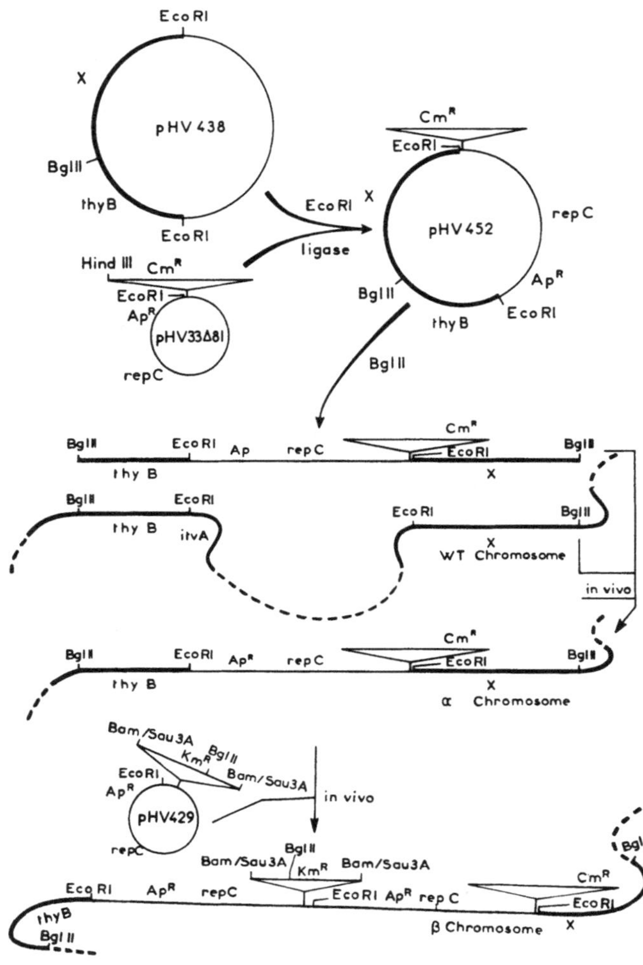

Fig. 2. Construction of a chromosome carrying repeated pBR322 sequences. Thin and thick lines represent sequences of plasmid and chromosome origin; interrupted line, chromosomal sequences replaced by plasmid sequences.

All the constructions made corresponded to those predicted, as judged by analysis of the plasmids (size, restriction patterns, genetic markers) and chromosomes (Southern hybridization, recloning in *E. coli* of the inserted plasmids and the flanking sequences, and analysis of their size, restriction pattern, and genetic markers).

RECOMBINATION FREQUENCIES

These were first determined by measuring the rate of appearance of cells sensitive to kanamycin (Km^S). In 12 independent clones harboring plasmid pHV467 (or pHV468) this rate was constant during at least 60 generations of growth and equaled 0.17 per cell generation. In the case of cells harboring β chromosome the rate was only $1.2.10^{-4}$ per generation. Since the Km^S cells appear as a consequence of recombination between the repeated pBR322-derived sequences, these results indicate that the recombination frequency in plasmids is 10^3 times higher than in the chromosome, provided that the segregation of the genomes devoid of the Km^R gene from those carrying it, is fast.

The plasmid recombination frequency was next determined by measuring the ratio of the replicating plasmid pHV467, carrying the repeated sequences and the nonreplicating plasmid pHV429, produced from pHV467 by recombination between the repeated sequences. This was achieved by analyzing in 2 ways the plasmid DNA extracted from cells harboring pHV467. (a) The DNA was used to transform _E. coli_ to kanamycin resistance. Ninety-six percent of the transformants harbored the parental plasmid pHV467, 4% the plasmid pHV429 generated by recombination; (b) the DNA was cleaved with _BglII_, and after gel electrophoresis hybridized with a probe specific for the Km^R gene (labeled pUB110 was used). Ninety percent of hybridizing DNA corresponded to pHV467, 10% to pHV429.

Nonreplicating plasmid pHV429 could only have been produced by recombination between repeats carried on pHV467. The ratio of the 2 plasmids can therefore be used as a measure of recombination frequency between the repeats. The values observed 0.04 and 0.1 should correspond to recombination frequency per cell generation. The difference between the 2 values may be due to integration of newly formed pHV429 into a pHV467 molecule present within the same cell. This would give rise to a plasmid having the structure pHV33Δ1 $(pHV429)_2$, which would not be distinguished from pHV467 (having the structure pHV33Δ1-pHV429, Fig. 1) by our transformation analysis, but would give rise to monomeric pHV429 in the restriction/hybridization analysis, since only pHV429 carries a _BglII_ site (Fig. 1). Both values were, nevertheless, close to that obtained by following the rate of appearance of Km^S cells (0.17/generation, see above), which indicates that plasmids devoid of the Km^R gene segregated rapidly from plasmids carrying that gene.

Free pHV429 was not detected by hybridization in the total DNA extracted from cells harboring the β chromosome. This indicates that the recombination frequency between the repeated sequences present in the chromosome was below 10^{-3}/generation, the limit of detection of our analysis.

Genetic and biochemical analysis gave values of about 10^{-1}/generation for the recombination frequency of plasmid-borne sequences. Genetic analysis (the only one having sufficient sensitivity) gave a value of 10^{-4}/generation for chromosome-borne sequences. Recombination in plasmids is therefore 10^3 times more frequent than in the chromosome. Could this difference be due to the presence of a higher number of copies of the plasmids than of the chromosome in the cell? To answer this question, we amplified the repeated structure carried by the chromosome and determined recombination frequency within the amplified structure.

AMPLIFICATION

Gene amplification has been well documented in prokaryotes (14-19). It occurs spontaneously in a number of bacteria provided that they carry in their genome an amplifiable structure of a general form RMR, where R and M stand for different DNA sequences (repeat and marker, respectively). Cells carrying the amplified structure $R(MR)_n$, where n represents the number of repeats of the amplification unit MR, can be isolated when their phenotype differs from that of the parental cells in which amplification did not occur.

The amplifiable structure RMR is present on the β chromosome, where pBR322 sequences and the Km^R gene stand for R and M, respectively (Fig. 2). Cells harboring such a chromosome were resistant to 1 μg/ml of kanamycin. Cells resistant to higher doses of the drug (2-10 μg/ml) appeared with frequencies of 10^{-4}-10^{-6}. They were enriched by growth in appropriate liquid medium, their DNA was extracted and analyzed by restriction enzyme cleavage and gel electrophoresis. Amplification of pBR322 sequences and the Km^R gene (the amplification unit) was detected in all cases, as judged by the appearance of visible restriction bands of the expected size over the background of the chromosomal restriction pattern. The degree of amplification, estimated by densitometry (and in some cases also by hybridization with an appropriate probe) is shown in Tab. 1. It increased with the kanamycin resistance up to 5 μg/ml of antibiotic and then leveled off, probably because of accumulation of mutations rendering the cells less permeable to aminoglycosides. The final value, about 30, is close to that reported for the pC194 copy number, 35 (19a).

RECOMBINATION WITHIN AMPLIFIED STRUCTURE

To measure the recombination frequency within the amplified structure we propagated the cells harboring 30 copies of the amplification unit (which is equivalent to pHV429, Fig. 1) for 145 generations in a medium devoid of antibiotic and (a) determined their plating efficiency on a medium containing 7.5 μg/ml of kanamycin; (b) extracted their DNA and measured the number of copies of the amplification unit.

Plating efficiency of the cells was equal to 1.0. Cells harboring 15 copies of the amplification unit plate on the same medium with an efficiency of 10^{-3}. Assuming that the precision of this

Tab. 1. Amplification of the Km^R gene in B. subtilis cells.

Resistance to kanamycine (µg/ml)	Degree of amplification
1	1
2	9
2.5	15
3	20
4	26
5	30
8	31
10	31

assay was such that we could not distinguish plating efficiency of 0.9 from that of 1.0, the frequency of the loss of 15 copies from the amplified structure could have been 7.10^{-4} per generation.

The copy number of the amplification unit, determined by analysis of DNA extracted from the same cells, was 28, which is probably not significantly different from the starting value of 30. Assuming that the difference was significant, the frequency of the loss of 2 copies of the amplification unit would be 5.10^{-4}/generation.

We also attempted to measure recombination frequency by determining the amount of free nonreplicative plasmid in the cells harboring the amplified structure. For this purpose, DNA from such cells was analyzed by Southern hybridization, using a probe specific for the Km^R gene. No free plasmid was seen under conditions in which about .03 copies/chromosome would have been detected, as judged from a control experiment in which a low amount of pHV429 was deposited on the gel. Recombination frequency was therefore lower than about 10^{-3}/generation.

None of the 3 methods used allowed us to measure with precision the recombination frequency within the chromosomally located amplified structure. The values obtained, ranging from 5.10^{-4} to 10^{-3}/generation are very likely too high, but are nevertheless close to that observed for a single duplication ($1.2.10^{-4}$/generation) and much lower than 10^{-1}/generation observed for plasmids pHV467 and pHV468.

DISCUSSION

Recombination between repeated pBR322 sequences carried on plasmid pC194 occurs 10^3 times more often than between the same sequences carried in the *thyB* region of *B. subtilis* chromosome. This observation raises several questions.

(A) Do plasmid-borne sequences generally recombine very frequently? It was previously observed that different pUB110-derived repeated sequences, replicated by pUB110 replicator, indeed recombine very frequently (20). In addition, pBR322 sequences carried on pE194 also appear to recombine frequently (unpub. results). Further experiments will, however, be needed to confirm and extend these observations.

(B) Do chromosomally located sequences recombine in general infrequently? We observed no recombination within a structure containing 20 copies of an 11-kb amplification unit, composed of the *thyB*, X, pBR322, and Km^R sequences, in experiments which should have detected events occurring with a frequency of 7.10^{-4}/generation. On the other hand, repeated and amplified sequences in the *pheA* (21) and *amy* (A. Albertini and S. Galizzi, pers. comm.) region of the chromosome recombined with an easily observable frequency. These results suggest that some sequences and/or chromosomal regions may be more prone to recombination than others.

(C) What is the molecular basis for the observed difference in recombination frequency? Several hypotheses may be considered; (i) Duplicated sequences may recombine by an intermolecular rather than intramolecular process, which could possibly occur more frequently between plasmids (22) than between chromosomes. This is ruled out by the fact that the *recE4* mutation, which reduces intermolecular plasmid recombination about 100 times (10,23) without affecting intramolecular recombination (24,25) did not change recombination frequency between plasmid-carried repeated sequences studied here; (ii) Recombination stimulating sequences (chi-like elements, Ref. 26) could be carried on plasmids and be absent from the *thyB* chromosomal region. They should be present not only within 400 bp of pC194 by which autonomously replicating pHV33Δ81 differs from nonreplicating pHV33Δ81 (Fig. 1 and 2), but also on 2 other plasmids, pUB110 and pE194 (see above, A), and should exert an effect on recombination much greater than that (5-8 times), previously described (26,27). We cannot rule out this hypothesis at present, but consider it unlikely; (iii) Recombination stimulating functions, such as *oriT* of plasmid F (28), may be present on plasmids and not on the chromosome; (iv) The proximity of a replication region may stimulate recombination of repeated sequences. This is supported by the preliminary evidence that initiation of replication in the chromosome induces recombination in its vicinity; (v) Plasmid structure (i.e., degree of supercoiling) may stimulate recombination. Any of the last 3 hypotheses could account for our results and for the fact that plasmids and the chromosome do not follow the same recombination pathways (i.e., do not use with the same efficiencies the enzymes capable of carrying out the process of recombination,

Ref. 1-10). Further experiments will be needed to support one over the others.

It is tempting to speculate about the possible biological significance of the observed difference in recombination rates. Frequent recombination may allow rapid evolution by generating novel genetic assemblies. The cost of this velocity, engendered by the loss of unsuccessful variants, may be acceptable in a polyploid system composed of small genetic units (the energy invested in synthesizing a nonviable copy of a 3-kb plasmid pC194 may not be too high) but intolerable in a predominantly haploid system containing a large genetic unit (the energy invested in creating not only the 4.10^3-kb chromosome of *B. subtilis* but also the totality of cell components must be very many orders of magnitude higher). Mechanisms ensuring high plasmid and low chromosomal recombination may thus have an evolutionary justification.

Finally, it may be worthwhile to draw attention to the fact that the foreign genetic information is likely to be maintained more stably in an infrequently recombining chromosome than in a frequently recombining plasmid. This consideration may be of importance in devising strategies to construct genetically modified organisms useful to humankind.

ACKNOWLEDGMENTS

We thank P. Noirot and D. Pujante for providing unpublished data and E. Pierre for her excellent technical help. This work was supported, in part, by a grant from the Centre National de la Recherche Scientifique (A.T.P. Biotechnologie).

REFERENCES

1. Fishel, R.A., A.A. James, and R. Kolodner (1981) *recA* independent general genetic recombination of plasmids. *Nature* 294:184-186.
2. Laban, D., and A. Cohen (1981) Interplasmidic and intraplasmidic recombination in *Escherichia coli* K12. *Mol. Gen. Genet.* 184:200-207.
3. James, A.J., P.T. Morrison, and R. Kolodner (1982) Genetic recombination of bacterial plasmid DNA. Analysis of the effect of recombination deficient mutations on plasmid recombination. *J. Mol. Biol.* 160:411-430.
4. Potter, M., and D. Dressler, (1979) DNA recombination: *In vivo* and *in vitro* studies. *Cold Spring Harbor Symp. Quant. Biol.* 43:969-985.
5. Clark, A.J. (1973) Recombination deficient mutants of *E. coli* and other bacteria. *Ann. Rev. Genet.* 7:67-86.

6. Clark, A.J. (1980) A view of the RecBC and RecF pathways of E. coli recombination. In Mechanistic Studies of DNA Replication and Genetic Recombination, B. Alberts and C.F. Fox, eds. Academic Press, New York, pp. 891-899.
7. Prozorov, A.A., V.I. Bashkirov, N.M. Lakomova, and N.N. Surikov (1982) Study of the phenomenon of marker rescue in plasmid transformation and transduction of intact cells, protoplasts and different rec mutants of Bacillus subtilis. Mol. Gen. Genet. 185:363-364.
8. Weinrauch, Y., and D. Dubnau (1983) Plasmid marker rescue transformation in Bacillus subtilis. J. Bact. 154:1077-1087.
9. Dubnau, D., R. Davidoff-Abelson, B. Scher, and C. Cirigliano (1973) Fate of transforming deoxyribonucleic acid after uptake by competent Bacillus subtilis: Phenotypic characterization of radiation-sensitive recombination-deficient mutants. J. Bact. 114:273-286.
10. Michel, B., B. Niaudet, and S.D. Ehrlich (1983) Intermolecular recombination during transformation of Bacillus subtilis competent cells by monomeric and dimeric plasmids. Plasmid 10:1-10.
11. Primrose, S.B., and S.D. Ehrlich (1981) Isolation of plasmid deletion mutants and study of their instability. Plasmid 6:193-201.
12. Dagert, M., I.M. Jones, A. Goze, S. Romac, B. Niaudet, and S.D. Ehrlich (1984) Replication functions of pC194 are necessary for efficient plasmid transduction by phage M13. EMBO J. 3:81-86.
13. Niaudet, B., A. Goze, and S.D. Ehrlich (1982) Insertional mutagenesis in Bacillus subtilis: Mechanism and use in gene cloning. Gene 19:277-284.
14. Rownd, R.H., and S. Mickel (1971) Dissociation of reassociation of RTF and r determinants of the R-factor NR1 in Proteus mirabilis. Nature New Biology 234:40-43.
15. Anderson, R.D., and J.R. Roth (1977) Tandem genetic duplications in phage and bacteria. Ann. Rev. Microbiol. 31:474-505.
16. Clewell, D.B., V. Yagi, and B. Bauer (1975) Plasmid-determined tetracycline resistance in Streptococcus faecalis: Evidence for gene amplification during growth in presence of tetracyline. Proc. Natl. Acad. Sci., USA 72:1720-1724.
17. Mattes, R., M.J. Burkardt, and R. Schmitt (1979) Repetition of tetracycline resistance determinant genes on R plasmids pRSD1 in Escherichia coli. Mol. Gen. Genet. 168:173-184.
18. Chandler, M., E. Boy de la Tour, D. Williams, and L. Caro (1979) Some properties of the chloramphenicol resistance transposon Tn9. Mol. Gen. Genet. 176:221-231.
19. Peterson, B., and R.H. Rownd (1983) Homologous sequences other than insertion elements can serve as recombination sites in plasmid drug resistance gene amplification. J. Bact. 156:177-185.
19a. Shivakumar, A.G., and D. Dubnau (1978) Plasmid replication in dnats mutants of Bacillus subtilis. Plasmid 1:405-416.
20. Michel, B. (1981) Recombination génétique homologue chez Bacillus subtilis. Thèse 3e cycle à l'Université de Paris VII.

21. Young, M. (1984) Gene amplification in *Bacillus subtilis*. *J. Gen. Microbiol.* (in press).
22. Bedbrook, J., and F. Ausubel (1976) Recombination between bacterial plasmids leading to the formation of plasmid multimers. *Cell* 9:707-716.
23. Keggins, K.M., E.J. Duval, and P.S. Lovett (1978) Recombination between compatible plasmids containing homologous segments requires the *Bacillus subtilis recE* gene product. *J. Bact.* 134:514-520.
24. Tanaka, T. (1979) *recE4* independent recombination between homologous deoxyribonucleic acid segments of *Bacillus subtilis* plasmid. *J. Bact.* 139:775-782.
25. Michel, B., B. Niaudet, and S.D. Ehrlich (1982) Intramolecular recombination during plasmid transformation of *Bacillus subtilis* competent cells. *EMBO J.* 1:1505-1511.
26. Stahl, F.W. (1979) Special sites in generalized recombination. *Ann Rev. Genet.* 13:7-24.
27. James, A.J., P.T. Morrison, and R. Kolodner (1983) Isolation of genetic elements that increase frequencies of plasmid recombination. *Nature* 303:256-259.
28. Porter, R.D. (1981) Enhanced recombination between F42lac and λplac: dependence on F42lac fertility functions. *Mol. Gen. Genet.* 184:355-358.

MERCURY RESISTANCE TRANSPOSON Tn813 MEDIATES CHROMOSOME TRANSFER IN RHODOPSEUDOMONAS SPHAEROIDES AND INTERGENERIC TRANSFER OF pBR322

A.R. St. G. Bowen and J.M. Pemberton

Department of Microbiology
University of Queensland
St. Lucia 4067, Australia

INTRODUCTION

Until relatively recent times little was known of the mechanisms of genetic exchange among photosynthetic bacteria (15). Although these organisms have been used as model systems for studying the biochemical and biophysical bases of photosynthesis, photochemical nitrogen fixation, and membrane formation, parallel studies involving genetic analysis have been, at best cursory. Because of the difficulties associated with the growth, mutant isolation and manipulation of obligate anaerobes, facultative photosynthetic bacteria, particularly members of the genus Rhodopseudomonas, provided the first detailed insights into mechanisms of genetic exchange among photosynthetic bacteria (14).

After the first discovery by Marrs (11) of genetic exchange in Rhodopseudomonas capsulata, mediated by phage-like particles known as Gene Transfer Agents, Sistrom (16) reported chromosome transfer in Rhodopseudomonas sphaeroides mediated by the broad host-range plasmid, pR68.45, a plasmid shown to mobilize the chromosomes of a wide variety of gram-negative bacteria (10). Pemberton and Bowen (13) then reported high frequency chromosome transfer in R. sphaeroides promoted by the broad host-range IncP-1 plasmid pRP1 carrying the mercury transposon Tn501. Using a variety of auxotrophic, antibiotic resistance and photopigment mutations, a map of the genome of R. sphaeroides was compiled. The map produced, using pRP1::Tn501, consisted of 2 linkage groups. In this paper we compare further mapping data obtained with pRP1::Tn501 to patterns of marker inheritance from chromosome mobilization promoted by plasmid pR751 carrying a mutant of the mercury transponson Tn21 deleted for its resolvase function (Tn813; Ref. 4). These data demonstrate that R. sphaeroides possess but a single linkage group carrying all mutations so far mapped. In addition to mobilizing the chromosome, plasmid pR751::Tn813 is able to mobilize small cloning vectors, such as pBR322, from Escherichia coli to a number of gram-negative bacteria including R. sphaeroides.

MATERIALS AND METHODS

Bacterial Strains, Plasmids, and Transposons

Details of the bacterial strains, plasmids, and transposons are listed in Tab. 1.

Media and Cultural Conditions

Z broth, Z agar, and minimal medium have been described previously (13). Unless otherwise specified, all cultures were incubated at 35°C.

Antibiotics, Amino Acids, and Mercuric Salts

Because of the sensitivity of certain auxotrophic mutants, particularly phenylalanine mutants, to the levels of amino acids in minimal medium, all amino acids, purines, and pyrimidines were added to a final concentration of 5 µg/ml. For E. coli, the contraselective levels of antibiotics were: carbenicillin, 100 µg/ml; tetracycline hydrochloride, 20 µg/ml; kanamycin sulfate, 50 µg/ml; streptomycin sulfate, 100 µg/ml; mercuric chloride, 12 µg/ml; trimethoprim, 50 µg/ml; rifampicin, 100 µg/ml; naladixic acid, 25 µg/ml. For R. sphaeroides, the antibiotic levels were; carbenicillin, 1 µg/ml; tretracycline hydrochloride, 1 µg/ml; kanamycin sulfate, 20 µg/ml; streptomycin sulfate, 30 µg/ml; mercuric chloride, 7 µg/ml; trimethoprim, 50 µg/ml; rifampicin, 30 µg/ml; naladixic acid, 10 µg/ml.

Isolation of Auxotrophic Mutants and Mating on Solid Medium

These techniques have been described previously (13).

RESULTS

Chromosome Mapping Using RP1::Tn501

Use of RP1::Tn501 produced polarized chromosome transfer in R. sphaeroides, and this discovery has allowed the construction of a genomic map of this bacterium consisting of 2 linkage groups, each possessing early and late markers (13). One of the priorities in extending mapping data obtainable with RP1::Tn501 was to determine if the large linkage group, containing the photopigment gene cluster, was in fact linked to the small linkage group, containing the histidine gene cluster. Based on the possibility that chromosome transfer was bidirectional from a single origin on a circular *R. sphaeroides* chromosome, we examined the linkage between later markers of both linkage groups.

The marker furthest from the origin of transfer on the large linkage group was *trp-3*$^+$ (Fig. 1); with RP1:: Tn*501* this marker showed occasional low linkage (1-2%) to *his-3*$^+$ on the small linkage group. By comparison *leu-2*$^+$, a late marker on the small linkage group showed more appreciable linkage (8%) to *str-14*, on the large linkage group. While these low linkage values were inherent in mapping this late region of the chromosome with RP1:: Tn*501*, there was some hesitation in concluding that the 2 linkage maps could have been unified using mapping data obtained with this plasmid. To provide more substantial linkage values between markers in this late region, a variety of new plasmid-transposon combinations were tested for chromosome mobilizing ability in *R. sphaeroides* (Bowen and Pemberton, manuscript in preparation). The one chosen for current study was pR751::Tn*813*.

Chromosome Mobilization and Mapping Using pR751::Tn813

Initial experiments with pR751::Tn*813* demonstrated that this plasmid was capable of chromosome mobilization in *R. sphaeroides*.

A comparison of the level of recovery of recombinants, for markers spanning the entire *R. sphaeroides* linkage group, suggested that pR751::Tn*813* produced chromosome transfer predominantly from an origin close to, and with high frequency transfer of, the photopigment gene cluster (Tab. 2). Using the same group of markers, plasmid pRP1::Tn*501* showed its characteristic polarized transfer from origins close to the photopigment gene cluster, and the histidine gene cluster (Tab. 2; Fig. 1; Ref. 13). A linkage analysis of recombinants derived from crosses involving pR751::Tn*813* showed substantial coinheritance between markers in the late region (*leu-2*$^+$, *trp-3*$^+$, and *str-14*) of the *R. sphaeroides* chromosome (Fig. 2). These data confirmed that there is but one linkage group, containing all known markers. The pattern of coinheritance was such as to suggest that pR751::Tn*813* transfers large segments of the main chromosome, perhaps occasionally the entire chromosome. Using both pR751::Tn*813* and pRP1::Tn*501* a map of the *R. sphaeroides* chromosome was produced (Fig. 3); the relative positions of each of the markers was determined by linkage to surrounding genes.

Intergeneric Transfer of Cloning Vectors

Photosynthesis is but one of many interesting characteristics of gram-negative bacteria, not expressed in *E. coli* (1). A number of solutions have been devised which allow cloning into genetically

Tab. 1. Bacterial strains, plasmids, and transposons.

Strain Number	Genotype/Phenotype	Reference
Bacterial strains		
Rhodopseudomonas sphaeroides		
RS6122	his-1 1	(13)
RS6124	his-2	(13)
RS6127	arg-1	(13)
RS6132	his-3	(13)
RS6152	ade-3,crtE1,str-14	(13)
RS6167	his-4	(13)
RS6168	phe-1	(13)
RS6169	arg-2	(13)
RS6177	met-2,arg-4	(13)
RS6178	met-2,phe-2	(13)
RS6181	his-3,phe-4	(13)
RS6182	his-3,aro-2	(13)
RS6189	met-2,trp-1,his-6	This paper
RS6192	met-2,his-5,gly-1	(13)
RS6194	his-3,trp-3,phe-5	(13)
RS6195	his-3,trp-3,leu-1	(13)
RS6196	his-3,trp-3,arg-6	This paper
RS6197	his-3,met-5,leu-2	(13)
RS6198	his-3,met-5,phe-6	(13)
RS6199	his-3,met-5,gly-3	This paper
RS6200	his-3,met-5,trp-5	(13)
RS6202	his-3,ile-2,phe-7	(13)
RS6203	his-3,ile-2,gly-4	This paper
RS6204	his-3,aro-1,leu-3	(13)
RS6272	ade-3,crtE1 str-14,rif-10	This paper
RS6304	nif-1	This paper
RS7009	ade-3,crtE1,str-14,rsrI-1,rif-11	This paper
RS7014	his-1,mmc-1	This paper
Escherichia coli		
CSH358	$hsdR^-_k$ $hsdM^+_k$ supF $\phi 80^r$	(5)
JM103	Δ(lac pro),thi,strA,supE, endA,sbcB,$hsdR^-$, F' traD36,proAB, $lacI^q$,ZΔM15	(18)
Plasmids		
pRP1	Tra^+ $Ap^rKm^rTc^r$	(10)
pR751	Tra^+ Tp^rSu^r	(10)
pUC8	Ap^r	(18)
pBR322	Ap^rTc^r	(12)

Tab. 1. Continued

Strain Number	Genotype/Phenotype	Reference
Transposons		
Tn501	$TnpA^+TnpR^+$ Hg^r	(17)
Tn813	$TnpA^+TnpR^-$ Hg^r	(4)

Abbreviations: ade,adenine; Ap^r,ampicillin resistance; aro, aromatic amino acid;arg,arginine;bch,bacteriochlorophyll. crt,carotenoid;end,endonuclease;gly,glycine;his,histidine; Hg^r,mercury resistance;hsdM,host-determined modification; hsdR,host determined restriction;ile,isoleucine;Km^r,kanamycin resistance;lac,lactose;leu,leucine;met,methionine; mmc,mitomycin sensitivity; nif,nitrogen fixation;phe,phenylalanine; pro,proline; rif^r,rifampicin resistance; rsrI,restriction endonuclease RsrI; sbc,suppressor for recBC; str,streptomycin resistance;nal^r,naladixic acid resistance; Su^r,sulphonamide resistance; thi,thiamine; Tc^r,tetracycline resistance;Tpr,trimethoprim resistance;tra,transfer proficiency;trp,tryptophan; Δ ,deletion.

well-characterized organisms such as E. coli and subsequent transfer of the cloned genes back into their original host for expression. On the one hand, there are the dual cloning vectors which can replicate and be maintained both in E. coli and the other organism of interest; many such dual cloning vectors have been constructed in vitro by combining pBR322 with a naturally-occurring plasmid isolated from the organism of interest (12).

On the other hand, advantage has been taken of the broad host-range of the small, multi-copy plasmids of the IncQ group; although nonconjugative, plasmids such as pRSF1010, pR300B, and pR1162 can be mobilized by pRP1 to a wide variety of gram-negative bacterial species; Timmis and his coworkers have developed a variety of cloning vectors derived from pRSF1010 by in vitro manipulation (1). Although these 2 approaches have proven useful cloning strategies, the narrow host-range plasmid pBR322 and its many derivatives still remain the most popular cloning vectors available; more recent derivatives, such as pUC8 and pUC12, have the added advantage of being able to be used directly in sequencing by the method of Maxam and Gilbert (18).

In experiments involving the transposon Tn21, Grinsted and his coworkers (4) observed that a mutant deleted for its resolvase function (Tn813) was able to mediate the formation of highly stable cointegrates between 2 plasmids. Using the broad host-range plasmid

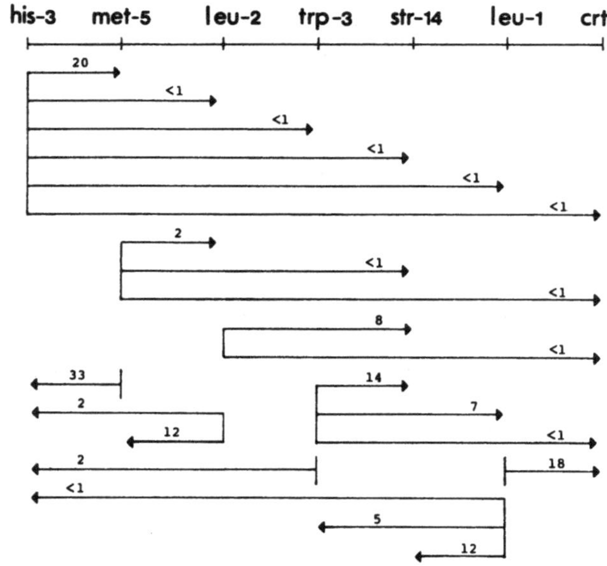

Fig. 1. Linkage data derived from matings between the donor strain RS6152 (pRP1::Tn501) and recipient strains RS6195 and RS6197. Selected marker (bar); unselected marker (arrowhead).

pR751 carrying this mutant transposon, we found pR751::Tn813 formed highly stable cointegrates with pBR322; if the level of cointegrate formation is measured by the level of cotransfer of pBR322 and pR751::Tn813, then approximately 1% of cells harboring both plasmids possess such stable cointegrates. The stability of these cointegrates (pR751::Tn813::pBR322) was such as to allow transfer of pBR322 from one strain of *E. coli* to another, and from *E. coli* to *R. sphaeroides*; the cointegrate containing pBR322 could be retransferred from *R. sphaeroides* to *E. coli* (Tab. 3). The presence of the cointegrate in the recipient strain conferred simultaneous resistance to trimethoprim (pR751), mercury (Tn813), carbenicillin, and tetracycline (pBR322). Other small cloning vectors such as pUC8 (Tab. 3), pHC79 (the cosmid derivative of pBR322), and pACYC184 could be transferred between strains of *E. coli* and from *E. coli* to *R. sphaeroides* (Bowen and Pemberton, unpub. data). In addition, the cointegrates containing pBR322 could be transferred from *E. coli* to such diverse gram-negative bacteria as the cellulolytic *Cellvibrio mixtus*, the amylolytic *Aeromonas punctata*, and the xenobiotic degrader *Alcaligenes eutrophus* (Pemberton, Wynne, and Gobius, unpub. data).

DISCUSSION

Following the discovery that RP1::Tn501 could be used for

Tab. 2. Comparison of the chromosome mobilizing abilities of pRP1::Tn501 and pR751::Tn813.

Selected Marker[b]	Frequency of recombinant formation[a]	
	Plasmids	
Group A	pRP1::Tn501	pR751::Tn813
his-3^{+}	230	16
met-5^{+}	120	64
leu-2^{+}	1.4	60
trp-3^{+}	5.6	30
str-14	0.4	42
leu-1^{+}	5.0	29
phe-5^{+}	6.0	18
Group B		
crt-1^{+}	175	210
ade-3+	190	255

[a] Frequency of recombinants calculated per 10^6 donor cells

[b] In crosses involving Group A markers the donor strain was RS6152 carrying pRP1::Tn501 or pR751::Tn813; for Group B markers, the donor strain was Rs6132 carrying the appropriate plasmid.

genomic mapping in R. sphaeroides, it became possible to locate a wide variety of genes (13). The preliminary map of this bacterium showed 2 linkage groups, with the important photopigment genes clustered together in a single region of the genome close to an origin of chromosome transfer. If it is assumed that R. sphaeroides has a single circular chromosome, then the pattern of recovery of recombinants and the linkage data obtained with RP1::Tn501 is consistent with bidirectional Hfr chromosome mobilization from a single origin on the main chromosome (Ref. 13, 14; Fig. 1; Tab. 2). Confirmation that R. sphaeroides does possess a single linkage group was obtained using pR751::Tn813; in this case, Hfr chromosome mobilization occurs predominantly from a single origin, in one orientation, and with the early transfer of the photopigment genes and the late transfer of the histidine biosynthetic genes (Fig. 1,2; Tab. 2). A contrasting characteristic of pR751::Tn813 compared to pRP:Tn501 is that the former plasmid-transposon combination transfers larger segments of the chromosome (Fig. 1 and 2); the reason for this difference remains unknown.

Evolutionarily related transposons such as Tn3, Tn501, Tn1721, and Tn21 mediate the cointegration of 2 replicons as an intermediate step in their transposition. The cointegrate is then resolved by a

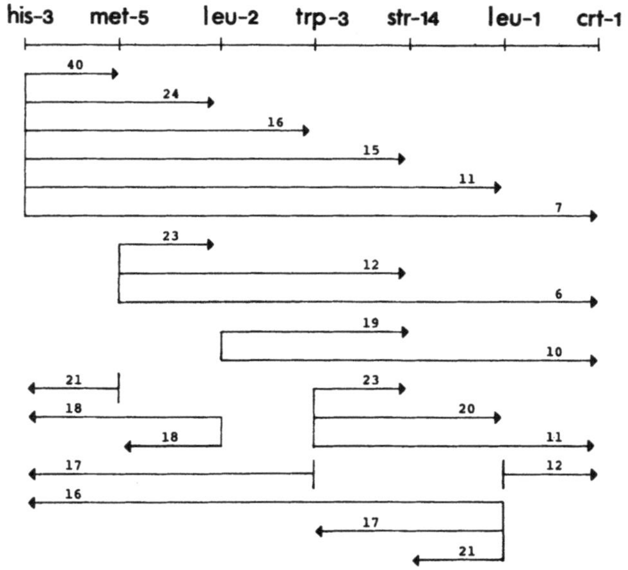

Fig. 2. Linkage data derived from matings between the donor strain RS6152 (pR751::Tn813) and recipient strains RS6195 and RS6197.

site-specific tnpR gene product (resolvase) acting at the res site (4). Usually, stable cointegrates are obtained using transposons carrying either a mutation in the res site or in the tnpR gene. However, Kitts et al. (6) reported that Hg^{++} induces readthrough

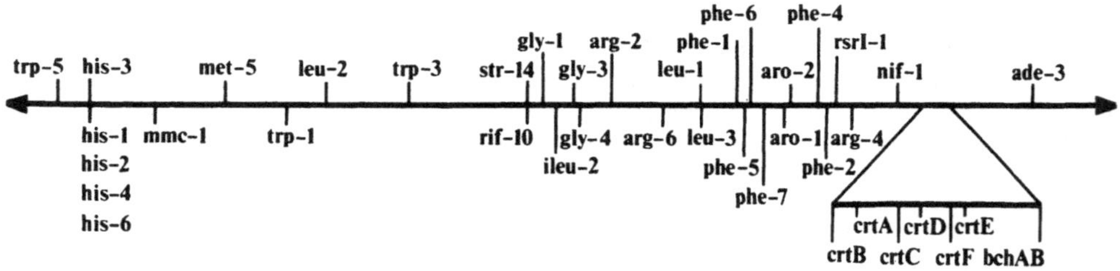

Fig. 3. Linkage map of the Rhodopseudomonas sphaeroides chromosome. The relative positions of markers were determined by patterns of coinheritance with surrounding markers using RP1::Tn501 and pR751::Tn813 promoted chromosome transfer. Gene symbol abbreviations as in Tab. 1. Arrowheads indicate position and orientation of Tn501 transfer origins; Tn813 promotes chromosome transfer from the origin close to ade-3+, with the same orientation as Tn501. The photopigment gene cluster is drawn from data in Ref. 13.

Tab. 3. Transfer of cointegrates between Escherichia coli and Rhodopseudomonas sphaeroides.

Donor	Recipient	Frequency of Transfer[a] Selected markers	
Escherichia coli	Rhodopseudomonas sphaeroides	Tp^rRif^r	$Tp^rCb^rRif^r$
CSH358(pR751::Tn813,pBR322)	RS7009 Res^-Rif^r	$1x10^{-1}$	$1x10^{-3}$
CSH358(pR751::Tn813,pBR322)	RS6152 Res^+Rif^r	$5x10^{-4}$	$1x10^{-6}$
JM103 (pR751::Tn813,pUC8)	RS7009 Res^-Rif^r	$2x10^{-2}$	$5x10^{-5}$
Rhodopseudomonas sphaeroides	Escherichia coli	Tp^rNal^r	$Tp^rCb^rNal^r$
RS7009(pR751::Tn813::pBR322)	CSH358 Res^-Nal^r	$1x10^{-1}$	$1x10^{-3}$
RS7009(pR751::Tn813::pUC8)	CSH358 Res^-Nal^r	$4x10^{-2}$	$5x10^{-4}$

[a]Frequency of transfer calculated per donor cell.

transcription from the mercury-resistance operon into the tnpR gene. These data support the suggestion that the tnpR gene of Tn501 may not have an independent promoter, while tnpA does; this provides an explanation for the efficient transposition and cointegrate formation by Tn501, and the highly inefficient resolution of these cointegrates. The chromosome transfer observed in R. sphaeroides appears to be due to cointegration of RP1::Tn501 and the main chromosome to form an Hfr; more recent reports of Tn501 mediated chromosome transfer in Pseudomonas putida and Bordetella pertussis support this view (3,19). Although Tn3 and Tn1721 do not promote chromosome transfer in R. sphaeroides, this may simply reflect the known efficiency of their resolvase functions, indicating that resolvase-deficient mutants of these 2 transposons could be used for replicon mobilization.

The generation of a linkage map of R. sphaeroides, the availability of a range of mutations in each region of the chromosome, and the ability to transfer small cloning vectors, such as pUC8, pBR322, and its cosmid derivative, pHC79, from E. coli to R. sphaeroides via a cointegrate with pR751::Tn813 provides the opportunity to produce a detailed physical and genetic map of this photosynthetic bacterium. It should be possible to isolate and identify sequences of the chromosome encoding such important biological processes as photosynthesis, photochemical nitrogen fixation, and membrane formation.

ACKNOWLEDGMENTS

The authors wish to thank Cathy Smith and Ann van Pham for their expert technical assistance. A.R.St.G.B. is supported by a Commonwealth Postgraduate Award from the Australian Government. Research funds were provided by the Australian Research Grants Commission, grant number D2/83/155701.

REFERENCES

1. Bagdasarian, M., and K.N. Timmis (1982) Host: Vector systems for gene cloning in Pseudomonas. In Current Topics in Microbiology and Immunology. Springer-Verlag, Berlin, 96:47-67.
2. De La Cruz, F. and J. Grinsted (1982) Genetic and molecular characterization of Tn21, a multiple-resistance transposon from R100-1. J. Bact. 151:222-228.
3. Dean, H.F., and A.F. Morgan (1983) Integration of R91-5::Tn501 into the Pseudomonas putida PPN chromosome and genetic circularity of the chromosomal map. J. Bact. 153:485-497.
4. Diver, W.P., J. Grinsted, D.C. Fritzinger, N.L. Brown, J. Altenbuchner, P. Rogowsky, and R. Schmitt (1983) DNA sequences of and complementation by the tnpR genes of Tn21, Tn501 and Tn1721. Mol. Gen. Genet. 191:189-193.
5. Karn, J., S. Brenner, L. Barnett, and G. Cesarini (1980) Novel bacteriophage cloning vector. Proc. Natl. Acad. Sci., USA 77:5172-5176.
6. Kitts, P., D. Symington, M. Burke, R. Reed, and D. Sherratt (1982) Transposon-specified site-specific recombination. Proc. Natl. Acad. Sci., USA 79:46-50.
7. Gest, H. (1966) Comparative biochemistry of photosynthetic processes. Nature (Lond.) 209:879-882.
8. Grinsted, J., F. De La Cruz, J. Altenbuchner, and R. Schmitt (1982) Complementation of transposition of tnpA mutants of Tn3, Tn21, Tn501 and Tn1721. Plasmid 8:276-286.
9. Hohn, B., and J. Collins (1980) A small cosmid for efficient cloning of large DNA fragments. Gene 11:291-298.
10. Holloway, B.W. (1978) Plasmids that mobilise the bacterial chromosome. Plasmid 2:1-19.
11. Marrs, B.L. (1974) Genetic recombination in Rhodopseudomonas capsulata. Proc. Natl. Acad. Sci., USA 71:971-974.
12. Old, R.W., and S.B. Primrose (1982) Principles of gene manipulation: An introduction to genetic engineering. Blackwell Scientific Publications (Lond).
13. Pemberton, J.M., and A.R.St.G. Bowen (1982) High frequency chromosome transfer in Rhodopseudomonas sphaeroides promoted by the broad host range plasmid RP1 carrying the mercury transposon Tn 501. J. Bact. 147:110-117.
14. Pemberton, J.M., S. Cooke, and A.R.St.G. Bowen (1983) Gene transfer mechanisms among members of the genus Rhodopseudomonas. Ann. Microbiol. (Inst. Pasteur) 134B:195-204.

15. Saunders, V.A. (1982) Genetics and molecular biology of photosynthetic bacteria and cyanobacteria. In Photosynthesis: Development, Carbon Metabolism and Plant Productivity Vol. III. Academic Press, Inc., New York.
16. Sistrom, W.R. (1977) Transfer of chromosomal genes mediated by plasmid R68.45 in Rhodopseudomonas sphaeroides. J. Bact. 131:526-532.
17. Stanisich, V.A., P.M. Bennett, and M.H. Richmond (1977) Characterisation of a translocation unit encoding resistance to mercuric ions that occurs on a non-conjugative plasmid in Pseudomonas aeruginosa. J. Bact. 129:1227-1233.
18. Vieira, J., and J. Messing (1982) The pUC plasmids, an M13mp7-derived system for insertion mutagenesis and sequencing with synthetic universal primers. Gene 19:259-268.
19. Weiss, A.A., and S. Falkow (1983) Transposon insertion and subsequent donor formation promoted by Tn501 in Bordetella pertussis. J. Bact. 153:304-309.

REPORT ON A WORKSHOP:

PLASMIDS IN UNUSUAL SYSTEMS

Simon Silver, John Pemberton[1], Kenji Sakaguchi[2]

Washington University
St. Louis, Missouri, 63110

The first problem for this workshop was to define "unusual systems" as more than a catch-all for topics not covered elsewhere at the meeting. This is related to the "specialized functions" sessions held later in the week, for functions not covered in special single topic sessions. Unusual systems are (i) plasmids in organisms other than *E. coli* and especially those that have not gained sufficient research activity to stand alone at the meeting. Thus, some "unusual systems" of 1984 will enjoy a different status in future Plasmid Symposia. (ii) Unusual systems also mean plasmid functions, and especially metabolic capacities, governed by plasmids that have not yet seen sufficient study to stand in sections by themselves.

An emerging trend in prokaryotic molecular genetics is the increasing interest in unusual microorganisms; "unusual" meaning organisms which possess biologically important properties (for example, photosynthesis) but for which there are few, if any, means of genetic analysis. There are a number of factors which are stimulating this interest. On the one hand, there is the increasing involvement and influence of biotechnology/genetic engineering companies which are attempting to obtain microorganisms with novel biological properties and which are examining ways of increasing the efficiency of known important medical, industrial, and agricultural processes. On the other hand, there is an increasing awareness by microbial geneticists that it is possible to blend and extend the range of genetic and molecular biological techniques developed in *E. coli* to many interesting and important species of both gram-positive and gram-negative bacteria.

---

[1]Dept. of Microbiology, University of Queensland, St. Lucia 4067, Australia. [2]Mitsubishi-Kasei Institute of Life Sciences, Minamiooya 11, Machida, Tokyo, Japan, 194.

The discussion included how systems of genetic exchange in well-known organisms such as *E. coli* and *P. aeruginosa* might be adapted for genetically less well characterized bacteria such as *Nitrobacter*, *Bacteroides*, *Clostridium* and *Methanobacterium*. For many of these organisms, the broad host-range plasmids and cloning vectors, used so successfully in members of the genus *Pseudomonas*, are either unstable or simply not transferrable. Another hurdle to overcome is the problem of strain improvement when a biologically important property is encoded in a large number of genes. A common strategy is to carry out cloning in *E. coli* and transfer the genes back into the bacterium of interest for expression and genetic analysis. This latter approach necessitates more than just a rudimentary understanding of the genetic topography of the unusual bacterium. Questions of gene transfer systems in *Xanthomonas* (for xantham gum production), *Bacteroides nodosus* (for production of antipilus vaccine against footrot), *Nitrobacter* (conversion of nitrite to nitrate), *Thermus* (heat-stable proteins and biological processes) and *Clostridium acetobutylicium* (industrial chemicals) were a few of those organisms discussed. Possibilities of setting up systems of conjugation and chromosome transfer (considered the most potentially useful), and transformation and transduction were canvassed. The most popular approach was to be cloning in *E. coli* and adapting naturally occurring plasmids as potential "shuttle" vectors. One major impediment was a lack of awareness of the range of cloning vectors and strategies that are available for developing gene transfer systems in unusual bacteria.

In addition to the interest in organisms and what they can do, there was an interest in the metabolic products and capacities that are "native" to the organisms and also in products that can be produced by genetic systems cloned into the organism of choice. Among the industrially useful products discussed were amino acids and heat-stable enzymes (as biocatalysts and marketable products). Solvent production might be improved by use of genetically altered organisms with greater solvent resistance. Calf renin and chymosin are bulk products that will be manufactured more economically from bacterial cells rather than isolated from animal tissues. Higher value, smaller market products include peptide hormones.

There was a ready exchange during the workshop and elsewhere during the meeting that showed that people functioning in university laboratories were greatly concerned with producing plasmids and strains with important economic values and, equally, that people working in biotechnology companies were interested in the basic scientific understanding of the systems being developed. It was considered that basic understanding was essential for development of useful strains and their maintenance and improvement.

CHAIRMAN'S INTRODUCTION:

REPLICATION, INCOMPATIBILITY, AND PARTITION

Kurt Nordström

Department of Microbiology
Biomedical Center
Uppsala University
Box 581
S-751 23 Uppsala, Sweden

The life cycle of plasmids comprises 2 different events, increase in number (replication) and distribution of the plasmid copies to the daughter cells (partition). Both these events are controlled by plasmid-borne genetic information. They were discussed in 16 talks, one informal workshop (on partition) and 37 posters.

Replication and partition are independent processes. Deletion of the _par_ region of a plasmid does not affect the control characteristics of plasmid replication but causes unstable inheritance of the plasmid. A _par_-deleted replicon can be stabilized by the _par_ region of any plasmid. Hence, it is possible to study replication and partition independently.

Replication

Plasmid replication is controlled such that a given plasmid in a given host under given conditions is present in a defined average number of copies per cell. Theoretically, this could be achieved by saturation kinetics where replication is controlled by a rate-limiting host-controlled reaction (Nordström). However, all plasmids studied so far control their own replication by genetic information causing negative control. The main inhibitor can either be protein (λ_dv_, Matsubara), RNA (CloDF13, Nijkamp; pT181, Novick; ColE1, Polisky; FII plasmids, Rownd) or series if direct repeats [RK2, Figurski; R6K, Helinski; P1, Yarmolinsky; and presumably R1162 (RSF1010), Meyer; pSC101, Churchward et al.]. The functioning of the λ_dv_ system is fairly well understood, although the biochemical events at initiation of replication still are poorly known;

Matsubara addressed the question as to where the first deoxyribonucleotide is introduced during initiation of replication of λdv (Tsurimoto et al.). Similarly, 3 replication control systems using small, unstable RNA species as replication inhibitors are fairly well understood. In all 3 (ColE1-CloDF13, Polisky, Nijkamp; FII plasmids, Rownd; pT181, Novick), the inhibitor has a high degree of secondary structure and interacts with the complementary region of the opposite long transcript. This presumably results in an RNA duplex (Tamm et al.), that changes the configuration of the longer transcript further downstream. This results in a decreased efficiency of primer activity (ColE1-CloDF13) or of expression of a rate-limiting Rep protein (FII and pT181); Wong et al. reported ColE1 mutations changing the secondary structure of the preprimer and affecting replication control. The hypothetical structure of RNAI (the inhibitor of plasmid ColE1) was verified by structural studies presented by Polisky. Similarly, Rownd presented computer studies showing that the secondary structure of the RepA-message of FII plasmids may change by the formation of an RNA duplex with CopA-RNA such that the Shine-Dalgarno region becomes hidden in a duplex region. This may explain why the inhibitory CopA-RNA acts as a translational inhibitor. Interestingly, the Rep protein of FII plasmids is mainly cis-acting, whereas that of plasmid pT181 is also acting in trans. This has some consequences for the incompatibility properties as discussed by Novick.

The large group of important plasmids that have series of DNA repeats as their replication inhibitors was discussed in several papers. As pointed out by Helinski, there are many similarities between the structure of these replicons: the size and spacing of the repeats, the location of at least some of the repeats in the vicinity of the origin, the size of the plasmid-coded Rep proteins (28-38 kb), the location of the rep genes near the origin, the autoregulation of the expression of the rep genes. The function of the repeats is still obscure, but titration of a positively needed function has been suggested, but has not yet been proven. The plasmid-coded proteins are naturally the prime suspects and Yarmolinsky reported that the RepA protein of plasmid P1 binds to the incA repeats. High concentrations of the R6K Rep protein (π) and the P1 Rep protein (RepA) actually inhibit replication (Helinski; Yarmolinsky). Titration of an autoregulated Rep protein maybe poses a paradox. Yarmolinsky tried to solve this paradox with his beautiful Clepsydra (water clock) analogy. The solution could be a trifunctional Rep protein (autoregulated, titrated, and active) as suggested by Kline and Trawick. Results reported by Yarmolinsky support this picture; the RepA protein inhibits transcription from and into the origin of plasmid P1. Furthermore, the P1 replication system works properly even if the origin is physically separated from the incA repeats and the autoregulated repA gene. Hence, the signal caused by binding of the RepA protein to the repeats has to travel through the cytoplasm.

Plasmid RK2 has got an even more complicated control system than the other plasmids with inc repeats. Two proteins, the korA and korB products, interfere with the expression of the Rep protein coded for by the trfA gene (Figurski; Thomas et al.).

A considerable fraction of the contributions dealt with the genetic organization of the various basic replicons. Hence, the knowledge about the structure of the replicons is advancing fast. Similarly, knowledge of the molecular biology is also expanding rapidly. Particularly the ColE1-CloDF13 systems are exploited in many ways: Dooley et al. aim at a total characterization of the replication control region of plasmid ColE1 by systematically mutating "every single base pair"; Elble et al. have isolated copy mutants of ColE1 where the stability of RNAI is changed; van Putten et al. reported that methylation in the control region is important for copy number control of plasmid CloDF13; Naito et al. reported that RNaseH does not seem to be needed for in vivo replication of plasmid ColE1 in contrast to in vitro results reported previously.

The biochemistry of replication is studied intensely in several laboratories. Koguni described a purified system for replication of oriC plasmids of *Escherichia coli* that fulfills all characteristics of the in vivo system. Similarly Haring et al. and Scholz et al. reported in vitro studies of replication of plasmid RSF1010. This plasmid codes for 3 rep genes and is independent of several of the *E. coli* DNA replication enzymes (e.g., RNA polymerase and the DnaB, C, and G proteins). This implies a unique replication system which may be the key to the broad host range expressed by plasmid RSF1010. The host range of plasmid RK2 requires specific plasmid-coded functions in some hosts but not in others (Schmidhauser and Helinski).

Less progress has been made in the physiological aspects of plasmid replication. Highlander and Novick studied replication of plasmid pT181 during copy-number-shift experiments. Rownd discussed measurements of the concentration of the various transcripts of the basic replicon of the FII plasmid NR1 at different concentrations of the corresponding genes. Furthermore, Rownd has performed an interesting theoretical analysis based upon the thermodynamics of the interaction between the loops of the CopA-RNA and the RepA-message. These measurements and calculations came to the same conclusions as the physiological experiments performed in other laboratories during the last 10 years, which is reassuring.

Finally, Pritchard discussed the differences between control of plasmid replication and chromosome replication. Perhaps the chromosome uses other principles than the plasmids do? It is striking that oriC plasmids do not express inhibitor activity or sensitivity to a possible inhibitor of replication. There are also differences with respect to the timing of chromosome and plasmid replication.

The picture that now emerges is one in which different plasmids use different approaches to control their replication - there are at least 3-4 different schemes known so far. The idea that plasmids are simple-minded organisms may have received a blow!

Partition

Most naturally occurring plasmids are maintained very stably in exponentially growing populations of host bacteria. However, deletion of regions of plasmids often causes destabilization of the plasmids. Deletion derivatives of low copy-number plasmids are often found to be lost in a frequency that fits with random distribution of the plasmid copies at cell division (P1, Austin; F, Hiraga). However, random distribution of a high copy-number plasmid would not cause any instability of inheritance. Kasner and Rownd reduced the copy number of plasmid ColE1 and observed unstable maintenance and concluded, as others have before, that ColE1 does not have any *par* function. However, deletion derivatives of high copy-number plasmids are often lost in a high frequency (pSC101, Cohen; *oriC*, Kunze and Messer), whereas the parent plasmids are stably maintained. This implies that under these circumstances, the unit of segregation consists of more than one plasmid genome, either by the formation of concatemers or by some sort of clustering or compartmentalization. Hence, the deleted region does not have anything to do with the process of partition. This matter was discussed in an informal workshop and it was agreed upon that *a par system secures that each daughter cell is guaranteed at least one plasmid molecule at cell division*. A system that breaks up oligomers (e.g., by site-specific recombination; Nijkamp; Grinter and Barth) or clustering/compartmentalization is not a *par* system.

Low copy-number plasmids (P1, Austin; Abeles et al.; F, Hiraga; R1, Gerdes et al.) possess *par* systems. These seem to code for 2 proteins and a centromer-like cis-acting region. There is randomization during partition; hence, *par* expresses incompatibility (P1, Austin; R1, Gerdes et al.). The synthesis of the ParA protein of plasmid P1 is autoregulated by the ParA protein in cooperation with the ParB protein.

Deletion of various parts of the stability region of plasmid pSC101 gave a series of derivatives (Cohen; Miller et al.). Some of these were stably maintained, but were preferentially lost from cells also containing the wild-type parent plasmid. This phenotype was denoted Cmp^- (competition-negative) and indicates a correlation between replication and stable maintenance. Maybe clustering (dimers, trimers, etc. were not observed in this case) removes the plasmids from the replication pool?

Kunze and Messer observed that *oriC* plasmids are unstably maintained, whereas the plasmid is stably inherited. Hybrids between pSC101 and *oriC* plasmids were found to be unstably maintained.

In order to completely stabilize a low copy-number plasmid, the par system is not enough. There is also a system that inhibits cell division in cells carrying only one plasmid copy - the ccd (Control of Cell Division) system of plasmid F (Hiraga). Plasmid R1 seems to code for an analogous system (Gerdes et al.). The ccd system may be responsible for the induction of the SOS system that is caused by the introduction of UV-damaged λminiF but not by UV-damaged phage λ (Bailone et al.).

In conclusion, stable maintenance of bacterial plasmids is affected by at least 4 different systems:

1. Replication
2. Partition
3. Ccd (Control of Cell Division)
4. Site-Specific Recombination.

In order to be stably maintained, low copy-number plasmids seem to need at least 3 of these. Systems from different plasmids can be combined to give stable inheritance--the cassette model of Hiraga.

ROLE OF THE π INITIATION PROTEIN AND DIRECT NUCLEOTIDE SEQUENCE REPEATS IN THE REGULATION OF PLASMID R6K REPLICATION

Marcin Filutowicz, Michael McEachern, Alan Greener
Pradip Mukhopadhyay, Elliott Uhlenhopp,*
Ross Durland and Donald Helinski

Department of Biology
University of California, San Diego
La Jolla, California 92093

INTRODUCTION

Bacterial plasmids vary considerably in size and copy number. A fundamental property shared by these elements, however, is their stable inheritance in a growing population of bacteria. This stable maintenance at a copy number characteristic of the element is determined by genetic regions on the plasmid that regulate the initiation of replication and subsequent partitioning of the plasmid. As in the case of bacterial and animal viruses that employ different mechanisms for their duplication in a particular host cell, plasmids have evolved different strategies to assure their maintenance at a specific copy number. In some cases both primary and secondary control circuits function to regulate the initiation of plasmid replication events and prevent loss of the element in an exponentially growing bacterial population. This paper will be concerned largely with the regulation of replication of the naturally occurring antibiotic resistance plasmid R6K.

Plasmids With Direct Nucleotide Sequence Repeats at Their Origin

The majority of plasmid elements of *Escherichia coli* analyzed to date contain direct nucleotide sequence repeats at their origin of replication. This list includes plasmids R6K (1,2), RK2 (3,4), F (5,6), pSC101 (7,8), Rts 1 (9), P1 (10), λdv (11,12) and R485 (D. Stalker, unpub. observ.) (Tab. 1). The direct repeats vary from 17

* Current Address: Department of Chemistry, Grinnell College, Grinnell, Iowa.

to 24 bp in length and are either tandemly joined or separated by one or more nucleotides. In each case, the plasmid also specifies a protein that is required for its replication (Rep protein). The start of the Rep protein structural gene is generally a short distance (25-200 bp) from 1 end of the direct nucleotide sequence repeats (Tab. 1). In the case of λdv (11), the origin sequence is located within the O initiation protein gene. The size of the Rep proteins that have been characterized varies from 29,000 to 38,000 d. In the case of plasmid R6K, the Rep protein (designated π protein) has been shown to be required for the initiation of R6K replication (13) and to bind to the direct nucleotide sequence repeats at the origin of replication (14,15).

The nucleotide sequences of the direct repeat units for the plasmids λdv, R6K, F, RK2, and P1 are displayed in Fig. 1. It is worth noting that with few exceptions, the hexanucleotide TGAGPuG is present in each of the direct repeats. While the role of this hexanucleotide sequence in plasmid DNA replication remains to be determined, it appears that for plasmid R6K, the G occupying the second and fourth positions is essential for binding of the π initiation protein and a functional replication origin (see section below). The important role of the direct repeats in the regulation of plasmid DNA replication is also indicated by the finding that the repeat sequences of plasmids F (16,17), P1(10), R6K (S. Yang, unpub. observ.) and RK2 (T. Schmidhauser, M. deSmet, and P. Reeves, unpub. observ.), when inserted into another plasmid replicon, express incompatibility properties characteristic of the parent plasmid.

General Features of R6K Replicon

The origins of replication, designated α, β, and γ, are contained within a 4-kb region of the 38-kb plasmid R6K (Fig. 2) (18-21). The α and β origins display in E. coli sequential, bi-directional replication toward an asymmetrically located terminus (18,19). An in vitro system, capable of replication of R6K or its

Tab. 1. Direct nucleotide sequence repeats at their origin of replication.

Plasmid	Direct Repeats at Origin (A)	Replication Protein (mol. wt.) (B)	Distance Between Genetic Regions A and B	Ref.
R6K	7-22 bp	35,000	207 bp	1, 2
pSC101	3-18 bp	37,500	95 bp	7, 8
Rts1	5-24 bp	33,000	26 bp	9
F	4-19 bp	29,000	98 bp	5, 6
P1	5-19 bp	32,000	56 bp	10
RK2	8-17 bp	32,000	~4000 bp	3, 4

```
   TCCCCCGTTT|TGAGGG|AT
   TCCCCCAATT|TGAGGG|AT                          LAMBDA
   TCCCTCGTTT|TAGGGG|AT
   TCCCTCGTTT|TGAGGG|AT

        AAACA|TGAGTG|GATAGTACGAA *
        AAACA|TGAGAG|CTTAGTACGAT
        AAACA|TGAGAG|CTTAGTACGAG
        AAACA|TGAGAG|CTTAGTACGAA
        AGCCA|TGAGGG|TTTAGTTCGAA                 R6K
        AGCCA|TGAGAG|CTTAGTACGAA
        AAACA|TGAGAG|CTTAGTACGAA
        AAACA|TGAGAG|CTTAGTACGAC

             |TGAGGG|TTATCTGTCACAGATT
             |TGAGGG|TGGTTCGTCACATTTG
             |TGAGGG|TAATTTGTCACAGTTT            MINI-F
             |TGAGGG|CAGTTTGTCACAGTTG
             |TGAGGG|TTGATTATCACAGTTT

      TGACAGA|TGAGGG|GCGG
      TGACACT|TGAGGG|GCCG
      TGACAGA|TGAGGG|GCAG
      TGACACT|TGAGGG|GCGC
      TGACAGA|TGAGGG|GCGC                        RK2
      TGACACT|TGAGGG|GCAG
      TGACAGA|TGAGGG|GCGC
      TGACATT|TGAGGG|GCTG

    GTGTGTGCT|GGAGGG|AAA
    ATGTGTGCT|GCCGGG|AAG
    ATGTGCGCT|GGAGGG|GGA
    ATGTGTGCT|GGAGGG|AAA
    ATGTGTGCT|GGAGGG|AAA                         PI
    ATGTGCACT|GGAGGG|AAA
    ATGTGTGCT|GGAGGG|AAA
    ATGTGCTGT|GGAGGG|TAC
    ATGTGCGAT|AGAGGG|AAG
    ATGTGTGCT|GGAGGG|AGA
```

Fig. 1 Families of direct repeat units from the origin region of 5 different plasmid replicon sequences are aligned to indicate the location of a 6-bp homologous region. The 8th direct repeat of plasmid R6K that is in the putative promoter-operator region of the π structural gene is indicated (*). The lambda direct repeats are present in the origin region of the λdv plasmid. Reference citations for these sequences are given in the text.

derivatives, has been developed from E. coli cells (13,22). A minimal replication region, consisting of 2 separable components: the γ origin and the structural gene for the π initiation protein (pir), has been defined (23). The γ origin contains 7 tandemly associated 22-bp direct repeats (1). The eighth copy of this repeat is located in the putative promoter-operator region of the pir gene (2,24) and is likely to play a role in the autoregulation of expression of the pir gene (2,25).

The π initiation protein is also required for the initiation of replication from the α and β origins (26). However, unlike the γ

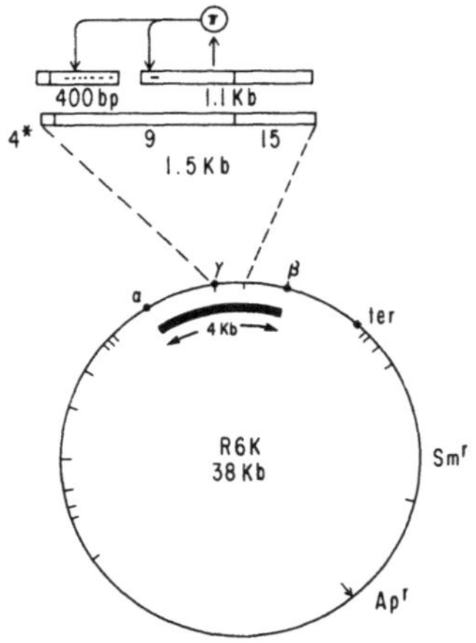

Fig. 2. Physical and genetic map of plasmid R6K. Hash marks on the circular map refer to the position of HindIII sites. 4*, 9 and 15 refer to HindIII fragments present in the origin region. α, γ and β indicate the locations of the 3 origins of replication. ter refers to the assymetric terminus of replication. The positions of the streptomycin and ampicillin resistance genes are indicated by Sm^r and Ap^r, respectively. The single Bam H1 restriction site is located by an arrow (↓). The relative locations of the 7 direct repeats in the λ origin and the 8 repeats in the putative operator-promoter region are indicated by dash marks.

origin it has not been possible to isolate the α or β origin as a short segment of DNA capable of replication when the π protein is provided in trans. As indicated in Fig. 3, the minimal size of the functional α and β origins obtained to date is approximately 2 kb (24,26). The complete nucleotide sequence of the region containing the γ and β origins of replication has been determined (1,2,24,26). This region also includes a functional promoter [(27) and P. Mukhopadhyay, unpub. observ.] specifying transcription in the direction of the repeats, the seven 22-bp direct nucleotide sequence repeats that express R6K incompatibility (designated incA, S. Yang, unpub. observ.), a second incompatibility region (designated IncB; S.Yang, unpub. observ.) the pir gene and a second open reading frame (ORF) (26).

The γ origin has been delineated by Tn5 transpositional analysis (28). It consists of a 260-bp region extending from a short distance to the left of the junction of HindIII fragments 4 and 9 (Fig. 2 and 3) to just before the blgII site in HindIII fragment 9. Removal of 3 or more of the direct repeats from the γ origin results in loss of the origin activity (28). As indicated in Fig. 4, several deletions and insertions have been produced in the β-origin

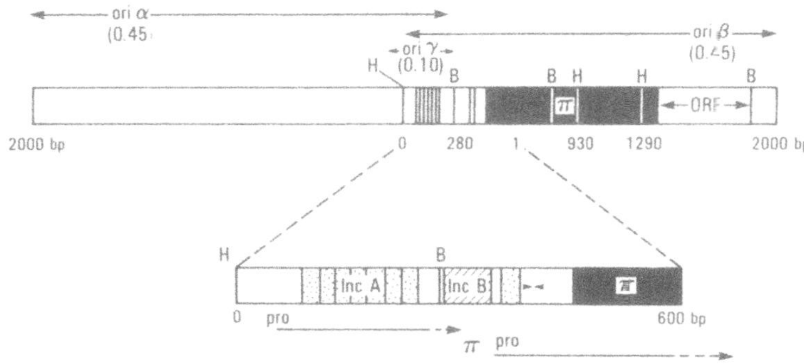

Fig. 3. Structure of the replication origin region of R6K. The relative frequency of initiation from the α, γ and β origins in vivo is indicated in parenthesis. The 7 direct repeats (designated the Inc A region) lie downstream from a functional promoter region. A 2nd incompatibility region (Inc B) lies between the 7 direct repeats and the 8th repeat. The putative promoter-operator region of the π structural gene contains an inverted repeat indicated by facing arrows. The π structural gene and a 2nd open reading frame (ORF) are present in the β-origin fragment. <u>Hind</u>III and <u>Bgl</u>II sites are indicated by H and B, respectively.

region. These analyses have established the essential role of both the <u>Hind</u>III-Bgl II fragment (bp positions 0 to 277) containing the 7 direct nucleotide sequence repeats and the π initiation protein in the activity of both the α and β-origins of replication. In addi-

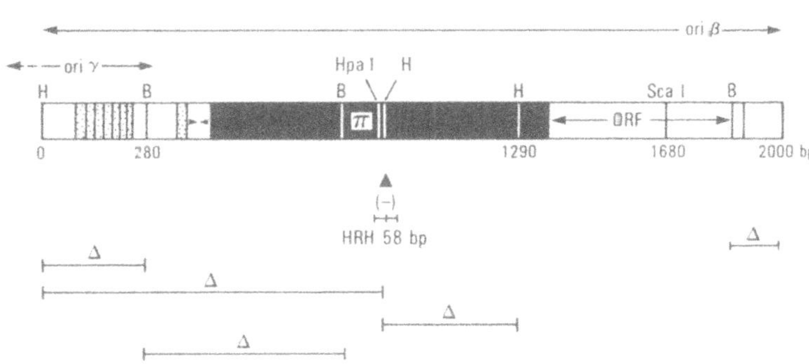

Fig. 4. The position of deletions and insertions in the β-origin fragment (26). The extent of each deletion is indicated by cross bars. HRH refers to a <u>Hind</u>III fragment of 58 bp that contains an internal <u>Eco</u> RI site. Other designations are described in the legend to Fig. 3.

tion, insertions of nucleotide sequences ranging from 12-bp to 340-bp (including sequences that maintain the translational reading frame) at the Hpa I (position 931-bp) and Sca I (position 1,680-bp) sites inactivate the β origin (P. Mukhopadhyay, unpub. observ.). Except for a 12-bp insertion at the Hpa I site, the insertional inactivation cannot be fully complemented by providing in trans the 2 Bgl II segments extending from the 275-bp to 1,869-bp positions which code for the π protein. These observations suggest a requirement for structural integrity in at least certain regions of this 2-kb segment for a functional β origin (26).

Expression of the π Structural Gene is Autoregulated

Plasmids carrying a fusion of the *pir*-promoter to the *lacZ* gene express β-galactosidase activity in *E. coli* at a substantially reduced level when the π protein is provided in trans [(25) and G. Davis and M. Filutowicz, unpub. observ.]. The activity of the π protein in repressing the expression of the *pir* gene has also been demonstrated in vitro utilizing purified π protein and *E coli* RNA polymerase (A. Greener, unpub. observ.). This activity of the π protein, illustrated in Fig. 5, has been examined also by the DNase I footprint technique (15) and M. Filutowicz, unpub. observ.]. It was found that purified π protein and RNA polymerase bind to the presumed operator-promoter region of the π structural gene. As shown in Fig. 6, the binding sites overlap for 3 helical turns that include 2 partial repeat sequences in an inverted orientation and adjacent to the eighth repeat. Competition binding studies further demonstrated that the π protein will displace prebound RNA polymerase. These results strongly support a mechanism for autoregulation of the *pir* gene that involves competition between the π protein and RNA polymerase for binding to the *pir* promoter region.

Mutations in the Direct Nucleotide Sequence Repeats Alter π Protein Binding and Inactivate the γ-origin

A fusion protein consisting of the π protein joined near its

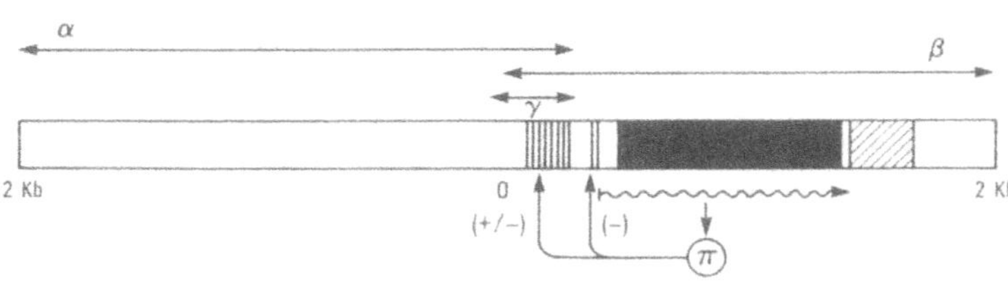

Fig. 5. Interaction of the π protein with the direct nucleotide sequence repeats. The relative positions of the *pir* gene (solid region) and the ORF (diagonal lines) are indicated with respect to the α, γ and β origins of replication. The narrow vertical boxes locate the positions of the direct repeats.

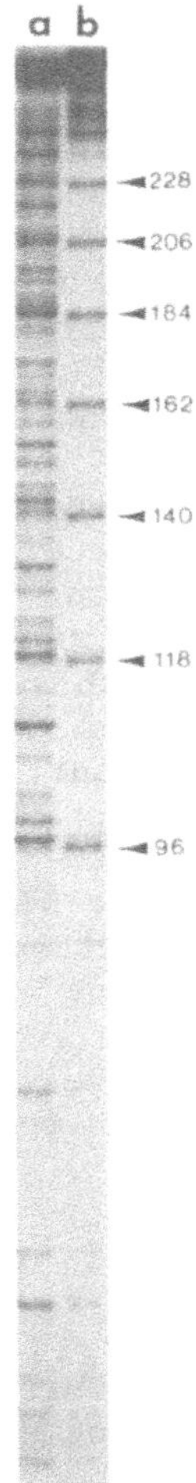

Fig. 6. Footprint pattern of a HindIII-Bgl II fragment containing the direct repeats. This fragment labeled at the 3' end was subjected to DNase I cleavage in the absence (a) and presence of purified π protein (b). Within the protected segment characteristic enhancements, each in 1 of the 7 direct repeats, are indicated by arrow. Positions of enhanced cleavages by DNase I were determined by comparison to sequencing samples run in parallel and numbered according to Stalker et al., (2). (Filutowicz, Uhlenhopp and Helinski, ms. in prep.).

C-terminal end to an N-terminal α-donor polypeptide of *E. coli* β-galactosidase has been shown to bind to the 7 tandem repeats within the γ-origin (14,15). DNase I footprint analysis of this specific binding using purified wild-type π protein exhibits a similar association with the tandem repeats (Fig. 7) as shown by the chimeric protein (M. Filutowicz, unpub. observ.). Sodium bisulfite in vitro mutagenesis of the direct repeat region within the γ-origin was carried out to assess the primary structural requirements for π binding and origin activity. The mutagenized γ-origin, joined to a pBR322 replicon, was analyzed for functionality of the γorigin in a <u>polA214</u> temperature-sensitive host strain. A number of partial or completely defective γ-origin mutants have been obtained (M. McEachern, unpub. observ.). The mutational changes, identified by nucleotide sequence analysis are shown for mutants γ117 and γ120 in Fig. 8. Mutant γ117, containing G to A transitions at 3 positions of the first nucleotide sequence repeat, exhibits defective γ-origin activity and loss of binding of the wild-type π protein specifically to this altered repeat (M. McEachern, unpub. observ.). Two of the 3 G to A substitutions are in the consensus hexanucleotide sequence shown in Fig. 1. A revertant of this γ-origin mutant also has been obtained (M. McEachern, unpub. observ.). This revertant of γ117 was deleted for the altered first repeat. It is of interest that this first repeat is immediately downstream from the active promoter in the γ-origin region (Fig. 3 and 8). Since <u>E. coli rpoB</u> mutants have been found to suppress temperature-sensitive π protein mutants (M. Filutowicz, unpub. observ.), defective binding of the π protein to the first repeat in γ117 may prevent a specific interaction of the π protein with RNA polymerase at the γ-origin promoter site which may be required for origin activity. Deletion of this mutant repeat may restore the structural requirements for this specific interaction of the π protein and RNA polymerase.

Mutant γ120 contains 5 C to T transitions. DNase I footprint analysis showed that π was unable to bind to the altered sixth repeat and bound less efficiently to the altered seventh repeat (M. McEachern, unpub. observ.). A revertant of γ120 contained a deletion of the sixth repeat, suggesting again that spacing requirements

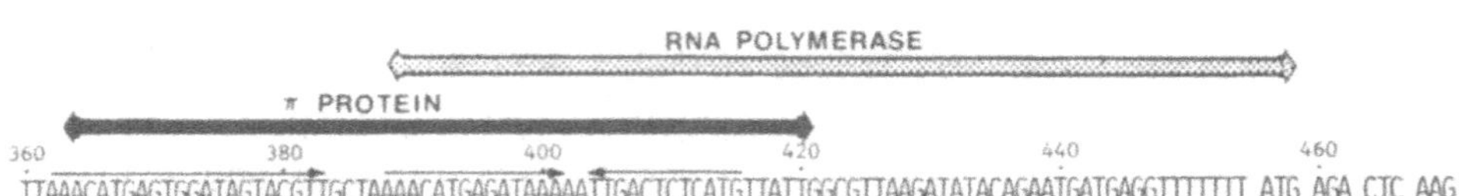

Fig. 7. π protein and RNA polymerase binding sites in the promoter-operator region of the <u>pir</u> gene. The coordinates are taken from Stalker et al., (2). The putative -10 and -35 segments of the <u>pir</u> are underlined. The horizontal arrows indicate the 8th direct and the inverted repeats which show partial homology with the repeat of 22 bp. The vertical arrows indicate position of enhanced DNase I cleavage frequencies observed in the presence of the π protein. (Filutowicz, Davis, Greener and Helinski, ms. in prep.).

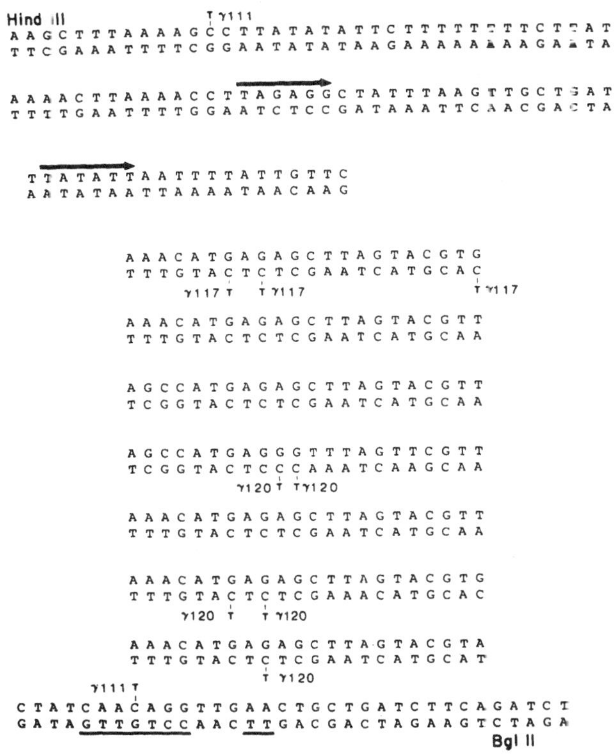

Fig. 8. Location of bisulfite-induced mutations resulting in an inactivation of the γ-origin. The sequence of the HindIII-Bgl II segment is taken from Stalker et al., (1). The horizontal arrows indicate putative -10 and -35 sequences of a promoter located upstream of the seven 22-bp repeats [(27) and P. Mukhopadhyay, unpub. observ.]. Solid bars indicate conserved sequences present in several plasmids containing direct repeats. (McEachern, Filutowicz and Helinski, ms. in prep.).

are critical to γ origin functionality. Thus, in the case of γ-origin mutants γ117 and γ120 base changes at the seventh and ninth positions within the 22-bp repeat (and within the consensus hexanucleotide sequence) result in a loss of π binding to the altered repeat. It is worth noting that the bases at these positions are among those that interact with the π-β-galactosidase fusion protein (15).

A γ-origin Mutant Contains a Base Substitution in a Consensus ACAGG Sequence

Analysis of plasmids containing direct nucleotide sequence repeats at their origins revealed that a majority of these plasmids contained the sequence ACAGG in the immediate proximity of the direct repeats (Tab. 2) (M. McEachern, unpub. observ.). One of the γ-origin replication defective mutants, obtained by sodium bisulfite mutagenesis and designated γ111, contained 2 G to A transitions with 1 of the base changes in the second position of the ACAGG sequence (Fig. 8) (M. McEachern, unpub. observ.). The frequency of occurrence of this pentanucleotide sequence in the vicinity of the direct nucleotide sequence repeats in a variety of plasmids and the mutational analysis of γ111 suggests an essential role for this sequence in R6K origin activity.

π Initiation Protein Acts Both Positive and Negatively at the Origin of Replication

Mutant π proteins have been produced that result in temperature-sensitivity (13), a change in copy number (30) or a change in incompatibility properties of the R6K replicon (M. Filutowicz, M. McEachern, unpub. observ.). One of these mutants, designated *pir405*-Cos (cold sensitive), contains a single base substitution resulting in a glycine to aspartic acid amino acid substitution at the 8first position of the π protein (30). This mutation was produced in a derivative of R6K, plasmid PRK 419, that contains the *pir* gene and the β and λ-origins of replication (Fig. 9). The *pir405*-Cos exhibits a substantially elevated plasmid copy number in *E. coli* cells growing at 42° and an even greater copy number at 30° (30). As shown in Fig. 9, the mutant phenotype is fully suppressed when the wild-type *pir* sequence is provided in trans. Complete suppression requires the N-terminal and at least part of the C-terminal sequence of the π protein (30). A second π structural gene mutant,

Tab. 2. Conserved pentanucleotide sequence in the origin region of plasmids containing direct repeats. The sequences were taken from Ref. 1 (R6K), 29 (pSa), 3 (RK2), 9 (Rts1), and 5 (F). The sequence for plasmid R485 was obtained by D. Stalker (unpub. observ.).

R6K	t a c t a t C a ACAGG t t g A A c t g c t
R485	g t a c a t C C ACAGG c t a A A a a a a t
pSa	a c c c g g a g ACAGG c a t A A t a a t a
RK2	g g c t g t C C ACAGG c a g A A a a t c c
Rts1	g c c a a a C C ACAGG c a c A c t t a t t
F	a a t t t g t C ACAGG g t t A A g g g c a

Conserved pentanucleotide sequence in the origin region of plasmids containing direct repeats. The sequences were taken from references 1 (R6K), 29 (pSa), 3 (RK2), 9 (Rts1), and 5 (F). The sequence for plasmid R485 was obtained by D. Stalker (unpublished observations).

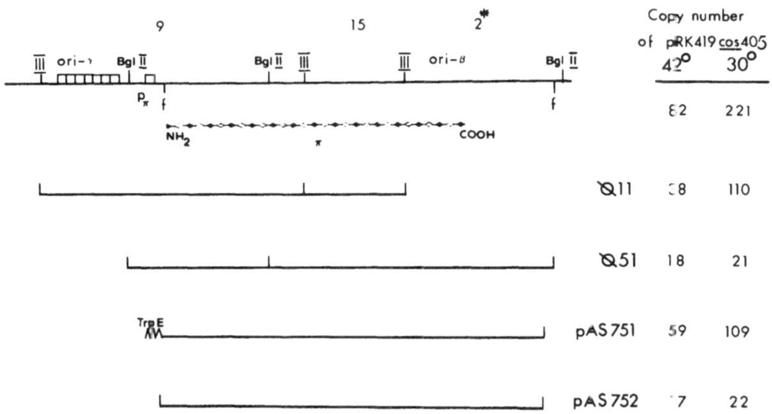

Fig. 9. Copy number of pRK419 pir405-cos in the presence of wild-type and altered π proteins supplied in trans. Relevant restriction sites are indicated within the pir structural gene. III and f refer to HindIII and Hinf I restriction sites, respectively. The HindIII fragments of R6K present in pRK419 and pir405-cos plasmids are designated 9, 15, and 2*. The location of the π promoter is designated p_{π}, whereas the eight 22bp repeats are indicated by boxes. Ø11 and Ø51 are π coding sequences integrated into the host chromosome. pAS751 and pAS752 contain π coding sequences inserted into the pBR322 derivative pAS620. TrpE refers to the 7 NH_2-terminal amino acid residues of the trpE gene and the tryptophan operon promoter. The copy number of pRK419 pir405-cos was measured in 10-ml cultures grown at 42°C and in cultures that have been shifted from 42°C to 30°C for 4 hr. Linearized pRK248 was the internal standard in strains Ø11 and Ø51. Data were presented in Ref. 30.

designated pir13-Cop and resulting in a leucine to arginine substitution at the 9 first amino acid position of the π protein, similarly exhibits a high copy-number phenotype that is suppressed by a wild-type pir sequence provided in trans (M. McEachern, unpub. observ.). The pir13-Cop or pir405-Cos mutation, present in plasmid pRK 419, also results in a decrease in R6K incompatibility (M. McEachern and M. Filutowicz, unpub. observ.).

The properties of the pir405-Cos and pir13-Cop mutants indicate that the π protein plays a role in the negative control of R6K replication in addition to its positive requirement for the initiation of plasmid replication. This negative activity of π is also supported by pulse-labeling experiments that measure R6K γ-origin activity in E. coli cells after induction of π synthesis. As shown in Fig. 10, a pBR322 hybrid plasmid containing a λ cl-ts gene and the R6K pir gene downstream from the λ P_R promoter was constructed and

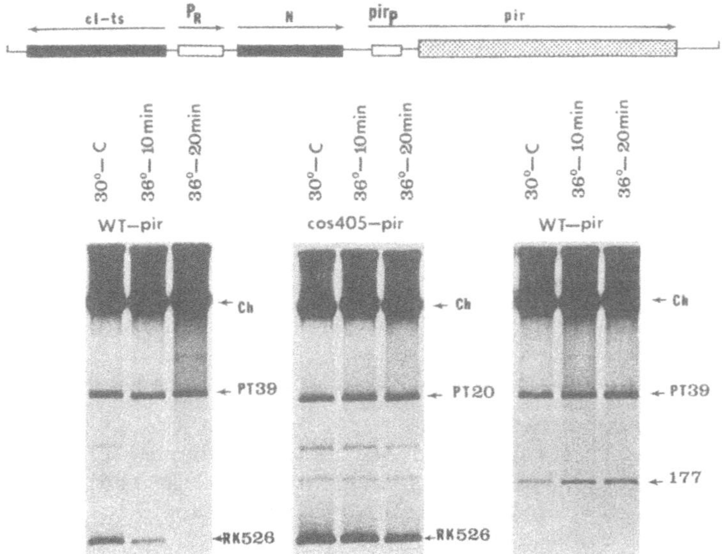

Fig. 10. The effect of increased intracellular π concentration on the replication of a γ origin plasmid. Plasmids pPT39 and pPT20 contain a FnuDII-pir fragment from wild-type pRK419 and the pir405-cos mutant, respectively, placed downstream of an inducible P_R-promoter of phage λ. The key element involved in control of the π protein expression in these constructs are schematically shown at the top of the figure. E. coli strain P678-54 (Thy-) was transformed at 30°C with appropriate combinations of plasmids. 2 ml of logarithmically growing cultures (2×10^8 cells ml^{-1}) supplemented with 5 µg ml^{-1} of thymidine were labeled with [$2^{14}C$]-thymidine (1 µci ml^{-1}) for 15 min. The cultures were subjected to a temperature shift (35°) and then transferred back to 30°C after the time indicated and before the addition of radioisotope. The cells were harvested and washed with ^{14}C-free medium. The plasmid DNA molecules were separated by agarose gel electrophoresis. RK526 and 177 refer to the R6K γ-origin plasmid and pACY§c177, respectively.

used to temperature-induce wild-type π protein synthesis in E. coli (P. Tooker and M. Filutowicz, unpub. observ.). Replication of the λ-origin plasmid pRK 526 in cells carrying the inducible pir plasmid was measured by pulse labeling DNA with C^{14} thymidine before and after temperature induction of π synthesis results in a substantial reduction of γ-origin replication (Fig. 10). As a control, the replication of plasmids pACYC177, R300B, PB§t322 and an RK2 derivative were not affected by the induction of π synthesis. Finally, using analogous constructs, the induction of mutant pir405-Cos π synthesis appeared to be considerably less effective in repressing γ-origin replication. These results indicate that excessive levels of π protein specifically repress R6K γ-origin replication.

π Protein Binds to the Nucleotide Sequence Repeats and Acts Positively and Negatively in the Regulation of R6K Replication

The bifunctional properties of the π protein and the analysis of γ-origin replication mutants suggest a model of regulation of R6K replication (Fig. 11) where the π protein at low concentrations initiates R6K replication by virtue of its binding to the direct nucleotide sequence repeats. Subsequent biochemical events in the region of the tandem repeats result in the initiation of replication at the α, β or γ origin of replication. The mechanism by which π protein association with the nucleotide sequence repeats results in the initiation of replication at a distant site (α or β origin) is unknown. Some structural integrity involving the nucleotide sequence between the direct repeats and the β-origin appears to be required for a functional β-origin. According to this model at higher π concentrations a negative activity of π is exerted at both regions containing repeats of 22 bp; the γ-origin and the operator region for the _pir_ gene. The range of π concentrations required for positive or negative activity is unknown. In an earlier study involving a functional π protein lacking several N-terminal amino acids, the results indicated a constant copy number of a γ-origin replicon over a broad range of π concentrations (25). The role of autoregulation of the _pir_ gene expression and the role of the promoter upstream from the direct nucleotide sequence repeats in the regulation of R6K replication are also unknown.

CONCLUDING REMARKS

The display of different strategies for the regulation of replication of plasmid elements underscores the versatility with which

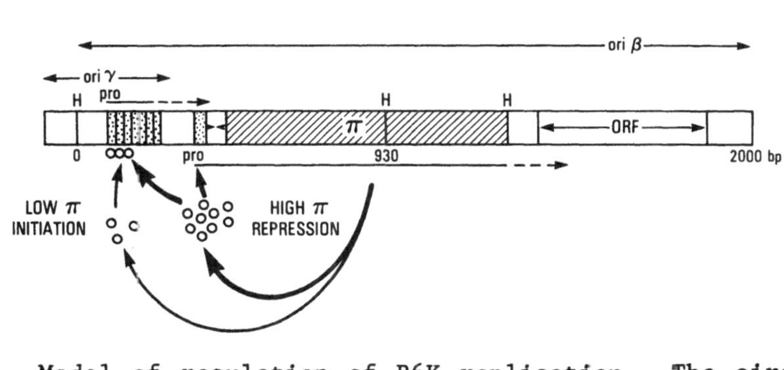

Fig. 11. Model of regulation of R6K replication. The circles represent π protein molecules. The various designations are described in the legend to Fig. 3.

plasmid elements have employed basic components (origin sequences, replication proteins, trans-acting repressors) in evolving regulatory circuits that provide for a fixed copy number in a bacterial host. In several of the plasmid systems examined, this copy number is the result of the interplay of separable positive- and negative-acting elements. In the case of plasmid R6K, both positive and negative activities reside in a single component, the π protein. This protein recognizes and binds to a multiple set of 22-bp nucleotide repeats. Since mutations that eliminate binding of the protein to a single repeat abolish replication activity at the origin sequence, this interaction appears to be critical for initiating a round of plasmid replication. The association of the π protein with the direct repeats most often results in replication origin activity at a considerable distance from the repeats. The structural gene for the π protein is autoregulated by a mechanism involving competition between the π protein and RNA polymerase for the putative promoter operator region of the <u>pir</u> gene. This region contains nucleotide sequences similar to the direct repeats at the origin of replication. The majority of plasmids in <u>E. coli</u> share with R6K the property of nucleotide sequence repeats at the origin of replication and a plasmid-specified replication protein. It will be of interest to determine if similarities exist among these plasmids in the nature of the interaction of the replication protein with these nucleotide sequence repeats.

ACKNOWLEDGMENT

This work was supported by grants from the National Institutes of Allergy and Infectious Diseases and the National Science Foundation. E.U. was supported by the National Science Foundation College Faculty Program.

REFERENCES

1. Stalker, D.M., R. Kolter, and D.R. Helinski (1979) Nucleotide sequence of the region of an origin of replication of the antibiotic resistance plasmid R6K. <u>Proc. Natl. Acad. Sci., USA</u> 76:1150-1154.
2. Stalker, D.M., R. Kolter, and D.R. Helinski (1982) Plasmid R6K DNA replication. I. Complete nucleotide sequence analysis of an autonomously replicating segment. <u>J. Mol. Biol.</u> 161:33-43.
3. Stalker, D.M., C.M. Thomas, and D.R. Helinski (1981) Nucleotide sequence of the origin of replication of the broad host range plasmid RK2. <u>Mol. Gen. Genet.</u> 181:8-12.
4. Smith, C.A., and C.M. Thomas (1984) Nucleotide sequence of the <u>trf</u>A gene of broad host-range plasmid RK2. <u>J. Mol. Biol.</u> <u>175</u>:000-000.
5. Murotsu, T., K. Matsubara, H. Sugisaki, and M. Takanami (1981) Nine unique repeating sequences in a region essential for rep-

lication and incompatibility of the mini-F plasmid. Gene 15:257-271.
6. Tolun, A., and D.R. Helinski (1982) Separation of the minimal replication region of the F plasmid into a replication origin segment and a trans-acting segment. Mol. Gen. Genet. 186:372-377.
7. Churchward, G., P. Linder, and L. Caro (1983) The nucleotide sequence of replication and maintenance functions encoded by plasmid pSC101. Nucl. Acids Res. 11:5645-5659.
8. Armstrong, K.A., R. Acosta, E. Ledner, Y. Machida, M. Pancotto, M. McCormick, H. Ohtsubo, and E. Ohtsubo (1984) A 37 kilodalton plasmid-encoded protein is required for replication and copy number control in pSC101 and Ts derivative pHS1. J. Mol. Biol. (in press).
9. Kamio, Y., A. Tabuchi, Y. Itoh, H. Katagiri, and Y. Terawaki (1984) Complete nucleotide sequence of mini-Rts1 and its copy mutant. J. Bact. 158:307-312.
10. Abeles, A., K.M. Snyder, and D. Chattoraj (1984) P1 plasmid replication: replicon structure. J. Mol. Biol. 173:307-324.
11. Scherer, G. (1978) Nucleotide sequence of the O gene and of the origin of replication in bacteriophage λ DNA. Nucl. Acid Res. 5:3141-3156.
12. Grosschedl, R., and G. Hobom (1979) DNA sequences and structural homologies of the replication origins of lambdoid bacteriophages. Nature. 277:621-627.
13. Inuzuka, M., and D.R. Helinski (1978) Replication of antibiotic resistance plasmid R6K DNA in vitro. Biochem. 17:2567-2573.
14. Germino, J., and D. Bastia (1983) The replication initiator protein of plasmid R6K tagged with β-galactosidase shows sequence-specific DNA binding. Cell 32:131-140.
15. Germino, J., and D. Bastia (1983) Interaction of the plasmid R6K-encoded replication initiator protein with its binding sites on DNA. Cell 34:125-134.
16. Tolun, A., and D.R. Helinski (1981) Direct repeats of the F plasmid IncC region expresses F incompatibility. Cell 24:687-694.
17. Tsutsui, H., A. Fujiyama, T. Murotsu, and K. Matsubara (1983) Role of 9 repeating sequences of the mini-F genome for expression of F-specific incompatibility phenotype and copy number control. J. Bact. 155:337-344.
18. Lovett, M.A., R.B. Sparks, and D.R. Helinski (1975) Bidirectional replication of plasmid R6K DNA in Escherichia coli: correspondence between origin of replication and position of single-strand break in relaxed complex. Proc. Natl. Acad. Sci., USA 72:2905-2909.
19. Crosa, J.H., L.K. Luttrop, F. Heffron, and S. Falkow (1975) Two replication initiation sites on R-plasmid DNA. Molec. Gen. Genet. 140:39-50.
20. Inuzuka, N., M. Inuzuka, and D.R. Helinski (1980) Activity in vitro of 3 replication origins of the antibiotic resestare plasmid RSF 1040. J. Biol. Chem. 255:11071-11074.

21. Crosa, J.H. (1980) Three origins of replication are active *in vivo* in the R-plasmid RSF 1040. *J. Biol. Chem.* 255:11075-11077.
22. Inuzuka, M., and D.R. Helinski (1978) Replication of antibiotic resistance plasmid R6K DNA *in vitro*. *Biochemistry* 17:2567-2573.
23. Kolter, R., M. Inuzuka, and D.R. Helinski (1978) Trans-complementation dependent replication of a low molecular weight origin fragment from plasmid R6K. Cell 15:1199-1208.
24. Shon, M., J. Germino, and D. Bastia (1982) The nucleotide sequence of the replication origin of the plasmid R6K. *J. Biol. Chem.* 257:13823-13827.
25. Shafferman, A., R. Kolter, D.M. Stalker and D.R. Helinski (1982) Plasmid R6K DNA replication. III. Regulatory properties of the π initiation protein. *J. Mol. Biol.* 161:57-76.
26. Shafferman, A., and D.R. Helinski (1983) Structural properties of the beta origin of replication of plasmid R6K. *J. Biol. Chem.* 258:4083-4090.
27. Shafferman, A.L., and D.R. Helinski (1984) Transcription signals in a region essential for replication of plasmid R6K. *Plasmid* (submitted for publication).
28. Kolter, R., and D.R. Helinski (1982) Plasmid R6K DNA replication. II. Direct nucleotide sequence repeats are required for an active gamma origin. *J. Mol. Biol.* 161:45-56.
29. Tait, R.C., C.I. Kado, and R.L. Rodriguez (1983) A comparison of the origin of replication of pSa with R6K. *Mol. Gen. Genet.* 192:32-38.
30. Stalker, D., M. Filutowicz, and D.R. Helinski (1983) Release of initiation control by a mutation in the plasmid encoded protein required for R6K DNA replication. *Proc. Natl. Acad. Sci., USA* 80:5500-5505.

INITIATION OF REPLICATION OF THE *ESCHERICHIA COLI* CHROMOSOMAL ORIGIN RECONSTITUTED WITH PURIFIED ENZYMES

Jon M. Kaguni,[1] LeRoy L. Bertsch, David Bramhill, James E. Flynn,[2] Robert S. Fuller,[3] Barbara Funnell, Satoko Maki, Tohru Ogawa, Kazuko Ogawa, Arie van der Ende, and Arthur Kornberg

Biochemistry Department of
Stanford University
Stanford, California 94306

ABSTRACT

A mixture of purified proteins has replaced a crude enzyme fraction capable of efficient replication of *oriC*-containing plasmids. The reconstituted enzyme system contains proteins which provide initiation, replication, and specificity functions required for dnaA-dependent replication specific for an *oriC* template. Replication can be separated into successive stages of RNA synthesis and DNA replication. Isolation of an intermediate no longer requiring RNA polymerase action requires the presence of dnaA protein, DNA gyrase, dnaB protein and dnaC protein. Intermediate formation likely involves binding of dnaA protein to a 9-bp sequence present 4 times as inverted repeats within the chromosomal origin sequence.

INTRODUCTION

The study of replication of single-stranded DNA bacteriophages M13, G4, and ϕX174 has contributed significantly to our current

---

[1]Present address: Biochemistry Dept., Michigan State University, East Lansing, Michigan 48824; [2]deceased; [3]present address: Dept. of Microbiology and Immunology, University of California, Berkeley, California 94720.

understanding of priming of nascent chains in *E. coli* (1). These bacteriophages rely almost entirely on host-encoded enzymes; their modes of replication represent 3 distinct mechanisms of priming DNA chains, all extended by DNA polymerase III holoenzyme, the major replicative enzyme of *E. coli* (2). Such studies, although useful in understanding nascent chain synthesis, do not explain how replication is initiated at the unique chromosomal origin, *oriC*, nor how this process is regulated and coupled to other cellular events.

Understanding the biochemical nature of the initiation process and its regulation has become accessible because of 2 advances. The first is the availability of recombinant DNAs containing the *E. coli* chromosomal origin, *oriC* (3,4). This unique sequence of 245 bp, highly conserved among many gram-negative bacteria (5), has been well characterized by deletion, mutation, and comparative DNA sequence analyses (6). The second advance is the discovery of a soluble enzyme system which efficiently replicates such recombinant DNAs in a bidirectional manner by initiating DNA synthesis from or near *oriC* (7,8).

Further studies, focused on resolving this enzyme system into its component parts, have identified 3 functional classes of proteins: (i) initiation proteins involved in recognizing the *oriC* sequence and transcribing it, (ii) replication proteins involved in extension of DNA chains, and (iii) specificity proteins which suppress DNA synthesis at extraneous sites throughout the DNA duplex while permitting it to occur at the *oriC* sequence. A key feature of the initiation process likely involves recognition of specific repeated sequences within *oriC* by dnaA protein.

A reconstituted mixture of purified proteins has been developed which performs initiation, replication and specificity functions (9). Replication of a variety of *oriC*-containing DNAs occurs efficiently and can be separated into distinct stages of RNA synthesis and DNA replication. A biochemical understanding of the initiation process and hopefully its regulation can now be approached.

dnaA PROTEIN BINDING SITES

The product of *dnaA* has been implicated in genetic and physiological studies at an early stage in initiating a cycle of chromosomal replication (10). With the exception of pSC101 (11,12), this *dnaA* requirement is unique for replication from the *E. coli* origin, *oriC*. The dnaA protein has been purified based on its ability to bind to restriction fragments containing *oriC* (13) and by a complementation assay for in vitro replication of *oriC*-containing templates in conjunction with its overproduction in an expression vector (14).

As judged by nitrocellulose filter retention assays, dnaA protein specifically binds to restriction fragments containing the oriC sequence. DNase I protection experiments indicate that dnaA protein specifically binds to and protects a 9-bp sequence found 4 times in oriC (15). Protection by dnaA protein appears to be cooperative. At levels of dnaA protein of less than 20 molecules per DNA fragment, little or no protection was observed; a level of 30 to 40 is effective in protecting the repeat sequences as well as adjacent sequences. Sites of enhanced cleavage are observed within some repeat sequences as well as outside these repeats (15).

This sequence of 9 bp is not unique to the chromosomal origin; homologous sequences are found in a variety of DNAs (Tab. 1). In addition to binding to oriC, dnaA protein also binds to restriction fragments containing: (i) the origin regions of pBR322, ColE1, and

Tab. 1. Consensus Sequence of dnaA Protein Binding Sites (15).

A. Sequences in known binding sites:	
oriC R1	TTG TTATCCACA GGC
oriC R2	GGG TTATACACA ACT
oriC R3	TAG TTATCCAAA GAA
oriC R4	GAG TTATCCACA GTA
ori-pBR322	CGG TTATCCACA GAA
ori-ColE1	CAG TTATCCACA GAA
ori-pSC101	GAG TTATACACA GGG
dnaA promoter region	GAT TTATCCACA GGA
dnaA structural gene	AGT TTTTCCACA CCT
"X-protein" promoter region	TAC TTTTCCACA GGT
IR_L-Tn5	CTC TTATACACA AGT
M13 site	GCG TTATACAAA TCT
Consensus:	gag $TTAT_{A}^{C}CA_{A}^{C}A$ gga
B. Other homologous sequences:	
Origins of replication:	
R100	TGC TTATCCACA TTA
R1	TGC TTATCCACA TTT
CloDF13	AAG TTACCCACA GAT
P1	CAC TTATCCACA TTA
P1	ACA TTATCCACT GGA
F	GGG TTATCCACT TAT
F	CAC TTATCCACG GGG
Regulatory regions:	
pyrBI (aspartate transcarbamylase)	ACC TTATCCAAA CTG
argF (ornithine transcarbamylase)	GAT TTATACAAA AAA
pclA	AAA TTATCCACA TGA
uvrB	TGG TTATCCACA GGA
dam	GCT TTCTCCACA GCC

pSC101; (ii) the promoter regions for the dnaA protein and the "X-protein"; (iii) the dnaA structural gene; (iv) the left inverted repeat of Tn5; and (v) a coding region of M13. Other homologous sequences presumably bound by dnaA protein are found in origin regions of phage P1 and plasmids R100, R1, CloDF13, and F, as well as in regulatory regions for pyrB1, argF, polA, uvrB and dam.

These numerous dnaA binding sites may be important in regulating replication or gene expression. Support for such roles comes from the following lines of evidence: dnaA protein is required for replication of the E. coli chromosome (7,10), replication of pSC101 (11,12) and transposition of Tn5 (Johnson and Reznikoff, pers. comm.); dnaA protein may autoregulate its own expression (16).

INITIATION OF REPLICATION

Reconstitution of oriC Replication

A reconstituted enzyme system has been developed which depends on dnaA protein and efficiently replicates a variety of oriC-containing DNAs which differ both in the orientation of the oriC-containing fragments relative to the vector and in the amount of chromosomal DNA flanking oriC contained in the DNA fragment.

Maximal DNA synthesis is observed with these recombinant DNAs upon addition of partially purified fractions (resolved by chromatography of an ammonium sulfate fraction on an Amicon dye affinity column) to a mixture of highly purified proteins. In the presence of topoisomerase I (17), RNase H (18), and HU protein (19), which ensure that DNA synthesis is specific for oriC-containing DNA templates, replication is efficient and depends on dnaA protein. The activities contained in these partially purified fractions are currently being further purified in order to understand how they stimulate replication.

Replication of M13oriC2LB5 RF Requires Many Purified Proteins

A recombinant DNA, M13oriC2LB5 (Fig. 1), containing the minimal oriC sequence of 245 bp within a 327-bp fragment replicates efficiently in the presence of purified enzymes (Tab. 2). Requirements are observed for RNA polymerase, the gyrase B subunit, dnaA protein, single-strand DNA binding protein (SSB), dnaB protein, dnaC protein and DNA polymerase III holoenzyme. The incomplete dependence on protein HU for DNA synthesis may be due to its presence in other enzyme fractions. Poor dependencies on primase and gyrase A subunit are also observed. Based on specific assays for primase, it seems unlikely that this activity is present in significant amounts in other enzyme fractions. One possibility is that RNA polymerase may replace the requirement for primase in priming DNA chains for lagging strand synthesis. Alternatively, this reaction may represent

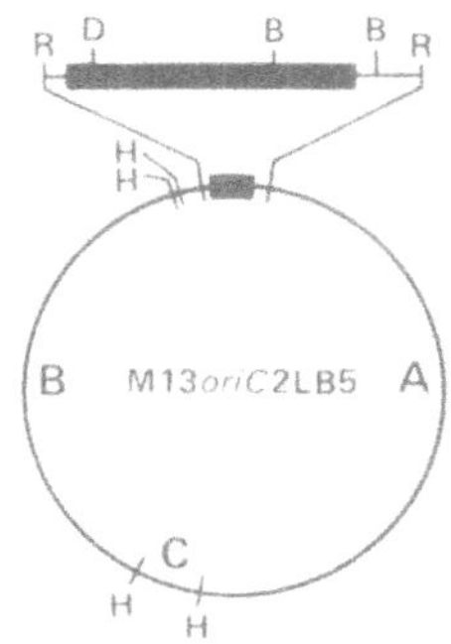

Fig. 1. Restriction map of M13oriC2LB5. This M13 chimera contains a 327-bp HaeIII-AccI fragment containing oriC with EcoRl linker ends from pTOA7 (T. Ogawa and A. Kornberg, unpub. data) inserted into the vector M13ΔE101. ■, oriC; H, HaeII; R, EcoRl; D, HindIII; B, BamHl.

an uncoupling of leading and lagging strand synthesis as has been observed for ColE1 replication (20).

Despite the poor dependence on the gyrase A subunit, its participation is inferred by inhibition by oxolinate, the specific

Tab. 2. Requirements for replication of M13oriC2LB5 RF in the presence of PVA.*

Component omitted	DNA synthesis (pmol)
Expt. 1:	
None	541
RNA polymerase	26
Gyrase A subunit	512
Gyrase B subunit	22
dnaA protein	64
Expt. 2:	
None	762
Protein HU	456
SSB	105
dnaB protein	23
dnaC protein	54
Primase	374
Pol III holoenzyme	5

*Topoisomerase I and RNase H were present to maintain oriC template specificity and dnaA dependence.

gyrase A inhibitor, and by the dependence on gyrase A under other reaction conditions described below.

Interestingly, restriction fragments containing the oriC sequence serve as templates for DNA replication (Tab. 3). Enzymatic requirements for DNA synthesis are similar to those of a reaction with supercoiled DNA, including the requirement for the gyrase B subunit (Tab. 4) and the gyrase A subunit (inferred by oxolinate-sensitivity).

Whereas replication of pSY317 and M13oriC26 RF started at or near the oriC sequence with bidirectional progress with a crude enzyme system (8), replication of M13oriC2LB5 RF is principally unidirectional from oriC in the purified enzyme system. It is not yet clear whether the lack of bidirectional progress is due to a lack of enzymes or certain DNA sequences, or both. The fewer priming events implied for a unidirectional mode of replication may simplify analysis of the initiation event.

The absolute need in oriC replication for a high concentration of a hydrophilic polymer such as polyvinyl alcohol or polyethylene

Tab. 3. Replication of restriction fragments containing oriC.

	Template (pmol)	DNA synthesis (pmol)	
		+dnaA	-dnaA
Expt. I:			
M13oriC2LB5 RF	600	342	18
HaeII-digested M13oriC2LB5 RF	600	45	9
M13ΔE101 RF (vector)	600	18	9
HaeII-digested M13ΔE101 RF	600	13	6
Exp. II:			
HaeII-digested M13oriC2LB5 RF	600	45	8
HaeII fragment A (oriC)	360*	45	7
HaeII fragment B	240*	12	6

*Equivalent to 600 pmol of M13oriC2LB5 digest.

Tab. 4. Requirements for replication of a restriction fragment containing oriC.*

Component omitted	DNA synthesis (pmol)
None	61
HU protein	51
SSB	18
RNA polymerase	6
Gyrase A subunit	38
Gyrase B subunit	19
dnaB protein	3
dnaC protein	13
Primase	21
dnaA protein	15
DNA pol III holoenzyme	3

*Topoisomerase I and RNase H were present to maintain oriC template specificity and dnaA dependence. Reactions contained 360 pmol of the HaeII fragment A containing oriC.

glycol (7) may be due, in part, to a macromolecular crowding effect (21) which concentrates the DNA template as well as scarce replication enzymes. Such polymers complicate the isolation of products of the reaction, inasmuch as the DNA becomes highly catenated by the action of DNA gyrase or by the combined action of topoisomerase I and exonuclease III (22).

By systematically altering the levels of each reaction component in the absence of such polymers, an acceptable level of replication was observed with an increased level of gyrase A. In the absence of polyvinyl alcohol, a strong dependence on the gyrase A subunit was observed. As in presence of polyvinyl alcohol, replication in its absence with the increased level of gyrase A depended on dnaA protein and is specific for oriC containing DNAs. The absence of polyvinyl alcohol minimizes catenation and, as shown below, permits isolation of an intermediate in replication.

Replication can be Separated into Successive Stages of RNA Synthesis and DNA Replication

The reaction could be staged by omission of DNA polymerase III holoenzyme from an initial reaction and by addition of specific inhibitors of RNA polyermase in a second incubation. Rifampicin, which inhibits initiation, and streptolydigin, which inhibits RNA elongation, when present at the outset inhibited DNA synthesis almost entirely (17 pmol). Addition of the inhibitors after the first stage of incubation concomitant with the addition of DNA polymerase III holoenzyme permitted DNA synthesis to occur (267 pmol). Thus,

RNA polymerase forms a product during the first stage of incubation which is required for subsequent DNA synthesis to occur.

Other reaction components, except for those which confer specificity, were tested to determine whether any of them were required prior to or during the action of RNA polymerase in the first stage. Except for RNA polymerase, omission of any one of these proteins from the first stage and its addition with DNA polymerase III holoenzyme, rifampicin, and streptolydigin at the outset of the second stage of incubation did not markedly reduce the amount of subsequent DNA synthesis. Thus, these proteins are presumed to act after RNA polymerase. Attempts to isolate by gel filtration an intermediate which no longer requires RNA polymerase action indicated that several of the proteins including dnaA, dnaB, dnaC, and DNA gyrase were required in the first stage. In their absence no functional intermediate could be isolated. However, primase and DNA polymerase III holoenzyme were dispensable in the first stage. These results indicate that dnaA, dnaB, dnaC, and DNA gyrase are needed to form an intermediate sufficiently stable for gel filtration and functional for DNA replication.

SUMMARY

A crude soluble enzyme system capable of authentic *oriC* replication has been replaced by a purified enzyme system which specifically sustains replication of a variety of *oriC*-containing DNAs. This reconstituted system contains proteins which fall into 3 groups: *initiation proteins* responsible for transcription at *oriC* to prime DNA synthesis, *specificity proteins* which ensure that DNA synthesis is specific to the chromosomal origin, and *replication proteins* responsible for advancement of the replication fork. A number of significant questions may now be answered with this reconstituted enzyme system: Where are the transcripts formed by RNA polymerase for *oriC* replication located and how are they utilized? What enzymes and DNA sequences are required for bidirectional replication from *oriC*? How does binding of dnaA protein to the chromosomal origin direct initiation of DNA synthesis? Does DNA gyrase have a role in replication beyond supercoiling of DNA? How closely do the mechanisms that operate in small phage replication resemble those at the elongating replication forks in the in vitro replication from *oriC*? What elements control the frequency of initiation at the chromosomal origin?

ACKNOWLEDGMENTS

We mourn the loss of Jim Flynn whose sudden death on May 8, 1984, deprived us of a gifted and devoted colleague and friend. Had he presented this paper as planned, it would have included a report of his extensive studies of the organization and replication of pSC101, as well as the studies on oriC to which he contributed.

REFERENCES

1. Dixon, N.E., L.L. Bertsch, S.B. Biswas, P.M.J. Burgers, J.E. Flynn, R.S. Fuller, J.M. Kaguni, M. Kodaira, M.M. Stayton, and A. Kornberg (1983) Single-stranded phages as probes of replication mechanisms. In Mechanisms of DNA Replication and Recombination, UCLA Symposia on Molecular and Cellular Biology, New Series, Vol. 10, N.R. Cozzarelli, ed. Alan R. Liss, Inc., New York, pp. 93-113.
2. Burgers, P.M.J., and A. Kornberg (1982) ATP activation of DNA polymerase III holoenzyme from Escherichia coli. II. Initiation complex: Stoichiometry and reactivity. J. Biol. Chem. 257:11474-11478.
3. Yasuda, S., and Y. Hirota (1977) Cloning and mapping of the replication origin of Escherichia coli. Proc. Natl. Acad. Sci., USA 74:5458-5462.
4. von Meyenburg, K., F. G. Hansen, L.C. Nielsen, and E. Riise (1978) Origin of replication, oriC, of the Escherichia coli chromosome on specialized transducing phages λasn. Mol. Gen. Genet. 160:287-295.
5. Zyskind, J.W., J.M. Cleary, W.S.A. Brusilow, N.E. Harding, and D.W. Smith (1983) Chromosomal replication origin from the marine bacterium Vibrio harveyi functions in Escherichia coli: oriC consensus sequence. Proc. Natl. Acad. Sci., USA 80:1164-1168.
6. Oka, A., K. Sugimoto, M. Takanami, and Y. Hirota (1980) Replication origin of the Escherichia coli K-12 chromosome: the size and structure of the minimum DNA segment carrying the information for autonomous replication. Mol. Gen. Genet. 178:9-20.
7. Fuller, R.S., J.M. Kaguni, and A. Kornberg (1981) Enzymatic replication of the origin of the Escherichia coli chromosome. Proc. Natl. Acad. Sci., USA 78:7370-7374.
8. Kaguni, J.M., R.S. Fuller, and A. Kornberg (1982) Enzymatic replication of the E. coli chromosomal origin is bidirectional. Nature (London) 296:623-627.
9. Kaguni, J.M., and A. Kornberg (1984) Replication initiated at the origin (oriC) of the E. coli chromosome reconstituted with purified enzymes. Cell, in press.
10. Hirota, Y., J. Mordoh, and F. Jacob (1970) On the process of cellular division in E. coli. III. Thermosensitive mutants of Escherichia coli altered in the process of DNA initiation. J. Mol. Biol. 53:369-387.
11. Hasunuma, K., and M. Sekiguchi (1977) Replication of plasmid pSC101 in Escherichia coli K12: requirement for dnaA function. Mol. Gen. Genet. 154:225-230.

12. Frey, J., M. Chandler, and L. Caro (1979) The effects of an Escherichia coli dnaAts mutation on the replication of the plasmids ColE1, pSC101, R100.1, and RTF-TC. Mol. Gen. Genet. 174:117-126.
13. Chakraborty, T., K. Yoshinaga, H. Lother, and W. Messer (1982) Purification of the E. coli dnaA gene product. EMBO J. 1:1545-1549.
14. Fuller, R.S., and A. Kornberg (1983) Purified dnaA protein in initiation of replication at the Escherichia coli chromosomal origin of replication. Proc. Natl. Acad. Sci., USA 80:5817-5821.
15. Fuller, R.S., B.E. Funnell, and A. Kornberg (1984) The dnaA protein complex with the E. coli chromosomal replication origin (oriC) and other DNA sites. Cell (submitted for publication).
16. Hansen, F.G., and K.V. Rasmussen (1977) Regulation of the dnaA product in Escherichia coli. Mol. Gen. Genet. 155:219-225.
17. Kaguni, J.M., and A. Kornberg (1984) Topoisomerase I confers specificity in enzymatic replication of the Escherichia coli chromosomal origin. J. Biol. Chem. 259:8578-8583.
18. Ogawa, T., G.G. Pickett, T. Kogoma, and A. Kornberg (1984) RNase H confers specificity in the dnaA-dependent initiation of replication at the unique origin of the Escherichia coli chromosome in vivo and in vitro. Proc. Natl. Acad. Sci., USA 81:1040-1044.
19. Dixon, N.E., and A. Kornberg (1984) Protein HU in the enzymatic replication of the chromosomal origin of Escherichia coli. Proc. Natl. Acad. Sci., USA 81:424-428.
20. Staudenbauer, W.L., E. Scherzinger, and E. Lanka (1979) Replication of the colicin El plasmid in extracts of Escherichia coli: uncoupling of leading strand from lagging strand synthesis. Mol. Gen. Genet. 177:113-120.
21. Zimmerman, S.B., and B.H. Pheiffer (1983) Macromolecular crowding allows blunt-end ligation by DNA ligases from rat liver or Escherichia coli. Proc. Natl. Acad. Sci., USA 80:5852-5856.
22. Low, R.L., J.M. Kaguni, and A. Kornberg (1984) Potent catenation of supercoiled and gapped DNA circles by topoisomerase I in the presence of a hydrophilic polymer. J. Biol. Chem. 259:4576-4581.

ORIGIN AND INITIATION SITES OF λdv DNA REPLICATION IN VITRO

Toshiki Tsurimoto, Haruhiko Kouhara, and Kenichi Matsubara

Institute for Molecular and Cellular Biology
Osaka University
Suita, Osaka 565, Japan

ABSTRACT

The sequence of λ DNA essential for the unique initiation of replication was analyzed in an in vitro replication system. Fragments of λ DNA were inserted into pBR322 and used as templates or were circularized in vitro in the absence of pBR322 and employed in the same tests. A 165-bp region left of the EcoRI site in the O gene of the λ genome was defined as the functional origin. This region, which we defined as the ori region, carries, in order, the 4 19-bp repeat sequences where the O protein binds (ori-repeats), an A+T-rich stretch, and a region that constitutes part of a large palindromic structure. Regions that have long been suspected to participate in λ DNA replication initiation, ice and oop were not required for the O, P-dependent λ-specific replication initiation. The λdv and the "ori region plus" recombinant plasmids initiated DNA synthesis at or around this region, and the reaction depended on the presence of λ-coded initiators, O and P proteins.

Early replicative intermediates of λdv were prepared in an in vitro replication system in the presence of ddCTP, an inhibitor of DNA chain elongation. This system allows only that DNA synthesis that is a result of replication initiation. Using this system, the initiation site(s) of the DNA synthesis was finely analyzed by mapping the transition sites from primer RNA to DNA synthesis. Short-chain DNAs produced from regions near the ori region were purified from the intermediates. A fraction of the short-chain DNAs was covalently linked to primer RNA. The 5'-ends that had been linked to RNA (transition sites) were exposed by alkaline hydrolysis, labeled with ^{32}P, and the transition sites were mapped along the nucleotide sequence of the genome. Two striking features emerged from this analysis: (i) The transition sites are located on both sides of the ori region, and no transition arose within the 165-bp ori region; (ii) The transition sites on both sides are not unique, but multiple, and are clustered in one of the 2 strands. Furthermore, their orientation demonstrates that the DNA synthesis in initiation of replication from the 2 sides of the _ori_ region converge. The frequency of the "leftward" DNA synthesis is several times higher than that of "rightward" synthesis. These results reflect asymmetric bidirectional replication of λdv DNA, and may also reflect replication of λ phage DNA.

INTRODUCTION

The initiation process of DNA replication, which controls the whole process of replication, is tightly regulated by multiple components. The key process in initiation has been intensively studied in several prokaryote replication systems, mostly in _Escherichia coli_, and many enzymes and various DNA signals that participate in the initiation reaction have been identified (1,2). The replication mechanism of bacteriophage λ and plasmid λdv (a minireplicon of λ phage) have also been intensively studied (for a review, see Ref. 3). A large number of genetic elements have been identified, reflecting the complicated processes that regulate the initiation of λ replication (4).

The replication initiation of λ phage or λdv plasmid depends on 2 λ-coded initiator proteins, the products of _O_ and _P_ genes (5). Replication initiates at or around a unique site located within the _O_ gene (about 81% of the λ phage genetic map) and proceeds bidirectionally (6,7). Prior to the initiation, the initiation region must be "activated" by transcription (transcriptional activation; Ref. 8,9), though the mechanism of this is unknown. The replication origin has been defined by several cis-dominant replication-defective mutations (10,11), which have been mapped at or around the origin defined by electron microscopic studies. The nucleotide sequence in this region shows 3 characteristic structures around the defined origin (12,13): (i) 4 19-bp repeating sequences arranged tandemly (named ori-repeats), (ii) an A+T-rich stretch consisting of about 40 bp, (iii) a 28-bp large palindromic sequence with an _EcoRI_ site (see Fig. 8). These structures together make up a region of about 200 bp that has been termed _ori_ (for replication _origin_).

So far, however, in vivo studies have failed to show if the defined _ori_ is the essential structure (functional origin) for replication of λ or λdv. Lusky and Hobom (14,15) have proposed that another signal, in addition to the 200-bp stretch of _ori_, is important for the replication of λ in vivo. They named the hypothetical functional region for replication "inceptor" (_ice_) and claimed that it is located far to the left of _ori_, in the middle of the _cII_ gene. They further postulated that it would be a site of switching from primer RNA to DNA synthesis. No further evidence to support this idea has been obtained, however, and another line of evidence obtained by Moore et al. (16) cast doubt on the notion that _ice_ is an indispensable site for λ replication.

Recently, biochemical studies have revealed further details of the replication mechanism. Many E. coli components participating in the elongation of DNA chains have been purified (17). In the case of λ, the initiation proteins O and P have been purified and characterized (18,19). The O protein binds specifically to the ori-repeats (20), and the P protein interacts with E. coli replication protein, dnaB (18,21). Purified dnaK protein interacts with both the O and P proteins (22). The O and P proteins were postulated to interact from a genetical study (23) but this has not been proved in in vitro studies with purified proteins. Despite the progress of interaction studies, nothing is yet known of how these proteins trigger the initiation of DNA replication.

An in vitro system has been developed that allows replication of λdv in the presence of the O and P proteins (24,25,26). This system allows de novo replication of an externally added mature λdv template, and is highly useful for analyzing the DNA signals that function in the initiation event and other molecular events associated with the initiation, such as location of initiation site(s). This paper describes the results of studies to define the functional origin of λ and map the initiation sites of replication, where the transition from primer RNA to DNA synthesis occurs.

RESULTS AND DISCUSSION

Defining the ori-Region Essential for Initiation of λdv Replication In vitro

Replication activity of pBR322-cloned λ ori fragments. We have developed an in vitro replication system for λdv DNA that depends on externally added O and P proteins (26). This system was used to analyze the minimal region of λ DNA essential for replication. As mentioned in the "Introduction," particular attention has been paid to ice and ori.

We constructed several pBR322-based recombinant plasmids carrying various regions around λ ori (Fig. 1). The replication ability of the recombinant DNAs in closed circular form was assayed in the presence or absence of O and P proteins (Tab. 1). pMY44, pOri-1, and pOA-4 carry respectively 165-bp, 354-bp, and 715-bp regions located to the left of the EcoRI site in the O gene (Fig. 1). The 165-bp region of pMY44 covers just the 3 unique structures of ori as mentioned above (see Fig. 8). The 715-bp region of pOA-4 covers ice and the oopRNA region, in addition to ori. pOri-D1 is similar to pOA-4 but lacks its rightmost portion, i.e., a 46-bp A+T-rich stretch in ori. pMY61-16 carries an additional 186-bp region to the right of pOri-1. In the absence of O and P proteins, λdv DNA does not replicate in the in vitro system. However, these pBR322-based plasmids allowed DNA synthesis to some extent, initiated from the pBR322-origin of replication. Addition of the O and P proteins

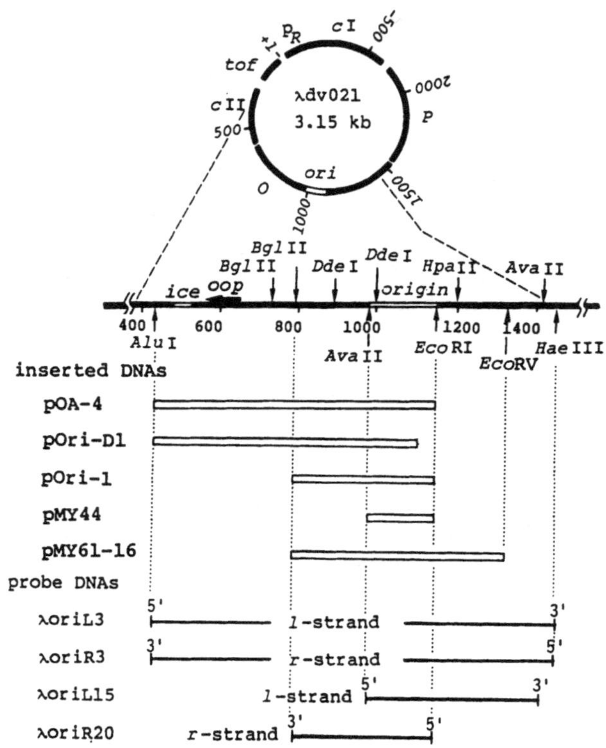

Fig. 1. λ fragments carried in the pBR322-based recombinant plasmids used as templates for the in vitro replication system. The genome of λdv021 (4,27) is shown with relevant restriction sites and the regions for ice and ori (open boxes). Genes cI, tof, cII, O, and P are shown in thick lines. oopRNA is indicated by an arrow. Numbers outside the circle indicate nucleotide numbers from mRNA starting point (+1) at p_R. The recombinant plasmids were constructed by inserting the fragments between EcoRI and BamHI sites of pBR322 in the same orientation. Constructions of pOA-4 (18), pOri-D1 (28), and pOri-1 (29) have been reported. pMY44 carries a 165-bp λ fragment isolated by AvaII (983) and EcoRI (1147) digestion. The AvaII end was converted to a BamHI end by end-filling by T4 DNA polymerase followed by ligation with BamHI linker. pMY61-16 carries the region between BglII (793) and EcoRV (1333). This plasmid was constructed by insertion of a 186-bp fragment covering EcoRI (1147) to EcoRV (1,333) into the EcoRI site of pOri-1. The EcoRV end was joined to the EcoRI end using EcoRI linker. Probe DNAs demonstrate regions of λ genome that were used, in single-stranded form, to purify short-chain DNAs (see text).

Fig. 1. (Continued)

DNAs covering these regions have been cloned into single-stranded M13 phage vectors, mp8 and mp9 (30). Lines represent the DNA regions and strands cloned in M13 phage DNAs. λoriL3 and λoriR3 carried the sequences of the l- and r-strands, respectively, covering the AluI (432) to HaeIII (1439) region. λoriL15 carried the l-strand spanning the 2 AvaII sites (983 to 1416). λoriR20 carried the r-strand of the region from BglII (793) to EcoRI (1147). The single-stranded phage DNAs were purified as described in (31), except that BND-cellulose chromatography was omitted.

stimulated DNA synthesis of pMY44, pOA-4, pOri-1, and pMY61-16, but not of pBR322 or pOri-D1. These results demonstrate that the functional origin of λ is located within the 165-bp region in pMY44. The A+T-rich stretch was indispensable for the λ replication, since pOri-D1 was inactive. All these DNAs, except for pMY61-16, lacked the 28-bp-palindromic sequence, because EcoRI truncated the palindrome. Since the activities of pMY61-16 and other active templates were the same, the complete palindromic sequence is dispensable to the replication. Whether a fraction of the palindromic sequence is needed is unknown.

Lusky and Hobom have argued that a region called ice takes part in the initiation of DNA synthesis. Our study has clearly shown that the ice region can be deleted without impairing the O,P-dependent replication ability in vitro. Recently, we have constructed a λdv genome that is completely devoid of the ice signal but which replicates autonomously (unpub. observation). Therefore, we conclude that ice is not essential for O,P-dependent λ replication, either in vitro or in vivo.

Replication of λ is sensitive to rifampicin, an inhibitor of RNA polymerase. The O,P-dependent DNA synthesis of pMY44, as well as λdv, was found to be sensitive to this drug (data not shown), showing that the minimized 165-bp λ ori sequence retained the original requirement for this enzyme. Since the 165-bp fragment lacks a normal promoter, the transcription needed for activating the ori may occur at a promoter(s) in the vector pBR322.

Replication activity of λ ori circles. In the experiments with pBR322-based recombinant plasmids, the replication activities of λ and pBR322 were superimposed. Though the results shown in Tab. 1 clearly demonstrated the presence of the functional λ origin within the 165-bp region, one could argue that it might be active only under the influence of the coexistent pBR322 replication system. To eliminate this possibility, we constructed the corresponding λ DNA fragment that had been circularized in vitro (λ ori circles). For

Tab. 1. Replication in vitro of pBR322-based recombinant plasmids carrying various fragments from around λ origin.

Plasmids	DNA synthesis (pmoles) Without O and P proteins	With O and Pproteins	differences
Experiment 1.			
pOA-1	74.4	150.4	76.0
pOri-D1	80.8	89.6	8.8
pOri-1	58.4	134.4	76.0
pBR322	72.0	80.8	8.8
λdv021	1.1	248.0	246.9
Experiment 2.			
pOri-1	162.2	226.2	64.0
pMY44	145.2	241.2	96.0
pMY61-16	115.2	209.5	94.3
pBR322	148.6	155.8	7.2
λdv021	2.8	198.8	196.0

Recombinant plasmids carrying various fragments of λ (see Fig. 1) were added as templates in the _in vitro_ replication system (26). Experiment 1 was carried out in the standard reaction mixture (25 μl) containing 40 mM Hepes (pH 8.0), 80 mM KCl, 2 mM ATP, 0.5 mM GTP, 0.5 mM CTP, 0.5 mM UTP, 50 μg/ml of bovine serum albumin, 100 μg/ml of creatine kinase, 21.6 mM creatine phosphate, 6 % (w/v) polyethylene glycol (PEG) 20,000, 11 mM $M_g(CH_3COO)_2$, 100 μM dNTP with 100 cpm/pmol of [α-^{32}P]dCTP, 8 μg/ml of template DNA and 120 μg of _E. coli_ YS-1 lysate prepared according to the method of Fuller et al (32). The mixture was incubated at 30°C for 30 min with or without O (20 μg/ml) and P (16 μg/ml) proteins. Experiment 2 was carried out similarly, except that 25 μM dNTP, 10 μg/ml of DNA, 30 μg/ml of O protein, and 10 μg/ml of P protein were employed in a volume of 20 μl. The results are expressed as total dNMP incorporated. Differences between DNA syntheses with and without O and P proteins are shown in the right column. As the lot numbers of _E. coli_ lysates used in experiments 1 and 2 were different, replication activity associated with pBR322 was somewhat higher in experiment 2 than in experiment 1.

this purpose, DNA fragments that had been cloned in pBR322 were extracted and circularized by DNA ligase under very dilute conditions in the presence of ethidium bromide to introduce negative turns in the DNA circle (Fig. 2). Since the λ _ori_ fragments in question are too small to use in such an assay, a 0.9-kb coding region for chloramphenicol-resistance gene (Cm^r) of Tn9 (33) was employed as a space DNA which showed no replication activity. The λ fragments carried in pMY44 and pMY61-16 were first inserted into an _EcoRI_ site located in the center of the Cm^r gene carried in pBR322. These Cm^r-λ _ori_ fragments were extracted from the recombinant plasmids and circularized. c53X and cLDX are λ _ori_ circles carrying 165-bp and 540-bp λ fragments within the Cm^r gene, respectively. The λ _ori_ in each circle was transcribed in the same orientation as in λdv from $\underline{p}_{lac}$, which was joined with the Cm^r gene (see legend to Fig. 2). cLB, which was used as a positive control in this assay system, carried a 1.9-kb λ fragment covering a region from $\underline{p}_R$ to ori. cV2 is a circle consisting only of the spacer Cm^r gene. The replication activities of these small template DNAs were assayed similarly as before. The results in Tab. 2 demonstrate that without O and P proteins, none of the small templates initiated DNA replication. In the presence of these proteins, cLB, c53X, and cLDX, but not cV2, initiated DNA synthesis. Thus, the in vitro assay system allows replication of minimized λ ori circles, in the absence of any influence from the replication of coexistent pBR322. The results for the small templates were in complete agreement with those obtained with pBR322-based recombinant plasmids. The 165-bp λ fragment (c53X), active in the O,P-dependent DNA replication, showed the same level of DNA synthesis as cLB and cLDX demonstrating that the 165-bp fragment carries all the signals necessary for the O,P-dependent DNA synthesis in the in vitro system. Thus, we defined the 165-bp region as the "ori region", indicating that it is the minimal functional origin of λ replication.

Identifying the Region Where λ DNA Replication is Initiated.

The initiation site of DNA replication on λdv021 and pMY44 was investigated using ddCTP, which blocks chain elongation at an early phase of replication (26). The template DNAs were replicated in the presence of [α-^{32}P]dCTP and ddCTP, and specific activity of the ^{32}P-count in each restriction fragment was analyzed (Fig. 3). Those fragments located in the O gene were found to be preferentially labeled when λdv021 was used as a template (Fig. 3A). Because DNA elongation is halted randomly by ddCTP, the relative specific activity in each fragment should reflect its distance from the initiation site of DNA synthesis, which, in this case, must lie within, or close to, the ori region. The gradual decrease of radioactivity on both sides of the ori region demonstrates that the DNA synthesis proceeds bidirectionally. However, the bidirectional synthesis seems to be asymmetric, with leftward synthesis stronger than rightward.

pMY44 was analyzed similarly. The result in Fig. 3B shows again that the initiation of DNA synthesis occurs within, or close to, the ori region. Asymmetric DNA synthesis was again observed, but in this case rightward synthesis was stronger. The cause of this difference between λdv and pMY44 is unknown. It could reflect a difference between the sequences flanking the ori region, which either promote or inhibit procession of the replication fork, or it could reflect a difference between transcription proceeding either across or away from the ori region in connection with primer RNA synthesis or transcriptional activation.

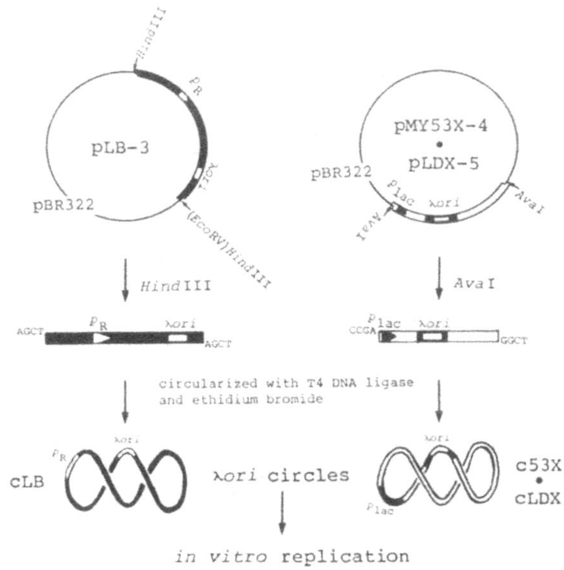

Fig. 2. Construction of λ ori circles. pLB-3 is a pBR322-based recombinant plasmid, carrying a 1.9-kb fragment of λdv covering from HindIII (-563) to EcoRV (1,333) inserted into the HindIII site of the vector. The EcoRV end had been converted to a HindIII end by ligation with HindIII linker. Thin and thick lines represent, respectively, regions of pBR322 and λdv sequence. The open arrow and box in the λ sequence are $\underline{p}_R$ (a right λ promoter) and λ ori region, respectively. pMY53X-4 and pLDX-5 are pBR322-based recombinant plasmids that carry λ fragments inserted into the Cm^r gene used as spacer (open bar). The spacer DNA, a 0.9 kb fragment consisting of a lactose operon promoter (plac: filled arrow) joined with 0.8 kb TaqI fragment coding for Cm^r gene of Tn9 (details will be described elsewhere), was inserted into the AvaI site of pBR322. The 165-bp and 540-bp λ fragments carried in pMY44 and pMY61-16, respectively, were inserted into a unique EcoRI site in the spacer DNA, resulting in pMY53X-4 and pLDX-5. The ori region in the inserted λ fragment was transcribed from the plac in the same direction as in λdv genome, where transcription is initiated at $\underline{P}_R$.

pLB-3 was digested with HindIII, and pMY53X-4 and pLDX-5 were digested with AvaI. DNA fragments of 1.9 kb from pLB-3, 1.2 kb from pMY53X-4 and 1.5 kb from pLDX-5 were purified by electrophoresis and self-ligated at a concentration of 1-2 μg DNA/ml in the presence of 5 μg/ml of ethidium bromide. The ligated DNA was extracted with

Fig. 2. (Continued)

phenol to remove ligase and ethidium bromide, and the resulting circles, cLB (from pLB-3), c53X (from pMY53X-4) and cLDX (from pLDX-5) were precipitated with ethanol. The majority of the DNAs in these preparations were in monomeric negatively twisted closed circular form.

The result in Fig. 3B alone does not allow us to conclude that the DNA synthesis of pMY44 is bidirectional. However, our mapping studies of the transition sites from primer RNA to DNA synthesis indicate that the DNA synthesis of this plasmid also proceeds bidirectionally, with several transition sites located on both strands around the _ori_ region (data not shown).

Determination of the Multiple Initiation Sites Flanking the ori Region

Analysis of transition sites from primer RNA to DNA synthesis. To study the first step in the initiation of λ DNA synthesis, we attempted to localize the site(s) of transition from primer RNA synthesis to DNA synthesis. For this purpose, λdv021 DNA was added as a template to the in vitro replication system. The reaction conditions were essentially the same as described in Tab. 1, except that 300 μl of reaction mixture contained 20 μg/ml of λdv DNA, 20 μM of each dNTP with 4.2 mCi/μmol of [methyl-^{3}H]TTP, and 40 μM ddCTP, and was incubated at 30°C for 10 min. A higher ddCTP/dCTP ratio (40/20) was chosen in this experiment than that adopted in Fig. 3 (100/100), in order to minimize the possibility of excessive chain

Tab. 2. Replication of λ _ori_ circles in vitro.

λ _ori_ circles	DNA synthesis (pmoles)	
	without O and P proteins	with O and P proteins
cLB	3.6	80.4
c53X	6.0	67.6
cLDX	4.4	63.6
cV2	4.4	10.8

Various fragments of λ from around _ori_ were joined with the $\underline{CM^r}$ gene and made into closed circles in an _in vitro_ ligation system (see text, and Fig. 2). The reactions were carried out at 30°C for 30 min in standard reaction mixture (20 μl) as described in Table 1, with 10 μg/ml of template λ _ori_ circles. The results are expressed as total dNMP incorporated. Under the condition used, 117 pmols of dNMP was incoporated into λdv021. cV2 is the control, a 0.9kb circle consisting solely of spacer $\underline{CM^r}$ DNA.

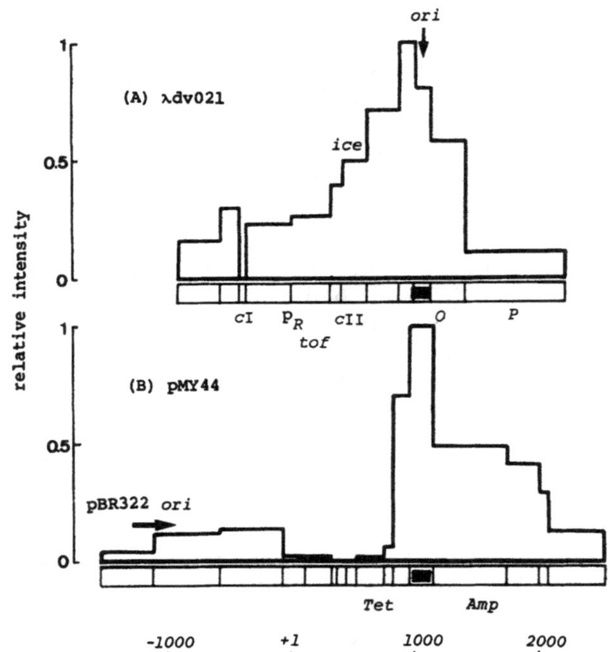

Fig. 3. Analysis of initiation sites of DNA replication in vitro in λdv021 and pMY44 genomes using ddCTP. (A) λdv021 DNA was replicated in the standard reaction mixture containing 100 μM of [α-^{32}P]dCTP and 100 μM ddCTP. The replicated DNA was purified and digested with DdeI and EcoRI. Distribution of newly incorporated radioactivity among the restriction fragments was analyzed by electrophoresis and autoradiography. The specific radioactivity of each fragment was measured by scanning with a densitometer and mapped along λdv genome (abscissa). The results are expressed as relative intensities (ordinate). The ori region is shown as a filled box in the map. For ori and ice, see text. (B) pMY44 DNA was employed similarly, except that 20 μM [α-^{32}P]dCTP and 40 μM ddCTP were used. The DNA was digested with HaeIII and analyzed by electrophoresis as before. Amp and Tet are the drug-resistance markers in pBR322. pBR322 origin and arrow are the replication origin of the vector and its direction of replication. The direction of λ ori region (filled box) in B is the same as in A. Numbers under the graphs represent nucleotide distance on λdv021 genome as shown in Fig. 1.

elongation that might lead to synthesis of new Okazaki fragments, i.e., general chain elongation. During this incubation, approximately 30% of the template DNAs initiated DNA synthesis (the amount of active templates was approximately 0.9 pmols), and short-chain DNAs of 60-160 nucleotides were synthesized (corresponding to approximately 1.9 pmols). Therefore, about 2 short-chain DNAs were formed per active template. The reaction was stopped and the replicative intermediates of λdv were purified by successive chromatographies on sepharose 2B and benzoylated-naphthoylated DEAE cellulose (31). Details of replicative intermediate purification will be described elsewhere.

The replicative intermediates were heated at 95°C for 3 min., and denatured short-piece DNA was purified with a sepharose CL-4B column. Fractions with kd values of 0.4 to 0.8, which contained 60-160 nucleotide short-piece DNAs, were pooled. Their 5' ends were masked with T4 polynucleotide kinase and 0.1 mM ATP at 37°C for 30 min., then the sample was incubated in 0.15 M NaOH and 4 mM EDTA at 37°C for 14 hr. to expose 5'-OH groups that had been covalently linked with primer RNA. The freshly exposed 5'-OH groups were specifically labeled with T4 polynucleotide kinase and 1.3 μM [γ-^{32}P]ATP at 0°C for 60 min. (34). Approximately 10% of total short-pieces were end-labeled. Note that this figure represents the fraction of DNA pieces that had been linked to primer RNA. The remaining 90% of the fragments had been freed from the primer RNA before or during sample preparation.

As shown in Fig. 3, when the chain terminator ddCTP was administered, practically all the DNA synthesis occurred in or around the <u>ori</u> region. Two cycles of hybridization with appropriate probes were carried out as shown below, to select those fragments carrying only the sequences of the <u>ori</u> region and its vicinity. Possible contamination by Okazaki fragments representing chain elongation that might have occurred in other regions of the plasmid DNA was eliminated by these hybridizations. Thus, the junction between primer RNA and short DNA in our sample can be taken to represent the transition site(s) in the earliest phases of replication.

The labeled DNA was divided into 2 equal portions, (I) and (II), which were mixed respectively with an excess (30 μg) of λoriL3 DNA and λoriR3 DNA (see Fig. 1). The DNA in each sample was denatured and renatured at 65°C for 6-14 hr. (first hybridization). The nonrenatured, end-labeled short DNA chains were purified by chromatography using CL-4B, and fractions with kd values of 0.4-0.8 were collected. Short pieces for <u>l</u>- or <u>r</u>-strand synthesis were enriched in these fractions. To these samples from (I) (II), was added, respectively, an excess (30 μg) of λoriR20 or λoriL15 (see Fig. 1), and denaturing and renaturing treatments were done as above (second hybridization). The renatured DNA was purified by applying the sample on a CL-4B column and collecting the DNAs in the excluded fractions. The labeled materials in (I) and (II) represent short-chain DNAs of <u>l</u>- and <u>r</u>-strands, respectively, at or around the <u>λ-ori</u> region.

The hybrids of the short-chain DNAs and probe DNAs were digested with S1 nuclease to remove single-stranded regions (35), and the mixture was passed through Sephadex G-100 columns. The excluded DNAs were pooled and digested with appropriate restriction enzymes. Through these series of treatments, about 3.0% and 10% of the initially incorporated ^{32}P counts ertr recovered in (I) and (II), respectively. The digested DNAs were analyzed by electrophoresis in denaturing gel.

Transition sites are located outside of the *ori* region. The S1-treated sample (I) was digested with *Eco*RI and electrophoresed. This experiment was designed to detect *l*-strand DNA that had undergone transition within the *ori* region, since such DNA would show a band(s) shorter than 165 nucleotides. However, no discrete bands were detected (Fig. 4, lane 1). The same sample exhibited several bands upon treatment with *Ava*II (lane 2), indicating that several transition sites are located to the left of the *ori* region. Further analysis of these sites will be described below.

We then examined the S1-treated sample (II) by digestion with *Dde*I to detect the transition sites on the *r*-strand within the *ori* region. A band of 152-155 nucleotides appeared (Fig. 4, lane 3), indicating that a transition site is located at around the 1155 position, to the right of the *ori* region (marked A in Fig. 7 and 8). Some shorter bands were also observed. These bands appeared reproducibly, irrespective of whether the sample had been treated with *Dde*I or *Hpa*II (lanes 3 and 4), or restriction enzyme treatment omitted (data not shown). Thus, they must be "background" bands of this assay system, representing preferential termination products that tend to terminate in cytosine clusters in the ori-repeats in the presence of ddCTP (see legend to Fig. 6). Thus, we conclude that all the transition sites are located outside of the *ori* region. The *Hpa*II digest exhibited additional bands representing other transition sites to the right of the *ori* region (see below).

To locate the transition sites outside the *ori* region, the S1-treated *l*-strand specific sample (I) was digested with *Dde*I or *Ava*II. The results in Fig. 5 demonstrate that transition sites are located to the left of the *ori* region, in a stretch from 793 to 1,000. Many bands of various intensities were observed in both samples. Since the region analyzed carries 2 *Dde*I sites (887 and 1,000), transition sites were determined precisely with the *Ava*II digested sample, and 3 major clusters of transition sites (Z, Y, and V) were identified within the region from 895 to 980. Z is composed of one strong and 2 weak bands; Y, of 2 strong bands; and V, of 4 strong and 7 weak bands. In addition, 5 clusters of weak bands (x, w, s, t, u) were noted. "s" and "x" overlap in lane 1.

Transition sites of the *r*-strand located to the right of the *ori* region were analyzed from DNAs in sample (II) that had been

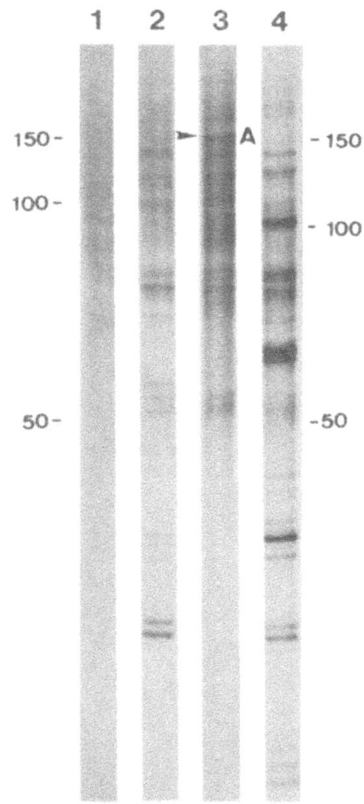

Fig. 4. Search for transition sites within or close to the ori region. Purified, end-labeled l-strand short-chain DNA that had been hybridized with λoriR20 followed by S1-digestion was digested with EcoRI (lane 1) or AvaII (lane 2). Purified, end-labeled short-chain DNA that had been hybridized with λoriL15 followed by S1-digestion was digested with DdeI (lane 3) and HpaII (lane 4). Numbers beside the figures represent the nucleotide length.

end-labeled and hybridized with λoriL15. The sample was digested with EcoRI or HpaII (Fig. 6). Transition sites between 1150 and 1416 can be located by this analysis. Seven major clusters (A, B, E, F, G, H, I) of transition sites were observed. Each cluster appeared in ladders of 4-8 bands. Weaker transition sites (c, d, j, k, l) were also visible. Weak ladders between G and H clusters and between H and I clusters (lanes 2 and 3) are part of the "background" that was also observed in Fig. 4 (for the background, see above, also legend to Fig. 6).

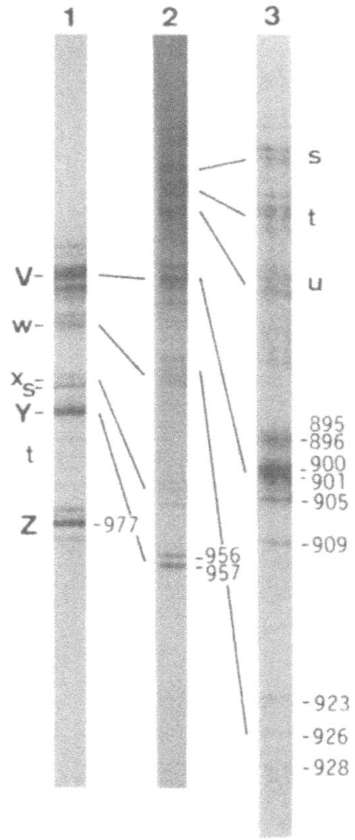

Fig. 5. Analysis of l-strand transition sites located to the left of the ori region. Purified, end-labeled l-strand short-chain DNA, hybridized with λoriR20 and treated with S1 nuclease, was digested with DdeI (lane 1) and AvaII (lanes 2 and 3). The data are assembled from 3 independent electrophoreses. Lane 3 is the result of long electrophoresis of the same sample as in lane 2. Z, Y, and V are the major transition clusters that consist of strong bands and some flanking weak bands. x, w, u, t, and s are the minor transition sites. Corresponding bands are connected by lines. Numbers indicate the locations of the 5' ends of the DNA pieces along the nucleotide sequence of λ (see Fig. 7 and 8). These locations were determined from the positions of restriction enzyme sites and the measured nucleotide lengths of bands obtained by comparing with size markers.

Next, the transition sites were examined in a wider stretch of DNA covering about 1 kb around the ori region, using λoriL3 and λoriR3 as probes. No further strong transition sites were detected, although weak sites were found on both strands (data not shown: the results are incorporated in Fig. 7). The weak transition sites in the _l_-strand were dispersed in the regions from 750 to 800 and from 1230 to 1310. The weak transition sites in the _r_-strand were clustered in the region from 910 to 930.

A Summary Map Showing RNA-DNA Transition Sites in λdv Replication

The foregoing analyses by electrophoresis enabled us to locate the transition sites within 2 nucleotides. The relative intensities of the major bands were assayed in 2 or 3 independent experiments. The results are assembled in Fig. 7. Most of the transitions for the rightward DNA synthesis occur to the left of the _ori_ region, while minor transitions occur to the right of the _ori_ region. Conversely, most of the transitions for the leftward DNA synthesis are located to the right of the _ori_ region. Therefore, the major DNA chain initiations occur on both sides of the _ori_ region and the syntheses converge. The result is bidirectional replication.

To locate the transition sites precisely, the ^{32}P-labeled mononucleotides obtained from the 5' ends of the short DNA chains (V, Y, Z and A, B, E, F, G, H, I) were analyzed. The results (data not shown) agreed well with those obtained from the analyses of chain length. Figure 8 summarizes the transition sites on the nucleotide sequence of the λ _ori_ region. So far, we have not been able to detect a consensus sequence for the transition sites, except for the preferential use of thymidine residues. Thus, 12 out of 19 strong transition sites were T, and only 4, 2, and 1 were A, C, and G, respectively.

No transition sites were discovered within the _ori_ region. This implies that the _ori_ region essential for replication initiation contains no site of initiation of primer RNA synthesis. Two possibilities can be considered concerning the synthesis of primer RNA. First, the primer RNA synthesis might be initiated at multiple sites that are somehow associated with the transition sites we observed. Each of the product RNAs would then be short molecules. Alternatively, the primer RNA might be made as a large transcript starting far upstream of the transition sites. Rightward RNA synthesis from p_R is one candidate for the long right primer. However, there is no known long leftward RNA synthesis that goes across the _ori_ region. Therefore, the leftward long primer must be made by a new mechanism; and this, of course, might also operate for rightward primer synthesis. These problems can only be solved by analysis of the 5' ends of the primer RNAs.

Our observations also demonstrate that the _ori_ region does not contain the site(s) of transition from primer RNA to DNA. At present, too little information is available to link these observations with the fact that O protein binds the ori-repeating sequences (20).

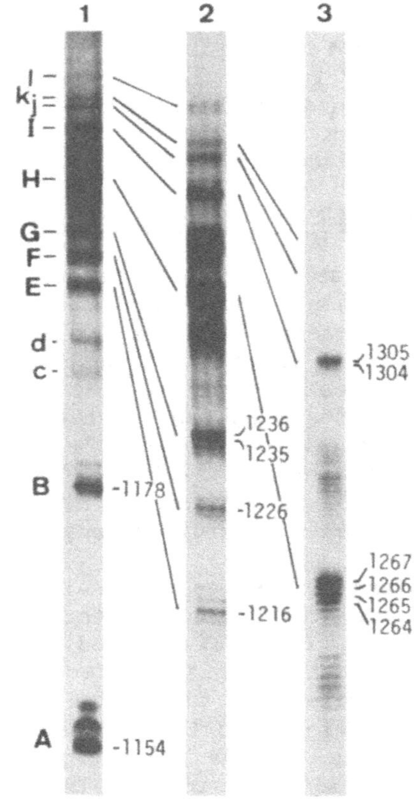

Fig. 6. Analysis of transition sites on r-strand to the right of the ori region. Purified, end-labeled r-strand short-chain DNA, hybridized with λoriL15 and treated with S1 nuclease, was digested with EcoRI (lane 1) or HpaII (lane 2). Lane 3 is the result of long electrophoresis of the same sample as in lane 2. Clusters of major transition sites are indicated as A, B, E, F, G, H, and I, and minor sites are indicated as c, d, j, k, and l. The HpaII digest (lanes 2 and 3 in b) shows 2 additional ladders which consist of background bands produced as a result of preferential chain termination of DNA synthesis of ddCTP at the cytosine cluster. Ladders between H and I, and between G and H may represent the DNA chains initiated respectively at sites B and A and terminated at the cytosine cluster at positions 1,092-1,096. We have three reasons to assume this: (i) The chain lengths of these bands, 57-61, and 76-84 nucleotides, match the distances from sites B and A to the termination. (ii) Nucleotides at the 5' ends of these bands are mainly C and T, suggesting that they share the same ends with the A and B bands,

Fig. 6. (Continued)

respectively (see Fig. 4). (iii) These bands did not appear in the sample digested with EcoRI, which cuts just downstream of the sites A and B. But they appeared at the same positions upon digestion with several other enzymes.

Several lines of evidence suggest that the O protein interacts with P protein (23), which in turn interacts with such host-coded proteins as dnaB (18,21) and dnaK (22). The ori sequence may thus be used as a signal that attracts the replication machinery complex. This complex could alter the structure of neighboring DNA regions to facilitate initiation of primer RNA synthesis. Alternatively, they could facilitate modification or processing of the incoming primer RNA molecules or could regulate their synthesis.

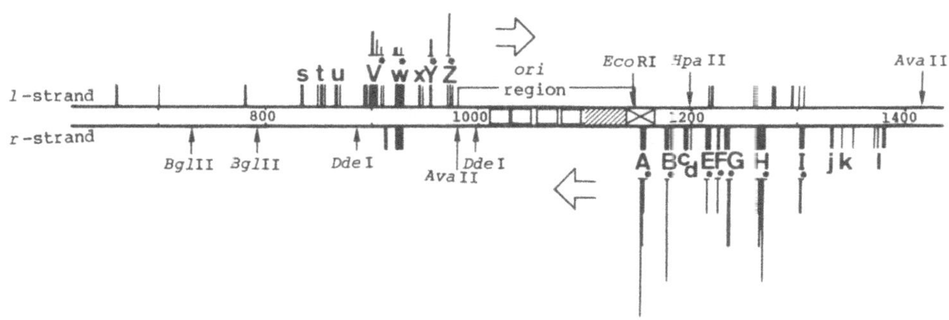

Fig. 7. Summary map of the transition sites on l- and r-strands of λ genome. Vertical bars on each strand indicate locations of the transition sites observed in this work. The letters correspond to those in Fig. 5 and 6. Additional bars above and below the asterisked letters are drawn to represent relative frequency of occurrence of the transition. These were obtained from densitomeric tracing of the autoradiographed bands. Results for minor bands that could not be accurately measured are not shown. The results for l- and r-strands can be compared with each other, since they were taken from the same autoradiography using ^{32}P-labeled DNAs that were prepared from 2 identical aliquots prepared for each strand (not shown). Arrows represent the direction of DNA synthesis. The ori region carries ori-repeats (4 boxes), an A+T-rich stretch (///), along with a part of a palindromic 28-bp sequence (▷◁). The cutting sites of restriction enzyme used in the analyses are also shown. It is possible that some weaker transition sites have been missed in these analyses.

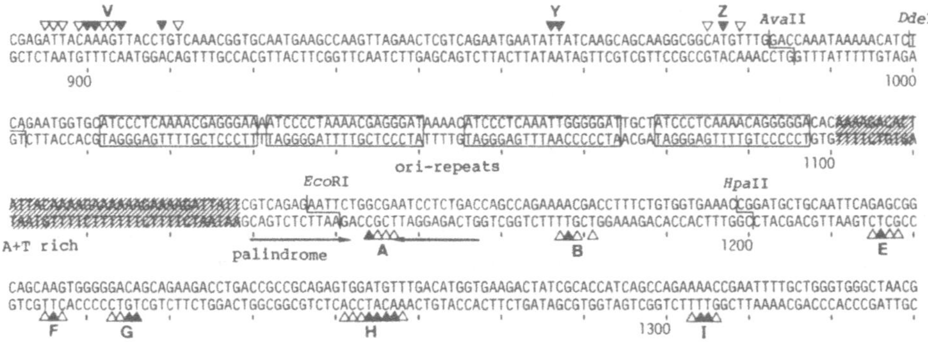

Fig. 8. Transition sites as represented in the nucleotide sequence of λ around the ori region. λ DNA sequence from 891 to 1,330 is shown. The "ori region" is the region from AvaII (983) to EcoRI (1,147). Ori-repeats (boxed), A+T-rich stretch (///), and palindromic sequence (arrows) are shown. Three clusters of strong transition sites on the l-strand (V, Y, Z) and 7 clusters of strong sites on the r-strand (A, B, E, F, G, H, I) are defined in Fig. 5 and 6. Mononucleotides at the 5' end of the DNA joined to primer RNA were analyzed as described in (36). Closed and open triangles show the sites of strong and weak transitions, respectively. In the A cluster, some weak sites located close to the EcoRI site might have been missed due to technical limitations.

Of the leftward and rightward DNA syntheses, the former is several times more frequent. This asymmetry was reflected in the higher recovery of r-strand than l-strand after S1 treatment. This fact agrees well with the results of Fig. 3, which show that leftward DNA synthesis occurs more strongly than rightward DNA synthesis in the process of bidirectional DNA replication of λdv in vitro. The implications of this asymmetry are unknown, but it could reflect a difference in the initiation mechanisms for leftward and rightward DNA synthesis.

The multiplicity of transition sites from primer RNA to DNA synthesis in λdv replication is in sharp contrast with the situation in ColE1 (37,38), which synthesizes a long primer RNA starting at a unique site, followed by transition at a unique site. This difference means that the "multiple site" mechanism must be different from that of the "unique site" mechanism observed with ColE1. It would be of interest to see, at least in procaryotes, how the "multiple site" and "unique site" mechanisms are distributed among different replicons.

Although multiple, the λ transition sites seem not to be distributed randomly: in fact, the replicon seems to have favored sites that appear reproducibly in independent experiments, although how these sites are selected is unclear. At this stage, we can only infer that combinations of local nucleotide sequences and higher order structures of the DNA are responsible for the selection of the transition sites, and that thymidine residues play an important role. The possibility remains that different transition sites might be selected when the experimental system is modified.

Since the in vitro system used exhibits a similar mode of DNA synthesis to that observed in in vivo replication, in the requirements for the reaction, the bidirectional nature of the replication, and the *ori*-dependent initiation, the initiation process of λ replication in vivo may also show transition at multiple sites. However, the possibility cannot be ruled out that the finding of multiple transition sites reflects a lack of a factor(s) necessary for selection of a unique site(s). In this connection, it will be of interest to analyze the in vivo transition sites of λ DNA replication.

CONCLUSION

Deletion studies have shown that the functional origin of λ DNA replication can be defined within a 165-bp region (*ori* region) that carries reiterated sequences, an A+T-rich stretch, and a portion of a long palindromic sequence, all located within the coding region of the *O* gene. This region agrees well with the previously identified site of initiation of λ DNA replication, from genetic studies as well as electron microscopic examinations. The minimized *ori* region replicates only in the presence of O and P proteins and the RNA polymerase-dependent RNA synthesis. The DNA synthesis that is promoted by the *ori* region, as examined by incorporation tests, proceeds bidirectionally.

Short-chain DNAs joined to primer RNAs were purified from earliest replicative intermediates of λdv, as prepared in an O,P-dependent in vitro replication system. The 5' ends of these short DNA chains (i.e., the transition sites from primer RNA to DNA synthesis) were freed from RNA, specifically labeled with ^{32}P, then mapped along the known nucleotide sequence of λ. The results showed that transition occurred at multiple sites located on both strands flanking the λ *ori* region. The orientation of the transition sites demonstrates that the chain elongation proceeds convergently. This, again, is in accord with the bidirectional nature of initiation in λ replication. The fact that no transition sites were observed within the *ori* region implies that the sequence essential for λ replication involves neither the transition sites from primer RNA to DNA nor the initiation sites of primer RNA synthesis. To understand the initiation mechanism of λ replication totally, it is necessary to elucidate the sites and mechanisms of primer RNA synthesis.

ACKNOWLEDGMENTS

We are grateful to Dr. Asao. Fujiyama for his valuable advice throughout this work. This work was supported by Scientific Research Grants from the Ministry of Education, Science, and Culture of Japan.

REFERENCES

1. Tomizawa, J.-I., and G. Selzer (1979) Initiation of DNA synthesis in Escherichia coli. Ann. Rev. Biochem. 48:999-1034.
2. Nossal, N.G. (1983) Prokaryotic DNA replication systems. Ann. Rev. Biochem. 53:581-615.
3. Matsubara, K. (1981) Replication control system in lambda dv. Plasmid 5:32-52.
4. Tsurimoto, T., and K. Matsubara (1983) Replication of bacteriophage λ DNA. Cold Spring Harbor Symp. Quant. Biol. 47:681-691.
5. Ogawa, T., and J. Tomizawa (1968) Replication of bacteriophage DNA. I. Replication of DNA of lambda phage defective in early function. J. Mol. Biol. 38:217-225.
6. Schnös, M., and R.B. Inman (1970) The position of branch point in replicating lambda DNA. J. Mol. Biol. 51:61-73.
7. Inman, R.B. (1981) DNA replication. In Proceedings of the 3rd Cleveland Symposium on Macromolecules, A.G. Walton, ed. Elsevier Publishing Co., Amsterdam, pp. 155-165.
8. Dove, W., E. Hargrove, M. Ohashi, F. Haugli, and A. Guha (1969) Replicator activation in lambda. Japan J. Genet. Vol. 44, Suppl. 1, pp. 11-22.
9. Furth, M.E., W.F. Dove, and B.J. Meyer (1982) Specificity determinants for bacteriophage λ DNA replication. III. Activation of replication in λric mutants by transcription outside of ori. J. Mol. Biol. 154:65-83.
10. Dove, W.F., H. Inokuchi, and W.F. Stevens (1971) Replication control in phage lambda. In The Bacteriophage Lambda, A.D. Hershey, Cold Spring Harbor Laboratory, New York, pp. 747-771.
11. Rambach, A. (1973) Replicator mutants of bacteriophage λ: Characterization of 2 subclasses. Virology 54:270-277.
12. Denniston-Thompson, K., D.D. Moore, K.E. Kruger, M.E. Furth, and F.R. Blattner (1977) Physical structure of replication origin of bacteriophage lambda. Science 198:1051-1056.
13. Grosschedl, R., and G. Hobom (1979) DNA sequence and structural homologies of the replication origins of lambdoid bacteriophages. Nature 277:621-627.
14. Lusky, M., and G. Hobom (1979) Inceptor and origin of DNA replication in lambdoid coliphages. I. The λ DNA minimal replication system. Gene 6:137-172.

15. Lusky, M., and G. Hobom (1979) Inceptor and origin of replication in lambdoid coliphages. II. The λ DNA maximal replication system. Gene 6:173-197.
16. Moore, D.D., K. Denniston-Thompson, K.E. Kruger, M.E. Furth, B.G. Williams, D.L. Daniels, and F.R. Blattner (1979) Dissection and comparative anatomy of the origins of replication of lambdoid phages. Cold Spring Harbor Symp. Quant. Biol. 43:155-163.
17. Kornberg, A. (1979) Aspects of DNA replication. Cold Spring Harbor Symp. Quant. Biol. 43:1-9.
18. Tsurimoto, T., T. Hase, H. Matsubara, and K. Matsubara (1982) Bacteriophage lambda initiator: Preparation from a strain that overproduces the O and P proteins. Mol. Gen. Genet. 187:79-86.
19. Roberts, J.D. and R. McMacken (1983) The bacteriophage λ O replication protein: Isolation and characterization of the amplified initiator. Nucl. Acids Res. 11:7435-7452.
20. Tsurmioto, T., and K. Matsubara (1981) Purified bacteriophage λ O protein binds to four repeating sequences at the λ replication origin. Nucl. Acids Res. 9:1798-1799.
21. Wickner, S.H. (1979) DNA replication proteins of Escherichia coli and phage λ. Cold Spring Harbor Symp. Quant. Biol. 43:303-310.
22. Zylicz, M., J.H. LeBowitz, R. McMacken, and C. Georgopoulos (1983) The dnaK protein of Escherichia coli possesses as ATPase and autophosphorylating activity and is essential in an in vitro DNA replication system. Proc. Natl. Acad. Sci., USA 80:6431-6435.
23. Tomizawa, J. (1971) Functional cooperation of genes O and P. In The Bacteriophage Lambda, A. Hershey, ed. Cold Spring Harbor Laboratory, New York, pp. 549-552.
24. Anderl, A., and A. Klein (1982) Replication of λdv DNA in vitro. Nucl. Acids Res. 10:1733-1740.
25. Wold, M.S., J.B. Mallory, J.D. Roberts, J.H. LeBowitz, and R. McMacken (1982) Initiation of bacteriophage λ DNA replication in vitro with purified λ replication proteins. Proc. Natl. Acad. Sci., USA 79:6176-6180.
26. Tsurimoto, T., and K. Matsubara (1982) Replication of λdv plasmid in vitro promoted by purified λ O and P proteins. Proc. Natl. Acad. Sci., USA 79:7639-7643.
27. Chow, L.T., N. Davidson, and D. Berg (1974) Electron microscopic study of the structure of λdv DNAs. J. Mol. Biol. 86:69-89.
28. Tsurimoto, T., and K. Matsubara (1981) Purification of bacteriophage λ O protein that specifically binds to the origin of replication. Mol. Gen. Genet. 181:325-331.
29. Tsurimoto, T., and K. Matsubara (1981) Interaction of bacteriophage lambda O protein with the lambda origin sequence. In The Initiation of DNA Replication, ICN-UCLA Symp. 22, D.S. Ray, ed. Academic Press, New York, pp. 263-275.

30. Messing, J., and J. Vieira (1982) A new pair of M13 vectors for selecting either DNA strand of double-digested restriction fragments. Gene 19:269-276.
31. Hay, R.T., and M.L. DePamphilis (1982) Initiation of SV40 DNA replication *in vivo*: Location and structure of 5' ends of DNA synthesized in the *ori* region. Cell 28:767-779.
32. Fuller, R.S., J.M. Kaguni, and A. Kornberg (1981) Enzymatic replication of the origin of the *E. coli* chromosome. Proc. Natl. Acad. Sci, USA 78:7370-7374.
33. Alton, N.K., and D. Vapnek (1979) Nucleotide sequence analysis of the chloramphenicol resistance transposon Tn9. Nature 282:864-869.
34. Okazaki, R., S. Hirose, T. Okazaki, T. Ogawa, and Y. Kurosawa (1975) Assay of RNA-linked nascent DNA pieces with polynucleotide kinase. Biochem. Biophys. Res. Commun. 62:1018-1024.
35. Fujiyama, A., Y. Kohara, and T. Okazaki (1981) Initiation sites for discontinuous DNA synthesis of bacteriophage T7. Proc. Natl. Acad. Sci., USA 78:903-907.
36. Ogawa, T., and T. Okazaki (1979) RNA-linked nascent DNA pieces in phage T7-infected *Escherichia coli*. III. Detection of intact primer RNA. Nucl. Acids Res. 7:1621-1633.
37. Tomizawa, J-I., H. Ohmori, and R.E. Bird (1977) Origin of replication of colicin El plasmid DNA. Proc. Natl. Acad. Sci., USA 74:1865-1869.
38. Itoh, T., and J-I. Tomizawa (1980) Formation of an RNA primer for initiation of replication of colEl DNA by ribonuclease H. Proc. Natl. Acad. Sci, USA 77:2450-2454.

BROAD HOST-RANGE PLASMID R1162: REPLICATION, INCOMPATIBILITY, AND COPY-NUMBER CONTROL

Richard J. Meyer, Lung-Shen Lin, Kyunghoon Kim, and Michael A. Brasch

The University of Texas at Austin
Department of Microbiology
Experimental Science Building, 226
Austin, Texas 78712-1095

INTRODUCTION

R1162 is a small [8.7 kilobases (kb)] plasmid encoding resistance to sulfonamides (Su^R) and streptomycin (Sm^R), and replicating with a high copy-number in Escherichia coli (at least 10 copies per chromosome) (1). The plasmid is not self-transmissible, but may be readily mobilized by certain other conjugative plasmids, such as those of the P-1 incompatibility group (2). Originally isolated from Pseudomonas aeruginosa (3), R1162 has a broad host-range (1,4-6). Plasmids which appear very similar to R1162 have been isolated from diverse bacteria (1,7); many of these may be identical representatives of the same far-ranging plasmid. At present there do not seem to be any clear differences between R1162 and the best-studied member of this group, RSF1010 (from E. coli) (1).

Genetic Organization of R1162

Figure 1 is a map of R1162 showing cleavage sites for several restriction enzymes, and also the approximate locations of coding sequences for known plasmid functions. The Su^R and Sm^R genes are cotranscribed, with Su^R promoter proximal (6,8). The single EcoRI cleavage site lies between the structural genes for Su^R and Sm^R, and insertion of DNA into this site can have a variable effect on expression of both drug resistances. Thus, a promoter cloned in one direction can activate Sm^R, and change the level of this resistance (Ref. 6 and unpub. observations), whereas in the other direction it can reduce the level of Su^R by transcriptional interference. These characteristics are useful for molecular cloning (for example, see Ref. 9).

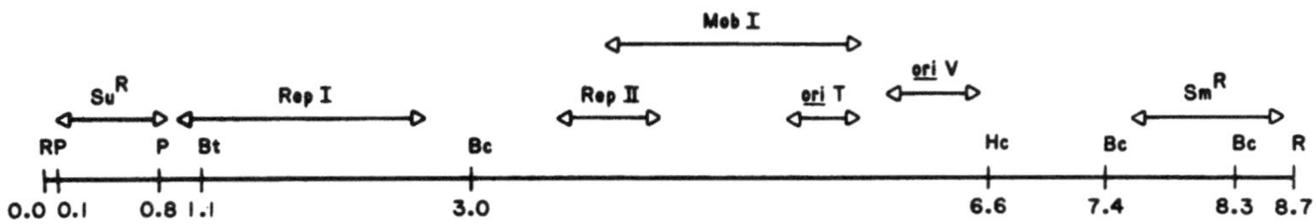

Fig. 1. Genetic organization of R1162. The cleavage sites for a few restriction enzymes are shown; numbers refer to distances, in kilobase pairs (kb), from the single EcoRI cleavage site. Abbreviations: Bc: BclI, Bt: BstEII, Hc: HincII, P: PstI, and R: EcoRI. Other abbreviations are defined in the text.

Two regions of the plasmid supply products necessary in trans for plasmid maintenance: sequences between 0.8 and 3.0 kb, and 3.8 and 4.3 kb (Rep I, II: Fig. 1). Initially, we believed that the latter DNA was dispensible, because pMS14, which lacks this DNA, can replicate (2). However, we have found that it does so by virtue of a second, low-copy number R1162 derivative, also present in the cell, which supplies RepII. Satellite plasmids consisting of oriV must have RepI and RepII functions supplied in order to be maintained in the cell.

DNA-encoding functions necessary in trans for conjugal mobilization of R1162 (Mob) are located between 3.9 and 5.7 kb. The origin for conjugal mobilization (oriT) and the origin for DNA replication (oriV), are located in the region 5.3-6.6 kb (2). The close physical association of oriT and oriV has raised the possibility of a functional interdependence (10). However, it is clear from cloning studies described below that both are capable of independent activity. No additional DNA required for replication or transfer is located to the right of the 6.6 kb position on the map.

To examine the RepI region in greater detail, we have cloned the 0.0 to 3.0 kb region containing RepI into pBR322, and generated deletions rightward and leftward into this DNA by Bal31 digestion (Fig. 2). We asked whether the resulting plasmid derivatives still contained the complete RepI region by testing their ability to support replication of a satellite plasmid, pMS125, which is $RepII^+$ $oriV^+$ but deleted for RepI. The results for some derivatives are shown in Fig. 2. Essential sequences are evidently located as far left as 1.07 kb, and as far right as 2.57 kb. We also tested whether any of the cloned DNAs would complement pMS88, a derivative of R1162 which is temperature-sensitive for maintenance due to mutation in RepI (2). The plasmids unable to provide RepI function for pMS125 were also unable to complement pMS88. The simplest interpretation is that there is a single transcriptional unit lying within the region 0.91-2.63 kb, which may encode one or more proteins.

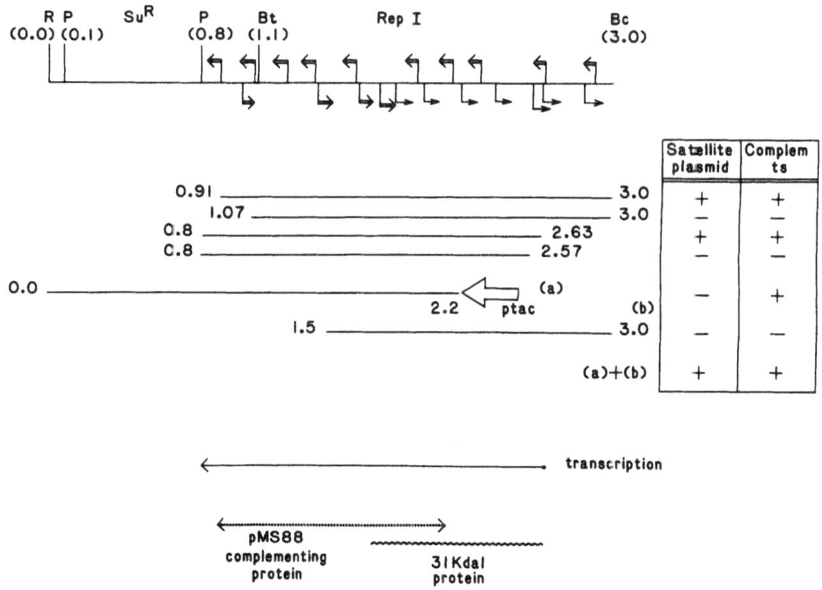

Fig. 2. Region of R1162 containing genes required in trans for plasmid maintenance. A physical map with some restriction enzyme cleavage sites (see Fig. 1) is at the top. The vertical lines indicate the endpoints of Bal31-generated deletions, and the attached arrows the direction of transcription within R1162 tested at that point. An open arrow means that transcription was detected. Below, fragments of R1162 DNA cloned into pBR322 are represented by the horizontal lines. Whether each fragment is able to support replication of an $oriV^+$ $RepII^+$ satellite plasmid, or to complement the R1162 derivative pMS88, which has a mutation in RepI, is shown at the right. At the bottom of the figure, the location of a gene encoding a 31 kd protein, and also the location and maximum size of DNA encoding the product complementing the defect in pMS88 are indicated; the direction of transcription of these genes is shown just above.

We fused the variable ends of the Bal31 deletions to a lacZ fragment lacking both the lacZ promoter and amino-terminal coding sequence (11). The resulting fusions can be used to determine the direction of transcription within R1162 DNA at each junction point. This is possible even when the fusions are out of phase, since the assay for β-galactosidase is very sensitive (11). The results are shown in Fig 2: the arrows indicate the direction of transcription being tested; open arrows mean that β-galactosidase is produced. There is transcription from right to left throughout the RepI region, and also transcription rightward in the left segment.

We looked for polypeptides that are encoded by the pBR322-RepI plasmids, both in vitro (12) and in minicells (13). Since these plasmids each contain a different, known deletion of the RepI region, the locations of the coding sequences can be determined for any proteins detected. A 31 kd protein is encoded within the region 1.7-2.6 kb (Fig. 2). The R1162 fragments 1.8-3.0 kb and 1.96-3.0 kb encode truncated fragments of approximately 29 and 22 kd respectively: thus, transcription of the gene is from right to left, in agreement with the lacZ fusion results.

The 31 kd protein will not complement the temperature-sensitive RepI protein encoded by pMS88. However, a plasmid derivative containing the 0.0-2.2 kb R1162 DNA fragment, fused at the 2.2 end to the tac promoter (14), will complement pMS88 (Fig. 2). Such a derivative will still not support replication of the satellite plasmid pMS125. If another plasmid, containing R1162 DNA between 1.5 and 3.0 kb, is also present, pMS125 is maintained (Fig. 2). Thus, it appears that both the putative pMS88 complementing protein and the 31 kd protein are required for replication of R1162 DNA. These products are made either from a single transcript, with the 31 kd gene promoter-proximal, or from 2 overlapping transcripts with closely spaced promoters. We have not yet clearly identified a protein corresponding to the pMS88 complementing activity. However, Scherzinger et al. (15) have shown that at least 2 proteins, required for replication of RSF1010 DNA in vitro, are encoded in the region which we call RepI. They have purified a 27 kd polypeptide which, from their mapping studies, is probably the pMS88-complementing protein. If this is so, the 1.07-2.57 kb RepI region is probably just sufficient to encode 27 and 31 kd polypeptides without overlap.

Genetic Organization of the oriT-OriV Region

We have used molecular cloning methods to characterize more fully the region of R1162 containing oriT and oriV. Some of the results of this analysis are given in Fig. 3. Specific fragments of DNA from R1162 were cloned into pBR322, and then tested for an active oriV as outlined in Fig. 4. Because the cloned R1162 DNA lacked the Rep functions, we tested whether this DNA could replicate as a satellite plasmid when Rep functions were provided in trans. Cloned R1162 DNA linked to DNA encoding kanamycin-resistance (Km^R) was released from cloning vector DNA by SalI digestion, and used to transform MV10 (R1162) to Km^R (Fig. 4). Plasmid DNA, extracted from transformants resistant to kanamycin but sensitive to carbenicillin (Cb^S) was then used to transform MV10 (pMS69) and MV10 (pBR322) to Km^R. The plasmid pMS69 is a chimera of pBR322 and an R1162 derivative that contains all Rep functions (2). If a $Km^R(Cb^S)$ satellite plasmid had been present in the transformed MV10 (R1162) cells, then the Km^R transformants of MV10 (pMS69) would not, in general, also be Sm^R, since pMS69 could provide any missing functions for plasmid maintenance. In contrast, the Km^R transformants of MV10 (pBR322)

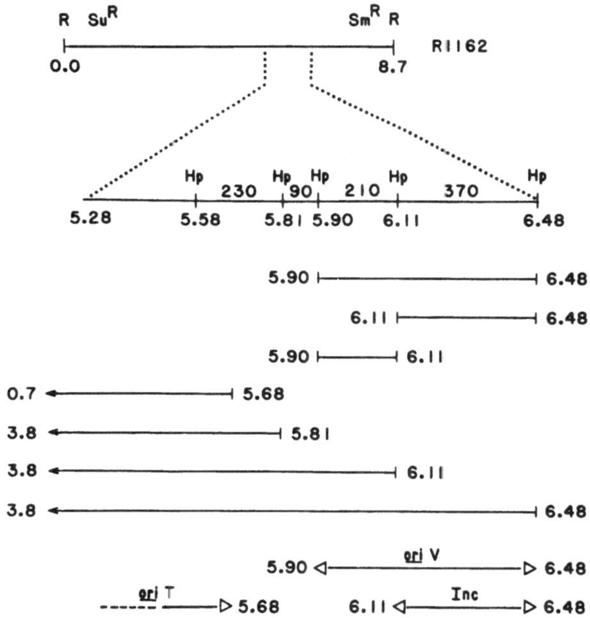

ori V	Inc	Mob
+	+	−
−	+	−
−	−	−
−	−	2×10^{-5}
N.T.	N.T.	9×10^{-7}
N.T.	N.T.	4×10^{-3}
N.T.	N.T.	4×10^{-3}

Fig. 3. The oriV region of R1162. R1162 DNA between 5.28 and 6.48 kb (Fig. 1), containing the origin of plasmid replication, is expanded and the location of several HpaII (Hp) cleavage sites are shown. Numbers between the HpaII cleavage sites give fragment sizes in base pairs (bp). Below are represented various fragments of DNA cloned into pBR322, and, to the right, the results for testing each of these for ability to form satellite plasmid (oriV), expression of incompatibility (Inc), and conjugal mobilization by R751 (Mob). If the plasmid is mobilizable, the frequency (transconjugants per donor cell) is given.

would all be Sm^R, since cotransformation of R1162 would be required for maintenance of the satellite. If the initial $Km^R(Cb^S)$ transformants arose by a mechanism not involving satellite plasmid formation (e.g., by deletion of the Cb^R gene in pBR322, or by insertion of the Km^R gene into R1162 by recombination), then this pattern of acquiring Sm^R in the second transformation would not be observed (Fig. 4).

We found that the $oriV^+$ function required the presence of 2 contiguous HpaII fragments of 210 and 370 bp (Fig. 3). Neither fragment by itself was capable of replication as a satellite plasmid. Different fragments of R1162 DNA, cloned into pBR322, were also tested for expression of incompatibility. DNA of each plasmid derivative was used to transform MV10 (R1162), and the resulting Cb^R

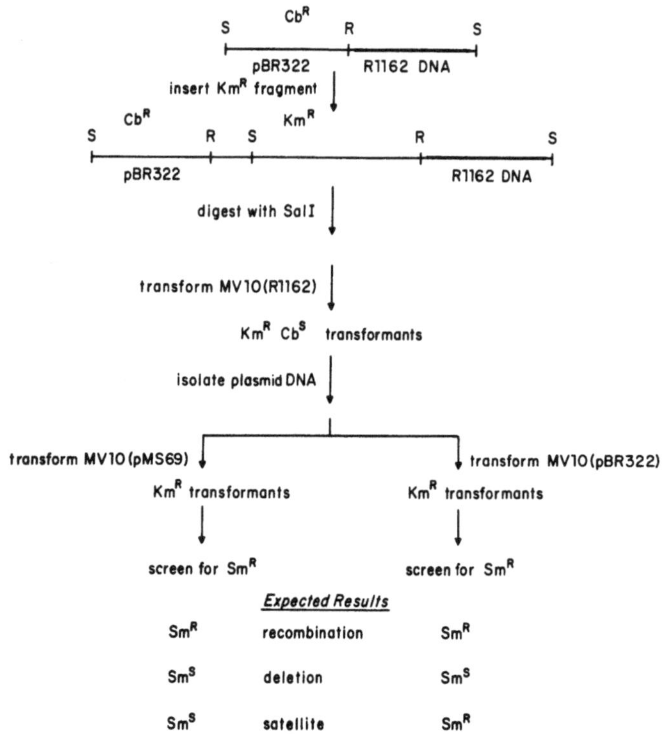

Fig. 4. Method for detection of satellite plasmids. The procedure is outlined in the text.

cells screened for retention of R1162 by scoring for Sm^R. The incompatibility determinant was localized to the 370 bp HpaII fragment. Thus, the incompatibility determinant and oriV are closely linked in R1162 (16).

We previously observed that pMS94, a pBR322 derivative containing R1162 DNA between 0.70 and 5.68 kb, was mobilizable by the self-transmissible IncP1 plasmid R751 (2). Thus, the oriV-Inc DNA lying between 5.90 and 6.48 kb is not required for mobilization, although oriV and oriT are closely linked on R1162 (2). However, we noticed that the mobilization frequency of pMS94 was typically about $2X10^{-5}$ transconjugants per donor cell, under conditions where R1162 was mobilized at a frequency of 0.1 (Fig. 3). It is possible, therefore, that sequences lying between oriV and oriT affect the efficiency of mobilization. We measured the mobilization frequencies for pBR322 derivatives containing R1162 DNA from 3.8 kb to a variable endpoint within the oriV region. In each case, the R1162 DNA was located in an identical position and orientation within pBR322. The results suggest that DNA sequences in the oriV region are important for mobilization. The mobilization, at intermediate frequencies, of the 0.70-5.68 bp DNA fragment, which had been cloned into pBR322 at a different location, confirms the importance of DNA sequences adjacent to oriT on transfer frequencies.

Properties of the Origin of Replication of R1162

The base sequence of DNA containing the origin of replication of R1162 is shown in Fig. 5. This DNA consists of 2 HpaII-generated fragments, and spans 5.90 to 6.48 kb on the R1162 map (Fig. 3). In addition, incompatibility is encoded within the 370 bp HpaII fragment (6.11-6.48 kb).

There are no sequences that correspond closely with those of E. coli promoters (17,18) and there is only one ATG-initiated open reading frame contained within the sequence and sufficient to encode a polypeptide of greater than 40 amino acids. This open reading frame, which could encode a polypeptide of about 5,000 daltons, extends from position 558 to 409. This sequence is preceded by a ribosome binding site (19) and an in-phase nonsense codon. There is no obvious promoter; if the DNA is transcribed, it is probably part of a larger transcriptional unit. This potential gene could be expressed in the cloned DNA, with transcription provided by an exogenous promoter. However, the putative polypeptide is not required for expression of incompatibility or an active oriV. We deleted 150 bp of DNA, from the RsaI cleavage site (position 435) to the end of the 370 bp fragment (position 587). The remaining DNA (positions 1-435) was able to form a satellite plasmid, and also continued to express incompatibility. This does not exclude a regulatory role for the polypeptide.

The result with the RsaI-generated deletion means that the incompatibility determinant must lie between positions 212 and 435. The outstanding feature of this region is the presence of 3 perfectly conserved direct repeats, part of which is repeated a fourth time (boxed sequences, Fig. 5). In view of the results with other plasmids (20-23), it is likely that these direct repeats are necessary for expression of incompatibility and copy-number control. This is considered further in the next section. Finally, the adjacent 210 bp HpaII fragment shows extensive 2-fold symmetry (Fig. 6).

Both the 210 and 370 bp HpaII fragments are required in cis for an active oriV (although at least 150 bp of the latter fragment are dispensible). We find that neither fragment will replicate as a satellite plasmid in a cell containing (a) pMS94 (Rep functions provided but other HpaII fragment missing); (b) R1162 (Rep functions and other HpaII fragment provided in trans as part of replicon); or (c) pMS94 plus cloned other fragment (Rep functions and other HpaII fragment provided in trans but not as part of active replicon). However, the 2 fragments need not be contiguous, or in the same relative orientation. The plasmids pMS320 and pMS321, which contain

```
    1 HpaII  10          20   HaeIII30         40          50          60
5' CCGGGCTGAATGATCGACCGAGACAGGCCCTGCGGGGCTGCACACGCGCCCCCACCCTTC
3' GGCCCGACTTACTAGCTGGCTCTGTCCGGGACGCCCCGACGTGTGCGCGGGGGTGGGAAG
   <----------Δ1----------
   <--------------------------------Δ2--------------------------
            70 HaeIII   80          90          100         110         120
5' GGGTAGGGGGAAAGGCCGCTAAAGCGGCTAAAAGCGCTCCAGCGTATTTCTGCGGGGTTT
3' CCCATCCCCCTTTCCGGCGATTTCGCCGATTTTCGCGAGGTCGCATAAAGACGCCCCAAA
   ------Δ2------
            130         140         150         160 FnuDII 170         180
5' GGTGTGGGGTTTAGCGGGCTTTGCCCGCCTTTCCCCCTGCCGCGCAGCGGTGGGGCGGTG
3' CCACACCCCAAATCGCCCGAAACGGGCGGAAAGGGGGACGGCGCGTCGCCACCCCGCCAC
                                              -------Δ3-------
            190         200         210 HpaII 220         230         240
5' TGTAGCCTAGCGCAGCGAATAGACCAGCTATCCGGCCTCTGGGCATATTGGGCAGGGCAG
3' ACATCGGATCGCGTCGCTTATCTGGTCGATAGGCCGGAGACCCGTATAACCCGTCCCGTC
   -----------Δ3------------->
            250         260         270         280         290         300
5' CAGCGCCCCACAGGGCGTGACTAACCGCGCCTAGTGGATTATTCTTAGATAATCATGGAT
3' GTCGCGGGGTGTCCCGCACTGATTGGCGCGGATCACCTAATAAGAATCTATTAGTACCTA

            310         320         330         340         350         360
5' GGATTTTTCCAACACCCCGCCAGCCCCCGCCCCTGCTGGGTTTGCAGGTTTGGGGGCGTG
3' CCTAAAAAGGTTGTGGGGCGGTCGGGGGCGGGGACGACCCAAACGTCCAAACCCCCGCAC

            370         380         390         400         410         420
5' ACAGTTATTGCAGGGGTTCGTGACAGTTATTGCAGGGGGGCGTGACAGTTATTGCAGGGG
3' TGTCAATAACGTCCCCAAGCACTGTCAATAACGTCCCCCCGCACTGTCAATAACGTCCCC

            430   RsaI 440         450         460         470         480
5' TTCGTGACAGTTAGTACGGGATGACGGGCACTGGCTGGCAATGTCTAGCAACGGCAGGCA
3' AAGCACTGTCAATCATGCCCTACTGCCCGTGACCGACCGTTACAGATCGTTGCCGTCCGT

            490         500         510         520         530         540
5' TGTCGGCTGAGGGTAAAAGAACTTTCCGCTAAGCGATAGACTGTATGTGAAACACAGTAT
3' ACAGCCGACTCCCATTTTCTTGAAAGGCGATTCGCTATCTGACATACACTTTGTGTCATA

            550         560         570         580 HpaII
5' TGCAAGGACGCGGAACATGCCTCATGTGGCGGCCAGGACGGCCAGCCGG
3' ACGTTCCTGCGCCTTGTACGGAGTACACCGCCGGTCCTGCCGGTCGGCC
```

Fig. 5. Base sequence of DNA containing the origin of replication and determinant for incompatibility of R1162. The recognition sequences for certain restriction enzymes are boxed, as are those sequences that make up the direct repeats. The locations of deletions Δ1, Δ2, and Δ3 are shown.

R1162 DNA organized as shown in Fig. 7, are both able to replicate as satellite plasmids. We have also generated in vitro deletions within the 210 bp HpaII fragment to determine which regions are essential (Fig. 5 and 6). The deletion Δ1 does not interfere with the

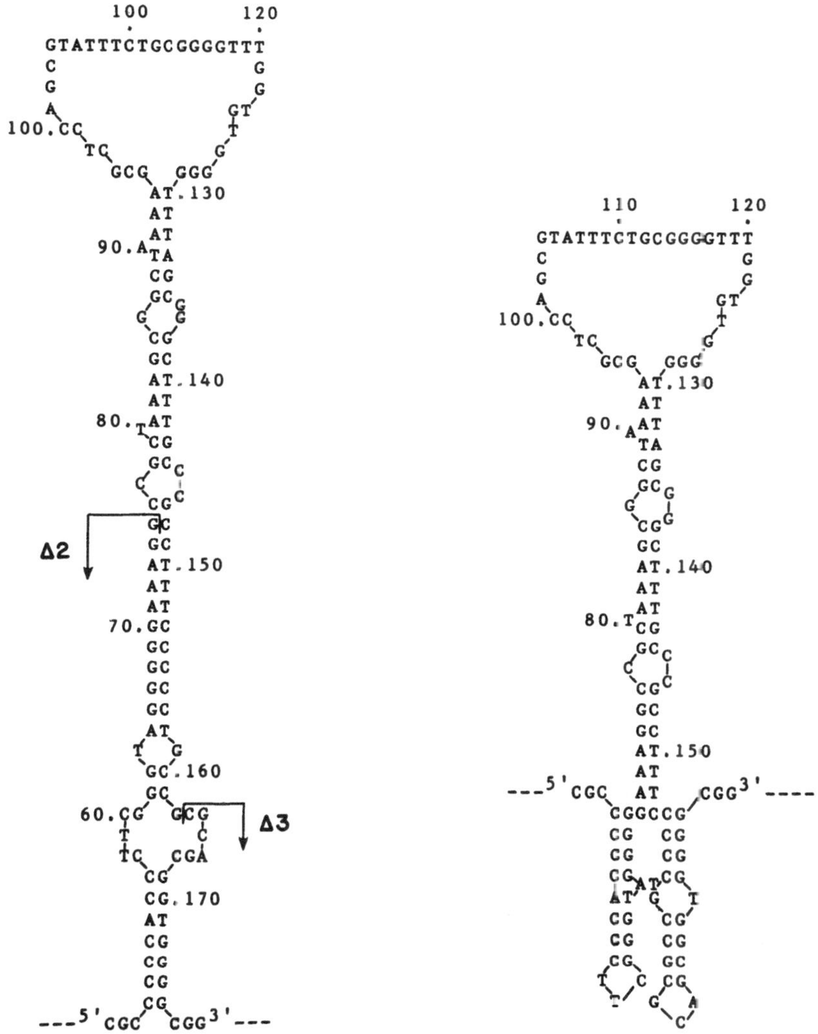

Fig. 6. Part of the base sequence of one strand of the DNA in Fig. 5, redrawn to emphasize the 2-fold symmetry. The starting points of deletions Δ2 and Δ3 are indicated.

ability of the <u>oriV</u> DNA to replicate as a satellite, whereas satellite plasmids cannot be obtained from DNA containing deletions Δ2 or Δ3 (Fig. 5 and 6). These results are consistent with the idea that the 2-fold symmetry of the 210-bp fragment is significant with respect to plasmid DNA replication.

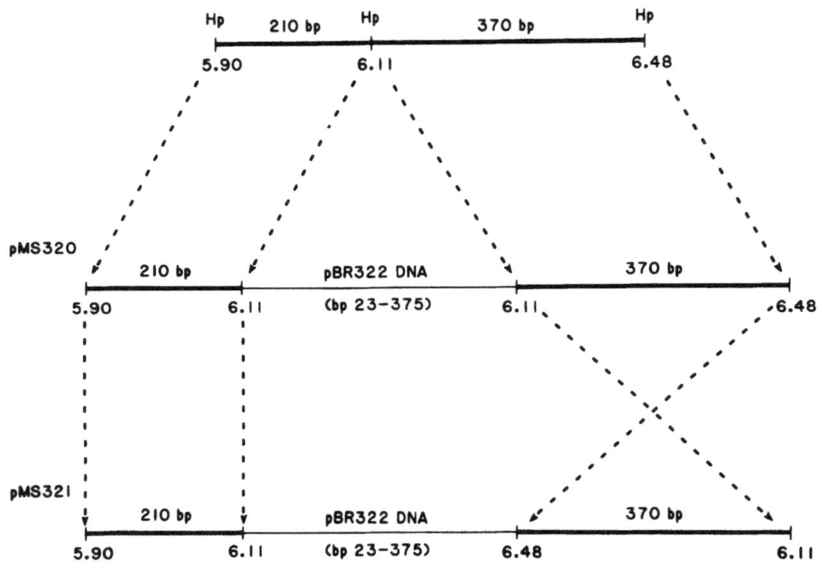

Fig. 7. Location and orientation of R1162 DNA in satellite plasmids pMS320 and pMS321. The arrangement of this DNA in R1162 is shown at the top.

Incompatibility and Copy-Number Control

The 370-bp HpaII-generated fragment of R1162 contains the determinant for incompatibility: derivatives of pBR322 containing this fragment eliminate R1162 from the cell (Fig. 3). Is this DNA also involved in copy number control? We joined the 370-bp Inc fragment to DNA encoding resistance to trimethoprim (Tp^R), and cloned this DNA into the EcoRI site of R1162 (Fig. 8, left). As a control we constructed another Tp^R fragment, identical to the first except that it lacks the Inc DNA, and also cloned this into R1162 (Fig. 8, right). Duplication of the Inc region, with the second copy at a novel location, resulted in a lower plasmid copy number (Fig. 9, lanes 4-6). The orientation of insertion of the DNA was not important. This decrease in copy number could come about because of duplication of a DNA sequence encoding a trans-acting inhibitor of plasmid replication. Alternatively, a positively acting product may be removed by binding to the DNA, perhaps at the directly repeated sequences within this DNA (Fig. 5) (2). If a constitutively produced, required plasmid product were not present in excess, then nonproductive binding to a second site could result in lower plasmid copy number, yet the product itself might not be involved in normal regulation of copy number. To exclude this possibility, we introduced the pBR322 derivative pMS94, which encodes the necessary gene products active in trans for replication of R1162, into strains containing R1162 with the duplicated Inc segment. No

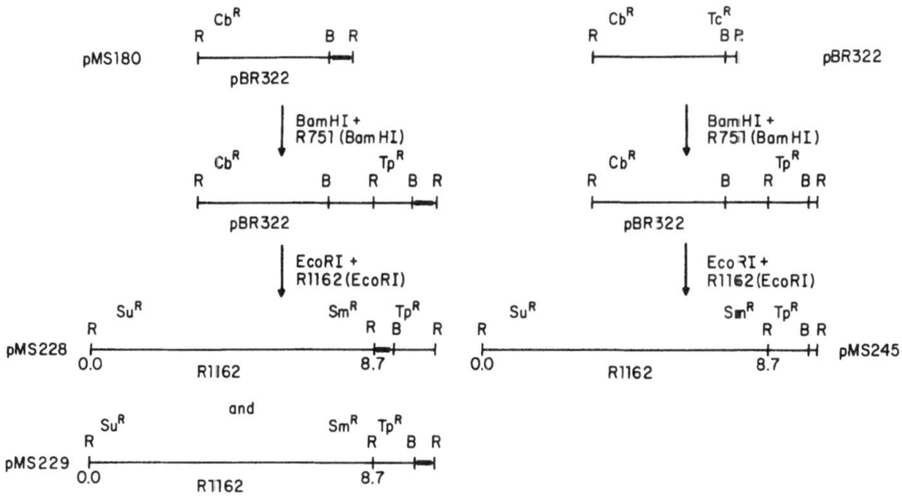

Fig. 8. Cloning DNA encoding the incompatibility determinant of R1162 (left), and control DNA (right) into R1162. The plasmid pMS180 is a pBR322 derivative in which R1162 DNA (coordinates 6.11-6.48 kb, Fig. 3) has been cloned into the ClaI site. This R1162 DNA, specifying Inc, is indicated by a thick horizontal line. Note that this DNA is present in opposite orientations in pMS228 and pMS229.

increase in copy number to normal levels was observed (Fig. 9, lanes 1-3). We conclude that the lower copy number of plasmids with the duplicated Inc DNA reflects the copy-control mechanism for R1162.

DISCUSSION

As a detailed picture of how R1162 DNA replicates becomes available, it will provide insight into how this plasmid achieves its broad host range. Several observations can already be made. First, it is clear that a large fraction of R1162 DNA encodes genes required for plasmid maintenance (Fig. 1). Moreover, 3 essential proteins, active in an in vitro system for plasmid DNA replication, have been identified by Scherzinger et al. for RSF1010 (15). In principle, broad host range can be the consequence of a plasmid providing many of its own requirements for replication or, alternatively, it can be due to universalized sites on its DNA, able to parasitize whatever host replicative functions are available. R1162 may have evolved in the former direction. Interestingly, another broad host-range plasmid, RK2, might require only one plasmid-encoded protein for replication (24,25), although other proteins are necessary for cellular maintenance (26,27).

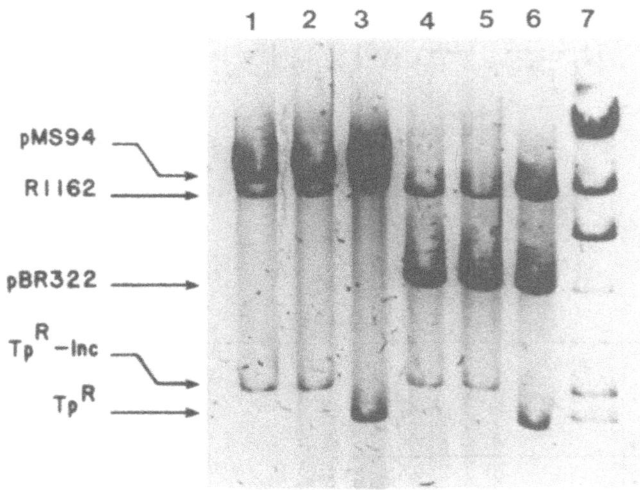

Fig. 9. Agarose gel electrophoresis of EcoRI-digested plasmid DNAs from strains containing plasmids (see Fig. 8): Lane 1 pMS94 + pMS228, lane 2 pMS94 + pMS229, lane 3 pMS94 + pMS245, lane 4 pBR322 + pMS228, lane 5 pBR322 + pM229, lane 6 pBR322 + pMS245. Lane 7 is λ DNA, HindIII-digested.

The origin of replication of R1162 is located between 5.90 and 6.48 kb on the R1162 map, and this is quite distant from those genes providing the proteins which presumably act on it. A similar situation is observed for RK2; in contrast, for many plasmids, coding sequences sufficient for replication lie close together (28-31). Broad host-range plasmids could arise by fusion of separate replicons, followed by a slow process of homogenization: loss of redundant genetic elements along with mutation of others to broader activity. The lack of clustering of coding sequences essential for replication could reflect this evolutionary history.

The origin region of R1162 contains no open reading frames of substantial size likely to encode protein. The amount of transcriptional activity within the oriV region cannot be determined from the sequence alone. There are no putative promoters with substantial homology to both the -35 and -10 consensus sequences for the promoters of *E. coli* (17,18), although the promoters of broad host-range plasmids might be expected to diverge from this consensus. However, if RNA polymerase is involved in replication, it probably does not provide a primer, since replication of RSF1010 DNA can take place in vitro in the presence of rifampicin (32). It has been suggested that differences among the RNA polymerases of gram-negative bacteria may be great enough to preclude their playing this role in replication of broad host-range plasmids (15).

Replication of RSF1010 proceeds bidirectionally or unidirectionally (33), and probably similarly so in R1162. This property might be reflected in the extensive 2-fold symmetry of the 210-bp fragment, although we have not ruled out the possibility that R1162 contains 2 closely linked origins, and that we have cloned only one of these. We considered that the 2-fold symmetry might indicate the presence of 2 origins, back-to-back. However, if this is so, they are not independent, since a deletion within either of these potential origins results in failure of the remaining DNA to form a satellite plasmid.

Direct repeats about the size of those present in the R1162 oriV region are found in several plasmids, and evidence has accumulated that these bind positively acting proteins (34,35). Perhaps the direct repeats serve as entry points for proteins required in formation of a plasmid DNA replication complex. The direct repeats are also associated with expression of incompatibility (20-23). This has led Tsutsui et al. to propose a model for control of copy number in which a positively acting product, present in constant amount in the cell, is titrated by dummy sites represented by some or all of the directly repeated sequences (20). Our results are consistent with this model, since increasing the dosages of the genes for required plasmid products does not affect the copy number of derivatives containing a duplicated Inc segment. However, the possibility of synthesis of an inhibitor, directed by the Inc DNA, has not been ruled out.

A comparison of the base sequence of the direct repeats of R1162 with those of other plasmids shows that they bear the greatest homology with those of another broad host-range plasmid, RK2 (Fig. 10) (23). Perhaps the direct repeats serve to bind host-encoded products, and the homology represents evolution toward a consensus sequence able to function in large variety of hosts. It is also conceivable that R1162 and RK2 shared a common ancestor.

Not all the sequences required in cis for replication of R1162 need be in their wild-type location or orientation with respect to

```
C G[T G A C A G T T]A T T G C[A G G G G]        R1162

[T G A C A G T T]G[A G G G G]C G G              RK2
           C A
            T
```

Fig. 10. Comparison of the direct repeats of R1162 and RK2 (23). The homologous regions are boxed. The sequence of one strand is shown, written 5' to 3'.

each other. This indicates that the oriV region is made up of a number of discrete functional domains, separated by nonessential segments. Possibly, replication components are assembled at one location, and then, in a primosome-like fashion, migrate along the DNA to the origin, where DNA synthesis is initiated (36).

ACKNOWLEDGMENTS

Parts of this work have been supported by the National Science Foundation (Grants PCM-8002628 and PCM-8219916), and by the Robert A. Welch Foundation (Grant F-964).

REFERENCES

1. Barth, P., and N. Grinter (1974) Comparison of the deoxyribonucleic acid molecular weights and homologies of plasmids conferring linked resistance to streptomycin and sulfonamides. J. Bact. 120:618-630.
2. Meyer, R., M. Hinds, and M. Brasch (1982) Properties of R1162, a broad host-range, high-copy number plasmid. J. Bact. 150:552-562.
3. Bryan, L.E., H.M. van den Elzen, and J.T. Tseng (1972) Transferable drug resistance in Pseudomonas aeruginosa. Antimicrob. Ag. Chemother. 1:22-29.
4. Gautier, F., and R. Bonewald (1980) The use of plasmid R1162 and derivatives for gene cloning in the methanol-utilizing Pseudomonas Aml. Mol. Gen. Genet. 178:375-380.
5. Quivey, Jr., R.G., R.J. Meyer, and F.R. Tabita (1981) Plasmid transfer into Rhodospirillum rubrum. Eighty-first Ann. Meeting, American Society for Microbiology (Abstract H113).
6. Meyer, R., R. Laux, G. Boch, M. Hinds, R. Bayley, and J.A. Shapiro (1982) Broad host-range Inc P-4 plasmid R1162: Effects of deletions and insertions on plasmid maintenance and host range. J. Bact. 152:140-150.
7. Grinter, N.J., and P. Barth (1976) Characterization of SmSu plasmids by restriction endonuclease cleavage and compatibility testing. J. Bact. 128:394-400.
8. Rubens, C., F. Heffron, and S. Falkow (1976) Transposition of a plasmid dexoyribonucleic acid sequence that mediates ampicillin resistance: Independence from host rec functions and orientation of insertion. J. Bact. 128:425-434.
9. Bagdasarian, M., E. Amann, R. Lurz, B. Ruckert, and M. Bagdasarian (1983) Activity of the hybrid tryp-lac (tac) promoter of Escherichia coli in Pseudomonas putida. Construction of broad host-range, controlled-expression vectors. Gene 26:273-282.
10. Nordheim, A., T. Hashimoto-Gotoh, and K.N. Timmis (1980) Location of two relaxation nick sites in R6K and single sites in pSC101 and RSF1010 close to origins of vegetative replication:

Implications for conjugal transfer of plasmid deoxyribonucleic acid. J. Bact. 144:923-932.

11. Casadaban, M.J., J. Chou, and S.N. Cohen (1980) In vitro gene fusions that join an enzymatically active β-galactosidase segment to amino-terminal fragments of exogenous proteins: Escherichia coli plasmid vectors for the detection and cloning of translational initiation signals. J. Bact. 143:971-980.
12. Pratt, J.M., G.J. Boulnois, V. Darby, E. Orr, E. Wahla, and I.B. Holland (1981) Identification of gene products programmed by restriction endonuclease DNA fragments using an E. coli in vitro system. Nucl. Acids. Res. 9:4459-4474.
13. Frazer, A.C., and R. Curtiss, III (1973) Derepression of anthranilate synthetase in purified minicells of Escherichia coli containing the Col-trp plasmid. J. Bact. 115:615-622.
14. Amann, E., J. Brosius, and M. Ptashne (1983) Vectors bearing a hybrid trp-lac promoter useful for regulated expression of cloned genes in Escherichia coli. II Gene 25:167-178.
15. Scherzinger, E., M.M. Bagdasarian, P. Scholz, R. Lurz, B. Ruckert, and M. Bagdasarian (1984) Replication of the broad host-range plasmid RSF1010: Requirement for three plasmid-encoded proteins. Proc. Natl. Acad. Sci., USA 81:654-658.
16. Lin, L.-S., and R.J. Meyer (1984) Nucleotide sequence and functional properties of DNA encoding incompatibility in the broad host-range plasmid R1162. Mol. Gen. Genet. (194:423-431).
17. Rosenberg, M. and D. Court (1979) Regulatory sequences involved in the promotion and termination of RNA transcription. Ann. Rev. Genet. 13:319-353.
18. Hawley, D.K., and W.R. McClure (1983) Compilation and analysis of Escherichia coli promoter DNA sequences. Nucl. Acids Res. 11:2237-2255.
19. Shine, J., and K. Dalgarno (1974) The 3' terminal sequence of E. coli 16S rRNA: Complementarity to nonsense triplets and ribosome binding sites. Proc. Natl. Acad. Sci., USA 71:1342-1346.
20. Tsutsui, H., A. Fujiyama, T. Murotsu, and K. Matsubara (1983) Role of nine repeating sequences of the mini-F genome for expression of F-specific incompatibility phenotype and copy number control. J. Bact. 155:337-344.
21. Kamio, Y., and Y. Terawaki (1983) Nucleotide sequence of an incompatibility region of mini-Rts1 that contains five direct repeats. J. Bact. 155:1185-1191.
22. Tolun, A., and D.R. Helinski (1981) Direct repeats of the F plasmid incC region express F incompatibility. Cell 24:687-694.
23. Stalker, D.M., C.M. Thomas, and D.R. Helinski (1981) Nucleotide sequence of the region of the origin of replication of the broad host-range plasmid RK2. Mol. Gen. Genet. 181:8-12.
24. Kornacki, J., A. West, and W. Firshein (1984) Proteins encoded by the trans-acting replication and maintenance regions of broad host-range plasmid RK2. Plasmid 11:48-57.

25. Schmidhauser, T., M. Filutowicz, and D.R. Helinski (1983) Replication of derivatives of the broad host-range plasmid RK2 in two distantly related bacteria. Plasmid 9:325-330.
26. Figurski, D.H., R.F. Pohlman, D.H. Bechhofer, A.S. Prince, and C.A. Kelton (1982) Broad host range plasmid RK2 encodes multiple Kil genes potentially lethal to *Escherichia coli*. Proc. Natl. Acad. Sci., USA 79:1935-1939.
27. Thomas, C.M., A.A.K. Hussain, and C.A. Smith (1982) Maintenance of broad host-range plasmid RK2 replicons in *Pseudomonas aeruginosa*. Nature 298:674-676.
28. Oka, A., N. Nomura, M. Morita, H. Sugisaki, K. Sugimoto, and M. Takanami (1979) Nucleotide sequence of small ColE1 derivatives: Structure of the regions essential for autonomous replication and colicin E1 immunity. Mol. Gen. Genet. 172:151-159.
29. Andres, I., P.M. Slocombe, F. Cabello, J.K. Timmis, R. Lurz, H.J. Burkardt, and K.N. Timmis (1979) Plasmid replication functions II. Cloning analysis of the RepA replication region of antibiotic resistance plasmid R6-5. Mol. Gen. Genet. 168:1-25.
30. Manis, J.J., and B.C. Kline (1977) Restriction endonuclease mapping and mutagenesis of the F sex factor replication region. Mol. Gen. Genet. 152:175-182.
31. Kolter, R., and D.R. Helinski (1978) Construction of plasmid R6K derivatives *in vitro*: Characterization of the R6K replication region. Plasmid 1:571-580.
32. Diaz, R., and W.L. Staudenbauer (1982) Replication of the broad host-range plasmid RSF1010 in cell-free extracts of *Escherichia coli* and *Pseudomonas aeruginosa*. Nucl. Acids Res. 10:4687-4702.
33. DeGraaff, J., J.H. Crosa, F. Heffron, and S. Falkow (1978) Replication of the nonconjugative plasmid RSF1010 in *Escherichia coli* K-12. J. Bact. 134:1117-1122.
34. Germino, J., and D. Bastia (1983) The replication initiator protein of plasmid R6K tagged with β-galactosidase shows sequence-specific DNA binding. Cell 32:131-140.
35. Vocke, C., and D. Bastia (1983) DNA-protein interaction at the origin of DNA replication of the plasmid pSC101. Cell 35:495-502.
36. Arai, K., and A. Kornberg (1981) Unique primed start of phage phiX 174 DNA replication and mobility of the primosome in a direction opposite chain synthesis. Proc. Natl. Acad. Sci., USA 78:69-73.

CONTROL OF PLASMID REPLICATION: THEORETICAL CONSIDERATIONS AND PRACTICAL SOLUTIONS

Kurt Nordström

Department of Microbiology
Biomedical Center
Uppsala University
Box 581
S-751 23 Uppsala, Sweden

INTRODUCTION

The life cycle of bacterial plasmids involves 2 different processes, replication (*rep*) and partitioning (*par*). Both these processes have to be controlled in order to secure stable inheritance of a plasmid in a growing population of bacteria. The crucial genetic information for (control of) replication (21) and partitioning (31,41) is present on the plasmids themselves. The 2 processes seem to be independent of each other since deletion of the *par* region of a plasmid does not affect the copy-number control although it results in unstable inheritance of the plasmid (41). Furthermore, the *par* region of 1 plasmid can be replaced by that of another, giving totally stable inheritance and the same copy number as that of the initial replicon (42). A third system adds to the stability of inheritance of plasmids, the *ccd* (control of cell division) system that inhibits cell division in cells containing only 1 copy of the plasmid (45).

Different plasmids are present in different copy numbers. I will discuss the mechanisms that control plasmid replication.

GENERAL PROPERTIES OF PLASMID REPLICATION

All plasmids seem to have a series of replication properties in common. They utilize host systems for their replication. Hence, the plasmid-borne genetic information needed for replication is fairly small, 1-3 kb (kilobase pairs). This genetic information is often denoted the basic replicon (20). A basic replicon consists of 1 (or sometimes 2-3) origin(s) for replication, and, in some cases, but not always, the structural gene(s) (rep) for 1-3 proteins needed for replication (see also Tab. 2). These proteins are presumably needed for initiation of replication (although the biochemical function has not yet been clarified for any of the rep proteins); initiation is the rate-limiting step in replication and, at least for some plasmids, replication is controlled by regulation of the expression of a rep protein (24; Novick et al., unpub. data).

The population biology as well as the replication control systems (cop for copy-number control) of all or most plasmids have a series of properties in common (Tab. 1):

1.) Plasmid replication is autonomous and not linked to any specific event in the chromosome replication cycle or in the cell cycle of the host. This conclusion is mainly based on the fact that the ratio between the number of plasmid copies and the number of chromosomal origins and termini varies with the growth rate (7,11).

2.) There is a widespread in inter-replication times in the plasmid population, i.e., the daughter of a plasmid molecule can replicate much earlier or much later than one generation after its formation. This conclusion is based on Meselson-Stahl density shift experiments (2,12,53). (This is often referred to as random selection.) The spread in inter-replication times has got a minimum value because plasmids under replication are not available for a

Tab. 1. General characteristics of plasmid replication.

Characteristics of plasmid replication	References
1. Autonomous, i.e. independent of chromosome replication and cell division cycles.	7,11
2. Wide spread in interreplication times	2,12,38,53
3. Replication spread over the entire cell cycle	1,12
4. Defined copy numbers defined by plasmid, host, and growth conditions	7,9,11,28,34, 40,66
5. Deviations from normal copy number adjusted.	
6. Copy mutants exist. Some are dominant others recessive to the wild type	3,34,40,54,66, 74
7. Cloned DNA may cause inhibition of replication, incompatibility (inc or cop)	65

second replication event (38); this eclipes period can be fairly long, e.g., more than 25-30 min for <u>oriC</u> plasmids (23).

The spread in inter-replication times causes randomization during replication which together with random assortment for partitioning at cell division forms the basis of plasmid incompatibility (Fig. 1) (44).

3.) Plasmid replication is spread over the whole cell cycle. This conclusion is based upon 2 kinds of experiments, Meselson-Stahl density shift experiments and analysis of plasmid replication in cells of different size (1,12).

4.) In exponentially growing populations of plasmid-carrying bacteria, plasmids are present in distinct copy numbers. These are defined by the plasmid (34,40,66), the host (9,28), and the growth conditions (7,11). It should be stressed that in the steady state each plasmid copy replicates, on the average, once during a doubling time; this is an average value, and does not apply to the individual plasmid copy (see point 2 above).

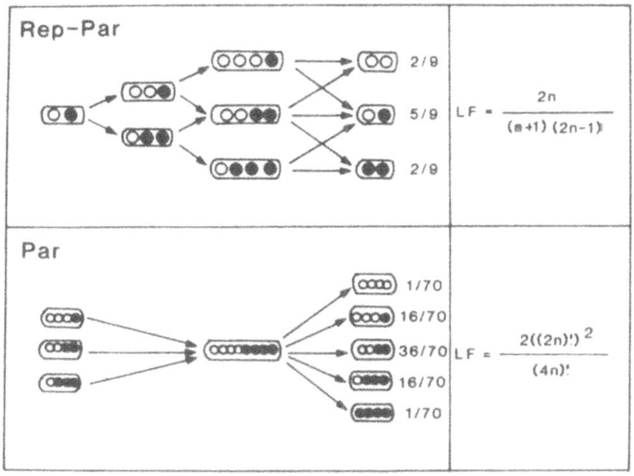

Fig. 1. Plasmid incompatibility as a consequence of randomization during replication and partitioning. The open and full circles represent 2 differently marked plasmids: In the Rep-Par case, the 2 plasmid species have the same replication and partitioning systems; in the Par case, 2 different replicons have the same Par system. The ratios to the right give the frequency of the different combinations of 2 plasmid species after cell division. LF represents the rate of reduction of the relative size of heteroplasmid population during each cell cycle.

5.) The same defined copy number is obtained after introduction (by transfection, transduction, or conjugation) of 1 plasmid copy into a plasmid-free cell. Similarly, perturbations giving average copy numbers higher or lower than the defined one are gradually corrected during growth eventually resulting in the defined average copy number. Hence, the number of replications per plasmid and cell cycle (which is in average exactly 1 in the steady state) is more than 1 at copy numbers lower than the defined one and less than 1 at copy numbers higher than the defined one (Fig. 2), i.e., the rate of replication per plasmid copy falls with increasing plasmid concentration (the hatched areas in Fig. 2).

6.) Mutations in the plasmid (34,40,66) and (in some cases) the host chromosome (9,28) may result in an increased plasmid copy number. Such plasmid mutants are called copy mutants (Cop, cop). Some of these are dominant, others recessive to the parent plasmid. This means that there is a preferential loss of the wild type from cells also carrying a dominant copy mutant, whereas the copy number of a recessive copy mutant is reduced by the simultaneous presence of the wild-type parent plasmid (3,74). There are also copy mutants with a lower copy number than that of the wild-type. Some copy mutants have become compatible with their wild-type parent (54,74).

7.) Cloning of parts of a plasmid onto a vector that is compatible with the plasmid may result in trans-acting inhibition of replication of the parent plasmid, often to such an extent that the plasmid is lost in the presence of the cloned fragment; this has been used to identify inc (incompatibility) genes, that more suitably should be called cop genes. [It should be noted that incompatibility can also be caused by the par (partitioning system)

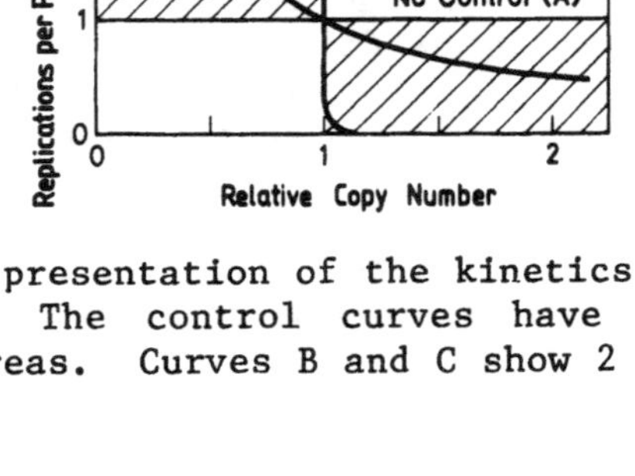

Fig. 2. Schematic presentation of the kinetics of control of replication. The control curves have to lie within the hatched areas. Curves B and C show 2 examples of control kinetics.

(42).] In some cases a cop^+ gene may reduce the copy number of a *cop* mutant without expressing incompatibility, i.e., causing loss of the mutant plasmid (5,33).

THEORETICAL MODELS FOR PLASMID REPLICATION AND ITS CONTROL

I will now take up a few simple theoretical systems for control of replication and discuss how they meet the conditions of Tab. 1. In doing this, I will treat the plasmid population in a culture as if it formed 1 pool; the cell is irrelevant for such a discussion.

Passive Control of Plasmid Replication

When bacteria are growing exponentially in a medium that contains all nutrients in saturating amounts, steady states are established that are characterized by constant ratios between all components of the cell, i.e., the concentration of each cell component does not change with time. The actual concentration of the various cell constituents at steady state growth is in some cases due to active control, but presumably the concentration of many cell components is merely a consequence of the rate constants of the numerous biochemical reactions that take place in a cell. Consequently, it cannot be excluded that the copy number of a plasmid could be passively maintained. The question then arises whether such a system might result in the replication properties discussed above and particularly whether it might allow adjustment of the copy number after perturbation.

The rate of plasmid replication will in such a system be dependent upon the concentrations of many different compounds. However, mathematically all of these may be summarized and replication may be schematically described as a simple enzyme reaction:

$$S \overset{E}{\rightleftharpoons} P \qquad (1.)$$

where E, S, and P are the enzyme, the substrate, and the product of the reaction. The rate of such a reaction ($v = dP/dt$, but since the product is the same molecular species as the substrate, $v = dS/dt$):

$$v = V \cdot S/(K + S) \qquad (2.)$$

V is the maximal rate of this reaction, S is the concentration of the substrate, K is the Michaelis constant.

At saturation, i.e., $S \gg K$ the rate becomes constant, independent of S. The rate of replication of each plasmid copy will drop as S increases:

$$v/S = V/(K + S) \xrightarrow[S \gg K]{} V/S \qquad (3.)$$

i.e., at higher S values, the rate function is an hyperbolic function (as is the rate curve for at least some plasmids, see below).

Let us briefly discuss whether such a system might fit conditions 1-7 in the preceding section and Tab. 1.

Replication of the plasmid would, of course, be a purely statistical event. Hence, conditions 1-3 would clearly be fulfilled. The system would also lead to stable copy numbers (n values). At steady state, the rate can be expressed as the formation of n molecules in a volume (a cell) that, at the start, has n molecules. Hence,

$$v = dS/dt = n \cdot \ln 2/T = N \cdot \ln 2 \cdot n/T(K' + n) \qquad (4.)$$

where T is the generation time and N is the maximum rate expressed as number of replications per cell and cell cycle. K' is the Michaelis constant expressed as number of plasmid molecules per baby cell. This gives the steady state copy number:

$$n = N - K' = \text{constant} \qquad (5.)$$

Depending on the rate constants, the n value can be at any part of a Michaelis-Menten curve (Fig. 3).

Since the rate of the reaction per plasmid copy decreases by increasing concentration of the substrate (Fig. 3), the system will also be able to adjust aberrations in copy number (Fig. 4). This

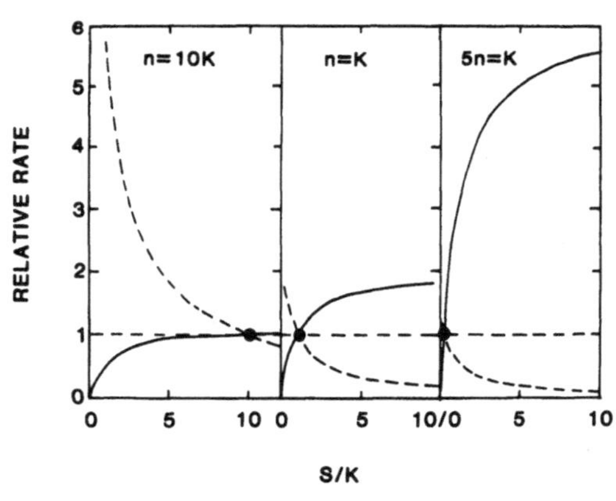

Fig. 3. Passive control of plasmid replication. The graph shows 3 examples of control kinetics (eq. 5.). The solid lines represent the rate of replication (eq. 2.); the dotted lines, the rate of replication per plasmid copy (eq. 3.); and the solid circles, the steady state values.

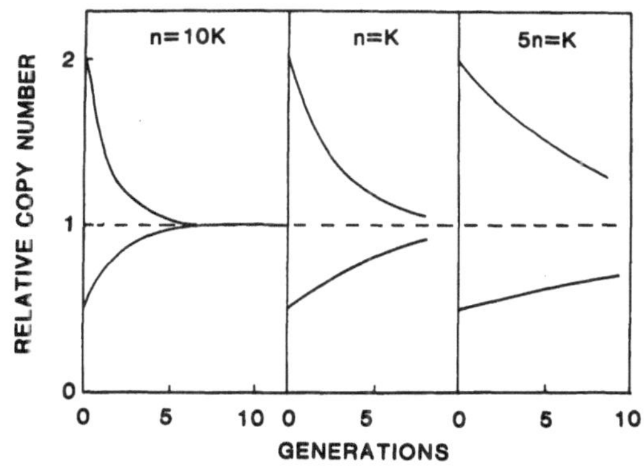

Fig. 4. Passive control of plasmid replication: kinetics of restoration of steady states after perturbations of the copy number. Conditions are as in Fig. 3. The dotted lines represent the steady state conditions.

adjustment will be fairly rapid, but the time required to regain the steady state will be different at different values of the enzyme constants (Fig. 4).

Plasmid replication is specific and the specificity has to reside in the nucleotide sequence of the plasmid. If 2 plasmids carry the same specificity region, they will compete for the reaction and, hence, they will show incompatibility.

The wide spread in inter-replication times causes variations in the ratio between the 2 plasmids, deviations that cannot be corrected (see Fig. 1).

Mutations in the specificity region could change $\underline{V}$ as well as $\underline{K}$, i.e., result in change in copy number (copy mutants). The simultaneous presence of a copy mutant plasmid and the wild-type parent plasmid in a cell would resemble an enzyme reaction with 2 substrates present. In such a case, the substrate will be competitive inhibitors of each other:

$$E + S_1 \rightleftharpoons C_1 \longrightarrow P_1 + E \qquad (6.)$$

$$E + S_2 \rightleftharpoons C_2 \longrightarrow P_2 + E \qquad (7.)$$

which gives:

$$\underline{v}_1 = \underline{N}_1 \cdot \underline{n}_1 / (\underline{K}_1 + \underline{n}_1 + \underline{n}_2 \cdot \underline{K}_1 / \underline{K}_2) \qquad (8.)$$

$$\underline{v}_2 = \underline{N}_2 \cdot \underline{n}_2 / (\underline{K}_2 + \underline{n}_2 + \underline{n}_1 \cdot \underline{K}_2 / \underline{K}_1) \qquad (9.)$$

Since $\underline{n} = \underline{N} - \underline{K}'$, increased copy number can be obtained by a decreased $\underline{K}'$, an increased $\underline{N}$, or a change in both (Fig. 5). Similarly, mutants with decreased copy number would also be possible. The cop mutants would behave differently when present in cells also carrying the wild-type plasmid (Fig. 5). If $\underline{K}'$ of the mutant is lower or $\underline{N}$ is higher than those of the wild-type, the cop mutant would be dominant (Fig. 5 a and b), whereas change in both $\underline{K}'$ and $\underline{N}$ may give recessiveness (Fig. 5c). Hence, also condition 6 is

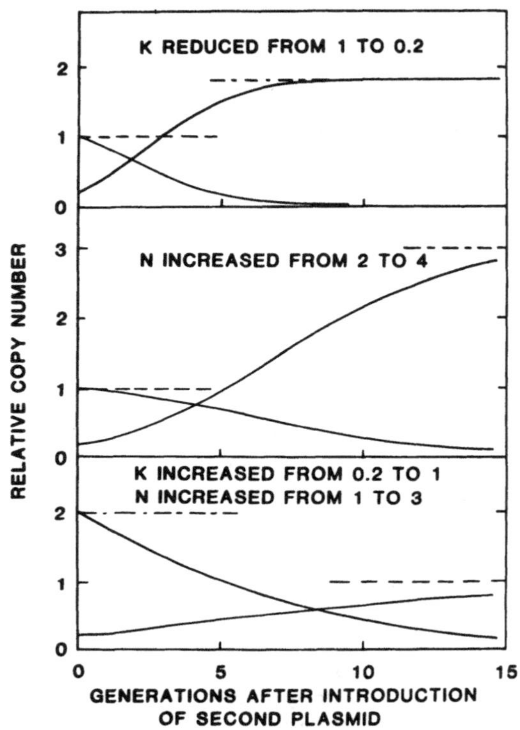

Fig. 5. Passive control of plasmid replication and incompatibility: kinetics of replication after creation of a heteroplasmid situation. a (top) and b (middle): the wild-type plasmid has been given the K' and N values of 1 and 2, respectively, whereas the mutant in the top case has a 5-fold lower K' value, and, in the middle case, a 2-fold increased N value. The mutant was introduced at a 5-fold lower concentration than the steady state concentration of the wild-type plasmid. c (bottom): the wild-type has a K' value of 0.2 and an N value of 1, whereas the mutant has the K' and N values 1 and 3, respectively. The wild-type plasmid was introduced at a 10-fold lower concentration than the steady-state concentration of the mutant.

filled. However, there seems to be one exception; it is hard to explain how mutation might cause an increase in copy number and also a total change of specificity, resulting in compatibility between the mutant and its parent.

Host mutations changing the copy number of a plasmid would be caused by changes in the concentration of the enzyme (*E*) or in the properties of the enzyme, changing *K'* and/or *N*.

The incompatibility often exerted by a cloned fragment from a plasmid will be a special case of the competitive inhibition discussed above. If the specificity region were cloned onto a compatible plasmid, the cloned fragment would cause a reduction in the replication frequency of its parent plasmid, i.e., the cloned fragment would express negative control of replication and express incompatibility. Hence, this passive control is an example of negative control.

In conclusion, the simple passive control would give replication properties as 1-7 discussed above. Furthermore, the discussion shows that negative control does not require the existence of specific inhibitors.

However, there is no real indication of the existence of saturation kinetics as a device for control of plasmid replication. It is very hard to imagine what the rate limitation could be, since the system has to be plasmid-specific; there are at least 20 compatibility groups in *Escherichia coli*. Compatible plasmids should use different rate-limiting reactions although, biochemically, replication of all plasmids is very similar.

Furthermore, there is a quantitative aspect: the wet weight of an *E. coli* cell is above 10^{-12} g corresponding to a volume of 10^{-15} l. Hence, 1 plasmid molecule per cell represents a plasmid concentration of 2×10^{-9} M. This is a very low concentration. During one generation (20 min or more) a low-copy-number plasmid (*n* = 2-4) replicates only very few copies per cell; hence, the turnover number is extremely small, <<1 per enzyme molecule and min. This argues for an active negative control rather than for a passive control. A hard argument is, of course, that in those cases where anything specific is known, the plasmid codes for a negative control element (see below). However, this does not exclude that there, somewhere, might be a plasmid that uses a very simple passive control of its replication. This might, for example, apply to some *cop* mutants where the copy number is very high which might result in host factors/systems being rate-limiting.

Negative Control of Plasmid Replication

During the vegetative cycle, phages replicate as fast as possible. Phage λ, for example, goes through an early phase of exponential replication during which the phage DNA population doubles every 2-3 min (4,77). Plasmids, on the other hand, replicate on average exactly once per cell cycle, i.e., much more rarely than vegetative phages do. Hence, plasmids are most of the time nonreplicating. Consequently, plasmid replication is inhibited most of the time. This requires some sort of negative control whose function is to titrate the concentration of the plasmid. The rate of replication per plasmid copy decreases with the concentration of the plasmid. This might be achieved by passive control (see the previous section). The alternative would be a system with a negative loop. Negative control was first proposed by Pritchard (46) and the so-called Inhibitor-Dilution-Model has been discussed in several reviews (47,48). Therefore, I will not discuss these models, but try to generalize the discussion even further. There seem to be 2 types of models proposed: one where the inhibitor is formed constitutively (47,48), the other where it is autoregulated (58). Some general properties of these models are discussed in the following sections.

Negative control by a constitutively formed repressor. The simplest system of this type would be one in which the constitutively formed repressor acts at the origin of replication (or at the site where the first event in replication occurs).

$$\text{Cop} + \text{ori} \rightleftharpoons \text{Cop-ori} \quad (10.)$$

$$\underline{dS}/\underline{dt} = \underline{k}_1 \cdot \underline{ori} \quad (11.)$$

$$\underline{ori} + \underline{Cop\text{-}ori} = \underline{S} \quad (12.)$$

$$\underline{Cop} \cdot \underline{ori}/\underline{Cop\text{-}ori} = \underline{K}. \quad (13.)$$

In the steady state,
$$\underline{Cop} = k_2 \cdot \underline{S} \quad (14.)$$

Hence,
$$\underline{ori} = \underline{K} \cdot \underline{S}/(\underline{K} + \underline{k}_2 \cdot \underline{S}) \quad (15.)$$

and
$$\underline{dS}/\underline{dt} = k_1 \cdot \underline{K} \cdot \underline{S}/(\underline{K} + k_2 \cdot \underline{S}) = S \cdot \ln 2/T, \quad (16.)$$

which gives
$$S = K(k_1 \cdot T/\ln 2 - 1)/k_2. \quad (17.)$$

Hence, conditions 1-4 are clearly met. Furthermore, the discussion of conditions 5-7 will be very much the same as for autoregulation (see next section).

Autorepressor control. Sompayrac and Maaløe (58) suggested an autorepressor model for control of replication of the *E. coli* chromosome. Murotsu and Matsubara (36) concluded that λ *dv* is an example of a plasmid that uses an analogous system for control of its replication (see below).

An autorepressor can either also be the positively needed Rep function or a separate one; in the latter case, the repressor and the Rep function form an operon; see plasmid $\underline{\lambda dv}$ in Fig. 6. The rate of replication is determined by the concentration of the Rep function. The rate of formation of Rep is determined by the concentration of operator-promoter (op) that has not bound the inhibitor Cop. The concentrations of Cop, op, and Rep are denoted $\underline{Cop}$, $\underline{op}$, and $\underline{Rep}$, respectively:

$$\mathrm{op} + \mathrm{Cop} \rightleftharpoons \mathrm{op\text{-}Cop}, \tag{18.}$$

which gives

$$\underline{op}\cdot\underline{Cop}/\underline{op}\text{-}\underline{Cop} = \underline{K}, \tag{19.}$$

and

$$\underline{op} + \underline{op}\text{-}\underline{Cop} = \underline{S}, \tag{20.}$$

where S is the substrate of the reaction, the replicon. This gives

$$\underline{op} = \underline{K}\cdot\underline{S}/(\underline{K} + \underline{Cop}) \tag{21.}$$

In the steady state, the concentrations of all components in the cells are constant. This also means that the rate of formation of any component is proportional to that of any other. Hence,

$$\underline{Rep} = \underline{k}_2\cdot\underline{S} \text{ and } \underline{dRep}/\underline{dt} = \underline{k}_2\cdot\underline{dS}/\underline{dt} \tag{22.}$$

and

$$\underline{Cop} = \underline{k}_3\cdot\underline{S} \text{ and } \underline{dCop}/\underline{dt} = \underline{k}_3\cdot\underline{dS}/\underline{dt}. \tag{23.}$$

Since

$$\underline{dRep}/\underline{dt} = \underline{k}_4\cdot\underline{op}, \tag{24.}$$

$$(\underline{dS}/\underline{dt})\underline{k}_2 = \underline{dRep}/\underline{dt} = \underline{k}_4\cdot\underline{op} \tag{25.}$$

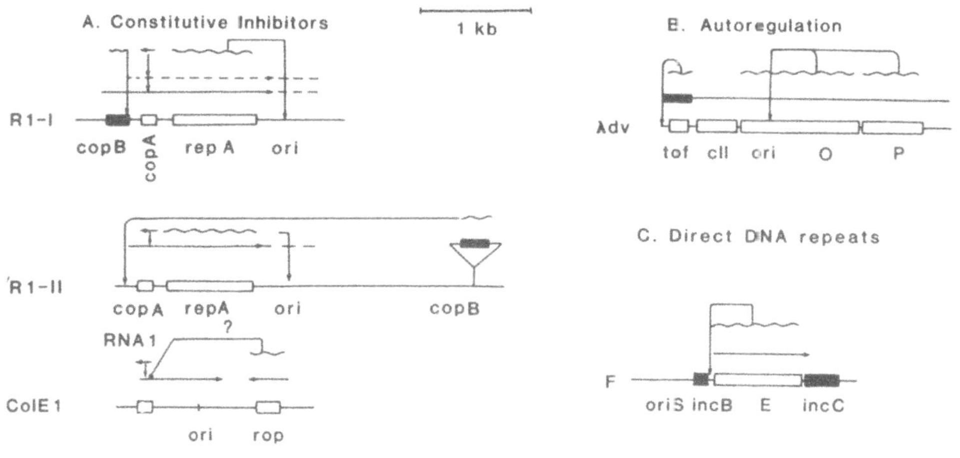

Fig. 6. Control circuits of the basic replicons of a few plasmids.

and $$\underline{dS}/\underline{dt} = \underline{k}_4 \cdot \underline{op}/\underline{k}_2 = \underline{k}_4 \cdot \underline{K} \cdot \underline{S}/\underline{k}_2(\underline{K} + \underline{k}_3 \cdot \underline{S}). \quad (26.)$$

The steady state so obtained will give a defined $\underline{n}$ value, which can be calculated as follows: The population of plasmids grows exponentially with a defined generation time, $\underline{T}$.

$$\underline{dS}/\underline{dt} = \underline{k}_1 \cdot \underline{S} = \underline{S} \cdot \underline{\ln 2}/\underline{T}. \quad (27.)$$

Hence, $$\underline{n} \cdot \underline{\ln 2}/\underline{T} = \underline{k}_4 \cdot \underline{K} \cdot \underline{n}/\underline{k}_2 \cdot (\underline{K} + \underline{k}_3 \cdot \underline{n}) \quad (28.)$$

and $$\underline{n} = \underline{K}(\underline{k}_4 \cdot \underline{T}/\underline{\ln 2} - \underline{k}_2)/\underline{k}_2 \cdot \underline{k}_3. \quad (29.)$$

The autorepressor system is of course truly statistical and conditions 1-3 in Tab. 1 are met, and so is condition 4, as was demonstrated in equation (29.).

What about condition 5, the ability to compensate for deviations from the steady state level? If a cell happens to get more copies of the replicon, e.g., by uneven distribution at cell division, the rate of replication in the cell will increase. However, the output of Cop will also increase and gradually slow down transcription from op, and decreasing the probability of replication to <1 per plasmid copy and cell division. The opposite will be true for cells that happen to get fewer copies than the average cell. The actual kinetics will be dependent upon a series of conditions, e.g., the stability of Rep and Cop.

Copy mutants may be obtained by mutations that change any of the constants $\underline{k}_2$, $\underline{k}_3$, $\underline{k}_4$, and $\underline{K}$; the mutations might be in the $\underline{op}$ region or the $\underline{cop}$ or $\underline{rep}$ genes or in the target for the Rep protein. Mutations in the $\underline{rep}$ gene or the Rep target as well as promoter-down mutations will decrease the copy number. Such mutants would be recessive to the wild-type, since they replicate less than the wild-type at a given concentration of the Cop function. Mutations in the $\underline{cop}$ gene resulting in a less efficient Cop function would also cause an increased copy number; such mutants would be recessive to the wild-type in the sense that they would affect the wild-type less than a normal plasmid would, and presumably also in the sense that they would be preferentially lost from heteroplasmid cells also containing the wild-type plasmid. Mutations affecting the operator might lead to decreased sensitivity to the Cop function and, hence, to an increased copy number. Such mutants would be dominant over the wild-type since the concentration of the Cop function would be higher in cells carrying such a mutant than in cells carrying the wild-type plasmid.

What would happen if a $\underline{cop}$ gene together with its operator-promoter were cloned onto a cloning vehicle? At high ratios between the copy number of the vector and the plasmid the concentration of op + op-Cop would remain constant and equal to the concentration of the vehicle; T; at lower ratios, the plasmid will also contribute to $\underline{op} + \underline{op}\text{-}\underline{Cop}$

$$\underline{op} + \underline{op}\text{-}\underline{Cop} = \underline{T} \tag{30.}$$

$$\underline{dCop}/\underline{dt} = \underline{k}_5 \cdot \underline{op} \tag{31.}$$

Equation (19.) then gives

$$\underline{op} = \underline{K} \cdot \underline{T}/(\underline{K} + \underline{Cop}) \tag{32.}$$

$$\underline{dCop}/\underline{dt} = \underline{k}_5 \cdot \underline{op} = \underline{k}_5 \cdot \underline{K} \cdot \underline{T}/(\underline{K} + \underline{Cop}). \tag{33.}$$

However, in the steady state

$$\underline{dCop}/\underline{dt} = \underline{ln2} \cdot \underline{Cop}/\underline{T} = \underline{k}_5 \cdot \underline{op}, \tag{34.}$$

hence

$$\underline{K} \cdot \underline{T}/(\underline{K} + \underline{Cop}) = \underline{ln2} \cdot \underline{Cop}/\underline{k}_5 \cdot \underline{T} \tag{35.}$$

and

$$\underline{Cop} = -\underline{K}/2 \pm \sqrt{\underline{K}^2/4 + \underline{K} \cdot \underline{k}_5 \cdot \underline{T} \cdot \underline{T}/\underline{ln2}.} \tag{36.}$$

Hence, a cloned cop gene will express inhibition of replication (Inc) at an extent that increases with the concentration of the gene.

Positive Control of Plasmid Replication

The membrane-site model of Jacob, Brenner, and Cuzin (18) has been used, not completely correctly, as the prime model of positive control. A lot of effort has been used to distinguish between positive and negative control of replication. However, all this has been a waste of time and effort, since nobody seems ever to have tried to define what positive control of replication means. A try to define positive control inevitably leads to the conclusion that it cannot exist. To get a decreasing rate of replication at increasing concentrations of the substrate (see Fig. 2), a negative control loop is necessary; the only alternative is saturation kinetics, where an outside (host) system sets the maximum level of replication by, for example, supplying a rate-limiting enzyme in constant amounts. But this is not positive control.

Hence, the distinction between positive and negative control is uninteresting. What matters is to find out whether individual control components act positively, negatively, or both.

CONTROL OF REPLICATION OF KNOWN PLASMIDS

Basic Replicons and Their Components

Basic replicons. Natural plasmids can be any size from a few kb to hundreds of kb. However, large plasmids carry a whole series of genes that are concerned with functions other than plasmid maintenance. The size of a replicon can be reduced by using restriction endonucleases in vitro (e.g., 65). Kollek et al. (20) introduced the concept basic replicon for the smallest piece of plasmid DNA that is able to replicate. I prefer a more limited definition, namely a basic replicon is the smallest piece of DNA that replicates with the wild-type copy number.

Basic replicons normally form a continuous piece of DNA the size of which in most cases is 1.0-2.5 kb (Tab. 2). Hence, a basic replicon is fairly small, with a potential of coding for 3-4 proteins, because plasmids use host functions for their replication. As a comparison, conjugal transfer requires one order of magnitude more DNA (76).

Some large plasmids have more than one basic replicon; presumably one of these is solely or preferentially used.

Basic replicons consist of origins, *cop* genes, and (in some cases) *rep* genes (Tab. 2). Fairly little is known about how replication is initiated; plasmid ColEl is an exception here (15).

Origins of replication. Replication does not start at random but at defined origins (15). Some plasmids have more than 1 origin. Normally, only 1 of the origins seems to be active; however, a word

Tab. 2. Functions coded for by some plasmid basic replicons.

Plasmid	Size of basic replicon (kb)	Origin	Cop function	*rep* genes
ColEI	1-1.5	1	RNAI Rop Protein	None
R1(FII)	2.5	1	CopA-RNA CopB Protein	*repA*
pT181		1	RNA	*repC*
λ *dv*	2	1	Cro protein	*O*,*P*
F	1.5	*oriV*,*oriS*		*repE*
R6K		α,β,γ	Direct repeats	*pir*
PI		1	of DNA + ?	
RK2		1		
RSF1010	5kb	1	?	*repA*,*B*,*C*

of caution is needed since very little direct experimentation has been devoted to investigate this. Plasmid R6K is an example of a plasmid in which 3 origins (α, β, and γ) are all used with about equal frequency (22).

Origins are not well-defined. In most cases they are identified by electron microscopy as bubbles on early replicative intermediates (ss Ref. 15,65). Plasmid ColE1 uses an RNA primer and the exact location of the shift from RNA to DNA synthesis has been identified to be any of 3 consecutive base pairs (68). Similarly, plasmid _oriC_ starts its replication at 3-4 closely linked base pairs (14).

Origins were once concluded to have a high degree of secondary structure with the possibility of forming stem-loop structures. However, the so identified origin of FII plasmids (51) was recently found to be outside of the functional origin [identified by in vitro replication (29)]. However, it is possible that the region of secondary structure might have a function in determining the activity of the origin.

Rep functions. Some plasmids (e.g., ColE1) do not require any plasmid-coded functions for their replication. However, it is fairly common that plasmids code for 1-3 Rep functions (Tab. 2). These are supposed to be involved in initiation of replication, although the exact function has not been identified for the Rep function of any plasmid. Since there are in vitro replication systems for plasmids R6K (16,17), R1 (10,29), and pT181 (19), progress can be be expected in the biochemical understanding of initiation of replication.

The formation of Rep protein is rate-limiting for replication of at least some plasmids [FII (24), and F (57)]. There are at least 2 cases where a Rep protein is mainly cis-acting [the λ O protein and the FII RepA protein (29)]; this means that they preferentially replicate the DNA molecule from which they were coded. They act in trans only if the _cis_-target is absent (29); the target is the origin (_ori_) of replication. The λ O protein is clearly very unstable (29) and the RepA protein seems to be "destroyed" by the presence of _oriR1_ or stoichiometrically used in replication (29).

De novo protein synthesis is needed for each round of replication of FII plasmids (10,75).

Cop functions. Deletion and cloning analysis of several plasmids have always led to the identification of DNA sequences that are acting in trans as inhibitors of replication. A careful analysis (genetic or, in a few instances, biochemical) have identified 3 types of inhibitors (Tab. 2), small proteins (5,30,33), small RNA molecules (3,35,62,69), and a series of direct repeats of DNA (22,60,61,64,67,73). In the last case, formation of RNA has not been completely ruled out. However, it appears that DNA itself may cause trans-inhibition of replication.

The small RNAs have a high degree of secondary structure allowing the formation of long stems with small loops (3,35,62,69). The specificity rests on the loops (3,35,62) and the inhibitory effect is specified by interaction between RNA and RNA (25).

Control kinetics. The rate of replication per plasmid copy decreases with increasing plasmid concentration (hatched areas in Fig. 2). The actual kinetics is not known in more than 2 cases; plasmids F and R1 (a member of the FII group). The control curve of the latter plasmid was recently discussed in detail (43) and shown to be an hyperbolic one, whereas it is more like a step function for the former plasmid (72).

Brief Presentation of a Few Systems

The presentation of plasmid replication control systems is not intended to be a complete review, but aims at showing the principles used by some plasmid systems (see Fig. 6).

Plasmids using small RNA molecules as main copy number control device. Three different plasmids clearly use small (80-100 nucleotides) RNA molecules as their main control device, namely the *E. coli* plasmids ColE1 (35,69) and those of the FII group (3,54,62), and the *Staphylococcus aureus* plasmid pT181 (Novick, pers. comm.) (Tab. 2).

There are several similarities between these systems. The main control system is an RNA molecule with a high degree of secondary structure; the specificity resides in a small loop at the end of a long stem (3,35,62,69). The target for the small RNA is a long transcript formed from the complementary strand. Hence, the primary interaction is one between 2 complementary RNA species (26,35,62, and Novick et al., unpub. data). The specificity (and the specific activity) of the interaction can be drastically affected by single base-pair substitutions, some of which completely change the incompatibility properties of the system (35,62). The biochemistry of the RNA-RNA interaction is not known, but the formation of an RNA duplex is an appealing possibility. In all 3 cases, the effect of the interaction is down-stream on the long transcript, reducing the ability to translate a *rep* gene (FII and pT181) (26) or reducing the ability to process the long transcript into a primer (ColE1) (35,69). The Rep protein of the FII group is (mainly) cis-acting (29), whereas that of plasmid pT181 is acting both in cis and in trans (Novick et al., unpub. data).

These 3 plasmids may be regarded as examples of plasmids using a simple negative control by a constitutively formed repressor (see above); the target is not the ori but an earlier step that is rate-limiting and that is post-transcriptional.

The kinetics of the replication control system are best known in the FII group. Plasmid R1 of this group forms an unstable RNA molecule (half life 1-2 min., Ref. 62). This means that the concentration of the copA-RNA always is proportional to the concentration of the plasmid and the target (43). The expression of the repA gene is proportional to the concentration of plasmid, but inversely proportional to the concentration of the inhibitor and, hence, the total rate of formation of the RepA protein in a given volume is constant. This was previously found experimentally in shifts between different copy numbers (11,13). Hence, the control curve is hyperbolic (Fig. 2).

Interestingly, another replication system, namely transposition of transposon Tn10 is negatively controlled by a small transcript that is complementary to the early region of the transposon message (56).

Since the RepA protein is cis-acting, a mutation increasing the output of RepA protein is a cis-acting mutation and the mutant plasmid is dominant over the wild-type. This is not the case for plasmid pT181 where the RepC protein is acting both in cis and in trans (Novick et al., unpub. data). Hence, the increased pool of RepC protein can be used by the mutant as well as the wild-type plasmid.

Plasmid ColE1 has got a second cop gene (rop; repressor of primer), coding of a small acidic protein (5). Deletion of the rop gene increases the copy number several-fold. Addition of extra rop genes does not decrease the copy number of a wild-type plasmid; hence the Rop protein does not exert incompatibility. The Rop protein seems to act together with the RNAI, maybe to terminate transcription (not at all clear at present) (5).

The secondary control system of FII plasmids, the CopB system. Plasmids of the FII group superficially resemble the ColE1 group (Fig. 6). They also contain the structural gene, copB, coding for a small protein (in this case basic, though) (33). Deletion of the copB gene increases the copy number several-fold; addition of extra copB copies does not change the copy number of the wild-type plasmid. However, mutants deleted of the copB gene are very sensitive to extra copB copies; hence, copB expresses incompatibility towards the copB mutants but not towards the wild-type (33,49). However, there is a difference between the FII and the ColE1 group. The target for the CopB protein is a promoter and the effect of the protein is to almost totally switch-off a second promoter (repAp) (25). Womble and Rownd (unpub. data) and Riise (unpub. data) have demonstrated that this promoter is derepressed at concentrations of the CopB protein below the wild-type level. Similarly, Riise et al. (49)

showed that insertion of the CopB gene and its promoter into an R1 replicon deleted of the CopB gene and its promoter gave a replicon with a functional copB gene at a new location (Fig. 6). This plasmid has a copy number 1/4-1/3 of that of the wild-type replicon. The control kinetics of this replicon has not been worked out (and is hard to study, because of the very low copy number) but a fair guess is that the control curve would be steeper than a hyperbolic one.

Finally, it should be stressed that a plasmid in which the active copB gene has been moved to a new location still is a plasmid of the same type as the parent plasmid, but the plasmid now has 2 independent constitutively formed repressors (Fig. 6).

Plasmid λdv using a simple autorepressor control system. Plasmid λdv controls its replication by an autorepressor system. The autorepressor is the Cro protein, which controls a transcript that, downstream of the cro message, contains 2 rep genes (O and P) (30). The O protein is cis-acting and the target is a region with 4 direct repeats in the O gene, which also contains the replication origin (71). The O protein is unstable and the replication rate is either determined by the output of the O protein, by transcriptional activation of ori, or both (15,70).

Murotsu and Matsubara (36) have shown that plasmid λdv fits well into the Sompayrac-Maaløe (58) autorepressor theory. They have isolated promoter and operator mutations and these behave in the predicted way with respect to copy number, incompatibility, and expression of the control gene (cro).

Lee and Bailey [Plasmid 11 (1984) 151-165 and 166-177] have presented a mathematical model for replication of plasmid λdv and its copy mutants. The model is based upon the wealth of quantitative information that is available for phage λ. The authors have been able able to predict the correct copy number as well as several other replication parameters of the plasmid. They also predict that the control curve is very close to an hyperbolic curve (see Fig. 2).

Plasmids with direct repeats as cop/inc genes. Several plasmids have cop-DNA that expresses its function without any sign of formation of gene products. The cop-DNA has got a series of carefully spaced direct repeats (18-22 bp). This applies, for example, to plasmids F (67,73), R6K (22,60), P1 (Chattoray and Yarmolinsky, pers. comm.), and RP1 (RK2) (61,64). On top of this they (at least some of them) also seem to have an autorepressor system where a rep gene is autoregulated (22,57,59). A lot is known about plasmid R6K, where the rep gene (pir) is autoregulated (22,59). Mutations in the pir gene may cause an increase in copy number (59) but there are also data that contradict the conclusion that the pir gene product (the π protein) is rate-limiting (22). The function of the repeats is obscure and very little is known about the control parameters.

Plasmid F is better known with respect to control kinetics. Tsutsui and Matsubara (72) showed that F replication is totally switched-off at plasmid concentrations twice that of the wild-type level. Hence, the control curve seems to be a step function. Expression of the E gene is autoregulated (57). Tolun and Helinski (67) have shown that a 55-bp fragment containing 2 of the 9 direct repeats of the basic F replicon expresses a weak inhibition of replication. Tsutsui et al. (72) have shown that the number of repeats in a cell controls replication: the location or orientation does not seem to matter.

What is the molecular basis for the trans-acting negative control of replication of plasmid F? There is no indication of the formation of any gene product from this region. The suggestion has then been that the direct repeats in the DNA nucleotide sequence titrates a positive factor [Tsutsui et al. (73)]. Some workers in the field seem to be hesitant to this as if DNA were of a different stuff than all other macromolecules. However, I can't see any difference between the binding of an inhibitory protein to a DNA sequence and the binding of a positively needed protein (Rep) to an inhibitory DNA segment (rpt for repeat).

If we, for simplicity, assume that $\underline{Rep}_o$ is constant, then:

$$Rep + rpt \rightleftharpoons C \qquad (37.)$$

$$\underline{Rep}\cdot\underline{rpt}/\underline{K} = \underline{C} \qquad (38.)$$

$$\underline{rpt} = \underline{a}\cdot\underline{S} \qquad (39.)$$

$$\underline{C} + \underline{Rep} = \underline{Rep}_o. \qquad (40.)$$

This gives

$$\underline{Rep} = \underline{K}\cdot\underline{Rep}_o/(\underline{K} + \underline{a}\cdot\underline{S}). \qquad (41.)$$

Hence, the frequency of replication ($\underline{dS}/\underline{dt} = \underline{k}_1\cdot\underline{Rep}$) drops with increasing substrate concentration ($\underline{S}$). However, the decrease is slow (at high $\underline{S}$ values $\underline{dS}/\underline{dt}$ is inversely proportional to $\underline{S}$). The equations are based on a high ratio $\underline{rpt}/\underline{Rep}_o$. If this ratio is closer to 1, a second condition has to be added.

$$\underline{rpt + C} = \underline{rpt}_o \qquad (42.)$$

If $\underline{K}$ is small, the situation will be similar to titration and almost every molecule of Rep will bind to rpt as long as these are in surplus. If there is a slow but steady production of Rep, suddenly all Rpt will be bound by Rep and free Rep molecules become available. Replication then supplies new Rpt that bind all free Rep molecules.

Hence, there will be a switch-on/switch-off of replication (a step function). However, this requires either that the number of Rep molecules is very small (maybe 20 per cell for plasmid F; some proteins are present in such low concentrations) but still synthesized steadily or that each repeat binds many Rep molecules (either as a cluster or by modifying them by binding and hence catalytically destroying Rep activity).

Nothing is known about what positive factor binds to the repeats, although the 29 k E protein of F is a prime suspect. Nothing seems to be known about the amount of 29 k protein formed. An intriguing fact is also that this and analogous proteins in other plasmid systems are autoregulated (22,57,59), which seems to be hard to understand; autoregulation of a titrated Rep function seems to be a contradiction. One possibility could be that the protein is bifunctional (as suggested by Molin et al., unpub. data) with one part being inhibitory and the other being a Rep function. This could keep autoregulation intact and allow destruction of the Rep function. Possibilities of establishing the identity of the Rep function and the protein that binds to the repeats are now clearly at hand for several of the plasmids that have repeats in their DNA nucleotide sequence as a main part of their replication control system.

CONTROL OF CHROMOSOME REPLICATION

Does chromosome replication obey the rules in Tab. 1? The answer is partly no. 1) Cooper and Helmsetter (8) have clearly shown that initiation of chromosome replication is not linked to a specific part of the cell division cycle. 2) At rapid growth, 2 or more initiations of chromosome replication take place per cell and cell cycle. However, the spread in inter-replication times is much less for the bacterial chromosome than for plasmids (8,12). 3) Similarly, multiple initiations essentially coincide in time and do not spread over the whole cell cycle (8). Requirements 4) and 5) are fulfilled. 6) Mutations exist that affect the copy number of the chromosome (52,63). However, these affect RNA polymerase and not specific <u>cop</u> genes. Finally, 7) has not been tested experimentally.

How do we explain that multiple initiations at chromosomal origins spread very little in time? One possibility would be that the control curve is a step function and that the <u>cop</u> gene is located some distance from the chromosomal origin. Multiple initiations at the same origin may be made impossible by steric reasons. Presumably, the newly replicated DNA is not in a form that is suitable for replication immediately after replication. This would resemble the slow replication cycle (from CCC-DNA to CCC-daughter-DNA) of (some) plasmids including <u>oriC</u> plasmids (23,75).

There is no direct evidence for the existence of a negative control element involved in chromosome replication. So far, work with oriC plasmids has not contributed much to the understanding of replication control (see brief discussion in Ref. 39).

Since there seem to be some differences between chromosome and plasmid replication, it might be a fruitful approach to turn the matter upside down. Can plasmids run the chromosome and, in such a case, what are the characteristics of the chromosome replication and the cell-division cycles? The answer to the first question is yes. Nishimura et al. (37) showed that plasmid F can integratively suppress a dnaA(ts) mutation. Chandler et al. (6) found that also plasmid R100 is able to perform integrative suppression; they also reported that the R100 origin is used at temperatures nonpermissive for chromosomal replication. Molin and Nordström (32) reported that an integrated plasmid R1 expresses CopA activity and is sensitive to CopA. This should make it possible to study the chromosome replication and cell-division cycles. However, these strains grow poorly in minimal medium at high temperature, perhaps due to effects on the membranes because dnaA expression is autoregulated. However, preliminary data in our laboratory (Luud Koppes) indicate that the size distribution of intR1 cells at 42°C is much broader than that of normal E. coli strains. We are presently engaged in the construction of $dnaA^{+}$ strains where oriC has been replaced by the R1 and F basic replicons.

FINAL COMMENTS

Studies of control of plasmid replication have now reached a state where the mechanisms involved gradually are being unraveled. Clearly, negative control is involved, but the inhibitors can be DNA, RNA, or protein. Control is exerted by regulating transcription and/or translation rather than by interfering with the origin of replication and the biochemical events that take place there. In some cases, in vitro and in vivo studies of the quantitative aspects of replication control are possible. This will require careful measurements of binding constants, rate constants, etc.

ACKNOWLEDGMENTS

This work was supported by the Swedish Natural Science Research Council and the Swedish Cancer Society. I am grateful for numerous discussions throughout the years with Bob Pritchard, Richard Novick, Michael Yarmolinsky, and many others, as well as with present and previous co-workers, especially Søren Molin, Luud Koopes, and Gerhart Wagner. However, none of these should be blamed for the content of this paper.

REFERENCES

1. Andresdottir, V., and M. Masters (1978) Evidence that F'lac replicates asynchronously during the cell cycle of Escherichia coli B/r. Mol. Gen. Genet. 163:205-212.
2. Bazaral, M., and D.R. Helinski (1970) Replication of a bacterial plasmid and an episome in Escherichia coli. Biochemistry 9:399-406.
3. Brady, G., J. Frey, H. Danbara, and K.N. Timmis (1983) Replication control mutations of plasmid R6-5 and their effects on interactions of the RNA-I control element and its target. J. Bact. 154:429-436.
4. Carter, B.J., and M.G. Smith (1970) Intracellular pools of bacteriophage λ deoxyribonucleic acid. J. Mol. Biol. 50:713-718.
5. Cesareni, G., M.A. Muesing, and B. Polisky (1982) Control of ColE1 DNA replication - The rop gene product negatively affects transcription from the replication primer promoter. Proc. Natl. Acad. Sci., USA 79:6513-6517.
6. Chandler, M.L., L. Silver, and L. Caro (1977) Suppression of an Escherichia coli dnaA mutation by the integrated R factor R100.1: Origin of chromosome replication during exponential growth. J. Bact. 131:421-430.
7. Collins, J., and R.H.Pritchard (1973) Relationship between chromosome replication and F'lac episome replication in Escherichia coli. J. Mol. Biol. 78:143-155.
8. Cooper, S., and C.E. Helmstetter (1968) Chromosome replication and the division cycle of Escherichia coli B/r. J. Mol. Biol. 31:519-540.
9. Cress, D.E., and B.C. Kline (1976) Isolation and characterization of Escherichia coli chromosomal mutants affecting plasmid copy number. J. Bact. 125:635-642.
10. Diaz, R., K. Nordström, and W.L. Staudenbauer (1981) Plasmid R1 DNA replication dependent on protein synthesis in cell-free extracts of E. coli. Nature 289:326-328.
11. Engberg, B., and K. Nordström (1975) Replication of R-factor R1 in Escherichia coli K-12 at different growth rates. J. Bact. 123:179-186.
12. Gustafsson, P., K. Nordström, and J.W. Perram (1978) Selection and timing of replication of plasmids R1drd-19 and F'lac in Escherichia coli. Plasmid 1:187-203.
13. Gustafsson, P., and K. Nordström (1980) Control of replication of plasmid R1. Kinetics of replication in shifts between different copy number levels. J. Bact. 141:106-110.
14. Hirose, S., S. Hiraga, and T. Okazaki (1983) Initiation site of deoxyribonucleotide polymerization at the replication origin of the Escherichia coli chromosome. Mol. Gen. Genet. 189:422-431.
15. Hobom, G. (1981) Replication signals in prokaryotic DNA. Curr. Topics Microbiol. Immunol. 94/95:93-142.
16. Inuzuka, M., and D.R. Helinski (1978) Replication of antibiotic resistance plasmid R6K DNA in vitro. Biochemistry 17:2567-2573.

17. Inuzuka, M., and D.R. Helinski (1978) Requirements of a plasmid-encoded protein for replication in vitro of plasmid R6K. Proc. Natl. Acad. Sci., USA 75:5381-5385.
18. Jacob, F., S. Brenner, and F. Cuzin (1963) On the regulation of DNA replication in bacteria. Cold Spring Harbor Symp. Quant. Biol. 28:329-348.
19. Khan, S.A., S.M. Carleton, and R.P. Novick (1981) Replication of plasmid pT181 DNA in vitro: Requirement for a plasmid-coded product. Proc. Natl. Acad. Sci., USA 78:4902-4906.
20. Kollek, R., W. Oertel, and W. Goebel (1978) Isolation and characterization of the minimal fragment required for autonomous replication of a copy mutant (pKN102) of the antibiotic resistance factor R1. Mol. Gen. Genet. 162:51-58.
21. Kolter, R., and D.R. Helinski (1979) Regulation of initiation of DNA replication. Ann. Rev. Genet. 13:355-392.
22. Kolter, R. (1981) Replication properties of plasmid R6K. Plasmid 5:2-9.
23. Leonard, A.C., J.A. Hucul, and C.E. Helmstetter (1982) Kinetics of minichromosome replication in Escherichia coli B/r. J. Bact. 149:499-507.
24. Light, J., and S. Molin (1981) Replication control functions of plasmid R1 act as inhibitors of expression of a gene required for replication. Mol. Gen. Genet. 184:56-61.
25. Light, J., and S. Molin (1982) The sites of action of the two copy number control functions of plasmid R1. Mol. Gen. Genet. 187:486-493.
26. Light, J., and S. Molin (1983) Post-transcriptional control of expression of the repA gene of plasmid R1 mediated by a small RNA molecule. EMBO J. 2:93-98.
27. Lipinska, B., A. Podhajska, and K. Taylor (1980) Synthesis and decay of λ DNA replication proteins in minicells. Biochem. Biophys. Res. Commun. 92:120-126.
28. Macrina, F.L., G.G. Weatherly, and R. Curtiss III (1974) R6K plasmid replication: Influence of chromosomal genotype in minicell-producing strains of Escherichia coli K-12. J. Bact. 120:1387-1400.
29. Masai, H., Y. Kaziro, and K.-I. Arai (1983) Definition of oriR, the minimum DNA segment essential for initiation of R1 plasmid replication in vitro. Proc. Natl. Acad. Sci., USA 80:6814-6818.
30. Matsubara, K. (1976) Genetic structure and regulation of replication of plasmid λdv. J. Mol. Biol. 102:427-439.
31. Meacock, P.A., and S.N. Cohen (1980) Partitioning of bacterial plasmid during cell division: A cis-acting locus that accomplishes stable plasmid inheritance. Cell 20:529-542.
32. Molin, S., and K. Nordström (1980) Control of plasmid R1 replication: Functions involved in replication, copy number control, incompatibility, and switch-off of replication. J. Bact. 141:111-120.
33. Molin, S., P. Stougaard, J. Light, M. Nordström, and K.

Nordström (1981) Isolation and characterization of new copy mutants of plasmid R1 and identification of a polypeptide involved in copy number control. Mol. Gen. Genet. 181:123-130.

34. Morris, C.F., H. Hashimoto, S. Mickel, and R. Rownd (1974) Round of replication mutant of a drug resistance factor. J. Bact. 118:855-866.

35. Muesing, M., J. Tamm, H.M. Shepard, and B. Polisky (1981) A single base-pair alteration is responsible for the DNA overproduction phenotype of a plasmid copy-number mutant. Cell 24:235-242.

36. Murotsu, T., and K. Matsubara (1980) Role of an autorepression system in the control of λdv plasmid copy number and incompatibility. Mol. Gen. Genet. 179:509-519.

37. Nishimura, Y., L. Caro, C.M. Berg, and Y. Hirota (1971) Chromosome replication in Escherichia coli. IV. Control of chromosome replication and cell division by an integrated episome. J. Mol. Biol. 55:441-456.

38. Nordström, K. (1983) Replication of plasmid R1: Meselson-Stahl density shift experiments revisited. Plasmid 9:218-221.

39. Nordström, K. (1983) Control of plasmid replication: A synthesis occasioned by the recent EMBO Workshop "Replication of prokaryotic DNA," held at de Eemhof, The Netherlands, May 1982. Plasmid 9:1-7.

40. Nordström, K., L.C. Ingram, and A. Lundbäck (1972) Mutations in R-factors of Escherichia coli causing an increased number of R-factor copies per chromosome. J. Bact. 110:562-569.

41. Nordström, K., S. Molin, and H. Aagaard-Hansen (1980) Partitioning of plasmid R1 in Escherichia coli. I. Kinetics of loss of plasmid derivatives deleted of the par region. Plasmid 4:215-227.

42. Nordström, K., S. Molin, and H. Aagaard-Hansen (1980) Partitioning of plasmid R1 in Escherichia coli. II. Incompatibility properties of the partitioning system. Plasmid 4:332-349.

43. Nordström, K., S. Molin, and J. Light (1984) Control of replication of bacteriol plasmids: Genetics, molecular biology, and physiology of the plasmid R1 system. Plasmid (in press).

44. Novick, R.P., and F.C. Hoppensteadt (1978) On plasmid incompatibility. Plasmid 1:421-434.

45. Ogura, T., and S. Hiraga (1983) Mini-F plasmid genes that couple host cell division to plasmid proliferation. Proc. Natl. Acad. Sci., USA 80:4784-4788.

46. Pritchard, R.H. (1968) Control of DNA synthesis in bacteria. Heredity 23:472.

47. Pritchard. R.H. (1969) Control of replication of genetic material in bacteria. In Bacterial Episomes and Plasmids. A Ciba Foundation Symposium, G.E.W. Wolstenholme and M. O'Conner, eds. J. & A. Churchill Ltd., London, pp. 65-74.

48. Pritchard, R.H., P.T. Barth, and J. Collins (1969) Control of DNA synthesis in bacteria. Symp. Soc. Gen. Microbiol. 19:263-297.

49. Riise, E., P. Stougaard, B. Bindslev, K. Nordström, and S. Molin (1982) Molecular cloning and functional characterization of the copy number control gene copB from plasmid R1. J. Bact. 151:1136-1145.
50. Rasmussen, K.V., T. Atlung, G. Kerszman, G.E. Hansen, and F.G. Hansen (1983) Conditional change of DNA replication control in an RNA polymerase mutant of Escherichia coli. J. Bact. 154:443-454.
51. Rosen, J., H. Ohtsubo, and E. Ohtsubo (1979) The nucleotide sequence of the region surrounding the replication origin of an R100 resistance factor derivative. Mol. Gen. Genet. 171:287-293.
52. Rosen, J., T. Ryder, H. Ohtsuba, and E. Ohtsuba (1981) Role of RNA transcripts in replication incompatibility and copy number control of antibiotic resistance plasmid derivatives. Nature 290:794-797.
53. Rownd, R.H. (1969) Replication of a bacterial episome under relaxed control. J. Mol. Biol. 44:387-402.
54. Rownd, R.H., A.M. Easton, and P. Sampathkumar (1981) Copy number control and incompatibility of incFII R plasmids. In Molecular Biology, Pathogenicity, and Ecology of Bacterial Plasmids, R.C. Clowes, S.B. Levy, and E.L. Koenig, eds. Plenum, New York, pp. 303-315.
55. Schafferman, A., R. Kolter, D. Stalker, and D. Helinski (1982) Plasmid R6K DNA replication. III. Regulatory properties of the initiation protein. J. Mol. Biol. 161:57-76.
56. Simons, R.W., and N. Kleckner (1983) Translational control of IS10 transposon. Cell 34:683-691.
57. Søgard-Andersen, L., L.A. Rokeach, and S. Molin (1984) Regulated expression of a gene important for replication of plasmid F in E. coli EMBO J. 3:257-262.
58. Sompayrac, L., and O. Maaløe (1973) Autorepressor model for control of DNA replication. Nature (London) New Biol. 241:133-135.
59. Stalker, D., M. Filutowicz, and D. Helinski (1983) Release of initiation control by a mutational alteration in the R6K π protein required for plasmid DNA replication. Proc. Nat. Acad. Sci., USA 80:5500-5504.
60. Stalker, D.M., R. Kolter, and D.R. Helinski (1979) Nucleotide sequence of an origin of replication of the antibodic resistance plasmid R6K. Proc. Natl. Acad. Sci., USA 76:1150-1154.
61. Stalker, D.M., C.M. Thomas, and D.R. Helinski (1981) Nucleotide sequence of the region of replication of the broad host range plasmid RK2. Mol. Gen. Genet. 181:8-12.
62. Stougaard, P., S. Molin, and K. Nordström (1981) RNAs involved in copy-number control and incompatibility of plasmid R1. Proc. Natl. Acad. Sci., USA 78:6008-6012.
63. Tanaka, M., H. Ohmori, and S. Hiraga (1983) A novel type of E. coli mutants with increased chromosomal copy number. Mol. Gen. Genet. 192:51-60.

64. Thomas, C.T. (1981) Molecular genetics of broad host range plasmid RK2. Plasmid 5:10-19.
65. Timmis, K.N., S.N. Cohen, and F.C. Cabello (1978) DNA cloning and the analysis of plasmid structure and function. Progr. Mol. Subcell Biol. 6:1-58.
66. Timmis, K., and U. Winkler (1973) Gene dosage studies with pleiotrophic mutants of Serratia marcescens superactive in the synthesis of marcescin A and certain other exocellular proteins. Mol. Gen. Genet. 124:207-217.
67. Tolun, A., and D.R. Helinski (1981) Direct repeats of the F plasmid incC region express F incompatibility. Cell 24:687-694.
68. Tomizawa, J., H. Ohmori, and R.E. Bird (1977) Origin of replication of colicin El plasmid DNA. Proc. Natl. Acad. Sci., USA 74:1865-1869.
69. Tomizawa, J.-L., T. Itoh, G. Selzer, and T. Som (1981) Inhibition of ColEl RNA primer formation by a plasmid-specified small RNA. Proc. Natl. Acad. Sci., USA 78:1421-1425.
70. Tsurimoto, T., and K. Matsubara (1981) Purification of bacteriophage λO protein that specifically binds to the origin of replication. Mol. Gen. Genet. 181:325-331.
71. Tsurimoto, T., and K. Matsubara (1981) Interaction of bacteriophage lambda O protein with the lambda origin sequence. ICN-UCLA Symp. Mol. Cell Biol. 22:263-275.
72. Tsutsui, H., and K. Matsubara (1981) Replication control and switch-off function as observed with a mini-F factor plasmid. J. Bact. 147:509-516.
73. Tsutsui, H., A. Fujiyama, T. Murotsu, and K. Matsubara (1983) Role of nine repeating sequences of the mini-F genome for expression of F-specific incompatibility phenotype and copy number control. J. Bact. 155:337-344.
74. Uhlin, B.E., and K. Nordström (1975) Plasmid incompatibility and control of replication: Copy mutants of the R-factor R1 in Escherichia coli Kd-12. J. Bact. 124:641-649.
75. Uhlin, B.E., and K. Nordström (1978) A runaway-replication mutant of plasmid R1drd-19: Temperature-dependent loss of copy number control. Mol. Gen. Genet. 165:167-179.
76. Willetts, N., and R. Skurrey (1980) The conjugation system of F-like plasmids. Ann. Rev. Genet. 14:41-76.
77. Young, E.T., and R.L. Sinsheimer (1967) Vegetative bacteriophage DNA. II. Physical characterization and replication. J. Mol. Biol. 30:165-200.

THE PARTITION FUNCTIONS OF P1, P7, AND F MINIPLASMIDS

Stuart Austin and Ann Abeles

Laboratory of Genetics and Recombinant DNA
Litton Bionetics, Inc., Basic Research Program
National Cancer Institute, Frederick Cancer Research Facility
P.O. Box B
Frederick, Maryland 21701

ABSTRACT

The partition regions of P1, P7, and F miniplasmids are discrete DNA sequences of about 3 kb in length that will promote accurate partition of hybrid plasmids independent of the source of replication functions or the position or orientation of the elements.

Each of the *par* regions seems to be very similarly organized, with open reading frames for essential proteins and a terminal site which appears to be analogous to the centromere of eukaryotic cells. When cloned, these terminal sites exert incompatibility against their respective parent plasmids presumably because they can compete with the parent plasmids as substrates for partition.

We have determined the complete DNA sequence of the P1 *par* region. In addition to the open reading frame for the essential *parA* protein (42-44 kd), the region contains a second open reading frame which could encode a 38-kd protein. The 2 large open reading frames appear to form an operon that is negatively regulated from a site adjacent to the promoter and responds to the *par* gene products in trans. Both this site and the downstream "centromere" site, *incB*, contain blocks of extremely AT-rich sequences, which are postulated to be binding sites for *par* proteins. The *incB* and upstream AT-rich regions both contain 20-bp imperfect inverted repeats. Further downstream from the minimal *incB* sequence (172 bp) lies an additional region which is essential for partition.

The further analysis of the P1 par region should be greatly facilitated by the finding that it can function in cis to stabilize pBR322 vectors under conditions where the copy number of pBR322 is reduced.

INTRODUCTION

Of the low copy-number plasmids of *Escherichia coli*, the sex-factor F and the plasmid prophage of bacteriophage P1 are among the most stringently controlled (9). At slow growth rates, the extreme case of having only 2 copies per dividing cell seems to be achieved. The plasmid prophage of bacteriophage P7 is closely related to P1, but differs in key plasmid maintenance functions. It is maintained at similar copy number to F and P1 as judged by the quantity of plasmid DNA detected by gel electrophoresis or hybridization techniques (Austin, unpub. data). Loss of these plasmids from growing cells is very rare; P1 being lost in less than 1 in 10^5 cell division events (5). In addition to a stringent, low-error system for replication control, these plasmids must be substrates for a highly accurate partition system which insures that each daughter cell receives a plasmid copy. It is generally assumed that the partition system promotes attachment of the daughter DNAs to some cellular structure, probably the cell envelope, which pushes or pulls the DNA copies into opposite halves of the dividing cell. These plasmids should, therefore, contain information for their own partitioning which is analogous to the centromere of eukaryotic chromosomes. Three important criteria have been used to establish the existence of such sequences. Firstly, mutation or deletion of the sequences allows replication to continue but causes instability of the plasmid due to random distribution of plasmid copies to daughter cells. Secondly, the necessary sequences can stabilize foreign replicons which, themselves, lack an autochthanous partition system. Thirdly, the essential cis-acting sequences which constitute the recognition sites for partition act as incompatibility determinants because, when cloned in a suitable vector, they compete for partitioning with a resident plasmid, displacing it from the growing population.

Random distribution of plasmids with such low copy number leads to a very high rate of loss, approaching 25% per cell division event at slow growth rates. Thus, plasmid mutants which are partition-defective have a very strong phenotype. Partition mutants can readily be distinguished from mutants defective in accessory mechanisms which aid partition such as those which resolve plasmid dimers (5) or block cell division when errors in replication or partitioning occur (8).

As the 3 above-mentioned unit-copy plasmids are all large and contain much information which is dispensable for plasmid maintenance, a necessary first step for their analysis is to construct miniplasmids which lack inessential information. Studies on F (2,7) and P1 (2) miniplasmids have shown that the information required for partition of these plasmids is contained in a region of approximately 3 kb which is adjacent to, but separable from, the replication functions. Figure 1 illustrates the general organization of the plasmid maintenance regions of F and P1 as cloned in a λΔ<u>att</u> vector. Each encodes proteins essential for partition, and contains a cis-

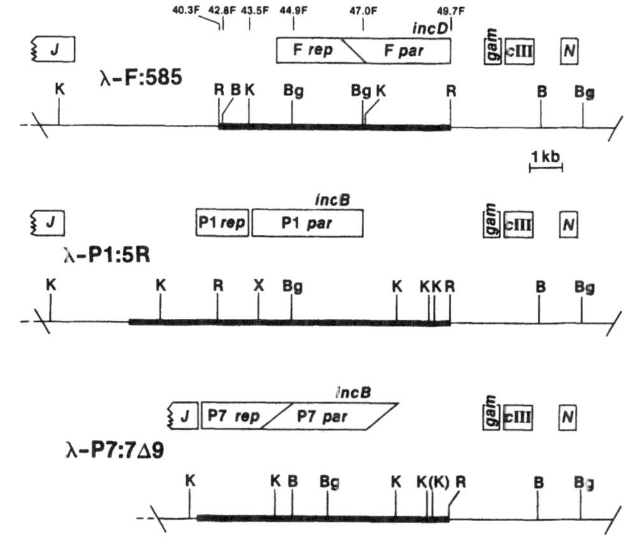

Fig. 1. Physical maps of λ clones containing F, P1, and P7 plasmid maintenance regions. The λ<u>imm</u> 21<u>b</u>538 phages lack the <u>att</u> attachment site, and therefore, are dependent on the plasmid maintenance functions to form lysogens. λ-F:585 and λ-P1:5R were constructed as previously described (2). λ-P7:7Δ9 was constructed by cloning <u>Eco</u>R1 fragments of P7 DNA into the unique <u>Eco</u>RI site of the λ vector. One clone, λ-P7:7, contained an 11.8-kb fragment which was substantially homologous to the <u>Eco</u>R1 fragments 8 and 5 of P1 and contained parts of the P7 <u>Bam</u>HI fragments 1 and 3 (6). Deletion mutants which delete into the inserted DNA were selected as previously described (5), and were mapped by restriction digest and gel electrophoresis. λ-P7:7Δ9 is one of these which was able to lysogenize as a stable unit-copy plasmid. The P7 insert is oriented in the opposite direction to the standard P7 map (6). The existence of the <u>Kpn</u>I site in parenthesis is not certain. Both the regions to the left and right of the P7 <u>par</u> region are substantially homologous to the corresponding P1 regions. R = <u>Eco</u>RI; B = <u>Bam</u>HI; Bg = <u>Bgl</u>II; K = <u>Kpn</u>I; X = <u>Xho</u>I.

acting site which is thought to be the recognition site for partition. When cloned in suitable vectors, these sites exert incompatibility against their parental plasmids by competing efficiently as substrates for partition. We have recently cloned the plasmid maintenance region from P7. Initial studies show that it contains a partition region which seems to be similarly organized to P1 and F par regions (Fig. 1). Each of the 3 par regions are different in specificity as none of the cloned incompatibility determinants (partition sites) is effective at competing for partition with plasmids other than those from which it was derived.

This paper describes the properties of the newly isolated P7 par region in comparison to F and P1 and describes the fine-structure and function of the P1 region.

RESULTS

A Comparison of the Properties of the P1 par Region with that of P1 and F

Table 1 summarizes the properties of the P7 par region and its associated incompatibility determinant P7incB, and compares them to those of the P1 and F regions. Each region is capable of stabilizing a foreign replicon, showing that each is sufficient to promote partition. Neither the position or orientation of the par region with respect to the origin of replication seems important. Note that one combination of rep and par regions (P1rep, P7par) could not be isolated despite a powerful selection for its formation. We feel that this may reflect some inappropriate juxtaposition of sequences in this one particular construct rather than a replicon specificity for P7par. Otherwise, all possible combinations give rise to stable composites. The par regions, although functionally equivalent, show a complete species specificity of mechanism as they do not compete with each other for partition, and the cloned incompatibility determinants are only effective against the plasmid from which they were derived (Tab. 2).

The Complete DNA Sequence of the P1 par Region

We have determined the DNA sequence of 6 kb of the P1 plasmid maintenance region which includes both the rep and par sequences. The open reading frame for the 42-44 kd par protein (3) is present in the position predicted (parA; Fig. 1). An additional large open reading frame is also present (parB) which could encode a 34-kd protein. Although we have not been able to detect a protein product for this open reading frame by the maxicell techniques (10), it seems likely that such a produce is made as the potential codon usage of parB is normal with respect to parA and repA genes of P1 (data not shown, Ref. 1). The parB open reading frame is read in

Tab. 1. Composite plasmids were formed by the recombination of deletion mutants of 2 of the λminiplasmids shown in Fig. 1, which retain only the rep or par regions of those plasmids (2). The composites contain the sequences from both parents, bringing different combinations of rep and par sequences together. The strains carrying the composites were then made recombination defective to prevent composite breakdown, and the stability of plasmid maintenance without selection determined as previously described (2).

Genotype of Composite Plasmid rep	par	% Loss of Plasmid in 25 Generations
P1	None	96
F	None	97
P1	F	6
F	P1	1
P1	P7	-*
F	P7	2

Composite plasmids were formed by the recombination of deletion mutants of two of the λminiplasmids shown in Figure 1, which retain only the rep or par regions of those plasmids (see Austin and Abeles, 1983a). The composites contain the sequences from both parents, bringing different combinations of rep and par sequences together. The strains carrying the composites were then made recombination defective to prevent composite breakdown, and the stability of plasmid maintenance without selection determined as previously described (Austin and Abeles, 1983a).

*No composites were formed.

the same direction as parA and lies between parA and incB. Thus, the general organization of the P1 par region much resembles that of F (Fig. 2; Ref. 7,12).

We have also sequenced the par regions of the par17 amber mutant. The par17 mutation, which blocks partition (3), is caused by a CAG to TAG change at codon 180 of the parA open reading frame. This generates the 16-kd nonsense fragment of the protein as found previously in maxicell experiments (3).

Tab. 2. Strain N100 recA carried incompatibility determinates from the par regions of F, P1, or P7 in a pBR322 vector. The λ-miniplasmids were introduced by infection, selecting lysogens with clear selector phages (2). The lysogenic colonies were picked and allowed to grow for 25 generations without selection. Isolated colonies formed by these cells were tested for the presence of the λ-miniplasmids by scoring for λ-immunity. No loss of the pBR322 derivatives was seen during the experiment.

Resident Plasmid	Loss of λ Plasmid After 25 Generations in Cells Transformed With			
	pBR322	pALA17 (P1incB)	pALA32 (FincD)	pALA152 (P7incB)
λ-P1:5R	8	98	5	2
λ-F:583	<1	<1	85	1
λ-P7:7Δ9	3	1	7	97

Strain N100 recA carried incompatibility determinates from the par regions of F, P1 or P7 in a pBR322 vector. The λ-miniplasmids were introduced by infection, selecting lysogens with clear selector phages (Austin and Abeles, 1983a). The lysogenic colonies were picked and allowed to grow for 25 generations without selection. Isolated colonies formed by these cells were tested for the presence of the λ-miniplasmids by scoring for λ-immunity. No loss of the pBR322 derivatives was seen during the experiment.

The Par Region of P1 Can Stabilize pBR322

High copy-number plasmids should not need a sophisticated partition mechanism, as random distribution would be effective in ensuring partition. However, if the copy number is reduced sufficiently, random distribution should be ineffective. We have found that the copy number of pBR322 is drastically reduced in a polAts strain at 30°C due to the relatively poor activity of the polA gene product which is essential for replication of plasmids of this type (11). pBR322 is very unstable under these conditions. Derivatives of pBR322 that carry the P1 par region are considerably more stable (Fig. 2), suggesting that pBR322 does not contain a partition system of its own, but can be segregated by the P1 system. The P1parA-mutation par17 blocks the stabilization. As par17 is an amber mutation in the parA open reading frame, this result confirms that the par system is being used by the plasmid under these conditions. Using this assay for par function in pBR322 constructs, we have been able to localize the minimal par region to a HindIII-HpaI DNA fragment of 3.3 kb. The necessary material extends

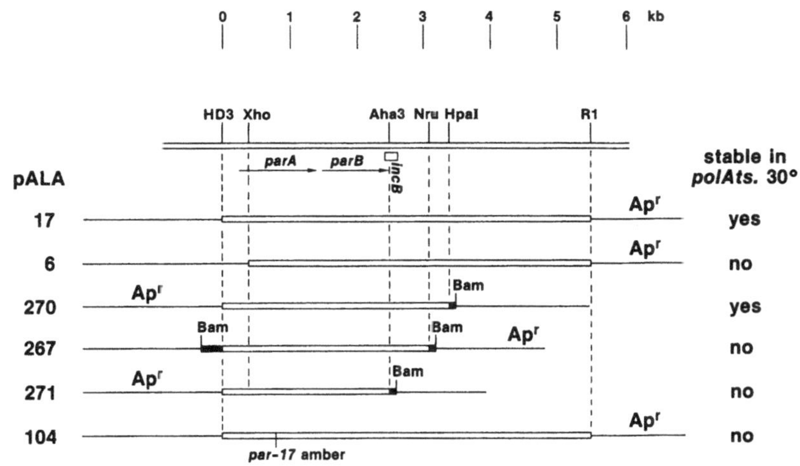

Fig. 2. All plasmids were constructed by cloning fragments into pBR322 vectors. The fragments were derived from λ-P1:5R or from λ-P1:5R fragments cloned into M13Mp10 or M13Mp11 vectors. Single lines = pBR322 DNA (part of). Open boxes = P1 DNA. Closed boxes = linkers derived from M13 or pBR322 vectors. The polA12ts strain was transformed with the plasmid DNA's at 30° selecting for ampicillin resistance. Stability was measured by scoring the ampicillin resistance of the cells after 10 generations of unselected growth in broth. Typically, stable plasmids are lost in each 5% of the cells, whereas unstable ones are lost in 95%. At 40°, all plasmids were lost for <98% of the cells in 10 generations. HD3 = HindIII; Xho = XhoI; Aha3 = AhaIII; Bam = BamIII.

further to the right than previously estimated (Fig. 1; Ref. 2) and includes the parA and parB open reading frames, the incB sequence, and some additional sequences to the right of incB. Plasmids which lack the amino-terminal of the parA gene, or incB, or the region to the right of it are no more stably maintained than pBR322 itself (Fig. 2).

Regulation of the Par Proteins

Although the parB open reading frame appears to be used, inspection of the DNA sequence shows no obvious promoter site immediately upstream of it, and the distance between parA and the probable translation start point for parB is only 16. It seems probable that parA and parB are contranscribed from a promoter upstream of parA. Figure 3 shows the DNA sequence adjacent to the -10 region of the putative promoter and occupying the space between this promoter and the likely Shine and Dalgarno (ribosome binding site) suggested that

```
                     1870                 1890                 1910                 1930                1950
AAGCTTGGCT GCCCATTGAA GCAGGAAATC ACCAAAATGA TTCAGGCTAC AACCTGAACG TAGAAGAAAT CCGCGTCCTT TATGCGTGGA GGATGCCAAA
                         ----       -----                                  ----         -----         ---
                                                -35?                  -10?

                     1970                 1990                 2010                 2030                2050
GCATGTTGTG ACACACTTGG CAAAGGAGTA AGCATGCAGA GAATGCTATG TACAAGCATC TACGCATACA TTATTATTTT ATGCAGCATT TTTAATTAAA
                                                                                    ------------->  <-
                                                                            ------------   -----------
               S-D?
                     2070                           2090                    2110                    2130
TTCAAAAATA CAGCATAAAG GATGACTTTC G ATG AGT GAT TCC AGC CAG CTT CAC AAG GTT GCT CAA AGA GCA AAC AGA ATG CTC
                                   met ser asp ser ser gln leu his lys val ala gln arg ala asn arg met leu
------------------------
---------   -----

                                                          ParA
```

Fig. 3. The DNA sequence of the left end of the P1 par region. The sequence begins at the Hind III site which defines the start of the par region and extends into the parA open reading frame. Although the start point for parA shown fits the available data best, a start point at the second ATG sequence in the frame cannot be completely excluded. The dashed underlines show runs of 4 or more A or T bases. The 20 bp inverted repeat sequence requires 2 mismatches and a single base loop-out to give maximum complementarity.

some regulatory element acts here. In order to test this, a plasmid was constructed which has the first 53 codons of parA fused in frame with the β-galactosidase gene. This plasmid produces a Lac^{+} fusion protein under the control of the par promoter. We have shown that the production of the hybrid β-galactosidase activity is strongly downregulated by the presence of an intact P1 par region in trans (data not shown). Initial experiments suggest that both parA and parB products are required for this downregulation. Thus, the par operon appears to be subject to a novel type of autoregulation for which both products are required to act as the repressor. Our observations on the properties of amber mutants with lesions in the genes for the two F par proteins suggested the existence of the same type of regulation in F (4).

The Nature of incB

The P1 incB locus exerts a strong incompatibility effect when cloned in a pBR322 vector (Tab. 2). By subcloning fragments from this region, we have located a sequence sufficient to exert this phenotype to a 175-bp TaqI-HincII fragment (Fig. 4). The sequence of this fragment is shown in Fig. 5. It contains the end of the parB open reading frame and a downstream region which contains an imperfect 20 inverted repeat sequence. This sequence is AT-rich. Bracketing this feature are 2 highly AT-rich sequences, the right, most of which has 17 contiguous AT pairs with 2 blocks of 7As in the same strand.

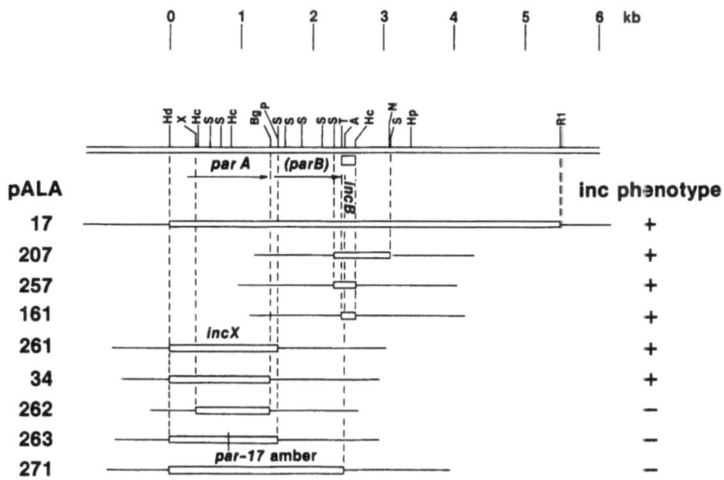

Fig. 4. All plasmids were derived from pBR322 vectors. The presence of incompatibility determinants were determined by: (a) introducing λ-P1:5R under selection to W100 recA cells carrying the plasmid and then measuring the loss of λP1:5R under nonselective conditions, or (b) transforming established λ-P1:5R lysogens with the pBR322 derived plasmids and scoring loss of λ-P1:5R under nonselective conditions. With inc+ clones, λ-P1:5R was very unstably maintained in both tests, whereas a λ-P7 or λ-F miniplasmid was stably maintained. Hd = HindIII; Hc = HincII; Hp = HpaI; A = AhaIII; Bg = BglII; RI = EcoRI; X = XhoI; P = PvuI; N = NruI; S = Sau3A; T = TaqI. (All other TaqI sites not shown. All 4 base recognitions between HpaI and EcoRI not shown).

A Second Incompatibility Locus Due to Overproduction of parA

During our efforts to locate incB, we discovered a second pBR322 cloned par fragment which could exert incompatibility specifically against P1 (Fig. 4). This incompatibility is due to parA protein synthesis from the fragment as it is entirely abolished by the parA17 amber mutation and requires an intact amino-terminus of the parA open reading frame (Fig. 4). It seems probable that such clones overproduce parA protein due to lack of a parB on the high copy plasmid, a condition which appears to release the autoregulation of the operon. An excess or imbalance of parA protein relative to other components appears to interfere with proper partition.

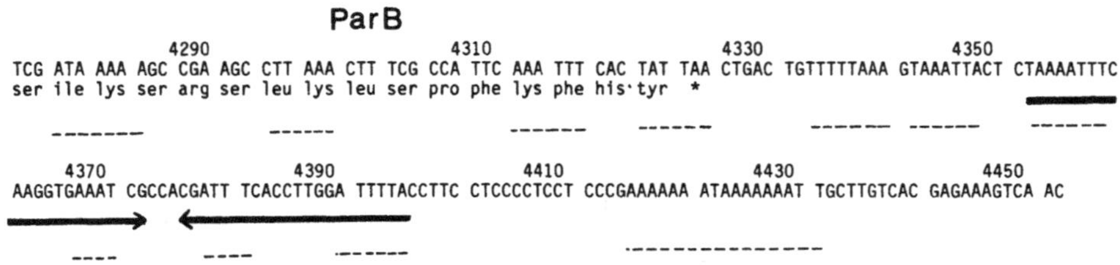

Fig. 5. The DNA sequence of the minimal incB locus. The 175 base pairs shown are those present in pALA161 which exerts incB incompatibility. The dashed underlines show runs of 4 or more A or T bases. The 20 inverted repeat sequence requires one mismatch and an adjacent one base loop-out to give maximum complementarity.

CONCLUSIONS

The partition regions of P1, F, and P7 seem to be very similar in their function and general organization. However, each system appears to specify a unique pathway for partition as none exerts incompatibility against the other types. The specificity is reflected in the specificity of the *par* region components. Neither F nor P7 encoded *par* proteins can complement P1 *parA* mutants (data not shown), and the *inc* regions of at least F and P1 show no homology as judged by Southern Blot hybridization (Austin, unpub. data). A ca. 2-kb region of nonhomology between P1 and P7 (nonhomology E, Ref. 13) appears to correspond approximately to the position of the P1 and P7 *par* regions (this paper; Ref. 6).

The individual *par* regions contain all the necessary plasmid-specified information to promote partition. This allows them to function in the context of foreign replicons. The observation that P1 *par* can stabilize pBR322 shows that *par* has no specificity with respect to replication strategy. P1 will, presumably, stabilize any replicon as long as copy number is reduced to the point where specific partition becomes limiting on stability.

It seems probable that *par* proteins promote partition by binding to the DNA at the *inc* site, and to the cell envelope to form a specific partition complex. It appears that the concentration of *par* proteins in the cell is critical for this to occur. Their synthesis appears to be tightly autoregulated in P1, and overproduction of *parA* protein blocks specific partition. Autoregulation of F par proteins also seems to occur. We would, therefore, expect *par* proteins to bind to 2 distant regions within the *par* sequences; in the vicinity of the promoter for *par* and at the putative centromere site. A striking feature of the DNA sequence in these 2 regions is the presence of highly AT-rich runs of bases with a strong tendency for blocks of As or Ts to be in the same strand. In both cases, the AT-rich regions include 20 bp imperfect inverted repeats spaced 1-3 apart. Neither the inverted repeat regions nor the other AT-rich regions show extensive homology between the upstream and downstream loci, although similarities abound. It seems probable that these features are important for par binding, as we have found no other features common to the 2 regions. We are currently attempting to purify the 2 P1 proteins and study their binding properties and the identity of the DNA sites involved.

ACKNOWLEDGMENTS

Research sponsored by the National Cancer Institute, DHHS, under contract No. N01-CO-23909 with Litton Bionetics, Inc. The contents of this publication do not necessarily reflect the views of policies of the Department of Health and Human Services, nor does mention of trade names, commercial products, or organizations imply endorsement by the U.S. Government.

REFERENCES

1. Abeles, A., K. Snyder, and D. Chattoraj (1984) P1 plasmid replication: Replicon Structure. J. Mol. Biol. 173:307-324.
2. Austin, S., and A. Abeles (1983a) The partition of unit-copy miniplasmids to daughter cells. I. P1 and F miniplasmids contain discrete, interchangeable sequences sufficient to promote equipartition. J. Mol. Biol. 169:353-372.
3. Austin, S., and A. Abeles (1983b) The partition of unit-copy miniplasmids to daughter cells. II. The partition region of miniplasmid P1 encodes an essential protein and a centromere-like site at which it acts. J. Mol. Biol. 169:373-387.
4. Austin, S., and A. Wierzbicki (1983) Two mini-F encoded proteins are essential for equipartition. Plasmid 10:73-81.
5. Austin, S., M. Ziese, and N. Sternberg (1981) a novel role for site-specific recombination in maintenance of bacterial replicons. Cell 25:729-736.
6. Iida, S., and W. Arber (1979) Multiple physical differences in the genome structure of functionally related bacteriohage P1 and P7. Mol. Gen. Genet. 173:249-261.
7. Ogura, T., and S. Hiraga (1983a) Partition mechanism of F plasmid: Two plasmid gene-encoded products and a cis-acting region are involved in partition. Cell 32:351-360.
8. Ogura, T., and S. Hiraga (1983b) Mini-F plasmid genes that couple host division to plasmid proliferation. Proc. Natl. Acad. Sci., USA 80:4784-4788.
9. Prentki, P., M. Chandler, and L. Caro (1977) Replication of the prophage P1 during the cell-cycle of *Escherichia coli*. Mol. Gen. Genet. 152:71-76.

10. Sancar, A., A. Hack, and W. Rupp (1979) Simple method for the identification of plasmid-coded proteins. J. Bact. 137:692-693.
11. Kinsbury, D., and D. Helinski (1970) DNA polymerase as a requirement for the maintenance of the bacterial plasmid colicinogenic factor E1. Biochem. Biophys. Res. Commun. 41:1538-1544.
12. Komai, N., T Nishizewa, Y. Hayakawa, P. Murotsu, and K. Matsubara (1982) Detection and mapping of six miniF-encoded proteins by cloning analysis of dissected miniF segments. Mol. Gen. Genet. 186:193-203.
13. Yun, T., and D. Vapnek (1977) Electron microscopic analysis of bacteriophage P1, phage P1Cm, and phage P7. Determination of genome sizes, sequences, homology, and location of antibiotic resistance determinants. Virology 77:376-385.

GENETIC INTERACTIONS OF BROAD HOST-RANGE PLASMID RK2: EVIDENCE FOR A COMPLEX REPLICATION REGULON

David H. Figurski, Calvin Young, Helen C. Schreiner, Robert F. Pohlman,[1] David H. Bechhofer,[2] Alice S. Prince,[3] and Thomas F. D'Amico

Department of Microbiology and Cancer Center
College of Physicians & Surgeons
Columbia University
701 West 168th Street
New York, New York 10032

INTRODUCTION

As first recognized by Jacob, Brenner, and Cuzin (13), a successful plasmid must have an efficient replication control system that ensures its stable maintenance by the host. It is now very apparent that bacterial plasmids have evolved an intriguing variety of regulatory strategies to accomplish this (23). However, it is important to realize that for any such system to work effectively, there are certain requirements which concern the plasmid-host interaction. (i) Plasmid genes for replication control must be expressed adequately by the host at the proper time in the cell cycle. (ii) For initiation of replication, the plasmid initiation complex must interact productively with the host's replication apparatus. (iii) Expression of the plasmid replication determinants must be tightly controlled to prevent an unnecessary metabolic burden or potentially lethal titration of host replication components. (iv) The plasmid must interact with host cell components to guarantee segregation of at least one plasmid copy to each daughter cell at division.

---

[1] Department of Biochemistry, University of Minnesota, Minneapolis, Minnesota 55108; [2] Department of Microbiology, The Public Health Research Institute of the City of New York, 455 First Avenue, New York, New York 10016; and [3] Departments of Medicine and Microbiology, College of Physicians and Surgeons, Columbia University.

For plasmids maintained only by closely related hosts, it is easy to imagine how such efficient plasmid-host relationships could evolve. However, when one superimposes on these requirements the additional property of an extremely broad host range, as seen for plasmids of incompatibility groups P (7,18) and Q (3), it is less clear how stable maintenance is achieved in each host. Obviously all hosts are not equal with respect to plasmid maintenance because no other plasmid group has a host range which includes virtually all gram-negative bacteria.

The molecular basis for this striking capability is not obvious. One may ask if these plasmids have uniquely capitalized on some common feature of gram-negative bacteria. For example, their initiation complexes may interact well with the replication apparatus of any host. Alternatively, these plasmids may encode different functions for different hosts. Another question concerns the ability to express the essential plasmid genes sufficiently well in each host. Have broad host-range plasmids evolved universal promoters which are reasonably functional in any host, or do important genes have multiple promoters? The answers to these basic questions about the mechanism of broad host range are not yet known and they should be interesting because the challenge to these plasmids is greater than that of their narrower host-range counterparts.

We have been studying the IncP plasmid RK2 with the idea that this broad host-range plasmid may reveal novel genetic determinants for replication control. This expectation seems to have been justified because we have found a new class of maintenance determinants. These are not absolutely required for the plasmid to replicate, but they can have a dramatic effect on plasmid replication control.

ESSENTIAL MAINTENANCE DETERMINANTS OF RK2

Plasmid RK2 (Fig. 1) is a 56 kb broad host-range plasmid of the IncP group (12,15). Two plasmid determinants (*oriV* and *trfA*) have been shown to be absolutely required for plasmid replication. *OriV* is the origin of replication, originally mapped to coordinate 12' of the physical map by electron microscopy of replicative intermediates (16). *OriV* was subsequently shown to behave as a cis-acting replication element which could be activated in trans by the product of another RK2 gene (8). The gene for this diffusible function was named *trfA* (29) and has been mapped to the 16'-18' region of RK2 (19,29). The ability to activate *oriV* in trans allowed strategies for precise localization and identification of the important nucleotide sequences (26). Those needed for replication involve several direct repeats, some of which exhibit incompatibility with RK2-based replicons (30).

OriV and *trfA* comprise the essential minimum for RK2 replication. Plasmids containing only these 2 determinants are able to

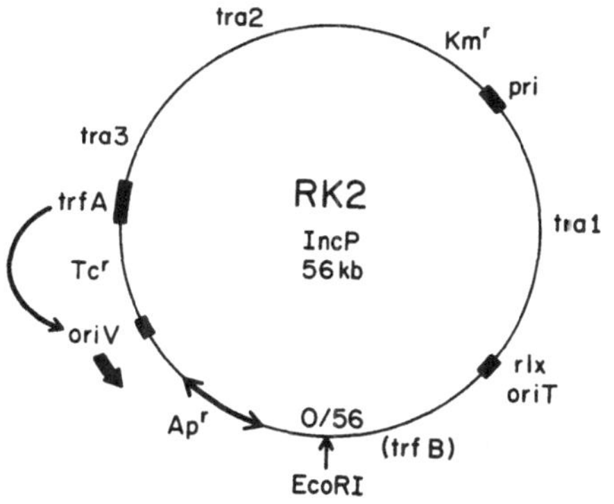

Fig. 1. Basic genetic map of RK2. Presented here are landmark determinants to show the basic organization of the RK2 genetic map. In the text, coordinates indicate the distance in kilobases, clockwise, from the unique EcoRI site and are designated with a prime (e.g., 12'). OriV and trfA are discussed in the text; the arrow over oriV indicates the direction of replication (16). Tra indicates regions of genes for conjugal transfer; pri is the primase gene (14); rlx, relaxation complex nick site (11); Ap^r, Tc^r, and Km^r are genes for resistance to ampicillin, tetracycline, and kanamycin, respectively. The arrowheads in the region of Ap^r show that this is a transposon.

replicate in *Escherichia coli*, *Pseudomonas*, and other gram-negative hosts (22; Schmidhauser and Helinski, pers. comm.). Thus, at least some broad host range capability is encoded within these functions.

QUASI-ESSENTIAL MAINTENANCE DETERMINANTS

We have found that RK2 encodes a second class of maintenance determinants which we call "quasi-essential." This class consists of genes lethal to the host (*kil* genes) and their controlling elements (*kor* genes). Whether one of these genes is essential depends on which other genes are present in the plasmid. Our studies have revealed the following: (i) *kor* genes are essential whenever the corresponding *kil* genes are present on the plasmid; and (ii) a *kil* gene may be essential for plasmid replication when the appropriate *kor* gene is present. Thus, loss of either a *kil* determinant *or* a *kor* determinant will have serious consequences for RK2 maintenance. However, it is possible to lose both because an *oriV*$^+$ *trfA*$^+$ basic replicon can be functional in the absence of *kil* and *kor* genes. Thus, these genes are considered essential only when the corresponding partner is still present in the plasmid.

The existence of kil determinants was first indicated by our inability to clone certain regions of RK2 (10). We reasoned that these regions encode functions toxic to the host cell (kil) and that other functions of RK2 (kor) are normally able to control them. By first cloning regions suspected of having kor genes, we were able to construct host strains capable of tolerating the cloned kil determinants. That the kil determinants are actually lethal to the host was demonstrated by curing cells of the kor^+ helper plasmid and showing that these cells failed to form colonies in the absence of selection for either plasmid (10).

Thus, new determinants of RK2 were revealed by 2 unambiguous phenotypes: (i) host cell killing (by kil determinants) and (ii) prevention of host death (by kor determinants). These phenotypes allowed us to locate three separate kil determinants (kilA, kilB, and kilC), each of which is controlled by the product of a specific kor gene (korA, korB, and korC). Their relative positions on the RK2 physical map are shown in Fig. 2.

kilB1 AND THE QUESTION OF trfB

Earlier studies with miniderivatives of RK2 suggested that, in addition to oriV and trfA, a third region (trfB) is required for replication (29). The trfB region was curious because in certain plasmids it seems to be essential (27,29), whereas in others it is clearly dispensable (27). After finding that kilB maps near trfA

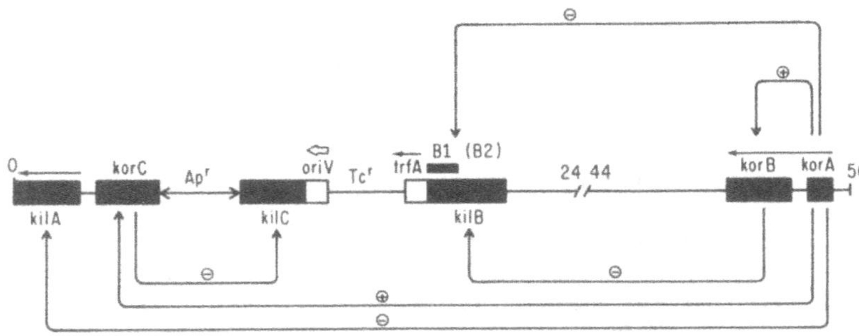

Fig. 2. The kil and kor determinants of RK2 and their interactions. This linear representation of the RK2 map begins at the EcoRI site. Only the relevant regions are presented; the 20-kb segment between 24' and 44' is not shown. Interactions are depicted with "+" indicating a positive effect or "-", a negative effect. Open arrow over oriV shows direction of replication. Direction of transcription for trfA (25), kilA (Young, Prince, and Figurski, in prep.) and korA-korB (4; Bechhofer and Figurski, in prep.) is indicated by an arrow overhead.

and that its control gene (korB) maps near trfB, we proposed that the trfB region actually encodes a kor gene required only when a kil gene near trfA is present (10). We proved this by demonstrating that a trfB-requiring plasmid encodes a kil-like determinant immediately adjacent to trfA (20). As predicted, this kil is controlled by a function from the trfB region. Furthermore, plasmids defective in this kil do replicate, but do not require trfB. Thus, trfB is not an essential replication function.

This kil determinant maps within the kilB region and very likely contributes to the phenotype conferred by kilB. However, it can be regarded as only a component of kilB because it is controlled by the trfB region, whereas the complete kilB is not. We named this component "kilB1." The corresponding control gene was originally named "korB1," but we have since found that korA is sufficient for control of kilB1 (see below). Clearly another determinant of the kilB region is responsible for the requirement for korB, and we have reserved the designation "kilB2" for this component.

[In another study, Smith and Thomas (24) found a kil-like determinant in the same region and named it "kilD," with "korD" as the control gene in trfB. KilB1 and kilD are very likely the same, as are korA and korD.]

In summary, the kilB region has 2 distinct kil determinants: kilB1 (whose existence is verified) and kilB2 (whose existence is deduced). KilB1 is adjacent to trfA and can be controlled by korA, which maps in the region designated trfB. KilB2 is on the other side of kilB1, and together they require more than korA. This additional determinant is korB, which maps next to korA, but extends beyond the trfB region. These relationships are shown in Fig. 2.

STRUCTURE OF korA

By subcloning and manipulation of DNA fragments from the korA-korB region, we mapped the korA gene to the region between coordinates 55.6' and 55.2' (4). Transcription proceeds from 55.6' to 55.2' as determined by promoter replacement experiments and by hybridization analysis with korA-specific RNAs synthesized in vivo. The nucleotide sequence of this region revealed one intact open reading frame preceded by a ribosome binding site and a promoter-like structure with strong homology to the -35 and -10 consensus sequences. The product of this open reading frame, as determined by translation of the sequence, should be a polypeptide of 13,000 d. Evidence that this is korA was provided by BAL31-generated deletion mutants which removed nucleotides from the 3' end of the open reading frame. One mutant which gives a KorA$^-$ phenotype (i.e., unable to control kilA) is missing nine carboxy-terminal residues from the putative product. Another mutant which confers a KorA$^-$ phenotype (i.e., fewer and smaller transformants with *kilA*) is missing 2 residues (4).

We have also found that this minimal *korA*-encoding region is able to control *kilB1*, and furthermore that the *korA* mutants are also defective in the control of *kilB1*. Thus, the same *kor* gene is responsible for negative control of 2 separate *kil* determinants (Bechhofer, D'Amico, and Figurski, unpub. results).

korA INHIBITS *kilA* EXPRESSION

The *kilA* gene is located in the 0'-2.3' region (32). Insertion of a DNA linker at coordinate 2.2' causes complete loss of *kilA* expression (Young, Prince, and Figurski, in prep.), but expression is restored if another promoter is inserted here (Fig. 3). With the promoter for tetracycline resistance (p_{tet}), the gene is constitutively expressed at a low level; i.e., cells with this plasmid form small colonies. With p_{lac}, *kilA* is inducible with IPTG. A host cell mutant partially resistant to *kilA* is also resistant to the killing activity from these promoter fusions.

In contrast to the wild-type *kilA* gene, *kilA* fused to p_{tet} or p_{lac} is not affected by *korA* (Fig. 3). Thus, replacement of the natural *kilA* expression region has removed sensitivity to *korA*. To

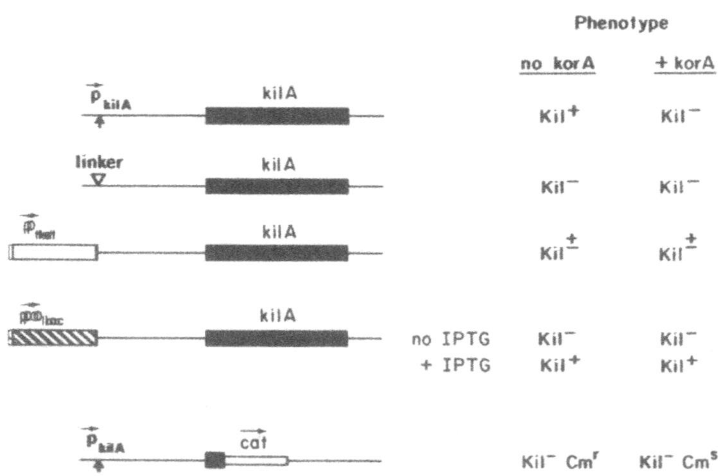

Fig. 3. Analysis of *korA* control of *kilA*. This schematic diagram presents constructed variations of the *kilA* region, as described in the text. The Kil phenotype was assayed by transformation of cells with and without *korA*. For the *cat* fusions, transformants were screened for resistance or sensitivity to 50 μg/ml chloramphenicol (Cm).

examine this further, we have fused the structural gene for chloramphenicol transacetylase (cat) to the wild-type kilA promoter region. Plasmids with this fusion confer Cm^r on the cells in the absence of korA, but the cells are Cm^s when korA is present. We conclude that korA negatively controls kilA at the level of gene expression. It is also reasonable to expect that korA controls kilB1 in the same way.

RELATIONSHIP OF korA TO EXPRESSION OF korB

KorA and korB are separate genes because interruption of the SstII site at 54.2' inactivates korB function but not korA (10). However, deletion of the 5' end of korA eliminates expression of both korA and korB. Hybridization analysis of korB-specific RNA showed (i) the direction of transcription is the same as that of korA, and (ii) the korB-specific probe hybridizes to the same RNA bands as does the korA-probe, while these probes show no homology to each other. This evidence suggests that korB is downstream of korA and within the same transcriptional unit (Bechhofer and Figurski, in prep.).

KorA product is not required for korB product to control kilB because it is possible to isolate $korB^+$ plasmids completely deleted for korA (2; D'Amico and Figurski, unpub. data). In one case, we have verified that expression depends on a new promoter fused to korB. This finding implies that, like korA, korB is also able to control the kilB1 component of kilB, and we have preliminary evidence that this is true (Schreiner and Figurski, unpub. data).

Our results suggest that the korA product might be a positive regulator for korB expression. This study involves a Tn76 insertion of the korA-korB region, originally isolated by P. Barth as a host range mutant of RP4 (1), an IncP plasmid identical to RK2 (5,31). Recent studies showed the Tn76 to have inserted between korA and korB (2). This mutation was transferred by marker rescue to a cloned korA-korB segment, and the mutant segment was found to express both $KorA^+$ and $KorB^+$ phenotypes. Because the 15-kb Tn76 insertion is not polar on korB, it seemed likely that a transposon promoter is expressing korB. Indeed, we found that deletion of the Tn76 segment adjacent to the 5' end of korB caused loss of $KorB^+$ phenotype.

We were therefore surprised to find the following results: (i) deletion of korA from this Tn76 derivative also caused a $KorB^-$ phenotype; and (ii) addition of korA in trans restored the $korB^+$ phenotype. Because we already established that korA product is not required for korB control of kilB, we hypothesized that korA product is needed to express the korB gene.

One possibility we considered is activation of a promoter for korB. However, we found that mutants deleted for korA promoter could not express korB even when korA was supplied in trans. Likewise, in the Tn76-korB fusion, removal of the Tn76 sequences results in a KorB$^-$ phenotype that is no longer reversed by korA in trans. Thus, it seems that korB expression requires a promoter and korA product.

Another possibility we considered is that a transcription termination signal occurs before korB and that korA is required to prevent termination to allow transcription of korB. Thus, regardless of the promoter, korA would be required for expression of korB. In the promoter fusions in which korB is expressed independent of korA, we proposed that the terminator was lost in the construction. Although we have no direct evidence for this model with korB, supporting results have emerged from studies on the interaction of korA and korC.

POSITIVE REGULATION BY korA IN korC EXPRESSION: EVIDENCE FOR ANTITERMINATION

The phenotypes of various insertion and deletion mutants suggested that korC maps in the 3.3'-5.0' region. However, we were surprised to find that this region, when cloned, does not express korC. A second determinant is also required for the KorC$^+$ phenotype, and we found this to be korA (32).

Because we thought korA is required for korB expression by functioning as an antiterminator of transcription, we reasoned that the same mechanism is involved in the korA-dependence of korC. It therefore seemed possible that the putative antitermination function might be bypassed in an E. coli host defective in the termination factor rho (21). Thus, the korC region was tested for its ability to control kilC without korA in rho$^+$ (termination proficient) and rho$^-$ (termination defective) hosts. The results showed clearly that the rho$^-$ mutation bypasses the requirement for korA in the expression of korC (Fig. 4) (32). This is consistent with the interpretation that korA acts as an antiterminator of transcription.

INVOLVEMENT OF kil AND kor DETERMINANTS IN REPLICATION CONTROL

What are the true functions of the kil and kor genes? An indication that they might be important to the IncP group of plasmids first came from our finding that all of 14 different IncP plasmids were positive when tested for korA- and korB-like functions (10). In contrast, plasmids of 19 other incompatibility groups expressed no functions able to control either RK2 kilA or kilB. Thus, the kor genes, and very likely their corresponding kil determinants, are unique features of IncP plasmids. The striking conservation of

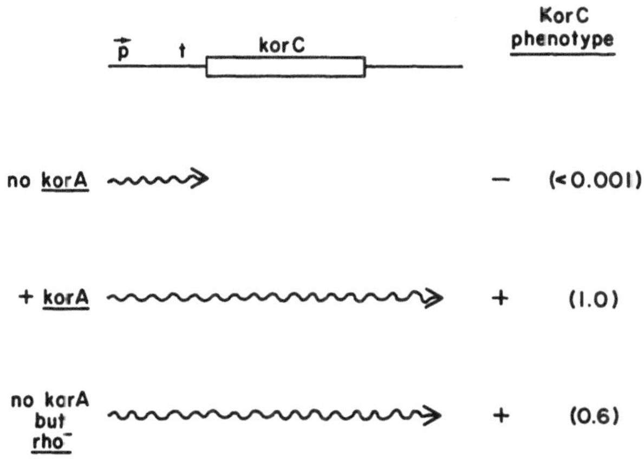

Fig. 4. Effect of a host *rho* mutation on the *korA*-dependence of *korC*. *KorC* function of this cloned segment was assayed by its ability to control a *kilC*$^+$ plasmid when *korA* was present or when the host was *rho*$^-$. The numbers show relative efficiencies of transformation. A possible mechanism is depicted in the drawing. "p" denotes a promoter; "t", a *rho*-dependent transcription termination signal.

these functions within the IncP group led us to propose that these genes might be important to plasmid maintenance. We suggested that they were important in non-*E. coli* hosts because they did not seem to be part of the essential minimum for RK2 replication in *E. coli*. However, we now have evidence that the action of *kil* and *kor* genes can have a profound effect on RK2 replication even in *E. coli*.

We have observed that both *korA* and *korB* exhibit a strong incompatibility towards certain RK2 replicons (Schreiner, Bechhofer, Pohlman, Young, Borden, and Figurski, in prep.). Our analysis of this phenomenon has led us to conclude the following. (i) The incompatibility results from inhibition of replication by the action of *korA* or *korB* on the expression of *trfA*. (ii) The presence or absence of *kilB1* in *cis* to *trfA* has a striking effect on the sensitivity of the plasmid to the *korA*- or *korB*-mediated inhibition of *trfA*.

These interactions are illustrated by the results in Tab. 1. Whenever a *korA*$^+$ *korB*$^+$ plasmid is present in a strain, transformation of that strain by certain RK2 replicons is inhibited at least 100-fold (compare lines 1 and 2; 4 and 5). Non-RK2 replicons are not affected, and in fact linkage of a non-RK2 replicon (e.g., P15A; line 3) to the RK2 plasmid prevents the inhibition. Thus, the interference is at the level of replication, and only plasmids dependent on RK2 *oriV* for replication are inhibited by the *korA*-*korB* region.

Tab. 1. Incompatibility by korA and korB.

	Incoming plasmid	Resident plasmids in host 1	Resident plasmids in host 2	Relative transformation efficiency[a]
1.	oriV$^+$	trfA$^+$ kilB1$^-$	---	1.0
2.	"	trfA$^+$ kilB1$^-$	korA$^+$B$^+$	<0.01
3.	oriV$^+$ P15A rep	trfA$^+$ kilB1$^-$	korA$^+$B$^+$	1.0
4.	oriV$^+$ trfA$^+$ kilB1$^-$	---	---	1.0
5.	"	---	korA$^+$B$^+$	<0.01
6.	"	---	korA$^+$B$^-$	<0.01
7.	"	---	korA$^-$B$^+$	<0.01
8.	"	---	korA$^-$B$^-$	0.5
9.	kor-resistant variant of oriV$^+$ trfA$^+$ kilB1$^-$	---	---	1.0
10.	"	---	korA$^+$B$^+$	1.1
11.	oriV$^+$ trfA$^+$ kilB1$^+$	---	korA$^+$B$^-$	0.3
12.	"	---	korA$^-$B$^+$	0.4

[a] The host strain in each case was MV10 (10). All transformation efficiencies are corrected for variations in competence as measured by transformation with saturating amounts of a non-RK2, compatible plasmid. Variations of less than three-fold are considered insignificant.

Which determinant(s) in the korA-korB region are actually responsible for this incompatibility? Meyer and Hinds (17) have reported an incompatibility determinant, IncP-1(II), in this region. However, the incompatibility observed here is clearly different. The most notable difference is that, unlike the effect of IncP-1(II), the interference by the korA-korB region is suppressed by the addition of another replicon. In fact, we have found that the effect is mediated by the korA and korB genes themselves. Surprisingly, either korA or korB alone will exhibit strong incompatibility (lines 6 and 7). Only when both are missing is the incompatibility lost (line 8). Therefore, korA and korB are each able to inhibit RK2-specific replication.

To begin to answer how this might be accomplished, we sought to identify the targets for korA and korB. The effect of these determinants on the oriV$^+$ trfA$^+$ kilB1$^-$ plasmid used in Tab. 1 is so strong that it was possible to select a plasmid mutant resistant to korA and korB (Tab. 1, line 9 and 10). Analysis of the structure of this mutant has revealed a likely mechanism for korA- and korB-mediated incompatibility (Fig. 5). Because this plasmid has suffered a deletion in the promoter region of trfA and because the plasmid is completely resistant to korA and korB, the deleted segment must encode the targets.

The simplest explanation is that korA and korB products interact with recognition signals in the promoter region of trfA to inhibit expression of the trfA gene. In this way plasmid replication is prevented. In the kor-resistant mutant, the recognition signals for korA and korB products are missing.

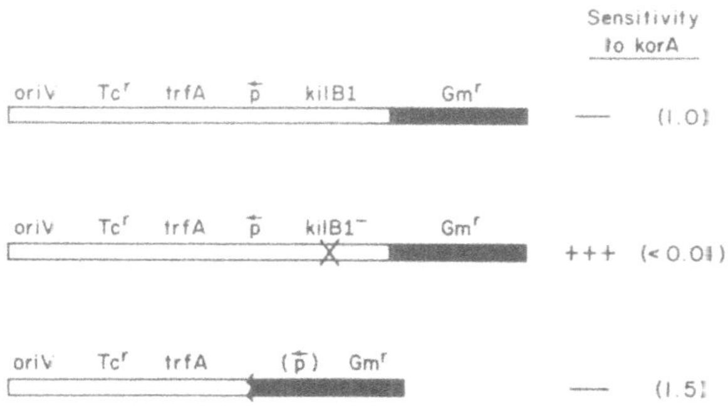

Fig. 5. Structure of plasmids resistant to *korA* incompatibility. These plasmids are an isogenic set which stems from the *kilB1*-derivative in the middle. Numbers show relative transformation efficiencies into a *korA*$^+$ host. In the top plasmid, the *kilB1*$^-$ allele is replaced with *kilB1*$^+$. At bottom is the structure of a *korA*- and *korB*-resistant mutant. Promoters are indicated with "p" and direction of transcription with an arrow. "(p)" in the bottom derivative suggests that *trfA* is expressed from a promoter in the Gm^r-encoding region because the known *trfA* promoter (25) is deleted.

We also noticed that *kilB1* has a dramatic effect on the ability of *korA* and *korB* to inhibit replication. The test plasmids used in the studies in Tab. 1 contain a *kilB1*$^-$ allele of a spontaneous Kil$^-$ mutant isolated previously (20). As shown above, these plasmids are highly sensitive to *korA* and *korB* incompatibility. However, if the *kilB1*$^-$ allele is replaced with the wild-type *kilB1*$^+$ allele, the new plasmid shows little if any sensitivity to *korA* and *korB* (Tab. 1, lines 11 and 12). This effect of *kilB1*$^+$ occurs only in cis; a *kilB1*$^+$ determinant alone in trans has no effect (data not shown). This is a remarkable result, for it implies that in a *korA*$^+$ or *korB*$^+$ plasmid *kilB1* should be considered an essential replication determinant. This is, of course, the case for parental RK2.

Thus, we have shown that *korA* and *korB* can inhibit RK2-specific replication. The results indicate that *korA* and *korB* products inhibit expression of the *trfA* gene and that this is accomplished by direct interaction with regulatory sequences in the promoter region of *trfA*. Furthermore, the inhibitory effects of *korA* and *korB* are somehow modulated by the *kilB1* determinant. These interactions are summarized in Fig. 6.

SUMMARY

The *kil* and *kor* genes of RK2 are novel genetic determinants

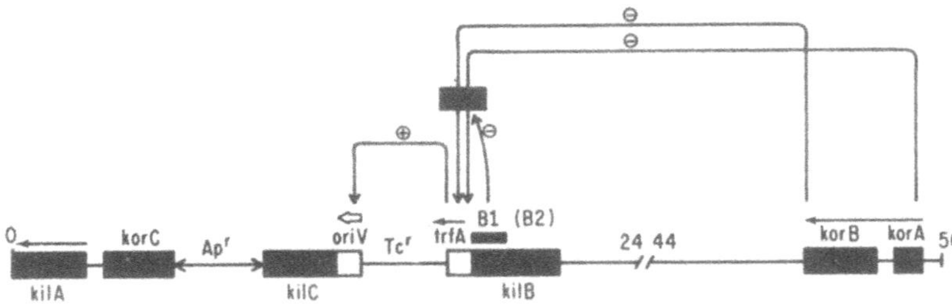

Fig. 6. Interactions of kil and kor functions with the replication determinants of RK2.

which seem to be common to all IncP plasmids. Their presence is easily detected by the strong phenotypes of host cell killing and its prevention. Thus, it has been possible to do extensive genetic analysis of these determinants, and this has revealed an intricate network of interactions. KorA has emerged as the central component in this network because of its multiple functions--both positive and negative.

It is now clear that the kil and kor genes constitute a new class of elements involved in plasmid maintenance. They cannot be considered absolutely essential for replication because it is possible to construct functional RK2 replicons which are devoid of any known kil or kor determinants. Nevertheless, the network of kil and kor interactions has now been shown to include interactions with essential replication determinants, and therefore there is a direct link between the action of these genes and control of plasmid replication (Fig. 7).

Originally we speculated that the kil and kor genes might be important to the broad host range of RK2 (10). We wish to propose

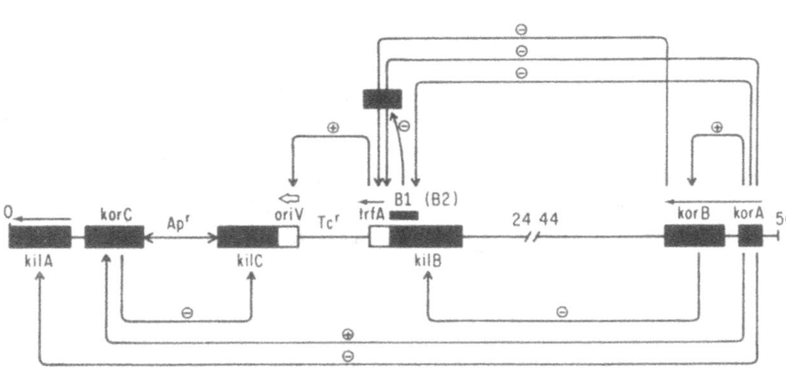

Fig. 7. Summary of the genetic interactions of RK2.

further that the *kil* and *kor* network constitutes a replication regulon, and that perhaps the function of this regulon is to ensure expression of *trfA* at appropriate levels. The complexity of this regulon may reflect an ability of the system to adapt to the intracellular environments of a variety of hosts. Indeed, there is tantalizing evidence that regions encoding *kil* or *kor* genes are important to host range (1,2,6,28; Schmidhauser and Helinski, pers. comm.). We are therefore hopeful that the study of these genes and the eventual determination of the molecular basis of their actions will lead to a complete understanding of the replication control and broad host range capability of IncP plasmids.

ACKNOWLEDGMENTS

We deeply appreciate the help of N. Scott, H. Spivak, B. Lynch and numerous students who spent summers or rotations in the lab: L. Babiss, S. Bessim, P. Borden, J. Davis, M. Dimitrov, M. Donnenberg, L. Epstein, P. Freimuth, I. Gelman, P. Gregor, B. Hill, O. Jovanovic, C. Kelton, N. LaMonica, S. Malik, N. Marder, L. Mielke, and T. Siegel. We also thank W. Maas for helpful discussions; and we are grateful to B. Hamilton for typing the manuscript. This work was supported by NSF Grant PCM78-08906 and NIH Grants GM26863 and GM29085 to D.F.

REFERENCES

1. Barth, P.T. (1979) RP4 and R300B as wide host-range plasmid cloning vehicles. In *Plasmids of Medical, Environmental and Commercial Importance*, K.N. Timmis, and A. Puhler, Eds. Elsevier/North Holland, Amsterdam/New York, pp. 399-410.
2. Barth, P.T., K. Ellis, D.H. Bechhofer, and D.H. Figurski (1984) Involvement of *kil* and *kor* genes in the phenotype of a host-range mutant of RP4. *Mol. Gen. Genet.* (in press).
3. Barth, P.T., and N.J. Grinter (1974) Comparison of the deoxyribonucleic acid molecular weights and homologies of plasmids conferring linked resistance to streptomycin and sulfonamides. *J. Bact.* 120:618-630.
4. Bechhofer, D.H., and D.H. Figurski (1983) Map location and nucleotide sequence of *korA*, a key regulatory gene of promiscuous plasmid RK2. *Nucl. Acids Res.* 11:7453-7469.
5. Burkardt, H.J., G. Reiss, and A. Puhler (1979) Relationship of group P1 plasmids revealed by heteroduplex experiments: RP1, RP4, R68, and RK2 are identical. *J. Gen. Microbiol.* 114:341-348.
6. Cowan, P., and V. Krishnapillai (1982) Tn7 insertion mutations affecting the host range of promiscuous IncP-1 plasmid R18. *Plasmid* 8:164-174.
7. Datta, N., and R.W. Hedges (1972) Host ranges of R-factors. *J. Gen. Microbiol.* 70:453-460.

8. Figurski, D.H., and D.R. Helinski (1979) Replication of an origin-containing derivative of plasmid RK2 dependent on a plasmid function provided in _trans_. _Proc. Natl. Acad. Sci., USA._ 76:1648-1652.
9. Figurski, D.H., R.J. Meyer, and D.R. Helinski (1979) Suppression of ColE1 replication properties by the IncP-1 plasmid RK2 in hybrid plasmids constructed _in vitro_. _J. Mol. Biol._ 133:295-318.
10. Figurski, D.H., R.F. Pohlman, D.H. Bechhofer, A.S. Prince, and C.A. Kelton (1982) The broad host range plasmid RK2 encodes multiple _kil_ genes potentially lethal to _Escherichia coli_ host cells. _Proc. Natl. Acad. Sci., USA_ 79:1935-1939.
11. Guiney, D.G., and E. Yakobson (1983) Location and nucleotide sequence of the transfer origin of the broad host range plasmid RK2. _Proc. Natl. Acad. Sci., USA_ 80:3595-3598.
12. Ingram, L., M.H. Richmond, and R.B. Sykes (1973) Molecular characterization of the R factors implicated in the carbenicillin resistance in a sequence of _Pseudomonas aeruginosa_ strains isolated from burns. _Antimicrob. Agents and Chemother_ 3:279-288.
13. Jacob, F., S. Brenner, and F. Cuzin (1963) On the regulation of DNA replication in bacteria. _Cold Spring Harbor Symp. Quant. Biol._ 28:329-348.
14. Lanka, E., R. Lurz, and J.P. Furste (1983) Molecular cloning and mapping of SphI restriction fragments of plasmid RP4. _Plasmid_ 10:303-307.
15. Meyer, R., D. Figurski, and D.R. Helinski (1977) Phyiscal and genetic studies with restriction endonucleases on the broad host range plasmid RK2. _Mol. Gen. Genet._ 152:129-135.
16. Meyer, R., and D.R. Helinski (1977) Unidirectional replication of the P-group plasmid RK2. _Biochem. Biophys. Acta_ 478:109-113.
17. Meyer, R., and M. Hinds (1982) Multiple mechanisms for expression of incompatibility by broad-host-range plasmid RK2. _J. Bact._ 152:1078-1090.
18. Olsen, R.H., and P. Shipley (1973) Host range and properties of the _Pseudomonas_ aeruginosa R factor R1822. _J. Bact._ 113:772-780.
19. Pohlman, R.F., and D.H. Figurski (1983) Conditional lethal mutants of the _kilB_ determinant of broad host range plasmid RK2. _Plasmid_ 10:82-95.
20. Pohlman, R.F., and D.H. Figurski (1983) Essential genes of plasmid RK2 in _Escherichia coli_: _trfB_ region controls a _kil_ gene near _trfA_. _J. Bact._ 156:584-591.
21. Roberts, J.W. (1969) Termination factor for RNA synthesis. _Nature_, London 224:1168-1174.
22. Schmidhauser, T.J., M. Filutowicz, and D.R. Helinski (1983) Replication of derivatives of the broad host range plasmid RK2 in two distantly related bacteria. _Plasmid_ 9:325-330.
23. Scott, J.R. (1984) Regulation of plasmid replication. _Microbiol. Rev._ 48:1-23.

24. Smith, C.A., and C.M. Thomas (1983) Deletion mapping of kil and kor functions in the trfA and trfB regions of broad host range plasmid RK2. Mol. Gen. Genet. 190:245-254.
25. Smith, C.A., and C.M. Thomas (1984) Nucleotide sequence of the trfA gene of broad host-range plasmid RK2. J. Mol. Biol. 175:251-262.
26. Stalker, D.M., C.M. Thomas, and D.R. Helinski (1981) Nucleotide sequence of the region of the origin of replication of the broad host range plasmid RK2. Mol. Gen. Genet. 181:8-12.
27. Thomas, C.M. (1981) Complementation analysis of replication and maintenance of broad host range plasmids RK2 and RP1. Plasmid 5:277-291.
28. Thomas, C.M., A.A.K. Hussain, and C.A. Smith (1982) Maintenance of broad host range plasmid RK2 replicons in Pseudomonas aeruginosa. Nature 298:674-676.
29. Thomas, C.M., R. Meyer, and D.R. Helinksi (1980) Regions of broad host range plasmid RK2 which are essential for replication and maintenance. J. Bact. 141:213-222.
30. Thomas, C.M., D.M. Stalker, and D.R. Helinski (1981) Replication and incompatibility properties of segments of the origin region of replication of the broad host range plasmid RK2. Mol. Gen. Genet. 181:1-7.
31. Villarroel, R., R.W. Hedges, R. Maenhaut, J. Leemans, G. Engler, M. Van Montagu, and J. Schell (1983) Heteroduplex analysis of P-plasmid evolution: The role of insertion and deletion of transposable elements. Mol. Gen. Genet. 189:390-399.
32. Young, C., D.H. Bechhofer, and D.H. Figurski (1984) Gene regulation in plasmid RK2: Positive control by korA in the expression of korC. J. Bact. 157:247-252.

REPLICATION DETERMINANTS OF THE BROAD HOST-RANGE PLASMID RSF1010

Peter Scholz, Volker Haring, Eberhard Scherzinger, Rudi Lurz, Miroslawa M. Bagdasarian,* Heinz Schuster, and Michael Bagdasarian*

Max-Planck-Institute of Molecular Genetics
Abteilung Schuster
1000 Berlin 33
Ihnestrasse 63-73, Germany

INTRODUCTION

RSF1010 is a small (8.7 kb), multicopy plasmid of the incompatibility group IncQ that confers resistances to streptomycin and sulfonamide to its host cells (1). It is very similar or identical with the 2 other representatives of the IncQ group, R300 (2) and R1162 (3). One of the most striking properties of this plasmid is its extraordinary broad host-range among gram-negative bacteria. This feature has made it attractive as a replicon for construction of vectors for gene cloning outside of *Escherichia coli* species. Both general- and special-purpose cloning vectors, derived from RSF1010, have proved very useful for gene manipulation in different species of soil bacteria (for review, see Ref. 4,5).

The molecular regulatory mechanisms that govern host-range specificity of plasmids are not well understood. This fact has prompted studies on the mechanism of replication, copy-number control, and maintenance of RSF1010 as a model plasmid that offers the convenience of small size and well-known physical structure (6,7). The picture emerging from these studies is that of a replicon encoding multiple essential replication genes that render the plasmid replication independent of several vital replication functions of the host cell.

*Unit for Applied Cell and Molecular Biology, Umea University, S-901 87 Umea, Sweden.

ESSENTIAL REGIONS OF RSF1010

The origin of vegetative replication (oriV) has been located at the coordinate of 2.6 kb on the physical map of RSF1010 (Fig. 1 and 2) by electron microscopic examination of replicating molecules linearized at the unique EcoRI-sensitive site of the plasmid (8). These measurements were performed on the replicative intermediates isolated from E. coli cells. They could not therefore provide adequate information on the mechanism of RSF1010 replication in other bacterial species. To answer the question whether in different host bacteria RSF1010 replication could be initiated specifically from other origins than in E. coli, we have examined, by electron microscopy, replicative intermediates of this plasmid isolated from Pseudomonas aeruginosa. Our measurements have shown (Fig. 1) that the replication of RSF1010 in this bacterium starts preferentially at the coordinate 2.65 kb. This location of oriV has been confirmed by measurements performed on replicating molecules linearized at the PvuII-sensitive sites (results not shown). From these results it is clear that in 2 organisms as different as E. coli and P. aeruginosa the same region of the plasmid, and therefore most probably the same origin, is used to initiate the replication of RSF1010. If this is indeed the case, other plasmid-coded replication or maintenance determinants must be responsible for the broad host-range specificity of this plasmid.

Restriction endonuclease cleavage analysis of RSF1010 indicated that the region comprising oriV should be contained within the stretch of the molecule spanned by the fragments HaeII (1.2-2.3 kb), HaeII (2.3-2.4 kb) and HaeII (2.4-3.3). However, attempts to generate self-replicating mini-plasmids from these 3 HaeII fragments ligated to different DNA fragments carrying antibiotic resistance genes were unsuccessful. This suggested the possibility that genes essential for replication or maintenance of RSF1010 are located in regions of the plasmid distant from oriV. To identify these regions, we have constructed hybrid plasmids carrying different DNA fragments of RSF1010 cloned into an unrelated plasmid vector and determined their ability to support the replication of the oriV-carrying HaeII fragments. One such helper plasmid was constructed by inserting a 5.0-kb SstI fragment from pKT228 (RSF1010 carrying Tn3 insertion at coordinate 4.2 kb) into the single SstI site of the vector pKT101 (7). The resulting recombinant plasmid, pMMB2, contains RSF1010 sequences located between coordinates 4.2 and 8.7 kb as well as a 0.3-kb segment of Tn3 (Fig. 2). E. coli strains carrying pMMB2 were used subsequently as recipients for recombinant DNA obtained by ligation of HaeII-digested RSF1010 DNA in the presence of nonself-replicating HaeII fragments carrying ampicillin- or chloramphenicol-resistance genes derived from a temperature-sensitive plasmid pHSG415 (9). Ap^r and Cm^r clones were selected and analyzed for the presence of satellite plasmids. All were found to contain, in addition to pMMB2, small plasmids. The smallest Ap^r plasmid, pMMB4 (4.5

kb), contained HaeII fragments [(1.2-2.3 kb), (2.3-2.4kb), and (2.4-3.3 kb)] of RSF1010 and the HaeII Ap^r fragment of pHSG415 (Fig. 2). The smallest one of the Cm^r plasmids, pMMB12, contained the same HaeII fragments of RSF1010 ligated to the HaeII Cm^r fragment derived from pHSG415 (Fig. 2). The structure of the later miniplasmid was confirmed by electron microscopic analysis of heteroduplex molecules formed with RSF1010 (7).

DNA of the miniplasmids, pMMB4 and pMMB12, separated from the DNA of the helper plasmids (pMMB2) by electrophoresis and extracted from the agarose gels, could be introduced by transformation into *E. coli* cells carrying the helper plasmid pMMB2 or the parental plasmid RSF1010. The miniplasmids could not, however, be introduced into plasmid-free *E. coli* strains (Tab. 1). It is clear, thus, that the replication or maintenance of the miniplasmids carrying *oriV* of RSF1010 is dependent upon trans-acting functions located in the region of RSF1010 between 4.2 and 8.7 kb that is present in pMMB2.

MUTATIONS IN THE ESSENTIAL GENES OF RSF1010

To obtain mutations in the essential genes coding for the trans-acting functions required for replication of RSF1010, we have treated purified DNA of RSF1010 with NH_2OH (10) and transformed the mutagenized DNA into *E. coli* strains C600 and RH2358, a strain carrying $supF_{ts}$ mutation (11). Transformants in which the plasmid was segregated out after a cycle of growth at 42°C were picked and the plasmids in them were analyzed for stability at nonpermissive conditions (42°C or in a $supF_{ts}$ host). Two mutant derivatives of RSF1010 have been selected for further analysis; a temperature-sensitive mutant, pMMB19; and an amber mutant, pMMB13. Both mutant plasmids were lost rapidly from the cells growing at nonpermissive conditions (Fig. 3) and this property could be used to study the complementation of the mutated gene by the wild-type function and for mapping of the affected locus (see next section). The fact that an amber mutation affects stability of RSF1010 indicated that at least one of the essential genes of this plasmid codes for a protein product.

LOCATION OF THE ESSENTIAL GENES ON THE PHYSICAL MAP OF RSF1010

In the plasmid pMMB2, the trans-acting, essential genes of RSF1010 were joined to the replicon of the pKT101 vector and could therefore be subjected to a deletion analysis. To do this, pMMB2 DNA was linearized at the single AccI site or by the excision of the small PstI fragment. Linearized DNA was digested with Bal31 nuclease, religated, and used to transform strain C600 to Km^r. Two series of deletions were thus obtained, one originating at the AccI site within RSF1010 sequences and one entering the RSF1010 segment

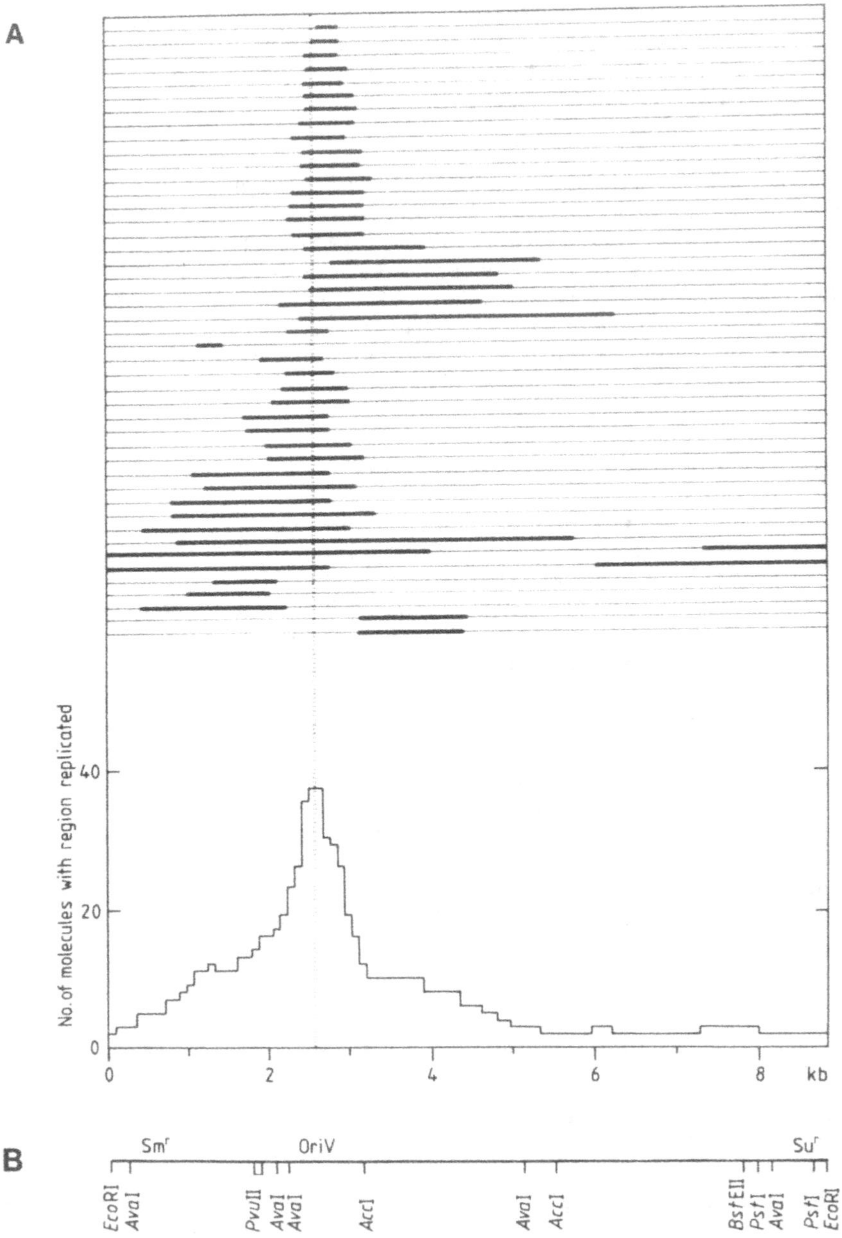
A
No. of molecules with region replicated
40
20
0
0
2
4
6
8
kb
B
Sm^r
OriV
Su^r
EcoRI
AvaI
PvuII
AvaI
AvaI
AccI
AvaI
AccI
BstEII
PstI
AvaI
PstI
EcoRI

C

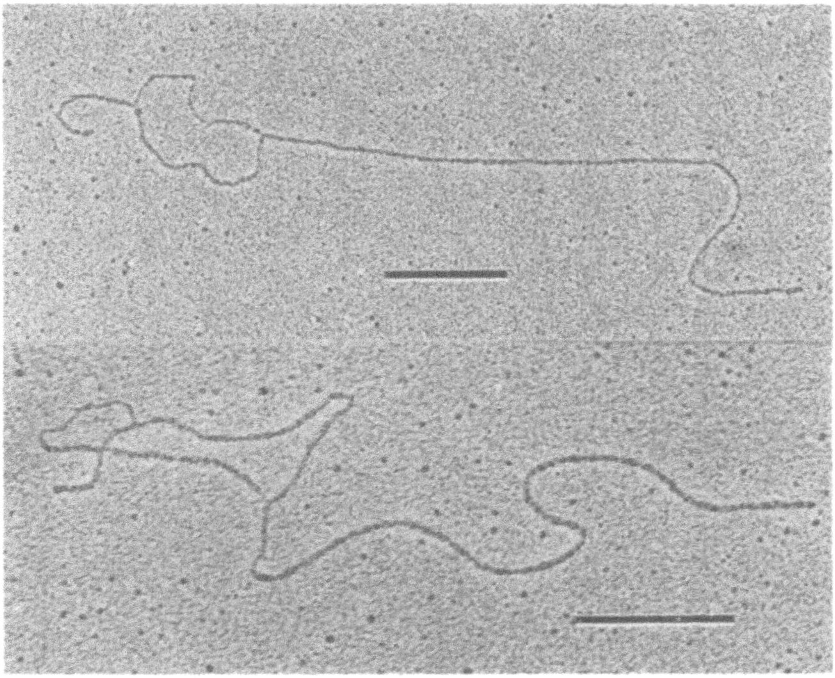

Fig. 1. Partially replicated molecules of RSF1010 isolated from Pseudomonas aeruginosa cleaved with EcoRI and visualized by electron miscroscopy. Upper part: Line drawings and histogram of RSF1010 replicative intermediates. Middle part: Restriction endonuclease cleavage map of RSF1010. Lower part: Electron micrographs of replicating molecules of RSF1010 from P. aeruginosa, linearized with EcoRI. The bar represents 1 kb.

from the PstI site (Fig. 4). A third series of deletions, entering the RSF1010 segment from the end opposite to the PstI end was obtained from DNA linearized with SstI. In this case, the parental plasmid used was a previously constructed derivative of pMMB2, Δ38, in which about 1 kb around the single EcoRI site had been deleted to remove the adjacent SstI site (Fig. 4). The end point of the deletions, mapped by electron microscopic analysis of heteroduplex molecules formed with parental plasmids, are shown in Fig. 4. All deletion derivatives of pMMB2 obtained this way were tested for their ability to provide the functions required for the replication and maintenance of the miniplasmid, pMMB12, and for the rescue of RSF1010 mutants pMMB19 and pMMB13. The results indicated that functions essential for the replication of the oriV miniplasmid are located on both sides of the single AccI site present in the plasmid pMMB2 (Fig. 4). Mutant rescue experiments have shown that pMMB2 de-

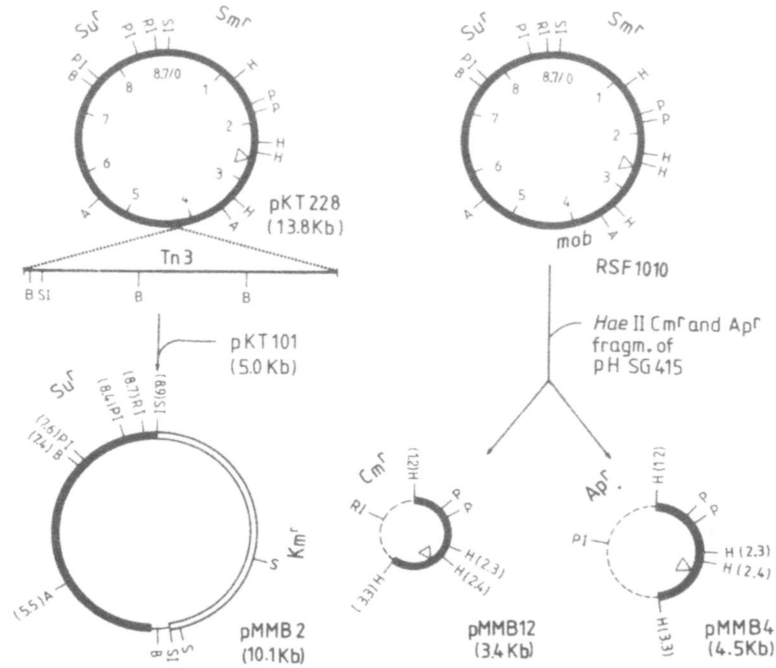

Fig. 2. Construction of plasmids carrying replication determinants of RSF1010: oriV and rep. The thick solid line represents sequences derived from RSF1010. Restriction endonuclease cleavage sites (Ref. 6, 7) are indicated by the code: A, AccI; B, BstEII; RI, EcoRI; H, HaeII; P, PvuII; PI, PstI; S, SmaI; SI, SstI. Of the numerous HaeII sites in RSF1010, only those present in the oriV region are indicated. The numbers in parenthesis refer to coordinates in parental RSF1010 molecule (in kb pairs from the unique SstI site). The open triangle indicates the origin of RSF1010 replication (8).

rivatives in which deletions entered the RSF1010 segment from the PstI site could rescue the mutation in pMMB19 but not that in pMMB13, while derivatives with deletions around the AccI site could rescue the mutation in pMMB13 but not in pMMB19. Moreover, deletions Δ63 and Δ67 entering the RSF1010 segment from the end opposite to the PstI end and unable to complement for the replication of the oriV miniplasmids were nevertheless able to rescue mutations in both pMMB19 and pMMB13 (Fig. 4). The region of RSF1010 comprised between the RSF1010 coordinates 4.2 - 8.7 may thus be subdivided into 3 segments, each encoding a function essential for replication or maintenance of plasmids with RSF1010 origin. We have proposed to call these essential genes repA, repB, and repC. These results have been confirmed by the demonstration that an in vitro DNA replication sys-

Tab. 1. Transformation frequency of miniplasmids carrying the origin of RSF1010 in E. coli.

Plasmid Resident	Plasmid Incoming	Transformants / 1 μg DNA
none	pMMB4	< 200
none	pMMB12	< 200
pKT101	pMMB4	< 200
pKT101	pMMB12	< 200
pMMB2	pMMB4	6.0×10^5
pMMB2	pMMB12	8.0×10^5
RSF1010	pMMB4	1.8×10^5

Approximately 10^9 competent cells of E.coli C600 were transformed with 5 ng of mini-plasmid DNA purified by agarose gel electrophoresis. Transformants were selected for Ap^r in the case of pMMB4 and for Cm^r in the case of pMMB12.

tem requires the addition of protein products of all 3 rep genes, encoded on the plasmid pMMB2, in order to replicate RSF1010 DNA. The products of genes repA, repB, and repC have been purified to homogeneity by using an in vitro complementation test as assay for their activity. Finally the positions of the individual rep genes, shown in Fig. 4, have been confirmed by complementation tests run with extracts, prepared from strains carrying deletion derivatives of pMMB2, to which purified rep proteins were added in the in vitro DNA replication assay (7).

NUCLEOTIDE SEQUENCE OF repA GENE

Cloning experiments have shown that the 1.24-kb BalI fragment located between AccI and PstI sites of RSF1010 (Fig. 4 and 5) contains a complete functional repA gene. Moreover, repA protein purified from cells carrying this 1.2-kb BalI fragment, isolated from the thermosensitive plasmid pMMB19 and cloned into an expression vector, was temperature-sensitive in an in vitro assay (unpub. data). These determinations provided direct evidence for the location of the structural gene for the repA protein (Fig. 4 and 5). They showed also that repA gene is transcribed in the direction from AccI towards the BssHII and BstEII site.

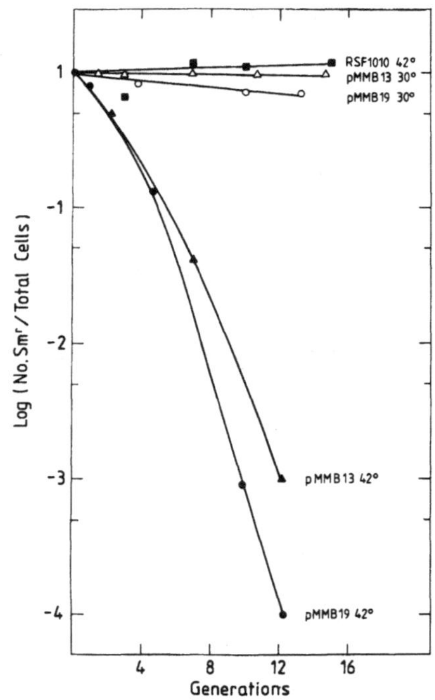

Fig. 3. Stability of RSF1010 derivatives at different temperatures. Logarithmically growing cultures of *E. coli* C600 carrying RSF1010 or pMMB19 and *E. coli* RH2358 [sup Fts (11)] carrying pMMB13 were transferred to the indicated temperature. At time intervals, suitable dilutions of the cultures were plated on Luria Broth agar with or without streptomycin. Colonies were scored after 24 hr at 30°C.

To carry on a more detailed analysis of the region containing the *repA* gene and to reveal possible signals for its regulation, we have determined the complete nucleotide sequence of the DNA fragment comprised between the *AccI* site at coordinate 5.5 kb and *BalI* site at coordinate 6.9 kb. The nonrandom sequencing strategy of Poncz et al. (12) was used to generate overlapping fragments that were sequenced by the dideoxy-chain termination method (13).

The sequence thus obtained contains several open frames reading in the direction from *AccI* towards *BssHII* site. The correct reading frame for the *repA* gene has been selected by taking into account the subunit size of the purified repA protein and of the sequence of 53 amino acids from its $-NH_2$ terminus (Scholz, Scherzinger, Haring, Wittmann, and Bagdasarian, to be published elsewhere).

The nucleotide sequence of the 1243-bp *BalI* fragment of RSF1010, containing the entire *repA* gene, is shown in Fig. 5. One

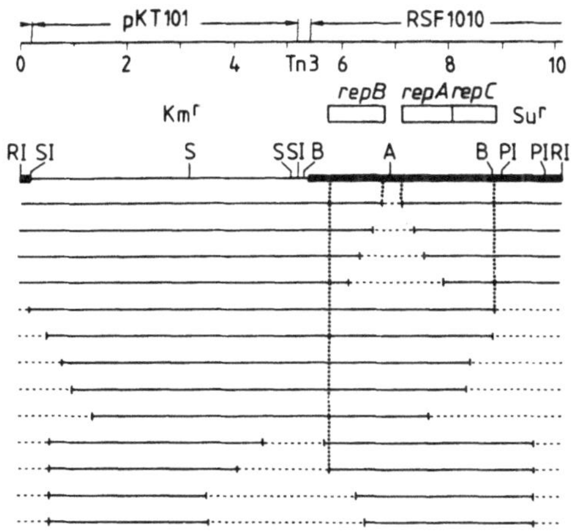

	COMPLEMENTATION OF		
	pMM312	pMMB19	pMMB13
pMMB2	+	+	+
Δ1	+	+	+
Δ5	-	-	+
Δ8	-	-	+
Δ10	-	-	+
Δ38	+	+	+
Δ15	-	+	-
Δ18	-	+	-
Δ20	-	+	-
Δ23	-	-	-
Δ71	-	+	+
Δ70	-	+	+
Δ63	-	+	+
Δ67	-	+	+

Fig. 4. Upper part: Physical and genetic map of pMMB2. Distances are given in kb pairs from the unique EcoRI site. Abbreviations for restriction endonucleases are explained in legend to Fig. 2. The thick solid line denotes sequences derived from RSF1010. Open boxes represent the positions of rep genes deduced from the data in this table, reference (7), and nucleotide sequence analysis.

Lower part: Diagrams of pMMB2 deletion derivatives used in this work. Broken lines represent the areas deleted, as determined by DNA-DNA heteroduplex analysis (7).

Table: Complementation of the replication ability of mutant RSF1010 derivatives by deletions of pMMB2. Strains containing the pairs of plasmids indicated were plated and grown at nonpermissive conditions without antibiotics. Individual colonies were then tested for the appropriate resistance markers. Positive complementation (+) was assigned to those pairs where at least 90% retained the marker of the RSF1010 derivative. Lack of complementation (-) indicates that 95% of the colonies have lost the marker of the RSF1010 derivative.

strand of DNA is displayed and the entire sequence is numbered, in both directions, beginning at the translational start codon, ATG, of the repA protein. The coding region of the gene is preceded, at positions -7 through -13, by a sequence -AAGGAGG- which is complementary to the end of the 16S ribosomal RNA and represents the ribosome binding site (14). We assume that the sequences -GTGACC-, at positions -49 through -44, and -CCACCT-, at positions -26 through -21, represent the RNA polymerase binding site (-35 region) (15) and the Pribnow box (-10 region) (15) of the repA gene promoter. This assumption, however, is based solely upon the location of these se-

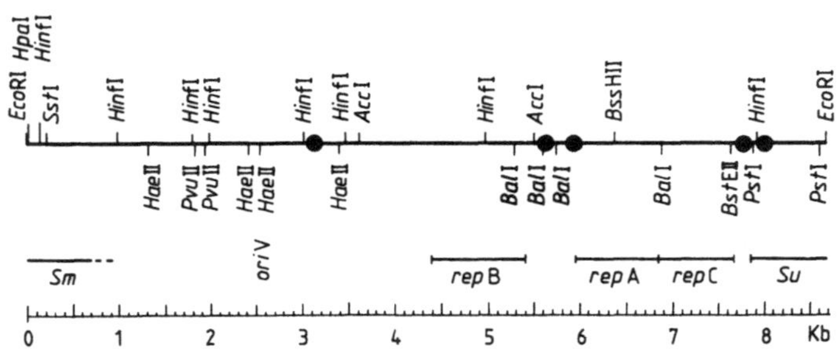

Fig. 5. Physical and genetic map of RSF1010. The map is linearized at the unique EcoRI site. Of the numerous HaeII and BalI sites of RSF1010 only the HaeII sites of the oriV region and the BalI sites of the repA-repC region are indicated. The filled circles represent RNA polymerase binding sites determined by electron microscopy (6).

quences close to and upstream of the Shine-Dalgarno sequence and the translational start codon of the gene.

The coding sequence of the repA gene contains 837 bp and terminates with GCC (Ala) followed by a TAA stop codon. Since the initiator Met residue is not present in the mature repA protein this sequence codes for 278 amino acids. The molecular weight of the polypeptide coded for by this sequence is 29.7 kd-- a value that is in good agreement with the subunit size of 29.0 kd determined for the purified repA protein by the SDS-polyacrylamide electrophoresis (unpub. results).

There is no recognizable transcription termination structure (15) following the coding region of the gene either in the sequence presented in Fig. 5 or downstream of it up to the region that represents coding sequences of the repC gene. We must assume, therefore, that the genes repA and repC are transcribed as a unit and, therefore, most probably, share at least some regulatory features.

REQUIREMENTS OF HOST FUNCTIONS FOR THE REPLICATION OF RSF1010

As shown in this and previous work (7), RSF1010 requires 3 proteins encoded by its own genome to start the replication from its origin. Both the number of essential genes and their arrangement on the physical map of this plasmid distinguishes it from the narrow host-range plasmids studied to date, such as ColE1, R1, R6-5, R6K, F, and pSC101, which have their replication determinants clustered around the origin of replication (16-22). The genetic organization of RSF1010 resembles rather that of another broad host-range plasmid, RK2/RP1/RP4, in which several essential replication and maintenance genes are scattered over a large portion of the genome (23).

The existence of several essential regions in the plasmid RK2 has led to the presumption that broad host-range plasmids encode more replication functions than do the narrow host-range plasmids. It is believed that this property makes them less dependent upon the replication functions of the host cell and allows thus a greater degree of promiscuity (23).

It has already been shown that rifampicin-sensitive RNA polymerase, one of the host proteins implicated in the initiation of replication of several narrow host-range plasmids (including dv), is not required for the replication of RSF1010 DNA in vitro (7,24). To test the requirement of other host DNA replication factors for the replication of RSF1010, we have introduced the plasmid into strains containing conditional mutations in the genes *dnaB*, *dnaC*, and *dnaG*. The cells of these strains were labeled with ^{3}H-thymidine under permissive and nonpermissive conditions, total DNA was extracted from these cells by lysis with Sarkosyl, and analyzed by CsCl-ethidium bromide gradient centrifugation.

Under permissive conditions (30°C) total radioactivity incorporated into the supercoiled DNA fraction, which represents plasmid DNA, amounts to 0.02 - 0.06 of the total radioactivity incorporated into the linear DNA fraction, composed predominantly of chromosomal DNA (Fig. 7, Tab. 2). This ratio corresponds to 9-28 copies of the plasmid per chromosome equivalent if the value of 3,750 kbp is taken as the size of the *E. coli* chromosome. These figures are in good agreement with the previous estimates of 10-50 copies of RSF1010 per cell (1). Under nonpermissive conditions (41°C) the incorporation of thymidine into the chromosomal DNA is drastically reduced whereas the incorporation into the supercoiled DNA fraction is reduced to a small extent (Fig. 7). In all 3 mutants tested there was considerably more thymidine incorporated into the plasmid DNA fraction than into the chromosome (Tab. 2). We conclude, therefore, that under conditions where either *dnaB*, *dnaC*, or *dnaG* function is inhibited, the replication of RSF1010 DNA continues. In a strain containing *dnaZ* mutation, RSF1010 seems to be present at such low copy number that its estimation by labeling and dye-buoyant density centrifugation of total cellular DNA was difficult even at permissive conditions. We consider this as an indication that DNA polymerase III, or at least some of its components, are required for the replication of RSF1010 in *E. coli* cells.

CONCLUDING REMARKS

Among plasmids that are able to replicate in different, distantly related, bacterial hosts the genetics of replication determinants has been investigated in only 2 cases, that of RP4/RK2 plasmid (23,25-27) and of RSF1010 (Ref. 7, this work). In both plasmids it was found that essential genes are scattered over a large portion

```
BalI -190        -180        -170        -160        -150        -140
TGGCCAGCCCTGACGCTGTACGCCAAGCGCGATATGCCGAGCGCATGAAGGCCAAAGGGA
     -130        -120        -110        -100         -90         -80
TGCGTCAGCGCAAGTTCTGGCTGACCGACGACGAATACGAGGCGCTGCGCGAGTGCCTGG
      -70         -60         -50         -40         -30         -20
AAGAACTCAGAGCGGCGCAGGGCGGGGTAGTGACCCCGCCAGCGCCTAACCACCAACTG
      -10           0               10                20          30
CCTGCAAAGGAGGCAATCA ATG GCT ACC CAT AAG CCT ATC AAT ATT CTG
                        Ala Thr His Lys Pro Ile Asn Ile Leu
            40              50              60              70
GAG GCG TTC GCA GCA GCG CCG CCA CCG CTG GAC TAC GTT TTG CCC
Glu Ala Phe Ala Ala Ala Pro Pro Pro Leu Asp Tyr Val Leu Pro
    80              90             100             110             120
AAC ATG GTG GCC GGT ACG GTC GGG GCG CTG GTG TCG CCC GGT GGT
Asn Met Val Ala Gly Thr Val Gly Ala Leu Val Ser Pro Gly Gly
           130             140             150             160
GCC GGT AAA TCC ATG CTG GCC CTG CAA CTG GCC GCA CAG ATT GCA
Ala Gly Lys Ser Met Leu Ala Leu Gln Leu Ala Ala Gln Ile Ala
   170             180             190             200             210
GGC GGG CCG GAT CTG CTG GAG GTG GGC GAA CTG CCC ACC GGC CCG
Gly Gly Pro Asp Leu Leu Glu Val Gly Glu Leu Pro Thr Gly Pro
           220             230             240             250
GTG ATC TAC CTG CCC GCC GAA GAC CCG CCC ACC GGC ATT CAT CAC
Val Ile Tyr Leu Pro Ala Glu Asp Pro Pro Thr Ala Ile His His
   260             270             280             290             300
CGC CTG CAC GCC CTT GGG GCG CAC CTC AGC GCC GAG GAA CGG CAA
Arg Leu His Ala Leu Gly Ala His Leu Ser Ala Glu Glu Arg Gln
           310             320             330             340
GCC GTG GCT GAC GGC CTG CTG ATC CAG CCG CTG ATC GGC AGC CTG
Ala Val Ala Asp Gly Leu Leu Ile Gln Pro Leu Ile Gly Ser Leu
   350             360             370             380       BssHII
CCC AAC ATC ATG GCC CCG GAG TGG TTC GAC GGC CTC AAG CGC GCC
Pro Asn Ile Met Ala Pro Glu Trp Phe Asp Gly Leu Lys Arg Ala
           400             410             420             430
GCC GAG GGC CGC CGC CTG ATG GTG CTG GAC ACG CTG CGC CGG TTC
Ala Glu Gly Arg Arg Leu Met Val Leu Asp Thr Leu Arg Arg Phe
```

```
 440       450        460        470        480
CAC ATC GAG GAA GAA AAC GCC AGC GGC CCC·ATG GCC CAG GTC ATC
His Ile Glu Glu Glu Asn Ala Ser Gly Pro Met Ala Gln Val Ile
         490        500        510        520
GGT CGC ATG GAG GCC ATC GCC GCC GAT ACC GGG TGC TCT ATC GTG
Gly Arg Met Glu Ala Ile Ala Ala Asp Thr Gly Cys Ser Ile Val
 530       540        550        560        570
TTC CTG CAC CAT GCC AGC AAG GGC GCG GCC ATG ATG GGC GCA GGC
Phe Leu His His Ala Ser Lys Gly Ala Ala Met Met Gly Ala Gly
         580        590        600        610
GAC CAG CAG CAG GCC AGC CGG GGC AGC TCG GTA CTG GTC GAT AAC
Asp Gln Gln Gln Ala Ser Arg Gly Ser Ser Val Leu Val Asp Asn
 620       630        640        650        660
ATC CGC TGG CAG TCC TAC CTG TCG AGC ATG ACC AGC GCC GAG GCC
Ile Arg Trp Gln Ser Tyr Leu Ser Ser Met Thr Ser Ala Glu Ala
         670        680        690        700
GAG GAA TGG GGT GTG GAC GAC GAC CAG CGC CGG TTC TTC GTC CGC
Glu Glu Trp Gly Val Asp Asp Asp Gln Arg Arg Phe Phe Val Arg
 710       720        730        740        750
TTC GGT GTG AGC AAG GCC AAC TAT GGC GCA CCG TTC GCT GAT CGG
Phe Gly Val Ser Lys Ala Asn Tyr Gly Ala Pro Phe Ala Asp Arg
         760        770        780        790
TGG TTC AGG CGG CAT GAC GGC GGG GTG CTC AAG CCC GCC GTG CTG
Trp Phe Arg Arg His Asp Gly Gly Val Leu Lys Pro Ala Val Leu
 800       810        820        830        840
GAG AGG CAG CGC AAG AGC AAG GGG GTG CCC CGT GGT GAA GCC TAA
Glu Arg Gln Arg Lys Ser Lys Gly Val Pro Arg Gly Glu Ala ***
       850       860       870       880       890       900
GAACAAGCACAGCCTCAGCCACGTCCGGCACGACCCGGCGCACTGTCTGGCCCCCGGCCT
       910       920       930       940       950       960
GTTCCGTGCCCTCAAGCGGGGCGAGCGCAAGCGCAGCAAGCTGGACGTGACGTATGACTA
       970       980       990      1000      1010      1020
CGGCGACGGCAAGCGGATCGAGTATCAGCGGCCCGGAGCCGCTGGGCGCTGATGATCTGC
      1030         BalI
GCATCCTGCAAGGGCTGGTGGCCA
```

Fig. 6. Nucleotide sequence of the BalI fragment of RSF1010 containing the repA gene. The putative promoter sequences of the gene are enclosed in solid line boxes and the ribosome binding site in broken line box. Nucleotides are numbered, in both directions, from the translation start codon, ATG of the repA coding sequence. Amino acids of repA protein are shown under their respective codon triplets. The stop codon of repA coding sequence is indicated by asterisks.

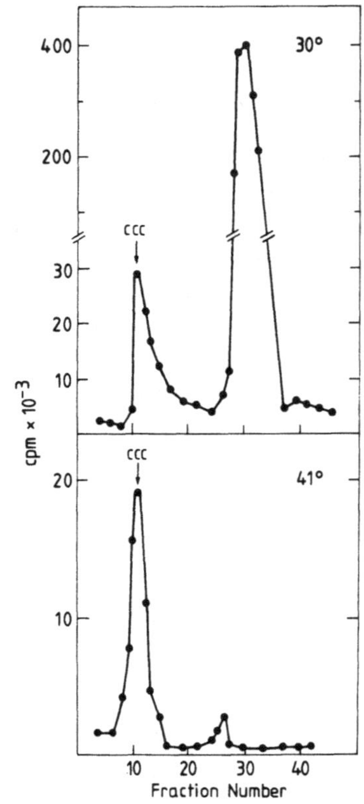

Fig. 7. Dye-buoyant density centrifugation profile of total DNA of a dnaG [RSF1010] strain labeled at permissive (30°C) and nonpermissive (41°C) temperature. One ml culture (5×10^8 cells) of the strain BW86 [RSF1010] (dnaG3 deoB thyA) was placed at temperatures indicated and labeled by the addition of ^{3}H-thymidine (0.04 mCi, 3.5 Ci per 1 mmol) for 60 min. The cultures were lysed with Sarkosyl (1.0%) and centrifuged to equilibrium in CsCl-Ethidium bromide density gradient. Fractions of the gradient were collected on Whatman GF/C discs, washed with 5% trichloroacetic acid, then ethanol, dried, and counted in a liquid scintillation counter.

of the genome. Moreover, the number of essential genes in broad host-range plasmids is greater than in the plasmids with narrow host specificity. Of the 2 broad host-range plasmids studied so far, RSF1010 represents a more dramatic example of this type of genetic structure. Despite the relatively small size, about 50% of its genome is taken up by essential genes.

The rep genes of RSF1010, described in this and previous papers (7), code for proteins that are specifically required for the repli-

Tab. 2. Incorporation of thymidine into RSF1010 DNA in *dna* mutants.

Chromosomal mutation	c p m ccc.DNA / linear DNA	
	30°	42°
*dna*B	0.020	5.00
*dna*G	0.065	12.02
*dna*C	0.170	3.20

Cultures were labelled and analysed as described in the legend to Fig. 7. The counts in the peaks were summed to determine the ratios given in the table. Strains were *dna*B165/70 *thy*A *thi* [RSF1010], *dna*C2 *thy*A *dra* [RSF1010] and *dna*G3 *deo*B *thy*A [RSF1010].

cation of plasmids carrying the replication origin of RSF1010 in vivo and in vitro. It could be expected, therefore, that their role would be either to replace the functions of some host cell DNA replication proteins in the replication of RSF1010 or to serve as a functional (adaptational) link between the replication factors of the host and the initiation factor of the plasmid. On the basis of the in vivo experiments reported in this work it cannot be ruled out that the replication factors of RSF1010 act by stabilizing the temperature-sensitive *dna*B, *dna*C, and *dna*G proteins and thus mimic the independence of RSF1010 replication of these host proteins. However, parallel experiments performed with the in vitro replication system indicated that, at least in *E. coli*, the 3 *dna* proteins of the host are replaced by the *rep* proteins of RSF1010. The 3 *rep* proteins are specific for the RSF1010 template; they are not able to substitute for the *dna* functions of the host either in the chromosome replication in vivo or in the replication of ColE1 DNA in an in vitro system (unpub. data).

Differences in properties of the *dna*B, *dna*C, and *dna*G proteins among gram-negative bacteria may constitute a barrier for those plasmids that rely on the functions of these host proteins for their replication. The example of RSF1010 shows that at least some of the broad host-range plasmids have evolved by acquiring their own genes that perform functions analogous to the *dna*B, *dna*C, and *dna*G genes of *E. coli* but are specific for the replication of plasmid DNA.

ACKNOWLEDGMENTS

This work was supported in part by the grant No. GI-3-060-D from the Biomolecular Engineering Program of the Commission of European Communities.

REFERENCES

1. Guerry, P., J. van Embden, and S. Falkow (1974) Molecular nature of two nonconjugative plasmids carrying drug resistance genes. J. Bact. 117:619-630.
2. Barth, P., and N.J. Grinter (1974) Comparison of the deoxyribonucleic acid molecular weights and homologies of plasmids conferring linked resistance to streptomycin and sulfonamides. J. Bact. 120:618-630.
3. Meyer, R., M. Hinds, and M. Brasch (1982) Properties of R1162, a broad host range, high copy number plasmid. J. Bact. 150:552-562.
4. Bagdasarian, M., and K.N. Timmis (1982) Host:vector systems for gene cloning in *Pseudomonas*. Curr. Topics Microb. Immun. 96:47-67.
5. Lehrbach, P.R., and K.N. Timmis (1983) Genetic analysis and manipulation of catabolic pathways in *Pseudomonas*. Biochem. Soc. Symp. 48:191-219.
6. Bagdasarian, M., R. Lurz, B. Rückert, F.C.H. Franklin, M.M. Bagdasarian, J. Frey, and K.N. Timmis (1981) Specific purpose plasmid cloning vectors. II. Broad host range, high copy number, RSF1010-derived vectors, and host-vector system for gene cloning in *Pseudomonas*. Gene 16:237-247.
7. Scherzinger, E., M.M. Bagdasarian, P. Scholz, R. Lurz, B. Rückert, and M. Bagdasarian (1984) Replication of the broad host range plasmid RSF1010: Requirement of three plasmid-encoded proteins. Proc. Natl. Acad. Sci., USA 81:654-658.
8. De Graaff, J., J.H. Crosa, F. Heffron, and S. Falkow (1978) Replication of the non-conjugative plasmid RSF1010 in *Escherichia coli* K12. J. Bact. 134:1117-1122.
9. Hashimoto-Gotoh, T., F.C.H. Franklin, A. Nordheim, and K.N. Timmis (1981) Specific-purpose plasmid cloning vectors. I. Low copy number, temperature-sensitive, mobilization-defective pSC101-derived containment vectors. Gene 16:227-235.
10. Humphreys, G.D., G.A. Willshaw, H.R. Smith, and E.S. Anderson (1976) Mutagenesis of plasmid DNA with hydroxylamine: Isolation of mutants of multi-copy plasmids. Mol. Gen. Genet. 145:101-108.
11. Delcuve, G., T. Cabezon, A. Ghysen, A. Herzog, and A. Bollen (1977) Amber mutations in *Escherichia coli* essential genes: Isolation of mutants affected in the ribosomes. Mol. Gen. Genet. 157:149-153.

12. Poncz, M., D. Solowiejczyk, M. Ballantine, E. Schwartz, and S. Surrey (1982) "Nonrandom" DNA sequence analysis in bacteriophage M13 by the dideoxy chain termination method. Proc. Natl. Acad. Sci., USA 79:4298-4302.
13. Messing, J. (1983) New M13 vectors for cloning. Meth. in Enzym. 101:20-77.
14. Shine, J., and L. Dalgarno (1975) Determinant of cistron specificity in bacterial ribosomes. Nature 254:34-38.
15. Rosenberg, M., and D. Court (1979) Regulatory sequences involved in the promotion and termination of RNA transcription. Ann. Rev. Genet. 13:319-353.
16. Backman, K., M. Betlach, H.W. Boyer, and C. Yanofsky (1978) Genetic and physical studies on the replication of ColEl-type plasmids. Cold Spring Harbor Symp. Quant. Biol. 43:69-76.
17. Leight, J., and S. Molin (1981) Replication control functions of plasmid Rl act as inhibitors of expression of a gene required for replication. Mol. Gen. Genet. 184:56-61.
18. Timmis, K.N., H. Danbara, G. Brady, and R. Lurz (1981) Inheritance functions of group IncFII transmissible antibiotic resistance plasmids. Plasmid 5:53-75.
19. Stalker, D.M., R. Kolter, and D.R. Helinski (1982) Plasmid R6K DNA replication. I. Complete nucleotide sequence of an autonomously replicating segment. J. Mol. Biol. 161:33-43.
20. Komai, N., T. Nishizawa, Y. Hayakawa, T. Murotsu, and K. Matsubara (1982) Detection and mapping of six mini F-encoded proteins. Mol. Gen. Genet. 186:193-203.
21. Lindner, P., G. Churchward, and L. Caro (1983) Plasmid pSC101 replication mutants generated by insertion of the transposon Tn1000. J. Mol. Biol. 170:287-303.
22. Thomas, C.M., R. Meyer, and D.R. Helinski (1980) Regions of broad host range plasmid RK2 which are essential for replication and maintenance. J. Bact. 141:213-222.
23. Diaz, R., and W.L. Staudenbauer (1982) Replication of the broad host range plasmid RSF1010 in cell-free extracts of Escherichia coli and Pseudomonas aeruginosa. Nucl. Acid Res. 10:4687-4702.
24. Figurski, D., and D.R. Helinski (1979) Replication of an origin-containing derivative of plasmid RK2 dependent on a plasmid function provided in trans. Proc. Natl. Acad. Sci., USA 76:1648-1652.
25. Thomas, C.M. (1981) Molecular genetics of broad host range plasmid RK2. Plasmid 5:10-19.
26. Thomas, C.M. (1981) Complementation analysis of replication and maintenance functions of broad host range plasmids RK2 and RP1. Plasmid 5:277-291.

REGULATION OF REPLICATION AND MAINTENANCE FUNCTIONS OF BROAD HOST-RANGE PLASMID RK2

C. M. Thomas, C. A. Smith, V. Shingler, M. A. Cross, A. A. K. Hussain,* and M. Pinkney

Department of Genetics
University of Birmingham
P. O. Box 363
Birmingham B 15 2TT, England

ABSTRACT

Replication of broad host-range plasmid RK2 depends on a cis-acting vegetative replication origin $oriV_{RK2}$ and the polypeptide product(s) of the trans-acting gene trfA as well as on host-specified products. The trfA gene is the second cistron in a polycistronic unit whose first cistron may be kilD, one of 4 known RK2-specified kil loci (kilA, B, C, and D) which are inhibitory for bacterial host or plasmid vector in the absence of kor functions which suppress in trans the effect of their respective kil genes. Transcription of the operon containing trfA is negatively regulated by the products of both the trfB locus (alias korD and korA) and korB. The loci, trfB and korB, are expressed from a single transcriptional unit which we propose to be negatively autoregulated by the products of both loci, although an additional, weaker and unregulated transcript may also express korB. While deletions in the $oriV_{RK2}$ region have indicated the presence of copy number control elements adjacent to and possibly overlapping with the minimal $oriV_{RK2}$ segment, the overriding control of copy number seems to reside in the trfB and korB lóci which in conjunction appear to reduce expression of the trfA gene to levels limiting for replication. Coregulation of trfA with kil genes may indicate that kil genes play a role in plasmid maintenance other than replication.

* Present address: Max-Planck Institut für Molekulare Genetik, Abteilung Wittman, D-1000 Berlin 33 (Dahlem), Federal Republic of Germany.

INTRODUCTION

Escherichia coli IncP-1 plasmid RK2 (similar and probably identical to RP1, RP4, R68, and R18) is capable of conjugal transfer between and stable maintenance in most gram-negative bacterial species (for a general review of this plasmid, see Ref. 1). Interestingly, it is reported elsewhere in this volume (2) that RK2 cannot replicate in gram-negative anaerobes such as *Bacteroides* spp., the first well documented example of a limit on the host range of RK2. A physical and genetic map of RK2 is shown in Fig. 1 using the recently revised size of 60 kb (3). The locations of plasmid genes are given by reference to these kb coordinates clockwise from the unique *EcoRI* site. In addition to genes (Pn^r, Tc^r, and Km^r) conferring resistance to penicillin (carried within the transposon Tn1 which is present in RK2), tetracycline, and kanamycin, respectively, about one third of the genome is composed of blocks of genes involved in conjugal DNA transfer (*tra1*, *tra2*, and *tra3*; Ref. 4, 5, 6; a DNA primase, *pri*, Ref. 7; and the transfer origin *oriT* Ref. 8, 9). The rest of the genome contains the essential replication functions as well as *kil* and *kor* genes whose possible significance in plasmid maintenance is discussed briefly at the end of this paper.

For replication RK2 encodes $oriV_{RK2}$ (12.3 to 13.0 kb) (10-13), the origin from which vegetative replication proceeds unidirectionally anticlockwise (14), and *trfA* (16.0 to 17.3 kb) which encodes a polypeptide(s) essential for $oriV_{RK2}$ function (10,11,15). Replicons whose essential components consist of just these 2 RK2-derived loci

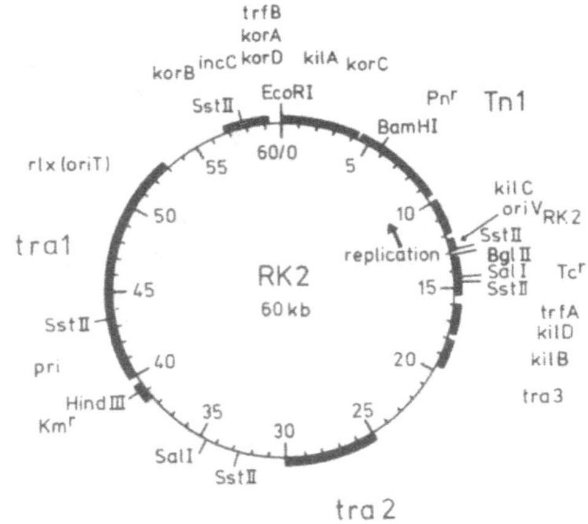

Fig. 1. Physical and genetic map of RK2. The loci shown are defined in the Introduction. The genome is calibrated in kb coordinates. The location of replication genes have been revised to the new 60 kb map by using *EcoRI* and *SalI* sites as reference points.

have at least part of the host range of parental RK2 (E. coli, Pseudomonas aeruginosa, P. putida, Azotobacter vinelandii, Acinetobacter calcoaceticus, Rhizobium meliloti, Agrobacterium tumefaciens, Rhodopseudomonas sphaeroides, Caulobacter crescentus; Ref. 16, 16a; this laboratory, unpub. data). Although only $oriV_{RK2}$ and trfA are essential for replication initial attempts to obtain small derivatives of RK2 by deletion analysis using partial digestion with the restriction endonuclease HaeII resulted in miniplasmids with DNA from 3 regions of the parental genome: 12.3 to 13.0 kb ($oriV_{RK2}$), 16.0 to 18.7 kb (trfA), and 57.6 to 59.6 kb (11). This latter region was designated the trfB region because it also appeared to provide a trans-acting function involved in plasmid maintenance (15). However, it now seems clear that at least part of the reason for the initial inability to delete this trfB region is that kilD lies adjacent to trfA (17,18), the 2 not being separable by HaeII digestion, so that kilD was present with trfA rendering it essential that korD (apparently the same locus as trfB and korA) be present in these minireplicons unless a mutation inactivates kilD. Since the kilD locus appears only to interfere with maintenance of the plasmid carrying it, it may be different from the other RK2 kil loci previously described (10,19) which are host lethal in the absence of their equivalent kor (or kil-override) functions. However, this apparent difference may result from the way the locus was mapped.

In addition to the trfB/korA/korD locus, the trfB region present in these mini RK2 replicons also encodes a distinct locus, designated incC by us, which has been reported (19; K. Ellis and P. Barth, pers. comm.) to express in trans incompatibility towards RK2 and certain large derivatives of RK2 but not towards miniplasmids like those described above.

In this chapter, we describe studies carried out in E. coli which have mapped the genetic loci known to lie in or adjacent to the trfA and trfB regions of RK2. In addition, we have investigated the circuits involved in regulated expression of these genes and we show that regulation of trfA gene expression appears to be involved in plasmid copy number control.

RESULTS

Analysis of the trfA Region

Analysis of the trfA region has been facilitated by the use of a plasmid, pCT87 (15), which carries a mutant trfA region containing a point mutation rendering kilD inactive and which was obtained after deletion of the trfB (korD) region. The nature of this mutation is discussed below, but suffice it to say at this stage that the overall gene organization of this region and the nature of gene products is not altered by this single base change. Analysis of the *trfA* region was carried out by transposon mutagenesis of pCT87 using *Tn1723*, a Km^r insertion derivative of *Tn1722*, the minor transposon of *Tn1721* (20). *Tn1723* has the useful property of having *EcoRI* sites 15 bp from each end of the transposon within its inverted repeats. This allows accurate mapping of the transposon's point of insertion as well as deletion of all but 30 bp of the whole transposon. Since *Tn1723* appears to have a polar effect on transcriptional units into which it inserts, it can inactivate genes either by disrupting their coding sequences or simply by reducing their transcription. In the latter case, however, reactivation would be expected on deletion of the internal transposon *EcoRI* fragment while the remaining 35 bp insertion (30 bp of *Tn1723* and a 5 bp of direct duplication of the target site) would be expected to give a frame shift in any coding sequence and so still to inactivate the coding sequence. Such analysis allowed us to define the limits of the region required for *trfA* activity and established that there is a substantial segment upstream of the *trfA* coding sequence where polar effects can abolish *trfA* expression (21). Using *galK* promoter probe vectors (pKO1 and derivatives, Ref. 22) to determine the direction of transcription at each point of the transposon, insertion indicated that there is only anticlockwise transcription in the region required for *trfA* activity (21) (Fig. 2). The promoter from which this transcription originates has been more accurately mapped to coordinate 17.65 kb approximately, first, by RNA polymerase binding studies using retention of RNA polymerase-DNA fragment complexes on nitro-cellulose filters and DNase I protection studies (23), and second, by reverse transcriptase mapping of the 5'end of *trfA* mRNA produced in vivo (M. Pinkney, unpub. data). This mRNA has the same 5' ends in *E. coli*, *Ps. aeruginosa* and *Ps. putida*, indicating that RNA polymerases from all 3 species can recognize a common DNA sequence which is interesting, since, apparently, standard *Pseudomonas* spp. promoters are not efficiently used by *E. coli* RNA polymerase.

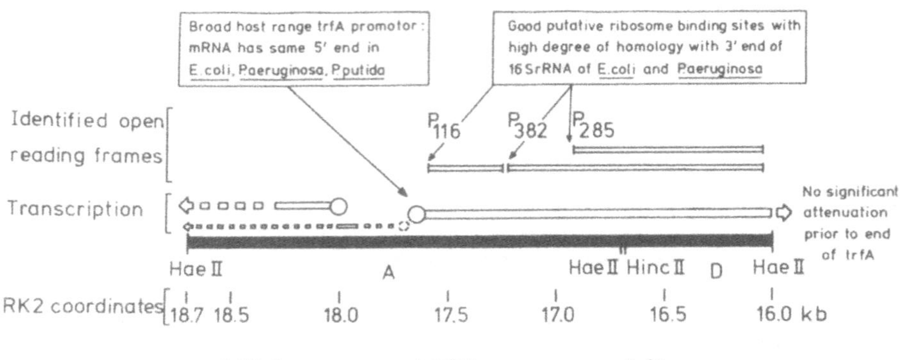

Fig. 2. Organization of the *trfA* region. See text for the experimental basis of these conclusions.

There is also clockwise transcription in the general trfA region as shown in Fig. 2. While weak clockwise transcription may originate from close to the trfA promoter, a second stronger transcription originates between coordinates 18.0 and 18.3 kb. As discussed below the promoter at this position may represent the transcriptional signal for the kilB gene.

Analysis of the gene products of the region required for trfA expression has been achieved by a combination of DNA sequence analysis (24) and identification of polypeptide products in maxicells (21). This region produces polypeptides of approximately 13 kd, 43 kd, and 32 kd, and since both the larger products are still produced by certain deletion derivatives which do not retain enough DNA to code for both separately, it seemed likely that their coding regions overlapped in some way. Analysis of polypeptide truncation products in maxicells containing internally deleted transposon insertion derivatives of pCT87 indicated that the 13 kd product must be the product of the region upstream from the trfA coding region itself in which transposon insertions have a polar effect on trfA activity. These analyses showed also that the coding region for the 43 dk polypeptide starts before that for the 32 kd polypeptide, and that a series of insertions which cause truncation of both the 32 kd and 43 kd polypeptides give products that consistently differ in size by 10-11 kd, suggesting that both polypeptides are translated from alternative translational starts within a single open reading frame. Only the 32 kd polypeptide is essential for trfA activity in *E. coli* but the 43 kd polypeptide may also have trfA activity.

This analysis of the organization of the trfA region is confirmed by DNA sequence analysis which reveals only one open reading frame (of 382 codons) large enough to code for either of the 32 or 43 kd polypeptides. This open reading frame is preceded by a good putative ribosome binding site and contains a second ribosome binding site prior to an internal ATG which would give a second translational product of 285 amino acids. Both the 382 and 285 amino acid polypeptides would be basic, consistent with a role as DNA binding proteins, and contain a sequence of amino acids which may be homologous to the family of α helix-turn-α helix DNA binding structures present in many dsDNA binding proteins (25).

Preceding the trfA open reading frame is another open reading frame of 116 codons whose location, with respect to transposon insertions, is consistent with its coding for the 13 kd polypeptide product observed in maxicells. For reasons discussed below we think that this may be the kilD gene product, so that, as summarized in Fig. 2 a polycistronic mRNA may be translated to give both a kil gene product and replication gene products. Interestingly the 13 kd polypeptide (P_{116}) is produced in much larger absolute amounts (rather than just by mole to mole ratio) than either the 43 kd (P_{382}) or 32 kd (P_{285}) polypeptides as estimated by both the amount of coomassie blue stained material and the ^{14}C-amino acid label incorporation. Since quantitative estimates for the transcription indicate no significant attenuation prior to the end of the trfA gene (26), this must be due to either differential stability of mRNA segments, differential rates of translation initiation, and/or elongation or protein stability. The Shine/Dalgarno sequence is 9 base before the AUG for P_{116} but only 4 and 3 bases (far from ideal) for P_{382} and P_{285}, respectively suggesting that translational initiation efficiency may be at least one of the factors involved in this difference. The degree of efficiency with which P_{116} is expressed suggests that it may provide a useful set of signals for directing gene expression in a wide range of gram-negative bacterial species.

An additional locus previously proposed to lie within the trfA region is tra3, defined by a Tn7 insertion (in pRP101) which abolishes RP4 conjugal transfer and surface exclusion (6). While it has been demonstrated that tra3 is different from kilB (26a) from published data it is still possible that tra3 lies in the 16.0 to 18.7 kb region of RK2. However, accurate restriction mapping of pRP101 and subcloned segments in this laboratory indicates that the transposon insertion lies clockwise from the end of kilB and, therefore, completely outside the trfA region.

Analysis of the trfB Region

Analysis of the trfB region of RK2 and the adjacent segment containing korB has been carried out by a combination of insertion and deletion analysis, by the fusion of various segments of this region to galK promoter-probe plasmids and by analysis of polypeptide products in maxicells (17,27). The korA/korD/trfB locus within this region has been sequenced (28) and we have almost completed the DNA sequence of the rest of the 57.6 to 60.0 kb region. We have identified a single transcriptional unit running anticlockwise from a promoter which overlaps the HincII site at 59.3 kb (23) and continuing across all 3 known loci in this region trfB/korA/korD, incC, and korB (27). (The incC locus expresses incompatibility towards parental RK2-like plasmids (Ref. 27,29; K. Ellis and Barth, unpub. data; but its mode of action is unknown). All 3 loci are encoded clockwise of coordinate 56.8 kb. A second weaker transcriptional unit originating at an undefined point somewhere near coordinate 58.6 kb appears to transcribe across korB (27), Fig. 3.

The lengths of the polypeptides associated with incC activity (P_{38kD}) and korB activity (P_{49kD}) are sufficient to fully occupy the space between the promoter at 39.3 kb and the end of the korB gene at 56.8 kb, Fig. 3. Thus, if the trfB/korA/korD locus codes for a polypeptide, then it must either be a (processed?) subsegment of the incC-associated polypeptide or be translated from an overlapping reading frame in different phase from the incC open reading frame. The DNA sequence of the korA/trfB/korD region has

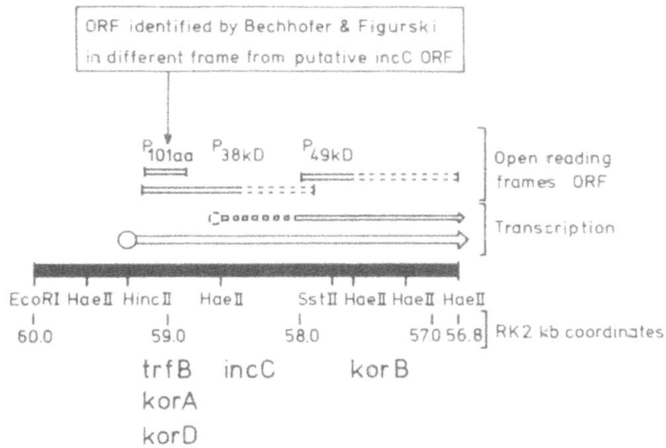

Fig. 3. Organization of the trfB, korB region. See text for the experimental basis of these conclusions.

been determined by Bechhofer and Figurski (28) and reveals both a 101 codon reading frame within the region defined as containing this locus and a second, out of phase reading frame starting just before this smaller open reading frame and continuing without termination across the end of the published sequence. We do indeed observe an approximately 14 kd polypeptide product which we proposed to be associated with trfB/korA/korD activity (27). Whether or not this corresponds to the product of the small reading frame in this region will only be resolved when we have point mutations which affect the product of only one of the two open reading frames. In view of the organization of transcription and translation signals in this region it seems highly unlikely that the trfB/korA/korD product could be a trans-acting RNA. The organization of the trfB region is summarized in Fig. 3.

Regulation of Genes in the trfA and the trfB Regions

In order to determine whether or not transcription in the trfA region is regulated by any of the products of the trfA or trfB regions various pKO plasmid derivatives with the trfA promoter transcribing across galK were used alone or in combination with compatible plasmids carrying various segments of these 2 RK2 regions (26). In all these experiments the level of pKO-derivative plasmid DNA was screened by rapid DNA isolation and gel electrophoresis and it was determined that the observed effects could not be explained simply by variations in plasmid copy number. Quantitative analysis of galactokinase production by fusions in the trfA transcript is such that the translation would produce: (1) only galactokinase; (2) both galactokinase and the 13 kd polypeptide; (3) galactokinase,

13 kd polypeptide and both 32 kd and 43 kd trfA polypeptides, suggested that the polypeptide products of the trfA region did not autoregulate their own expression. This conclusion was strengthened with other experiments where the products of the trfA region were provided in trans (26).

When trfB/korA/korD was placed in trans an approximately 10-fold decrease in galactokinase production and hence trfA transcription was observed consistently at every point assayed within the trfA operon. However, no effect on the divergent clockwise transcription in this region (Fig. 2) was observed. This result suggested that kor genes may act by regulating transcription of kil genes. Since of the 2 transcriptional units identified in this region only the trfA operon is regulated by the korD (korA/trfB) locus on its own, we concluded that kilD is probably part of this operon and may be associated with the 13 kd polypeptide product. Consistent with transcription from this promoter being responsible for kilD expression we have found that the only difference in DNA sequence between the wild-type trfA region (cloned always along with the trfB/korA/korD locus) and the mutant trfA region in pCT87 (which was cloned in the absence of the trfB region and which is effectively $kilD^-$) is a point mutation in the trfA promoter which reduces its homology with the E. coli concensus promoter sequence and which appears to reduce its strength in transcriptional initiation.

Since the kilB locus could at least partly overlap with the trfA region (17) it seemed possible that the clockwise transcription in this region might be responsible for the expression of kilB. The effect of plasmids carrying trfB, incC and korB in trans on galactokinase production by the trfA region transcriptional fusions was therefore determined and it was found that indeed the major clockwise transcription was reduced 10- to 15-fold when korB was present. Surprisingly, however, transcription of the trfA operon itself was also further reduced when korB was present, the combination of trfB and korB reducing transcription 100-fold (26). Since the clockwise transcription is negatively regulated when korB is present, as we proposed, we have tentatively assigned the major promoter in this region to an operon involved in kilB expression. Thus we find that the kilB and trfA promoters are both qualitatively and quantitatively different in their patterns of regulation of gene expression.

While it is possible that regulation of transcription might occur by means of premature termination of transcription it seems more likely that regulation occurs by interference with initiation of transcription, most likely by repressor protein binding to operator sites. The DNA sequence of the trfA promoter (identified both by RNA polymerase binding studies and reverse transcriptase mapping of mRNA 5' end) was compared to that of the putative kilB promoter (identified as the only region of good homology to the E. coli concensus promoter sequence between 2 Tn1723 insertion sites between which the level of clockwise transcription increases significantly in a restriction fragment known to bind RNA polymerase) and also the trfB promoter (identified as a region of strong homology to the E. coli concensus promoter sequence lying across the trfB region HincII site which was predicted to lie within the promoter from studies with galK fusions and deletion analysis (27) and confirmed by protection from HincII restriction endonuclease digestion by bound RNA polymerase) (23). These promoter sequences are aligned in Fig. 4 with respect to putative -10 and -35 sequences. A feature of all 3 promoters is the presence of inverted repeat sequences which suggest the presence of protein binding sites. Both the trfA and trfB promoters are identical in this respect and have 2 related but different inverted repeats, one overlapping the -10 sequence, the other upstream from but close to the -35 sequence. The kilB promoter has an identical copy of this latter repeat but shifted to overlap the -10 sequence, Fig. 4. The inverted repeat common to all three promoters we propose as being the site which provides sensitivity to repression by korB while the inverted repeat found only in the trfA and trfB promoters we propose to be associated with trfB/korA/ korD sensitivity. While we do not yet have a good data for regulation of transcription from the trfB promoter, the striking similarities between the trfA and trfB promoters suggest that their regulatory properties are probably similar. So we propose that transcription from the trfB promoter is probably doubly autoregulated by the products of both the trfB/korA/korD locus and the korB locus. These regulatory circuits are summarized in Fig. 5.

Control of RK2 Plasmid Copy Number

Of major interest is the means by which RK2 copy number is regulated. Plasmid copy number is regulated by the frequency of initiation of plasmid replication and, in general, elements that

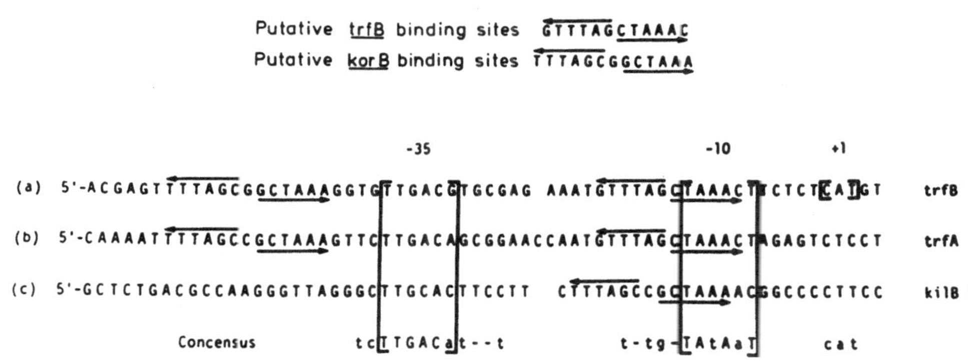

Fig. 4. Sequence of trfA, trfB and kilB promoters showing shared palindromic sequences which may be repressor binding sites.

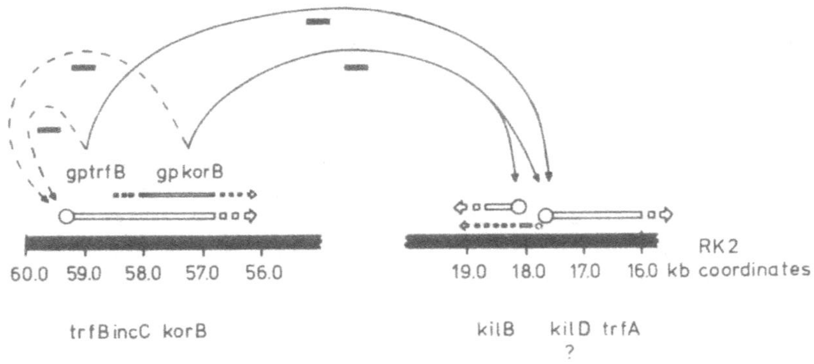

Fig. 5. Summary of established and proposed regulatory circuits in the trfA and trfB regions.

control plasmid copy number have been identified by their ability to interfere with plasmid replication in trans (incompatibility), and by the fact that mutations that alter them in cis may alter plasmid copy number. Since the region of the RK2 origin of vegetative replication contained the first major incompatibility determinant(s) located (30,11) initial efforts to understand copy number control concentrated on this region. The region containing the vegetative replication origin has been sequenced (13,31) and while the elements in and around the replication origin are currently under investigation the significance (and in some cases the reality) of elements inferred from the DNA sequence will become clear only after further work. A summary of this region is shown in Fig. 6. Obvious features in this region are: (1) a series of direct repeats ($\mathrm{TGACA}^{\mathrm{CT}}_{\mathrm{GA}}\mathrm{TGAGGGGCN}^{\mathrm{C}}_{\mathrm{G}}$) of about 17 conserved base pairs which are arranged as a single copy, a group of three and a group of five - in the group of five the repeat unit is 22 or 23 bp; (2) an A+T rich region of about 60 bp which may represent the initiation region and whose function may be enhanced by the adjacent 65 bp G+C rich region; (3) putative transcription and translation signals. So far the information essential to allow initiation to occur has been mapped to within a 393 bp HpaII fragment (12) which contains the group of 5 direct repeats but still lacks much of the potentially significant information in this region. That this additional information is involved in determining the frequency at which replication occurs from oriV$_{RK2}$ is indicated by studies with mini RK2 plasmids which are isogenic except for deletions adjacent to, but outside, the minimal oriV$_{RK2}$ region (32). The deletions studied differ from the wild-type region either by a 57 bp deletion (defining copA and incA) removing 2 of the 3 direct repeats or by a deletion (defining copB and incB) up to the end of the HaeII site at the end of the HaeII oriV$_{RK2}$ fragment close to the BglII fragment and thus removing the single copy of the direct repeat and the beginning of the puta-

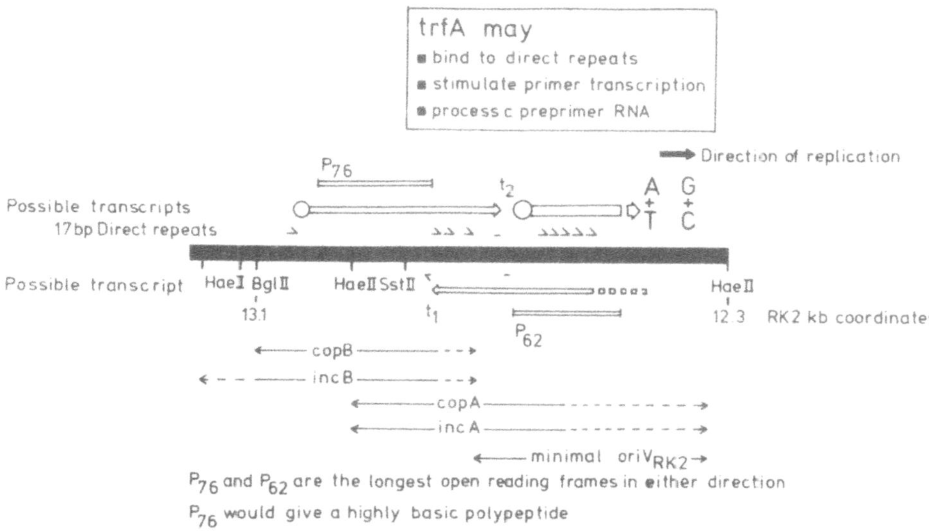

Fig. 6. Organization of $\underline{oriV}_{RK2}$ region.

tive transcript and open reading frame shown as P_{76}. The minireplicons studied contained the loci trfB, incC, kilD, and trfA, but not kilB or korB. The structure of these plasmids (pCT539, pCT460, pCT543 and pCT461) and their copy numbers, estimated by comparison with pBR322 as an internal control, are shown in Fig. 7. The re-

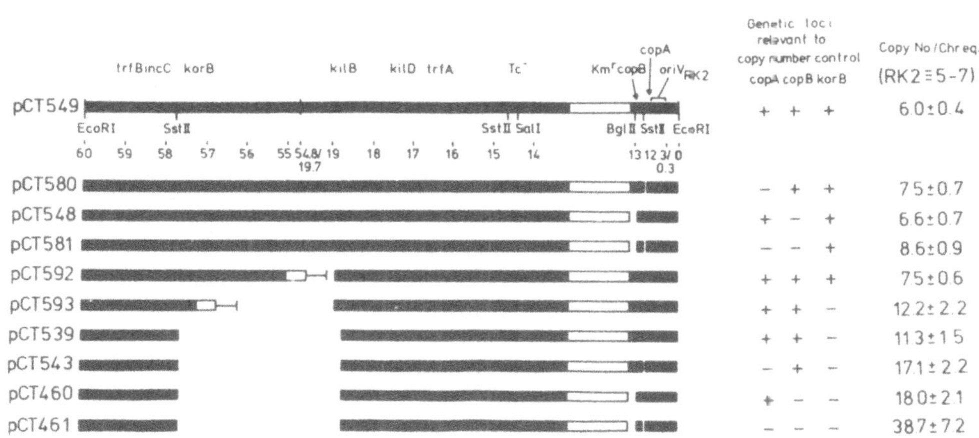

Fig. 7. Copy number control in RK2. Structure and copy number of plasmids used to investigate the importance of copA, copB and korB. RK2 DNA is shown as a thick solid line whose RK2 kb coordinates are shown at the top for pCT549.

sults showed that with a wild-type $\underline{oriV}_{RK2}$ region these minireplicons have a copy number of about 11 while either one of the 2 deletions adjacent to the minimal $\underline{oriV}_{RK2}$ raises copy number to approximately 17 to 18 copies per chromosome equivalent. With both deletions present together plasmid copy number is approximately 35 to 40 copies per chromosome equivalent. Thus the information adjacent to the minimal $\underline{oriV}_{RK2}$ region can play a role in copy number control. While the loci involved do also express incompatibility, we do not know at present what the nature of the control mechanism is.

One additional fact that did emerge from the above experiments was that the minireplicons being studied had a copy number higher than that of RK2 (estimated as 5-7 copies per chromosome equivalent, Ref. 33). It appeared, therefore, that these mini RK2 plasmids may lack some regulatory element that in the parent plasmid reduces the frequency of initiation of replication even further. From the studies on regulation of trfA operon transcription described above, it appeared possible that korB may represent such an element which is absent from these minireplicons. Consistent with this possibility was the fact that korB in trans was found to express incompatibility towards these minireplicons (34). When korB was reintroduced into these minireplicons the copy number was found to drop to a level within or close to the range found for RK2 itself (see pCT549 and pCT592, Fig. 7) (34) while similar manipulations on the plasmid without introducing a functional korB gene (pCT593, Fig. 7) had no significant effect on plasmid copy number. It thus appears that in these minireplicons korB may be further reducing trfA gene expression to a level where trfA gene product concentration has become limiting for replication.

To determine what effect the copA and copB deletions described above have on copy number when korB is present in minireplicons these mutant $\underline{oriV}_{RK2}$ regions were introduced into plasmids which are otherwise isogenic with pCT549. Copy number determinations showed that neither the copA deletion, the copB deletion, nor both of these together have a large effect on plasmid copy number when trfA gene expression is apparently reduced to rate limiting levels (see plasmids pCT580, pCT548 adn pCT581, Fig. 7). It appears, therefore, that not only may limitation of trfA gene expression play a role in copy number control but it may in fact be the overriding copy number control mechanism. Models for situations where copy number is negatively controlled require that the concentration of an inhibitor should be proportional to the copy number so that under given conditions plasmid replication without cell volume increase should lead to increased inhibitor concentration and therefore a reduced chance of initiation of replication occurring. If korB expression depended solely on the autogenously regulated circuit proposed for transcription from the trfB promoter then its concentration would not be expected to vary in this way. It is worthwhile recalling, therefore, that there seems to be a secondary promoter in the trfB region which should lack the regulatory properties of the trfB promoter and which should give expression of *korB* in proportion to copy number (Fig. 3). Clearly a knowledge of the exact way in which these regulatory circuits work awaits much more detailed analysis.

DISCUSSION

The results described in the previous sections provide a reasonably comprehensive map of the loci so far identified in the small plasmids derived from RK2. The finding that *kor* genes negatively regulate the expression of *kil* genes provides an understanding of how the lethal effect of these *kil* genes may be overridden. Of even greater interest is the organization of the positively acting *trfA* gene in an operon with *kilD* and its coregulation with *kilD* and *kilB* by the products of the *trfB/korA/korD* locus and the *korB* locus and the apparent role of this negative regulatory circuit in copy number control. The detailed mechanism of these control circuits still remains to be elucidated however. For example, it remains unclear whether the *trfB* and *korB* products act independently or in some concerted way. In addition we still do not understand the role of *incC* in these circuits (see Fig. 3 and 4). Recently we have found that the miniplasmids containing *korB* are sensitive to *incC* in trans suggesting that limiting levels of *trfA* gene expression make replicons sensitive to *incC* incompatibility. Thus, *incC* may play an additional role in this replication and maintenance regulatory circuit.

Another intriguing question is what role do the *kil* genes of RK2 play? Since they are coregulated with *trfA* it is tempting to suggest that they play a role in the stable maintenance of RK2. The coupled cell division genes of plasmid F (35,36) provide an example of pair of genes, one potentially host lethal and the other a suppressor of this lethality which show close parallels to *kil* and *kor*. It is possible also that while the F and P1 partitioning systems do not seem to consist of any components which are host lethal, RK2 may code for a membrane protein involved in partitioning which might interfere with host membrane functions if overproduced. Some current experiments are designed to test the role of *kil* and *kor* in RK2 maintenance.

It seems likely therefore that the studies described in this paper, rather than solving all the puzzles about RK2, provide a more rational basis for studies of the way in which RK2 maintains itself stably, not only in *E. coli* but also in a wide range of other gram-negative bacterial species.

ACKNOWLEDGMENTS

We thank K. Ellis and P. Barth for communication of results prior to publication; M. Rosenberg for pKO plasmids; R. Schmitt for Tn*1722* derivatives; H. Howell for technical assistance; E. Badger and B.S. Price for help in preparing the manuscript. V. Shingler was supported by a Medical Research Council (MRC) Studentship for training in research methods. M.A. Cross was supported by a Science and Engineering Research Council (SERC) Quota Award; SERC also provided assistance in his attending this Conference. C.A. Smith was supported by MRC Project Grant No. G80/0374/9CB; M. Pinkney is supported by MRC Project Grant No. G83/0983/8CB; both awarded to C.M. Thomas. The Wellcome Trust and the Royal Society provided assistance for C.M. Thomas to attend this Conference.

REFERENCES

1. Thomas, C.M. (1981) Molecular genetics of broad host range plasmid RK2. Plasmid 5:10-19.
2. Guiney, D.G. (1984) In Plasmids in Bacteria, D.R. Helinski et al., eds. Plenum, New York.
3. Lanka, E., R. Lurz, and J.P. Furste (1983) Molecular cloning and mapping of SphI restriction fragments of RP4. Plasmid 10:303-307.
4. Barth, P.T., and N.J. Grinter (1977) Map of plasmid RP4 derived by insertion of transposon C. J. Mol. Biol. 113:455-474.
5. Barth, P.T., N.J. Grinter, and D.E. Bradley (1978) Conjugal transfer system of plasmid RP4: Analysis by transposon 7 insertion. J. Bact. 133:43-52.
6. Barth, P.T. (1979) PR4 and R300B as wide host-range plasmid cloning vehicles. In Plasmids of Medical, Environmental and Commercial Importance, K.N. Timmis and A. Puhler, eds. Elsevier, North Holland, Amsterdam/New York, pp. 399-410.
7. Lanka, E., and P.T. Barth (1981) Plasmid RP4 specifies a deoxyribonucleic acid primase involved in its conjugal transfer and maintenance. J. Bact. 148:769-781.
8. Guiney, D.G., and D.R. Helinski (1979) The DNA-protein relaxation complex of the plasmid RK2: Location of the site specific nick in the region of the proposed origin of transfer. Mol. Gen. Genet. 176:183-189.
9. Guiney, D.G., and E. Yakobsen (1983) Location and nucleotide sequence of the transfer origin of the broad host range plasmid RK2. Proc. Natl. Acad. Sci., USA 80:3595-3598.
10. Figurski, D., and D.R. Helinski (1979) Replication of an origin-containing derivative of plasmid RK2 dependent upon a plasmid function provided in trans. Proc. Natl. Acad. Sci., USA 76:1648-1652.
11. Thomas, C.M. R. Meyer, and D.R. Helinski (1980) Regions of broad host range plasmid RK2 which are essential for replication and maintenance. J. Bact. 141:213-222.
12. Thomas, C.M., D.M. Stalker, and D.R. Helinski (1981) Replication and incompatibility properties of segments of the origin region of replication of broad host range plasmid RK2. Mol. Gen. Genet. 181:1-7.

13. Stalker, D.M., C.M. Thomas, and D.R. Helinski (1981) Nucleotide sequence of the region of the origin of replication of the broad host range plasmid RK2. Mol. Gen. Genet. 181:8-12.
14. Meyer, R., and D.R. Helinski (1977) Unidirectional replication of the P-group plasmid RK2. Biochim. Biophys. Acta 478:109-113.
15. Thomas, C.M. (1981) Complementation analysis of replication and maintenance functions of broad host range plasmids RK2 and RP1. Plasmid 5:277-291.
16. Schmidhauser, T., M. Filutowicz, and D.R. Helinski (1983) Replication of derivatives of the broad host range plasmid RK2 in two distantly related bacteria. Plasmid 9:325-330.
16a. Schmidhauser, T., and D.R. Helinski (1984) In Plasmids in Bacteria, D.R. Helinski et al., eds. Plenum, New York.
17. Smith, C.A., and C.M. Thomas (1983) Deletion mapping of kil and kor functions in the trfA and trfB regions of broad host range plasmid RK2. Mol. Gen. Genet. 190:245-254.
18. Pohlman, R.F., and D.H. Figurski (1983) Essential genes of plasmid RK2 in Escherichia coli: trfB region controls a kil gene near trfA. J. Bact. 156:584-591.
19. Figurski, D.H., R.F. Pohlman, D.H. Bechhofer, A.S. Prince, and C.A. Kelton (1982) The broad host range plasmid RK2 encodes multiple kil genes potentially lethal to Escherichia coli host cells. Proc. Natl. Acad. Sci., USA 79:1935-1939.
20. Schmitt, R., E. Bernhard, and R. Mattes (1979) Characterization of Tn1721, a new transposon containing tetracycline resistance genes capable of amplification. Mol. Gen. Genet. 172:53-65.
21. Shingler, V., and C.M. Thomas (1984) Analysis of the trfA region of broad host range plasmid RK2 by transposon mutagenesis and identification of polypeptide products. J. Mol. Biol. 175:229-250.
22. McKenney, K., H. Shimatake, D. Court, U. Schmeissher, C. Brady, and M. Rosenberg (1981) A system to study promoter and terminator signals recognized by Escherichia coli RNA polymerase. In Gene Amplification and Analysis. Volume II: Analysis of Nucleic Acids by Enzymatic Methods, J.C. Chirikjian and T.S. Papas, eds. Elsevier/North Holland, Amsterdam, pp. 282-415.
23. Smith, C.A., V. Shingler, and C.M. Thomas (1984) The trfA and trfB promoter regions of broad host range plasmid RK2 share common potential regulatory sequences. Nucl. Acids Res. 12:3619-3630.
24. Smith, C.A., and C.M. Thomas (1984) Nucleotide sequence of the trfA gene of broad host range plasmid RK2. J. Mol. Biol. 175:251-262.
25. Ohlendorf, D.H., W.F. Anderson, and B.W. Matthews (1983) Many gene-regulatory proteins appear to have a similar α-helical fold that binds DNA and evolved from a common precursor. J. Mol. Evol. 19:109-114.
26. Shingler, V., and C.M. Thomas (1984) Transcription in the trfA region of broad host range plasmid RK2 is regulated by trfB and korB. Mol. Gen. Genet. (in press).

26a. Pohlman, R.F., and D.H. Figurski (1983) Conditional lethal mutants of the kilB determinant of broad host range plasmid RK2. Plasmid 10:82-95.

27. Smith, C.A., and C.M. Thomas (1984) Molecular genetic analysis of the trfB and korB region of broad host range plasmid RK2. J. Gen. Microbiol. 130:1651-166.

28. Bechhofer, D.H., and D.H. Figurski (1983) Map location and nucleotide sequence of korA, a key regulatory gene of promiscuous plasmid RK2. Nucl. Acids Res. 11:7452-7469.

29. Meyer, R., and M. Hinds (1982) Multiple mechanisms for expression of incompatibility by broad host range plasmid RK2. J. Bact. 152:1078-1090.

30. Meyer, R. (1979) Expression of incompatibility by derivatives of the broad host range IncP-1 plasmid RK2. Mol. Gen. Genet. 177:155-161.

31. Waters, S.H., P. Rogowsky, J. Grinsted, J. Altenbuchner and R. Schmitt (1983) The tetracycline resistance determinants of RP1 and Tn1721: nucleotide sequence analysis. Nucl. Acids Res. 11:6089-6015.

32. Thomas, C.M., M.A. Cross, A.A.K. Hussain and C.A. Smith (1984) Analysis of copy number control elements in the region of the vegetative replication origin of the broad host range plasmid RK2. EMBO 3:75-63.

33. Figurski, D., R. Meyer and D.R. Helinski (1979) Suppression of ColE1 replication properties by the IncP-1 plasmid RK2 in hybrid plasmids constructed in vitro. J. Mol. Biol. 133:295-318.

34. Thomas, C.M. and A.A.K. Hussain (1984) The korB gene of broad host range plasmid RK2 is a major copy number control element which may act together with trfB by limiting trfA expression. EMBO 3: (in press).

35. Ogura, T., and S. Hiraga (1983) Mini-F plasmid genes that couple host cell division to plasmid proliferation. Proc. Natl. Acad. Sci., USA 80:4784-4788.

36. Karoui, H., F. Bex, P. Dréze and M. Couturier (1983) Ham22, a mini-F mutation which is lethal to host cell and promotes recA-dependent induction of lambdoid prophage. EMBO 2:1863-1868.

CONTROL OF CHROMOSOME REPLICATION IN BACTERIA

R.H. Pritchard

Department of Genetics
The University of Leicester
Leicester, England

INTRODUCTION

The exciting progress that has recently been made in unravelling the replication control mechanisms that determine the copy numbers of several plasmids has been reviewed and updated by several contributors to this volume. In contrast, the control of chromosome replication in bacteria remains largely a closed book.

As with plasmids, three principal approaches have been used to investigate the control of initiation of chromosome replication in *E. coli*.

Isolation of Minireplicons

It at one time seemed likely that the isolation of oriC plasmids (1) would open up chromosome initiation control to investigation, but these plasmids have not been as useful as they had promised. One of their more puzzling properties is their failure to express strong incompatibility against the chromosome itself. Thus (2) a stable pBR322 derivative with a copy number of about 30 containing a functional 1.4-kb oriC plasmid as an insert did not significantly alter host-cell size or growth rate. This means that this segment of chromosomal DNA does not code for a trans-acting rate determining effector of chromosome initiation, and makes it unlikely that it contains a target capable of titrating out such an effector coded for by the host.

Interpretation of similar data obtained using larger oriC plasmids (3) is less straightforward because, although the average copy number is 5-10, a significant proportion of the cell population is plasmid-free. Nevertheless, the survival and growth of the plasmid-containing component of the population surely leads to the same conclusion.

The fact that oriC plasmids do not code for a trans-acting inhibitor (or controlling initiator) of replication is not surprising. There is no a priori reason why the coding sequence should be in close proximity to the initiation site. On the contrary, there are good reasons (4,5) for supposing that such a gene should be located some distance from the initiation site in order to permit synchronous replication of origins in cells with more than one. The apparent absence of a target close to the site of initiation is less easy to understand and suggests at least three possibilities:

(i) The regulatory system which controls the timing of initiation is fundamentally different from corresponding systems controling plasmid replication. The difficulty is to imagine a control system which does not involve a trans-acting effector and a target site for that effector near the site of initiation, especially since initiation of replication in cells with more than one origin is essentially synchronous.

(ii) A target site for the effector is located on oriC plasmids but the titrating activity of these sites is lethal to the host cell. On this hypothesis oriC plasmids would all be target-site mutants resulting from the demands of the selection technique used to isolate them.

(iii) A third and interesting possibility, which came out of discussions at this conference with G. Churchward, is that after initiation has taken place replication forks stall at a site some distance away and that the primary control of initiation is at this site. OriC plasmids would lack the normal trans-acting initiation control functions and target sites but would, nevertheless, behave as replicons.

Isolation of Initiation-Defective Mutants

So far, none of the genes which mutate to inhibit initiation appear to be the primary control genes which determine the timing of initiation. Thus, in the case of dnaA, the most fully studied gene, defective mutants are recessive, indicating a positive function for the gene product, which is consistent with in vitro studies (6). In addition, the concentration of the gene product appears not to be rate-limiting for replication in nonmutant strains (7). These properties are not consistent with the gene product being the primary control gene.

There is also evidence that the primary control functions which determine the timing of initiation are intact in dnaA strains.

Thus, at intermediate temperatures, initiation of replication is delayed so that the DNA:mass ratio in the culture is maintained at a lower level than in wild-type strains (8,9). Addition of chloramphenicol to such a culture results in an immediate release from the initiation delay but the origin:mass ratio only rises approximately to the wild-type level. This result suggests that the normal initiation control which sets the DNA:mass ratio is still fully functional in *dnaA* strains.

Incidentally, the release of the *dnaA* block does not occur if protein synthesis is inhibited by amino-acid starvation instead of chloramphenicol, indicating that it is not inhibition of protein synthesis per se which overcomes this block (8,10). However, if amino-acid starvation is performed in a relaxed (*rel*) background the result mimics that obtained with chloramphenicol. This striking result was interpreted in the following way: in *dnaA* mutants there is a failure to synthesize a stable RNA which is necessary for initiation but is not normally rate-limiting. Addition of chloramphenicol, and amino-acid starvation of a *rel* strain, stimulates stable RNA synthesis (11). At intermediate temperatures, this allows a sufficient rate of synthesis of the stable RNA needed for initiation to permit a normal rate of initiation despite the deficiency of *dnaA* gene product.

Isolation of *cop* Mutants

This has been less rewarding in the case of chromosome replication than it has been in studies on copy-number control in plasmids.

A strain of *E. coli* B/r with a DNA:mass ratio which is half that of typical *E. coli* strains has been found (12). The difference was attributed to a mutation close to, and possibly in the *dnaA* gene.

There have recently been 2 reports of the isolation of several mutants with a 2- to 4-fold increase in DNA:mass ratio (13, 14). All these mutants were found to carry mutations in either *rpoB* or *rpoC* which specify subunits of RNA polymerase. E. Gutzman and A. Jimenez-Sanchez in my laboratory have also isolated a number of thermosensitive *cop* mutants with elevated DNA:mass ratios at high temperatures and these, likewise, appear to carry mutations in *rpoC*.

These are curious findings. Is it conceivable that such a precisely timed discontinuity in the cell cycle as initiation would be controlled by the level of RNA polymerase or a polymerase cofactor? Do these mutations alter RNA polymerase specificity or is the DNA:mass ratio perturbed simply by an alteration in RNA polymerase activity?

To investigate these questions we asked if rifampicin could mimic the effect of these mutations, which have a slightly reduced

RNA polymerase activity. The result is shown in Fig. 1. This shows that sufficient rifampicin to cause a slight reduction in growth rate has an immediate stimulatory effect on DNA synthesis leading to a 2-fold enhancement in the DNA:mass ratio which was maintained over several mass doublings.

We conclude that mutations in RNA polymerase which increase the DNA:mass ratio do so simply as a result of a reduction in the activity of RNA polymerase. Our interpretation of this rather general effect is as follows.

Replication in some plasmids (such as, ColE1, see Ref. 5) is controlled by a negative-feedback loop in which the rate of initiation of each plasmid is inversely proportional to the concentration of an inhibitor, which is a small unstable RNA. Since this is synthesized constitutively there will be a simple proportionality between the concentration of the plasmid and the concentration of the inhibitor and an inverse proportionality between the concentration of the plasmid and its rate of replication.

If chromosome replication were mediated by a similar negative loop, with constitutive synthesis of inhibitor, it would be necessary to assume a much stronger dependence of the probability of initiation on inhibitor concentration than is the case with plasmids like ColE1 in order to account for the very strong dependence of the timing of initiation on cell volume.

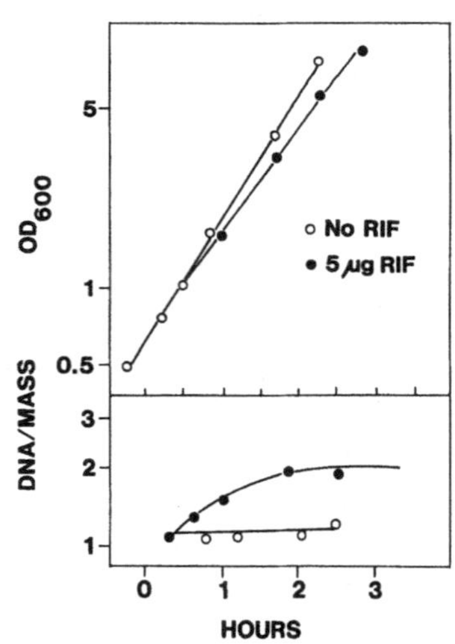

Fig. 1. Mass (absorbance) increase and relative DNA:mass ratio in E. coli LE234 (metB, argH, ilv, rpsl, tna, xyl, mtl, supE) incubated at 42°C in the presence and absence of rifampicin. say 10%.

Consider now the effect of an inhibition of RNA polymerase sufficient to reduce the overall rate of RNA synthesis and growth by say 10%. There would also be a 10% reduction in the rate of inhibitor synthesis. Because of the stronger dependence of initiation on inhibitor concentration this would have a disproportionate effect on the rate of initiation.

If the inhibitor of chromosome replication were a small unstable RNA this might also account for the failure, so far, to isolate thermosensitive copy effect mutants of it.

REFERENCES

1. Yasuda, S., and Y. Hirota (1977) Cloning and mapping of the replication origin of Escherichia coli. Proc. Natl. Acad. Sci., USA 74:5458-5462.
2. Yamaguchi, K., M. Yamaguchi, and J. Tomizawa (1982) Incompatibility of plasmids containing the replication origin of the Escherichia coli chromosome. Proc. Natl. Acad. Sci., USA 79:5347-5351.
3. von Meyenburg, K., F.G. Hansen, E. Riise, H.E.N. Bergmans, M. Meijer, and W. Messer (1978) Origin of replication, oriC, of the Escherichia coli chromosome: Genetic mapping and minichromosome replication. Cold Spring Harbor Symp. Quant. Biol. 43:121-128.
4. Pritchard, R.H. (1978) Control of DNA replication in bacteria. In DNA Synthesis: Present and Future, I. Molineux and M. Kohijama, eds. Plenum, New York.
5. Pritchard, R.H. (1984) Control of DNA replication in bacteria. In The Microbial Cell Cycle, P. Nurse, and E. Streiblova, eds. CRC Press.
6. Fuller, R.S., and A. Kornberg (1983) Purified dnaA protein in initiation of replication at the Escherichia coli chromosomal origin of replication. Proc. Natl. Acad. Sci., USA. 80:5817-5821.
7. Churchward, G., P. Holmans, and H. Bremer (1983) Increased expression of the dnaA gene has no effect on DNA replication in a $dnaA^{+}$ strain of Escherichia coli. Mol. Gen. Genet. 192:506-508.
8. Orr, E., P.A. Meacock, and R.H. Pritchard (1978) Genetic and physiological properties of an Escherichia coli strain carrying the dnaA gene mutation T46. In DNA Synthesis: Present and Future, I. Molineaux, and M. Kohimaja, eds. Plenum, New York.
9. Hansen, F.G. and K.V. Rasmussen (1977) Regulation of the dnaA product in Escherichia coli. Mol. Gen. Genet. 155:219-225.
10. Lycett, G.W., E. Orr, and R.H. Pritchard (1980) Chloramphenicol releases the block in initiation of chromosome replication in a dnaA strain of Escherichia coli K12. Mil. Gen. Genet. 178:329-336.

11. Shen, V., and H. Bremer (1977) Chloramphenicol induces changes in the synthesis of ribosomal transfer and messenger ribonucleic acids in Escherichia coliB/r. J. Bact. 130:1098-1108.
12. Choung, K.K., E. Estiva, and H. Bremer (1981) Genetic and physiological characterization of a spontaneous mutant of Escherichia coliB/r with aberrant control of deoxyribonucleic acid replication. J. Bact. 145:1239-1246.
13. Rasmussen, K.V., T. Atlung, G. Kersman, G.E. Hansen, and F.G. Hansen (1983) Conditional change of DNA replication control in an RNA polymerase mutant of Escherichia coli. J. Bact. 1564:443-451.
14. Tanaka, M., H. Ohmori, and S. Hiraga (1983) A novel type of E. coli mutants with increased chromosomal copy number. Mol. Gen. Genet. 192:51-60.

STABLE MAINTENANCE OF PLASMID CLO DF13: STRUCTURAL AND FUNCTIONAL RELATIONSHIPS BETWEEN REPLICATION CONTROL, PARTITIONING, AND INCOMPATIBILITY

H. John J. Nijkamp, Bob van Gemen, Marcel J.J. Hakkaart, Arnold J. van Putten, and Eduard Veltkamp

Department of Genetics, Biological Laboratory
Vrije Universiteit
de Boelelaan 1087
1081 HV Amsterdam, The Netherlands

INTRODUCTION

Nonconjugative plasmids are a very attractive model system for the study of the structural and functional aspects of stable maintenance of replicons in a growing bacterial population. Biochemical and molecular genetic studies have resulted in more insight into the genes and sites involved in the processes responsible for plasmid maintenance such as plasmid replication and its control, accurate partitioning of plasmids over daughter cells, and incompatibility between replicons.

In this chapter, the significance of plasmid-determined processes in the stable maintenance of replicons will be exemplified by the nonconjugative bacteriocinogenic plasmid Clo DF13. We will summarize the structural and functional aspects of Clo DF13 replication control, partitioning, and incompatibility, and we will discuss their inter-relationships.

Plasmid Clo DF13 (5.75 MD) originates from *Enterobacter cloacae* DF13, but is also stably maintained in *Escherichia coli*. In the latter host, the plasmid is present to the extent of about 10 copies per cell (1). A genetic and functional map of Clo DF13 (for reviews, see Ref. 2,3) is given in Fig. 1, which includes the relevant genes and sites involved in replication control, partitioning, and incompatibility.

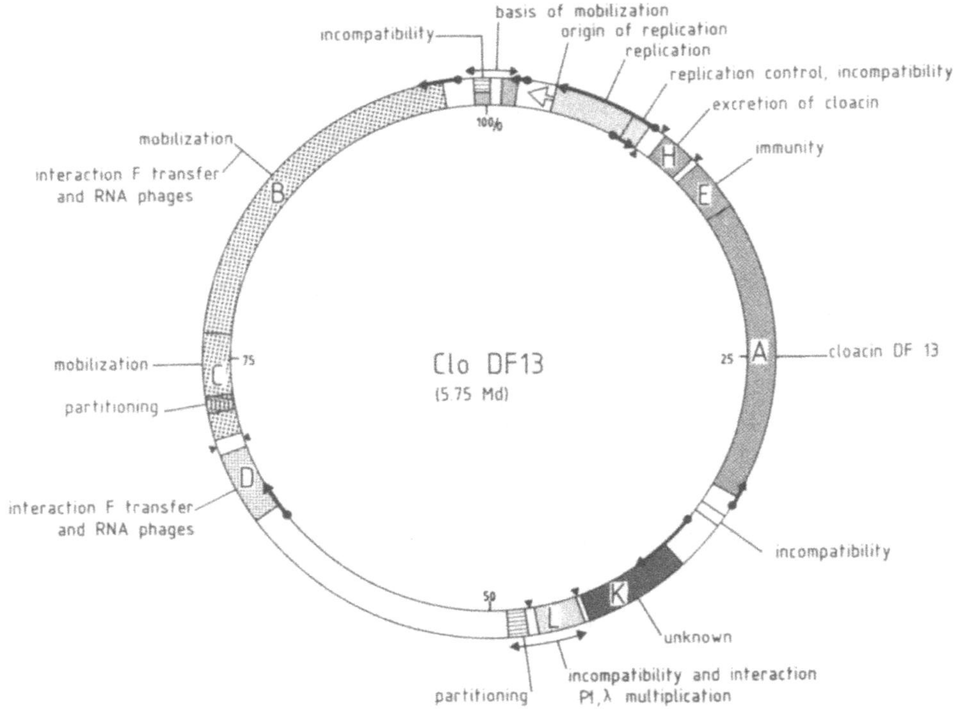

Fig. 1. Genetic and functional map of plasmid Clo DF13. Identified promoters (●) and terminators (▼) as well as the direction of transcription (→) are indicated in the map.

REPLICATION CONTROL

The autonomous replication of nonconjugative plasmids takes place in a controlled manner so that the concentration of a given plasmid is maintained at a constant level in a particular host bacterium grown under defined conditions. Nonconjugative plasmids like Clo DF13 and Col E1 do not code for proteins involved in the replication machinery but they have sequences that meet the requirements for autonomous and controlled replication in dividing cells. They possess determinants that constitute the so-called minimal replicon. The determinants include (i) the origin of replication, (ii) a sequence for the synthesis of RNA involved in priming of DNA synthesis, (iii) overlapping with this latter sequence a gene coding for RNA molecules that control the formation of primer RNA, and (iv) because a minimal replicon is stably maintained, genetic information for accurate partitioning can be expected in that stretch of DNA. In the case of Col E1, the product of the rop gene located outside the minimal replicon negatively affects the synthesis of primer RNA (5). Both replication control and partitioning are presumed to be the basis of phenomena known as incompatibility (see next sections).

Figure 2 summarizes the relevant features of the Clo DF13 minimal replicon located between 1.8% and 11.5% on the plasmid genome (3). Replication of Clo DF13 starts at a unique sequence, designated oriV and proceeds unidirectionally counter clockwise. Two species of RNA are encoded in the region preceding the replication origin. One species, the preprimer RNA (RNA II), is transcribed from a promoter (P2) about 550 basepairs upstream the origin and terminating beyond it. RNase H processes RNA II at the origin to form a primer upon which DNA polymerase I acts to initiate replication (6). The second RNA species, RNA I, is smaller (105 nucleotides) and is transcribed in the opposite direction to RNA II, starting from promotor P1 and terminating at terminator t1 (7). The coding sequences for RNA I and RNA II overlap and are thus complementary. RNA I negatively controls DNA replication by preventing the processing of RNA II by RNase H (8).

Clo DF13 Copy Control Mutants

A number of plasmid copy control (*cop*) mutants have been isolated and characterized (9,10,11,12,13). With respect to Clo DF13, 2 different types of replication control mutants can be distinguished. Mutants of the first type were shown to have a nonconditional

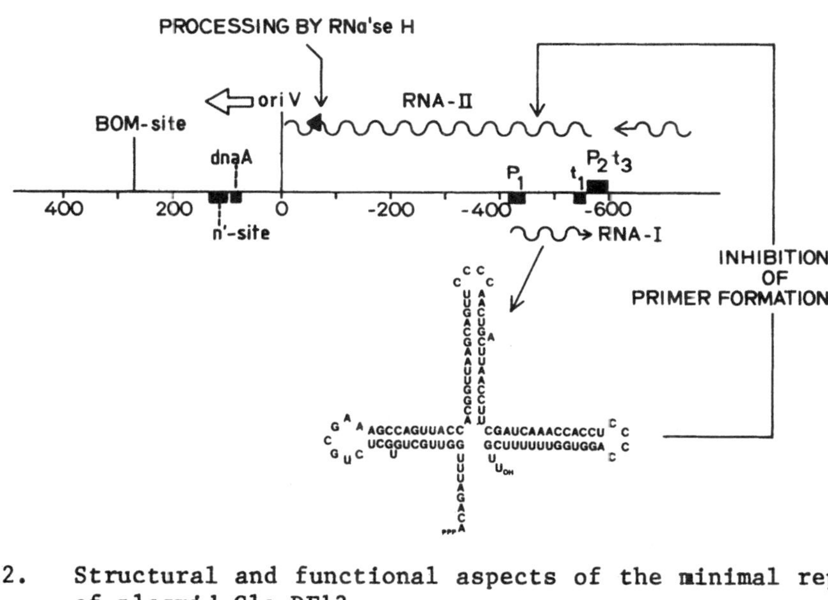

Fig. 2. Structural and functional aspects of the minimal replicon of plasmid Clo DF13.

increase in copy number (9). Sequence analysis revealed that these mutations are located within the region encoding RNA I (11). Based on the location of this type of cop mutations it has been proposed that the target of RNA I is the complementary sequence of RNA II. As shown in Fig. 2, RNA I may adapt a tRNA-like structure with 3 stem-loops. It is assumed that the single-stranded loop structures are involved in the inhibitory activity of RNA I by basepairing with the complementary sequence of RNA II. The second type of replication control mutants is represented by Clo DF13 cop1 (TS), which is a conditional copy mutant (10). At 30°C the copy number of this mutant is, in comparison with that of the wild-type plasmid, only slightly increased; at 42°C the copy number increases drastically. At this latter temperature the mutant plasmid shows a "runaway" behavior. Here, we report recent data on the nature of this mutation, which is located within a region involved in the initiation of RNA II (P2) and the termination (t3) of the cloacin operon transcript (see Fig. 2). From experiments in which we cloned this region into promoter-study respectively terminator-study systems, we concluded that the Clo DF13 cop1 (TS) mutation results in a pleiotropic effect, namely, a nonconditional increase in promoter P2 activity and a temperature-dependent decrease in termination efficiency both resulting in an increased synthesis of preprimer RNA. (A. van Putten et al., manuscript in prep.).

Methylation Regulated Clo DF13 Replication

Recently, we obtained evidence that the absence of Dam or Dcm methylation leads to a loss of the runaway behavior of the cop1 (TS) mutant. In the sequence of terminator region t3, overlapping the preprimer promoter region (P2), dam and dcm methylation sites are present (Fig. 3). One of these sites is altered by the cop1 (TS)

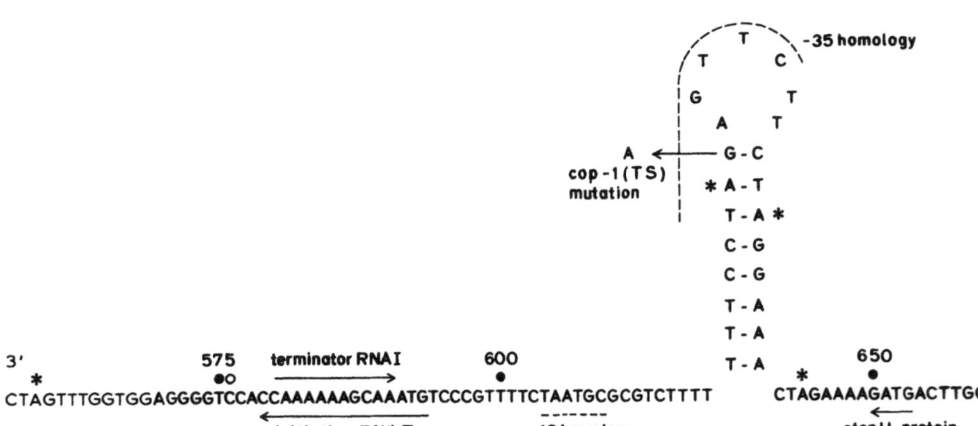

Fig. 3. Nucleotide sequence of the terminator region (t3) of the cloacin operon. Putative dam (*) and dcm (o)methylation sites are indicated.

Tab. 1. Galactokinase (galK) activity of cells harboring the plasmids pKG1800 and pVU402. Cells were grown at 42°C. The galactokinase activity observed in the absence of transcription is approximately 10 U. Pgal: galactose promoter of plasmid pKG1800; Tcop1 (TS): fragment containing the mutant cloacin operon terminator (t3).

Plasmid construction		Strain C600 *dam⁺ dcm⁺*	Strain GM48 *dam⁻ dcm⁻*
pKG1800	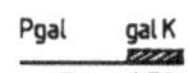	613	505
pVU402	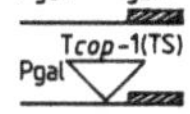	196	31

Cells were grown at 42°C. The galactokinase activity observed in the absence of transcription is approximately 10 U.

Pgal: galactose promoter of plasmid pKG1800;

Tcop1 (TS): fragment containing the mutant cloacin operon terminator (t3).

mutation. In order to study the molecular basis of this phenomenon, we determined the efficiency of termination of the cloacin operon terminator structure in the presence and the absence of methylation. For this purpose, we used plasmid pVU 402 harboring the mutant terminator (t3) in terminator-study plasmid pKG1800. The data obtained are shown in Tab. 1. From these results, we conclude that the absence of Dam or Dcm methylation leads to an increased termination at t3 in the cop1 (TS) mutant, abolishing read-through transcription and resulting in a reduced plasmid copy number at elevated temperatures. These results suggest a regulatory role of methylation in Clo DF13 replication. However, the biological function of this methylation regulation is not yet clear. No function has yet been found for the occurrence of many methylation sites in the origin of replication of *E. coli* (31), nor has a phenotypic effect been observed for the Dam methylation on DNA replication initiation. In the case of Clo DF13, however, we have observed an effect of methylation on the initiation of Clo DF13 replication. In the cop1 (TS) mutant, the increased initiation of replication at 37°–42°C is brought back to the level of wild-type frequency of initiation by preventing read-through transcription in the absence of methylation. A methylation-regulated system might be important in the situation that the plasmid DNA is undermethylated, e.g., in case of hemimethylated DNA, occurring directly after synthesis of the new DNA strand. Whether undermethylation, improving termination of transcription at t3, triggers immediate and proper initiation (at P2) of synthesis of RNA II which is sensitive to RNA I control, or affects interaction of this region with the cell membrane or regulatory proteins, is not clear at the moment. An accurate termination of transcription in the Clo DF13 minimal replicon in case of hemimethylated DNA may be important anyway because the terminator of the RNA I gene, t1, also harbors methylation sites.

PARTITIONING

DNA replication is an absolute requirement for the stable maintenance of a plasmid in a growing cell population. However, plasmid replication alone does not ensure stable inheritance of the plasmid over a great number of cell divisions. To ensure stable inheritance there has to exist, in addition to a proper, controlled replication system, a specific mechanism for accurate partitioning of plasmid molecules at all cell divisions. When there is no specific mechanism for partitioning, and partitioning is an at random process, then it can be demonstrated by mathematical calculations that plasmidless daughter cells rapidly will be generated (14).

It was found for a number of plasmids, including Clo DF13, that certain plasmid DNA regions are involved in the partitioning process. It was shown already in 1978 (15) that certain deletion mutants of Clo DF13 <u>cop</u> plasmids are not stably inherited in the progeny and that the instability of these <u>cop</u> mutants is accompanied by the formation of high amounts of multimeric plasmid DNA molecules. Analysis of these instable Clo DF13 <u>cop</u> mutants revealed that 2 re-

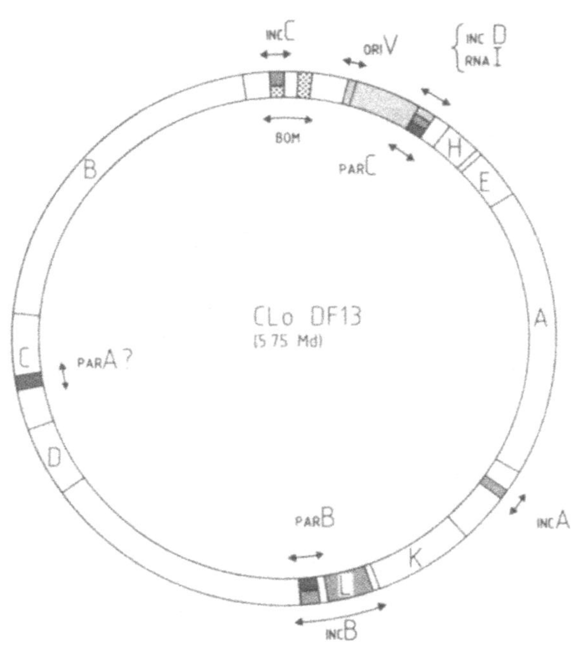

Fig. 4. Genetic information involved in stable maintenance of plasmid Clo DF13: replication control (RNAI), partitioning (<u>par</u>), and incompatibility (<u>inc</u>).

gions on the plasmid genome, designated parA and parB (see Fig. 4) contribute to plasmid maintenance (16). These loci map respectively between 71%-72% and 47%-49% of the Clo DF13 genome. While parA contributes only weakly to plasmid stability, it was found that deletion of the parB region resulted in a strong increase in the generation of multimeric molecules together with an impaired segregation of the plasmids into daughter cells.

Cloning of the site-specific resolution functions of the Tn3-like transposon Tn901 into unstable Clo DF13 derivatives abolished the multimeric forms and reestablished the stable maintenance of these plasmids (16). Apparently, the plasmid par function deleted can be replaced by the resolution functions of transposon TN901. Based on these observations, it was postulated that parB functions in a recombination process that generates the conversion of multimeric DNA, possibly formed during termination of replication, into monomeric DNA (16). Recently, we showed that a recA-independent site-specific resolution process at parB is involved in the maintenance of Clo DF13 cop mutants (17). We found that site-specific resolution occurs if a Clo DF13 plasmid contains 2 parB loci in an intramolecular direct orientation. A specific Clo DF13 sequence within the parB region acts as a site for a site-specific recombination process involved in the resolution of multimeric molecules. This sequence is capable of forming an intra-strand secondary structure (see Fig. 5) because of extensive internal symmetry. A similar system has been described for the stringently replicating phage P1 (18). For a comparison of the site-specific recombination functions of Clo DF13 (parB), P1 (loxP), λ (att site), and of tn3/$\gamma\delta$ (res site) see Hakkaart et al. (17).

Clo DF13 gene L, located in the direct vicinity of parB (19) is not essential for parB functioning. However, our genetic data indicate that the transcription from the gene L containing operon into parB is required (17). We have suggested that termination of transcription in the parB site at the stem-loop structure might be a crucial step (20).

The parB resolution system has a number of surprising features: (i) the parB resolution site is located outside the minimal replicon of Clo DF13, (ii) the parB resolution system is required for stable maintenance of Clo DF13 cop mutants, but not for Clo DF13 plasmids with a wild-type number of plasmid copies, and (iii) there is no evidence at all for a crucial and exclusive role of a plasmid-specified gene product required for site-specific recombination at the parB site.

A Clo DF13 cop^{+} minimal replicon is stably maintained for at least 100 cell generations, indicating that an active partitioning mechanism is involved. Thus far, a partitioning system exclusively for the Clo DF13 minimal replicon has not been identified.

The difference between wild-type plasmids and *cop* mutants, with respect to their dependence on *parB* for stable maintenance, can be explained by assuming that dissociation of multmeric DNA can be performed by 2 systems, a *parB*-dependent and a *parB*-independent system. The *parB*-independent system, likely located in the minimal replicon, should be either a recA-independent, intramolecular recombination process or a site-specific recombination process occurring at an as yet unknown resolution site(s). The *parB*-independent system might only be capable of resolving a limited number of multimeric molecules during cell growth. Due to this limited capacity, an increase in the frequency of initiation of replication due to a *cop* mutation will result in the accumulation of multimeric DNA unless the *parB* resolution function is present on the plasmid. From all our studies we could not obtain any evidence for the crucial and exclusive role of a plasmid-specified gene product, involved in the resolution process. Therefore, we postulate that recombination enzymes encoded by the host are involved in this process. In this respect it should be mentioned that in a *recE* activated host, both *parB*$^+$ and *parB*$^-$ plasmids are stably maintained (16). It is known that in *recE*-activated cells the interconversion between multimeric and monomeric plasmid molecules is drastically stimulated due to the action of exonuclease VIII (22). In addition, all Clo DF13 *cop*$^-$ *par*$^+$ plasmids regenerate multimeric plasmid DNA molecules and are, therefore, not stably maintained in the *E. coli* host mutant, W5445 ZH37, isolated by P. Derbyshire.

Further observations (21) that may be relevant to Clo DF13 stability are the following: (i) 2 Clo DF13 replicons of exactly the same size but different in their antibiotic-resistance markers have an equal chance to colonize the daughter cells upon cell division, resulting in an equal distribution of the 2 plasmids in the cell population, (ii) in case of cells harboring Clo DF13 replicons that differ in size, the smaller plasmid is gradually lost from the cell population, (iii) in the case of cells harboring Clo DF13 replicons that differ in copy number, the low copy number plasmid is also lost from the cell population, and (iv) the effect of plasmid copy number is dominant over the effect of plasmid size.

INCOMPATIBILITY

Plasmids that are unable to coexist in the same bacterial cell are designated incompatible plasmids. Naturally occurring incompatible plasmids invariably have some sequence homology. This sequence homology is structurally and functionally related at least to essential features of plasmid maintenance: replication control and partitioning (see next section).

For several plasmids it has been possible to map loci for incompatibility. We have examined the incompatibility properties of plasmid Clo DF13 and we were able to identify and localize 4 different incompatibility (inc) regions on the Clo DF13 genome (23). The map positions of these regions designated incA, incB, incC, and incD are shown in Fig. 4. We have studied the contribution of each of the 4 inc regions, separately as well as in combination with each other, to the incompatibility of 2 plasmid replicons. We found that: (i) plasmid incompatibility can be evoked by the incD region independently of the other inc regions; (ii) the incA, incB, and incC regions do not function on their own; (iii) the function of both incA and incC in plasmid incompatibility is dependent on the presence of the incB region (gene L) (base sequence analysis indicates that incA as well as incC is located in regions not coding for a gene product); and (iv) the incC region only appears to function when the incD region is present in cis.

Based on these observations we propose a distinction between the incompatibility evoked by the incD region (arbitrarily designated type I incompatibility) and the incompatibility evoked by the other Clo DF13 inc regions, incA, incB, and incC (designated type II incompatibility) (23). We will discuss the 2 incompatibility systems in the context of replication control and partitioning in the next section.

RELATIONSHIPS BETWEEN REPLICATION CONTROL, PARTITIONING, AND INCOMPATIBILITY

Incompatibility is a description of a phenomenon: the noncoexistence of 2 replicons in the same cell. This noncoexistence can be regarded as the consequence of fundamental mechanisms for the stable maintenance of plasmids in dividing cells, e.g., replication control and partitioning. We will discuss the structural and functional relationships between the 2 Clo DF13 incompatibility systems, Clo DF13 replication control, and partitioning.

Type I Incompatibility (incD)

The incD region overlaps the Clo DF13 minimal replicon, including the copy control region. The incD exerts the so-called type I incompatibility (23). The favored model for this type of incompatibility is that of Pritchard (24), further developed by others (14, 25), which states that the initiation of replication is controlled by a cytoplasmic repressor substance, the specific activity of which is linked to cell growth (24). Two incompatible plasmids have identical repressors and repressor target sites. Nonconjugative plasmids produce RNA I as replication repressor and their target sites are likely to be the complementary RNA II molecules (8). Copy control mutants that have been located in the RNA I gene (9,10,12, 13) alter repressor and target site in a complementary manner and both have properties different from the wild-type repressor and target. This might explain why these types of copy control mutants have altered incompatibility properties (8,13,23). These findings fit the model in which the incD region evokes incompatibility at the level of replication control by the action of the trans-acting RNA I repressor, as predicted by the repressor dilution model of Pritchard (24).

A Clo DF13 minimal replicon is stably maintained in dividing cells and therefore this region may also contain information for accurate partitioning. No partitioning system has been identified yet for the Clo DF13 minimal replicon, but we observed that part of the RNA I gene (tentatively designated parC) has a significant nucleotide sequence homology with the parA locus, which is located in one of the mobilization genes, gene C (21). The extent of homology is illustrated in Fig. 5. In contrast to parB, parA contributes only weakly to the stable maintenance of Clo DF13. Therefore, the question is justified whether or not parA is a real partitioning site. It might be a secondary site just because of some sequence homology with a primary par site. If it turned out that parC within the RNA I gene were a partitioning site then incD incompatibility could be the result of 2 mechanisms: replication control and partitioning, 2 mechanisms that may overlap or interact not only structurally but also functionally. The minimal replicon possibly contains also a membrane binding site for replication. This site and a partitioning site are not necessarily the same.

Type II Incompatibility (incA, incB, and incC)

The functioning of incA as well as incC depends on the presence of the incB region (23). From previous studies (27,30) it is known that the gene coding for protein L is located within the 45%-50% region of Clo DF13. Protein L is an inner membrane protein (28) and has a MW of 9.9 Kd (20). Sequence analysis of the incA (29) and the incC (20) region as well as analysis of the proteins produced by Clo DF13 (27) suggests that these inc loci are located within a DNA region not coding for any protein. Taking these observations in account, we have proposed that the incA, incB, and incC form an additional incompatibility system (designated type II incompatibility) that can be explained by a model which is similar to the membrane attachment site model of Jacob et al. (26). We postulated (23) that protein L, product of the gene L in the incB region, is required for the attachment of the plasmid to the cell membrane. This attachment protein facilitates the binding between specific sites, as incA and incC, and the appropriate membrane sites. The inhibition of the multiplication of double-stranded DNA phages by the gene L protein (27) can be the result of the same type II incompatibility mechanism.

The underlying mechanism for the type II incompatibility is likely, at least in part, the partitioning mechanism. The parB site is part of the incB locus. A nucleotide sequence in parB (Fig. 6) located next to gene L (see Fig. 4) is involved in the site-specific recombination involved in the resolution of multimeric plasmid molecules. One of the possible explanations for the generation of plasmidless cells in case of parB⁻ Clo DF13 cop mutants could be that the proper partitioning of plasmids at cell division is disturbed in case of multimeric molecules; in other words, the multimeric molecules "clog up" the segregation system. At the moment we don't know whether the plasmid instability results from an impaired active partitioning process or from a strongly reduced copy number in an ad random partitioning system. Protein L is not required for multimer resolution. However, transcription from the KL operon into the parB site is necessary. It is not fully excluded that the L protein plays a role in resolution. In Clo DF13 L⁻ mutants the function of the L protein could be replaced by a host protein with a similar function.

The incA locus is located at about 32% of the Clo DF13 genome. This region is strongly involved in gene expression. Both the promoter for the cloacine operon and the promoter for the KL operon are located in that area (20,30). binding sites for the cAMP-CRP complex and lexA protein can be found there as well (30). A nucleotide sequence in the incA region permits the formation of a stem-loop structure as is illustrated in Fig. 6.

In our incompatibility type II model, incA is attached to the membrane with the help of protein L. The natural function of this membrane attachment is unknown. It may play a role in the control of the expression of the KL region, or it may play a more direct role in the site specific recombination. Recently, we have obtained evidence that other sites than parB are involved in recA-independent site-specific recombination (A. van Putten, unpub. results). The nature of this recA independent recombination process is subject to present investigation.

The incC locus, located at 97% of the Clo DF13 genome, overlaps a part of the cis-acting bom site (basis of mobilization). Inactivation of incC by transposon insertion prevents mobilization of Clo DF13 (23). The incC locus does not code for any obvious polypeptides (29) but it contains a nucleotide sequence that permits the formation of a stem-loop structure, very similar to that of the incA locus (see Fig. 6).

About the natural function of incC, one can only speculate at the moment. The incC site may have, after attachment to the membrane by the L protein, either a regulatory function in the initiation or termination of transfer replication or a partitioning function. As partitioning site it may be involved in partitioning of plasmid molecules over daughter-cells or may be involved in partitioning of single-stranded DNA molecules over the conjugating acceptor and donor cell during the initial phase of transfer replication. Under nonconjugative conditions the incC site will not be

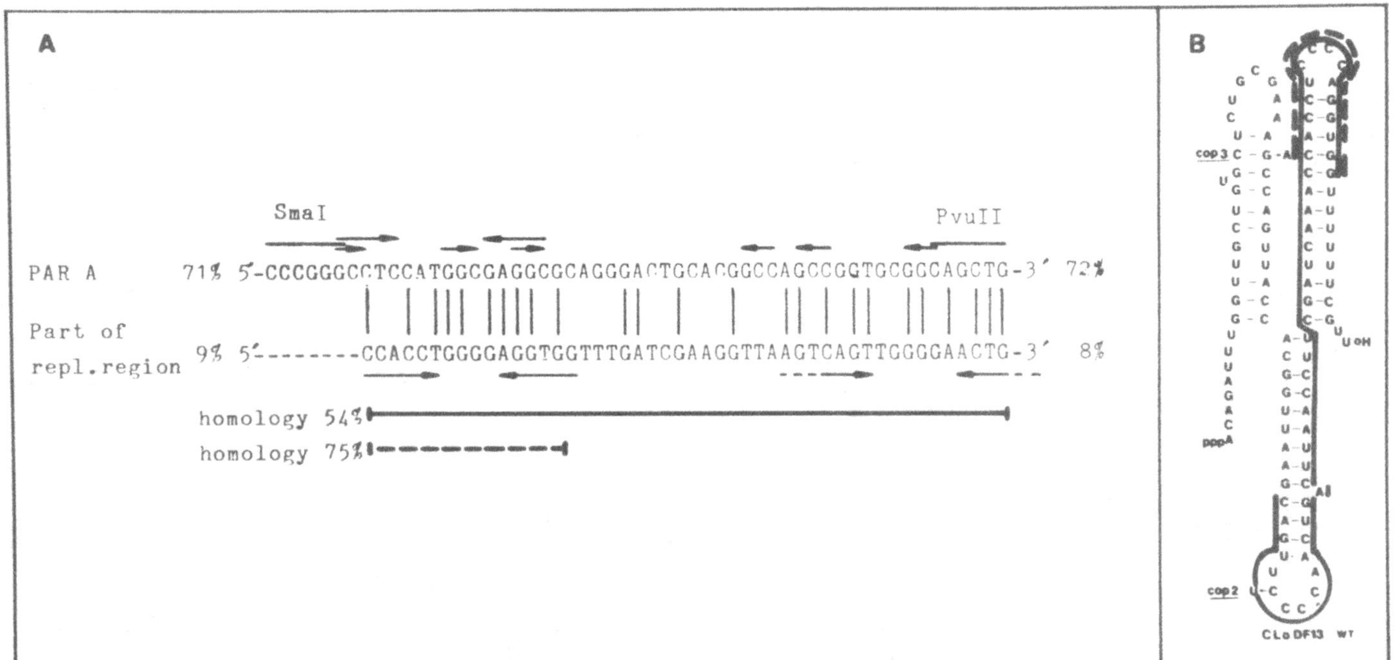

Fig. 5. Nucleotide sequence homology study. A. Nucleotide sequence (5'-3') of part of the parA region of Clo DF13 as determined, and part of the minimal Clo DF13 replicon (11). Marked are: SmaI and PVUII restriction sites. Arrows indicate regions of internal symmetry. Indicated is the region with 54% (-) and 75% (---) sequence homology. B. Nucleotide sequence of the RNA I repressor molecule (11) of Clo DF13. The regions of homology with parA are indicated. (see Fig. 5A).

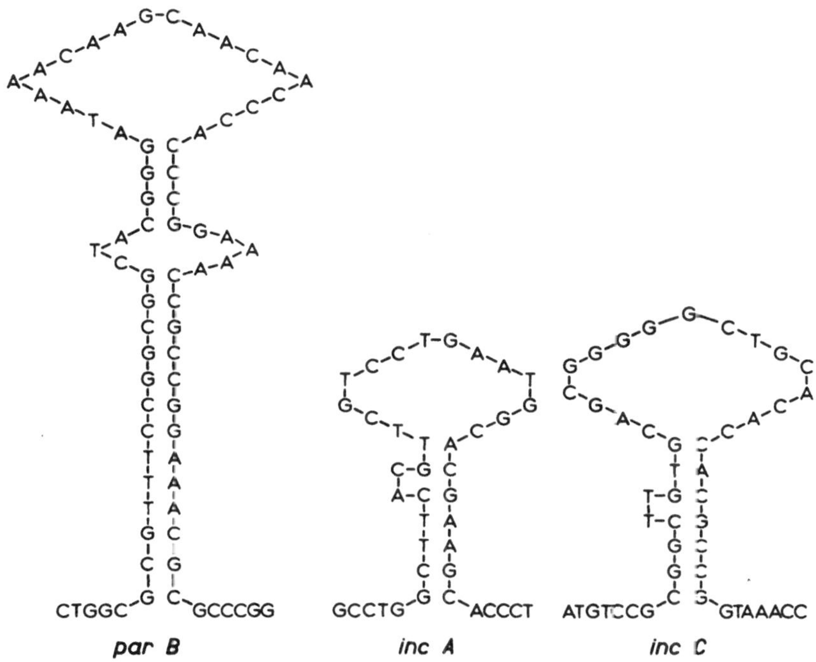

Fig. 6. Possible stem-loop structures within respectively the parB, incA, and incC region of plasmid Clo DF13.

functional in partitioning of single-stranded molecules because of the absence of signals for the initiation of transfer replication, but may be still susceptible for certain partitioning proteins.

A requirement for the incC functioning in incompatibility is the presence in cis of the minimal replicon. Likely, transcription (or replication) from this region into the incC locus is necessary. The stem-loop structure in incC, together with transcription, might be necessary for incC functioning.

In this chapter we have summarized the present knowledge of factors determining the stable maintenance of plasmid Clo DF13 in a growing bacterial cell population. We have discussed the structural and functional aspects of Clo DF13 replication control, partitioning, and incompatibility. In addition, attention is given to the structural and functional relationships between replication control, partitioning, and incompatibility.

Further studies are necessary to unravel the molecular basis of stable inheritance of plasmid DNA not only in E. coli but also in other microorganisms. It might be, however, that the elements determining plasmid maintenance in *E. coli* reflect a more general system operating in the stable maintenance of replicons in microorganisms.

REFERENCES

1. Veltkamp, E., and H.J.J. Nijkamp (1976) Characterization of a replication mutant of the bacteriocinogenic plasmid Clo DF13. Biochim. Biophys. Acta 425:356-367.
2. Nijkamp, H.J.J., and E. Veltkamp (1981) Genetic organization and expression of non-conjugative plasmids. In Molecular Biology, Pathogenicity and Oecology of Bacterial Plasmids, S.B. Levy, R.C. Clowes, and E.L. Koenig, eds. Plenum Press, New York, pp. 247-258.
3. Veltkamp, E., and A.R. Stuitje (1981) Replication and structure of the bacteriocinogenic plasmids Clo DF13 and Col E1. Plasmid 5:76-99.
4. Veltkamp, E., M.J.J. Hakkaart, A.J. van Putten, B. van Gemen, and H.J.J. Nijkamp (1984) Application of plasmids in gene technology: Factors determining maintenance of plasmids in bacterial cells. In Progress in Industrial Microbiology, Elsevier (in press).
5. Cesareni, G., M.A. Muesing, and B. Polisky (1982) Control of Col E1 DNA replication: The *cop* gene product negatively affects transcription from the replication primer promoter. Proc. Natl. Acad. Sci., USA 79:6313-6317.
6. Itoh, T., and J. Tomizawa (1980) Formation of an RNA primer for initiation of replication of Col E1 DNA by ribonuclease H. Proc. Natl. Acad. Sci., USA 77:2450-2454.
7. Stuitje, A.R., E. Veltkamp, J. Maat, and H.L. Heyneker (1980) The nucleotide sequence surrounding the replication origin of the *cop*3 mutant of the bacteriocinogenic plasmid Clo DF13. Nucl. Acids Res. 8:1459-1474.
8. Tomizawa, J., T. Itoh, G. Selzer, and T. Som (1981) Inhibition of Col E1 RN primer formation by a plasmid-specified small RNA. Proc. Natl. Acad. Sci., USA 78:1421-1425.
9. Kool, A.J., and H.J.J. Nijkamp (1974) Isolation and characterization of a copy mutant of the bacteriocinogeic plasmid Clo DF13. J. Bact. 120:569-578.
10. Andreoli, P.M., R.A. Brandsma, E. Veltkamp, and H.J.J. Nijkamp (1978) Isolation and characterization of a Clo DF13::Tn901 plasmid mutant with a thermosensitive control of DNA replication. J. Bact. 135:612-621.
11. Stuitje, A.R., C.E. Spelt, E. Veltkamp, and H.J.J. Nijkamp (1981) Identification of mutations affecting replication control of plasmid Clo DF13. Nature 290:264-267.
12. Muesing M., J. Tamm, H.M. Shepard, and B. Polisky (1981) A single base-pair alternation is responsible for the DNA overproduction phenotype of a plasmid copy-number mutant. Cell 24:235-242.

13. Lacatena, R.M., and G. Cesareni (1981) Base pairing of RNA I with its complementary sequence in the primer transcript inhibits Col El replication. Nature 294:623-626.
14. Novick, R.P., and F.C. Hoppensteadt (1978) On plasmid incompatibility. Plasmid 1:421-434.
15. Stuitje, A.R., E. Veltkamp, P.J.M. van den Elzen, and H.J.J. Nijkamp (1978) In vitro construction of deletion mutants of the bacteriocinogenic plasmid Clo DF13. Nucl. Acids Res. 5:1801-1820.
16. Hakkaart, M.J.J., E. Veltkamp, and H.J.J. Nijkamp (1982) Maintenance of the bacteriocinogenic plasmid Clo DF13 in Escherichia coli cells. II. Specific recombination functions involved in plasmid maintenance. Mol. Gen. Genet. 188:338-344.
17. Hakkaart, M.J.J., P.J.M. van den Elzen, E. Veltkamp, and H.J.J. Nikjamp (1984) Maintenance of multicopy plasmid Clo DF13 in E. coli cells: Evidence for site-specific recombination at parB. Cell 36:203-209.
18. Austin, S., M. Ziese, and N. Steinberg (1981) A novel role for site-specific recombination in the maintenance of bacterial replicons. Cell 25:729-736.
19. Andreoli, P.M., N. Overbeeke, E. Veltkamp, J.D.A. van Embden, and H.J.J. Nijkamp (1978) Genetic map of the bacteriocinogenic plasmid Clo DF13 derived by insertion of the transposon Tn901. Mol. Gen. Genet. 160:1-11.
20. Van den Elzen, P.J.M., M.J.J. Hakkaart, A.J. van Putten, H.H.B. Walters, E. Veltkamp, and H.J.J. Nijkamp (1983) Structure and regulation of gene expression of a Clo DF13 plasmid DNA region involved in plasmid segregation and incompatibility. Nucl. Acids Res. 11:8791-8808.
21. Hakkaart, M.J.J., B. van Gemen, E. Veltkamp, and H.J.J. Nijkamp (1984) Maintenance of multicopy plasmid Clo DF13: III. Role of plasmid size and copy number in partitioning. Mol. Gen. Genet. (in press).
22. Fishel, R.A., A.A. James, and R. Kolodner (1981) RecA-independent general genetic recombination of plasmids. Nature 294:184-186.
23. Hakkaart, M.J.J., J.G. Wesseling, E. Veltkamp, and H.J.J. Nijkamp (1982) Maintenance of the bacteriocinogenic plasmid Clo DF13 in Escherichia coli cells: I. Localization and mutual interactions of four Clo DF13 incompatibility regions. Mol. Gen. Genet. 186:531-539.
24. Pritchard, R.H., P.T. Barth, and J. Collins (1969) Control of DNA synthesis in bacteria. Symp. Soc. Gen. Microbiol. 19:263-297.
25. Hashimoto-Gotoh, T., and K.N. Timmis (1981) Incompatibility properties of Col El and pMBl derivative plasmids: Random replication of multicopy replicons. Cell 23:299-238.
26. Jacob, F., S. Brenner, and F. Cuzin (1963) On the regulation of DNA replication in bacteria. Cold Spring Harbor Symp. Quant. Biol. 28:329-348.

27. Van de Pol, H., E. Veltkamp, and H.J.J. Nijkamp (1980) Genetic information of the bacteriocinogenic plasmid Clo DF13 involved in the inhibition of the multiplication of double stranded DNA phages. Mol. Gen. Genet. 178:535-540.
28. Van de Pol, H. (1980) The bacteriocinogenic plasmid Clo DF13: Mobilization and phage interaction. Thesis, Vrije Universiteit, Amsterdam.
29. Snijders, A., A.J. van Putten, E. Veltkamp, and H.J.J. Nijkamp (1983) Localization and nucleotide sequence of the *bom* region of Clo DF13. Mol. Gen. Genet. 192:444-451.
30. Van den Elzen, P.J.M., J. Maat, H.H.B. Walters, E. Veltkamp, and H.J.J. Nijkamp (1982) The nucleotide sequence of the bacteriocin promoters of plasmid Clo DF13 and Col E1: Role of *lexA* repressor and cAMP in the regulation of promotor activity. Nucl. Acids Res. 10:1913-1927.
31. Sugimoto, K., A. Oka, H. Sugisaki, M. Takanami, A. Nishimura, Y. Yasuda, and Y. Hirota (1979) Nucleotide sequence of *E. coli* K12 replication origin. Proc. Natl. Acad. Sci., USA 76:575-579.

REPLICATION CONTROL FOR PT181, AN INDIRECTLY REGULATED PLASMID

Richard P. Novick, Steven J. Projan, C. Chandra Kumar, Stephen Carleton, Alexandra Gruss, Sarah K. Highlander, and John Kornblum

Public Health Research Institute
of the City of New York, Inc.
New York, New York 10016

SUMMARY

PT181 is a fully sequenced Staphylococcus aureus plasmid whose size is 4,437 bp. It specifies tetracycline resistance and has a copy number of about 22 per cell in exponentially growing cultures. The functional organization of the pT181 replicon is centered around the coding sequence for a 35-kd protein, RepC, that is absolutely required for replication of the plasmid. The replication origin is contained within the repC coding sequence and the region immediately 5' to the RepC start is involved in control of the plasmid replication rate. PT181 replication is controlled at the level of RepC synthesis by a negative regulatory system that is functionally similar to that of the ColE1 and IncFII plasmids of Escherichia coli. The pT181 control circuit involves 2 short transcripts, RNA I and RNA II, that are transcribed from the region specifying the 5' end of the untranslated repC mRNA leader and in the opposite direction. These are referred to as countertranscripts. The countertranscripts regulate RepC synthesis by a mechanism that probably involves interaction with the repC mRNA leader in a manner that interferes with translation. Both of the countertranscripts seem to be necessary for normal replication control; their separate roles remain unclear.

Unlike plasmids of the ColE1 and IncFII groups, plasmids such as ColE1 are considered to have direct regulation of replication because the inhibitory element of the copy control circuit directly inhibits the initiation of replication. Plasmids such as pT181 are considered to have indirect regulation of replication because the product of the regulated step, RepC, is trans-active. Plasmids of the IncFII type are considered to have direct regulation of replication because the product of the regulated step, RepA is cis-active. The analysis of pT181 replication physiology has illustrated 2 important differences between directly and indirectly regulated plasmids: a) for directly regulated plasmids, copy mutants specifying a normal inhibitor substance but an inactive target site exclude the wild-type or recessive mutants by directly interfering with their replication. Analogous mutants of indirectly regulated plasmids coexist readily with the wild-type and all mutants (although they do manifest segregational incompatibility) because the Rep protein is always shared by all plasmids in the cell, regardless of its source. b) Mutations of directly regulated plasmids in the region where target transcript and countertranscript overlap may give rise to totally new incompatibility groups because they engender independently self-correcting copy pools. Such mutants for indirectly regulated plasmids never give rise to new incompatibility groups, even if they have totally different inhibitor-target recognition specificities because they do not engender independently self-correcting copy pools due to the mutual availability of the Rep protein.

These considerations lead to a general view of the basis of plasmid incompatibility that is somewhat broader than can be gained from a consideration of directly regulated plasmids alone.

INTRODUCTION

The details of replication control systems are quite variable from plasmid to plasmid and their comparative analysis has revealed certain basis principles. One of these principles is that the characteristics of the control system are dependent in an important way on the point in the initiation pathway at which the controls act. If the inhibitor acts directly at the initiation step, the regulation is referred to as <u>direct</u>; if the inhibition is of the synthesis of a trans-acting factor required for replication, the regulation is referred to as <u>indirect</u>. Of the 3 groups of plasmids whose replication control mechanisms are best understood, namely the ColE1 and IncFII groups of *Escherichia coli* and the Inc3 group of *Staphylococcus aereus*, the first 2 have direct regulation, the third, indirect. All 3 groups utilize regulatory circuits of the inhibitor-target type in which the inhibitor is an RNA molecule referred to as a countertranscript. Other well-studied plasmids such as F, R6K, P1, and RK2 utilize a regulatory mechanism that involves binding of a trans-acting replication factor (the initiator) to sets of directly repeated nucleotide sequences (iterons) at or near the replication origin. These <u>initiator-binding</u> mechanisms are probably of the direct variety; however, from the available data, one cannot yet be certain. In this report, we describe the characteristics of the indirect type of regulation, as exemplified by pT181, and contrast these to the direct type as exemplified by the ColE1 family and by R1/NR1/R6-5.

THE pT181 REGULATORY CIRCUIT

pT181 is the prototype of a large and widely dispersed family of small, multicopy staphylococcal plasmids. It specifies inducible tetracycline resistance, has a molecular length of 4,437 bp, and is maintained at a copy number of about 22 in exponential cells. It is inherited stably in *S. aureus* but unstably in *Bacillus subtilis*, although it is maintained at approximately the same copy number in both species. The entire nucleotide sequence of the plasmid has been determined (2) and the origin of the numbering sequence is the unique PvuI site (see Fig. 1). The essential replication functions are located in a region of approximately 1.5 kb, between positions 3500 and 500. A region that is required for stability has recently been identified between positions 500 and 700 (A. Gruss and R. Novick, unpub. data). PT181 specifies a trans-active protein, RepC, that is absolutely required for the replication of the plasmid and is specific for plasmids of the Inc^{3} group (3). The determinant for this protein, *repC*, has been shown by genetic means to correspond to a 939-bp open reading frame extending from bases 3646 to 151 and reading counterclockwise as the map is conventionally drawn (3). We

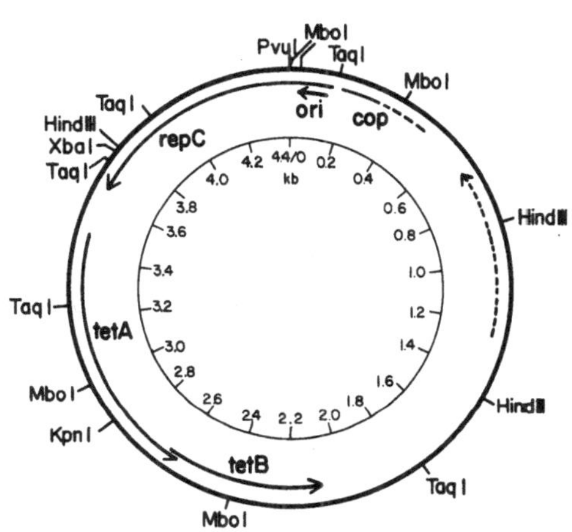

Fig. 1. pT181 map. Kilobase coordinates originate at the Pvu I site at 12:00 and are read in the clockwise direction. Nucleotide positions are based upon the known sequence as determined by Khan and Novick (2). Solid arrows labeled repC, tetA, and tetB indicate open reading frames based on the sequence and known to encode functional proteins. The dashed arrow indicates an open reading frame that is not known to encode any polypeptide. "Ori" indicates the approximate outer limits of the replication origin and the direction of replication (5). "Cop" indicates the region in which the known elements of the replication control system are located (8).

have shown previously that pT181 is capable of semiconservative replication in cell-free extracts of *S. aureus* strains containing the plasmid but not in extracts of strains containing no plasmid or any other plasmid (4). The activity of pT181-containing extracts was shown to be due to a plasmid-encoded protein which is probably RepC.

PT181 replicates unidirectionally from a single origin located between nucleotides 15 and 157 (5). Replication is in the counterclockwise direction (see Fig. 1); additionally, all of the known and potential polypeptide coding sequences are also transcribed in the counterclockwise direction. Thus, there is a strong counterclockwise directionality to the general flow of information around the plasmid, which may reflect some degree of coupling between replication and transcription.

Elements of the replication control system for pT181 are located in the region immediately 5' to the RepC coding sequence. As currently understood, this region occupies at least 300 bp and contains the promoters for 4 transcripts: 2 left tandem starts at positions 339 and 410 are used for the *repC* mRNA and the 2 transcripts are referred to as RNA III and RNA IV, respectively. One right start at position 246 is used for 2 countertranscripts referred to as RNA I and RNA II. The former terminates at about 340, the latter at about 400 (6). The functional organization of the essential region is diagrammed in Fig. 2.

The CopA function

As is the case with most plasmids, when pT181 is cloned to a vector that is maintained at a copy number higher than that of pT181, the composite plasmid prevents the establishment of autonomous pT181. This inhibition is due to the activity of a negative regulator of replication, referred to as CopA, whose activity is gene-dosage dependent. We have demonstrated (7) that CopA activity directly blocks pT181 replication in an experiment in which pT181 was cloned to a thermosensitive replication-defective (Tsr) mutant of pE194. It was possible to establish a heteroplasmid containing pT181 and this latter cointegrate by performing the cross at an elevated temperature. As soon as the temperature was lowered, however, the Tsr replicon recovered its normal copy number and replication of the pT181 replicon was blocked. This experiment is shown in Fig. 3. Using the pT181::pE194 cointegrate, we mapped the position of the inhibitory activity by generating deletions by exonuclease III digestion of Pvu I-linearized DNA in vitro. These results were clear: all of the deletion-bearing plasmids were either as effective as wild-type in excluding pT181 or totally ineffective. The map positions of the deletions are shown in Fig. 4 and indicate that CopA is encoded in the region 5' to the *repC* start. The inhibitor gene was further localized to the region between positions 158 and

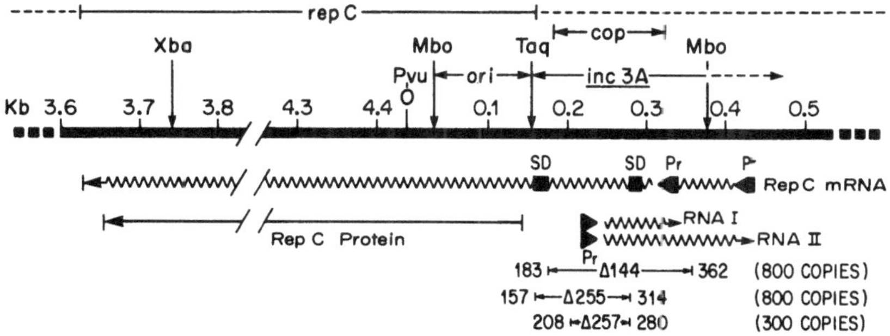

Fig. 2. pT181 regulatory region. Figures are in kilobases, referring to the coordinate system indicated in Fig. 1. Wavy lines indicate known transcripts. For all 4 transcripts shown, the start sites are known and the termination sites are approximated (4). "repC" indicates the extent of the repC gene based on its coding sequence. "Inc3A" refers to the map location of the negative regulatory element of the copy control system, copA, that is responsible for interference in trans with replication of sensitive plasmids. SD = Shine-Dalgarno sequence; Pr = promoter. Below the map are shown the extents of 3 spontaneous deletions, given as the nucleotide positions of their ends, along with the copy number of the deletion-bearing plasmids.

570 by subcloning of pT181 Taq I fragments to a derivative vector as shown in Fig. 5 (7).

Copy mutants

Evidence implicating the above-mentioned inhibitor in pT181 replication control was provided by the isolation and analysis of copy mutants (8). These were obtained by selection for increased resistance to tetracycline and their plasmid linkage was verified by outcrossing. A screening gel showing a set of typical copy mutants is presented in Fig. 6. Samples are sheared, whole-cell minilysates (9) that are loaded in carefully calibrated amounts; gels of this type are scanned by fluorescence densitometry to determine copy numbers directly (10).

The copy mutants could be divided into 2 classes on the basis of resistance or sensitivity to CopA inhibition. Resistant mutants are referred to as dominant, sensitive as recessive. Included among the copy mutants are 3 short spontaneous deletions, 144, 256 and 257. All 3 are dominant and map in the copA region. One of them, Δ144, was cloned to pE194 and shown not to express CopA activity, indicating that the determinants of the inhibitor and its target are very closely linked and probably overlapping (8). Sequence analysis

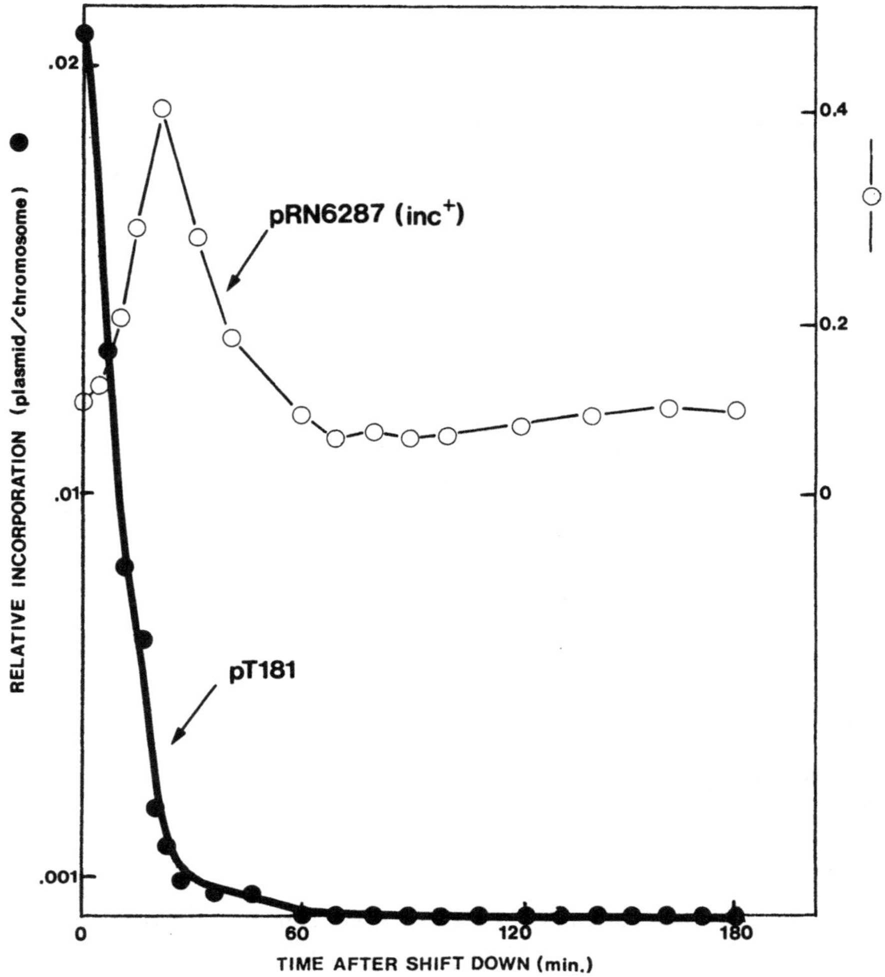

Fig. 3. Inhibition of the replication of pT181 by copA. pRN6287 is a derivative of pE194 (Tsr) to which has been cloned the Taq I-A fragment of pT181, indicated by "(inc$^+$)". Replication of pT181 (solid circles) and pRN6287 (open circles) was monitored following a temperature shift from 42°C to 32°C by successive 5 min pulses with ^{3}H thymidine. Lysates were prepared and separated by agarose gel electrophoresis and the resulting bands were excised and counted. The plasmid counts were normalized by dividing by the counts in the corresponding chromosomal bands. Note the difference in scale for the 2 plasmids.

showed that all the copy mutants are located in the copA region. The recessive copy mutants synthesized RNA I and RNA II either in reduced amounts or in abnormal patterns--thus cop-619 and 609 syn-

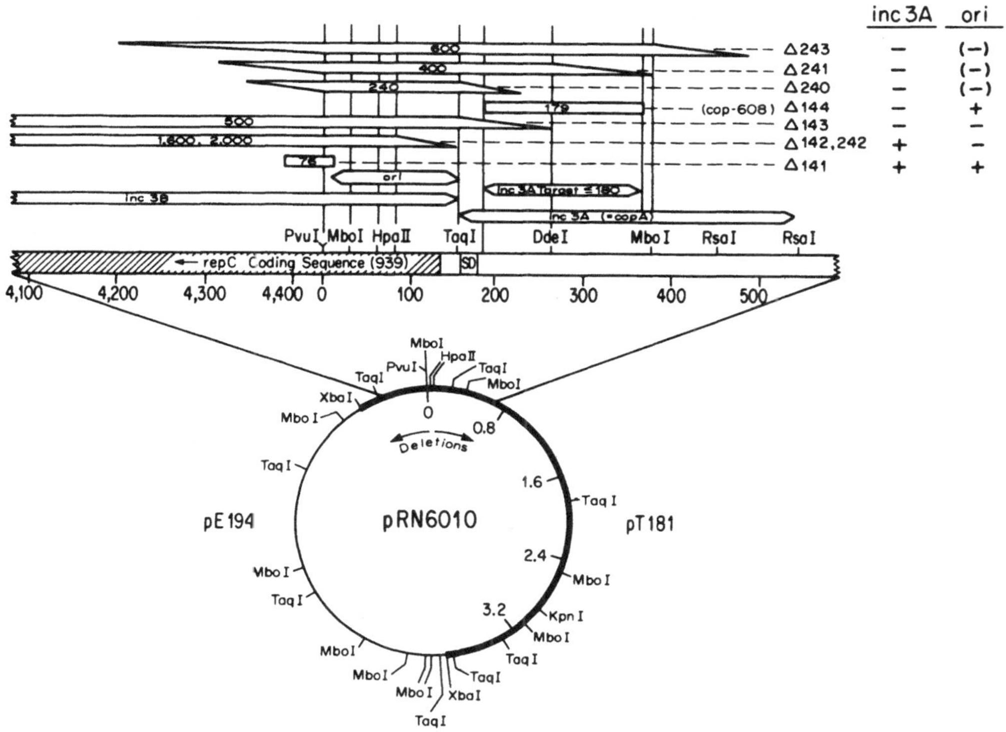

Fig. 4. Deletion map of the repC control region. Deletions 141-143 and 240-243 were generated by enzymatic digestion of Pvu I-linearized pRN6010 (see Fig. 5) DNA and were mapped by restriction analysis and scored for Inc3A activity (see text) and origin activity (5). Numbers indicate estimated sizes of the deletions and tapering portions indicate the limits of uncertainty of the locations of the ends. The locations of Δ141 and Δ144 have been determined precisely by sequencing (2,4). The locations of inc3A and inc3B were defined by cloning (see text). The Inc3A target was defined by Δ144.

thesized much less of the 2 than the wild-type; cop-621, 622, and 629 synthesized only RNA I or RNA II but not both (see Fig. 7). The dominant mutants, on the other hand, always synthesized greatly elevated amounts of these 2 RNA species. These results indicate that the inhibitory activity expressed by the copA gene is due to the countertranscripts and that both may be required for normal copy control.

Regulatory Levels

Preliminary measurements of the relative countertranscript

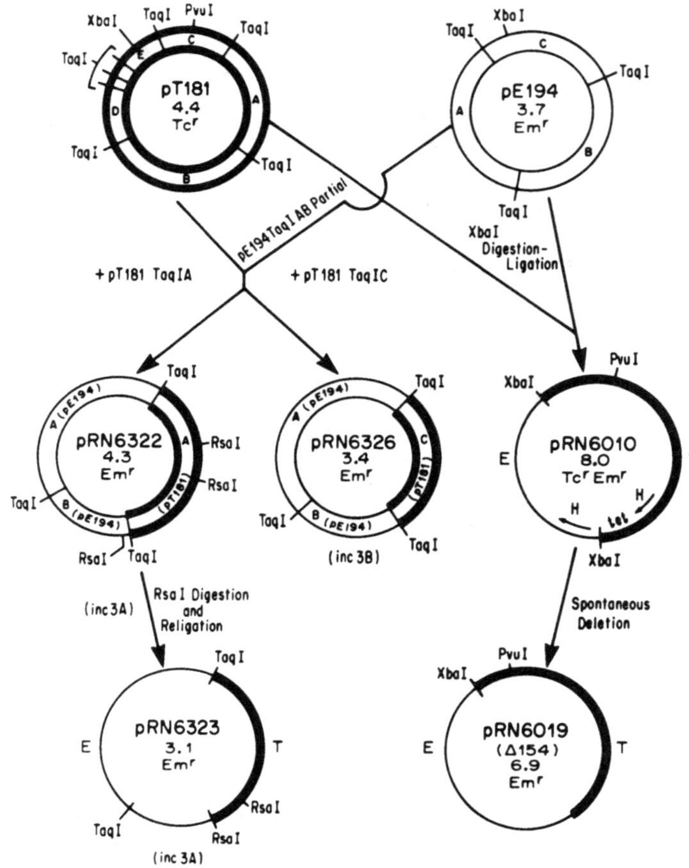

Fig. 5. Plasmid constructions. Only restriction sites involved in the cloning are shown. pT181 and its fragments are shown with heavy lines; pE194 and its fragments are drawn lightly. At right is shown the construction of pRN6010 by Xba I digestion and ligation. pRN6019 arose by the spontaneous deletion of 2 kb between the arrows marked "H". At left is shown the cloning inc3A (copA) and inc3B. For inc3A, the largest Taq I fragment (A) of pT181 was ligated to a partial Taq I digest of pE194 and the resulting chimera was digested with Rsa I and re-ligated. The final subclone, pRN6323, lacks the Rsa I fragment between position 570 of pT181 and 611 of pE194 and has therefore lost a 23-bp stretch of pE194 DNA between the Rsa I (611) and Taq I (588) sites (22) at 6 o'clock as shown. For inc3B, the third largest Taq I fragment of pT181 (C) was ligated to the same partial digest of pE194 (pRN6326). Both of these clones lack the non-essential Taq I-C fragment of pE194.

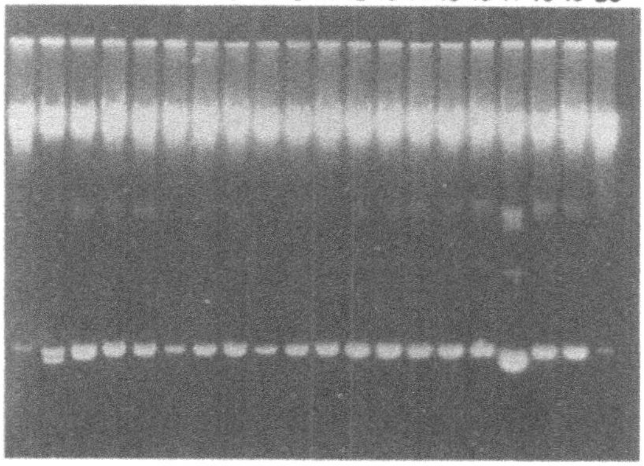

Fig. 6. Agarose gel electrophoretic analysis of copy mutants. Whole cell sheared minilysates were separated on 1% agarose in tris-borate buffer for 18 hr at 2.5 volts/cm, stained with ethidium bromide and photographed. Samples corresponding to equivalent numbers of starting cells were loaded (about 10^8 cells per sample) in order to permit comparison. Gels of this type were used for copy number determination. Tracks 1,20: pT181; 2: cop-626; 3: 623; 4: 622; 5: 621; 6: 620; 7: 619; 8: 618 9: 617; 10: 615; 11: 614; 12: 613; 13: 612; 14: 611; 15: 610; 16: 609; 17: 608; 18: 607; 19: 601.

The heavy upper and lower bands represent sheared chromosomal and supercoiled plasmid monomeric DNA, respectively. The uppermost band represents a small amount of material that remains in the sample well. The intermediate bands represent plasmid isomers including oligomers and relaxed monomers.

levels in copy mutants suggests that copy number is determined by the countertranscript concentrations per se. For example, in promoter-down mutants such as cop-619, the countertranscript levels are very close to wild-type despite a 10-fold increase in copy number; in other words, a 10-fold increase in countertranscript gene dosage is required to establish the standard regulatory inhibitor level. On the other hand, with dominant mutants, the countertranscript levels are increased in proportion to the copy number; because the mutant plasmid is less sensitive to the inhibitor, a higher concentration of the latter is required to achieve a regulatory level. Note that the target concentration seems not to be important.

Analysis of the repC mRNA Leader Sequence

A possible mechanism by which such countertranscripts might

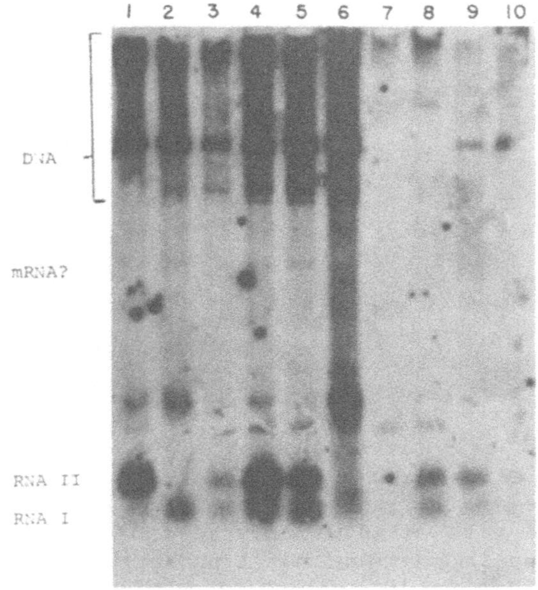

Fig. 7. Northern blot hybridization analysis of pT181 countertranscripts. Whole-cell minilysates were prepared by standard techniques, separated on 1.5% agarose and blotted to nitrocellulose. Blots were hybridized to a nick-translated sample of pT181 Mbo I-D (31-376; see Fig. 2) and autoradiographed. Samples corresponding to equivalent numbers of cells were run. In track 1 is a lysate prepared from a strain containing pT181 cop-622; track 2: cop-621; 3: cop-619; 4: cop-618; 5: cop-611; 6: cop-607; 7: pRN6050; 8: pRN6010; 9: pT181; 10: pC221. The following features of this analysis are noteworthy: the mutants in tracks 1 and 2 are defective in termination of RNA I and RNA II, respectively; these 2 and that in track 3 are recessive copy mutants. The mutants in tracks 5 and 6 are dominant copy mutants and, typically, overproduce the countertranscripts. The mutant in track 6 is also dominant but is evidently defective in countertranscript production as well as sensitivity. The extracts in tracks 7 and 8 are from strains containing the 2 orientations of the pT181::pE194 cointegrate; the difference in countertranscript levels suggests the possibility of transcriptional interference in 1 of the 2 orientations. Finally, the extract in track 10 is from a strain containing pC221, a plasmid that is compatible with but related to pT181.

regulate pT181 replication was suggested by analysis of the sequence of the repC mRNA leader region and of the copy mutations as shown in Fig. 8. Note that there is a strong cluster of dominant mutations in the short GC-rich region from 308-320 and that all but one of some 20 mutants of this type had a GC to AT change in this region. Note also that the 2 mutants with reduced amount of RNA I and II are in the -10 and -35 regions of the countertranscript promoter. The mutants defective in the synthesis of either but not both of the countertranscripts are evidently terminator mutants. A possible secondary structure that could be implicated in the termination of RNA I is shown in Fig. 9. The 2 mutations that affect RNA I termination both affect the penultimate G in the 3' arm of the RNA I stem. A possible hairpin that could be involved in RNA II termination involves nucleotides 358-399 (see Fig. 8). Note that this hairpin has a run of 5' A residues (358-361) as well as a run of 3' U's (396-400) so that it could also serve as a terminator for transcription in the opposite direction. Thus cop-621 (367 G-A) could actually owe its increased copy number to increased left transcription, as well as, or rather than, to decreased RNA II levels.

Given the striking parallel between this pattern of mutations and that observed for ColE1 (11,12,13) and the IncFII plasmids (14,15), it seemed likely that a similar control mechanism was operative. Accordingly, we analyzed the secondary structures of RNA I and II by the computer method of Zuker and Stiegler (16) and found that when optimally folded, the 2 countertranscripts displayed the region containing the cluster of dominant mutations as a single-stranded loop. The complement of this structure is displayed in the leader transcripts only if the folding program is run on a part of the leader sequence (Fig. 9). In the optimal secondary structure predicted for the entire leader, the target region is mostly base-paired and presumably unavailable for interaction with the inhibitory RNA molecule. This structure, shown in Fig. 10, reveals several features of the leader region that we consider to be important in the operation of the copy control system. The Shine-Dalgarno sequence for RepC translation is located in the distal arm of a strong potential hairpin involving nucleotides 151-192. We refer to the proximal arm of this dyad as the anti-SD sequence. This hairpin is terminated by a run of 6 U residues followed by the start codon. Upstream, at positions 320-328, is a 9 base segment that is a precise direct repeat of the Shine-Dalgarno sequence and flanking bases. In the computer-generated secondary structure of the leader, the upstream segment is paired with the anti-SD segment which would pre-empt the formation of the downstream hairpin, leaving the SD segment single-stranded and in a configuration that would be accessible for translation. This is referred to as the active configuration. We propose that the early configuration of the leader is the actual target for RNA I/II, that the interaction would involve the target loop as a nucleation site, and that an RNA-RNA duplex would form, as is the case with ColE1 (17) and, presumably, with the IncFII plasmids. The consequence of the RNA-RNA interaction would be the formation of the downstream stem in which the SD sequence is fully base paired and unavailable for the initiation of RepC translation. Additionally, this hairpin could serve as an early termina-

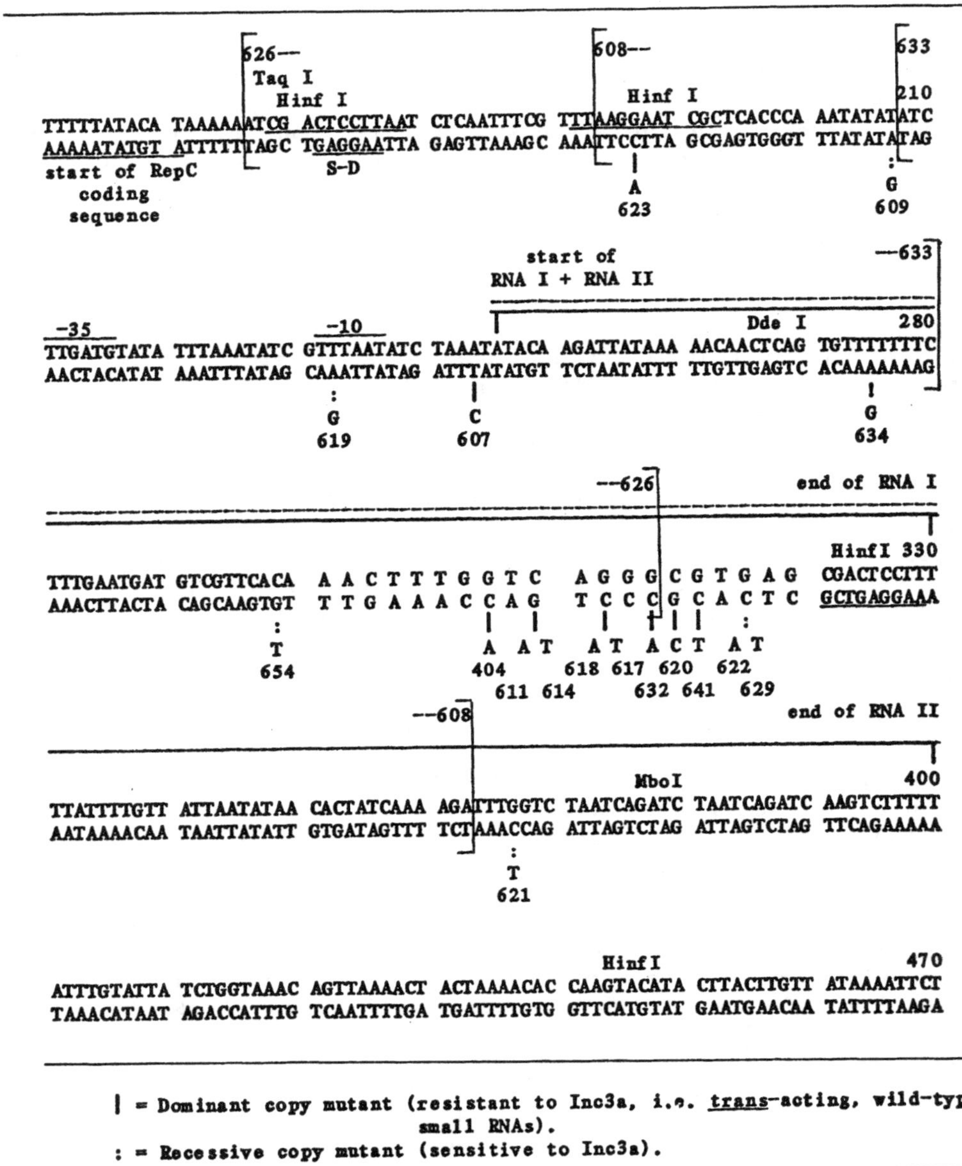

Fig. 8. Sequence of the pT181 copy control region between bases 141 and 470. Restriction enzyme recognition sites are indicated as well as a putative ribosome binding site and start site for the RepC coding sequence; read from right to left. Base sequence changes in several independently isolated copy mutants of pT181 are as follows:

Fig. 8 (Continued)

| = Dominant mutation

: = Recessive

! = Temperature sensitive for replication

tion site (attenuator) for repC mRNA transcripts. We refer to this as the inactive configuration. The model is thus a dynamic one in which the critical inhibitory interaction occurs before transcription of the leader has progressed beyond nucleotide 140. Any leader transcripts that progress a further 10 bases are predicted to refold into the active configuration and to be insensitive to the inhibitory interaction. Any transcripts that are "caught" by RNA I/II before refolding would be irreversibly inactivated. The level of RNA I/II, therefore, would determine the probability that any transcript would progress to an active form and by this means would set the rate of RepC production and therefore the replication rate. Note that this process is regarded as a steady state system rather than one in which fluctuations in the inhibitor concentration have an all-or- none effect on replication as was suggested in the original formulation of the inhibitor-dilution model (18). Available evidence in support of this model is mostly circumstantial and is as follows:

1. We have replaced the G at position 168 (see Fig. 8) with a C by synthetic oligonucleotide-directed mutagenesis and found that the mutant sequence is translationally inactive in E. coli. This supports the identification of the SD sequence as indicated and therefore of the RepC start (J. Kornblum et al., unpub. data).

2. A spontaneous mutation (cop-623) at position 189 (C to A) has a dominant high-copy phenotype and is completely indifferent to RNA I/II inhibition. This mutation changes the calculated free energy of the SD-anti SD loop from -14 to -6 kcal, supporting the conclusion that the anti-SD segment is critical for the regulatory interaction.

3. Sequence analysis of pC221, a closely related S. aureus plasmid that has a high degree of homology with pT181 in the leader region, has revealed a hairpin similar to that seen in pT181 positions 151-192, as well as a direct repeat of the putative SD sequence upstream (unpub. data). As shown in Fig. 11, however, the pC221 anti-SD sequence differs at 2 positions from the corresponding pT181 sequence and these 2 differences are mirrored by compensating differences in the putative pre-emptor sequence.

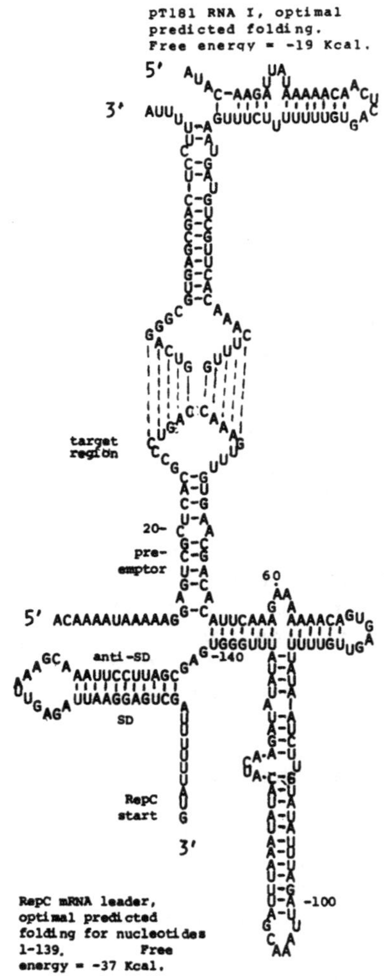

Fig. 9. Possible interaction between computer-predicted secondary structures for RNA I and the repC leader. In this model, the folding program has been run only on the first 140 nucleotides of the leader, as indicated. The secondary structure from 140 to the 3' end was predicted independently. Note that the target region is displayed as a single-stranded loop and that the Shine-Dalgarno sequence is fully base-paired.

4. Deletion of nucleotides 208-280 results in a high copy-number phenotype in which the mutant plasmid is insensitive to CopA inhibition. A T to C mutation at position

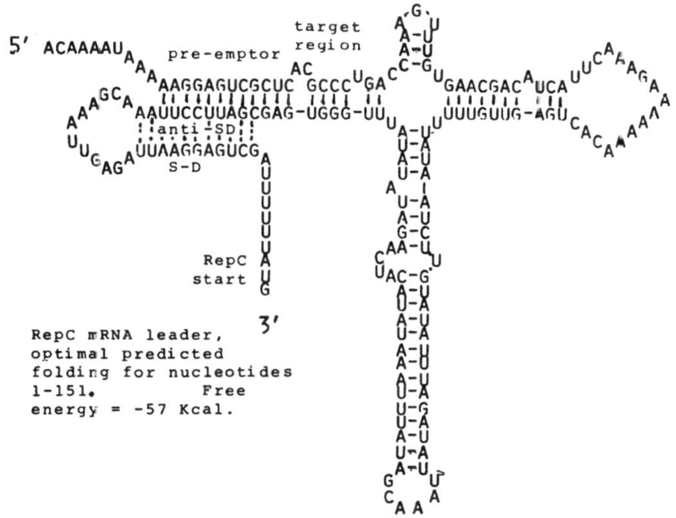

Fig. 10. Predicted secondary structure for the repC mRNA leader. (See text).

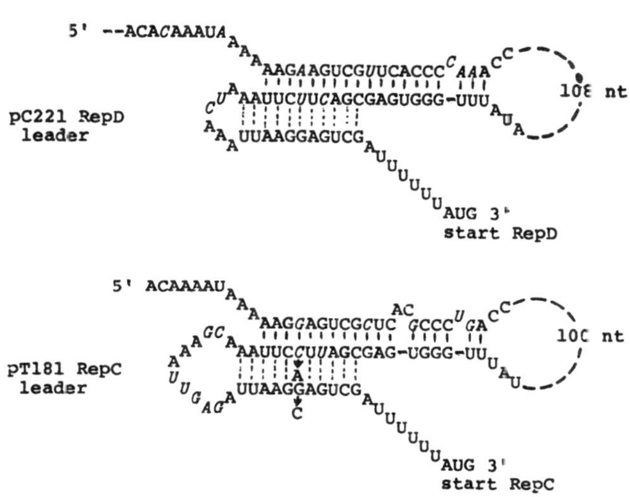

Fig. 11. Partial representation of the repC mRNA leader secondary structure for pT181 compared to the corresponding region of pC221. Note that the 5' end of the putative <u>repD</u> leader is uncertain. The nucleotides in the 2 corresponding regions of the 2 plasmids that do not match are shown in italics.

244 (cop-607) gives a similar phenotype. Note that these mutations involve none of the critical interacting sequences described above. However, the deletion eliminates a strong potential hairpin involving nucleotides 203-258, and the mutation occurs within this hairpin, lowering its free energy. It is suggested that this hairpin may serve as a strong RNA polymerase pause site which would ensure that the early section of the leader remains in the CopA-sensitive configuration for sufficiently long to interact with the inhibitor; elimination of this putative pause site would result in very rapid refolding of the leader to the late, CopA-insensitive configuration. The predicted secondary structures for the deleted leader are shown in Fig. 12.

At present, this model remains speculative; however, it has the virtue of being readily testable.

INDIRECT REGULATION

Non-exclusion

For ColE1, copy mutants that produce wild-type RNA I but are insensitive to it exclude the wild-type (as well as any other mutants that are sensitive to the inhibitor) because the elevated RNA I levels directly block replication in trans (B. Polisky, pers. comm.). For pT181, there is no such exclusion. In all cases thus

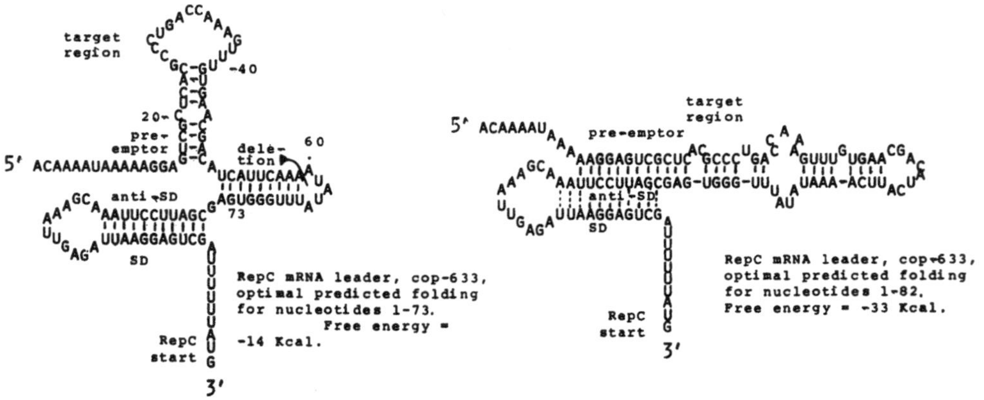

Fig. 12. Predicted secondary structures for the partially and completely folded cop-633 repC mRNA leader. Nucleotides are numbered from the 5' end of the leader in this diagram and the numbers do not correspond to the primary nucleotide numbers as shown in Fig. 8.

far tested, it was possible to co-establish the wild-type plasmid or any of the recessive mutants with any other mutant, including one dominant mutant that has been shown to overproduce wild-type RNA I and RNA II, namely cop-623 (see Fig. 8). These results are presented in Tab. 1, which lists the mutants that have been tested in this manner, and gives their copy numbers alone or in the presence of the wild-type or another mutant plasmid. Note that in the presence of cop-623, the wild-type plasmid is amplified to the extent that the copy pool is approximately split between the 2 plasmids. On the basis of previous findings that RepC synthesis by the wild-type plasmid is blocked when the countertranscript concentration is elevated to approximately 2.5x its normal level, it is inferred that the same is true in the WT/ cop-623 heteroplasmid, so that all of the RepC activity in this strain would be due to cop-623. The situation for other dominant mutants is similar; as these have mutations in the target region and therefore do not make wild-type countertranscripts, RepC synthesis by the wild-type plasmid would be blocked by its own countertranscripts so long as its copy number is elevated above normal. Note that in most of the examples, the copy pool is approximately split, as would be expected if the constituent plasmids can compete equally for all of the RepC protein present. In a few cases, the copy pools are asymmetric; the plasmid with the higher copy number evidently has a competitive advantage over the other; the basis of such advantage is not presently understood. An additional feature of these heteroplasmids is that one can predict

Tab. 1. Interactions between pT181 copy mutants and pSA5000.

PLASMID	COP ALLELE	COPY NO. IN HETEROPLASMID STATE pT181 REPLICON	pSA5000	SUM OF WILD-TYPE PLUS MUTANT	HOMOPLASMID COPY NUMBER
pT181	wild-type	—	—	—	22
pSA5000	wild-type	—	—	—	22
pT181	wild-type	12	10	22	22
pRN8009	cop-609(R)*	22	32	52	74
pRN8011	cop-611(D)*	25	110	140	200
pRN8020	cop-615(D)	70	63	130	150
pRN8023	cop-618(D)	46	170	210	210
pRN8024	cop-619(R)	20	21	41	140
pRN8026	cop-621(R)	85	23	110	150
pRN8027	cop-622(R)	32	22	54	120
pRN8061	cop-623(D)	110	130	240	400

*R = recessive; D = dominant or overlap mutant.

the approximate total pools on the basis of the type of mutant plasmid that is present. Thus, for recessive mutants, the pool is approximately twice the size of that for the wild-type alone; here, both plasmids can synthesize RepC but only one can synthesize active inhibitor. In order to achieve the regulatory inhibitor level, the inhibitor-synthesizing plasmid must reach its normal (wild-type) copy number. Since the inhibitor-negative plasmid can compete equally for RepC, it also reaches the same level; hence, the double-sized pool. For dominant mutants, only the mutant can synthesize RepC but both plasmids utilize it; thus, the overall pool should be somewhat less than that of the mutant alone, which is roughly true in most cases.

Incompatibility

Although indirectly regulated plasmids do not show exclusion in tests such as those just described, they do show segregational incompatibility. The segregation rates for heteroplasmids containing isogenic replicons can be predicted on the basis of copy numbers according to a Markov Chain analysis for random replication plus random assortment with equipartitioning (19). In practice, determination of segregation rates for heteroplasmids with high copy numbers is inconvenient and we have adopted a different strategy for evaluating incompatibility. This strategy involves a phenomenon referred to as reciprocal intrapool variation (20), in which selection for one of the plasmids in a heteroplasmid pool causes amplification of that plasmid at the expense of the other if the 2 are segregationally incompatible but not if they are compatible. A typical example for wild-type incompatible (Inc3) plasmids pT181 and pSA5000, and for compatible plasmids pT181 and pC194 is shown in Fig. 13 and 14. Note in particular that the reversal of relative copy numbers is virtually complete before there is any detectable segregation for the incompatible pair and that there is no detectable variation for the compatible pair. Note also that the overall copy pools remain constant in both cases. Similar results are observed for heteroplasmid strains containing wild-type and mutant or pairs of mutant plasmids, except that where only one of the plasmids can synthesize RepC, the overall pool does not remain constant. Selection for the RepC-producing plasmid elevates the overall pool and selection for the other lowers the overall pool, very much as one would expect (20).

Inhibitor-target specificity and incompatibility

As has been shown for ColE1 (13) and for several IncFII plasmids (21), mutations affecting the specificity of inhibitor-target recognition for directly regulated plasmids have profound effects on segregational incompatibility; if the mutant inhibitor is less active against the target of another mutant or the wild-type, then the strength of the observed segregational incompatibility is decreased

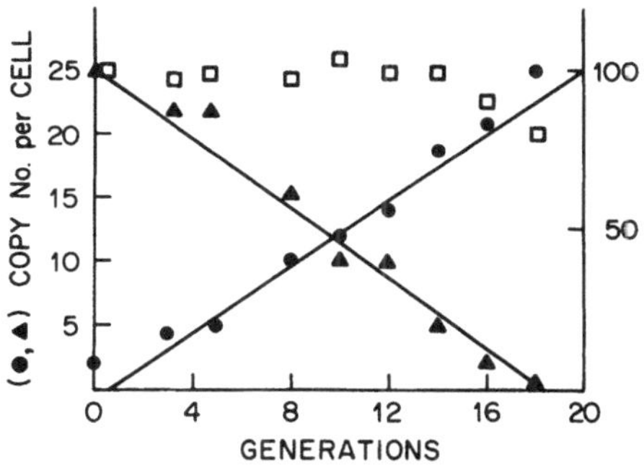

Fig. 13. Reciprocal intrapool variation for isogenic wild-type plasmid replicons. A culture of a strain containing pT181 and pSA5000 was grown exponentially for 20 generations in CY broth containing Cm and Tc at 20 and 1 μg/ml, respectively, then shifted into CY broth containing Cm and Tc at 1 and 10 μg/ml, respectively, and followed for a further 18 generations of exponential growth. The culture was sampled periodically and analyzed by gel electrophoresis for the relative amounts of the 2 plasmids and simultaneously plated to score for the heteroplasmid population. Solid triangles: copy number of pSA5000; solid circles: copy number of pT181; open squares: percentage of heteroplasmids.

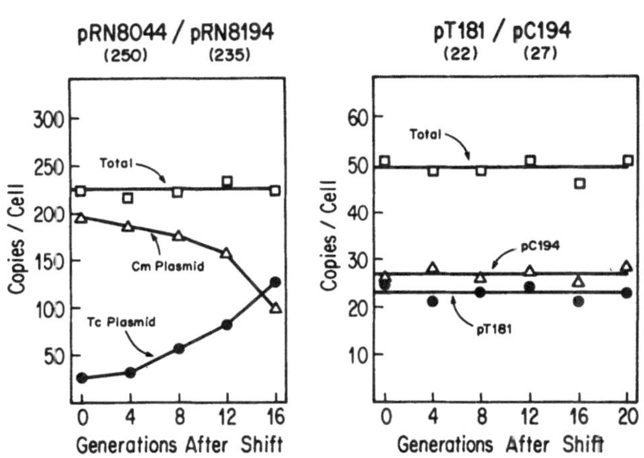

Fig. 14. Effect of antibiotic selection on heteroplasmids containing pT181 copy mutants (left) and compatible plasmids (right). Experiments performed as in Fig. 13. The 2 mutants whose intrapool data are shown at left have non-cross-reacting inhibitors and targets. The compatible plasmids used in the experiment shown at right are pT181 and pC194. The cultures were assayed periodically for plasmid content and for heteroplasmid frequency.

(13); if the recognition specificities for the contralateral inhibitors and targets are asymmetric, then so is the segregational incompatibility; the plasmid that is more sensitive to the other's inhibitor is lost more rapidly from the heteroplasmid strain. In extreme cases, where the inhibitor-target recognition specificities are changed completely, the 2 plasmids become compatible. This establishment of new incompatibility types is understandable in terms of the stochastic nature of segregational incompatibility; in particular, segregational incompatibility is a consequence of random fluctuations in the individual copy pools of the component plasmids, due to random selection for replication and random assortment for partitioning. Since the self-corrective mechanism does not distinguish the component plasmids, it cannot correct the disproportions and the eventual result is segregation at rates that are predictably determined by the overall copy number of the heteroplasmid pool. The mutations that change inhibitor-target recognition serve to establish independent self-correcting copy pools for any 2 or more such plasmids in the heteroplasmid state; this has the effect of making the 2 plasmids segregationally compatible. It is also obvious that in such cases, if one were to clone the inhibitor determinant to an independent replicon, it would have an effect on other plasmids in trans that was a reflection of its target recognition specificity. This is amply demonstrated by the experiments of Lacatena and Cesareni (11,12) with ColE1 mutants; conservely, it is predicted that mutants selected for insensitivity to the wild-type (or mutant) inhibitor would show the same type of incompatibility changes.

For indirectly regulated plasmids, it is predicted that in no case will changes in inhibitor-target specificity alter the segregational incompatibility properties of the plasmid, despite the fact that the mutant plasmid may be totally indifferent to the passively cloned inhibitor gene of the wild-type or of another mutant. This is because the product of the regulated step is mutually available; any attempt by one plasmid to correct fluctuation in its copy number will also affect the other and, moreover, will make matters worse. For example, if one of the pair happens to suffer a relative diminution of copy number, the concentration of its own inhibitor will also decrease and it will then increase its rate of rep protein synthesis. However, the other plasmid will not only utilize this new product, it will utilize it with higher probability due to its elevated copy number; in other words, as soon as any significant fluctuation occurs, the plasmid with the lower copy number is doomed to a stochastic death.

This prediction has been verified with pT181. The inhibitor region (nucleotides 158-570) has been cloned to pE194 for several of the dominant pT181 copy mutants and the cloned inhibitor has been found to have lower activity than the wild-type inhibitor towards either wild-type or other dominant mutants. In some combination, no inhibitory activity could be demonstrated, the target plasmid having the same copy number in the presence as in the absence of the cloned inhibitor determinant. Heteroplasmids were constructed for several pairs of plasmids whose inhibitors and targets were mutually indifferent to one another by this test and these were tested for segregational incompatibility both by growth in non-selective medium and by the reciprocal intrapool variation test. Because of the relatively high copy numbers of these mutants, segregational incompatibility was difficult to measure accurately, although it was always demonstrable. As shown in Fig. 14, reciprocal intrapool variation was readily demonstrable for one of these pairs. The rate of reversal of the pool was somewhat slower than that for the wild-type pair; this difference is tentatively attributed to the larger size of the mutant copy pool; experiments to test this by means of isogenic plasmids with the same copy pool are in progress.

ACKNOWLEDGMENTS

This work was supported by research grants from the American Cancer Society (MV-8D) and National Institutes of Health (GM14372) to R.P.N. S.J.P. held a National Research Service Award from the National Institutes of Health (F32-GM07899); C.C.K. was a postdoctoral fellow of the Leukemia Society of America, Inc.; S.C., A.G., and S.K.H. were trainees in the Graduate Assistants Program of New York University Medical School. Expert technical and secretarial assistance of Soraya Moghazeh and Paula Offricht, respectively, are gratefully acknowledged.

REFERENCES

1. Grindley, N.D.F., J.N. Grindley, and W.S. Kelley (1978) Mutant plasmid with altered replication control. In Microbiology - 1978, D. Schlessinger, ed. Am. Soc. Microbiol., Washington, D.C., pp. 71-73.
2. Khan, S.A., and R.P. Novick (1983) Complete nucleotide sequence of pT181, a tetracycline resistance plasmid from staphylococcus aureus. Plasmid 10:251-259.
3. Novick, R.P., G.K. Adler, S. Majumder, S.A. Khan, S. Carleton, W. Rosenblum, and S. Iordanescu (1982) Coding sequence for the pT181 repC product: A plasmid-coded protein uniquely required for replication. Proc. Natl. Acad. Sci., USA 79:4108-4112.
4. Khan, S., S. Carleton, and R.P Novick (1981) Replication of plasmid pT181 DNA in vitro: Requirement for a plasmid coded product. Proc. Natl. Acad. Sci., USA 78:4902-4906.
5. Khan, S.A., G.K. Adler, and R.P. Novick (1982) Functional origin of replication of pT181 plasmid DNA is contained within a 168-base-pair segment. Proc. Natl. Acad. Sci., USA 79:4580-4584.
6. Kumar, C., and R.P. Novick (1984) Plasmid pT181 replication is regulated by two countertranscripts. Submitted to Proc. Natl. Acad. Sci., USA.

7. Novick, R.P., G.K. Adler, S.J. Projan, S. Carleton, S.K. Highlander, A. Gruss, S.A. Khan, and S. Iordanescu (1984) Control of pT181 Replication I. The pT181 copy control function acts by inhibiting the synthesis of a replication protein. Submitted to Embo J.
8. Carleton S., S.J. Projan, S.K. Highland, S.M. Moghazeh, and R.P. Novick (1984) Control of pT181 Replication II. Mutational analysis. Submitted to Embo J.

9. Eckhardt, T. (1978) A rapid method for the identification of plasmid deoxyribonucleic acid in bacteria. Plasmid 1:584-588.
10. Projan, S.J., S. Carleton, and R.P. Novick (1983) Determination of plasmid copy number by fluorescence densitometry. Plasmid 9:182-190.
11. Lacatena, R.M., and G. Cesareni (1981) Base pairing of RNA I with its complementary sequence in the primer precursor inhibits ColE1 replication. Nature 294:623-626.
12. Lacatena, R.M., and G. Cesareni (1983) Anatomy of the target of the inhibitor of ColE1 replication. In Abstracts of the 1982 EMBO Conference, "Replication of Prokaryotic DNA," The Netherlands, pp.62-65.
13. Tomizawa, J.-I., and T. Itoh (1981) Plasmid ColE1 incompatibility determined by interaction of RNA I with primer transcript. Proc. Natl. Acad. Sci., USA 78:6096-6100.
14. Brady, G., J. Frey, H. Danbara, and K. Timmis (1983) Replication control mutations of plasmid R6-5 and their consequences for interactions of the RNA-I control element with its target. J. Bact. 154:429-436.
15. Nordstrom, K., S. Molin, and H. Aagaard-Hansen (1981) Plasmid R1 incompatibility. In Molecular Biology, Pathogenicity, and Ecology of Bacterial Plasmids, R.C. Clowes, S.B. Levy, and E.L. Koening, eds. Plenum Press, New York, pp.291-301.
16. Zuker, M., and P. Stiegler (1981) Optimal computer folding of large RNA sequences using thermodynamics and auxiliary information. Nucleic Acids Res. 9:133-148.
17. Tomizawa, J.-I., and T. Itoh (1982) The importance of RNA secondary structure in ColE1 primer formation. Cell 31:575-583.
18. Pritchard, R.H., P.B. Barth, and J. Collins (1969) Control of DNA synthesis in bacteria. Symp. Soc. Gen. Microb. 19:263-297.
19. Novick, R.P., and F. Hoppensteadt (1978) On plasmid incompatibility. Plasmid 1:421-434.
20. Projan, S.J., and R.P. Novick (1984) Reciprocal intrapool variation in plasmid copy numbers: A characteristic of segregational incompatibility. Plasmid (in press).
21. Easton, A.M., and R.H. Rownd (1982) The incompatibility product of IncFII R plasmid NR1 controls gene expression in the plasmid replication region. J. Bact. 152:829-839.
22. Horinouchi, S., and B. Weisblum (1982) Nucleotide sequence and functional map of pE194, a plasmid that specifies inducible resistance to macrolide, linosamide and streptogramin type B antibiotics. J. Bact. 150:815-825.

CONSTRUCTION OF ColEl RNAl MUTANTS AND ANALYSIS OF THEIR FUNCTION IN VIVO

Barry Polisky, Joseph Tamm, and Tim Fitzwater

Program in Molecular, Cellular, and Developmental Biology
Department of Biology
Indiana University
Bloomington, Indiana 47405

ABSTRACT

We have carried out experiments designed to investigate the relationship between structure and function for the ColEl RNAl species. RNAl is a small RNA (108 nucleotides) that has been implicated in copy number control of the multicopy plasmid ColEl. In vitro, RNAl inhibits the processing of the primer precursor required for initiation of DNA replication. The RNAl gene is entirely complementary to the 5'-terminal region of the primer. We have functionally separated these 2 RNA species by cloning the RNAl gene downstream from the _S. marscesens trp_ promoter. When cloned in a ColEl-compatible plasmid, a _trp_-RNAl fusion has been shown to mediate ColEl-type incompatibility _in vivo_. The construction scheme described here also generates mutant RNAl species with altered sequences at the 5' terminus of RNAl which have been assayed for function in vivo. These experiments have indicated that sequences at the 5' terminus play a critical role in RNAl function.

INTRODUCTION

ColEl is a small plasmid of _Escherichia coli_ that has proven to be a valuable model for studies of DNA replication control. By biochemical and genetic dissection of the sequence and replication machinery requirements of this plasmid, it has been possible to isolate and characterize the major plasmid-encoded elements involved in the initiation and regulation of plasmid replication. ColEl exists at about 20-30 copies per chromosome and is totally dependent on host replication enzymes (1,2). Its replication does not require any plasmid-encoded information that must function in trans in a "positive" genetic sense (1,2). Indeed, a variety of genetic experiments involving copy-number mutants have indicated that the major mode of control was via the synthesis of plasmid-specific, negatively acting replication control elements (3,4).

The development of an in vitro replication system using supercoiled plasmid DNA and purified enzymes has revealed a detailed picture of the minimal requirements for high-fidelity replication initiation. The enzymatic components are RNA polymerase, DNA polymerase I and ribonuclease H (5). The ColE1 replication primer is initiated 555 nucleotides upstream from the origin by RNA polymerase and is endonucleolytically processed by RNase H to create a functional 3'-OH onto which DNA polymerase I adds deoxyribonucleotides (5).

It is obvious that ColE1 replication must be regulated at some level to account for its constant copy number and its genetic stability. A number of lines of evidence indicate that primer formation is the locus of control in vitro and that a central regulatory factor is a small RNA species encoded by ColE1 (6). This small RNA is transcribed from the DNA strand opposite to that encoding the primer and is entirely complementary to 5' terminal 108 nucleotides of the primer (7). Addition of purified ColE1 RNA1 to an in vitro ColE1 replication system results in inhibition of RNase H processing of the nascent primer and an inhibition of addition of deoxyribonucleotide incorporation by DNA polymerase I (6).

Like other plasmids, ColE1 displays incompatibility in vivo towards related replicons. One way to tie the in vivo incompatibility behavior with the in vitro replication results is provided by the incompatibility model proposed by Pritchard and colleagues (8,9). In its broadest sense this model suggests that incompatibility is a consequence of negative replication control mediated by trans-acting repressors which have specificity for replicons of a given incompatibility group. In conjunction with a random replication and/or segregation scheme this proposal can account for the incompatibility behavior of ColE1-type plasmids. In this model RNA1 is the molecular basis of incompatibility, acting in trans to lower the probability of DNA initiation as a consequence of its interaction with the replication primer (10,11).

We are interested in the structure and function of RNA1 for several reasons. First, we wanted to investigate whether RNA1 is sufficient to mediate ColE1 incompatibility in vivo. Second, RNA1 represents a novel type of repressor molecule in that its activity seemed likely to be mediated largely through its secondary and tertiary structural features and those of its target, the replication primer (10,11). Our long-term goal was to establish a genetic system with which we could introduce mutations specifically in RNA1, test the effects of these changes in vivo on ColE1 replication, and then investigate the structural features of RNA1 that had been altered by these changes.

RESULTS AND DISCUSSION

Because of the physical overlap between RNA1 and the primer, point mutations in one create complementary changes in the other. We isolated a DNA fragment containing the RNA1 structural gene, but lacking functional elements of both the RNA1 promoter as well as primer promoter. This fragment, a 148 bp HaeIII-FNUdIII segment, was isolated from the cop^{+} ColE1 plasmid pNOP42 (4). The scheme by which it was placed under the transcriptional control of the *Serratia marcesens* tryptophan promoter is depicted in Fig. 1a. The RNA1-containing fragment has blunt ends to which synthetic *Hind*III linkers were ligated. The resulting fragment was cloned into the unique *Hind*III site immediately downstream of the *S. marcesens trp* promoter present in the pBR322-derivative plasmid pM59. The resulting plasmid was designated pTF114. pTF114 also carries the pBR322-derived RNA1 gene and is ColE1-incompatible. To create a plasmid whose ColE1-incompatibility determinants were derived from the *trp*-RNA1 moiety, we replaced the pBR322 origin region with a 2055 bp *Pst*I-*Sma*I fragment containing the replication origin of the ColE1-compatible plasmid pACYC177 (12). These manipulations created the plasmid designated pTF275 (Fig. 1a). To provide sufficient quantities of the *trp* repressor to control the multicopy *trp* operator we introduced a 1250 bp *Bam*HI fragment that contains the *E. coli trp* repressor structural gene (13). This fragment was introduced into the unique *Bam*HI site of pTF114. The *Hind*III downstream from the 3' end of the RNA1 gene in pTF275 was deleted after partial *Hind*III digestion and *Bal*31 treatment. These manipulations created a plasmid with a unique *Hind*III site located between the *trp* promoter and the ColE1 RNA1 gene. This plasmid was designated pTF325 (Fig. 1a).

Subsequent experimental steps were designed to create fusions of various types between the *trp* promoter and the RNA1 gene. Features of the DNA sequence relevant to these steps are shown in Fig. 1b. The overall strategy was to open the plasmid at the *Hind*III site and remove nucleotides from both termini with limited *Bal*31 nuclease digestion. Following ligation of the resulting ends the altered molecules were introduced into recipient *E. coli* by transformation. It should be noted that in pTF325 the ColE1 RNA1 gene transcript would be expected to contain an extra 49 residues at its 5' terminus if transcriptional initiation occurred at the *trp* start (see leftmost wavy line in Fig. 1b). In in vitro transcription experiments carried out with supercoiled PTF325 DNA and *E. coli* RNA polymerase a prominent band with the expected mobility was observed (results not shown). In addition, we have observed that

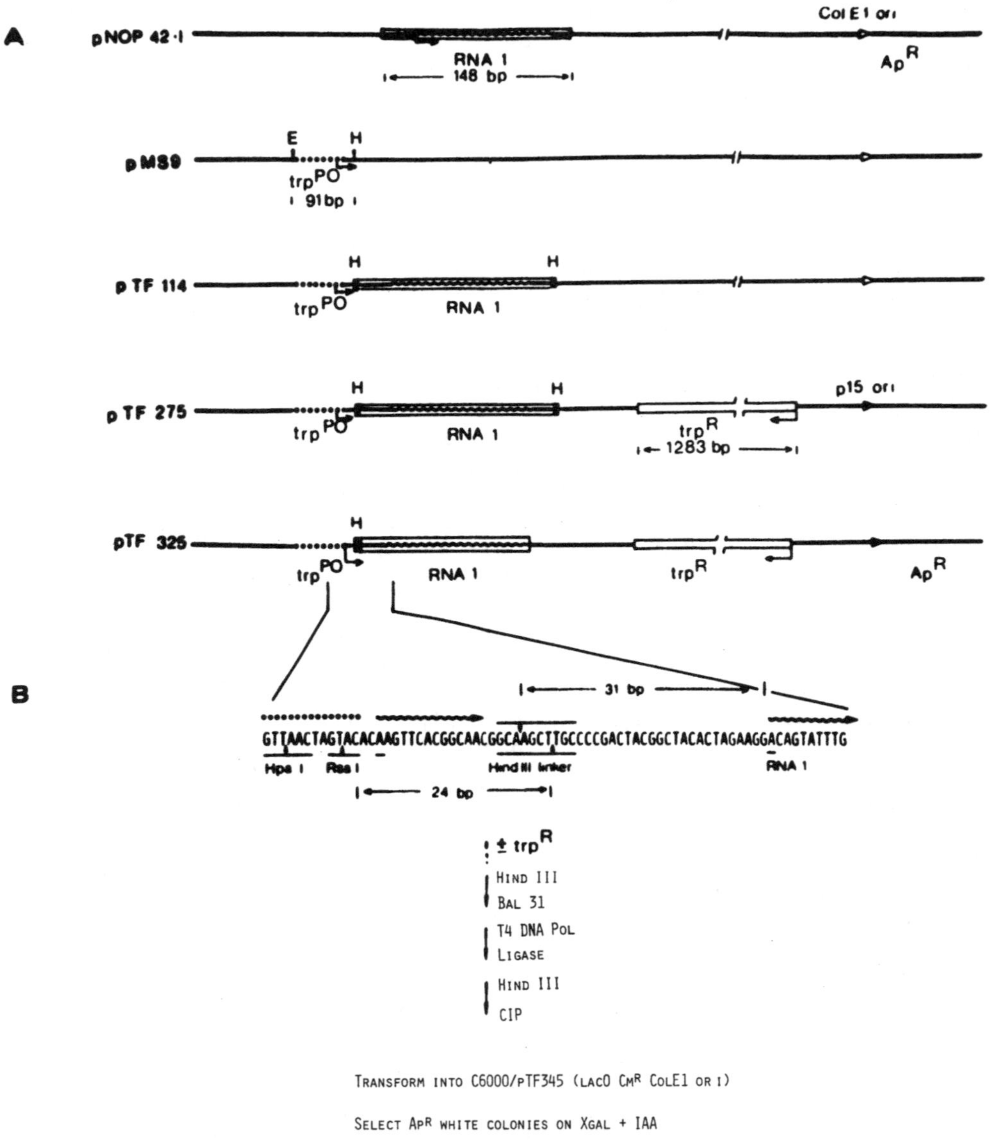

Fig. 1. (a) Scheme for the construction of trp promoter-RNA1 fusion plasmids; pNOP42·I is a 4.2x10^3 bp cop$^+$ ColE1 plasmid. A 148 bp HaeIII-FnudII fragment was isolated from this plasmid by electroelution from a 6% (w/v) polyacrylamid gel. This fragment contains the coding region for RNA1 but lacks the RNA1 promoter and the replication primer promoter. Synthetic decameric HindIII linkers (BRL) were attached to this fragment which was then cloned into the unique HindIII site of pMS9. A plasmid with the proper orientation of insertion of the RNA1 fragment with re-

Fig. 1. (Continued)

spect to the *trp* promoter was selected and designated pTF114. pTF114 contains a ColE1-type replication origin, is incompatible with ColE1, and encodes Ap^R. A 2055 bp *PstI*-*SmaI* fragment containing the replication origin from the ColE1-compatible plasmid pACYC177 was inserted to replace the *PstI*-*PruII* fragment of pTF114 which contains the replication origin. The insertion re-creates the *bla* gene and the resulting plasmid encodes Ap^R. In addition, a 1250-bp *BamHI* fragment which encodes *trpR* from *E. coli* was isolated from ptrpR3 (13). This fragment was inserted into the unique *BamHI* site of pTF114. These manipulations resulted in pTF275 which was treated with *HindIII* under conditions designed to partially cleave at one of the 2 *HindIII* sites. One-cut linear molecules were isolated from a 1% (w/v) agarose gel and treated with 0.1 unit of *Bal31* nuclease/μg DNA for 25 min at 30°C. Following extraction with phenol and precipitation with ethanol, 100 units of DNA ligase were added and the mixture was used to transform *E. coli* C600 cells with selection for Ap^R. Transformats were analyzed for a plasmid that retained the *trp* promoter-proximal *HindIII* site but lacked the distal *HindIII* site. A plasmid designated pTF325 was isolated by this procedure. pTF325 is deleted for approximately 75 bp and lacks the distal *HindIII* site.

(b) The nucleotide sequence (hyphens omitted for clarity) of pTF325 in the region containing the *trp* operator, the unique *HindIII* site and the 5'-proximal nucleotides of ColE1 RNA1. The dots over the sequence designate sequences relevant for *trp* repressor binding. The underlined A residue adjacent to the *RsaI* site is the initiating nucleotide of the truncated *trp* leader sequence. The wavy line designates the start of RNA1 transcription and the initiating A residue of RNA1 is also underlined. The decameric *HindIII* site is shown with lines above and below the sequence. Arrows designate the sizes of particular regions or the direction of transcription.

Deletions were introduced into pTF325 as follows. 20 μg of pTF325 DNA were cut with *HindIII* (Biolabs) in TA buffer (33 mM-Tris-acetate, pH 7.9, 66 mM-potassium acetate, 10mM-magnesium acetate, 100 μg bovine serum albumin/ml, 0.5 mM-dithiothreitol) to completion, extracted with phenol, precipitated with ethanol, and resuspended at 1 μg/μl in TE buffer (10 mM-Tris·HCl, 1 mM-EDTA, pH 8). 1 μg portions in 25 μg 1 x *Bal31* salts (200 mM-NaCl, 12.5 mM-$CaCl_2$, 12.5 mM-$MgSO_4$, 20 mM-Tris·HCl, pH 8.1, 1 mM-EDTA) were treated with 0.1 unit *Bal31* (BRL) at 30°C for 25 min. Pilot experiments indicated that these conditions removed an average 30 bp from each end. After *Bal31* treatment,

Fig. 1. (Continued)

portions were pooled, extracted with phenol, and precipitated with ethanol. The DNA was then treated with 2.5 units T4 DNA polymerase/μg (BRL) in TA buffer at 37°C for 5 min. 2 mM-dNTPs were added and repair proceeded for 60 min at 37°C, followed by 70°C heat-inactivation for 10 min. 4 μg of DNA were ligated with 100 units T4 DNA ligase in TA buffer plus 1 mM-ATP at 14°C overnight, and heat-inactivated at 70°C for 10 min, followed by *HindIII* digestion at 37°C for 2 hr to eliminate any residual supercoiled DNA. DNA was then extracted with phenol precipitated with ethanol, and resuspended in 50 μl of CIP buffer (50 mM-Tris·HCl, pH 9.0, 1 mM-$MgCl_2$, 0.1 mM-$ZnCl_2$, 1 mM-spermidine) and treated with 1 unit calf intestinal phosphatase (CIP; Boehringer Mannheim) at 37°C for 30 min. to prevent in vivo ligation of *HindIII* ends that had escaped *Bal31* digestion. An additional unit of CIP was then added and incubated as before. CIP was inactivated by the addition of 40 μl water, 10 μl 10 x TNE buffer (100 mM-Tris·HCl, pH 8.0, 1m-NaCl, 10 mM-EDTA), and 5 μl of 10% (w/v) sodium dodecyl sulfate followed by incubation at 68°C for 15 min. DNA was then extracted with phenol and precipitated with ethanol prior to resuspension in TE buffer (pH 8.0) for transformation. IAA, indoleacrylic acid.

pTF325 is compatible with ColE1-derived plasmids in vivo under conditions in which *trp*-directed transcription is expected to be constitutive. These results indicated that the ColE1 RNA1 produced by pTF325 was not capable of mediating ColE1-type incompatibility in vivo, most plausibly as a consequence of structural alterations in the fusion RNA1. Consequently, it was possible to select for functional *trp*-RNA1 fusions by introducing the *Bal31*-treated pTF325 molecules into an *E. coli* strain harboring a ColE1 plasmid and screening recipients for those which, upon subsequent growth, had lost the ColE1 resident. For this purpose we used a ColE1 plasmid designated pTF345. This plasmid confers chloramphenicol resistance (CM^R) and also carries a fragment containing the *lac* operator of *E. coli*. As a consequence of its multicopy nature, this plasmid titrates the endogenous *lac* repressor making the host constitutive for β-galactosidase expression (14). Consequently, colonies harboring pTF345 are blue on *X-gal* plates. We expected that a functional *trp*-RNA1 fusion would displace pTF345 resulting in a white Cm^S colony on *X-gal* plates. Indoleacrylic acid, an inducer of the *trp* operon (15), was also included to ensure that the *trp* promoter was active.

Using this scheme we isolated a fusion derivative designated pTF421 that had the expected properties. DNA sequence analysis across the fusion joint indicated that a 50 bp deletion occurred in pTF421. *In vitro*, this plasmid produces an RNA1 with mobility identical to that of *cop*$^+$ pNOP42. Finally, analysis of the 5'-terminal sequences of pTF421-encoded ColE1 RNA1 indicated that it is identical to wild-type ColE1 RNA1 (results not shown). Subsequent transformation experiments revealed that pTF421 could displace resident ColE1 plasmids even in the presence of excess tryptophan, suggesting that RNA1 synthesis was constitutive. DNA sequence analysis demonstrated that the basis for this expression was the extent of the deletion on the *trp* side which resulted in a defective operator sequence (results not shown).

The characteristics of pTF421 demonstrated that the ColE1 RNA1 structural gene provides sufficient information to convert a ColE1-compatible plasmid into a ColE1-incompatible plasmid. The results indicate that RNA1 is an important incompatibility determinant in vivo but do not prove that it is the only ColE1-encoded component involved in this phenomenon.

In addition to pTF421 we have isolated other *Bal*31 fusion derivatives that are capable of displacing a resident ColE1 plasmid. Some of these plasmids displace ColE1 residents with apparently slower kinetics than pTF421. Two possible explanations could account for this behavior. The spacing between the *trp* promoter and the 5 'terminus of wild-type RNA1 might have been altered resulting in less efficient expression of a functional RNA1 species in vivo. Alternatively the *Bal*31 deletion step might have resulted in efficiently expressed altered RNA1 species that contained sequence changes at its 5' terminus such as 5' extensions, deletions, or fusions with upstream sequences.

As a qualitative measure of the *trp* promoter strength we have transcribed candidate plasmids in vitro with *E. coli* RNA polymerase and assessed the extent of the transcription of the *trp*-RNA1 species. For the plasmids described in this report we detected no differences between the deletion derivatives and the parent plasmid pTF325 (results not shown).

To obtain a quantitative measure of differences on ColE1-type incompatibility mediated by different deletion derivatives we carried out segregation tests in vivo. Over a period of 80 generations of growth in the absence of selection a wide variety of segregation kinetics were observed (Fig. 3). pTF421, encoding wild-type RNA1 segregated essentially completely from pTF345 within 8 generations. As noted previously, pTF325, in which 49 nucleotides are appended to the 5' terminus of RNA1, displays no incompatibility with pTF345. The sequence of several other derivatives are schematically depicted in Fig. 2b. pTF370 encodes an RNA1 with 5 additional residues at its 5' terminus and displaces pTF345 with slower kinetics than pTF421. pTF365 has 15 additional residues and does not detectably displace pTF345.

Two additional derivatives are worthy of mention. These were picked from a collection of colonies in which no segregation between the incoming *Bal*31 plasmid and the resident occurred. We examined

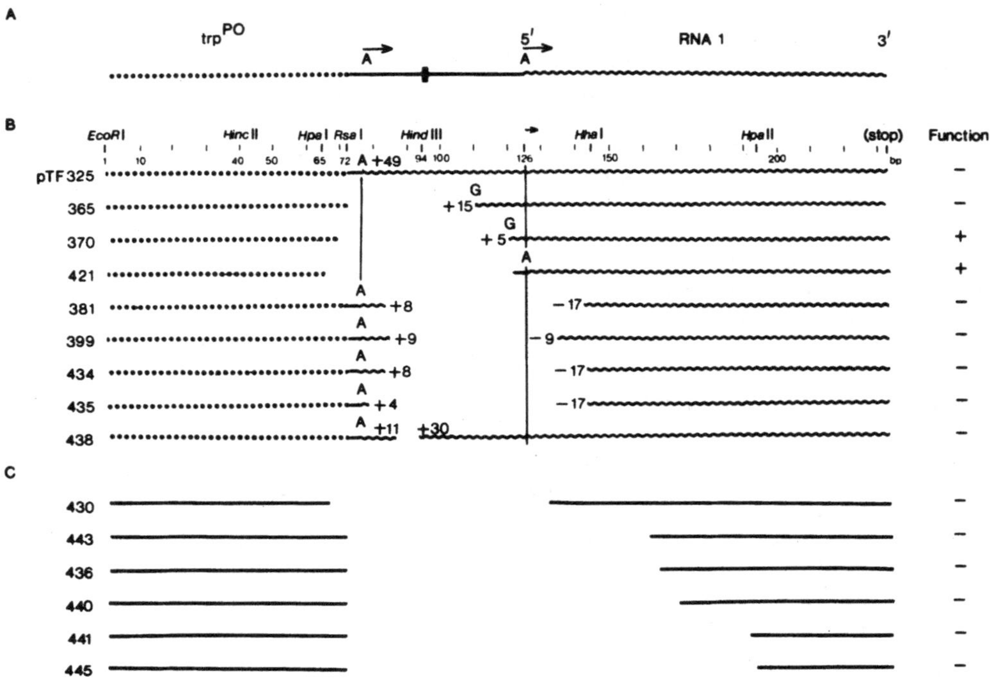

Fig. 2. Schematic diagram of the sequences of trp-RNA1 fusions. (a) trp-RNA1 fusions plasmid before Bal31 deletion. Dotted line designates trp promoter-operator sequences. The A adjacent to the promoter indicates the trp leader A. Arrows designate the direction of transcription which is capable of initiation either from the trp leader A start or the RNa1 A start depending on the extent of Bal31 digestion. The dark vertical block represents the location of the HindIII site in pTF325.

(b) Wavy lines designate transcription (left to right). For each plasmid shown, the initiating nucleotide and relative position of initiation is depicted. Plus and minus signs refer to nucleotide positions relative to the trp leader A or the correct A start of wild-type RNA1. Thus, pTF365-encoded ColE1 RNA starts with a G residue 15 nucleotides upstream from the correct A start. Several of the plasmids shown initiate RNA1 transcription from the trp leader A start. Deletion of 5' portions of RNA 1 in some plasmids has resulted in fusions of trp sequences to remaining RNA1 sequences.

Sequences of the trp-RNA1 fusion sequences were determined by analysis of transcripts produced in vitro with E. coli RNA polymerase or by M13 DNA sequencing. RNA transcripts were prepared from supercoiled DNA, purified by CsCl cen-

Fig. 2. (Continued)

trifugation, using [$\gamma^{-32}P$] GTP or ATP as labels. The labeled trp-RNA1 bands were isolated and digested with T1 ribonuclease under standard sequencing conditions (21) or subjected to limited alkaline hydrolosis. The position of each G or A residue was then determined by comparison to the known DNA sequence of the region. This approach permitted unambiguous assignment of the 5' end.

The column labeled "Function" refers to the ability of RNA1 produced by the various plasmids to mediate ColE1 incompatibility in vivo.

(c) Deletion size estimates were made on the basis of restriction fragment analysis. Plasmid DNA was cut with EcoRV, HindII, and BamH1 in 1XTA buffer and run on a 6% (w/v) polyacrylamide gel.

fusion structures to isolate constructs that lacked biological activity as a result of inappropriate 5'-terminal sequences. Two plasmids, pTF381 and pTF399, were shown by restriction analysis to contain 58 bp and 48 bp deletions, respectively. Both plasmids have a functional trp promoter in vitro (results not shown). We have determined the RNA sequence of the RNA1 species produced by these plasmids. Each produce a hybrid RNA1 that contains sequences derived from the trp leader transcription initiation point fused to a short deletion of 5'-proximal RNA1 sequences (depicted schematically in Fig. 2b). In pTF381 the first 8 residues of the trp transcript are fused to an RNA1 body that is deleted for the first 17 nucleotides. In pTF399 9 nucleotides of the trp leader sequence are fused to an RNA1 body that is deleted for the first 9 residues. Consequently, pTF399 RNA1 is the same length as wild-type RNA1 but differs in 7 of 9 positions. The first 9 nucleotides of wild-type RNA1 are known to be in an "accessible" configuration as measured by nuclease and chemical probes of secondary structure (16). The inability of pTF399 RNA1 to mediate ColE1 incompatibility in vivo suggests that the 5'-tail region of RNA1 is critical for RNA1 function and/or stability in vivo.

In addition to these plasmids several other trp-RNA1 fusion joints have been examined at the RNA sequence level. These RNA1 structures are depicted schematically in Fig. 2b. The derivatives designated pTF434 and 435 encode hybrid RNA1 species which initiate at the trp leader A residue and contain 8 or 4 nucleotides from the trp sequence, respectively. These leaders are fused to an RNA1 body deleted for 17 nucleotides at the 5' terminus of RNA1. Like pTF381 these RNA1 species are non-functional in mediating ColE1 incompati-

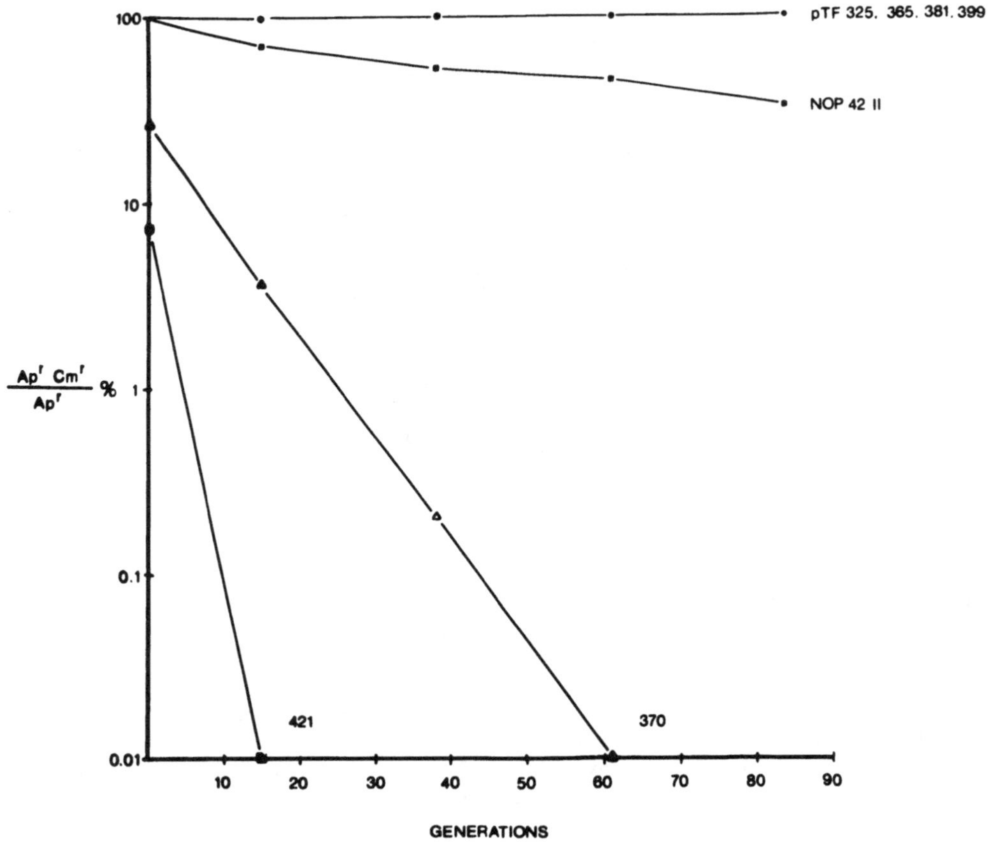

Fig. 3. Kinetics of segregation of trp-RNA1 fusion plasmids and a ColE1 resident plasmid. Plasmid DNA was transformed into the C600 strain of *E. coli* containing the ColE1 resident plasmid pTF345 (Cm^r), and plated on VB minimal medium E plates supplemented with 0.2% (w/v) glucose, 50 μg threonine/ml, 50 μg leucine/ml, 12 μg vitamin B1/ml, 100 μg ampicillin/ml and 30 μg chloramphenicol/ml. Zero hour points were obtained by diluting colonies in 0.85% (w/v) NaCl and plating 100 μl on VB soft agar containing 10 μg tetrazolium chloride/ml and 100 μg ampicillin/ml or 100 μg ampicillin/ml plus 30 μg chloramphenicol/ml to determine the percentage of Ap^r Cm^r/Ap^r cells. Diluted cells were also inoculated into 3 ml of VB medium containing 100 μg ampicillin/ml and allowed to grow for approximately 18 generations prior to plating. Subsequent time points were obtained by periodic dilution of cells grown into stationary phase followed by plating as above.

bility in vivo (results not shown), presumably because of alterations in secondary structure. However, effects on RNA1 stability in vivo have not been ruled out for these derivatives. Finally, pTF438 has 11 trp residues fused to a +30 RNA1 gene, i.e., an RNA1 gene is generated with an additional 41 residues. Like pTF324, pTF438 is compatible with ColE1 derivatives in vivo. A number of additional constructs with varying sized deletions in the RNA1 gene have also been isolated (Fig. 2c). None of the derivatives encodes a biologically active RNA1 species in vivo (results not shown).

One unexpected result which emerges from these experiments concerns the importance of the 5' terminal region of RNA1 for its ability to function as a replication repressor (20). Mutations affecting a limited number of residues only in the 5' "tail" region of RNA1 lead to a complete loss of activity (pTF399). More recent results using a different method of selecting RNA1 defective mutations has revealed that a single base-pair alteration in the 5' tail region can inactivate RNA1 (T. Dooley, pers. comm.). Preliminary secondary structural analysis of these mutant RNA1 species has indicated that no alterations are detectable in the 3 stems and loops. These results demonstrate that the 5'-tail of RNA1 is critical for RNA1 function. One possible role for this region is in providing a nucleation region for interaction with the primer precursor. Alternatively this region may be involved in tertiary intramolecular interactions in RNA1.

These experiments demonstrate that a small plasmid-encoded RNA molecule plays a key role in plasmid incompatibility in vivo. The basic result is consistent with the behavior of RNA1 in a purified replication system in vitro where it has been shown to negatively modulate proper processing of the replication primer (6,10). An important question concerns whether RNA1 is the sole ColE1-mediated incompatibility determinant. It should be noted that ColE1 encodes a second negative, trans-acting copy number determinant, designated the rop function (17). The rop gene encodes a small polypeptide of 63 amino acid residues which is capable of decreasing the extent of transcription initiated from the primer promoter in vivo (17,18). Recent experiments of R. Lacatena and G. Cesareni (pers. comm.) indicate that the effects of the rop gene product on primer transcription require ColE1 RNA1. Although a molecular mechanism for the action of the rop gene product is not yet at hand, several possibilities are apparent. RNA1 and the rop protein may interact directly and the resulting RNA-protein complex may be a more potent inhibition of primer processing than either alone. Alternatively, RNA1 and rop may have separate targets on the primer precursor. It is clear from deletion studies that RNA1 can function in the absence of rop (19), raising the question of whether the rop protein serves as a secondary, fine-tuning copy number control mechanism. Additional experiments using purified components in an in vitro replication system should permit their roles to be more precisely defined.

ACKNOWLEDGMENTS

We thank C. Yanofsky, D. Gelfand, and R. Somerville for strains and plasmids. This work was supported by National Institutes of Health grant GM 24212. B.P. was supported by an NIH Research Cancer Development Award.

REFERENCES

1. Donoghue, D.J., and P.A. Sharp (1978) Replication of colicin E1 plasmid DNA in vivo requires no plasmid-encoded proteins. J. Bact. 133:1287-1294.
2. Kahn, M., and D. Helinski (1978) Construction of a novel plasmid-phage hybrid: Use of the hybrid to demonstrate ColE1 DNA replication in vivo in the absence of a ColE1-specified protein. Proc. Natl. Acad. Sci., USA 75:2200-2204.
3. Conrad, S.E., and J.L. Campbell (1979) Role of plasmid-coded RNA and ribonuclease III in plasmid DNA replication. Cell 18:61-71.
4. Shepard, H.M., D.H.Gelfand, and B. Polisky (1979) Analysis of a recessive copy number mutant of the plasmid ColE1: Evidence for negative control of plasmid replication. Cell 18:267-276.
5. Itoh, T., and J. Tomizawa (1980) Formation of an RNA primer to initiation of replication of ColE1 DNA by ribonuclease H. Proc. Natl. Acad. Sci., USA 77:2450-2454.
6. Tomizawa, J., T. Itoh, G. Selzer, and T. Som (1981) Inhibition of ColE1 RNA primer formation by a plasmid-specified small RNA. Proc. Natl. Acad. Sci., USA 78:1421-1425.
7. Morita, M., and A. Oka (1979) The structure of a transcriptional unit of Colicin E1 plasmid. Eur. J. Biochem. 97:435-443.
8. Pritchard, R.H., P.T. Barth, and J. Collins (1969) Control of DNA synthesis in bacteria. Symp. Soc. Gen. Microbiol. 19:163-297.
9. Pritchard, R.H. (1978) Control of DNA replication in bacteria. In DNA Synthesis, I. Molineaux and M. Kohiyama, eds. Plenum Publishing Co. New York, pp. 1-26.
10. Tomizawa, J., and T. Itoh (1981) Plasmid ColE1 incompatibility determined by interaction of RNA1 with primer transcript. Proc. Natl. Acad. Sci., USA 78:6096-6100.
11. Lacatena, R.M., and G. Cesareni (1981) Base pairing of RNAI with its complementary sequence in the primer precursor inhibits ColE1 replication. Nature 294:623-626.
12. Chang, A.C.Y., and S.N. Cohen (1978) Construction and characterization of amplifiable mutlticopy DNA cloning vehicles derived from the p15A cryptic miniplasmid. J. Bact. 134:1141-1156.
13. Roeder, W., and R.L. Somerville (1979) Cloning the trpR gene. Mol. Gen. Genet. 176:361-368.

14. O'Farrell, P.H., B. Polisky, and D.H. Gelfand (1978) Regulated expression by read through translation from a plasmid-encoded β-galactosidase. J. Bact. 134:645-654.
15. Morse, D., R.D. Mosteller, and C. Yanofsky (1969) Dynamics of synthesis, translation and degredation of trp operon messenger RNA in E. coli. Cold Spring Harbor Symp. Quant. Biol. 34:725-740.
16. Tamm, J., and B. Polisky (1983) Structural analysis of RNA molecules involved in plasmid copy number control. Nucl. Acids Res. 11:6381-6397.
17. Cesareni, G., M.A. Muesing, and B. Polisky (1982) Control of ColE1 DNA replication: The rop gene product negatively affects transcription from the replication primer promoter. Proc. Natl. Acad. Sci., USA 79:6313-6317.
18. Som, T., and J. Tomizawa (1983) Regulatory regions of ColE1 that are involved in determination of plasmid copy number. Proc. Natl. Acad. Sci., USA 80:3232-3236.
19. Twigg, A.J., and D. Sherratt (1980) Trans-complementable copy-number mutants of plasmid ColE1. Nature 283:216-218.
20. Lacatena, R.M., and G. Cesareni (1983) Interaction between RNA1 and the primer precursor in the regulation of ColE1 replication. J. Mol. Biol. 170:635-650.
21. Wrede, P., R. Wurst, J. Vournakis, and A. Rich (1979) Conformational changes of yeast tRNA phe and E. coli tRNA glu as indicated by different nuclease digestion patterns. J. Biol. Chem. 254:9608-9616.

INCOMPATIBILITY AND INCFII PLASMID REPLICATION CONTROL

Robert H. Rownd, David D. Womble, Xin-nian Dong, Verne A. Luckow, and Ru Ping Wu*

Department of Molecular Biology
The Medical and Dental Schools
Northwestern University
Chicago, Illinois 60611

ABSTRACT

The DNA coding for replication control and incompatibility of the plasmid NR1 serves as a template in vivo and in vitro for RNA transcription in both directions. In the rightward direction, RNA synthesis begins from 2 different promoters, one of which is regulated and the other constitutive. In vivo, each of these transcripts is more than 1,000 nucleotides long, terminating near the estimated site for the origin of replication. These transcripts serve as messenger RNA for several proteins. One protein (repA1) is required for replication and another (repA2) serves as the repressor for the regulated rightward promoter. RNA synthesis in the leftward direction is constitutive and produces a single transcript of 91 nucleotides which is complementary in sequence to the rightward transcripts. This small transcript is the incompatibility product which regulates the replication of the plasmid. When the intracellular concentration of the small transcript is experimentally varied, the rate of translation of the rightward transcripts and the rate of initiation of replication (plasmid copy number) vary inversely to its concentration. The mode of action of this inhibitor RNA is likely to be formation of an RNA-RNA duplex with the rightward transcripts, thereby inhibiting the translation which would produce the required replication protein. The probability that a rightward transcript will escape interaction with the small RNA molecules and thus allow replication to initiate can be predicted from the laws of

* Present address: Shanghai Institute of Materia Medica, Chinese Academy of Sciences, 319 Yue-Yang Road, Shanghai 20031, China.

mass action based on base-stacking free energies for the likely sequences of initial contact. The estimated intracellular RNA concentrations, based on quantitative hybridization experiments, are in agreement with those predicted from the calculated equilibrium constants for duplex formation.

INTRODUCTION

Plasmid incompatibility is the inability of 2 closely related plasmids which share the same replication control mechanism to stably coexist in the descendants of a bacterial cell. The 90-kb antibiotic resistance plasmid NR1 belongs to the FII incompatibility group and is normally maintained at about 2 copies per chromosome or 4 copies per cell (1,2). The genes required for autonomous replication of NR1 and the related IncFII plasmids R6 and R1 are located within a 2-kb region of the plasmid DNA (3-6) which retains all of the replication and incompatibility properties of the 90 -kb NR1 plasmid (3,7). The nucleotide sequences of the replication regions of NR1, R6, and R1 and of several copy number (*cop*$^-$) or incompatibility (*inc*$^-$) mutants have been determined (8-15).

The replication genes of NR1 are located within 2 contiguous *Pst*I restriction fragments. The origin of replication is in a *Pst*I 1.6-kb fragment (Ref. 16,17; J. Kasner and R. H. Rownd, unpub. data), while incompatibility (*inc*) and copy number (*cop*) control are in a *Pst*I 1.1-kb fragment (3,5-7) (Fig. 1). Rightward transcription toward the replication origin is initiated in vitro at 2 promoters in the *Pst*I 1.1-kb fragment more than 1,000 bp from the replication origin (12-14). The promoter for the 380 base RNA-C is located within the 60-bp *Sau*3A fragment, while the promoter for the 1,150 base RNA-A is located near the rightward end of the 332-bp *Sau*3A fragment (Fig. 1). RNA-C is terminated in vitro at a site about 50 bases into the 262-bp *Sau*3A fragment (12). Evidence presented here (and unpub. data) suggests that transcription reads through this site in vivo to form RNA-CX, which contains all the sequences in RNA-C and RNA-A (Fig. 1). The rightward transcripts could code for several proteins, called repA1, repA2, and repA3 (9), which are produced by minireplicators in minicell systems (9) or in in vitro coupled transcription/translation systems (18). The repA1 protein appears to be required for replication (3,6,9). The rightward transcripts might also serve as primers for initiation of replication at the origin (12,13). Transcription in the leftward direction produces the 91 base RNA-E (12-14,19) which is coded by the opposite strand of the 262-bp *Sau*3A fragment (Fig.1). It is thus complementary in sequence to the rightward transcripts. Most evidence suggests that plasmid incompatibility and replication control are mediated by this small untranslated RNA (7,11,13,19,20).

In this communication we show that the target for the RNA-E incompatibility inhibitor is within a complementary region of the

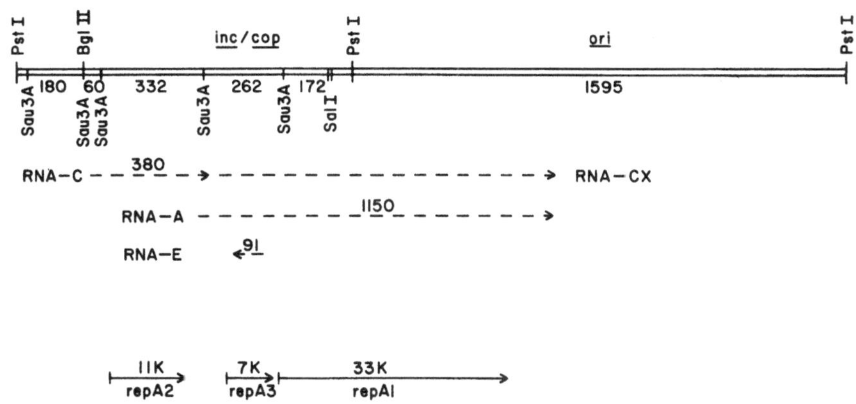

Fig. 1. Linear representation of the DNA from the replication control region of plasmid NR1. This region consists of 2 PstI restriction fragments of 1.1 kb (inc/cop) and 1.6 kb (ori). The location and direction of transcription for RNA-C, RNA-CX, RNA-A, and RNA-E are indicated by dashed arrows (12-15). The protein coding regions (9,10) are shown by the solid arrows for repA1 (33,000 d), repA2 (11,000 d), and repA3 (7,000 d). Relevant restriction sites are indicated. Minireplicators used in this study consist of these 2 PstI fragments plus a 2.2 kb PstI fragment coding for resistance to chloramphenicol (3,7). Fusion phages were constructed using the left-most SalI site in the 1.1 kb fragment, fusing RNA-CX and RNA-A to the β-galactosidase coding region of λRS205 (7). The 1.1 kb fragment and the various Sau3A fragments were also cloned into plasmid pBR322 or into the single-stranded phage M13mp10 (25).

rightward transcripts RNA-A and RNA-CX. We have examined the rates of in vivo transcription and translation from the replication region of NR1 by measuring the amount of RNA synthesis using quantitative hybridization and by assaying β-galactosidase activity after fusing the replication sequences to the lacZ gene in the lambda phage λRS205 (7). Minireplicators and λ-lacZ fusion phages from NR1 and its copy number or incompatibility mutants were also examined. The results suggest a mechanism for regulation of IncFII plasmid DNA replication. Copy number is proportional to the rate of rightward gene expression from the replication control region. This expression is regulated at the level of transcription by the repA2 protein and at the level of translation by interaction of the incompatibility inhibitor RNA-E with complementary sequences in the rightward transcripts.

RESULTS

Plasmid and Phage Derivatives of NRl and its Cop⁻ and Inc⁻ Mutants

The plasmid and λ-*lacZ* phage derivatives of NRl and several of its copy number mutants are listed in Tab.1. pRR12, pRR21, and pRR39 are Cop⁻ mutants of the 90-kb NRl plasmid. pRR933 and pRR949 are minireplicator derivatives of NRl which consist of the *PstI* 1.1 (*inc*/*cop*) and 1.6 (*ori*) kb fragments ligated to a 2.2-kb *cat* fragment which confers chloramphenicol resistance. These plasmids have the *cat* fragment in opposite orientations. pRR942 and pRR1919 are *PstI* minireplicator derivatives of pRR12 and pRR21, respectively, in which the 2.2 -kb *cat* fragment is in the same orientation as in pRR933. pRR5406 has a structure similar to pRR949 except that the RNA-C/CX promoter has been inactivated owing to insertion of *Tn3* in the 60-bp *Sau3A* fragment of the parental plasmid pRR39. pRR1909 and pRR1910 were derived from pRR949 by deletion of certain *Sau3A* fragments. pRR1909 has deleted the RNA-C/CX promoter in the 60-bp *Sau3A* fragment and thus has only the RNA-A promoter for rightward transcription, whereas pRR1910 has both the RNA-C/CX and RNA-A promoters.

The *PstI* 1.1-kb (*inc*/*cop*) fragments from NRl, pRR12, and pRR21 have been cloned on the high copy number vector pBR322 to form pRR935 (from NRl), pRR939 (from pRR12), and pDPT261 from pRR21), respectively. pWNRR3 is a mutant of pRR935 which no longer expresses incompatibility toward NRl and does not mediate synthesis of RNA-E owing to a "down-promoter" mutation (see below). The *Sau3A* 262-bp fragment of NRl and pWNRR3, which contains only the RNA-E (leftward) promoter, has been cloned into the *BamHI* site of pBR322 to form pDXRRl, pDXRR2 and pDXRR4. pVLRR7 contains the *Sau3A* 332-bp fragment of NRl cloned into the *BamHI* site of the high copy number vector pUC8. This plasmid can mediate the synthesis of the repA2 protein under the control of the *lac* promoter/operator. The properties of these plasmid derivatives will be discussed in further detail in the following sections.

The λ*lacZ* phages, constructed as described previously (7), contain the part of the *PstI* 1.1 kb fragment to the leftward of the *SalI* site (Fig. 1) fused to a *lacZ* (β-galactosidase) coding region on the phage λRS205. In lysogens of these fusion phages β-galactosidase synthesis is under the control of the rightward NRlx replication transcripts and the prophages are named after the source of the plasmid DNA insert. For example, λ933 was derived from plasmid pRR933.

Incompatibility Properties of Plasmid Derivatives

The incompatibility properties of the minireplicators (and others not included in Tab.1) were found to be the same as those of the parent plasmids from which they were derived (Ref. 3,7; unpub.

Tab. 1. Plasmids used in this study.

PLASMID	DESCRIPTION	REFERENCE
Mutants of 90 kb NR1 plasmid		
NR1	natural isolate, IncFII resistance plasmid	21
pRR12	inc$^-$ cop$^-$ mutant of NR1	22
pRR21	inc$^+$ cop$^-$ mutant of NR1	1,6
pRR39	inc$^+$ cop$^-$ mutant of NR1, Tn3 inserted into the 60 bp Sau3A Fragment	C.R. Barton
Minireplicator derivatives		
pRR933	minireplicator from NR1, PstI 1.1 kb + 1.6 kb + 2.2 kb cat	3,7
pRR949	minireplicator from NR1, PstI 1.1 kb + 1.6 kb + 2.2 kb cat (reverse orientation)	7
pRR942	minireplicator from pRR12, PstI 1.1 kb + 1.6 kb + 2.2 kb cat	3,7
pRR1919	minireplicator from pRR21, PstI 1.1 kb + 1.6 kb + 2.2 kb cat	7
pRR5406	minireplicator from pRR39, similar structure to pRR949	C.R. Barton
pRR1909	Sau3A deletion of pRR949, retains the 332 bp and 262 bp fragments	A.M. Easton
pRR1910	Sau3A deletion of pRR949, retains the 60 bp, 332 bp, and 262 bp fragments	A.M. Easton
PstI fragments cloned on high copy number vectors		
pBR322	cloning vector	23
pRR935	pBR322 plus the PstI 1.1 kb inc fragment of NR1	7
pRR939	pBR322 plus the PstI 1.1 kb inc fragment of pRR12	7
pDPT261	pBR322 plus the PstI 1.1 kb inc fragment of pRR21	6,7
pWNRR3	inc$^-$ mutant of pRR935	R.P. Wu
Sau3A fragments cloned on high copy number vectors		
pDXRR1	pBR322 plus the Sau3A 262 bp fragment of pWNRR3, + orientation	X. Dong
pDXRR2	pBR322 plus the Sau3A 262 bp fragment of pWNRR3, - orientation	X. Dong
pDXRR4	pBR322 plus the Sau3A 262 bp fragment of NR1, - orientation	X. Dong
pUC8	cloning vector	24
pVLRR7	pUC8 plus the Sau3A 332 bp fragment of NR1, + orientation	V.A. Luckow
λ-lacZ phage derivatives		
λ933	λRS205 derivative containing RNA-C and RNA-A promoters of pRR933 fused to lacZ	7
λ942	λRS205 derivative containing RNA-C and RNA-A promoters of pRR942 fused to lacZ	7
λ1919	λRS205 derivative containing RNA-C and RNA-A promoters of pRR1919 fused to lacZ	7
λ1909	λRS205 derivative containing RNA-A promoter of pRR1909 fused to lacZ	X. Dong
λ1910	λRS205 derivative containing RNA-C and RNA-A promoters of pRR1910 fused to lacZ	X. Dong

data). All of the NR1, pRR21, and pRR39 derivatives were incompatible with other derivatives of these plasmids in all possible combinations. Various pRR12 (inc$^-$ cop$^-$) minireplicators were compatible with NR1 (inc$^+$ cop$^+$) and pRR21 (inc$^+$ cop$^-$) derivatives and had only weak incompatibility against other pRR12 minireplicators. However, when cloned on a high copy-number vector, the PstI 1.1-kb or Sau3A 262-bp fragments from pRR12 displayed strong incompatibility against pRR12 minireplicators, but were compatible with NR1 Inc$^+$ derivatives. The corresponding derivatives of NR1 or pRR21 displayed strong incompatibility against NR1, pRR21, or pRR39 derivatives, but were compatible with pRR12 derivatives even though the cloned inc fragments were present on high copy-number vectors. Thus, pRR12 is mutant simultaneously in both the incompatibility inhibitor and its target (3,7).

Measurement of In Vivo Transcription Rates

In order to examine the rates of expression of the NR1 replication control genes (Fig. 1) quantitative hybridization of (^{3}H)-labeled RNA was used to measure transcription rates in either the rightward or leftward direction independently. The DNA "probes" were composed of the single-stranded phage M13mp10 (25) containing specific Sau3A fragments from the replication region of NR1. The 332-bp fragment in the "+" orientation will hybridize with RNA-C (see Fig. 1), the 262-bp fragment in the "+" orientation will hybridize with RNA-CX and RNA-A, the 262-bp fragment in the "-" orientation will hybridize with RNA-E, and the 332-bp fragment in the "-" orientation measures only nonspecific binding (unpub. data). In vivo transcription rates were measured in cells harboring derivatives of NR1 or its copy mutants pRR12 (inc$^-$ cop$^-$) and pRR21 (inc$^+$ cop$^-$) (Tab.2). λ933, λ942, and λ1919 are λ-lac derivatives of the minireplicators pRR933, pRR942, and pRR1919, respectively, while pRR935, pRR939, and pDPT261 are pBR322 derivatives into which the corresponding PstI 1.1-kb fragments have been cloned. This series allows measurement of in vivo transcription when the templates are all maintained at the same low copy number (λ-lacZ lysogens), when they are replicating at their natural copy numbers (minireplicators) or when they are maintained at the same high copy number (pBR322 derivatives).

The corresponding λ-lacZ (λ933 and λ942), minireplicator (pRR933 and pRR942), and pBR322 (pRR935 and pRR939) derivatives of NR1 and the inc$^-$ cop$^-$ pRR12 are transcribed at equivalent levels per gene copy. For example, the relative transcription rate per copy of λ933 and λ942 for RNA-E was 2.20 and 2.03, for RNA-C was 0.22 and 0.24, and for RNA-CX plus RNA-A was 1.00 and 1.10, respectively (Tab.2). Except for the λ-lacZ lysogens, the inc$^+$ cop$^-$ pRR21 derivatives have a 10-fold increase in the level of both RNA-C and RNA-CX plus RNA-A sequences whereas they have approximately the same relative transcription rate per copy for RNA-E as the NR1 and pRR12 derivatives. RNA-C transcription is also elevated 10-fold in the

Tab. 2. In vivo transcription rates.[a]

Template	inc	cop	Copies per Chromosome[c]	Relative Transcription Rate per Copy[b] RNA-E	RNA-C	RNA-CX + RNA-A	(RNA-CX + RNA-A) / RNA-C
λ933	+	+	1.0	2.20	0.22	1.00	4.6
pRR933	+	+	2.2	2.77	0.25	0.19	0.76
pRR935	+	+	40	4.94	0.24	0.14	0.60
λ942	-	-	1.0	2.03	0.24	1.10	4.6
pRR942	-	-	7.3	3.21	0.26	0.17	0.64
pRR939	-	-	40	3.41	0.24	0.14	0.59
λ1919	+	-	1.0	6.02	2.50	2.20	0.88
pRR1919	+	-	9.9	1.14	2.44	1.68	0.69
pDPT261	+	-	40	1.52	2.52	1.63	0.65
pWNRR3	-	d	40	0.08	0.27	0.30	1.11

[a] Exponential phase cells were pulse-labeled for 2.0 min with 5-[^{3}H]-uridine at 20 μCi/ml, 25 Ci/mmole, followed by quick killing with a frozen mixture containing chloramphenicol and sodium azide (26). RNA was extracted as described (26) yielding about 250 μg at greater than 10^5 cpm/μg. RNA (about 5 x 10^6 cpm) was quantitatively hybridized to a vast excess of single-stranded DNA fixed to nitrocellulose using published methods (26, 27). The DNA was from the single-stranded phage M13mp10 (25) carrying the 332bp or 262bp Sau3A fragments from the NR1 replication segment (Fig. 1) inserted in either orientation relative to the viral strand (unpublished data). The filters were treated with RNase prior to measuring the bound cpm. Non-specific binding (about 100 cpm) was assayed for each RNA sample using single-stranded M13mp10 DNA alone. The method is quantitative, directly proportional to RNA input, and specific (unpublished data).

[b] The fraction of ^{3}H-labeled RNA which bound to specific DNA probes was normalized to the number of uracil residues in each transcript in the portion of the transcript complementary to the probe and divided by the copy number of the template. This gives the relative transcription rate for each type of RNA, with the value for (RNA-CX + RNA-A) in λ933 called 1.00. The actual fraction bound for (RNA-CX + RNA-A) in λ933 was 1.56 x 10^{-4} ± 0.30 x 10^{-4}.

[c] The relative copy numbers of plasmids were estimated from the covalently closed circular plasmid DNA to chromosome DNA in CsCl - ethidium bromide gradients (1) or from the specific activity of chloramphenicol acetyltransferase in cell extracts (7, 28).

[d] Cannot form a mir replicator.

pRR21 λ-lacZ lysogen (λ1919). The smaller relative increase in RNA-CX plus RNA-A sequences of λ1919 is explained below. DNA sequence analysis has shown that pRR21 has a mutation in the RNA-C promoter (unpublished data), which explains the 10-fold increase in the rate of transcription of RNA-C and RNA-CX. In most cases, RNA-E is produced in excess compared to RNA-CX plus RNA-A (Tab.2). However, compared to its parent pRR935 (pBR322 plus the 1.1-kb PstI fragment of NR1), RNA-E is reduced more than 50-fold from the mutant plasmid pWNRR3, which does not express incompatibility (exclusion) toward either wild-type plasmids or other incompatibility mutants (unpub. data). DNA sequence analysis has revealed a "down" mutation in the RNA-E promoter of pWNRR3. Thus, RNA-E is the incompatibility inhibitor of IncFII plasmids.

The RNA-A Promoter is Repressed by the RepA2 Protein

The levels of transcription and translation of the rightward transcripts can be monitored using the λ-*lacZ* lysogens. λ-*lacZ* fusion phages were constructed from the *Sau*3A deletion derivatives pRR1909 and pRR1910 (Tab.1). The λ1909 derivative lacks the *Sau*3A 60-bp fragment containing the RNA-C promoter and thus does not produce RNA-C, RNA-CX, or repA2 protein. RNA-A is the only rightward NR1 transcript from λ1909. λ1910 has the *Sau*3A 60-bp fragment and therefore makes RNA-C, RNA-CX, and repA2. Lysogens with λ1909 or λ1910 were compared to the λ1919 lysogen, which has an increased level of RNA-C and RNA-CX transcription. When repA2 protein is provided in trans by plasmid pVLRR7, which has the repA2 gene cloned into the vector pUC8 (Tab.1), β-galactosidase activity from the RNA-A promoter of λ1909 is completely repressed (Tab.3). In the absence of extra repA2 protein β-galactosidase activity from all lysogens is proportional to the sum of RNA-CX plus RNA-A, while in the presence of excess repA2 protein it is proportional to RNA-CX alone. Therefore, when RNA-A is repressed, RNA-CX provides a basal level of rightward transcription.

When the repA2 gene dosage is high enough to repress completely RNA-A transcription, the ratio of (RNA-CX + RNA-A)/RNA-C should reflect the frequency of extension of RNA-C to form RNA-CX. This ratio is 0.6-0.7 for all the minireplicators and pBR322 derivatives (Tab.2), suggesting that sufficient repA2 protein is present to repress RNA-A transcription for all these plasmids, including the minireplicator derivative pRR933 from the wild-type, low copy-number plasmid NR1. The 6- to 7-fold increase in this ratio for the λ933 and λ942 lysogens is due to the derepression of RNA-A synthesis at the low copy number of one per chromosome. This increase is much smaller in the λ1919 lysogen because most of the rightward transcription comes from synthesis of RNA-C and RNA-CX owing to the mutation in the RNA-C promoter. At the low copy number in the lysogens, the endogenous amount of repA2 protein produced by λ933 and

Tab. 3. Repression of the RNA-A promoter by trans repA2 protein.

	Relative Transcription Rate[a]		β-galactosidase Activity[b]	
Lysogen	RNA-CX + RNA-A	RNA-C	-repA2	+repA2
λ1909	1.02	0.00	377	0
λ1910	0.66	0.36	163	58
λ1919	2.20	2.50	965	1102

[a] Determined in the absence of additional repA2 protein as described in Table 2.

[b] Miller Units (ref. no. 29). The additional repA2 protein was supplied in *trans* by coresident plasmid pVLRR7.

λ942 is apparently insufficient to repress completely the RNA-A promoter. However, repression of RNA-A synthesis is established when the repA2 gene dosage is doubled, as in the case of the minireplicator pRR933. Thus, the transcription data in Tab. 2 suggests that almost complete repression of RNA-A synthesis is achieved at the normal copy number of the minireplicators. The fully derepressed level of RNA-A transcription could be estimated from the λ1909 lysogen (Tab.3) and from the minireplicator pRR5406 (data not shown) which do not make any repA2 protein. In the wild-type plasmid derivatives, the strengths of the promoters can be ranked as RNA-E > derepressed RNA-A > RNA-C > repressed RNA-A.

RNA-E Inhibits and RNA-A Stimulates repA1 Gene Expression and Plasmid Copy Number

In order to examine the effects of RNA-A, RNA-E, and repA2 protein on gene expression and copy number in trans, plasmids which synthesize these products were introduced into either a λ933 lysogen or a strain harboring the pRR933 minireplicator. The derivations of the coresident plasmids are presented in Tab.1, and the trans products they produce were assayed independently (Tab.2 and unpub. data). All are derivatives of pBR322 carrying the PstI 1.1-kb or Sau3A 262-bp or 332-bp fragments from the replication region of NR1 or its incompatibility mutants. The RNA-E promoter, which is nonfunctional in pWNRR3 derivatives, is the only promoter on the Sau3A 262-bp fragment (12,13). Therefore, when the 262-bp fragment of pWNRR3 is cloned into the BamHI site of pBR322, "RNA-A" or "RNA-E" hybrid RNA sequences are synthesized by transcription from the tetracycline resistance gene (tet) promoter (Fig. 2). The results in Tab.4 show that excess RNA-E both inhibited β-galactosidase synthesis from λ933 and reduced the copy number of a coresident pRR933 plasmid. RNA-E produced from an incompatibility mutant (pRR939) or as a hybrid from the tetracycline resistance promoter (pDXRR2) was less effective than wild-type RNA-E (pRR935, pDXRR4). Excess RNA-A both stimulated β-galactosidase activity in the lysogen and raised the copy number of a coresident minireplicator. Presumably this resulted from titration of the RNA-E synthesized from the lysogenic phage or the minireplicator by the trans RNA-A. Thus, the rightward transcript sequences from the 262-bp Sau3A fragment (RNA-A and RNA-CX sequences) must be the target for the incompatibility repressor RNA-E. The plasmid pDXRR5 (Fig. 2) had little or no transcription from either strand of the 262-bp Sau3A fragment owing to deletion of the tet gene promoter after cleavage of pDXRR1 with HindIII, S1 digestion, and ligation. pDXRR5 had little effect on either β-galactosidase synthesis or pRR933 copy number. Thus, the presence of the DNA sequences alone had no effect in trans, showing that it is the rightward RNA transcripts and not the DNA which is the target of RNA-E inhibition. An excess of repA2 protein from plasmid pVLRR7 reduced β-galactosidase activity in the lysogen (Tab.2). However, in the presence of excess repA2 protein, excess RNA-E or RNA-A had their expected inhibitory or stimulatory effects, respectively.

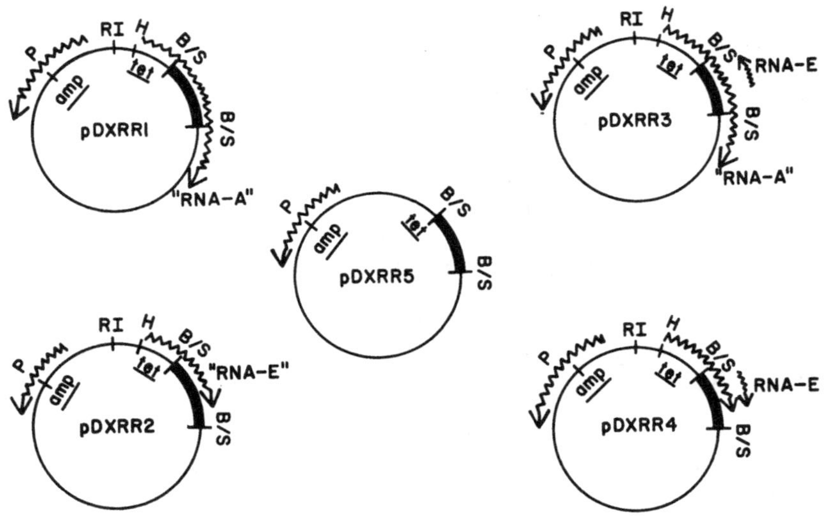

Fig. 2. Maps of pBR322 clones carrying the 262 bp Sau3A fragments from the replication control region of pWNRR3 (pDXRR1, pDXRR2, and pDXRR5) or NR1 (pDXRR3 and pDXRR4). Insertions were in either orientation at the BamHI site in the tetracycline resistance gene. Transcription products are indicated as wavy arrows. Restriction sites are: P, PstI; RI, EcoRI; H, HindIII; B/S, BamHI-Sau3A fusion.

Intracellular RNA Concentrations

The stabilities of RNA-C, RNA-A, and RNA-E were also examined by hybridization after a pulse-chase experiment. Each was unstable with a half-life of less than 10% of a generation (data not shown). If it is assumed that these transcripts behave similarly to normal mRNA (36), their intracellular concentrations can be estimated from the hybridization data (Fig. 3). The levels of RNA-C and RNA-E were constant per plasmid copy over the entire range of copy numbers. The level of derepressed RNA-A per plasmid copy was also constant in the case of plasmids which have deleted or inactivated the RNA-C promoter and do not synthesize the repA2 repressor protein. The level of RNA-CX plus RNA-A per copy was also constant above about 5 copies per cell. At lower copy numbers the level of RNA-CX plus RNA-A increased owing to a lower level of synthesis of repA2 and approached that expected for the constant level of RNA-CX plus the fully derepressed level of RNA-A. The normal copy number of NR1 is about 4.4 copies per cell. The dotted lines extrapolated to the ordinate represent the unregulated, or fully derepressed, levels of each transcript.

Tab. 4. Regulation of gene expression and copy-number.

		Relative β-galactosidase Activity	Relative Copy Number
Coresident	*trans* Products[a]	λ933	pRR933
none	none	0.83	1.0
pBR322	none	1.00	1.0
pDXRR1	A_{tet}	3.00	2.5
pDXRR2	E_{tet}	0.30	0.7
pDXRR4	E	0.30	0.0[b]
pDXRR5	none	1.21	1.1
pVLRR7	repA2	0.29	0.8
pWNRR3	repA2, A	0.98	3.2
pRR935	repA2, A, E	0.02	0.0[b]
pRR939	repA2, A_{12}, E_{12}	0.19	0.7

[a] A=sequences of RNA-A, A_{tet}=*tet*-A hybrid RNA, E=RNA-E, E_{tet}=*tet*-E hybrid RNA, A_{12} and E_{12} are the transcripts from the *inc*⁻ pRR12 derivative, and repA2 is the protein coded by RNA-C (Fig. 1).

[b] The coresident plasmid expressed incompatibility against pRR933, effectively reducing its copy number to 0.

DISCUSSION

Derivatives of the *inc*$^-$ *cop*$^-$ mutant pRR12 are transcriptionally indistinguishable from the wild-type NR1, but have stimulated gene expression (lysogen) (Ref.7; unpub. data) and increased copy number (minireplicator) (3,7). This suggests that the pRR12 incompatibility mutation affects a post-transcriptional step in replication control. There is a single base substitution in pRR12 (Ref.9; unpub. data) which results in a complementary change in sequence to both RNA-E and the rightward transcripts RNA-CX and RNA-A.

Derivatives of the *inc*$^+$ *cop*$^-$ pRR21 have a 10-fold increase in rightward transcription and the λ1919 lysogen also mediates increased β-galactosidase synthesis. The copy-number increase for this mutant is likely a result of the increased rate of rightward transcription from the RNA-C promoter, which increases the level of RNA-C and RNA-CX.

Increased levels of trans RNA-E inhibit rightward gene expression and lower the copy number of a coresident minireplicator. This inhibition shows incompatibility group specificity as seen by the difference between the wild-type NR1-derivative pRR935 and the mutant pRR12-derivative pRR939 (Tab.4). Increased levels of trans RNA-A stimulate both rightward gene expression and increase the copy number of a coresident minireplicator, probably by binding the endogenous RNA-E synthesized by the lysogen or the minireplicator. Other experiments have shown that the transcription rates themselves are not affected by excess trans RNA-A or RNA-E (unpub. data). Together with the results for the pRR12 derivatives, which show no alteration in transcription rates, these observations are indicative of post-transcriptional regulation. It is likely that RNA-E inhibits translation of the rightward transcripts by forming a complementary RNA-RNA duplex. The repA2 protein also reduces rightward gene expression by repression of transcription from the RNA-A promoter. These interactions suggest that IncFII plasmids have 2 basic mechanisms of replication control.

One mechanism of control is through derepression of RNA-A synthesis. This will occur if the plasmid copy number falls below the wild-type level and the repA2 concentration is decreased owing to gene dosage effects. The active form of the repA2 repressor appears to be a tetramer (unpub. data). Thus, a decrease in repA2 concentration would result in a considerably larger decrease in its repressor activity. The derepressed promoter for RNA-A is stronger than the RNA-CX promoter and the increased level of total rightward transcription under these conditions would increase the probability of replication. The copy mutant pRR21 has a higher level of transcription illustrating how additional rightward transcription can increase the rate of replication. Other experiments have shown that plasmids with derepressed RNA-A transcription, such as pRR39, or increased transcription owing to fusion to some other strong promoter also have high copy numbers (unpub. data).

A second mechanism of control is mediated by the interaction of the incompatibility inhibitor with the rightward transcripts to inhibit translation of repA1. This would occur at all plasmid copy numbers but would be most effective at normal copy numbers or above when the RNA-E and rightward transcript concentrations are higher. In this situation the RNA-A promoter is repressed and a basal level of rightward transcription is provided by RNA-CX (Fig.3). Additional repA2 protein has little effect on plasmid copy number (Tab.4).

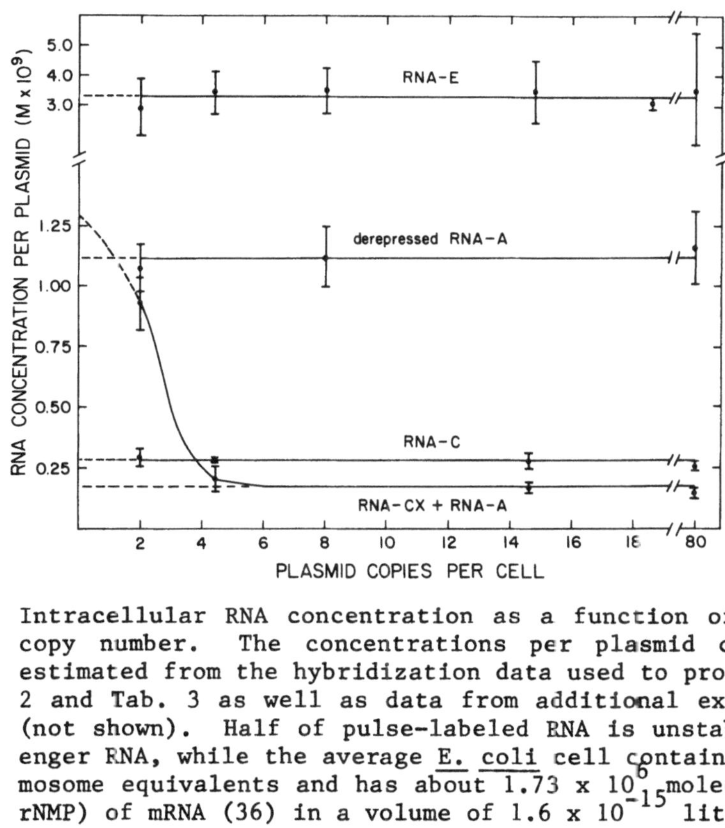

Fig. 3. Intracellular RNA concentration as a function of plasmid copy number. The concentrations per plasmid copy were estimated from the hybridization data used to produce Tab. 2 and Tab. 3 as well as data from additional experiments (not shown). Half of pulse-labeled RNA is unstable messenger RNA, while the average E. coli cell contains 2 chromosome equivalents and has about 1.73×10^{6} molecules (as rNMP) of mRNA (36) in a volume of 1.6×10^{-15} liters (31). The intracellular concentration of any unstable RNA molecule can be estimated by: $(F) \times 2 \times 1.73 \times 10^{6} \div (H) \div 6.023 \times 10^{23} \div 1.6 \times 10^{-15}$ = mole/liter where F is the fraction of cpm bound to a probe with H bp of homology to the transcript. The solid lines connect the points for the measured values (mean ± standard deviation) while the dashed lines extrapolate the data to the Y-axis to represent the unregulated level of synthesis. For reference, one molecule per cell gives a concentration of 1.04×10^{-9} M.

Replication control is provided by the interaction of RNA-E with the rightward transcripts which would be primarily RNA-CX. Copy number mutations which affect incompatibility simultaneously change the sequence of RNA-E and the rightward transcripts. The base changes in these mutants are located in or next to possible 6-base loops of proposed secondary structures for both RNA-E and the rightward transcripts (9-11,13). A comparison of the loop sequences of NR1 and pRR12 is shown in Tab.5 If it is assumed that the initial interaction between RNA-E and the rightward transcripts occurs by base pairing of the 6-base loops (as suggested by the mutations), then base-stacking free energies (30) and equilibrium constants for this pairing, corrected to in vivo ionic conditions (31), can be calculated (Tab.5). The probability that a rightward transcript will escape interaction with RNA-E and thus allow the initiation of plasmid replication can be predicted from the laws of mass action assuming the likely sequences of initial contact. Free RNA-E (E6) binds to free transcript [A6], forming a complex base paired by the 6-base loops (A6/E6). The complex then unfolds and forms a complete 91 bp duplex (A91/E91).

$$A6 + E6 \rightleftharpoons A6/E6 \longrightarrow A91/E91$$

If the second step is faster, its rate will be directly dependent on the equilibrium established by the first step. This suggests that the mutant pRR12 will have a lower amount of duplex formation, owing to a lower base-stacking energy (Tab.5).

Tab. 5. Equilibrium constants for incompatibility interactions.[a]

	NR1	pRR12
Loop Sequence	UUGGCG	UUUGCG
ΔG° (kcal/mole)	-13.6	-9.80
Kd (<u>M</u>)	1.10×10^{-10}	6.67×10^{-8}
[RNA-E] (<u>M</u>)	1.51×10^{-8}	5.01×10^{-8}
[RNA-CX + RNA-A] (<u>M</u>)	9.10×10^{-10}	2.65×10^{-9}

[a] The RNA sequence shown is the 6-base loop in the proposed secondary structure for the rightward transcripts (Fig. 5A) (9-11, 13). The proposed structure for RNA-E has a complementary loop sequence which can pair with the loop of the rightward transcripts (Fig. 5B). Base-stacking free energies (ΔG°) for this pairing (30) corrected to <u>in vivo</u> ionic conditions (0.23 <u>M</u> NaCl, + 0.5 kcal/mole per base-pair; ref. no. 30, 31) are presented and standard dissociation constants (Kd) are calculated:

$$Ka = \frac{[A6/E6]}{[A6]\,[E6]} = 1/Kd = e^{-\Delta G^\circ/RT}$$

See text for definition of A6, E6, and A6/E6. The RNA concentrations per cell are taken from Fig. 3 for pRR933 (NR1 derivative, 4.4 copies/cell) and pRR942 (pRR12 derivative, 14.6 copies/cell).

From the data in Fig. 3 the intracellular RNA concentrations expected for the wild-type plasmid NR1 can be calculated as a function of plasmid copies per cell. The fraction of rightward RNA not bound by RNA-E can then be calculated using the NR1 equilibrium constant in Tab. 5 (Fig. 4, dashed line). The total free rightward RNA concentration [A6], can then be calculated as a function of copy number (Fig. 4, solid line). The repression of RNA-A transcription above 4 copies per cell would result in the lower level of [A6] indicated by the alternate dashed-dotted line. The estimated free rightward RNA concentration can be considered to reflect the probability that a rightward transcript will escape regulation by RNA-E, resulting in translation and the initiation of replication. At normal NR1 copy numbers or higher, the level of [A6] is constant and independent of copy number. This would result in a copy number-independent rate of initiation of plasmid replication, thus automatically correcting for random fluctuations in pool size due to un-

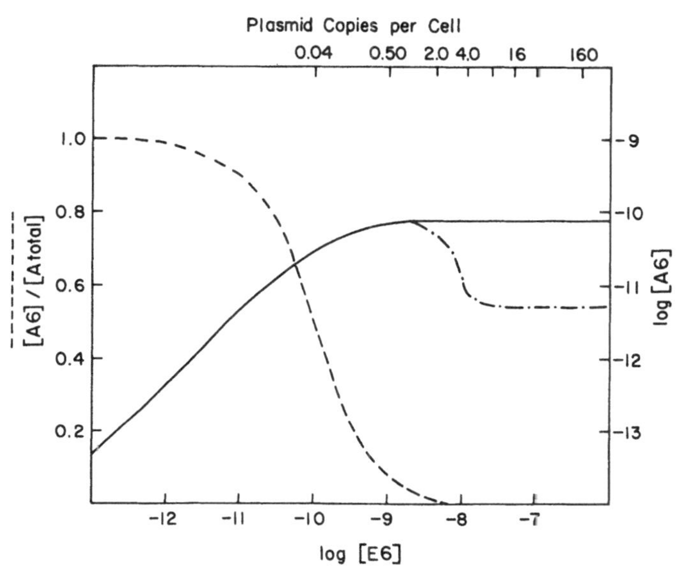

Fig. 4. Calculated interaction of RNA-E with the rightward transcripts RNA-CX and RNA-A as a function of plasmid NR1 copy number per cell. Using the RNA concentration estimates in Fig. 3 and the dissociation constant for NR1 in Tab. 5, the unbound fraction of rightward RNA, [A6]/[A total] (dashed line), and the molar values for the free rightward RNA concentration, [A6], were calculated as a function of plasmid copy number. The solid line was calculated assuming a constant ratio of rightward and leftward transcription. The alternate dashed-dotted line takes into account the repression of RNA-A transcription as the plasmid copy number increases above the NR1 copy number of 4 per cell.

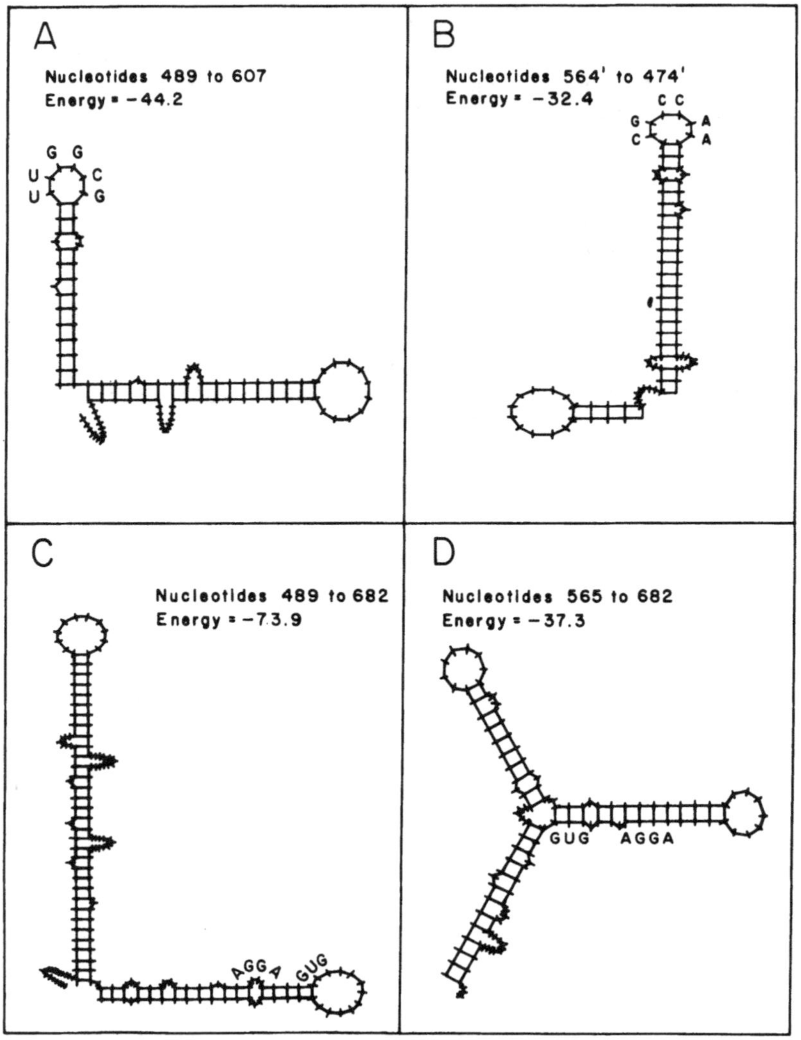

Fig. 5. Predicted secondary structures for various segments of the rightward transcripts and RNA-E. The target segments of rightward RNA and RNA-E which are complementary are shown in A and B, respectively. The 6-base loops are the presumed sites of initial interaction. The translation start site for the repA1 coding sequence is shown in the absence (C) or presence (D) of RNA-E. The base numbering is that of Rosen et al. (9), which starts at the leftward end of the 60 bp Sau3A fragment in Fig. 1 and increases to the rightward. The structures were predicted with the computer program of Zuker and Steigler (37), which uses the base-stacking and destabilizing rules of Salser (38).

equal plasmid partitioning at cell division or premature initiation of replication. Similar calculations for the copy mutants pRR12, pRR21, and pRR39 predict higher levels of [A6] than for NR1 which would result in a higher steady-state plasmid copy number (not shown). The higher level of [A6] for pRR12 is the result of a weaker RNA-RNA interaction. The higher levels for pRR21 and pRR39 are the result of a higher level of rightward transcription.

How might the formation of a duplex between RNA-E and the rightward transcripts inhibit translation and replication? If the repA1 protein is a required initiator protein, then preventing translation of its mRNA would inhibit replication. Possible secondary structures for the rightward transcripts and RNA-E were examined by computer analysis (Fig. 5). If a full 91-base duplex is formed between RNA-E and the rightward transcript, then 91 bases of the rightward RNA would be prevented from further secondary structure interactions with other regions of its own sequence. The predicted secondary structures for rightward RNA in the absence or presence of RNA-E are different. The free rightward transcripts could fold such that the ribosome binding site (32) for repA1 is single-stranded, allowing efficient translation. The complexed RNA results in the ribosome binding site for repA1 folded into a region of double-stranded secondary structure, thus preventing translation (33).

This model for regulation of plasmid DNA replication includes elements of the negative model of Pritchard et al. (34), where RNA-E would be the unstable inhibitor. Positive regulation is also suggested from the requirement for a positively acting initiator protein, repA1, similar to the replicon model (35). A distinct difference from these earlier models is that the present one does not predict or require any defined periodicity in the probability of replication initiation. Instead, a constant probability per cell would duplicate the pool of plasmids each generation, and random fluctuations would automatically be corrected. The RNA-RNA interactions described here could be an important element in many regulated DNA replication systems, including those for bacterial chromosomes and multiple origins of replication in eukaryotic organisms. Such a novel regulation system could also be involved with many other kinds of cellular regulatory processes, where post-transcriptional regulation might be desirable.

ACKNOWLEDGMENTS

We would like to thank Malcolm E. Winkler for advice on RNA extraction and hybridization; Steven C. Kowalczykowski for helpful discussions of RNA-RNA base-stacking interactions, Alan M. Easton, Padmini Sampathkumar, and Carolyn R. Barton for permission to cite their unpublished data; and Joseph E. Gerwin, for his excellent technical assistance. This work was supported in part by U.S. Public Health Services research grants GM14398 and GM30731, and by grants from the Otho S.A. Sprague Memorial Institute and the Arlo M. Bane fund. The secondary structure analysis was performed with funds provided by the Vogelback Computing Center, Northwestern University.

REFERENCES

1. Womble, D.D., D.P. Taylor, and R.H. Rownd (1977) A method for obtaining more-accurate covalently closed circular plasmid-to-chromosome ratios from bacterial lysates by dye-buoyant density centrifugation. J. Bact. 130:148-153.
2. Rownd, R.H., and D.D. Womble (1978) Molecular nature and replication of R factors. In R factor, Drug Resistance Plasmid, S. Misuhashi ed. University of Tokyo Press, Tokyo, pp. 161-193.
3. Miki, T., A.M. Easton, and R.H. Rownd (1980) Cloning of replication, incompatibility, and stability functions of R plasmid NR1. J. Bact. 141: 87-99.
4. Molin, S., and K. Nordstrom (1980) Control of plasmid R1 replication: Functions involved in replication, copy number control, incompatibility, and switch-off of replication. J. Bact. 141:111-120.
5. Timmis, K.N., I. Andres, and P.M. Slocombe (1978) Plasmid incompatibility: cloning analysis of an incFII determinant of R6-5. Nature (London) 273: 27-32.
6. Taylor, D.P., and S.N. Cohen (1979) Structural and functional analysis of cloned DNA segments containing the replication and incompatibility regions of a miniplasmid derived from a copy number mutant of NR1. J. Bact. 137:92-104.
7. Easton, A.M., and R.H. Rownd (1982) The incompatibility product of IncFII R Plasmid NR1 controls gene expression in the plasmid replication region. J. Bact. 152:829-839.
8. Rosen, J., H. Ohtsubo, and E. Ohtsubo (1978) The nucleotide sequence of the region surrounding the replication origin of an R100 resistance factor derivative. Mol. Gen. Genet. 171:287-293.
9. Rosen, J., T. Ryder, H. Inokuchi, H. Ohtsubo, and E. Ohtsubo (1980) Genes and sites involved in replication and incompatibility of an R100 plasmid derivative based on nucleotide sequence analysis. Mol. Gen. Genet. 179:527-537.
10. Stougaard, P., S. Molin, K. Nordstrom, and G. Hansen (1981) The nucleotide sequence of the replication control of the resistance plasmid R1drd-19 Mol. Gen. Genet. 181:116-122.
11. Brady, G., J. Frey, H. Danbara, and K.N. Timmis (1983) Replication control mutations of plasmid R6-5 and their effects on interactions of the RNA-I control element with its target. J. Bact. 154:429-436.
12. Easton, A.M., P. Sampathkumar, and R.H. Rownd (1981) Incompatibility of IncFII R plasmid NR1. In The Initiation of DNA Replication, D.S. Ray, ed. Academic Press, Inc., New York, pp. 125-141.

13. Rosen, J., T. Ryder, H. Ohtsubo, and E. Ohtsubo (1981) Role of RNA transcripts in replication incompatibility and copy number control in antibiotic resistance plasmid derivatives. Nature (London) 290:794-797.
14. Lurz, R., H. Danbara, B. Ruckert, and K.N. Timmis (1981) Plasmid replication functions. VII. Electron microscopic localization of RNA polymerase binding sites in the replication control region of plasmid R6-5. Mol. Gen. Genet. 183:490-496.
15. Chan, P.T., and J. Lebowitz (1982) Mapping of RNA plymerase binding sites in R12 derived plasmids carrying the replication-incompatibility region and the insertion element IS1. Nucleic Acids Res. 10:7295-7311.
16. Ohtsubo, E., J. Feingold, H. Ohtsubo, S. Mickel, and W. Bauer (1977) Unidirectional replication in Escherichia coli of three small plasmids derived from R factor R12. Plasmid 1:8-18.
17. Synenki, R.M., A. Nordheim, and K.N. Timmis (1979) Plasmid replication functions. III. Origin and direction of replication of a "mini" plasmid derivative from R6-5. Mol. Gen. Genet. 168:27-36.
18. Brawner, M.E., and S.R. Jaskunas (1982) Identification of polypeptides encoded by the replication region of resistance factor R100. J. Mol. Biol. 159:35-55.
19. Stougaard, P., S. Molin, and K. Nordstrom (1981) RNAs involved in copy-number control and incompatibility of plasmid R1. Proc. Natl. Acad Sci., USA 78:6008-6012.
20. Light, J., and S. Molin (1983) Post-transcriptional control of expression of the repA gene of plasmid R1 mediated by a small RNA molecule. EMBO J. 2:93-98.
21. Rownd, R., R. Nakaya, and A. Namamura (1966) Molecular nature of the drug resistance factors of the Enterobacteriaceae. J. Bact. Biol. 17:376-393.
22. Morris, C.F., H. Hashimoto, S. Mickel, and R. H. Rownd (1974) Round of replication mutant of a drug resistance factor. J. Bact. 118:855-866.
23. Bolivar, F., R.L. Rodriquez, P.J. Green, M.C. Betlach, H.L. Heynecker, H.W. Boyer, J.H. Crosa, and S. Falkow (1977) Construction and characterization of new cloning vehicles. II. A multi-purpose cloning system. Gene 2:95-113.
24. Vieira, J., and J. Messing (1982) The pUC plasmids, an M13mp7-derived system for insertion mutagenesis and sequencing with synthetic universal primers. Gene 19:259-268.
25. Messing, J., and J. Vieira (1982) A new pair of M13 vectors for selecting either DNA strand of double-digest restriction fragments. Gene 19:269-276.
26. Forchhammer, J., E.N. Jackson, and C. Yanofsky (1972) Different half-lives of messenger RNA corresponding to different segments of the tryptophan operon of Escherichia coli. J. Mol. Biol. 71: 687-699.
27. Gillespie, D., and S. Spieglman (1965) A quantitative assay for DNA-RNA hybrids with DNA immobilized on a membrane. J. Mol. Biol. 12:829-842

28. Shaw, W.V. (1975) Chloramphenicol acetyltransferase from chloramphenicol-resistant bacteria. Methods Enzymol. 43:737-755.
29. Miller, J.H. (1972) Experiments in Molecular Genetics. Cold Spring Harbor Press, Cold Spring Harbor, New York.
30. Tinoco, I.P., N. Borer, B. Dengler, M.D. Levine, O.C. Uhlenbeck, D.M. Crothers, and D. Gralla (1973) Improved estimation of secondary structure in ribonucleic acids. Nature (London) New Biol. 246:40-41.
31. von Hippel, P.H., S.C. Kowalcyzkowski, N. Lonberg, J.W. Newport, and L.S. Paul 91982) Autoregulation of gene expression. Quantitative evaluation of the expression and function of the bacteriophage T4 gene 32 (single-stranded DNA binding) protein system. J. Mol. Biol. 162:795-818.
32. Shine, J., and L. Dalgarno (1974) The 3'-terminal sequence of Escherichia coli 16S ribosomal RNA: Complementarity to nonsense triplets and ribosome binding sites. Proc. Natl. Acad. Sci., USA 71:1342-1346.
33. Horinouchi, S., and B. Weisblum (1982) Nucleotide sequence and functional map of PE194, a plasmid that specifies inducible chloramphenicol resistance. J. Bact. 150:815-825.
34. Pritchard, R.H., P.T. Barth, and J. Collins (1969) Control of DNA synthesis in bacteria. Symp. Soc. Gen. Microbiol. 19:263-297.
35. Jacob, F., S. Brenner, and F. Cuzin (1963) On the regulation of DNA replication in bacteria. Cold Spring Harbor Symp. Quant. Biol. 28:329-348.
36. Watson, J.D. (1976) Molecular Biology of the Gene, 3rd Edition. W.A. Benjamin, Inc., Reading, Massachusetts, p. 69.
37. Zuker, M., and P. Steigler (1981) Optimal computer folding of large RNA sequences using thermodynamics and auxillary information. Nucleic Acids Res. 98:133-148.
38. Salser, W. (1977) Globin mRNA sequences: Analysis of base pairing and evolutionary implications. Cold Spring Harbor Symp. Quant. Biol. 42:985-1102.

P1 PLASMID MAINTENANCE: A PARADIGM OF PRECISE CONTROL

Dhruba K. Chattoraj, Ann L. Abeles,
and Michael B. Yarmolinsky

Laboratory of Genetics and Recombinant DNA
LBI-Basic Research Program
NCI-Frederick Cancer Research Facility
Frederick, Maryland 21701

OVERVIEW

Certain bacterial plasmids, P1 prophage among them, are stably inherited despite their very low copy number (59). In each lysogenic cell, the mechanisms that guarantee a P1 replication prior to cell division and an equipartition of the daughter plasmids at cell division operate with error frequencies comparable to rates of mutation (7,61). In reviewing what we know about how such fidelity is achieved, this article will concentrate on plasmid replication and its control; the accompanying one by Austin and Abeles will consider partitioning.

A number of plasmid-encoded functions that contribute to stable plasmid maintenance have been distinguished. Discrete functions, however, are not necessarily the work of physically separable determinants. A replication origin may be internal to an initiator gene, as in λdv (77) or pT181 (34). An incompatibility determinant may be coincident with the determinant of its target, as in colE1 (76) or R1 (43). For several years, the hypothesis was widely accepted that the origin of plasmid replication and the site of partition are, in general, one and the same (32). We are fortunate that in the case of P1 it has been possible to dissect the determinants of the principal plasmid maintenance functions from each other and study them

---

in isolation as well as variously combined. Dissection has revealed an origin of replication (oriR) and its associated specific initiator (RepA) and control element (incA) (1,6); a partitioning apparatus that consists of a centromere analog with which an incompatibility element (incB) is associated and 2 proteins (ParA and ParB) implicated in its function (2,4); a site (loxP) (28) at which a protein (Cre) (2) efficiently resolves dimeric plasmids that might otherwise interfere with plasmid equipartitioning (7,71); very likely, a mechanism that delays cell division when the plasmid copy number falls below a critical threshold value (see later); and at least one alternative origin under separate control (oriL) (70). The elements of P1 mentioned above, partition elements aside, will be considered below.

Our understanding of P1 maintenance has been aided by comparative studies. For example, the concept of independent replication and partition modules, made explicit with the demonstration of the functionality of experimentally rearranged elements from the unrelated plasmids P1 and F (3), had been adumbrated by comparison of the closely related plasmid prophages P1 and P7. Much of the partition regions of these largely homologous members of incompatibility group Y appears to lie within a 2.5-kb region of P1/P7 nonhomology (29,83,84).

In addition to coliphages P1 and P7, the members of incompatibility group Y (26) include the closely related plasmid p15B (26,30), presumably a defective prophage; pIP231, an H_2S-producing plasmid (14); and the Salmonella phage j2 (52). P1 shows extensive base-sequence homology, at least in the ori-repA-incA region, with plasmid Rts1, a member of incompatibility group T (35). Plasmids that do not have a DNA sequence homology to P1 may, nevertheless, exhibit evidence of relatedness. The most striking example is the fertility factor F. Over those regions of F and P1 for which DNA sequence comparisons have been made by computer analysis, hardly any significant homology is apparent (1), but in overall organization, the 2 plasmids are seen to exhibit a remarkable similarity (Fig. 1). We shall have more than one occasion to refer to implications of this relationship.

An aspect of plasmid replication that has been more thoroughly studied in F than in P1 is worth mentioning at the outset because of its importance; that is the manner of restoration of normal plasmid copy number in response to the introduction of excess copies of the plasmid. The added copies cause a total inhibition of plasmid replication until such time as cell division has reduced the plasmid copy number to its normal level (79). A very different restoration kinetics is characteristic of R1 (23). We assume that the manner by which P1 restores its copy number is essentially similar to that which has been reported for F.

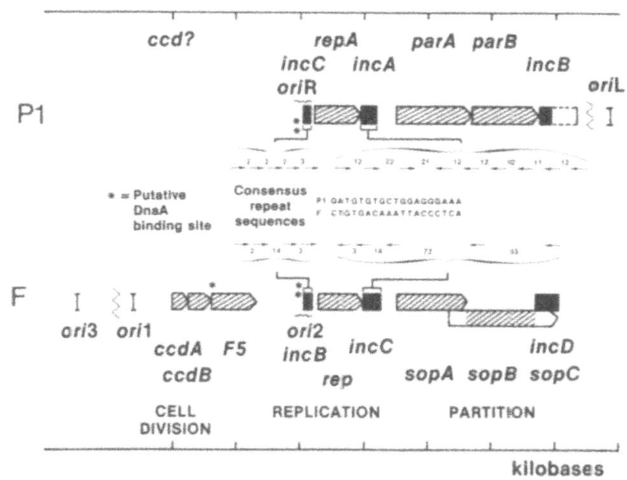

Fig. 1. The shaded arrows represent open reading frames for proteins that have been identified; black boxes represent segments that can exert incompatibility. Small arrows indicate the arrangements of the 19-bp repeat sequences. The distance between oriR and oriL is 10 ± 2 kb (70). The P1 coordinates of the replication region are from (1), and of the partition region from (5). The F coordinates of the cell division region are from (11,56), of the replication region are from (40,42) and of the partition region are from (57). There is evidence that ccdA and ccdB may each produce 2 small proteins (11). The pair of asterisks represent the sequence TTATCCACATTATCCAC in oriR of P1 and GTTATCCACTTATCCACG in ori2 of F; the single asterisk just within the putative open reading frame of the F5 protein represents the sequence GTTATACACAG (50).

ORIGINS

A Choice of Origins

Two separate and independent plasmid replicons have been isolated from a region of P1 DNA that represents about one-eighth of the genome and that contains, between them, the partition module. Because these replicons were detected in λ clones of rightward and leftward extensions of P1 fragment EcoRI-5, the original hybrids were designated λ-P1:5R and λP1:5L (70). The R and L replicons are subject to different controls. As will be discussed in a separate section, the R replicon is under control of the adjacent incA element, the determinant of group Y incompatibility; the L replicon is not (Ref. 73, N. Sternberg, unpub. data). The existence of an incA-insensitive plasmid replicon in a complete but chromosomally integrated P1 is indicated by the experiments of Fig. 2. Plasmid rep-

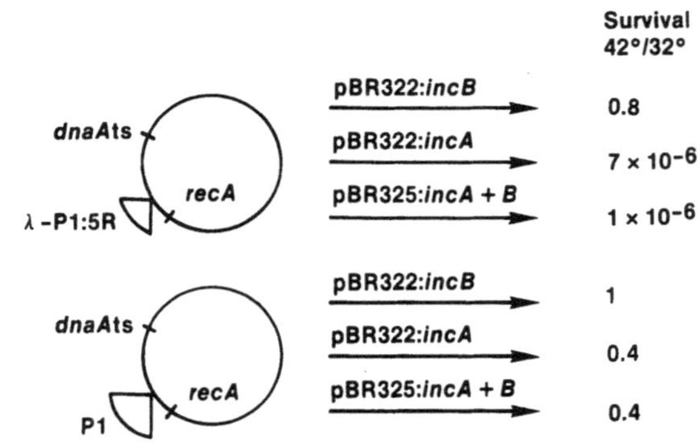

Fig. 2. Top. A λ-P1:5R that also carries the locus of site-specific recombination, loxP, was inserted into the bacterial locus loxB of a recA56 dnaA46(ts) E. coli by the method of prophage displacement (72). Bottom. P1Cm was similarly integrated.

The bacteria were transformed with the indicated pBR322 derivatives, spread on nutrient agar, and incubated 32°C and 42°C to determine the surviving fraction. Interference with survival at 42°C by an introduced inc element is taken as measure of interference with replication initiated from the P1·DNA.

lication initiating from λ-P1:5R or from P1 is seen to permit survival at 42°C of a dnaAts host into which these plasmids have been suitably integrated. Although inhibition of λ-P1:5R replication can be accomplished by additional copies of a cloned incA element, they are insufficient to inhibit markedly reduced plasmid replication initiated from P1. Preliminary experiments suggest that components of the P1 immunity system are required in addition to incA if P1-mediated plasmid replication is to be effectively blocked (D. Chattoraj and N. Sternberg, unpub. data).

The sensitivity of free P1 to destabilization by incA suggests that, during normal maintenance conditions, P1 is driven by the R replicon predominantly. Any alternative replicon of P1, if functional in the prophage state, is therefore insufficiently active to keep pace with cell division. However, an alternative replicon might be important in certain bacterial hosts other than E. coli or quite possibly during the establishment of lysogeny.

Alternative origins are not uncommon in bacterial plasmids. For example, a cluster of 3 origins has been noted in F (19,42,75). The central one of the 3, ori-1, is not expressed unless the nearby ori-2 is present, whereas the functioning of the ori-2 and ori-3 are independent of the presence of ori-1. The origin of the P1 R replicon, to which we have devoted most of our attention, appears to be analogous to ori-2 of F (Fig. 1). It remains to be determined whether there exist in P1 other origins with properties similar to those of F, and, if so, what advantages this extravagance might confer.

An Isolated Functional Origin

We have previously reported that a 2.1-kb P1 DNA segment determines a functional R replicon of unit copy (1). This mini-P1 includes the essential replication gene, repA flanked by the sets of 19-bp sequences indicated in Fig. 1. The set upstream of repA (designated incC) consists of 5 repeats; the downstream set (designated incA) consists of 9 repeats. Deletions that reduce the segment to 1.5 kb have recently been introduced without significant effect on plasmid replication. Additional deletions that eliminate all but ine and one-half repeats from incA result in a plasmid (pALA317) with elevated copy number (D. Chattoraj, K. Cordes, and A. Abeles, submitted for publication). In pALA317, there is only 1.2 kb of P1 DNA. Recently, it has been possible to dissect further this mini-P1 into a repA gene (without its own promoter) and a 245-bp region upstream of repA that is capable of acting as an origin when supplied in trans with a source of RepA protein. The construct pALA159 is essentially a P1 origin that depends entirely on the presence of pALA69, which furnishes RepA protein to it. The structures are given in Fig. 3, and the DNA sequence of the 245 bp of P1 present in pALA159 is shown in capital letters in Fig. 4. How much of this DNA is essential to origin function remains to be seen. Some hints are provided below.

Comparisons Among ori Sequences

A comparison of the 245-bp origin region of P1 with several other plasmid origins reveals a conserved sequence, GTTATCCACAGG, that, with minor variations, has been recognized as occurring 4 times in oriC of enterobacteria that belong to various genera (85) and once in the promoter area of a gene adjacent to oriC in *Escherichia coli* (24,25), as well as between the 2 promoters of the *E. coli* dnaA gene (24) and near the ends of the Tn5 (33). This sequence is generally found in association with several GATC sequences. In F, as in P1, variants of this sequence occur as a tandem pair (Fig. 4). An additional copy occurs near the beginning of the open reading frame (50) for the F5 protein of F (Fig. 1). Those that occur in oriC of *E. coli* bind dnaA protein specifically (20). Why dnaA protein binding sites should be conserved in plasmids such as P1, F, and pBR322 (46), that appear independent of dnaA, remains to be determined. Localized mutagenesis may be expected to answer this question.

We have included in Fig. 4 the DNA sequence of a primosome assembly site in F that has been identified as a segment of the oriS

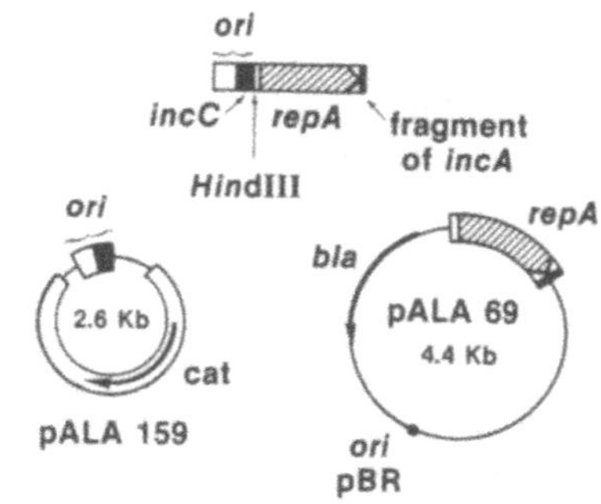

Fig. 3. The linear P1 fragment of the diagram carries a functional origin, repA, and 1-1/2 of the 9 repeat sequences of incA. The P1 DNA in pALA159 and pALA69 was derived from the linear fragment by HindIII cleavage at a site that borders on incC. Genes cat and bla determine chloramphenicol acetyl transferase and β-lactamase, respectively. Transcription of the repA gene of pALA69 is from an unidentified pBR322 promoter. The RepA protein furnished by pALA69 permits replication of pALA159. This complementation permits the 245-bp ori region to establish a copy number similar to that of plasmids in which ori and repA have not been separated.

region that might be essential for initiation. A scan of the 245-bp origin region of P1 reveals no analogous site, nor do we note such a site in P1 beyond the minimal origin within about 100 bp from the region of local P1/F homology described above. (In Fig. 4, P1 DNA exterior to the 245-bp region is distinguished by lower case letters.) It will, therefore, be of some interest to determine whether the primosome binding site plays a significant role in replication initiated at oriS. Paradoxically, a promising candidate for an essential site is present in only 1 of 2 similarly organized replicons, whereas sites deemed probably superfluous for both are uniquely conserved.

The most striking feature of the origin region compared in Fig. 4 has yet to be mentioned. That is the presence of from 3 to 5 copies of closely spaced, 18- to 21-bp sequences. These replicon-specific repeated sequences are so distributed as to present similar domains on a given side of the DNA helix, suggesting that their role is in the interactive binding of a specific protein characteristic of each replicon. In each of the cases that have been tested, the isolated set of repeated sequences exerts group-specific imcompatibility when present on a high copy-number vector (P1: Ref. 1, D. Chattoraj, K. Cordes, and A. Abeles, submitted for publication; Rts1: Ref. 36; F: Ref. 75,78; pSC101: Ref. 44). This incompatibility is most simply interpreted as a consequence of competition for a

```
P1      cattcctgtacgagggcgatgaccttaaatgctatgaacatggtcttctcatcgaaagcgtagaataataccactccctTCCAACCGGCTACGTTGGCCGGTTTTTCAC
Rts1    TAGTTCCGGGAACCGTTGTCGGGTAATGAATGGGTTTACCTTTCATTGCCCGGCAGCGGTTGGACAAGCGTAGCGCGTCAGATATTCCTCCACATTACCATCATTTAAA
F       GCTCGCCGCAGTCGAGCGACGGCGAAGCCCTCGAGTGAGCGAGGAAGCACCAGGGAACAGCACTTATATATTCTGCTTACACACGATGCCTGAAAAAACTTCCCTTGGG
pSC101  GCCCGAGCTTGCGAGGGTGCTACTTAAGCCTTTAGGGTTTTAAGGTCTGTTTTGTAGAGGAGCAAACAGCGTTTGCGACATCCTTTTGTAATACTGCGGAACTGACTAA

P1      TTATCCACATTATCCACTGGATAGATCCAATAATCAGGTCCATACAGATCCCAATTAGATCCATATAGATCCCTGATCGTTGCAGGCCGCGCCACGTCTGGCTTAGAAGTGT
Rts1    GTTATCCACATATCCACCGTGTAGATCCAATAATAGATCCATAGAGAGATCCAGATAAAACCAAAAAGATCCCCGTGGTCTGTAGCCTTACGCCACAAGGCTTACAACGTTT
F       GTTATCCACTTATCCACGGGGATATTTTTATAATTATTTTTTTTATAGTTTTTAGATCTTCTTTTTTTAGAGCGCCTTGTAGGCCTTTATCCATGCTGGTTCTAGAGAAGGT
pSC101  AGTAGTGAGTTATACACAGGGCTGGGATCTATTCTTTTTATCTTTTTTTATTCTTTATTCTATAAATTATAACCACTTGAATATAAACAAAAAAAACACACAAAGGTCTAGC

P1      ATCGCGATGTGTGCTGGAGGGAAAACGATGTGTGCTGGAGGGATAAAAATGTGTGCTGACGGGTTGCTAATGTGTGCTGGCGGGATATAGGATGTGTGTTGACGGGAAAGCT
Rts1    TTCGGTGTGTGCTGAGGGGAAAAAGGTGTGTGCTGAGGGGAAGAAAGTGTGTGTTACGGGGATTTGGGTGTGTGCTGAGGGGAAAAAAGGTGGGCGTCACGGGGAAAATGTG
F       GTTGTGACAAATTGCCCTTTCAGTGTGACAAATCACCCTCAAAATGACAGTCCTGTCTGTGACAAATTGCCCTTAACCCTGTGACAAATTGCCCTCAGAAGAAGCTGTTTTT
pSC101  GGAATTTACAGAGGGTCTAGCAGAATTTACAAGTTTTCCAGCAAAGGTCTAGCAGAATTTACAGATACCCACAACTCAAAGGAAAAGGACTAGTAATTATCATTGACTAGCC
```

Fig. 4. The sequence of the 245-bp origin region of P1 is in capitals. Major repeat sequences are indicated by arrows, putative DnaA protein binding sites (20) by solid underlining, GATC sequences by overlining, and regions of homology to a primosome loading site in colE1 (31) by dashed lines. The pSC101 sequence given contains a region with some homology to F (marked by dashed lines and dots). The sequence of P1 is taken from (1), of Rts1 from (35), of F from (53), and of pSC101 from (16). See Fig. 5 for continuation of sequences.

limited supply of the essential initiator that the replicon determines. Both in vivo and in vitro experiments to be discussed below support this view.

Transcription Initiated in the Origin Region under RepA Control

Transcription initiated from the 245-bp origin region of P1 has been detected by galK operon fusion (Tab. 1). Transcription proceeds in both directions, although with unequal intensities. Both the strong promoter (prepA), which is pointed towards repA, and the weak promoter (pI), which is oppositely oriented, are repressed by RepA protein. The extent of this repression is most dramatic when the RepA protein is supplied from repA cloned in the high-copy-number vector pST52 (Tab. 1, line 3). Even the low, presumably physiological, level of RepA furnished by λ-P1:5R reduced by more than 90% the galK expression that is dependent upon prepA (Tab. 1, line 4); RepA has a considerable, possibly comparable, effect on transcription from pI, although the effect is difficult to measure accurately because the promoter is so weak.

These results lend support to our supposition that the repeat units here, as in the origin regions of other replicons (22,77,80), are binding sites for a replicon-specific initiator. The supposition is further strengthened by recent experiments that indicate that both promoters lie within a 135-bp fragment, 104 bp of which are occupied by the incC set of repeats (data not shown). The observation that the initiator represses 2 differently oriented promoters suggests the possibility that it has more than one role to play. We consider next what these roles may be.

Tab. 1. Effect of incA and repA on transcription from the P1 origin region. Operon fusions were made between the 245-bp origin fragment of P1 and the galK gene of plasmids pK06 and pK04 (48). In pK06 (data column 1), the orientation of the fragment with respect to galK is the same as its orientation with respect to repA in P1; in pK04 (data column 2), the orientation is reversed. The plasmids were used to transform a galK2 recA3 strain of E. coli and galactokinase units were measured according to Ref. 48. The denominators of the tabulated ratios, i.e., the basal enzyme levels in bacteria harboring pK06 or pK04, varied between 3.5 and 3.8 units in different experiments. The pST52 vector is a RSF1030 derivative (66). The repA103 mutation is an amber mutation described in (1).

Source of incA & repA	Ratio of galactokinase units with/without origin fragment: Orientation 1 (prepA)	Orientation 2 (pI)
pST52 (vector)	48	2.0
pST52 incA	43	2.1
pST52 repA	0.3	0.4
λ-P1:5RrepA$^+$	1.7	0.4
λ-P1:5RrepA103	40	2

THE RepA INITIATOR

Role of RepA in Replication

The capacity of pALA69 to permit the replication of the isolated P1 origin in pALA159 (Fig. 3) is presumably conferred by the *repA* gene. Detectable RepA protein is made in maxicells from the replication region of P1 cloned into pBR322 (pALA33), although the amount of protein made is low (1). By placing the *repA* gene under the λp_L promoter, it has been possible to synthesize the protein in high yield. After successive purifications on columns of afigel blue and phosphocellulose, the protein is about 90% pure. The relative electrophoretic mobility of *repA* indicates a protein of about 32 kd, in good agreement with the value of 32,224 predicted from the 286-codon open reading frame (1). RepA is typical of several plasmid initiator proteins that have been characterized with respect to such features as size and net positive charge (Tab. 2).

We would like to know how plasmid copy number varies as a function of *repA* expression and whether RepA protein is rate-limiting for replication under normal maintenance conditions. The data available, although minimal, reveal that the relationship is not a simple monotonic one. A high copy-number plasmid (pALA162) that supplies RepA in large excess, owing to the presence of a strong *bla-p2* promoter, blocks the replication of various mini-P1 plasmids including λ-P1:5R and an *incA*-deleted plasmid, pALA139. Insertion

Tab. 2. Comparison of plasmid-determined initiator proteins.

Plasmid	Initiator	Mol. Wt. (Kd)	Net Charge	Autoregulation	Refs.
P1	RepA	32	+14	Yes	(1)*
Rts1	RepA	33	+13	Not determined	(35)
F	Rep (or E)	29	+16	Yes	(40, 58, 65, 82)
pSC101	Rep	37	+13	Yes	(16, 81)
R6K	π	35	+11	Yes	(21, 39)
pT181	RepC	37.5	+10	Not determined	(55)
λdv	O	34	+ 9	(Yes)†	(17, 47)

* See Table 3

† Autoregulator gene *cro* in same operon as initiator gene *O*.

of a moderately effective transcription terminator downstream from the strong promoter relieves this inhibitory effect. We conclude that RepA in considerable excess is directly inhibitory to RepA-mediated replication. P1 appears not to be unique in possessing an initiator protein that can exert a negative as well as a positive effect on the initiation of replication. Studies of copy-number mutations in the initiator proteins of F (58) and R6K (68) indicate a similar complexity of functions. Inhibitory effects on replication of initiator overproduction have been noted in R6K (27) and pSC101 (G. Churchward, P. Linder, X. Guixan, and L. Caro, submitted for publication).

Indirect evidence that the normal concentration of initiator protein is such as to limit, rather than saturate or inhibit, the replication machinery will be presented in subsequent sections.

RepA Regulation of *repA* Transcription

An autoregulatory role of RepA is shown with the aid of *repA-lacZ* fusion (Tab. 3). When RepA is furnished from a high-copy-number plasmid, synthesis of the fusion protein is completely eliminated. The promoter that is affected is the p*repA* in *incC* that we previously noted to be markedly repressed by RepA. There does not appear to be a more proximal promoter that is sufficiently active to be responsible for significant *repA* expression. By appropriate insertion of an efficient terminator, *repA* expression from a high copy-number plasmid could be made to depend solely on whatever promoters exist in the P1 DNA between *incC* and *repA*. The resultant plasmid failed to permit maintenance of the isolated origin plasmid, pALA159. Normal transcription of *repA* must, therefore, depend upon the highly RepA-sensitive promoter, p*repA*, that is in the set of *incC* repeats.

The organization of P1 and F are not the same in the region between the clustered origin repeats and the start of the initiator gene (Fig. 5). The transcription of the initiator gene of F has been found to originate primarily from a promoter in this region. The initiator is also autoregulated, but in this case, the proposed autoregulatory binding site, although recognizably related to the repeat sequences, is separated from the cluster (65). A similar separation has been noted in R6K (69) and pSC101 (16). The autoregulatory binding site is also significantly different in sequence from any of the clustered sites. These structural differences are likely to impart a different form to the dose-response curves for the activity of the initiator in replication and in the regulation of its own synthesis.

One implication of this statement is that it might be possible to mutationally alter the autoregulated initiator protein so as to differentially affect 1 of its 2 roles. A mutation in the initiator

Tab. 3. Autoregulation of repA transcription. Fusions were made that join in frame the first 66 codons of repA DNA to the lacZ gene (missing codons 1-8) of pMLB1034 (64). The prepA- containing P1 DNA of the plasmid used in obtaining the data of column 1 included 297 bp upstream of repA (the 245-bp origin region and 52 bp proximal to repA). The other insert (data of column 2) included only 56 bp upstream of repA. By providing a ribosome-binding site, this 56-bp insert allows β-galactosidase expression from a weak RepA-insensitive promoter (p?) that presumably lies within the vector itself. The plasmids were used to transform a Δlac recA56 strain of E. coli, and β-galactosidase units were measured according to (51). The basal enzyme level in bacteria harboring pMLB1034 was <1 unit.

	β-galactosidase units from repA-lacZ protein fusion	
Source of repA	prepA	p?
pST52 (vector)	2700	42
pST52 repA	< 1	43

that specifically diminishes its autoregulatory role would cause an increase in plasmid copy number, provided that replication was initiator-limited. At least some cop mutations that have been identified in P1 (9) and in F (58,63) have been localized to the respective initiator genes and possibly belong to this category.

Specific Binding of RepA to Repeat Sequences

The genetic arguments for RepA binding to the DNA repeats are supported by a direct physical demonstration of sequence-specific binding (Fig. 6). We look forward to obtaining quantitative data with which to answer questions about the stoichiometry and cooperativity of this binding. Cooperative interactions presumably lie at the heart of precise replication control.

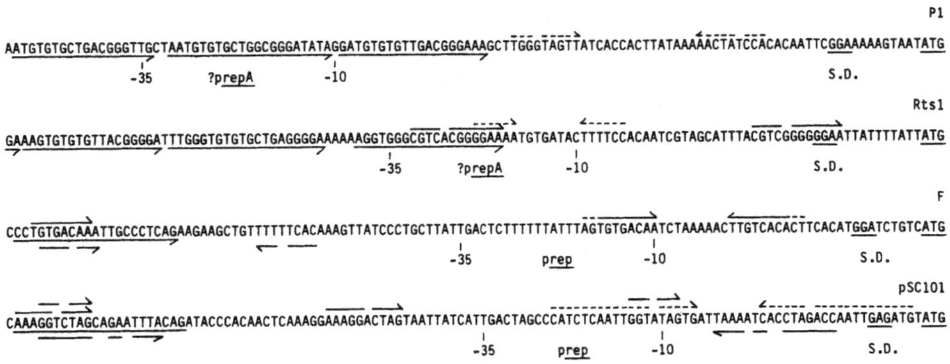

Fig. 5. Characteristic -10 and -35 sequences (60) of likely promoters for transcription of the initiator proteins of P1 and Rts1 are the choices of the authors based on a TARGSEARCH analysis (54). The corresponding choices for F and pSC101 are those of (65) and (16), respectively. The P1 sequence is taken from (1), the Rts1 sequence from (35). Arrows without dashes represent major repeats of repeated segments thereof. Certain of these are postulated to be autoregulatory sites. Dashed arrows represent other (inverted) repeats.

REPLICATION CONTROL ELEMENT *incA*

Dispensability of incA

In P1, Rts1, and F, the repeats that bind (or presumably bind) the plasmid-determined initiators are found flanking the respective initiator genes. The downstream repeats are more numerous than the upstream repeats, are less densely distributed, and appear to play a different role. Whereas the upstream repeats (of F, and presumably of P1) appear to be an essential component of the origin (M. Couturier, pers. comm.), the downstream repeats are dispensable. Elimination from a mini-P1 of 7 1/2 of the 9 downstream (*incA*) repeats results in a 6-fold increase in plasmid copy number (D. Chattoraj, K. Cordes, and A. Abeles, submitted for publication). In F, the deletion of all repeat units downstream of the initiator gene has been achieved and increases the copy number about 6- to 8-fold (10,63,78). Loss of these repeats does not result in runaway replication.

Reinsertion of *incA* into a mini-P1 that has suffered a substantial deletion of *incA* repeats (pALA139) restores the plasmid to a

Fig. 6. In vitro binding of RepA protein to repeat sequences.

A fragment of DNA containing the 9 incA repeats and a control piece of pBR322 DNA were excised from a pBR322 clone carrying incA and end-labeled with ^{32}P dNTP. The fragments were then mixed with the indicated increasing amounts of purified RepA and electrophoresed in a 5% acrylamide gel. The autoradiogram shows that the incA fragment formed several slow-moving bands, apparently by complexing with the protein, whereas the mobility of the control fragment was not affected. Binding of RepA to incC has been demonstrated in similar experiments (data not shown).

low copy number. The incA can be variously positioned within the plasmid or placed in trans within a chromosomally integrated λ prophage. The effects on copy number of such reinsertions are comparable, at least to a first approximation. Therefore, neither the position nor the orientation of this set of repeats is of major significance in determining copy number. The results of these experiments also imply that the incA repeats are dispensable in the target replicon for sensitivity to the destabilizing action of incA (Tab. 4). The latter deduction is further strengthened by an experiment (not shown) in which a single copy of incA per bacterial chromosome destabilizes the isolated origin plasmid pALA159 when the source of RepA is from pALA69, a plasmid in which repA is transcribed from an unidentified vector promoter, rather than from a P1 promoter.

Tab. 4. Destabilization of an incA-deleted mini-P1, pALA139, by incA in trans. pALA139 retains only 1 1/2 incA repeats (as little of incA as is present in the chimera of line 3), a functional repA gene, a P1 origin, and a chloramphenicol acetyl transferase gene. The plasmid is stably maintained in the absence of equipartition functions because of its elevated copy number relative to its $incA^+$ parent. It can be maintained in the presence of 1 copy of incA by selection with chloramphenicol. The tabulated percentages are based on a minimum of 64 colonies tested for drug resistance after 25 generations of growth with selection for the presence of pBR322 alone or λ alone. The bacterium was recA3 strain N100.

Vector and incA repeats	Percent bacteria retaining pALA139 after 25 generations
1) pBR322: all nine	<2
2) pBR322: part of #7, #8, #9	<2
3) pBR322: part of #8, #9	94
4) pBR322: none	>99
5) λ prophage: all nine	<1
6) λ prophage: none	>99

We conclude that the incA present in unit-copy mini-P1 is dispensable for incA sensitivity, as it is for the capacity to replicate. It is an element in replication control that is capable of inhibiting a P1 replicon when present as a single extra copy.

Replication Control Exerted by Subsets of incA

To determine the mechanism(s) by which incA exerts its control, we have asked whether destruction of the integrity of incA influences its activity. The efficacies of subsets of the incA repeats cloned into pBR322 are presented in Tab. 4 and 5. Like the complete set of 9 repeats, each set of 3 repeats tested causes more than 98% loss of λ-P1:5R within 25 generations. A clone of less than 2 repeats failed to exert significant imcompatibility by this criterion.

Tab. 5. Destabilization of the mini-P1 λ-P1:5R by fragments of incA. The experiments are similar to those of Tab. 4 except that loss of the mini-P1 was followed by loss of the λ immunity marker.

incA repeats carried by pBR322 vector	Percent bacteria retaining λ-P1:5R after 25 generations
1) all nine	<3
2) #1, #2, #3	<1
3) #4, #5, #6	<1
4) #7, #8, #9	<3
5) part of #7, #8, #9	3
6) part of #8, #9	96
7) none	>99

By a less sensitive test, based on dnaAts integrative suppression, none of the set of 3 repeats was effective in expressing incompatibility, but 2 different sets of 6 repeats were as effective as intact incA (data not shown). These results demonstrate that no unique site within incA is required for incA activity (Tab. 4 and 5). Since even 2 1/2 repeats exert marked incompatibility when present in high copy number, we consider it unlikely that incA encodes small RNA or oligopeptide mediators of incompatibility. However, we do not imply that the repeats act independently of one another.

Relief by RepA of Replication Block by incA Repeats

There is a simple way to explain how the subsets of incA can cause specific replication inhibition. We suggest that the DNA repeats inhibit initiator function by uniting with the specific initiator. Provided that an appreciable fraction of the initiator can be removed by binding to incA, nothing more elaborate than a mass action effect need be postulated. According to this hypothesis, incompatibility expression by incA should be relieved by an increased supply of RepA. The results presented in Tab. 6 show that a source of RepA overcomes the blockage by incA of integrative dnaAts suppression. This relief can be overcome, in turn, by additional copies of incA. We have already seen in Tab. 1 that incA by itself does not affect transcription from the origin. Both these experi-

Tab. 6. Replication block by incA relieved by repA. At least 2 λ incA prophages are required to block integrative suppression of the dnaAts defect afforded by the integrated mini-P1, λ-P1:5R. One λincA prophage is insufficient. RepA is overproduced from the pBR322 bla-p2 attenuated by a transcription terminator promoter in the plasmid (pA1A176) of lines 5 and 6. The relief afforded by this overproduction of RepA, in turn, overcome by furnishing incA from the high-copy-number plasmid pST52 incA. A small fragment of incA is associated with the cloned repA gene and is the same fragment that is shown in line 3 of Tab. 4 to exert negligible incompatibility even when in high copy number. The pBR322 vector used here contained an inserted chloramphenicol acetyl transferase gene.

Episomes carried by recA56 dnaA46(ts) E. coli			Surviving fraction 42°/32°
Integrated at loxB	Prophages at attλ	Plasmids	
1) none	none	none	4×10^{-6}
2) λ-P1:5R	none	none	1
3) λ-P1:5R	(λincA)	none	5×10^{-4}
4) λ-P1:5R	(λincA)	pBR322	5×10^{-4}
5) λ-P1:5R	(λincA)	pBR322 repA	1
6) λ-P1:5R	(λincA)	pBR322 repA; pST52 incA	2×10^{-6}

ments support the contention that incA controls copy number by sequestering the rate-limiting RepA protein.

A direct demonstration that incA binds repA protein is shown in Fig. 6. Quantitative measurements of affinities of the initiator protein for the repeats before and after the initiator gene have not yet been made. We expect that the incA repeats are loaded with RepA before the putative site that triggers replication is filled. We also expect that the relative affinities of RepA for repeat groups of various sizes are correlated with their relative strengths as incompatibility elements.

REPLICATION CONTROL

How are the Antagonistic Control Circuits to be Reconciled?

The arguments presented for a control of RepA concentration by autoregulation and by binding to incA may appear separately compelling, but these 2 possibilities are also mutually exclusive. A responsive autoregulatory circuit would nullify any attempt to regulate the concentration of free RepA by varying the concentration of RepA binding sites. If the 2 modes of control are to have any biological meaning, their operation must not be such that one control inevitably nullifies the other.

A reconciliation might be achieved if there existed a concentration range of RepA over which autoregulation is effective and a different concentration range over which incA is effective. Unfortunately, the evidence suggests that both kinds of controls operate at the same RepA concentration, i.e., when a single copy of a mini-P1 is present.

A more satisfactory resolution of the problem may be to invoke different kinetics for the 2 control circuits. The following formulation of the "over shoot model" is, in part, the fruit of discussions with Kurt Nordström and Robert Pritchard, among others, but the component ideas have originated independently in several laboratories. The model depends on the existence of a refractory period in the autoregulatory circuit that is due to the finite duration of transcription, translation, and message decay. Let us suppose that the refractory period in the autoregulatory circuit is sufficiently long that, after the incA repeats have filled with initiator protein, free initiator can transiently accumulate up to a threshold concentration or amount necessary for replication to be initiated. Although replication will double the initiator gene dosage, the concentration of free initiator will fall sharply rather than rise. At the time of replication, the delayed autoregulatory shut-off of initiator synthesis will have occurred and, as has been proposed for the analogous case of F (78), existing initiator molecules will bind to the newly created repeats. Until these repeats fill and free initiator accumulates again, there will be no reinitiation of replication. In this model, there is a hierarchy of affinities for initiator protein. The autoregulatory site will have the highest affinity, the copy-number control repeats a lesser affinity, and the site(s) that triggers replication the lowest. The model also assumes a concentration of free initiator protein sufficiently low that titration by a few repeats can be significant. Titration must also be slow enough to account for the observed interreplication times.

An alternative resolution of the problem can be made if the autoregulator and the initiator of replication are assumed to be different molecular species that are determined by the same operon. A model of this kind was proposed for *E. coli* chromosome replication in 1973 (67) and was shown to operate in the case of λdv, which is controlled by the cotranscribed *cro* autoregulator and the *O* initiator (47). The applicability of this model to F is suggested by J.D. Trawick and B.C. Kline (submitted for publication; and Ref. 38). These authors propose, for simplicity, that the initiator is a modified form of the autoregulator, perhaps derived from it by oligomerization. If the rate of this modification is so slow that only a very low concentration of initiator forms, its titration by repeats can significantly alter the availability of initiator at the site that triggers replication. Initiator titration need not affect autoregulation. The relative concentration of autoregulator is not specified by the model and the affinity of the autoregulatory site for the autoregulator need not be particularly high. Although this and the preceding model have certain features in common, there are testable differences between them. Both these models adequately account for the abrupt adjustment of plasmid replication rates that is characteristic of F (79) and presumably of P1.

Autoregulated initiators are characteristic of a number of different plasmids (Tab. 2). A rationale for this autoregulation is apparent. In general, replacement of an autoregulated rate-limiting initiator by a constitutive one would cause the plasmid copy number to increase out of control or, in some cases, only to the point where further initiator synthesis would inhibit plasmid replication.

The separate set of copy-control repeats that are found in P1, Rts1, and F play a less obvious role in plasmid maintenance. Runaway replication does not occur when these repeats are deleted, although, as previously mentioned, elevated copy numbers do result. To appreciate the possible utility of these repeats in plasmid maintenance, we ask what disadvantage might their loss confer on a plasmid that correspondingly reduced the synthesis (or efficacy) of the initiator protein to compensate for its increased availability for triggering replication. Such a compensated loss would leave the average plasmid copy number unaltered. However, the longer the set of binding sites, the more reproducible will be the length of the period needed for initiator resynthesis that will follow each DNA duplication. Without *incA*, the refractory periods between successive rounds of replication would be more variable. A broadened distribution of plasmid copy numbers in the bacterial population would cause an increased proportion of the bacterial population to be disadvantaged in the sense of carrying an excessive plasmid burden or none at all.

A way of formally analyzing the situation has been suggested to us by Dr. Luud Koppes (University of Uppsala). He proposes that the interval between successive replications can be regarded as the waiting time for successive random associations of initiator with a particular number of repeats that must be filled before DNA synthesis initiates. It is assumed that the occupation of a particular repeat unit at the origin region triggers the initiation of DNA synthesis and occurs only after an orderly filling of the other repeats. Koppes calculates, in an idealized case, that the coefficient of variation of the interreplication time is proportional to the reciprocal of the square root of the total number of repeats. The important conclusion is that the larger the number of repeats, the smaller the fluctuation.

It is not excluded that an adequate explanation of P1 plasmid maintenance will require the invocation of an additional control loop whose existence has been overlooked. Evidence for the existence of 1 additional stabilizing circuit that we believe is present in intact P1, but not in the mini-P1 plasmids so far considered, will be presented next.

COORDINATION WITH CELL DIVISION

For a plasmid to be maintained as stably as is P1 or F, fluctuation in interreplication time that could result in plasmid loss cannot be tolerated. It is not surprising that a special mechanism should have evolved to ensure that emergency measures are invoked when there is an immediate threat of such loss.

Any of a variety of unit-copy plasmids that have been impaired in their capacity to replicate and then introduced into healthy bacteria cause a temporary cessation of cell division. The bacteria filament and subsequently exhibit other SOS responses as if they had been damaged themselves (8,12,13,18,41,45,62). These responses can be brought about by UV irradiation of the plasmid or by more specific means of reducing plasmid copy number. Repair of damage to the plasmid is favored, and cell division prior to repair is inhibited. The genetic basis for this coupling of cell division to the well-being of the plasmid has been determined in F (8,11,13,37,49,50,56). The location of the coordination with cell division (ccd) genes of F is shown in Fig. 1. The ccdB gene product(s) inhibits cell division; the ccdA gene product(s) normally prevents this inhibition but cannot do so at gene dosages that are so low as to present a threat to plasmid stability.

Although ccd genes have not been identified in P1, there are indications that a ccd function is determined within EcoRI-8 of P1, in a region situated similarly to that of F. EcoRI-8 is a 6-kb fragment that carries the P1 origin near its junction with EcoRI-5. Mini-P1 plasmids that carry EcoRI-8 have been shown to have the potential for causing SOS induction (15), in contrast to the diminutive mini-P1 plasmids that we have studied (M. Yarmolinsky, unpub. data). In view of the several similarities between P1 and F that we have already noted, it would be surprising to find that both plasmids have adopted similar strategies for addicting the host to their presence.

CONCLUSIONS: THE CLEPSYDRA ANALOGY

The mechanisms of plasmid tenacity discussed in this and the accompanying article are of some antiquity because they appear to be common to diverse replicons that, by normal criteria, are nonhomologous. In several plasmids, a modular arrangement of DNA regions that are involved in replication and its control, coordination with cell division, partitioning, and dimer resolution enables us to perform a variety of mix-and-match experiments. Comparative studies make it clear that Nature was doing this long before our arrival on the scene. The process has been named reticulate (as opposed to arboreal) evolution.

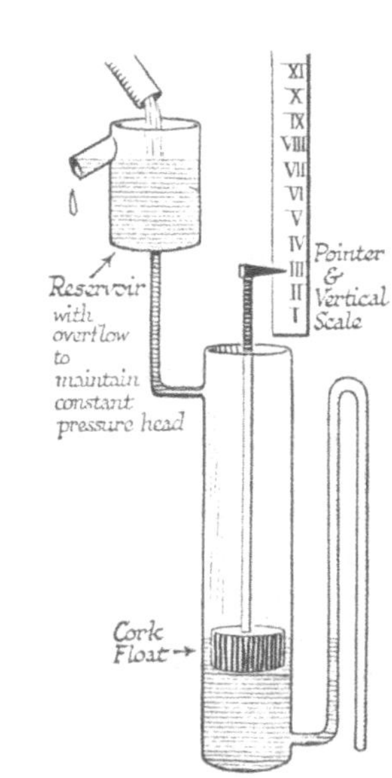

Fig. 7. The upper reservoir, with its means of maintaining a constant pressure head, resembles the autoregulated source of RepA. The lower vessel resembles the set of repeats that must be filled before a putative critical repeat can fill and trigger the sudden creation by replication of a set of empty repeats. Alterations in the length of the siphon tube correspond to alterations in the number of incA repeat sequences and affect the interval between cycles.

The P1 control circuit to which we have paid most attention involves the regulation of DNA synthesis by a stretch of DNA that acts as a sink or chambered reservoir for an essential initiator protein. We have suggested that filling the chambers triggers a round of DNA synthesis that generates a new empty reservoir. The action is much like that of a pipet washer that fills slowly up to the point where a siphon empties the vessel suddenly, whereupon it recommences the cycle. The principle was used more than 2,000 years ago in the construction of water clocks (clepsydras) (Fig. 7). In some clepsydras, constancy of the cycles was assured by having the reservoir filled from a tank that was maintained at the level of an overflow. This constant influx of water is analogous to a synthesis of initiator that, due to autoregulation, is at least partially buffered against changes in gene dosage. We would like to believe that the repeated DNA sequences and the control circuits associated with them form a molecular clock by which to regulate plasmid generation times.

We are attracted to plasmids as objects of study because they are dispensable elements. Ironically, we are most fascinated by the ingenuity of the mechanisms that make a resident plasmid almost impossible to disown, mechanisms that appear to be among Nature's early inventions. It is appropriate that one of these mechanisms may represent a molecular analog of an early human invention.

ACKNOWLEDGMENTS

We thank Martin Mulligan and William McClure of Carnegie-Mellon University for providing us with a TARGSEARCH computer analysis of potential promoters in the region upstream of the _repA_ gene of P1; Bruce Kline, Luud Koppes, Kurt Nordström, and Robert Pritchard for helpful discussions of regulatory circuitry; and Julie Ratliff for cheerful preparation of the manuscript, sequences included.

Research sponsored by the National Cancer Institute, DHHS, under contract No. N01-CO-23909 with Litton Bionetics, Inc. The contents of this publication do not necessarily reflect the views or policies of the Department of Health and Human Services, nor does mention of trade names, commercial products, or organizations imply endorsement by the U.S. Government.

REFERENCES

1. Abeles, A.L., K.M. Snyder, and D.K. Chattoraj (1984) P1 plasmid replication: Replicon structure. _J. Mol. Biol._ 173:307-324.
2. Abremski, K., and R. Hoess (1983) Bacteriophage P1 site-specific recombination: Purification and properties of the Cre recombinase protein. _J. Biol. Chem._ 259:1509-1519.

3. Austin, S., and A. Abeles (1983) Partition of unit-copy mini-plasmids to daughter cells. I. P1 and F miniplasmids contain discrete interchangeable sequences sufficient to promote equipartition. J. Mol. Biol. 173:353-372.
4. Austin, S., and A. Abeles (1983) Partition of unit-copy mini-plasmids to daughter cells. II. The partition region of mini-plasmid P1 encodes an essential protein and a centromere-like site at which it acts. J. Mol. Biol. 169:373-387.
5. Austin, S., and A. Abeles (1984) The partition functions of P1, P7 and F miniplasmids. This volume.
6. Austin, S., F. Hart, A. Abeles, and N. Sternberg (1982) Genetic and physical map of a P1 miniplasmid. J. Bact. 152:63-71.
7. Austin, S., M. Ziese, and N. Sternberg (1981) A novel role for site-specific recombination in maintenance of bacterial replicons. Cell 25:729-736.
8. Bailone, A., R. Devoret, A. Brandenburger, A. Lévine, M. Dutreix, and M. Pierre (1984) Aborted installation or maintenance of mini-F induces SOS functions. Abstract, this volume.
9. Baumstark, B.R., L. Lowery, and J.R. Scott (1984) Location by DNA sequence analysis of cop mutations affecting the number of plasmid copies of prophage P1. Mol. Gen. Genet. (in press).
10. Bergquist, P.L., R.A. Downard, P.A. Caughey, R.C. Gardner, and H.E.D. Lane (1981) Analysis of mini-F plasmid replication by transposition mutagenesis. J. Bact. 147:888-899.
11. Bex, F., H. Karoui, L. Rokeach, P. Drèze, L. Garcia, and M. Couturier (1983) Mini-F encoded proteins: Identification of a new 10.5 kilodalton species. EMBO J. 2:1853-1861.
12. Borek, E., and A. Ryan (1958) The transfer of irradiation-elicited induction in a lysogenic organism. Proc. Natl. Acad. Sci., USA 44:374-377.
13. Brandenburger, A., A. Bailone, A. Lévine, and R. Devoret (1984) The SOS signal produced by miscarriage of an intact mini-F plasmid. J. Mol. Biol. (in press).
14. Briaux S., G. Gerbaud, and A. Jaffé-Brachet (1979) Studies of a plasmid coding tetracycline resistance and hydrogen sulfide production incompatible with the prophage P1. Mol. Gen. Genet. 170:319-325.
15. Capage, M.A., and J.R. Scott (1983) SOS induction by P1Km mini-plasmids. J. Bact. 155:473-480.
16. Churchward, G., P. Linder, and L. Caro (1983) The nucleotide sequence of replication and maintenance functions encoded by plasmid pSC101. Nucleic Acids Res. 11:5645-5659.
17. Daniels, D.L., J.S. Schröeder, W. Szybalski, F. Sanger, A.R. Coulson, G.F. Huong, D.F. Hill, G.B. Petersen, and F.R. Blattner (1983) Complete annotated lambda sequence. In Lambda II, R.W. Hendrix, J.W. Roberts, F.W. Stahl and R.A. Weisberg, eds. Cold Spring Harbor Laboratory, Cold Spring Harbor, New York, pp. 519-676.
18. Devoret, R., and J. George (1967) Induction indirecte du prophage λ par le rayonment untraviolet. Mutat. Res. 4:713-725.

19. Eichenlaub, R., D. Figurski, and D. Helinski (1977) Bidirectional replication from a unique origin in a mini-F plasmid. Proc. Natl. Acad. Sci., USA 74:1138-1141.
20. Fuller, R.S., and A. Kornberg (1983) Purified dnaA protein in initiation of replication at the Escherichia coli chromosomal origin of replication. Proc. Natl. Acad. Sci., USA 80:5817-5821.
21. Germino, J., and D. Bastia (1982) Primary structure of the replication initiation protein of plasmid R6K. Proc. Natl. Acad. Sci., USA 79:5474-5479.
22. Germino, J., and D. Bastia (1983) Interaction of the plasmid R6K-encoded replication initiator protein with its binding sites on DNA. Cell 34:125-134.
23. Gustafsson, P., and K. Nordström (1980) Control of replication of plasmid R1. Kinetics of replication in shifts between different copy number levels. J. Bact. 141:106-110.
24. Hansen, E.B., F.G. Hansen, and K. von Meyenburg (1982) The nucleotide sequence of the dnaA gene and the first part of the dnaN gene of Escherichia coli K-12. Nucl. Acids Res. 10:7373-7385.
25. Hansen, F.G., S. Koefoel, K. von Meyenburg, and T. Atlung (1981) In The Initiation of DNA Replication, ICN-UCLA Symposia on Molecular and Cellular Biology, D.S. Ray and C.F. Fox, eds. Academic Press, New York, Vol. 21, pp. 37-55.
26. Hedges, R.W., A.E. Jacob, P.T. Barth, and N.J. Grinter (1975) Compatibility properties of P1 and ΦAMP prophages. Mol. Gen. Genet. 141:263-267.
27. Helinski, D. (1984) This volume.
28. Hoess, R., M. Ziese, and N. Sternberg (1982) P1 site-specific recombination: Nucleotide sequences of the recombining sites. Proc. Natl. Acad. Sci., USA 79:3398-3402.
29. Iida, S., and W. Arber (1979) Multiple physical differences in the genome structure of functionally related bacteriophages P1 and P7. Mol. Gen. Genet. 173:249-261.
30. Ikeda, H., M. Inuzuka, and J-I. Tomizawa (1970) P1-like plasmid in Escherichia coli 15. J. Mol. Biol. 50:457-470.
31. Imber R., R.L. Low, and D.S. Ray (1983) Identification of a primosome assembly site in the region of the ori2 replication origin of the Escherichia coli mini-F plasmid. Proc. Natl. Acad. Sci., USA 80:7132-7136.
32. Jacob, F., S. Brenner, and F. Cuzin (1963) On the regulation of DNA replication in bacteria. Cold Spring Harbor Symp. Quant. Biol. 20:329-348.
33. Johnson, R.C., and W.S. Reznikoff (1983) DNA sequences at the ends of transposon Tn5 required for transposition. Nature 304:280-282.
34. Kahn, S.A., G.K. Adler, and R.P. Novick (1982) Functional origin of replication of plasmid pT181 DNA is contained within a 168 base pair segment. Proc. Natl. Acad. Sci., USA 79:4580-4584.

35. Kamio, Y., A. Tabuchi, Y. Itoh, H. Katagiri, and Y. Terawaki (1984) Complete nucleotide sequence of mini-Rts1 and its copy mutant. J. Bact. 158:307-312.
36. Kamio, Y., and Y. Terawaki (1983) Nucleotide sequence of an incompatibility region of mini-Rts1 that contains five direct repeats. J. Bact. 155:1185-1191.
37. Karoui, H., F. Bex, P. Drèze, and M. Couturier (1983) Ham22, a mini-F mutation which is lethal to host cell and promotes recA-dependent induction of lambdoid prophage. EMBO J. 2:1863-1868.
38. Kline, B.C., and J.D. Trawick (1984) A proposed model for control of mini-F replication. Abstract, this volume.
39. Kolter, R., and D. Helinski (1982) Plasmid R6K DNA replication II. Direct nucleotide sequence repeats are required for an active γ-origin. J. Mol. Biol. 161:45-56.
40. Komai, N., T. Nishizawa, Y. Hayakawa, T. Murotsu, and K. Matsubara (1982) Detection and mapping of six mini F-encoded proteins by cloning analysis of dissected mini-F segments. Mol. Gen. Genet. 186:193-203.
41. Koyama, A.H., and T. Yura (1975) Plasmid mutations affecting self-maintenance and host growth in Escherichia coli. J. Bact. 122:80-88.
42. Lane, D., and R.C. Gardner (1979) Second EcoRI fragment of F capable of self-replication. J. Bact. 139:141-151.
43. Light, J., and S. Molin (1982) The sites of action of the two copy number control functions of plasmid R1. Mol. Gen. Genet. 187:486-493.
44. Linder, P., G. Churchward, and L. Caro (1983) Plasmid pSC101 replication mutants generated by insertion of the transposon Tn1000. J. Mol. Biol. 170:287-303.
45. MacQueen, H.A., and W.D. Donachie (1977) Intracellular localization and effects on cell division of a plasmid blocked in deoxyribonucleic acid replication. J. Bact. 132:392-397.
46. Sutcliffe, J.G. (1978) The complete nucleotide sequence of the Escherichia coli plasmid pBR322. Cold Spring Harbor Symp. Quant. Biol. 43:77-90/
47. Matsubara, K. (1981) Replication control system in lambda dv. Plasmid 5:32-52.
48. McKenney, K., H. Shimatake, D. Court, U. Schmeissner, C. Brady, and M. Rosenberg (1981) In Gene Amplification and Analysis, J.G. Chirikjian and T.S. Papas, eds. Elsevier-North Holland, Amsterdam, Vol. 2, pp. 384-417.
49. Miki, T., Z-T. Chang, and T. Horiuchi (1984) Control of cell division by sex factor F in Escherichia coli II. Identification of genes for inhibitor proteins and trigger protein on the 42.84-43.6 F segment. J. Mol. Biol. 174 (in press).
50. Miki, T., K. Yoshioka, and T. Horiuchi (1984) Control of cell division by sex factor F in Escherichia coli. I. The 42.84-43.6 F segment couples cell division of the host bacteria with replication of plasmid DNA. J. Mol. Biol. 174 (in press).
51. Miller, J.H. (1972) Experiments in Molecular Genetics. Cold Spring Harbor Laboratory, Cold Spring Harbor, New York.

52. Mise, K., Y. Yoshida, and M. Kawai (1983) Generalized transduction between Salmonella typhi and Salmonella typhimurium by phage j2 and characterization of the j2 plasmid in Escherichia coli. J. Gen. Microbiol. 129:3395-3400.
53. Murotsu, T., K. Matsubara, H. Sugisaki, and M. Takanami (1981) Nine unique repeating sequences in a region essential for replication and incompatibility of the mini-F plasmid. Gene 15:257-271.
54. Mulligan, M.E., D.K. Hawley, R. Entriken, and W.R. McClure (1984) Escherichia coli promoter sequences predict in vitro RNA polymerase selectivity. Nucl. Acids Res., (in press).
55. Novick, R.P., G.K. Adler, S. Majumder, S.A. Khan, S. Carleton, W.D. Rosenblum, and S. Iordanescu (1982) Coding sequence for the pT181 repC product: A plasmid-coded protein uniquely required replication. Proc. Natl. Acad. Sci. 79:4108-4112.
56. Ogura, T., and S. Hiraga (1983) Mini-F plasmid genes that couple host cell division to plasmid proliferation. Proc. Natl. Acad. Sci., USA 80:4784-4788.
57. Ogura, T., S. Hiraga (1983) Partition mechanism of F plasmid: 2 plasmid gene-encoded product and a cis-acting region are involved in partition. Cell 32:351-360.
58. Piérard, P., J. Janssens, L. Garcia, P. Drèze, M. Couturier, and F. Bex (1984) Mini-F copy number mutations affecting the primary structure of the E protein. Arch. Int. de Phys. et de Bioch. 92:B52-B53.
59. Prentki, P., M. Chandler, and L. Caro (1977) Replication of the prophage Pl during the cell cycle of Escherichia coli. Mol. Gen. Genet. 152:71-76.
60. Rosenberg, M., and D. Court (1979) Regulatory sequences involved in the promotion and termination of RNA transcription. Annu. Rev. Genet. 13:319-353.
61. Rosner, J.L. (1972) Formation, induction and curing of bacteriophage Pl lysogens. Virology 48:679-689.
62. Rosner, J.L., L.R. Kass, and M.B. Yarmolinsky (1968) Parallel behavior of F and Pl in causing indirect induction of lysogenic bacteria. Cold Spring Harbor Symp. Quant. Biol. 33:784-789.
63. Seelke, R.W., B.C. Kline, J.D. Trawick, and G.D. Ritts (1982) Genetic studies of F plasmid maintenance genes involved in copy number control, incompatibility and partitioning. Plasmid 7:163-179.
64. Silhavy, T.J., M.L. Berman, and L.W. Enquist (1984) Experiments with Gene Fusions. Cold Spring Harbor Laboratory, Cold Spring Harbor, New York.
65. Søgaard-Anderson, L., L.A. Rokeach, and S. Molin (1984) Regulated expression of a gene important for replication of plasmid F in E. coli. EMBO J. 3:257-262.
66. Som, T., and J-I. Tomizawa (1982) Origin of replication of Escherichia coli plasmid RSF1030. Mol. Gen. Genet. 187:375-383.
67. Sompayrac, L., and O. Maaløe (1973) Autorepressor model for control DNA replication. Nature New Biol. 241:133-135.

68. Stalker, D.M., R. Kolter, and D.R. Helinski (1982) Plasmid R6K replication I. Complete nucleotide sequence of an autonomously replicating segment. J. Mol. Biol. 161:33-43.
69. Stalker, D.M., M. Filutowicz, and D.R. Helinski (1983) Release of initiation control by a mutational alteration in the R6K π protein required for plasmid replication. Proc. Natl. Acad. Sci., USA 80:5500-5504.
70. Sternberg, N., and S. Austin (1983) Isolation and characterization of P1 minireplicons, λ-P1:5R and λ-P1:5L. J. Bact. 153:800-812.
71. Sternberg, N., and D. Hamilton (1981) Bacteriophage P1 site-specific recombination. I. Recombination between 2 loxP sites. J. Mol. Biol. 150:467-486.
72. Sternberg, N., D. Hamilton, and R. Hoess (1981) P1 site-specific recombination. II. Recombination between loxP and the bacterial chromosome. J. Mol. Biol. 150:487-507.
73. Sternberg, N., and R. Hoess (1983) The molecular genetics of bacteriophage P1. Annu. Rev. Genet. 17:123-154.
74. Tolun, A., and D. Helinski (1981) Direct repeats of the F plasmid incC region express F incompatibility. Cell 24:687-694.
75. Tolun A., and D. Helinski (1982) Separation of the minimal replication region of the F plasmid into a replication origin segment and a transacting segment. Mol. Gen. Genet. 186:372-377.
76. Tomizawa, J-I., and T. Itoh (1981) Plasmid colE1 incompatibility determined by interaction of RNAI with primer transcript. Proc. Natl. Acad. Sci., USA 78:6096-6100.
77. Tsurimoto, T., and K. Matsubara (1981) Purified bacteriophage λO protein binds to four repeating sequences at the λ replication origin. Nucl. Acids Res. 9:1789-1799.
78. Tsutsui, H., A. Fujiyama, T. Murotsu, and K. Matsubara (1983) Role of 9 repeating sequences of the mini-F genome for expression of F-specific incompatibility phenotype and copy number control. J. Bact. 155:337-344.
79. Tsutsui, H., and K. Matsubara (1981) Replication control and switch-off function as observed with a mini-F factor plasmid. J. Bact. 147:509-516.
80. Vocke, C., and D. Bastia (1983) DNA-protein interaction at the origin of DNA replication of the plasmid pSC101. Cell 35:495-502.
81. Vocke, C., and D. Bastia (1983) Primary structure of the essential replicon of the plasmid pSC101. Proc. Natl. Acad. Sci., USA 80:6557-6561.
82. Wehlmann, H., and R. Eichenlaub (1980) Plasmid mini-F encoded proteins. Mol. Gen. Genet. 180:205-211.
83. Yarmolinsky, M. (1982) Bacteriophage P1. In Genetic Maps, 2. S.J. O'Brien, ed. National Cancer Institute, National Institutes of Health, Frederick, Maryland, pp. 34-43.
84. Yun, T., and D. Vapnek (1977) Electron microscopic analysis of bacteriophage P1, P1Cm and P7. Determination of genome size,

sequence homology, and location of antibiotic-resistance determinants. *Virology* 77:376-385.

85. Zyskind, J.W., J.M. Cleary, W.S.A. Brusilow, N.E. Harding, and D.W. Smith (1983) Chromosomal replication origin from the marine bacterium *Vibrio harveyi* functions in *Escherichia coli*: *oriC* consensus sequence. *Proc. Natl. Acad. Sci., USA* 80:1164-1168.

PARTITIONING OF THE pSC101 PLASMID DURING CELL DIVISION

Stanley N. Cohen,[1] Christine A. Miller,[1] William T. Tucker,[2] Peter A. Meacock,[3] and Peter Gustafsson[4]

[1]Department of Genetics
Stanford University School of Medicine
Stanford, California 94305

INTRODUCTION

It is generally accepted that a mechanism must exist for ensuring that each of the products of a bacterial cell division receives a copy of the chromosome. However, until recently, the existence of a specific mechanism for actively distributing extrachromosomal elements such as plasmids to daughter cells was less clear. Certainly, for plasmids that normally exist in high copy-number within the cell, random distribution alone should be adequate to assure stability in a growing cell population (6,9). However, when the number of individual copies of such plasmids is decreased as a consequence of concatenation (i.e., multimer formation), instability results (1,4,5,12). Addition of a DNA segment that resolves the multimers into monomeric units raises the number of individual copies to a level where random distribution of the copies can accomplish stable inheritance; loci involved in the resolution of multimers, and consequently affecting the extent of plasmid stability, have been identified on the plasmids P1 (1), Clo DF13 (4,5), and ColEI, (12). Similarly, introduction of a strong promoter onto certain high copy-number plasmids is known to interfere with plasmid propagation, decrease copy number, and consequently reduce stability; addition of a terminator downstream from the promoter reverses these effects (13).

---

[2] Research School of Biological Sciences, Australian National University, Canberra City, A.C.T. 2601, Australia; [3] Department of Biochemistry, University of Leicester, Leicester E1 7RH, England; [4] Department of Microbiology, University of Umeå, S-901 87, Umeå, Sweden.

Plasmids having a copy-number too low to accomplish stable inheritance by a random distribution process presumably must employ an active mechanism for this purpose. Loci that encode such a mechanism have been termed par or stb, and have in recent years been identified on the plasmids pSC101 (6), NR1 (7), R1 (9), and F (11). Our investigations of plasmid partitioning have involved pSC101.

Structural Properties of the par Locus of the pSC101 Plasmid

In the initial studies of the par locus of plasmid pSC101, par was assigned to a 370-bp EcoRI/AvaI-generated DNA fragment adjacent to the essential replication region of the plasmid (Fig. 1). Deletion of the DNA segment between the EcoRI site and the proximal HaeII site resulted in plasmid loss from about 75% of cells growing for 100 generations in the absence of selection for the antibiotic-resistance trait encoded by the plasmid (Fig. 2A). Mutants of Par$^-$ pSC101 plasmids that were maintained at a copy number approaching 50 per cell (6) were still found to be unstable, although less so than wild-type pSC101 replicons (Fig. 2B). Thus, even when Par$^-$ pSC101 derivatives were maintained at a copy number that would be expected to fully ensure stable plasmid maintenance by a random distribution model, some instability was nevertheless observed.

The key properties of the pSC101 par locus, as described by Meacock and Cohen (6) are summarized in Tab. 1. The par locus functions normally when placed at sites on pSC101 other than its native location--and it functions equally well when placed in either orientation. However, the presence of the pSC101 par region on a coexisting compatible plasmid does not stabilize a Par$^-$ pSC101 replicon in trans. Furthermore, Par$^-$ high copy-number mutants of pSC101 are unstable despite their elevated copy number, as noted above.

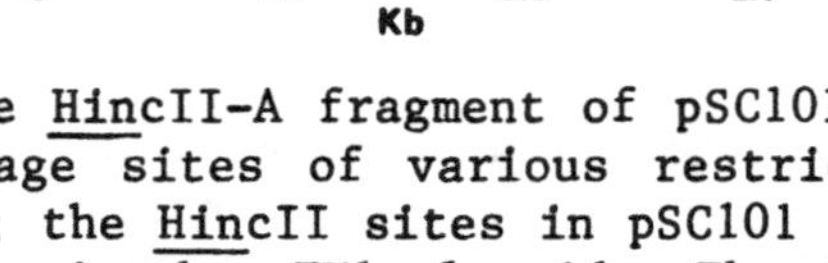

Fig. 1. Map of the HincII-A fragment of pSC101. The location of the cleavage sites of various restriction endonucleases are shown; the HincII sites in pSC101 have been made into EcoRI sites in the pTU1 plasmid. The position of the partition (Par) and replication (Rep) regions are shown. Diagram from Meacock and Cohen (6).

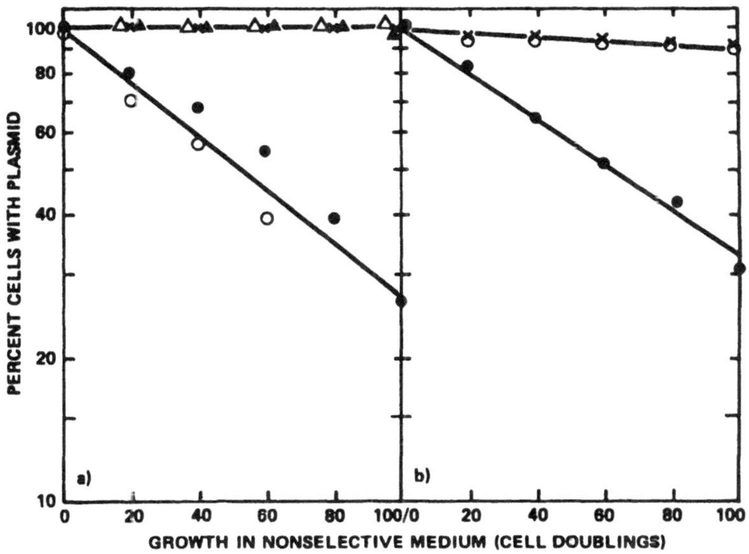

Fig. 2. Stability of autonomously replicating plasmids derived from pSC101 and the effect of an increase in plasmid copy number. Cells of E. coli C600 containing various plasmids were examined for loss of plasmids during growth in non-selective medium by scoring for markers on the plasmids. Various markers on various plasmids were used: Tc^r, pSC101; Km^r, pPM20; Lac^+, pPM21; Ap^r, pPM22, pPM23, pPM25, and pPM29. In panel A the rate of loss of plasmids pSC101(×), pPM20(△), pPM21(▲), pPM22(●), and pPM23 (○) is shown. In panel B the rate of loss of plasmid pPM22 (●) is compared to the rates of loss of 2 mutants of pPM22 that have a copy number of 5 times pPM22, pPM25(○), and pPM29(×). Control experiments show that C600 and C600 containing pSC101 show no detectable difference in growth rate. See Meacock and Cohen (6) for experimental details.

Tab. 1. Summary of major properties of PSC101 par locus (6).

1. Works at different locations in cis; non-functional in trans.
2. Works in either orientation.
3. Par^- phenotype not eliminated by elevation in copy number.
4. Stabilizes unrelated plasmids.
5. No instability of pSC101 results from multiple copies of pSC101 par locus on compatible coexisting plasmid.

When the pSC101 *par* region was placed on an unrelated unstable replicon, such as the pACYC184 plasmid, which is a multicopy, chloramphenicol-amplifiable, polA-dependent p15A-derived replicon, stabilization occurs; thus, the partitioning system of pSC101 is not replicon specific, or even specific for replicons of a particular type. Similarly, later studies have shown that pSC101 *par* region can stabilize Par$^-$ derivatives of the large, cointegrate, oligocopy plasmid R1 (10). Finally, no competition between copies of the *par* region was observed when the multicopy pACYC184 plasmid carrying the pSC101 *par* region was introduced into a cell carrying a wild-type pSC101 plasmid. In analogous experiments carried out earlier (2), placement of the pSC101 *incompatibility* region onto a multicopy *ColE1* plasmid resulted in "super incompatibility" (i.e., loss of a coexisting pSC101 plasmid at an extremely rapid rate).

Taken together, the above findings suggested that the determinant necessary for stable inheritance of the pSC101 plasmid is physically and functionally separate from the replication, incompatibility, and copy number control functions of the plasmid. Moreover, we concluded from our results that the pSC101 *par* locus does not encode a trans-acting protein; instead, our findings suggested that the locus contains a site that accomplishes even distribution of plasmids to daughter cells in growing bacterial cell populations by interacting with cellularly encoded components of the partitioning system.

DNA Sequence Analysis

DNA sequence analysis has been carried out for the 375-bp *EcoRI/AvaI* DNA fragment that contains the pSC101 *par* locus. The resulting sequence shows a number of structurally interesting features. Within the *par* region are 3 segments that potentially can participate in the formation of hairpin loop structures (8,14). These partition-related (PR) segments have been designated *a* (location bp 170-185), *b* (location bp 194-204), and *a'* (location bp 236-251) in Fig. 3, where the relevant sequences are underscored. The *a* and *a'* are direct repeats (14 of 16 bases identical), and *b* is an inverted repeat of both *a* and *a'* (8 of 9 and 9 of 12 bases paired). Thus, hairpin-loop structures can be drawn from both *a* and *b* and *a'* and *b*. As summarized in Tab. 2, plasmids that have deleted *a* and *a'*, *a* and *b*, or *b* and *a'* are phenotypically Par$^-$, but are less unstable than those that have deleted all 3 of the PR segments; the latter deletion mutants show a phenotype that we have termed "Super-par$^-$" (14).

The Super-Par$^-$ Phenotype

Historically, the Par$^-$ phenotype of pSC101 was defined in terms of the plasmid instability displayed by the *HaeII* replication region fragment shown in Fig. 1. Plasmids containing only this region are

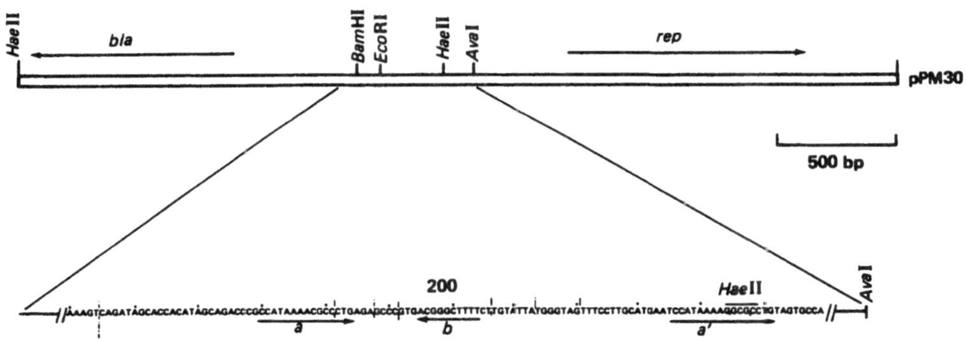

Fig. 3. Map of pPM30 and the sequence of the par region. The upper diagram depicts linearization of the par+ pSC101 derivative pPM30 at one of its HaeII sites. The bla (β-lactamase) gene and rep (replication) region are indicated. The EcoRI/AvaI fragment containing the par locus was sequenced (8), and by deletion analysis (14) the PR segments were defined. They are indicated in this figure by a, b, and a'. Figure modified from Tucker et al. (14).

maintained in 25% of cells grown for 100 generations of the absence of selective pressure (6). However, replicons that lack the entire EcoRI/AvaI par region fragment show a rate of loss that far exceeds that seen with plasmids that lack only the par region segment between the EcoRI (HincII) and HaeII sites. When the entire EcoRI/AvaI fragment is missing, less than 20% of the cells retain the plasmid after 40 generations of growth in nonselective medium, and less than 1% of cells (the limit of detection in the procedure used) contain the plasmid after 60 generations (14). A similar "Super-par$^-$" phenotype was observed when all 3 of the PR segments, a, b, and a', were deleted by BAL31 digestion. Plasmids that have deleted only a or a' are perfectly stable in the absence of a coexisting plasmid derived from the same replicon, and plasmids that have deleted either a and a', a and b, or b and a' display the Par$^-$ phenotype as noted above and previously (14).

The Cmp$^-$ Phenotype

In the course of our studies, we observed that all Par$^-$ plasmids (Gustafsson, Meacock and Cohen, unpub. data), as well as certain pSC101 par deletion mutants having a Par$^+$ phenotype (14), showed a diminished ability to coexist with pSC101 derivatives that contain an intact par region. We have termed this phenotype, which is characterized by an increased sensitivity to incompatibility, Cmp for competition; plasmids that are unable to compete on equal terms with another replicon of the same incompatibility group are Cmp$^-$. Cmp$^-$ Par$^+$ plasmids are completely stable for more than 1,000 genera-

Tab. 2. Phenotypes of par deletion mutants.

Plasmid	Sequences[1] a	b	a'	Phenotype Cmp[2]	Par[3]	Super-par[4]	Stability[5]
pPM30	+	+	+	+	+	+	100
pPM22	-	-	p	-	-	+	25
pWTT310	+	+	+	+	+	+	100
pWTT314	+	+	+	+	+	+	100
pWTT315	-	-	-	-	-	-	16/40
pWTT316	+	+	p	-	+	+	100
pWTT317	+	+	+	+	+	+	100
pWTT318	-	-	p	-	-	-	14/40
pWTT319	+	+	p	-	+	+	100
pWTT301	+	+	+	+	+	+	100
pWTT303	-	+	+	-	+	+	100
pWTT304	-	-	+	-	-	+	32
pWTT460	+	+	p	+	+	+	100
pWTT461	+	+	p	-	+	+	100
pWTT462	+	+	-	-	+	+	100
pWTT463	+	+	-	-	+	+	100
pWTT464	+	-	-	-	-	+	29
pWTT465	p	-	-	-	-	+	10
pWTT466	-	-	-	-	-	-	20/40
pJGB11	-	+	p	-	-	+	79
pJGB12	-	+	+	-	+	+	100
pJGB111	-	+	p	-	-	+	15

[1]The sequence of the par region is shown in Fig. 3. In it the 3 PR (partition related) sequences are underlined and marked a, b, and a'. To show how much of these sequences is present in various plasmids, a "+" indicates the sequence is present, a "-" indicates it is deleted, and a "p" indicates it is partially present.

[2]Cmp phenotype is measured by a plasmid's ability to transform the strain PM191 carrying the incompatible wild-type pSC101 plasmid compared to its ability to transform the host strain PM191. Normal incompatibility reduces this number by approximately 10-fold; in cases where the transforming ability of a plasmid is reduced more than 1,000-fold, that plasmid is called Cmp$^-$.

[3]Par phenotype is an indication of the plasmid's stability over 100 generations of unselected growth.

Tab. 3. Characteristics of the Cmp^- phenotype.

1. Plasmid is totally stable in absence of coexisting Cmp^+ plasmid of same incompatibility type.
2. Reduces ability to compete with Cmp^+ coexisting plasmid of same incompatibility type.
3. Normal incompatibility is expressed against other Cmp^- incompatible plasmids.
4. Copy number is normal.

tions of growth in nonselective media. However, they show a reduction of transformation frequency that exceeds by 3 orders of magnitude the reduction normally observed when a wild-type plasmid is introduced into cells already harboring an incompatible wild-type plasmid. The characteristics of the Cmp^- phenotype are summarized in Tab. 3.

The Cmp phenotype of pSC101 is determined by sequences within its *par* region and appears to represent an intermediate stage of expression of the *par* function (14). When only one PR segment is deleted, a Cmp^- plasmid is formed; however, when 2 of the PR segments

---

[4]The Super-par^- phenotype reflects a degree of instability so great that after 100 generations, less than 1% of the cells growing in the absence of selection contain the plasmid.

[5]Stability is the percent of cells retaining the plasmid, measured by percent of Ap^r cells after 100 generations of unselected growth; each is an average of 3 separate experiments in which less than 10% deviation was observed. When the Par value would be 0 in 100 generations, we took the last time point that yielded any ampicillin-resistant cells and expressed the value in Ap^r cells in that number of generations of unselected growth (e.g., 16/40 is 16% Ap^r cells after 40 generations).

are deleted, a Par$^-$ plasmid results. However, the Cmp phenotype also appears to be associated in some way with the replication/incompatibility system of the plasmid, since a Cmp$^-$ pSC101 derivative coexists stably with pACYC184-derived replicon that carries the par region of pSC101.

Apparent Plasmid Pool Size vs. Actual Pool Size in par$^-$ Mutants

As indicated in Fig. 2 and Tab. 1, when a par$^-$ derivative of pSC101 that is maintained at 5 times the number of copies as the wild type was examined, some instability was observed--despite the calculation that random distribution of the multiple copies of the plasmid should not have resulted in its loss from any of the cells in the population (6). Super-par$^-$ derivatives of pSC101 that carry a wild-type replication region are lost at a rate consistent with random distribution of a replicon having a copy number of 4 per cell at division, whereas the actual copy number of the plasmid was measured at several times this value (14). Thus, the number of copies in the intracellular pool of plasmids does not seem to be taken into account in the segregation of Super-par$^-$ mutants. Similarly, direct examination of the effects of copy number on the stability of Par$^-$ and Super-par$^-$ mutants confirmed that high copy-number mutations have only a small effect on the rate of loss of a Super-par$^-$ mutant, supporting the view that the separate plasmids in the intracellular pool are not individually "counted" in the absence of par; an increase in plasmids from 15 copies per cell at division to 5 times this number of copies does not yield a correspondingly slower rate of loss (14).

Studies with a Super-par$^-$ derivative of pSC101 that also carries a temperature-sensitive mutation in the replication region support this interpretation; following abrupt inhibition of plasmid replication by shift to the nonpermissive temperature, half of the cell population loses the Super-par$^-$ ts plasmid in the first generation following the temperature shift (Fig. 4). In contrast, a par$^+$ ts derivative of pSC101 was lost from cells growing at the nonpermissive temperature at a rate consistent with a plasmid copy number averaging about 10 per cell; for more than 2 doublings after the temperature shift, most cells retained a copy of this plasmid. During each subsequent generation, the fraction of cells retaining the plasmid decreased by half.

It has already been noted that the loss of ColE1, pACYC184, and other high copy-number plasmids from growing cell populations is often associated with the formation of multimers; such multimerization apparently reduces the number of separate plasmid molecules to a level where random distribution can no longer ensure stable inheritance of these plasmids. Introduction of a plasmid locus that accomplishes resolution of the multimers into unit-length molecules restores stability (4,5,12). However, as shown in Fig. 5, stabili-

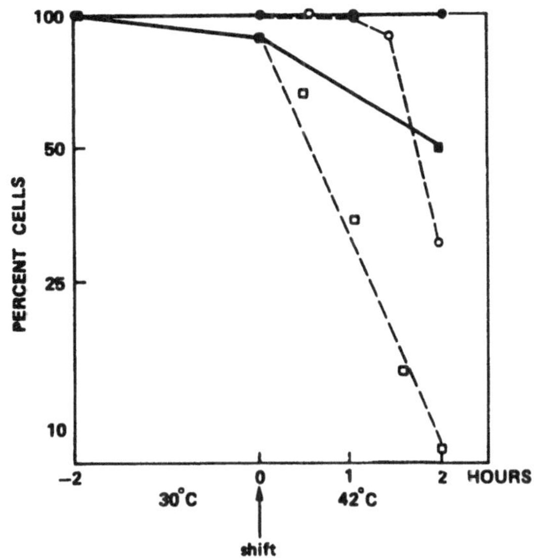

Fig. 4. Rate of loss of Par^+ vs. $Super-par^-$ ts pSC101-derivative plasmids upon shift to the nonpermissive temperature. PM191 (a $recA^-$ derivative of C600) carrying plasmids pCM301 (Par^+) or pCM302 ($Super-par^-$) which are ts for replication, were grown at 30°C without antibiotics. After 2 hours, half of each culture was shifted to 42°C. Samples were taken at the time points shown, plated on nonselective media, and the resulting colonies were tested for resistance to Ap to determine the presence of the plasmid. Circles are pCM301, squares are pCM302. Solid lines and closed symbols represent growth at 30°C; dotted lines and open symbols represent growth at 42°C. The generation time at 42°C was approximately 30 min. Figure from Tucker et al. (14).

zation of the pACYC184 plasmid by the pSC101 par region does not occur by resolution of pACYC184 multimers to monomeric molecules. Thus, plasmid stabilization by the pSC101 par region does not appear to involve a mechanism that regulates the number of individual plasmid molecules in the cell. This conclusion is consistent with our failure to observe multimers of pSC101 derivatives that lack the par region (14).

DISCUSSION

It is now apparent that at least 3 discrete segments of DNA within the par region of pSC101 are involved in partitioning of the

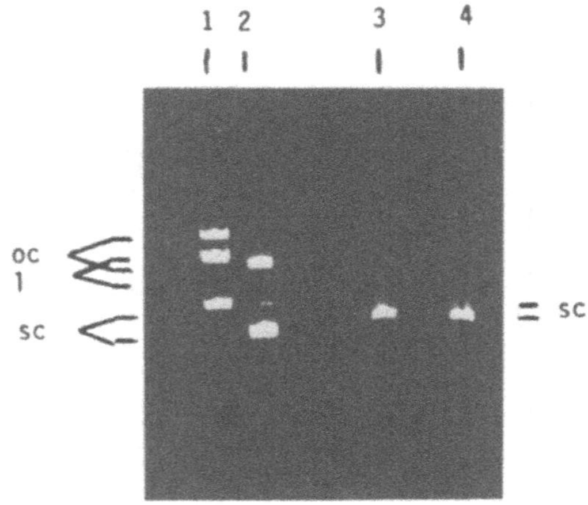

Fig. 5. Multimer formation of par$^+$ and Par$^-$ derivatives of p15A replicons and pSC101 replicons. Lane 1 contains plasmid DNA of pACYC184, a p15A replicon that is unstable, and Lane 2 contains plasmid DNA of pPM31, a derivative of pACYC184 that has the pSC101 par region and is stable. The prominent bands are the supercoiled, open circular, and linear DNA, as indicated, while the less prominent bands indicate various multimers. Details about these plasmids are reported in Chang and Cohen (3) and Meacock and Cohen (6). Lane 3 contains plasmid DNA of pPM30, a par$^+$ pSC101 replicon, and Lane 4 contains plasmid DNA of pWTT315, a Super-par$^-$ pSC101 replicon. The prominent band is the supercoiled DNA and no multimers of either plasmid are detectable. These plasmids are described in Meacock and Cohen (6) and Tucker et al. (14).

plasmid to daughter cells. Deletion of all 3 segments results in the Super-par$^-$ phenotype, which is also observed with plasmids that lack the entire EcoRI/AvaI segment. Deletion of 2 of the PR segments yields the simple Par$^-$ phenotype. Plasmids that have only one of the PR segments deleted are perfectly stable when in the absence of coexisting wild-type plasmid derived from the same replicon, but show a decreased ability to compete with incompatible coexisting plasmids (the Cmp phenotype). The structural and functional interrelationships of the phenotypes associated with par region deletions are summarized in Tab. 4.

Our results suggest a model in which the par function of the pSC101 plasmid enables the nonreplicating plasmids of the intracellular pool to be: 1) counted as individual molecules, and 2) distributed evenly to daughter cells. Moreover, our data with par$^-$ ts plasmids suggest that the plasmid being replicated at the time of

Tab. 4. Summary of structural and function interrelationships of phenotypes associated with *par* region deletions.

Deletion*	Phenotype
a, *b*, and *a'*	Super-par⁻
a and *b*	Par⁻
a' and *b*	Par⁻
a and *a'*	Par⁻
a	Cmp⁻
a'	Cmp⁻

*Complete deletion of PR segment yields greater effects than partial deletion. Extending a deletion between PR segments has no detectable effect (see Tucker *et al.* (1984) for details).

cell division may be distributed separately from the pool and randomly with regard to the pool.

We view the Super-par$^-$ phenotype as the result of total ablation of *par* function, so that plasmids having this phenotype behave entirely and quantitatively according to the proposed model. The simple Par$^-$ phenotype represents a lesser degree of loss of function; there is some capacity for counting and partitioning the individual plasmids in the nonreplicating pool Presumably, the presence of a single intact PR segment in simple Par$^-$ mutants is sufficient to impart some degree of *par* function.

While the above model is speculative, it is consistent with our observations. However, the applicability of the model to the Cmp$^-$ phenotype is less direct. While previously it has been shown that the *par* region of the pSC101 plasmid is able to stabilize Par$^-$ plasmids that utilize an entirely different mode of replication, the discovery of the Cmp$^-$ phenotype suggests that the partition and the replication/incompatibility mechanisms for the pSC101 plasmid nevertheless have some degree of functional linkage. When a Cmp$^-$ plasmid is present in the same cell as a wild-type derivative of the same replicon, the Cmp$^-$ plasmid is at a competitive disadvantage. However, when 2 replicatively similar Cmp$^-$ plasmids coexist in the same cell, they behave identically to 2 coexisting Cmp$^+$ plasmids and show normal incompatibility properties (14). The exact nature of the replicative disadvantage of the Cmp$^-$ plasmid remains to be elucidated, as does the mechanism by which the PR segments *a*, *b*, and *a'* accomplish the partitioning of the pSC101 plasmid during the active cell growth.

ACKNOWLEDGMENTS

These studies were supported by Grant GM26355 to S.N.C. During part of this study, P.A.M. was the recipient of a NATO Fellowship from the Science Research Council of the United Kingdom, and P.G. was the recipient of a Postdoctoral Fellowship Award from the Medical Research Council of Denmark.

REFERENCES

1. Austin, S., M. Ziese, and N. Sternberg (1981) A novel role for site-specific recombination in maintenance of bacterial replicons. Cell 25:729-736.
2. Cabello, F., K. Timmis, and S.N. Cohen (1976) Replication control in a composite plasmid constructed by *in vitro* linkage of 2 distinct replicons. Nature 259:285-290.
3. Chang, A.C.Y., and S.N. Cohen (1978) Construction and characterization of amplifiable multicopy DNA cloning vehicles derived from the P15A cryptic miniplasmid. J. Bact. 134:1141-1156.
4. Hakkaart, M.J.J., P.J.M. vanden Elzen, E. Veltkamp, and H.J.J. Nijkamp (1984) Maintenance of the bacteriocinogenic plasmid Clo DF13 in *E. coli* cells: Evidence for Site-Specific Recombination at parB. Cell 36:203-209.
5. Hakkaart, M.J.J., E. Veltkamp, and H.J.J. Nijkamp (1982) Maintenance of the bacteriocinogenic plasmid Clo DF13 in *Escherichia coli* cells. II. Specific recombination functions involved in plasmid maintenance. Mol. Gen. Genet. 188:338-344.
6. Meacock, P.A., and S.N. Cohen (1980) Partitioning of bacterial plasmids during cell division. A *cis*-acting locus that accomplishes stable plasmid inheritance in cell populations. Cell 20:529-542.
7. Miki, T., A.M. Easton, and R.H. Rownd (1980) Cloning of replication, incompatibility and stability functions of R plasmid NR1. J. Bact. 141:87-99.
8. Miller, C.A., W.T. Tucker, P.A. Meacock, P. Gustafsson, and S.N. Cohen (1983) Nucleotide sequence of the partition locus of *Escherichia coli* plasmid pSC101. Gene 24:309-315.
9. Nordström, K., S. Molin, and H. Aagard-Hansen (1980) Partitioning of plasmid R1 in *Escherichia coli*: I. Kinetics of loss of plasmid derivatives deleted of the *par* region. Plasmid 4:215-227.
10. Nordström, K., S. Molin, and H. Aagard-Hansen (1981) Plasmid R1 incompatibility. Contribution from the *cop/rep* and from the *par* systems. In *Molecular Biology, Pathogenicity, and Ecology of Bacterial Plasmids*, R.C. Clowes, S.B. Levy, and E.L. Koenig eds. Plenum, New York, pp. 291-301.
11. Ogura, T., and S. Hiraga (1983) Partition mechanism of F plasmid: Two plasmid gene-encoded products and a *cis*-acting region are involved in partition. Cell 32:351-360.

12. Summers, D.K., and D.J. Sherratt (1984) Multimerization of high copy number plasmids causes instability: ColE1 encodes a determinant essential for plasmid monomerization and stability. Cell 36:1097-1103.
13. Stueber, D., and H. Bujard (1982) Transcription from efficient promoters can interfere with plasmid replication and diminish expression of plasmid specified genes. EMBO Vol. 1, No. 11, pp. 1399-1404.
14. Tucker, W.T., C.A. Miller, and S.N. Cohen (1984) Structural and functional analysis of the *par* region of the pSC101 plasmid. Cell 38:191-201.

DNA-PROTEIN INTERACTION AT THE REPLICATION ORIGINS OF PLASMID CHROMOSOMES

D. Bastia, C. Vocke, J. Germino, and J. Gray

Department of Microbiology and Immunology
Duke University Medical Center
Durham, North Carolina 27710

ABSTRACT

Novel techniques have been developed to purify replication initiator proteins of the plasmids R6K and pSC101. The techniques consist of tagging the initiator cistrons at the C-terminus with β-galactosidase-encoding DNA of Escherichia coli in the correct translational phase. The hybrid proteins are then rapidly purified by adsorption to and elution from a β-galactosidase- specific affinity column. Two procedures have been devised to isolate the nonfused initiator proteins using the fused protein as a handle. The first procedure, called subunit association chromatography, exploits the association of a monomer of nontagged protein with that of β-galactosidase-tagged protein in isolating both types of proteins by β-galactosidase specific affinity column chromatography. The second procedure involves the fusion of the initiator protein to β-galactosidase via a specific linker DNA. The linker DNA encodes a protein which is readily and specifically hydrolyzed by a sequence specific protease, thus releasing the initiator protein from β-galactosidase. Using purified or partially purified initiator protein, we have demonstrated that the R6K encoded initiator protein (Pi protein) binds to a consensus 22 bp sequence at 2 regions of the plasmid chromosome. The pSC101-encoded initiator protein binds to sequences at or near the plasmid replication origin. At low concentrations the protein binds to a nucleation site and upon raising the concentrations of the protein binding is promoted at 4 adjacent sequences that have partial homologies with the nucleation sequence. Deletion of the binding site leads to a nonfunctional replication origin.

INTRODUCTION

The mechanism of initiation of Cairns-type DNA replication continues to be an interesting unsolved problem in molecular biology. In order to study this problem we have concentrated on the purification of initiator proteins from 2 plasmid systems namely R6K (1) and pSC101 (2,3).

The plasmid R6K is 26 x 10^6 daltons in molecular weight and contains 3 replication origins (4,5,6), a unique replication terminus (7,8,9,10), and encodes a 35-kd initiator protein also known as Pi protein (11,12,13).

The plasmid pSC101 discovered by Cohen and co-worker (2,3) is a low copy-number replicon that is dependent on the *E. coli* dnaA protein (14) and a plasmid-coded initiator protein (15,16) for its replication. The primary structure of the plasmid-coded initiator has been determined from its nucleotide sequence and the protein has been partially purified (15). The minimal replicon of pSC101 was previously localized in a 1800-bp autonomously replicating fragment (17). The topology of this fragment has been determined both structurally and functionally by nucleotide sequence analysis and in vitro deletion analysis (15).

In this chapter, we present novel procedures for the purification of plasmid-encoded initiator proteins and summarize results pertaining to the interaction of these proteins with their respective binding sites on the plasmid chromosomes.

RESULTS

Purification of the R6K Initiator Protein (Pi Protein) by Subunit Association Chromatography

The principle of this purification procedure is to express 2 copies of the R6K initiator cistron, one of which is fused to the *lac* Z cistron, in tandem under the transcriptional control of the *lac* promoter. Since the R6K initiator is polymeric in structure we expected that the nonfused initiator protein would associate by subunit interaction with the β-galactosidase-fused initiator subunit. Because of this association, both types of protein should be recoverable in a relatively pure form by β-galactosidase specific affinity column chromatography. With this approach in mind, we constructed the plasmid pJG11 which contains 2 tandem copies of the R6K initiator cistron, one of which is fused to β-galactosidase (18) in the correct translational phase (Fig. 1). Both the fused and nonfused cistrons are under the transcriptional control of the lac promoter. The analysis of the proteins isolated from cells that carry the plasmid pJG11 shows that affinity chromatography in pNH_2-Phe-S-gal. (p-amino-phenyl-thio-galactoside-sepharose) column indeed purifies both the β-galactosidase tagged and the nontagged initiator proteins (Fig. 1). Amino-terminal amino acid sequence analysis confirms that the 35-kd molecular weight protein has the same first 10 amino acids as the β-galactosidase tagged 160-kd protein (18).

Purification of the R6K Initiator by Gene Fusion and Site-Specific Proteolysis

We wished to develop a more general procedure for the rapid purification of proteins from recombinant DNA clones. Our approach was to fuse the R6K initiator cistron to the β-galactosidase cistron via a piece of DNA encoding the pro α-2 collagen of chicken. The tripartite fusion cistron was in the correct translational phase and under the transcriptional control of the λP_R promoter in the plasmid pJG201 (Fig. 2).

The λP_R promoter was activated by raising the temperature to 42°C and thereby inactivating the thermosensitive λ CI857 repressor. The tripartite fusion protein was purified by β-galactosidase-specific affinity column chromatography and then cleaved with collagenase of *Achromobactor iophagus*. Microbial collagenase are known to cleave specifically the sequence pro-X-gly-pro-Y (19). The collagenase cleavage separated the initiator protein moiety from β-galactosidase. Column chromatography on a TSK 4000 column using high performance liquid chromatography (HPLC) enabled us to purify the R6K initiator in a homogenous form (Fig. 3 and Ref. 20).

Biological Activity of the β-Galactosidase-Tagged R6K-Encoded Initiator Protein

The biological activity of the β-galactosidase-tagged initiator protein was examined by in vitro DNA replication assay and by sequence-specific DNA binding assay described in a subsequent section. In vitro replication of R6K DNA was previously demonstrated in crude cell extracts (21). We wished to determine if the β-galactosidase-tagged initiator protein of R6K was competent to initiate DNA replication. The conditions of the in vitro replication reaction are shown in the legend to Fig. 4. The results show that the β-galactosidase-tagged initiator protein promotes rifampicin-sensitive and novobiocin-sensitive DNA synthesis. We have analyzed the reaction products by 2-dimensional neutral-neutral and neutral-alkaline agarose gel electrophoresis and confirmed that a full range of replication intermediates are formed and these appear to be Cairns-type intermediates (18).

Sequence-Specific DNA-Binding Activity of the R6K Initiator Protein and Location of the DNA-Binding Domain

We have previously demonstrated that the β-galactosidase-tagged initiator protein of R6K binds to 2 specific regions located at the 5' untranslated region of the initiator cistron (19,22). In this

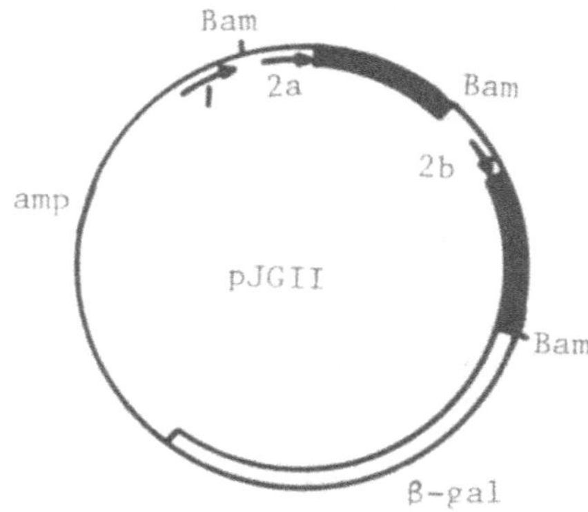

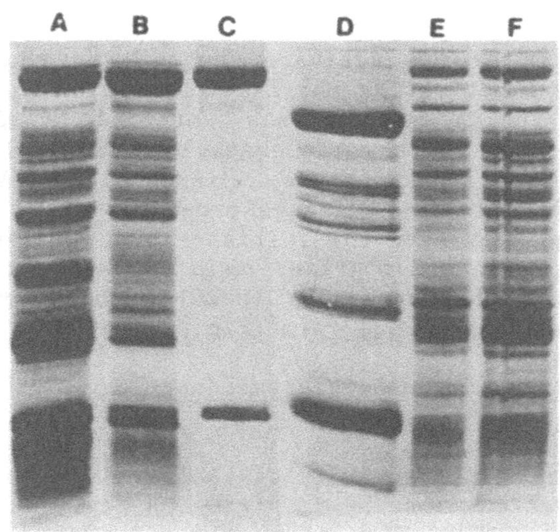

chapter, we summarize the DNA-protein interaction studies performed with both the β-galactosidase-tagged protein and the collagenase - cleaved purified initiator protein. We further demonstrate that the DNA binding domain of the R6K initiator is located in the C-terminal 2/3rd of the protein. The amino terminal region of the initiator protein, unlike the lac and λ repressor proteins (23,24), did not show consistent sequence-specific DNA-binding.

In order to compare the DNA-binding characteristics of the β-galactosidase-tagged protein and the nontagged purified initiator we

Fig. 1. *Upper* - Structure of the plasmid pJG11. The plasmid contains 2 copies of the initiator protein cistron of R6K in the correct transcriptional phase with the *lac* promoter [1]. The second copy of the initiator cistron is fused to the lac Z cistron in the correct translational phase at the 9th codon from the N terminus of the lac Z gene. Each initiator cistron also contains its own natural promoter [2a, 2b].

Lower, SDS-polyacrylamide gel analysis of proteins produced by *E. coli* (*lac* Z) cells carrying the plasmid pJG11. A, total protein; B, a 40% $(NH_4)_2SO_4$ precipitate; C, protein purified by affinity column chromatography through a pNH_2PheS gal column (Ref. 18). Note the simultaneous isolation of the 160-kd β-galactosidase tagged protein and the 35-kd R6K initiator which is not tagged with β-galactosidase, presumably because of subunit association between the 2 types of protein monomers.

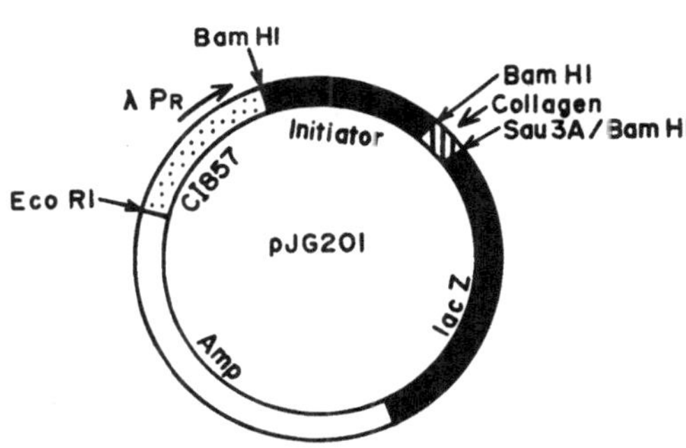

Fig. 2. Structure of the plasmid pJG201. The plasmid contains the initiator cistron of R6K fused at the Bam H1 site in the correct translational phase with the linker DNA sequence coding for 60 amino acids from the triple helical region of chicken pro-α-2 collagen. The linker has several units of sequence of the type pro-X-gly-pro-y, which is specifically cleaved by microbial collagenase. The collagen linker coding sequence is fused to the *lac* Z gene at the 9th codon from the N-terminus and in the correct translational phase.

Thus pJG201 encodes a tripartite protein under the transcriptional control of the inducible λ P_R promoter. The plasmid carries the thermosensitive λ CI857 repressor gene, the ribosome binding site of *Cro* and the ATG of *Cro* which is not in phase with the tripartite hybrid protein.

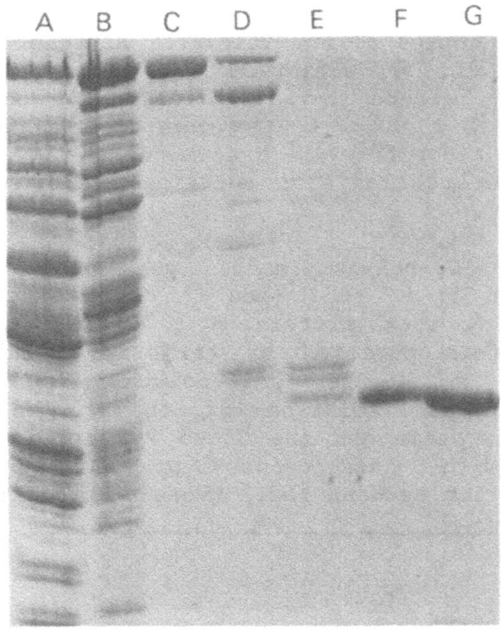

Fig. 3. Purification of the R6K initiator protein by collagenase cleavage. 10% polyadrylamide-SDS gel (stained with Coomassie blue) profiles of A, total protein from E. coli cells carrying pJG201 after induction of λ P_R; B, 40% $(NH_4)SO_4$ pellet of the total protein; C, the tripartite hybrid protein after elution from a p-NH_2-Phe-S-gal affinity column; D, protein shown in C after cleaverage with collagenase, note multiple ∿ 35-kd bands because of multiple collagenase cleavage sites in the linker; E, partially purified protein after a 40% (NH_4) SO_4 precipitation of the protein shown in E, the protein in the supernatant is shown; F, G, 15 μg and 30 μg of purified R6K initiator after fractionating the protein shown in B through a TSK 3000 HPLC gel filtration column.

performed nitrocellulose filter binding experiments (22,25). The results of these experiments depicted in Fig. 5 show that both the β-galactosidase-tagged and nontagged initiator bind to identical restriction fragments belonging to the region called Site I and Site II (22,26). Therefore, as previously noted by us (26), the presence of the β-galactosidase has no detectable influence on the sequence-specific DNA binding of the R6K initiator protein.

We have performed footprinting experiments (Fig. 6) and methylation protection experiments to further analyze the interaction of

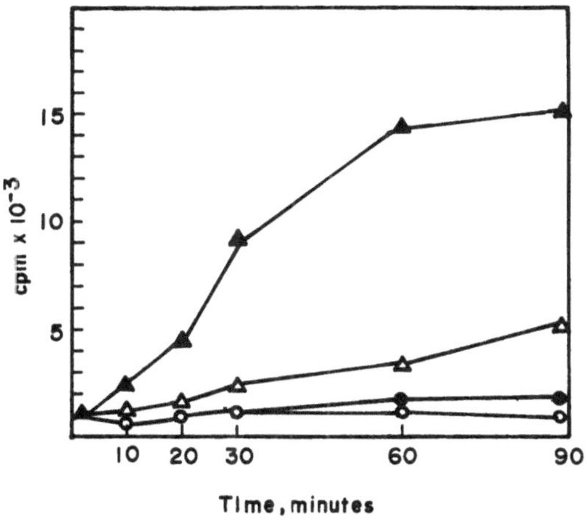

Fig. 4. In vitro replication of the plasmid pJG14 (ori γ cloned in pUC9). Cell lysates were prepared by the lysozyme-freeze-thaw method from the strain D110 (E. coli Pol A^-) as described before (18). The 50 μl reaction mixture contained [α-^{32}P] dATP; 200 μm each of dCTP, TTP, and dGTP; 500 μm each of GTP, CTP, UTP; 2 mM ATP; 40 mM Hepes (pH 7.6); 11 mM $(CH_3COO)_2Mg$; 100 μg/ml creatine kinase; 21.6 mM creatine phosphate; 50 μg/ml of bovine serum albumin; 6% polyethylene gylcol 8,000; 200 μg of crude protein from cell lysate; 8.6 μg of pJG14 DNA; 420 μg of initiator (β-gal.-tagged; ▲, with initiator; △, without initiator; ○, rifampicin plus; and ●, navobiocin plus.

the R6K-initiator protein with the binding Sites I and II. The results are summarized in Fig. 7. The DNase footprinting experiments demonstrate that the protein interacts with the consensus sequence APuNCATGAGPyTTAGTNCGTN. The binding Site I contains 7 tandem copies of this sequence (27) while the Site II contains one complete copy and 2 partial copies of these sequences (Fig. 7; Ref. 22). Interaction with the initiator protein makes the 10th residue from the left of the top strand and 3rd residue from the right of the bottom strand of each 22 bp repeat specifically "hypersensitive" to digestion by DNase I thus suggesting some structural alterations such as bending or kinking of the binding site.

Methylation protection studies show that the protein specifically makes contact with purine residues located mostly on the top strand of the DNA (Fig. 7; Ref. 22).

In order to localize the DNA-binding domain of the protein, we digested the initiator cistron at the single Bgl II site (Fig. 8) and separately cloned the N-terminal and C-terminal pieces in the

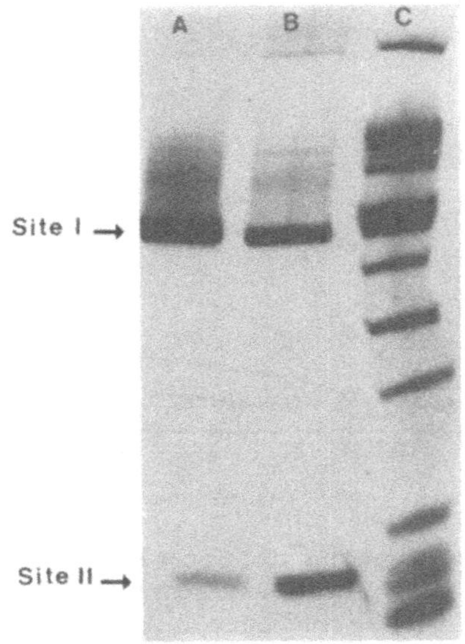

Fig. 5. Interaction of R6K initiator protein with R6K DNA. Autoradiogram of a 5% polyacrylamide gel of 3' end labeled, Hind III-Hinf I restriction fragments of the plasmid pJG4 (which contains the β-replicon of R6K cloned in pBR322, see Ref. 26). A, DNA fragments bound to purified R6K initiator, after collagenase cleavage and retained on a nitrocellulose filter; B, DNA fragments bound by the tripartite hybrid protein (Fig. 3C); C, Input 3'-end labeled restriction fragments. Thus both the β-galactosidase tagged initiator and purified nontagged initiator both bind to identical restriction fragments (Site I and Site II).

vector pJG200 (20) in the correct translational phases with the collagen and β-galactosidase sequences. Both types of the protein were purified using the procedures described before (20) and were used to perform nitrocellulose filter-binding experiments. The results depicted in Fig. 9 demonstrates that the C-terminal fragment, but not the N-terminal fragment, shows sequence specific DNA binding. Thus, the R6K-initiator (Pi protein) contains at least one sequence-specific DNA-binding domain in the two-thirds of the peptide from the carboxy-terminus.

The Replication Origin and Initiation Protein of Plasmid pSC101

The replication origin of the plasmid pSC101 was previously localized in a 1,800-bp autonomously replicating fragment by Cohen and co-workers (17). This fragment lacked the partition locus

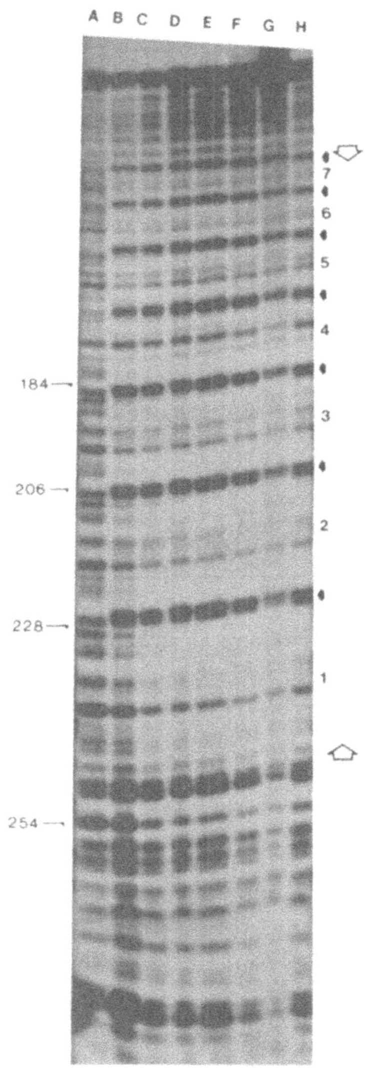

Fig. 6. DNase I footprint of the β-galactosidase tagged R6K initiator at binding Site I. The restriction fragment containing Site I was 5'-end labeled and digested with DNase I (panel A) or was first complexed with increasing amounts of initiator protein and then cleaved with DNase I (B-H). Note, 7 tandem protected regions in B-H each of which are bounded by a residue that is hypersensitive to DNase I (→). See Fig. 7 for more details. The numbers on the left correspond to the coordinates shown in Fig. 7.

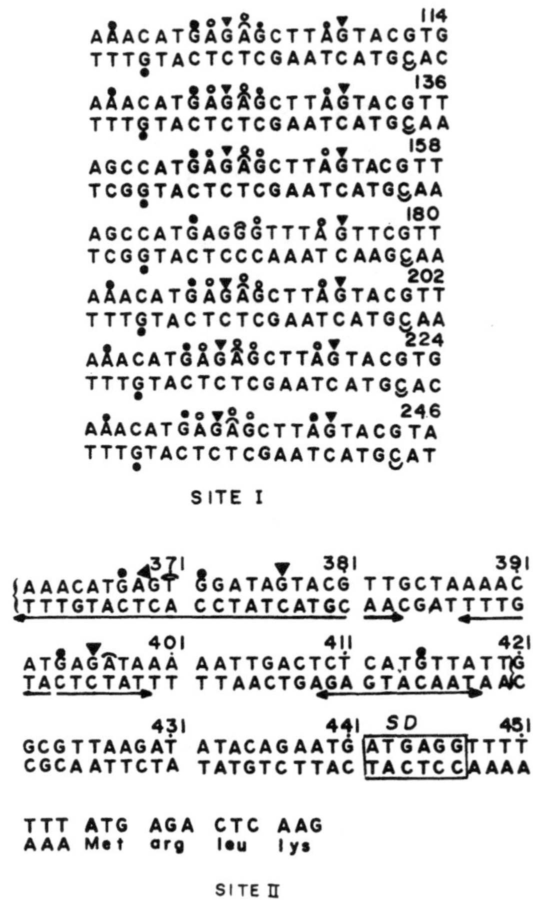

Fig. 7. Binding Sites I and II of R6K showing the summary of footprinting and methylation protection experiments. υ, η = enhanced sensitivity to DNase I, ● = strongly blocked from methylation; ○ = weakly blocked from methylation; ▼ = enhanced methylation; SD = Shine-Delgarno sequence. The region protected from DNase digestion at Site II is shown within brackets.

(17,28). We have further dissected the replication origin by nucleotide sequence analysis and in vitro mutagenesis. The functional replicon of pSC101 consists of a trans-acting initiator cistron (Fig. 10; and Ref. 16) and a cis-acting origin (Fig. 10; Ref. 15, 16). The origin region consists of an AT-rich region, a putative dnaA binding site and the binding sites for the pSC101-encoded initiator protein (Fig. 10).

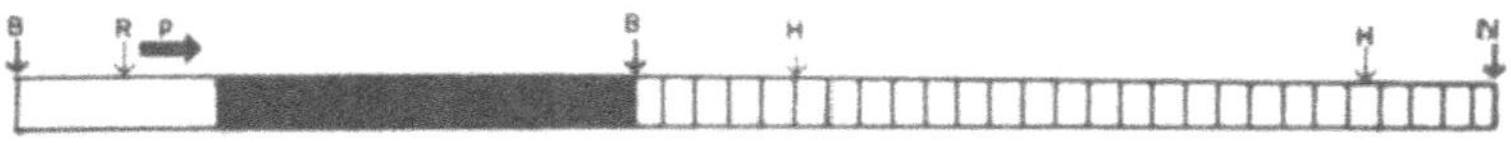

Fig. 8. Restriction map of the R6K initiator cistron. The N-terminal fragment ( ) and the C-terminal DNA fragment ( ) were separately cloned into the vector pJG200 which contains the collagen linker (Ref. 20). The N- and C- terminal peptides were purified as described in the text. R_1 = Eco R_I, Bl = Bam HI, B = Bgl I, S = Sau 3A, p = promoter of the initiator cistron.

In vitro deletion analyses showed that all sequences beyond the cistron for the initiator are dispensable. However, deletions extending into the coding region corresponding to the C-terminus give nonfunctional replicon. Similarly, up to approximately 100 bp from the left end of the 1,800-bp region are dispensable. However, when deletions extend into the AT-rich region and beyond of the untranslated sequences, inactivation of the origin of replication (Fig. 10) takes place.

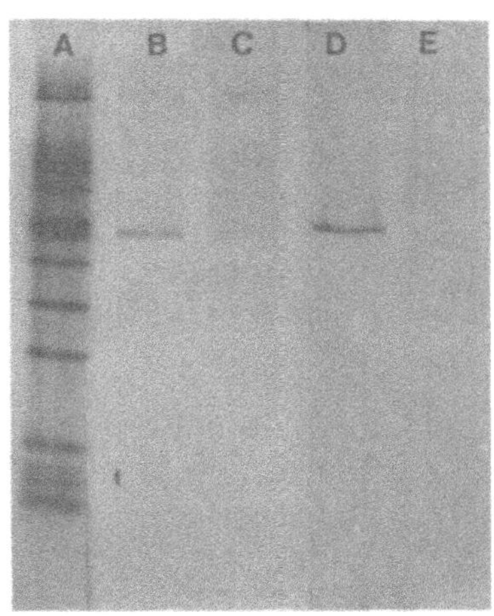

Fig. 9. Interaction of the C-terminal and N-terminal peptides of R6K initiator (see Fig. 8) with DNA. Autoradiograms of a 5% polyacrylamide gel showing A, input DNA; B, D, restriction fragment bound to C-terminal peptide; and C, E, restriction fragment bound to the N-terminal peptide. Note that only the C-terminal peptide shows sequence-specific DNA-binding.

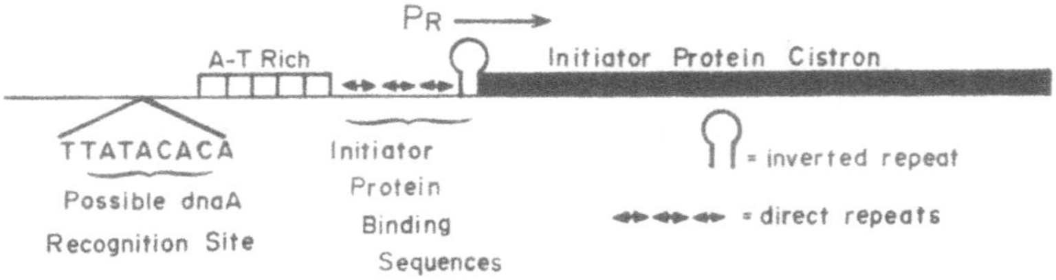

Fig. 10. The replication origin region of the plasmid pSC101. Note the AT-rich region, putative dnaA binding sites, the plasmid-encoded protein binding site, and the 3 direct repeats (→).

The Initiator Protein of pSC101

Using the techniques mentioned earlier, we have partially purified the replication initiator (37.5 kd) protein of pSC101 and studied its interaction with the replication origin.

We constructed the plasmid pCV2 (Fig. 11; Ref. 29) which contains the 1,800-bp autonomously replicating fragment of pSC101 cloned into the vector pUC9 (30). The HinfI-generated restriction sites of pCV2 are shown (Fig. 11). The pCV2 DNA was digested with HinfI and 3'-end labeled with [α-p^{32}] dATP and the labeled DNA was used to study DNA-protein interactions with the pSC101-encoded in-

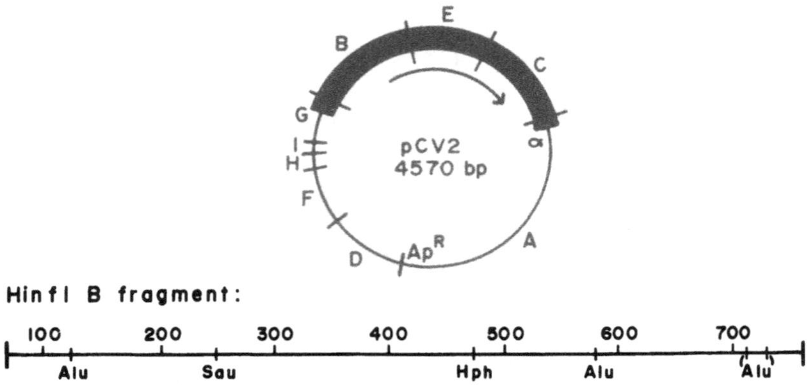

Fig. 11 Restriction (HinfI) map of the plasmid pCV2. pSC101 replicon (■), pUC9 vector DNA (□). Alu = Alu I, Sau = Sau 3A, Hph = HphI, Ap^R = β-lactamase gene, α = α-donor (peptide) gene of β-galactosidase. The Hinf 1B fragment specifically binds to the pSC101-encoded 37.5-kd initiator protein (Ref. 29).

itiator protein. We performed both nitrocellulose filter-binding and anti-β-galactosidase antiserum promoted immunoprecipitation of β-galactosidase-tagged initiator DNA complex to localize the binding region on the pSC101 DNA. The results showed that the Hinf 1B fragment specifically binds to the plasmid-encoded initiator protein (29).

In order to study the interaction of the protein with DNA as a function of the input protein concentrations, we performed exonuclease III-footprinting as described in the legends to Fig. 12. The results show that at the lowest protein concentration, the protein occupies a nucleation site. Upon increasing the protein concentration the protein binds to 4 adjacent sites.

We also performed methylation protection and interference experiments (29) to delineate the pattern of interaction of the protein with DNA. The results are summarized in Fig. 13.

The first 3 direct repeats are homologous to each other but they share partial homologies with the nucleation site and the intervening binding sites.

DISCUSSION

The protein purification procedure described in this chapter does not depend on the biological activity of the protein being purified but is based on the assay and specific affinity procedure for β-galactosidase which is used as a tag. Therefore, the procedures should be applicable to other types of proteins. Since β-galactosidase of E. coli is expressed in yeast cells, in mammalian cells, and in Drosophila (31, 32, 33), with appropriate modifications of the vector (selectable marker and proper replication origins), the techniques could be used to purify proteins from these eukaryotic cells.

The apparent limitation of the subunit association chromatography is that it would only be effective if the target protein is polymeric in structure. This limitation has been overcome to a large extent by the use of the second approach that uses the collagen linker. Previous work, using Clostridial collagenase and synthetic peptides of defined sequences, has shown that collagenase fractions were found that selectively hydrolyzed the sequence pro-X-gly-pro-y (19). Very few proteins other than collagen have these sequences (34). Therefore, the collagenase-linker technique should be useful for the purification of a wide range of proteins provided that the addition of the pro-X to the C-terminus, resulting from collagenase cleavage did not impair the biological activity of the target protein.

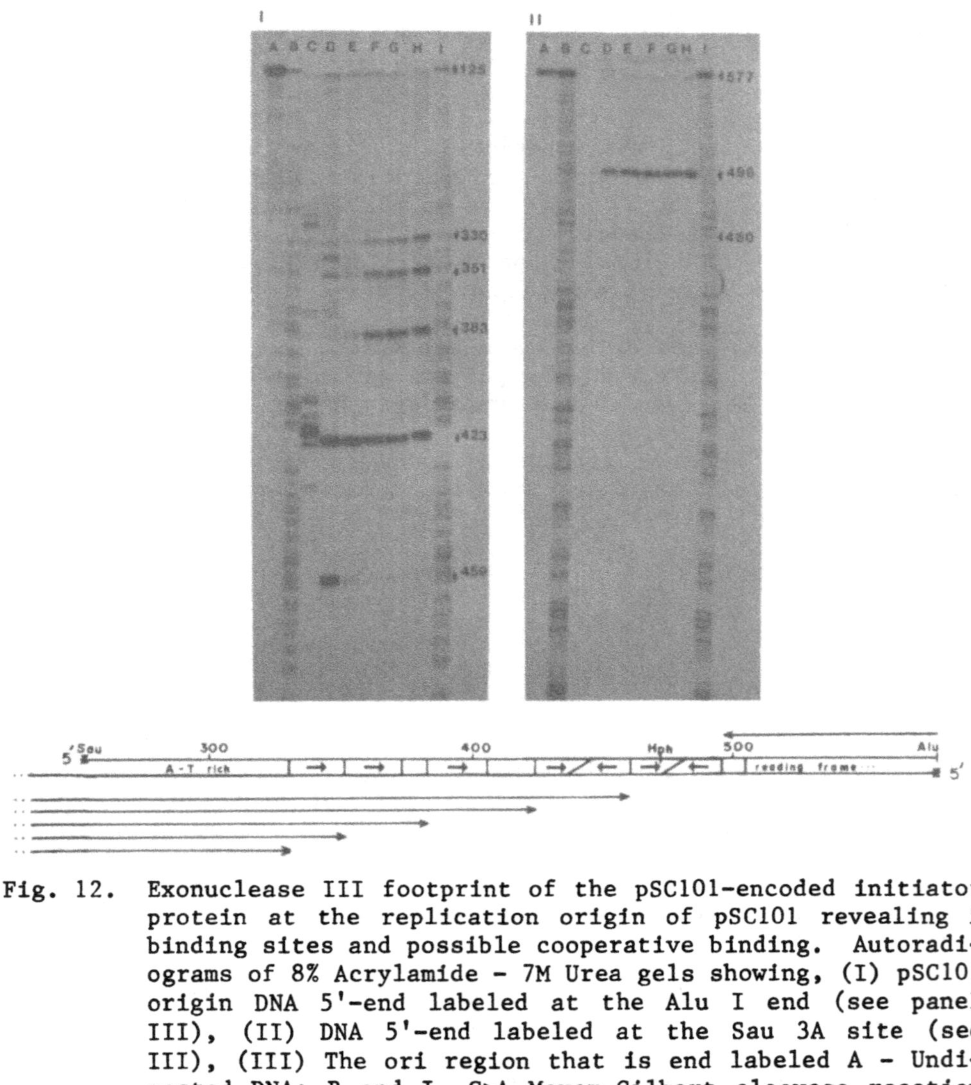

Fig. 12. Exonuclease III footprint of the pSC101-encoded initiator protein at the replication origin of pSC101 revealing 5 binding sites and possible cooperative binding. Autoradiograms of 8% Acrylamide - 7M Urea gels showing, (I) pSC101 origin DNA 5'-end labeled at the Alu I end (see panel III), (II) DNA 5'-end labeled at the Sau 3A site (see III), (III) The ori region that is end labeled A - Undigested DNA; B and I, G>A Maxam-Gilbert cleavage reaction supplying molecular weight markers; C - exo-III digestion of DNA without added initiator; D-H, exo-III digestion in the presence of increasing amounts of initiator protein. Note that in panel I the band (corresponding to coordinate 459) protected at the lowest protein concentration is the nucleation site. Note, 4 additional sites protected with increasing protein concentrations.

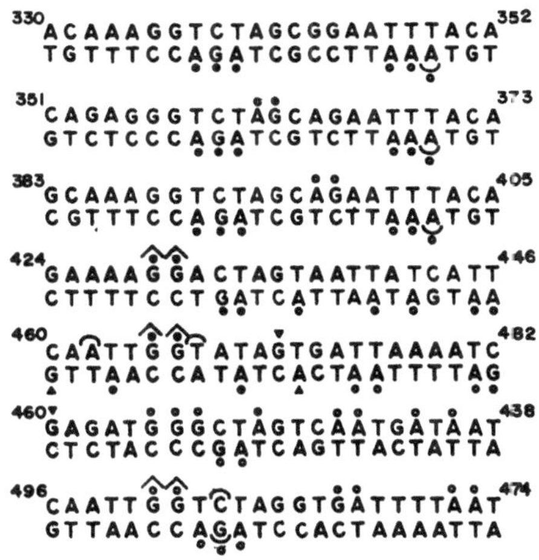

Fig. 13. Summary of the DNA-protein interaction studies between the pSC101 origin and the plasmid encoded 37.5-kd initiator protein. The sequences between the coordinates 496-474 and 460-482 constitute the nucleation site to which the protein binds at the lowest concentration. The sequences between 460-438 and 424-446 constitute the second binding domain. The sequence between 330-352 and the 3 direct repeat 5-330) constitute the remainder 3rd, 4th, and 5th binding sites. ●, strongly blocked from methylation; o, weakly blocked; ▾, enhanced methylation; ∪,∩, enhanced DNase cleavage; ^, methylation interference.

The results presented here demonstrate that fusion of β-galactosidase does not affect the sequence-specific DNA-binding activities of the R6K and pSC101-encoded initiator proteins. This is not surprising in view of the fact that the DNA-binding domains of many proteins are located at a small region of the molecule and that the DNA-binding domain can be isolated without the loss of ligand binding activity by controlled proteolysis (23, 24). Previous work has shown that both the <u>lac</u> and the λ repressors can be fused to β-galactosidase without the loss of biological activities. Hence β-galactosidase fusions, in principle, could be used for rapid screening for sequence-specific DNA-binding proteins from a variety of systems.

The purified β-galactosidase tagged and nontagged proteins bind to 2 regions of the R6K chromosome. The binding Site I is located at or near ori γ but is separated by approximately 2,000 bp from ori α and 1,000 bp from ori β. The Site I is essential for DNA replication from all 3 origins (35). We have suggested that the binding Site I may serve as an "initiator loading site."

The binding Site II of R6K on the other hand is not essential for DNA replication (36). However deletion of this site leads to loss of autoregulation of the initiator synthesis (36). Previous work had shown that R6K initiator (Pi protein) autoregulated its own synthesis (35).

The techniques described above have also been used to partially purify the 37.5 kd pSC101-encoded initiator protein. This protein has sequence-specific binding activity (29). Footprinting experiments using exonuclease III show that the protein binds in a sequential manner to 4 sites after binding to a nucleation site at the lowest effective protein concentration. Either the binding is cooperative or else the protein has a gradient of affinities for the binding sites at the replication origin of pSC101.

Future experiments, using purified proteins and in vitro replication system, should reveal the role of the plasmid-encoded initiators in initiation of replication and plasmid incompatibility.

ACKNOWLEDGMENT

This work was supported by the grants AI-19881 and P01 CA30246 awarded by the NIH. D. B. is an Established Investigator of the American Heart Association. We thank Hilda Smith for the preparation of this manuscript. The Fig. 6, 7, 11, 12, and 13 have been reproduced with the permission of the publishers of _Cell_ (MIT Press).

REFERENCES

1. Kontomichalou, P., M. Mitani, and R. Clowes (1970) Circular R-factor molecules controlling penicillinase synthesis, replicating in _Escherichia coli_ under either relaxed or stringent control. _J. Bact._ 104:34-44.
2. Cohen, S.N., and A.C.Y. Chang (1973) Recircularization and autonomous replication of a sheared R-factor DNA segment in _Escherichia coli_ transformants. _Proc. Natl. Acad. Sci., USA_ 70:1293-1297.
3, Cohen, S.N., and A.C.Y. Chang (1977) Revised interpretation of the origin of pSC101 plasmid. _J. Bact._ 132:734-737.
4. Crosa, J. (1980) Three origins of replication are active _in vivo_ in the R plasmid RSF 1040. _J. Biol. Chem._ 255:11075-11077.
5. Inuzuka, N., M. Inuzuka, and D. Helinski (1980) Activity _in vitro_ of three origins of replication of the antibiotic resistance plasmid RSF 1040. _J. Biol. Chem._ 255:11071-11074.
6. Shon, M., J. Germino, and D. Bastia (1982) The nucleotide sequence of the replication origin of the plasmid R6K. _J. Biol. Chem._ 257:13823-13827.

7. Crosa, J., L.K. Luttrop, and S. Falkow (1976) Mode of replication of the conjugative R plasmid RSF 1040 of E. coli. J. Bact. 129-454-466.
8. Lovett, M., R.A. Sparks, and D. Helinski (1975) Bidirectional replication of the plasmid R6K in E. coli: Correspondence between origin of relication and position of single strand break in relaxation complex. Proc. Natl. Acad. Sci., USA 72:2905-2909.
9. Kolter, R., and D. Henlinski (1978) Activity of the replication terminus of the plasmid R6K in hybrid replicons in E. coli. J. Mol. Biol. 124:425-441.
10. Bastia, D., J. Germino, J. Crosa, and J. Ram (1981) The nucleotide sequence surrounding the replication terminus of R6K. Proc. Natl. Acad. Sci., USA 78:2095-2099.
11. Germino, J., and D. Bastia (1982) Primary structure of the replication initiation protein of plasmid R6K. Proc. Natl. Acad. Sci., USA 79:5475-5479.
12. Stalker, D.M., R. Kolter, and D. Helinksi (1982) Plasmid R6K replication: Complete nucleotide sequence of an autonomously replicating segment. J. Mol. Biol. 161:33-43.
13. Kolter, R., M. Inuzuka, and D. Helinski (1978) Transcomplementation dependent replication of a low molecular weight origin fragment from plasmid R6K. Cell 15:1199-1208.
14. Hasanuma, K., and M. Sakiguchi (1977) Replication of plasmid pSC101 in Escherichia coli K12: Requirement for dnaA function. Mol. Gen. Genet. 154:225-230.
15. Vocke, C., and D. Bastia (1983) Primary structure of the essential replicon of the plasmid pSC101. Proc. Natl. Acad. Sci., USA 80:6557-6561.
16. Churchward, G., P. Linder, and L. Caro (1983) The nucleotide sequence of replication and maintenance functions encoded by plasmid pSC101. Nucl. Acids Res. 11:5645-5659.
17. Meacock, P., and S.N. Cohen (1980) Partitioning of bacterial plasmids during cell division: A cis-acting locus that accomplishes stable plasmid maintenance. Cell 20:529-542.
18. Germino, J., J.G. Gray, H. Charbonneau, T. Vanaman, and D. Bastia (1983) Use of gene fusion and protein-protein interaction in the isolation of a biologically active regulatory protein: The replication initiator protein of plasmid R6K. Proc. Natl. Acad. Sci., USA 80:6848-6852.
19. Harper, E. (1970) Studies on the mechanism of action of bacterial collagenase. In Collagenase, I. Mandel, ed. Gordon and Breach, New York, pp. 19-24.
20. Germino, J., and D. Bastia (1984) Rapid purification of a cloned gene product by genetic fusion and site-specific proteolysis. Proc. Natl. Acad. Sci., USA Vol. 81 (in press).
21. Inuzuka, M., and D. Helinski (1978) Requirement of a plasmid-encoded protein for replication in vitro of plasmid R6K. Proc. Natl. Acad. Sci., USA 75:5381-5385.
22. Germino, J., and D. Bastia (1983) Interaction of the plasmid R6K-encoded replication initiator protein with its binding sites on DNA. Cell 34:125-134.

23. Ogata, R., and W. Gilbert (1978) An amino-terminal fragment of lac repressor binds specifically to lac operator. Proc. Natl. Acad. Sci., USA 75:5851-5854.
24. Pabo, C.O., W. Krovatin, A. Jeffrey, and R.T. Sauer (1982) The N-terminal arms of λ repressor wrap around the operator DNA. Nature (London) 298:441-443.
25. Riggs, A.D., H. Suzuki, and S. Bourgeoise (1970) Lac repressor operator interaction I: Equilibrium studies. J. Mol. Biol. 48:67-83.
26. Germino, J., and D. Bastia (1983) The replication initiator protein of plasmid R6K tagged with β-galactosidase shows sequence-specific DNA-binding. Cell 32:131-140.
27. Stalker, D.M., R. Kolter, and D. Helinski (1979) Nucleotide sequence of an origin of replication of the antibiotic resistance plasmid R6K. Proc. Natl. Acad. Sci., USA 76:1150-1154.
28. Miller, C.A., W.T. Tucker, F. Meacock, and S.N. Cohen (1983) Nucleotide sequence of the partition locus of pSC101. Gene 24:309-315.
29. Vocke, C., and D. Bastia (1983) DNA-protein interaction at the origin of DNA replication of the plasmid pSC101. Cell 35:495-502.
30. Vieira, J., and J. Messing (1982) The pUC plasmids, an M13 mp7 derived system for insertion mutagenesis and sequencing with synthetic universal primers. Gene 19:259-268.
31. Lis, J.T., J.A. Simon, and C.A. Sutton (1983) New heat shock puffs and β-galactosidase activity resulting from transformation of Drosophila with an hep 70-lac Z hybrid gene. Cell 35:403-410.
32. Rose, M., M.J. Casadaban, and D. Botstein (1981) Yeast genes fused to β-galactosidase in Escherichia coli can be expressed normally in yeast. Proc. Natl. Acad. Sci., USA 78:2460-2464.
33. Nielsen, D.A., J. Chou, A. Mackrell, M.J. Casadaban, and D.F. Steiner (1983) Expression of a preproinsulin-β-galactosidase gene fusion in mammalian cells. Proc. Natl. Acad. Sci., USA 80:5198-5202.
34. Anglister, L., and I. Silman (1979) Molecular structure of elongated forms of electric eel acetylcholine sterase. J. Mol. Biol. 125:293-309.
35. Shafferman, A., R. Kolter, and D. Helinski (1982) Plasmid R6K DNA replication: Regulatory properties of the initiation protein. J. Mol. Biol. 161:67-72.
36. Germino, J. (1983) Studies on the initiation and termination of replication of plasmid R6K. Ph.D. thesis, Department of Microbiology, Duke University Medical Center.

POSITIVE REGULATION AND TRANSCRIPTION INITIATION OF *XYL* OPERONS ON TOL PLASMID

Teruko Nakazawa,* Sachiye Inouye, and Atsushi Nakazawa

*School of Allied Health Sciences
Yamaguchi University
and Department of Biochemistry
Yamaguchi University School of Medicine
Ube 755, Japan

INTRODUCTION

Degradative Pathway Specified by TOL

Members of the bacterial genus *Pseudomonas* have a capacity to degrade a wide range of organic compounds. Many degradative pathways are controlled by chromosomal genes, while some catabolic activities are specified by genes borne on plasmids.

A number of plasmids have been isolated which determine degradation of aromatic hydrocarbons and their products. These include NAH, encoding enzymes for naphthalene oxidation (1); SAL, controlling salicylate degradation (2); and TOL, specifying enzymes for degradation of toluene and xylenes (3,4). The plasmids carry a whole array of functional genes responsible for successive catabolism of substrates. Therefore, such plasmids would provide a useful experimental system for detailed analyses of genetic organization and of expression of catabolic pathways in nonenteric bacteria.

The natural host of TOL is *Pseudomonas putida* (*arvilla*) mt-2 (ATCC23973) that was isolated in 1959 by K. Hosokawa in Japan from enrichment culture of *m*-toluate. The reactions applied to toluene as encoded by the TOL plasmid are controlled in the 2 reaction sequences by 2 gene clusters (Fig. 1). The upper pathway sequence is responsible for oxidation of toluene to benzoate through benzyl alcohol and benzaldehyde. The lower pathway sequence includes the formation of catechol followed by the meta-cleavage of catechol and successive reactions to yield pyruvate and acetaldehyde. A whole

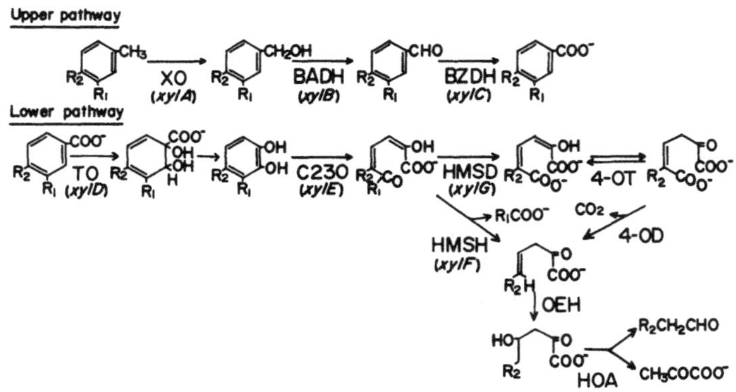

Fig. 1. The degradative pathway encoded by TOL. Enzyme abbreviations: XO, xylene oxygenase; BADH, benzyl alcohol dehydrogenase; BZDH, benzaldehyde dehydrogenase; TO, toluate oxygenase; C230, catechol 2,3-dioxygenase; HMSD, 2-hydroxymuconic semialdehyde dehydrogenase; HMSH, 2-hydroxymuconic semialdehyde hydrolase; 4-OD, 4-oxalocrotonate decarboxylase; OEH, 2-oxopent-4-enoate hydratase; HOA, 4-hydroxy-2-oxovalerate aldolase. xylA to xylG, genes for the degradative pathway. Compounds: R1, H; R2, H; toluene. R1, CH_3; R2, H; *m*-xylene. R1, H; R2, CH_3; *p*-xylene.

set of enzymes responsible for these reactions also catalyzes the metabolism of methylated derivatives of the aromatic compounds. The meta-fission of catechol is characteristic of plasmid-encoded aromatic catabolism of TOL, NAH, and SAL.

A Model for Regulation of Gene Expression

Based on the mutant analysis, Worsey et al. (5) proposed a model for the regulation of the early enzymes. The model is modified by recent studies (6,7) as shown in Fig. 2. The enzymes are organized in 2 regulatory blocks, i.e., operons. The first operon (xylABC) consists of xylA, xylB, and xylC for the upper pathway enzymes responsible for the conversion of aromatic hydrocarbons to carboxylic acids. The second operon (xylDEGF) consists of xylD, xylE, xylG, xylF and so forth for the lower pathway enzymes catalyzing further oxidation and complete degradation of carboxylic acids.

Two regulatory genes, xylR and xylS, are involved; xylR plays a role in the induction of both operons by the hydrocarbons and the alcohol metabolites, whereas xylS is responsible for the induction of only the second operon by the carboxylic acid. Both xylR and xylS are necessary for the induction of the second operon by *m*-xylene or *m*-methylbenzyl alcohol.

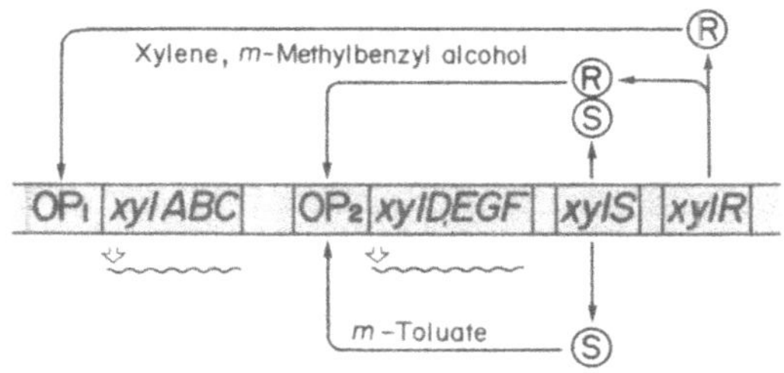

Fig. 2. A model for the positive regulation of xyl operons. The xylR product combines with the operator of xylABC (OP1) in the presence of m-xylene or m-methylbenzyl alcohol and activates the xylABC transcription. The xylS product combines with the operator of xylDEGF (OP2) in the presence of m-toluate and activates the xylDEGF transcription. Both xylR and xylS products are required for the m-xylene- or m-methylbenzyl alcohol-dependent activation of the xylDEGF transcription.

In the following sections, evidence will be presented that the xylR and xylS products are trans-acting in the xylABC and xylDEGF operons to activate the gene expression.

POSITIVE REGULATION OF XYL OPERONS

Isolation of RP4-TOL Recombinants

Since the *Pseudomonas* system for gene manipulation had not been established at the early stage of our work on TOL, we looked for a system that could be analyzed in *E. coli*.

The TOL plasmid in a strain of *Pseudomonas aeruginosa* PAO is unstable at high temperature, while a broad host-range IncP1 plasmid, RP4, is temperature-resistant (8). Plasmid RP4 (54-kb) was introduced into the PAO strain carrying TOL (117-kb), and clones were selected on m-toluate-M9 agar at 42°C. Thus, a recombinant plasmid pTN1 (114-kb) was obtained that was stable even at high temperature (9). The pTN1 plasmid in *P. putida* specified a constitutive low expression of meta-pathway enzymes, whereas the upper pathway enzymes remained inducible (Tab. 1). *P. putida* cells carrying pTN1 grew poorly on m-toluate-M9 agar and produced 2 types of segregants at a frequency of about 10^{-4} per cell (10). One of the segregants carried pTN2 (111-kb) that specified inducible synthesis of TOL enzymes similar to that of the wild-type TOL plasmid. Another segregant carried pTN8 (114-kb) which specified a constitutive high level of the meta-pathway enzymes (Tab. 1).

Tab. 1. Enzyme activities in crude extracts of P. putida carrying RP4-TOL recombinants.

Strain	Inducer[a]	Sp act (mU/mg) BADH	C23O	HMSH	β-Lactamase[b]
TN2100(pTN1)	None	2	240	3	410
	m-Xylene	550	200	4	480
	m-Methylbenzyl alcohol	110	250	3	370
	m-Toluate	2	250	3	420
TN2100(pTN2)	None	2	270	3	420
	m-Xylene	350	8900	120	400
	m-Methylbenzyl alcohol	70	7800	80	340
	m-Toluate	1	8000	90	470
TN2100(pTN8)	None	4	2340	23	460
	m-Xylene	520	1950	22	410
	m-Methylbenzyl alcohol	330	2200	21	490
	m-Toluate	3	2410	22	430

[a] Cells were grown at 27°C overnight in L broth in the absence or presence of m-xylene vapor, 5mM m-methylbenzyl alcohol, or 5 mM m-toluate.

[b] β-Lactamase encoded by RP4 was determined to check the copy number of recombinants.

Inefficient and Temperature-sensitive Expression of xyl Operons in E. coli

RP4-TOL recombinants were transferred to E. coli, and the expression of xyl genes in E. coli was determined at various temperatures (Tab. 2). C23O and HMSH levels in cells of E. coli carrying either pTN1, pTN2, or pTN8 at 30°C are substantially lower than those in P. putida grown at 27°C (Tab. 1) under both induced and noninduced conditions. Induced levels of BADH are slightly lower in E. coli than in P. putida. The constitutive expression of C23O and HMSH encoded by pTN1 and pTN8 in E. coli is temperature resistant, while the inducible synthesis of BADH encoded by pTN1, pTN2, and TN8 and the induction of C23O and HMSH encoded by pTN2 is temperature sensitive. The thermosensitivity of enzyme induction was also observed in P. aeruginosa.

Structure of RP4-TOL Recombinants

Plasmid pTN2 has the entire length of RP4 and a part of TOL (56-kb) that contains all the degradative genes as well as the regulatory genes required for enzyme induction (10). The physical map of the TOL region of pTN2 (Fig. 3) is indistinguishable from the corresponding region of TOL (pWWO) (11,12). The structural genes of the operons were mapped on pTN2 by using the molecular cloning technique in E. coli (13,14). The genetic map of pTN2 is consistent

Tab 2. Enzyme activities in crude extracts of E. coli carrying RP4-TOL recombinants grown at various temperatures. [a] Cells were grown overnight in L broth in the absence or presence of m-xylene vapor, 5mM m-methylbenzyl alcohol, or 5mM m-toluate.

Strain	Temp.	Inducer[a]	Sp act (mU/mg) BADH	C230	HMSH	β-Lactamase
20SO(pTN1)	30°C	None	10	9	0.1	1390
		m-Xylene	119	3	<0.1	790
		m-Methylbenzyl alcohol	62	5	<0.1	1380
		m-Toluate	9	7	0.1	1470
	37°C	None	8	29	0.3	1420
		m-Methylbenzyl alcohol	22	28	0.3	1590
		m-Toluate	13	37	0.6	1620
	42°C	None	21	110	0.5	450
		m-Methylbenzyl alcohol	18	110	0.6	400
		m-Toluate	31	120	0.7	430
20SO(pTN2)	30°C	None	10	6	<0.1	1910
		m-Xylene	215	1670	14.8	1540
		m-Methylbenzyl alcohol	60	750	7.4	1570
		m-Toluate	12	150	2.1	1520
	37°C	None	13	3	<0.1	1950
		m-Methylbenzyl alcohol	24	290	2.9	2160
		m-Toluate	15	61	0.7	2110
	42°C	None	27	3	<0.1	550
		m-Methylbenzyl alcohol	24	5	<0.1	570
		m-Toluate	31	6	<0.1	590
20SO(pTN8)	30°C	None	6	320	0.1	1390
		m-Methylbenzyl alcohol	60	300	3.0	1390
		m-Toluate	11	310	2.8	2000
	37°C	None	12	630	3.6	1510
		m-Methylbenzyl alcohol	19	830	5.0	1700
		m-Toluate	13	630	4.5	1340
	42°C	None	20	740	6.9	540
		m-Methylbenzyl alcohol	14	810	6.3	570
		m-Toluate	21	760	6.0	560

[a] Cells were grown overnight in L broth in the absence or presence of m-xylene vapor, 5mM m-methylbenzyl alcohol, or 5mM m-toluate.

with those for pWWO mapped by transposon mutagenesis, molecular cloning, and analysis of insertion mutants (7,12,15). We have determined the complete nucleotide sequence of xylE (921 bp) and its flanking sequences (16). The xylE sequence of pWWO reported by Zukowski et al. (17) is coincided well but not identical to that of pTN2.

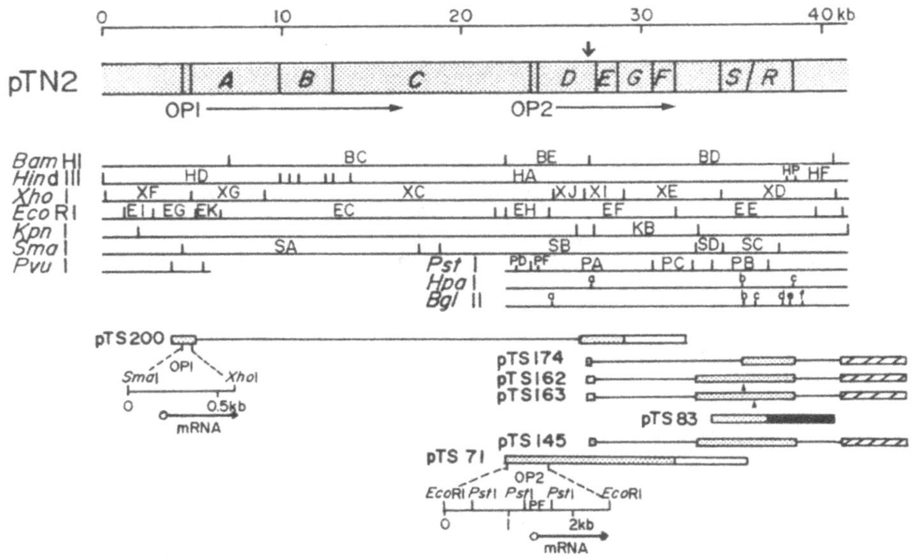

Fig. 3. Physical and functional map of the TOL region of pTN2. The notations of xyl genes and operator-promoter regions are the same as described in the legend to Fig. 2. A vertical arrow indicates the insertion site of a 3.6-kb segment in pTN1. The horizontal arrows indicate the direction of transcription. The lower part of the figure shows the structures of hybrid plasmids. Stippled, open, hatched, and closed boxes indicate the TOL, pBR322, pACYC 184, and pACYC177 regions, respectively. The fine mapping of transcription start sites in OP1 and OP2 are also indicated.

In addition to pTN2, several recombinants of a IncP1 plasmid and TOL were reported (18-22). A transposition model involving a unique segment of the TOL plasmid is unlikely, since the TOL moieties present in the different RP4-TOL plasmids are not identical (21). The pTN1 plasmid has an insertion of 3.6-kb fragment at the right end of the BamHI BE fragment in Fig. 3, which leads a constitutive low expression of xylE, xylG, and xylF.

During in vivo formation of pTN1 in *P. aeruginosa*, 2 events should have occurred at relatively high frequencies. One was the insertion of the 3.6-kb segment, and the other was the recombination of the 56-kb TOL segment and the temperature-resistant RP4 replicon. Thus, the cells carrying pTN1 could grow on m-toluate at high temperature because of the temperature-resistant replication and expression. Although the xylD expression was inducible and temperature-sensitive in pTN1, the chromosomal isofunctional enzyme, benzoate oxygenase might be able to complement the activity.

The 3.6-kb insertion found in pTN1 is very similar to an insertion (3.4-kb) described by Lehrbach et al. (15). Both insertions lead to inactivation of TOL genes and have no restriction sites for HindIII, EcoRI, SalI, PvuII, XhoI, PstI, BamHI, and BglII. The parent plasmid used to obtain insertion mutants by Lehrbach et al. is an RP4-TOL plasmid, pED3300, containing a 69-kb segment of TOL. From restriction, DNA hybridization, or heteroduplex analysis, the segment was shown to originate from within pWWO fragment HindIII-E (15) that is involved in the 69-kb TOL segment of pED3300, but not in the 56-kb segment of pTN2. The 3.6-kb insert can excise precisely from pTN1 to produce pTN2 and also can cause deletions of nearby DNA segments (10).

Plasmid pTN8 is a spontaneous mutant of pTN1 which was found in a segregant selected as a good grower on m-toluate. C230 is synthesized constitutively at a high level in cells carrying pTN8 (Tab. 1 and 2), and the plasmid still contains the 3.6-kb insert at the same site and orientation as in pTN1. It appears that the DNA sequence of 3.6-kb insert in pTN1 changed to produce a strong promoter in pTN8. We are currently investigating the DNA sequence surrounding the transcription start site within the insertion of pTN8.

Molecular Cloning of Operator-Promoter Regions of xyl Operons and Regulatory Genes xylR and xylS

As described in a previous section, pTN2 in E. coli specifies inducible synthesis of TOL enzymes similar to that in P. putida. These observations prompted us to analyze the regulation of xyl expression by molecular cloning in E. coli.

The lower part of Fig. 3 shows the structures of hybrid plasmids constructed to map the regulatory region of the xylABC and the xylDEGF operons as well as the regulatory genes xylR and xylS (6,14).

Plasmid pTS200 has a 1.7-kb PvuI-EcoRI fragment (OP1) joined to XhoI fragment XI (xylE) on pBR322 replicon. Plasmid pTS174 contains a 2.7-kb HpaI fragment (xylR) on pACYC184 replicon, pTS83 contains the PstI-PB fragment (xylS) on pACYC177 replicon, and pTS145 contains a 5.4-kb KpnI-HindIII fragment (xylR and xylS). The HindIII fragment HP is contained in PTS145. Plasmids pTS162 and pTS163 are insertion derivatives of pTS145 that were constructed by insertion of the 1.5-kb kan fragment of Tn5 into the BglII b site and c site, respectively. Plasmid pTS71 is a TOL-pBR322 hybrid containing xylD, xylE, xylG and xylF (13).

The activities of C230 in cells carrying these plasmids are shown in Fig. 4 (6). Cells carrying pTS200 alone do not show induction of the enzyme. In the presence of either pTS174 or pTS162, the enzyme is induced by m-xylene or m-methylbenzyl alcohol. pTS163 has

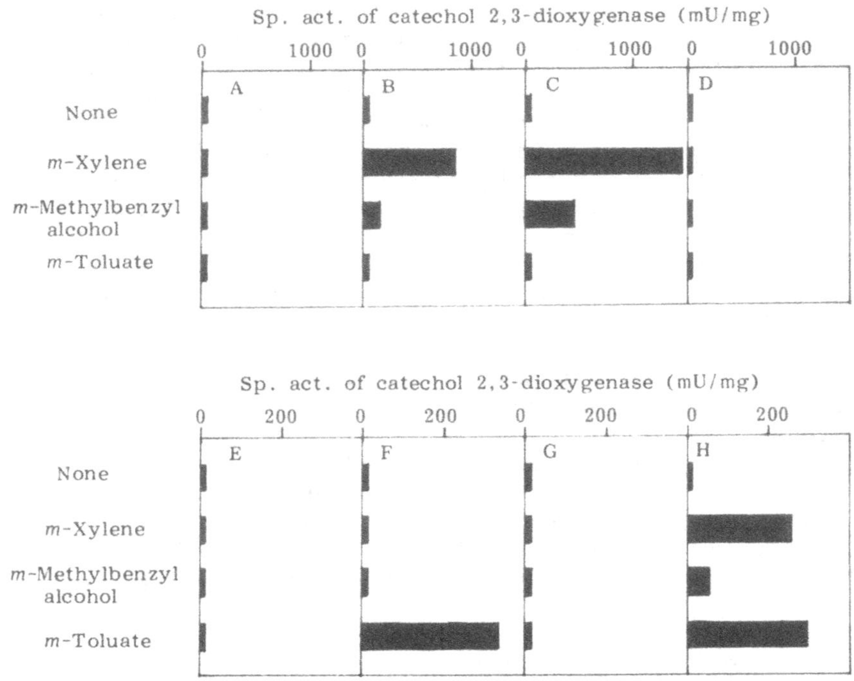

Fig. 4. Activation of *xylABC* and *xylDEGF* operons by *xylR* and *xylS*. Cells of *E. coli* carrying hybrid plasmids were grown in L broth at 30°C in the absence (none) or presence of *m*-xylene, *m*-methylbenzyl alcohol, or *m*-toluate. Plasmids carried: A, pTS200 (OP1-*xylE*$^+$); B, pTS200 and pTS174 (*xylR*$^+$); C, pTS200 and pTS162 (*xylR*$^+$); D, pTS200 and pTS163 (*xylS*$^+$); E, pTS71 (OP2-*xylDEGF*$^+$); F, pTS71 and pTS83 (*xylS*$^+$); G, pTS71 and pTS174; H, pTS71 and pTS145 (*xylR*$^+$ *xylS*$^+$).

no effect on the activation of OP1. These results indicate that pTS200 contains OP1 and that the *xylABC* operon is activated by the gene product of *xylR* that is present in the 2.8-kb region from the *Bgl*II b site to the *Hpa*I c site. OP2 was further mapped on the 0.6-kb fragment ranging from the left end of the *Sma*I fragment SA to the right end of the *Xho*I fragment XF (6).

Cells carrying pTS71 and either pTS83 or pTS145 show induction of C230 in the presence of *m*-toluate, while pTS174 has no effect on the induction. In addition, pTS163 but not pTS162 gave *m*-toluate -dependent induction. Therefore, pTS71 contains OP2, and pTS83, pTS145 and pTS163 contain *xylS* that was mapped on the *Pst*I PB fragment (Fig. 3). OP2 was further analyzed and mapped on the *Pst*I PF fragment (Fig. 3) (6).

The m-xylene- or m-methylbenzyl alcohol-dependent induction of C230 encoded by pTS71 occurs only in the presence of pTS145. As evidenced by the activation of OP1, both pTS174 and pTS162 contain the functional xylR gene. If the same xylR gene is involved in the activation of both operons, the xylR gene alone is not sufficient for the induction of xylDEGF. It was concluded that both xylR and xylS are essential to the m-xylene- or m-methylbenzyl alcohol-dependent activation of xylDEGF. The same conclusion was reached by Franklin et al. from the analysis of transposon mutations (7).

In the above experiments, the operator-promoters were cloned on the pBR322 replicon, and the regulatory genes were cloned on either the pACYC184 or pACYC177 replicon. Since pBR322 is compatible with pACYC replicon, the trans-acting effect of xylR or xylS could be determined.

In summary, we have analyzed the regulation of xyl gene expression by molecular cloning in E. coli, and the results obtained so far are consistent with the positive-regulation model shown in Fig. 2.

The xylE Gene Product C230 as a Promoter Probe

The precise mapping of OP1 and OP2 as well as xylR and xylS described above was carried out by monitoring the expression of xylE. The xylE product, C230 catalyzes the conversion of catechol to a yellow compound, 2-hydroxymuconic semialdehyde which has an absorption peak at 375 nm with the molar extinction coefficient of 14,700 at pH7.5 (23). The activity within cells is easily detected by spraying 0.1 M catechol solution on colonies and observing development of a yellow color. Catechol may be transported into cells by a simple diffusion and attacked by C230 which has a low Km value for catechol (3 μM) (23).

C230 has the molecular weight of 140,000 and consists of 4 identical subunits consisting of 307 amino acid residues (16). Several base pairs upstream from the initiation codon, the ribosome binding site has been predicted, that is complementary to the 16S rRNA of E. coli and P. aeruginosa (24). There are stop codons in front of the initiation codon so that a fused product of xylE is not synthesized upon ligation with the coding sequence upstream from the gene. In addition, since xylE is neither the first nor the last cistron of the operon, there is no proper promoter or terminator around the gene.

When xylE was joined to the tetracycline promoter of pBR322 by inserting the BamHI-BD fragment into the BamHI site, C230 was synthesized by a read-through transcription from the vector promoter in E. coli that was more than 300-fold of that encoded by pTN2 under noninduced conditions (13). Therefore, the low expression of TOL genes in E. coli is not due to a translational barrier, but is a reflection of poor transcription efficiency.

The gene expression of colicin E1 was studied by following the induction of C230 encoded by a ColE1-xylE fusion plasmid (25). The operator-promoter region of the colicin E1 gene was joined to the SstII fragment (1.2-kb) containing xylE. It was found that the C230 synthesis from the hybrid plasmid was cyclic AMP-dependent and controlled by the recA-lexA system (25). The xylE gene, therefore, is a useful probe for detection and quantitative determination of promoters in a wide range of gram-negative bacteria.

The usefulness of xylE as a promoter probe was also demonstrated in a Bacillus subtilis system (17).

TRANSCRIPTION INITIATION OF XYL OPERONS

Localization of Transcription Start Site of xylABC Operon

The induction pattern of xyl operons in E. coli is quite similar to that in P. putida, suggesting that the same operator-promoter is functioning in both bacteria. To see whether this is the case, we determined the in vivo transcription start site of the xylABC operon by S1 nuclease and reverse transcriptase mapping (26).

The detailed mapping revealed that the OP1 region is in the 0.6-kb fragment ranging from the left end of the SmaI fragment SA to the right end of the XhoI fragment XF (Fig. 3) (6). A fragment containing the OP1 region was labeled at the XhoI site of the sense strand and used as a probe for S1 mapping analysis that was performed as described (27). Crude RNA preparations were obtained from cells of P. putida or E. coli carrying pTN2 grown under induced or noninduced conditions. RNA was hybridized with the probe DNA, treated with S1 nuclease, and the protected fragments were subjected to polyacrylamide-urea gel. It was found that the transcription started at 390 nucleotides upstream from the labeled XhoI site when RNA from m-methylbenzyl alcohol-induced cells of P. putida carrying pTN2 was subjected to the assay.

The transcription start site was further analyzed by reverse transcriptase mapping (28). The sense strand of a small DNA fragment was labeled at the 5' ends, hybridized with mRNA and extended with reverse transcriptase to the transcription start site. The extended product was analyzed by coelectrophoresis on a sequence gel with chemically cleaved reaction products by the procedure of Maxam and Gilbert. As a result, a single start site was found with RNA from induced cells of P. putida carrying pTN2. The same start site was also found with RNA from induced cells of E. coli carrying pTN2, but the amount of mRNA was only 10% of that of P. putida. The mRNA levels in these bacteria determined by S1 mapping coincided with the enzyme levels in vivo (see Tab. 1 and 2). No transcription was found with RNA preparations from noninduced cells of P. putida or E. coli carrying pTN2 or cells free from the plasmid.

These results indicate that the m-methylbenzyl alcohol-dependent transcription of the xylABC operon in E. coli starts at exactly the same site as in P. putida. The low level of mRNA in E. coli appears to be a reflection of inefficient transcription of xyABC by E. coli RNA polymerase.

Localization of Transcription Start Site of xylDEGF Operon

The transcription start site of xylDEGF was also determined by S1 nuclease and reverse-transcriptase mapping (29). We have previously mapped OP2 on the PF fragment (Fig. 3) (6). S1 nuclease mapping revealed that the transcription started in the PF fragment in the induced cells of P. putida or E. coli carrying pTN2.

The precise initiation site for transcription of xylDEGF was further determined by reverse-transcriptase mapping. As for the xylABC operon, a single start site was found in both P. putida and E. coli cells under both m-methylbenzyl alcohol- and m-toluate-induced conditions.

```
   SmaI
   1 HpaII                                     TaqI
5' CCCGGGAAAGCGCGATGAACCTTTTTTATCGCTGCCTTGATCAAATCGAC
   GGGCCCTTTCGCGCTACTTGGAAAAAATAGCGACGGAACTAGTTTAGCTG

51
   AGGTGGTTATGCGCGATTGATGATTTGCTCAAATACAGCCAGCGTGCTGT
   TCCACCAATACGCGCTAACTACTAAACGAGTTTATGTCGGTCGCACGACA

101
   AGATTTTCTCTCATACCCCCCCTTTCTTTTTTACAAAGAAAATCAATAAT
   TCTAAAAGAGAGTATGGGGGGGAAAGAAAAAATGTTTCTTTTAGTTATTA

151                                                AluI
   TTAGATGAAATAAGGGGATCGGTATAAGCAATGGCATGGCGGTTGCTAGC
   AATCTACTTTATTCCCCTAGCCATATTCGTTACCGTACCGCCAATGATCG

201                          BstEII
   TATACGAGACTTAAAATAAAAATAGTGGTGACCCTTCAATGTTGTATTTT
   ATATGCTCTGAATTTTATTTTTATCACCACTGGGAAGTTACAACATAAAA
                                                MetLeuTyrPhe
251
   CTCAACTCTGTTCAGATTGGTTGCTTTCGCCATGTATATCCTCAAAGCGG
   GAGTTGAGACAACTCTAACCAACGAAAGCGGTACATATATGAGTTTCGCC
   LeuAsnSerValGlnIleGlyCysPheArgHisValTyrProGlnSerGly
301 HaeIII                  HinfI        HaeIII
   GCCAGCCGTAGCCGTTACGCCTGGAGTCTGTGACGCTGGCC 3'
   CGGTCGGCATCGGCAATGCGGACCTCAGACACTGCGACCGG
    ProAlaValAlaValThrProGlyValCysAspAlaGly
```

Fig. 5. Nucleotide sequence of the OP1 region of the TOL plasmid (26). Nucleotides are numbered from the SmaI site. A vertical arrow indicates the transcription start site of the xylABC operon. Inverted repeats are indicated by solid horizontal arrows and direct repeats, by dashed horizontal arrows. The underlined bases are complementary to the 3' end of the 16S rRNA of P. aeruginosa. Amino acids predicted from the nucleotide sequence are shown in italics.

The transcription in *E. coli* cells was poor especially when induced by *m*-toluate. It is not known whether *m*-toluate uptake in *E. coli* is lower than in *P. putida*.

Nucleotide Sequence of OP1 and OP2

The sequences surrounding the transcription start sites of *xylABC* (340 bp) (Fig. 5) and *xylDEGF* (336 bp) (Fig. 6) were determined (26) by the Maxam and Gilbert procedure.

In the region downstream from the transcription start site, there is one AUG in *xylABC* and 3 AUGs in *xylDEGF*, from which a continuous reading frame exists in each sequence at least up to the end of the sequences determined. The predicted amino acid sequences probably correspond to the NH_2-terminal region of the enzyme of the first cistron of each operon. Several base pairs upstream from the initiation codon of each operon, there is a typical ribosome binding site.

Preceding the transcription initiation site of *xylABC*, there are two pallindromic AT-rich sequences as candidates for operators,

```
   1 HpaII
5' CCGGTTTGATAGGGATAAGTCCAGCCTTGCAAGAAGCGGATACAGGAGTG
   GGCCAAACTATCCCTATTCAGGTCGGAACGTTCTTCGCCTATGTCCTCAC

 51               XbaI
   CAAAAAATGGCTATCTCTAGAAAGGCCTACCCCTTAGGCTTTATGCAACA
   GTTTTTTACCGATAGAGATCTTTCCGGATGGGGAATCCGAAATACGTTGT

101               HinfI                     BstNI   TaqI
   GAAACAATAATAATGGAGTCATGACCATGACAATGCACCTGGGGCTCGAC
   CTTTGTTATTATTACCTCAGTACTGGTACTGTTACGTGGACCCCGAGCTG
                              MetThrMetThrMetHisLeuGlyLeuAsp
151
   TATATAGATAGTCTCGTTGAAGAAGATGAGAACGAGGGCATCTACCGCTG
   ATATATCTATCAGAGCAACTTCTTCTACTCTTGCTCCCGTAGATGGCGAC
   TyrIleAspSerLeuValGluGluAspGluAsnGluGlyIleTyrArgCys
201                                    TaqI
   CAAGCGCGAGATGTTCACCGACCCTCGGCTGTTCGATTTAGAGATGAAAC
   GTTCGCGCTCTACAAGTGGCTGGGAGCCGACAAGCTAAATCTCTACTTTG
   LysArgGluMetPheThrAspProArgLeuPheAspLeuGluMetLys
251                                                AvaI
   ACATCTTTGAGGGCAACTGGATTTATCTCGCCCACGAGAGCCAGATTCCC
   TGTAGAAACTCCCGTTGACCTAAATAGAGCGGGTGCTCTCGGTCTAAGGG
   HisIlePheGluGlyAsnTrpIleThrLeuAlaHisGluSerGlnIlePro
301                                    HpaII
   GAGAAGAACGACTATTACACCACGCAGATGGGCCGG 3'
   CTCTTCTTGCTGATAATGTGGTGCGTCTACCCGGCC
   GluLysAsnAspTyrTyrThrThrGlnMetGlyArg
```

Fig. 6. Nucleotide sequence of the OP2 region of the TOL plasmid (29). Nucleotides are numbered from the *HpaI* site. A vertical arrow indicates the transcription start site of the *xylDEGF* operon. An inverted repeat is indicated by horizontal arrows. The underlined bases are complementary to the 3' end of the 16S rRNA of *P. aeruginosa*. Amino acids predicted from the nucleotide sequence are shown in italics.

which have the axis of diad symmetry at -85 and -76 bp (Fig. 5). Such a sequence, however, is absent from OP2. On the other hand, a pallindromic sequence is found at -15 bp (Fig. 6). The xylABC operon is activated by xylR, whereas the xylDEGF operon is activated either by xylS alone or xylS together with xylR. Thus, the sequences of the 2 operators could be different from each other.

In the region upstream from the transcription start sites in both sequences, the consensus sequences of E. coli promoters were not found with either the -10 region (Pribnow box) or the -35 region (RNA polymerase recognition site). The sequences of the 2 operons, however, are homologous with each other around at the -35 and -10 regions. The homologous sequences may represent signals for transcription initiation of positively controlled operons in Pseudomonas.

ACKNOWLEDGMENTS

This work was partly supported by Grants-in-Aid for Scientific Research from the Ministry of Education, Science, and Culture, Japan. A part of this work was done under contract with the Science and Technology Agency of the Japanese Government, and was supported by the Special Fund for Multiministrial Projects.

REFERENCES

1. Dunn, N.W., and I.C. Gunsalus (1973) Transmissible plasmid coding early enzymes of naphthalene oxidation in Pseudomonas putida. J. Bact. 144:974-979.
2. Chakrabarty, A.M. (1972) Genetic basis of the biodegradation of salicylate in Pseudomonas. J. Bact. 112:815-823.
3. Williams, P.A., and K. Murray (1974) Metabolism of benzoate and methylbenzoates by Pseudomonas putida (arvilla) mt-2: Evidence for the existence of a TOL plasmid. J. Bact. 120:415-423.
4. Worsey, M.J., and P.A. Williams (1975) Metabolism of toluene and xylenes by Pseudomonas putida (arvilla) mt-2: Evidence for a new function of the TOL plasmid. J. Bact. 124:7-13.
5. Worsey, M.J., F.C.H. Franklin, and P.A. Williams (1978) Regulation of the degradative pathway coded for by the TOL plasmid (pWWO) for Pseudomonas putida mt-2. J. Bact. 134:757-764.
6. Inouye, S., A. Nakazawa, and T. Nakazawa (1983) Molecular cloning of the regulatory gene xylR and the operator-promoter regions of xylABC and xylDEGF operons of TOL plasmid. J. Bact. 155:1192-1199.
7. Franklin, F.C.H., P.R. Lehrbach, R. Lurz, B. Rueckert, M. Bagdasarian, and K.N. Timmis (1983) Localization and functional analysis of transposon mutations in regulatory genes of the TOL catabolic pathway. J. Bact. 154:676-685.

8. Nakazawa. T. (1978) TOL plasmid in Pseudomonas aeruginosa PAO: thermosensitivity of self-maintenance and inhibition of host cell growth. J. Bact. 133:527-535.
9. Nakazawa, T., E. Hayashi, T. Yokota, T. Ebina, and A. Nakazawa (1978) Isolation of TOL and RP4 recombinants by integrative suppression. J. Bact. 134:270-277.
10. Nakazawa, T., S. Inouye, and A. Nakazawa (1980) Physical and functional mapping of RP4-TOL plasmid recombinants: Analysis of insertion and deletion mutants. J. Bact. 144:222-231.
11. Downing, R., and P. Broda (1979) A cleavage map of the TOL plasmid of Pseudomonas putida mt-2. Mol. Gen. Genet. 177:189-191.
12. Franklin, F.C.H., M. Bagdasarian, M.M. Bagdasarian, and K.N. Timmis (1981) Molecular and functional analysis of the TOL plasmid pWWO from Pseudomonas putida and cloning of genes for the entire regulated aromatic ring meta cleavage pathway. Proc. Natl. Acad. Sci., USA 78:7458-7462.
13. Inouye, S., A. Nakazawa, and T. Nakazawa (1981) Molecular cloning of TOL genes xylB and xylE in Escherichia coli. J. Bact. 145:1137-1143.
14. Inouye, S., A. Nakazawa, and T. Nakazawa (1981) Molecular cloning of gene xylS of the TOL plasmid: Evidence for positive regulation of the xylDEGF operon by xylS. J. Bact. 148:413-418.
15. Lehrbach, P.R., D.J. Jeenes, and P. Broda (1983) Characterization by molecular cloning of insertion mutants in TOL catabolic functions. Plasmid 9:112-125.
16. Nakai, C., H. Kagamiyama, M. Nozaki, T. Nakazawa, S. Inouye, Y. Ebina, and A. Nakazawa (1983) Complete nucleotide sequence of the metapyrocatechase gene on the TOL plasmid of Pseudomonas putida mt-2. J. Biol. Chem. 258:2923-2928.
17. Zukowski, M.M., F.F. Gaffney, D. Speck, M. Kauffmann, A. Findeli, A. Wisecup, and J.P. Lecocq (1983) Chromogenic identification of genetic regulatory signals in Bacillus subtilis based on expression of cloned Pseudomonas genes. Proc. Natl. Acad. Sci., USA 80:1101-1105.
18. Jacoby, G.A., J.E. Rogers, A.E. Jacob, and R.W. Hedges (1978) Transposition of toluene degrading genes and expression in E. coli. Nature 274:179-180.
19. Chakrabarty, A.M., D.A. Friello, and L.H. Bopp (1978) Transposition of plasmid DNA segments specifying hydrocarbon degradation and their expression in various organisms. Proc. Natl. Acad. Sci., USA 75:3109-3122.
20. White, G.P., and N.W. Dunn (1977) Apparent fusion of the TOL plasmid with the R91 drug resistance plasmid in Pseudomonas aeruginosa. Aust. J. Biol. Sci. 30:345-355.
21. Muelien, P., R.G. Downing, and P. Broda (1981) Excision of the 40-kb segment of the TOL plasmid from P. putida mt-2 involves direct repeats. Mol. Gen. Genet. 184:97-101.
22. Jeenes, D.J., and P.A. Williams (1982) Excision and integration of the degradative pathway genes from the TOL plasmid pWWO. J. Bact. 150:188-194.

23. Nozaki, M. (1979) Oxygenases and dioxygenases. Topics of Current Chemistry. 78:145-186.
24. Shine, J., and L. Dalgarno (1975) Determination of cistron specificity in bacterial ribosomes. Nature (Lond.) 254:34-38.
25. Ebina, Y., Y. Takahara, K. Shirabe, M. Yamada, T. Nakazawa, and A. Nakazawa (1983) Plasmid encoded regulation of colicin El gene expression. J. Bact. 156:487-492.
26. Inouye, S., Y. Ebina, A. Nakazawa, and T. Nakazawa (1984) Nucleotide sequence surrounding transcription initiation site of xylABC operon on TOL plasmid of Pseudomonas putida. Proc. Natl. Acad. Sci., USA 81:1688-1691.
27. Ebina, Y., and A. Nakazawa (1983) Cyclic AMP-dependent initiation and ρ-dependent termination of colicin El gene transcription. J. Biol. Chem. 258:7072-7078.
28. Débarbouillé, M., and O. Raiboud (1983) Expression of the Escherichia coli malPQ operon remains unaffected after drastic alteration of its promoter. J. Bact. 153:1221-1227.
29. Inouye, S., A. Nakazawa, and T. Nakazawa (1984) Nucleotide sequence of the operator-promoter region of the xylDEGF operon on TOL plasmid in Pseudomonas putida. Gene 29:323-330.

CHAIRMEN'S INTRODUCTION: PLASMID TRANSFER

Neil Willetts and Don B. Clewell[1]

Biotechnology Australia Pty. Ltd.
Roseville, NSW 2069, Australia

Bacterial plasmids from a wide range of genera are now being studied in numerous laboratories around the world, and from a variety of perspectives. Plasmids provide systems for the study of DNA replication and its control; they are useful in studies of bacterial genetics and are invaluable as vectors for molecular cloning. From the standpoint of the bacteria, however, what is it about these molecules that makes them so ubiquitous?

It now seems clear that plasmids provide bacterial populations with a greatly enhanced flexibility for rapid change. Bacteria are among the most variable organisms in nature, and the role of plasmids in this capacity is of major importance. One of the most significant features of plasmids is their ability to transfer from one bacterium to another. This most frequently occurs by a conjugative process, although in some situations plasmids can transfer by transduction or transformation. The prevalance of conjugation as a plasmid phenotype is perhaps an indication of its importance.

While many plasmids are conjugative, others can transfer only if mobilized by a conjugative plasmid. Such nonconjugative plasmids often supply genes contributing to the mobilization process, while others may need to be co-integrated with the mobilizing plasmid; even the bacterial chromosome can be transferred by the latter mechanism. Cointegrate structures are frequently intermediates in the transposition of an insertion sequence or transposon from one replicon to another.

---

[1]Department of Oral Biology and Microbiology/Immunology, Schools of Dentistry and Medicine and The Dental Research Institute, The University of Michigan, Ann Arbor, Michigan 48109.

The ultimate advantage of the resultant DNA mobility is that when a given bacterial cell has a gene that is uniquely protective or provides a growth advantage in a certain environment, a mechanism is available for distributing this information to other members of the bacterial population. As the information is gained and more cells become potential donors, the speed at which the gene can spread through the population increases; and selective processes aid its establishment.

Recent studies of conjugation in a number of different systems has revealed a rich variety of means whereby DNA transfer can be accomplished. The formation of mating pairs in gram-negative bacteria via an extracellular organelle, the pilus, can be contrasted to the pheromone-induced aggregation of certain gram-positive species; and the 33-kb DNA segment required for conjugation by IncF plasmids is distinguished from the 2 kb required by some _Streptomyces_ plasmids. In the papers which follow, we shall learn more of this diversity.

LINEAR PLASMIDS WITH TERMINAL INVERTED REPEATS OBTAINED FROM *STREPTOMYCES ROCHEI* AND *KLUYVEROMYCES LACTIS*

Kenji Sakaguchi, Hirohiko Hirochika, and Norio Gunge

Mitsubishi-Kasei Institute of Life Sciences
Minamiooya 11, Machida, Tokyo, Japan, 194

SUMMARY

Linear plasmids with inverted terminal repeats of 614 bp were obtained from *Streptomyces rochei* which produced lankacidin. The 5' ends were blocked by the association of a terminal protein. A DNA model of racket frame-like structure is presented implying the juxtaposition of 2 double-stranded DNAs of the same sequence through the binding of cohesive proteins which recognize and bind to the DNA.

Two linear plasmids with the inverted terminal repeats of 202 and 184 bp were obtained from a yeast, *Kluyveromyces lactis*. A killer toxin was produced from the shorter plasmid. The protein toxin inhibited the adenylatecyclase activity of the yeast membrane.

INTRODUCTION

Most procaryotic and eucaryotic plasmids reported are composed of covalently-closed circular DNAs. Linear plasmids were first found in our laboratory from microbes. From *Steptomyces rochei* producing the antibiotics lankacidin and lankamycin, 2 linear plasmids of 17 kb designated as pSLA 1 and pSLA 2 (1,2,3,4) were obtained in 60 copies each per chromosome, as well as a circular plasmid, pSLA 3, which had the abilities of self-transmissibility and pock formation when the strain was costreaked with other strains.

From the yeast *Kluyveromyces lactis* which grows in cheese whey utilizing lactose as a carbon source, 2 linear plasmids of different

Tab. 1. Actinomycetes strains tested for the presence of plasmid.

Antibiotics or β-lactamase produced	Strain	Antibiotics or β-lactamase produced	Strain
Blasticidin S	*Streptomyces griseochromogenes* 2A-327	Streptothricin	*Streptomyces lavendulae* NIHJ-E-2
Lankacidin	*Streptomyces* sp. 6642-GC_1		*Streptomyces microflavus* NIHJ-13-A
	Streptomyces sp. 7434-AN_4	Chlortetracycline	*Streptomyces aureofaciens* NRRL-B-2209
Lonomycin	*Streptomyces ribosidificus* TM-481	Polyoxin	*Streptomyces cacaoi* var. *asoensis* POM-23-1
Viomycin	*Streptomyces diastatochromogenes* KCC-S-0119	Polyoxin + lysocellin	*Streptomyces cacaoi* var. *asoensis* POM-23-1 mutant
	Streptomyces vinaceus ICC-S-0091	Methylsalinomycin	*Streptomyces albus* ATCC-21838
	Streptomyces californicus KCC-S-0143	Xanthomycin	*Streptomyces* sp. 1134-7
	Streptomyces puniceus KCC-S-0406	Geldanamycin	*Streptomyces hygroscopicus* 434
	Streptomyces sp. KCC-U-0197	Cephamycin	*Streptomyces clavuligerus* KCC-X-0710
	Streptoverticillium olivoreticuli KCC-S-0197		*Streptomyces clavuligerus* IFO 13307
Mikamycin	*Streptomyces mitakaensis* 74-4		*Streptomyces lipmanii* KCC-S-0711
Cycloheximide	*Streptomyces* sp. 2217-G_1		*Streptomyces lipmanii* IFO 13306
	Streptomyces sp. 8465-MC_1		*Streptomyces rochei* IFO 12908
	Streptomyces sp. 8482-CC_1		*Streptomyces fimbriatus* KCC-S-0190
Neomycin	*Streptomyces albogriseolus* NRRL-B-1305	Gentamicin	*Micromonospora echinospora* NRRL-2985
	Streptomyces sp. 7068-CC_1	β-Lactamase	*Streptomyces* sp. E-750-3
	Streptomyces sp. 8104-MC_2		*Streptomyces* sp. E-756-1

sizes were found in 50-100 copies each per haploid cell. They were designated as pGKl 1 and pGKl 2; 8, 9 kb and 13, 4 kb, respectively (5,6).

A protein toxin against *Saccharomyces*, *Kluyveromyces*, and several other yeast species was produced from the shorter plasmid, pGKl 1. The pGKl 1 DNA was totally sequenced and it contained 4 open reading frames.

The existence of linear plasmids in maize (*Zea mays*) mitochondria has been reported by Pring et al. (7). Their nucleotide sequences with inverted terminal repeats were recently reported (8) to possess a DNA structure similar to our *Streptomyces* and *Kluyveromyces* plasmids. From a fungus, *Ascobolus inversus*, a linear plasmid was found by Francou (9) which may have possessed the inverted terminal repeats.

LINEAR PLASMIDS FROM *STREPTOMYCES ROCHEI*

Isolation of Circular and Linear DNA Plasmids from Various Streptomyces Species

Thirty-four *Streptomyces* strains producing various antibiotics,

Tab. 2. Summary of plasmids from four strains of Streptomyces.

Strain	Antibiotics produced	Plasmid	S-value	Mol.Wt. ($\times 10^{-6}$)
S. puniceus KCC-S-0406	viomycin	pSPU1 pSPU2	45.8 69.0	17.3 39.1
Streptomyces sp. 2217-G1	cycloheximide	pSCY1 pSCY2	21.9 46.7	3.2 18.0
Streptomyces sp. 7068-CC1	neomycin	pSNE1 pSNE2	51.7 75.9	22.2 47.0
S. hygroscopicus 434	geldanamycin	pSHY1	67.9	37.9

including viomycin, blasticidin-S, cycloheximide, cephamycin, and others were isolated and generously supplied by Professor N. Otake, Institute of Applied Microbiology, University of Tokyo (Tab. 1 and 2). Through CsCl-EtBr gradient centrifugation, 7 circular plasmids were isolated from 4 strains of Streptomyces. The plasmids obtained from cycloheximide- or viomycin-producing strains were small enough for possible use as vectors.

From S. rochei producing lankacidin and lankamycin, 3 kinds of plasmids were isolated using the electrophoretic method, not by CsCl-EtBR centrifugation. Two linear and 1 circular plasmids were obtained from different variant strains and designated as pSLA 1,2, and 3 (Tab. 3).

pSLA 3, Circular, Transferable, and Pock-forming Plasmid

A circular plasmid 9, 2 kb was obtained from a S. rochei strain 7434 AN4 (3) with a high copy-number of 40 per chromosome. Its electrophoretic pattern and restriction map are presented in Fig. 1 and 2. There are 6 cleaving sites by SalI. There is no cut by EcoRI, HindIII, BglII, HpaI, SstI, XbaI, or XhoI. In a mixed culture of many spores of the 7434 AN 4 (2) strain with fewer spores of the 7434 AN 4 (3) strain, pocks were formed on the lawn of the former strain by lethal zygosis. The colony appearing in the clear zone

Tab. 3. Linear and circular plasmids isolated from S. rochei. LC: Lankacidin; LM: Lamkamycin.

strain	plasmid		antibiotics	resistance to LC
6642 GC1	pSLA 1	linear	LC, LM	+
7434 AN4(2)	pSLA 2	linear	LC, LM	+
7434 AN4(3)	pSLA 3	circular	LM	+
7343 LZ4(2,3)	pSLA 2 + 3		LC, LM	+

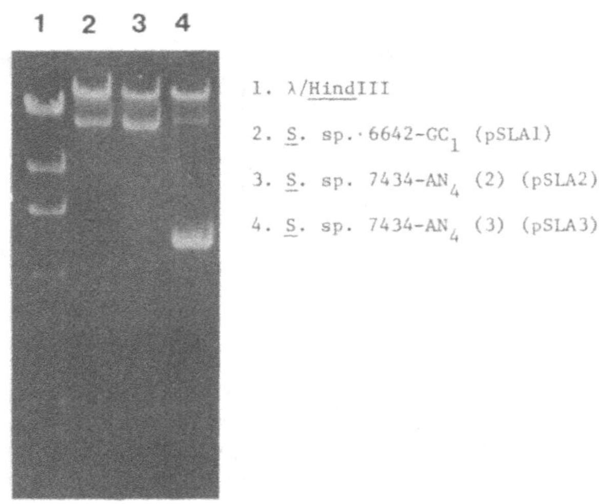

Fig. 1. Presence of 2 linear plasmids, pSLA 1 and pSLA 2, and circular transferable plasmid pSLA 3 in 3 strains of Streptomyces rochei.

was transferred to the pSLA 3 plasmid, giving rise to the strain 7434 LZ 4 (2,3) which contained both pSLA 2 and 3. The labeled pSLA 3 DNA had a weak homology with the left part of pSLA 2 linear plasmid DNA, as proven by the Southern hybridization technique.

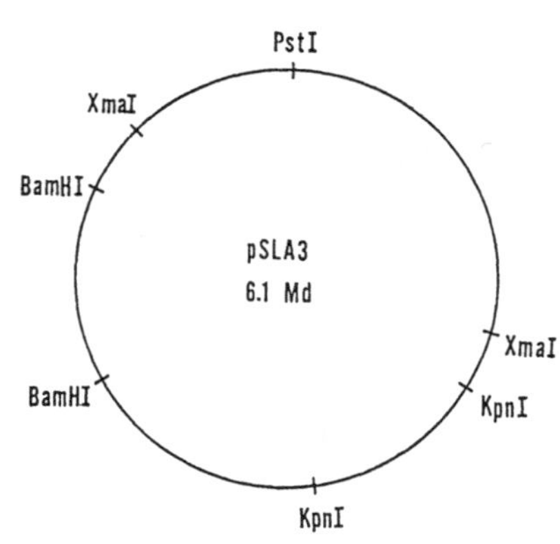

Fig. 2. Restriction map of pSLA 3.

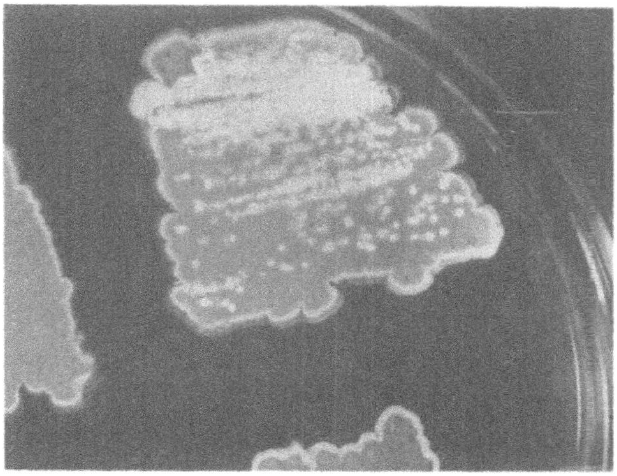

Fig. 3. Pock formation by the conjugative transfer of pSLA 3 plasmid from S. rochei 7434-AN4(3) to 7434-AN4(2).

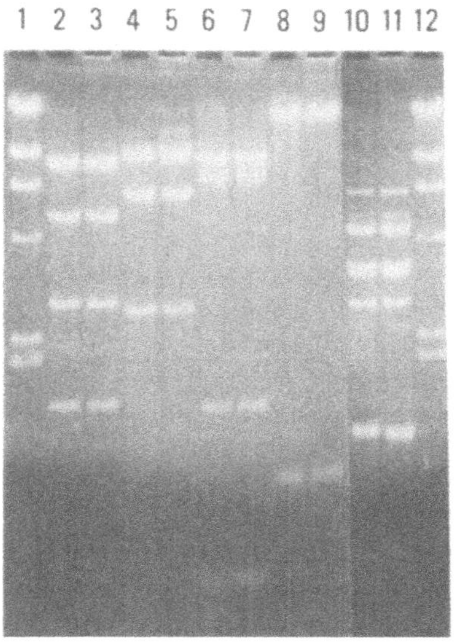

Fig. 4. Agarose gel electrophoresis of pSLA1 and pSLA2 treated with different restriction endonucleases. The arrow indicates the position of XmaI-E fragment of pSLA1. 1, 12: λ DNA treated with HindIII; 2, 3: BamHI; 4, 5: SalI: 6, 7: BglI; 8,9: EcoRI; 10, 11: XmaI; 2,4,6,8,10: pSLA1; 3,5,7,9,11: pSLA2.

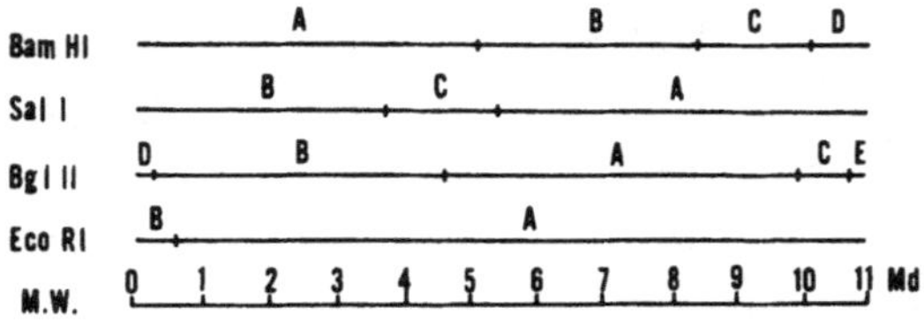

Fig. 5. Cleavage maps of pSLA2.

This transferable plasmid can be a good candidate as the vehicle in *Streptomyces*, for the following reasons:

a) Small molecular weight of 6.1 Md.

b) Unique cleaving site by *PstI*. Through a partial digestion technique, other cleaving sites can be used for cloning.

c) High copy number of 40 per chromosome which can result in the gene dosage effect for the greater expression of genes. Also, the pSLA 3 plasmid can easily be detected and isolated from a small quantity of cell cultures of even 0.5 ml.

d) Capability of "pock formation," which enables the isolation of a transferred clone with a high efficiency.

e) Transferability by mixed culture.

Restriction Pattern of Linear Plasmids pSLA 1 and pSLA 2 (3)

The 2 linear plasmids pSLA 1 and pSLA 2 DNA were isolated and their electrophoretic patterns were very similar (Fig. 4). The difference in size and number of the fragments generated by *BamHI*, *SalI*, and *EcoRI* was not detectable between the 2 plasmids. The only detectable difference was the *XmaIE* fragment where pSLA 1 was 60 bp longer. The cleavage map of the pSLA 2 DNA is presented in Fig. 5.

Association of Terminal Protein with the 5'-termini of Linear Plasmid DNA

Bacteriophages and animal viruses such as *Bacillus subtilis* ϕ 29 (10,11) and adenoviruses (12) contain linear DNAs. They have a unique structure at the ends of 5'-termini of their DNA, inverted repetitive sequences covalently attached to terminal proteins. The protein serves for the priming of DNA replication. We examined the possibility of the presence of such structures in pSLA 1 and pSLA 2. The attachment of protein to DNA can greatly reduce the electrophoretic mobility of DNA in the gel (10). We electrophoresed the

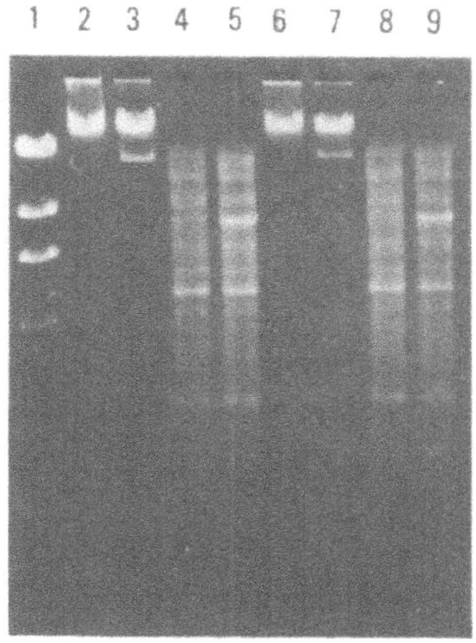

Fig. 6. Agarose gell electrophoresis of untreated or Pronase-treated pSLA1 and pSLA2. Cleared lysates prepared from pSLA1-harboring (2-5) or pSLA2-harboring (6-9) strains without Pronase treatment were subject to agarose gel electrophoresis after the following treatment. 2,6: No treatment; 3,7: treated with Pronase; 4,8: treated with BamHI; 5,9: treated with Pronase after BamHI digestion. 1: λ DNA treated with HindIII.

cleared lysates of *S. rochei* cells with or without the pronase treatment. The linear DNA bands were only detectable with the pronase treatment of the lysates. Furthermore, the same results were obtained in the case of BamHI-treated fragments of the linear DNAs; the 2 terminal fragments appeared only when they were treated with pronase (Fig. 6). These results indicate the firm binding of proteins to the 2 ends of the linear plasmid DNA.

We further examined the location of protein on pSLA 1 and pSLA 2. If a protein is associated with the end of the plasmid DNA, the DNA may be resistant to exonuclease, such as exonuclease III (3'-exonuclease) or λ-exonuclease (5'-exonuclease). The susceptibility of both plasmid DNAs to the 2 exonucleases was examined. As shown in Fig. 7, pSLA 2-protein complex was sensitive to exonuclease III, but resistant to λ-exonuclease. The same results were obtained in pSLA 1. These results strongly suggest that the 3' ends of these plasmid DNAs are free and the 5' ends are blocked by a protein covalently-bound.

The presence of this 5' end block is also assured by a 5' end-labeling experiment with polynucleotide kinase and [γ-^{32}P] ATP. The ^{32}P labeling of terminal DNA was doubled when the bound protein was removed by the alkali treatment of 0.1^{N} NaOH 30 Min (4).

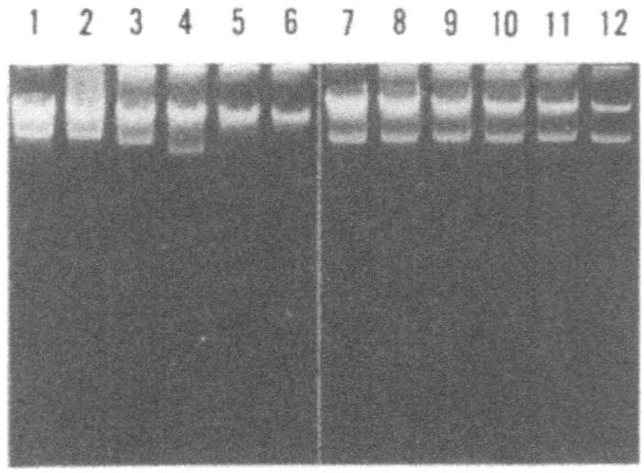

Fig. 7. Agarose gel electrophoresis of pSLA2 treated with exonucleases. Cleared lysates were treated with exonuclease III (1-6) or λ-exonuclease (7-12) for 0 min (1,7), 2 min (2,8), 5 min (3,9), 10 min (4,10), 15 min (5,11), or 40 min (6,12) at 37°C. After Pronase treatment, samples were subjected to agarose gel electrophoresis.

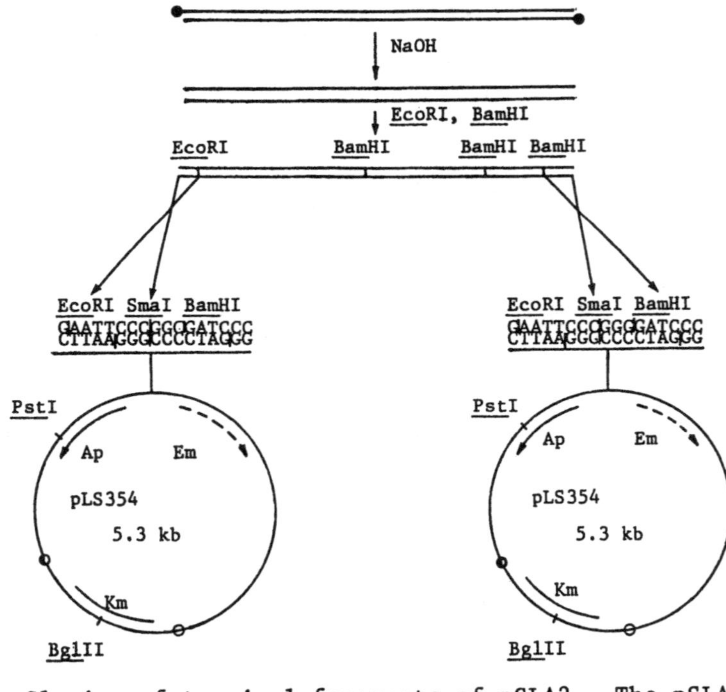

Fig. 8. Cloning of terminal fragments of pSLA2. The pSLA2 DNA was treated with NaOH in order to remove 5' end block (o).

Inverted Terminal Repetition Sequences (4)

The terminal fragments of the linear plasmid pSLA 2 were cloned in *Escherichia coli* cells using the plasmid pLS 354 (Fig. 8) The structure and the exact length of the inverted repetition were examined by determining the nucleotide sequence using the method of Maxam and Gilbert (13). Fig. 9 indicates the presence of an inverted terminal repetition of 614 bp. In the region beyond the 614 bp

```
Left
CCCGCGGAGCGGTCCCTGCGGGCTTCGCCCGCGCAGAGACACGCCCCGCTGCGCGGGGCGTGTGAACGG   70
AGGGTCACCGCCCGCCCTGCGGGCGGGCGTTCGGCCGCTGCGCGGCCTGCCGGCTCGCTCGGCCTGCGGC  140
CCGAGCGGCGCGGGCCACTGCGCTCCCCGTGCGAATCTCGCGCTCCGCGCGTACACGGGCCGCAGGGCGC  210
CGAGGGGCCCGTCCGCGCGCCCGGACCCGAGGCCCCTGCGGGGCCTGGCTGGCTACGAGGGGAGGGGTCG  280
GGGCGGGTGGGACAACTCGGGGGATCATCACGGTGAGTTGATGCAGCATGTGCTGTTGGCAGATCTGCAT  350
CACCCCCGTGCTTCAAATGTGACGGCTAGACCGACTCTGGAGCCGCGCCCTGCCACCCCTGCGGGGTGGA  420
CAGGCCCTGTCAACACCGGAACAAGAGCCAGGCGCGGCGCGGCACCGCTGTCCCCGGAGCACTCGGGCGC  490
GAGGGCGAAGCATGGGGAGCAGCGCTGTGACCACGGTAAACACCGCCAGAGACCCACTCCGGCACGTGT  560
TGCGCACCCTCACGGCCCGCCGCCGAGCGGACCCGCACGGACCGCCGGCCACCCGCACTCCGAGCCCCGG  630
ACGAGCTGCACGGGACACTTCTGACGCCTCATCAGCGCCACAGGCGAATACTATCCGTCGCCCCTAACAA  700
GGTTGACACCCCGAAAAAAACCAGCCCTAAACACGAAACGCCCCCGGAACAAAAAGCATCCCGAGGCGT  770
TCCGCAGACTCCCGAATTACGCGACGGCTTTGGCTTCAGCAGATACGGTGTTCCGATTCCGGGAATTC   840

Right
TGCGCACCCTCACGGCCCGCCGCCGAGCGGACCCGCACGGACCGCCGGCCACCCCAGGCGCGATGCTCCG  630
ACCGCCACGCACGGGCAGGTTCTGACGGATCATCAGCCCTCACAGACGAATACACTCAGTCGCCCCAAAC  700
AAGGTTGACACCCTCCCGGAAATCCGGCCGTAAACACAGAACGCCCCCGGAACAAAAAGCATCCCGAGGG  770
CGTCCGCAGGCTTCCGAATTACGCGACCCTCCCCCACTCCCTATTCCGCCGATTAACAGTCATGGAGGT   840
TCTGCGGGCGTGGCTGGAGCAGCGCTGCCCTTGGCGCTCGCGGCCTTGGCCTCCCGTAGGTAGCGCTCCA  910

CCGAGTC
```

Fig. 9. Terminal nucleotide sequences of pSLA2. The inverted terminal repetition extends to nucleotide 614. The sequences homologous between the 2 ends are boxed. The palindromic sequences are indicated by the arrows. The initiation and termination codons of a possible coding sequence are indicated by the broken underlining.

of complete terminal homology, there are still 11 sites of interrupted homologous sequences which are indicated by boxes in Fig. 9. The total length of homologous sequences in the region from 615 to 800 bp from both ends reached 65%.

Juxtaposing 4 Strands of DNA and the Racket Frame Model

We postulate the presence of a juxtaposing structure of 4 strands of DNA. Such a structure cannot be considered from the physico-chemical character of pure DNA; however, the existence of such a structure in nature is implicitly understood in many aspects of the functions of DNA in organisms. The juxtaposition of 2 double-stranded DNAs of identical or similar sequences is a prerequisite to the recombination reactions. Another visible proof of the presence of such a structure is the allele arrangement of chromosomes, for instance, in the salivary chromosomes of *Drosophila*, and in the case of mitosis of eucaryotic cells.

This juxtaposition of identical double-stranded DNA sequences is probably supported by the binding of various specific and nonspecific binding proteins to the DNA. These proteins are called juxtaposition proteins. In pSLA 2, 9 palindromic structures are detected at the terminal half of the repetition sequences (see Fig. 9). One of their functions could be the binding of specific proteins, which might act to make the 2 double-stranded DNAs stick together. An example of such cohesive proteins is observed in the terminal protein of the adenovirus DNA. The proteins are reported to have the characteristic of sticking to each other (12).

If the postulation of such a 4-stranded structure can be admitted, racket frame-like DNA structure which is presented in Fig. 10 is deduced for the DNA possessing the sequences of terminal inverted repeats. The function of such a DNA is not known. However, the inverted terminal repetition sequences are common in transposable elements, in adenoviruses, in the proviral structure of some viruses (14), in the transposable element of the foldback type in *Drosophila* cells (15), and in the maize S1 and S2 linear plasmids (8). A common feature of these DNA elements is their capability to insert themselves into the chromosomal or mitochondrial DNA. The sharing of such a common nature suggests that the racket-frame structure of double-stranded DNA is such that it inserts and integrates into other double-stranded DNAs. In the case of pSLA 2, attempts to detect such integration are underway.

Hairpin Structures at Termini of Inverted Repeats

From computer data analysis of the possible secondary structure of pSLA 2 DNA termini from 1 to 70 bp, several hairpin structures can be depicted on the basis of a low free-energy requirement, which can easily be formed, perhaps by the binding of proteins. This was accomplished by the analysis of matrices constructed to pair each

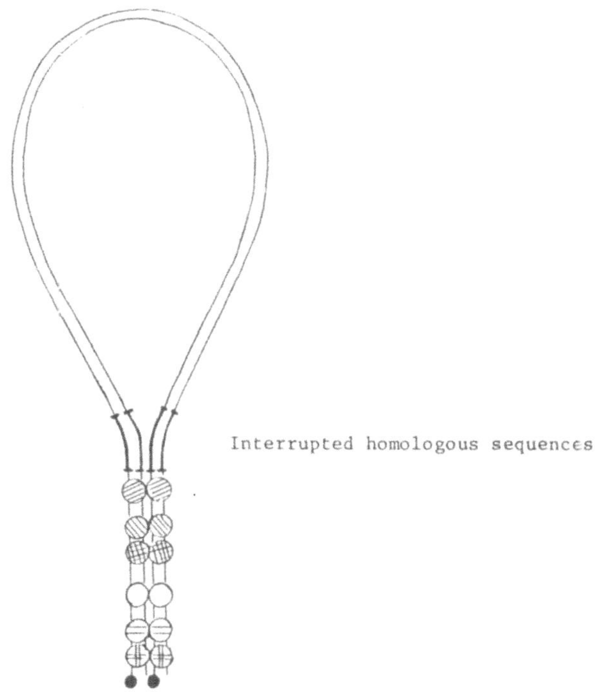

Fig. 10. Racket frame model.

● The terminal protein which initiates DNA replication and has itself the cohensive nature.

Juxtaposition proteins.

○ The proteins which recognize and bind to the specific DNA sequences and stock together.

nucleotide of a sequence in order with every other nucleotide. To calculate free energies of base pairing, the rules of Tinoco et al. (16) were followed, as modified to account for the effects of neighboring pairs (17). Two secondary structures formed from the 5'-terminal repeats are depicted in Fig. 11. Fig. 11A and 11B show the 2 most energetically favorable structures. Relative to an unpaired single DNA strand, the free energies of formation of these base paired structures are -52.1 Kcal/mole and -43.2 Kcal/mole, respectively. Given the number of nucleotides in the repeat sequences and the numbers of G-C pairs within the folded structures, 2 energies indicate a high degree of stability of the structures shown, and they could be expected to form spontaneously in single-stranded sequences at the termini.

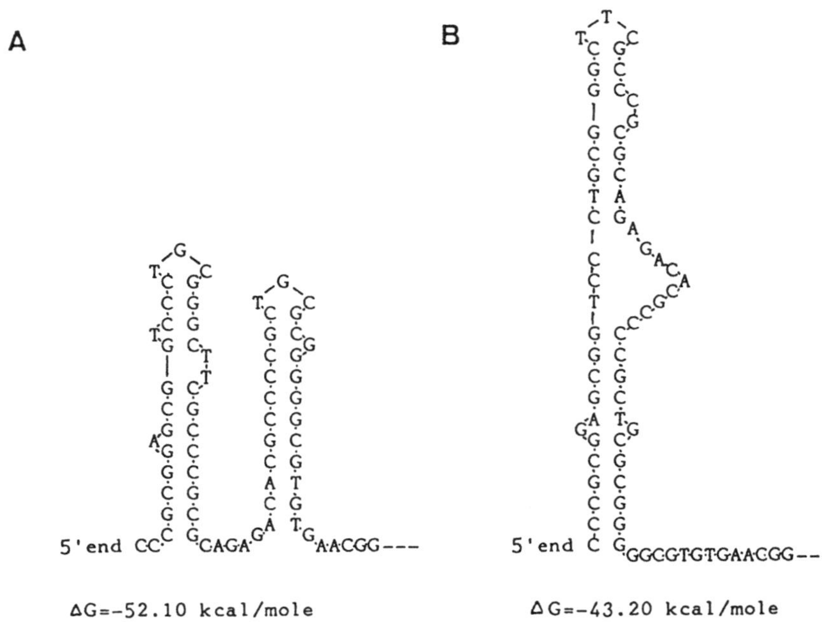

Fig. 11. Hairpin structures at termini of inverted repeats.

DISCUSSION

Various specific structures such as cohesive ends, direct or inverted repeats, or hairpins are usually present at the terminal region of linear DNA genomes (18). We have shown that the structure of the ends of pSLA 2 is similar to that of adenovirus DNA and Bacillus subtilis phage ϕ29 DNA: (a) these genomes have the inverted terminal repetitions and the terminal protein covalently linked to each 5' end; (b) the linkage between the genome and the terminal protein is alkali-labile, and 5' phosphate is released after alkali-treatment; (c) the ends of the genomes are blunt and contain free 3' OH. The terminal protein appears to function as a primer for DNA polymerase in adenovirus (12,19) and ϕ29 (20,21,22). This type of priming is called protein priming.

We have cloned intact termini of pSLA 2 after removal of the terminal protein with alkaline treatment. Using the same procedure, we have also succeeded in cloning the termini of pSLA 1 DNA, which showed an almost identical restriction pattern to that of pSLA 2, but had a small insertion or deletion of less than 100 bp in the nonterminal area. The nucleotide sequence of the cloned termini of pSLA 2 has been determined and an inverted terminal repetition of 614 bp long has been found. This is longer than that found in adenovirus, ϕ 29, and maize S1 or S2 plasmids (Lonsdale et al., 1981). The terminal sequence of pSLA 2 plasmid is highly homologous up to about 800.

LINEAR PLASMIDS FROM KLUYVEROMYCES LACTIS

Most strains of Saccharomyces cerevisiae contain the plasmid 2-μm DNA, a class of covalently-closed circular DNA with a monomeric contour length of approximately 2 μm (23). A similar type of plasmid has been found in S. italicus (24) and recently in other yeasts [see review by Gunge (6)]. We changed the procedure of detecting plasmids from routinely employed CsCl-EtBR gradient centrifugation to the electrophoretic method, as applied for the discovery of linear plasmids from Streptomyces rochei.

In looking among 70 yeast strains of various genera and species, in the petite negative yeast Kluyveromyces lactis IFO 1267, synonymous to K. lactis CBS 2359 and K. Lactis ATC 5356, we found 2 unique double-stranded linear DNA species, pGKl 1 of 8.9 kb and pGKl 2 of 13.4 kb, in 50-100 copies per haploid cell (5,6).

General Properties and Sequence Analysis of pGKl Plasmid DNA

Both pGKl plasmids stably coexist in K. lactis cells and show cytoplasmic inheritance. The densities of both plasmids are identical (1,687 g/cm^3), lower than those of the mitochondrial (1,692 g/cm^3) and nuclear (1,699 g/cm^3) DNAs.

The terminal structures of linear plasmids are of considerable interest, especially as to the DNA replication since DNA polymerase requires a primer to initiate replication. Several lines of evidence show that linear DNA molecules have unique terminal structures such as hairpin termini, palindromic termini, repetitive sequences, and protein association. Yeast chromosome ends (telomeres) are cross-linked and have repetitive sequences. Sequence analysis showed that pGKl 1 and pGKl 2 have repetitive unique ends. The termini consist of A-T rich inverted repeats of 202 bp in pGKl1 and 184 bp in pGKl 2, as shown in Fig. 12 (F. Hishinuma, pers. comm.). These inverted terminal sequences were also reported by Sor et al. (25). There are several short dispersed repeats, either direct or inverted, but no internal palindromic sequence in the terminal area. Also, there is no obvious sequence homology between pGKl 1 and pGKl 2 in the area.

The total sequence of the shorter pGKl 1 plasmid which is responsible for the production of the killer toxin was found to be 8,876 bp with 5 open reading frames (H. Hishinuma et al., pers. comm.) (Fig. 13).

Protein Toxin Produced from pGKl 1 Plasmid

K. lactis cells harboring pGKl plasmids secrete a toxic protein to kill sensitive strains over a wide range of yeasts, including Kluyveromyces, Saccharomyces, Candida, and Torulopsis, whereas the plasmid-containing cells are immune to the toxin. The purified toxin is composed of 2 subunits of proteins, one 27 kd and one over

```
pGKl 2
                                                  AZuI
Left                                               ↓                                60
        AAAAGGTATA  TATATAAAAA  ATAAAGTTGG  GTTTTTAAGC  TAATAAAAGT  TGAAATCAGG
Right
                                                                         Taq I
Left                                                                  Xho I ↓↓
        TCAAAAAAAA  GCAACCCAAA  GTATGTTTTA  CGTGTTTTTT  TATGTTTAGG  TCATATCTCG
Right

Left      Rsa I       Hpa II   Hinf I                                          180
        AGTACGCCCT  TTTTCGCCGG  AGTCAATTAG  GTCATACTTT  TCTATATAAT  CCAAATCCCA
Right
                                                                               240
Left        ATCAGT  TATAGTAGCA  TACAGTGTTA  AATGTATATT  TAATGTTACA  TAATAATTTA
        AAAA
Right       TCAATT  GAATGATTCT  TAATATGATT  TAATAGTTTA  TGATTATAAA  TGTCTTATAT
                      Hinf I                                              Cla I

pGKl 1

Left                          Rsa I                          Rsa I  Mbo II  60
        ACACATAACA  TAGGGGAGAG  TACTAAAAGT  GAGATTATTG  GAAGATTAGT  ACGTCTCCAT
Right

Left                                                                           120
        TTTTTTCTGT  TTTTTTGTTT  TTATATATTA  GGTTATTTTT  TTTCAGTTTT  ATATCAACTC
Right

Left                                                                           180
        TGTATAACAA  GTCTATTTTT  TTATATTTTA  AGTCTATTTT  ACACTTTTGA  CCTATAAGTC
Right
                                                                               240
Left                              ACTATAAT  ATATGAATTA  CATTATTAAT  TTAAAAATGG
        ATTTTATTAT  ACACATTTTC  CA
Right                             TTCTGTAG  ATTATTCATA  CTATCTATTT  TTTCAGTACA
                                                                        Rsa I
```

Fig. 12. Inverted terminal repetitive sequences of pGKl 1 and pGKl 2.

80 kd (26). The subunits are further characterized as containing another polypeptide of 24 kd (T. Fujimura, pers. comm.). The purified toxin protein blocks the growth of the sensitive cells of *S. cerevisiae* by inhibiting the adenylate cyclase activity in the membrane fraction (27) (Fig. 14). This is the first indication among the bacteriocin- type protein toxins of blockage of adenylate cyclase in the membrane. It is to be noted that the dsRNA encoded *S. cerevisiae* toxin is a simple protein of 11,740 d and distorts the ion transport of yeast membranes (28).

Genetic Analyses

Intergeneric transfer of both pGKl 1 and pGKl 2 into *S. cerevisiae* was achieved either by protoplast fusion (29) or by

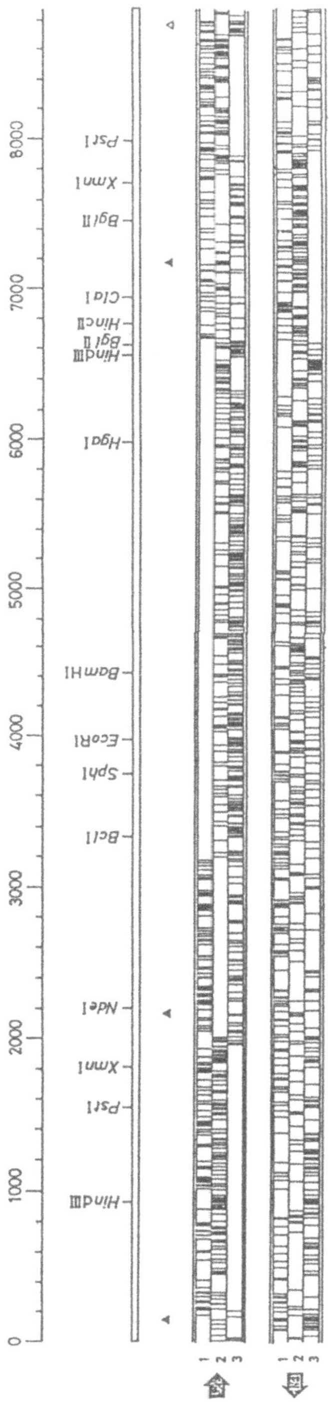

Fig. 13. Open reading frames present in pGK1 linear plasmid DNA.

transformation (30). The transferred plasmids exist in the cytoplasm and confer the killer and immunity characteristics on the S. cerevisiae host cells. Cytoplasmic nonkiller mutants were isolated from K. lactis and S. cerevisiae by spontaneous or induced mutation.

The deletion mutants we have obtained are shown in Fig. 15. The pGKl 1S which deleted the central part of pGKl 1 lost the killer activity, but the immunity against the killer protein was present. The other 2 mutants which had only left part of the plasmid, singly or doubled, lost both activities. Therefore, the immunity gene is located on the right side.

The pGKl 2 is likely to be necessary for the replication of both plasmids because the non-killer mutants which deleted only pKGl 1 was found in K. lactis, but never vice versa. Thus, the pGKl killer system resembles the dsRNA killer system of S. cerevisiae where the smaller (M)dsRNA encodes for both killer and immunity phenotypes, whereas the larger (L) dsRNA is necessary for replication or maintenance of both RNA plasmids.

Neither pGKl 1 nor pGKl 2 hybridizes with the nuclear and mitochondrial DNAs of K. lactis (31).

Incompatibility with Mitochondrial DNA

In the experiment of intergeneric transfer, the pGKl plasmids are stably maintained in mitochondria deficient (ρ^-) cells of S. cerevisiae (29,30). If, however, wild-type mtDNA (ρ^+) is introduced from a wild-type strain by cytoduction, the killer DNAs become unstable and are frequently lost during cell growth. This phenomenon appears to be ascribable to the incompatibility of pGKl plasmids with the incoming mtDNA, because the plasmid stability could be restored in some mitochondria-deficient cells induced by ethidium bromide treatment (29). Several lines of evidence suggest that the incompatibility could be due either to the competition between mitochondria and the plasmids for their DNA replication, or to interference of mitochondria with plasmid transmission during cell division. The pGKl plasmids were able to be introduced from K. lactis into Candida pseudotropicalis and Kluyveromyces fragilis by protoplast fusion or rare mating. In these experiments, the plasmid instability was again observed in the taxonomically distant Candida yeast, but not in the more related yeast K. fragilis. This suggests a relationship between mitochondrial phylogeny and incompatibility.

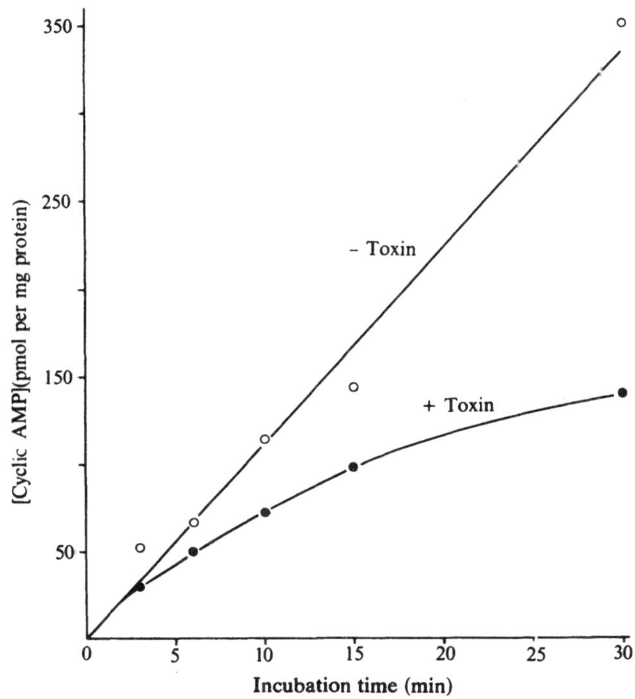

Fig. 14. Effect of K. lactis toxin on the activity of adenylate cyclase in a membrane fraction prepared from a sensitive strain of S. cerevisiae.

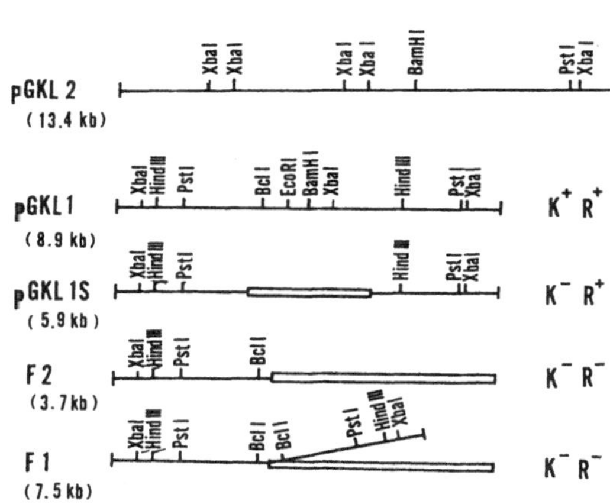

Fig. 15. Maps of pGKl 2, pGKl 1, and deletion mutants. The boxed areas are deleted.

REFERENCES

1. Hayakawa, T., N. Otake, H. Yonehara, T. Tanaka, and K. Sakaguchi (1979) Isolation and characterization of plasmids from _Streptomyces._ _J. Antibiotics_ 32:1348-1350.
2. Hayakawa, T., T. Tanaka, K. Sakaguchi, N. Otake, and H. Yonehara (1979) A linear plasmid-like DNA in _Streptomyces_ sp. producing lankacidin group antibiotics. _J. Gen. Appl. Microb._ 25:255-260.
3. Hirochika, H., and K. Sakaguchi (1982) Analysis of linear plasmids isolated from _Streptomyces_: Association of protein with the ends of the plasmid DNA. _Plasmid_ 7:59-65.
4. Hirochika, H., and K. Sakaguchi (1984) A linear DNA plasmid from _Streptomyces rochei_ with an inverted terminal repetition of 614 base pairs. _EMBO J._, Vol. 3, No. 4:761-766.
5. Gunge, N., A. Tamaru, F. Ozawa, and K. Sakaguchi (1981) Isolation and characterization of linear deoxyribonucleic acid plasmids from _Kluyveromyces lactis_ and the plasmid associated killer character. _J. Bact._ 145:382-390.
6. Gunge, N. (1983) Yeast DNA plasmids. _Ann. Rev. Microb._ 37:253-276.
7. Pring, D.R., C.S. Levings III, W.W.L. Hu, and D.H. Timothy (1982) Unique DNA associated with mitochondria in the "S" type cytoplasm of male-sterile maize. _Proc. Natl Acad. Sci., USA_ 74:2904-2908.
8. Levings, C.S., and R.R. Sederoff (1983) Nucleotide sequence of the S2 mitochondrial DNA from the S-cytoplasm of maize. _Proc. Natl. Acad. Sci., USA_ 80:4055-4059.
9. Francou, F. (1981) Isolation and characterization of a linear DNA molecule in the fungus _Ascobolus immersus_. _Mol. Gen. Genet._ 184:440-444.
10. Ito, J. (1978) Bacteriophage ϕ29 terminal protein: Its association with the 5' termini of the ϕ29 genome. _J. Virol._ 28:895-904.
11. Mellado, R.P., and M. Salas (1983) Initiation of phage ϕ29 DNA replication by the terminal protein modified at the carboxyl end. _Nucleic Acid Res._ 11:7397-7407.
12. Chalberg, M.D., S.V. Desiderio, and T.J. Kelly, Jr. (1980) Adenovirus DNA replication _in vitro_. _Proc. Natl. Acad. Sci., USA_ 77:5105-5109.
13. Maxam, A.M., and W. Gilbert (1980) Sequencing end-labelled DNA with base-specific chemical cleavages. _Meth. Enzymol._ 65:499-560.
14. Varmus, H.A. (1982) Form and functions of retroviral proviruses. _Science_ 216:812-827.
15. Potter, S.S., M. Truen, M. Phillips, and A. Mahler (1980) _Cell_ 20:639-647.
16. Tinoco, Jr., I., O.C. Uhlenbeck, and M.D. Levine (1971) _Nature_ 230:362-367.
17. Tinoco, Jr., I. et al. (1973) Improved estimation of secondary structure in ribonucleic acid. _Nature_ 246:40-41.

18. Kornberg, A. (1980) DNA replication (Freeman, San Francisco)
19. Rekosh, D.M.K., W.C. Russel, A.J.D. Bellet, and A.J. Robinson (1977) Identification of a protein linked to the ends of adenovirus DNA. Cell 11:283-295.
20. Watabe, K. et al. (1982) In vitro replication of bacteriophage ϕ29 DNA. Proc. Natl. Acad. Sci., USA 79:5245-5248.
21. Penalva, M., and M. Salas (1982) Initiation of phage ϕ29 DNA replication in vitro: Formation of a covalent complex between the terminal protein, p^3, and 5'-dAMP. Proc. Natl. Acad. Sci., USA 79:5522-5526.
22. Blanco, L., et al. (1983) Factors involved in the initiation of phage ϕ29 DNA replication in vitro: Requirement of the gene 2 product for the formation of the protein P^3-DNA complex. Nucleic Acid Res. 11:1309-1323.
23. Clark-Walker, G.D. (1972) Isolation of circular DNA from a mitochondrial fraction from yeast. Proc. Natl. Acad. Sci., USA 69:388-392.
24. Jazwinski, S.M., and G.M. Edelman (1979) Replication in vitro of the 2μm DNA plasmid of yeast. Proc. Natl. Acad. Sci., USA 76:1223-1227.
25. Sor, F., M. Wesolowski, and H. Fukuhara (1983) Inverted terminal repetitions of the 2 linear DNA associated with the killer character of the yeast Kluyveromyces lactis. Nucleic Acids Res. 11:5037-5044.
26. Sugisaki, Y. et al. (1984) Characterization of novel killer toxin encoded by a double stranded linear DNA plasmid of Kluyveromyces lactis. Eur. J. Biochem. (in press).
27. Sugisaki, Y., N. Gunge, K. Sakaguchi, M. Yamasaki, and G. Tamura (1983) Kluyveromyces lactis killer toxin inhibits adenylate cyclase of sensitive yeast cells. Nature 304:464-466.
28. Bostian, K.A., J.E. Hopper, D.T. Rogers, and D.J. Tipper (1980) Translational analysis of the killer-associated virus-like particle dsRNA genome of S. cerevisiae: M dsRNA encodes toxin. Cell 19:403-414.
29. Gunge, N., and K. Sakaguchi (1981) Intergeneric transfer of DNA killer plasmids, pKGl 1 and pKGl 2, from Kluyveromyces lactis into Saccharomyces cerevisiae by cell fusion. J. Bact. 147:155-160.
30. Gunge, N., K. Murata, and K. Sakaguchi (1982) Transformation of Saccharomyces cerevisiae with linear DNA killer plasmids from Kluyveromyces lactis. J. Bact. 151:462-464.
31. Wesolowski, M., P. Dumazert, and H. Fukuhara (1982) Killer DNA plasmids of the yeast Kluyveromyces lactis. II. Restriction endonuclease maps. Curr. Genet. 5:199-203.

CONJUGAL PLASMID TRANSFER IN *BACILLUS THURINGIENSIS*

John S. Chapman and Bruce C. Carlton

Department of Molecular and Population Genetics
University of Georgia
Athens, Georgia 30602

INTRODUCTION

Bacillus thuringiensis (BT) is a gram-positive, sporeforming bacterium distinguished by its ability to produce parasporal protein crystals. These proteins, collectively known as δ-endotoxins, are toxic to a number of lepidopteran insects of economic importance as well as to certain dipterans which are important vectors of human disease (7). Over 20 different varieties of *B. thuringiensis* are currently recognized on the basis of flagellar antigens (6) and immunological (19) and toxicity (8) properties of the parasporal crystals.

Strains of *B. thuringiensis* typically have complex arrays of extrachromosomal plasmids, in some instances up to 11 plasmids in a single strain (11). Genetic analysis of several parental strains and cured derivatives has demonstrated a correlation between specific plasmids and δ-endotoxin production (12), the implicated plasmid being a different size in different varieties. The discovery that toxin plasmids could be transmitted at high frequencies between crystal-producing (Cry^+) and acrystalliferous (Cry^-) strains during growth in mixed culture (13) confirmed the localization of genes for toxin production on the same specific plasmids implicated from curing studies (14,15).

Conjugal transfer of plasmids is not limited to *Escherichia coli* and other gram-negative bacteria. It is a well-documented occurrence in such gram-positive genera as *Streptococcus* (9), *Clostridium* (2), and *Streptomyces* (16), and there are indications that it may occur in *Bacillus subtilis* (18,20). Conjugal proficiency and cell clumping are induced in the *Streptococcus faecalis* system by exposure of donor cells to small peptides ("sex pheromones") produced by recipient cells (10). Several peptides are produced by recipient cells, each being specific for induction of transfer of a particular plasmid.

Preliminary experiments in our laboratory (13) had suggested that the natural mechanism for plasmid transfer among *B. thuringiensis* strains was due to a conjugation-like process. This work confirms and extends those findings using as a model system a 75-Mdal toxin-coding plasmid from strain HD2, a serotype 1 prototype strain of *B. thuringiensis*.

METHODS

Strains

The strains used in this study are listed in Tab. 1, together with their relevant phenotypes and plasmid complements. Strains were either obtained from J.M. González of our laboratory, or derived in this work. Strains resistant to 50 μg/ml nalidixic acid or 12 μg/ml rifampicin were generated by ethylmethanesulfonate mutagenesis (3) followed by selection on antibiotic-supplemented nutrient agar. Strain HD2 encodes the toxin genes on a transmissible 75-Mdal plasmid.

Media and Materials

Strains were maintained on slants of nutrient salts agar and routinely grown in nutrient broth. Nutrient salts medium consisted

Tab. 1. Strains used in these studies and their properties.

Strain[a]	Phenotype; plasmid complement[b]	Reference
HD2 *B. thuringiensis* var. *thuringiensis*; serotype 1	wild-type; 150, 75, 57, 54, 37, 32, 7.6, 7.2, 6.2, 5.2	González et al. (14)
HD2-13	crystal-positive; 75, 54, 32, 7.6, 7.2, 6.2, 5.2	González et al. (14)
HD2-62	$strep^r$, crystal-negative; 54, 32, 7.6, 7.2, 6.2, 5.2	González et al. (14)
HD2-106	nal^r, $strep^r$, crystal-negative; 54, 32, 7.6, 7.2, 6.2, 5.2	This work
HD2-107	rif^r, $strep^r$, crystal-negative; 54, 32, 7.6, 7.2, 6.2, 5.2	This work
HD2-108	nal^r, crystal-positive; 75, 54, 32, 7.6, 7.2, 6.2, 5.2	This work

[a] Strains HD2-13 and HD2-62 are derived from HD2 by curing spontaneously and with $NaDODSO_4$; the antibiotic resistance markers were generated via ethyl methane-sulfonate mutagenesis.

[b] Plasmid complement is indicated in megadaltons. Abbreviations: $strep^r$, streptomycin-resistant; nal^r, nalidixic acid-resistant; rif^r, rifampicin-resistant

of 8.0 g/l nutrient broth, 15.0 g/l Bacto-agar, 0.1 g/l $CaCl_2$, 10.0 mg/ml $MnCl_2 \cdot 4H_2O$, 1.0 ml/l 1.0 M $MgCl_2 \cdot 6H_2O$; SCG medium, used for growing cells for agarose gel electrophoresis was as previously described (12), as were the SP I and SP II media used for transformation (1).

Bacto-agar, nutrient broth, and casamino acids were obtained from Difco. Antibiotics and enzymes were purchased from Sigma.

Plasmid Transfer Conditions

The basic protocol for plasmid transfer was that of González et al. (14), modified for specific experiments. A Cry^- streptomycin-sensitive donor strain and a Cry^- streptomycin-resistant recipient were grown separately in nutrient broth in a shaker incubator for 14-20 hr at 30°C. These stationary phase cultures were diluted 1:50 into fresh nutrient broth and incubated with shaking for 3-4 hr. Matings were then performed in nutrient broth by mixing 10 µl/ml of donor and recipient cultures, usually in a final volume of 2.0 ml, for 8-20 hr. The recipient population was selected by streaking onto streptomycin-supplemented nutrient salts agar.

The DNase I experiments were performed as above with 2 modifications. First, mixed cultures received 40 µl/ml DNase I or an equal volume of distilled water, and second, matings were performed in SP II medium whenever the used medium was to be employed in a *B. subtilis* transformation experiment to monitor DNase I activity. Transformation was performed according to Anagnostopoulos and Spizizen (1). The transforming DNA was pBC16, a 3-Mdal plasmid coding for tetracycline resistance.

U-tube experiments were done in a commercially available apparatus purchased from Bellco Biological Glassware, Vineland, NJ. The final volume of the mating mix was 10 ml, with the donor and recipient strains being inoculated at a concentration of 10 µl/ml each. Filters with a 0.45 µ pore size (Millipore Corp.) were used to separate the chambers containing donor and recipient cultures.

In matings where the effects of inhibitors were studied, the inhibitor-sensitive member of the mating pair was inoculated at a 10-fold higher concentration than the inhibitor-resistant mating pair member and the matings performed for 20 hr. Streptomycin sulfate was used at a concentration of 100 µg/ml; nalidixic acid, at 50 µg/ml; and rifampicin, at 12 µg/ml. Mating cultures received 8 µl/ml of inhibitor solution or distilled water.

To determine the effects of initial cell density on mating, 1.0 ml and 100 µl aliquots of donor and recipient cultures were centrifuged for 1.0 min in a microcentrifuge (Model 235A, Fisher Scientific Co.) and resuspended to the original volume in fresh nutrient broth. This wash was designed to eliminate media carryover effects during mating. Donor and recipient strains were inoculated into nutrient broth (final volume 10.0 ml) at initial concentrations of 10 µl/ml and 100 µl/ml each, respectively.

Conditioned media for use in mating experiments were prepared by growing mixed or axenic cultures for 8 hr, removal of the cells by centrifugation, and sterilization of the media by filtration. In some of these experiments, overnight cultures were treated by diluting them into media consisting of 50% conditioned media and 50% fresh nutrient broth. Donor and recipient inocula were centrifuged in a microcentrifuge, resuspended in fresh nutrient broth, and inoculated to the appropriate medium at 100 µl/ml each in a final volume of 10.0 ml. In some experiments the conditioned media were supplemented with 5X nutrient broth.

To determine donor mating capability, logarithmically growing cultures of donor and recipient cells were diluted 1:10 into distilled water and were vortexed vigorously. Cell counts were performed microscopically using a hemacytometer, and replicated 3 times. The cultures were then further diluted to obtain donor concentrations of less than 10 cells/10 µl and recipient concentrations of 50 cells/10 µl. Recipient cells were present at a higher density to avoid the possibility of a recipient conjugating with more than one donor cell. Actual donor and recipient cell numbers were determined by spotting 10 fractions of 10 µl each onto nutrient agar, incubating for 8-12 hr, and counting with the aid of a dissecting microscope. Parallel matings were performed using 10 µl inocula of the diluted donor and recipient cultures in a volume of 1.0 ml and incubating for 18-20 hr. The mating cultures were diluted and plated onto nutrient salts medium with and without streptomycin to determine final donor and recipient cell numbers. After 48 hr of incubation at 30°C, colonies on the plates were counted and the frequency of Cry^+ transconjugants determined.

Determination of Plasmid Transfer Frequencies, Cell Numbers, and Growth

At the termination of each mating experiment, samples were streaked to antibiotic-supplemented nutrient salts plates and incubated at 30°C for 48 hr. Frequencies of Cry^+ transconjugants were determined by microscopy of randomly selected colonies or by visual comparison of colony morphologies. Cry^+ colonies are whiter and more glossy than adjacent acrystalliferous colonies. The random screening method generally results in higher frequencies of Cry^+ transconjugants, perhaps due to an unconscious bias in selecting colonies.

Growth of mixed and axenic cultures was monitored in 25-ml side arm flasks by following changes in culture turbidity with a Klett-

Summerson colorimeter fitted with a red (660 nm) filter. Cell numbers were determined by direct microscopic counting using a hemacytometer or by serial dilution and plating on appropriate media.

RESULTS

Nature of the Plasmid Transfer System

The plasmid transfer process displays a requirement for cell-cell contact, as shown by the results of U-tube experiments (Tab. 2). Separation of the 2 chambers of the U-tube with a 0.45 μ filter eliminated transfer of the 75-Mdal plasmid, while in U-tubes without filters approximately 50% of the streptomycin-resistant recipients (strain HD2-62) acquired the crystal-producing phenotype associated with the 75-Mdal plasmid.

Transfer of the 75-Mdal plasmid between crystal-producing and acrystalliferous strains of HD2 was insensitive to the action of 40 μg/ml DNase I in 20 hr matings (Tab. 2). DNase I in filtered, used media was still active after this period of time, as shown by its ability to eliminate transformation of *Bacillus subtilis* MI 112 by pBC16 to tetracycline-resistance.

Plasmid transfer experiments using mutagenized derivatives of strains HD2-13 and HD2-62 resistant to inhibitors of DNA, RNA, and protein synthesis (nalidixic acid, rifampicin, and streptomycin, respectively) were conducted to determine the possible roles of these

Tab. 2. Effects of DNase I and cell-cell contact on plasmid transfer.

Mating pair[a]	Condition	Frequency of cry^+ transconjugants	n[c]
HD2-13 x HD2-62	control	48.7 ± 1.32	6
HD2-13 x HD2-62	+DNase I	47.2 ± 3.9	4
HD2-13 x HD2-62	u-tube without filter	49.8 ± 2.9	4
HD2-13 x HD2-62	u-tube with filter	0	3

[a] HD2-13 is a crystal-producing, streptomycin sensitive derivative of HD2; HD2-62 is crystal-negative, streptomycin resistant derivative.

[b] Frequency was determined by microscopic examination of 30-50 colonies per determination and are presented as the means plus and minus the standard error of the mean.

[c] n = number of determinations.

processes in plasmid transfer. If the role of the donor strain is only to provide DNA (via cell lysis) for a transformation-like process, the appearance of Cry$^+$ phenotype in the inhibitor-resistant recipients should be insensitive to inhibition of macromolecular synthesis in the donor strain. In these experiments, the donor strains were initially present at a 10-fold higher concentration than the recipient strain to counter the lack of growth of the sensitive donor strains in the presence of the inhibitors. Table 3 indicates that nalidixic acid, rifampicin, and streptomycin eliminated transfer of the Cry$^+$ phenotype. Thus, it appears the donor must continue to synthesize DNA, RNA, and protein in order for transfer to occur. Reciprocal experiments using a nalidixic acid-resistant donor (line 8) or when both donor and recipient were resistant (line 10) indicate that it is necessary for the recipient to synthesize DNA (as expected if repliconation is to occur) and that elimination of transfer by nalidixic acid was due to the inhibitor-sensitive phenotype of the nalidixic acid-sensitive member of the mating pair. The insensitivity of plasmid transfer to DNase I, the requirement for cell-cell contact, and the necessity for DNA, RNA, and protein synthesis are all consistent with the hypothesis that transfer of the 75-Mdal plasmid occurs via conjugation.

Tab. 3. Effects of inhibitors of macromolecular synthesis on plasmid transfer.

Mating pair[a]	Condition	Frequency of cry$^+$ transconjugants	n[c]
HD2-13 x HD2-62	control	48.7 ± 1.32	6
HD2-13 x HD2-62	+ streptomycin	0	3
HD2-13 x HD2-106	control	33.0 ± 4.0	3
HD2-13 x HD2-106	+ nalidixic acid	0	3
HD2-13 x HD2-107	control	36.4 ± 2.2	3
HD2-13 x HD2-107	+ rifampicin	0	3
HD2-108 x HD2-62	control	39.0 ± 3.2	3
HD2-108 x HD2-62	+ nalidixic acid	0	3
HD2-108 x HD2-106	control	36.01 ± 1.8	3
HD2-108 x HD2-106	+ nalidixic acid	37.4 ± 2.2	3

[a] HD2-13 is crystal$^+$, streptomycin sensitive; HD2-62 is crystal-negative, streptomycin resistant; HD2-106 is crystal-negative, nalidixic acid resistant; HD2-107 is crystal-negative, rifampicin resistant; HD2-108 is crystal-positive, nalidixic acid resistant.

[b] Frequency was determined by microscopic examination of 30-50 colonies per determination and are presented as means plus and minus the standard error of the mean.

[c] n = number of determinations.

Characterization of the Conjugal System

The kinetics of the appearance of Cry^+ transconjugants parallels that of the logarithmic growth of a mixed population of donor and recipient cells. Occasional Cry^+ transconjugants could be found in the lawn of a sample streaked from broth after 2 hr of mixed growth. The frequencies of Cry^+ transconjugants became measurable after 3-4 hr of mixed growth and attained final frequencies of 30-40%.

Conjugal transfer of the 75-Mdal plasmid was independent of the initial cell density. The appearance and final frequency of transconjugants was identical in cultures receiving either the usual inocula (2-4 x 10^4 cells/10µl of both donor and recipient) or 10 times the usual inocula (Fig. 1). Two implications may be drawn from these data: (i) that cell-cell contact is necessary but not sufficient for conjugation, and (ii) that one or both members of the mating pair must undergo a growth-related alteration in order to become conjugation proficient.

To lend credence to the idea that a growth-phase related change must take place for conjugation to occur, short matings (4 hr) were performed by mixing donor and recipient cells initially at different phases of the growth cycle. Strains HD2-13 and HD2-62 have identical growth rates in nutrient broth. Logarithmically growing cells show little or no lag when transferred to fresh nutrient broth, while stationary phase cells exhibit an extended lag of 2-3 hr before commencing logarithmic growth (Fig. 2). Thus, a culture established from logarithmically growing inocula can be considered to be in mid-log at the end of a 4-hr mating, while one derived from a stationary inoculum will be in early log phase after 4 hr.

When both the recipient and donor inocula were in log phase, transconjugants began to appear at 2 hr and their frequency continued to increase throughout the course of the experiment (Tab. 4). When both donor and recipient inocula were in stationary phase, no Cry^+ transconjugants were found after 4 hr. If the donor inoculum was in the stationary phase and the recipient inoculum was in log phase, again, no transconjugants were found after 4 hr. However, if the donor inoculum was in log phase and the recipient inoculum was in stationary phase, transconjugants began to appear after 3 hr. At this point the recipient cells were entering log phase and the donor cells were in late log phase. These results are interpreted as indicating that recipient cells can receive plasmids while in early log phase, while donor cells must be in mid- to late-log phase before they become conjugation proficient.

Formation of cell aggregates of 2 to several hundred cells (cell clumping) occurs during the growth of mixed or axenic cultures

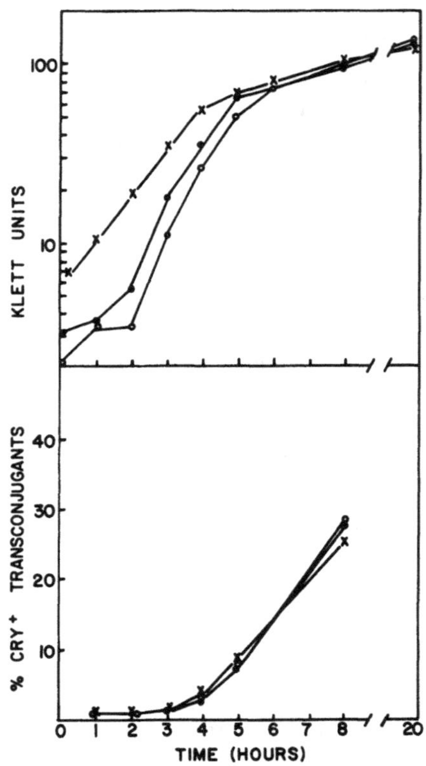

Fig. 1. Effect of initial cell density on growth of mixed cultures and transfer of crystal-producing phenotype. The top panel details the growth of mixed cultures of strains HD2-13 and HD2-62. Cells were centrifuged and resuspended in fresh nutrient broth before being used as inocula. Symbols: (○), 10 µl/ml inoculum of both donor and recipient; (×) 100 µl/ml inoculum of both donor and recipient; (●), uncentrifuged control, 10 µl/ml inoculum of donor and recipient. The lower panel describes the kinetics of appearance of crystal-producing transconjugants during growth of mixed cultures, as described in Panel A.

of *B. thuringiensis*. The fraction of the population in clumps increases as the culture progresses through logarithmic growth to stationary phase, although the clumping is not as extensive as that seen in the *S. faecalis* system. The data in Tab. 5 show that clumping in *B. thuringiensis* is nonspecific. Media used for the growth of any of the strains of *B. thuringiensis* examined will induce cell clumping in any of the other *B. thuringiensis* strains examined, including *B. thuringiensis* var. *israelensis* which does not conjugate with any of the strains listed in the table. Suspending cells in fresh nutrient broth produces no clumping.

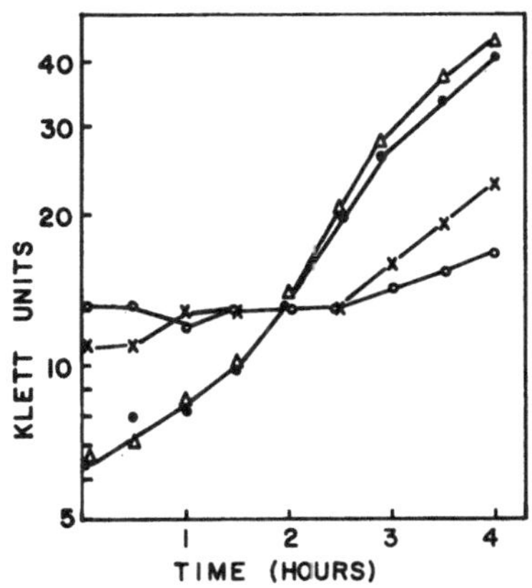

Fig. 2. Growth of strains HD2-13 and HD2-62 following inoculation from precultures at different growth stages. Symbols: (●), HD2-13, log inoculum; (○), HD2-13, stationary inoculum; (Δ), HD2-62, log inoculum; (×), HD2-62, stationary inoculum.

Pretreatment of either donor or recipient, or mixture of them with used media from donor, recipient, or mixed cultures prior to mating decreased transfer of the 75-Mdal plasmid to the recipient strain (Fig. 3). This effect is probably related to the increase in the length of the lag phase of the culture when transferred to the fresh broth used for the mating, most notably when both strains were treated.

Tab. 4. Effect of growth phase of donor and recipient inocula on plasmid transfer.

Growth Phase		Presence of cry$^+$ transconjugants[a]			
HD2-13	HD2-62	1 hr	2 hr	3 hr	4 hr
log	log	-	+	+	+
stationary	log	-	-	-	-
stationary	stationary	-	-	-	-
log	stationary	-	-	+/-	+

[a] Determined by microscopic examination of lawns of cells streaked onto nutrient salts medium supplemented with 100 μg/ml streptomycin sulfate. Several thousand cells were examined in each case; the experiment was repeated four times.

Tab. 5. Clumping of *Bacillus thuringiensis* cells by conditioned media.

Strain tested	Conditioned media from:				
	HD2	HD2-13	HD2-62	HD567-61	Control[b]
HD2	+	+	+	+	-
HD2-13	+	+	+	+	-
HD2-62	+	+	+	+	-
HD567-61	+	+	+	+	-

[a] Conditioned media were prepared by growing the indicated strains in nutrient broth for eight hours; followed by centrifugation to remove the cells and filtration to resterilize the media.

[b] Control was fresh nutrient broth.

When cells grown in fresh broth were mated in media used in previous matings, or in media in which either donor or recipient strains alone had been grown, growth of the mixed culture proceeded at a lower rate although final cell densities were comparable to those obtained in fresh broth (Fig. 3). Conjugal plasmid transfer was severely inhibited in conditioned media from both mixed and axenic cultures of strain HD2. Frequencies of Cry^+ transconjugants in these experiments rarely exceeded 2%. Addition of nutrients (in the form of concentrated nutrient broth) did not alleviate the inhibition of conjugal transfer.

The addition of trypsin, pepsin, α-chymotrypsin, carboxypeptidase, protease K, or *Staphylococcal* V8 protease to mating mixtures had no effect on conjugation. These data, combined with those from the conditioned media experiments, would seem to preclude the possibility of involvement of peptides analogous to the sex pheromones of *Streptococcus faecalis*.

Mating Capability

If the ability to conjugate is limited to a specific subset of the donor cell population, dilution of the donor culture to obtain very small inocula sizes (i.e., <10 cells) would increase the chance that some donor inocula in parallel matings would not contain conjugation-proficient cells. Ideally, the donor inocula size would be one cell, however, the tendency of *B. thuringiensis* towards filamentous growth and cell clumping make it difficult to attain this ideal. The 3 experiments presented in Tab. 6 were selected from a larger number of experiments because they involved the smallest donor inocula (approximately 10 cells). The range of progeny:donor

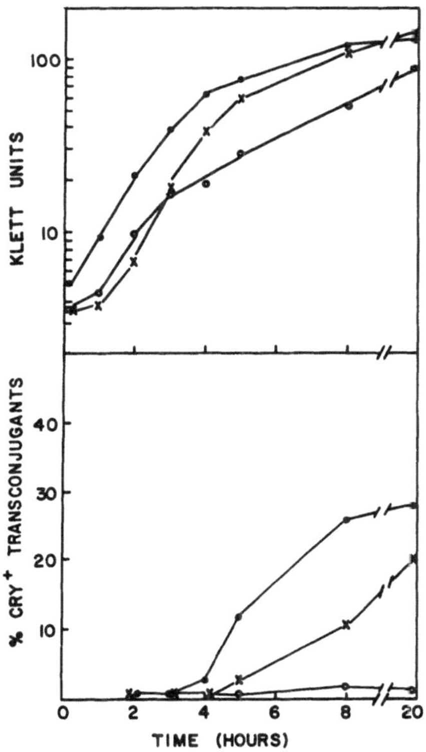

Fig. 3. Effect of conditioned media on growth of mixed cultures and transfer of crystal-producing phenotype. The top panel shows growth of mixed cultures of strains HD2-13 and HD2-62 under different conditions. Symbols: (●) derived from overnight cultures diluted into fresh broth, then mated in fresh broth; (×) derived from overnight cultures diluted into conditioned media, then mated in fresh broth; and (○) derived from overnight cultures diluted into fresh broth, then mated in conditioned broth. The lower panel describes the appearance of crystal-producing transconjugants under the same conditions described in Panel A.

ratios obtained in these experiments were similar to those obtained when using much larger inocula sizes (2-4 x 10^4 cells). In all experiments, progeny:donor ratios of less than one indicate that not all members of the donor population conjugated. Within experiments, the variance (a statistical measure of distribution about the mean) of progeny:donor ratios was very small compared to the mean, indicating the small segments of the population examined are not very different from the population as a whole. Thus, it seems likely that conjugal ability is not limited to a specific subset of the

Tab. 6. Variance of progeny:donor ratios.

Donor Inoculum[a]	Recipient Inoculum	Progeny:donor ratios[b]	Mean	Variance
11.5	40.2	6.7×10^{-2}, 6.6×10^{-2}, 8.3×10^{-2}, 1.1×10^{-1}, 7.2×10^{-2}, 8.2×10^{-2}, 5.5×10^{-2}	8.9×10^{-2}	5×10^{-4}
15	26	8.8×10^{-2}, 3.4×10^{-2}, 4.0×10^{-2}, 4.9×10^{-2}	5.2×10^{-2}	5.6×10^{-4}
9.6	40	3.1×10^{-1}, 2.2×10^{-1}, 2.1×10^{-1}, 3.0×10^{-1}	2.6×10^{-1}	8.2×10^{-3}

[a] Number of cells.

[b] Progeny:donor ratio is defined as the number of crystal positive transconjugants per donor cell after 20 hour matings.

donor population. Similar reasoning would apply to the recipient cells; experiments to determine recipient mating capability are in progress.

Transfer of Chromosomal Markers

In some conjugal systems (e.g., the F system of *E. coli*) mobilization of chromosomal genes by conduction occurs. In *B. thuringiensis* there are several possible instances of transmissible plasmids spontaneously integrating into the chromosome, and in some cases dissociating from the chromosome with and without size changes (Ref. 15 and J.M. González, pers. comm.). It is therefore reasonable to assume that chromosomal markers could be mobilized by the *B. thuringiensis* conjugal system. However, in numerous experiments designed to detect transfer of chromosomal markers at frequencies as low as 10^{-9}, no transmission of antibiotic-resistance markers (streptomycin, nalidixic acid, and rifampicin) or auxotrophic markers (histidine, tryptophan, and riboflavin) has yet been observed.

DISCUSSION

The data presented in this report lead to the conclusion that plasmid transfer in *B. thuringiensis* fits the classical definition of conjugation. It must be emphasized that this report has used, as a model system, the 75-Mdal toxin plasmid of strain HD2. However, transfer of many other plasmids in this strain and in other strains of B. thuringiensis has been shown to occur (13,14) and we have no evidence that transmission of other plasmids occurs by a mechanism different than that defined here. Transfer of the 75-Mdal toxin plasmid is insensitive to the action of DNase I and requires cell-cell contact. In addition, studies with inhibitors have indicated an active metabolic role for donor cells. Protein synthesis in the donor is required as indicated by the sensitivity of the process to streptomycin; the sensitivity of plasmid transfer to rifampicin and nalidixic acid imply that donor conjugal RNA and DNA synthesis is necessary (4). Likewise, the involvement of recipient conjugal DNA synthesis is implied by inhibition of plasmid transfer by nalidixic acid when a sensitive recipient strain is used.

Aggregates of cells are formed during both mixed and axenic growth of B. thuringiensis, which, in mixed culture, presumably facilitates conjugation. This cell clumping appears to be nonspecific, however. Unlike the conjugal system of S. faecalis (10), there seems to be no involvement of pheromone-like peptides. In fact, preexposure of cells to conditioned media or mating the cells in conditioned media inhibits conjugation, rather than inducing it as in the Streptoccocus system. Since the inhibition also occurs when the conditioned media is supplemented with concentrated nutrient broth, it seems unlikely the inhibition is due simply to reduced growth. Thus, the inhibition is most likely due to a compound or class of compounds produced during growth. Further investigations of the nature of this phenomenon are in progress.

The frequency of crystal-producing transconjugants achieved in this mating system is slightly higher than those observed in other conjugation systems. The cointegrate plasmid pAD1::Tn 917 transfers to up to 10% of the recipient cells in the S. faecalis conjugal system (10), while in the gram-negative bacterium Pseudomonas stutzeri, plasmid markers transfer to up to 23% of the recipient population (21). The high frequencies seen in the B. thuringiensis conjugal system could be the result of many donor cells undergoing infrequent conjugations, or a few promiscuous cells conjugating many times. If the ability to conjugate is limited to a specific subset of the donor population, differences in progeny:donor ratios should be amplified in parallel matings where the donor inoculum size is very small (approximately 10 cells). Progeny:donor ratios of less than one indicate that not all donor cells conjugate; however, the variances observed within experiments is very small compared to the mean. This evidence implies, but does not unequivocally demonstrate, that conjugal proficiency is not limited to a subset of the donor population. Thus, while it seems that all donor cells may be able to conjugate, not all of them do. Factors affecting progeny numbers have been shown to include collision rate between donor and recipient cells, the abilities of both donor and recipient to mate, segregation of the plasmid in dividing recipient cells, and retransfer of the plasmid by recipients (5).

Although considerable progress has been made in the investigation of the *B. thuringiensis* conjugal system, much remains to be done. Mobilization of chromosomal markers has not yet been observed by us despite numerous attempts; work is in progress with different markers and different conditions to determine if this can be detected. The location of the genes responsible for conjugation is not certain, although the 75-Mdal toxin plasmid is an excellent candidate. Determination of the range of strains able to exchange plasmids conjugatively, and of which *B. thuringiensis* plasmids are mobilizable is another topic currently being investigated. A more complete understanding of the detailed mechanism of this genetic exchange system and the resultant ability to manipulate it will increase the importance of *B. thuringiensis* as a vehicle for the genetic engineering of biopesticides.

REFERENCES

1. Anagnostopoulos, C., and J. Spizizen (1961) Requirements for transformation in *Bacillus subtilis*. *J. Bact.* 81:741-746.
2. Brefort, G., M. Magot, H. Ionesco, and M. Sebald (1977) Characterization and transferability of *Clostridium perfringens* plasmids. *Plasmid* 1:52-66.
3. Carlton, B.C., and B.J. Brown (1981) Gene mutation. In *Manual of Methods for General Bacteriology*, P. Gerhardt, ed. Am. Soc. Microbiol., Washington, D.C.
4. Clark, A.J., and G.J. Warren (1979) Conjugal transmission of plasmids. *An. Rev. Genet.* 13:99-125.
5. Cullum, J., J.F. Collins, and P. Broda (1978) Factors affecting the kinetics of progeny formation with F'lac in *Escherichia coli* K12. *Plasmid* 1:536-544.
6. de Barjac, H. (1981) Identification of H-serotypes of *Bacillus thuringiensis*. In *Microbial Control of Pests and Plant Diseases, 1970-1980*, H.D. Burges, ed. Academic Press, New York, pp. 35-43.
7. Dulmage, H.T. (1979) Genetic manipulation of pathogens: Selection of different strains. In *Genetics in Relation to Insect Management*, M.A. Hoy and J.J. McKelvey, Jr., eds. Rockefeller Foundation, New York, pp. 116-127.
8. Dulmage, H.T. and cooperators. (1981) Insecticidal activity of isolates of *Bacillus thuringiensis* and their potential for pest control. In *Microbial Control of Pests and Plant Diseases, 1970-1980*, H.D. Burges, ed. Academic Press, New York, pp. 193-222.
9. Dunny, G.M., B.L. Brown, and D.B. Clewell (1978) Induced cell aggregation and mating in *Streptococcus faecalis*: Evidence for a bacterial sex pheromone. *Proc. Natl. Acad. Sci., USA* 75:3479-3483.
10. Dunny, G.M., R.A. Craig, R.L. Carron, and D.B. Clewell (1979) Plasmid transfer in *Streptococcus faecalis*: Production of multiple sex pheromones by recipients. *Plasmid* 2:454-465.

11. González, Jr., J.M., and B.C. Carlton (1980) Patterns of plasmid DNA in crystalliferous and acrystalliferous strains of Bacillus thuringiensis. Plasmid 3:92-98.
12. González, Jr., J.M., H.T. Dulmage, and B.C. Carlton (1981) Correlation between specific plasmids and δ-endotoxin production in Bacillus thuringiensis. Plasmid 5:351-365.
13. González, Jr., J.M., and B.C. Carlton (1982) Plasmid transfer in Bacillus thuringiensis. In Genetic Exchange: A Celebration and a New Generation, U.N. Streips, S.H. Goodgal, W.R. Guild, and G.A. Wilson, eds. Dekker, New York. pp. 85-95.
14. González, Jr., J.M., B.J. Brown, and B.C. Carlton (1982) Transfer of Bacillus thuringiensis plasmids coding for δ-endotoxin among strains of B. thuringiensis and B. cereus. Proc. Natl. Acad. Sci., USA 79:6951-6955.
15. González, Jr., J.M., and B.C. Carlton (1984) A large transmissible plasmid is required for crystal toxin production in Bacillus thuringiensis variety israelensis. Plasmid 11:28-38.
16. Hopwood, D.A., and M.J. Merrick (1977) Genetics of antibiotic production. Bacteriol. Rev. 41:595-635.
17. Jacob, A.E., and S.J. Hobbs (1974) Conjugal transfer of plasmid-borne multiple antibiotic resistance in Streptococcus faecalis var. zymogenes. J. Bact. 117:360-372.
18. Klier, A., C. Bourgouin, and G. Rapoport (1983) Mating between Bacillus subtilis and Bacillus thuringiensis and transfer of cloned crystal genes. Mol. Gen. Genet. 191:257-262.
19. Krywienczyk, J., and T.A. Angus (1967) A serological comparison of several crystalliferous insect pathogens. J. Invert. Pathol. 9:126-128.
20. Orrego, C., M. Arnaud, and H.O. Halvorson (1978) Bacillus subtilis 168. Genetic transformation mediated by outgrowing spores: Necessity for cell contact. J. Bact. 134:973-981.
21. Stewart, G.J., C.A. Carlson, and J.L. Ingraham (1983) Evidence for an active role of donor cells in natural transformation of Pseudomonas stutzeri. J. Bact. 156:30-35.

MECHANISMS ESSENTIAL FOR STABLE INHERITANCE OF MINI-F PLASMID

Sota Hiraga, Teru Ogura,* Hirotada Mori, and Masafumi Tanaka

Institute for Virus Research
Kyoto University
Kyoto 606, Japan

INTRODUCTION

F plasmid of *Escherichia coli* is known to be stably maintained in dividing cells, despite of low copy number per cell. Mini-F plasmids, which were constructed in vitro from the 9.2 kb *Eco*RI-generated f5 fragment of the F plasmid, are also stably inherited, even though the copy number is still low (for review, see Ref. 1). We have sought to understand the molecular mechanisms involved in the stable maintenance of mini-F.

We have found that there are 2 different mechanisms for mini-F, besides the control mechanism of replication. The first mechanism presumably acts for equipartition of plasmid DNA molecules to daughter cells (2). This mechanism of partition depends on 2 trans-acting genes (*sopA* and *sopB*: *sop* stands for stability of plasmid) and cis-acting DNA site (*sopC*) of the plasmid (Fig. 1). The second mechanism acts to maintain stably plasmid by coupling host cell division to plasmid proliferation (3). This mechanism depends on 2 plasmid genes (*ccdA* and *ccdB*: *ccd* stands for coupled cell division). When replication of a plasmid carrying the *ccdA* and *ccdB* genes is prevented and the plasmid copy number decreases through subsequent cell divisions to a critical threshold, as few as 1 per cell, increase of colony formers is inhibited, and appearance of plasmid-free segregants is effectively prevented (Fig. 2). We have proposed a hypothesis that the *ccdB* gene product inhibits cell division via an unknown mechanism, and the *ccdA* gene product suppresses the *ccdB* function, except when the copy number decreases to the critical threshold (3).

* Department of Technology, National Institute of Health, Tokyo 141, Japan.

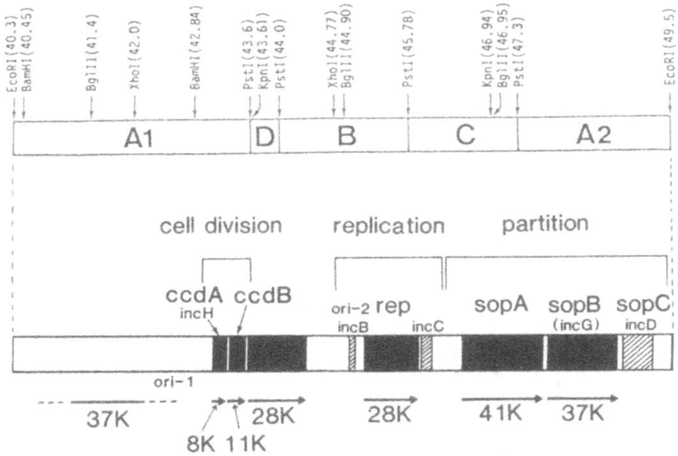

Fig. 1. Physical and functional map of EcoRI-generated f5 fragment of F plasmid. The numbers in parentheses denote kb coordinates on the F genome (1,3). The PstI segments are called A1, D, B, C, and A2 according to Murotsu et al. (4). The ori-1 and ori-2 denote replication origins of the mini-F plasmid (1). The rep codes for a protein essential for replication from ori-2 (5,6,7). The sop region essential for plasmid partitioning contains sopA, sopB, and sopC (2). The ccd region (coupled cell division) essential for coupling of host cell division to plasmid proliferation contains ccdA and ccdB (3). The incC (1), incB (1), incD (1,2,8), incG (2), and incH (Ref. 3, and see text) regions show incompatibilities. Filled regions on the map denote coding regions for proteins. Hatched regions denote direct repeats; 19 or 22 base-pair repeats in incB and incC (4,9,10), 43 base-pair repeats in incD (Takeyoshi Miki, pers. comm.). Arrows denote proteins (K, kilodalton) and their transcriptional direction (Ref. 1 and 11 and Takeyoshi Miki, pers. comm.).

MECHANISM OF PARTITION OF MINI-F PLASMID

Stabilization of oriC Plasmids by Joining with Mini-F Segments

We have described that F' plasmids carrying the replication origin (oriC) of the E. coli chromosome (12) or Salmonella thyphimurium (13) can replicate in Hfr strains of E. coli, in which F plasmid can not replicate. This suggests that a circular DNA molecule containing oriC can replicate as a plasmid in the cells using oriC. We have further isolated oriC plasmids from a transducing λ phage carrying oriC of E. coli by in vitro recombination tech-

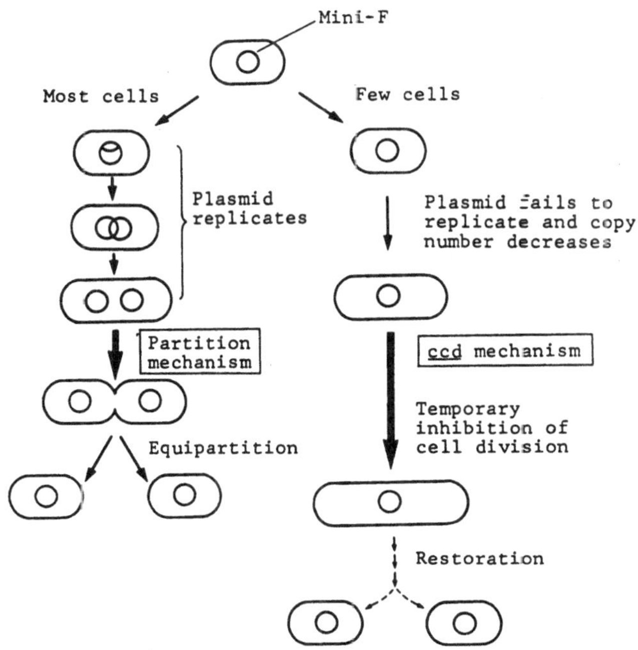

Fig. 2. Scheme showing mechanisms of stable inheritance of mini-F plasmid.

niques (14,15). We found that oriC plasmids are not stably maintained, and plasmid-free segregants appears at high frequencies in a nonselective medium. We assumed that the instability of oriC plasmids was due to the lack of function essential for equipartition.

To test this possibility, we constructed a composite plasmid (pXX11) consisting of an oriC plasmid and the mini-F plasmid (pSC138). The composite plasmid was stably maintained in Hfr strains, in which mini-F could not replicate. The oriC plasmid might be stabilized by the partition mechanism of mini-F (15).

We further constructed oriC plasmids carrying various segments of mini-F and examined the instability of these plasmids in F^- cells (recA1) to determine the minimal region necessary for stabilization of the oriC plasmid. These analyses indicate that the HaeII-SalI segment (46.4-49.4 kb in the F coordinate map) of mini-F is sufficient for the stabilization of the oriC plasmid (3). The stabilization of the oriC plasmid is achieved without any detectable increase in plasmid copy number per plasmid-carrying cell. This suggests that the HaeII-SalI segment is essential for the partition of the plasmid.

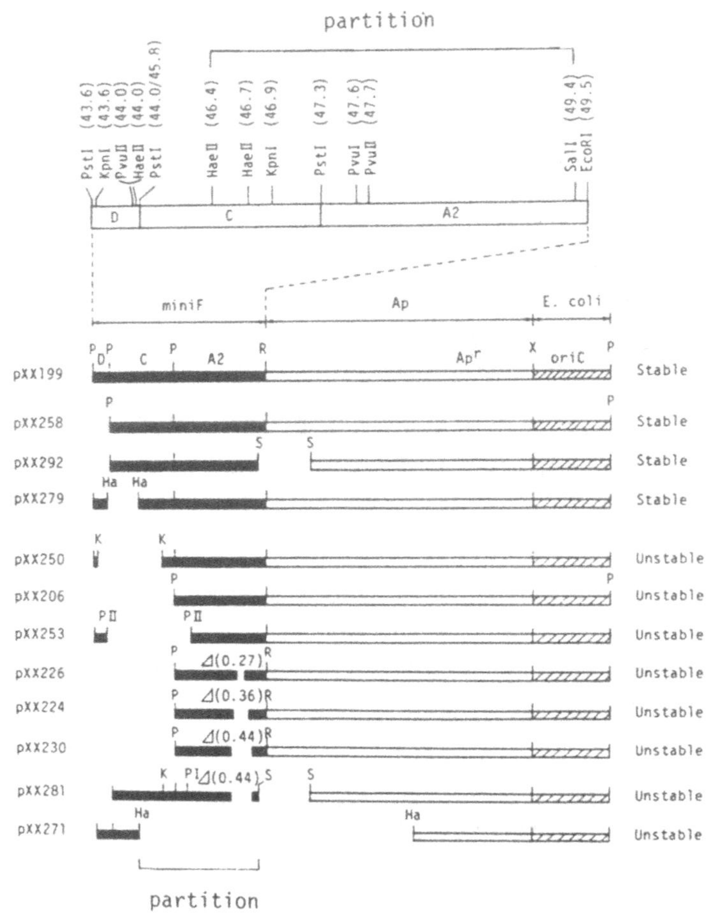

Fig. 3. Structure of deletion derivatives of an oriC plasmid carrying mini-F segments. The cleavage sites for EcoRI (R), PstI (P), and XhoI (X) are indicated. Only the sites for restriction endonucleases used to construct the deletion derivatives are shown on the map of each plasmid. S = SalI; Ha = HaeII; K = KpnI; PII = PvuII; PI = PvuI. The HaeII sites used to construct pXX279 are 44.0 kb and 46.4 kb on the F coordinate map, respectively. (Δ) A spontaneous deletion. (Number in parentheses = size of deletion in kb.)

Trans-Complementation Analyses of Functional Regions in the Partition Segment

To analyze the HaeII-SalI segment, we constructed 2 series of plasmids carrying various subsegments of the partition segment, us-

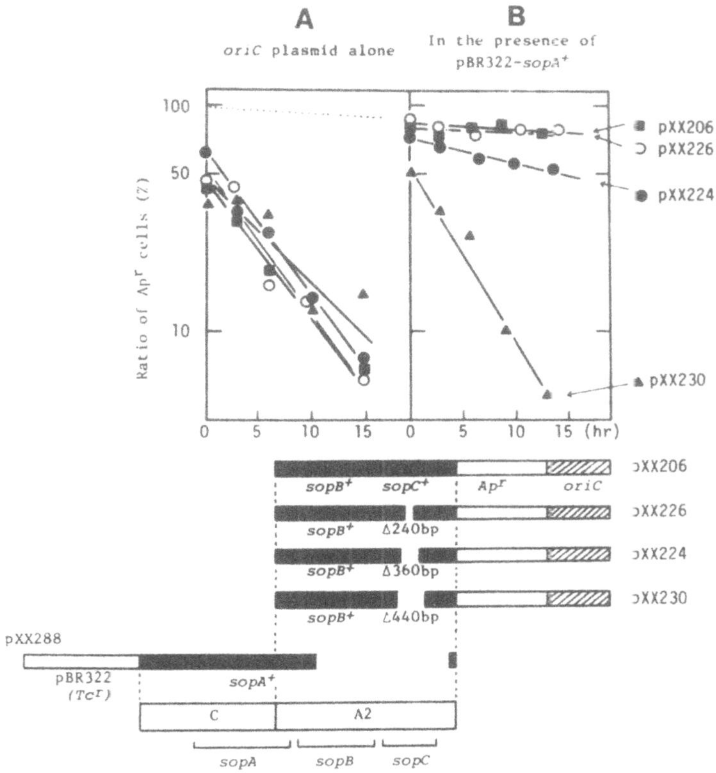

Fig. 4. Trans-complementation test for stability of oriC plasmids carrying a mini-F segment in the presence of a pBR322 derivative carrying a segment of mini-F. Samples were removed at intervals, diluted, and spread on L-agar plates. Colonies grown on the plates were tested for antibiotic-resistance. Dotted line in A shows the stability of pXX199 or pXX258 (see Fig. 3).

ing 2 vectors, i.e., an oriC plasmid and pBR322 (2). Stability of an oriC plasmid carrying a subsegment was measured in the presence or absence of a pBR322 derivative carrying another subsegment. These types of experiments were performed in various pairs of plasmids. Results from these complementation tests are shown in Fig. 4 and 5. Figure 4A shows stability of oriC plasmids carrying segments of mini-F in the absence of coexisting plasmids. All the oriC plasmids shown here are unstable. Figure 4B shows the stability of these oriC plasmids in the presence of pXX288, a pBR322 derivative carrying the PstI-HincII and HincII-EcoRI segments (47.3-47.9 kb and 49.4-49.5 kb, respectively). The oriC plasmid pXX206 was maintained stably in the presence of this pBR322 derivative. In contrast, the

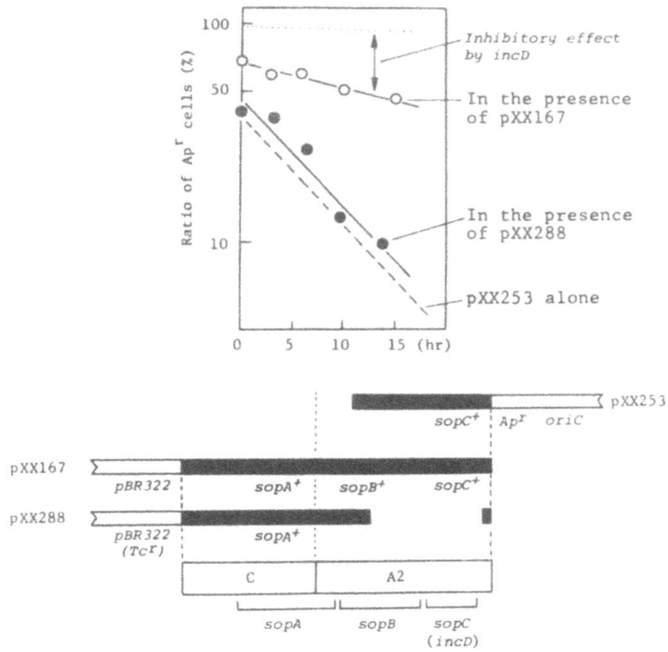

Fig. 5. Stability of an oriC plasmid pXX253 carrying sopC in the simultaneous presence of a pBR322 derivative carrying a mini-F segment. Cells carrying plasmids are grown in a nonselective medium, L broth.

pXX230 with a 440 base-pair (bp) deletion was still unstable. These results suggest that a trans-acting gene, designated sopA, essential for the stabilization is located in the mini-F segment carrying by pXX288. A cis-acting functional site, called sopC, which is located in the region deleted in oriC plasmid pXX230 is also involved in the stabilization.

The oriC plasmid pXX230 with a 360 bp deletion in the sopC region was partially stabilized by pXX288. With a 240 bp deletion, pXX226 was fully stabilized as pXX206. The activity of the sopC region, therefore, may be related to the size of the sopC segment.

As shown in Fig. 5, oriC plasmid pXX253 (see Fig. 3) is not stabilized in the presence of the pBR322 derivative carrying the sopA gene (pXX288), but can be partially stabilized by pXX167, which carries the entire segment essential for partition. This suggests that another trans-acting gene, termed sopB, exists between sopA and sopC. The oriC plasmid pXX253 is not fully stabilized by pXX167. This may be due to the incompatibility (incD) directed by the sopC region existing in both the plasmids (discussed later). These and further results suggest that sopA, sopB, and sopC are located in mini-F as shown in Fig. 6.

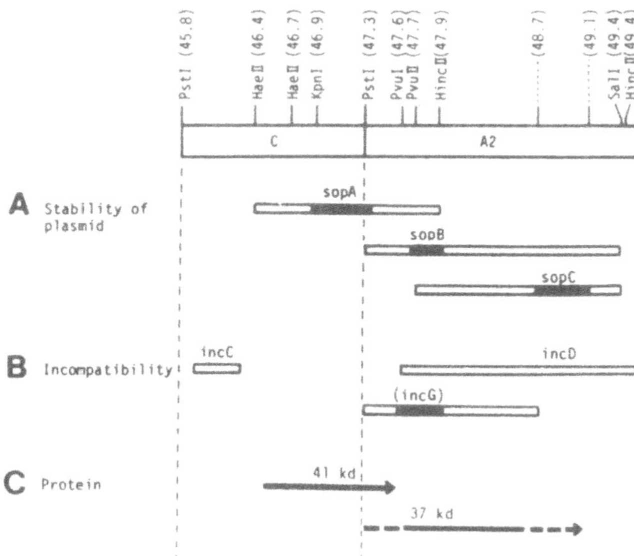

Fig. 6. The location of the sopA, sopB, and sopC genes in the partition segment.

A pBR322 derivative carrying the sopA gene, but neither sopB nor sopC, codes for a 41-kd protein. Another pBR322 derivative carrying the sopB gene, but neither sopA nor sopC, codes for a 37-kd protein. SopA and SopB proteins were recovered in cytoplasmic fraction and membrane fraction, respectively, in the presence of 10 mM $MgCl_2$ (K. Nagai et al., unpub. data). Austin and Wierzbicki (16) have isolated 2 amber mutants of mini-F which are not stably maintained. These mutant plasmids have a mutation in the genes coding for 41-kd and 37-kd proteins, respectively, consistent with our conclusion on sopA and sopB.

Isolation of Host Mutants Defective in Functions Essential for Partition of Mini-F

To investigate host chromosomal genes which are involved in the partition of mini-F, we isolated many host mutants in which an F' plasmid is unstable, and plasmid-free segregants appear at high frequencies under nonselective conditions. Then, we introduced a mini-F plasmid carrying the partition segment (pXX325, see Fig. 11) or that lacked the segment (pXX327, see Fig. 11) into these host mutants by transformation. Resulting transformants were analyzed for the stability of these mini-F plasmids in a nonselective medium. In one of the host mutants (hop-18, hop stands for host function of plasmid maintenance), the Sop^+ plasmid pXX325 was as unstable as the Sop^- plasmid pXX327 (Fig. 7). This host mutant may be defective in a function essential for the partition of mini-F.

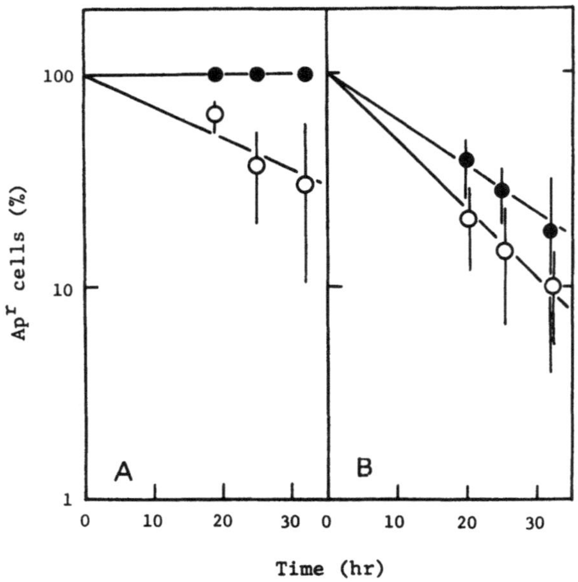

Fig. 7. Stability of mini-F plasmids in a host mutant (hop-18) defective in the essential function for partition of mini-F. Cells carrying a mini-F plasmid were spread on nonselective plates and incubated at 37°C for various times, indicated. To test for percent of plasmid-free segregants to total cells in each colony grown on the plates, colonies were independently removed, suspended in saline, diluted, and spread on nonselective plates. Colonies grown on the plates were removed with toothpicks and tested for ampicillin-resistance. Average ratio of ampicillin- resistant colonies to total colonies is shown in the figure. A vertical line at each circle denotes confidence limits ($\alpha = 0.05$). A, parental host (KZ344). B, hop-18 mutant. Solid circle, pXX325, a mini-F plasmid carrying the partition segment. Open circle, pXX327, mini-F plasmid lacking the partition segment. (see Fig. 11). Doubling time of cells was 50-55 min in all the strains.

In other host mutants tested, the Sop+ plasmid pXX325 was unstable at various levels, but the Sop$^-$ plasmid pXX327 was extremely less stable, more than 10-fold, than pXX325. We, therefore, judged that this type of mutants was partially defective in the host function essential for the replication of mini-F.

Hypothetical Apparatus of Partition

We have described a model for the partition mechanism of mini-F plasmid (2). The sopC region acts as a specific DNA site necessary for the actual partitioning, interacting with the cellular components consisting of SopB protein and host proteins. SopA protein may also interact directly or indirectly with the _sopC_ region, and/or SopA protein may regulate the expression of the _sopB_ gene (2). The expression of the _sopA_ gene may be regulated by SopB protein (17; our unpub. data). The hypothetical partition apparatus is assumed to be related to the cell membrane (2).

We propose to call generally the hypothetical partition apparatus "partisome," and the cis-acting specific DNA site, like _sopC_ of mini-F, "partor site" (Fig. 8). Each plasmid DNA molecule carrying the partition segment forms a partisome at the partor site. The partisome consists of the partor site and proteins specified by plasmid and host chromosomal genes, and associates with cytoplasmic membrane. Partisomes make pairs at random. Each pair of partisomes splits simultaneously at a time in cell cycle and the released partisomes move each other to opposite directions of both sides of a plane, where the septum will be formed later to separate daughter cells.

In the partisome model, when 2 kinds of plasmids having homologous partition mechanisms, but heterologous replication mechanisms, exist in a cell, a weak incompatibility between both plasmids should be observed, because random assortment of DNA molecules of the plasmids to daughter cells results from the partition. IncD phenotype directed by the _sopC_ region of mini-F, therefore, may be due to the random assortment of plasmid molecules. This model is consistent with the speculation that random assortment in equipartition results in a weak incompatibility (18,19).

MECHANISM OF COUPLING OF HOST CELL DIVISION TO PLASMID PROLIFERATION

Novel Mechanism of Plasmid Maintenance

We have also described an additional novel mechanism which acts to stably maintain a plasmid (3). This mechanism is specified by 2 plasmid genes, _ccdA_ and _ccdB_, which are located within the _HpaI-Pst_-I segment (42.9-43.6 kb; _ccd_ region) of mini-F (see Fig. 1). A plasmid (e.g., _oriC_ plasmid) carrying this segment is more stably maintained than that lacking the segment. An _oriC_ plasmid carrying both the _ccd_ segment and the partition segment of mini-F is extremely stably maintained as the parental mini-F pSC138.

When temperature-sensitive _polA_ mutant cells (km1213), which harbor a pBR322 derivative carrying the _ccd_ segment, pXX306 (see Fig. 10), were grown at 30°C in a nonselective medium and transferred to 42°C, colony formers (viable cells) continued to increase exponentially for 3-3.5 hr (3.5-4 generations) after transfer

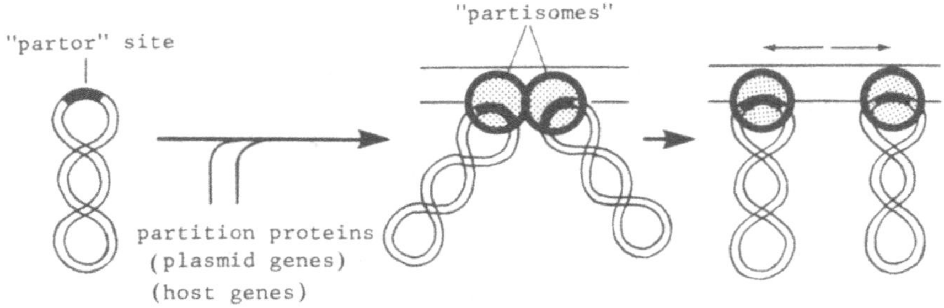

Fig. 8. Hypothetical apparatus for partition of plasmid in prokaryotes: "Partisome" model.

to 42°C, and then stopped increasing (Fig. 9B). Turbidity increased at least for 6-7 hr and filamentous cells with many nucleoids were observed. Plasmid-free segregants appeared only infrequently after 7 hr incubation at 42°C. In contrast, the inhibition of cell division was not observed in a strain carrying pBR322, and segregants began to appear at high frequencies after about 3 hr (Fig. 9A). It

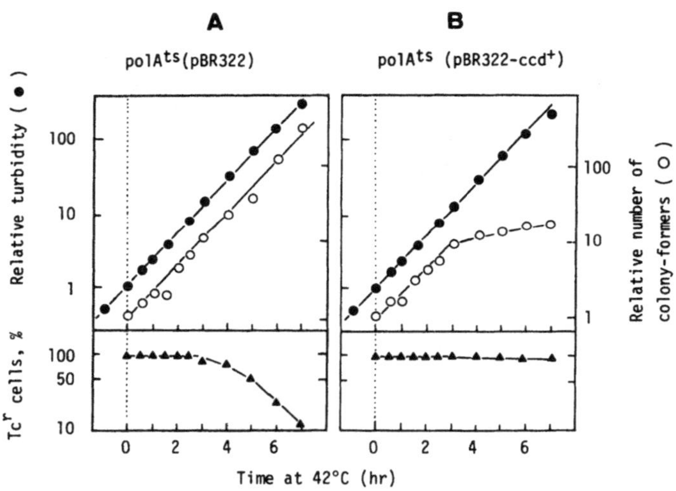

Fig. 9. Inhibition of host cell division in <u>polA</u>^{ts} cells which harbor a pBR322 derivative carrying the <u>ccd</u> segment, pXX306. Cells (<u>polA</u>^{ts}) carrying a plasmid grown at 30°C were transferred to 42°C, and then turbidity of the culture, colony formers and ratio of plasmid-carrying cells to total cells were measured at intervals. (A) <u>polA</u>^{ts} cells harboring pBR322. (B) <u>polA</u>^{ts} cells which harbors a pBR322 derivative carrying the <u>ccd</u> segment of mini-F, pXX306.

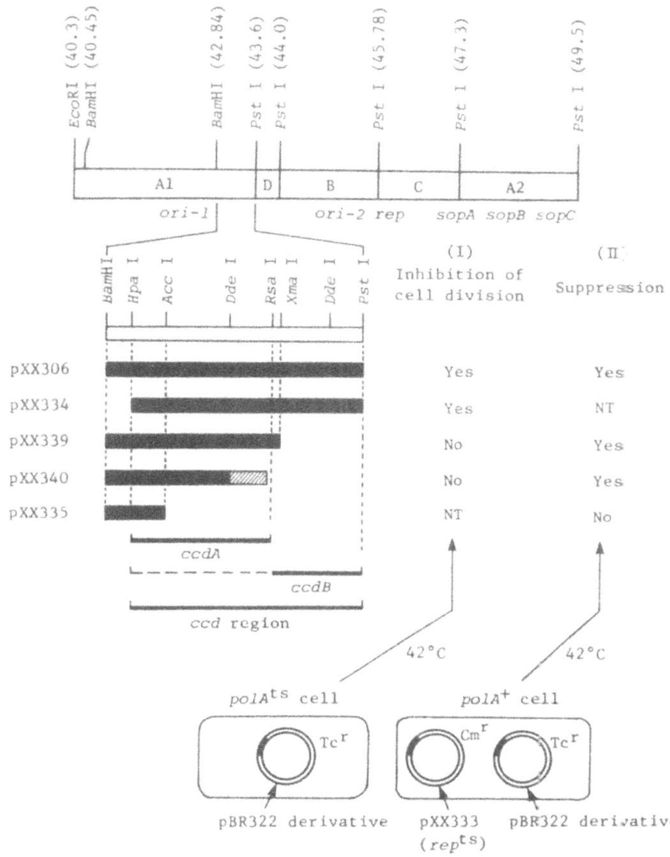

Fig. 10. Properties of pBR322 derivatives carrying mini-F segments. Column I, inhibition of cell division in polA^{ts} cells (km1213) carrying a pBR322 derivative carrying the indicated segment of mini-F (solid bar) after transfer to 42°C. Column II, suppression effect of the indicated plasmid on the inhibition of cell division which is caused at 42°C by a thermosensitive plasmid carrying the ccd region (PXX333) (see text). KH802 cells (polA^{+}) harboring both the pXX333 plasmid and the indicated pBR322 derivative were transferred to 42°C, and samples were removed and tested for colony formers. NT = not tested.

should be noted that the time when colony formers stop increasing in cells harboring the ccd-carrying pBR322 derivative nearly coincides with the time when plasmid-free segregants begin to appear in control pBR322-harboring cells. DNA synthesis of the cells harboring the ccd-carrying pBR322 derivative was not inhibited at least for 6 hr (3).

It seems unlikely that the inhibition of plasmid replication per se inhibits cell division, because the strain harboring the ccd-carrying pBR322 derivative grows normally and the number of colony formers increases for 3-3.5 hr after transfer to 42°C. Replication of ColE1-type plasmids in the $polA^{ts}$ strain (km1213) is inhibited immediately after transfer to high temperature. Therefore, the inhibition of cell division may be caused by reduction of plasmid copy number (3).

To confirm the effect of plasmids carrying the ccd segment on host cell division, we recloned the ccd segment of mini-F on a vector plasmid pHSG415, which has a thermosensitive replication mutation and can not replicate at 42°C. The resulting ccd-plasmid pXX333 was introduced in $recA^+$ cells, and transformants were tested for colony formers after transfer to 42°C. In this case, host cell division was also inhibited at 2-2.5 hr (about 3 generations) after transfer to 42°C, whereas, turbidity continued to increase for at least 4.5 hr. Plasmid-free segregants appeared only at low frequencies at 4.5 hr after transfer. By contrast, no detectable inhibition of cell division was observed in control $recA^+$ cells carrying the parental pHSG415, and plasmid-free segregants began to appear at 2-2.5 hr (3,20). These results are consistent with the above result in $polA^{ts}$ cells harboring the ccd-carrying pBR322 derivative.

Inhibition of host-cell division was also observed in recA1 mutant cells harboring pXX333 (3,20).

Suppression Effect of a Coexisting Replication-Proficient ccd-carrying Plasmid on the Inhibition of Cell Division

If it is true that the reduction of copy number of a ccd-carrying plasmid triggers a signal for inhibition of cell division, it could be expected that the inhibition is suppressed by the presence of replication-proficient plasmid carrying the homologous ccd segment. Accordingly, we introduced mini-F (pSC138) into the polA-ts strain which harbors the ccd-carrying pBR322 derivative of mini-F, and examined cell division after transfer to 42°C. As expected, cell division was no longer inhibited.

Similar suppression of the inhibition of cell division was observed in a $polA^+$ strain which harbored both the ccd-carrying thermosensitive plasmid pXX333 and the ccd-carrying pBR322 derivative. The inhibition of cell division caused by reduction of copy number of the ccd-carrying plasmid is thus suppressed by the replication-proficient plasmid carrying the homologous ccd segment.

Two Genes Located in the ccd Segment

To analyze the genes located within the ccd region, we constructed pBR322 derivatives carrying various subsegments of the ccd region, introduced them into *polA*^{ts} cells, and tested for the inhibition of host cell division after transfer to 42°C. As shown in Fig. 10 (column I), the functional region sufficient for the inhibition of cell division is located within the *HpaI*-*PstI* segment (42.9-43.6 kb) (3).

We introduced 1 of these pBR322 derivatives into *polA*$^{+}$ cells harboring the thermosensitive plasmid, pXX333, which carried whole the *ccd* segment, and tested for the inhibition of host cell division at 42°C. Results shown in Fig. 10 (column II) indicate that the suppression function on the inhibition of cell division is specified by a gene, named *ccdA*, which is located within the *HpaI*-*RsaI* segment (42.9-43.35 kb). Another gene, *ccdB*, which causes inhibition of cell division, may be located at the right side of the *ccdA* gene. The *ccdA* gene product suppresses the inhibitory function of the *ccdB* gene, except when the copy number of *ccd*-plasmid reduces to the critical threshold (3).

Similar phenomena have been described by other groups (21-24). Two proteins coded by the *ccd* segment were detected (21,22). A protein corresponding to CcdB protein inhibits cell division and another corresponding to CcdA protein suppresses the inhibitory function of the former. There are 2 open frames coding for 8 and 11 kd proteins, respectively, in the *ccd* region (24).

Prophage Induction by Functions of the ccd Genes

We found that weak induction of prophage λ was observed, besides inhibition of cell division, in a *recA*$^{+}$ λ-lysogen harboring pXX333 after incubation at 42°C (20). The induction of prophage in the pXX333-harboring λ-lysogen occurs only in a few percent of cells in the culture.

Both the λ induction and the inhibition of cell division caused by blocking of replication of pXX333 were suppressed in the presence of a replication-proficient pBR322 derivative carrying the *ccdA* gene. The λ induction, but not the inhibition of cell division, depends on the *recA*$^{+}$ gene function. The signal caused by the *ccdB* gene function may inhibit cell division via an unknown *recA*-independent pathway. The weak induction of prophage may be due to the side effect of the *ccdB* function via a *recA*-dependent pathway.

Weak induction of synthesis of RecA protein was observed in the *recA*$^{+}$ strain harboring pXX333 after incubation for 3 hr at 42°C, but not in the *recA1* mutant harboring this plasmid (20). Karoui et al. (22) have described induction of prophage λ and SfiA protein synthesis caused by inactivation of the *ccdA* gene function of mini- F. Our recent data (A. Jaffe and S. Hiraga, in preparation) indicate that the inhibition of cell division by the *ccd* function is independent of *sfiA* and *sfiC* (25) functions.

It is also unknown what is the molecular mechanism which monitors directly or indirectly the decreased copy number to one copy of a *ccd*-carrying plasmid. One possible explanation is that the rate of synthesis of CcdA protein depends on the copy number of the plasmid and, when CcdA protein is reduced to a critical concentration in the cell, the inhibitory function of CcdB protein appears and cell division is inhibited via a *recA*-independent pathway. If only oligomer of CcdA protein, but not monomer, had the suppression function against the inhibitory function of CcdB protein, the dependency of this phenomenon on the concentration of CcdA protein could be very clear, because, for example, the concentration of the dimer is in proportion to the 2 power of that of the monomer.

We have proposed, in addition, a hypothesis (20) that the *ccd* mechanism of F plasmid contributes to indirect induction (26) of prophage λ caused by an F plasmid transferred into a λ lysogen via conjugation.

STABILITY OF MINI-F PLASMID

Evidence of the Two Mechanisms Acting for Stable Inheritance of Mini-F

We have demonstrated that the partition mechanism and the *ccd* mechanism of mini-F act actually for stable inheritance of mini-F plasmid (2,3). We constructed deletion mutants of mini-F pSC138 and tested their stability. As shown in Fig. 12A, pXX9 and pXX317, which have both the Sop$^+$ and Ccd$^+$ functions, are stably maintained as the parental plasmid pSC138. By contrast, pXX318 and pXX344, which have the Ccd$^+$ function but not the Sop$^+$ function, are unstable (Fig. 12B). The Sop$^+$ function, therefore, acts for the stable maintenance of mini-F. Plasmid pXX327, which has neither the Sop$^+$ function nor the Ccd$^+$ function, is less stable than pXX318 and pXX344 (Fig. 12C). Therefore, the Ccd$^+$ function acts for stable inheritance of mini-F plasmid. Resulting from impaired equipartition of the Sop$^-$ plasmid, pXX318 or pXX344, cells which have only 1 copy of the plasmid per cell may appear at a high frequency in cell population. In these cells, cell division may be inhibited by the *ccd* mechanism. The Ccd$^+$ Sop$^-$ plasmids pXX318 and pXX344 are, therefore, more stably maintained than the Ccd$^-$ Sop$^-$ plasmid pXX327.

pXX325 and pXX326, which have the Sop$^+$ function but not the Ccd$^+$ function, are as stable as pSC138 (Fig. 12D). The effect of the *ccd* mechanism can not be clearly detected in Sop$^+$ mini-F plasmids as expected (3).

Incompatibilities of Mini-F

Mini-F plasmid pSC138 is known to be destabilized by the presence of another heterologous plasmid (for example, pBR322) carrying a segment of the mini-F plasmid (1). Incompatibilities of incB and incC, located in the essential region for replication of mini-F, are mediated by clusters of directly repeated nucleotide sequences of 19 or 22 bp (4,9,10) (see Fig. 1). As described above, the cis-acting sopC region shows an incompatibility. The incompatibility may correspond to the incD-mediated incompatibility in the IncFI group of plasmids (8) (Fig. 1 and 6). We have described another incompatibility located in the A2 segment. The functional region for the incompatibility, named incG, is located within the 47.3-48.7 kb segment (Fig. 1 and 6). The incG incompatibility was observed only under particular conditions as follows. When the A2 segment was cloned on pBR322, most of the plasmids carried the A2 segment with a deletion at the sopC region. When one of these pBR322 derivatives carrying the sopC-deleted A2 segment coexisted with pSC138, pSC138 was unstable and pSC138-derived plasmids with a deletion at sopC were selected during subsequent growth. The incG incompatibility may be caused by overproduction of SopB protein (Ref. 2; unpub. data).

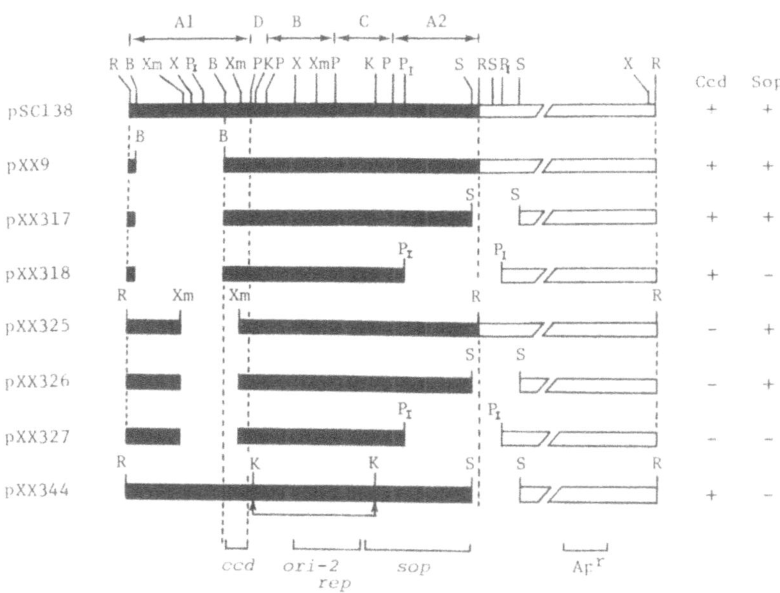

Fig. 11. Deletion mutants of mini-F plasmid pSC138. Restriction enzymes are shown in the legend of Fig. 3. B = BamHI; Xm = XmaI. In pXX344, the KpnI segment is inserted in the opposite direction.

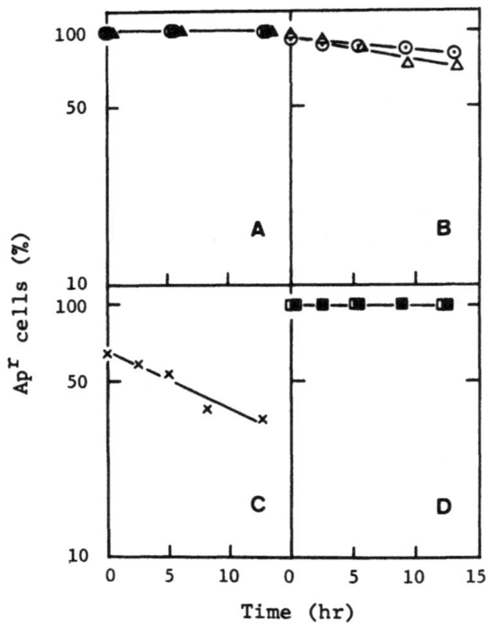

Fig. 12. Stability of deletion mutants of mini-F plasmid. (A) ●, pSC138; ○, pXX9; ▲, pXX317. (B) △, pXX318; ⊙, pX344. (C) ×, pXX327. (D) ■, pXX325; □, pXX326.

The stability of the Ccd$^+$ Sop$^-$ mini-F plasmid pXX318 (see Fig. 11) is reduced by the presence of a pBR322 derivative carrying the ccd segment (pXX306, see Fig. 10), but not by the presence of a pBR322 derivative lacking the ccd segment (pXX256), as shown in Fig. 13B. This destabilization of pXX318 by the extra copies of the ccd-carrying plasmid pXX306 is proper, as expected from our hypothesis on the ccd mechanism described above. This incompatibility mediated by the ccdA gene product is named the incH incompatibility. The incH and incG incompatibilities are thus novel types of incompatibilities of mini-F which are mediated by gene products.

Since incompatibilities are caused by different maintenance mechanisms of mini-F, analyses of incompatibilities are also important to study maintenance mechanisms of other plasmids.

SUMMARY

The F plasmid has its own partition mechanism and ccd mechanism (coupled cell division), besides its own replication mechanism, in order to be stably inherited into daughter cells through cell division. These 3 mechanisms are independent of one another. There-

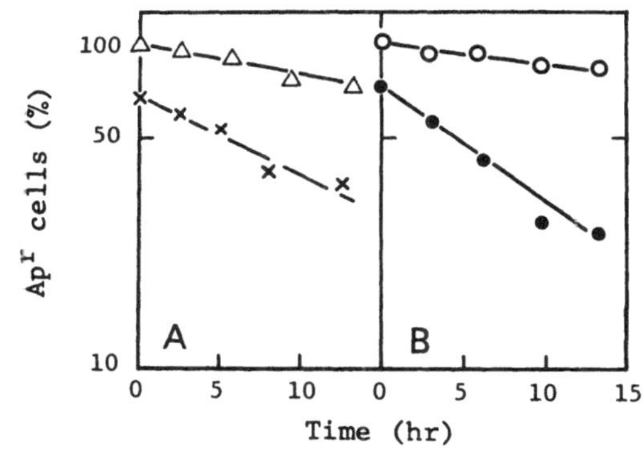

Fig. 13. Stability of mini-F plasmids. (A) Δ, mini-F pXX318, which carries both the ccdA and ccdB genes, but not the partition segment. x, mini-F pXX327, which carries neither the ccd genes nor the partition segment. (B) •, stability of mini-F pXX318 in the presence of a pBR322 derivative carrying both the ccdA and ccdB genes (pXX306). o, stability of mini-F pXX318 in the presence of a pBR322 derivative lacking the ccd genes (pXX256).

fore, when a DNA segment essential and sufficient for a mechanism is jointed to other heterologous plasmids, the segment is also functional. Most of natural low copy number plasmids might also have their own replication, partition, and ccd mechanisms. These 3 mechanisms may be fundamental to ensure stable inheritance for low copy-number replicons in prokaryotes.

REFERENCES

1. Lane, H.E.D. (1981) Replication and incompatibility of F and plasmids in the incFI group. Plasmid 5:100-126.
2. Ogura, T., and S. Hiraga (1983a) Partition mechanism of F plasmid: Two plasmid gene-encoded products and a cis-acting region are involved in partition. Cell 32:351-360.
3. Ogura, T., and S. Hiraga (1983b) Mini-F plasmid genes that couple host cell division to plasmid proliferation. Proc. Natl. Acad. Sci., USA 80:4784-4788.
4. Murotsu, T., K. Matsubara, H. Sugisaki, and M. Takanami (1981) Nine unique repeating sequences in a region essential for replication and incompatibility of the mini-F plasmid. Gene 15:257-271.
5. Watson, L.A., S.H. Phua, P.L. Bergquist, and H.E.D. Lane (1982) An Mr 29,000 protein is essential for mini-F maintenance in E. coli. Gene 19:173-178.
6. Maki, S., M. Kuribayashi, T. Miki, and T. Horiuchi (1983) An amber replication mutant of F plasmid mapped in the minimal replication region. Mol. Gen. Genet. 191:231-237.

7. Kline, B.C., and J. Trawick (1983) Identification and characterization of a second copy number control gene in mini-F plasmids. Mol. Gen. Genet. 192:408-415.
8. Gardner, R.C., L. Malcolm, P.L. Bergquist, and H.E.D. Lane (1982) IncD, a genetic locus in F responsible for incompatibility with several plasmids of the incFI group. Mol. Gen. Genet. 188:345-352.
9. Tolun, A., and D.R. Helinski (1981) Direct repeats of the F plasmid incC region express F incompatibility. Cell 24: 687-694.
10. Komai, N., T. Nishizawa, Y. Hayakawa, T. Murotsu, and K. Matsubara (1982) Detection and mapping of six mini-F-encoded proteins by cloning analysis of dissected mini-F segments. Mol. Gen. Genet. 186:193-203.
11. Tsutsui, H., A. Fujiyama, T. Murotsu, and K. Matsubara (1983) Role of nine repeating sequences of the mini-F genome for expression of F-specific incompatibility phenotype and copy number control. J. Bact. 155:337-344.
12. Hiraga, S. (1976) Novel F prime factors able to replicate in Escherichia coli Hfr strains. Proc. Natl. Acad. Sci., USA 73:198-202.
13. Joh, K., and S. Hiraga (1979) Genetic mapping of the chromosomal replication origin of Salmonella typhimurium. J. Bact. 138:297-304.
14. Miki, T., S. Hiraga, T. Nagata, and T. Yura (1978) Bacteriophage λ carrying the Escherichia coli chromosomal region of the replication origin. Proc. Natl. Acad. Sci., USA 75:5099-5103.
15. Ogura, T., T. Miki, and S. Hiraga (1980) Copy-number mutants of the plasmid carrying the replication origin of the Escherichia coli chromosome: Evidence for a control region of replication. Proc. Natl. Acad. Sci., USA 77:3993-3997.
16. Austin, S., and A. Wierzbicki (1983) Two mini-F-encoded proteins are essential for equipartition. Plasmid 10:73-81.
17. Phua, S.H., P.L. Bergquist, and H.E.D. Lane (1982) Effects of Tn5 insertion in the incD region on mini-F maintenance and polypeptide synthesis. Mol. Gen. Genet. 188:353-355.
18. Nordström, K., S. Molin, and H. Aagaard-Hansen (1980) Partitioning of plasmid R1 in Escherichia coli. II. Incompatibility properties of the partitioning system. Plasmid 4:332-349.
19. Austin, S., and A. Abeles (1983) Partition of unit-copy miniplasmids to daughter cells. J. Mol. Biol. 169:373-387.
20. Mori, H., T. Ogura, and S. Hiraga (1984) Prophage induction caused by mini-F plasmid genes. Mol. Gen. Genet. (in press).
21. Bex, F., H. Karoui, L. Rokeach, P. Drèze, L. Garcia, and M. Couturier (1983) Mini-F encoded proteins: Identification of a new 10.5 kilodalton species. EMBO J. 2:1853-1861.
22. Karoui, H., F. Bex, P. Drèze, and M. Couturier (1983) Ham22, a mini-F mutation which is lethal to host cell and promotes recA-dependent induction of lambdoid prophage. EMBO J. 2:1863-1868.

23. Miki, T., K. Yoshioka, and T. Horiuchi (1984) Control of cell division by sex factor F in Escherichia coli. I. The 42.84-43.6 F segment couples cell division of the host bacteria with replication of plasmid DNA. J. Mol. Biol. 174 (in press).
24. Miki, T., Z.T. Chang, and T. Horiuchi (1984) Control of cell division by sex factor F in Escherichia coli. II. Identification of genes for inhibitor protein and trigger protein on the 42.84-43.6 F segment J. Mol. Biol. 174 (in press).
25. D'Ari, R., and O. Huisman (1983) Novel mechanism of cell division inhibition associated with the SOS response in Escherichia coli. J. Bact. 156:243-250.
26. Borek, E., and A. Ryan (1960) The transfer of a biologically active irradiation product from cell to cell. Biochim. Biophys. Acta 41:67-73.

SEX PHEROMONES AND PLASMID TRANSFER IN STREPTOCOCCUS FAECALIS: A PHEROMONE, cAM373, WHICH IS ALSO EXCRETED BY STAPHYLOCOCCUS AUREUS

Don B. Clewell, Florence Y. An, Bryan A. White, and Cynthia Gawron-Burke

Departments of Oral Biology and Microbiology/Immunology
and The Dental Research Institute
Schools of Dentistry and Medicine
The University of Michigan
Ann Arbor, Michigan 48109

INTRODUCTION

Recipient strains of *Streptococcus faecalis* excrete small, peptide sex pheromones which induce donors harboring certain conjugative plasmids to synthesize a proteinaceous adhesin that facilitates the formation of mating aggregates. (For a recent review of this phenomenon, see Ref. 2.) Referred to as "aggregation substance," the adhesin has been revealed as a unique antigen that uniformly coats the donor cell surface (14,25). Self aggregation of donors also occurs; for this reason the pheromones are also referred to as clumping inducing agents (CIAs). It has been proposed that "aggregation substance" binds specifically to a "binding substance" present on the surface of both donor and recipient cells (6,7).

Upon acquiring a copy of the plasmid, the recipient ceases to excrete the related pheromone, but pheromones specific for donors harboring unrelated classes of conjugative plasmids continue to be excreted (5,7). A specific sex pheromone, therefore, is defined by the plasmid system being used to resolve it. The activity designated cAD1 represents the pheromone to which strains harboring pAD1 respond; similarly, donors harboring pPD1, pAMγ2, etc., respond, respectively, to cPD1, cAMγ2, etc.

The "shut-off" of pheromone synthesis in donor cells is accomplished by a plasmid-encoded inactivation of the peptide via a chemical addition, the nature of which is unknown (12). A phosphodiester bond is generated in the process, and the modified pheromone is capable of inhibiting exogenous pheromone activity--probably by competition for a pheromone receptor site. On the basis of gel filtration chromatography, the pheromones cAD1 and cPD1 were roughly sized at 1,050 and 800 daltons, respectively, and the corresponding modified forms (designated mcAD1 and mcPD1, respectively) appear to be 300-400 d larger (12). Purification of the peptides has been difficult because of the extremely low concentrations in filtrates. However, Suzuki and coworkers have recently been successful in the purification of cPD1; the peptide was found to consist of 8 amino acids and is very hydrophobic (A. Suzuki, pers. comm.). The molecular weight is 912.

Mutants significantly reduced in the production of cPD1 (i.e., to 3-4% that of the wild-type) have been derived and found to also be reduced in cAD1 production and probably additional pheromones (2,12). These variants were poor recipients; however, those which acquired plasmid DNA (e.g., pAM714, a derivative of pAD1) were good donors. Indeed, since the latter also failed to excrete the modified pheromone (i.e., mcAD1), with its attendant inhibitory effects, donor sensitivity to exogenous pheromone was greater than normal (2). The reduction of more than one pheromone activity in a single mutation raises the possibility that synthesis or subsequent processing or transport of these substances is linked.

The genetics of the pAD1 pheromone response have been partially analyzed using the erythromycin (Em)-resistance transposon Tn*917* (11). Two regions on the plasmid, designated *traA* (1.5 kb) and *traB* (1.3 kb) and separated by 1.7 kb, appear to be involved in a negative regulation of the synthesis of "aggregation substance"; insertions into these regions give rise to constitutive aggregation and the ability to donate at high frequency in short (10 min) matings. Other mutants, exhibiting a significantly reduced mating potential, were also derived (11).

CIA ACTIVITIES PRODUCED BY SPECIES OTHER THAN *STREPTOCOCCUS FAECALIS*

Although extracellular compounds that affect competence for transformation are well known in *Streptococcus pneumoniae* and *Streptococcus sanguis* (15), substantive evidence for bacterial sex pheromones which promote conjugative DNA transfer has been reported only in *Streptococcus faecalis*. To determine if different species of bacteria might excrete substances similar to the *S. faecalis* sex pheromones, 10 *S. faecalis* strains exhibiting high sensitivity to induced clumping by plasmid-free *S. faecalis* filtrates were used to screen filtrates from a variety of other bacterial sources. (Relevant strains and plasmids used in this study are listed in Tab. 1 and 2.) Table 3A shows results for 4 of these responder strains, 3 of which (39-5, DS16, and DS5) have been previously characterized

Tab. 1. Bacterial strains used in this study.

Strain	Relevant Chromosome markers	Plasmid Content	Source
Strep. faecalis			
JH2-2	rif,fus	none	(13)
FA2-2	rif,fus	none	(3)
JH2SS	str,spc	none	(22)
FA1001	rif,fus	none	This study; (derivative of JH2-2)
Y11	str,spc,tet(Tn916)	none	(12)
39-5		pPD1,pPD2,pPD3, pPD4,pPD5,pPD6	(25)
DS16	tet(Tn916)	pAD1,pAD2	(3, 23)
DS5	tet	pAMγ1,pAMγ2,pAMγ3 pAMβ1,pAMα1	(4, 5)
RC73	tet(Tn918) (probable chromosomal location)	pAM373, at least four additional plasmids	This study; University Hospital, Ann Arbor
FA373	rif,fus,tet(Tn918)	pAM373	This study
FA381	str,spc,tet(Tn918)	none	This study
CG110	rif,fus,tet(Tn916)	none	(9)
Strep. faecium			
9790(ATCC)		cryptic plasmids present	G. Shockman
Strep. sanguis			
Challis (NCTC 7868)(V288)		none	F. Macrina
Challis RF	rif,fus	none	This study
G9B		none	(18), Garth Jones
Staph. aureus			
879R4	(restriction negative)	12.5 kb cryptic plasmid	(20) P.Pattee
879R4S	str (derivative of 879R4)	" "	This study
879R4RF	rif,fus (derivative of 879R4)	"	This study
RN450		none	(17), R. Novick
Escherichia coli			
CG170	DH1 background	pAM170	(10)
FA390	DH1 background	pAM390	This study

Tab. 2. Relevant plasmids used in this study.

Plasmid	Marker	Comment
pAD1	Hem-Bac	57 kb, conjugative plasmid from S. faecalis DS16
pAM373	none	36 kb, conjugative plasmid from S. faecalis RC73
pAM377	Em^r (Tn917)	pAM373 with Tn917 insertion
pAM378	Tc^r (Tn918)	pAM373 with Tn918 insertion
pAM380	Em^r (Tn917); Tc^r (Tn916)	pAM377 with Tn916 insertion
pAMα1::Tn917	Tc^r, Em^r (Tn917)	non-conjugative; mobilized easily; see reference 1
pAM180	Em^r, Tc^r (Tn916)	a derivative of the 16 kb, conjugative, Em^r plasmid pAM81 with Tn916 insertion; see reference 9
pAM170	Ap^r, Tc^r (Tn916)	E. coli plasmid vector pGL101 with a cloned fragment carrying Tn916; see reference 10
pAM390	Ap^r, Tc^r (Tn918)	E. coli plasmid vector pGL101 with a cloned fragment carrying Tn918

with respect to their plasmid content (3,4,5,23,25). The 3 latter strains, as well as the 6 S. faecalis strains not shown (DU9, RC11, RC26, RC48, RC56, and RC87), all failed to respond to any of the filtrates. Interestingly, strain RC73 responded to filtrates from Strep. faecium (strain 9790), Strep. sanguis (strains Challis and G9B), and Staphylococcus aureus (strains 879R4S and RN450). Filtrates from the following bacteria (not shown in Tab. 3a) exhibited no CIA activity for any of the 10 responders: Strep. sanguis A1, Strep. pyogenes AC1, Strep. agalactiae T-2, Strep. lactis 4027, Strep. cremoris 4158, Strep. milleri CRS8507, Staph. epidermidis UM899, Bacillus subtilis BR151, and Escherichia coli DH1. It was also of interest that 2 mutants of S. faecalis (FA1001 and Y11) that were defective in cPD1 and cAD1 activity, and whose filtrates generated little or no response by 39-5, DS16, or DS5, produced essentially unaltered levels of CIA activity to which RC73 responded (Tab. 3B).

The response of RC73 to the 2 Staph. aureus filtrates prompted an examination of additional clinical isolates of Staph. aureus as

Tab. 3. CIA titers of culture filtrates from different species.

			S. faecalis responders			
			39-5	DS16	DS5	RC73
A.						
	Strep. faecalis	JH2-2	64	32	64	128
	Strep. faecium	9790	<2	<2	<2	8
	Strep. sanguis	Challis	<2	<2	<2	16
		G9B	<2	<2	<2	8
	Staph. aureus	879R4S	<2	<2	<2	32
		RN450	<2	<2	<2	32
B.						
	Strep. faecalis	JH2-2	64	32	64	128
		FA1001	2	<2	<2	64
		Y11	2	<2	2	128

a Titers were measured using a microtiter dilution assay as described in reference 7. The values represent the reciprocal of the highest serial (two fold) dilution which still exhibited a detectable clumping response using the indicated responder strain.

A. Filtrates of selected strains of different species containing CIA activity detected by at least one *S. faecalis* responder strain.

B. Comparisons of filtrates from *S. faecalis* variants reduced in production of cAD1 and cPD1. FA1001 was generated by mutagenesis of JH2-2 with "nitrosoguanidine" using the same screening method used to obtain Y11 (see reference 12).

well as its less pathogenic, coagulase negative counterparts (e.g., Staph. epidermidis). All of 21 additional strains of Staph. aureus examined excreted activities (titers of 16-64) to which RC73 responded (data not shown). This was in contrast to 22 isolates of coagulase negative staphylococci among which only 2 excreted such an activity. (In the 2 positive cases, the titers were only 2.)

The production of CIA activity by Strep. sanguis is not as prevalent as was found for Staph. aureus. Whereas strains Challis and G9B excreted significant activity, several other Strep. sanguis strains tested did not.

IDENTIFICATION OF THE PLASMID IN S. FAECALIS RC73 ENCODING THE RELATED CIA RESPONSE

The apparent correlation of a CIA activity with staphylococcal pathogenicity stimulated further analyses of the responding system. S. faecalis RC73 is hemolytic, bacteriocinogenic, and resistant to tetracycline; sedimentation and electrophoretic analyses revealed at least 5 plasmids -- the largest being about 57 kb. In order to derive strains harboring only one or a few of these plasmids, matings were performed between RC73 and the plasmid-free S. faecalis strain FA2-2. Selection for Tc-resistant transconjugants from overnight filter matings (8) yielded transconjugants at a frequency of about 10^{-6} per donor (about 10% were hemolytic), and these contained a variety of plasmid combinations. It became evident that only strains harboring a 36-kb plasmid to be designated pAM373 exhibited a response to filtrates of Staph. aureus and Strep. sanguis; indeed, one transconjugant (FA373) harbored only this plasmid. A restriction map of pAM373 is shown in Fig. 1. The plasmid is maintained at about 4 copies per chromosome and does not bear the RC73 determinants for hemolysin, bacteriocin, or Tc-resistance. (The nature of the Tc-resistance determinant selected for in the mating experiments is discussed below.)

To serve as a useful marker, the 5 kb Em-resistance transposon Tn917 (3,21,22) was introduced onto pAM373 using the following approach. Strain FA373, which harbors pAM373 as its only plasmid, was mated (overnight on filters at a donor to recipient ratio of 1:1) with S. faecalis JH2SS (pAMα1::Tn917). [pAMα1 is a 9-kb, nonconjugative Tc-resistance plasmid (1), known to be readily mobilizable by several different conjugative plasmids.] The mating mixture was plated on media containing rifampin, fusidic acid (for donor selec-

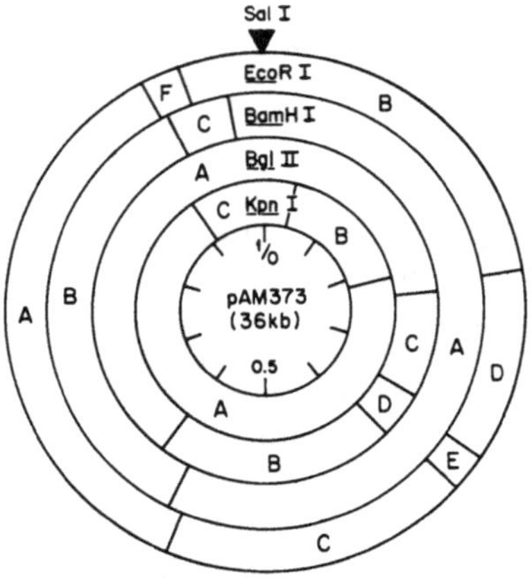

Fig. 1. Restriction map of pAM373. Mapping was based on analyses of double and partial digestions with appropriate enzymes.

tion) and Em in order to select for transconjugants in which pAM373 transferred to the recipient and mobilized back the pAMα1::Tn917 plasmid to donors. Transconjugants arose at approximately 10^{-8} per donor; of these 1 out of 3 tested transferred Em-resistance at high frequency without cotransfer of Tc-resistance. Subsequent analyses showed that Tn917 had inserted into pAM373, probably during the initial mobilization event. The resulting plasmid was designated pAM377, and it transferred at high frequency (10^{-3}) in 4-hr broth matings. As shown in Fig. 2A (lane 2), an insertion (5 kb) into the EcoRl A fragment had occurred.

An additional derivative of pAM373, having both Em- and Tc-resistance determinants, was generated by inserting the conjugative transposon Tn916 (Tc) onto pAM377 (Em) using an approach previously described (8,9). Briefly, pAM377 was introduced into *S. faecalis* CG110 (harbors Tn916 on the chromosome); use of the resulting derivative as a donor in subsequent matings that selected for transfer of Tc-resistance gave rise to a pAM377 derivative carrying Tn916. Designated pAM380, the plasmid acquired the insertion in its EcoRl A fragment, which already contained Tn917 (Fig. 2A, lane 4). Further analyses (data not shown) have shown that Tn916 is actually inserted into a Bgl II fragment within Tn917.

Interestingly, the Tc-resistance determinant selected for in the initial RC73 x FA2-2 matings proved to be chromosome-borne; indeed, subsequent matings yielded a few Tc-resistant transconjugants found to be plasmid-free. Matings using strain FA373 (Tc-resistant, yet harboring only pAM373) as a donor and JH2SS as a recipient yielded Tc-resistant transconjugants at a frequency of 10^{-5}; and of 14 isolates tested for donor potential in cross-streak matings (3), 8 were found to transfer Tc-resistance at high frequency. Plasmid DNA from 3 of these strains was examined and found to contain a segment of DNA inserted into the EcoRl fragment C of pAM373. A restriction digest of one of these, designated pAM378, is shown in Fig. 2A (lane 3). (The inserted DNA did not contain an EcoRl site.)

The Tc-resistance determinant was also capable of inserting into other plasmids. When the hemolysin plasmid pAD1 was introduced into a plasmid-free Tc-resistant derivative (FA381; a derivative of JH2SS) and subsequent matings were performed using FA2-2 as a recipient, Tc-resistant transconjugants (of which 60% were hemolytic) with the resistance marker linked to pAD1 could be generated at a frequency of about 10^{-8} (about 10-20% of the hemolytic transconjugants). Figure 2B (lanes 2 and 3) shows examples of insertions into 2 different EcoRl restriction fragments (B and D), one of which (D) gave rise to a hyperhemolytic phenotype. The enhanced hemolysin expression resembled a phenomenon known to occur in the case of some Tn916 insertions into pAD1 (8).

Finally, it is noted that among the transconjugants of the original mating between RC73 and FA2-2, a hemolytic derivative subsequently used as a donor in matings with JH2SS gave rise to Tc-

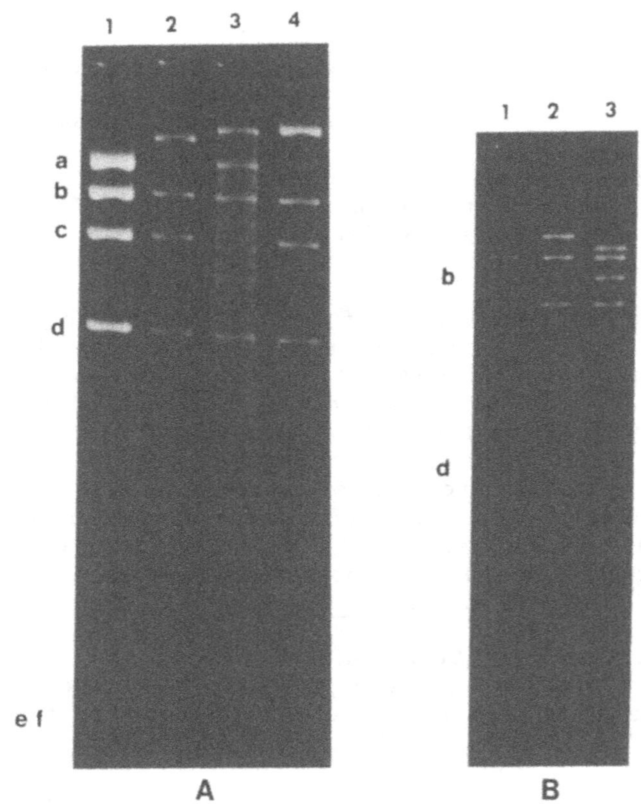

Fig. 2. Agarose gel electrophoresis of plasmid DNA digested with EcoR1. The procedure was as described elsewhere (8). A. pAM373 (lane 1); pAM377 (lane 2); pAM378 (lane 3); pAM380 (lane 4). The lower case letter on the left corresponds to the EcoR1 fragment of pAM373 (fragments e and f migrate together). B. pAD1 (lane 1); pAD1::Tn918 (lane 2) (insertion in b fragment); pAD1::Tn918 (lane 3) (insertion in d fragment). The b and d fragments of pAD1 are indicated on the left.

resistant, hemolytic strains of which about 10% were hyperhemolytic. In these cases the Tc-resistance determinant was linked to a 57-kb hemolysin plasmid originally present in RC73.

In all of the cases where it was measured (e.g., those shown in Fig. 2), the size of the inserted DNA associated with Tc-resistance was estimated at ∿15kb (similar to that of Tn916). The element will henceforth be designated Tn918.

ON THE TRANSFERABILITY OF pAM373

As shown in Tab. 4, the pAM373 derivatives pAM377(Em) and pAM378(Tc) transferred efficiently in 4-hr broth matings between strains of S. faecalis; under the same conditions, however, no transfer could be observed from S. faecalis to either Strep. sanguis Challis or the restriction negative (20) Staph. aureus 879R4RF. In further efforts to transfer plasmid DNA to these species, overnight filter matings were performed. In the case of pAM377 (Em), transfer of Em-resistance to Staph. aureus or Strep. sanguis could not be detected; however, when pAM378(Tc) was used, Tc-resistant transconjugants appeared at 3×10^{-5} among Staph. aureus and at a very low frequency (4×10^{-9}) among Strep. sanguis. When several of these transconjugants were examined for plasmid content, pAM378 could not

Tab. 4. Mating experiments α.

Donor	Recipient	Selection	Frequency. Transconjugants per donor
A. Broth matings (4 hrs)			
JH2SS(pAM377)	FA2-2	Rif,Fa,Em	5.9×10^{-3}
"	879R4RF	Rif,Fa,Em	$<3 \times 10^{-9}$
"	Challis RF	Rif,Fa,Em	$<5 \times 10^{-9}$
JH2SS(pAM378)	FA2-2	Rif,Fa,Tc	3.0×10^{-2}
"	879R4RF	Rif,Fa,Tc	$<1 \times 10^{-9}$
"	Challis RF	Rif,Fa,Tc	$<1 \times 10^{-9}$
B. Filter matings (overnight)			
JH2SS(pAM377)	FA2-2	Rif,Fa,Em	4.6×10^{-2}
"	879R4RF	Rif,Fa,Em	$<2 \times 10^{-9}$
"	Challis RF	Rif,Fa,Em	$<2 \times 10^{-9}$
JH2SS(pAM378)	FA2-2	Rif,Fa,Tc	3.3×10^{-2}
"	879R4RF	Rif,Fa,Tc	3.0×10^{-5}
"	Challis RF	Rif,Fa,Tc	4×10^{-9}
FA2-2(pAM380)	879R4S	Em,Sm	$<1 \times 10^{-9}$
"	"	Tc,Sm	1×10^{-7}
JH2-2(pAM180)	879R4S	Em,Sm,	1×10^{-7} (6% Tc^r)
"	"	Tc,Sm	1×10^{-7} (83% Em^r)

a Mating and plating procedures were as described elsewhere (7, 8). Rif, rifampin; Fa, fusidic acid.

be found. (Only a 13.5-kb cryptic plasmid already present in the staphylococcal recipient was evident.) In the case of pAM380 (pAM373::Tn917::Tn916), Tc-resistant transconjugants arose at 10^{-8}; whereas Em-resistant transconjugants could not be obtained. Those Tc-resistant transconjugants that were obtained were sensitive to Em, and no new plasmid could be detected.

The establishment of the Tn918 and Tn916 markers in *Staph. aureus* is consistent with a behavior previously observed for Tn916 in *S. faecalis* (9). When located on a conjugative plasmid such as pAM81 (24 kb, Em-resistance), Tn916 appears to undergo a "zygotic induction" when introduced into a new strain. That is, following transfer of pAM180 (pAM81::Tn916), Tn916 excises from the plasmid and is either lost or inserts into the chromosome. When pAM180 was used in matings with *Staph. aureus* (Tab. 4), Tc-resistant transconjugants as well as Em-resistant transconjugants could be obtained; however, not all Em-resistant transconjugants were Tc-resistant and vice versa. This is similar to the behavior previously observed (9) in matings between *S. faecalis* strains. In view of the multiple similarities between Tn918 and Tn916, it seemed likely that transfer of Tc-resistance from pAM378 donors resulted in insertions of Tn918 into the recipient chromosome. Since the frequency of Tc-resistant transconjugants was relatively high ($\sim 10^{-5}$), it is conceivable that the plasmid actually entered the staphylococcal recipient but could not be maintained and was subsequently segregated.

FURTHER ANALYSES OF TN918 AND COMPARISONS TO TN916

It has been possible to clone Tn918 in *Escherichia coli* DH1 by ligating a pAD1 *Eco*R1 D::Tn918 fragment into the single *Eco*R1 site of pGL101 [ampicillin resistance (Ap), 2.4 kb] (16). *E. coli* transformants resistant to both Ap and Tc could be obtained; however, Tc-resistance was lost at high frequency (e.g., over 90% of the cells became sensitive to Tc during unselected overnight growth) without loss of Ap-resistance. Analyses of the sensitive derivatives have shown that Tn918 was lost and that there was a regeneration of the *Eco*R1 D fragment (i.e., the pAD1 fragment into which Tn918 had been inserted). Exactly the same phenomenon had been previously observed for Tn916 (10). When a Tn918 clone designated pAM390 was compared with a similar Tn916 clone (pAM170), restriction digests revealed 5 similar *Hinc*II fragments (within the transposons) and single recognition sites for *Hind*III and *Kpn*I (data not shown). As expected, strong homology was evident between Tn916 and Tn918 in Southern blot hybridizations (19) using Tn916 as a probe.

ANALYSIS OF THE TN918 AND TN916 INSERTS IN *STAPH. AUREUS*

Tn916 has been previously shown to insert into a number of different sites in the *S. faecalis* chromosome; this could be demonstrated by hybridizing a ^{32}P-labeled restriction fragment containing Tn*916* to chromosomal DNA cleaved with *Hind*III (9). In the case of each insert, 2 fragments hybridized; these represented the 2 transposon-host DNA junction fragments extending from the single *Hind*III site within Tn*916*. Because the size of these fragments differed in the various isolates, insertions into different sites must have occurred. Figure 3 shows the results of similar experiments in which ^{32}P-labeled pAM170 (a pGL101 chimera containing Tn*916*) was hybridized to *Hind*III digested chromosomal DNA isolated from several staphylococcal strains containing Tn*918* or Tn*916* inserts. The transposons had been introduced from their positions on plasmid DNA (i.e., from pAM378 or pAM380; see above). It is readily apparent that insertions of both Tn*918* and Tn*916* occurred into different sites on the recipient chromosome; and in a few cases more than one transposon insertion was evident.

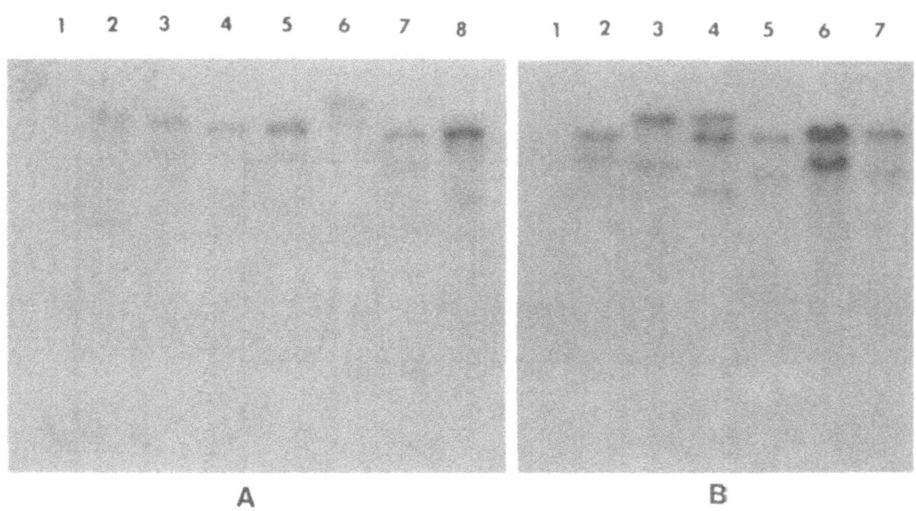

Fig. 3. Autoradiograms obtained from filter blot of *Hind*III-digested chromosomal DNA of Tc-resistant transconjugants of *Staph. aureus* and hybridization to ^{32}P-labeled pAM170 (contains Tn*916*). The method was that of Southern (19) as modified by Wahl et al. (24). A. Lanes 2 through 8 represent 7 different transconjugants from matings between *S. faecalis* JH2SS(pAM378) and *Staph. aureus* 879R4RF. Lane 1 represents a Tc-sensitive recipient (879R4RF). B. Lanes 2 through 7 represent transconjugants acquiring Tn*916* as a result of matings between *S. faecalis* FA2-2 (pAM380) and *Staph. aureus* 879R4S. Lane 1 is a Tc-sensitive recipient (879R4S).

PROPERTIES OF THE PHEROMONE cAM373

Similar to the cases of cAD1 and cPD1, the pheromone that induces pAM373 transfer, now referred to as cAM373, was heat stable, resistant to trypsin, and sensitive to chymotrypsin. The results of gel filtration chromatography are shown in Fig. 4A, where cAM373 is observed to elute between cAD1 and cPD1. On the basis of the previously estimated sizes of cAD1 and cPD1 (12), cAM373 was approximated at 850 d.

It was also found that strains harboring pAM373 or related derivatives excreted low levels of an inhibitor activity presumed to represent a modified form of the pheromone. Designated mcAM373, this activity was observed to have a mass about 300-400 d larger than cAM373 (Fig. 4B).

CONCLUDING REMARKS

The observation that members of a few other bacterial species can excrete a substance resembling a specific sex pheromone in *S. faecalis* does not necessarily imply a pheromonal role in those

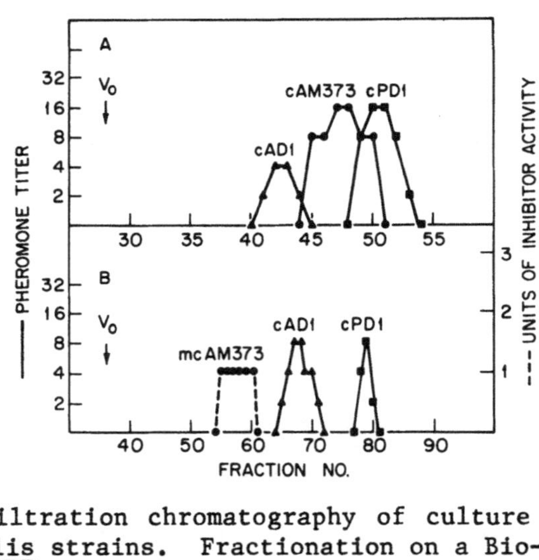

Fig. 4. Gel filtration chromatography of culture filtrates of *S. faecalis* strains. Fractionation on a Bio-Rad P2 polyacrylamide column (200-400 mesh; 1.6 x 100 cm) and the corresponding assays were as described elsewhere (12). Vo refers to the void volume. Strain FA373 was used as a responder to measure cAM373 activity. (A) Filtrate of JH2-2. (B) Filtrate of FA2-2(pAM378). (Note: a unit of inhibitor is defined as the amount capable of giving rise to a 50% reduction of pheromone activity.

species. Indeed, the fact that pAM373 derivatives could not be established in *Staph. aureus* or *Strep. sanguis* suggests the possible absence of a direct relationship between cAM373 and mating in these organisms. As previously noted (1), it is conceivable that bacterial production of the substances which donor strains recognize as sex pheromones could have preceded the evolution of the related conjugative systems, with the latter simply taking advantage of these available, extracellular compounds as mating signals. If this is true, the question arises as to the nature of the original, perhaps continuing, function of these peptides. The production of cAM373 activity by all *Staph. aureus* strains tested, and a general absence of the activity in most coagulase-negative staphylococci, raises the possibility of a role in virulence. The derivation of mutants defective in cAM373 production and their examination in virulence assays could shed light on this interesting question.

Although pAM373 derivatives failed to establish in *Staph. aureus*, the fact that they could facilitate insertions of Tn*918* and Tn*916* at multiple sites in the staphylococcal chromosome suggests their potential as delivery vehicles for these transposons in this species. After targeting a particular gene by mutagenic insertion, restriction fragments containing either of these transposons could be cloned in *E. coli* (10) where excision then gives rise to a regenerated segment of DNA containing the specific staphylococcal locus of interest.

ACKNOWLEDGMENTS

We wish to thank M. Smith, Y. Yagi, E. Ehrenfeld, G. Fitzgerald, Y. Ike, R. Wirth, J. Shaw, and R. Kessler for helpful discussions. We also thank M. O'Connor-Scarlet and our many colleagues for their generous donations of bacterial strains. This work was supported by National Institutes of Health grants DE02731 and AI10318. C.G.-b is the recipient of a Junior Faculty Research Award from the American Cancer Society, and B.W. is supported by a National Institute of Health Postdoctoral Fellowship.

REFERENCES

1. Clewell, D. (1981) Plasmids, drug resistance, and gene transfer in the genus *Streptococcus*. *Microbiol. Rev.* 45:409-436.
2. Clewell, D.B., B.A. White, Y. Ike, and F. An (1984) Sex pheromones and plasmid transfer in *Streptococcus faecalis*. In *Microbial Development*, R. Losick and L. Shapiro, eds. Cold Spring Harbor Press, Cold Spring Harbor, New York, pp. 122-149.
3. Clewell, D., P. Tomich, C. Gawron-Burke, A. Franke, Y. Yagi, and F. An (1982) Mapping of *Streptococcus faecalis* plasmids pAD1 and pAD2 and studies relating to transposition of Tn*917*. *J. Bact.* 152:1220-1230.

4. Clewell, D., Y. Yagi, G. Dunny, and S. Schultz (1974) Characterization of three plasmid DNA molecules in a strain of Streptococcus faecalis. Identification of a plasmid determining erythromycin resistance. J. Bact. 117:283-289.
5. Clewell, D., Y. Yagi, Y. Ike, R. Craig, B. Brown, and F. An (1982) Sex pheromones in Streptococcus faecalis: Multiple pheromone systems in strain DS5, similarities of pAD1 and pAMγ1, and mutants of pAD1 altered in conjugative properties. In Microbiology-1982, D. Schlessinger, ed. American Soc. Microbiol., Washington, D.C. pp. 97-100.
6. Dunny, G., B. Brown, and D. Clewell (1978) Induced cell aggregation and mating in Streptococcus faecalis. Evidence for a bacterial sex pheromone. Proc. Natl. Acad. Sci., USA 75:3479-3483.
7. Dunny, G., R. Craig, R. Carron, and D. Clewell (1979) Plasmid transfer in Streptococcus faecalis. Production of multiple sex pheromones by recipients. Plasmid 2:454-465.
8. Franke, A., and D. Clewell (1981) Evidence for a chromosome-borne resistance transposon in Streptococcus faecalis capable of "conjugal" transfer in the absence of a conjugative plasmid. J. Bact. 145:494-502.
9. Gawron-Burke, C., and D. Clewell (1982) A transposon in Streptococcus faecalis with fertility properties. Nature 300:281-284.
10. Gawron-Burke, C., and D. Clewell (1984) Regeneration of insertionally inactivated streptococcal DNA fragments following excision of Tn916 in Escherichia coli: Strategy for targeting and cloning of genes from gram-positive bacteria. J. Bact. 159:214-221.
11. Ike, Y., and D. Clewell (1984) Genetic analysis of the pAD1 pheromone response in Streptococcus faecalis using Tn917 as an insertional mutagen. J. Bact. 158:777-783.
12. Ike, Y., R. Craig, B. White, Y. Yagi, and D. Clewell (1983) Modification of Streptococcus faecalis sex pheromones after acquisition of plasmid DNA. Proc. Natl. Acad. Sci., USA 80:5369-5373.
13. Jacob, A., and S. Hobbs (1974) Conjugal transfer of plasmid-borne multiple antibiotic resistance in Streptococcus faecalis var zymogenes. J. Bact. 117:360-372.
14. Kessler, R., and Y. Yagi (1983) Identification and partial characterization of a pheromone-induced adhesive surface antigen of Streptococcus faecalis. J. Bact. 155:714-721.
15. Lacks, S. (1977) Binding and entry of DNA in bacterial transformation. In Microbial Interactions, Series B, Receptors and Recognition, Vol. 3. J.L. Reissig, ed. Chapman and Hall, London, pp. 177-232.
16. Lauer, G., R. Pastrana, J. Sherley, and M. Ptashne (1981) Construction of overproducers of the bacteriophage 434 repressor and cro proteins. J. Molec. and Appl. Genet. 1:139-147.

17. Novick, R., and R. Brodsky (1972) Studies on plasmid replication. 1. Plasmid incompatibility and establishment in Staphylococcus aureus. J. Molec. Biol. 68:285-302.
18. Rosan, B. (1976), Relationship of the cell wall composition of group H streptococci and Streptococcus sanguis to their serological properties. Infect. and Immun. 13:1144-1153.
19. Southern, E. (1975) Detection of specific sequences among DNA fragments separated by gel electrophoresis. J. Molec. Biol. 98:503-517.
20. Stobberingh, E., and K. Winkler (1977) Restriction-deficient mutants of Staphylococcus aureus. J. Gen. Microbiol. 99:359-367.
21. Tomich, P., F. An, and D. Clewell (1978) A transposon (Tn917) in Streptococcus faecalis which exhibits enhanced transposition during induction of drug resistance. Cold Spring Harbor Symp. Quant. Biol. 43:1217-1221.
22. Tomich, P., F. An, and D. Clewell (1980) Properties of erythromycin-inducible transposon Tn917 in Streptococcus faecalis. J. Bact. 141:1366-1374.
23. Tomich, P., F. An, S. Damle, and D. Clewell (1979) Plasmid related transmissibility and multiple drug resistance in Streptococcus faecalis subsp. zymogenes strain DS16. Antimicrob. Ag. Chemother. 15:828-830.
24. Wahl, G., M. Stern, and G. Stark (1979) Efficient transfer of large DNA fragments from agarose gels to diazobenzyloxymethyl paper and rapid hybridization by using dextran sulfate. Proc. Natl. Acad. Sci., USA 76:3683-3687.
25. Yagi, Y., R. Kessler, J. Shaw, D. Lopatin, F. An, and D. Clewell (1983) Plasmid content of Streptococcus faecalis strain 39-5 and identification of pheromone (cPD1)-induced surface antigen. J. Gen. Microbiol. 129:1207-1215.

GENERAL GENETIC RECOMBINATION OF BACTERIAL PLASMIDS

Amikam Cohen, Zipora Silberstein,
Shimshon Broido, and Avraham Laban

Department of Molecular Genetics
Hadassah Medical School
The Hebrew University
P.O. Box 1172
Jerusalem 91-101, Israel

PLASMID RECOMBINATION

Analysis of genetic recombination between 2 chromosomes, which differ from each other by at least 2 genetic markers, is based mainly on the production of recombinants which are phenotypically distinguishable from their progenitors. The use of small DNA molecules such as bacterial plasmids as recombination substrates provides an additional handle for the investigation of recombination. It also allows the description of recombination events in molecular terms. Recombination of 2 plasmids, genetically distinguishable from each other, can be followed as recombination reactions in other systems by monitoring changes in the phenotypic expression of the plasmid genes (1,2,3). But in addition, plasmidic recombination may be investigated by analyzing changes of their molecular structures. Interplasmidic recombination leads to the formation of plasmid oligomers and intraplasmidic recombination between repeated sequences can lead to deletion of the region between the crossing-over sites, and, when the recombination substrate is an oligomer, to monomerization. When mutations affect restriction endonuclease sites, recombination products may be compared to recombination substrates by restriction endonculease analysis (2,4,5). The small size of bacterial plasmids also allows the isolation and characterization of molecular intermediates of the recombination pathway (6,7).

These features of the plasmid recombination system allow determination of the dependency of plasmid recombination on host gene functions (1,2,3,6,7,8), isolation of recombinogenic elements that increase frequency of recombination (9), determination of the length of homology which is involved in recombination (10,11), and investigation of the molecular nature of the recombination pathway (4,5,6).

While plasmids are preferred probes for the investigation of bacterial recombination, their susceptibility to the bacterial system is of major concern to those who are interested in molecular cloning of DNA. Recombination reactions which affect plasmid molecular structure may also interfere with the integrity of cloned DNA fragments, especially when these fragments contain repetitive sequences. Loss of DNA sequences by intraplasmidic recombination has affected several cloned DNA fragments, including highly repeated satellite Drosophila DNA (12), yeast ribosomal DNA (13,14) adenovirus DNA (15), balbini ring DNA (16), retrovirus-like repeated mouse gene family (17), and Moloney sarcoma virus DNA (18). Attempts have been made to overcome the problem of the instability of cloned DNA by using _recA_ mutants as recipient bacteria. However, since plasmid recombination reactions are at least partially _recA_-independent (2,3,4,11,14,15) this approach has limited success.

The length of the repeated DNA sequence and the degree of homology determine recombination frequency. Frequencies of 10^{-3} to 10^{-4} recombinants per cell have been observed when recombination occurred between homologous sequences of several hundred nucleotides in rec^+ hosts (2,3). When very short direct repeats are involved (5 to 7 nucleotides), frequencies of 10^{-7} to 10^{-8} deletions per cell have been determined (10,11). It has not yet been established whether the same mechanism is involved in the 2 classes of recombination events.

Plasmid maintenance is affected by recombination activity. Since plasmid copy number is determined through the control of the number of replication origins in the cell (19), plasmid oligomerization reduces the ratio of plasmid molecules to plasmid origins and, therefore, reduces also the number of plasmid molecules in the cell. This reduction may interfere with partition of plasmids to daughter cells, especially when unit copy-number plasmids, or high copy-number plasmids which lack a partition function, are involved.

The problems of oligomerization is overcome by the activity of specific recombination systems which resolve oligomers to monomeric products. Recombinogenic elements which are involved in site-specific recombination or activate _recA_-dependent intraplasmidic recombination have been described both in unit copy-number and in high copy-number plasmids (20,21,22).

THE MOLECULAR NATURE OF PLASMID RECOMBINATION

The Role of Strand Exchange

Two classes of genetic exchange can lead to the reconstruction of a functional gene by recombination of 2 genes mutated at different sites: one involves a double-strand exchange between the mutation sites, and the other a single-strand transfer, which is followed by branch migration across one or both mutations and processing of the heteroduplex intermediate to homoduplex by replication or mismatch repair (23). While the first mechanism contributes to recombination between distal markers, the second is involved mainly in recombination between close markers (24,25,26).

A system which facilitates the study of the molecular nature of plasmid recombination has been constructed (2,3). This system consists of 2 copies of the tet gene, each of which is mutated at a different restriction endonuclease site. The mutated genes are located on 2 compatible plasmids, for the analysis of intermolecular recombination (Fig. 1A), or, as a tandem repeat on one plasmid, for the analysis of intramolecular recombination (Fig. 1B). Recombination events are scored by determining the frequency of Tet^r recombinants, and the molecular structure of plasmid recombination products, which confer tetracycline-resistance, is determined by restriction endonuclease analysis. Mutations are induced in the tet gene by plasmid linearization with the appropriate restriction endonuclease, subsequent conversion of the single-strand ends to blunt ends by S1 nuclease digestion (deletions), or DNA polymerase 1-induced synthesis of the complementary strand (additions), and religation. Alternatively, molecular linkers or DNA sequences of various lengths are inserted at the restriction sites (2,3,4,5). In an interplasmidic recombination system, derivatives of pACYC184 (27) and pBR322 (28) are used as recombination substrates. These 2 plasmids share a sequence of 1.4 kb which is bound at one side by the EcoRI site on pBR322. Each of these plasmids confers antibiotic-resistance, in addition to the resistance to tetracycline; pACYC184 confers resistance to chloramphenicol, and pBR322, to ampicillin.

The molecular mechanism which yields tet^+ recombination products determines their structure. Structural analysis of these products can, therefore, contribute to the understanding of the recombination pathway (2,4,5). Examples of alternative interplasmidic recombination reactions in the BamHI-SalI interval of the tet gene, which yield distinct classes of recombination products, are presented in Fig. 2 and Fig. 3. Recombination of a pACYC184 derivative mutated at the SalI site, with a pBR322 derivative mutated at the BamHI site, can yield tet^+ products through a reciprocal double-strand exchange between 2 sites (Fig. 2), or by a process of single-strand exchange across one or both restriction sites and repair of the mismatched sequences (Fig. 3). The first possible pathway would

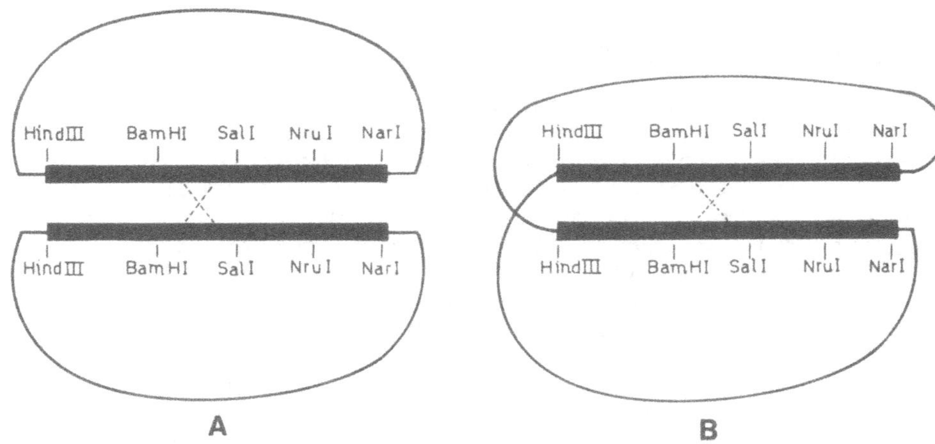

Fig. 1. Intermolecular and intramolecular recombination of the mutated tet gene. For intermolecular recombination, tet genes on 2 compatible plasmids (pBR322 and pACYC184) are mutated, each at a different restriction endonuclease site (A). In intramolecular recombination (B), both mutated genes are located as a tandem repeat on one plasmid. Recombination proficiencies are determined by scoring Tet^r cells (2,3). Recombination products are subjected to restriction endonuclease analysis (4,5).

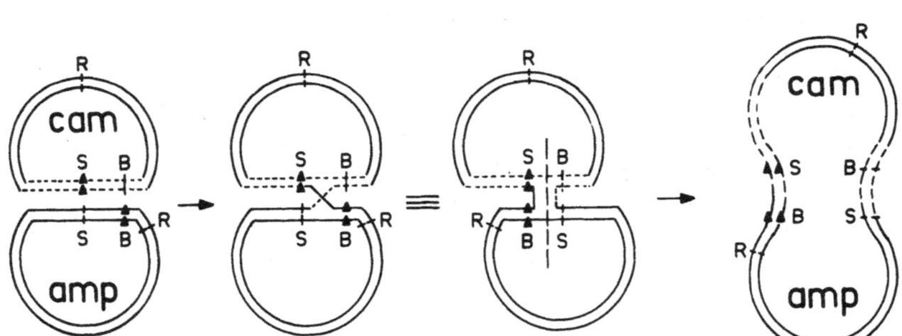

Fig. 2. The anticipated tet^+ interplasmidic recombination products of a reciprocal double-strand exchange between mutated BamHI and SalI restrictions endonuclease sites. A pBR322 derivative conferring ampicillin resistance (amp) and mutated at the BamHI site (B) was recombined with a pACYC184 derivative conferring chloramphenicol resistance (cam) and mutated at the SalI site (S), to yield a tet^+ amp^+ cam^+ heterodimer with one site for each restriction enconuclease.

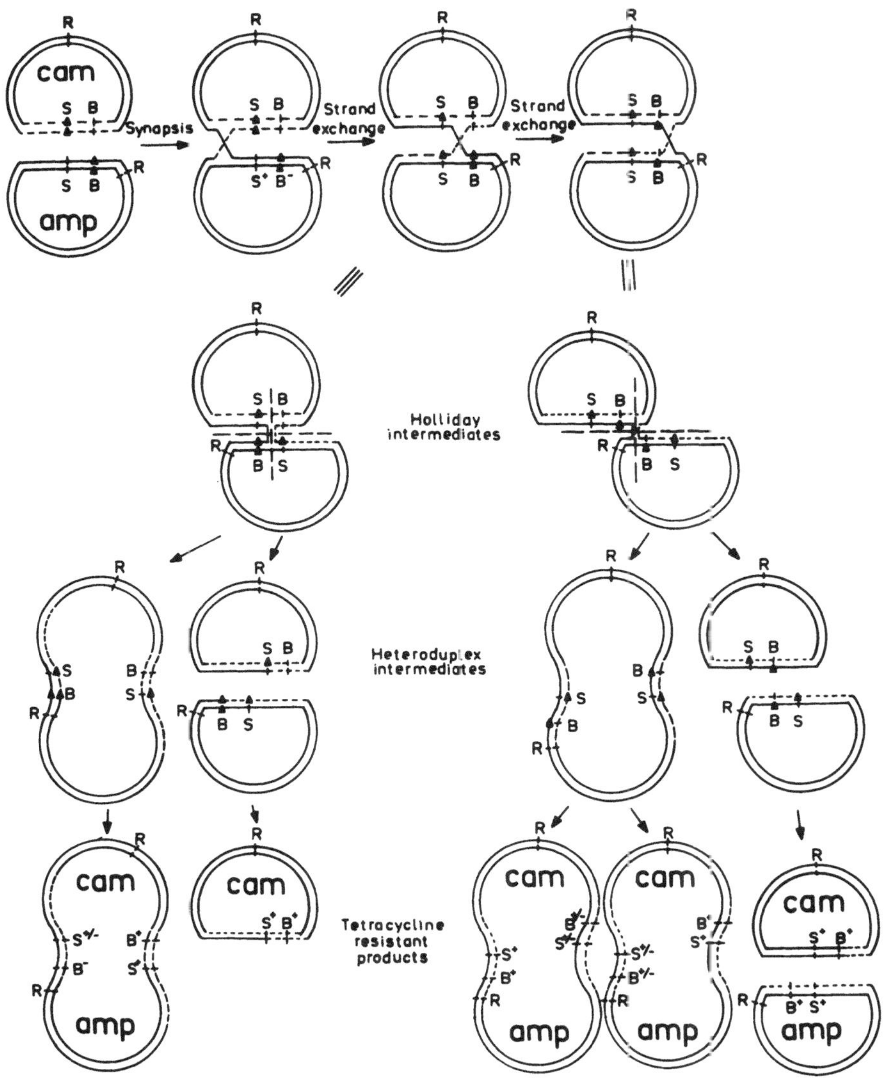

Fig. 3. Products of recombination reactions which involved heteroduplex formation and mismatch repair. In a recombination reaction between the substrates presented in Fig. 2, strand exchange across the SalI site and repair of the mismatched sequence can yield 3 possible tet^+ products; pACYC184 or one of 2 heterodimers (bottom left). Strand exchange across both mutation sites and independent repair of each mismatched sequence can yield 9 possible products (bottom right).

yield a single tet$^+$ product: a heterodimer with one functional tet gene and one gene carrying both mutations (Fig. 2). The second pathway can yield different classes of recombination products, depending on the path of strand exchange and the nature of the mismatch repair reaction (Fig. 3). By this mechanism, a synaptic complex which migrates across the SalI site of the BamHI-SalI interval forms a heteroduplex at the SalI site. Endonucleolitic resolution of the recombination intermediate and repair of the heteroduplex can yield 3 possible tet$^+$ plasmid products (Fig. 3, Tab. 1): a pACYC184 monomer or either one of 2 possible heterodimers. Both dimers would have one functional tet gene, and a tet gene mutated at the BamHI site. The SalI site of the mutated tet gene may either be mutated or repaired. Since in this case the heteroduplex does not include the mutated BamHI site, the functional tet gene which is produced includes the BamHI site of pACYC184. In an alternative pathway, migration of the branch point across both mutations forms mismatched sequences at the 2 restriction sites. Since each strand has a mu-

Tab. 1. Recombination products of crosses in the BamHI-SalI genetic interval.

Box	Recombining products	Recombining plasmids pAL231xpAL242	pAL234xpAL246
A, B, C	R — R — B S — R	21.1%	100%
B, C	R S — R — B S — R	30.3%	0
B, C	B S — R	44.9%	0
C	R B — R — B S — R	2.8%	0
C	R B S — R — B S — R	0	0
C	R B S — R — B — R	0	0
C	R B S — R	0.9%	0
C	R B S — R — S — R	0	0
C	R B S — R — R	0	0

Clones of E. coli K-12 (AB1157) cells harboring tet$^+$ recombination products were selected on tetracycline plates. Plasmid DNA preparations of 2,000-4,000 Tet^r colonies were used to transform recA (DR100) cells, under conditions of limiting DNA concentration, and Tet^r transformants were analyzed for resistance to ampicillin and chloramphenicol (about 500 transformants for each cross). Plasmids of 59 Tet^r Amp^r Cam^r transformants of the cross pAL231xpAL242 and 67 of the cross pAL234xpAL246 were subjected to restriction endonuclease analysis. R, B, and S designate restriction endonuclease sites of EcoRI, BamBI, and SalI, respectively. The molecular structure which is indicated by box A is the anticipated product of a reciprocal double-strand exchange between the mutation sites (Fig. 2). Box B designates the anticipated products of a recombination reaction which involves a single-strand exchange across the SalI site and mismatch repair. The products which were marked by box C are anticipated from a recombination reaction which involves independent repair of mismatched sequences at both restriction sites (Fig. 3).

tation at one of the sites and a wild allele at the other, a functional tet gene cannot be formed by plasmid replication or by co-repair of the 2 mutations. In such an event only independent mismatch repair at each site would yield a functional gene. A recombination intermediate with mismatched sequences at each of the restriction sites can yield 9 different tet^+ products by independent repair: pACYC184 or pBR322 monomers, a dimer carrying 2 functional tet genes, 2 possible dimers carrying one functional gene and one gene with 2 mutations, and 4 possible dimers, each with one functional gene and one gene with one of the 2 mutations (see Tab. 1). Using a similar approach, one can predict the class of tet^+ products if recombination involves strand exchange across the BamHI but not across the SalI site (not shown); this class includes a pBR322 monomer and 2 possible dimers, each with one SalI site and either one or 2 BamHI sites.

In order to test the predictions of the schemes which are presented in Fig. 2 and Fig. 3, a pBR322 derivative carrying a SI-induced deletion at the BamHI site (pAL231) was co-transformed with a pACYC184 derivative carrying a SI-induced deletion at the SalI site (pAL242) into an *E. coli* K-12 cell (AB1157). Following growth in the absence of antibiotics, clones harboring tet^+ recombination products were selected by plating samples on tetracycline plates and plasmids conferring tetracycline-resistance were isolated and subjected to structural analysis. The differences between the antibiotics-resistance markers on the 2 recombining plasmids allows discrimination between the 2 possible monomeric tet^+ products and the heterodimers. Heterodimers, which confer resistance to both chloramphenicol and ampicillin, are subjected to restriction endonuclease analysis. Since each of the 2 recombining plasmids has a unique EcoRI site at a different location with respect to the tet gene, detailed analysis of the heterodimer products is accomplished by size determination of restriction endonuclease products of SalI, BamHI, and either one of these endonucleases in combination with EcoRI. Results of structural analysis of recombination products of the pAL231xpAL242 cross are presented in Tab. 1. The presence of monomeric products and the high incidence of conversion rule out the possibility that most recombinations in this system involve a reciprocal double-strand exchange between the mutation sites. The location of the functional tet gene on most dimers, the observation that most tet^+ monomers are pACYC184 and that the incidence of conversion at the SalI site is exceedingly higher than that at the BamHI site is consistent with the notion that most recombination events which yield tet^+ products involve a single-strand exchange across the SalI site, but not across the BamHI site. The high incidence of gene conversion at the SalI site may suggest an asymmetric strand-exchange mechanism as proposed by Meselson and Radding (29).

Results of a recombination experiment in the HindIII-BamHI interval of the tet gene are presented in Tab. 2. A pBR322 derivative carrying an SI-induced deletion at the HindIII site (pAL230) was

Tab. 2. Recombination products of crosses in the HindIII-BamHI interval.

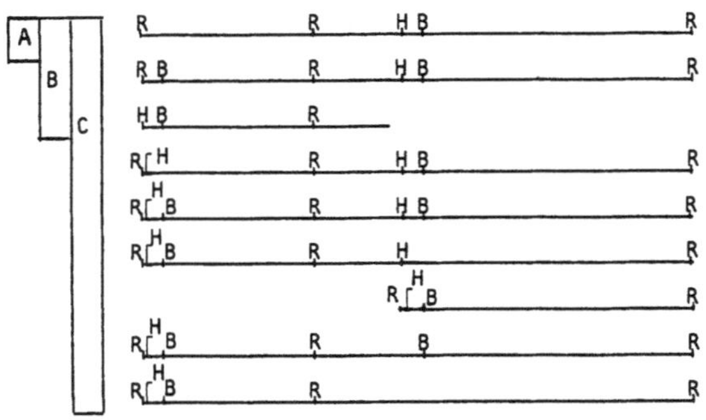

Recombining products (rows as drawn, top to bottom)	Recombining plasmids: pAL230xpAL241
1	36.7%
2	46.4%
3	16.2%
4	0
5	0
6	0
7	0.7%
8	0
9	0

Recombination products were analyzed as in Tab. 1. Except that HindIII, BamHI, and EcoRI restriction endonucleases were used. Forty-three Tet^r Cam^r Amp^r transformants were analyzed this way. R, B, and H indicate restriction sites of EcoRI, BamHI, and HindIII, respectively. Six hundred fifteen transformants were analyzed for resistance to ampicillin and chloramphenicol. Box A indicates the anticipated products of a reciprocal double-strand exchange between the mutation sites. Box B, the products of a recombination reaction which involves strand exchange across the BamHI site, and box C the products of a recombination reaction which involves strand exchange across both sites.

crossed with a pACYC184 derivative carrying an SI-induced deletion at the BamHI site (pAL241). Most tet^+ monomers obtained by this cross are also resistant to chloramphenicol, as occurs with the recombination substrate with the BamHI mutation. All dimers have a single HindIII site and most have 2 BamHI sites. The location of a single HindIII site in all products is the same as in pACYC184, 1.4 kb away from an EcoRI site.

Thus, in both crosses, most recombinations involved a conversion at the mutated restriction endonuclease site which is located at the distal position with respect to the HindIII site of the tet gene.

The results are consistent with a proposal that most recombination events which yield tet^+ recombinants are initiated outside the HindIII-SalI interval of the tet gene on the SalI side, and strand exchange is in the SalI → HindIII direction. This conclusion is also supported by the distribution of recombination products of crosses in the NruI-SalI and NruI-BamHI interval of the tet gene (data not shown).

The Effect of Nonhomologous Sequences on Strand Exchange

Do major nonhomologies block strand exchange? Studies of λ phage recombinations indicate that, in this system, major nonhomologies are included in the heteroduplex region but their resolution to functional products depends on DNA replication (Ref. 30, M. Lichten and M.S. Fox, in prep.). In an in vitro system, RecA-promoted strand exchange across major nonhomologies depends on ATP regeneration and on the presence of single-strand DNA binding protein (31). In the plasmid recombination system, major nonhomologies block strand exchange. A cross between a pBR322 derivative with a major insertion at the BamHI site (pAL234) and a pACYC184 derivative with a major insertion at the SalI site (pAL246) yields only one class of tet^+ recombination products -- heterodimers with one functional gene and one gene with both mutations (Tab. 1). The difference between the product of this cross and the product of the pAL231xpAL242 cross suggests that major nonhomologies are not included in a heteroduplex region, or, if included, are not repaired to yield a functional tet gene.

If, as suggested above, most recombination events in the pAL231xpAL242 cross are initiated outside the BamHI-SalI interval on the SalI side, and major nonhomologies block strand exchange, then a major insertion at the SalI site, but not at the BamHI site, should lower recombination frequency in this interval. Results which are presented in Tab. 3 are consistent with this prediction. Bracketing

Tab. 3. The effect of nonhomology on recombinant frequency in short genetic intervals.

Genetic interval	Recombining plasmids	Site of major non-homology	Recombinant frequency
BamHI-SalI	pAL231xpAL242	-	1×10^{-3}
BamHI-SalI	pAL234xpAL246	BamHI, SalI	4×10^{-5}
BamHI-SalI	pAL231xpAL246	SalI	9×10^{-5}
BamHI-SalI	pAL234xpAL242	BamHI	3.8×10^{-3}
HindIII-BamHI	pAL234xpAL240	BamHI	1.9×10^{-4}
HindIII-BamHI	pAL231xpAL240	-	2×10^{-3}

Recombinant frequencies were determined as described (2,5,8). The major nonhomology at the BamHI site is a 650 nucleotide sequence of λ phage CII region and at SalI site a 1,300 nucleotide sequence of *Streptomyces griseus* plasmid pSC1 (5).

the BamHI-SalI interval by inserting major nonhomologies at both restriction sites (pAL234xpAL246) leads to a major reduction in recombinant frequency. A major insertion at the SalI site (pAL231x pAL246) has a similar effect, but a major insertion at the BamHI site (pAL234xPAL242) does not lower recombination frequency. In fact, a major insertion at the BamHI site increases recombination frequency in the BamHI-SalI interval, possibly because of interference with strand exchange across the entire interval and, thereby, promoting endonucleolitic resolution of recombination intermediates within the interval. Heteroduplexes which are formed by such resolution can yield functional homoduplexes by replication or corepair. While a major insertion at the BamHI site does not decrease recombination proficiency at the BamHI-SalI interval, it decreases recombination proficiency in the HindIII-BamHI interval. Recombination frequencies of the pAL234xpAL240 are 10 times lower than recombination frequencies of the pAL231xpAL240 cross. This result also supports the proposal of initiation of recombination events at a site or a region outside the interval, on the SalI side, and a polar strand exchange in the SalI → HindIII direction. The nature of the hypothetical recombinogenic element which is involved in initiation of recombination events in this system, the mechanism of strand exchange and resolution of recombination intermediates, and the dependency of these processes on bacterial gene functions are under investigation.

THE DEPENDENCY OF PLASMID RECOMBINATION ON HOST FUNCTIONS

Two pathways are involved in postconjugational recombination of chromosomal DNA in *E. coli* K-12: the RecBC pathway which is responsible for most recombination events in the cell and the RecF pathway which is involved in about 1% of chromosome recombination (32). A third pathway, the RecE, can be activated in strains of *E. coli* which carry the *rac* prophage (33). Like the RecF pathway, the RecE pathway does not play a significant role in chromosomal recombination in wild-type cells. Mutations in the *recBrecC* genes, which inactivate the RecBC pathway and thereby reduce chromosomal recombination activity, can be suppressed by an *sbcA* or an *sbcB* mutation. The *sbcA* mutation activates the RecE pathway by depression of the synthesis of the *recE* gene product, exonuclease VIII (34). The *sbcB* mutation inactivates exonuclease I and thereby increases recombination activity via the RecF pathway (35). All 3 chromosomal recombination pathways have an absolute dependency on the *recA* gene product (32). The RecBC pathway also depends on the *recBrecC* gene product, exonuclease V. The *RecE* pathway depends on the *recE* gene product, exonuclease VIII, and on *recF* gene activity (36); the RecF pathway depends on functional *recF* and *recJ* activities (37) and on a *lexA*-controlled function which is coded by a gene located at 57.5 min on the *E. coli* chromosome (38).

Differences between the dependency of plasmid recombination and the dependency of chromosomal recombination on bacterial gene functions have been reported in studies involving recombination between 2 *Flac tra*$^-$ elements. Recombination in this system has a *recA* dependency, but is completely independent of the *recBrecC* gene product. On the other hand, this system has partial dependency on the *recF* function (1). Further characterization of the role of host gene functions in the process of plasmid recombination was accomplished by employing the interplasmidic and intraplasmidic recombination system which is described above (2,3,7,8).

These studies indicate several differences between the activities which are involved in postconjugational recombination and in plasmid recombination (Tab. 4). Like chromosomal recombination, plasmid recombination in wild-type cells depends on a functional *recA* gene product. However, while chromosomal recombination has an absolute dependency on the *recA* function, a residual plasmid recombination activity (1-4% of that in wild type cells) is observed in *recA* mutants (2,3). The dependency of intraplasmidic recombination on the *recA* function is completely alleviated when recombination

Tab. 4. The dependency of plasmid recombination on bacterial gene functions.

Bacterial relevant genotype (representative strain)	Relative recombinant frequency		
	Intraplasmidic	Interplasmidic	Postconjugational
rec$^+$ (AB1157)	1.0	1.0	1.0
recA (JC2926)	1.5×10^{-2}	2.7×10^{-2}	10^{-5}
recBrecC (JC5519)	1.0	1.0	10^{-2}
recF (JC9239)	2.6×10^{-2}	2.7×10^{-2}	1.0
sbcB (JC11451)	1.0	1.0	1.0
recBsbcA (JC5183)	32	4.8	0.5
RecArecBsbcA (DR107)	33	0.[illegible]3	10^{-5}

Relative recombinant frequencies are determined in relation to frequencies in *rec*$^+$ (AB1157) cells. Relative plasmid recombinant frequencies are from Ref. 2, 8, and postconjugational recombination frequencies are of Ref. 32, 34, 36, 37. Recombinant frequencies of *recBsbcB* strains are not determined due to the instability of plasmids in this genetic background.

proceeds via the RecE pathway. Intraplasmidic recombination proficiency in *recArecBsbcA* cells is similar to that in *recBsbcA* cells. In both genetic backgrounds this proficiency is significantly higher than that in wild-type cells (2,3).

Interplasmidic recombination via the RecE pathway is partially dependent on the *recA* activity. Interplasmidic recombination proficiencies in *recArecBsbcA* cells is about 40-fold lower than that in *recBsbcA* (2).

RecA-independent recombination via the RecE pathway has also been observed with replicating λ DNA molecules as a substrate (39). It is conceivable that in both systems DNA replication intermediates serve as substrates for *recA*-independent recombination reaction. The activity which suppresses *recA* deficiency in the *recBsbcA* genetic background has not yet been characterized. An attractive possibility is the functioning of an activity, similar to that of λ phage β protein (40), which is associated with exonuclease VIII or with another prophage *rac* product.

Plasmid recombination in wild-type cells does not depend on the *recBrecC* gene product. It does however depend on the *recF* gene activity (7,8). Unlike chromosomal recombination, recombination of plasmids via a *recF*-dependent pathway is not enhanced by inactivation of exonuclease I by a *sbcB* mutation (8).

The differences between postconjugational chromosomal recombination and plasmid recombination in the degree of the dependency of these processes on *recBrecC* and *recF* functions suggest that the role of different gene products in recombination is determined by the molecular structure of the DNA substrate. It is possible that in vivo, like in vitro (41,42), circular DNA molecules do not serve as substrates for the *recBrecC* gene product, exonuclease V. Therefore, recombination of plasmids and other circular molecules depends on another enzymatic activity, possibly on the *recF* gene product. This may be related to the proposal that postconjugational recombination via the *recF* pathway involves circular intermediates (43). Understanding the nature of the enzymatic activity associated with the *recF* gene product and with the product of other genes which are involved in recombination via the RecF pathway will contribute to the investigation of the molecular nature of plasmid recombination.

REFERENCES

1. Willetts, N.S. (1975) Recombination and the *Escherichia coli* K-12 sex factor F. *J. Bact.* 121:36-43.
2. Laban, A., and A. Cohen (1981) Interplasmidic and intraplasmidic recombination in *Escherichia coli* K-12. *Mol. Gen. Genet.* 184:200-207.

3. Fishel, R.A., A.A. James, and R. Kolodner (1981) RecA independent general recombination of plasmids. Nature 294:184-186.
4. Doherty, M.J., P.T. Morrison, and R. Kolodner (1983) Genetic recombination of bacterial plasmid DNA. Physical and genetic analysis of the products of plasmid recombination in Escherichia coli. J. Mol. Biol. 167:539-560.
5. Laban, A., Z. Silberstein, and A. Cohen (1984) The effect of non-homologous DNA sequences on interplasmidic recombination. Genetics (in press).
6. Potter, H., and D. Dressler (1976) On the mechanism of genetic recombination: Electron microscopic observation of recombination intermediates. Proc. Natl. Acad. Sci., USA 73:3000-3004.
7. James, A.A., P.T. Morrison, and R. Kolodner (1982) Genetic recombination of bacterial plasmid DNA. Analysis of the effect of recombination-deficient mutations on plasmid recombination. J. Mol. Biol. 160:411-430.
8. Cohen, A., and A. Laban (1983) Plasmidic recombination in Escherichia coli K-12: The role of recF gene function. Mol. Gen. Genet. 189:471-474.
9. James, A.A., P.T. Morrison, and R. Kolodner (1982) Isolation of genetic elements that increase frequencies of plasmid recombinants. Nature 303:256-259.
10. Jones, I.M., S.B. Primrose, and S.D. Ehrlich (1982) Recombination between short direct repeats in a recA host. Mol. Gen. Genet. 188:486-489.
11. Albertini, A.M., M. Hofer, M.P. Calos, and J. Miller (1982) On the formation of spontaneous deletions: The importance of short sequence homologies in the generation of large deletions. Cell 29:319-328.
12. Brutlag, D., K. Fry, T. Nelson, and P. Hung (1977) Synthesis of hybrid bacterial plasmids containing highly repeated satellite DNA. Cell 10:509-519.
13. Cramer, J.H., F.W. Farrelly, J.T. Barnitz, and R.H. Rownd (1977) Construction and restriction endonuclease mapping of hybrid plasmids containg S. cerevisiae ribosomal DNA. Mol. Gen. Genet. 151:229-244.
14. Cohen, A., D. Ram., H.O. Halverson, and P.L. Wensink (1978) Deletions within E. coli plasmids carrying yeast rDNA. Gene 3:135-147.
15. Perricauset, M., A. Fritsch, U. Pattersson, L. Philipson, and P. Thiollais (1977) Excision and recombination of adenovirus DNA fragments in Escherichia coli. Science 196:208-210.
16. Case, S.T. (1982) Selective deletion of large segments of Balbiani ring DNA during molecular cloning. Gene 20:169-176.
17. Keshet, E., and Y. Shaul (1981) Terminal direct repeats in a retrovirus-like repeated mouse gene family. Nature 289:83-85.
18. McClements, W.L., L.W. Enquist, M. Oskarsson, M. Sullivan, and G.F. Vande Woude (1980) Frequent site-specific deletion of Coliphage λ murine sarcoma virus recombinants and its use in the identification of a retrovirus integration site. J. Virol. 35:488-497.

19. Prichard, R.H., and N.B. Grover (1981) Control of Plasmid Replication and its Relationship to Incompatibility. In Molecular Biology, Pathogenicity, and Ecology of Bacterial Plasmids, S.B. Levi, R.C. Clowes, and E.L. Koenig, eds. Plenum Press, New York, pp. 271-278.
20. Austin, S., M. Ziese, and N. Sternberg (1981) A novel role for site-specific recombination in maintenance of bacterial replicons. Cell 25:729-736.
21. Hakkaart, M.J.J., E. Veltkamp, and H.J.J. Nijkamp (1982) Maintenance of the bacteriocinogenic plasmid Clo DF13 in Escherichia coli cells. II. Specific recombination functions involved in plasmid maintenance. Mol. Gen. Genet. 188:338-344.
22. Hakkaart, M.J.J., P.J.M. van den Elzen, E. Veltkamp, and H.J.J. Nijkamp (1984) Maintenance of multicopy plasmid Clo DF13 in E. coli cells: Evidence for site-specific recombination at parB. Cell 36:203-209.
23. Holliday, R. (1964) A mechanism for gene conversion in fungi. Genet. Res. 5:282-304.
24. Amati, P., and M. Meselson (1965) Localized negative interference in bacteriolophage λ. Genetics 51:369-379.
25. Meselson, M. (1967) The Molecular Basis of Genetic Recombination. In Heritage from Mendel, R.A. Brink, ed. The University of Wisconsin Press, Madison, pp. 81-104.
26. Fox, M.S. (1978) Some features of genetic recombination in procaryotes. Ann. Rev. Genet. 12:47-68.
27. Chang, A.C.Y., and S.N. Cohen (1978) Construction and characterization of amplifiable multicopy DNA cloning vehicles derived from the P15A cryptic miniplasmid. J. Bact. 134:1141-1156.
28. Bolivar, F., R.L. Rodriguez, P.J. Green, M.C. Betlach, H.L. Heyneker, H.W. Boyer, J.H. Crosa, and S. Falkow (1977) Construction and characterization of new cloning vehicles. II. A multi-purpose cloning system. Gene 2:95-113.
29. Meselson, M.S., and C.M. Radding (1975) A general model for genetic recombination. Proc. Natl. Acad. Sci., USA 72:358-361.
30. Lichten, M., and M.S. Fox (1983) Effect of non-homology on bacteriophage lambda recombination. Genetics 103:5-22.
31. Bianchi, M.E., C. DasGupta, and C.M. Radding (1983) Insertions, deletions and mismatches in heteroduplex DNA made by RecA protein. Cell 34:931-939.
32. Clark, A.J. (1973) Recombination deficient mutants of E. coli and other bacteria. Ann. Rev. Genet. 7:67-86.
33. Kaiser, K., and N.E. Murray (1979) Physical characterization of the "rac prophage." Mol. Gen. Genet. 175:159-174.
34. Barbour, S.D., H. Nagaishi, A. Templin, and A.J. Clark (1970) Biochemical and genetic studies of recombination proficiency in Escherichia coli. II. Rec^+ revertants caused by indirect suppression of Rec^- mutations. Proc. Natl. Acad. Sci., USA 67:128-135.
35. Kushner, S.R., H. Nagaishi, and A.J. Clark (1972) Indirect suppression of recB and recC mutations by exonuclease I deficiency. Proc. Natl. Acad. Sci., USA 69:1366-1370.

36. Gillen, J.R., D.K. Willis, and A.J. Clark (1981) Genetic analysis of the RecE pathway of genetic recombination in Escherichia coli K-12. J. Bact. 145:521-532.
37. Horii, Z.I., and A.J. Clark (1973) Genetic analysis of the RecF pathway of genetic recombination in Echerichia coli K-12. Isolation and characterization of mutants. J. Mol. Biol. 80:327-344.
38. Lloyd, R.G., S.M. Picksley, and C. Prescott (1983) Inducible expression of a gene specific to the RecF pathway. Mol. Gen. Genet. 190:162-167.
39. Gillen, J., and A.J. Clark (1974) The recF pathway of bacterial recombination. In Mechanism in Recombination, R.F. Grell, ed. Plenum Press, New York, pp. 123-136.
40. Kmiec, E., and W.K. Hollman (1981) β protein of bacteriophage λ promotes renaturation of DNA. J. Biol. Chem. 256:12636-12639.
41. Karu, A.E., V. Mackay, P.S. Goldmark, and S. Linn (1973) The recBC deoxyribonuclease of Escherichia coli K-12. Substrate specificity and reaction intermediates. J. Biol. Chem. 248:4874-4884.
42. Taylor, A., and G.R. Smith (1980) Unwinding and rewinding of DNA by the RecBC enzyme. Cell 22:447-457.
43. Bresler, S.E., I.Y. Goryshin, and V.A. Lanzov (1981) The process of general recombination in Escherichia coli: Structure of intermediate products. Mol. Gen. Genet. 183:139-143.

THE ORIGIN OF PLASMID DNA TRANSFER DURING BACTERIAL CONJUGATION

Donald G. Guiney, Gary Chikami, Cornelia Deiss, and Emanuel Yakobson

Department of Medicine H811F
University of California Medical Center
225 Dickinson Street
San Diego, California 92103

INTRODUCTION

Bacterial conjugation is the process of gene transfer requiring cell-to-cell contact. Such mating systems are widespread in bacteria and have been described in many gram-positive and gram-negative organisms (1-3). However, detailed studies on the mechanism of DNA transfer during conjugation have been carried out only in gram-negative bacteria (4). The conjugation systems studied in gram-negatives are encoded by plasmids, and mediate the transfer of plasmid and, occasionally, chromosomal DNA from the donor to the recipient cell. Although transformation and transduction of plasmids occur, conjugation appears to be the major mechanism of plasmid DNA transfer in nature. In addition, conjugation is the only means of genetic exchange between different groups of bacteria, since transformation and transduction are generally limited to exchange between closely related members of the same genus or species. The promiscuous gene transfer mediated by broad host-range plasmids means that the gene pool available to bacteria is immensely larger than if genetic exchange was restricted by species barriers. Therefore, broad host-range conjugation enables bacteria to draw upon a very large genetic reservoir when subjected to evolutionary pressures. The most dramatic clinical example of this exceptional adaptability is the development and spread of antibiotic resistance in pathogenic bacteria.

Current knowledge of the molecular mechanisms of plasmid DNA transfer during conjugation is based heavily on studies of the F plasmid in *Escherichia coli* (reviewed in Ref. 1, 2, and 4). The general features of F plasmid transfer are shown in Fig. 1. The F

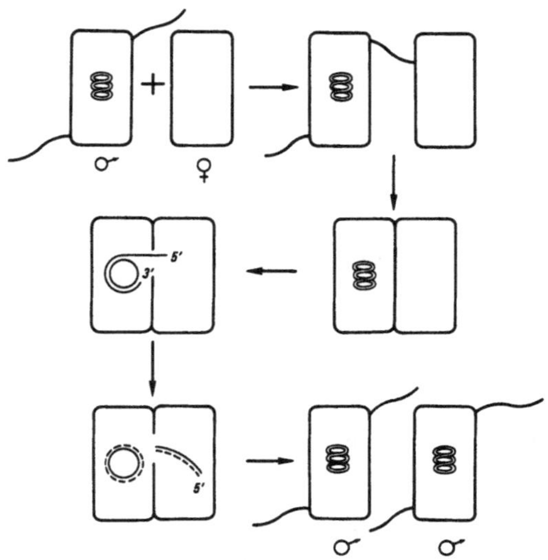

Fig. 1. Scheme of bacterial conjugation, based on the F plasmid of E. coli. For simplicity, the chromosomal DNA is not shown. Plasmid DNA synthesized in the donor and recipient during conjugal DNA replication is shown by the dashed lines.

sex pili are required to establish early mating contact between the donor and recipient cells (5). The donor and recipient then form stable mating pairs or aggregates with extensive wall-to-wall contacts (6). A mating bridge is formed, and a mating signal triggers conjugal DNA replication and transfer in the donor cell. A single specific strand is transferred to the recipient with the 5' end leading (7,8). Conjugal DNA synthesis in the donor replaces the transferred strand, while a complement to this strand is made in the recipient. The transferred strand is recircularized and ligated in the recipient to form the intact plasmid. The extent to which this scheme of F plasmid transfer can be generalized to other plasmids is not known. Single strand-specific transfer with 5' orientation has not been demonstrated for other plasmids. Nevertheless, the model shown in Fig. 1 forms the basis for studies of other plasmid conjugation systems.

The origin of transfer (oriT) is the region of particular interest in studying the molecular mechanism of DNA transfer. It is at oriT that the major events of conjugative DNA metabolism occur: nicking of the transferred strand, conduction from donor to recipient, initiation of conjugal DNA synthesis, and recircularization in the recipient. Because of this central role in conjugation, the oriT region has received detailed study.

PLASMID DNA-PROTEIN RELAXATION COMPLEXES

Properties of Relaxation Complexes

The discovery of plasmid DNA-protein relaxation complexes has provided a key to the mechanism of initiation of DNA transfer at oriT. Many bacterial plasmids, including resistance plasmids and colicinogenic factors, can be isolated from gently lysed cells in the form of a superhelical DNA-protein complex, which can be induced in vitro to nick the DNA and convert the plasmid to the open circular form (Fig. 2) (9,10). In the case of the ColE1 plasmid for *E. coli*, the nick occurs at a specific site in the heavy strand, and a 60,000 d protein remains covalently attached to the 5' end of the DNA (11-13). These results led to the proposal that the relaxation complex initiates plasmid transfer in vivo by nicking at the origin of transfer, mediates movement of the 5'end of the DNA into the recipient, and promotes recircularlization and ligation of the plasmid (13,14). This hypothesis was substantiated by Sherratt and coworkers who showed that the ColE1 relaxation complex is required for mobilization and that the oriT and relaxation nick sites coincide (14,15).

Triggering of Relaxation Complex Nicking

The in vivo signal that triggers relaxation complex nicking is unknown. The initial in vitro studies found that relaxation was induced by agents which disrupted protein structure: pronase, ionic

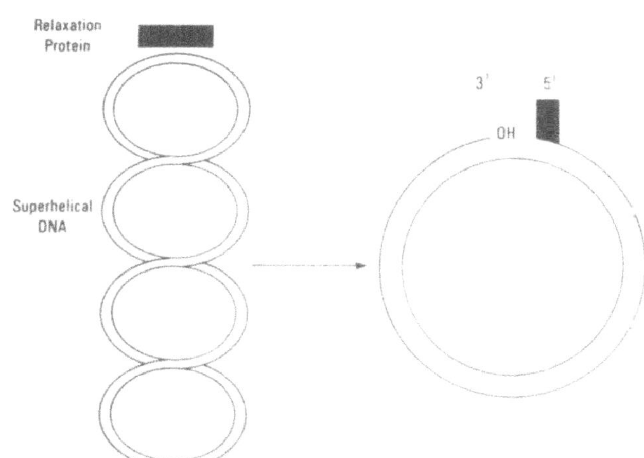

Fig. 2. Reaction mediated by plasmid relaxation complexes. The superhelical DNA is tightly bound to one or more relaxation proteins. Relaxation induces a strand- and site-specific nick in the DNA. In the open circular product, the 3' end of the nicked strand is free, but the 5' end is covalently linked to one of the relaxation proteins (13).

detergents such as sodium dodecyl sulfate, and heat (9). We have found that changes in the DNA structure also trigger the nicking event. Ethidium bromide, which unwinds the DNA helix, is a very efficient inducer of relaxation. Furthermore, as shown in Fig. 3, the nicking event occurs at an ethidium bromide concentration which just removes all the negative superhelical turns in the native plasmid DNA. In this experiment, the supercoiled plasmid DNA-protein complex is sedimented through sucrose gradients containing different concentrations of ethidium bromide. As ethidium bromide binds to superhelical DNA, it removes the negative superhelical turns and decreases the sedimentation rate (16). At an ethidium bromide concentration of about 2 μg/ml, there are no superhelical turns remaining, and the DNA sediments as an open circle (see Panel C, Fig. 3). At higher concentrations, positive superhelical turns are introduced, and the sedimentation rate increases again. Figure 3 shows that ethidium bromide concentrations above 2 μg/ml induce nicking by the relaxation complex, so that most of the plasmid DNA continues to sediment as open circles, even in higher concentrations of ethidium (see Panels D, E, F of Fig. 3). These results show that DNA unwind-

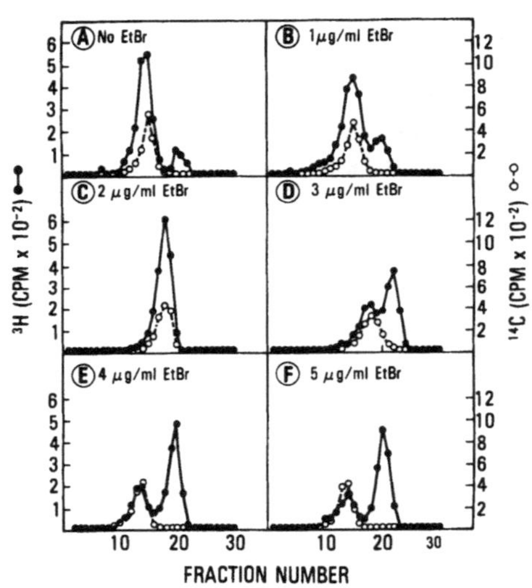

Fig. 3. Relaxation of the ColE1 complex by ethidium bromide (EtBr). The ColE1 relaxation complex (labeled with ^{3}H-thymine) and uncomplexed supercoiled ColE1 DNA (labeled with ^{14}C-thymine) were purified as described previously (13). The ColE1 relaxation complex and supercoiled DNA marker were sedimented through 5-20% sucrose gradients containing EtBr at the indicated concentrations. Methods for centrifugation, fractionation, and radioisotope counting are described in Ref. 13. This experiment was also done using ^{14}C-labeled ØX174 single-strand DNA as a marker to show that the concentration of EtBr needed to remove the negative superhelical turns is 2 μg/ml.

ing triggers the nicking event, and suggest that physiological processes such as DNA replication and transcription, which also unwind DNA, may trigger nicking in vivo. There may be a continuous equilibrium in the cell between nicked and covalently closed forms of the relaxation complex. For some plasmids, nicking may not be triggered by a specific mating signal, but rather this continuous nicking process could supply open circular plasmid molecules ready for transfer. Such a continuous nicking and ligation process would explain the results of Everett and Willetts, who found that phages containing the F transfer origin were nicked even in cells containing F mutants unable to make sex pili (17). Nicking and ligation by the ColE1 relaxation complex during replication would also account for the properties of a temperature-sensitive relaxation complex mutation isolated by Collins and Helinski (18). The relaxation complex of this plasmid was unstable at elevated temperatures. The mutant plasmid was also temperature sensitive for replication, yet other studies showed that the relaxation complex is not required for replication (14,15,19). However, if the nicking/ligation function was disrupted at the nonpermissive temperature, this plasmid could be "stuck" in the open circular state and be unable to complete replication.

THE TRANSFER ORIGIN OF RK2

The conjugation system of IncP plasmid RK2 is of particular interest because of the broad host range properties of this plasmid. The RK2 transfer system is particularly well adapted to overcome species barriers and can mediate gene transfer between a wide variety of gram-negative bacteria (20). The physiology of the RK2 conjugation system differs in several important aspects from the F plasmid. RK2 sex pili are thin and rigid while F pili are thick and flexible (21). RK2 transfers efficiently only on solid medium while F transfers equally well in the liquid or solid phase (22). RK2 is well suited for broad host range conjugation, being 10^4 to 10^5 times more efficient than F at promoting plasmid transfer from *E. coli* to *Pseudomonas* (23). Finally, RK2 and other IncP plasmids are able to mobilize a wide variety of small, nonconjugative plasmids, including ColE1, pSC101, RSF1010 (R300B), and the gonococcal penicillinase plasmid pWD2 (24,25). In contrast, mobilization of pSC101, RSF1010, and pWD2 by F occurs at low to undetectable frequencies. These studies indicate that the transfer system of RK2 has special characteristics which contribute to the broad host range properties of the plasmid.

Mapping the Relaxation Nick Site and oriT of RK2

Figure 4 shows a map of RK2 with the location of the relaxation nick site (*rlx*) and the transfer origin (*oriT*) which coincide at the 48 kb coordinate. The *rlx* site was located by sizing the single-stranded DNA after nicking, restriction enzyme digestion, and alkaline denaturation (26). *oriT* was defined by genetic criteria:

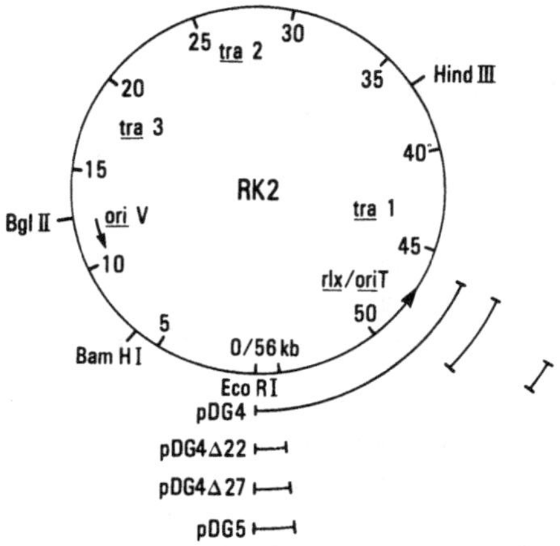

Fig. 4. Map of RK2 (taken from Ref. 27). tra 1,2, and 3 are the 3 transfer regions defined by Barth (34,35) on the basis of Tn7 insertions. rlx is the relaxation nick site (26) and oriT is the transfer origin (27). The RK2 DNA cloned into the pDGV1 vector in the derivatives pDG4, pDG4Δ22, pDG4Δ27, and pDG5 is shown by the solid bars.

1) the oriT site is required in cis for transfer, 2) oriT confers high frequency, recA independent mobilization when the conjugation system is supplied by a helper plasmid in trans, and 3) cointegrate formation is not involved in mobilization (27). The functional oriT region was subcloned as a 760 bp HaeII fragment on pDG5 (Fig. 5).

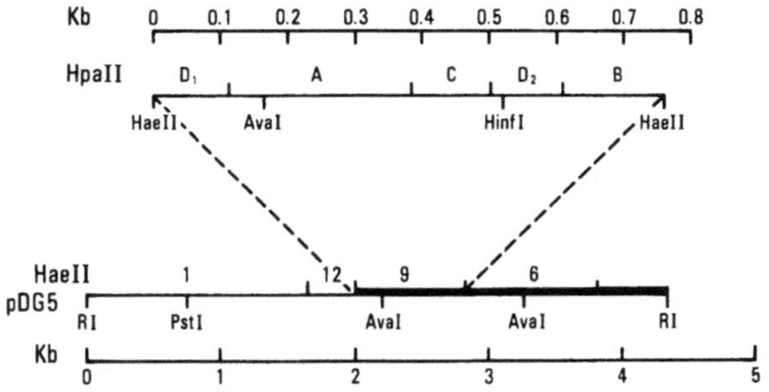

Fig. 5. Map of pDG5. The RK2 DNA present in pDG5 is shown by the heavy line. pDG5 was generated in several steps by partial HaeII digests of pDG4. The HaeII numbers refer to fragments from pDG4 that are present in pDG5 (see Fig. 4). Fragment 9 is a 760 bp sequence spanning oriT, and fragment 6 is from the region just counterclockwise from the EcoRI site of RK2. HpaII digestion of fragment 9 yields 5 subfragments labeled A,B, C, D_1, and D_2 according to decreasing size.

This HaeII fragment contains 4 HpaII sites producing 5 subfragments, designated A, B, C, D_1, and D_2 according to size. The D_2 and B HpaII fragments were cloned into the EcoRI/ClaI region of the vector pDGV1. As shown in Tab. 1, the fully functional oriT is contained in the D_2B region, since pDG16 is transferred nearly as efficiently as pDG5 by the pRK231 helper plasmid. The D_2 fragment alone functions as the transfer origin, but the transfer frequency of pEYDG59 containing D_2 is reduced to about 1% of the frequency of pDG16. Since pDG17 (containing B alone) is not mobilized, the B region by itself is not a transfer origin, but B is required for full function of the oriT activity contained in D_2.

The DNA sequence of the 112 bp HpaII D_2 fragment has been determined and is shown in Fig. 6 (27). The most striking feature of the region is a 19 bp inverted repeat in the right-hand half of the sequence. This inverted repeat could form a hairpin or cruciform structure, as shown in Fig. 7. We have postulated that this inverted repeat may be the binding site for the RK2 relaxation complex. This suggestion is strengthened by our subsequent finding that deletion of 22 bp from the right end of the sequence (forming plasmid pDG11, see Fig. 6) completely abolishes oriT activity (see transfer data in Tab. 1). The deletion in pDG11 removes the outside 10 bp from the right-hand inverted repeat, and indicates that the intact inverted repeat is required for oriT activity. This deletion also

Tab. 1. Mobilization of oriT containing plasmids.

Plasmid[a]	oriT HpaII Fragments[b]	Transfer Frequency[c]
pDG5	D_1ACD_2B	1.4
pDGV1	none	<0.00001
pDG16	D_2B	0.7
pDG17	B	<0.00001
pEYDG59	D_2	0.003
pDG11	D_2 deletion	<0.00001

[a]All plasmids were transformed into E. coli HB101 containing the RK2 helper plasmid pRK231 (tra^+, Tc^R, Km^R).

[b]The HpaII D_2,B, and D_2B fragments were cloned into the EcoRI/ClaI sites of the vector pDGV1. All fragments are in the same orientation with respect to the vector.

[c]The donors (HB101 containing pRK231 and the indicated plasmid) were mated with the HB101 nal^R recipient on filters as described previously (27). Transfer frequency is the number of transconjugants containing the mobilized plasmid divided by the number of transconjugants containing pRK231.

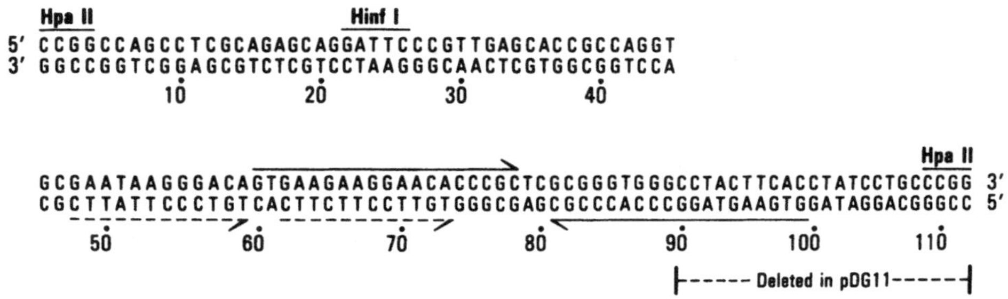

Fig. 6. Nucleotide sequence of HpaII D_2 fragment containing oriT. The inverted repeat is shown by solid arrows, and the direct repeat is seen with dashed arrows. The 22 bp deleted in pDG11 is also shown, and extends from a HaeIII site at base pair 90 to the end of the sequence.

shows that the two direct repeats do not function as a transfer origin.

Homology in the Transfer Origins of IncP Plasmids

RK2 is similar or identical to RP1, RP4, R18, and R68 (28). All these plasmids were isolated in Birmingham, England, and are epidemiologically related (29). A number of other IncP plasmids have been isolated from around the world and carry different drug resistance markers. We have examined the DNA homology between the RK2 oriT region and 9 other IncP plasmids isolated from different geographic locations (30). Figure 8 shows a representative Southern hybridization between the 760-bp HaeII oriT fragment of RK2 and a HaeII digest of 5 other IncP plasmids. R996 and pUZ8 contain strongly hybridizing HaeII fragments of identical size to the HaeII oriT fragment of RK2, while for R772, R906, and R751, the homologous HaeII fragment is smaller and the hybridization is much weaker. The other IncP plasmids tested, R702, R839, R938, and R1033, all had homologous HaeII fragments identical to RK2. We have extended these homology studies to examine 3 other important regions of RK2: 1) the primase gene, 2) the trfA region, which produces a trans-acting product required for replication, and 3) the oriV region. For each of these regions, we find that 6 IncP plasmids are identical to RK2: R702, R839, R938, R995, pU28 and R1033. The 3 other IncP plasmids, R751, R906, and R772, are similar to each other but quite different in the pattern of hybridization from RK2. Based on these hybridization studies, we have designated the RK2-like plasmids as IncPα and R751-like plasmids as IncPβ (30,31). The recent heteroduplex results of Villarroel et al. show that the IncPα plasmids all share a common core structure (32). The homology studies indicate that the IncP plasmids diverged some time ago into 2 distinct evolutionary groups that continue to share incompatibility and broad host-range properties.

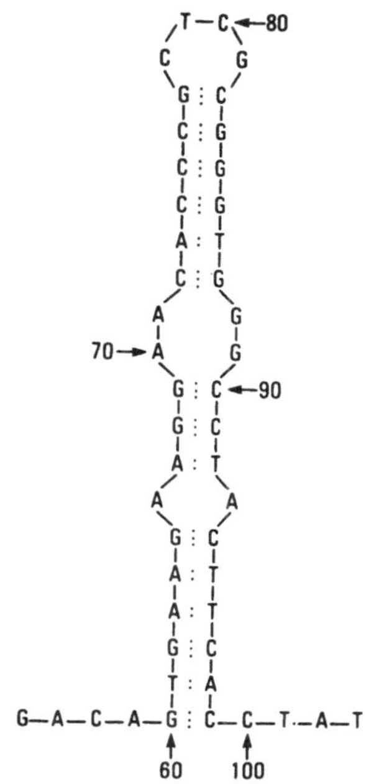

Fig. 7. Potential hairpin structure involving the inverted repeats in oriT. There are 3 bp mismatches, and the first base pair of the loop probably does not form due to steric hindrance. The free energy of formation from single-stranded DNA is -10.8 kcal mol^{-1} (calculated by the method of Tinoco et al., 36). The deletion in pDG11 reduces the free energy of formation to -6.0kcal mol^{-1}.

Specificity of the RK2 oriT Region

Since the homology data indicated substantial differences between the oriT regions of IncPα and IncPβ plasmids, we tested the functional relationships between the transfer systems of these plasmids. Table 2 shows the mobilization frequency of plasmids containing the RK2 oriT region using R702 (IncPα) and R751 (IncPβ) as the tra^+ helper plasmids. R702 is able to transfer pDG5 (760 bp oriT) and pEYDG59 (112 bp oriT) as efficiently as RK2. However, R751 was not able to mobilize either pDG5 or pEYDG59, but could transfer pDG4Δ22 and pDG4 as efficiently as RK2 (see Fig. 4 for maps of pDG4Δ22 and pDG4). Similar results were obtained with other IncPα and IncPβ plasmids. These studies show that the IncPβ transfer system cannot utilize the oriT of RK2, probably because the IncPβ relaxation proteins are unable to recognize the IncPα oriT sequence. However, since pDG4 forms a functional RK2 relaxation complex, this can be recognized by the rest of the IncPβ conjugation system and efficient mobilization occurs. Since pDG4Δ22 is also transferred by R751, the RK2-specific gene product which interacts with oriT is

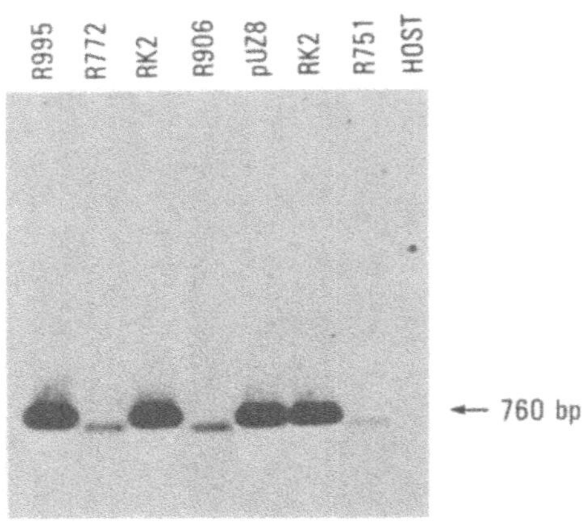

Fig. 8. Southern hybridization between the oriT region of RK2 and 5 other IncP plasmids. The 760 bp HaeII fragment 9 of pDG5 was labeled by nick translation and hybridized under low strigency conditions to a complete HaeII digest of the IncP plasmids.

contained within a 3 kb region surrounding oriT (see Fig. 4). These studies show that the functional specificity of the IncP conjugation system is at the level of recognition of the oriT sequence by the relaxation complex. Once this binding is accomplished, transfer can occur by either the IncPα or IncPβ systems. However, the interaction of conjugation systems with the RK2 relaxation complex is specific for the IncP group, since pDG4 cannot be mobilized by Flac or R64drd11.

Tab. 2. Mobilization of the RK2 oriT region by IncPα and IncPβ plasmids.

Transfer Plasmid[a]	Mobilization of RK2 oriT Plasmid[b]			
	pEYDG59	pDG5	pDG4 Δ22	pDG4
R702 (IncPα)	1.3	1.0	—	—
R751 (IncPβ)	0.001	0.0001	0.1	15

[a]The RK2 oriT plasmids were transformed into E. coli HB101 containing either R702 or R751.

[b]Matings were done as described in Table 1. Mobilization is expressed as the ratio: transfer of the RK2 oriT plasmid by the indicated IncP plasmid divided by transfer of the RK2 oriT plasmid by pRK231.

Host Range of the RK2 Conjugation System

RK2 and related IncP plasmids can transfer to and be maintained in a wide variety of aerobic and facultative gram-negative bacteria. This broad host-range requires a conjugation system capable of transfer between unrelated bacteria, and a replication system able to maintain the plasmid in the new host. In a direct comparison of the RK2 and F plasmid transfer systems, RK2 was 10^4 to 10^5 times more efficient than F at transfers between E. coli and Pseudomonas aeruginosa (23). Despite the extensive host range of the IncP plasmids, we have been unable to transfer RK2 into the gram-negative anaerobe, Bacteroides. These organisms are the most numerous bacteria in the human colon, and thus are in close contact with E. coli. However, systematic attempts to transfer both narrow and broad host-range plasmids from E. coli to Bacteroides fragilis have been unsuccessful (20). We postulated that this barrier to successful gene transfer from E. coli to Bacteroides could be due to 1) lack of plasmid DNA transfer, 2) lack of RK2 replication in Bacteroides, or 3) lack of drug-resistance gene expression in Bacteroides. To examine this problem, we constructed a novel shuttle vector consisting of pDG5 and the Bacteroides clindamycin resistance plasmid pCP1. This hybrid plasmid, designated pDP1 (Fig. 9) replicates in both E. coli and B. fragilis, and contains the RK2 oriT sequence (33). As seen in Tab. 3, pDP1 can be transferred into B.

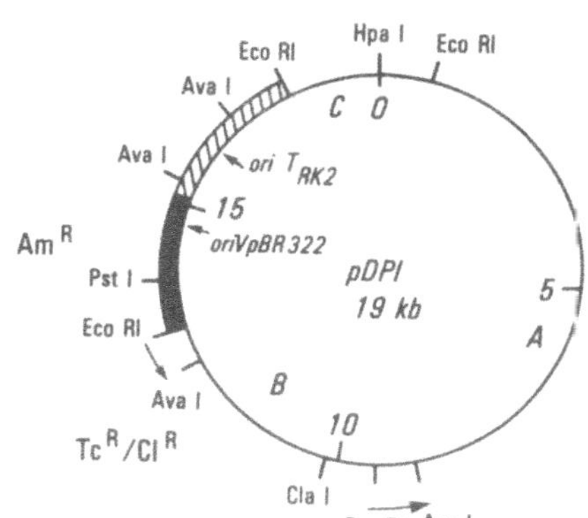

Fig. 9. Map of pDP1, showing pDG5 (thick line) linked to the Bacteroides plasmid pCP1 (thin line). The region of pBR322 containing oriV and ampicillin resistance (Am^R) is shown by the thick solid line, and the RK2 DNA containing oriT is seen by the thick hatched line. The three EcoRI fragments of pCP1 are labeled A, B, and C. The arrows show the location of the direct repeats of pCP1. The EcoRI B fragment of pCP1 encodes tetracycline resistance (Tc^R) in E. coli and clindamycin resistance (Cl^R) in Bacteroides (37).

Tab. 3. Transfer of pDP1 into Bacteroides fragilis.

Plasmids in E. coli Donor[a]	Transfer Frequency to B. fragilis[b] pDP1	pRK231
pDP1	$<10^{-8}$	—
pDP1+pRK231	$3x10^{-6}$	$<10^{-8}$

[a]E. coli donor was JA221 leuB, recA, $hsdR^-$, $hsdM^+$.

[b]The B. fragilis recipient was strain 638 rif^R. Transfers were performed by a plate mating technique (33). Transfer frequency is the number of transconjugants containing the selected plasmid divided by the total number of B. fragilis recipients present at the end of the overnight matings. pDP1 transfer was selected with 5 ug/ml of clindamycin and pRK231 was selected with 1 ug/ml of tetracycline.

fragilis by the RK2 helper plasmid, pRK231. However, pRK231 DNA was never found in the B. fragilis transconjugants selected for pDP1 transfer. Since pRK231 is almost always cotransferred in E. coli matings, we interpret these results to indicate that pRK231 cannot be maintained in Bacteroides. Furthermore, the TEM β-lactamase gene from pDG5 is not expressed in Bacteroides, while the erythromycin-clindamycim resistance determinant from pCP1 is not expressed in E. coli. This novel gene transfer system mediated by RK2 has allowed us to determine that differences in both plasmid replication and gene expression form substantial barriers to gene exchange between aerobic and anaerobic bacteria.

These studies show that the conjugation system of RK2 has a broader host-range than the replication system. This may be a general property of transmissible plasmids since the E. coli F plasmid is able to mediate low frequency transfer to Pseudomonas but cannot replicate in that host (23). Recently, RK2-mediated plasmid transfer from E. coli to cyanobacteria has been reported (38). Our results further extend the host range of RK2 conjugation to include anaerobic bacteria. The RK2 conjugation system is an extremely versatile genetic exchange mechanism for the study and manipulation of gram-negative bacteria.

ACKNOWLEDGMENTS

This work was supported by grants from the National Institute of General Medical Sciences and the National Institute of Allergy and Infectious Diseases. D.G.G. is a Teaching and Research Scholar of the American College of Physicians.

REFERENCES

1. Clark, A., and G. Warren (1979) Ann. Rev. Genet. 13:99-125.
2. Willetts, N., and R. Skurray (1980) Ann. Rev. Genet. 14:41-76.
3. Clewell, D.B. (1981) Microbiol. Rev. 45:409-436.
4. Willetts, N., and B. Wilkins (1984). Microbiol. Rev. 48:24-41.
5. Achtman, M., and R. Skurray (1977) In Microbial Interactions, Receptors and Recognition, Ser. B. V3, J.L. Reissig, ed. Chapman and Hall, London, pp. 233-279.
6. Achtman, M., S. Schwuchow, R. Helmuth, G. Morelli, and P.A. Manning (1978) Molec. Gen. Genet. 164:171-183.
7. Rupp, W., and G. Ihler (1968) Cold Spring Harbor Symp. Quant. Biol. 33:647-650.
8. Ohki, M., and J. Timizawa (1968) Cold Spring Harbor Symp. Quant. Biol. 33:651-657.
9. Clewell, D.B., and D.R. Helinski (1969) Proc. Natl. Acad. Sci., USA 62:1159-1166.
10. Kupersztoch-Portnoy, Y.M., G. Miklos, and D.R. Helinski (1974). J. Bacteriol. 120:545-548.
11. Blair, D.G., and D.R. Helinski (1975) J. Biol. Chem. 250:8785-8789.
12. Lovett, M.A., and D.R. Helinski (1975) J. Biol. Chem. 250:8790-8795.
13. Guiney, D.G., and D.R. Helinski (1975) J. Biol. Chem. 250:8796-8803.
14. Warren, G.J., A.J. Twigg, and D.J. Sherratt (1978) Nature 274:259-261.
15. Dougan, G., and D.J. Sherratt (1977) Mol. Gen. Genet. 151:151-160.
16. Helinski, D.R., and D.B. Clewell (1971) Ann. Rev. Biochem. 40:899-942.
17. Everett, R., and Willetts (1980) J. Mol. Biol. 136:129-150.
18. Collins, J., S. Yanofsky, and D.R. Helinski (1978) Mol. Gen. Genet. 167:21-28.
19. Hershfield, V., H.W. Boyer, L. Chow, and D.R. Helinski (1976) J. Bacteriol. 126:447-453.
20. Guiney, D.G. (1984) J. Infect. Dis. 149:320-329.
21. Bradley, D. (1980) Plasmid 4:155-169.
22. Bradley, D., D. Taylor, and D. Cohen (1980) J. Bacteriol. 143:1466-1470.
23. Guiney, D.G. (1982) J. Mol. Biol. 162:699-703.
24. Willetts, N. (1980) In Molecular Biology, Pathogenicity, and Ecology and Bacterial Plasmids, S. Levy, R. Clowes, and E. Koenig, eds. Plenum Press, New York, pp. 207-216.
25. Guiney, D.G., and J. Ito (1982) J. Bacteriol. 150:298-302.
26. Guiney, D.G., and D.R. Helinski (1979) Mol. Gen. Genet. 176:183-189.
27. Guiney, D.G., and E. Yakobson (1983) Proc. Natl. Acad. Sci., USA 80:3595-3598.
28. Burkardt, J., G. Riess, and A. Puhler (1979) J. Gen. Microbiol. 114:341-348.

29. Ingram, L., M.H. Richmond, and R.B. Sykes (1973) Antimicrob. Ag. Chemother. 3:279-288.
30. Yakobson, E., and D.G. Guiney (1983) Mol. Gen. Genet. 192:436-438.
31. Chikami, G., D.G. Guiney, T. Schmidthauser, and D.R. Helinski (submitted for publication).
32. Villarroel, R., R.W. Hedges, R. Maenhaut, J. Leemans, G. Engler, M. Van Montagu, and J. Schell (1983) Mol. Gen. Genet. 189:390-399.
33. Guiney, D.G., P. Hasegawa, and C.E. Davis (submitted for publication).
34. Barth, P.T., N. Grinter, and D.E. Bradley (1978) J. Bacteriol. 133:43-52.
35. Barth, P.T. (1978) In Plasmids of Medical, Environmental, and Commercial Importance, K. Timmis and A. Puhler, eds. Elsevier/North-Holland, Amsterdam, pp. 399-410.
36. Tinoco, I., O. Uhlenbeck, and H. Levine (1971) Nature 230:362-367.
37. Guiney, D.G., P. Hasegawa, and C.E. Davis. Plasmid (in press).
38. Wolk, C., A. Vonshak, P. Kehoe, J. Elhai (1984) Proc. Natl. Acad. Sci., USA 81:1561-1656.

GENES AND GENE PRODUCTS INVOLVED IN THE SYNTHESIS OF F-PILI

Susan Laine, Deanna Moore, Pushpa Kathir, and Karin Ippen-Ihler

Department of Medical Microbiology and Immunology
Texas A & M University
College Station, Texas 77840

ABSTRACT

Membrane fractions containing [35-S]methionine labeled proteins synthesized by F*lac* and F*lac tra* mutant strains or by λ*tra* transducing phages expressed in such strains have been analyzed in order to investigate the pathway for synthesis of the F-pilin subunit and the gene products involved in synthesis of F-pili. Our data indicate that the synthesis of a mature F-pilin subunit requires the expression of at least 2 *tra* operon genes in addition to the structural gene for F-pilin, *traA*. In the absence of these activities, *traA* expression results primarily in the synthesis of a polypeptide, Ap14, with an apparent molecular weight of approximately 14,000. We assume this polypeptide corresponds to the direct product of the *traA* gene. In the presence of *traQ* activity, the major detectable product of *traA* is a polypeptide, Ap7(Q), which migrates with an apparent molecular weight of 7,000, suggesting that *traQ* product may process or assist in the processing of Ap14. Polypeptide Ap7(Q) is not, however, mature F-pilin, since it reacts poorly with anti-F-pilus-serum. Synthesis of a polypeptide which appears to be antigenically equivalent to F-pilin and which we assume requires a modification of the F-pilin N-terminus, is detected as synthesis of a polypeptide, Ap7*. This protein migrates slightly more slowly than Ap7(Q) on our polyacrylamide gels. Polypeptide Ap7*, can be efficiently precipitated with F-pilus antiserum, and can be detected in both inner membrane and outer membrane fractions under conditions where assembly of F-pili can occur. These data suggest that Ap7* is the mature F-pilin subunit and is assembled from an inner membrane pool. Synthesis of Ap7* appears to require *traG* activity, but may also be dependent upon additional *tra* activities.

INTRODUCTION

F-pili are filmentous structures elaborated from the surface of cells carrying the conjugative plasmid F. They are required for adsorption of a number of male-specific bacteriophage and can be most easily differentiated from other surface structures by this property. Similar fimbrial appendages are found on the surface of cells which carry a variety of conjugal plasmids, and pili are therefore thought to play an important role in plasmid directed transfer systems. Despite the fact that F-pili appear to be composed of a single subunit (F-pilin), analysis of the activities required to synthesize an F-pilus has proved complex. The genes required to elaborate F-pili are encoded together with other F transfer (tra) genes within a 33 kilobase (kb) region of F DNA. (for review, see Ref. 17 and 18). Expression of at least 14 tra genes is required for F-piliation, and it is likely that additional genes within the cluster remain to be described (3,17). Although the structural gene for F-pilin was determined to be traA, (7), assignment of other functions has been complicated by the difficulty of assaying for intermediates in F-pilin synthesis and assembly. However, the detection of unassembled F-pilin polypeptide in inner membrane preparations of cells harboring F, together with analysis of λtra transducing phage products in Ftra mutant backgrounds permitted us to determine that a gene product (traQ), encoded in a tra operon region 14 kb distal to traA, was required in synthesis of the F-pilin subunit (3,10). In the work described below, additional analyses have been performed to examine the membrane location of unassembled and assembled F-pilin, and to more completely characterize the activity of traQ. We report that an F activity in addition to traQ is required to obtain synthesis of mature F-pilin from traA.

DETECTION OF F-PILIN IN LABELED MEMBRANE PREPARATIONS

Comparison of [35-S] labeled membrane preparations derived from Flac and F- cells and fractionated by SDS-polyacrylamide gel electrophoresis (SDS-PAGE) permits the detection of only a few F tra operon products. These include the products of traG, traT, and F-pilin which migrate in our gel system at apparent relative molecular weights (M_R) of 100,000, 25,000, and 7,000 respectively (8,9). Since, however, F-pilin subunits are readily detected by this means, we have used such preparations to determine both the membrane location of F-pilin, and the effect of F tra operon mutations on its synthesis and location. In earlier work, we reported that a large number of F-pilin subunits are found in inner membrane preparations, and suggested that F-pili are anchored in the inner membrane and assembled from an inner membrane pool of F-pilin (8). An F-pilin polypeptide could still be detected in membranes of cells carrying Flac tra nonsense or frameshift mutations in any one of the genes traL, traE, traK, traB, traV, traW, traC, traU, traF, or traH, or in the promoter proximal region of traG. Since these mutations prevent expression of F-pili, this result suggested that these gene products were required for assembly of the pilus structure rather than synthesis of its subunit, although the possibility that one or more of these products effected a change which was not detected on our gels was not excluded (9).

Recent analysis of sucrose density fractionated membranes confirms that such mutations do indeed affect the location of membrane F-pilin. Figure 1 shows a typical density profile of F*lac* [35-S] methionine labeled membrane proteins after fractionation by SDS-PAGE. As can be seen in this autoradiogram, analysis across the sucrose density gradient revealed significant variations in protein content in samples of different density. Some proteins are remarkably well concentrated in one or 2 fractions, while others are more broadly distributed. Since the preparation and collection of the gradient permits some variation in density of given fractions we distinguish 4 general regions: We label as "OM," fractions of density 1.20 - 1.23 gm/cc which contain the highest concentration of the major host outer membrane proteins. "IM" fractions are density 1.14 - 1.18 gm/cc and contain proteins and enzymatic activities representative of the inner membrane. Intermediate between the OM and IM regions, outer and inner membrane proteins overlap considerably; membranes which appear to represent adhesion zones have been reported at this density (2). Fractions of density 1.14 gm/cc or less contain a different set of proteins, possibly those bound to the cytoplasmic side of the inner membrane. It is interesting to note that we observe these proteins to be lost if membranes are pelleted prior to running the sucrose gradient. They are recovered together with the other membrane proteins after such spins only if the membrane pellet is resuspended by boiling in SDS sample buffer.

Examination of Fig. 1 shows that, as expected, F *traT* product is concentrated in the OM fractions, while the product of *traG* is primarily concentrated in the IM fractions (1,9). F-pilin is, however, recovered in significant amounts from fractions spanning the OM, intermediate, and IM regions. In contrast, when membranes of F *tra* mutants defective in F-pilus assembly are examined (Fig. 2A and 2B) F-pilin is very obviously concentrated in the IM fractions. F-pilin in membranes from an F*lac* *traG24* strain which is defective in F-pilus production exhibits a similar density distribution (Fig. 2C), while much more F-pilin is found in the OM fractions when *traG79*, an amber mutation which does not prevent F-pilus production is tested (Fig. 2D). Such experiments clearly demonstrate that F-pilin subunits which cannot be assembled accumulate in an inner membrane pool. Since assembled F-pili were not removed from the cell surface of the F*lac* or F*lac* *traG79* strain prior to preparation of the membrane samples analyzed in these experiments, we assume that the presence of F-pilin in OM samples indicates that subunits which have been completely assembled or which are in the process of assembly are in these fractions. Therefore we can obtain an indicator of F-pilus assembly with such gradients.

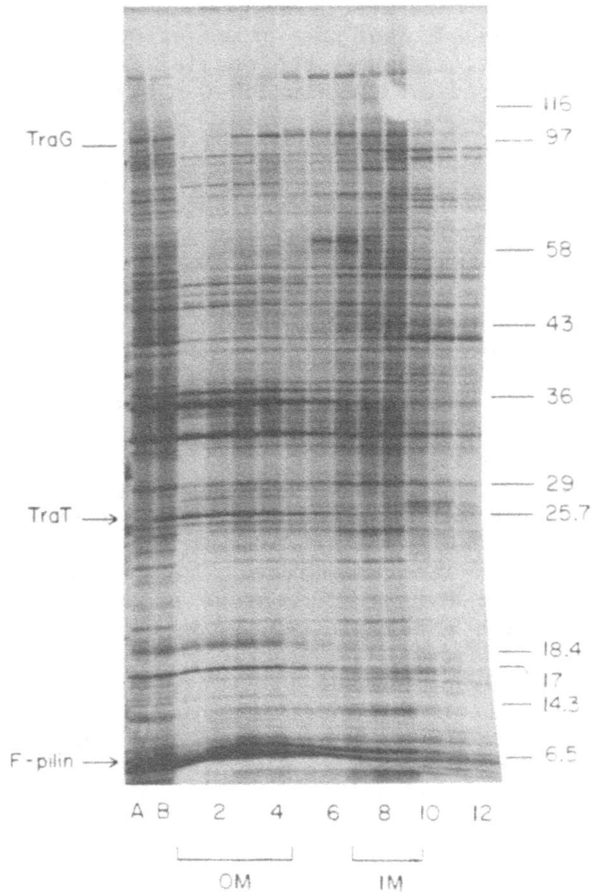

Fig. 1. Autoradiogram of a polyacrylamide gel on which [35-S] methionine labeled membrane proteins from an Flac strain were fractionated. Samples analyzed were whole membranes prepared from the female strain, JC3272 (lane A) and its Flac derivative, JC3273 (lane B), as well as proteins in membrane pellets derived from each of 12 fractions (labeled in order of decreasing density) of an isopycnic sucrose density gradient of proteins from the Flac strain. Fractions corresponding to outer membrane (OM) and inner membrane (IM) densities are indicated. Arrows on the left denote positions of the products of *traG*, *traT*, and F-pilin. Numbers on the right indicate the positions and molecular weights ($x10^{-3}$) of marker proteins used, as ascertained from the corresponding stained gel.

Strains, growth media, and procedures were as previously described (8) except that a 10 ml culture of the Flac strain was labeled by adding a total of 100 µci [35-S]

Fig. 1. Continued

methionine for a 5 min interval during log phase. The labeled cells were pelleted by centrifugation, resuspended in 1 ml Tris-EDTA buffer [10 mM Tris-HCl, 5 mM EDTA, 1 mM dithiothreitol, pH 7.8], disrupted by passage through a French pressure cell (16,000 psi), and centrifuged for 3 min in an Eppendorf microcentrifuge to remove unlysed cells and large vesicles. A 0.7 ml portion of the supernatant was then placed on top of a sucrose gradient which had been formed by adding successive layers of 0.5 ml 55% and 0.8 ml each of 50, 45, 40, 35, and 30% sucrose in 5 mM EDTA pH 7.5 to an ultracentrifuge tube. After centrifugation for 18 hr at 45,000 rpm in an SW 50.1 rotor, equal fractions of decreasing density were collected from the bottom of the tube. A 50 µl sample was used to determine the refractive index in each aliquot. The remainder was diluted with Tris-EDTA buffer and each fraction was then centrifuged for 18 min at 50,000 rpm in a 70.1 rotor to pellet membrane material. The drained pellets were subsequently resuspended in 50 µl sample buffer and analyzed immediately by SDS polyacrylamide gel electrophoresis using 10-16% exponential gradient gels as described previously except that electrophoresis was at 10 W constant power with cooling (3,8).

EXPRESSION OF *traA*, THE STRUCTURAL GENE FOR F-PILIN

Evidence has accumulated for some time, that the F *traA* gene is the structural gene for F-pilin. The ability of RNA phages to bind to F-pilin is affected by mutation of *traA* (16) or complementation of *traA* mutants with R-100-1 (14). F-pilin purified from a *supF* strain carrying the *Flac traA1* (amber) plasmid contained an additional tyrosine (7), and *Flac traA* amber strains were unable to synthesize membrane F-pilin (9). However, the direct product of *traA*, a polypeptide which neither comigrated with F-pilin nor appeared to react with F-pilus antiserum was reported to migrate at 13,700 daltons (5,6). This suggested that if *traA* was the structural gene for F-pilin, alteration of its product must be necessary.

The existence and location of an additional *tra* operon activity which participates in F-pilin biosynthesis became evident when it was found that *Flac tra* polar mutations or Hfr deletion strains which were defective in expression of the *tra* region between *traF* and *traH* were also deficient in synthesis of membrane F-pilin (8,9). The presence of this function, named *traQ*, could be assayed in UV-irradiated hosts by examining the products of lambda transducing phages carrying the *traA* gene (9). When a female host was used, the major *traA* product was an M_R 14,000 polypeptide which accumulated in the inner membrane. In an *Flac* host, the major *traA* product was an

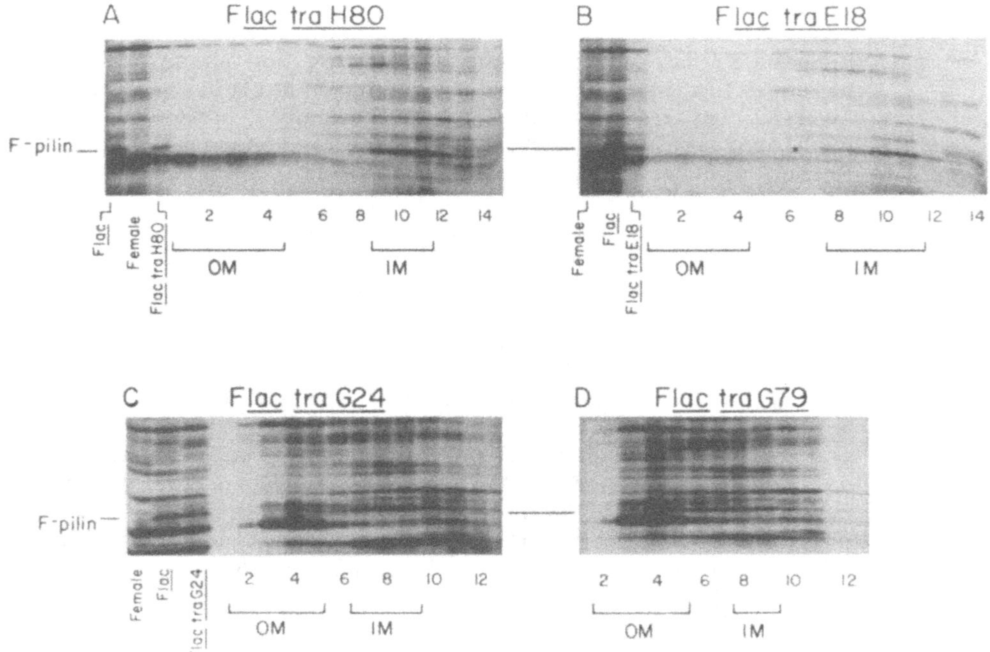

Fig. 2. Lower portions of autoradiograms showing [35-S]methionine labeled proteins present in sucrose gradient fractionated membrane preparations obtained and analyzed as described for Fig. 1. Strains analyzed were derivatives of JC3272 carrying: A. Flac traH80; B. Flac traE18; C. Flac traG24; and D. Flac traG79. Whole membrane preparations used as controls were from JC3272 (female) or its Flac or Flac tra derivatives as indicated below the autoradiogram. Numbered lanes correspond to samples from sucrose fractions (increase in number indicates decrease in density) from which density fractionated membrane samples were derived. The position at which F-pilin migrates is indicated on the left.

M_R 7,000 protein which comigrated with F-pilin. Similarly, an M_R 7,000 polypeptide was a prominant phage product in experiments in which a phage carrying the entire traA-traH region was used or when both a $TraA^{+}$ phage and a phage carrying the traQ region were used to coinfect a female host (3).

We have found that examination of density fractionated membrane preparations is also a very useful technique for visualizing tra products synthesized after transducing phage infection of UV-irradiated hosts. The background of labeled lambda phage products

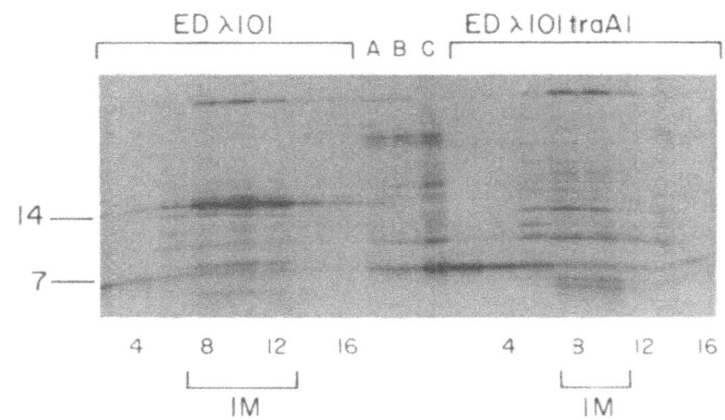

Fig. 3. Lower portion of an autoradiogram of a gel on which membrane proteins labeled during infection of the UV-irradiated female host XK1800 with the λtraALE transducing phage EDλ101 (left side) or its traA1 mutant derivative (right side) were analyzed after fractionation by sucrose density. Preparation, fractionation, and analysis of samples was as described in Fig. 1 except that, in this case, labeling of the transducing phage products in the UV-irradiated host was as described in Ref. 10, and only even-numbered sucrose gradient fractions were analyzed. Also shown are whole cell samples labeled after infection with EDλ101 (lane A), the parental phage, EDλ4 (lane B), or the EDλ101 traA1 mutant derivative (lane C). Numbers on the left indicate the positions and apparent molecular weight (x 10^{-3}) of the labeled polypeptides present in samples obtained after EDλ101 infection and absent in samples obtained after infection with the EDλ101 traA1 derivative.

is greatly reduced in these experiments. In Fig. 3 the membrane associated products of EDλ101 [λ ptraALE (4)] and those synthesized by a traA1 amber derivative of this phage are compared after infection of the UV-sensitive female host, XK1800. Labeled traE product is prominent in the inner membrane preparations from both phage infections, while traL product is not obvious, possibly because of the low methionine content of this protein (Frost, Paranchych, and Willetts, in press). It can be seen that the traA1 mutation prevents the synthesis of a prominent M_R 14,000 EDλ101 encoded inner membrane band. It also prevents the appearance of an M_R 7,000 inner membrane polypeptide. The latter is found in small quantity relative to the M_R 14,000 product, and is only detectable in fractionated membrane samples such as are shown (3). Our interpretation of this result was that, under conditions of over-production and accumulation of the traA M_R 14,000 product, (Ap14) the F^- host might inefficiently process this protein to an M_R 7,000 polypeptide (Ap7) which resembles F-pilin. The relationship of these products to those synthesized in an Flac host is shown in Fig. 4, where inner membrane fractions from the EDλ101 and EDλ101 traA gradients analyzed in Fig. 3 are compared with membrane proteins synthesized by EDλ101 in an Flac traA1 strain. (Background synthesis of F-pilin is avoided in this host since it cannot itself synthesize F-pilin, but can synthesize all other F products prior to the UV treatment.) In the Flac traA1

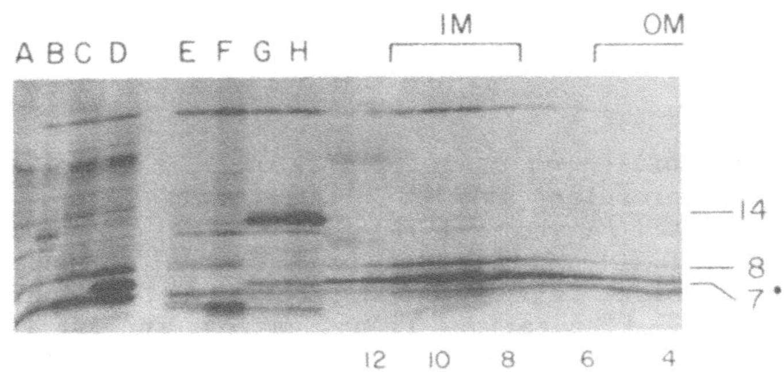

Fig. 4. Proteins labeled during infection with the traALE phage EDλ101 in the UV-irradiated Flac traA1 host, EM 1801. Shown, is the lower portion of the autoradiorgram of a gel on which membrane proteins in fractions 4-13 of a sucrose density gradient were analyzed. In this case, the fractions were placed on the gel such that the density of the fractions increases from left (highest-numbered fractions) to right.

Also analyzed on the same gel, for comparison, are the inner membrane proteins labeled during infection of the female host XK1800 with EDλ101 (lanes G and H) or the EDλ101 traA1 derivative (lanes E and F). These lanes contained fractions 9 and 11 from each of the 2 gradients analyzed in Fig. 3. Labeled proteins detectable in whole cell samples after infection of the UV-irradiated female strain with EDλ101traA1 or EDλ101 are shown in lanes A and B, respectively. Corresponding whole cell samples obtained using the Flac traA1 host are in lanes C and D.

Numbers at the right indicate the position and approximate molecular weight ($x\ 10^{-3}$) of the M_R 14,000 traA product readily detected after infection of the female host (lanes B, G, and H), and of the M_R8,000 and M_R7000 products detected after infection of the Flac traA1 host. Since this M_R7000 product can be seen to migrate slightly more slowly than the similar polypeptide detectable in lanes G and H, it is indicated as 7*.

host, the traA products of EDλ101 are very different. Polypeptide Ap14 is no longer detected. A band which comigrates with F-pilin is prominent and can clearly be seen to migrate at a position slightly above the Ap7 polypeptide observed in female hosts. Since it appears to differ from and have a fractionally higher molecular weight than Ap7, we refer to it as Ap7*. In addition, an M_R 8,000 polypeptide (Ap8) appears.

PRECIPITATION OF TraA PRODUCTS WITH ANTISERUM

In order to confirm the identity of Ap7* as F-pilin and to investigate the relationship between F-pilin and the additional products of traA which could be detected in our system we raised rabbit antiserum to purified F-pilin and examined the membrane products of EDλ101 which it precipitated. When the traA gene of EDλ101 was expressed in an Flac traA1 host, the serum could be used to efficiently immunoprecipitate both Ap7* (F-pilin) and Ap8. However, we were unable to detect either Ap14 or Ap7 in anti-F-pilin precipitates of EDλ101 expressed in female cells (3). Since the traA1 amber mutation affected the synthesis of all of these polypeptides, and suppression of traA1 results in F-pilin with an altered amino acid content (7), these proteins must contain a common amino acid sequence. Our result could be explained then, only if this amino acid sequence is insufficient to comprise the major antigen of F-pilin. Evidence that, in fact, the major antigen of F-pilin might derive from a modification of its amino acid sequence had been provided by the work of Worobec et al. (19) who demonstrated that the major antigen of EDP208 pilin is the blocked N-terminus of this polypeptide. We surmised that F-pilin might have a similarly blocked N-terminus.

Nevertheless, it seemed reasonable to predict that a common amino acid sequence shared by all traA products should be detectable with anti-F-pilus serum if our system could be made sufficiently sensitive. Therefore, we constructed strains which would make these products in greater quantity by cloning a PstI fragment containing traALE from pBE268 into the PstI site of pLA2311 (12,13). In the plasmid (pKI301) thus constructed, genes traALE are located downstream from the lambda promoter P_L. A strain expressing λCI857 repressor and carrying pKI301 grows normally at 30°C. When it is shifted to 42°C, an immediate increase in pKI301 products and concommitant decrease in host protein synthesis is observable. Within 30 min, synthesis of the products of pKI301 including the products of traA and traE, predominates completely over host protein synthesis; during this time, however, total protein synthesis diminishes. Cellular viability is eventually lost (D. Moore, unpub. data). Figure 5 shows [35-S]methionine membrane proteins obtained when an Flac traA1 (lane 6) or a female (lane 7) host carrying pKI301 were labeled for 5 min in the interval 10-15 min after induction at 42°C. Whole-cell samples of EDλ101 (traALE) products are shown in lanes 1

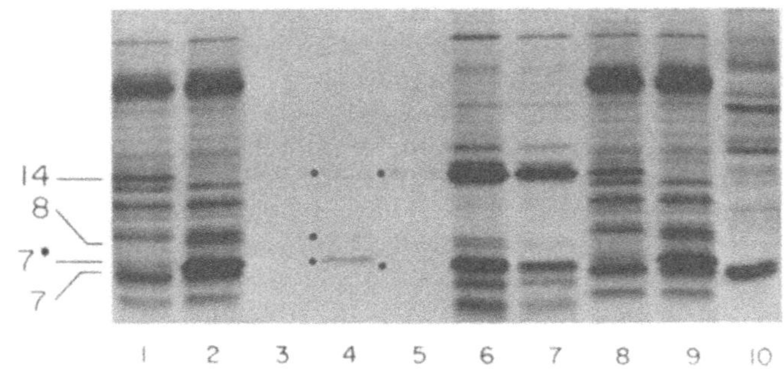

Fig. 5. Analysis of antibody precipitable proteins synthesized by pKI301 in a F^- and in an Flac traA1 host. pKI301 is a derivative of the vector pLa2311 (12) into which a PstI fragment carrying genes traALE has been inserted. Expression of the tra segment occurs from lambda P_L, at 42°C. Analyses of whole membrane samples of [35-S]methionine labeled proteins synthesized in the interval 10-15 min after induction at 42°C are shown for a pKI301/F^- strain (lane 7), a pKI301/Flac traA1 strain (lane 6) and for the pLa2311/Flac traA1 control (lane 10). Analyses of the proteins precipitated from these samples with anti-F-pilus serum and IgSorb are also shown: pKI301/F^-, lane 5; pKI301/Flac traA1, lane 4; pLa2311/Flac traA1, lane 3. Whole cell samples of labeled EDλ101 products obtained after infection of a female host (lanes 1 and 8) or an Flac traA1 host (lanes 2 and 9) were also analyzed for reference. Numbers on the left indicate the apparent molecular weight of the corresponding antibody precipitable bands (marked by dots) in lanes 4 and 5.

Membrane preparation and antibody precipitation was performed as described previously, except that only 1 ml samples of the plasmid-containing strains were used, and cells were disrupted by sonication (3).

and 8 (F^-host) and lanes 2 and 9 (Flac traA1 host) for comparison. It is clear that Ap14 is synthesized from pKI301 in abundance in both host backgrounds. The Ap7 polypeptide is also made in the female host, while Ap7* together with Ap8 are synthesized in the Flac traA1 strain. It is further demonstrated (lanes 4 and 5) that anti-F-pilus serum can be used to immunoprecipitate all of these proteins, confirming a relationship in their structure which we assume derives from the common amino acid sequence we predicted they must have. As we had observed in the λtra system, Ap7* was clearly the most efficiently precipitated polypeptide in keeping with our identification of this polypeptide as the antigenically mature F-pilin subunit.

PRODUCTS OF EDλ134 [tra(Y)ALEKBPVWCUNFQH(G)]

As indicated above, we have reported previously that no M_R 7,000 band resembling F-pilin could be detected in membranes of Hfr mutants defective in expression of the region between traF and traH. We postulated that an activity stemming from a tra gene (traQ) in the traF-traH region was required for F-pilin synthesis (10). When the products of a $TraA^+$ phage coinfected with a $TraQ^+$ phage, or the products of a large transducing phage (EDλ134) which expresses the entire traA - traH region were examined, an M_R 7,000 polypeptide became a major product of traA (3). These results are illustrated by experiments such as that shown in Fig. 6 in which inner membrane associated polypeptides synthesized by EDλ134 during infection of F^- cells were analyzed. The lambda products synthesized by the parental phage, EDλ4, which were analyzed simultaneously, are also shown. Comparison of the products of EDλ4 and EDλ134 permits the detection of many tra operon proteins. Some of these have been provisionally identified (Moore and Ippen-Ihler, unpub. data). Further analysis of the products of cloned subsegments of DNA from the tra operon as well as the products of additional transducing phages is currently in progress to determine the exact tra regions specifying others. Examination of the region in which F-pilin migrates shows clearly, that an M_R 7,000 band is synthesized by EDλ134 and that lesser quantities of an M_R 8,000 protein are also readily detectable. These same polypeptide bands appeared to be synthesized whether an Flac traA1 or an F^- host was used, and to comigrate with the products of the $TraA^+$ phage EDλ101 which are synthesized in an Flac traA1 host. Thus, such results initially suggested that the tra genes expressed from the $TraA^+TraQ^+$ phage EDλ134 could accomplish synthesis of F-pilin (3). However, a close examination of well-resolved gels on which EDλ134 products made in male and female hosts were compared, as well as analysis of anti-F-pilus serum precipitated proteins has shown that they are not identical.

As can be seen in Fig. 7, the anti-F-pilus serum worked efficiently in immunoprecipitation of $AP7^*$ (F-pilin) together with Ap8 when the male host was used for the EDλ134 infection. In contrast, the M_R 7,000 and 8,000 polypeptides made in F^- cells were very poorly precipitated. Again, when good resolution is obtained, the M_R 7,000 EDλ134 product can be seen to migrate slightly faster than F-pilin, resembling Ap7 rather than Ap7*. Since we do not known if it is exactly identical to the Ap7 polypeptide observed in A^+Q^- infections we refer to it as Ap7(Q). It is obvious that the tra gene products expressed from EDλ134 cannot, alone, accomplish addition of the major antigen of F-pilin. That is, EDλ134 products appear to accomplish efficient synthesis of Ap7(Q) and AP8, but not Ap7*.

This is intriguing because EDλ134 carries tra(Y)ALEKBPVWCUNFQH(G), and, therefore, expresses almost all of the genes required for F-pilus assembly. It thus seemed possible that addition of the

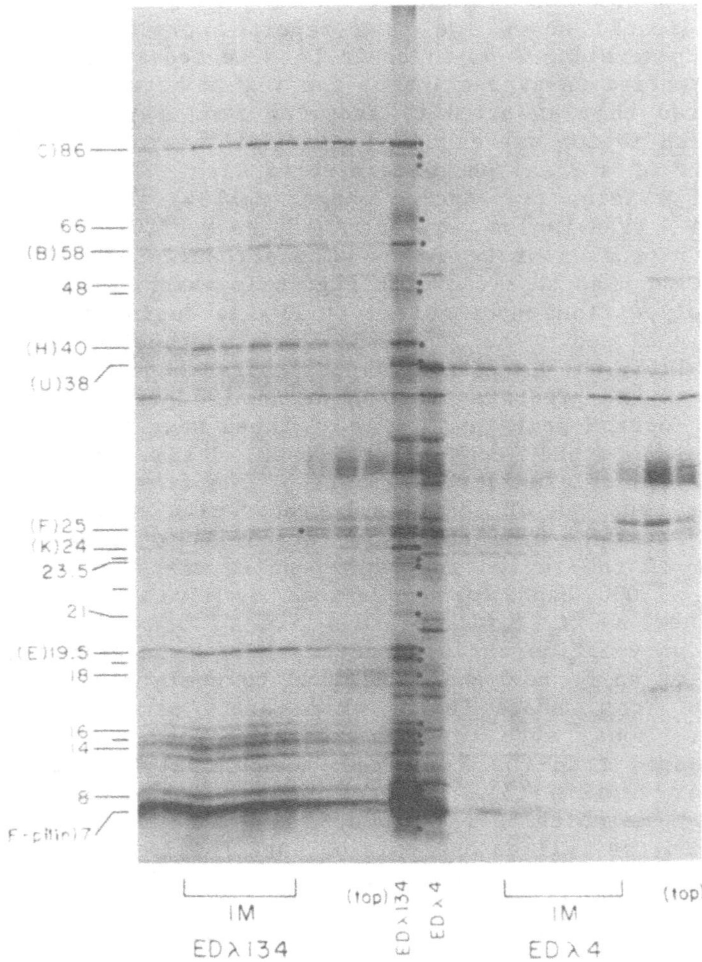

Fig. 6. Analysis of [35-S]methionine labeled proteins synthesized after infection of UV-irradiated XK1800 with either EDλ134, which carries a tra(Y)ALEKBPVWCUNFH(G) segment or its parental phage, EDλ4 (3,4). Whole membrane samples are shown in the center lanes; membrane samples obtained after sucrose density fractionation are shown on the left for EDλ134, on the right, for EDλ4. Only fractions of intermediate to light density were analyzed on this gradient. Black dots indicate bands detected in membranes from EDλ134 infected cells which were not present after EDλ4 infection. Numbers on the left indicate the apparent molecular weight (x 10^{-3}) of these bands; the tra cistrons thought to encode some of these products are also indicated.

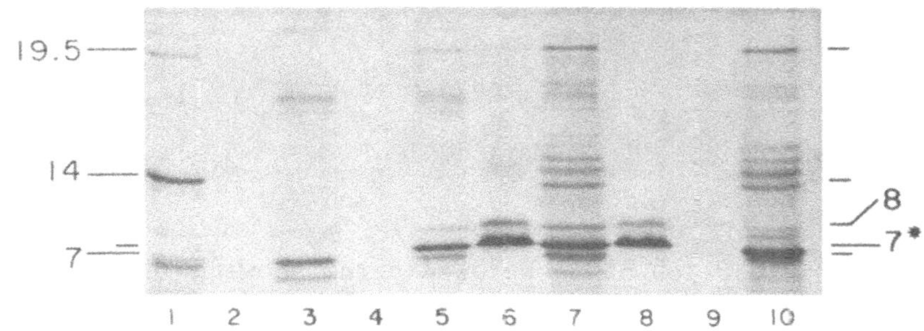

Fig. 7. Lower portion of an autoradiogram showing analysis of membrane proteins labeled during transducing phage infection and the proteins immunoprecipitated from these preparations after adsorption to anti-F-pilus serum. Samples analyzed were: membranes (lane 1) and immunoprecipitate (lane 2) from EDλ101 infected F⁻ cells; membranes (lanes 3,5, and 7) and immunoprecipitates (lanes, 4,6 and 8) from Flac traA1 cells infected with EDλ4 (lanes 3 and 4), EDλ101 (lanes 5 and 6) and EDλ134 (lanes 7 and 8); and the immunoprecipitate (lane 9) and membrane preparation (lane 10) from EDλ134 infected F⁻ cells. Experimental procedures were as described previously (3). Numbers at the side of the figure indicate the apparent molecular weight (x 10^{-3}) and positions of the product of traE (19.5) and polypeptides associated with traA expression.

F-pilin major antigenic group might be associated with the assembly of F-pilin into F-pili, a process which could occur in the Flac traA1 host, but not in the F⁻ host. Examination of EDλ134 membrane products on sucrose gradients indicated that while most of the F-pilin polypeptide synthesized in either the F⁻ or Flac traA1 host could be found in the inner membrane, a significant amount could also be detected at densities associated with intermediate and outer membrane when the Flac traA1 host was used, that is, in samples we consider to have contained assembled F-pilin (P. Kathir, unpub. data). Therefore, we tested the capacity of our anti-F-pilus serum to precipitate F-pilin from membrane preparations of different density. The results are shown in Fig. 8. When the Flac traA1 host was used, the antiserum precipitated F-pilin from all portions of the gradient. That is Ap7* was found not only in membrane samples of outer membrane density, but also, indeed most heavily, was concentrated in samples of inner membrane density. Therefore, while it is still possible that addition of the antigenic group occurs as a condition of assembly, it appears to occur in the inner membrane, and the unassembled inner membrane pool of F-pilin in male cells carries this antigenic modification.

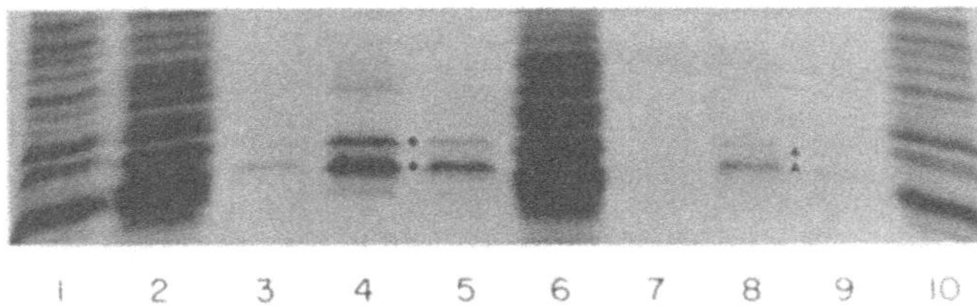

Fig. 8. Lower portion of an autoradiogram showing labeled proteins present in immunoprecipitates obtained after adsorption of sucrose density fractionated membranes with anti-F-pilus serum. Shown are the immunoprecipitates from (outer) membranes of density 1.21-1.23 gm/cc (lanes 3 and 7), (inner) membranes of density 1.14-1.18 gm/cc (lanes 4 and 8), and light fractions of density 1.07-1.11 gm/cc (lanes 5 and 9). Samples were labeled during EDλ134 infection of either the *Flac traAl* host (lanes 3-5) or the F^- host (lanes 7-9). Dots indicate the polypeptides Ap8 and Ap7* synthesized in the *Flac traAl* host; ▲, the polypeptides Ap8 and Ap7(Q) synthesized in the female host. Also shown are whole cell preparations of proteins labeled after EDλ4 or EDλ134 infection of the *Flac traAl* host (lanes 1 and 2, respectively) or of the female host (lanes 10 and 6, respectively).

DISCUSSION

Our data allow us to conclude that biosynthesis of F-pilin requires expression of at least 2 *tra* activities in addition to expression of the structural gene, *traA*. An M_R 14,000 product (Ap14) is the major product of *traA* when other *tra* activities are not present. A small amount of an M_R 7,000 polypeptide which we presume stems from inefficient host processing of Ap14 is also observable. Ap14 is efficiently replaced by an M_R 7,000 polypeptide [Ap7(Q)] in $TraA^+TraQ^+$ systems, while yet another F activity must also be present in order to synthesize the modified polypeptide (Ap7*) which is antigenically identical to F-pilin. An M_R 8,000 polypeptide (Ap8) is also observed in $TraA^+TraQ^+$ systems and may be an intermediate or byproduct of the *traQ* reaction. These results are summarized in Fig. 9.

The Nature of the traQ Reaction

We have previously suggested that *traQ* product functions in a processing reaction in which the direct product of *traA* (Ap14) is cleaved (3,10). Although the exact nature of the *traQ*-mediated reaction is still unclear, this model remains the most straightforward explanation of our data. The DNA sequence of *traA*, recently determined by Frost, Parenchych, and Willetts (*J. Bact.*, in press), can be interpreted quite consistently with our hypothesis. We assume

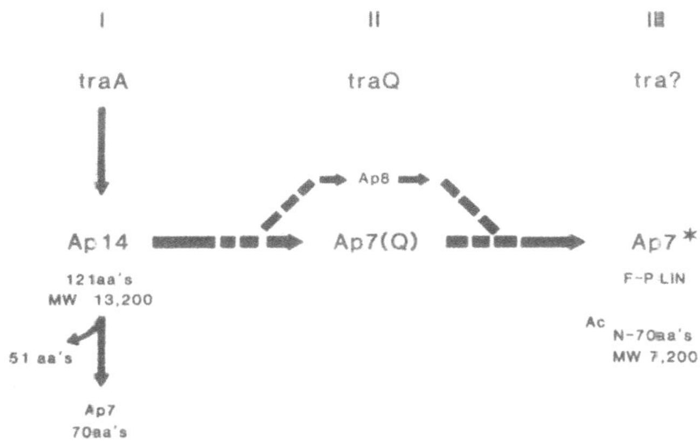

Fig. 9. Diagram showing the possible pathway of F-pilin synthesis. Expression of traA in a female host leads to synthesis of Ap14 from which small amounts of Ap7 can be obtained. When traQ activity is present Ap14 appears to be converted to Ap7(Q) and small amounts of Ap8 are also detected. If additional tra activities are also present, a polypeptide, Ap7*, which appears to be antigenically equivalent to F-pilin is synthesized. Numbers of amino acids per polypeptide, and the acetylated N terminus for F-pilin are surmised for these products based on the data of Frost, Paranchych, and Willetts (J. Bact., in press). Dotted lines indicate that the relationship of Ap8 and Ap7(Q) to the pathway is not yet clear (see text).

polypeptide Ap14 corresponds to the 121 amino acid polypeptide these authors predict to be encoded by the traA reading frame. Polypeptides Ap7 and Ap7(Q) correspond well in size to the 70 amino acid C terminal portion of the traA gene coded product they have ascribed to F-pilin. Since both a core sequence of hydrophobic amino acids and a typical signal peptidase recognition site (11) appear in the sequence immediately preceding those amino acids which match the F-pilin N-terminus, Frost, Paranchych, and Willetts have suggested that a signal peptidase cleavage can occur at the F-pilin N terminus, i.e., between amino acids 51 and 52 of the traA product. We assume that our ability to detect Ap7 reflects this occurrence. Since, however, we observed very little Ap7 polypeptide to be made when traA was expressed in an F$^-$ host, and Ap14 always predominated under such conditions, it would appear that host enzymes can accomplish this processing only very inefficiently. Indeed, it seems likely that the appearance of Ap7 is a consequence of overproduction of Ap14 in the absence of the tra products with which it would normally associate.

It is possible to imagine a number of different mechanisms through which traQ product might accomplish the efficient production of Ap7(Q). It seems likely, however, that *traQ* product either: 1) is a signal peptidase which efficiently recognizes the direct product of *traA* and cleaves it, or 2) interacts with or acts on the *traA* product in such a way as to effect efficient subsequent recognition by a host or *tra* operon-specified signal peptidase. In light of the latter hypothesis, it is interesting to note that under TraA$^+$TraQ$^+$ conditions, small quantities of an M_R 8,000 polypeptide (Ap8) which also appeared to derive from *traA* were detected in concert with the appearance of Ap7(Q). It is conceivable that Ap8 results from a *traQ* processing event which shortens the 51 amino acid "leader" portion of the *traA* product to permit proper orientation and recognition of its signal sequence. If Ap8 is such an intermediate, we would expect its amino terminal sequence to correspond to an internal sequence in the leader, and we are currently examining this possibility. Since we cannot presently discount other explanations for Ap8, such as aberrent migration or host modification of Ap7(Q), and it is possible that either one or both of these proteins are intermediates in synthesis of Ap7*, the dotted lines in Fig. 9 reflect this ambiguity. Our present assumption is that Ap7(Q), like Ap7, results from a cleavage between amino acids 51 and 52. Since *traQ* mediated processing could differ from host processing, the N-terminal sequence of Ap7(Q) is also being analyzed.

It is important to note that, at this point, we have not been able to demonstrate precursor relationships among the *traA* products. Ap14, for example, is typically detectable only when synthesized in excess to *traQ* expression, and under these circumstances appears stable in chase experiments. Similarly, we do not yet know if Ap8 or Ap7(Q) are actual precursors to Ap7*. It is possible that one or both are end products made only when appropriate interactions with other *tra* proteins are precluded. Attempts to reversibly inhibit the synthesis of F-pilin using ethanol or a *secA*(ts) host were unsuccessful (Laine, unpub. data). It seems likely that translation of *traA* and *tra* mediated processing of its product are tightly coupled reactions. We are approaching this problem by constructing sets of compatible plasmids on which expression of various *tra* genes can be induced.

The *traQ* product itself has not yet been identified. Our most recent experiments indicate that *traQ* activity can be expressed from a 6.45 kb *SmaI* fragment which includes F DNA between coordinates 77.45 and 83.9 on the map of Skurray and Willetts (15,17). Although the activities of *traU*, *traN* and *traF* which are carried in this segment do not appear to be involved in processing the *traA* product, our clone expresses at least 2 additional *tra* polypeptides (Moore, Wu, and Ippen-Ihler, unpub. data). Further experiments are therefore necessary to determine whether the *traQ*-mediated reaction requires one or both of these polypeptides.

The nature of the modification reaction. It is apparent from our data that when all *tra* genes are expressed, the final product of the traA gene is not Ap7(Q), but Ap7*. Only Ap7* appears antigenically equivalent to F-pilin. Ap7* appears to be a modified form of Ap7(Q) which is synthesized only if yet another F activity in addition to that stemming from the traQ region is present. Thus, Ap7* may be derived from Ap7(Q) through the activity of a modifying enzyme. Alternatively, Ap7(Q) may be an aberrent product which is made when the traQ product acts in the absence of the modifying enzyme. In any case, our assay for the latter activity is the synthesis of Ap7* rather than Ap7(Q).

Frost, Paranchych, and Willetts have also recently established that the N-terminal amino acid of F-pilin is acetylated, (J. Bact., in press). This finding, together with the work of Worobec et al. (19) which demonstrated that a similar N terminus is the major antigen of EDP208 pilin, strongly suggests that the antigenic difference we detect in comparison of Ap7(Q) and Ap7* reflects the acetylation of the F-pilin N terminus. As we are able to find Ap7* in the inner membrane, our data further suggest that the acetylation occurs during the synthesis of the F-pilin inner membrane pool, prior to or in concert with assembly of F-pili. Although the system in which we detected the addition of antigen also permitted assembly of F-pili, it seems likely that a complete assembly apparatus is not an absolute requirement for completion of the modification. We have previously shown that the M_R 7,000 product of a TraA$^+$ phage expressed in Flac hosts carrying mutations in traB, traV, traW, traC, traU, or traF migrates more slowly than the polypeptide we now call Ap7. We presume, therefore, that synthesis of Ap7* occurs despite such mutations. Nevertheless, we have not yet completed antibody assays for Ap7* in such mutant strains, and our earlier experiments in which we reported the presence of F-pilin polypeptide in the membranes of other Flac tra mutant hosts would not necessarily have distinguished Ap7* from Ap7(Q). Additional experiments are, therefore, needed to determine the minimal number of tra gene products necessary to accomplish the modification reaction. However, the gene products expressed by EDλ134, which carries a very large segment of the tra operon, were clearly insufficient. It is obvious that some Flac product, not expressed by EDλ134 is required. Yet, mutation or deletion of most of the tra genes which are not expressed from EDλ134 does not affect F-piliation. If we assume that only Ap7* can be assembled into F-pili, the gene products of traM, traS, traT, traD, traI, or traZ cannot be involved. The most obvious remaining candidates are traG and traJ. Our preliminary experiments suggest that it is the N-terminal portion of traG product which has been shown to be required for piliation that is also necessary for synthesis of Ap7*. The modification appears to progress in an Flac traJ90 background (unpub. results). However, synthesis of competing unlabeled F-pilin has complicated interpretation of antiserum precipitates of EDλ134 products synthesized in Flac traG, and other Flac hosts. Therefore, although traG product appears to be required in the synthesis of Ap7*, further experiments are required to establish its exact role. If, indeed, one function of traG product is to acetylate the F-pilin N terminus, it is interesting to speculate that the C-terminal portion of the traG polypeptide, [which is not required for expression of F-pilin, but is essential for mating pair stabilization and DNA transfer (See Ref. 17 for summary of evidence)] might modulate a mating-stimulated alteration of this reaction.

F-pilus assembly. In addition to the products of traA, traQ, and traG, at least 10 additional tra genes are known to be required in order to accomplish expression of an F-pilus on the cell surface. Since these genes span a region of the tra operon which has not yet been genetically saturated and appears to encode more gene products than could be encoded by known genes (3), it seems probable that the number of gene products involved in F-pilus assembly, in fact, exceeds this number. As a working model, we propose that these proteins interact in a trans-membrane complex which creates an F-specific adhesion between the inner and outer membrane of the host. We imagine that this adhesion complex functions in assembly and disassembly of F-pilin from the inner membrane and in conduction of nucleic acid during conjugation and infection by F-pilus specific bacteriophages. Although we find it experimentally convenient to determine the associations of these proteins with host inner and outer membrane material, it is important to comment that we imagine that these associations reflect orientations of tra proteins which are associated with each other in a complex which spans both. Although, as yet, we know very little about the individual functions of the numerous tra products associated with F-piliation, we expect that the techniques we have developed for analysis of the pathway of F-pilin synthesis will prove useful in identifying these products and analyzing the interactions which occur between these proteins.

ACKNOWLEDGMENT

We are grateful to Laura Frost, William Paranchych and Neil Willetts for sharing their unpublished results, and providing us with the DNA sequence of traA prior to its publication. We thank David McMurray for his assistance in the preparation of anti-F-pilus serum. Our work was supported by Public Health Service Grant AI 14426 from the National Institute of Allergy and Infectious Diseases.

REFERENCES

1. Achtman, M., N. Kennedy, and R. Skurray (1977) Cell-cell interactions in conjugating E. coli: Role of traT protein in surface exclusion. Proc. Natl. Acad. Sci., USA 74:5104-5108.
2. Bayer, M., G. Costello, and M. Bayer (1982) Isolation and partial characterization of membrane vesicles carrying markers of the membrane adhesion sites. J. Bact. 149:758-767.

3. Ippen-Ihler, K., D. Moore, S. Laine, D.A. Johnson, and N.S. Willetts (1984) Synthesis of F-pilin polypeptide in the absence of F traJ product. Plasmid 11:116-129.
4. Johnson, D.A., and N.S. Willetts (1983) λ-Tranducing phages carrying transfer genes isolated from an abnormal prophage insertion into the traY gene of F. Plasmid 9:71-85.
5. Kennedy, N., L. Beutin, M. Achtman, R. Skurray, W. Rahmsdorf, and P. Herrlich (1977) Conjugation proteins encloded by the F sex factor. Nature 270:580-585.
6. Manning, P.A., and M. Achtman (1974). Cell to cell interactions in conjugating E. coli: The involvement of the cell envelope. In Bacterial Outer Membranes, M. Inouye, ed. Wiley, New York, pp. 409-447.
7. Minkley, Jr., E.G., S. Polen, C.C. Brinton, Jr., and K. Ippen-Ihler (1976) Identification of the structural gene of F-pilin. J. Mol. Biol. 108:111-121.
8. Moore, D., B.A. Sowa, and K. Ippen-Ihler (1981) Location of an F-pilin pool in the inner membrane,. J. Bact. 146:251-252.
9. Moore, D., B.A. Sowa, and K. Ippen-Ihler (1981) The effect of tra mutations on the synthesis of the F-pilin membrane polypeptide. Mol. Gen. Genet. 184:260-264.
10. Moore, D., B.A. Sowa, and K. Ippen-Ihler (1982) A new activity in the F tra operon which is required for F-pilin synthesis. Mol. Gen. Genet. 188:459-464.
11. Perlman, D., and H.O. Halvorson (1983) A putative signal peptidase recognition site and sequence in eukaryotic and prokaryotic signal peptides. J. Mol. Biol. 167:391-409.
12. Remaut, E., P. Stanssens, and W. Fiers (1981) Plasmid vectors for high efficiency expression controlled by the P_L promoter of coliphage lambda. Gene 15:81-83.
13. Thompson, R., and M. Achtman (1978) The control region of the F Sex Factor DNA transfer cistrons: Restriction mapping and DNA cloning. Mol. Gen. Genet. 165:295-304.
14. Willetts, N.S. (1971) Plasmid specificity of two proteins required for conjugation in E. coli K-12. Nature New Biol. 230:183-85.
15. Willetts, N., and G. Johnson (1979) R. SmaI cleavage map of the transfer operon of the E. Coli K12 sex factor F. Genet. Res. Camb. 34:195-202.
16. Willetts, N.S., P.H. Moore, and W. Paranchych (1980) Variant pili produced by mutants of the Flac plasmid. J. Gen. Microbiol. 117:445-464.
17. Willetts, N., and R.A. Skurray (1980) The conjugation system of F-like plasmids. Ann. Rev. Genet. 14:41-76.
18. Willetts, N., and B. Wilkins (1984) Processing of plasmid DNA during bacterial conjugation. Microbiol. Rev. 48:24-41.
19. Worobec, E.A., A.K. Tanja, R.S. Hodges, and W. Paranchych (1983). Localization of the major antigenic determinant of EDP208 pili at the N-terminus of the pilus protein. J. Bact. 153:955-961.

GENETICS OF CLINDAMYCIN RESISTANCE IN BACTEROIDES

C. Jeffrey Smith,[1] Melissa C. Liechty,[2]
Jeanette L. Rasmussen,[2] and Francis L. Macrina[2]

[1]Laboratory of Molecular Microbiology
National Institute of Allergy and Infectious Diseases
Fort Detrick
Frederick, Maryland 21701

INTRODUCTION

The colonic Bacteroides species are numerically predominant members of the human intestinal tract microflora (2,6). In addition to their presence as indigenous flora, these organisms are significant opportunistic pathogens implicated in a variety of infectious processes (1). Clindamycin, a lincosamide antibiotic, is used extensively as a first line drug for the treatment of anaerobic infections caused by Bacteroides species. However, the increasing emergence of lincosamide-macrolide (LM) (e.g., clindamycin-erythromycin) resistant Bacteroides in the clinical setting (18) has severely compromised routine antimicrobial chemotherapy.

Following the appearance of clindamycin-resistant Bacteroides strains, there were several reports demonstrating conjugal transfer of antibiotic-resistance among Bacteroides species, and the first Bacteroides R-plasmids were described (12,14,20,21,23,24). Thus far, 3 Bacteroides R-plasmids have been described, and a summary of their relevant properties is presented in Tab. 1. The 3 plasmids vary widely in size and represent distinct, self-transmissible replicons; however, they each encode high-level resistance (> 100 μg per ml) to the antibiotics clindamycin, erythromycin, and lincomycin. Based on DNA-DNA hybridization studies, a related LM-

---

[2] Department of Microbiology and Immunology, Virginia Commonwealth University, Richmond, Virginia 23298.

Tab. 1. Summary of properties of Bacteroides R-plasmids.

Host Strain	Plasmid	Molecular Size (kb)	Phenotype[a]	Geographic Origin	Ref.
B. fragilis TMP10	pBFTM10	14.6	LM resistance conjugal transfer	Los Angeles, CA	(21)
B. fragilis V479-1	pBF4	41	LM resistance conjugal transfer	Paris, France	(23)
B. ovatus IB106	pBI136	82	LM resistance conjugal transfer	Paris, France	(15)

[a]The phenotype of LM resistance indicates resistance to more than 100 μg per ml of clindamycin or erythromycin.

resistance determinant has been found associated with all 3 plasmids (14,15). Although apparently widespread, there is a lack of information regarding the mode of dissemination of this antibiotic-resistance determinant.

In addition to the R-plasmids found in Bacteroides, several laboratories have described strains that conjugally transfer an LM-resistance phenotype by a mechanism which does not appear to involve detectable plasmid DNA (8,10,19). The cellular location of these "nonplasmid associated" resistance determinants is not yet known but they may be chromosomally located. Hybridization studies with these strains indicate that the LM-resistance determinant is similar to the one associated with the Bacteroides R-plasmids described in Tab. 1 (5,9,10).

By way of example, using the recently described R-plasmid pBI136 (15), this paper will summarize much of the current information regarding transmissible LM-resistance in the Bacteroides. Data will be presented on the molecular structure of both the plasmid-mediated and nonplasmid associated LM-resistance determinants.

RESULTS AND DISCUSSION

The Antibiotic Resistance Transfer Event

Bacteroides ovatus IB106 is a clinical isolate resistant to more than 100 μg per ml of clindamycin and erythromycin. Transfer of this LM-resistance phenotype to 2 suitably marked recipient strains was demonstrated by cocultivation of cells on nitrocellulose filters placed on nonselective agar plates in an anaerobic chamber. As shown in Tab. 2, the transfer frequency ranged from 10^{-7} to 10^{-5} transconjugants per input donor cell. The transconjugant clones obtained from these matings were also transfer proficient at comparable frequencies (Tab. 2, expt. 3 and 4). The transfer frequencies observed were in the range reported for other Bacteroides systems (10,21,23). The frequency of transfer was somewhat dependent upon the Bacteroides species participating in the matings (Tab. 2; compare expt. 1 and 2, and expt. 3 and 4). Although the basis for this difference is not understood, other reports also have shown differences in transfer frequencies of interspecies matings (23).

Available evidence suggests that the transfer of LM resistance from IB106 occurs by a mechanism that fits the criteria for a conjugation-like event. Using previously published methods (16,23), the transfer process was found to be insensitive to the action of DNase I and it required cell-to-cell contact between donor and recipient strains. Moreover, neither chloroform-treated donor cells nor donor cell-free filtrates could mediate the antibiotic resistance transfer event. In addition, transfer did not occur in liquid broth cultures of donor and recipient strains. These results are characteristics of all Bacteroides transfer systems described to date (5,16,19,21,23). There are no known reports of genetic exchange mediated by transformation or transduction in Bacteroides.

Plasmid Analysis of IB106

B. ovatus IB106 contains 2 plasmid species: pBI106, 44 kilobases (kb) and pBI136, 82 kb (Fig. 1). Detection of pBI136 was initially complicated by a relatively high concentration of pBI106 in the plasmid preparations. Estimates based on the fluorescence of

Tab. 2. Conjugal transfer of LM resistance among Bacteroides strains.

Expt.[a]	Donor Strain	Recipient Strain	Selective[b] medium	Transfer[c] frequency
1.	B. ovatus IB106	B. fragilis 638	Cc Rf	5×10^{-7}
2.	B. ovatus IB106	B. uniformis V528	Cc Rf	1×10^{-5}
3.	B. fragilis IB131	B. ovatus V211	Cc Rha	2×10^{-7}
4.	B. uniformis IB128	B. ovatus V211	Cc Rha	4×10^{-5}

[a]The transconjugants retained for further study from each experiment were as follows: Expt. 1, B. fragilis IB126 and IB131; Expt. 2, B. uniformis IB128 and IB129; Expt.3, B. ovatus IB136; and Expt. 4 B. ovatus IB137.

[b]The selective medium for experiments 1 and 2 was supplemented Brain Heart Infusion agar (16) containing 5 μg per ml clindamycin (Cc) and 25 μg per ml rifampicin (Rf). For experiments 3 and 4 a defined medium (16) containing 10 μg per ml clindamycin, and rhamnose (0.2%; Rha) as sole energy source was used for selection.

[c]Transfer frequency was estimated by averaging results from at least three independently performed experiments. Transfer frequency is the number of LM resistant transconjugants per viable input donor cell.

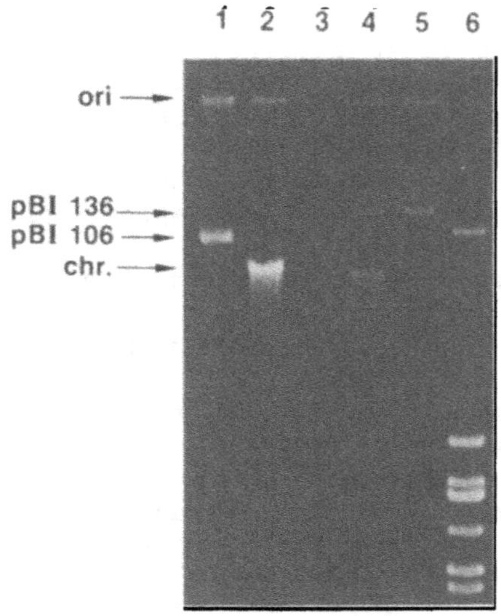

Fig. 1. Plasmid DNA isolated from IB106 and LM resistant transconjugants. IB106 (lane 1), IB128 (lane 2), IB131 (lane 3), IB137 (lane 4), IB136 (lane 5), and V517 (lane 6) plasmid size reference molecules (7). For derivation of transconjugant strains see the footnotes to Tab. 2. Plasmid DNA was purified by CsCl-ethidium bromide ultra-centrifugation of cleared lysates (4). (Ori), origin of the gel; and (chr), region corresponding to chromosomal DNA.

diluted plasmid samples electrophoresed on agarose gels and stained with ethidium bromide indicated that pBI106 was present in at least 20-fold higher concentration than pBI136 (data not shown).

Examination of LM-resistant transconjugants for plasmid DNA generally resulted in a plasmid species similar in size to pBI136 (Fig. 1). In most strain backgrounds (e.g., B. uniformis IB128 and B. fragilis IB131) the presence of pBI136 was difficult to detect and sometimes impossible to visualize by ethidium bromide staining. Samples from the equivalent of 100 ml or more of culture were usually required for analysis on agarose gels. Plasmid pBI136 DNA more routinely was obtained in higher amounts when it was present in the B. ovatus V211 strain (e.g., strains IB136 or IB137). The yields of plasmid DNA were still significantly lower than those seen for pBF4 and pBFTM10 in the same genetic background.

Due to this, inconsistency of plasmid isolation from transconjugant strains, a role for pBI136 in LM-resistance transfer, was demonstrated by filter blot hybridization studies. EcoRI restricted total cellular DNA from IB106 and transconjugants was probed with either pBI106 or pBI136 plasmid DNA and the resulting autoradiographs are shown in Fig. 2. These results clearly show that pBI106 was not involved in either LM resistance or its transfer since the original donor IB106 was the only strain that possessed sequences homologous to pBI106 DNA (Fig. 2A). When pBI136 was used as the probe, strain IB106 and all LM-resistant transconjugants exhibited homologous sequences which comigrated with purified, EcoRI-digested pBI136 (Fig. 2B). No such hybridizing sequences were observed in the original recipient strains. These data suggested a stable inheritance of pBI136 among all transconjugants including those in which the plasmid was not readily detected.

Besides the 7 EcoRI fragments of pBI136, each of the transconjugants contained an additional fragment which hybridized weakly to the pBI136 probe and migrated between the EcoRI F and G fragments. The origin of these additional sequences is not known but their migration patterns varied consistently depending on the strain background suggesting these are not partial EcoRI digestion products (compare Fig. 2B, lanes 1 and 2 to lanes 3 and 5).

Properties of pBI106 and pBI136. During the course of this investigation, it became apparent that restriction endonuclease digest patterns of the LM-resistance plasmid pBF4 were remarkably similar

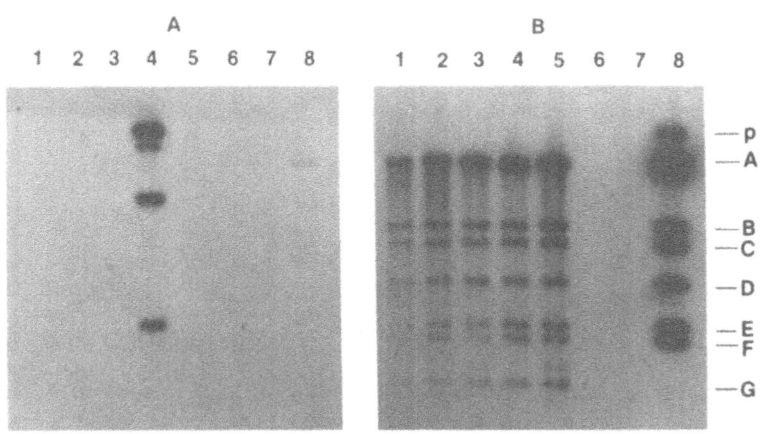

Fig. 2. Autoradigraphs of EcoRI-digested total cellular DNA from IB106 and LM-resistant transconjugants following hybridization with ^{32}P-labeled pBI106 (panel A) or pBI136 (panel B). The total cellular DNAs used were as follows: IB129 (lane 1), IB128 (lane 2), IB126 (lane 3), IB106 (lane 4), IB131 (lane 5), strain 638 (lane 6), V528 (lane 7), and purified pBI136 from IB136 (lane 8). For derivation of strains see the footnotes to Tab. 2. The pBI136 EcoRI fragments (A to G) are shown at the right of the photograph. Note that an artifact in the blotting procedure prevented hybridization of the pBI136 EcoRI-G fragment to itself in Lane 8. (p), Partial digestion product.

to those of pBI106 depending upon the enzyme used (e.g., *Hind*III). This observation coupled with the fact that pBI136 (not pBI106), was implicated in IB106 LM resistance, prompted a comparative analysis of the 3 plasmids. Using filter blot hybridization techniques, each plasmid was used to detect homologous sequences in *Eco*RI digests of each of the other plasmids; the resulting autoradiographs are shown in Fig. 3. The pBI106 probe did not hybridize to pBI136 but showed strong homology to most of the pBF4 molecule. Only the 3.7 kb pBF4 *Eco*RI-D fragment failed to hybridize with pBI106 (Fig. 3B). This is the region that is implicated in LM resistance (24). The pBF4 probe showed homology to both pBI106 and pBI136. The hybridization to pBI106 occurred at all 3 of its *Eco*RI fragments but homology to pBI136 was restricted to the *Eco*RI-C and -G fragments (Fig. 3C). The reciprocal blots with pBI136 as the probe indicate that the cross-hybridization between pBF4 and pBI136 was primarily mediated by the pBF4 *Eco*RI-D fragment but some homology to the pBF4 *Eco*RI-B fragment was also observed (Fig. 3D). Weak homology between the pBI136 probe and DNA present in the pBI106 preparation also occurred. However, these sequences comigrated with the pBI136 *Eco*RI

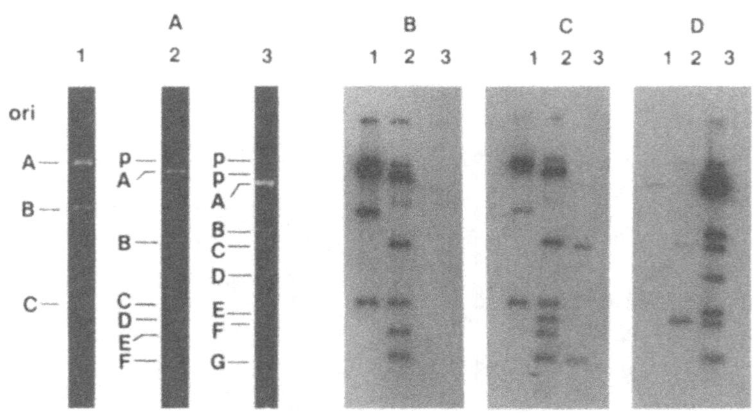

Fig. 3. Sequence homology between pBF4, pBI106, and pBI136. Purified plasmid DNA was digested with *Eco*RI and then samples were electrophoresed on the same agarose gel. DNA was then transferred to nitrocellulose filters (17) and then hybridized at 67°C (22) with the ^{32}P-labeled (13) plasmid DNA indicated below. Panel A, Agarose gel of digested plasmid DNAs. pBI106 (lane 1), pBF4 (lane 2), and pBI136 (lane 3). Panels B, C, and D are autoradiographs of blotted DNA from gel slices identical to panel A. The following ^{32}P-labeled probes were used: panel B pBI106, panel C pBF4, and panel D pBI136. (ori), Origin of gel; (p), partial digestion product; and the various *Eco*RI fragments of each plasmid are shown to the left of the appropriate gel slice in panel A.

fragments indicating that this hybridization was due to small amounts of pBI136 present in all samples of pBI106.

Endonuclease restriction maps of pBF4 and pBI106 revealed a number of interesting similarities (Fig. 4). There is a highly conserved region of these plasmids that extends for a few kilobases on either side of the EcoRI site which serves as the 0 reference point on the maps. This region includes the EcoRI-C fragments which are identical in size in both plasmids. A 3.4 kb BstEII fragment is also present in both pBF4 and pBI106. Insertion into this fragment of pBF4 inactivates the transfer function of the plasmid (D. Guiney, pers. comm.). There presently is no evidence that pBI106 is transmissible.

The relationship between pBF4 and the 2 plasmids of IB106 is quite intriguing. pBF4 and pBI106 may have evolved from a common ancester, and preliminary studies indicate that these plasmids are in fact incompatible. With the exception of the exact region implicated in LM resistance a high degree of homology extends over the entire length of these 2 molecules. The pBF4 EcoRI-D fragment is, on the other hand, the only region which shares homology with the LM-resistance plasmid pBI136. These results cannot be taken as direct evidence for transposition of the LM-resistance determinant between pBI106 and pBI136. However, these results do indicate that this antibiotic-resistance determinant is maintained on a highly mobile DNA sequence which is capable of widespread dissemination.

The LM-Resistance Determinant

Preliminary evidence based on homology with pBF4, suggested that the pBI136 EcoRI-C and -G fragments were involved in the LM-resistance phenotype. Use of the pBR325::pBF4 EcoRI-D chimeric plasmid pVA831 (8) as a hybridization probe against pBI136 proved that this pBF4 fragment was responsible for the homology between the 2 plasmids. The probe hybridized strongly to the pBF4 EcoRI-D fragment and to the EcoRI-C and -G fragments of 2 pBI136 molecules (Fig. 5 lanes 1, 2, and 3). In addition, the probe hybridized to the pBF4 EcoRI-B fragment. Homology to this fragment is due to a proposed directly repeated sequence present on both the -B and -D fragments (14) and will be discussed below. Previous studies (8,10,14) have shown that the pBF4 EcoRI-D fragment has homology to many of the LM-resistant *Bacteroides* strains including those with the nonplasmid associated determinants.

Further evidence implicating the pBI136 EcoRI-C and -G fragments in LM resistance was obtained from analysis of a spontaneously occurring, antibiotic sensitive, pBI136 deletion derivative. Restriction analysis of pBI136Δ1 showed a nearly precise deletion of the 7.2 kb EcoRI-C fragment. When this plasmid was probed with the pVA831 clone, the results confirmed that this fragment was totally absent and no fusion fragment was present, but the EcoRI-G fragment

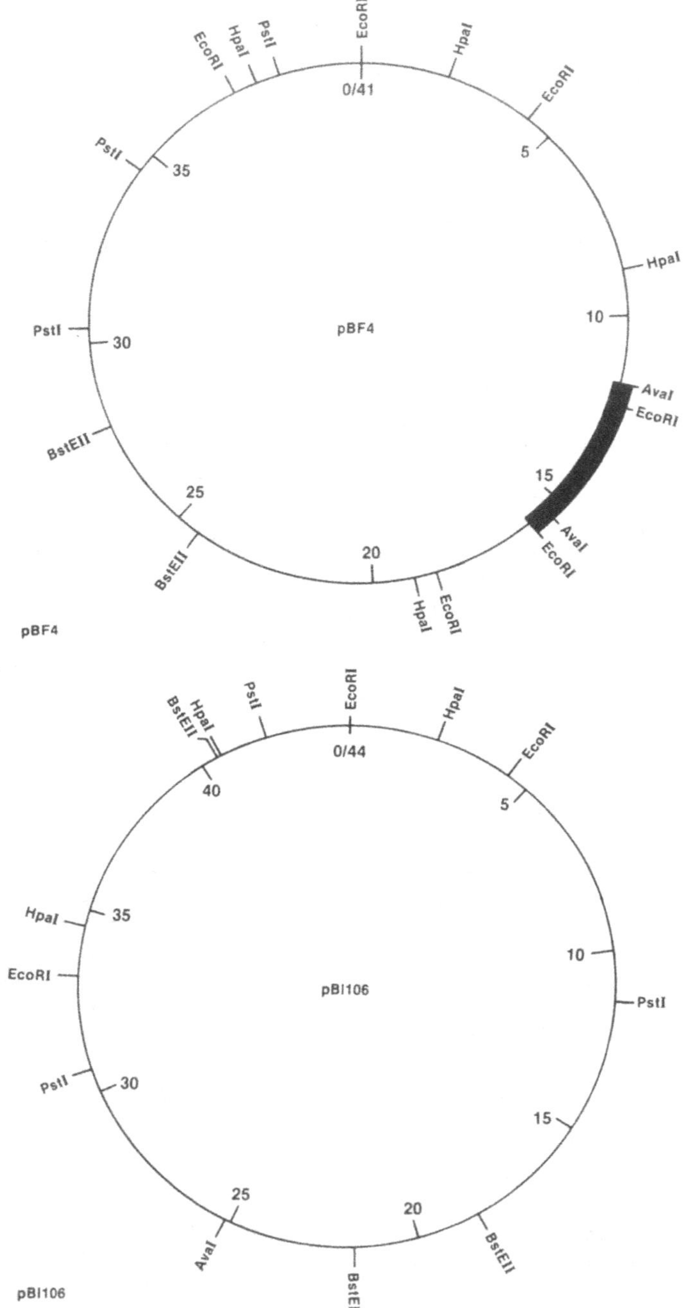

Fig. 4. Comparative endonuclease restriction site maps of pBF4 (left) and pBI106 (right). The heavy line on the pBF4 map indicates the region implicated in LM resistance.

remained intact (Fig. 5). Similar studies with pBF4 (EcoRI-D, 24) and pBFTM10 (EcoRI-B, 21) have also noted seemingly precise deletions of specific EcoRI fragments associated with the loss of LM-resistance.

Comparison of plasmid mediated determinants. Restriction endonuclease mapping (21,24) and hybridization studies (14) with pBF4 and pBFTM10 have suggested that their LM-resistance determinants are bounded on both sides by a directly repeated sequence. The repeated sequence is characterized by restriction enzyme recognition sites for EcoRI and AvaI which are separated by about 500 base pairs (bp). To further characterize the organization of the pBI136 resistance determinant, EcoRI plus AvaI double digests of pBFTM10, pBF4, and pBI136 were analyzed by filter blot hybridization using the pVA831 probe. Presumably, 500 bp of the repeated sequence is contained within the pBF4 EcoRI-D fragment of this probe. The results in Fig. 6 clearly show the presence of the 500-bp EcoRI/AvaI fragments in all 3 plasmids. The pBI136 EcoRI-C and -G fragments each have decreased in size by about 500 bp indicating that each fragment contained a copy of the repeated sequence (data not shown). As expected the pVA831 probe hybridized strongly to the pBFTM10 EcoRI-B and pBF4 EcoRI-D fragments but the probe also retained strong homology

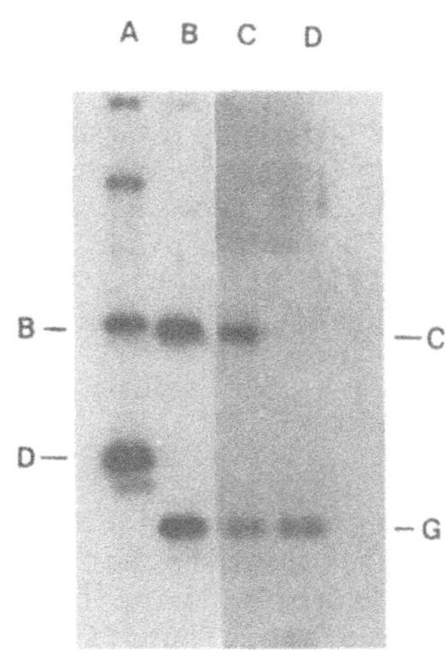

Fig. 5. Homology of pBI136 and pBI136Δ1 to the pBF4 EcoRI-D fragment cloned into pBR325. Purified plasmid DNAs were digested with EcoRI, electrophoresed on an agarose gel, blotted to nitrocellulose filters, and probed with the chimeric plasmid pVA831 (pBR325::pBF4 EcoRI-D;14). pBF4 (lane A), pBI136 from IB106 (lane B), pBI136 from IB128 (lane C), and pBI136Δ1 from IB178 (lane D). The EcoRI-B and -D fragments of pBF4 are shown to the left of the photograph and the EcoRI-C and -G fragments of pBI136 are shown to the right of the photograph.

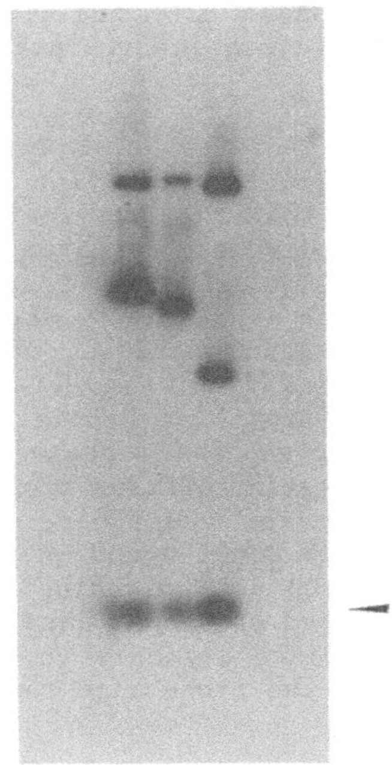

Fig. 6. Evidence for a directly repeated sequence present on the *Bacteroides* R-plasmids pBFTM10, pBF4, and pBI136. *Eco*RI plus *Ava*I double digests of the plasmids were electrophoresed on a 0.8% agarose slab gel, blotted to a nitrocellulose filter, and probed with the pBF4 *Eco*RI-D containing chimeric plasmid, pVA831. pBFTM10 (lane A), pBF4 (lane B), and pBI136 (lane C). The common 500 bp *Eco*RI - *Ava*I fragment is indicated by an arrow to the right of the photograph.

to the *Eco*RI-A and -B fragments of pBFTM10 and pBF4 respectively (Fig. 6). This result would suggest that the repeated sequence extends well beyond the *Ava*I site in these fragments. In order to accurately define the limits of this putative repeated sequence, primary DNA sequence analysis of pBF4 has been initiated. Preliminary results from these studies have so far verified the presence of the directly repeated sequence on both the *Eco*RI-D and -B fragments (Rasmussen and Macrina, in prep.).

Cloning of the pBI136 *Eco*RI-C and -G fragments into *E. coli* JM83 has facilitated restriction mapping of these 2 important DNA segments. The pUC19 vector system (11) was used for standard shotgun cloning of the pBI136 *Eco*RI fragments, and clones containing the C, D, E, F, and G fragments were obtained. Restriction digest studies together with additional hybridization results (data not shown) permitted the construction of the restriction map in Fig. 7. Typical of the results with pBF4 and pBFTM10, the region of pBI136 implicated in LM resistance (EcoRI-C) is flanked on both sides by restriction sites for EcoRI and AvaI in a direct orientation. The intervening DNA sequence of the pBI136 element is much larger in size than seen in pBF4 or pBFTM10 (Fig. 7). A comparison of the restriction sites within the region does not reveal any highly conserved sites except for the repeated sequence bounded by EcoRI and AvaI sites. This apparent heterogeneity makes impossible even speculative placement of the LM-resistance genes within these elements.

In addition to the LM-resistance determinant, the pBF4 EcoRI-D and pBFTM10 EcoRI-B fragments contain a cryptic tetracycline-resistance determinant (D. Guiney, pers. comm.). These determinants are expressed in aerobically grown *E. coli* but do not express tetracycline resistance in *Bacteroides*. *E. coli* strains containing either the pBI136 EcoRI-C or -G fragments cloned into the vector pUC19, do not express tetracycline resistance in either orientation. These results and the obvious structural differences in the 3 LM-resistance elements, imply that there has been a broad evolutionary divergence of these sequences among the *Bacteroides*.

Nonplasmid associated determinants. The widespread dissemination of a common LM-resistance determinant among *Bacteroides* includes not only the plasmid mediated resistance (14,15) but the nonplasmid associated resistance as well (5,9,10). In order to better understand the molecular organization of the nonplasmid determinant

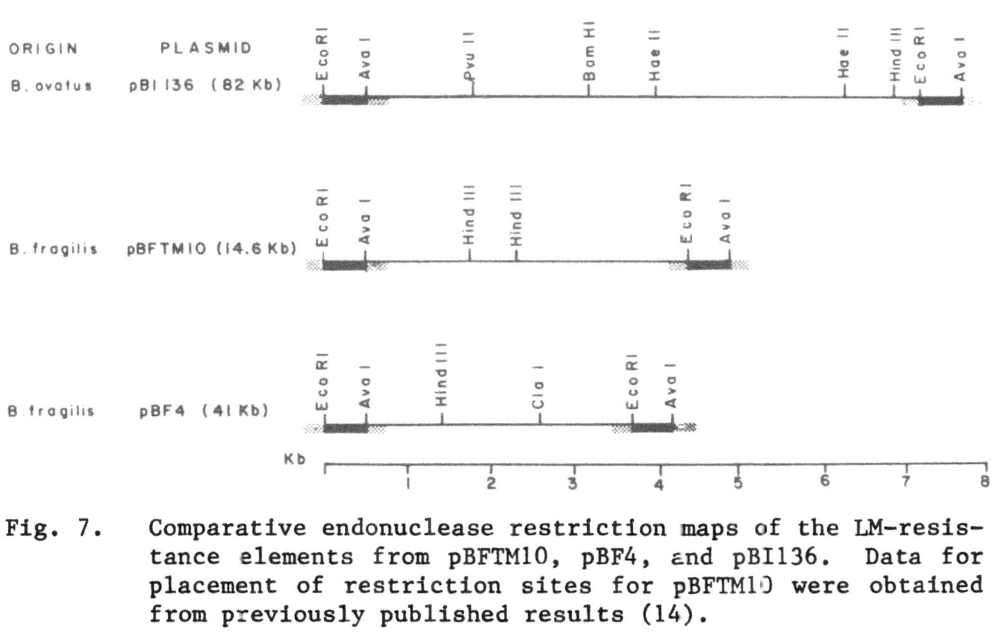

Fig. 7. Comparative endonuclease restriction maps of the LM-resistance elements from pBFTM10, pBF4, and pBI136. Data for placement of restriction sites for pBFTM10 were obtained from previously published results (14).

we have continued our studies with B. fragilis V503 (10). This previous work has shown that the pVA831 chimeric plasmid is homologous to 2 HindIII fragments present in the total cellular DNA of V503 and its LM-resistant transconjugants. Further hybridization data using the LM-sensitive deletion plasmid pBF4Δ1 (24) as a hybridization probe, failed to show hybridization with V503 (10). This plasmid probe apparently does not contain the LM-resistance determinant but does carry at least 500 bp of the directly repeated sequence (14,24). Thus these results suggested that the V503 determinant does not carry the directly repeated sequence common to the R-plasmids. Evidence supporting this conclusion was obtained by filter blot hybridization analysis of EcoRI and AvaI double digests of V503 and its transconjugants. The radiolabled probes were pVA831 or a small EcoRI/HpaII fragment from within the directly repeated sequence on the pBF4 EcoRI-D fragment. The autoradiographs in Fig. 8A show that the positive control, pBF4, hybridized strongly to the pVA831 probe at the EcoRI-B and -D fragments and at the 500 bp EcoRI/AvaI fragment. Neither V503 nor its transconjugants seem to contain the small 500-bp fragment but they do show homology with 2 fragments of about 15 kb and 6.4 kb in size. When the small pBF4 EcoRI/HpaII fragment was used as the probe (Fig. 8B), the only hy-

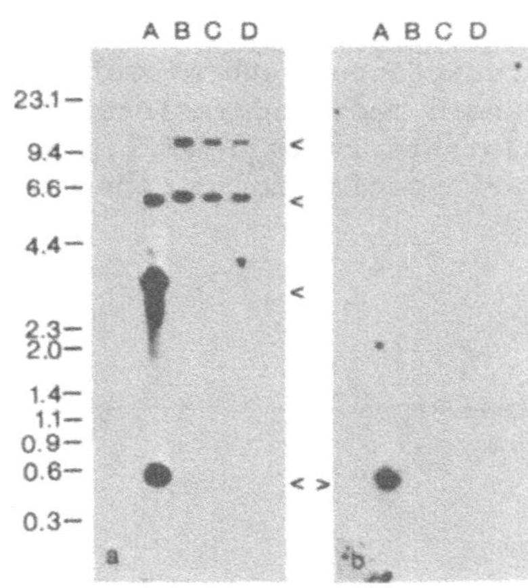

Fig. 8. Homology between V503 and the pBF4 EcoRI-D fragment is restricted to those pBF4 sequences which lie between the direct repeat sequences. EcoRI plus AvaI double digests of the plasmid pBF4 (lane A), and of total cellular DNA from V503 (lane B) and its transconjugants V622 (lane C) and V654 (lane D) were subjected to filter blot hybridization analysis. The following probes were used: Panel a, pVA831 (pBR325::pBF4 EcoRI-D, Ref. 10); panel b, pVA200 and EcoRI/HpaII fragment cloned from within the EcoRI/AvaI 500-bp fragment of pVA831. Size reference markers in kb are given to the left of the photograph and areas of hybridization are indicated by the arrows in the center of the photograph.

bridization observed was to the positive control pBF4 in lane A. No hybridizing sequences within V503 or its transconjugants were seen even upon prolonged exposure of the autoradiograph.

SUMMARY AND CONCLUSIONS

Results presented in this paper have shown that a widely distributed LM-resistance determinant is present on at least 3 distinct Bacteroides R-plasmids. In fact these plasmids bear no homology outside of the defined regions implicated in the LM resistance. This resistance determinant on pBF4, pBFTM10, and pBI136 is located within DNA segments bounded on each side by a directly repeated sequence of more than 500 bp. The intervening sequences of these 3 elements are variable, and range in size from about 3.7 kb to 7.2 kb (Fig. 7). Apart from the EcoRI/AvaI restriction sites which characterize the repeated sequence, there is a notable lack of common restriction sites within these elements. These results suggest that the elements do possess a certain degree of structural similarity but significant evolutionary divergence has occurred.

The presence and location of the directly repeated sequences, their association with specific deletions, and their association with an antibiotic-resistance determinant, are features common to many antibiotic-resistance transposons described for other prokaryotes. In addition, these elements are highly mobile, being found on a number of R-plasmids. The unique relationship between pBF4, pBI106, and pBI136 described here is a clear indication of the potential for these DNA sequences to move from one molecule to another. Given the extensive dissemination and the genetic and structural characteristics described above, it seems likely that the LM-resistance determinant is carried on transposon-like elements present in pBF4, pBFTM10, and pBI136. However, further experimentation will be necessary to document the transposition event.

Bacteroides strains such as B. fragilis V503, possess a transmissible LM-resistance determinant which does not appear to be associated with detectable extrachromosomal elements (5,9,10). Presently, a number of strains of this type have been found over a wide geographic area. The LM-resistance genes associated with these strains are apparently similar to the one carried on the Bacteroides R-plasmids because homology between the 2 has been observed. However, it is important to note that within the limits of Southern filter blot hybridization, neither V503 nor its transconjugants possess the directly repeated sequence found on the LM-resistance plasmids (Fig. 8).

The elusive nature of the V503 LM-resistance elements presents an intriguing problem. One model that has been proposed is that these resistance determinants reside on a conjugative transposon similar to Tn916 of Streptococcus faecalis (3). Such a mechanism would account for the transmissibility and lack of extrachromosomal detection, and it would be consistent with the structural differences between the LM-resistance elements of the plasmids and that of V503. However, the recent discovery of pBI136 has provided direct evidence for the existence of difficult to detect, large molecular weight R-plasmids in the *Bacteroides*. Thus there is a possibility that the resistance determinants of V503 and similar strains reside on large plasmids.

Future studies with pBI136 should result in an understanding of the problems associated with its detection in many cellular backgrounds. These studies may be useful in understanding the molecular nature of the nonplasmid-associated antibiotic-resistance determinants present in *Bacteroides* species.

ACKNOWLEDGMENTS

This research was in part supported by Public Health Service grants DE04224 and AI/DE20153 and by a grant from the Jeffress Trust to F.L.M. We thank G. Harbaugh, Donna Bandy, and Sonja Blowe for assistance in manuscript preparation.

REFERENCES

1. Finegold, S.M. (1977) *Anaerobic Bacteria in Human Disease*. Academic Press, Inc., New York.
2. Finegold, S.M., H.R. Atteberg, and V.L. Sutter (1974) Effect of diet on human fecal flora: comparison of Japanese and American diets. *Am. J. Clin. Nutr.* 27:1456-1469.
3. Franke, A.E., and D.B. Clewell (1981) Evidence for a chromosome-borne resistance transposon (Tn*916*) in *Streptococcus faecalis* that is capable of "conjugal" transfer in the absence of a conjugative plasmid. *J. Bact.* 145:494-502.
4. Guerry, P., D.J. LeBlanc, and S. Falkow (1973) General method for the isolation of plasmid deoxyribonucleic acid. *J. Bact.* 116:1064-1066.
5. Guiney, D.G., P. Hasegawa, D. Stalker, and C.E. Davis (1983) Genetic analysis of clindamycin resistance in *Bacteroides* species. *J. Infect. Dis.* 147:551-558.
6. Holdeman, L.V., E.P. Cato, and W.E.C. Moore, eds. (1977) *Anaerobe Laboratory Manual*, Virginia Polytechnic Institute and State University Anaerobe Laboratory, Blacksburg, VA.
7. Macrina, F.L., D.J. Kopecko, K.R. Jones, D.J. Ayers, and S.M. McCowen (1978) A multiple plasmid-containing *Escherichia coli* strain: Convenient source of size reference plasmid molecules. *Plasmid* 1:417-420.
8. Macrina, F.L., T.D. Mays, C.J. Smith, and R.A. Welch (1981) Nonplasmid associated transfer of antibiotic resistance in

Bacteroides. J. Antimicrob. Chemother. Vol. 8, Suppl. D, pp. 77-86.
9. Marsh, P.K., M.H. Malamy, M.J. Shimell, and F.P. Tally (1983) Sequence homology of clindamycin resistance determinants in clinical isolates of Bacteroides spp. Antimicrob. Agents Chemother 23:726-730.
10. Mays, T.D., C.J. Smith, R.A. Welch, C. Delfini, and F.L. Macrina (1982) Novel antibiotic resistance transfer in Bacteroides. Antimicrob. Agents Chemother. 21:110-118.
11. Norrander, J., T. Kempe, and J. Messing (1983) Construction of improved M13 vectors using oligodeoxynucleotide-directed mutagenesis. Gene 26:101-106.
12. Privitera, G., A. Dublanchet, and M. Sebald (1979) Transfer of multiple antibiotic resistance between subspecies of Bacteroides fragilis. J. Inf. Dis. 139:83-87.
13. Rigby, P.W.J., M. Diechmann, C. Rhodes, and P. Berg (1977) Labeling deoxyribonucleic acid to high specific activity in vitro by nick translation with DNA polymerase I. J. Mol. Biol. 113:237-251.
14. Shimell, M.J., C.J. Smith, F.P. Tally, F.L. Macrina, and M.H. Malamy (1982) Hybridization studies reveal homologies between pBF4 and pBFTM10, two clindamycin-erythromycin resistance transfer plasmids of Bacteroides fragilis. J. Bact. 152:950-953.
15. Smith, C.J., and F.L. Macrina (1984) Large transmissible clindamycin resistance plasmid in Bacteroides ovatus. J. Bact. 158:739-741.
16. Smith, C.J., R.A. Welch, and F.L. Macrina (1982) Two independent conjugal transfer systems operating in Bacteroides fragilis V479-1. J. Bact. 151:281-287.
17. Southern, E.M. (1975) Detection of specific sequences among DNA fragments separated by gel electrophoresis. J. Mol. Biol. 98:503-517.
18. Tally, F.P., G.J. Cuchural, N.V. Jacobus, S.L. Gorbach, K.E. Aldridge, T.J. Cloary, S.M. Finegold, G.B. Hill, P.B. Iannini, R.V. McCloskey, J.P. O'Keefe, and C.L. Pierson (1983) Susceptibility of the Bacteroides fragilis group in the United States in 1981. Antimicrob. Agents Chemother. 23:536-540.
19. Tally, F.P., M.J. Shimell, G.R. Carson, and M.H. Malamy (1981) Chromosomal and plasmid mediated transfer of clindamycin resistance in Bacteroides fragilis. In Molecular Biology, Pathogenicity and Ecology of Bacterial Plasmids, S. Levy, R.C. Clowes, and E. Koenig, eds. Plenum Press, New York, pp. 51-59.
20. Tally, F.P., D.R. Snydman, S.L. Gorbach, and M.H. Malamy (1979) Plasmid mediated transferable resistance to clindamycin and erythromycin in Bacteroides fragilis. J. Infect. Dis. 139:83-88.
21. Tally, F.P., D.R. Snydman, M.J. Shimell, and M.H. Malamy (1982) Characterization of pBFTM10, a clindamycin-erythromycin resistance transfer factor from Bacteroides fragilis. J. Bact. 151:686-691.11.

22. Thayer, R.E. (1980) An improved method for detecting foreign DNA in plasmids of *Escherichia coli*. *Anl. Biochem.* 98:60-63.
23. Welch, R.A., K.R. Jones, and F.L. Macrina (1979) Transferable lincosamide-macrolide resistance in *Bacteroides*. *Plasmid* 2:261-268.
24. Welch, R.A., and F.L. Macrina (1981) Physical characterization of the *Bacteroides* R plasmid pBF4. *J. Bact.* 145:867-872.

EFFECT OF CHROMOSOME HOMOLOGY OF PLASMID TRANSFORMATION AND PLASMID CONJUGAL TRANSFER IN *HAEMOPHILUS INFLUENZAE*

M. Balganesh and J. K. Setlow

Biology Department
Brookhaven National Laboratory
Upton, New York 11973

INTRODUCTION

Homology between a resident plasmid in *H. influenzae* and a plasmid entering the cell has been shown to be very important for the acquisition by the cell of genetic markers from the incoming plasmid. For example, the frequency of transformation of recombination-proficient *H. influenzae* by two 51-kbp plasmids increases by more than 3 orders of magnitude when the recipient cells contain a partially homologous plasmid (1). This increase is not simply the result of increased uptake of plasmid DNA into the cell, since no transformation could be observed with Rec^- recipients which take up DNA as well as the wild type. Furthermore, one of the plasmids has been shown to contain the DNA uptake sites necessary for efficient uptake into competent *H. influenzae* cells (Setlow and Albritton, unpublished experiments), and the data suggest that the other plasmid also contains such sites. Thus, the plasmids are able to transform Rec^+ cells containing homologous DNA at a higher frequency because of events occurring following DNA uptake.

A study of transformation by a hybrid plasmid, pNov1, constructed by inserting a 9 kbp piece of chromosomal DNA coding for resistance to novobiocin into the 5.7-kpb plasmid RSF0885, showed that there is a similar effect of homology on the establishment of this plasmid (16). Efficient transformation by pNov1 is also dependent on the Rec^+ genes of the recipient. There is considerable recombination between homologous regions of the plasmid pNov1 and the chromosome in Rec^+ recipients, resulting in around 75% substitution of homologous chromosomal genetic information for plasmid genes in the established plasmid. Even more probable than plasmid establishment following uptake is the integration of homologous parts of the plasmids into the chromosome, causing loss of the plasmid from the cell. This interaction takes place at the time of the transformation rather than after the plasmid becomes established (16).

Competence in H. influenzae is a physiological state in which DNA may be taken up efficiently from the medium (7). However, the competent state involves more than ability to take up DNA. There is also a competence-induced increase in the ability of DNA to recombine. For example, the frequency of H. influenzae phage recombination is around 10 times higher in competent cells than in exponentially growing or stationary phase cells (4). The chromosomal DNA of competent cells contains regions of single-strand gaps and tails (10,11). These regions are apparently necessary for recombination between transforming DNA and the recipient chromosome. A recombination-defective mutant, rec2, that is transformed with a frequency 10^{-7} that of the wild type, does not contain such regions, although when it is in the competent state it takes up DNA from the medium as well as does the competent wild type. A possible explanation of the interaction between plasmids and chromosomes during transformation, but not after the plasmids become established, was that the competent state was required for such interaction. However, no interaction was observed when wild-type cells with an established plasmid were made competent (16). Thus, the interaction apparently only takes place upon entrance of the plasmid into the cell. We present evidence here that when a plasmid enters the cell by a route other than transformation, namely conjugal transfer, there is also no interaction between homologous regions of plasmid and chromosome. We also show that the ability of a plasmid to become established by transformation is dependent upon the degree of homology between plasmid and chromosome, but such homology does not increase plasmid establishment following conjugal transfer.

METHODS

In order to make an adequate comparison of the effect of chromosomal homology on transformation and conjugal transfer, we needed plasmids with and without homology that could both transform and be transconjugated. Since the smaller plasmids in H. influenzae are not conjugally transferred, we used the 35-kbp mobilizing plasmid pHD147 (6). Unfortunately, we could not detect mobilization of pNov1 (Tab. 1), although the parent plasmid RSF0885 was mobilized at a reasonable frequency, as previously reported (6). These data suggest that the PvuII site in RSF0885, used for insertion of the chromosomal DNA of pNov1, is involved with mobilizability. Therefore, we have used another cloning vehicle, pDM2, previously constructed in this laboratory (12), and have ligated into it chromosomal portions from pNov1. Figure 1 shows a simplified restriction

Tab. 1. Mobilization of RSF0885 and pNovl.

Mating pairs	Transconjugation frequency
rec1(pNovl)(pHD147) x rec2	$<8 \times 10^{6}$
rec1(pNovl)(pHD147) x wild type	$<8 \times 10^{-7}$
rec1(RSF0885)(pHD147) x rec2	10^{-3}
rec1(RSF0885)(pHD147) x wild type	10^{-3}

Cells were mixed and incubated overnight on hemoglobin agar, a condition that fosters conjugal transfer of plasmids (Albritton, Setlow and Slaney, 1982). The recipients (rec2 or wild type) were resistant to streptomycin, and RSF0885 and pNovl in the donors conferred resistance to ampicillin. The transconjugation frequency was measured as the ratio of viable cells doubly resistant to ampicillin and streptomycin to the total number of recipients (the streptomycin-resistant cells).

map of pNovl as well as drawings of the other plasmids used. None of these plasmids contains an active transposon. The 9.8-kbp parent plasmid pDM2, coding for ampicillin and chloramphenicol resistance, contains 2 EcoRI sites separated by 1.1 kbp. Following digestion by EcoRI, the large (8.7-kbp) fragment was then ligated together with itself, or in the case of pMB1 and pMB2, ligated to 1 (pMB1) or 2 (pMB2) copies of the 1.7 kbp EcoRI fragment of pNovl. The large plasmid pMB3 was constructed by first digesting the protruding single-strand ends of EcoRI-cut pDM2s with S1 nuclease, and then blunt-end ligating the whole of the chromosomal portion of pNovl, removed by PvuII digestion. The SmaI-cut vector was also blunt-end ligated to the 9.0-kbp pNovl fragment. The latter plasmid, pMB4, no longer coded for chloramphenicol resistance. The other plasmids, however, conferred both ampicillin and chloramphenicol resistance. The 1.7-kbp EcoRI fragment of pNovl confers resistance to 25 μg/ml novobiocin, but is not expressed in cells carrying pMB1 or pMB2, unless the resistance gene has been transferred to the chromosome. In cells carrying pMB3 or pMB4, which include from pNovl the 1.7-kbp fragment as well as chromosomal DNA on either side of that fragment, the novobiocin resistance is expressed even without recombination with the chromosome. Figure 2 shows evidence that the orientation of the single EcoRI fragment in pMB1 is the same as that of the 2 EcoRI fragments in pMB2.

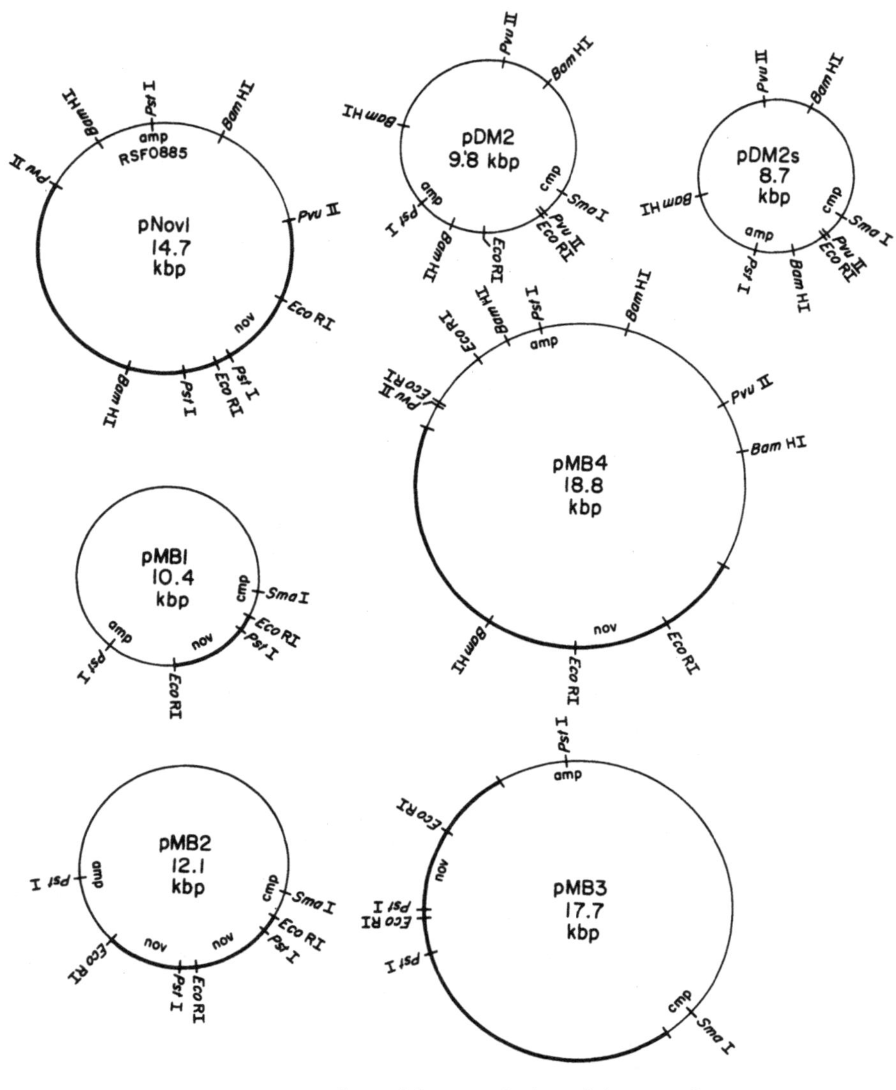

Fig. 1. Plasmids used in this study.

Transformation was carried out as previously described (15) in cells made competent in MIV medium (17). All of the plasmids we have constructed, except pNovl, were mobilizable by pHD147 in three-way matings. Mating was usually done on membrane filters on top of hemoglobin agar for 2 to 3 hr (with little or no observable difference in transconjugation frequency between these times). Because of the phenomenon of Rec^+-dependent transfer of chromosomal markers from cell to cell (2), we used Rec^- strains for the donors, to avoid the possibility that chromosomal markers in the recipients could en-

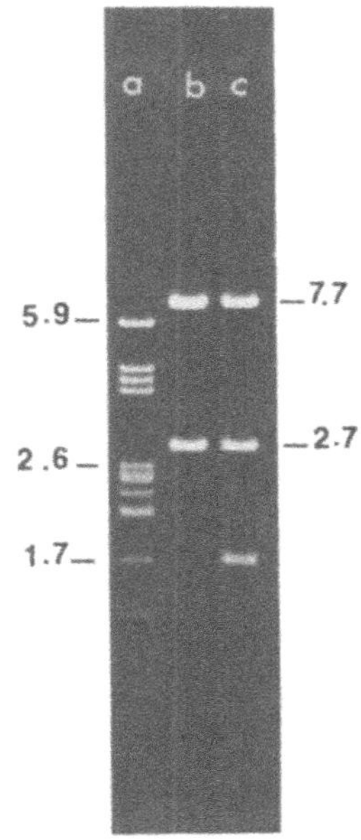

Fig. 2. Gel electrophoresis of pMB1 (b) and pMB2 (c) cut with PstI, along with molecular weight standards (HpaI digested T7 DNA) (a). The numbers refer to kbp. Since the digestion of pMB1 yields 2 fragments the same size as 2 of the 3 fragments of digested pMB2, the orientation of the 2 EcoRI chromosomal fragments in pMB2 must be the same as the orientation of the single chromosomal fragment of pMB1.

ter donor cells and give spurious numbers for transconjugant frequency. Transconjugation was completely dependent on the presence of the mobilizing plasmid, as shown in the example of Tab. 2. The frequencies were independent of the recombination proficiency of the recipient, as in the two-way mating shown in Tab. 1, and were also independent of the state of competency of the recipient, and whether deoxyribonuclease at 125 μg/ml was added to the cells at the time of mixing them and adding them to the filter (data not shown).

RESULTS

Transformation and Transconjugation Frequencies

Figure 3 shows a comparison of frequency of plasmid establishment following transformation or conjugation as a function of size of the chromosomal insert in wild-type or Rec^- recipients. The plasmid pDM2s with its EcoRI piece removed is more comparable to the larger plasmids than the untreated pDM2. Conjugal transfer but not

Tab. 2. A typical mating experiment.

	Total Number of Cells in the Suspension			
Mating cells	Donors	Containing the mobilizing plasmid	Recipients	Transconjugants
rec2(pHD147) rec1(pMB3) Rd(wild type)	$1.6x10^9$	$2.4x10^9$	$1.2x10^9$	$1.2x10^5$
rec1(pMB3) Rd	$2.2x10^9$	-	$8.8x10^8$	0 (<1 in 4×10^6 recipients)

After incubation of the cells on a filter for two hours, they were suspended in a volume of 2 ml, and plated with appropriate dilutions and antibiotics. The donors, rec1(pMB3), were resistant to novobiocin and chloramphenicol, the cells containing the mobilizing plasmid, rec2(pHD147), were resistant to kanamycin, and the recipients, Rd, were resistant to streptomycin. Transconjugants were resistant to both streptomycin and chloramphenicol.

transformation was depressed in pDM2s relative to that of pDM2. There was an exponential increase in the number of wild-type transformants with increasing size. The data for transformation of the Rec$^-$ strain show that even the smallest insert increased the transformation several orders of magnitude over that of pDM2 without an insert. This increase is presumably the result of the presence of uptake sites on the inserts. However, the larger inserts did not cause a higher frequency of transformation in the Rec$^-$ cells; on the contrary, the frequency decreased slightly with an increase in insert size. The transformation of the rec2 strain by the plasmids with inserts was much lower than in the wild-type. Similar results were obtained with rec1 recipients (data not shown).

The conjugal transfer results were not different for the wild-type and Rec$^-$ recipients, and there was relatively little effect of insert size, except that there was a decrease in frequency of transfer of the largest plasmid.

Figure 4 shows transformation to novobiocin resistance in the same experiment presented in Fig. 3. The data for plasmid establishment (transformation to chloramphenicol resistance) from Fig. 3 are also shown for comparison. The plasmid with 2 inserts of the novobiocin resistance marker, pMB2, did not transform to novobiocin resistance as well as the plasmid with only 1, pMB1, although pMB2 became established with a frequency around 5 times higher. However, the transformation to novobiocin resistance by the largest plasmid was considerably higher than by the other 2 plasmids.

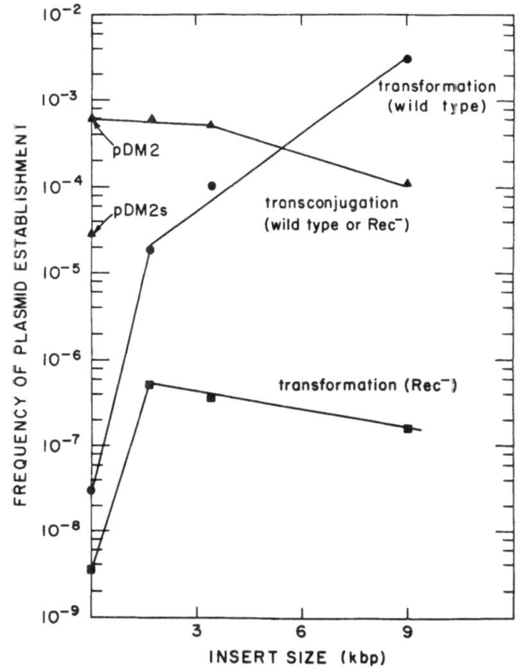

Fig. 3. Frequency of plasmid establishment by transformation and transconjugation as a function of size of the chromosomal insert in the EcoRI site of pDM2. Final concentration of purified plasmid DNA for transformation was 0.3 μg/ml. Transformation by pDM2 and pDM2s yielded the same frequency.

Recombination Between Plasmids and Chromosomes

Table 3 shows measurements of plasmid-chromosome recombination in transformants and transconjugants. Transformants or transconjugants containing pMB3 or pMB4, were selected by resistance to chloramphenicol or ampicillin. They were picked, grown up in chloramphenicol or ampicillin (the parent plasmid markers) and spotted on plates containing parent plasmid marker or this marker plus novobiocin., A clone was considered to be a recombinant if it had become novobiocin-sensitive. Since the novobiocin-resistance markers of pMB1 and pMB2 are not expressed unless they are integrated into the chromosome, we tested for transfer of novobiocin sensitivity from chromosome to plasmid by determining transformation to novobiocin resistance from lysates of chloramphenicol transformants of these plasmids. When a lysate failed to show any evidence of novobiocin resistance markers, the plasmids must have acquired novobiocin sensitivity from the chromosome.

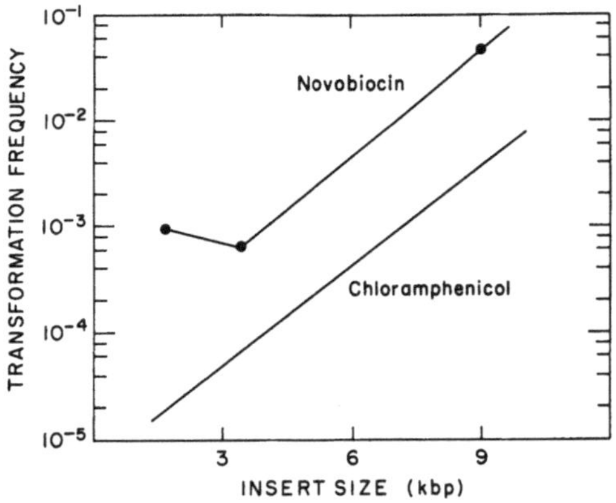

Fig. 4. Transformation to novobiocin resistance by pDM2 derivatives with the same cells and DNA as used for Fig. 3. The chloramphenicol transformation of Fig. 3 is also shown for comparison.

There was considerably more recombination following transformation with the larger plasmids, and there seemed to be an effect of the position of the chromosomal insert on the amount of recombination. While the inserts were the same size in pMB3 and pMB4, there was about twice as much recombination when the insert was in the SmaI site (pMB4) as when it was in the EcoRI site (pMB3). The difference between the observed recombination in pMB1 and pMB2 may not be significant. However, there was clearly less recombination in these than in the larger plasmids, indicating that decrease in size of the homologous region results in decrease in plasmid-chromosome interaction resulting in recombination.

Table 3 also shows that while all the plasmids interacted with the chromosome in transformation, there was no recombination observed among transconjugants, even when the recipient was competent. Following mating, novobiocin-resistant recipient colonies were picked, grown in novobiocin, and spotted on plates containing novobiocin or novobiocin plus chloramphenicol or ampicillin. The fact that all the clones tested were resistant to both drugs indicated that there had been no integration of the novobiocin marker into the chromosome, a procedure that would be expected to cause loss of the plasmid and the parent plasmid resistance marker. In accord with this conclusion was the fact that when some of the pMB4-containing transconjugants were treated so as to lose the plasmid, in all 18 cured clones tested both novobiocin and ampicillin resistance had been lost.

Further evidence for lack of interaction between conjugally transferred plasmids and chromosome is that we failed to find any novobiocin-resistant recipient cells following conjugation that re-

Tab. 3. Recombination between plasmids and chromosomes.

Plasmid	Number of plasmid-containing clones tested	% Recombination[a]	Number of novobiocin resistant clones tested	Number of ampicillin or chloramphenicol sensitive clones[b]
Transformation				
pMB1	20	15		
pMB2	20	5		
pMB3	84	38		
pMB4	83	71		
Transconjugation				
pMB4 (noncompetent recipient)	82	0	61	0
pMB4 (competent recipient)	48	0	44	0
pMB3 (noncompetent recipient)			60	0

[a] % of novobiocin-sensitive plasmid-containing clones (for pMB1 and pMB2 this was the % of clones not containing DNA capable of transformation to novobiocin resistance).

[b] among the novobiocin-resistant clones.

The recipients in the matings were all wild type and originally streptomycin-resistant and novobiocin-sensitive. Donors and mobilizing plasmid strains were Rec^- and streptomycin-sensitive.

sulted in transfer of pMB1 or pMB2. Although these plasmids do not express the novobiocin marker, the marker is expressed when there is recombination resulting in transfer of the resistance gene into the chromosome. The results of Tab. 4 show that there was no such recombination detectable, even though the wild type (strain Rd) was used as recipient.

DISCUSSION

The data of Fig. 4 suggest that the pairing between the plasmid and homologous parts of the chromosome that increases plasmid establishment may not be the same as the pairing that results in integra-

Tab. 4. Mating involving plasmids which do not express the novobiocin-resistance marker unless there is integration of plasmid DNA into the chromosome.

Mating cells	Number of cells/ml of suspension			Novobiocin-resistant recipients
	Donors	Recipients	Recipients containing plasmids	
rec2(pHD147) / rec1(pMB1) / Rd	$4x10^8$	$3x10^8$	$5x10^5$	0 (<1 in $3x10^7$ recipients
rec2(pHD147) / rec1(pMB2) / Rd	$6x10^8$	$5x10^8$	$6x10^5$	0 (< in $5x10^7$ recipients

The cells were mixed and incubated overnight on hemoglobin agar. Some were then scraped off, resuspended and plated, after appropriate dilution, with antibiotics to distinguish between the strains. Resistance of the strains was the same as in Table 2, except that rec1 (pMB1) and rec1(pMB2) were resistant to chloramphenicol only.

tion of a homologous region of the plasmid into the chromosome. Thus, novobiocin transformation decreases with the duplication of the chromosomal portion in pMB2, but establishment of the plasmid increases with this duplication. A model to explain these data is shown in Fig. 5. The homologous region of the plasmids is labeled A B C, to be paired with A'B'C' of the chromosome. The smaller plasmid, pMB1, is likely to pair with the chromosome in a contiguous manner. However, the pairing of the larger plasmid could also involve 1 portion of each of the 2 inserts, with a looped portion between. The model suggests that transformants of pMB2 should sometimes have lost the second copy of the EcoRI fragment, and thus be converted to pMB1, because of the possibility of elimination of the unpaired homologous region. Twenty chloramphenicol transformants of pMB2 were picked, grown up and cleared lysates made and run on a gel. Sixteen of them were found to contain plasmids the size of pMB1, and 4 contained plasmids of the original size of pMB2. These data are in accord with the model.

Recently, a model has been proposed for transformation in *H. influenzae* (3), in which DNA is first taken up into extensions of the membrane on the cell surface, which have been called "transformasomes." Following uptake of a linear DNA and its partial degradation, a single strand of the DNA can enter the cytoplasm of the cell. This single strand can then pair with a homologous single-strand region of the recipient cell chromosome, leading to integration. There is some evidence for this model. The membrane exten-

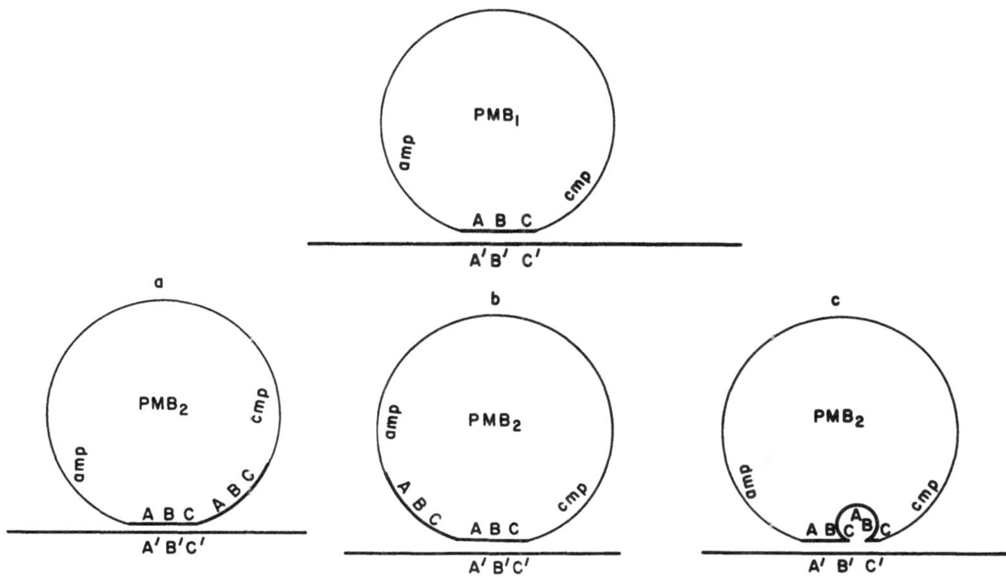

Fig. 5. A model for pairing of homologous parts of pMB1 and pMB2 with the recipient chromosome.

sions have been visualized, and shown to be more prevalent on competent cells (5, 9). A pore has been seen between the transformasome and the cytoplasm, through which the single strand of DNA could pass (9). Covalently closed circles do not readily exit from the transformasomes, nor do molecules with hairpin ends, and both of these are protected from degradation in the transformasome (3,8). Linear DNA, however, is degraded so that it has a single-stranded end (3,14). The DNA of competent H. influenzae has been shown to have one strand containing gaps and tails (10,11) that are necessary for pairing between transforming DNA and recipient DNA. Finally, it has been known for a long time that at least the overwhelming majority of the transforming DNA is integrated as a single strand in the double-stranded recipient molecule (13).

It is of interest to consider how the present results may be interpreted in terms of the model. The consideration is why the plasmids we have used transform at all. The cells were exposed to closed circular, supercoiled DNA (the form isolated from CsCl gradients). Since open circular or linear molecules are the only ones that exit rapidly from the transformasomes (3), either our plasmids become nicked inside the cell or their entrance into the cytoplasm takes a very long time.

We may also ask why the size of the homologous insert in the plasmids has such an enormous effect on the transformation frequency

(Fig. 3). It seems reasonable to suppose that the pairing of homologous regions of plasmid and chromosome not only can lead to integration of some of the plasmid into the chromosome, but also can stabilize the plasmid so that it can replicate and become established as an independent entity, as we postulated previously (16).

The lack of increase in transformation frequency of *rec2* recipients with increase in insert size reflects the inability of this mutant to promote pairing between homologous regions (11) that stabilizes the plasmid and increases the chances of its replication. Although the *rec1* mutant can promote such pairing, there is local DNA breakdown at pairing sites (11), and thus in this strain an increase in probability of pairing resulting from an increase in size of the homologous region does not increase the probability of establishment of the plasmid.

When a plasmid enters a cell by conjugal transfer, it need not become part of a transformasome, as indicated by the fact that competence does not affect such transfer. We postulate that the reason why there is no interaction between plasmid and chromosome following conjugal transfer is because the plasmids enter by a route other than transformation. The interaction may require single-strand ends to be formed, and this process may only take place when DNA is to go through the transformasome pore into the cytoplasm. The single-strand ends in the DNA of recipient cells alone do not result in interaction between plasmid and chromosome, since there was little or no interaction upon conjugal transfer of plasmids into competent cells (Tab. 3), and we observed earlier that when cells containing an established plasmid, pNov1, were made competent, there was no plasmid-chromosome interaction (16).

Antibiotic resistance in *H. influenzae* is undoubtedly spread in nature by conjugal transfer of plasmids. Our present observations suggest that at least in the absence of transposons there is rarely or never exchange of antibiotic resistance markers between plasmid and chromosome following such transfer.

ACKNOWLEDGMENTS

We thank William L. Albritton for the gift of the mobilizing plasmid, pHD147, and for help with the experiment showing lack of mobilization of pNov1. Research was carried out at Brookhaven National Laboratory under the auspices of the United States Department of Energy.

REFERENCES

1. Albritton, W.L., J.W. Bendler, and J.K. Setlow (1981) Plasmid

transformation in Haemophilus influenzae. J. Bact. 145:1099-1101.
2. Albritton, W.L., J.K. Setlow, and L. Slaney (1982) Transfer of Haemophilus influenzae chromosomal genes by cell-to-cell contact. J. Bact. 152:1066-1070.
3. Barany, F., M.E. Kahn, and H.O. Smith (1983) Directional transport and integration of donor DNA in Haemophilus influenzae transformation. Proc. Natl. Acad. Sci., USA 80:7274-7278.
4. Boling, M.E., J.K. Setlow, and D.P. Allison (1972) Bacteriophage of Haemophilus influenzae I. Differences between infection by whole phage, extracted DNA and prophage DNA extracted from lysogenic cells. J. Mol. Biol. 63:335-348.
5. Deich, R.A., and L.C. Hoyer (1982) Generation and release of DNA-binding vesicles by Haemophilus influenzae during induction and loss of competence. J. Bact. 152:855-864.
6. Deneer, H.G., L. Slaney, I.W. Maclean, and W.L. Albritton (1982) Mobilization of nonconjugative antibiotic resistance plasmids in Haemophilus ducreyi. J. Bact. 149:726-732.
7. Goodgal, S.H. (1982) DNA uptake in Haemophilus transformation. Ann. Rev. Genet. 16:169-192.
8. Kahn, M.E., F. Barany, and H.O. Smith (1983) Transformasomes: Specialized membranous structures that protect DNA during Haemophilus transformation. Proc. Natl. Acad. Sci., USA 80:6927-6931.
9. Kahn, M.E., G. Maul, and S.H. Goodgal (1982) Possible mechanism for donor DNA binding and transport in Haemophilus. Proc. Natl. Acad. Sci., USA 79:6370-6374.
10. LeClerc, J.E., and J.K. Setlow (1975) Single-strand regions in the deoxyribonucleic acid of competent Haemophilus influenzae. J. Bact. 122:1091-1102.
11. LeClerc, J.E., and J.K. Setlow (1975) Transformation in Haemophilus influenzae. In Mechanisms in Recombination, R.F. Grell, ed. Plenum Press, New York, pp. 187-207.
12. McCarthy, D., N.L. Clayton, and J.K. Setlow (1982) A plasmid cloning vehicle for Haemophilus influenzae and Escherichia coli. J. Bact. 151-1605-1607.
13. Notani, N., and S.H. Goodgal (1966) On the nature of recombinants formed during transformation in Haemophilus influenzae. J. Gen. Physiol. 49 (Part 2):197-209.
14. Sedgwick, B., and J.K. Setlow (1976) Single-stranded regions in transforming deoxyribonucleic acid after uptake by competent Haemophilus influenzae. J. Bact. 125:588-594.
15. Setlow, J.K., D.C. Brown, M.E. Boling, A. Mattingly, and M.P. Gordon (1968) Repair of deoxyribonucleic acid in Haemophilus influenzae. I. X-ray sensitivity of ultraviolet-sensitive mutants and their behavior as hosts to ultraviolet-irradiated bacteriophage and transforming deoxyribonucleic acid. J. Bact. 95:546-558.
16. Setlow, J.K., N.K. Notani, D. McCarthy and N.-L. Clayton (1981) Transformation of Haemophilus plasmid RSF0885 containing a

cloned segment of chromosomal deoxyribonucleic acid. J. Bact. 148:804-811.

17. Steinhart, W.L., and R.M. Herriott (1968) Fate of recipient deoxyribonucleic acid during transformation in *Haemophilus influenzae*. J. Bact. 96:1718-1724.

PLASMID DNA PRIMASES AND THEIR ROLE IN BACTERIAL CONJUGATION

B.M. Wilkins, L.K. Chatfield, C.C. Wymbs,
and A. Merryweather

Department of Genetics
University of Leicester
Leicester LE1 7RH, England

INTRODUCTION

RNA primers necessary for starting DNA chain growth are generally made by the transcriptional activity of a group of enzymes called DNA primases. The primase of Escherichia coli is the 60,000 dalton (60 kd) product of the dnaG gene, which functions to generate primers for discontinuous replication of the bacterial chromosome and of at least some plasmids, as well as for complementary strand-synthesis on certain single-stranded genomes (24). Recently a variety of conjugative plasmids have been shown to encode DNA primases active in an assay requiring primer synthesis on single-stranded phage fd DNA. The list includes members of the B, C, I_1 (= Iα), I_1+B, I_2, K, M, P, and U incompatibility groups (44). Representative plasmids of most of the other groups, including F, have been screened in this assay and are currently classified as primase negative (26), but it is possible that some may specify DNA primases that are unable to use fd DNA as a template. In this article, which concentrates on the primases specified by ColIb-P9 (IncI_1), RP4 (IncP), and R16 (IncB), we review the biochemistry of these enzymes and the organization of the appropriate structural genes, and discuss the physiological role of their products in plasmid DNA metabolism during bacterial conjugation.

BIOCHEMISTRY OF PLASMID DNA PRIMASES

A Perspective of E. Coli Primer-Generating Systems

The biochemistry of plasmid DNA primases needs to be considered in context with E. coli primer-generating systems. Three such bacterial mechanisms have been defined through in vitro studies of complementary strand-synthesis on single-stranded phage genomes, coated with bacterial single-stranded DNA binding protein (SSB). This area has been reviewed recently by Kornberg (24,25) and Nossal (33). SSB is required for bacterial DNA replication, and it acts in these in vitro systems to direct initiation and promote chain elongation by DNA polymerase III holoenzyme, the primary replicative DNA polymerase of *E. coli*.

The simplest system involving *dnaG* protein works with phage G4 DNA; the enzyme recognizes a region containing 2 potential hairpin structures and synthesizes, in a rifampin-resistant process, a specific RNA primer of 26 to 29 nucleotides. In a reaction coupled to DNA synthesis, primers are shorter and contain deoxyribonucleotides as well as ribonucleotides.

A more complex system, thought to reflect primer generation during the replication of some plasmids (22,32,39,45) and of bacterial DNA, has been defined using SSB-coated ϕX174 DNA. This process requires at least 6 prepriming proteins (*dnaB* and *dnaC* proteins, together with proteins i, n, n' and n") to allow primer generation by *dnaG* primase. These proteins are required for assembly and function of a multiprotein complex, termed the primosome, that is inferred to move progressively in the 5' to 3' direction from its assembly site to allow synthesis of multiple primers. The *dnaB* protein in this complex is envisaged to alter the configuration of single-stranded DNA so that it may be utilized by *dnaG* protein for primer synthesis.

The third primer-generating system is sensitive to rifampin and involves RNA polymerase itself: the enzyme recognizes a region rich in potential secondary structure within SSB-coated DNA of phages fd and M13, and synthesizes a specific RNA primer.

$IncI_1$ Plasmid Primase

The first plasmid-specified DNA primase to be purified and characterized was the encoded by R64*drd-11* (29). Since other plasmids of this group, such as ColIb-P9*drd-1* and R144*drd-3*, apparently carry homologous primase genes (15,16), the properties of the enzymes will be considered as a group. The $IncI_1$ plasmid primase gene (*sog*) specifies 2 antigenically related polypeptides with apparent molecular masses of 240 and 180 kd but only part of the larger protein has primase activity (see a later section). While R64*drd-11* primase activity was originally found to be associated with a 140 kd polypeptide (29), a protein of this size is undetectable in crude cell extracts using immunoassay, and it may have been selected during purification as a degradation product of a larger molecule (E. Lanka, pers. comm.). $IncI_1$ plasmid primase has a remarkable ability to generate primers on a range of different single-stranded phage genomes in vitro, since it can substitute for RNA polymerase, *dnaG* protein, and at least the *dnaB*-, *dnaC*-, and dnaG- protein components of the primosome in synthesizing functional primers on the DNA of phages fd, G4, and ϕX174, respectively. The enzyme generates short RNA primers in a rifampin-resistant process and these are elongated specifically by DNA polymerase III holoenzyme (29). Primers made in vitro range in size from 2 to about 10 nucleotides and, unlike primers made by dnaG protein, which contain a purine in the initiating nucleotide (3,34), they begin with cytidine-5'-monophosphate or cytidine with AMP in second position (26a).

In vivo, $IncI_1$ plasmid primase can apparently substitute for mutant dnaG protein during replication of the bacterial chromosome and other replicons, including $IncI_1$ and ColE1-like plasmids and phage λ DNA (42). This substitution was originally inferred from the ability of $IncI_1$ plasmids to suppress the temperature- sensitivity of DNA replication in dnaG mutants of *E. coli* (41). The plasmid gene responsible was designated sog (suppression of dnaG; Ref. 9) and shown through DNA cloning experiments to be the structural gene for the DNA primase (42).

Efficient suppression of the dnaG3 mutation by an $IncI_1$ plasmid requires derepression of the conjugative-transfer genes, suggesting that the primase functions during bacterial conjugation (41). Suppression has also been reported to require recA protein (35), but any involvement of this protein in the process must be indirect because a recombinant plasmid (pLG214; Ref. 42) specifying a truncated sog primase of 87 kd allowed dnaG3 recA1 cells to form colonies at 40°C with almost 100% efficiency (J.J. Overell and B.M. Wilkins, unpub. data). Since ColIb-P9drd-1, which carries a mutation derepressing expression of transfer genes, failed to suppress the temperature-sensitivity of DNA replication in dnaB70 mutants at 42°C, and in dnaC1 and dnaC2 strains at 41°C (unpub. data, and see Fig. 2 in Ref. 7), sog primase apparently cooperates in bacterial DNA replication with at least some primosome proteins. This contrasts the ability of the enzyme to generate primers on ϕX174 DNA in dnaB and dnaC cell extracts, and could reflect involvement of primosome proteins in movement of the replication fork along double-stranded DNA.

RP4 Plasmid Primase

The DNA primase specified by RP4 is the product of a gene designated pri and, like sog primase, it substitutes for the normal bacterial primer-generating mechanisms in initiating complementary strand-synthesis on the genomes of phages fd, G4, and ϕX174 (26). RP4 suppresses the dnaG3 mutation weakly but Pri^+ recombinant plasmids containing a mutation(s) that causes overproduction of the enzyme and a change in its specific activity allow dnaG3 cells to form colonies at 40°C with 100% efficiency (28). The pri gene specifies 2 antigenically related polypeptides of 118 and 80 kd, both of which possess primase activity (see a later section). The

RP4 enzyme synthesizes primers of similar composition and size distribution to those made by ColIb-P9 primase (26a), but the 2 enzymes are immunologically unrelated (26). Furthermore, sog and pri are not homologous since a fragment of ColIb-P9 containing sog did not hybridize with RP4 (14).

R16 Plasmid Primase

While the primase specified by R16, an IncB plasmid, has been less extensively characterized than the primases of $IncI_1$ plasmids and RP4, current evidence suggests that all 3 enzymes are functionally similar. The R16 enzyme initiates DNA synthesis on viral strands of phage M13 in cell-free extracts containing rifampin, and it promotes bacterial DNA replication in dnaG3 cells at 40°C. The gene, called pri, is different from sog and the RP4 pri determinant since a cloned primase-specifying fragment of R16 did not hybridize with RP4 or $IncI_1$ plasmids. Moreover, the R16 enzyme is immunologically distinct from sog primase and it has a larger apparent molecular weight (14,15).

ORGANIZATION OF PLASMID PRIMASE GENES

The $IncI_1$ plasmid sog gene and RP4 pri gene have been cloned and analysis of recombinant plasmids indicates that each locus specifies synthesis of 2 polypeptides that are apparently separate in-phase translation products.

The sog gene is located in an 8.1 kilobase pair (kb) EcoRI restriction fragment of ColIb-P9drd-1 (E-3 according to the numbering system of Uemura and Mizobuchi, Ref. 38) and this was cloned in pBR325 to give recombinant plasmid pLG215 (Fig. 1). The DNA fragment specifies 2 polypeptides, of 240 and 180 kd, which are also detectable in extracts of cells harboring the ColIb-P9 plasmid. The genetic determinants of these proteins overlap, since both are truncated by deletions in the cloned DNA, and the polypeptides apparently share a common amino acid sequence because both react with antiserum raised against a purified sog primase. Efficient synthesis of these polypeptides requires a promoter outside the 8.1 kb fragment, and since amber mutations prevented synthesis of the larger but not the smaller molecule, they are inferred to be separate translation products (10,42).

An important functional difference between the 2 polypeptides specified by sog is that only the 240 kd polypeptide has DNA primase activity. This activity requires a domain of between 36 and 75 kd at the N-terminal region of this large polypeptide, as estimated from sizes of proteins specified by deletion and amber mutants (10). Such a conclusion is consistent with the purification of an 87 kd

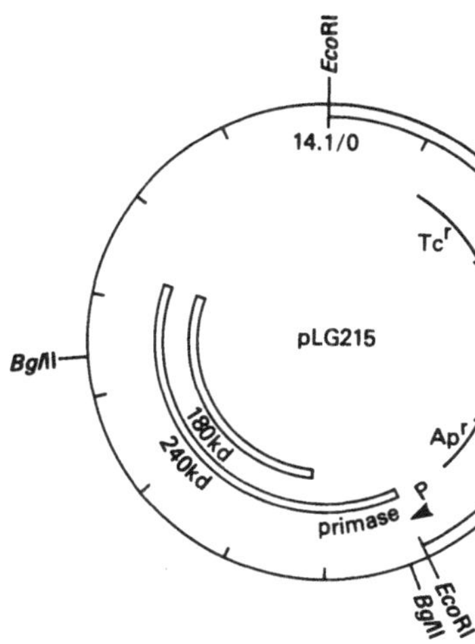

Fig. 1. Outline map of pLG215. The approximate locations of determinants of the 240 and 180 kd polypeptides are indicated, as deduced from the effect of deletion mutations that extend from the BamHI site without removing SalI. Efficient synthesis of both polypeptides requires transcription from the cat promoter of the vector (arrowhead labeled "P"). Further details are given elsewhere (10,42).

truncated polypeptide as a DNA primase with slightly greater specific activity than normal 240 kd sog product in the phage fd DNA assay system (42).

Overlapping genes with in-phase coding sequences have been detected in other genomes including those of phage λ and *E. coli* (36,37). The possible advantage of this type of gene arrangement is to allow formation of multifunctional protein complexes; regions common to both polypeptides may interact and carry out one function, while the domain that is unique to the larger molecule would participate in a separate reaction. As described in the next section, the primase moiety of the larger sog polypeptide acts to generate primers for conjugative DNA metabolism of $IncI_1$ plasmids but the role of the rest of this protein and the related 180 kd polypeptide is unknown. Elucidation will presumably require isolation and characterization of appropriate $IncI_1$ plasmid mutants.

The organization of the R16 pri gene has yet to be described. R16 belongs to the "I_1 complex" that includes plasmids of the I_1, I_1+B, B, and K incompatibility groups (12). These plasmids show significant DNA homologies and determine similar, thin, flexible pili that are serologically related (see Ref. 12 for a summary). Clearly members of the I_1 complex are heterogenous with regard to primase genes since $IncI_1$ and R16 DNA primases are antigenically distinct. A survey (Ref. 15; Tab. 1) of a number of these plasmids showed that R64, R144, R621a, R864a, and RIP72 all contain nucleotide sequences homologous with the cloned ColIb-P9 sog gene. However, TP125, R724, pLG101 (IncB), R805, and R387 showed sequence homology with the R16 pri determinant but not with sog.

It is likely that many genetically different conjugation systems exist among plasmids of gram-negative bacteria (44) and since sog is apparently a transfer gene, it is possible that plasmids containing sog-like sequences specify the same system. If so, such plasmids should encode similar pili, which are required for cellular interactions in conjugation, and mutually interchangeable proteins for DNA transfer itself. Members of the I_1 complex can be arranged into different sets using pilus-specific phages Iα and PR64FS; the former adsorbs to the shafts of pili encoded by $IncI_1$ plasmids, R144 and R864a ($IncI_1$+B), and R621a (IncB), whereas PR64FS has a wider range since it also recognizes the tips of pili specified by R805a ($IncI_1$+B), and R387 (IncK) (Ref. 12; D.E. Bradley, pers. comm.). Hence, plasmids determining sensitivity to phage Iα possess sog-like sequences but the converse does not hold, as shown by RIP72.

At least 2 different DNA transfer mechanisms are recognizable among I_1 complex plasmids through the ability of representative plasmids to mobilize a small recombinant plasmid (pLG2009) carrying the origin of transfer (oriT) of ColIb-P9. Table 1 shows that $IncI_1$ plasmids, R144, R864a, and R621a mobilized pLG2009 with high efficiency. Thus these plasmids determine functionally similar DNA

Tab. 1. Mobilization of a recombinant plasmid carrying oriT of ColIb-P9 by conjugative plasmids of the I_1 complex.

Conjugative plasmid	Inc group	Primase gene type	Mobilization of pLG2009 [a]
ColIb-P9*drd-1* [b]	I_1	*sog*	82
R64*drd-11*	I_1	*sog*	513
R144*drd-3*	I_1+B	*sog*	50
R864a*drd-1*	I_1+B	*sog*	128
R621a	B	*sog*	231
RIP72	B	*sog*	0.12
TP125	B	*pri*	0.16
R724	B	*pri*	0.24
R805a	I_1+B	*pri*	0.078
R387	K	*pri*	2.5

[a] pLG2009 is a derivative of the nonmobilizable vector plasmid pED825 (2.2 kb; Ref. 17) containing *oriT* of ColIb-P9*drd-1* on a 0.8 kb fragment. Values show yield of pLG2009-containing transconjugants expressed as a percentage of transconjugants harboring the conjugative plasmid. Donor strains were mutant at *recA*.

[b] pLG221 (ColIb-P9*drd-1 cib*::Tn*5*) was used to allow selection for kanamycin-resistance.

transfer systems; they also contain sog-like sequences and comprise the set determining sensitivity to the RNA phage Iα. The other plasmids in Tab. 1 mobilized pLG2009 with frequencies of at least 30-fold lower than ColIb-P9, indicating that they specify DNA transfer mechanisms differing from that of the $IncI_1$ plasmid. With the exception of RIP72, this second group contains sequences homologous with the R16 pri gene. Thus, this survey indicates that plasmids with sog-like sequences specify the same or closely related conjugation systems, but the association is not absolute.

The RP4 pri gene (3.2 kb) is located between coordinates 40.3 and 43.5 in the recent SphI map of the plasmid (27), and the locus specifies 2 antigenically related polypeptides of 118 and 80 kd. The putative promoter is about 5 kb upstream of the beginning of the gene. Results of deletion mapping, together with immunological cross-reaction of proteins, indicates that the 2 polypeptides are separate in-phase translation products. Unlike sog products, both the 118 and 80 kd pri polypeptides possess DNA primase activity and this is located in a region towards the C-terminus of the products (26,28).

ROLE OF $IncI_1$ PLASMID PRIMASE IN CONJUGATION

Sog^- Mutants and Conjugative DNA Synthesis

We recently reported the isolation of a Sog^- $IncI_1$ plasmid called pLG250 and its use to identify the physiological role of the DNA primase (11). This mutant is a derivative of pLG221, which is ColIb-P9drd-1 conveniently marked by the presence of Tn5 in the colicin Ib gene, and it contains a hydroxylamine-induced mutation(s) that eliminates primase activity without affecting the apparent molecular masses of the sog polypeptides. Plasmid pLG250 was stably maintained in *E. coli*, implying that sog primase is not required for vegetative replication of the plasmid. Presumably the primers necessary for this process are generated by bacterial primase. However, the mutant was approximately 80% deficient in generating transconjugants in *E. coli* and *Salmonella typhimurium* matings, suggesting that the enzyme functions during conjugative synthesis of plasmid DNA. Consistent with this are the observations that sog is coordinately expressed with transfer genes, as judged by the effect of drd mutations in derepressing expression of these genes (41), and that sog activity and transfer proficiency were reduced by insertion of Tn5 in the E-2 fragment of ColIb-P9drd-1 (unpub. data).

During conjugation, a specific $IncI_1$ plasmid strand is transferred to the recipient cell (40) where the complementary strand is synthesized by DNA polymerase III holoenzyme in a process independent of RNA polymerase and *E. coli* primase (9,43). Involvement of sog primase in this process was investigated at the molecular level by measuring conjugative DNA synthesis in recipient cells (RCDS).

This was detected by measuring [^{14}C]thymine incorporation when thymidine kinase-deficient (tdk) donors were mated with thymine-requiring recipients under conditions that prevented the latter cells from replicating the resident DNA. This inhibition was achieved by exposing the cells to a large dose of ultraviolet light before mating, thereby inducing sufficient pyrimidine dimers to render the DNA unusable as a template for DNA synthesis (11).

Figure 2 shows the amount of RCDS detected when pLG221 (Sog$^+$) and pLG250 (Sog$^-$) donors were mated with irradiated dnaG3 recipients that were treated with rifampin to inhibit transcription of transferred DNA and thus to prevent synthesis of sog primase in transconjugant cells. Clearly RCDS in this mating system is dependent on plasmid primase which, in view of the rifampin treatment, is presumably supplied by the donor cell. The primary defect in the pLG250-mediated mating is not failure of DNA transfer itself, because the deficiency of RCDS was alleviated partially when the recipient cells harbored the pBR325-sog$^+$ recombinant pLG215 (Tab. 2) and thus contained plasmid primase before mating. We therefore infer that sog

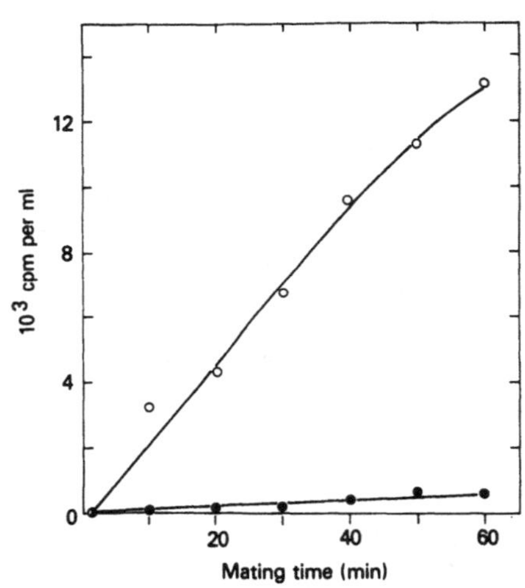

Fig. 2. Conjugative DNA synthesis in UV-irradiated, rifampin-treated dnaG3 recipients. Donor strain BW96 (dna$^+$ tdk rpoB) containing pLG221 (o) or the Sog$^-$ derivative pLG250 (●) was mated at 41°C with recipient strain BW86 (dnaG3 uvrB thyA), UV-irradiated with 400 J/m^2 before use. Mating was in the presence of rifampin (100 µg/ml) and [^{14}C]thymine (63 mCi/mmol); TCA-precipitable radioactivity was determined at the times indicated. Further details are given elsewhere (11). Data have been corrected by subtraction of counts incorporated by unmated bacteria (maximum of 400 cpm per ml).

Tab. 2. Contribution of dnaG and sog primases to conjugative DNA synthesis in recipient cells.

Strains		Relative amount
Donor	Recipient	of RCDS
dna$^+$(pLG221)	*dnaG3*	1.0
dna$^+$(pLG250)	*dnaG3*	0.05
dna$^+$(pLG250)	*dna*$^+$	0.15
dna$^+$(pLG250)	*dnaG3*(pLG215)	0.37
dna$^+$(pLG250,pLG215)	*dnaG3*	0.64

BW96 was donor strain. BW86 and the homogenic *dna*$^+$ strain BW39 were recipients. pLG221 (Sog$^+$) and pLG250 (Sog$^-$) are Inclα plasmids while pLG215 is a pBR325-*sog*$^+$ recombinant. Matings were as for Fig. 2. Values are expressed as the amount of RCDS detected at 60 min relative to that when BW96(pLG221) was mated with BW86 (line 1).

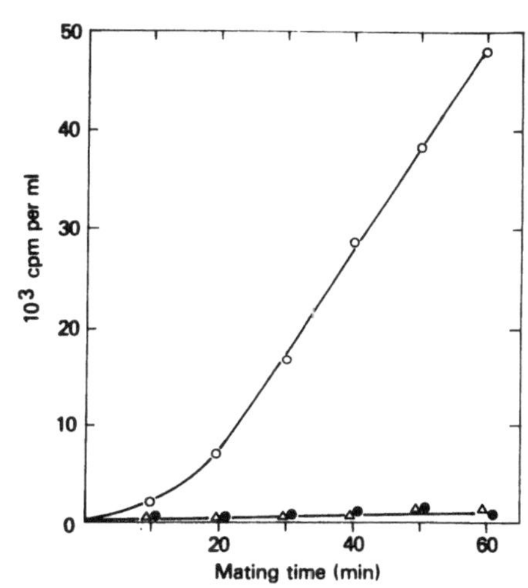

Fig. 3. Conjugative DNA synthesis in unirradiated, rifampin-treated dnaG3 recipients. Donor strain BW96 containing pLG221 (o) or pLG250 (•) was mated with BW86 recipients at 41°C. Counts incorporated when BW96(pLG221) was mated with BW86(pLG252) are also shown (Δ). pLG252 is a pBR325-eex$^+$ recombinant that excludes entry of pLG221. Other details are as for Fig. 2.

primase of the donor generates primers that are used after DNA transfer to initiate complementary strand-synthesis in the recipient; in a normal mating, primers might be made in the donor cell on the strand destined for transfer to be transmitted in a complex with the DNA, or they might be synthesized in the recipient cell following DNA and enzyme transfer.

The donor strain used in these experiments was $\underline{dna}^+$ and it is apparent from Fig. 2 that dnaG protein of donor cells cannot substitute for sog primase in RCDS. However, dnaG protein supplied by recipient cells can utilize the transferred DNA but with lower efficiency than sog primase (Tab. 2). This substitution of bacterial for plasmid enzyme presumably explains the partial transfer-proficiency of pLG250 in genetic experiments.

Further evidence that primers for sog-dependent RCDS are made by enzyme supplied by the donor cell is the observation that pLG250 was complemented by the presence of pLG215 (Sog^+) in the donor cell (Tab. 1). Since pLG215 is nonmobilizable, cells harboring pLG250 and pLG215 transmit a functional sog gene to recipients with only low frequency.

It is significant that pLG215 complemented pLG250 more effectively if present in the donor rather than recipient cell (Tab. 2). This difference suggests that plasmid primase acts positively in plasmid DNA metabolism in the donor cell, possibly to promote efficient synthesis of DNA to replace the transferred plasmid strand. We examined this by measuring conjugative DNA synthesis in donor cells (DCDS), using a strain containing the temperature-sensitive dnaB70 mutation and mating this at 43°C with tdk recipients. DCDS, which reflects synthesis of replacement strands, can be monitored in this system by thymine incorporation because, in contrast to vegetative DNA replication, conjugative DNA synthesis is a dnaB- independent process. The results, described elsewhere (11), showed that dnaB70 donors of pLG250 supported only 5% of the normal amount of DCDS although they gave rise to 35% of the normal yield of transconjugants. We therefore surmise that sog primase is inessential for transfer of at least pre-existing plasmid strands, as indicated by transconjugant formation, but it functions in the donor to allow efficient synthesis of replacement strands. Assuming, by analogy with F, that the transferred $IncI_1$ plasmid strand is transmitted with its 5' terminus leading, RNA primer would be required for DCDS if the 3' terminus were unavailable for extension because of its association with, for example, some component of the cell membrane (44). Alternatively, DCDS may be a discontinuous process needing multiple priming events.

In summary, we suggest that $IncI_1$ plasmid primase plays 2 roles in conjugation: it acts in the donor cell to promote synthesis of

DNA to replace the transferred plasmid strand, and it generates primers that are utilized for complementary strand-synthesis in the recipient. This contrasts models in which DCDS of R64drd-11 requires primers generated by RNA polymerase, with RNA synthesis initiating DNA transfer itself (13). These models stemmed from the observation that transfer and DCDS were more sensitive to rifampin than to chloramphenicol during mating of dnaB donors and minicells at 42°C (19). However, in another study (7), rifampin did not block early transfer of ColIb-P9drd-1 between dna^+ bacteria at 37°C and, since the antibiotic was found to cause a general disruption of DNA metabolism in bacteria harboring $IncI_1$ drd plasmids it was suggested that this effect of rifampin-treatment was responsible for curtailment of DNA transfer.

The more recent finding that the yield of R64drd-11 transconjugants was reduced by about 90% when cytidine-utilization was prevented in dnaB donors provides alternative, although indirect, evidence that transfer requires RNA synthesis in the donor cell (30). The enzyme responsible might be $IncI_1$ plasmid primase, since this generates primers initiated with cytidine, but it is stressed that our studies with pLG250 do not indicate requirement of this enzyme for transfer of pre-existing ColIb-P9 DNA. Clearly more remains to be learned about priming of DCDS and its relationship to DNA transfer. Further progress is likely to require better understanding of the nature of the 3' teminus created by nicking at oriT, and of enzymes that unwind plasmid strands in the donor cell. The latter aspect of conjugative DNA metabolism has been reviewed elsewhere (44).

Participation of a primer-generating enzyme in synthesis of replacement strands in the donor cell contradicts a central feature of the rolling circle model of DNA transfer which predicts continuous elongation of the plasmid strand undergoing transfer (20). Other contradictory evidence includes delay between first-round transfer of R64drd-11 and second-round transfer of replacement DNA (18), requirement of protein synthesis in the donor cell for each round of transfer (19), and accumulation of transferred ColIb-P9drd-1 DNA in the form of circular and linear monomeric units, even under conditions that impeded termination of the conjugation process and allowed transfer of about 5 genome equivalents to each cell (8).

Transfer of Plasmid DNA Primase to Recipients

Participation of sog primase of donor cells in RCDS suggests that the enzyme may be transmitted between conjugating cells. Since the primase can utilize a variety of DNA templates for primer synthesis, including the bacterial chromosome of dnaG mutants, 1 functional test for primase transfer asks whether unirradiated dnaG3 recipient cells recover ability to synthesize chromosomal DNA when mated with donors that are the sole source of the enzyme. This test assumes that RNA primers function on the template that serves for their synthesis, and that if RNA primers were transmitted during conjugation, they would be hydrogen-bonded to plasmid DNA and thus unable to act in trans to promote replication of bacterial DNA native to the recipient cell.

Comparison of the data in Fig. 2 and 3 shows that 3.6-fold more RCDS was detected in rifampin-treated dnaG3 recipients of pLG221 when the native DNA was unirradiated rather than rendered defective as a template for DNA synthesis by irradiation. Variation in the amount of DNA transfer in the 2 matings does not explain the difference, because irradiation of recipients only marginally reduced the amount of DNA transferred from this pedigree of donor cells. On the basis that incorporation of 1,200 counts per min per ml in this experimental system reflects synthesis of 31 x 10^6 daltons of single-stranded DNA (9), and that pLG221 (ColIb-P9::Tn5) has a molecular weight of about 69 x10^6, it is estimated that the average recipient in Fig. 3 supported synthesis of DNA equivalent to 36 single strands of pLG221 in 60 min. This high value itself suggests that the dnaG recipients were recovering their ability to replicate DNA during mating.

Recovery of bacterial DNA synthesis in mated dnaG3 recipient cells was confirmed directly by mating pLG221 donors with unirradiated, rifampin-treated, dnaG3 recipients for 60 min, as in Fig. 3, and then measuring the proportions of labeled DNA that were homologous with the plasmid and bacterial chromosome. This approach showed that 69% of the DNA labeled in the mated dnaG recipients was homologous with bacterial DNA and 24% with pLG221. Since the method showed that *E. coli* DNA and pLG221 share only about 2% homology, these results prove that the dnaG3 recipient cells recovered the ability to synthesize chromosomal DNA during mating, synthesizing almost 3 times more bacterial than plasmid DNA.

This recovery was dependent on plasmid DNA primase because few counts were incorporated if the donor strain harbored pLG250 (Sog^-) rather than pLG221 (Fig. 3). Furthermore, it was presumably mediated by sog protein derived from donor cells, because the recipient cells were treated with rifampin throughout mating to inhibit gene expression in plasmid transconjugants. The possibility that this treatment did not fully inhibit RNA polymerase native to the recipient cell, or that antibiotic-resistant RNA polymerase was transmitted from the donor cells, is refuted by the observation that rifampin-treated recipients also recovered the ability to synthesize bacterial DNA when mated with donors harboring both pLG250 and the small Sog^+ recombinant plasmid pLG215 (data not shown). Such cells contain functional plasmid primase but cannot donate a sog^+ gene during conjugation with significant frequency because pLG215 is not mobilizable.

These several results, to be described in detail elsewhere, lead to the inference that *sog* primase is transferred from donor to recipient during mating. The alternative explanation, that the donors transmit multiple mobile RNA primers that can function in trans to prime bacterial DNA synthesis in the recipient cell, is inconsistent with the current concept that a short RNA primer promotes DNA synthesis on the template that served for its own formation.

Mode of Transfer of Sog Primase

The phenomenon of entry exclusion has been exploited to determine whether *sog* primase is transferred between cells by a conjugation-dependent process. Entry exclusion is a barrier that prevents DNA transfer between cells harboring the same or a closely related plasmid. In F-mediated conjugation, *eex* genes may operate at 2 different stages: to reduce the formation of stable mating aggregates and to prevent DNA transfer between those aggregates that are formed (1). An entry exclusion gene(s) of ColIb-P9 has been cloned in pBR325: the recombinant plasmid is designated pLG252 and it contains a 3.5 kb *Eco*R1-generated fragment of the $IncI_1$ plasmid, which is E-9 according to the numbering system of Uemura and Mizobuchi (38). When present in a recipient strain, pLG252 decreased transfer of pLG221 by more than 98%, as measured by yield of transconjugants and by transmission of radioactively labeled DNA. The ColIb-P9 fragment present in pLG252 resembles the exclusion-determining fragment of R144 (21), both with regard to conferred phenotype and in containing a 1.15 kb *Eco*R1-*Hind*III fragment adjacent to a 0.51 kb *Hind*III-*Sal*I fragment. Since the cloned R144 *eex* gene(s) was shown to prevent conjugation at a stage subsequent to the formation of mating aggregates, we postulate that the corresponding gene in pLG252 functions likewise, possibly to inhibit DNA transfer between aggregated cells.

Figure 3 shows that donors of the Sog^+ conjugative plasmid pLG221 were unable to promote the rescue of bacterial DNA synthesis in *dnaG* recipient cells when the latter contained pLG252. This implies that *sog* primase is transmitted between mating cells by a conjugation-dependent process and, since the enzyme binds to single-stranded DNA, as revealed by its purification properties (29), we suggest that it is transmitted in a complex with the transferred plasmid strand. While it is simpler to assume that primers for RCDS are made following protein and DNA transfer, it is possible that they are generated in the donor cell and transmitted to the recipient bonded to the transferred DNA. If primers synthesized in vivo were as short as those made in vitro, they would require stabilization on the transferred strand and plasmid primase might fulfill this role, thus explaining the need for co-transfer of primers and enzyme.

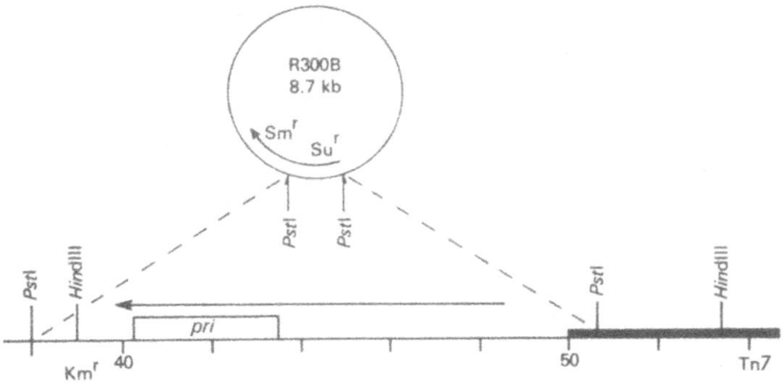

Fig. 4. Construction of pLG261: an R300B derivative carrying the pri region of the IncP plasmid pRP26. The latter is a RP4::Tn7 mutant specifying overproduction of plasmid primase (26). Map coordinates of RP4, location of pri and the putative transcript (line with arrowhead) are taken from Lanka et al. (27,28). Details of the Tn7 region are unpublished data of P.T. Barth and K. Ellis. Plasmid pRP1 is RP4 containing Tn7 inserted at about coordinate 41 (26,27).

ROLE OF RP4 PLASMID PRIMASE

The role of this enzyme was first studied by Lanka and Barth (26) using RP4 mutants carrying Tn7 insertions that inactivated primase activity. The insertions map within the operon that contains pri and the location of 1 (pRP1) is defined in the legend to Fig. 4. These mutants were lost from E. coli and S. typhimurium cells with a frequency of about 3% during 20 generations of unselected growth, indicating that the DNA primase plays a minor role in the maintenance of RP4. However, the requirement appears to be for resolution of recombination intermediates rather than for priming of vegetative replication itself (4).

Compared with RP4, E. coli donors of the Pri$^-$ mutants were defective in conjugation in some interspecific matings (26). Normal yields of transconjugants were detected when 6 bacterial species, including E. coli and Psuedomonas spp., were recipient strains but a reduced yield was recorded in 5 other matings, the decrease being about 25-fold in the case of S. typhimurium or Proteus mirabilis recipients. Back transfer from S. typhimurium donors established that the variation depended on the recipient rather than the donor strain. Assuming that a single strand of RP4 is transferred during conjugation, it was suggested that the plasmid primase acts in conjugation to generate primers for complementary strand-synthesis in the recipient cell, and that this activity is necessary for transconjugation formation in those species, such as *S. typhimurium*, that produce a primase which does not efficiently recognize the transferred DNA (26).

In collaboration with Dr. P.T. Barth, we have examined further the role of RP4 primase in conjugation, using pRP1 as a representative Pri⁻ plasmid (26) and pRP2 as a control. The latter plasmid contains a Tn*7* insertion near the RP4 tetracycline-resistance gene, which is outside the known transfer genes (5). To allow introduction of the *pri*⁺ gene into different recipient species, the gene was cloned in R300B, giving pLG261 (Fig. 4). R300B is a broad host-range plasmid of the IncQ group that is indistinguishable from RSF1010, R1162, and NTP2 (6).

Table 3 shows the relative yields of transconjugants formed when *E. coli* donors were mated with *E. coli* or *S. typhimurium* LT2 recipients. Transconjugant formation by the Pri⁻ plasmid was reduced by about 95% in the interspecific mating, in accordance with previous data (26), but it was also diminished, by about 30%, when *E. coli* recipients were used. Since these deficiencies were partially rectified when the recipient strains contained the Pri⁺ recombinant plasmid pLG261, it is apparent that RP4 primase is inessential for DNA transfer itself, but its products of reaction are needed in the recipient cell, presumably to initiate efficient conjugative DNA synthesis on transferred DNA. Clearly, *dnaG* protein of donor cells cannot bypass this requirement for RP4 primase, and even primer-generating enzymes specified by *E. coli* recipients are unable to substitute completely.

Tab. 3. Relative yield of transconjugants generated by RP4 Pri⁺ and Pri⁻ plasmids.

Recipient strain	Transconjugants
E. coli K-12	
C600	69
C600(pLG261)	98
S. typhimurium LT2	
SL329	4
SL329(pLG261)	61

Values show the yield of kanamycin-resistant, thymine nonrequiring transconjugants generated by W3110T⁻(pRP1) donors, expressed as a percentage of that obtained with the W3110T⁻(pRP2) control strain. Mating occurred on nitrocellulose filters placed on nonselective medium at 37°C for 3 hours. The construction of pLG261 is summarized in the legend of Fig. 4.

CONCLUDING REMARKS

A variety of conjugative plasmids encode DNA primases, the best studied being those of ColIb-P9 and RP4. These plasmids specify different but enzymatically similar primases that act in conjugative DNA metabolism. The ColIb-P9 enzyme is inferred to promote synthesis of DNA in the donor cell to replace the transferred strand, and it is apparently transmitted by a conjugation-dependent process to the recipient where its products initiate complementary strand-synthesis on the transferred DNA. The role of RP4 primase has not been studied at the molecular level, but genetic data suggest a function analogous to that of the ColIb-P9 enzyme.

While these primases are needed for efficient conjugative DNA synthesis in the recipient cell, they are apparently redundant for plasmid replication. Presumably primers for the latter process are generated by bacterial enzymes, but these are unable to substitute fully for plasmid primases during conjugation. This may be explained if DNA replication is unidirectional, with only 1 strand supporting synthesis of multiple primers for discontinuous DNA synthesis. The second strand might be deficient in recognition sites for bacterial primer-generating enzymes, since it acts as template for leading strand synthesis. If this second strand were transferred, efficient complementary strand-synthesis in the recipient might require a conjugation-specific priming mechanism and plasmid primase may contribute to 1 such process. RP4 replication is unidirectional and proceeds with opposite orientation to DNA transfer (2,31); if it is assumed that a distinct strand of RP4 is transferred with 5' terminus foremost, the transmitted strand will be the one that serves as template for leading strand synthesis. It is unknown whether the same applies to ColIb-P9 because the directions of replication and transfer of this plasmid have not been established.

Since plasmid primases render initiation of complementary strand-synthesis independent of bacterial primer-generating enzymes, they may contribute to host range by facilitating conversion of transferred DNA into double-stranded molecules in diverse bacterial species. The broad host-range of RP4 is well documented and, while ColIb-P9 is reported to be maintained in only a few enteric bacterial species (23) its conjugation system apparently allows DNA transfer between a much wider range of bacteria, as indicated by the ability of *E. coli* donors of pLG221 to denote Tn*5* to DNA of *P. putida*, *Methylophilus methylotrophus*, and *Alcaligenes eutrophus* (G. J. Boulnois et al., submitted for publication).

ACKNOWLEDGMENTS

C.C.W. holds a SERC research studentship and A.M. holds a CASE studentship in collaboration with Dr. P.T. Barth, ICI Corporate Bioscience Group, Runcorn. We thank Claire Corker for technical assistance.

REFERENCES

1. Achtman, M., N. Kennedy, and R. Skurray (1977) Cell-cell interactions in conjugating Escherichia coli: Role of traT protein in surface exclusion. Proc. Natl. Acad. Sci., USA 74:5104-5108.
2. Al-Doori, Z., M. Watson, and J. Scaife (1982) The orientation of transfer of the plasmid RP4. Genet. Res. 39:99-103.
3. Arai, K., and A. Kornberg (1981) Mechanism of dnaB protein action. IV. General priming of DNA replication by dnaB protein and primase compared with RNA polymerase. J. Biol. Chem. 256:5267-5272.
4. Barth, P.T., and N.J. Grinter (1983) A possible site-specific recombination and resolution system in plasmid RP4. J. Cell. Biochem. Suppl. 7B Abstract 0891, p. 149.
5. Barth, P.T., N.J. Grinter, and D.E. Bradley (1978) Conjugal transfer system of plasmid RP4: Analysis by transposon 7 insertion. J. Bact. 133:43-52.
6. Barth, P.T., L. Tobin, and G.S. Sharpe (1981) Development of broad host-range plasmid vectors. In Molecular Biology, Pathogenicity and Ecology of Bacterial Plasmids, S.B. Levy, R.C. Clowes, and E.L. Koenig, eds. Plenum Press, New York, pp. 439-448.
7. Boulnois, G.J., M.J. Beddoes, and B.M. Wilkins (1979) Rifampin disrupts conjugal and chromosomal deoxyribonucleic acid metabolism in Escherichia coli K-12 carrying some IncIα plasmids. J. Bact. 138:324-332.
8. Boulnois, G.J., and B.M. Wilkins (1978) A ColI-specified product, synthesized in newly infected recipients, limits the amount of DNA transferred during conjugation of Escherichia coli K-12. J. Bact. 133:1-9.
9. Boulnois, G.J., and B.M. Wilkins (1979) A novel priming system for conjugal synthesis of an IncIα plasmid in recipients. Mol. and Gen. Genet. 175:275-279.
10. Boulnois, G.J., B.M. Wilkins, and E. Lanka (1982) Overlapping genes at the DNA primase locus of the large plasmid ColI. Nucl. Acids Res. 10:855-869.
11. Chatfield, L.K., E. Orr, G.J. Boulnois, and B.M. Wilkins (1982) DNA primase of plasmid ColIb is involved in conjugal DNA synthesis in donor and recipient bacteria. J. Bact. 152:1188-1195.
12. Coetzee, J.N., D.E. Bradley, and R.W. Hedges (1982) Phages Iα and I_2-2: IncI plasmid-dependent bacteriophages. J. Gen. Microb. 128:2797-2804.
13. Curtiss, III, R., and R.G. Fenwick, Jr. (1975) Mechanism of conjugal plasmid transfer. In Microbiology - 1974, D. Schlessinger, ed. American Society for Microbiology, Washington, D.C., pp.156-165.

14. Dalrymple, B.P. (1982) Plasmid encoded DNA primases. Ph.D. thesis, University of Leicester, England.
15. Dalrymple, B.P., G.J. Boulnois, B.M. Wilkins, E. Orr, and P.H. Williams (1982) Evidence for two genetically distinct DNA primase activities specified by plasmids of the B and I incompatibility groups. J. Bact. 151:1-7.
16. Dalrymple, B.P., and P.H. Williams (1982) Detection of primase specified by IncB plasmid R864a. J. Bact. 152:901-903.
17. Everett, R., and N. Willetts (1982) Cloning, mutation, and location of the F origin of conjugal transfer. EMBO J. 1:747-753.
18. Fenwick, Jr., R.G., and R. Curtiss, III (1973) Conjugal deoxyribonucleic acid replication by *Escherichia coli* K-12: Stimulation in *dnaB*(ts) donors by minicells. J. Bact. 116:1212-1223.
19. Fenwick, Jr., R.G., and R. Curtiss, III (1973) Conjugal deoxyribonucleic acid replication by *Escherichia coli* K-12: Effect of chloramphenicol and rifampin. J. Bact. 116:1224-1235.
20. Gilbert, W., and D. Dressler (1968) DNA replication: The rolling circle model. Cold Spring Harbor Symposia on Quantitative Biology. Cold Spring Harbor, New York 33:473-484.
21. Hartskeerl, R.A., J.E.N. Bergmans, M.C. Kamp, and W.P.M. Hoekstra (1983) Cloning of an exclusion-determining fragment of the IncI plasmid, R144. Plasmid 10:11-20.
22. Imber, R., R.L. Low, and D.S. Ray (1983) Identification of a primosome assembly site in the region of the *ori2* replication origin of the *Escherichia coli* mini-F plasmid. Proc. Natl. Acad. Sci., USA 80:7132-7136.
23. Jacob, A.E., J.A. Shapiro, L. Yamamoto, D.I. Smith, S.N. Cohen, and D. Berg (1977) Plasmids studied in *Escherichia coli* and other enteric bacteria. In DNA insertion elements, plasmids, and episomes, A.I. Bukhari, J.A. Shapiro, and S.L. Adhya, eds. Cold Spring Harbor Laboratory, Cold Spring Harbor, N.Y., pp. 607-638.
24. Kornberg, A. (1980) DNA replication. W.H. Freeman and Co., San Francisco.
25. Kornberg, A. (1982) 1982 Supplement to DNA replication. W.H. Freeman and Co., San Francisco.
26. Lanka, E., and P.T. Barth (1981) Plasmid RP4 specifies a deoxyribonucleic acid primase involved in its conjugal transfer and maintenance. J. Bact. 148:769-781.
26a. Lanka, E., and J.P. Fürste (1984) Function and properties of RP4 DNA primase. In Proteins involved in DNA replication, U. Hübscher and S. Spadari, eds. Plenum Press, New York and London, pp. 265-280.
27. Lanka, E., R. Lurz, and J.P. Fürste (1983) Molecular cloning and mapping of *SphI* restriction fragments of plasmid RP4. Plasmid 10:303-307.
28. Lanka, E., R. Lurz, M. Kröger, and J.P. Fürste (1984) Plasmid RP4 encodes two forms of a DNA primase. Molecular and General Genetics 194:65-72.

29. Lanka, E., E. Scherzinger, E. Günther, and H. Schuster (1979) A DNA primase specified by I-like plasmids. Proc. Natl. Acad. Sci., USA 76:3632-3636.
30. Maturin, Sr., L.J., and R. Curtiss III (1981) Role of ribonucleic acid synthesis in conjugational transfer of chromosomal and plasmid deoxyribonucleic acids. J. Bact. 146:552-563.
31. Meyer, R., and D.R. Helinski (1977) Unidirectional replication of the P-group plasmid RK2. Biochim. et Biophys. Acta 478:109-113.
32. Nomura, N., R.L. Low, and D.S. Ray (1982) Identification of ColE1 DNA sequences that direct single strand-to-double strand conversion by a ϕX174 type mechanism. Proc. Natl. Acad. Sci., USA 79:3153-3157.
33. Nossal, N.G. (1983) Prokaryotic DNA replication systems. Ann. Rev. of Biochem. 53:581-615.
34. Rowen, L., and A. Kornberg (1978) A ribo-deoxyribonucleotide primer synthesized by primase. J. Biol. Chem. 253:770-774.
35. Sasakawa, C., and M. Yoshikawa (1978) Requirements for suppression of a dnaG mutation by an I-type plasmid. J. Bact. 133:485-491
36. Shaw, J.E., and H. Murialdo (1980) Morphogenetic genes C and Nu3 overlap in bacteriophage λ. Nature (London) 283:30-35.
37. Smith, R.A., and J.S. Parkinson (1980) Overlapping genes at the cheA locus of Escherichia coli. Proc. Natl. Acad. Sci., USA 77:5370-5374.
38. Uemura, H., and K. Mizobuchi (1982) Genetic and physical characterization of the ColIb plasmid using ColIb-R222 hybrids. Mol. Gen. Genet. 185:1-12.
39. Van der Ende, A., R. Teertstra, H.G.A.M. van der Avoort, and P.J. Weisbeek (1983) Initiation signals for complementary strand DNA synthesis on single-stranded plasmid DNA. Nucl. Acids Res. 11:4957-4975.
40. Vapnek, D., M.B. Lipman, and W.D. Rupp (1971) Physical properties and mechanism of transfer of R factors in Escherichia coli. J. Bact. 108:508-514.
41. Wilkins, B.M. (1975) Partial suppression of the phenotype of Escherichia coli K-12 dnaG mutants by some I-like conjugative plasmids. J. Bact. 122:899-904.
42. Wilkins, B.M., G.J. Boulnois, and E. Lanka (1981) A plasmid DNA primase active in discontinuous bacterial DNA replication. Nature (London) 290:217-221.
43. Wilkins, B.M., and S.E. Hollom (1974) Conjugational synthesis of $Flac^{+}$ and ColI DNA in the presence of rifampicin and in Escherichia coli K12 mutants defective in DNA synthesis. Mol. Gen. Genet. 134:143-156.
44. Willetts, N., and B. Wilkins (1984) Processing of plasmid DNA during bacterial conjugation. Microbiol. Rev. 48:24-41.
45. Zipursky, S.L., and K.J. Marians (1980) Identification of two Escherichia coli factor Y effector sites near the origins of replication of the plasmids ColE1 and pBR322. Proc. Natl. Acad. Sci., USA 77:6521-6525.

PROMOTERS IN THE TRANSFER REGION OF PLASMID F

Phil Mullineaux[1] and Neil Willetts[2]

Department of Molecular Biology
University of Edinburgh, Scotland

INTRODUCTION

Regulation of the conjugation (*tra*) genes of IncF plasmids occurs by a complex, 2-stage process. Most IncF plasmids carry 2 fertility inhibition genes, *finO* and *finP*, the products of which act in concert to prevent transcription of *traJ*, a gene which has its own promoter (6,22). F itself lacks *finO* but carries both a *finP* gene and the site at which FinOP regulation occurs, *fisO*. In turn, the *traJ* protein (pTraJ) is required for transcription of the major 32-kb *traY→Z* operon (10,22). The transfer region contains 2 further promoters, that for *traM* (21,23), and a weak promoter that also transcribes *traI* and *traZ* (1,23). There have also been suggestions that the F *traT* surface exclusion gene has its own promoter (18), as does the *traT* gene of R100 (16), although the polarity on surface exclusion and *traD* of Mu insertions into the *traYZ* operon argues against this (8,10). These regulatory stages were reviewed by Willetts and Skurray (24).

We have constructed fusions of all the *tra* promoters, and of the *finP* promoter, to the *galK* gene using the pKO plasmids described by McKenney et al. (14) or other plasmids derived from these. Measurement of the galactokinase levels produced by these plasmids under induced or repressed conditions has given further information about the regulatory systems, and allowed the *tra* promoter strengths to be compared to each other and to other prokaryotic promoters. We have paid particular attention to the promoter for the *traYZ* operon

Present address: [1] John Innes Institute, Colney Lane, Norwich, U.K.; [2] Biotechnology Australia Pty. Ltd., P.O. Box 20, Roseville, N.S.W., 2069, Australia.

since this is positively controlled, and such systems are relatively rare.

RESULTS AND DISCUSSION

Fusion of the tra Promoters to galK

In vitro recombination was used to fuse all of the potential tra promoters, and the putative finP promoter, to the galK gene. Vector plasmids were pKO1 and pKO4 (14), pHR11 (15), and 2 derivative plasmids constructed in our laboratory, pED101 and pED115. The latter contain the cloning sites BamI HindIII ClaI EcoRI and EcoRI ClaI HindIII BamI prior to galK, respectively, and are particularly useful since TaqI or HpaII/MspI termini can be readily cloned into ClaI termini by virtue of their CG single-strand extensions.

A variety of previously cloned segments of the F transfer region served as sources of the promoter fragments. Those fragments that were cloned into pKO-derived vectors are shown in Fig. 1. Details of the constructions will be presented elsewhere (P. Mullineaux and N. Willetts, in preparation). The galactokinase levels determined by these plasmids were measured in the presence or absence of other plasmids providing finO, finP or traJ, as appropriate (Tab. 1).

Two plasmids containing traJ-galK fusions were constructed, one (pED106) with a SalI-BglII fragment, and the second (pED121) with a TaqI fragment that by comparison was missing the 81 bp between the TaqI site in the traJ leader mRNA and the BglII site in traJ (Fig. 2). Genetic complementation tests showed that pED106, but not pED121, carried the finP gene: a part of this gene must therefore be located in the 81-bp TaqI-BglII region. This location is consistent with the sequence mapping of 2 finP⁻ mutations in this region (Fig. 2), and suggests that finP is a small gene corresponding to the DNA strand complementary to that encoding the traJ leader mRNA.

The galactokinase level of pED106, but not of pED121, was reduced by the FinOP system (Tab. 1), indicating that all or part of the site of action, fisO, also lies in the 81-bp TaqI-BglII region. pED121 includes the entire striking 12-bp perfect inverted repeat sequence located immediately prior to the TaqI site (Fig. 2), but not the F DNA sequence immediately following this. We hypothesize by comparison with transcriptional termination sites (19), that the fisO site includes both the inverted repeat sequence and the following DNA sequence where the usual run of 4 to 8 T's is replaced by specific binding of the FinOP repressor. Regulation of traJ would therefore be by premature termination of transcription rather than by prevention of initiation. Confirmation of this proposed location for the fisO site awaits the sequencing of fisO mutations such as those derived by Finnegan and Willetts (4).

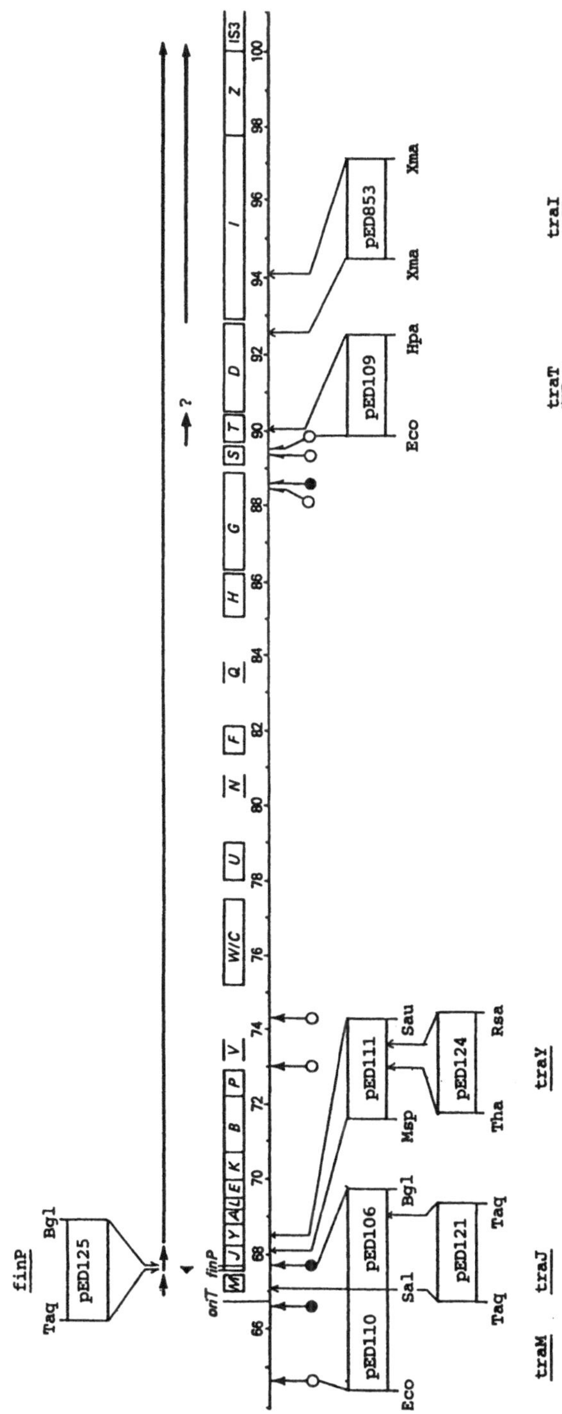

Fig. 1. Segments of the F transfer region carrying potential promoters that have been cloned into galK fusion plasmids. Restriction sites are marked for EcoRI (○) and BglII (●).

Tab. 1. Galactokinase levels of *tra-galK* fusion plasmids

Plasmid	Promoter	Induced level	Repressed level
pED110	*traM*	220	170
pED106	*traJ*	580	200
pED121	*traJ*	550	520
pED111	*traY*	1230	80
pED109	*traT*	1	1
pED853	*traI*	110	100
pED125	*finP*	12	12

Galactokinase assays were performed as described by McKenney *et al.* (14), and the number of units calculated according to Newman *et al.* (15). The host strain was JC3272 (*galK*$^-$). In our assays, pKG1800 gave 360 units and pKO1 gave 18 units (which has been subtracted from all values). Accessory plasmids were: R100 (*finO*$^+$), pNK46 (*finP*$^+$ *traJ*$^-$; 12); F*lac* (*traJ*$^+$).

Since pED106 carries *finP* but pED121 does not, it follows that the promoter for *finP* must also lie in the 81-bp *TaqI-BglII* fragment. Fusion of this to *galK* in the correct orientation gave a plasmid, pED125, encoding a low level of galactokinase activity that we presume is expressed from the *finP* promoter (Tab. 1). Potential -35 and -10 recognition sequences are marked in Fig. 2, together with a potential mRNA start site. There is a ribosome binding sequence downstream, prior to a 72-bp open reading frame, and the potential polypeptide product could be the novel 4-kd protein synthesized by a pBR*finP*$^+$ plasmid in minicells (11), although this might instead be a fusion protein created by the construction (11,21). A strong transcriptional termination sequence (12 bp inverted repeat, 5 GC pairs, and a run of 4 T's; Fig. 2) is present in the strand complementary to that encoding the *traJ* leader mRNA, and this would reduce the open reading frame to 48 bp, with no termination codon. We therefore propose that the *finP* product is an RNA molecule that together with the nonspecific *finO* product interacts with its complementary sequence to prevent *traJ* transcription and/or translation. This would account much more readily for the several plasmid-specific alleles of *finP* that have been described (5).

The promoter for the major *traYZ* operon was fused to *galK* in pED111, and as expected galactokinase expression by this plasmid was induced by *traJ* protein (Tab. 1). In contrast, the promoter for *traM* and the weak promoter for *traIZ* functioned constitutively.

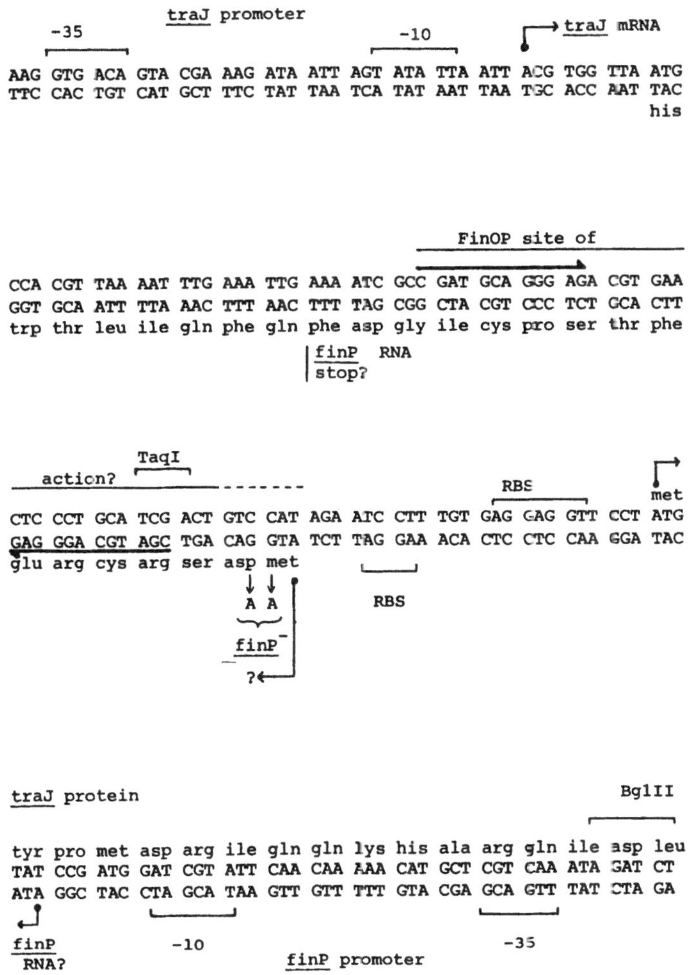

Fig. 2. Sequence of the traJ mRNA leader sequence (21). The likely location of the finP gene and promoter, and its potential 24 amino acid protein product, are shown. Two G→A base changes resulting in finP- mutations (N. Willetts, J. Maule, and R. Thompson, unpub. data) are indicated. RBS denotes ribosome binding site.

This result for the traM promoter differs from the pTraJ induction of expression of a traM-lacZ fusion found by Gaffney et al. (8) that might be correlated with the sequence similarity between the traM and traYZ promoter regions (Ref. 21; see Fig. 4 below). The reason for the discrepancy between the lacZ and galK fusion data remains to be elucidated.

The traYZ promoter was particularly powerful, giving about 4-fold more galactokinase than the gal promoter, and approaching the strength of the tac promoter (20). Cells containing pED111 together with single copy Flac traJ$^+$ grew poorly, and cells containing pED111 together with a multicopy pACYCtraJ$^+$ plasmid were unstable, so that the strength of the traYZ promoter in the presence of a high level of traJ product might be even greater. The traJ promoter was also unexpectedly powerful for a regulatory gene, and this together with the location of pTraJ in the outer membrane (12) may indicate that this protein additionally has a structural role in pilus formation and/or DNA transfer (reviewed by Willetts and Wilkins, Ref. 25). In accordance with the polarity data, we found that traT does not have its own promoter: the high levels of traT product produced by a repressed F plasmid may therefore be due to the long life or efficient translation of the basal uninduced level of traYZ mRNA.

The traY Promoter and Operator

The promoter and operator sites for traY were initially located on a 395-bp MspI-Sau3a fragment in pED111 (Fig. 3). Since this fragment lacks the traJ promoter, this finding confirms that the

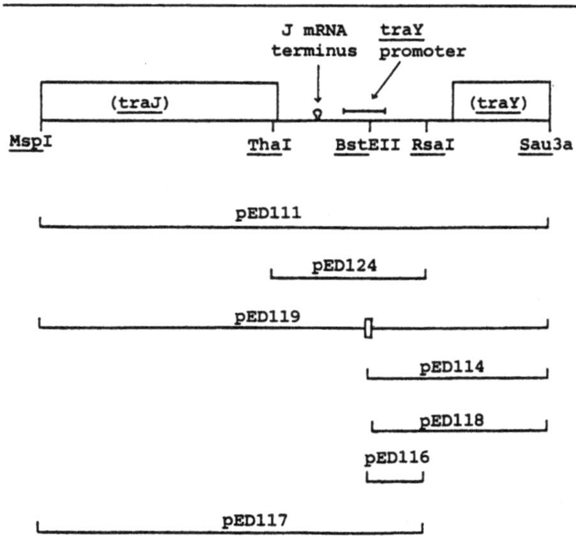

	GALACTOKINASE	
	Induced	Uninduced
pED111	1230	80
pED124	630	140
pED119	8	7
pED114	100	100
pED118	1	1
pED116	130	90
pED117	1340	110

Fig. 3. The traJ-traY intercistronic region, showing segments fused to galK and the galactokinase levels produced by the resultant plasmids in the presence or absence of Flac traJ$^+$. The termination site for traJ transcription (7), and the potential traY promoter, are marked.

```
ThaI
         10         20         30         40         50
CGCGTTAATA AGGTGTTAAT AAAATATAGA CTTTCCGTCT ATTTACCTTT

     J protein
       stop

                              BstEII

         60         70         80         90        100
TCTGATTATT CTGCAAACAT AAGTGGTAAC CAGAAGATAA ACAGCGGGAG

J mRNA        -35         |< pED114    -10
  stop                     |< pED118

                      RsaI
        110        120        130        140
GTGTATTGAA AAGATTTGGT ACACGTTCTG CAACAGGTAA GATG
                                     RBS     met
                                              |
                                              └→ pTraY
```

Fig. 4. Sequence of the traY promoter-operator region.

traYZ operon is not transcribed from the traJ promoter, with pTraJ functioning to prevent termination of such transcription. In the absence of the pTraJ inducer, a basal level of galactokinase was synthesized equivalent to about 6% of the induced level. This compares to <5% as measured by mRNA hybridization (22), and <0.3% measured in Flac traY-(MuAp) lac fusions (8): the reason for the higher level in the present case is unknown.

The promoter was located more accurately in 2 ways. Firstly, the central 118-bp ThaI-RsaI fragment was fused to galK in pED124, and this was shown to retain the inducible promoter (Fig. 3). Secondly, pED111 was digested with BstEII followed by brief BAL-31 treatment prior to religation, to give short deletion mutants. These are currently being sequenced, but polyacrylamide gel determination of the MspI-Sau3a fragment size showed that <20 bp had been deleted. One such plasmid was pED119, and this had totally lost ability to produce galactokinase. The traY promoter therefore covers or is very close to the BstEII site, and the uninduced level of galactokinase produced by pED111 is most likely due to the same traY promoter.

The sequence of the ThaI-Sau3a segment was determined (Fig. 4) and is identical to that found by Fowler et al. (7). Given the proximity of the traY promoter to the BstEII site, and the locations of the sites of termination of traJ mRNA and of the start of the traY gene (7), potential -35 and -10 sequences are marked that bracket the BstEII site. As for other positively controlled promoters such as those for the ara and mal operons (3,13), homology to the consensus recognition sequence (9) is poor. However, in contrast to the mal promoters, the -45 region is not unusually GC-rich. Our location for the traY promoter is distinct from those about 50 bp upstream suggested by Fowler et al. (7) on the basis of in vitro run-off transcription experiments. Since the traY promoter is positively controlled by pTraJ and may therefore have an unusual sequence, it is likely that such experiments would give misleading results (17).

The MspI-BstEII segment of pED111 was deleted to give pED114, which contained a noninducible promoter expressing galactokinase at a level similar to the uninduced level of pED111. However, the vector plasmid carried the tet promoter's -35 sequence 10 bp upstream from the ClaI/MspI site, and sequence data suggest that the promoter in pED114 may have been produced by fusion of this to the traY promoter's -10 sequence with a spacing of 18 bp. An adventitious Gal⁻ mutant obtained in this experiment, pED118, carried a deletion of 7 bp of vector DNA together with the first G of the BstEII recognition sequence. It gave no galactokinase activity, confirming that a part of the traY promoter must lie to the left of the BstEII site, and that the galactokinase synthesized by pED114 is therefore probably not derived from an uninduced traY promoter.

A further deletion of the RsaI-Sau3a fragment of pED114 gave pED116, which also carried the putative fusion promoter. In an in vitro reconstruction experiment, the MspI-BstEII segment of pED111 was recombined with the BstEII-RsaI segment of pED116 to give pED117. This recovered inducibility by pTraJ, suggesting that the traY operator site is upstream of the BstEII site. More precise location of this site awaits the sequencing of mutations leading to pTraJ independence (2) or pTraJ noninducibility, or isolation of functional pTraJ from the outer cell membrane followed by "footprinting" experiments to locate its binding sequence.

ACKNOWLEDGMENTS

We are grateful to John Maule and Elisabeth Marson for their expert assistance. This research was funded by MRC Programme Grant G978 88 to N. S. Willetts.

REFERENCES

1. Achtman, M., R. Skurray, R. Thompson, R. Helmuth, S. Hall, L. Beutin, and A.J. Clark (1978) Assignment of tra cistrons to Eco RI fragments of F sex factor DNA. J. Bact. 133:1383-92.
2. Achtman, M. (1973) Transfer-positive J-independent revertants of the F factor in E. coli. Genet. Res. 21:67-77.
3. Bedouelle, H., V. Schmeissner, M. Hofnung, and M. Rosenburg, (1982) Promoters of the malEFG and malK-lamB operons in E. coli K12. J. Mol. Biol. 161:519-531.
4. Finnegan, D.J., and N.S. Willetts, (1971) Two classes of Flac mutants insensitive to transfer inhibition by an F-like R factor. Mol. Gen. Genet. 111:256-264.
5. Finnegan, D.J., and N.S. Willetts (1972) The nature of the transfer inhibitor of several F-like plasmids. Mol. Gen. Genet. 119:57-66.
6. Finnegan, D.J., and N.S. Willetts (1973) The site of action of the F transfer inhibitor. Mol. Gen. Genet. 127:307-316.
7. Fowler, T., T. Taylor, and R. Thompson (1983) The control region of the F plasmid transfer operon: DNA sequence of the traJ and traY genes and characterization of the traYZ promoter. Gene 26:79-89.
8. Gaffney, D., R. Skurray, and N.S. Willetts (1983) Regulation of the F conjugation genes studied by hybridisation and tra-lacZ fusion. J. Mol. Biol. 168:103-122.
9. Hawley, D.K., and W.R. McClure (1983) Compilation and analysis of E. coli promoter DNA sequences. Nucl. Acids Res. 11:2237-2255.
10. Helmuth, R., and M. Achtman (1975) Operon structure of DNA transfer cistrons on the F sex factor. Nature 257, 652-656.
11. Johnson, D., R. Everett, and N.S Willetts (1981) Cloning of F DNA fragments carrying the origin of transfer oriT and the fertility inhibition gene finP. J. Mol. Biol. 153, 187-202.
12. Kennedy, N., L. Beutin, M. Achtman, R. Skurray, V. Rahmsdorf, and P. Herrlich (1977) Conjugation proteins encoded by the F sex factor. Nature 270:580-585.
13. Lee, N.L., W.O. Gielow, and R.G. Wallace, (1982) Mechanism of araC autoregulation and the domains of 2 overlapping promoters, P_c and P_{BAD}, in the L-arabinose regulatory region of E. coli. Proc. Natl. Acad. Sci. USA 78:752-756.
14. McKenney K., H. Shimatake, D. Court, U. Schmeissner, C. Brody, and M. Rosenberg (1981) A system to study promoter and terminator signals recognized by E. coli RNA polymerase. In Gene Amplification and Analysis, Vol. II J.G. Chirikjian & T. Papas, eds. Elsevier/North-Holland, New York, pp. 383-415.
15. Newman, A.J., J.C. Ma, K.M. Howe, I. Garner, and R.S. Hayward, (1982) Evidence that rifampicin can stimulate readthrough of transcriptional terminators in E. coli, including the attenuator of the rpoBC operon. Nucl. Acids Res. 10:7409-7424.

16. Ogata, R.T., C. Winters, and R.P. Levine, (1982) Nucleotide sequence analysis of the complement resistance gene from plasmid R100 J. Bact. 151:819-827.
17. Queen, C., and M. Rosenberg (1981) A promoter of pBR322 activated by cAMP receptor protein. Nucl. Acids Res. 9:3365-3377.
18. Rashtchian, A., J.H. Crooks, and S.B. Levy, (1983) traJ independence in expression of traT on F. J. Bact. 154:1009-1012.
19. Rosenberg, M., and D. Court (1979) Regulatory sequences involved in the promotion and termination of RNA transcription. Ann. Rev. Genet. 13:319-353.
20. Russell, D.R., and G.N. Bennett (1982) Construction and analysis of *in vivo* activity of *E. coli* promoter hybrids and promoter mutants that alter the -35 to -10 spacing. Gene 20:231-243.
21. Thompson, R., and L. Taylor (1982) Promoter mapping and DNA sequencing of the F plasmid transfer genes traM and traJ. Mol. Gen. Genet. 188:513-518.
22. Willetts, N.S. (1977) The transcriptional control of fertility in F-like plasmids. J. Mol. Biol. 112:141-148.
23. Willetts, N.S., and S. McIntire (1978) Isolation and characterization of λtra transducing phages from EDFL223 (FlactraB::EDλ4). J. Mol. Biol. 126:525-549.
24. Willetts, N.S., and R. Skurray (1980) The conjugation system of F-like plasmids. Ann Rev. Genet. 14:41-76.
25. Willetts, N.S., and B.M. Wilkins (1984) Processing of plasmid DNA during bacterial conjugation. Micro. Rev. 26:24-41.

CONJUGATIVE SEX PLASMIDS OF *STREPTOMYCES*

D. A. Hopwood, D. J. Lydiate, F. Malpartida, and H. M. Wright

John Innes Institute
Colney Lane
Norwich NR4 7UH, England

INTRODUCTION

Much is known about the ways in which the conjugative plasmids of gram-negative bacteria cause their hosts to mate (1). Detailed knowledge has been obtained about how the transfer of plasmid DNA occurs (2). The mechanisms by which such plasmids promote the mobilization of chromosomal genes have also been extensively studied (3-6). In contrast, very little is known about such phenomena for *Streptomyces* plasmids (7). However, now that many *Steptomyces* plasmids can be isolated with ease (8), genetically manipulated in vitro, and returned to the host cells by transformation of protoplasts (9,10) we can hope for more rapid progress in the understanding of the molecular biology of these plasmids. Here we review some facts and ideas about *Streptomyces* conjugation which may serve as a basis for future experimentation. As we see, conjugative *Streptomyces* plasmids are likely to possess properties interestingly different from those of plasmids of gram-negative bacteria and so they should repay a detailed investigation.

GENERAL FEATURES OF *STREPTOMYCES* PLASMIDS

The first conjugative plasmid to be identified in a gram-positive bacterium was SCP1 in the genetically studied *Streptomyces coelicolor* A3(2)(11). SCP1 was recognized through its influence on the recombination of host chromosomal genes ("fertility") and because it determines production of and resistance to a diffusible inhibitor of growth and sporulation, later identified as the antibiotic methylenomycin (12). Various derivatives of SCP1 have been characterized genetically, including those mediating high-frequency donation of chromosomal markers in which SCP1 seems to be stably integrated into the host chromosome, and autonomous SCP1-prime plasmids carrying segments of host DNA (13). Complete molecules of SCP1 DNA have not been isolated but segments of SCP1 have been cloned, including genes coding for methylenomycin resistance (14) and production (15); uncharacterized segments of the plasmid (15); and a region capable of autonomous replication when ligated with DNA carrying a selectable marker to form circular molecules (M.J. Bibb and J.M. Ward, pers. comm.). The size of SCP1 has been estimated to be in the range of 150 kb to at least 200 kb (15,16).

The first *Streptomyces* plasmid to be isolated physically was an initially cryptic CCC molecule of 31 kb (17) which was distinct from SCP1 and was later shown to be identical to a second sex plasmid, SCP2, of *S. coelicolor*. This plasmid had been identified through the occurrence of spontaneous variants (SCP2*) capable of promoting increased fertility in *S. coelicolor* A3(2)(18). The copy number of SCP2 in *S. coelicolor* is 1-4 per chromosome. This plasmid has not been shown to interact stably with the chromosome-indeed, an exhaustive search for high-frequency donors of chromosomal markers and for SCP2-prime plasmids was unsuccessful (J.A. Ewing, pers. comm.). However, the recent development of stable cloning vectors derived from SCP2* (19) has enabled segments of the host chromosome to be inserted into the plasmid, to provide artificial SCP2*-prime derivatives. Some of these, carrying segments of DNA which include genes for biosynthesis of the antibiotic actinorhodin (20), have been used in some genetic experiments described herein.

Many other plasmids have now been identified as CCC DNA (or occasionally as linear molecules; Ref. 21) in a wide variety of *Streptomyces* species (reviewed in Ref. 7). In contrast to SCP2 and (presumably) SCP1, several have a high copy number. Thus, pIJ101, a plasmid of 8.9 kb found naturally in *S. lividans* ISP5434, exists in 40-300 copies per chromosome depending on the age of the culture (22). A copy number as high as 800 has been reported for pNM100 in *S. virginiae* (23). A further interesting series of plasmids is represented by SLP1, pIJ110, and pIJ408. These are found as autonomous plasmids in *S. lividans* 66 after mating this strain with *S. coelicolor* A3(2), *S. parvulus* ATCC 12434, or *S. glaucescens* ETH 22794 respectively (24). They arise by the "looping out" of DNA sequences which occur as part of larger replicons (in the case of SLP1 in *S. coelicolor* this has been shown to be the chromosome; Ref. 25) and their conjugal transfer to *S. lividans*.

Most *Streptomyces* plasmids that have been studied give rise to "pocks" when a plasmid-carrying spore (or other plating unit) develops within a lawn of plasmid-free individuals. Pocks are circular areas, up to a few millimeters in diameter, in which the gross appearance of the culture is altered; usually the development of aerial mycelium and spores is delayed or prevented, and there may be precocious production of secondary metabolites, such as actinorhodin and undecylprodigiosin in S. coelicolor and S. lividans. Pock-formation (also called the Ltz^+ phenotype; Ref. 25) is discussed in more detail below.

STREPTOMYCES PLASMIDS AS TRANSFERABLE FERTILITY FACTORS

Minimum Transfer Regions

Table 1 lists some examples of comparatively small plasmids of Streptomyces that have been shown to be capable of conjugal transfer. For 3 plasmids, the segment of the plasmid required for transfer and pock formation has been partially defined. This has not been done for the other 4 plasmids. However, assuming that the essential replication functions of the plasmids occupy at least 2 kb (Tab. 2), it is apparent that not more than 3-10 kb of DNA is required for transfer and pock-formation by any of the 7 plasmids (and the upper end of this range is probably a considerable overestimate). This contrasts with the 15-35 kb transfer regions of several plasmids that have been analyzed in gram-negative bacteria (26-30) and indicates a simpler genetic determination of mating in Streptomyces (22,25).

"Fertility"

A "cross" in Streptomyces does not lend itself to kinetic analysis. It is a mixed culture of 2 parental strains growing on agar for a period of several days, during which a new generation of spores is produced. Among these spores, new genotypes are detected by inoculating samples on selective plating media. Most of the

Tab. 1. Small self-transmissible Streptomyces plasmids.

Plasmid	Size (kb)	Upper limit of transfer and Ltz region (kb)	Reference
pIJ101	8.9	3.1	22
SLP1.2	14.5	5	Kieser (unpublished)
SCP2*	31	10	19
pSF765	7.2	7.2	47
pNM100	9.3	9.3	23
pJV1	10.8	10.8	48
pVE1	11	11	49

Tab. 2. Basic replicons of Streptomyces plasmids.

Plasmid	Copy no.	Size (kb)	Upper limit of replicon (kb)[a]	References
pIJ101	40-300	8.9	2.05	22
SLP1.2	4-5	14.5	2.7 + 1.4[b]	Kieser (unpublished)
SCP2*	1-2	31	5.5	19, 50
pFJ103	(High)	20	2.9	51
pVE1	(High)	11	3.3	49
pNM100	c. 600	9.3	4.0	23
pSK1	(High)	10.3	3.8	52
pSK2	> 20	10.3	2.9	52

[a]In most cases the results are quite peliminary; therefore these values may be considerable over-estimates

[b]Two discontinuous regions of essential DNA

self-transmissible Streptomyces plasmids that have been examined promote the recombination of host chromosomal markers in such crosses. Several plasmids have been tested in a common background: genetically marked strains of S. lividans 66 carrying no other known plasmids (31). In crosses of a plasmid-carrying with a plasmid-free strain, the frequency of recombinants (selecting for one marker from each parent) as a fraction of the total spores varied reproducibly with the plasmid from > 10^{-3} for pIJ101 to < 10^{-5} for SCP1 and pJV1, with other plasmids having an intermediate fertility of around 10^{-4} (Tab. 3). On the other hand, SLP3 and SLP4 do not appear to act as fertility factors. The variation in fertility did not appear to reflect differences in the frequency with which the plasmids were acquired by the initially plasmid-free strain, since all the plasmids were transferred to at least 50% and usually close to 100% of the other parent. This frequency is not, however, a measure of the primary transfer of the plasmids but also depends on secondary transfer within the recipient culture (see below). For SLP2 and pJV1 in Tab. 3 (as well as SCP2 in S. coelicolor A3(2) and SRP1 in S. rimosus; Ref. 32, 33) fertility was significantly reduced when the same plasmid was present in both parents. This effect may be related to "entry disadvantage" (34) or "exclusion" (19), terms describing the failure of plasmid markers to reassort freely with host chromosomal genotypes in crosses between strains carrying marked plasmids. The

Tab. 3. Chromosomal recombination frequencies in crosses between *S. lividans* strains caused by various plasmids.

Plasmid	Recombinant fraction	
	$Plasmid^{+}$ X $Plasmid^{-}$	$Plasmid^{+}$ X $Plasmid^{+}$
SLP3	$< 2 \times 10^{-9}$	-
SLP4	$< 3 \times 10^{-9}$	-
SCP1	5.1×10^{-6}	3.2×10^{-6}
SCP2*	1.2×10^{-4}	1.0×10^{-4}
SLP2	2.5×10^{-4}	1.8×10^{-6}
pIJ6 (from SLP1.2)	1.4×10^{-4}	1.9×10^{-4}
pIJ303 (from pIJ101)	4.3×10^{-3}	4.3×10^{-3}
pIJ110	1.0×10^{-4}	1.2×10^{-4}
pIJ408	3.0×10^{-4}	2.2×10^{-4}
pIJ928 (from pIJ107)	5.3×10^{-5}	4.3×10^{-5}
pWOR191 (from pJV1)	5.1×10^{-6}	$e.1 \times 10^{-3}$

Data from ref. 31 except pIJ928 (19) and pWOR191 (48).

explanation of such "exclusion" is not known, but it might include the effects of any one, or a combination, of several factors, such as processes analogous to the surface exclusion and incompatibility shown by plasmids of gram-negative bacteria, as well as phenomena that may be peculiar to *Streptomyces* (7).

Patterns of Marker Inheritance in Crosses Involving SCP1

In crosses between $SCP1^{+}$ strains of *S. coelicolor* (presumed to carry autonomous SCP1) and $SCP1^{-}$ strains, there was a tendency for recombinants to inherit markers preferentially from the $SCP1^{+}$ parent (35). This result is compatible with the presence, within an $SCP1^{+}$ culture, of high-frequency donors which contribute markers preferentially to the progeny (rather as the fertility of F^{+} cultures of *E. coli* depends partially on a minority of Hfr bacteria in the F^{+} population; Ref. 36). Various types of SCP1-donor strains have been isolated from $SCP1^{+}$ cultures; these are listed in Tab. 4. For most of the donor types, the donor character was stable during subculture, but in a few cases it was highly unstable. In crosses of any of the donor types with an $SCP1^{-}$ strain, essentially 100% of the previously $SCP1^{-}$ genotype (taking account of the instability of the donor character where appropriate) were found to acquire donor properties apparently identical with those of the donor parent, but the patterns of inheritance of donor chromosomal markers varies with the strain (Fig. 1).

Tab. 4. Donor strains of S. coelicolor A3(2) involving SCP1.

Donor type	Origin	Characteristics	Reference
NF	SCP1$^+$ X SCP1$^+$ mating	Bidirectional donor from 9 o'clock position; integration of SCP1 into chromosome inactivated dagA	37, 53
NF-like donor (A634)	Unstable donor A620 crossed with SCP1$^-$	Bidirectional donor from near 9 o'clock position (but dag$^+$)	38
Bidirectional donor	SCP1'-cysB	Bidirectional donor from near cysB	39
Unstable "pabA" donors (e.g. A607, A620) Unstable "uraA" donor (A610) Stable "pabA" donor (A608) Stable "cysD" donor (2106)	SCP1$^+$ strain	Unidirectional donors from three different sites	38
SCP1'-cysB SCP1'-argA, uraB	SCP1$^+$	Primes carrying chromosomal genes	39

Markers were inherited from 2 of the stable donors (A608 and 2106) with a gradient of frequencies falling with increasing distance, in one direction only, from a fixed point on the linkage map. Such "unidirectional" donation of markers is reminiscent of the behavior of Hfr donors in E. coli which result from covalent integration of a plasmid into the chromosome; the main difference is the inheritance of the donor character by all recombinants, including those acquiring only markers close to the "integration" point. Three other stable donors--the original NF type and 2 donors derived secondarily from apparently autonomous plasmid variants [the NF-like donor A634 (Ref. 38) and the bidirectional donor derived from SCP1'-cysB (Ref. 39)]--showed a "bidirectional" gradient of marker donation. All these stable donors are likely to have originated by integration of SCP1 into the chromosome, and this idea is supported by 2 specific observations. 1) NF strains lack the activity of a chromosomal gene, dagA, which determines a diffusible agarase and which maps at the 9 o'clock position on the chromosome (37), the same point as the center of the bidirectional gradient of marker donation; it is postulated that dagA was inactivated by integration of SCP1. 2) The bidirectional donor derived from an SCP1-prime strain in which the plasmid carries the cysB gene donated

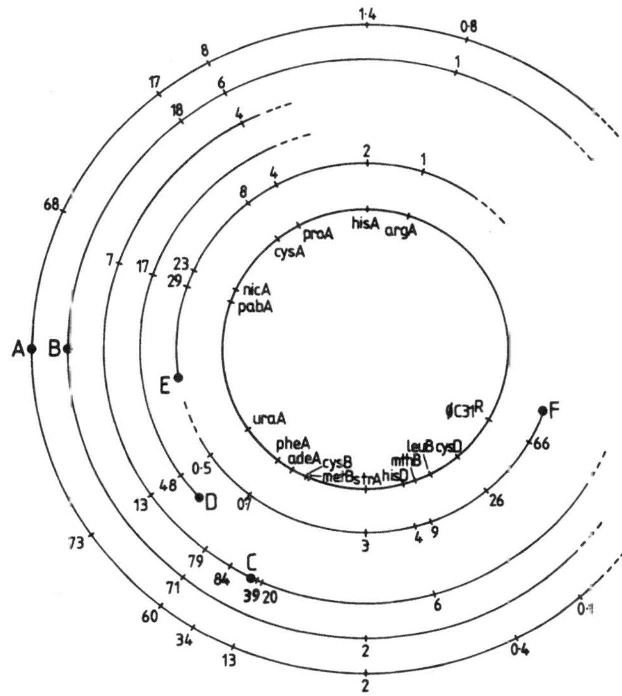

Fig. 1. Percentage frequencies of inheritance of donor markers by recombination in matings of various types of *S. coelicolor* SCP1-donors with $SCP1^-$ strains. A: NF (from Ref. 53); B: NF-like donor A634 (from Ref. 39); C: bidirectional donor derived from SCP1'-*cysB* (from Ref. 38); D: unstable "*uraA*" donor A610 (from Ref. 39); E: unstable "*pabA*" donor A607 and stable "*pabA*" donor A608 (from Ref. 39); F: stable "*cysD*" donor 2106 (from Ref. 13). The ● symbol indicates the presumed position of SCP1, inherited by approximately 100% of the recombinant progeny.

chromosomal markers with a gradient centered on this locus (39); integration via homologous recombination is the most likely explanation for the origin of this donor, and the slight instability of the donor property (∿ 0.1%) is compatible with re-excision of the plasmid by recombination between the tandemly repeated copies of the chromosomal segment which would be present.

Protoplast Fusions Between NF and $SCP1^-$ Strains

In protoplast fusions between pairs of *S. coelicolor* strains, recombinants originate at a high frequency. It can be deduced that they arise by repeated rounds of recombination within fusion bodies that, at least initially, contain the complete set of chromosomal genes of both parents (40,41). The published fusions were between pairs of SCP1⁻ strains. We later analyzed fusions between NF and SCP1⁻ strains (Hopwood and Wright, unpub. data). The fused protoplasts were allowed to regenerate nonselectively at varying dilutions, and spores were taken for analysis only from plates carrying subconfluent populations of colonies to avoid distortions of the results by mating on the regeneration plates. Otherwise, the technical details were as previously described (40).

Some of the results of these fusions are summarized in Fig. 2(a). In striking contrast to the situation in NF X SCP1⁻ matings the frequencies of all the donor and recipient markers (except at *uraA*) among nonparental progeny approached 50% (and were probably not significantly different from this value, allowing for the effects of a slight viability disadvantage of some of the alleles; Ref. 40). However, the NF character was inherited by up to 100% of the nonparental progeny, as in NF X SCP1⁻ matings (and this presumably explains the inheritance of *uraA1* by significantly more than 50% of the nonparental progeny because of its linkage to NF). That this obligate inheritance of NF was not due (primarily) to the effect of methylenomycin-production by NF strains inhibiting the

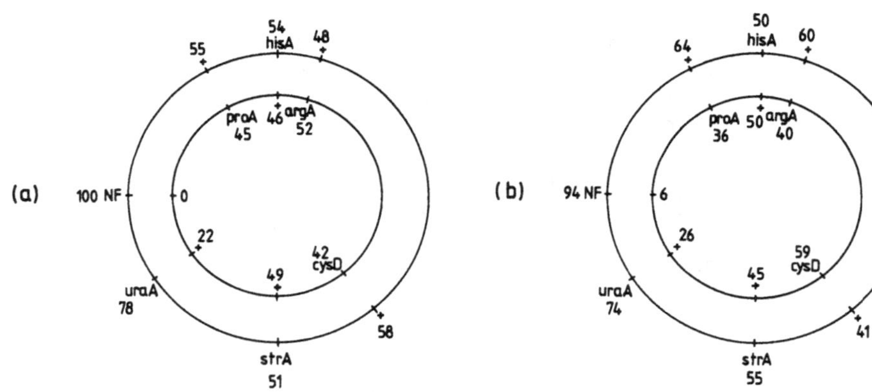

Fig. 2. Percentage inheritance of markers and of the NF character by recombinants in protoplast fusions of NF and SCP1⁻ strains of *S. coelicolor*. (a) NF strain A317 *hisA1 uraA1 strA1*; SCP1⁻ strain A700 *proA1 argA1 cysD18* (b) NF strain 2439 *hisA1 uraA1 strA1* (methylenomycin nonproducing mutant R11: Ref.42); SCP1⁻ strain A700 *proA1 argA1 cysD18*. In (a), there were 146 recombinants, representing 27.4% of the total progeny (which also included 316 of the NF parental genotype and 51 of the SCP1⁻ parental genotype); in (b) there were 102 recombinants, representing 27.3% of the total progeny (225 NF parentals and 46 SCP1⁻ parentals).

growth of $SCP1^-$ (and therefore methylenomycin-sensitive) strains was shown by the use of an NF methylenomycin nonproducing mutant (mmy) (42). The results of fusing such a strain with an $SCP1^-$ strain [Fig. 2(b)] were almost identical to those of the normal NF X $SCP1^-$ fusion in Fig. 2(a).

In an NF X $SCP1^+$ fusion (not shown) alleles from each parent were again approximately equal in frequency (this time including those at uraA) but NF progeny made up only about 50% of the total, the remainder being $SCP1^+$.

Patterns of Marker Inheritance in Crosses Involving SCP2*-Primes

In contrast to the situation in $SCP1^+$ X $SCP1^-$ crosses, $SCP2^+$ X $SCP2^-$ (or SCP2* X $SCP2^-$) crosses showed no tendency for progeny to inherit markers preferentially from the plasmid-carrying parent (32). This result is compatible with the observed absence (or perhaps extreme rarity) of high frequency donors of specific groups of chromosomal markers involving SCP2. Such donors have, however, recently been constructed by in vitro recombination, and include some containing DNA involved in actinorhodin biosynthesis (20).

The act cluster lies between the markers guaA and hisD on the S. coelicolor chromosome and closer to the latter (43). The entire set of act genes, and subsets of them, were cloned on to a derivative of SCP2* called pIJ922 (Fig. 3) which appears to be identical to SCP2* in its transfer and fertility properties (20). These clones, and subclones derived from them, provide a series of artificial SCP2*-primes with chromosomal segments of varying length inserted in either direction into the plasmid, all at the same site. Each clone, as well as pIJ922 itself, was introduced into an $SCP1^-$ $SCP2^-$ strain of S. coelicolor and the resulting strains were each crossed with an $SCP1^-$ $SCP2^-$ strain carrying a complementary set of chromosomal markers. The markers were chosen in such a way that the donation by the plasmid-carrying strain of alleles at loci (hisD, guaA, strA, and argA) situated at varying distances from the act region could be studied, with counter-selection at a distant locus, proA. The frequencies of donation of the markers are plotted in Fig. 4 as a function of the distance of the markers from act.

The main features of the results are as follows: 1) Whereas pIJ922 promoted the transfer of all 4 markers at a frequency of about 10^{-3}, characteristic of SCP2* X $SCP2^-$ crosses in S. coelicolor (32), the primes carrying inserts of chromosomal DNA at least 9.5 kb long caused increased donation of markers, at a frequency that fell with increasing distance of the marker from act. 2) The orientation of the inserts into the plasmid did not affect the pattern of marker donation (thus pIJ2300 and pIJ2302 have one orientation and pIJ2301 and pIJ2303 have the other); this is particularly telling for $hisD^+$ and $guaA^+$ which lie close to act and on opposite sides of

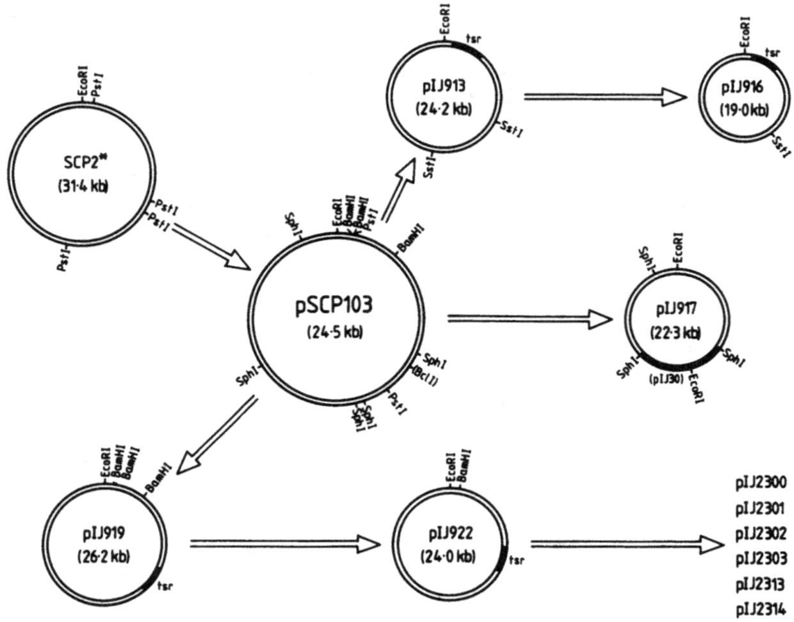

Fig. 3. Derivation from SCP2*, by in vitro deletions and additions of DNA, of the plasmids referred to in the text. Note that pSCP103 is drawn on a larger scale and that each plasmid includes only those restriction sites needed to explain the steps in the construction. The PstI C + D fragments of SCP2* were deleted to form pSCP103 (14). The BamHI B fragment of pSCP103 was replaced by the BamHI fragment of pIJ30 that carries tsr (thiostrepton resistance) to form pIJ913. The SstI B fragment of pIJ913 was deleted to form pIJ916. The SphI C + D + E fragments of pSCP103 were replaced by pIJ30 to form pIJ917. The BamHI tsr fragment of pIJ30 was inserted into BclI/ site 12 of pSCP103 (1 of 8 BclI/ sites in that plasmid: Ref. 19) to form pIJ919. The BamHI B + C fragments of pIJ919 were deleted to form pIJ922. Fragments of S. coelicolor chromosomal DNA, carrying various parts of the act gene cluster (20), were inserted into the BamHI site of pIJ922 to form pIJ2300, pIJ2301, pIJ2302, pIJ2303, pIJ2313, and pIJ2314.

it. 3) Recombinants selected to carry $hisD^+$ (or $guaA^+$), on one side of act, also inherited $guaA^+$ (or $hisD^+$) on the other side of act at a high frequency (25-75%) irrespective of the orientation of the chromosomal insert in the plasmid. 4) In all the crosses, the plasmids were transferred to 40-100% of the recipient (initially $SCP2^-$) strain.

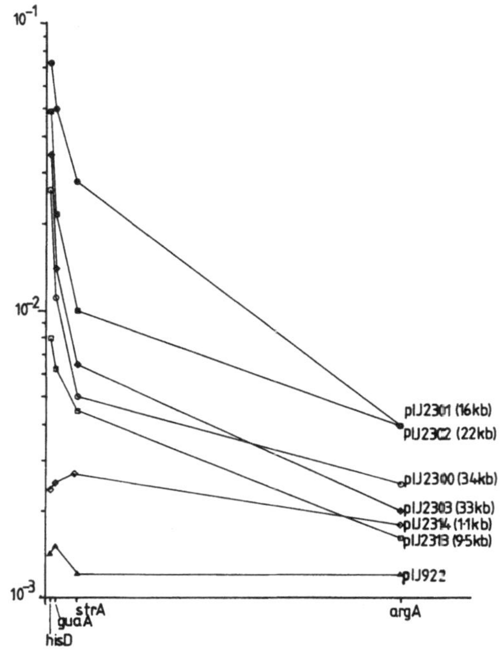

Fig. 4. Frequencies of inheritance of markers in matings of donor S. coelicolor strains carrying clones with chromosomal segments of varying length inserted into pIJ922 (and pIJ922 itself) crossed with SCP1⁻ strain B341. Ordinate: frequency of each donor marker among recombinants, expressed relative to the SCP1⁻ parental genotype. Abcissa: relative distance of each of the four markers from the act cluster. Numbers in parentheses are the sizes of the chromosomal inserts in each clone in kilobase pairs (the inserts in pIJ2300 and pIJ2302 had one orientation in the vector, while those in pIJ2301 and pIJ2303 had the opposite orientation). See text for further details.

These results indicate that homology between the plasmid and the chromosome can lead to an enhanced frequency of transfer of chromosomal markers, specifically of those close to the region of homology. This probably arises from an integration of the plasmid via homologous recombination to give a high frequency donor state. If so, the orientation of the integrated plasmid does not impose a polarity on the pattern of marker donation; the same, bidirectional, gradient results irrespective of orientation. (This is unlikely to arise from a combination of 2 unidirectional gradients, since markers on either side of the region of homology are often inherited together.) These are the same features as those observed in crosses involving the physically uncharacterized bidirectional SCP1 donors mentioned above. The minimum length of homologous DNA required for a significant interaction between $SCP2^*$ and the chromosome is certainly less than 9.5 kb (pIJ2313) while the insert of 1.1 kb in pIJ2314 appears not to give a significantly enhanced frequency of recombination over the $SCP2^*$ level.

POCK FORMATION: THE CONCEPT OF A PLASMID "SPREADING" FUNCTION

Pock formation is a common, perhaps universal, phenotype conferred by conjugative *Streptomyces* plasmids. Since pock formation depends on plasmid transfer, rather than on the action of diffusible inhibitors (18), the phenomenon was likened to lethal zygosis in *E. coli* (44); hence the phenotypic designation Ltz^+ (25). Copies of the plasmid that was present in the spore that initiated the pock can be found within the entire area of the pock, as shown for $SCP2^*$ by Bibb and Hopwood (32) and for pIJ101 by Kieser et al. (22). For pIJ101 (22), and also for SLP1.2 (T. Kieser, pers. comm.), a class of plasmid variants, constructed by in vitro recombination, gives rise to pocks of a characteristically reduced size, designated $Ltz^{+/-}$, in which the plasmid is present over a much smaller area than is the case for the wild-type, Ltz^+, plasmids (Fig. 5). The discovery of such $Ltz^{+/-}$ plasmids for pIJ101 gave rise to the concept of an active plasmid-determined spreading function, distinct from the ability to promote mating since the ability of $Ltz^{+/-}$ plas-

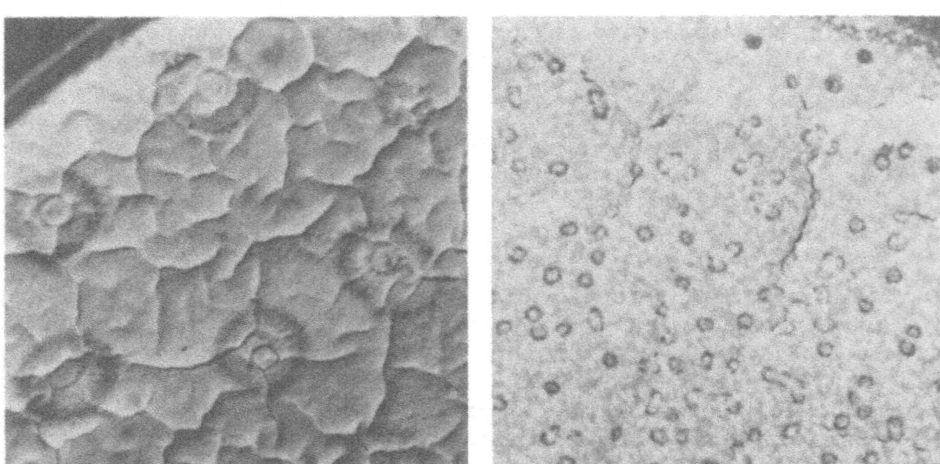

Fig. 5. Pocks produced by derivatives of pIJ101 in lawns of plasmid-free *S. lividans*. Left: normal sized pocks (Ltz^+) produced by pIJ303; right: small pocks ($Ltz^{+/-}$) produced by pIJ224. See Ref. 22 for details.

mids to be transferred in a mass mating appeared to be unimpaired. In vitro deletion or insertion of DNA into a region of the plasmid distinct from that involved in the $Ltz^{+/-}$ phenotype abolished pock formation completely and also rendered the plasmids incapable of transfer in matings.

Deletion of certain segments of $SCP2^*$ resulted in plasmids (e.g., pIJ917, Fig. 3) that were nonpock-forming, transfer-deficient (10^{-5} compared with 1), and incapable of promoting chromosomal recombination. Another derivative, pIJ916 (Fig. 3), was also nonpock-forming and transfer-deficient, but had only a slightly reduced fertility (10^{-5} compared with 10^{-4}) in otherwise plasmid-free _S. lividans_. Thus, the fertility promoted by pIJ916 was close to unity _per transfer_, compared with 10^{-4} for $SCP2^*$. Lydiate (19) suggested that pIJ916 was totally defective in the normal spreading function. As a consequence, one observed a high value of recombination per transfer because all transfer events were primary intermycelial events, rather than many of them being secondary events caused by plasmid spread within the genetically uniform $SCP2^-$ recipient culture and, therefore, incapable of generating recombinants. It appears, then, that primary transfer events have a very high probability of being associated with chromosomal recombination. [The very high frequency of recombination that was observed in the center of $SCP2^*$ pocks studied by Bibb and Hopwood (32) may also be a reflection of this. However, the strains in this study also carried SCP1 and the pocks may not be typical; further work will be needed to test the generality of this result.]

DISCUSSION: TRANSFER, SPREADING, AND FERTILITY OF _STREPTOMYCES_ PLASMIDS

One of the most striking characteristics of conjugative _Streptomyces_ plasmids is the small size of the segment of plasmid DNA required for the expression of conjugal properties; in the most studied example, pIJ101, this is not more than about 2 kb. Of course, the chromosome of the host (or perhaps other unrecognized plasmids) may provide functions in trans that are involved in conjugation; in that case, such functions are likely to be generalized in _Streptomyces_ since pIJ101 has been found to be conjugative in several host species including strains of _S. lividans_ which do not reveal evidence of the presence of any other plasmid. Plasmid transfer is, however, not entirely passive, since transfer-defective plasmid variants occur. Wild-type pIJ101 and several unrelated plasmids promote the transfer of transfer-defective pIJ101 derivatives at the same rather low level as they promote the transfer of chromosomal markers (22). This argues for a cis-acting function required for transfer of plasmid DNA and defective or absent in the nontransferable derivatives, all except one of which were known to have deletions of DNA including the transfer region (and the exception may well have had a small unrecognized deletion or insertion into the cis-acting site). The requirement of one or more (but not many) trans-acting plasmid-coded gene products for transfer is not excluded, but probably host gene products predominate in establishing the mating unions between hyphae; a purely host-determined hyphal fusion may perhaps provide the channel through which DNA is transferred.

In the cases where we have knowledge of the plasmid status of the host strains -- primarily _S. coelicolor_ and S. _lividans_ -- the transfer of chromosomal DNA (as measured by recombination of markers) does not occur, or occurs only at an extremely low level, in the absence of plasmids. The evidence is clearest for _S. lividans_ in which we were unable to detect definitive recombinants in SLP2$^-$ SLP3$^-$ strains (22,31); a low, variable level of nonparental colonies arose, but those tested had the genotypes that would have been expected to result from occasional mutation of a marker. In _S. coelicolor_, SCP1$^-$ SCP2$^-$ strains do yield a low level of true recombinants ($\sim 10^{-8}$; Ref. 32) but, since these strains harbor the integrated form of SLP1, which certainly promotes fertility in its autonomous form in _S. lividans_ (31,32), such recombinants could be plasmid-dependent. A wide variety of plasmids stimulate fertility, and to levels characteristic of each plasmid, the highest values reaching several percent (22). Moreover, some pairs of strains are less fertile when the plasmid is present in both strains than when it is present in only one (Tab. 3). These observations indicate that plasmids are involved in the transfer and/or recombination of chromosomal DNA. [The high frequency of recombination in protoplast fusions between plasmid-free strains (40) might argue against a role for plasmids in recombination itself, but it may be that protoplast formation or fusion disrupts a normal compartmentation of chromosomes which limits recombination.] However, a minority of plasmids, currently represented by SLP3 and SLP4 (31), does not promote the transfer of chromosomal DNA. Thus, chromosome transfer is not an inevitable consequence of plasmid transfer. Specific plasmid functions leading to fertility should therefore exist, but have not yet been demonstrated. In the case of SCP1, chromosome transfer could result from covalent association between plasmid and chromosomal DNA in various donor states, which presumably arise via recombination within special DNA sequences. This seems unlikely for SCP2. For other plasmids, no attempts to identify such donor states have been reported.

The analysis of genotypes within pocks, and the existence of the Ltz$^{+/-}$ class of plasmid variants (which are still fully transferable and promote normal high levels of fertility in mass matings), points to the existence of a "spreading" ability possessed by pock-forming conjugative plasmids. The mechanism of such spreading is unknown, but it has been suggested (22) that it might occur via intramycelial migration of plasmid DNA copies after primary intermycelial transfer by mating. We have postulated that, after primary transfer between mating individuals, the plasmid migrates away from the pock center, within mycelium exclusively containing chromosomes of recipient genotype. Thus, even if chromosomal DNA were mobilized during such migration, few further recombinants would be generated because the situation would be equivalent to constant "backcrossing" (45). A variant such as pIJ916, showing an abnormally high recombination frequency per transfer, is believed to be unable to spread; therefore, all transfers would be primary events and no such backcrossing would occur. While conduction of chromosomal DNA may be possible via some process not requiring covalent association between plasmid and chromosomal DNA, the high frequency of donor markers that results specifically from the presence of DNA homology between chromosome and plasmid in the SCP2*-primes suggests that this can be an important factor in the process.

The gradients of marker inheritance by recombinants in crosses involving some of the donor states of SCP1, and the artificial SCP2*-primes, might result from backcrossing. Thus, even if complete chromosomes were transferred via the mating reaction, continued conduction of chromosomal DNA during plasmid migration, and recombination with recipient chromosomes en route, would progressively dilute donor markers; and the more so, the looser their linkage with the point of association of donor DNA with the plasmid responsible for its conduction. The steepness of the gradient for a given set of markers would depend on the extent of migration compared with the frequency of primary plasmid transfer. The absence of a gradient in protoplast fusions between NF and SCP1 strains (except for markers close to the SCP1 integration site) would follow from the bringing together of statistically equal numbers of entire genomes of each parent into each fusion body; therefore, there would be no "backcrossing."

The donor state is inherited by almost all progeny of crosses of highly fertile strains carrying SCP1 variants with plasmid-free strains -- and also of fusions of NF with SCP1$^-$ strains -- whereas, crosses and fusions with SCP1$^+$ strains lead to a segregation of donor and SCP1$^+$ types. We have no unique explanation for this finding. However, it may result in part from the efficient spreading of plasmid copies, including those integrated into the chromosome, through the substrate mycelium, combined, perhaps, with an efficient mechanism that tends to ensure the inclusion of a plasmid copy in every spore. Possibly the development of mycelium and/or spores is inhibited in regions in which too few copies of SCP1 exist to allow their inheritance by every cellular compartment or spore, a phenomenon that would be analogous to the coupling of host-cell division to plasmid proliferation exhibited by F (46). When both variant and normal forms of SCP1 are present together, a segregation of plasmid copies would result in the observed inheritance of either plasmid form.

There remains the problem of explaining the contrast between the bidirectional gradients of marker inheritance seen in crosses involving the SCP2*-primes and some of the SCP1 donors, and the unidirectional gradients shown by other SCP1 donors. The bidirectional donation of markers could be explained by a unidirectional single-stranded DNA transfer mechanism similar to that revealed for Hfr forms in _E. coli_, assuming selection for the integrated plasmid and transfer of a DNA strand longer than the entire chromosome. This explanation can be extended to the unidirectional donor types if a hot spot for recombination is postulated to exist close to one side of the integrated plasmid. An attractive, alternative model for the bidirectional donation of markers involves the possible transfer of _Streptomyces_ plasmids as double-stranded circular molecules. Transfer of a plasmid integrated into the chromosome would then result in the passive conduction of the whole chromosome. Repeated rounds of recombination and plasmid spread would result in recombinants where donor alleles predominate at loci close to the site of plasmid integration, and recipient alleles would predominate at loci that are distant from the site of plasmid integration. Again, a hot spot for recombination close to one end of the integrated plasmid would explain the unidirectional donation of markers. At present, it is impossible to choose between these alternatives but it is to be hoped that future work, largely made possible by the improved techniques now available for analysis and manipulation of _Streptomyces_ DNA, will shortly throw more light on the biology of the interesting conjugative plasmids of this group of bacteria.

ACKNOWLEDGMENTS

We are grateful to Keith Chater and Tobias Kieser for helpful comments on the manuscript.

REFERENCES

1. Bradley, D.E. (1983) Morphological and serological relationships of conjugative pili. _Plasmid_ 4:155-169.
2. Willets, N., and B. Wilkins (1984) Processing of plasmid DNA during bacterial conjugation. _Microbiol. Rev._ 48:24-41.
3. Holloway, B.W. (1979) Plasmids that mobilize bacterial chromosome. _Plasmid_ 2:1-19.
4. Deonier, R.C., and R.G. Hadley (1980) IS2-IS2 and IS3-IS3 relative recombination frequencies in F integration. _Plasmid_ 3:48-64.
5. Guyer, M.S., R.R. Reed, J.A. Steitz, and K.B. Low (1980) Identification of a sex factor-affinity site in _E. coli_ as γδ. _Cold Spring Harbor Symp. Quant. Biol._ 45:135-140.

6. Krishnapillai, V., P. Royle, and J. Lehrer (1981) Insertions of the transposon Tn1 into the *Pseudomonas aeruginosa* chromosome. *Genetics* 97:495-511.
7. Hopwood, D.A., D.J. Lydiate, T. Kieser, and M.J. Bibb (1984) *Streptomyces* plasmids: Their biology and use as cloning vectors. In *Antibiotic Producing Streptomyces: A Treatise on Structure and Function*, L.E. Day and S.W. Queener, eds. Academic Press, New York (in press).
8. Kieser, T. (1984) Studies on the isolation of CCC DNA from *Streptomyces lividans* and *Escherichia coli*. *Plasmid* (in press).
9. Bibb, M.J., J.M. Ward, and D.A. Hopwood (1978) Transformation of plasmid DNA into *Streptomyces* at high frequency. *Nature* 274:398-400.
10. Bibb, M.J., K.F. Chater, and D.A. Hopwood (1983) Developments in *Streptomyces* cloning. In *Experimental Manipulation of Gene Expression*, M. Inouye, ed. Academic Press, New York and London, pp. 53-82.
11. Vivian, A. (1971) Genetic control of fertility in *Streptomyces coelicolor* A3(2): Plasmid involvement in the interconversion of UF and IF strains. *J. Gen. Microbiol.* 69:353-364.
12. Wright, L.F., and D.A. Hopwood (1976) Identification of the antibiotic determined by the SCP1 plasmid of *Streptomyces coelicolor* A3(2). *J. Gen. Microbiol.* 95:96-106.
13. Hopwood, D.A., and H.M. Wright (1976) Interactions of the plasmid SCP1 with the chromosome of *Streptomyces coelicolor* A3(2). In *Proceedings of the Second International Symposium on Genetics of Industrial Microorganisms*, K.D. MacDonald, ed. Academic Press, Lond., pp. 607-619.
14. Bibb, M., J.L. Schottel, and S.N. Cohen (1980) A DNA cloning system for interspecies gene transfer in antibiotic-producing *Streptomyces*. *Nature* 284:526-531.
15. Chater, K.F., and C.J. Bruton (1983) Mutational cloning in *Streptomyces* and the isolation of antibiotic production genes. *Gene* 26:67-78.
16. Hopwood, D.A., M.J. Bibb, J.M. Ward, and J. Westpheling (1979) Plasmids in *Streptomyces coelicolor* and related species. In *Plasmids of Medical, Environmental and Commercial Importance*, K.N. Timmis and A. Pühler, eds. Elsevier/North Holland, Amsterdam, pp. 245-258.
17. Schrempf, H., H. Bujard, D.A. Hopwood, and W. Goebel (1975) Isolation of covalently closed circular deoxyribonucleic acid from *Streptomyces coelicolor*. *J. Bact.* 121:416-421.
18. Bibb, M.J., R.F. Freeman, and D.A. Hopwood (1977) Physical and genetical characterization of a second sex factor, SCP2, for *Streptomyces coelicolor* A3(2). *Mol. Gen. Genet.* 154:155-166.
19. Lydiate, D.J. (1984) Molecular genetics of the *Streptomyces* plasmid SCP2. Ph.D. Thesis, University of East Anglia, Norwich.

20. Malpartida, F., and D.A. Hopwood (1984) Molecular cloning of the whole biosynthetic pathway of a Streptomyces antibiotic and its expression in a heterologous host. Nature 309:462-464.
21. Hayakawa, T., T. Tanaka, K. Sakaguchi, N. Otake, and H. Yonehara (1979) A linear plasmid-like DNA in Streptomyces sp. producing lankacidin group antibiotics. J. Gen. Appl. Microbiol. 25:255260.
22. Kieser, T., D.A. Hopwood, H.M. Wright, and C.J. Thompson (1982) pIJ101, a multi-copy broad host-range Streptomyces plasmid: Functional analysis and development of DNA cloning vectors. Mol. Gen. Genet. 185:223-238.
23. Jones, M.D., J.A. Mabe, W.M. Nakatsukasa, and J.T. Fayerman (1982) pFJ265, a new cloning vehicle for Streptomyces. Plasmid 11:92-95.
24. Hopwood, D.A., G. Hintermann, T. Kieser, and H.M. Wright (1984) Integrated DNA sequences in three streptomycetes form related autonomous plasmids after transfer to Streptomyces lividans. Plasmid 11:1-16.
25. Bibb, M.J., J.M. Ward, T. Kieser, S.N. Cohen, and D.A. Hopwood (1981) Excision of chromosomal DNA sequences from Streptomyces coelicolor forms a novel family of plasmids detectable in Streptomyces lividans. Mol. Gen. Genet. 184:230-240.
26. Willetts, N., and R. Skurray (1980) The conjugation system of F-like plasmids. Ann. Rev. Genet. 14:41-76.
27. Moore, R.J., and V. Krishnapillai (1982) Tn7 and Tn501 insertions into Pseudomonas aeruginosa plasmid R91-5: Mapping of two transfer regions. J. Bact. 149:276-283.
28. Moore, R.J., and V. Krishnapillai (1982) Physical and genetic analysis of deletion mutants of plasmid R91-5 and the cloning of transfer genes in Pseudomonas aeruginosa. J. Bact. 149:284-293.
29. Brown, A.M.C., and N.S. Willetts (1981) A physical and genetic map of the IncN plasmid R46. Plasmid 5:188-201.
30. Thatte, V., and V.N. Iyer (1983) Cloning of a plasmid region specifying the N transfer system of bacterial conjugation in E. coli. Gene 21:227-236.
31. Hopwood, D.A., T. Kieser, H.M. Wright, and M.J. Bibb (1983) Plasmids, recombination and chromosome mapping in Streptomyces lividans 66. J. Gen. Microbiol. 129:2257-2269.
32. Bibb, M.J. and D.A. Hopwood (1981) Genetic studies of the fertility plasmid SCP2 and its SCP2* variants in Streptomyces coelicolor A3(2). J. Gen. Microbiol. 126:427-442.
33. Friend, E.J., M. Warren, and D.A. Hopwood (1978) Genetic evidence for a plasmid controlling fertility in an industrial strain of Streptomyces rimosus. J. Gen. Microbiol. 106:201-206.
34. Kirby, R. (1976) Genetic studies on Streptomyces coelicolor plasmid one. Ph.D. Thesis, University of East Anglia, Norwich.
35. Hopwood, D.A., K.F. Chater, J.E. Dowding, and A. Vivian (1973) Advances in Streptomyces coelicolor genetics. Bact. Revs. 37:371-405.

36. Curtiss, R., and D.R. Stallions (1969) Probability of F integration and frequency of stable Hfr donors in F+ populations of E. coli K-12. Genetics 63:27-38.
37. Hodgson, D.A., and K.F. Chater (1981) A chromosomal locus controlling extracellular agarase production by Streptomyces coelicolor A3(2), and its inactivation by chromosomal integration of plasmid SCP1. J. Gen. Microbiol. 124:339-348.
38. Vivian, A., and D.A. Hopwood (1973) Genetic control of fertility in Streptomyces coelicolor A3(2): New kinds of donor strains. J. Gen. Microbiol. 76:147-162.
39. Hopwood, D.A., and H.M. Wright (1976) Genetic studies on SCP1-prime strains of Streptomyces coelicolor A3(2). J. Gen. Microbiol. 95:107-120.
40. Hopwood, D.A., and H.M. Wright (1978) Bacterial protoplast fusion: Recombination in fused protoplasts of Streptomyces coelicolor. Molec. Gen. Genet. 162:307-317.
41. Hopwood, D.A. (1983) Genetic studies with bacterial protoplasts. Ann. Rev. Microbiol. 35:237-272.
42. Kirby, R., and D.A. Hopwood (1977) Genetic determination of methylenomycin synthesis by the SCP1 plasmid of Streptomyces coelicolor A3(2). J. Gen. Microbiol. 98:239-252.
43. Rudd, B.A.M., and D.A. Hopwood (1979) Genetics of actinorhodin biosynthesis by Streptomyces coelicolor A3(2). J. Gen. Microbiol. 114:35-43.
44. Skurray, R.A., and P. Reeves (1973) Characterization of lethal zygosis associated with conjugation in Escherichia coli K-12. J. Bact. 113:58-70.
45. Chater, K.F. (1984) Morphological and physiological differentiation in Streptomyces. In Microbial Development, R. Losick and L. Shapiro, eds. Cold Spring Harbor Press (in press).
46. Ogura, T., and S. Hiraga (1983) Mini-F plasmid genes that couple host cell divison to plasmid proliferation. Proc. Natl. Acad. Sci., USA 80:4784-4788.
47. Murakami, T., C. Nojiri, H. Toyama, E. Hayashi, Y. Yamada, and K. Nagaoka (1983) Pock forming plasmids from antibiotic-producing Streptomyces. J. Antibiot. 36:429-434.
48. Bailey, C.J., M.J. Butler, J.E. Harris, C.J. Bruton, H.M. Wright, D.A. Hopwood, and K.F. Chater (1984) (manuscript in preparation).
49. Gibbons, P., and T. MacNeil (1983) Isolation and characterization of a plasmid from Streptomyces venezuelae useful as a Streptomyces cloning vector. Abstracts, Ann. Meeting of A.S.M., Washington, D.C.
50. Larson, J.L., and C.L. Hershberger (1984) Shuttle vectors for cloning recombinant DNA in Escherichia coli and Streptomyces griseofuscus C581. J. Bact. 157:314-317.
51. Richardson, M.A., J.A. Mabe, N.E. Beerman, W.M. Nakatsukasa, and J.T. Fayerman (1982) Development of cloning vehicles from the Streptomyces plasmid pFJ103. Gene 20:451-457.

52. Nabeshima, S., Y. Hotta, and M. Okanishi (1984) Construction of plasmid vectors from *Streptomyces kasugaensis* plasmids, pSK1 and pSK2. *J. Antibiot.* (submitted for publication).
53. Hopwood, D.A., R.J. Harold, A. Vivian, and H.M. Ferguson (1969) A new kind of fertility variant in *Streptomyces coelicolor*. *Genetics* 62:461-477.

THE STRUCTURE AND SOURCE OF PLASMID DNA DETERMINE THE CLONING PROPERTIES OF VECTORS FOR *BACILLUS SUBTILIS*

P.M. Rabinovich, M.Ya. Haykinson, L.S. Arutyunova, Yu.V. Yomantas, and A.I. Stepanov

Institute of Genetics and Selection of Industrial Microorganisms
Moscow 113545, U.S.S.R.

INTRODUCTION

The elaboration of a system for cloning in bacilli began with a search for plasmids capable of replicating in *Bacillus subtilis* cells. Various strains of bacilli often carry such plasmids, but most of them are either cryptic or their markers are not selective enough (for review, see Ref. 1,2). Although these plasmids are usable as vectors (2), major progress in gene engineering with bacilli is associated with the use of vectors originating from a phylogenetically different group of organisms. It was found that staphylococcal plasmids can transform *B. subtilis* cells and render them drug-resistant (3). The plasmids are maintained in a large number of copies in *B. subtilis* and carry some unique restriction sites suitable for cloning (4,5,6). Several genes were successfully cloned on staphylococcal plasmids as vectors, although as experimentation was extended, it was found that the efficiency of cloning in *B. subtilis* is much lower than in the traditional *Escherichia coli* system (4,7,8). Part of the reason for that has become clear after it was demonstrated that *B. subtilis*, unlike *E. coli*, cannot be transformed with plasmid monomers which constitute the bulk of plasmid populations isolated from cells or produced in routine in vitro ligation of DNA fragments. Accordingly, staphylococcal plasmids were found to transform *B. subtilis* only as oligomers (9,10), although both oligomers and monomers penetrate into the cell equally well. The plasmid enters the *B. subtilis* cell as a linear, single-stranded molecule (11), and the possibility of transformation is largely dependent on whether recycling of the strand can occur. The latter, in turn, is believed to depend on the presence of genome redundancy in the plasmid. Thus, the transforming efficiency of plasmid oligomers is associated with the existence of repeated sequences in them needed for the closure of the penetrated strand (11,12). This implies that transformation of B. subtilis with recombinant plasmids obtained by in vitro ligation could be accomplished through the use of special methods yielding oligomers of these plasmids. Success along these lines came with the creation of shuttle vectors capable of propagation both in B. subtilis and in E. coli (4,13,14). Recombinant plasmids based on these vectors are first accumulated in E. coli, and the DNA preparation obtained already containing plasmid oligomers, is used to introduce the plasmids into B. subtilis. The employment of E. coli as an intermediate host has greatly extended the potential of gene engineering in bacilli with regard to cloning individual DNA fragments (15,16,17,18) and creating a library of genes (19,20).

Another line of research was concerned with devising methods for transformation of bacilli with monomers of recombinant plasmids. One such method is "marker rescue", which involves recombination of a monomeric plasmid with a homologous region of a resident plasmid (21,22). It was found that the rescue could also occur on a homologous region of the host chromosome (23,24). Plasmid monomers lacking a region homologous to the recipient DNA can also be transformationally active if they carry direct repeats (25). Yet another approach to introducing plasmid monomers into bacilli consists in using PEG-dependent transformation of protoplasts (26). The development of the various techniques for cloning in B. subtilis has made this system more competitive with regard to the E. coli system, and it is now widely used for cloning homologous and heterologous genes (for review, see Ref. 27,28).

Some problems remain, however, and one of them is cloning of large (over 3 Mdal) fragments on staphylococcal plasmids or their derivatives. Cloned fragments in gene libraries obtained from B. subtilis are usually smaller than 3 Mdal (8,17,29), and when a mixture of fragments is subject to cloning, the heavier ones become underrepresented (30). We also experienced a problem when we attempted to propagate in B. subtilis a riboflavin operon within a 9.5-kb (6.3-Mdal) EcoRI fragment. A search for ways to overcome this problem has revealed some peculiarities of B. subtilis transformation with plasmid DNA, and vectors have eventually been constructed that offer effective cloning of relatively large DNA fragments.

Section 1. Characteristics of the Rib Fragment

Genes of riboflavin biosynthesis in B. subtilis are organized into an operon located between lys and ser markers (210°) on the chromosome map (31). The operon is repressed by the product of the regulator gene situated outside of this linkage group (32). The operon was found to contain a ribO operator locus whose mutations may cause derepression of the operon and overproduction of riboflavin (32,33). As a source of the rib operon we used 2 B. subtilis strains: SHgw with the wild-type operon($ribO^+$) and CHP335 with a derepressed rib operon ($ribO335 = ribO^c$) (Tab. 1).

Both chromosomal DNAs were treated with EcoRI, and the fragments obtained were cloned in E. coli $r^-_k m^+_k$ on the RSF2124 and pBR322 vectors. The resulting plasmids were designated pPR2 (RSF2124-Rib(O^c)), pLP95 (pBR322-Rib(O^c)) and pLP102 (pBR322-Rib (O^+)) (35,36). The plasmid pLP102 carries the wild-type operon and pPR2 and pLP95 carry the derepressed operon (Tab. 2). The Rib fragments in all the plasmids have the same size, 9.5 kb, and produce a similar restriction pattern (36). The rib genes of B. subtilis within all 3 plasmids are expressed in E. coli and complement the appropriate auxotrophic mutations in this host (17,39).

Section 2. Cloning of Rib Fragments with the Aid of the pUB110 Replicon

The plasmid pUB110 (4.5 kb), originating from Staphylococcus aureus (5), proved to be a good vector for B. subtilis in a number of cases (7,12,40). It codes for kanamycin resistance and has unique EcoRI and BamI sites suitable for cloning.

We made several attempts to propagate the Rib fragment on pUB110 in shotgun experiments, as well as to reclone it from plasmid pPR2. The chromosomal DNA of B. subtilis or the pPR2 DNA were treated with EcoRI and ligated with pUB110 linearized with the same enzyme. The ligation product was used for transformation of B. subtilis ribD107recE4 cells (Tab. 1). The recE4 mutation served to

Tab. 1. Bacteria.

Strain	Genotype	Source
Bacillus subtilis		
SHgw ($ribO^+$)	prototroph	T.P.Chernik
CHP335 ($ribO^c$)	ribO335	- " -
107 (rec^+)	ribD107	- " -
PG260 (recE4)	ribD107 recE4	(18)
Escherichia coli		
BSV82 (rib82)	thi rib82 strA hsdR $hsdM^+$	(34)
BSV821 (rib82 recA)	the same as BSV82 and recA srl::Tn10	(38)

Tab. 2. Plasmids.

Plasmid	Phenotype	Size (kb)	Source
pBR322	Ap Tc	4.3	G.Boyer
pPR2	Ap Rib(0^c)	20.3	(35)
pLP95	Ap Tc Rib (0^c)	13.8	(36)
pLP102	Ap Tc Rib(0^+)	13.8	- " -
pUB110	Km	4.5	D.Dubnau
pJJ1	Ap Km	8.7	(14)
pMX1	Ap Km Rib(0^+)	22.2	(37)
pMX9	Ap Km Rib(0^+)	18.3	- " -
pMX10	Km	4.1	- " -
pMX30	Em	18.3	- " -
pMX33	Em Rib(0^+)	27.8	(38)
pMX49	Em Rib(0^c)	27.8	This study
pMX39	Ap Em	22.8	(38)
pMX121	Ap Em	6.1	This study
pPIE10	Ap Em	9.5	A.S.Boitsov
pLA12	Ap Em	12.8	This study
pLA32	Ap Em	8.8	- " -
pLA110	Ap Em	9.1	- " -
pLA1718	Ap Em	9.4	- " -
pLA44	Ap Em Rib(0^c)	18.6	- " -

prevent recombination between homologous regions in the plasmid and chromosome. The transformants selected by the $Km^r Rib^+$ phenotype were used to isolate plasmid DNA which was tested for presence of the recombinant plasmids. No such plasmids were found, unfortunately, but the transformants, instead, contained small plasmids, similar in size to pUB110 and carrying only the Km^r marker. The Rib^+ phenotype of the transformants at first seemed mysterious to us. However, the evidence to be presented below allows the conclusion that this phenotype is due to a pUB110-Rib plasmid present in very few copies in the selected population and thus undetectable by the methods employed in the study.

In the next series of experiments, we attempted to clone the Rib fragment using *E. coli* as an intermediate host. We began with construction of a few plasmids each of which contained pUB110 DNA and the Rib fragment within the RSF2124 or pBR322 replicon. Neither of these plasmids was able to transform B. subtilis (17,37,38). Then we decided to expand the search for recombinant plasmid configurations in E. coli. The scheme of the experiment is depicted in Fig. 1.

The shuttle vector pJJ1 (pBR322-pUB110) (14) was chosen as the parental plasmid. It determines Ap^rKm^r properties in E. coli and the Km^r phenotype in B. subtilis and is cleaved by EcoRI into 2 fragments: fragment A (4.7 kb) which contains mainly the pBR322 DNA, and fragment B (4.1 kb) which contains mainly the pUB110 DNA. The A and B fragments are genetically marked by the Ap^r and Km^r determinants, respectively. The Rib fragment obtained by preparative fractionation of EcoRl digest of plasmid pLP102 was added to the mixture of A and B fragments (Fig. 1).

The mixture of A, B, and Rib was ligated and used to transform E. coli rib82 cells, with subsequent selection for the Ap^r, Km^r and Rib^+ characters. Plasmids from 11 independent E. coli transformants were checked for the ability to transform B. subtilis ribD107recE4 cells and render them Km^rRib^+. As Tab. 3 shows, 10 plasmid preparations out of the 11 failed to do so, but one (no. 7) did show transformational activity. Electrophoretic and restriction analyses of the plasmid preparations showed that all of them contained an 18.2-kb plasmid, while preparation no. 7 contained an additional 22.2-kb plasmid. Each of the 18.2-kb plasmids was composed of the 3 EcoRI fragments employed (A, B and Rib), their mutual orientation

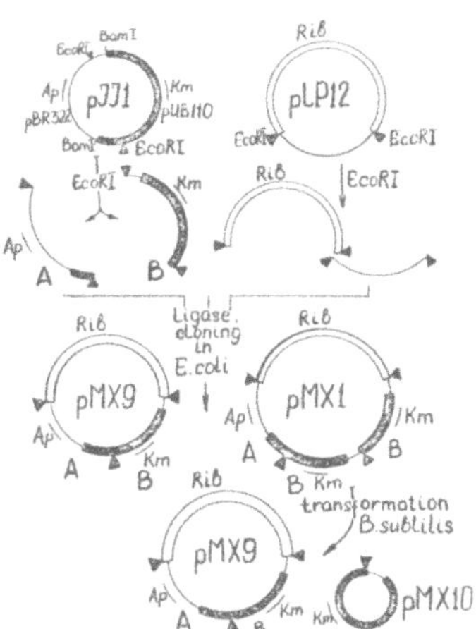

Fig. 1. Construction of plasmid pMX1 containing direct DNA repeats. (————) - pBR322, (▬▬▬) - pUB110, (▭) - Rib, (▼) - EcoRI sites; A, B - EcoRI fragments of pJJ1 plasmid.

Tab. 3. Transforming activity of plasmids containing EcoRI fragments A and B of plasmid pJJ1 and the Rib fragment.

Recipient	Selective markers	Number of transformants per μg DNA	
		Preparation no.1[a]	Preparation no.7
E. coli BSV82 (rib82)	Ap^rRib^+	$7.2\cdot10^4$	$9.7\cdot10^4$
B. subtilis PG260 (ribD107 recE4)	Km^rRib^+	0	$2.6\cdot10^3$

Note: a) Plasmid preparations no.2 through no.6 and no.8 through no.11 transform E. coli and B. subtilis similarly to preparation no.1

varying as could be expected, from the random ligation of the fragments in vitro. In order to analyze the heterogeneous preparation No. 7, we separated it into the heavier and lighter plasmids by means of transformation and cloning in E. coli rib82recA. The individual plasmids thus obtained were designated pMX1 and pMX9, respectively. Preparations containing pMX1, pMX9, or a mixture thereof, as well as the initial preparation no. 7, were all used to transform B. subtilis ribD107recE4 cells. It was found, as a result, that the transforming activity of preparation no. 7 is due entirely to plasmid pMX1 (Tab. 4). Schemes of pMX1 and pMX9 are shown in Fig. 1. The only difference between them is a duplication of segment B in pMX1, which suggests an essential role of this duplication for the transformation of B. subtilis.

Section 3. The Fate of pMX1 in B. subtilis Transformants

To our surprise, the plasmid DNA isolated from Km^rRib^+ clones of B. subtilis transformed with pMX1 produced only 1 electrophoretic

Tab. 4. Transforming activity of plasmids pMX1 and pMX9.

Plasmid	Number of Km^rRib^+ transformants per μg DNA in B. subtilis PG260 (ribD107 recE4)
pMX1	$3.1\cdot10^3$
pMX1 + pMX9	$2.1\cdot10^3$
Preparation no.7	$2.2\cdot10^3$
pMX9	0
pMX1 (monomer)	$2.2\cdot10^3$

band corresponding to a 4.1-kb plasmid. Restriction analysis suggested that this plasmid, designated pMX10, is a circular form of the B segment of pMX1 (Fig. 1). It could be assumed that pMX10, which carried the Km^r determinant, appeared in the cell as a result of intramolecular recombination of pMX1. The other product of the recombination then must be identical to pMX9 which carried the other selective marker, Rib^+. The absence of the appropriate heavy band on the electrophoregram could be due to a low copy number of pMX9 in bacilli. In order to demonstrate its presence in the *B. subtilis* transformants, we recloned the plasmid in *E. coli*. This was done by transforming *E. coli rib82recA* cells with the DNA preparation isolated from *B. subtilis* transformants, followed by selection for $Ap^r Km^r Rib^+$ determinants. All the 3 markers were transferred together, and the plasmid isolated from the *E. coli* transformants was shown by restriction analysis to be identical with pMX9. The pMX10 DNA, which was the major component of the transforming mixture, could not be maintained in *E. coli* as it lacked the pBR322 replicator.

Therefore, the plasmid pMX1 which is able to transform bacilli owing to the duplication of the B segment is dissociated in these cells into pMX9 and the circular form of the B segment, pMX10.

Then we isolated a monomer of pMX1 by means of preparative gel electrophoresis. The monomer was as effective for transformation of *B. subtilis ribD107 recE4* by $Km^r Rib^+$ markers as the unfractionated preparation (Tab. 5). In both cases the transformants contained pMX10 and pMX9, the latter in a small quantity detectable by recloning in *E. coli*. Thus, the presence of a repeat in pMX1 ensures transforming activity of the plasmid monomer and allows us to introduce a 9.5-kb *Rib* fragment into *B. subtilis* cells. Restriction analysis has indicated that the duplication of the B segment in pMX1 is a direct repeat. This is in agreement with the findings of Michel and coworkers (25) who studied the role of direct repeats in *B. subtilis* transformation and showed that they conferred transforming activity on plasmid monomers.

Tab. 5. Effectiveness of *B. subtilis* transformation by pMX30 plasmid DNA in the unfractionated, monomer and linear forms.

Form of plasmid DNA	Number of Em^r transformants per μg DNA	
	B. subtilis 107 (rec^+)	*B. subtilis* EG260(*recE4*)
pMX30 unfractionated	$1.2 \cdot 10^3$	$1.5 \cdot 10^4$
pMX30 monomer	$1.2 \cdot 10^3$	$2.9 \cdot 10^3$
pMX30 linearized with *EcoRI*	0	0

It should be noted that no pMX9 could be visualized in the DNA preparations isolated from B. subtilis transformants. In our opinion, this is due to the fact that the pUB110 vector fails to provide a sufficient copy number of the plasmid in B. subtilis. This assumption may also explain the failure to clone the Rib fragment on this vector. Indeed, the inability of pMX9 and the like plasmids to transform bacilli could well result from their low copy number and thus, the inability to sustain the Km^r phenotype, one of the traits used for transformant selection. The level of kanamycin resistance determined by pUB110 is apparently proportional to the plasmid copy number. As to the Rib^+ trait, it is expressed regardless of rib gene dosage since bacilli cells need but very little riboflavin for their growth. The possibility of selecting pMX1-mediated transformants is due to segregation of a high copy-number plasmid pMX10 which, together with a low copy-number plasmid pMX9, ensures selection of the Km^rRib^+ phenotype.

This segregation of pMX1 induced us to look for vectors able to transform bacilli in the monomeric form and retaining their initial molecular structure at the same time. Studies on the mechanism of B. subtilis transformation with plasmid DNAs (41) led us to assume that plasmids containing long enough inverted repeats would meet this requirement. No such plasmids were found in the above experiments, and we therefore decided to investigate transforming properties of a streptococcal plasmid pSM19035 which was known to carry long inverted repeats.

Section 4. Isolation and Properties of Plasmid pMX30 Containing Inverted Repeats

The plasmid pSM19035 has the size of 27 kb and carries an inducible determinant of erythromycin resistance, Em^r (42). A distinguishing feature of the plasmid are long inverted repeats comprising 80% of the contour length (43). It appears to be identical to another streptococcal plasmid, pSM22095, which is able to transform B. subtilis cells and make them erythromycin-resistant (42,44). We isolated pSM19035 from Streptococcus sanguis (Challis) and tested it in transformation of a B. subtilis recE4 strain. The Em^r transformants obtained had a very low content of pSM19035. We succeeded in isolating a spontaneous deletion derivative of pSM19035, designated pMX30, which had the size of 18.3 kb and was present at 3 copies per genome of bacilli. Based on the restriction map of pSM19035 (42, 45) and on a restriction analysis of pMX30, we constructed a restriction map of pMX30 which is shown in Fig. 2. Electron microscopy indicated that pMX30 retained a portion of the original inverted repeat of pSM19035 which made up about 35% of the pMX30 contour length (Fig. 2).

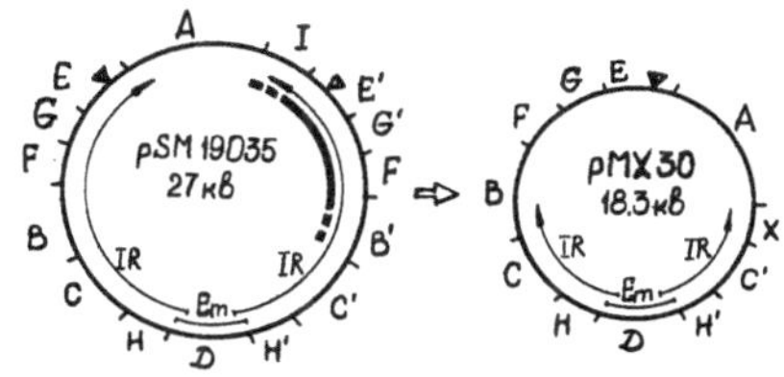

Fig. 2. Structural features of plasmid pMX30. Large letters - HindIII fragments (42,45), (▼)- EcoRI sites. (▬) - deletion, (——►IR)-inverted repeats.

Tab. 6. Characterization of vector molecules for B. subtilis.

Vector	Insert	Recombinant plasmid	Copy number[a)]	Plasmid stability(%)[b)] 30°	37°	42°
pMX30 18.3 kb	-		3	100	100	100
	Rib(O$^+$) 9.5 kb	pMX33 27.8 kb	3	100	100	100
	Rib(O^c) 9.5 kb	pMX49 27.8 kb	3	100	100	100
	pBR322 4.3 kb	pMX39 22.8 kb	3	100	100	100
pLA110 9.1 kb	-		35	100	90	<1
	Rib(O^c) 9.5 kb	pLA44 18.6 kb	15	50	30	<1
pLA32 8.8 kb	-		20	80	60	<1
pUB110 4.5 kb	-		40	100	100	100
	Rib(O$^+$)-pBR322 13.8 kb	pMX9	<1	<1	<1	<1

Notes: a) The copy number was measured as desribed (48). The cells were grown at 37°C.

b) The plasmid stability was determined as described elsewhere (49). The figures correspond to the percentage of clones retaining the plasmid after cultivation for 20 generations under non-selective conditions.

We isolated the pMX30 monomer from the agarose gel and tested its transforming activity with *rec*$^{+}$ and *rec*$^{-}$ *B. subtilis* strains. As is evident from Tab. 6, the monomer of pMX30 transforms *B. subtilis* *rec*$^{+}$ as effectively as does the unfractionated plasmid preparation. Linearization of pMX30 at its single *Eco*RI site outside of the repeat region abolishes the transforming activity completely. The pMX30 monomer also transforms *rec*$^{-}$ *B. subtilis* cells, although 10 times less effectively than the unfractionated preparation (Tab. 6).

We isolated plasmid DNA from *rec*$^{+}$ or *rec*$^{-}$ *B. subtilis* cells transformed with either monomers or unfractionated pMX30 preparations and subjected it to electrophoresis and homoduplex analysis. Within the experimental error of these methods, it can be stated that the structure of the plasmids isolated from any of the cell types is the same and identical to that of the initial donor DNA.

Section 5. Vector Properties of pMX30

The ability of the pMX30 monomer to transform *B. subtilis* cells and to retain its initial structure in the course of transformation makes it a suitable vector for cloning. The pMX30 DNA contains one *Eco*RI restriction site (Fig. 2). The site is situated outside of the replicative region (45) and may be employed for integration of foreign DNA. Our next experiment, whose schematic representation is shown in Fig. 3, was aimed at finding out whether the pMX30 vector could maintain the *Rib* fragment in bacilli (see Sec. 1), hoping that the fragment could serve as a probe for testing the vector proficiency of pMX30.

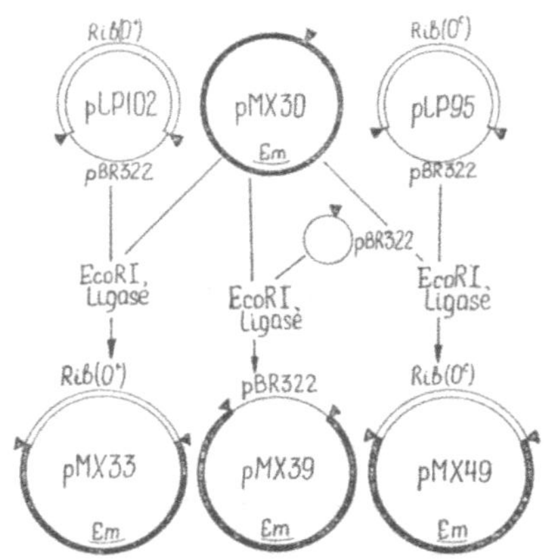

Fig. 3. Hibrid plasmids based on pMX30. (————)-pBR322, (▬▬)-pMX30, (▭)-*Rib*, (▼)- *Eco*RI sites.

The source of the Rib fragment was plasmid pLP102 (pBR322-Rib(0^+)) (see Sec. 1, Tab. 2). The pMX30 and pLP102 DNAs were treated with EcoRI, ligated, and the mixture was used to transform B. subtilis ribD107 recE4. The transformants which were selected by the $Em^r Rib^+$ markers contained 27-kb pMX30-Rib plasmids, in agreement with the sum of the vector and the Rib fragment masses. One of the recombinant plasmids was designated pMX33 (Fig. 3). A restriction analysis of this plasmid indicated that it contained the pMX30 vector and the entire Rib fragment.

In the next experiment we employed pMX30 to introduce a derepressed rib operon into B. subtilis. The design of the experiment was the same as the above, except the pLP95 plasmid (pBR322-Rib(0^c)) (Sec. 1) was taken as the source of the Rib fragment. The pMX30 and pLP95 plasmids were cleaved with EcoRI, the mixture ligated and used to transform bacilli. The $Em^r Rib^+$ transformants displayed 27-kb plasmids, one of which was designated pMX49 (Fig. 3). The restriction pattern produced by pMX49 is indistinguishable from that of pMX33. Phenotypically, however, the 2 plasmids are quite different, since pMX49 determines overproduction of riboflavin in bacilli. Both the plasmids are easily visualized in routine electrophoresis of plasmid DNA, which implies that cloning of the Rib fragment on the pMX30 vector does not lead to a dramatic decrease of the replicon copy number as is the case with the pUB110 vector (Sec. 2,3). In spite of the large insert (9.5 kb), both pMX33 and pMX49 are stably maintained in B. subtilis cells. No loss of the plasmids (< 0.1%) was observed after growing cells harboring them for 20 generations on a medium without erythromycin and supplemented with riboflavin (i.e., without selection for the determinants of the vector and the insert).

Section 6. Construction of the pMX39 Vector on the Basis of Plasmid pMX30

The basic vector properties of pMX30 described above were used in construction of a more convenient shuttle vector. The DNAs of pMX30 and pBR322 were restricted with EcoRI, treated with ligase, and used in the transformation of B. subtilis recE4. Several dozen clones selected by the Em^r phenotype were pooled, and the total DNA isolated from them was used in transformation of E. coli rib82 recA cells, followed by selection for the Ap^r marker of pBR322. The clones obtained carried 22-kb plasmids consisting of pMX30 and pBR322 joined by their EcoRI sites. One of these plasmids was designated pMX39 (Fig. 3). The pBR322 replicon of pMX39 allows the latter to accumulate in an E. coli culture. Experience has shown that preparative isolation of the plasmid from E. coli is more convenient than from B. subtilis, with regard to both yield and purity.

The basic characteristics of the pMX39 vector are as follows: i) unique BamI, PstI, SalGI, and EcoRV sites, and 2 EcoRI sites usable for cloning in bacilli; ii) a long inverted repeat important for the transforming properties of the plasmid; iii) stable maintenance of large DNA fragments within plasmids in bacilli.

These properties ensure usefulness of this vector for creating gene banks and in shotgun experiments. In a model experiment we employed pMX39 for cloning fragments of *B. amyloliquefaciens* DNA in *B. subtilis* cells. The donor DNA was subjected to limited cleavage with *Sau3A* restriction endonuclease (average fragment size 15 kb) and then ligated with the DNA of pMX39 linearized at the *BamI* site. After treatment with ligase, the preparation was used in transformation of *B. subtilis amy⁻ recE4*. The frequency of Em^r transformants was 10^5 per microgram DNA, and Em^r Amy^+ clones appeared among Em^r clones at the frequency of $5 \cdot 10^{-3}$. We isolated plasmid DNA from 10 $Em^r Amy^+$ clones, and the restriction and genetic analyses showed that all of them contained the α-amylase gene of *B. amyloliquefaciens* within inserts ranging from 5 to 15 kb. We also used pMX39 for shotgun cloning of *B. subtilis* α-amylase and lysine biosynthesis genes. All the genes were cloned within 5 to 30 kb inserts.

The experiments described above demonstrated that a treatment of pMX39 with BamI and ligase under conditions favoring the formation of oligomers increased the plasmid-transforming activity by nearly 2 orders of magnitude as compared to the native preparation. This is in accord with our results on the transforming properties of the pMX30 monomer (Sec. 4), which transforms *B. subtilis recE4* cells 10 times worse than does the unfractionated preparation (Tab. 6). Consequently, while long inverted repeats do impart transforming properties to a plasmid, its oligomerization enhances these properties further still. We believe that the presence of inverted repeats in plasmid DNA may determine a specific mechanism of *B. subtilis* transformation with plasmid monomers as well as oligomers.

Section 7. Models of B. subtilis Transformation by Plasmid DNA

Schemes of transformation are depicted in Fig. 4. In regard to plasmids without inverted repeats, the schemes are in accordance with the models proposed by Canosi and coworkers (23), Michel and coworkers (25), and by other investigators (41,46). The process of transformation is arbitrarily divided into 4 stages: linearization of the plasmid molecule, penetration of a single-stranded DNA into the cell, recyclization of this strand, and duplication of the strand. Although linearization may occur at random along the plasmid, for monomer transformation it is essential that it occur within the repeat region. This is shown on the scheme. For oligomer transformation, the site of linearization is unimportant since any DNA sequence in an oligomer is a repeat.

The key stage in the model is recyclization which is accomplished through synapsis between the ends of the penetrated molecule

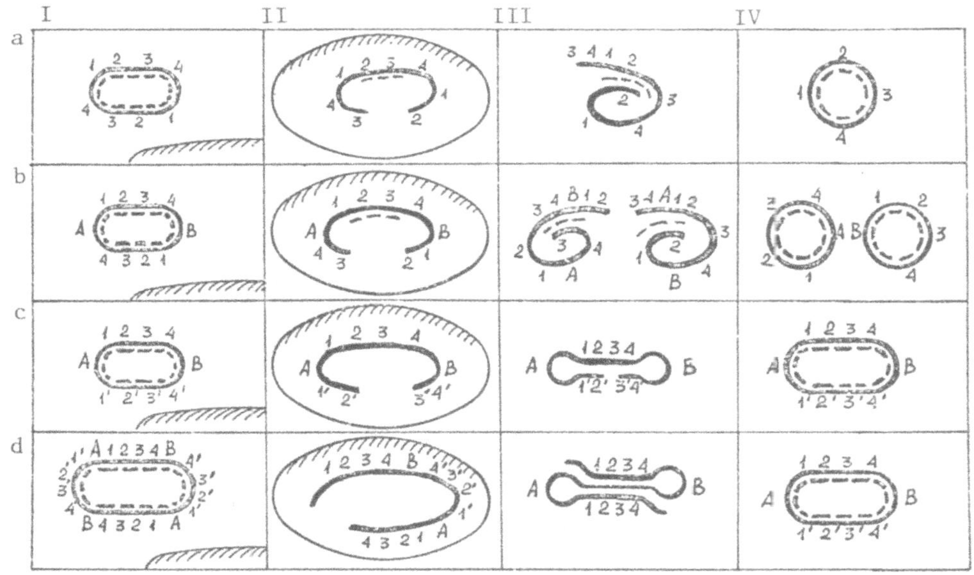

Fig. 4. Models for B. subtilis cells transformation. a) - plasmid as a dimer; b) - plasmid containing a direct repeat: c) - plasmid containing an inverted repeat; d) dimer of the plasmid with an inverted repeat. Transformation stages: I - DNA adsorbtion; II - DNA entry into cell; III-synapsis; IV - completion of transformation. Designation on DNA: repeated regions marked by numbers: 1,2,3,4, and 1',2'3',4'; unique DNA regions marked by letters; A,B.

and its inner site of homology. As can be seen from Fig. 4 (A,B), in transformation by an oligomer or by a monomer with direct repeats, the synapsis region is a triple-stranded structure. Two strands of this structure belong to 1 strand of the plasmid, and the third one, to the complementary strand. Since the plasmid as well as chromosomal DNA possibly penetrate into the cell as a single strand (11), the origin of the third strand remains a matter of speculation. It is hypothesized that small fragments of the complementary strand may "contaminate" the penetrating strand or be synthesized on it in the cell de novo (46).

The synapsis in case of plasmids carrying inverted repeats does not require any complementary DNA strand (Fig. 4:C,D). The synapsis structure may be double- or triple-stranded, and in both cases it is formed by a single DNA strand.

As is implied by the model, oligomers of a plasmid containing long inverted repeats have the best chance of transforming B.

subtilis cells since they can be recycled by either of the 2 mechanisms, i.e., involving direct (Fig. 4,A) or inverted (Fig. 4,D) repeats. As we have discussed in Sec. 6, the ligated mixture containing the pMX39 vector has a high transforming activity indeed.

Section 8. Cloning of the pMX39 Replication Region

Behnke and coworkers (47) studied several deletion derivatives of pSM19035 and established the region involved in the autonomous replication (rep) and the copy-number control (cop) in streptococci. Both the loci are mainly in the HindIII-F fragment of pSM19035. The rep locus also includes a small portion of the HindIII-B fragment while the cop locus slightly extends into the HindIII-G fragment (Fig. 2,5). The deletion of pSM19035 that gave rise to pMX30 touched a part of the inverted repeat, including the duplication of the HindIII-F fragment, and hence in pMX30 and pMX39, the replication region is present as a unique sequence (Fig. 2,5).

In order to isolate the pMX39 replication region we had to have auxiliary plasmids unable to reproduce themselves in bacilli but containing determinants expressible in these cells. One such plasmid was obtained by introducing the Em^r determinant of pMX39 into pBR322. This determinant is within the HindIII-D fragment (47). A mixture of pMX39 and pBR322 was treated with HindIII, ligated, and

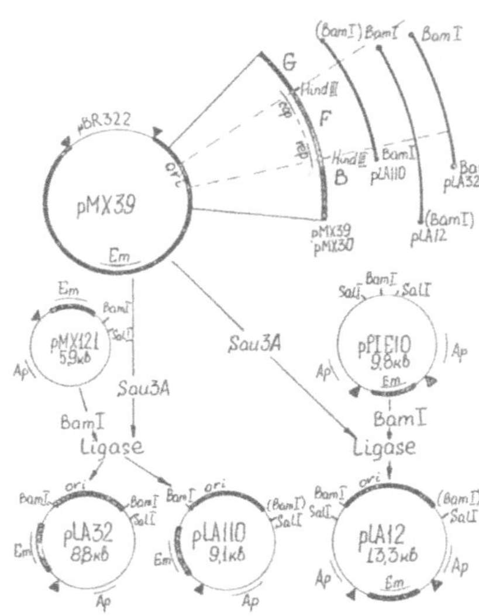

Fig. 5. Plasmids containing replication region of pMX30. (———) - pBR322; (▬▬▬) - pMX30; (▼) - EcoRI sites; The cop and rep determinants are indicated according to D. Behnke and S. Klaus (47).

used to transform E. coli rib82 recA cells. From the Ap^rEm^r transformants obtained, we isolated a 5.5-kb plasmid, pMX121, containing the pBR322 vector and the Em^r fragment joined by their HindIII sites (Fig. 5).

Another auxiliary plasmid, pPIE10, was kindly provided by A. Boitsov. The pPIE10, like pMX121, is a derivative of pBR322 at whose HindIII site the Em^r determinant of pSM22095 has been integrated. As a result of a spontaneous genome rearrangement of the plasmid during propagation in E. coli, the bulk of the pBR322 became duplicated and inverted in pPIE10. Both pPIE10 and pMX121 have unique BamI sites which were used for cloning the replicator region of pMX39.

The pMX39 DNA was subjected to a limited cleavage by Sau3A, and the resulting preparation was divided into 2 portions, to each of which either pPIE10 or pMX121 DNA, linearized at their BamI sites, were added. After ligation, the preparations were used to transform B. subtilis recE4 with selection for the Em^r phenotype. This yielded a collection of plasmids containing the pMX39 replication region within DNA fragments of varied length. Three of these plasmids having the smallest size of the cloned fragment were taken for further investigation. The plasmids were designated pLA12 (a pPIE10 derivative), pLA32 and pLA110 (pMX121 derivatives). The joining of the Sau3A and BamI cohesive ends during formation of these plasmids completely or partially restored the BamI sites at the insert ends. The DNA of pLA32 has 2 BamI sites, and pLA12 and pLA110 each have 1 BamI site (Fig. 5.). The restriction analysis of the 3 plasmids enabled us to measure the length of the "replicator" Sau3A fragments as 3.3, 2.7 and 3.0 kb for pLA12, pLA32 and pLA110, respectively, as well as to find out their coordinates on the pMX39 map. All the 3 fragments turned out to contain the rep locus, and the pLA110 fragment, in addition, contained the cop locus (Fig. 5.)

Section 9. Measuring Plasmid Copy Number

The plasmid copy number was determined as described (48) after growing plasmid-harboring B. subtilis under selective conditions at 37°C to $5 \cdot 10^8$ cells/ml. As follows from Tab. 6, pMX39 is maintained at 3 copies per genome. The reduction of the replicon size that accompanies the change from pMX39 to pLA110 entails a proportional growth of the copy number. The Sau3A fragments of pMX39 are similar to one another, as is suggested by properties of the respective plasmids. Indeed, pLA12 and pLA32 which lack the cop locus (Tab. 6 and Fig. 5) have the same copy number as pLA110 which carries the locus. This may indicate that the cop locus, which controls the copy number of pSM19035 derivatives in streptococci (Behnke et al. 47), has little if any effect on copy number determination in B. subtilis.

Section 10. Transformation of B. subtilis by a pLA12 Monomer

Electron microscopy and restriction data indicate that pLA12 carries an inverted repeat measuring 5 kb and comprising 40% of the plasmid genome (Fig. 5). This repeat is a duplication of the pBR322 segment (see Sec. 8), i.e., has an entirely different origin than the inverted repeat of pMX30. Therefore, it seemed important to find out whether the pLA12 repeat can provide for transforming activity of the plasmid monomer in *B. subtilis*. We isolated the pLA12 monomer by means of preparative electrophoresis and tested its transforming activity with rec^+ and rec^- cells of *B. subtilis*. As Tab. 7 shows, the monomer of pLA12 is an effective transforming agent. Using a double digestion with *BamI* and *SalGI*, we excised the replicator-containing fragment from pLA12 and recloned it on pMX121 (Fig. 6). The resulting plasmid, pLA1718, has the same replicator and the same Em^r determinant as pLA12, but lacks the latter's inverted repeat. As is evident from Tab. 7, the pLA1718 monomer is unable to transform *B. subtilis* cells. This result confirms our hypothesis on the role of inverted repeats in *B. subtilis* transformation (Sec. 7) and demonstrates that the transformation mechanism is essentially independent of the origin of the repeat.

Section 11. Vector Properties of pLA110

The plasmid pLA110 is a copy-number derivative of pMX39 and has some unique restriction endonuclease sites suitable for cloning (Fig. 5). We were interested in whether this plasmid retained the replicative property of its parent pMX30, namely, the ability to sustain replication of large DNA fragments. As a probe to check this, we have chosen the *Rib* fragment of pLP95 (see Sec. 1). The cloning scheme was identical to the one discussed in Sec. 5 (Fig. 3). The DNA of pLP95 was digested with *EcoRI* and joined with the pLA110 vector linearized with the same enzyme. After ligation, the preparation was used to transform *B. subtilis ribD107 recE4* with

Tab. 7. Effectiveness of *B. subtilis* transformation by unfractionated and monomer forms of pLA12 and pLA1718 plasmid DNAs.

Form of plasmid DNA	Number of Em^r transformants in *B. subtilis* PG260 (*recE4*) per μg DNA
pLA12 unfractionated	$9 \cdot 10^3$
pLA12 monomer	$2 \cdot 10^3$
pLA1718 unfractionated	$5 \cdot 10^3$
pLA1718 monomer	0

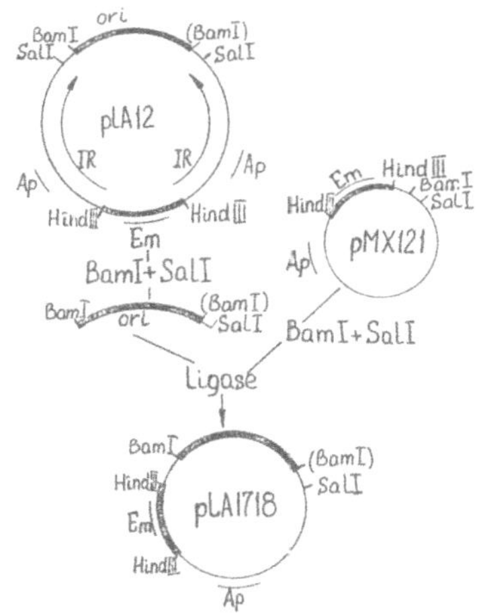

Fig. 6. Construction of plasmid pLA1718. (———) - pBR322; (▬▬) - pMX30; (—IR→) - inverted repeats.

subsequent selection for Em^rRib^+ markers. The plasmid pLA44 (pLA110-Rib) isolated from the transformants contained the vector and the Rib fragment joined at their EcoRI sites. The plasmid is stably maintained in *B. subtilis* grown in presence of erythromycin at 15 copies per genome, i.e., its copy number is greater than those of plasmids pMX33 and pMX45 constructed on the basis of the pMX30 vector (Tab. 6). It must be concluded, therefore, that the ability of pMX30 to propagate large DNA fragments in *B. subtilis* is due to its replication region and is retained in the pLA110 vector.

When the cells are grown under non-selective conditions, in the absence of erythromycin, pLA44 is eliminated while pMX49 persists (Tab. 6). The elimination of pMX30 and pMX39 derivatives is enhanced at higher cultivation temperatures. The observed difference in stability of the plasmids may be related to the difference in copy number. On the other hand, it cannot be ruled out that the stability determined by pMX30 is due to a certain genetic element lost during construction of the pLA32 and pLA110 plasmids.

CONCLUSIONS

In this study we attempted to estimate 2 most important characteristics of the *B. subtilis* vectors: their transforming activity and the ability to ensure maintenance of cloned genes. The tester used to characterize the vectors was a 9.5-kb Rib fragment of *B. subtilis* carrying a wild-type or a derepressed *rib* operon. The relatively large size of the fragment allows us to estimate the "strength" of a vector replicon, while expression of the *rib* genes offers a convenient way of monitoring the fate of recombinant plasmids without causing gross inhibitory effects in the vector-host system.

Our attempts to clone the *Rib* fragments on the pUB110 vector were unsuccessful. We attribute this result to the inability of pUB110 to sustain replication of large cloned fragments. Cloning with the aid of plasmid pJJ1 and the intermediate host, *E. coli*, resulted in construction of plasmid pMX1 which contained the *Rib* fragment and a direct repeat of DNA. The plasmid demonstrated the ability to transform *B. subtilis* in the form of monomer, which is in agreement with the finding of Michel and coworkers (25). The cleavage of pMX1 at its direct repeat seems to bear a relation to the mechanism of general recombination. As was shown for *E. coli*, excision of direct repeats from plasmids is a *recA*-dependent process (50). Indeed, in our experiments pMX1 was stably maintained in *E. coli rib82 recA* cells. The breakdown of pMX1 in the course of *B. subtilis* transformation proceeds quite effectively. At any rate, our method used to detect pMX9 in the transformants failed to reveal pMX1 in them.

An investigation of plasmid pMX30 implied that long inverted repeats ensure the ability of monomers of the plasmid to transform *B. subtilis*. As is discussed in Sec. 10, this ability is independent of the origin of the repeats.

Cloning of *Rib* fragments on pMX30 demonstrated a fundamental difference between this vector and pUB110, namely, its ability to provide for stable maintenance of large cloned DNA fragments in *B. subtilis*. On the basis of pMX30, the pMX39 plasmid was constructed. It contains a number of restriction sites suitable for cloning, and a ligated mixture containing pMX39 possesses a high transforming activity which is essential for shotgun experiments. In our opinion, this property of pMX39 owes to the presence in it of inverted repeats.

Construction of pMX39 derivatives on the basis of the plasmid's replication region yielded copy-number variants which preserved the ability to replicate large DNA inserts. The plasmids may prove useful in the construction of vectors for gene expression. Further studies of pMX39 derivatives should specify the role of various structural elements of the plasmid in its replication and maintenance in bacilli.

REFERENCES

1. Lovett, P.S., and M.G. Bramucci (1976) Plasmid DNA in bacilli. In *Microbilogy-1976*, D. Schlessinger, ed. Amer. Soc. Microbiol., Washington, D.C., pp. 388-393.

2. Kreft, J., and C. Hughes (1982) Cloning vectors derived from plasmids and phage of Bacillus. Curr. Top. Microbiol. Immunol. 96:1-17.
3. Ehrlich, S.D. (1977) Replication and expression of plasmids from Staphylococcus aureus in Bacillus subtilis. Proc. Natl. Acad. Sci., USA 74:1680-1683.
4. Ehrlich, S.D. (1978) DNA cloning in Bacillus subtilis. Proc. Natl. Acad. Sci., USA 75:1433-1436.
5. Gryczan, T.J., S. Contente, and D. Dubnau (1978) Characterization of Staphylococcus aureus plasmids introduced by transformation into Bacillus subtilis. J. Bact. 134:318-329.
6. Ehrlich, S.D., B. Niaudet, and B. Michel (1982) Use of plasmids from Staphylococcus aureus for cloning of DNA in Bacillus subtilis. Curr. Top. Microbiol. Immunol. 96:19-29.
7. Keggins, K.M., P.S. Lovett, and E.J. Duvall (1978) Molecular cloning of genetically active fragments of Bacillus DNA in Bacillus subtilis and properties of the vector plasmid pUB110. Proc. Natl. Acad. Sci., USA 75:1423-1427.
8. Michel, B., E. Palla, B. Niaudet, and S.D. Ehrlich (1980) DNA cloning in B. subtilis. III. Efficiency of random-segment cloning and insertional inactivation vectors. Gene 12:147-154.
9. Canosi, U., G. Morelli, and T.A. Trautner (1978) The relationship between molecular structure and transformation efficiency of some S. aureus plasmids isolated from B. subtilis. Mol. Gen. Genet. 166:259-267.
10. Mottes, M., G. Grandi, V. Sgaramella, U. Canosi, G. Morelli, and T.A. Trautner (1979) Different specific activities of the monomeric and oligomeric forms of plasmid DNA in transformation of B. subtilis and E. coli. Mol. Gen. Genet. 174:281-286.
11. deVos, W.M., G. Venema, U. Canosi, and T.A. Trautner (1981) Plasmid transformation in Bacillus subtilis: fate of plasmid DNA. Mol. Gen. Genet. 181:424-433.
12. Dubnau, D., S. Contente, and T.J. Gryczan (1980) On the use of plasmids for the study of genetic transformation in Bacillus subtilis. In DNA Recombination, Interaction and Repair, S. Zadrazil, and J. Sponar, eds. Pergamon Press, Oxford, pp. 365-386.
13. Kreft, J., K. Bernhard, and W. Goebel (1978) Recombinant plasmids capable of replication in B. subtilis and E. coli. Mol. Gen. Genet. 162:59-67.
14. Jomantas, J.V., P.M. Rabinovich, S.V. Bandrin, N.G. Demyanova, Y.I. Kozlov, and A.I. Stepanov (1979) Bireplicon vector plasmids for Escherichia coli and Bacillus subtilis. Dokl. Akad. Nauk SSSR 224:993-996.
15. Tanaka, T., and K. Sakaguchi (1978) Construction of a recombinant plasmid composed of B. subtilis leucine genes and a B. subtilis (natto) plasmid: Its use as a cloning vehicle in B. subtilis 168. Mol. Gen. Genet 165:269-276.
16. Gray, O., and S. Chang (1981) Molecular cloning and expression of Bacillus licheniformis β-lactamase gene in Escherichia coli and Bacillus subtilis. J. Bact. 145:422-428.

17. Jomantas, J.V., P.M. Rabinovich, and A.I. Stepanov (1980) Genetic engineering experiments in bacilli. Dokl. Akad. Nauk SSSR 254:493-495.
18. Jomantas, J.V., P.M. Rabinovich, and A.I. Stepanov (1982). Method for step-by-step cloning of the linked fragments of bacilli chromosome. Dokl. Acad. Nauk SSSR 264:482-484.
19. Rapoport, G., A. Klier, A. Billault, F. Fargette, and R. Dedonder (1979) Construction of a colony bank of E. coli containing hybrid plasmids representative of the Bacillus subtilis 168 genome. Mol. Gen. Genet. 176:239-245.
20. Ferrari, E., D.J. Henner, and J.A. Hoch (1981) Isolation of Bacillus subtilis genes from a Charon 4A library. J. Bact. 146:430-432.
21. Contente, S., and D. Dubnau (1979) Marker rescue transformation by linear plasmid DNA in Bacillus subtilis. Plasmid 2:555-571.
22. Weinrauch, Y., and D. Dubnau (1983) Plasmid marker rescue transformation in Bacillus subtilis. J. Bact. 154:1077-1087.
23. Canosi, U., A. Iglesias, and T.A. Trautner (1981) Plasmid transformation in Bacillus subtilis: Effects of insertion of Bacillus subtilis DNA into plasmid pC194. Mol. Gen. Genet. 181:434-440.
24. Iglesias, A., and T.A. Trautner (1983) Plasmid transformation in Bacillus subtilis: Symmetry of gene conversion in transformation with a hybrid plasmid containing chromosomal DNA. Mol. Gen. Genet. 189:73-76.
25. Michel, B., B. Niaudet, and S.D. Ehrlich (1982) Intramolecular recombination during plasmid transformation of Bacillus subtilis competent cells. EMBO J. 1:1565-1571.
26. Chang, S., and S.N. Cohen (1979) High frequency transformation of Bacillus subtilis protoplasts by plasmid DNA. Mol. Gen. Genet. 168:111-115.
27. Gryczan, T.J. (1982) Molecular cloning in Bacillus subtilis. In The Molecular Biology of the Bacilli, D. Dubnau, ed. Academic Press, New York, pp. 307-329.
28. Dubnau, D. (1983) Molecular cloning in Bacillus subtilis. In Experimental Manipulation of Gene Expression, M. Inouye, ed. Academic Press, Orlando, pp. 33-52.
29. Gryczan, T.J., S. Contente, and D. Dubnau (1980) Molecular cloning of heterologous chromosomal DNA by recombination between a plasmid vector and a homologous resident plasmid in Bacillus subtilis. Mol. Gen. Genet. 177:459-467.
30. Gryczan, T.J., and D. Dubnau (1982) Direct selection of recombinant plasmids in Bacillus subtilis. Gene 20:459-469.
31. Henner, D.J., and J.A. Hoch (1980) The Bacillus subtilis chromosome. Microbiol. Rev. 44:57-82.
32. Bresler, S.E., E.I. Cherepenko, T.P. Chernik, V.L. Kalinin, and D.A. Perumov (1970) Riboflavin operon of Bacillus subtilis. I. Genetic mapping of a linkage group. Genetika 6:116-134.
33. Bresler, S.E., L.G. Vyacheslavov, V.L. Kalinin, R.A. Kreneva, V.V. Kushev, V.A. Lanzov, M.I. Mosevitski, D.A. Perumov, and V.N. Rybchin (1973) Riboflavin operon. In Principal Genetical Processes, S.E. Bresler, ed. Nauka, Leningrad, pp. 211-233.

34. Bandrin, S.V., M.Y. Beburov, P.M. Rabinovich, and A.I. Stepanov (1979) Riboflavin auxotrophs of Escherichia coli. Genetika 15:2063-2065.
35. Rabinovich, P.M., M.Y. Beburov, Z.K. Linevich, and A.I. Stepanov (1978) Amplification of the riboflavin operon genes of Bacillus subtilis in Escherichia coli cells. Genetika 14:1696-1705.
36. Panina, L.I., J.V. Jomantas, M.J. Haykinson, and P.M. Rabinovich (1983) Cloning of the riboflavin biosynthesis genes of Bacillus subtilis in pBR322 plasmid vector in Escherichia coli. Genetika 19:174-176.
37. Haykinson, M.J., P.M. Rabinovich, and A.I. Stepanov (1982) The role of direct and inverted repeats in transformation of Bacillus subtilis by plasmid DNA. Dokl. Akad. Nauk SSSR 265:975-978.
38. Rabinovich, P.M., J.V. Jomantas, M.J. Haykinson, and A.I. Stepanov (1984) Cloning of genetic material in bacilli. Mol. Biol. 18:189-196.
39. Rabinovich, P.M., M.Y. Beburov, Z.K. Linevich, S.V. Bandrin, and A.I. Stepanov (1978) The expression of riboflavin operon genes of Bacillus subtilis cloned into pPR1 and pPR2 plasmids in Escherichia coli. Dokl. Akad. Nauk SSSR 238:1459-1462.
40. Palva, I. (1982) Molecular cloning of α-amylase gene from Bacillus amyloliquefaciens and its expression in B. subtilis. Gene 19:81-87.
41. Haykinson, M.J., P.M. Rabinovich, and A.I. Stepanov (1982) Models for plasmid transformation in Bacillus subtilis. Dokl. Akad. Nauk SSSR 265:1264-1267.
42. Behnke, D., V.I. Golubkov, H. Malke, A.S. Boitsov, and A.A. Totolian(1979) Restriction endonuclease analysis of group A streptococcal plasmids determining resistance to macrolids, lincosamids and streptogramin B antibiotics. FEMS Microbiol. Lett. 6:5-9.
43. Boitsov, A.S., V.I. Golubkov, I.M. Iontova, E.N. Zaitsev, H. Malke, and A.A. Totolian (1979) Inverted repeats on plasmids determining resistance to MLS antibiotics in group A streptococci. FEMS Microbiol. Lett. 6:11-14.
44. Boitsov, A.S., and V.I. Golubkov (1980) Transformation of Bacillus subtilis by streptococcal plasmids. Dokl. Akad. Nauk SSSR 254:490-492.
45. Behnke, D., and J.J. Ferretti (1980) Physical mapping of plasmid pDB101: A potential vector plasmid for molecular cloning in Streptococcus. Plasmid 4:130-138.
46. Stassi, D.L., P. Lopez, M. Espinosa, and S.A. Lacks (1981) Cloning of chromosomal genes in Streptococcus pneumoniae. Proc. Natl. Acad. Sci., USA 78:7028-7032.
47. Behnke, D., and S. Klaus (1983) Double or triple sets of replication functions as inverted and direct repeats on in vitro reconstructed streptococcal MLS resistance plasmids. Zeit. Allgem. Mikrobiol. 23:539-547.

48. Projan, S.J., S. Carleton, and R.P. Novick (1983) Determination of plasmid copy number by fluorescence densitometry. Plasmid 9:182-190.
49. Macrina, F.L., C.L. Keeler, Jr., K.R. Jones, and P.H. Wood (1980) Molecular characterization of unique deletion mutants of the streptococcal plasmid, pAM β1. Plasmid 4:8-16.
50. James, A.A., P.T. Morrison, and R. Kolodner (1982) Genetic recombination of bacterial plasmid DNA. Analysis of the effect of recombination-deficient mutations on plasmid recombination. J. Mol. Biol. 160:411-430.

SHUTTLE VECTOR FOR *ESCHERICHIA COLI*, *PSEUDOMONAS PUTIDA*, AND *PSEUDOMONAS AERUGINOSA*

A. A. Lushnikov, E. A. Andreevskaya, L. I. Lukashova, J. B. Lobach, and I. V. Domaradskij

All-Union Research Institute for Protein Biosynthesis
Main Department of Microbiological Industry
USSR Council of Ministers
109004, Moscow, B. Kommunisticheskaya, 27 USSR

SUMMARY

A hybrid plasmid capable of replication in 2 different genera, *Escherichia* and *Pseudomonas*, was constructed. This plasmid DNA can be used as a cloning vector in *E. coli* and pseudomonades cells.

The described hybrid plasmid pLD411 has been constructed on the basis of 2 small *E. coli* vector R-plasmids used in our laboratory and cryptic plasmid pWW2 or *P. putida* MT1. Plasmid pLD411 DNA was mapped with restrictases; its biological activity in transformations of different bacterial strains was studied, and the characteristics of transformed cells were also described.

INTRODUCTION

We could hardly be mistaken to state that more than 90% of all bacterial plasmid vector systems constructed by this time are proposed for *E. coli* as the most studied representative of *Procariotae*. As for *Pseudomonas*, the situation is different, and only some vector systems based on natural or in vitro constructed plasmid DNA are known now (1,2,3).

While studying some wild-type *E. coli* strains with plasmids that have drug resistance, we found polyplasmid strains of *E. coli*. By means of serial sequence steps of *E. coli* transformation by plasmid DNA from these strains and by selection of the clones obtained, we succeeded in isolating 2 small plasmids determining the ampicillin- or streptomycin-resistance of E. coli (4). These multicopy plasmids, as well as hybrid structures constructed on their bases, have high transformation activity of different E. coli strains but are incapable of transforming Pseudomonas strains. In this respect they differ from RP1(RP4) or RSF1010 plasmids which can exist in E. coli and Pseudomonas strains with the same progress probably due to the peculiarities of their transcription regulation (5).

Using a cryptic plasmid of P. putida and a plasmid of E. coli, we attempted to construct a hybrid plasmid capable of replication in these bacterial cells.

MATERIALS AND METHODS

Bacterial Strains and Plasmids

The strains and plasmids used in this study are listed in Tab. 1.

Isolation of Plasmid DNA, Restriction, and Ligation of its Fragments; Determination of Contour Length, Electrophoresis, and Transformation

DNA plasmid was isolated as described in Ref. 6; cleavage by specific endonucleases and ligation of plasmid DNA fragments was carried out as described in Ref. 7; determination of contour length of plasmid DNAs was made according to procedures found in Ref. 8; the transformation of E. coli cells was carried out as described in Ref. 9; transformation of P. putida and P. aeruginosa was performed according to Ref. 10. The technique used for rapid plasmid detection in transformed clones was conducted according to Ref. 11.

RESULTS AND DISCUSSION

One of the small nontransmissive R-plasmids, isolated from a wild polyplasmid E. coli strain, was used in our study. This plasmid, designated as pLD201 of 6.14 kb, conferred E. coli strains to grow on streptomycin-containing medium (4). For our purposes, this plasmid was minimized with EcoRV restrictase and the variant obtained was designated as pLD401. The pLD401 DNA of 3.82 kb had a limited number of restriction sites for some restrictases (see Fig. 1) and it is not cleaved by BamHI, BglII, HindIII, KpnI, or SalI endonucleases. While maintaining all other characteristics inherent in parental plasmids, pLD401, with the loss of the 2.32-kb EcoRV-fragment, obtained the valuable special feature of existing in cells in more copies increased from 10 to 80 per cell. This fact influenced the level of plasmid DNA yield during its preparative isolation. One must note that the addition of chloramphenicol in the

Tab. 1. Bacterial strains and plasmids.

Strain or plasmid	Source*
Strains	
E.coli C	Dr. Aleshkin A.G., Gamaleja's Institute for Epidemiology and Microbiology, Academy of Medical Sciences, USSR
P.aeruginosa PAO ML4262	Dr. Boronin A.M., Institute for Biochemistry and Physiology of Microorganisms, Academy of Sciences, USSR
P.aeruginosa PAO ML4600	same
P.putida BS250	same
P.putida mt-2 PaW85	Dr. Heinaru A.L., Tartu State University, Estonian SSR, USSR
P.putida MT1	same
P.putida MT1(Tol$^-$)	see text
Plasmids	
RSF1010	Dr. Boronin A.M., see above
pLD201	see text
pLD401	see text
pLD411	see text
pWW2	see text

* - Bacterial strains obtained from Dr. Heinaru A.L. while collaborating on biodegradated plasmids.

nutrient medium did not result in an increase of copy-number of the plasmids described, and their DNAs could not transform P. putida or P. aeruginosa strains.

Plasmid pWW2 of P. putida MT1 (12) of molecular mass 4.8±0.1 Md was another plasmid used in our study. This plasmid DNA was spontaneously isolated from the MT1 strain with the loss of another plasmid pWW1 of 52.0±0.4 Md in nonselective conditions of cultivation (L-broth). With the loss of plasmid pWW1, the strain cells lost their ability to use toluate as the only source of carbon. The properties of the strain (Tol$^-$) were similar to those of P. putida MT1-B1 isolated during the growing of parental strain MT1 on benzoate (13).

The study of pWW2 DNA showed that it had the only unique restriction site for EcoRV and SalI, 3 sites for PstI, but was not cleaved by EcoRI, BamHI, HindIII, and BglII endonucleases.

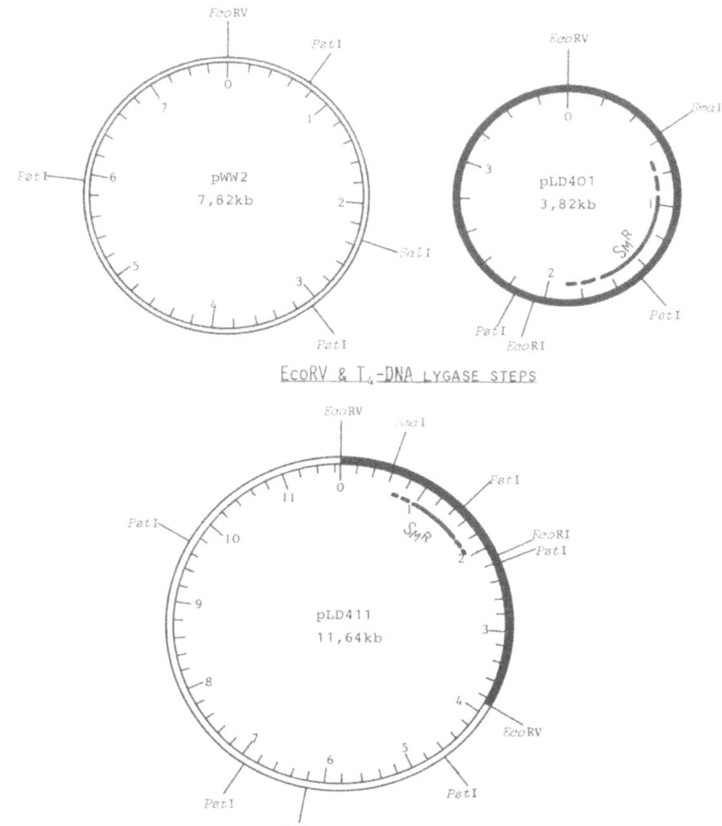

Fig. 1. Construction of the pLD411 plasmid on the basis of pLD401 and pWW2 plasmids. Plasmid pLD401 and pWW2 DNA were mixed in equimolar concentrations in restriction buffer containing 100 mM Tris•HCl (pH 7.5), 50 mM NaCl, 10 mM $MgCl_2$. Then the required quantity of <u>Eco</u>RV was added and the reaction maintained at 37°C for 60 min. DNA fragments were subjected to heating up to 65°C for 10 min and dialyzed with 66 mM Tris•HCl (pH 7.6), 6.6 mM $MgCl_2$. DTE and ATP were added with concentrations of 10 and 0.5 mM accordingly with required quantity of T_4-DNA ligase, and the reaction maintained at 4°C for 16-18 hr.

To obtain hybrid plasmid molecules, pLD401 and pWW2, their DNAs were mixed in equimolar quantities, cleaved by <u>Eco</u>RV endonuclease and the fragments obtained were ligated by T_4-DNA ligase. The <u>E. coli</u> C strain was transformed by the mixture of ligated fragments. The clones carrying hybrid plasmids were selected among the transformants resistant to 100 μg/ml of streptomycin using the technique

Tab. 2. Frequency of transformation (cells per μg DNA).

Plasmids	Bacterial strains				
	E.coli C	*P.putida*		*P.aeruginosa* PAO ML	
		BS250	mt-2 PaW85	4262	4600
pLD401	1.3×10^4	0	0	0	0
pLD411	1.0×10^3	4.7×10^2	5.8×10^2	1.1×10^3	1.3×10^3
RSF1010	2.2×10^4	3.3×10^2	7.1×10^2	1.8×10^3	2.1×10^3

described in Ref. 11. The clones of E. coli C, P. putida mt-2 PaW85 and BS250, and P. aeruginosa PAO ML 4262 and 4600 were transformed by hybrid plasmid DNA isolated from those clones and designated as pLD411. Plasmid RSF1010 DNA was used as a positive control and pLD401 as a negative one. The results of transformation are presented in Tab. 2, and the general scheme of obtaining of the hybrid plasmid pLD411 and its restriction map are shown in Fig. 1. Table 2 shows that plasmid pLD411 DNA transforms E. coli C as well as both Pseudomonas species. Moreover, in both cases the transformation frequency could be compared to that when plasmid RSF1010 DNA was used. The presence of the integral hybrid structure in transformed cells of different strains was confirmed by using the technique described in Ref. 11; plasmid DNAs of those strains we analyzed by means of restrictases and the data obtained enabled us to make a restriction map of pLD411.

The maximum streptomycin-resistance level of transformed E. coli strains was 1 mg/ml and the level of both species of Pseudomonas was more than 2 mg/ml of medium. The strain containing hybrid plasmid maintained streptomycin-resistance in nonselective conditions during 12 passages; the plasmid was not undergoing any structural alteration in this situation.

Plasmid pLD411 DNA can be used as a bifunctional vector in suitable Escherichia or Pseudomonas strains inserting the necessary genetic material in unique restriction sites. We believe that the plasmid functioning in cells of different genera was connected with separate use of its 2 parts for full replication of the whole structure. With regard to this fact, the plasmid is similar to bifunctional vector molecules used for E. coli and Saccharomyces yeast.

ACKNOWLEDGMENTS

We are thankful to S.V. Solovjeva for her excellent contribution in the construction and mapping of pLD401 vector DNA.

REFERENCES

1. Jacob, A.E., and N.J. Grinter (1975) Nature 255:504-506.
2. Sakaguchi, K. (1982) Gene cloning organ than E. coli. Berlin e.a., pp. 31-35.
3. Bagdasarian, M., and K.N. Timmis (1982) Gene cloning organ than E. coli. Berlin e.a.
4. Lushnikov, A.A., L.I. Koganitzkaja, and I.V. Domaradskij (1982) Metabolic Plasmids, A. Heinaru, ed. Tartu State University, Tallin, pp. 149-151.
5. Diaz, R., and W.L. Staudenbauer (1982) Nucl. Acid Res. 10:4687-4702.
6. Clewell, D.B., and D.R. Helinski (1970) Biochem. Biophys. Res. Comm. 41:150-156.
7. Bolivar, F., R.L. Rodriges, M.C. Betlach, and H.W. Boyer (1977) Gene 2:75-93.
8. Lang, D. (1970) J. Mol. Biol. 54:557-565.
9. Cohen, S.N., A.C.Y. Chang, and L. Hsu (1972) Proc. Natl. Acad. Sci., USA 69:2110-2114.
10. Chakrabarty, A.M., J.R. Milroie, D.A. Friello, and J.I. Vacca (1975) Proc. Natl. Acad. Sci., USA 72:3647-3651.
11. Kado, C.I., and L.N. Lee (1979) J. Bact. 140:227-238.
12. Duggleby, C.J., S.A. Bayley, M.J. Worsey, P.A. Williams, and P. Broda (1977) J. Bact. 130:1274-1280.
13. Williams, P.A., and M.J. Worsey (1976) J. Bact. 125:818-828.

REPORT ON A WORKSHOP:

PLASMID CLONING VEHICLES

J. Collins and C. Hershberger[1]

Gesellschaft für Biotechnologische Forschung mbH
Federal Republic of Germany

The first three contributions addressed the problem of designing plasmid vectors for the expression of foreign proteins in *E. coli*. All three used the principle that, to reduce the genetic and metabolic load to the cell, low or moderate copy-number plasmids should be used, and in two cases higher copy numbers were induced late in the growth phase. Yarraiton (Celltech) reported on experiments to express calf met-prochymosin. Initial experiments with the high copy ampicillin-resistance plasmid (based on pAT153) using a *trp* promoter and T7 terminator gave 5% loss/generation which could not be stabilized by ampicillin selection during scale-up. Repressed levels of production even in the presence of the pSC101 *par* locus caused instability because the *trp* promoter was not fully repressed when it was on the high copy plasmid. Vectors were therefore constructed in which the primer RNAII transcription could be brought under control of various inducible promoters, in particular, bacteriophage P_L under the control of the λcI857 gene product. In the presence of the pSC101 *par* region these plasmids carrying the *trp*-controlled prochymosin gene were stable at 30°C in the absence of antibiotic selection, having a copy number of 5/genome. This stability was high enough to meet the stringent industrial requirements with maintenance of excellent cell growth properties (final O.D.=60). On induction (37°C), a copy number of 400/cell was reached in which 10 to 5% of total protein was prochymosin. For the production of Insulin A and B chains (as β-galactosidase-Insulin A/B are *trp* E-Insulin A/B fusions) high expression from the *lac* or *trp* promoters on high copy plasmids gave 50% loss by the twentieth generation (Hershberger, Eli Lilly and Company). Using a down copy pBR322 derivative and transcriptional fusion of the expressed gene and Tc^R combined with the presence of the CI_{857} repressor on the plasmid in a CI^- lysogenic strain (i.e., suppression of a lethal

[1]Eli Lilly and Company, Indianapolis, Indiana.

phenotype forces plasmid maintenance), stability was achieved and at the end of production the cells could be lysed by lytic induction at 42°C for product release. For another interesting presentation on the use of fused peptides for gene product formation see Bastia (this volume), in which a collagenase-sensitive peptide is used as a digestible fusion bridge allowing later release of the mature peptide.

A third vector system which should have similar properties to those described above is based on the naturally low copy plasmid CloDF13 (Van Putten, Free University, Amsterdam) and uses a mutant temperature inducible for higher copy number. A hybrid was described containing the amp^r and tet^R genes of pBR322, the trp E gene and promoter and a strong downstream transcription terminator (see Bujard, this volume for the importance of strong terminators to prevent interference with replication) and the temperature-inducible mutant replicon of CloDF13.

A second main theme introduced by S. Ehrlich (Paris) and Hershberger was the formation of multimeric DNA segments (gene amplification) in the chromosomes of *Bacillus subtilis* and *Streptomyces* species. Ehrlich (this volume and *Pseudomonas* workshop) presented his proposal of integrating recombinant DNA more stably into the chromosome of *B. subtilis* ("chromosomal vectors"). The amplification of inserts located between direct repeats could be induced for the Neo^R-, Amp^R-, and Cm^R- but not for Ery^R- resistance genes. Coselection for Neo^R and Amp^R gave up to 50 copies/genome. Rownd (Northwestern University Medical School) reported that such amplification (a few copies) could be induced by antibiotic selection in a rec A ts *E. coli* mutant which then remained stable at the restrictive temperature (Hoffman, Rownd, in press).

Hershberger has identified a 10.5-kb region carrying an amplifiable unit of DNA and bounded by 2.2-kb direct repeats of the *Streptomyces fradiae* genome. Amplification of this 10.5-kb unit to give 500 tandem repeats (30% of the total final DNA) was obtained. Other naturally amplified sequences have been reported for different *Streptomyces* species by R. Hutter, H. Schrempf, M. Robinson, and U. Hornamann. At least 3 of the amplified DNAs are different as judged by a lack of homology detectable by hybridization.

The formation of concatamers (integrated "picolosomes": after Wigler) in the genomes of eukaryotic cells is used in the cosmid shuttle system to identify genes expressing in heterologous cells and to reisolate them in *E. coli* as cosmid hybrids via in vitro packaging (λ in vitro packaging system) of the crude genomic DNA from a eukaryotic cell clone (Collins, Braunschweig; see Lindenmaier, et al. 1983, *Nucl. Acids Res.*). It was proposed that this system might be effectively applied to any organism in which tandem repeats can be stably maintained. Recent advances in cosmid cloning methodology were reviewed in which the use of double *cos* vectors can facilitate the use of an Ish-Horowitz and Berg type method to prevent this formation of multimers of the vector and the use of dephosphorylated target DNA to prevent the cloning of non-contiguous fragments in a single hybrid. The use of an in vivo repackaging system allows the maintenance of the cosmid hybrid gene bank as a lysate of packaged particles (as for λ) which ensures stable maintenance of the bank over periods of several years. The development of a recombination screening method for cosmids analogous to that developed by Brian Seed for λ the π Vx-System) was reviewed [Collins, Braunschweig (in press), Lehrach et al., (in preparation), and Lindemaier and Houser]. In this system a small plasmid having no homology to the cosmid vector carries a DNA "probe." On transduction of the cosmid library into a λ lysogenic cell carrying this hybrid, the hybrid can cointegrate into a cosmid hybrid having homology in its insert. Subsequent in vivo repackaging and selection for antibiotic resistance markers carried by the cosmid and the small "probe" plasmid allows isolation of the co-integrate. This method is an alternative to colony hybridization for the isolation of long genomic fragments homologous to the DNA probe, and has the advantage that, due to the short residence time of the cosmid hybrid in *E. coli* (a few hours), even relatively unstable clones (e.g., human interleukin 2 gene) can be isolated.

Two further shuttle-vector systems were presented, one by Bagdasarian (Umia) for a broad host-range cosmid based on RSF1010 [see Pseudomonas workshop and Frey et al. *Gene* (1983)] and by Hershberger for *E. coli*/*Streptomyces*, based on the pock genes and ORI from SCP2. One variant of the latter can give single integration events leading to small "pocks." The presence of Neo^R and $thiostrepton^R$ genes on one vector allows insertional inactivation to be used to identify hybrid molecule-containing clones.

Although the development of small modifications to the huge range of existing vector systems to meet "local needs" is within the capability of most molecular geneticists, a few radical innovations can be seen to be constantly in development. For the further rational developments of vectors a better understanding of plasmid molecular biology, DNA replication, transposition, and recombination is required. This symposium provided useful additions to our understanding of these areas.

PLASMIDS IN THE DEGRADATION OF CHLORINATED AROMATIC COMPOUNDS

D. Ghosal, I.-S. You, D.K. Chatterjee,
and A.M. Chakrabarty

Department of Microbiology and Immunology
University of Illinois Medical Center
Chicago, Illinois 60612

INTRODUCTION

The soil-borne microbial community takes the "lion's share" of the mineralization process of the ever-increasing input of man-made organic compounds in the biosphere, converting them to biomass or maintaining the balance of various elements in the soil (1). In the broad spectrum of stability of these compounds, halogen-substituted aromatic pollutants are of major concern because of their recalcitrance and magnitude of use (2). Haloaliphatic compounds possess shorter half-life in nature, most probably because of the presence of broad-specificity enzymes which can dehalogenate such compounds and are known to be active in *Pseudomonas*, *Flavobacterium* sp., and *Moraxella* sp. (3,4,5,6). In the past few years, isolation of pure cultures capable of degrading various compounds that are toxic and classified as recalcitrant, such as 2,4,5-trichlorophenoxyacetic acid (2,4,5-T), 2,4-dichlorophenoxyacetic acid (2,4-D), pentachlorophenol, etc., boosted up research in both the fields of basic understanding of the metabolic pathways of these compounds as well as in their potential use in decontamination technology (7,8,9,10,11).

The present state of research increasingly strengthens the emerging concept that, like the multiple drug-resistance markers both in enteric and other pathogenic bacteria, the metabolic versatility of soil bacteria, with few exceptions, is determined totally or partly by resident plasmids (12).

On the basis of the nature of plasmid involvement in the biodegradation of haloaromatic compounds, 3 functional categorizations may be formulated.

(a) Plasmid material is mostly independent of the host because it carries at least all the structural genes necessary for the degradation of a specific compound to simple organic acids which are subsequently incorporated into the general metabolic pool. For instance, plasmid pWR1 or pAC25 of P. putida codes for all the enzymes necessary for the conversion of 3-chlorobenzoic acid (3Cba) to maleylacetic acid (13,24), but in these cases, host-coded enzymes can also complement part of the functional machinery because of the presence of an extra set of chromosomally coded enzymes (13,14).

(b) In the second category, some of the chromosomally coded enzymes, because of their broad specificity, can attack both halogenated and nonhalogenated substrates, leading to dead-end products like halocatechols in case of haloaromatic compounds (15,16). This block in the catabolic pathway is removed by the plasmid-encoded enzymes which have high specificities towards halogenated substrates (13,14,17,18,19). This case is best exemplified by Alcaligenes eutrophus containing plasmid pJP4 which can catabolize chlorocatechols, the dead-end products of 3-chlorobenzoate degradation, by the 2-step, chromosomally coded enzymes, benzoate-1-2 dioxygenase and dihydrodihydroxybenzoate dehydrogenase. Therefore, dissimulation of a haloaromatic compound is carried out by joint action of host and plasmid (13,14).

(c) In the third category, presence of side chain on the halogen-substituted aromatic ring makes them difficult substrates to degrade for host enzymes. The general assumption that cleavage of the side chain in 2,4-D and 4-chlorophenoxyacetic acid (4Cpa) is a plasmid-coded function (9,14) has not been challenged. For example, in case of 2,4-D, hydrolytic cleavage of the ether bond to remove the side chain, thus converting this compound to the corresponding chlorophenol, is believed to be a plasmid-specified function. The chlorophenol formed could then be degraded either by plasmid- or host-coded enzymes (9). A. paradoxus, containing plasmid pJP1 capable of degrading 2,4-D, is a well-known example of such a mechanism (9).

Knackmuss and his associates worked out the metabolic pathway for degradation of chlorobenzoates by Pseudomonas sp. B13 (19,20), identified the enzymes involved, and described the range of specificity of enzymes encoded by either chromosome or plasmid (20,21). Weightman et al., (14) recently proposed converging pathways for monochlorobenzoate and chlorophenoxyacetic acid by A. eutrophus harboring pJP4 via chlorocatechol. In order to study the role of plasmid and host determinants in the degradation of haloaromatic compounds, we have selected 2 of these well-investigated compounds, 3Cba and 2,4-D, and another compound, 2,4,5-T, which is least understood in terms of both catabolic pathway and degradative genes and enzymes involved.

Our laboratory is engaged in dealing with the following questions in relation to haloaromatic catabolism:

(a) The minimum segment of the plasmid DNA necessary for efficient expression of $3Cba^+$ or $2,4-D^+$ phenotype.

(b) Localization of the catabolic genes on the physical map of a catabolic plasmid.

(c) The nature of regulation or expression of a catabolic pathway and the regulatory genes involved.

(d) Do genes encoding similar catabolic steps but borne on different plasmids share any structural and functional homology as reminiscent of their common evolutionary origin?

(e) Is the evolution or spread of these catabolic genes similar to that of the multiple drug-resistance genes carried by various drug-resistance plasmids?

(f) To what extent are various legitimate and illegitimate recombinational events, such as amplification, deletion-fusion, insertion etc., which are basic mechanisms in evolution, operational in the evolution and dissemination of such catabolic plasmids?

(g) Do these catabolic plasmids possess unique structural and genetic features that provide them with raw materials for a unique mode of regulation and rapid evolution with dramatic increase in their catabolic voracity in response to selection pressure?

(h) What are the genetic and biochemical bases for specific association and expression of a specific catabolic pathway shared between a host bacterium and a plasmid?

Some of the answers to the above questions are discussed in the next few sections.

Plasmid-specified Genes in the Biodegradation of 3-Chlorobenzoic Acid (3Cba)

The transmissible nature of the 3Cba phenotype, both in *Pseudomonas* and *Alcaligenes* and other biochemical studies, indicates the essential participation of plasmids in total degradation of 3Cba (13,23,24,25). All the 5 enzymes necessary for the conversion of 3Cba to maleylacetic acid were shown to be present on plasmid pWR1 (13) which has been shown to be structurally homologous to pAC25 or pAC27, except a 6-kb deletion in pWR1 (26). Unlike pAC27, *Alcaligenes* plasmid pJP4 has been shown to encode only 3 enzymes for chlorocatechol catabolism and these appear to be similar in function with *Pseudomonas* sp. B13 plasmid pWR1-coded enzymes (14). The first 2 enzymes for the conversion of 3Cba to 3-chlorocatechol could be provided by chromosome both in Pseudomonas sp. B13 and A. eutrophus (13,14).

We have previously demonstrated the operation of the modified ortho pathway for 3Cba degradation encoded by pAC25 or pAC27. pAC27 is identical to pAC25 with regard to all known functions, but has an 11-kb segment deleted from pAC25 (27). To localize the genes for 3Cba degradation on plasmid pAC27 and pJP4, we took into consideration the fact that significant homology is expected between 2 plasmids only in the region carrying the 3Cba structural genes. Subsequent cloning of the regions in a broad host-range vector will undoubtedly localize the 3Cba genes. Physical maps for these 2 plasmids are given in Fig. 1A.

Following this rationale, EcoRl fragments A, B, C/D, E, F, and G of pJP4 were purified, nick-translated, and used as probes for hybridization with EcoRl, HindIII, and BglII digested restriction fragments of pAC25 transferred from agarose gels onto nitrocellulose filters. Homology was detected only with EcoRl-B and BglII-E fragments of pAC27 when EcoRl fragments B and E of pJP4 were used as probes (Fig. 1B and 1C). No hybridization was detected with any of the pAC25 fragments when EA, EC/ED, EF, and EG fragments of pJP4 were used as probes (data not shown). In an analogous experiment with BglII-E fragment (Fig. 1A) of the plasmid pAC25 was purified, nick-translated, and used for hybridization with EcoRl and EcoRl/BamHl digested pJP4 DNA fragments blotted to a nitrocellulose filter. As expected, only the EcoRl fragments B and E and EcoRl/BamHl double digested fragment C (10 kb) and F (3 kb) of pJP4 showed homology with the probe DNA fragment (Fig. 2). The region of homology between these 2 plasmids is diagrammatically presented in Fig. 1A.

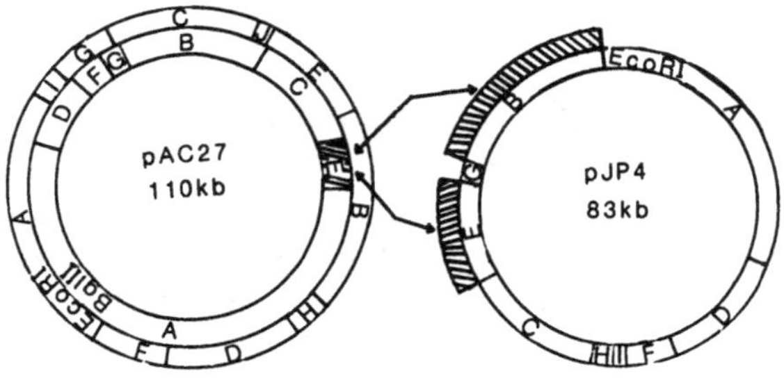

Fig. 1. A. Endonuclease cleavage map of plasmid pAC27 for EcoRl and BglII sites and pJP4 for EcoRl sites. Hatched areas on both the plasmids show the location of the chlorocatechol degradative genes. No homologous counterpart of EcoRl fragment EG of pJP4 is found in pAC27. Area encompassed by 2 arrows on pAC27 (BglII E fragment) and pJP4 (EcoRl fragment EB and EE) denote the region of homology between the 2 plasmids.

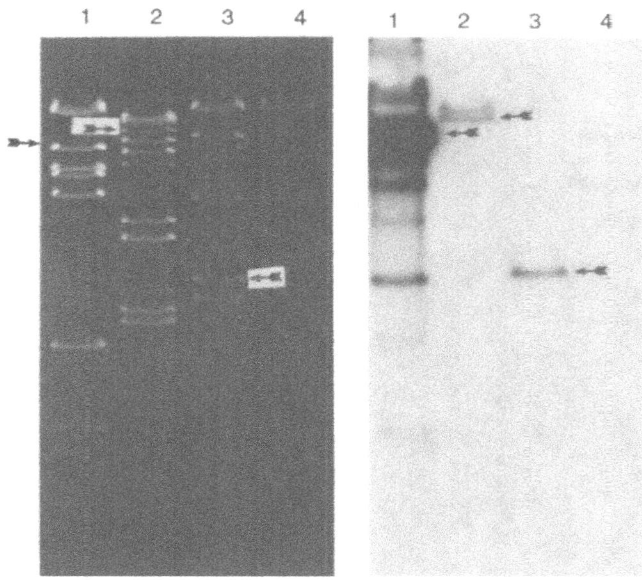

Fig. 1. B. Agarose gel electrophoresis of restriction fragments pJP4, EcoRl (lane 1), pAC25 EcoRl (lane 2), pAC25 BglII (lane 3), and pAC25 HindIII (lane 4). Corresponding Southern blot of the gel, using EcoRl fragment B of pJP4 (lane 1) as a nick-translated P^{32}-labeled probe, is shown on the right. Arrow shows the specific bands hybridized to such probe.

It is clear from the hybridization data that the segment representing sequence homology between the 2 plasmids is clustered in pAC27 on the 4.2-kb BglII-E fragment, whereas, the fragment EG of plasmid pJP4 with no detectable homology to pAC27 divides this region in pJP4 into 2 segments (Fig. 1A).

Cloning of Restriction Fragments of pAC27 and pJP4

Broad host-range cosmid vector pLAFR1 (28), specifying tetracycline-resistance (Tet^r), was used to construct EcoRl gene libraries of plasmid pAC27 and pJP4. The size of insert cloned in this phage lambda-derived vector is 20-30 kb (28). Partially EcoRl-digested plasmid DNAs were ligated to the EcoRl-cleaved vector, packaged in phage-lambda heads used for infection in *Escherichia coli* AC80 (Thr^- Leu^- Met^-) and selected for Tet^r. The plasmid pool was mobilized en masse from *E. coli* to *P. putida* PpG1900 (Kan^r Trp^-) or AC10 (Met^-) in a triparental mating using *E. coli* HB101/pRK2013 as a mobilizer (29). Conjugation mixtures were either plated on minimal glucose containing Tet or on 3Cba plates with recipient amino acid supplements. When plated on glucose Tet plate, colonies

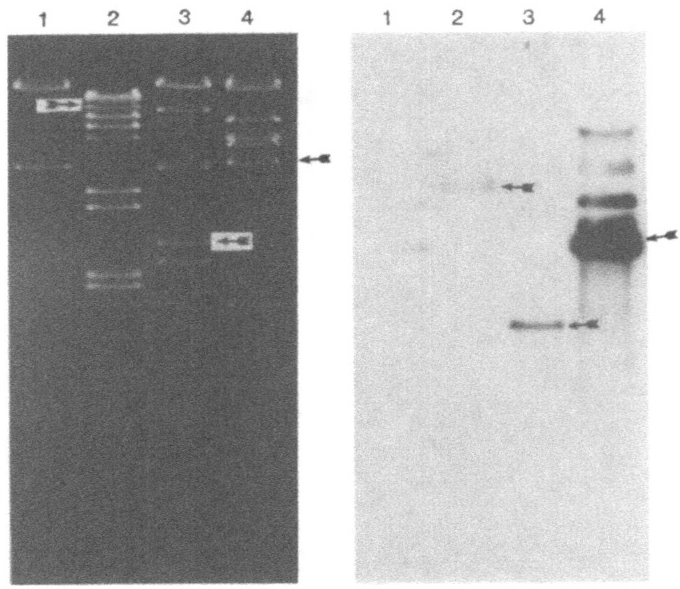

Fig. 1. C. Same as Fig. 1B; pAC25 HindIII (lane 1), pAC25 EcoRl (lane 2), pAC25 BglII (lane 3), and pJP4 EcoRl (lane 4). EcoRl fragment EE (lane 4 arrow) of pJP4 was used as a probe.

from such plates were tested for their 3Cba$^+$ phenotype. Restriction analysis of such 3Cba$^+$ clones is shown in Fig. 3A and 3B.

A 3Cba$^+$ clone obtained from pAC27 revealed the presence of pAC27 EcoRl fragment B and J only in the plasmid designated as pDC10 (Fig. 3A), whereas pJP4 derived clones such as YG410, and YG419 showed the presence of pJP4 EcoRl B and H, and only pJP4 EcoRl B fragment, respectively, besides pLAFR1. Unlike pDC10 which allowed rapid growth with 3Cba, the plasmids such as pYG410, pYG419, etc., present in YG410 or YG419 allowed only slow growth on 3Cba with accumulation of brown-colored products. The restriction analysis of such plasmids (e.g., pYG419) demonstrated the presence of the pJP4 EcoRl B fragment in relatively high intensity as compared to pLAFR1 (Fig. 3A) indicating the presence of multiple copies of EcoRl-B fragment on this plasmid. Electron microscopic visualization revealed a large plasmid of about 120 kb (Fig. 4). This plasmid on denaturation and renaturation did not show any "snap back" structure suggesting that the multiple EcoRl-B fragments are presumably present in tandem array. Slow growth of this strain in 3Cba suggested the presence of all the structural genes for chlorocatechol degradation on the EcoRl-B fragment of pJP4. These genes have subsequently been located more precisely on the 10-kb BamHl/EcoRl-C fragment. It

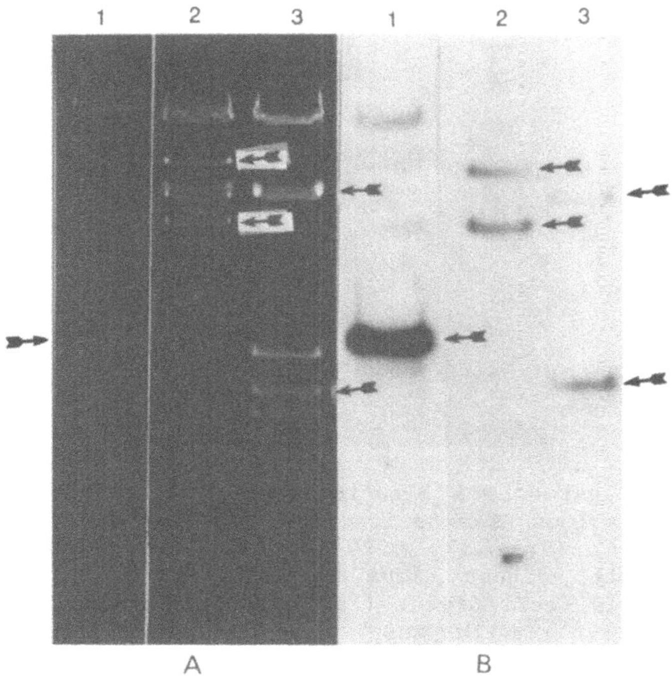

Fig. 2. Agarose gel electrophoresis (2A) of BglII digested pAC25 (lane 1), EcoRl digested pJP4 (lane 2), and BamHl/EcoRl digested pJP4 (lane 3). Southern blot (2B) of the same gel shows the homology of BglII E fragment (P^{32}-labeled probe) of pAC25 with EcoRl fragments B and E of pJP4 (lane 2) and 10-kb and 3-kb segments of pJP4 (lane 3).

was, however, apparent that an additional segment of pJP4 is necessary for the efficient expression of chlorocatechol genes. Absence of this regulatory gene is partly compensated for by amplification of the segment containing the structural genes, i.e., a gene dosage effect with pYG410 and pYG419. Since amplification of the structural genes allowed slow growth in absence of the regulatory gene product, it appears that the 3Cba degradative genes on pJP4 are, presumably, positively regulated. Additional evidence for this assumption will be provided in the next section.

The EcoRl-B fragment of pAC27 cloned on pDC10 is a 20-kb segment. To further localize the chlorocatechol-degradative genes in a shorter segment, pAC27 DNA was digested with BglII and ligated to a plasmid vector pRK290 (29) and a gene bank was constructed in E. coli AC80. Screening of $3Cba^+$ clones from this gene bank after mobilization into P. putida PpG1900 revealed the presence of 3 classes of clones (Fig. 3B). Class I contains 2 continuous BglII fragments, E and C (pDC15) and shows the best phenotypic expression; Class II

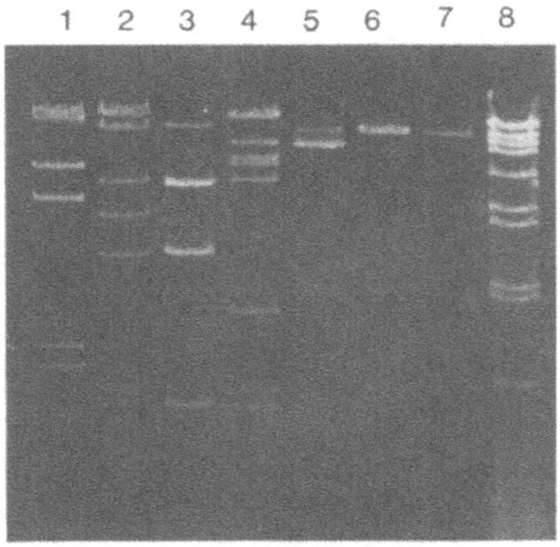

Fig. 3. A. Agarose gel electrophoresis of restriction fragments of various DNAs derived from pJP4 and pDC10. λ DNA EcoRl digest (lane 1); pJP4 HindIII digest (lane 2); pYG419 HindIII digest (lane 3); pJP4 EcoRl digest (lane 4); pYG419 EcoRl digest (lane 5); pLAFRI vector EcoRl digest (lane 6); pDC10 EcoRl digest (lane 7); and pAC27 EcoRl digest (lane 8). Lane 5 shows the presence of only EcoRl B fragment of pJP4 in multiple copies in pYG419 (higher intensity of the insert compared to vector) and lane 7 shows EcoRl B and J fragments of pAC27 in pDC10.

contains fragments E and F (pDC20) and demonstrates poor growth ability in minimal 3Cba medium. One Class III clone (DC25) was found to contain only a BglII-E fragment (pDC25) and also grows slowly on 3Cba. These data, in combination with previous results on hybridization, clearly suggest that chlorocatechol-degradative genes are homologous in plasmids pAC27 and pJP4 and are located on BglII-E fragment (4.2 kb) of pAC27 and in a 10-kb segment (BamHl/EcoRl fragment C) of pJP4.

To localize the putative regulatory gene(s) on plasmid pJP4, 15 individual AC80 Tet^r clones from the *E. coli* gene bank were randomly screened for the composition of recombinant plasmids by miniscale plasmid preparation (30). Seven clones all of which contain the pJP4 EcoRl-B fragment plus one or 2 of the EC, ED, EE, EF, EG, and EH fragments were selected, the plasmid from each clone was individually mobilized into various *P. putida* strains and tested for their $3Cba^+$ phenotype. Two $3Cba^+$ clones (Y43, Y52) were obtained from such transfer. Both of them grew well in minimal 3Cba medium, and restriction analysis of the plasmids from these 2 strains, grown in 3Cba-medium, showed them to be identical, i.e., both contained plasmids with EcoRl fragments EB, EE, and EF (Fig. 3B). In subsequent

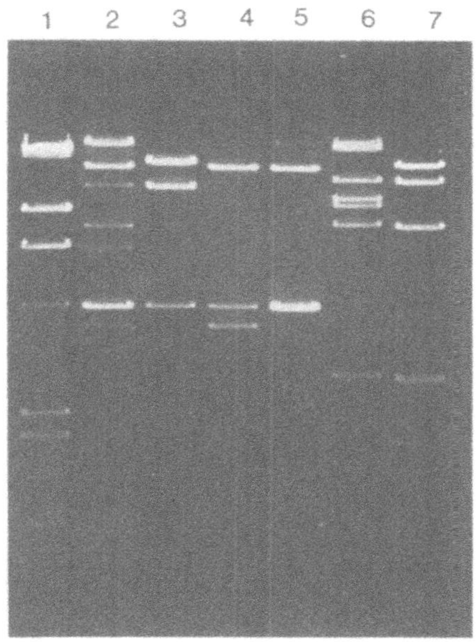

Fig. 3. B. Agarose gel electrophoresis of restriction fragments of various plasmids. Lane 1, λ (HindIII); lane 2, pAC27 (BglII); lane 3, pDC15 (BglII); lane 4, pDC20 (BglII); lane 5, pDC25 (BglII); lane 6, pJP4 (EcoR1); and lane 7, pY43 (EcoR1).

transformation experiments, this plasmid pY43 also conferred Tet^r $3Cba^+$ phenotype to various P. putida strains. Since EB alone or combinations of EB and other fragments besides EE and EF did not produce $3Cba^+$ phenotype readily, it appears that fragment EE and possibly EF are essential for rapid growth with 3Cba. The homology of the pJP4 EE fragment with the pAC25 BglII-E fragment (Fig. 2) further indicates that the EE fragment may contain the regulatory gene(s) needed for efficient expression of the chlorocatechol-degradative genes present on the pJP4 EB fragment.

Structural and Functional Features of the Plasmids Involved in Chlorocatechol Degradation

It is safe to assume from the results already presented that though the structural genes for chlorocatechol degradation are clustered and apparently homologous in both pAC27 and pJP4, some interesting structural and functional features indicate their remarkable evolutionary divergence.

Functionally, the plasmid pDC25 harboring the pAC27 4.2-kb BglII-E fragment or pDC10 containing the EcoR1-B fragment of pAC27

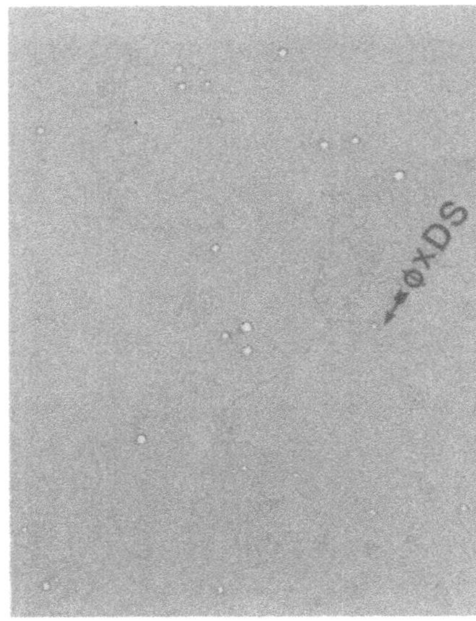

Fig. 4. Electron micrograph of covalently closed circular molecule of pYG419.

when transferred to various strains of P. putida or P. aeruginosa readily expresses the $3Cba^+$ phenotype. The plasmid pJP4 belongs to the P1 incompatibility group and can be transferred from its original host (A. eutrophus) to P. putida at a frequency of 10^{-2} to 10^{-3} when selected on PIA (Pseudomonas Isolation Agar) medium containing merbromin (100 μg/ml) or mercuric chloride (40-50 μg/ml); however, the $3Cba^+$ phenotype is recovered from the recipients with a frequency of 10^{-6} or less. Intact plasmid pJP4 was recovered from $3Cba^-$ Hg^r P. putida exconjugants (You and Ghosal, unpub. results). It has not been possible to grow P. putida harboring pJP4 in liquid minimal 2,4-D medium, though poor growth could be seen on 2,4-D plates.

Plasmid pJP4, like other 2,4-D-degradative plasmids such as pJP1, pJP2, etc., was originally isolated from Alcaligenes sp., whereas both pAC25 and pWR1 were originally isolated from Pseudomonas sp. A pertinent question is whether the difference in expression is due to an inherent variation in the structural and regulatory genes on the 2 plasmids or due to a mismatch of the signal sequences present on plasmid pJP4 with the transcription-translation machinery of the new host, i.e., Pseudomonas sp. The second explanation is plausible since Pseudomonas plasmids are known to be expressed poorly in other hosts, particularly the enteric bacteria. The presence of unique structural features present in the plasmid pJP4, however, appears to indicate that a variation in the continuity of the structural and regulatory genes pertinent to 3Cba degradation is the major difference between pAC25 and pJP4. The EcoR1-E fragment (EE) of pJP4 which demonstrates homology with pAC25 BglII-E fragment possesses a unique structural feature, a "stem-loop" structure (1.8-kb stem and 1-kb loop) not present in pAC25 DNA (Fig. 5). It has already been mentioned that the presence of intact plasmid pJP4 in *P. putida* does not confer $3Cba^+$ phenotype readily. Continued selection does allow emergence of $3Cba^+$ colonies. Plasmid DNA isolated from such $3Cba^+$ exconjugants were found to show unique structural rearrangements involving EcoR1 fragment E (EE) in the region of the above mentioned "step-loop" structure and most likely fragment B (EB) and F (EF). The type of rearrangements was found to be identical in the plasmid profiles isolated from 5 independent isolates of $3Cba^+$ *P. putida*. A typical plasmid with such rearrangement was designated as pYG3, which on EcoR1 digestion produced only 5 fragments instead of 10 for pJP4 (Fig. 6). *Alcaligenes eutrophus* harboring pJP4 (JMP134) on being maintained in 2,4-D media for many generations does not express $3Cba^+$ phenotype readily. Similar to *P. putida*, $3Cba^+$ *Alcaligenes* colonies can be obtained after prolonged selection on 3Cba plates. Restriction analysis of the plasmid DNA isolated from 3Cba grown *Alcaligenes* sp. demonstrated the presence of both pJP4 as well as the pYG3 type of plasmids with similar genetic rearrangements (Fig. 6). The structural changes of plasmid pJP4 which resulted in pYG3 forms were investigated by electron microscopic analysis, restriction enzyme digestion, and Southern hybridization. All 5 EcoR1 fragments of pYG3 were purified and separately

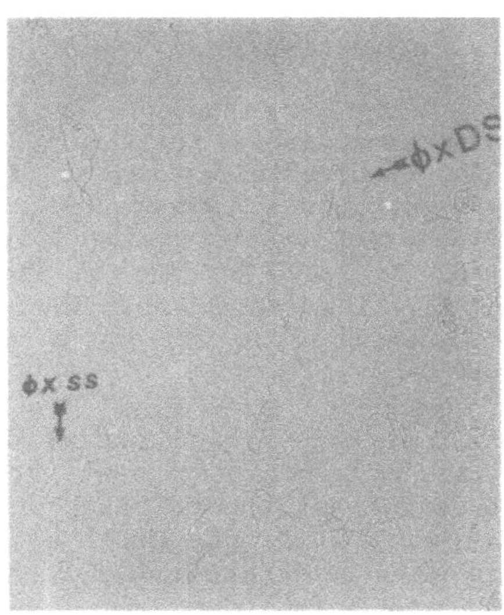

Fig. 5. Electron micrograph of "snap back" structure of EcoR1 fragment EE of pJP4 showing a 1.8-kb stem and a 1-kb loop.

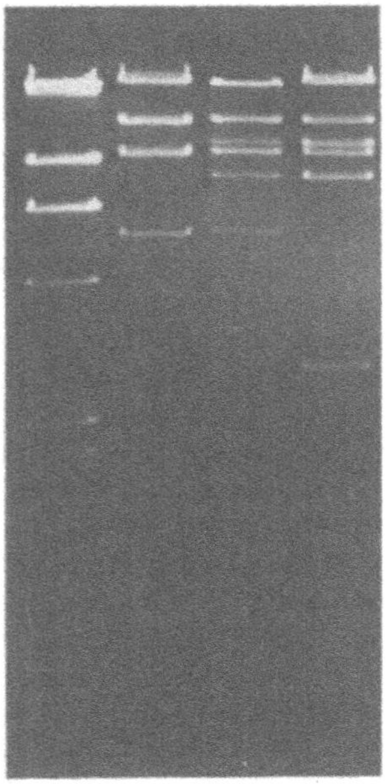

Fig. 6. Agarose gel electrophoresis of λ digested with HindIII (lane 1); EcoRl digested pYG3 (lane 2), and pJP4 DNA isolated from 3Cba-grown *Alcaligenes eutrophus* (lane 3) showing deletion. Lane 3 demonstrates the presence of 2 plasmids (both pJP4 and pYG3); pJP4 digested with EcoRl (lane 4).

used as probes for hybridization with EcoRl fragments of pJP4. Such hybridization results demonstrated that in pYG3, a deletion of approximately 15-kb DNA occurred with loss of pJP4 EcoRl fragments, EC, EH, EI, and 1.6-kb portions of fragment EE, and unknown amounts of EF and EG. Such deletion events created 3 new fragments, pYG3 EA' is the same as pJP4 EA, pYG3 EB' is the same as pJP4 EB, pYG3 EC' is derived from pJP4 fragments ED and EE, pYG3 ED' is derived from pJP4 fragment EF while pYG3 EE' is a very short deletion derivative of fragment EG.

Although restriction and hybridization data indicated the complete or partial loss of several fragments from pJP4 to generate the plasmid pYG3, electron microscopic contour-length measurement demonstrated the size of pYG3 of the order of 93 kb, compared to 83 kb of pJP4. An inverted duplication of a 25-kb segment involving EcoRl fragment E (EE) and G (EG) and most likely part of B and F occurred in pYG3. Such molecules of pYG3 on denaturation and fast renaturation give rise to a long-stem loop structure with a 25-kb stem and 1.3-kb and 43-kb loops as shown in Fig. 7.

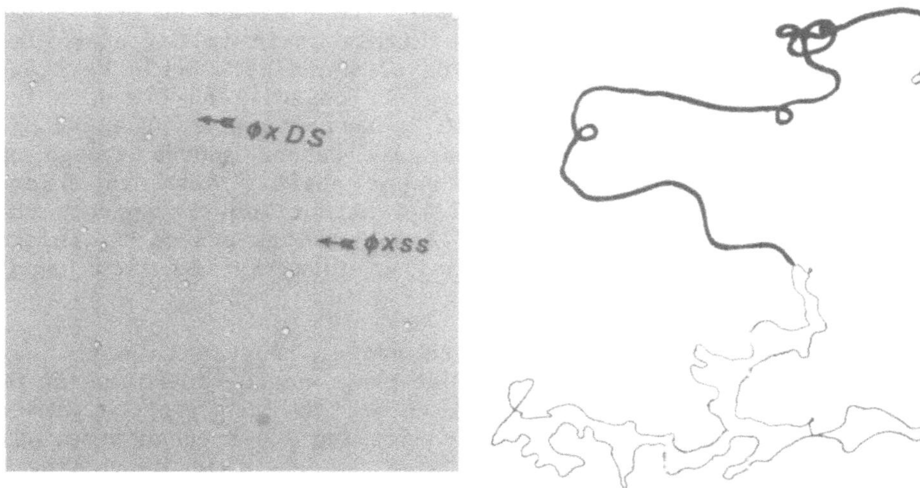

Fig. 7. Electron micrograph of "snap back" structure of the whole molecule of pYG3 showing 25-kb inverted duplication. Diagrammatic representation of the molecule is also shown.

The significance of this sequence rearrangement in the activation of 3Cba pathway is not known. Cells harboring pYG3 plasmid grow efficiently on 3Cba. It will be interesting to see whether the pYG3 form becomes sole plasmid population after prolonged cultivation of Alcaligenes on 3Cba. Such genetic rearrangements leading to $3Cba^+$ phenotype appear to make the cells unable to utilize 2,4-D. The precise map of deletion fusion points and duplicated segments is essential to have an understanding of the role of this unique structural rearrangement in functional expression of 3Cba genes.

Structural and Functional Interrelationship Among the Plasmids Isolated from Chlorophenoxyacetate-Degrading Bacteria

Cloning of the 2,4-D degradative genes. In the case of 2,4-D degradation, although all the enzymes for 2,4-D degradation are believed to be plasmid coded, apparent failure to get the expression of this phenotype in P. putida beyond a reasonable doubt, and failure to localize the genes for the first enzyme(s) of the pathway on a plasmid (14) do not exclude the serious question of any essential role of host determinants in 2,4-D catabolism. In our hand, though poor growth has been observed on the 2,4-D plate (600 μg/ml) of some P. putida strains harboring plasmid pJP4, after weeks of incubation (with $pJP4^+$, PpG1900, or PRS2015) all efforts to grow them in liquid culture have failed (I-S. You, unpub. observations). The adaptive status of the host (lack of tolerance to toxic intermediates such as chlorophenols generated during a 2,4-D catabolism) might be a major determinant in such inability of P. putida cells to utilize 2,4-D.

Plasmidless Alcaligenes sp. can support their growth on both phenol and dichlorophenol (9,22) which are highly toxic to organisms like P. putida. Both conjugative plasmids pJP4 and pJP2 could be transferred from Alcaligenes to a wild-type P. cepacia 383 (7) at a frequency of approximately 10^{-6} and 10^{-5}, respectively. The exconjugants grew efficiently on 2,4-D as sole carbon source though one could smell chlorophenols in the vapor phase. Although chlorophenols seem to accumulate during 2,4-D metabolism, it appears that P. cepacia 383 possesses a higher level of resistance to chlorophenols compared to other Pseudomonas as shown by the exconjugants and other reports (7).

The above observation led us to use P. cepacia as a host for cloning experiments using a broad host-range vector pLAFR1 which was found to propagate normally in this host, to identify the minimum fragment size of pJP4 required for conferring 2,4-D^+ phenotype. The cloning strategy was basically the same as described in cloning of the 3Cba genes. However, in this case we used both plasmid pJP4 and pJP2 to clone the 2,4-D genes.

Although we did not get any 2,4-D^+ clone from the EcoR1 library of pJP4, from the pJP2 EcoR1 library we succeeded in cloning a 20-kb segment containing 2 EcoR1 fragments (EA and EF of pJP2). This clone was designated as Y21. Further analysis of this clone and detailed biochemical characterization will be reported elsewhere (I.-S. You, in prep.). Recently, several 2,4-D^+ clones have also been obtained from a HindIII gene bank of pJP4. Their characterization is in progress.

Structural and functional interrelationships between plasmids pJP4 and pJP2. Plasmid pJP2 is about 20 kb smaller than pJP4 (25). Phenotypically, A. eutrophus harboring pJP2 can degrade both 2,4-D and its nonhalogenated analogue phenoxyacetate but not 3Cba, whereas, isogenic strains of Alcaligenes harboring pJP4 can utilize 2,4-D, 3Cba, and 4Cpa but not phenoxyacetate (14,22,25).

Our success in cloning 2,4-D catabolic genes in a 20-kb segment from pJP2 and initial failure to get such clones from pJP4 led us to speculate that 2,4-D genes including regulatory sequences may be present in a more clustered fashion on pJP2 compared to pJP4.

Supporting evidence in favor of this hypothesis has been obtained from a series of hybridization experiments in which individual EcoR1 fragments of pJP4 (EA, EB, EC/ED, EE, EF, EG, and EH) were purified and nick-translated. Each of this probe was hybridized to EcoR1 digested or SmaI digested pJP2 restriction fragments blotted to nitrocellulose filters. In complementary experiments, EcoR1 fragments EA, EF, and EG of pJP2 were purified and hybridized to various restriction fragments of pJP4. Such experiments in combination with cloning data stengthened the conclusion that most of the genes involved in chlorocatechol and substituted phenoxyacetic acid degradation are clustered in a 16-kb EcoRl-A (EA) fragment of pJP2.

Among the 8 EcoRl fragments of pJP4 tested, EB, EE, EF, and EG showed strong homology with the EA fragment of pJP2 while a small segment of EA near the vicinity of EB and another short stretch of EC in the vicinity of EE of pJP4 were found to be present in pJP2 EcoRl-F fragment (EF). Fragment ED of pJP4 is apparently not present in pJP2. Interestingly, these 2 EcoRl fragments A and F were recovered from 2 of the 3 2,4-D clones of pJP2 tested. Thus, it is logical to assume that about 40 kb of material of pJP2 is involved in other functions besides 2,4-D degradation. Construction of a restriction map of pY21 and its alignment with pJP2 are in progress.

It is interesting to note the evolutionary divergence of the functional aspects of these 2 plasmids. In plasmid pJP4, both 2,4-D and 3Cba, seem to be catabolized through a convergent pathway in which monochlorocatechol is a common intermediate, as discussed by Weightman et al., (14) and others (25). In support of the above view, 4Cpa was found to serve as a growth substrate for *Alcaligenes* sp. harboring pJP4 (14). However, the degradation of 2,4-D through a dichlorocatechol intermediate is a rather common pathway in nature (31), in which the removal of chlorine occurs after ring cleavage.

Since *A. eutrophus* harboring pJP2 is $3Cba^-$, it is likely that 2,4-D degradation pathway encoded by plasmid pJP2 proceeds via 3,5-dichlorocatechol. It was repeatedly observed that *A. eutrophus* harboring pJP2 grows faster in 2,4-D medium than that containing pJP4. This phenomenon could be explained by the fact that dehalogenation by lactonization, followed by HCl elimination, is energetically the most favored mechanism (31). In summary, the divergence of 2,4-D pathways and the convergence of 2,4-D and 3Cba pathways may occur at dichlorophenol and chlorocatechol stages respectively.

Structural and Functional Interrelationships Between 2,4-D Plasmid pJP4 and Plasmids Isolated from 2,4,5-T-Degrading Strain of *P. Cepacia*

Unlike 3Cba and 2,4-D degradation, proof for the functional involvement of plasmids in 2,4,5-T degradation is lacking. Recent evidence from the work done in this laboratory, however, indicates a probable role of resident plasmids of *P. cepacia* strain AC1100 in 2,4,5-T degradation.

Total plasmid DNA has been isolated from this strain and physical characterization has revealed some heterogeneity. The most predominant population (80% or more) consists of a single size plasmid of 170 kb ± 5 kb. Approximately 10% of the population had a size of 40 kb. These 2 plasmids were designated pDG3 and pDG4. The rest of the molecules were of sizes varying from 2 kb to 30 kb and were least predominant as judged from electron microscopic analysis. Curing of strain AC1100 of the resident plasmids using conventional techniques of mitomycin C treatment has met with failure. The instability of the 2,4,5-T phenotype especially in absence of selection pressure is a major problem in isolating plasmid-free 2,4,5-T derivatives.

In our preliminary study, we introduced RP4 or R68.45 into a strain derived from AC1100 (DC102, 2,4,5-T$^+$ Nal^r) and repeatedly subcultured under strong antibiotic selection. From such populations, we isolated derivatives of strain AC1100 that appear to have lost some resident plasmids as judged from the restriction analysis of plasmid DNAs isolated from 2,4,5-T$^-$, Kan^r, Tet^r, and Amp^r mutants. RP4 harboring DC102, 2,4,5-T$^-$ derivatives lacking plasmid DNA can grow well on salicylate indicating that salicylate-degrading genetic markers are present on the chromosome. A wild-type *P. cepacia* 383 (7) was also found to be Sal$^+$ and did not harbor any detectable plasmid. No homology was detected between a known SAL plasmid and AC1100 plasmids in Southern blotting experiment.

Since 2,4-D and 2,4,5-T are structurally analogous, it seems reasonable to assume that the genes encoding 2,4,5-T related catabolic functions may have some homology with 2,4-D degradative genes as part of their common evolutionary ancestry. In agreement with this assumption, a region of homology has been detected (∿ 4 kb) between pAC1100 (a composite of AC1100 plasmids) and a segment of the *EcoRl* fragment (EB) of pJP4 which harbors the structural genes for both the 2,4-D and halocatechol degradation. Additional regions of homology between these 2 plasmids have been detected, especially in the region of putative regulatory functions for halocatechol catabolism on the pJP4 physical map. Because of the large size of pAC1100, lack of restriction map, and other recognizable physical and genetic markers, analysis of the precise functional role of the plasmids in 2,4,5-T degradation has not yet been possible. Furthermore, plasmids from AC1100 are isolated in poor yields from batch cultures grown on 2,4,5-T, presumably because of a low copy number of these plasmids. Presence of extracellular DNAase in *P. cepacia* may also greatly reduce the yield of the plasmid DNA.

In spite of the significant homology detected between pJP4 and resident plasmids of AC1100, the fact that 2 catabolic pathways seem to have undergone drastic evolutionary divergence was suggested from an induction experiment. Induction of 2,4,5-T catabolic pathway with 200 μg/ml of 2,4,5-T did not support any detectable growth of *P. cepacia* DC102 in minimal medium containing 800 μg/ml 2,4-D. 2,4,5-T grown AC1100 cells are known to completely dechlorinate 2,4-D, but are unable to grow with this compound since the dechlorinating enzymes are not induced by 2,4-D.

Both pJP4 and pJP2 could be transferred at low frequency (10^{-8} and 10^{-7}, respectively) to the strain DC102 resulting in new hybrid strains, designated as IY43 (2,4-D$^+$, 2,4,5-T$^+$, Hg^r, Nal^r) and IY25 (2,4-D$^+$, 2,4,5-T$^+$, Nal^r), respectively, capable of growing with both 2,4-D and 2,4,5-T. In strain IY25, both plasmid pJP2 and pAC1100 exist in unmodified form whereas exconjugants containing pJP4 (6, including IY43, tested) showed a deletion of a 10-kb fragment of pJP4 but pAC1100 appeared to be unmodified (I.-S. You, in prep.).

Because of the significant homology of the 4-kb segment between the plasmids, pJP4 and pAC1100, it was anticipated that recombinational events might generate hybrid plasmids in vivo. Mobilization of these 2 plasmids in different genetic background may help in physical and functional analysis of _P. cepacia_ plasmids. Experiments along this line are presently underway in our laboratory.

CONCLUDING REMARKS

The extensive homology observed between pAC27 and pJP4 concerning the chlorocatechol-degradative genes and that between the same regions of pJP4 and the resident plasmids of the 2,4,5-T-degrading strain of _P. cepacia_ AC1100 is not only interesting, but it also points to the role of plasmid genes in the evolution and spread of degradative characters against toxic chemical compounds. Although this phenomenon is well-known in the evolution and dissemination of antibiotic-resistance genes, no clear evidence for the occurrence of a similar trait in soil bacteria was available until now. The structural studies presented in this paper also point to the important role of direct and inverted repeats in the genetic rearrangements necessary for rapid transition from growth on one substrate to that of a different substrate. The amplification of the structural genes in absence of a putative positive regulatory element is analogous to that observed in case of antibiotic resistance to allow the host microorganisms to withstand high and toxic concentrations of such antibiotics (32,33). Many toxic chemicals resemble antibiotics in terms of their inherent toxicity, both toward animals as well as microorganisms, and it is little wonder that microorganisms employ essentially identical tools, i.e., deletion, fusion, insertion, and other modes of genetic rearrangements, to help evolve new degradative functions. Continued studies of such processes are expected to provide unique insights regarding the mode of evolution of new biological functions in bacteria.

ACKNOWLEDGEMENTS

The investigation has been supported by a grant from the National Science Foundation (PCM 81-13558) and by a cooperative program grant (CR80-9666-01-0) from the U.S. Environmental Protection Agency.

REFERENCES

1. Alexander, M. (1969) Microbial degradation and biological effects of pesticides in soil iv soil biology. Rev. Res. Nat. Resources Res., UNESCO 9:209-226.
2. Alexander, M. (1981) Biodegradation of chemicals of environmental concern. Science 211:132-138.
3. Kawasaki, H., N. Tone, and K. Tonomura (1981) Purification and properties of haloacetate halidohydrolase specified by plasmid from Moraxella sp. strain B. Agric. Biol. Chem. 45:35-42.
4. Goldman, P., G.W.A. Milne, and D.B. Keister (1968) Carbon-halogen bond cleavage. III. Studies on bacterial halidohydrolases. J. Biol. Chem. 243:428-434.
5. Weightman, A.J., A.L. Weightman, and J.H. Slater (1982) Stereospecificity of 2-monochloropropionate dehalogenation by the two dehalogenases of P. putida PP3: Evidence for two different dehalogenation mechanisms. J. Gen. Microb. 124:433-437.
6. Chapman, P.J. (1979) Degradation mechanisms. In Microbial Degradation of Pollutants in Marine Environments, Environmental Protection Agency, Washington, D.C. pp. 28-66.
7. Kilbane, J.J., D.K. Chatterjee, J.S. Karns, S.T. Kellogg, and A.M. Chakrabarty (1982) Biodegradation of 2,4,5-trichlorophenoxyacetic acid by a pure culture of Pseudomonas cepacia. Appl. Env. Microb. 44:72-78.
8. Karns, J.S., J.J. Kilbane, D.K. Chatterjee, and A.M. Chakrabarty (1984) Microbial biodegradation of 2,4,5-trichlorophenoxyacetic acid and chlorophenols. In Genetic Control of Environmental Pollutants, G.S. Omenn and A. Hollaender, eds. Plenum Press, New York, pp. 3-21.
9. Fisher, P.R., J. Appleton, and J.M. Pemberton (1978) Isolation and characterization of the pesticide degrading plasmid pJP1 from Alcaligenes paradoxus. J. Bact. 135:798-804.
10. Stanlake, G.J., and R.K. Finn (1982) Isolation and characterization of pentachlorophenol degrading bacterium. Appl. Env. Microb. 44:421-427.
11. Pignatello, J.J., M.M. Martinson, J.G. Steiert, R.E. Carlson, and R.L. Crawford (1983) Biodegradation and photolysis of pentachlorophenol in artificial freshwater streams. Appl. Env. Microb. 46:1024-1031.
12. Chakrabarty, A.M. (1976) Plasmids in Pseudomonas. Ann. Rev. of Genet. 10:7-30.
13. Reineke, W., S.W. Wessels, M.A. Rubio, J. Latorre, V. Schwien, E. Schmidt, M. Schlomann, and H.-J. Knackmuss (1982) Degradation of monochlorinated aromatics following transfer of genes encoding chlorocatechol catabolism. FEMS Microb. Lett. 14:291-294.
14. Weightman, A.J., R.H. Don, P.R. Lehrbach, and K.N. Timmis (1984) The identification and cloning of genes encoding haloaromatic catabolic enzymes and the construction of hybrid pathways for substrate mineralization. In Genetic Control of Environmental Pollutants, G.S. Omenn and A. Hollaender, eds., Plenum Press, New York pp. 47-80.

15. Reineke, W., and H.-J. Knackmuss (1978) Chemical structure and biodegradability of halogenated aromatic compounds: Substitutent effects on 1,2-dioxygenation of benzoic acid. Biochim. Bioph. Acta. 542:412-423.
16. Reineke, W., and H.-J. Knackmuss (1978) Chemical structure and biodegradability of halogenated aromatic compounds: Substituent effects on dehydrogenation of 3,4-cyclohexadiene-1,2-diol-1-carboxylic acid. Biochim. Bioph. Acta. 542:424-429.
17. Knackmuss, H.-J., and M. Hellwig (1978) Utilization and cooxidation of chlorinated phenols by Pseudomonas sp. B13. Arch. Microb. 117:1-7.
18. Dorn, E., and H.-J. Knackmuss (1978) Chemical structure and biodegradability of halogenated aromatic compounds: Two catechol 1,2-dioxygenases from a 3-chlorobenzoate grown Pseudomonad. Biochem J. 174:73-84.
19. Schmidt, E., and H.-J. Knackmuss (1980) Chemical structure and biodegradability of halogenated aromatic compounds: Conversion of chlorinated muconic acids into maleoylacetic acid. Biochem J. 192:339-347.
20. Hartmann, J., W. Reineke, and H.-J., Knackmuss (1979) Metabolism of 3-chloro, 4-chloro, and 3.5 dichlorobenzoate by a Pseudomonad. Appl. Env. Microb. 37:421-428.
21. Reineke, W., and H.-J. Knackmuss (1980) Hybrid pathway for chlorobenzoate metabolism in Pseudomonas sp. B13 derivatives. J. Bact. 142:467-473.
22. Pemberton, J.M., B. Corney, and R.H. Don (1979) Evolution and spread of pesticide degrading ability among soil microorganisms. In Plasmids of Medical Environmental and Commercial Importance, K.N. Timmis and A. Puhler, eds. Elsevier/North Holland, Biomedical Press, Amsterdam.
23. Schwien, U., and E. Schmidt (1980) Improved degradation of monochlorophenols by a constructed strain. Appl Env. Microb. 44:33-39.
24. Chatterjee, D.K., S.T. Kellogg, S. Hamada, and A.M. Chakrabarty (1981) Plasmid specifying total degradation of 3-chlorobenzoate by a modified ortho pathway. J. Bact. 146:639-646.
25. Don, R.H., and J.M. Pemberton (1981) Properties of six pesticide degradation plasmids isolated from Alcaligenes paradoxus and Alcaligenes eutrophus. J. Bact. 145:681-686.
26. Chatterjee, D.K., and A.M. Chakrabarty (1983) Genetic homology between independently isolated chlorobenzoate-degradative plasmids. J. Bact. 153:532-534.
27. Chatterjee, D.K., and A.M. Chakrabarty (1982) Genetic rearrangement in plasmids specifying total degradation of chlorinated benzoic acid. Mol. Gen. Genet. 188:279-285.
28. Friedman, A.M., S.R. Long, S.E. Brown, W.J. Buikema, and F.M. Ausubel (1982) Construction of a broad host range cosmid cloning vector and its use in the genetic analysis of Rhibozium mutants. Gene 18:289-296.

29. Ditta, G., S. Stanfield, D. Cobin, and D. Helinski (1980) Broad host range DNA cloning system for gram-negative bacteria: Construction of a gene bank of Rhizobium meliloti. Proc. Natl. Acad. Sci., USA 77:7347-7351.

30. Birnboim, H.C., and J. Doly (1970) A rapid alkaline extraction procedure for screening recombinant DNA. Nucl. Acid. Res. 7:1513-1523.

31. Wood, J.M. (1982) Chlorinated hydrocarbons-oxidation in the biosphere. Environ. Sci. Tech. 16:291A-297A.

32. Perlamn, D., and R.H. Rownd (1975) Transition of R factor NR1 in Protens mirabilis: Molecular structure and replication of NR1 deoxyribonucleic acid. J. Bact. 123:1013-1034.

33. Ptashne, K, and S.N. Cohen (1975) Occurrence of insertion sequence (Is) region on plasmid deoxyribonucleic acid as direct and inverted nucleotide sequence duplications. J. Bact. 122:776-781.

NOTES ON METABOLIC PLASMID ORGANIZATION AND EXPRESSION

I. C. Gunsalus

School of Chemical Sciences
University of Illinois
Urbana, Illinois 61801

METABOLIC VERSATILITY AND PLASMIDS

The term metabolic plasmid, as opposed to resistance plasmid, refers to the encoding of multistep reaction sequences leading from noncellular structures to anaplerotic intermediates. The alternate term, degradative or catabolic plasmid, also denotes the control of these predominately energy coupled processes yielding principally the 2 to 4 carbon acids whose metabolism is chromosome encoded. The aromatic and terpene oxidative pathways, for example, each comprise more than a dozen reactions, several catalyzed by multiprotein enzyme systems. The genes are organized in 2 or more clusters under positive pleiotropic and secondary control. Figure 1 illustrates the phenotypes encoded by the "terpene" plasmid CAM, the "aromatic" ones NAH and SAL, and the "heterocyclic" NIC plasmid. For illustration we shall consider the state, structure, and control of CAM and of NAH/SAL. A related aromatic plasmid, TOL, will be considered by participants who have made the primary contributions.

Many, perhaps most, metabolic plasmids are self-fertile and range in size from ca. 100 to $\geq$ 200 kilobases (kb). They are scored on minimal media as positive-growth phenotypes. In contrast, the smaller resistance plasmids are scored on complete media and usually carry one, or a few, genes that encode enzymes to modify the procaryote toxic agents. The fertility regions of about 40-50 kb contribute at least in part, along with the multigene, multi-regulatory units, to the large sizes.

Metabolic plasmids were first recognized during our search for the genetic basis for the conversion in nature of organic structures to biosynthetic intermediates and to atmospheric CO_2. The first hint of an extra-chromosomal gene array came from lack of transductional linkages between the control of terpene oxygenation and auxotrophic loci in fluorescent pseudomonads, known to be prime partici-

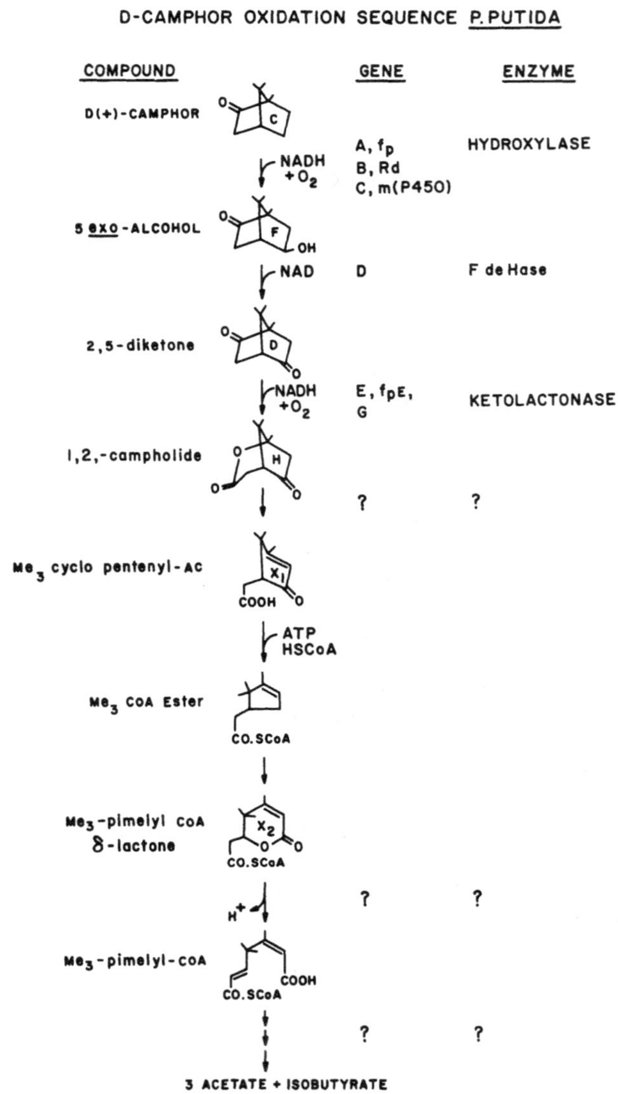

Fig. 1. Examples of phenotypes encoded by metabolic plasmids.

A. CAM, D-camphor to $3C_2 + C_4$ products.

pants in cycling carbon in the biosphere (1). The second signal came from the spontaneous, then induced, loss of entire pathways-so-called "curing" or Seg mutants. As the chemical intermediates became available for genotypic scoring, linkages were soon found for sequential steps along the metabolic pathways. Plasmid fertility was established, an additional role in chromosome mobilization docu-

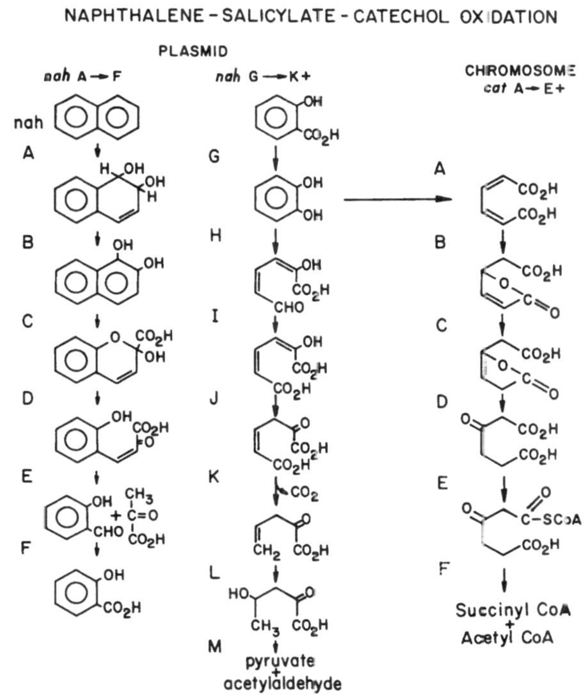

Fig. 1. (Continued)

B. NAH, naphthalene/salicylate to $C_1 + C_2 + C_3$, or to $C_2 + C_4$.

mented, and plasmid chromosome gene redundancy found at the isobutyrate level as we observed co-induction with the metabolism of the branched chain amino acid valine as a carbon source.

To inquire of a general, versatile role of plasmids, we selected the aromatic structures whose metabolic intermediates and pathways had been established, mostly by the Dagley and Evans schools (2). Noel Dunn arrived from the Holloway laboratory, examined the genetic base of naphthalene oxidation and validated a NAH plasmid, pDG7, in a stock culture strain, ATCC 17485. Two phenotypes, Nah^+ and Sal^+, were established and the plasmid implicated in the catechol meta cleavage. The ortho fission pathway was shown to be chromosome-encoded, and under inductive control at the level of catechol an early product thereof (3). Subsequently, general occurrence of plasmids carrying the genes for aromatic oxidative processes was recognized. The naphthalene-positive strains from the Stanier et al. collection (4), whether curable with mitomycin or not, have been shown by Patty Harder to carry plasmids with restrictase patterns compatible with pDG7. Today, TOL (5) is the best

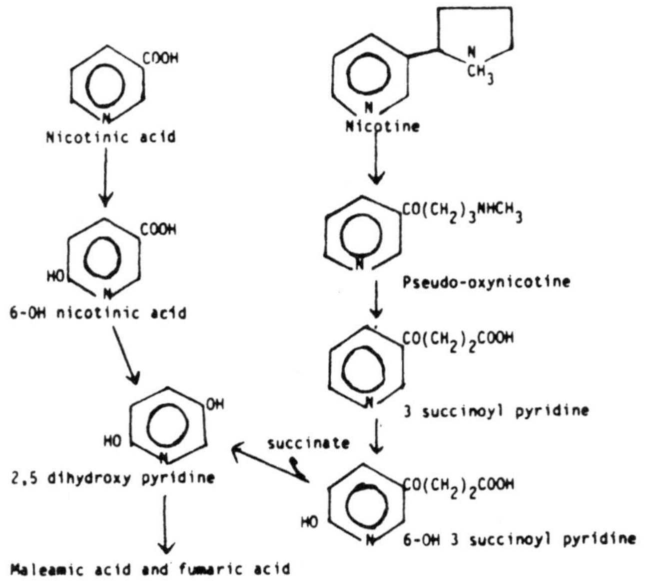

Fig. 1. (Continued)

C. NIC, nicotine to $C_1 + 2C_4 + NH_3 + CH_3NH_2$.

documented plasmid. In host strain mt-2 it specifies the genetic organization of the toluene-xylene oxidative patterns. The relevant data appear in this volume from the 3 principle groups, Williams in Bangor (5), Timmis (6), and Bagdasarian (7) in Berlin--now Geneva and Umea, respectively--and the Nakazawa laboratory (8) at the University of Yamaguchi in Japan.

The abundance and variety of organic structures released into the biosphere challenge the natural carbon cycling processes of nature in breadth of genetic and catalytic questions, many of whose solutions remain obscure (9).

Genotype Scoring

The mammalian liver and kidney microsomes and the microbes of the soil and water serve similar, prominent roles in the removal of these structures. Both have been shown to a major extent to activate molecular dioxygen at the expense of redox energy with direct incorporation of [O] or [O-O] into substrate at otherwise inert loci. The primary oxidative process occurs at a heme iron atom bearing a cysteinyl thiolate axial ligand which can serve as a 2 electron reservoir prior to H_2O release. Examples of plasmid versatility and the relevance of the catalytic centers can be illustrated by their role in encoding the oxygenases that induct hydrocarbon structures to the more familiar, or general, pathways. Two solutions to the selectivity have been experienced, with control in both gene and enzyme expression. The more selective, stereospecific, systems have been found in the mammalian mitochondria among the biosynthetic processes, e.g., for the formation of regulatory steroids, prostaglandins, bile acids, etc., and in the microbic processes. A contrasting example occurs in the relaxed control of steric requirements in the hepatic microsomes (9). The usual hepatic and renal processes convert to soluble derivatives and excrete the normal products of tissue metabolism, those ingested as pallative or therapeutic chemicals (drugs, antibiotics, etc.) or as residues of food constituents and flavors, and those introduced from the atmosphere as respiratory insult. The microbes recycle most carbon compounds to CO_2 aerobically, or anaerobically, form reduced products, e.g., methane, acetate, etc., and most generally derive also therefrom the intermediates of cell synthesis. Thus, the reaction sequences are more extensive, enjoy a larger modulation range, and are under more clustered and coordinated genetic control.

The hepatic P450 heme sulfur "cytochromes" of the microsomes have received the most extensive study from their implication in human and animal health, whereas the aerobic microorganisms have provided the bulk of genetic and chemical data as a result of amenability to genetic manipulation and of enzyme abundance and activity. The microbial systems contrast additionally with the hepatic microsomal ones. The hepatic enzymes were found to be of low affinity for a broad range of substrates with relatively low turnover, whereas the microbial enzyme systems are more selective with higher substrate affinities and turnover.

The gene-enzyme components and reactions of 2 microbial systems for the terpenes are illustrated in Fig. 2. The electron flow does not pass to the heme in the absence of substrate and thus does not dissipate cellular energy. Three genes, camABC, under coordinate regulation specify the 2 redox effector auxiliary proteins, and the substrate-selective monoxygenase $P450_{CAM}$. The 45 kd flavoprotein and a 12 kd iron sulfide protein are encoded, respectively, by camA and B; the camC product, a 45 kd $P450_{CAM}$ by camC, the following camD locus product is, an alcohol dehydrogenase composed of 2 equal size ∿40 kd subunits.

GENETIC DISTRIBUTION

The information among procaryote strains is carried in the population. Cells selected by enrichment usually express but part of the total metabolic potential. The general distribution of traits, the homology among genomes and related pathways follow from the broad host-range plasmid fertility and their ability to initiate chromosome transfer. Relatedness is also indicated by the plasmid incompatibility groups, for example, CAM and OCT in P2 and NAH and

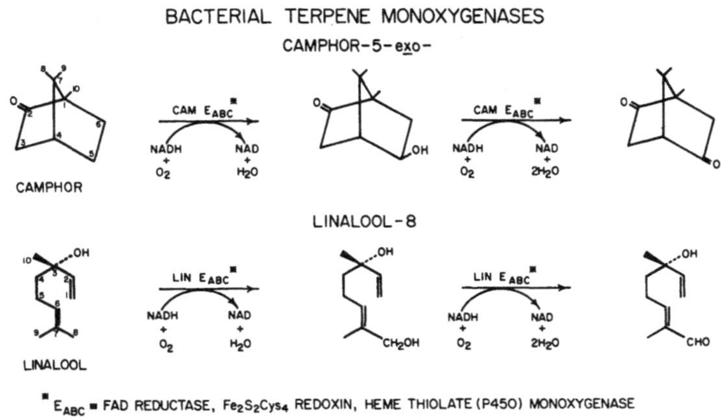

Fig. 2. Terpene P450 hydroxylase components - CAM/LIN.

SAL, P9 (28). In our hands convenient metabolic plasmid examples have included those that control terpene and aromatic oxidative pathways (5,28). More recent evidence suggests other terpene pathways under plasmid control, namely linalool, LIN. Preliminary reports have characterized a plasmid encoding heterocyclic metabolism of nicotine, NIC and nicotinate (29).

Although the CAM plasmid emerged first from our explorations, the aromatic plasmids, which followed, have provided the better method development and permitted more rapid progress in the analysis of structure and regulation (10-14). They are thus preferable examples.

Naphthalene/Salicylate

The naphthalene-positive phenotypes may utilize also salicylate, an intermediate of the first ring fission and pyruvate release, as shown in Fig. 1B. Thus, Nah^+ is a growth phenotype with cell yield equivalent to pyruvate, and Sal^+, a second one that provides through meta fission pyruvate and acetaldehyde or by the ortho pathway the succinyl- and acetyl-CoA derivatives. Aromatic structures with aliphatic substituents may yield ketones rather than aldehyde. These generalities apply to most aerobic, saprophytic cells of soil and water; of these we have used the fluorescent pseudomonads as biological prototypes (1,12,15).

A *Pseudomonas putida* strain PpG7 bearing the plasmid pDG7 expresses the inducible phenotypes Nah^+, Sal^+, with ring fission via the meta pathway (3). Chakrabarty (16) reported a plasmid, pAC1, of phenotype Nah^-Sal^+. The 2 are homologous except for an altered *nahA* gene locus (11-13), Fig. 3B. Among recent isolants, and culture collection strains, including several from the Rothamstead and Evans laboratories (4), were observed to carry one or more plasmids resembling pDG7 and pAC1.

Clues to the evolution of convergent and alternate genetic patterns among plasmids have been provided by the numbers of strains, the array of chemical intermediates, reactions and pathways observed with substrates from fossil fuels and products of current plant, animal, and synthetic chemistry. Among the convergent trends one observes the fission of catechol, and analogs, and the homology among genes for the meta pathway irrespective of the initial substrate, e.g., naphthalene/salicylate vs. xylene/toluene. Degeneracy may appear in the gene regulation with some retention in the enzyme complement. The range of structural tolerance among substrates is accompanied by mutation, altered gene array and regulation. For example, Yen's selection of Sal^+ revertants from Sal^- strains were found to delete plasmid genes which resulted in action of chromosomally encoded pathway, i.e., the ortho fission of catechol (24).

Alkane and Isoprenoid Gene Analyses

The plasmid-encoded hydrocarbon oxygenation processes of normal, branched, and cyclic carbon structures arose as we extended to genetic analyses the chemical inquiry of the terpene, camphor, and alkane metabolism (30-32). The 10 carbons of the terpene 2-bornanone are returned to CO_2 via 3 acetate and the debranching of isobutyrate to succinate. The bicyclic ones with ketone oxygen function gain solubility with less toxicity and hydrophobic rejection from the enzyme centers than the alcohols. Add the interesting stereochemical riddle presented by the examples of a hydroxylation of both the D- and L- enantomer of camphor by the single $P450_{CAM}$ enzyme center, indicated in Fig. 2, and a wedge to stereochemical permissiveness and group affinity is presented for refinement in a stringent enzyme model.

The acyclic monoterpenes are also prevalent plant products. Linalool, for example, with an assymetric (chiral) essential oxygen center at carbon 3 and 2 olefin centers exhibits configuration and binding requirements. Linalool-competent strains produce, under induction, a $P450_{LIN}$ system with stringent substrate requirements in a fluorescent pseudomonad. This curable phenotype presents an added access to genetic fine structure, some 60-odd analogs of linalool have been prepared to explore the heme-substrate site. From the exploration of chemical variety, features of homology emerge. For example, the system parallels of the camphor-5-exo-methylene and linalool-8 hydroxylases are indicated in Fig. 2. The enzyme systems differ only in the detailed primary and tertiary structure around the active center and in component interaction domains. The $P450_{LIN}$ active center probes extended to analogs modified at each of the 10 carbons and the essential oxygen function define first-order effects of spatial and affinity properties of the heme pocket (19).

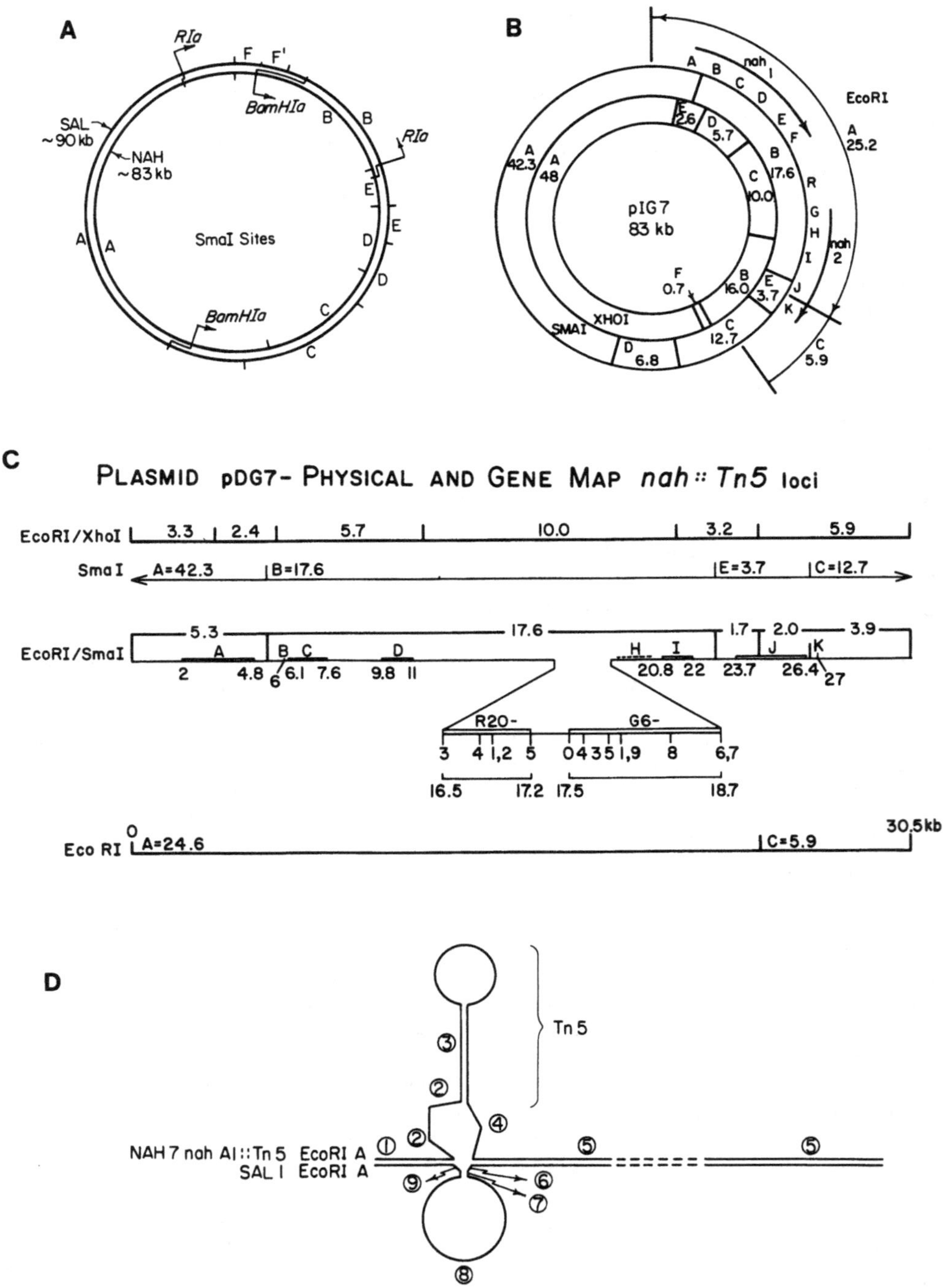

Fig. 3. Aromatic metabolic plasmid DNA: NAH(pDG7) + SAL(pAC1).
A. Circular SmaI restrictase maps.

Fig. 3. (Continued).

B. Structural and regulatory gene clusters of pDG7.

C. Nah gene organization: three operons, nahAF, GK, and nahR.

D. Heteroduplex of insertion/deletions, in nahA locus of pAC1/pDG7.

A third large class of terpenoid structures of the monocyclic group has been probed. Through para-cymene a bridge is offered to saturated and aromatized structures from a p-methyl and p-isopropyl base. Bhattacharyya et al. have developed the p-cymene system (20) and to a lesser extent the P. aeruginosa strains reported to metabolize geraniol (4); the redox and aliphatic vs. aromatic variety are added by the P450 para-cymene components. A $P450_{PCM}$ has been p-isolated; it responds to the inducing substrate p-cymene and numerous related saturated structures. The active center does not accommodate camphor nor linalool. The enzyme components in heterologous recombination exhibit affinities and limited catalysis.

The structure-function parameters of $P450_{CAM}$ have provided a prime model to physical, chemical, and genetic probes. The microbial origin affords control, quality, and abundance (15,21). The advances to the oxygenase field include primary structures of 2 components, the third well advanced. The carbon and O_2 center of $P450_{CAM}$ provides 2 regions with Cys residues with amino acid homology to other P450 systems including the hepatic microsomal enzymes (22,23). The current, working model targets 1 of the Cys residues as the essential sulfhydryl axial ligand to the heme (21-23) to provide the P450 character of the heme sulfur proteins. The unique 2 electron redox center is attributed to a retained sequence (26). A tertiary structure of $P450_{CAM}$ is now nearing completion with much anticipation of fine structure disclosure of binding, selection, and motion in catalysis (27).

HOST RANGE OF A PLASMID DNA EXPRESSION

As the methods for small plasmid DNA isolation were developed particularly in E. coli, we attempted their application to the larger metabolic plasmids of genus Pseudomonas. Early variations met with limited success prior to James Johnston's joining our laboratory with the primary desire to solve the DNA isolation problem and to document the genetic components by Pseudomonas transformation. Sporatic successes in a learning period blended into a predictable routine to yield circular plasmid DNA from metabolic plasmids of less than or equal to 100 kb. The primary focus on the NAH pDG7 plasmid was soon extended to the related TOL and SAL molecules with restrictase characterization and visualization by electrophoresis (10).

With the help of Jean Ito, Irving Crawford and I (11) found, by the first restriction analysis of NAH and SAL DNA preparations furnished by Johnston, similar fragments with the exception of a single larger component from SAL. *SmaI* was soon found to provide 5 near-identical size of fragments, and SAL 2 additional ca. 3 kb fragments. In *EcoRI* digests the latter arose in *SmaI* redigests of the *EcoRI* fragment A (11,12). Subsequently, Farrell modified the methods and carried out a broad spectrum of restrictase characterizations. These demonstrated NAH and SAL equivalence in Southern heteroduplex blots. The 117 kb TOL DNA anneals at about 70% homology with the 80 kb NAH. These data followed by transformation, with methods but slightly modified from the *E. coli* procedures of Helinski, confirmed and extended the earlier analyses by transduction and conjugation. The interstrain transfer, via transduction or conjugation, of the complete Nah^+ and Sal^+ phenotypes to the PpG1 cell line in strain 277 (*trpB615ΔCAM*) provided the a basis fof the genetic analysis of the organization and regulation of the plasmid-encoded aromatic metabolism.

The very large size of the terpene-encoded plasmids, for example, the CAM plasmid, has not yielded to routine isolation. Thus, the breakthrough of Koga to clone from a CAM plasmid DNA library with expression in *Pseudomonas* and *E. coli* has opened a route to full genetic participation. In *EcoRI* the pKT240 vector is amply expressed presumably from the Ap^r promoter; in vector pBR322 similar expression ensues from the same promoter. The *camC* gene product $P450_{CAM}$ 412 of amino acid residues is formed constitutively from a 2.2 kb insert; the genetic organization of these systems on the plasmids will be discussed. The *Pseudomonas* hosts express the *camC* insert of vector pKT240 constitutently from an unidentified promoter at levels equivalent to about twice the expression in *E. coli* from the vector pBR322.

PLASMID ORGANIZATION

Aromatic/Naphthalene

The circular restriction maps of the DNA from the NAH, SAL, and TOL plasmids, respectively pDG7, pAC1, and pWWO, were prepared. The intact DNA was shown to transform plasmid cured strains of *Pseudomonas putida* from both the homologous and a variety of heterologous cell lines (6,16). The pDG7 plasmid, first analyzed by Dunn, was subjected to Tn5 insertion mutagenesis by Yen (12). The insertion loci were determined by restrictase digestion and the genotypes scored by enzyme assay to define gene order and locus as shown in Fig. 3B. Two gene clusters were found, each of about 10 kb, bracketing a regulatory region of about 1 kb. These are indicated in more detail in Fig. 3C. The 2 gene clusters, under pleiotropic control by salicylate indicates 2 operons mediated by the locus nahR. The nahR locus maps just prior to the nahG locus of the second operon about 5 kb downstream from nahD. The nahEF loci specifying an aldolase and aldehyde dehydrogenase could reside in this region. The more detailed map, prepared by scoring the gene products of some 50 Tn5 inserts suggests the gene boundaries. They may be extended as additional mutant loci are identified. Especially to be noted is the rather extended region assigned to genes nahA and J. Gibson et al. have defined 3 protein components of the naphthalene hydroxylase; thus 3 genesare encoded within the 3 kb informally denoted as "nahA."

Homology of the NAH and SAL plasmids is apparent as indicated by the EcoRI and SmaI restrictase fragments, Fig. 3A, and by heteroduplex microscopy annealing and electron measurement. The EcoRI A fragment from pDG7 mutant nahA1::Tn5 vs. the wild-type pAC1 as shown in Fig. 3D with the Tn5 inverted repeat mapped as marker at 2.6 kb from the left margin of the EcoR1 fragment A, the SAL plasmid pAC1 shows a short inverted repeat at about 100 kb, and the longer insert, indicated by 8 which has not annealed with the pDG7 nahA gene locus. The Tn5 insert also makes visible small segments of nonannealed nahA as indicated by 2, 2', and 4 in the figure. The nature of the SAL insert and deletion will require added DNA analysis. These lesions and the pleiotropic control result in a full Nah$^-$ operon.

In summary, the Nah$^+$ and Sal$^+$ phenotypes of pDG7 result from 2 operons under pleiotropic control of salicylate or anthranilate (2-aminobenzoate). They account for about 30 kb of the 80 kb plasmid. Should the fertility functions require 40 kb of DNA, one would anticipate few additional genes in this aromatic plasmid.

VECTORS AND PROMOTERS IN PSEUDOMONAS AND E. COLI

The cell lines which have proved most useful in our analysis of the aromatic and terpene plasmids were derived from the wild-type CAM isolant PpG1. The CAM cured (Δcam) plasmid-free hosts PpG277 (trpB615Δcam) and PpG1343 (met-616Δcam) are preferred stable auxotrophs for transformation and expression. They serve also classical genetics for pf16 transduction, conjugation, and introduction of plasmid fertility. The preferred E. coli strains carry RR1 or are LE392 or HB101.

Broad host-range, multicopy vectors differing in compatibility have been prepared respectively from the vectors RSF1010, IncP4 and pRK2, IncP1. The derivatives used are pKT240, from RSF1010 and pRK290 from pRK2, kindly supplied, respectively, by Timmis and Bagdasarian, and Helinski. Figure 4A indicates their structures and sizes, ∿12 and 20 kb respectively, each suitably expressed in the P.

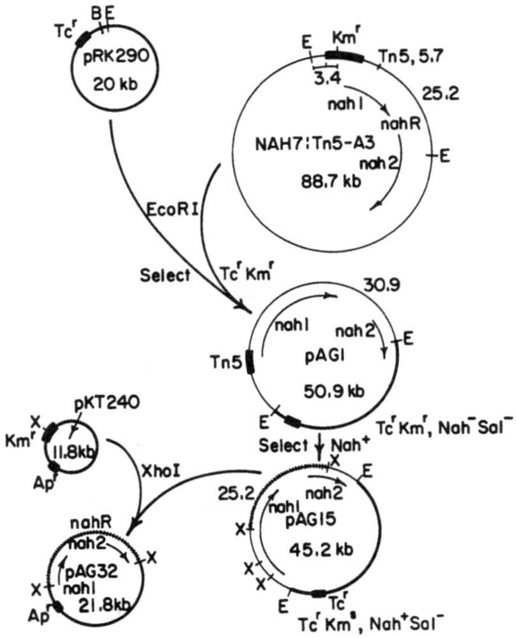

Fig. 4. Aromatic plasmid cloning and expression.

A. pDG7 subclones in compatible vectors pRK290/pKT240.

putida cell line; the vector pBR322 also expresses Pseudomonas genes in E. coli at similar enzyme levels.

The pKT240 vector expresses pseudomonad genes in the Pseudomonas and E. coli hosts, at approximately equal level, and in the latter pBR322, is an equally effective vector. Both pKT240 and pBR322 carry ampicillin, an Ap^r promoter upstream in the PstI insertion site from which expression occurs, whereas in the pseudomonad the pKT240 expression of pseudomonad insertions occurs from a promoter of unknown structure. In our experience, the difficulties encountered by some authors in obtaining expression of pseudomonad genes in E. coli, derives at least in part from attempts to use Pseudomonas promoters rather than from differences in the DNA expression per se (5,7,13,16).

DNA libraries have been prepared routinely from the aromatic plasmid DNAs in the case of NAH, SAL, and TOL. The constructs from the Tn5 insertion mutant, in nahA, were prepared first in pRK290 from an EcoRI library with Tc^r or Km^r selection. The Nah$^-$Sal$^-$ clones resulting from the Tn5 insertion were selected for Nah$^+$ indicating an effective promoter and subclone into pKT240 and an XhoI site by selecting for Ap^{r}Km^s clones. The inserts were mapped as shown in Fig. 4B. The subclones and their genotypes are indicated in Tab. 1.

Evidence of Positive Control

The subclones from pAG15 shown in Fig. 4B and Tab. 1 indicate expression by complementation with the nahR region carried in pKT240 in trans array. Complementation with pAG26 and 28 are indicated in Tab. 1 to occur at wild-type level. These data are most readily interpreted as a positive control mechanism.

EXPRESSION CLONES OF THE camC GENE: THE $P450_{CAM}$ MODEL

The CAM plasmid, pRG1 > 200 kb has eluded thus far routine isolation. Occasional success and selection for transcomplementation of the plasmid genes (1) from PstI in the restrictase digest or the use of whole-cell plus plasmid DNA have provided a route to suitable $P450_{CAM}$ and the related expression clones. The selection method is for growth on camphor in strain PpG543 (camC100); selection for Cam or Km^r markers provided a series of Cam^+ strains. Following for selection for rapid growth, plasmid DNA was isolated by the usual procedure (11) and the DNA used to transform strain PpG277 (trpB615Δcam). Clones were grown in PAS-Cam medium with a monosodium glutamate as a secondary substrate and screened for P450 production, supplemented with tryptophan. The ferrous-CO - ferrous spectra provided the expression levels of $P450_{CAM}$ indicated in Fig. 5D.

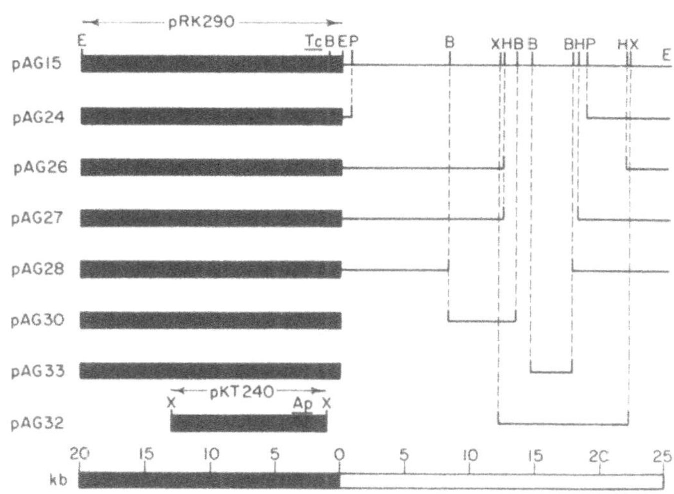

Fig. 4. (Continued).

B. pAG15 subclone expression with nahR in trans.

nah OPERON HYBRID AND DELETION PLASMIDS – pRK290-pDG7

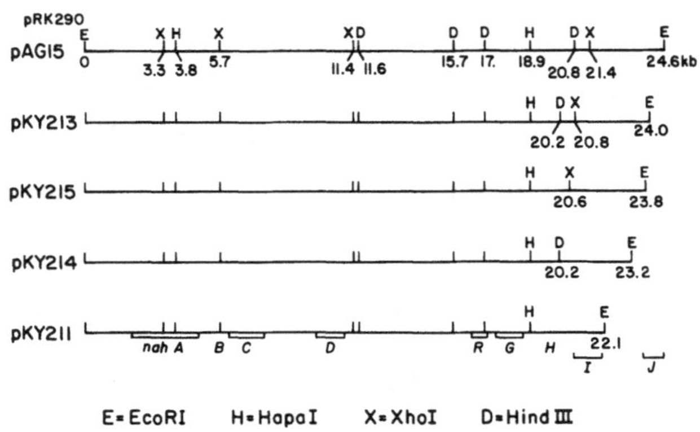

Fig. 4. (Continued)

C. Phenotypic revertants express aromatic ortho fission.

Defective strains were subcloned into the Pst site of pBR322, selected for Tc^rAp^s, then screened for P450 production. As indicated in Fig. 5D, the plasmid pKG300 produced ample P450 and E. coli RR under inducing conditions. As indicated by the data in Fig. 5A, the expression of both vectors in E. coli as from the ampicillin-resistant promoter, upstream to the PstI insertion site, i.e., plasmids pKG202 and 300, whereas in the pseudomonad, expression from vector pKT240 showed an unidentified promoter in the BstEII region

Tab. 1. Expression in trans of nah genes.

Plasmid	Enzyme Activity Expressed - nkat[a]								
	A	B	C	D	E	F	G	H	I
pDG7	100[b]	400	900	20	300	200	90	560	50
pAG15	85	500	1100	25	450	250	150	1000	50
pAG26 + 32	60	300	900	20			40	400	2
pAG28 + 32	65	300	nil				26	190	17
" -- 32					6	20	50	400	5
" 26 --	-	-	-	-			3	4	3
" 28 --	-	-					3	5	4

a) nmoles product/min/mg protein; - = nil; blank = not tested.

b) Induced with 0.3 mM anthrinilate; all noninduced are <.1 to .01 of induced.

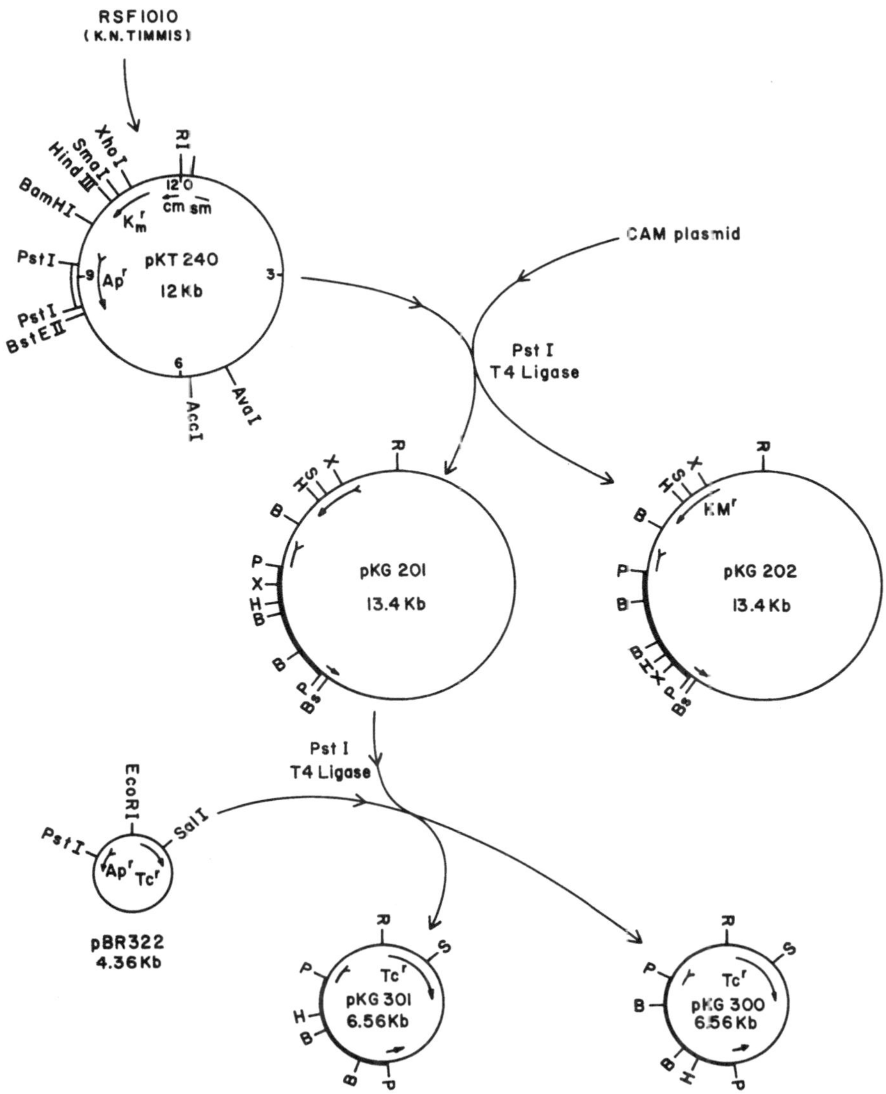

Fig. 5. Expression of the camC gene in Pseudomonas and in E. coli.

A. pKG201, a 2.2 kb PstI insert of camC in vector pKT240.

as indicated in pKG201. These differences are also expressed in Fig. 5D. Figures 5B and 5C concern the 2.2 kb insertion carried in vectors pBR322 and pKT240 whose expression is shown in Fig. 5D together with restriction maps of this region. The 2.2 kb insert expressed in pKG201 was also expressed at the same level in an XhoI

DNA SEQUENCE DATA FROM 2.2 kb FRAGMENT
by S. M. LEE

```
                                        -143                                                                    -100
                                        C G C T G G C C G A T C T G A T C A C C C A G C G T T T T T C C A T C G A C G A

                                                                                                                 -40
G G C C A G C A A G G C A C T T G A A C T G G T C A A G G C A G G A G C A C T G A T C A A A C C C G C G A T C G A C C C

                                                                              1                                   21
C A C T C T T T A G C C A A C C C G C G T T C C A G G A G A A C A A C A A C A A T G A C G A C T G A A A C C A T A C A A
                                                                               MET   THR   THR   GLU   THR   ILE   GLN

                                                                        61          171                          182
A G C A A C G C C A A T C T C G C C C C T C T G C C A C C C C A T G T G C C A G A G--      --C A A C G G C G G A C A C
 SER   ASN   ALA   ASN   LEU   ALA   PRO   LEU   PRO   PRO   HIS   VAL   PRO   GLU                 ASN   GLY   GLY   HIS

                                                                                                                 246
T G G A T C G C C A C T C G C G G C C A A C T G A T C C G T G A G G C C T A T G A G G A T T A C C G C C A C T T T T C C
 TRP   ILE   ALA   THR   ARG   GLY   GLN   LEU   ILE   ARG   GLU   ALA   TYR   GLU   ASP   TYR   ARG   HIS   PHE   SER

                                                                                                           297
T C C A G C G A G T G C C C G T T C A T C C C T C G T G A A G C C G G C G A A G C C T A C G A C T T C N T T C C C--
 SER   SER   GLU   CYS   PRO   PHE   ILE   PRO   ARG   GLU   ALA   GLY   GLU   ALA   TYR   ASP   PHE   ???   PRO

386                                                                                                              444
--G G A T C C A G G A G C T G G C C T G C T C G C T G A T C G A G A G C C T G C G C C C G C A A G G A C A G T G C A A C
     ILE   GLN   GLU   LEU   ALA   CYS   SER   LEU   ILE   GLU   SER   LEU   ARG   PRO   GLN   GLY   GLN   CYS   ASN

                                                                                                                 554
T T C A C C G A G G A C T A C G C C G A A C C C T T C C C G A T A C G C A T C T T C A T G C T G C T C G C A G G T C T A
 PHE   THR   GLU   ASP   TYR   ALA   GLU   PRO   PHE   PRO   ILE   ARG   ILE   PHE   MET   LEU   LEU   ALA   GLY   LEU

                                                                                                                 564
C C G G A A G A A G A T A T C C C G C A C T T G A A A T A C C T A A C G G A C C A G A T G A C C C G T C C G G A T G G C--
 PRO   GLU   GLU   ASP   ILE   PRO   HIS   LEU   LYS   TYR   LEU   THR   ASP   GLN   MET   THR   ARG   PRO   ASP   GLY

   1080                                                                                                         1137
   --C T G G C C C G C C G G G A A A T C A T C G T C A C C C T C A A G G A A T G G C T G A C C A G G A T T C C T G A C
       LEU   ALA   ARG   ARG   GLU   ILE   ILE   VAL   THR   LEU   LYS   GLU   TRP   LEU   THR   ARG   ILE   PRO   ASP

                                                                                                                1197
T T C T C C A T T G C C C C G G G T G C C C A G A T T C A G C A C A A G A G C G G C A T C G T C A G C C G C G T G C A G
 PHE   SER   ILE   ALA   PRO   GLY   ALA   GLN   ILE   GLN   HIS   LYS   SER   GLY   ILE   VAL   SER   GLY   VAL   GLN

                        1211                 1225                                                               1260
G C A C T C C C T C T G G T--                --A C T A C C A A A G C G G N A T A A A C A C A T G G G A G T G C G T G C
 ALA   LEU   PRO   LEU                           THR   THR   LYS   ALA         ***

                                                                                                                1320
T A A G T G A A C G C A A A C G A C A A C G T G G T A A A C G T C G G T A C C G G A C C G G A C T G G C T G G C G T T G
 ***

                                                                                                                1380
A G G T C G C C T T C G G C C T G C G C G C C A G C G G C T G G G A A G G C A A T A T C C G G T T G G T G G G G G A T G

                                                          1413                              1430                1440
C G A C G G T A A T T C C C C A T C A C T T A C C A C C C T T C C--                          --G G C A A A G C C A C
             ***

                                                                                                                1450
A G C G G A A A G C C T G T A C C T G A G A A C C C C A G A T G C C T A T G C A G C G C A G A A C A C C A A C A C C G G

                                                                                                                1560
A G G C C C A G G T A A C G G C T A T C A A C C G C G C C G A C A G C A A G T A T C C T A T C G G A T C G C C G G G C A

                                                                                                                1620
C T G G A T T A C G A C C G G C T G G T A T T G G C T A C C G G A G G G C G T C C A A G A C C C T A C C G G T G C C A G

                                                                                                                1680
T G G C G C A G T T G G A A A G G C G A A C A A C T T T C G A T A C C C C C C T C G A G T G C A T T C C C C G G C A G C

                                                                                                                1740
T G A T T G C G G A T A A C C G T C T G G T G G T G A T T G G T G G C G G C C A C A T T G G C C T T G A A G T G G C T G

                                                                                                                1800
C C A C C G C C A T C A A G G C G A T A C A T G C A C G T C A C C G T G C T T G A T A C G G C A G C C C G G T C T G G A

                                                    1829
G C G G G T T A C C G C C C G C C G G A G G C T A C G A G
```

Standard three letter amino acid code used.
*** = termination sequence

Fig. 5. (Continued)

B. Partial amino acid and DNA sequences of plasmid pKG201.

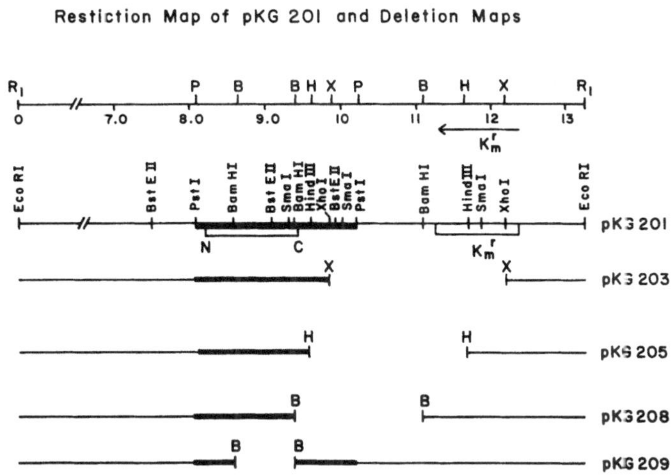

Fig. 5. (Continued)

C. Restrictase map and camC gene locus in pKG201.

and a HindIII subclone, indicated as pKG203 and 205, thus indicating that the structural gene lies between PstI and the HindIII site, about 1.6 kb.

Partial sequences from the restriction sites remaining in the

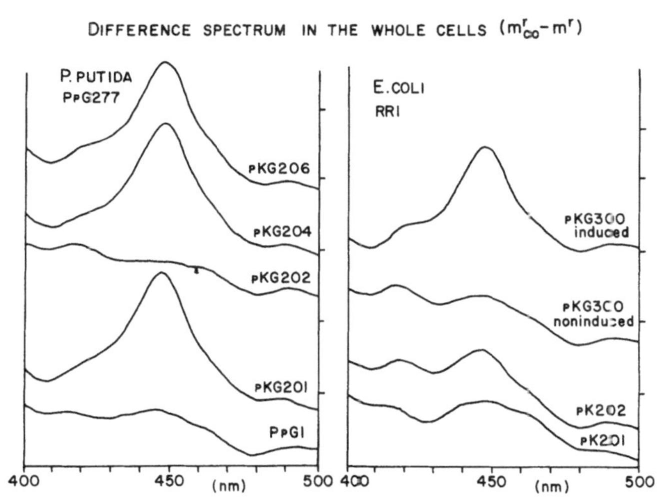

Fig. 5. (Continued)

D. $P450_{CAM}$ spectral assay.

1.6 kb insert are shown in Fig. 5B. The start signal is indicated as nucleotide 1, the known sequence of $P450_{CAM}$ starts with the N-terminal 3, proceeds 412 amino acid residues. At the terminal lies Ala at base 1239. The nucleotides preceding the start signal at about 150 bases to the PstI site are numbered negatively, the last base determined being -143. The DNA sequences stop signals at TAA beginning at nucleotide 1243, the second stop signal at 1261 coincide with the 45,000 molecular weight protein produced by the cloned insert in the wild-type plasmid.

Additional clones from a HindIII library by Dr. Koga express the first 4 genes of the Cam hydroxylase and alcohol dehydrogenase, see Fig. 1A. In summary, these advances provide a vector-promoter system with appropriate expression, 5-10 times the wild-type induced in *Pseudomonas putida* and in *E. coli*. They provide a wedge for genetic analysis, projection of P450 protein, further analysis of the control of the metabolic plasmids encoding terpene genes and pathway.

REFERENCES

1. Rheinwald, J.G., A.M. Chakrabarty, and I.C. Gunsalus (1973) A transmissible plasmid controlling camphor oxidation in *Pseudomonas putida*. Proc. Natl. Acad. Sci., USA 70:885-889.
2. Dagley, S., W.C. Evans, and D.W. Ribbons (1960) New pathways in the oxidative metabolism of aromatic compounds by micro-organisms. Nature 188:560.
3. Dunn, N.W., and I.C. Gunsalus (1973) A transmissible plasmid coding early enzymes of naphthalene oxidation in *Pseudomonas putida*. J. Bact. 114:974-979.
4. Stanier, R.Y., N.J. Palleroni, and M. Doudoroff (1966) The aerobic pseudomonads: A taxonomic study. J. Gen. Microbiol. 43:159-271.
5. Williams, P.A. (1981) Catabolic plasmids. Trends in Biochem. Sci. 6:23-26.
6. Franklin, F.C.H., M. Bagdasarian, M.M. Bagdasarian, and K.N. Timmis (1981) Molecular and functional analysis of the TOL plasmid pWWO from *Pseudomonas putida* and cloning of genes for the entire regulated aromatic ring meta cleavage pathway. Proc. Natl. Acad. Sci., USA 78:7458-7462.
7. Scherzinger, E., M.M. Bagdasarian, P. Scholz, R. Lurz, B. Rückert, and M. Bagdasarian (1984) Replication of the broad host range plasmid RSF1010: Requirement for three plasmid-encoded proteins. Proc. Natl. Acad. Sci., USA 81:654-658.
8. Inouye, S., Y. Ebina, A. Nakazawa, and T. Nakazawa (1984) Nucleotide sequence surrounding transcription initiation site of xylABC operon on TOL plasmid of *Pseudomonas putida*. Proc. Natl. Acad. Sci., USA 81:1688-1691.
9. Sato, R., and T. Omura (1978) Cytochrome P-450. Academic Press.

10. Johnston, J.B., and I.C. Gunsalus (1977) Isolation of metabolic plasmid DNA from Pseudomonas putida. Biochem. Biophys. Res. Commun. 75:13-19.
11. Farrell, R., I.C. Gunsalus, I.P. Crawford, J.B. Johnston, and J. Ito (1978) Restriction endonuclease sites and aromatic metabolic plasmid structure. Biochem. Biophys. Res. Commun. 82:411-416.
12. Yen, K.-M., and I.C. Gunsalus (1982) Plasmid gene organization: Naphthalene/salicylate oxidation. Proc. Natl. Acad. Sci., USA 79:874-878.
13. Yen, K.-M., M. Sullivan, and I.C. Gunsalus (1983) Electron microscope heteroduplex mapping of naphthalene oxidation genes on the NAH7 and SAL1 plasmids. Plasmid 9:105-111.
14. Grund, A.D., and I.C. Gunsalus (1983) Cloning of genes for naphthalene metabolism in Pseudomonas putida. J. Bact. 156:89-94.
15. Koga, H., B. Rauchfuss, and I.C. Gunsalus (1984) Cloning and expression of monoxygenase $P450_{CAM}$. J. Bact., submitted.
16. Chakrabarty, A.M. (1976) Plasmids in Pseudomonas. Ann. Rev. Genetics 10:7-30.
17. Lee, S.-M. Personal communication.
18. Ullah, A.H.J., P.K. Bhattacharyya, J. Bakthavachalam, G.C. Wagner, and I.C. Gunsalus (1983) Linalool 8-hydroxylase: A heme thiolate monoxygenase. Fed. Proc. 42:1897.
19. Samanta, T.B., A.H.J. Ullah, P.K. Bhattacharyya, and I.C. Gunsalus (1984) Chemical probe into the active site of linalool-8-hydroxylase. Fed. Proc. 42:1477.
20. Madhyastha, K.M., and P.K. Bhattacharyya (1968) Microbiological transformations of terpenes: Part XIII-Pathways for degradation of p-cymene in a soil pseudomonad (PL-strain). Ind. J. Biochem. 5:161-166.
21. Haniu, M., K.T. Yasunobu, and I.C. Gunsalus (1983) Heme binding and substrate-protected cysteine residues in $P450_{cam}$. Biochem. Biophys. Res. Commun. 116:30-38.
22. Fujita, V.S., S.D. Black, G.E. Tarr, D.R. Koop, and M.J. Coon (1984) On the amino acid sequence of cytochrome P-450 isozyme 4 from rabbit liver microsomes. Proc. Natl. Acad. Sci, USA (in press).
23. Kawajiri, K., O. Gotoh, K. Sogawa, Y. Tagashira, M. Muramatsu, and Y. Fujii-Kuriyama (1984) Coding nucleotide sequence of 3-methylcholanthrene-inducible cytochrome P-450d from rat liver. Proc. Natl. Acad. Sci., USA 81:1649-1653.
24. Yen, K.-M., and I.C. Gunsalus (1984) Gene nahR and regulation of plasmid-encoded naphthalene oxidation pathway. J. Bact. (in preparation).
25. Simonsen, J.L. (1949) The Terpene. Vol. II. The Dicyclic Terpenes and their Derivatives. Cambridge University Press.
26. Champion, P.M., B.R. Stallard, G.C. Wagner, and I.C. Gunsalus (1982) Resonance Raman detection of an Fe-S bond in cytochrome $P450_{CAM}$. J. Am. Chem. Soc. 104:5469-5472.

27. Poulos, T.L., M. Perez, and G.C. Wagner (1982) Preliminary crystallographic data on cytochrome $P450_{CAM}$. J. Biol. Chem. 257:10427-10429.
28. Pemberton, J.M. (1983) Degradative Plasmids. Intl Rev. of Cytology 84:155-183.
29. Thacker, R., O. Rorvig, P. Kahlon, and I.C. Gunsalus (1978) NIC, a conjugative nicotine-nicotinate degradative plasmid in *Pseudomonas convexa*. J. Bact. 135:289-290.
30. Gunsalus, I.C., H.E. Conrad, P.W. Trudgill, and L.A. Jacobson (1965) Regulation of catabolic metabolism. Israel J. Med. Sciences 1(6):1099-1119.
31. Chou, G.I.N., D. Katz, and I.C. Gunsalus (1974) Fusion and compatibility of camphor and octane plasmids in *Pseudomonas*. Proc. Natl. Acad. Sci., USA 71:2675-2678.
32. Shapiro, J., and M. Fennewald (1977) Regulatory mutations of the *Pseudomonas* plasmid *alk* regulon. J. Bact. 132:622-627.

THE STRUCTURE OF THE mer OPERON

Patricia Barrineau, Pearce Gilbert, W. James Jackson, Cheryl S. Jones, Anne O. Summers, and Sonya Wisdom

Department of Microbiology
The University of Georgia
Athens, Georgia 30602

ABSTRACT

The DNA sequence has been determined for a 3.8-kb region which encodes the mercury-resistance (mer) operon of the IncFII plasmid NR1. The sequence reveals 4 open reading frames which could encode proteins of 12,522, 9,429, 14,965, and 58,912 d corresponding to the 4 previously described Hg-inducible proteins detected in minicells carrying mer^+ plasmids. The Hg(II) reductase protein sequence is about 90% homologous to that of Tn501, but the DNA sequence shows a homology of 60-70% to that of Tn501 except for short regions of very high homology. The entire mer region is 63.4% G-C overall. The region encoding the merR (positive regulatory) function has 3 possible open reading frames, 2 of which overlap in one direction and the third of which reads in the opposite direction. Attempts to visualize the polypeptide(s) encoded by the merR cistron were unsuccessful.

INTRODUCTION

Resistance to mercury compounds is a commonly found plasmid-determined phenotype in both gram-positive and gram-negative bacterial genera (9,30). It occurs in 2 forms. The simpler form of the phenotype ("narrow spectrum") confers resistance only to inorganic mercury compounds, such as, $HgCl_2$ and to a few organomercurials, such as, merbromin and fluorescein mercuric acetate. The more complex phenotype ("broad spectrum") confers resistance to the above compounds as well as to a larger number of organomercurial compounds including phenyl mercuric acetate and merthiolate. There is evidence that these phenotypes are selected in nature by industrial and clinical use of mercury compounds (17,21). Broad-spectrum resistance occurs less frequently than narrow spectrum and both types of resistance have been found on transposable elements in gram-negative bacteria (4,23,29,31). Broad-spectrum resistance involves another enzyme, the organomercurial lyase (13,27), in addition to those functions described below.

Narrow-spectrum resistance to inorganic mercury compounds (mer) is effected by the conversion of toxic ionic Hg [Hg(II)] to the less toxic, relatively inert, volatile elemental Hg [Hg(0)]. This detoxification is effected by the Hg(II) reductase (HR), an enzyme characterized in gram-negative bacteria as a heat-stable, soluble, intracellular, NADPH-dependent, FAD-containing, disulfide oxidoreductase. This inducible enzyme is the product of the merA cistron (10,14,15). Previous biochemical studies have estimated the monomer molecular mass of this enzyme to be from 60-69,000 d. The recently published sequence of the HR encoded by Tn501 indicates a molecular mass of 58,660 (6) for that version of the enzyme. The HR apparently functions as a dimer and shows considerable homology, especially in the active site, to another FAD-containing disulfide oxidoreductase, human erythrocyte glutathione reductase (11,12). The HR is processed proteolytically rather slowly in vivo by the removal of between 3-7,000 d from its amino terminus. Presently, the agent and the functional correlate(s) of this processing are unknown. The HR has also been isolated from a limited number of gram-positive genera and, though less well characterized than the gram-negative enzyme, the gram-positive HR is similar except for heat stability (9,30)

The mer operon also encodes a Hg(II) uptake system, originally designated as merT which is only detectable in mutant strains which have lost the HR and so cannot volatilize Hg(II) (19). Since minicell polypeptide analysis of mer has associated 3 small proteins of approximately 15,000; 14,000; and 13,000 d with this function, the genotype designation merT (10) is inappropriate. However, for simplicity, we retain its use, in quotes ("merT") pending adoption of appropriate designations for the 3 peptides. The larger 2 are inner membrane proteins and the smallest protein is soluble and very rapidly processed to a 12,000-d form in intact minicells (14,15). All 3 proteins are very basic, detectable only by nonequilibrium isoelectric focusing in 2-dimensional polyacrylamide gel analysis.

Genetic studies indicate that both the HR and the Hg(II) uptake system are under positive regulation. As studied on monocopy plasmids, the operon consists of 2 complementation groups: Group A contains the positive regulatory function with apparently only one cistron, merR; and Group B contains both the Hg(II) uptake function ("merT") and the HR (merA). Recent studies of Tn5 mutants of mer in the monocopy plasmid R100-1 indicate that there is a second function (merC), mapping in or very near the amino terminus of the HR which is necessary for expression of the Hg(II) uptake function (3).

Other recent studies (20) with monocopy/multicopy mutant heteroplasmid strains also indicate that regulation may be somewhat more complex than initially posited (16). The positive regulatory function merR appears to be negatively autoregulated (20).

We report here the structure of the operon as revealed by DNA sequence analysis. The DNA sequence itself and a detailed analysis of the polypeptides which it encodes will be presented elsewhere (2).

MATERIALS AND METHODS

Bacterial Strains and Plasmids

DNA for sequencing was obtained from the plasmid pDB7 which carries the mer operon of NR1 cloned directly on a PstI fragment into pBR322. The regulated high-expression vehicle pλ8 (24) maintained in the host strain N4830 was used for subcloning and expression of the merR locus. The recombinant plasmids pRR130 and pRR134 carry, respectively, the entire mer operon and the EcoRI-H fragment of NR1 (18)

DNA Sequencing

The dideoxy chain termination method of Sanger (26) was used as modified by Messing (17). The enzymes Sau3A, RsaI, HaeIII, AluI, NaeI, HinPI, HpaII, and FnuDII were used to generate "shotgun" templates from a 4,000-bp NcoI-StuI fragment of pDB7 which confers the Hg-resistance phenotype when cloned alone into pBR322. Directed clones were also generated using the HincII and EcoRI sites within the mer operon templates. Data were recorded and analyzed using the GENED, GEL, and SEQ programs from Intelligenetics, Inc., Palo Alto, California.

Assessment of HgR Phenotype

Zone of inhibition, efficiency of plating, and Hg(II) reductase assays have been described (3).

Polypeptide Analysis

Labeling of plasmid-encoded polypeptides with ^{35}S-methionine and their analysis by gel electrophoresis and autoradiography has been described (14,15).

Reagents and Enzymes

The Klenow fragment of DNA polymerase was obtained from Bethesda Research Laboratories. Radioisotopically labeled compounds were obtained from Amersham, Inc. Nucleotides, and DNA sequencing primer were obtained from P-L Biochemicals. Restriction enzymes were obtained from New England Biolabs, Bethesda Research Laboratories, Promega Biotech, and Boehringer-Mannheim. All enzymes were used according to manufacturer's specifications.

RESULTS

DNA Sequence

The 3.4-kb region containing the mer operon begins at the left 38-bp inverted repeat which brackets one end of Tn21 and extends to the termination codon for the HR structural gene. The HindII site and the EcoRI site provided "landmarks" and were used for construction of directed templates. The sequence upon which the operon structure reported here is based is 96% complete on both strands. The single-strand regions include 3 short segments in the HR-structural gene and 1 segment of 30 bp in the merR region. The sequence of the HR of Tn501 reported last year by N. Brown et al. (6) was useful in determining the position of the HR coding region of NR1. The other reading frames described below were chosen on the basis of their being preceded by a reasonably good Shine-Dalgarno sequence (28), the similarity of their codon utilization patterns to that of the HR, and their correspondence in size, amino acid composition, and hydropathicity (16) to polypeptides previously shown to be specified by the operon (14,15). Since Hg(II) interacts with the sulfhydryl group of cysteine (the HR has 8, Ref. 6), we have also been concerned with the number of cysteines in each predicted protein.

There are 3 possible open reading frames in the left region of the operon (Fig. 1). This is the region previously shown to encode

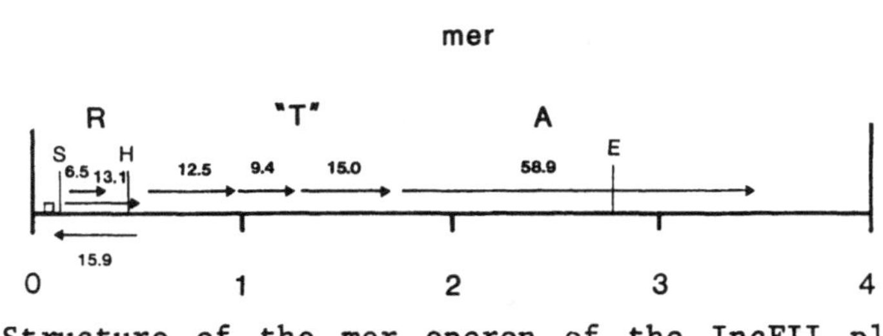

Fig. 1. Structure of the mer operon of the IncFII plasmid NR1. Distances are in kilobase pairs (kb). The small open box near the left end represents the left inverted repeat of Tn21 (8). Restriction sites indicated are EcoRI (E), HincII (H), and StuI (S). Sizes of open reading frames are indicated in kilodaltons (kd). Direction of reading frame is indicated by arrowhead.

the merR function. Two open reading frames read from left to right. The shorter of these would encode a peptide of 6,457 d and is not preceded by a Shine-Dalgarno sequence. The larger reading frame would encode a peptide of 13,139 d. There is a good Shine-Dalgarno sequence preceding this peptide, but it is 19 bp before the AUG, whereas effective translational start signals are typically found 8-9 bp before the AUG. The third possible open reading frame in the merR region lies on the complementary strand and reads towards IS1. This polypeptide would be 15,907 d and is preceded by a good Shine-Dalgarno sequence. Unlike the other 2 possible merR proteins, this one is also preceded by several good Pribnow-box-like sequences in the AT-rich region surrounding the HincII site. All 3 proteins are relatively basic and hydrophilic. Each of the 2 smaller predicted peptides would have 2 cysteines and the largest would have 4 cysteines (2).

We have previously shown that there is a merR-dependent, $HgCl_2$-inducible promoter (5) which maps in the vicinity of the HincII site. This is the least GC-rich region of the operon and contains several possible Pribnow-box (22) sequences. Restriction enzyme analysis (Ref. 5 and data not shown) indicates that this promoter directs transcription in a right direction, consistent with its being the promoter for the structural genes of the operon. The possibility that it also directs transcription to the left, i.e., that it also contains a promoter for the putative merR protein of 15,907 d encoded on the complementary strand (described above) has not been eliminated.

A transcript beginning in the promoter region and reading right would encode 4 polypeptides which we believe are the structural genes for the Hg(II) uptake system and the HR. The first open reading frame is preceded by a good Shine-Dalgarno sequence and would specify a peptide of 12,391 d. The protein has a very basic estimated isoelectric point and also has 4 cysteines. Hydropathicity analysis shows it is extensively hydrophobic (2). The second open reading frame in the Hg(II) uptake system region begins almost immediately after the first and is preceded by a Shine-Dalgarno sequence. It specifies a peptide of 9,429 d with 2 cysteines and a similarly basic isoelectric point. This peptide is largely hydrophilic except for its amino-terminus which consists of a short positively charged region followed by a sizeable (∿ 15 residues) hydrophobic region (2). The third open reading frame in this region is preceded by a Shine-Dalgarno sequence and specifies a peptide of 14,965 d which has 4 cysteines and 6 methionines. It also has a basic estimated isoelectric point and is extensively hydrophobic (2).

The fourth open reading frame specifies the Hg(II) reductase of molecular mass 58,912 d. It is preceded by a Shine-Dalgarno sequence. Its estimated isoelectric point is slightly higher than the measured value (15). It is a moderately hydrophilic protein with no extensive runs of hydrophobicity (data not shown). The protein sequence is 90% homologous to that of the HR from Tn501, with the only major discontinuity between the 2 proteins occurring in residues 70-77. However, the DNA sequences of the 2 HR genes are 60-70% homologous except for short stretches of 100 bp or less (2).

The overall GC content of the region sequenced is 63.4% with a marked gradient of GC content. The left end, up to the HincII site is 58.3% GC. The region encoding the Hg(II) uptake proteins is 61.5% and the HR structural gene is 65.2%.

Visualization of the merR Gene Product

We wished to determine which of the possible reading frames in the merR region is in fact the correct one. Since we have never seen a protein encoded by the merR cistron using the minicell system (3,14,15), to optimize expression of proteins encoded by this cistron we used the plasmid pRR134 (18) and recloned the NR1 EcoRI fragment which it carries into the regulated, high expression vehicle, pλ8 (Fig. 2). This EcoRI fragment of NR1 contains, in addition to merR, the "merT" peptides and the amino-terminal two-thirds of the HR.

The construction of the plasmids pJJ77 and pJJ80 is outlined in Fig. 2. They were designed to place the merR region in both orientations relative to the λP_L carried by pλ8. The advantage of this particular vehicle is that P_L is oriented such that it directs transcription of the β-lactamase gene (bla) of pBR322 (24). Thus the appearance of β-lactamase can be used as a diagnostic for the effective transcription of genes cloned into the sites between P_L and the bla locus. Each of these derivatives complements merR mutations as assessed by inhibition zone, efficiency of plating, and HR enzyme assay (data not shown). When expression from P_L is induced by incubation at 42°C neither pJJ77 nor pJJ80 directs the synthesis of any polypeptides (other than an N-IS1 fusion peptide, Fig. 3, fourth lane) which are not also seen in a strain carrying the vehicle pλ8 alone (Fig. 3). Thermoinducible synthesis of β-lactamase and its precursor is evident, indicating that the temperature induction worked. We discuss the implication of these results below.

DISCUSSION

1.) The HR (merA) is the largest predicted polypeptide encoded by the operon in NR1 and it contains 564 amino acids compared to 561 for the HR of Tn501 (6). While the polypeptide sequence of these 2 proteins is quite similar, the DNA sequence shows less homology. This protein appears to be the fourth in a transcript which initi-

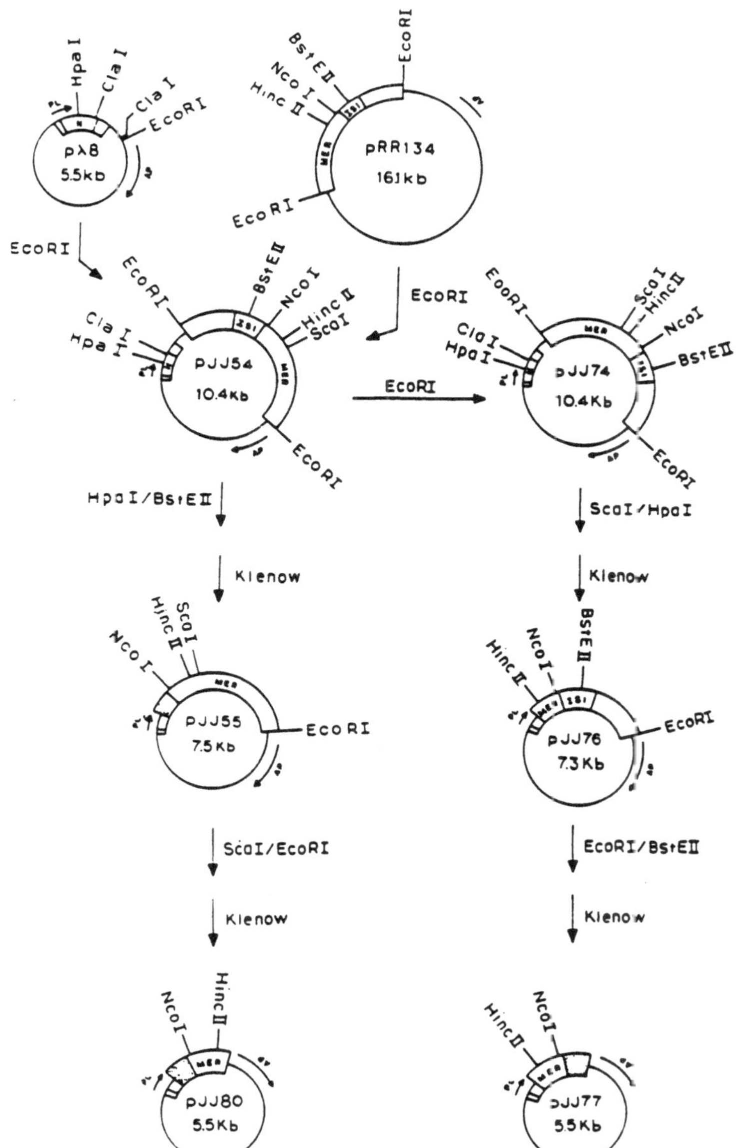

Fig. 2. Construction of the merR containing plasmids pJJ77 and pJJ80. The plasmid pRR134 carries the entire EcoRI fragment of NR1 cloned into the vehicle RSF2124 (18). This EcoRI fragment includes the merR cistron, the "merT" region and part of the HR structural gene (merA). The high expression vehicle pλ8 is described briefly in the text and in detail in Reed (24).

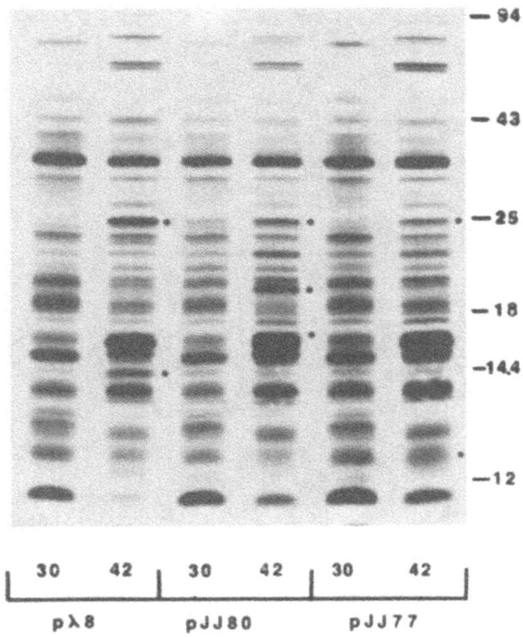

Fig. 3. An autoradiogram of a 16.5% SDS/urea-polyacrylamide gel analysis of the ^{3}H-amino acid-labeled polypeptides synthesized by N4830 carrying either the vehicle pλ8 alone, or the *merR* containing derivatives, pJJ77 or pJJ80. Each strain was labeled for 5 min at either 30°C or 42°C. Each lane was loaded with approximately 250,000 TCA precipitable counts. Molecular mass standards are indicated in the right margin. Dots mark thermoinducible proteins seen in pλ8 alone. The pJJ80 encoded thermoinducible peptide of ∿ 19,000 d is a fusion λ N of (39) with a heading frame in IS1b (35).

ates roughly 1,240 bp closer to IS1b, though in the absence of definitive information on the cis- and trans-acting elements of *mer* regulation, the existence of other promoters cannot be ruled out.

2.) Three polypeptides are predicted by the sequence in the region previously referred to as "*merT*." The 2 larger reading frames correspond quite well in size to the 15,000 and 14,000 d *mer* peptides previously described and their extensive hydrophobicity is consistent with their being intrinsic membrane proteins. Given the limits of precision of gel electrophoresis in the low molecular mass range, the smallest of these 3 predicted peptides is very likely the 13(12),000 d peptide. The amino-terminus of this predicted peptide has many features of a signal sequence, consistent with the processing and at least partial periplasmic location of the 13,000 d mer protein. The predicted amino acid sequences of the proteins all indicate rather basic isoelectric points, also consistent with previous observations (14,15). Until we have well-mapped mutations in each of these Hg(II) uptake proteins we will refrain from giving them genotypic designations and refer to them as ORF12.4, ORF9.4, and ORF15, respectively.

3.) Polypeptide analysis has not allowed us to decide which of the 3 predicted polypeptides in the merR region is the correct one. Presently, our bias is in favor of one or the other of the larger predicted peptides because, in addition to the presence of Shine-Dalgarno sequences, either of the larger peptides would include the HincII site. Ni'Bhriain et al. have shown that deletion of DNA from the HincII site right destroys merR function (20). The smaller open frame which reads from IS1b toward the operon structural genes does not extend as far as the HincII site. Moreover, there would be a very long intercistronic distance (266 bp) between the end of this predicted peptide and the beginning of the first of the "merT" peptides (2).

4.) While there are other open reading frames of lengths ranging from about 7,700 to 37,000 d (primarily on the complementary strand), there is none which can be unequivocally assigned to the merC function which we have mapped in or near the amino-terminus of the HR. Specifically, there is not an overlapping reading frame in the amino-terminus of the HR which would correspond to this function. Therefore, the loss of inducibility of the Hg(II) uptake proteins seen in merC insertion mutations of pDU202 (3) must be associated with either the HR itself or with the peptide immediately adjacent to its amino-terminus.

5.) A long open reading frame begins just beyond the 3' end of the HR structural gene. In Tn501 this frame apparently specifies a peptide of about 54,000 d (N. Brown, cited in Ref. 9). Ni'Bhriain et al. (20) and Foster (9) have suggested on the basis of Tn5 insertions in a multicopy mer operon, that this is an additional cistron of the operon which they have called merD. The proximity of this coding region to the HR structural gene almost insures that read-through transcription from the mer operon would effect synthesis of the mRNA encoding this polypeptide. However, the relationship of the putative merD to the Hg-resistance phenotype remains questionable since reduction of the copy number of the Tn5 merD mutants restored the resistant phenotype (20). In addition, deletions which remove everything to the right of a HincII site (approximately 80% of the reading frame) have no effect on the Hg-resistance phenotype of the multicopy plasmid (20). We also find that removal of the DNA right from a StuI site ($\sim$ 50% of the 54,000 d predicted peptide) has no effect on the expression of Hg resistance by a multicopy derivative. We suspect that the Hg^s phenotype associated with Tn5 insertions in this region is the result of Hg-inducible overexpression of Tn5-encoded proteins which leads to instability of the multicopy plasmid. The mer region is known to have strong Hg-inducible transcriptional activity (1).

We are currently making use of the DNA sequence information for mRNA mapping, and for in vitro, synthetic linker-directed mutagenesis and translational fusion construction to confirm and extend the observations reported here.

ACKNOWLEDGEMENTS

We are especially grateful to Susan Hollingshead and Barbara Rutledge for instruction in dideoxy sequencing. We also acknowledge the patient and enthusiastic support of the staff of Intelligenetics, Inc., Palo Alto, California, and Georgia colleague Rich Meagher with whose software and hardware the sequence data were reduced and analyzed. We also thank Paul Totis and Janice Kerr for competent and cheerful technical assistance. Our work is supported by NIGMS 28211 and NIAID-K04-00377.

REFERENCES

1. Alton, N.K., and D. Vapnek (1979) Transcription and translation of R-plasmid R538-1 DNA. Plasmid 3:366-376.
2. Barrineau, P., P. Gilbert, W. James Jackson, C.S. Jones, A.O. Summers, and S. Wisdom (1984) The DNA sequence of the plasmid-determined mer operon (in preparation).
3. Barrineau, P.J., and A.O. Summers (1983) A second positive regulatory function in the mer (mercury resistance) operon. Gene 25:209-221.
4. Bennett, P.M., J. Grinsted, C.L. Choi, and M.H. Richmond (1978) Characterization of Tn501, a transposon determining resistance to mercuric ions. Mol. Gen. Genet. 159:101-106.
5. Bohlander, F.A., A.O. Summers, and R.B. Meagher (1981) Cloning a promoter that puts the expression of tetracycline resistance under the control of the regulatory elements of the mer operon. Gene 15:395-403.
6. Brown, N.L., S.J. Ford, R.D. Pridmore, and D.C. Fritzinger (1983) DNA sequence of a gene from the Pseudomonas transposon Tn501 encoding mercuric reductase. Biochemistry 22:4089-4095.
7. Colwell, R.R., G.S. Sayler, J.D. Nelson, Jr., and A. Justice (1976) Microbial mobilization of mercury in the aquatic environment. In Environmental Biogeochemistry: Metal Transfer and Ecological Mass Balances, Vol. 2, J.D. Nriagv, ed. Ann Arbor Science, Ann Arbor, Michigan, pp. 473-486.
8. de la Cruz, F., and J. Grinsted (1982) Genetic and molecular characterization of Tn21, a multiple resistance transposon from R100.1. J. Bact. 151:222-228.

9. Foster, T.J. (1983) Plasmid-determined resistance to antimicrobial drugs and toxic metal ions in bacteria. Microbiol. Rev. 47:361-409.
10. Foster, T.J., H. Nakahara, A.A. Weiss, and S. Silver (1979) Transposon A-generated mutations in the mercuric resistance genes of plasmid R100.1. J. Bact. 140:167-181.
11. Fox, B.S., and C.T. Walsh (1982) Mercuric reductase: Purification and characterization of a transposon-encoded flavoprotein containing an oxidation-reduction active disulfide. J. Biol. Chem. 257:2498-2503.
12. Fox, B.S., and C.T. Walsh (1983) Mercuric reductase: Homology to glutathione reductase and lipoamide dehydrogenase. Iodoacetamide alkylation and the sequence of the active site peptide. Biochem. 22:4082-4088.
13. Furukawa, K., and K. Tonomura (1972) Induction of metallic mercury-releasing enzyme in mercury-resistant Pseudomonas. Agric. Biol. Chem. 36:2441-2448.
14. Jackson, W.J., and A.O. Summers (1982) Polypeptides encoded by the mer operon. J. Bact. 149:479-487.
15. Jackson, W.J., and A.O. Summers (1982) Biochemical characterization of the $HgCl_2$-inducible polypeptides encoded by the mer operon of R100. J. Bact. 151:962-970.
16. Kyte, J., and R. Doolittle (1982) A method for displaying the hydropathic character of a protein. J. Mol. Biol. 157:105-132.
17. Messing, J. (1983) New M13 vectors for cloning. Meth. Enzymol. 101:20-89.
18. Miki, T., A.M. Easton, and R.H. Rownd (1976) Mapping of the resistance genes of the R plasmid NR1. Mol. Gen. Genet. 158:217-224.
19. Nakahara, H., S. Silver, T. Miki, and R.H. Rownd (1979) Hypersensitivity to Hg^{2+} and hyperbinding activity associated with cloned fragments of the mercurial resistance operon of plasmid NR1. J. Bact. 140:161-166.
20. Ni'Bhriain, N.N., S. Silver, and T.J. Foster (1983) Tn5 insertion mutations in the mercuric ion resistance genes derived from plasmid R100. J. Bact. 155:690-703.
21. Porter, F., S. Silver, C. Ong, and H. Nakahara (1982) Selection for mercurial resistance in hospital settings. Antimicrob. Agents. Chemother. 22:852-858.
22. Pribnow, D. (1975) Nucleotide sequence of an RNA polymerase binding site at an early T7 promoter. Proc. Natl. Acad. Sci., USA 72:784-789.
23. Radford, A.J., J. Oliver, W.J. Kelly, and D.C. Reaney (1981) Translocatable resistance to mercuric and phenylmercuric ions in soil bacteria. J. Bact. 147:1110-1112.
24. Reed, R.R. (1981) Transposon-mediated site-specific recombination: a defined in vitro system. Cell 25:713-719.
25. Rinderle, S.J., J.E. Booth, and J.W. Williams (1983) Mercuric reductase from R-plasmid NR1: Characterization and mechanistic study. Biochem 22:869-876.

26. Sanger, F., S. Nicklen, and A.R. Coulsen (1977) DNA sequencing with chain terminating inhibitors. Proc. Natl. Acad. Sci., USA 74:5463-5467.
27. Schottel, J.L. (1978) The mercuric and organomercurial detoxifying enzymes from a plasmid-bearing strain of Escherichia coli. J. Biol. Chem. 253:4341-4349.
28. Shine, J., and L. Dalgarno (1974) The 3'-terminal sequence of Escherichia coli 16S ribosomal RNA: Complementarity to nonsense triplets and ribosomal binding sites. Proc. Natl. Acad. Sci, USA 71:1342-1346.
29. Stanisich, V.A., P.M. Bennett, and M.H. Richmond (1977) Characterization of a translocation unit encoding resistance to mercuric ions that occurs on a nonconjugative plasmid in Pseudomonas aeruginosa. J. Bact. 129:1227-1233.
30. Summers, A.O., and S. Silver (1978) Microbial transformations of metals. Ann. Rev. Microbiol. 32:637-672.
31. Summers, A.O., R.B. Weiss, and G.A. Jacoby (1980) Transposition of mercury resistance from a transferrable R plasmid of Escherichia coli. Plasmid 3:35-47.
32. Tanaka, N., J.H. Cramer, and R.H. Rownd (1976) EcoRI restriction endonuclease map of the composite R plasmid NR1. J. Bact. 127:619-636.

ANALYSIS AND MANIPULATION OF PLASMID-ENCODED PATHWAYS FOR THE CATABOLISM OF AROMATIC COMPOUNDS BY SOIL BACTERIA

K.N. Timmis,[1] P.R. Lehrbach,[1,a] S. Harayama,[1] R.H. Don,[1] N. Mermod,[1] S. Bas,[1] R. Leppik,[1,b] A.J. Weightman,[1,c] W. Reineke,[2] and H.J. Knackmuss[2]

[1]Department of Biochemistry
University of Geneva, Geneva, Switzerland

INTRODUCTION

Soil and water bacteria in general, and members of the genus *Pseudomonas* in particular, exhibit a fascinating wealth of exotic properties not found in commensal and parasitic bacteria, such as *Escherichia coli*. They are, for example, able to degrade and use as sources of carbon and energy a wide range of organic compounds, including some that are quite noxious and otherwise biocidal (e.g., phenol). Moreover, soil bacteria are able to evolve rapidly new metabolic activities in response to changes in environmental conditions, e.g., to evolve new pathways for the degradation of synthetic industrial compounds (xenobiotics) such as pesticides, that are newly applied to the environment (1). The rapidity of evolution of new pathways is particularly surprising because many pathways, e.g. those for the catabolism of aromatic hydrocarbons, involve a large number of enzymatic steps and often more than twice that number of polypeptides (2). It is hardly surprising, therefore, that many catabolic pathways are specified by plasmids (3), elements known to be instrumental for rapid genetic change in bacteria (4).

---

[2] Lehrstuhl für Chemische Mikrobiologie, Universität-Gesamthochschule Wuppertal, Federal Republic of Germany Present addresses: [a]Biotechnology Australia Pty Ltd., Roseville, NSW, Australia (P.R.L.); [b]Meat Research Laboratory, CSIRO, Cannon Hill, QLD, Australia (R.L); and [c]Dept. of Applied Biology, University of Wales Institute of Science and Technology, Cardiff, Wales (A.J.W.).

Soil bacteria generally live and grow under adverse conditions and nutrient concentrations are frequently extremely low; such bacteria are therefore expected to dispose of a highly regulated metabolism and an efficient use of energy. The complexity of the catabolic pathways implies, therefore, a sophisticated system of regulation to control the expression and to accomplish coordination of catabolic activities. Moreover, although catabolic activities of soil bacteria are very diverse, degradative routes are convergent and lead to a rather limited number of common intermediates that are then degraded by central pathways. The existence of pathway networks thus implies a high degree of integration and a subtle interplay of regulators and inducers for segments of pathways.

Our objectives are to analyze the genetic and biochemical organization of selected plasmid-encoded pathways and to characterize genetic and biochemical strategies utilized by soil bacteria to evolve new catabolic pathways. This, in turn, involves a study of the regulatory networks that control the expression of catabolic pathways. It is our ultimate goal to utilize this knowledge in order to effect evolution in vitro, i.e., to evolve new catabolic pathways, and to increase the activities of existing pathways. We are particularly interested in evolving pathways for the degradation of recalcitrant aromatics, especially xenobiotics, because this will provide important information on the biochemical basis of recalcitrance and on microbial strategies to overcome recalcitrance. An important aspect of this program will be the exploration of physiological limits to the evolution of new catabolic routes and the increase in global activities of catabolic pathways.

We have chosen as model systems for our studies the TOL plasmid specified pathway for the degradation of toluene, xylenes and related aromatic hydrocarbons, because of the considerable amount of biochemical data available on the pathway and the particular interest of certain pathway enzymes, and the pJP4 plasmid-specified pathway for the degradation of 3-chlorobenzoate and 2,4-dichlorophenoxyacetic acid, because halogenated organics constitute one of the largest group of xenobiotic compounds released into the biosphere and contain a range of both toxic and recalcitrant compounds (5,6).

CATABOLIC PATHWAY SPECIFIED BY THE TOL PLASMID pWWO

TOL plasmid pWWO (7,8), and its Tn*401* transposon derivative pWWO-161 (9), specify a pathway for the catabolism of toluene, xylenes and related hydrocarbons (10; Fig. 1). The pathway involves successive oxidations of the side chain to give benzoate or methyl benzoate (upper pathway), followed by dioxygenation of the aromatic ring to catechol or methyl catechol, *meta* (extradiol) cleavage of the ring to produce a semialdehyde, and degradation of the semialde-

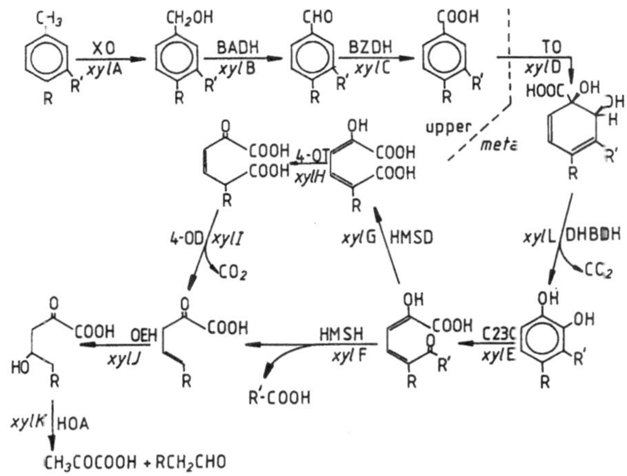

Fig. 1. TOL plasmid-specified pathway for the degradation of toluene and xylenes. Enzyme abbreviations: XO, xylene oxygenase; BADH, benzylalcohol dehydrogenase; BZDH, benzaldehyde dehydrogenase; TO, toluate/benzoate dioxygenase; DHBDH, dihydrodihydroxybenzoate dehydrogenase; C230, catechol 2,3-dioxygenase; HMSH, hydroxymuconic semialdehyde hydrolase; HMSD, hydroxymuconic semialdehyde dehydrogenase; 4-OT, 4-oxalocrotonate tautomerase; 4-OD, 4-oxalocrotonate decarboxylase; OEH, 2-oxopent-4-enoate hydratase; HOA, 2-oxo-4-hydroxy-pentonate aldolase. Initial compounds: R,R' = H: toluene; R = H, R' = CH_3: m-xylene; R = CH_3, R' = H: p-xylene. xylA to xylK: genes for the pathway enzymes (13,14).

hyde to central metabolites (meta-pathway or lower pathway). The pathway branches at the ring cleavage product and reconverges at 2-oxopent-4-enoate; the ring cleavage products formed from benzoate and p-toluate are channelled into the 4-oxalocrotonate branch whereas the product formed from m-toluate is metabolized by the hydrolytic branch (11).

By means of transposition mutagenesis of the TOL plasmid, and the cloning of TOL plasmid DNA segments, the catabolic genes have been localized to 2 distinct blocks, one for upper pathway enzymes and one for lower pathway enzymes (9,12; Fig. 2a). The 2 gene blocks are separated by a DNA segment about 10 kb in length. The genes of the upper pathway enzymes are organized in an operon, having the order promoter-xylC-A-B (13; Fig. 2b), as are the genes of the lower pathway, which have the order: promoter-xylD-L-E-G-F-J-I-H (14; Fig. 2c). Two gaps exist in the current lower pathway gene map and at least 1 gene, xylK, remains to be localized.

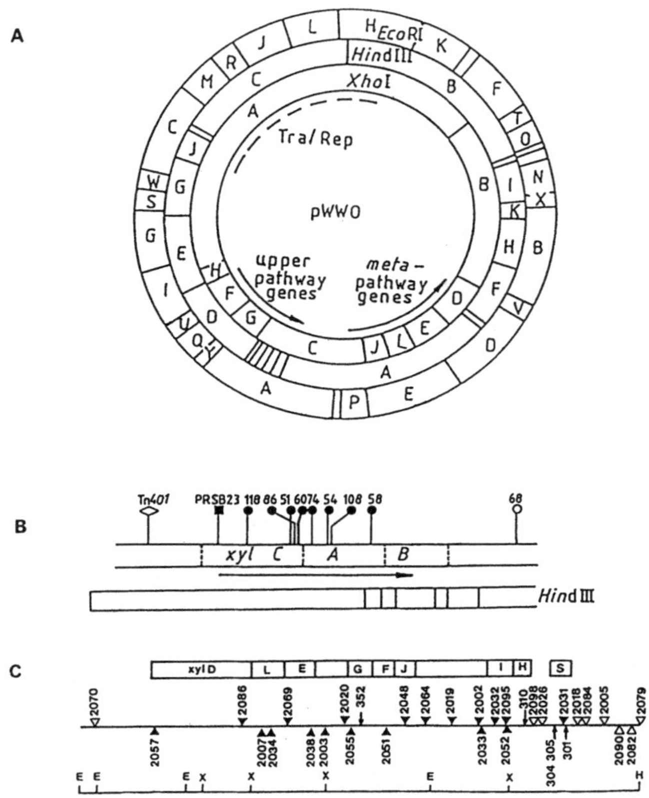

Fig. 2. The TOL plasmid pWWO and maps of the catabolic genes.

A: Restriction endonuclease cleavage map of pWWO. Tra/Rep indicates the region of the plasmid essential for its autonomous replication and conjugal transfer ability.

B: Map of the upper pathway operon. The Tn401 and the Tn5-68 mutations do not influence expression of the operon, whereas the filled square and circles indicate insertion mutations that prevent expression of one or more structural genes. This map is taken from Ref. 13.

C: Map of the lower pathway operon. Open triangles indicate insertions that do not influence expression of the operon whereas filled triangles indicate insertions that prevent expression of 1 or more structural genes. This map is taken from Ref. 14. E, H, X: cleavage sites for EcoRI, HindIII, and XhoI, respectively.

The designations of genes of the TOL plasmid catabolic operons are explained in Fig. 1.

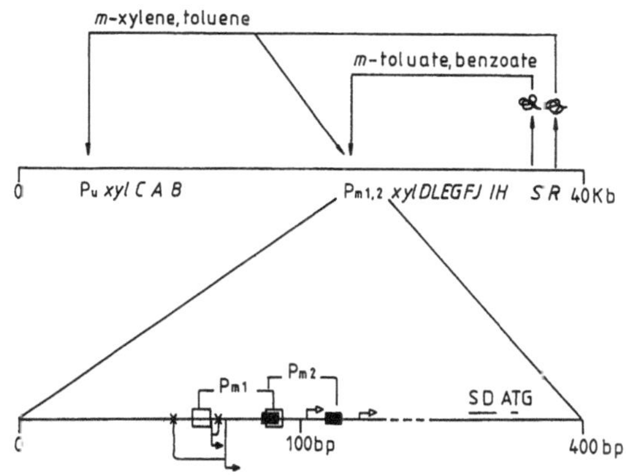

Fig. 3. Model for the regulation of the 2 catabolic operons of the TOL plasmid pWWO and structure of the meta-cleavage pathway operon promoters.

The proposed regulatory circuits are taken from references 15 and 17. P_u, P_{m1} and P_{m2} are the promoter of the upper pathway operon and the 2 overlapping promoters of the lower pathway operon, respectively. In the expansion, the -35 and -10 regions of P_{m1} (open boxes) and P_{m2} (filled boxes) are shown, as are their corresponding transcription initiation start sites (open arrows). SD and ATG indicate the locations of the ribosome binding site and the initiation codon of the first gene of the operon.

Point mutations leading to constitutive expression of the operon are indicated by crosses, and the new transcriptional start sites associated with these mutations are shown by filled arrows.

Toluene and xylenes, and their alcohols, act as inducers and provoke synthesis of all enzymes of the pathway, whereas benzoate and toluates induce the synthesis of the enzymes of the lower pathway. The pathway is therefore organized into 2 functional parts, the conversion of the hydrocarbons to the carboxylic acids, and the conversion of the carboxylic acids to central metabolites, and this biochemical organization is reflected in the genetic organization of the pathway. It was proposed some years ago that a regulator, the product of the xylS gene, is responsible for controlling expression of genes for enzymes of the lower pathway, whereas another regulator, the product of the xylR gene, is responsible for controlling expression of genes of both parts of the pathway (15; Fig. 3). The xylS and xylR genes have now been localized downstream of the lower pathway operon (16,17; Fig. 3). Mutations in xylR prevent expression of genes of upper pathway enzymes, but not those of lower pathway enzymes, whereas mutations in xylS prevent expression of genes of lower pathway enzymes but not those of upper pathway enzymes. Although upper pathway substrates ordinarily induce expression of both operons, they fail to induce expression of the meta-cleavage operon in xylS mutant bacteria (17); the xylS product is therefore an essential component of xylR-mediated induction of the meta-operon.

Nuclease S1 mapping studies of transcription initiation sites just upstream of the xylD gene, combined with deletion and polynucleotide sequence analysis of this region, have identified 2 overlapping promoters, P_{m1} and P_{m2}, that are responsible for xylS product-mediated regulated expression of the meta-cleavage pathway operon genes (18; Fig. 3). Mutant plasmids that express constitutively the meta-pathway operon have been isolated and some of their mutations characterized; most were shown to have created new, non-regulated promoters within or just upstream of the native promoters (18). The native promoters are structurally distinct from typical E. coli promoters, which is consistent with the poor expression of TOL plasmid genes in E. coli (9,18,19, 20,21). Moreover, a range of levels of activity in E coli was observed with the constitutive TOL plasmid promoters (18), which is further evidence that promoter structure is the major factor determining differences in efficiency of expression of TOL genes in Pseudomonas and E. coli.

These studies strongly suggest that the xylS gene product is a positive regulator which acts at least partially at the level of transcription, and whose activity is essential for expression of the lower pathway enzyme genes. The xylR product would also seem to be an essential positive regulator.

Several of the TOL plasmid-specified enzymes have useful properties. For example, the ring-cleavage enzyme, catechol 2,3-dioxygenase (C23O), which converts the colorless substrate catechol to the yellow-colored product 2-hydroxymuconic semialdehyde, is of considerable utility as an indicator enzyme. Its activity is readily and precisely measured by a simple photometric assay (22) and, most importantly, bacterial colonies containing the enzyme turn bright yellow when sprayed with catechol (9,19). The gene for C23O, xylE, whose polynucleotide sequence has been determined (23,24), is expressed in a wide range of bacteria, including Gram positives (23), and may therefore be used for a variety of genetic manipulations, such as the isolation of promoter-containing DNA fragments (23). The xylE gene thus represents a valuable addition to the limited number of currently available indicator genes such as lacz, whose products can be detected by simple color changes.

The xylD and xylL genes encode toluate 1,2-dioxygenase and dihydrodihydroxybenzoate dehydrogenase, respectively, which convert benzoate and toluates first to the dihydrodiol and then to catechol or methyl catechol (25,26). Unlike a number of analogous enzymes, they have a relaxed substrate specificity and form catechols from a variety of methyl- and halo-substituted benzoates (27). They are therefore useful for the construction of new strains exhibiting wide spectrum catabolic activities (28). TOL plasmid DNA segments carrying xylD or xylD,L, have been cloned and characterized; their features and their use in strain construction are described below.

CATABOLIC PATHWAY SPECIFIED BY THE pJP4 PLASMID

Two pathways for the degradation of haloaromatics have received considerable attention in recent years. The biochemistry of the modified ortho-cleavage pathway of Pseudomonas sp. B13 (WR1) for the degradation of 3-chlorobenzoate (3CB) is particularly well characterized (29,30) but its genetics have proven to be somewhat refractory. The strain contains a large plasmid, pWR1, which when transferred to other strains of Pseudomonas confers upon them the ability to utilize 3CB as sole source of carbon and energy. A closely-related plasmid, designated pAC25, has been identified in a $3CB^+$ isolate of P. putida (31). A number of $3CB^-$ transposon mutant derivatives of WR1 have been isolated and partially characterized (A.J. Weightman and M. Weisshaar, unpub. data), but the difficulty of isolation of pWR1 plasmid DNA has impeded their analysis.

A related pathway for the degradation of chloroaromatics that is less well characterized biochemically, but more amenable to genetic analysis, is the pathway for the catabolism of 3CB and 2,4-dichlorophenoxyacetic acid (24D) specified by Alcaligenes eutrophus JMP134 (1,32). In this case, crucial catabolic enzymes are encoded by the readily-isolated, broad host range P1 group plasmid pJP4.

The pathway for 24D degradation (Fig. 4) consists of removal of glyoxylic acid to form 2,4-dichlorophenol, followed by hydroxylation of the ring to form 3,5-dichlorocatechol, dioxygenation of the dichlorocatechol to form 2,4-dichloromuconate, cycloisomerization and spontaneous elimination of the chloro group at carbon atom 4, to form 4-carboxymethylene-2-chlorbut-2-ene-4-olide (2-chlorodienelactone), and hydrolysis to form 2-chloromaleylacetate. Chloromaleylacetate is further metabolized to 3-oxoadipate which is then metabolized to Krebs cycle intermediates (33,34).

A number of Tn5 and Tn1771 transposon insertion mutant derivatives of plasmid pJP4 have been obtained and characterized and 5 plasmid catabolic genes tfdB-E have been identified (35; Fig. 5). Genes tfdC-F form 1 operon whereas tfdB forms a separate operon. The enzymes encoded by four of these genes have been identified, i.e.: tfdB, 2,4-dichlorophenol hydroxylase, tfdC, chlorocatechol

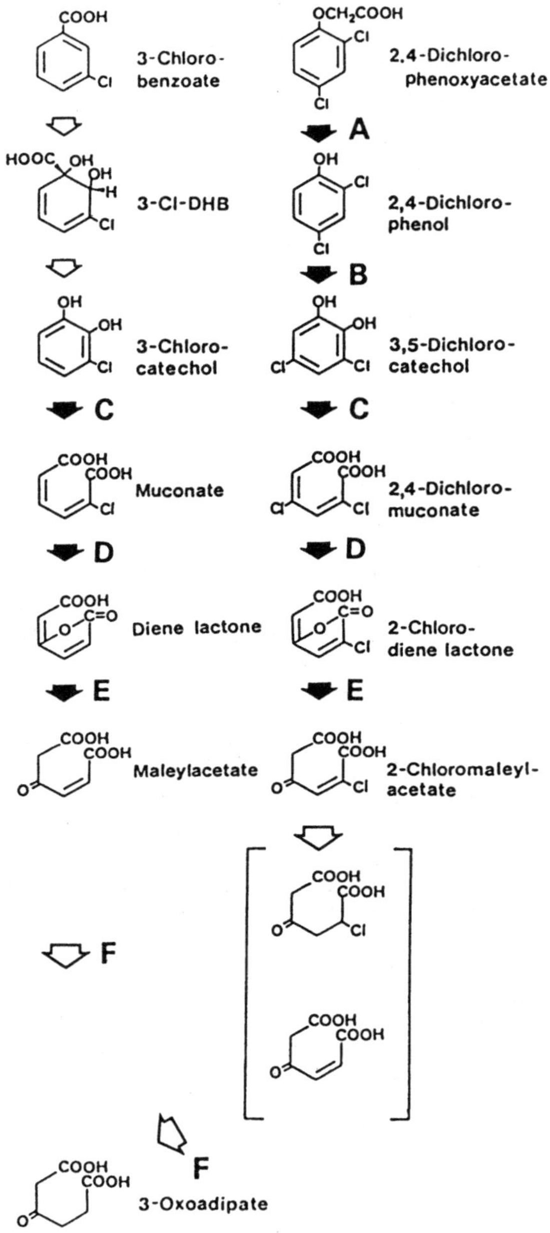

COOH
3-Chloro-benzoate
OCH2COOH
2,4-Dichloro-phenoxyacetate
A
HOOC OH
3-Cl-DHB
2,4-Dichloro-phenol
B
3-Chloro-catechol
3,5-Dichloro-catechol
C
C
Muconate
2,4-Dichloro-muconate
D
D
Diene lactone
2-Chloro-diene lactone
E
E
Maleylacetate
2-Chloromaleyl-acetate
F
F
3-Oxoadipate

Fig. 4. Proposed pathways for the degradation of 3-chlorobenzoate (3-chlorocatechol branch) and 2,4-dichlorophenoxyacetic acid in *Alcaligenes eutrophus* JMP134.

The letters in bold type indicate the pJP4 plasmid-carried catabolic genes *tfdA*,*B*,*C*,*D*,*E*, and *F*, which specify the enzymatic steps indicated by filled arrows. The two initial steps in the degradation of 3-chlorobenzoate are specified by chromosomal genes. The nature of the *tfdF* gene product is not yet established but suspected of being a trans-chloridiene lactone isomerase (see text).

1,2-dioxygenase; *tfdD*, chloromuconate cycloisomerase; *tfdE*, dienelactone hydrolase. The inability of plasmid-free derivatives of JMP134 to attack 24D suggests that the *tfdA* gene, which encodes the first enzyme of the pathway 24D monooxygenase, is plasmid borne. However, no mutants defective in this enzyme have thus far been obtained, and hence its location remains to be established. On the other hand, a fifth plasmid gene, *tfdE*, that encodes a step in 24D degradation (perhaps trans-chlorodiene-lactone isomerase) not thus far defined, has been localized on pJP4 (Fig. 5). Mutations in *tfdC*, *D* and *E* prevent growth of JMP134 on 3CB. Mutations in *tfdF* severely retard growth of JMP134 on 24D but not on 3CB. Introduction of the cloned *tfdC*, *D*, *E* genes into a pJP4-cured derivative of JMP134 enables the derivative to grow on 3CB, but not 24D. These results indicate that 24D is degraded to 3-oxoadipate by plasmid-encoded enzymes, whereas 3CB is transformed by enzymes encoded by the chromosome to chlorocatechol which is then degraded to 3-oxoadipate by the plasmid-encoded 24D pathway.

All identified catabolic genes of plasmid pJP4 have been localized to *Eco*RI fragment B. This fragment has been cloned in the pKT231 broad host range cloning vector plasmid in *E. coli* and subsequently transferred to a pJP4-cured derivative of JMP134 by RP4-mediated mobilization (35). Transconjugants containing pKT231-pJP4 *Eco*RI-B hybrid plasmids were able to grow on 24D and 3CB as sources of carbon and energy. The $24D^+$ and $3CB^+$ phenotypes could not, however, be maintained in the absence of continued selection pressure.

MANIPULATION OF CATABOLIC PATHWAYS BY MEANS OF CLONED GENES

An important potential use of a detailed biochemical and genetic characterization of a catabolic pathaway is the experimental manipulation of its activities. Manipulations of a catabolic pathway may be effected in order to accomplish a number of objectives, including:

(1) a vertical expansion of the pathway, i.e., an increase in

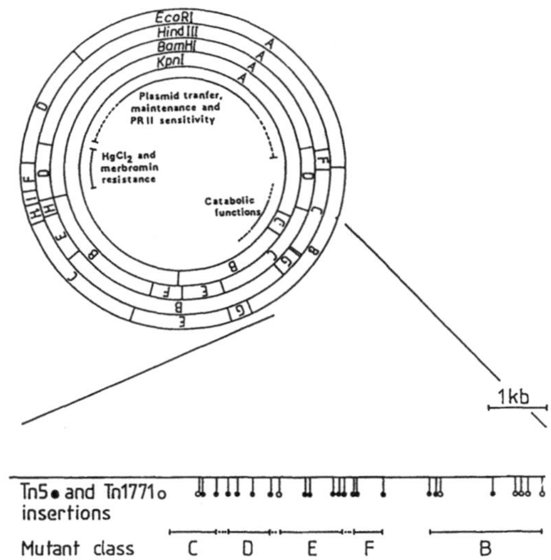

Fig. 5. Physical and genetic map of the pJP4 plasmid.

its length, accompanied by a corresponding increase in the number of compounds it can process (e.g., addition of genes encoding the conversion of chloronaphthalene to chlorocatechol to genes of a chlorocatechol pathway; see below);

(2) a horizontal expansion of a pathway, i.e., effecting a reduction in its specificity to enable degradation of a larger series of substrate isomers and analogues (Ref. 21, 25; see below);

(3) an increase in its activity;

(4) an alteration of the end product of the pathway, to provoke accumulation of a specific intermediate or to channel it into another pathway (26);

(5) an alteration in its regulation, e.g., in order that it be expressed constitutively (see above) or that new compounds act as inducers.

Gene cloning procedures are ideally suited for effecting the first four types of manipulation (21), and may be very useful for the last, although here an in vivo procedure may be the preferred choice initially. The manipulation of catabolic pathways by gene cloning methods has the major advantage over traditional methods of in vivo manipulation in that defined segments of DNA are used to effect predictable changes. Moreover, where supplementary genetic adjustments occur in vitro during a manipulation, they are more readily characterized at the molecular level. Perhaps most importantly, multiple changes can be effected either stepwise with high efficiencies or in a single step; classical procedures for effecting multiple genetic changes are invariably inefficient and require complicated and lengthy selection regimes.

We are currently exploring possibilities for the directed manipulation in vitro of catabolic pathways and are focusing our efforts on pathways for the degradation of aromatics and haloaromatics. Of crucial importance for these manipulations is the fact that many such pathways converge at a few common intermediates, such as catechol, or derivatives of catechol, which are the substrates for cleavage of the aromatic ring. Catechol may be subjected to ortho (intradiol) or meta (extradiol) cleavage (Fig. 6). The nature of the mode of ring fission and the corresponding pathway to tricarboxylic acid cycle intermediates depends upon the nature of the catechol derivative: catechol itself can be degraded by either pathway; on the other hand, methyl catechols are generally degraded by the meta pathway whereas chlorocatechols are usually degraded by a modified ortho-pathway. Chlorocatechols are in fact readily cleaved at the meta position but the resulting ring cleavage product either inhibits irreversibly the ring cleavage enzyme or is further metabolized to chlorosubstituted dead-end metabolites (36). Pathway

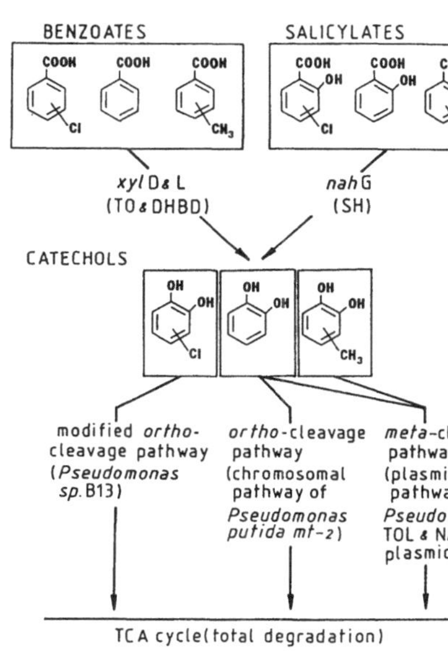

Fig. 6. Conversion of substituted benzoates and salicylates to catechols by broad specificity enzymes of cloned genes, and subsequent pathway for dissimilation of the catechols.

manipulation strategies must therefore take into account the need to metabolize a substrate to one of a few central intermediates, the need for the availability of an appropriate pathway for the further degradation of the selected intermediate, and the need to eliminate any pre-existing non-productive pathway for that compound.

Horizontal expansions of a 3CB pathway

Pseudomonas sp. B13 (29) degrades 3-chlorobenzoate, but not 4-chlorobenzoate (4CB) or 3,5-dichlorobenzoate (35DCB), owing to the fact that the first enzyme of the pathway, benzoate 1,2-dioxygenase, has a narrow substrate specificity and thus a low affinity for the latter 2 analogues (27). As indicated above, the substrate specificity of the equivalent enzyme encoded by the TOL plasmid of *P. putida* mt-2 is broader and the enzyme is able to transform 3CB, 4CB and 35DCB.

In an early experiment (28) to examine the feasibility of recruiting the TOL plasmid enzyme into the 3CB pathway, the TOL plasmid was transferred into strain B13. Transconjugants selected for their ability to grow on *m*-toluate were not able to utilize either 4CB or 35DCB as growth substrates. Mutant derivatives able to grow on 4CB could, however, be isolated from such transconjugants, as indeed could derivatives able to degrade 35DCB, although at lower frequencies. Apparently, therefore, recruitment of the TOL plasmid broad specificity enzymes into the 3CB *ortho*-pathway required at least one genetic change for acquisition of the $4CB^+$ phenotype and at least one more for acquisition of the $35DCB^+$ phenotype. As indicated above, the precise nature of these changes could not be readily analyzed although inactivation of the *meta*-cleavage pathway was observed (28,37).

We have now effected the same enzyme recruitment using cloned genes (25). A DNA segment carrying the *meta*-cleavage pathway operon promoters and *xylD,L*, the genes of toluate dioxygenase and dihydrodihydroxybenzoate dehydrogenase, but lacking genes for the *meta*-cleavage enzyme C23O and subsequent enzymes of the *meta* pathway, was inserted into the *Bam*HI site of pBR322 to give plasmid pPL401 (Fig. 7). In parallel, a DNA segment carrying *xylS*, the gene encoding the positive regulator of the *meta*-pathway operon, was cloned in the *Xho*I site of vector pKT231 (38), to produce plasmid pKT570. The large *Hin*dIII fragment of this plasmid, which contains most of the vector sequences plus a *xylS* containing sub-segment of the insert, was then inserted into the unique *Hin*dIII site of pPL401, to give pPL403, a broad host range hybrid plasmid carrying regulated genes encoding enzymes for the conversion of substituted benzoates to catechols. This plasmid was transferred to strain B13 but was unstable in this host. Stable derivatives were, however, readily selected and structural analysis of a hybrid plasmid, pPL403-1, from one derivative suggested that stability was gained as a result of a

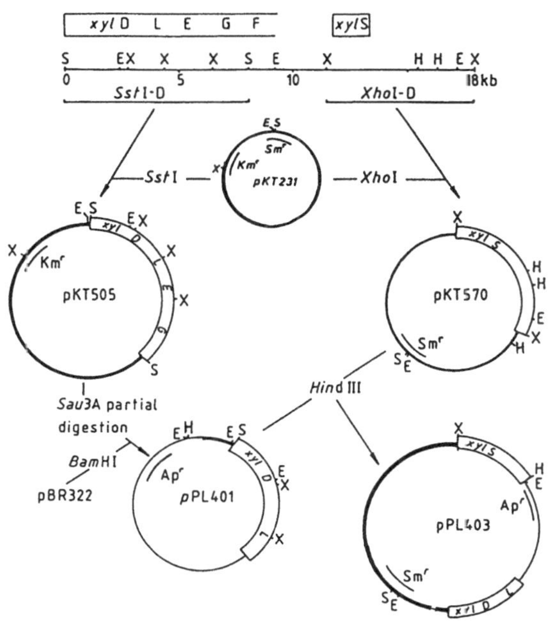

Fig. 7. Construction of a broad host range plasmid carrying the TOL plasmid xylD, L, S genes.

Abbreviations: S,E,X,H, cleavage sites for restriction endonucleases SstI, EcoRI, XhoI, HindIII, respectively.

deletion event that removed a small segment of vector DNA, located just upstream of the xylD,L operon.

When transferred to strain B13, the pPL403-1 plasmid extended the catabolic range of transconjugants to include 4CB, but not 35DCB (Tab. 1). However, 35DCB$^+$ derivatives of such transconjugants were readily isolated.

Analysis of the ability of mutants of pPL403, deleted of various segments of insert DNA, to give rise to 35DCB$^+$ derivatives showed that, whereas acquisition of the xylD gene alone permitted B13 to catabolize 4CB, acquisition of both xylD and xylL were necessary for the isolation of 35DCB$^+$ derivatives of B13. These results tentatively suggest that the dihydrodihydroxybenzoate dehydrogenase of the 3CB pathway of strain B13 is able to process the 4CB metabolite but not the 35DCB metabolite.

Thus expansion of the 3CB pathway to include 4CB by acquisition of the cloned xylD gene required no supplementary genetic change; expansion of the pathway to include 35DCB by acquisition of the

Tab. 1. Substrates utilized by Pseudomonas strains containing the cloned *xylD*, *L* and *nahG* genes[a]

Pseudomonas strain	ben	3CB	4CB	SAL	3CSAL	4CSAL	5CSAL
P.putida KT2442	+	-	-	-	-	-	-
KT2442 (pPL300-1)	+	-	-	+	-	-	-
P.sp B13	+	+	-	-	-	-	-
B13 (pPL300-1)	+	+	-	+	+	+	+
B13 (pPL403-1)	+	+	+	-	-	-	-

(Substrates[b])

[a] Growth determinations were carried out on solid and in liquid media; + indicates growth, - indicates no growth.

[b] Abbreviations: ben, benzoate; 3CB, 3-chlorobenzoate; 4CB, 4-chlorobenzoate; SAL, salicylate; 3,4,5CSAL, 3-, 4-, 5-chlorosalicylate.

cloned *xylD,L* genes did require such a change. 35DCB is not an effective inducer of the *meta*-pathway operon and we are therefore currently investigating possible modifications of the *meta*-pathway operon regulatory system in 35DCB$^+$ derivatives, particularly the molecular changes that may have occurred in the cloned *xylS* gene-containing DNA fragment.

A second type of expansion of the chlorocatechol pathway of strain B13, this time to include chlorosalicylates, was accomplished by introduction of the cloned gene for a salicylate hydroxylase (25). The NAH7 plasmid of *P. putida* encodes all enzymes required

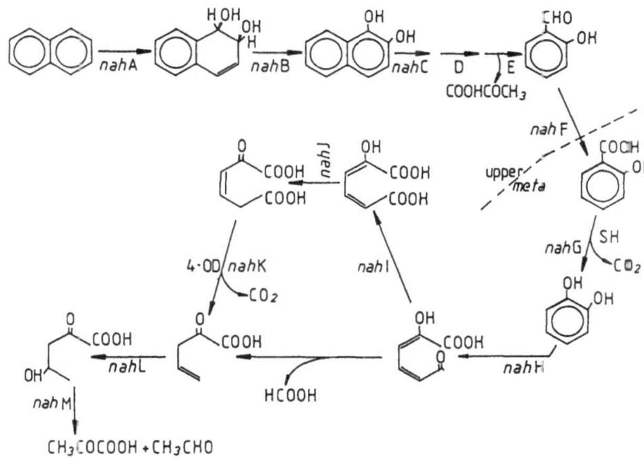

Fig. 8. NAH7 plasmid-specified pathway for the degradation of naphthalene.

This pathway is taken from Ref. 39.

for the catabolism of naphthalene via salicylate, catechol and a meta-cleavage pathway (39; Fig. 8). The salicylate hydroxylase encoded by the nahG gene of NAH7 exhibits a broad substate specificity and metabolizes salicylate, methyl- and chlorosalicylates to their corresponding catechol derivatives (25). A HindIII-generated fragment of the NAH7 plasmid that contains the nahG gene was inserted into pBR322 and the hybrid plasmid thereby constructed, pPL299, was fused to the pKT231 vector through their unique EcoRI cleavage sites, to produce pPL300. This hybrid was transferred to Pseudomonas bacteria but was found to be unstable. Stable derivatives were readily isolated and found to have suffered a small deletion of vector sequences just upstream of the cloned nahG gene. One such deletion derivative plasmid, pPL300-1, was transferred to strain B13 which, as a result, acquired the ability to degrade and use as sole source of carbon 3-, 4- and 5-chlorosalicylate. Thus, as was the case for expansion of the 3CB pathway to include 4CB, expansion of the same pathway to include chlorosalicylates required no genetic change other than that of acquisition of the gene of the enzyme to be recruited. This is almost certainly due to an absence on the cloned DNA segments of genes for enzymes of the non-productive meta-cleavage pathway.

Synthesis of salicyate hydroxylase in Pseudomonas bacteria containing pPL300 or pPL300-1 was inducible by salicylate, indicating that the cloned HindIII fragment contains the determinant of a regulator for the nahG gene. Thus, in both types of construction, inducible expression of the cloned genes was preserved. Clearly, in order for enzyme recruitment to be effectively accomplished by the genetic manipulations undertaken, it is necessary to assure balanced expression of cloned genes, either by native regulatory circuits (preferably) or by unnatural constitutive or regulated promoters.

Vertical Expansion of a Chloracatechol Pathway

An important potential use of gene cloning in pathway manipulation is the simultaneous accomplishment of multiple genetic changes, and the construction of strains that cannot be easily isolated by classical procedures. An example of this is the construction of chloronaphthalene-degrading organisms, which have thus far proven to be exceptionally difficult to obtain from natural sources.

A pseudomonad, SB104, able to degrade 2-methyl naphthalene was isolated from soil and was shown to carry an 80-kb plasmid which specifies a pathway similar to that of NAH7 for the degradation of naphthalene via salicylate, catechol and a meta-cleavage pathway. This plasmid, pSB104, is structurally similar though not identical to NAH7, and its catabolic pathway exhibits minor differences to that of NAH7 in terms of regulation of the pathway and the specificities and activities of certain enzymes (S. Bas, unpub. data).

Plasmid pSB104 catabolizes 2-chloronaphthalene to chlorocatechol. In order to explore the possibility of combining the chloronaphthalene to chlorocatechol pathway of SB104 with a chlorocatechol degradation pathway, an insertion mutant derivative of pSB104, containing a Tn5 insertion element in its C230 gene (i.e., a derivative

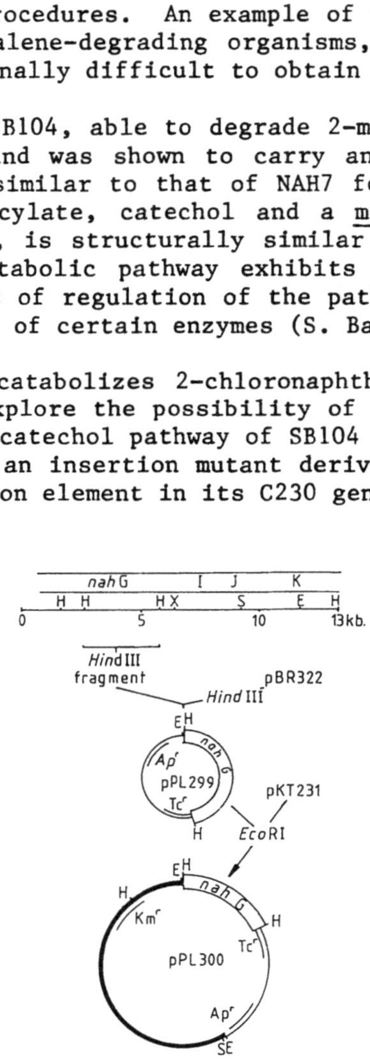

Fig. 9. Construction of a broad host range plasmid carrying the nahG gene of plasmid NAH7.

Abbreviations used are those described in Fig. 7.

blocked in the *meta*-cleavage pathway), was transferred by conjugation to *Pseudomonas* *sp*. B13. Transconjugants grew on 2-chloronaphthalene and released chloride.

As a first step in utilizing cloned DNA segments to construct an effective chloronaphthalene catabolic pathway, we have introduced the pSB104 plasmid into the laboratory strain of *P. putida*, KT2442, containing a pKT231 hybrid plasmid carrying the pJP4 genes for chlorocatechol degradation, and have isolated transconjugants able to grow on 2-chloronaphthalene as sole source of carbon and energy.

CONCLUDING REMARKS

The possibilities for the controlled, selective manipulation of microbial metabolic pathways by genetic engineering methods are enormous. Indeed, classical methods to effect such manipulations have been exploited for years to great effect in the area of secondary metabolism. Where a particular change can be selected or readily recognized, where the molecular details of the manipulation are not required, and in particular where little biochemical and genetic information on the pathway exists, a classical approach may still in most cases be the one of choice. However, where multiple changes are necessary, where no effective selection exists for the desired change, and particularly for those manipulations for which an understanding of the molecular events of the change is desired, the use of gene cloning procedures to introduce defined segments of DNA encoding just those enzymes of relevance into selected microorganisms is indicated.

The availability of broad host range cloning vectors and transposon donors (e.g., 21, 40) now permits genetic analysis and manipulation of a wide range of microbes, including those exhibiting a variety of important and interesting properties. It should be emphasized, however, that manipulation of complex pathways is most effective where extensive biochemical and genetic information already exists, as was the case for the experiments discussed here.

In this report, we have described the characterization of 2 model catabolic pathways, those of the TOL and pJP4 plasmids, and experiments in which cloned genes encoding broad specificity enzymes have been used to expand existing pathways for the degradation of chlorocatechols, and a cloned operon encoding chlorocatechol degradation has been used to create a hybrid pathway that catabolizes chloronaphthalene. These experiments have defined which enzymes were necessary for expansion for the 3CB pathway to include 4CB and 35DCB, and have indicated that a further mutation necessary for acquisition of the 35DCB phenotype may involve the regulation system for expression of the *xylD,L* genes.

It is to be anticipated that during the analysis of additional, in some instances unexpected, changes necessary for certain manipulations, important information on pathway functioning in vivo will be obtained, particularly on physiological and cellular parameters that constrain such manipulations, and on the regulatory circuits that assure optimal integration of the pathway under investigation with other other cellular metabolic activities.

ACKNOWLEDGMENTS

This work has been supported in part by grants from Transgène S.A., Strasbourg, the Bundesministerium für Forschung und Technologie, Bonn, and the Deutsche Forschungsgemeinschaft, Bonn-Bad Godesberg. We thank our catabolic colleagues in Göttingen for stimulating discussions, M. Rekik for technical assistance and F. Rey for secretarial assistance.

REFERENCES

1. Pemberton, J.M., B. Corney, and R.H. Don (1979) Evolution and spread of pesticide degrading ability among soil micro-organisms. In Plasmids of Medical, Environmental and Commercial Importance, K.N. Timmis and A. Pühler, eds. Elsevier/North-Holland, Amsterdam, pp. 287-299.
2. Ribbons, D.W., and R.W. Eaton (1982) Chemical transformations of aromatic hydrocarbons that support the growth of micro-organisms. In Biodegradation and Detoxification of Environmental Pollutants, A.M. Chakraborty, ed. CRC Press, Boca Raton, Florida, pp. 59-84.
3. Franklin, F.C.H., M. Bagdasarian, and K.N. Timmis (1981) Manipulation of degradative genes of soil bacteria. In Microbial Degradation of Xenobiotic and Recalcitrant Compounds, T. Leisinger, R. Hütter, A.M. Cook, and J. Nuesch, eds. Academic Press, London, pp. 111-132.
4. Farrell, R., and A.M. Chakrabarty (1979). Degradative plasmids: Molecular nature and mode of evolution. In Plasmids of Medical, Environmental and Commercial Importance, K.N. Timmis and A. Puhler, eds. Elsevier/North Holland, Amsterdam, pp. 97-109.
5. Keith, L.H., and W.A. Teilliard (1979) Priority pollutants I - a perspective view. Env. Sci. and Techn. 13:416-423.
6. Weightman, A.J., R.H. Don, P.R. Lehrback, and K.N. Timmis (1984) The identification and cloning of genes encoding haloaromatic catabolic enzymes and the construction of hybrid pathways for substrate mineralization. In Genetic Control of Environmental Pollutants, G.S. Ommen and A. Hollaender, eds. Plenum, pp. 47-80.
7. Williams, P.A., and K. Murray (1974) Metabolism of benzoate and

the methyl benzoates by Pseudomonas putida (arvilla) mt-2: evidence for the existence of a TOL plasmid. J. Bact. 120:416-423.

8. Wong, C.L., and N.W. Dunn (1974) Transmissible plasmid coding for the degradation of benzoate and m-toluate in Pseudomonas arvilla mt-2. Genet. Res. 23:227-232.
9. Franklin, F.C.H., M. Bagdasarian, M.M. Bagdasarian, and K.N. Timmis (1981) Molecular and functional analysis of the TOL plasmid pWWO from Pseudomonas putida and cloning of genes for the entire regulated aromatic ring meta cleavage pathway. Proc. Natl. Acad. Sci., USA 78:7458-7462.
10. Worsey, M.J., and P.A. Williams (1975) Metabolism of toluene and xylenes by Pseudomonas putida (arvilla) mt-2: evidence for a new function of the TOL plasmid. J. Bact. 124:7-13.
11. S. Harayama, N. Mermod, P.R. Lehrbach, and K.N. Timmis. Roles of the divergent branches of the meta-cleavage pathway for the degradation of benzoate and substituted benzoates. Submitted for publication.
12. Nakazawa, T., S. Inouye, and A. Nakazawa (1980) Physical and functional mapping of RP4-TOL plasmid recombinants: analysis of insertion and deletion mutations. J. Bact. 144:222-231
13. Leppik, R., P.R. Lehrbach, and K.N. Timmis. Genetic analysis of the TOL plasmid-specified upper pathway for the degradation of toluene and xylenes. Submitted for publication.
14. Harayama, S., P.R. Lehrbach, and K.N. Timmis (1984) Transposon mutagenesis analysis of the meta-cleavage pathway operon genes of the TOL plasmid of Pseudomonas putida mt-2. J. Bact. (in press).
15. Worsey, M.J., F.C.H. Franklin, and P.A. Williams (1978) Regulation of the degradative pathway enzymes coded for by the TOL plasmid (pWWO) from Pseudomonas putida mt-2. J. Bact. 134:757-764.
16. Inouye, S., A. Nakazawa, and T. Nakazawa (1981) Molecular cloning of gene xylS of the TOL plasmid: evidence for positive regulation of the xylDEGF operon by xylS. J. Bact. 148:413-418.
17. Franklin, F.C.H., P.R. Lehrbach, R. Lurz, B. Rueckert, M. Bagdasarian, and K.N. Timmis (1983) Localization and functional analysis of transposon mutations in regulatory genes of the TOL catabolic pathway. J. Bact. 154:676-685.
18. Mermod, N., P.R. Lehrbach, W. Reineke, H.-J. Knackmuss, and K.N. Timmis. Characterization of Pseudomonas putida TOL plasmid meta cleavage pathway operon promoters and mutant constitutive promoters. Submitted for publication.
19. Inouye, S., A. Nakazawa and T. Nakazawa (1981) Molecular cloning of TOL genes xylB and xylE in Escherichia coli. J. Bact. 145:1137-1143.
20. Jacoby, G.A., J.E. Rogers, A.E. Jacob, and R.W. Hedges (1978) Transposition of Pseudomonas toluene degrading genes and expression in E. coli. Nature (Lond.) 274:179-180.

21. Lehrbach, P.R., and K.N. Timmis (1983) Genetic analysis and manipulation of catabolic pathways in Pseudomonas. Biochem. Soc. Symp. 48:191-219.
22. Sala-Trepat, J.M., and W.C. Evans (1971) The meta-cleavage of catechol by Azotobacter species. 4-oxalocrotonate pathway. Eur. J. Biochem. 20:440-413.
23. Zukowski, M.M., D.F. Gaffney, D. Speck, M. Kauffmann, A. Findeli, A. Wisecup, and J.-P. LeCocq (1983) Chromogenic identification of genetic regulatory signals in Bacillus subtilis based on expression of a cloned Pseudomonas gene. Proc. Nat. Acad. Sci., USA 80:1101-1105.
24. Nakai, C., H. Kagamiyama, M. Nozaki, T. Nakazawa, S. Inouye, Y. Ebina, and A. Nakazawa (1983) Complete nucleotide sequence of the meta pyrocatechase gene on the TOL plasmid of Pseudomonas putida mt-2. J. Biol. Chem. 258:2923-2928.
25. Lehrbach, P.R., J. Zeyer, W. Reineke, H.-J. Knackmuss, and K.N. Timmis (1984) Enzyme recruitment in vitro: use of cloned genes to extend the range of haloaromatics degraded by Pseudomonas sp. B13: J. Bact., 158:1025-1032.
26. Zeyer, J., P.R. Lehrbach, and K.N. Timmis. Use of bacteria containing the cloned xylD gene of TOL plasmid pWWO to transform benzoate and substituted benzoates to the corresponding carboxyhydrodiols. Submitted for publication.
27. Reineke, W., and H.-J. Knackmuss (1978) Chemical structure and biodegradability of halogenated aromatic compounds. Substituent effects on 1,2-dioxygenation of benzoic acid. Biochim. Biophys. Acta 542:412-423.
28. Reineke, W., and H.-J. Knackmuss (1979) Construction of haloaromatics utilizing bacteria. Nature (Lond.) 277:385-386.
29. Dorn, E., M. Hellwig, W. Reineke, and H.-J. Knackmuss (1974) Isolation and characterization of a 3-chlorobenzoate degrading Pseudomonad. Archs. Microbiol. 99:61-70.
30. Schmidt, E., and H.-J. Knackmuss (1980) Chemical structure and biodegradability of halogenated aromatic compounds. Conversion of chlorinated muconic acids into maleylacetic acid. Biochem. J. 192:339-347.
31. Chatterjee, D.K., S.T. Kellogg, S. Hamada and A.M. Chakrabarty (1981) Plasmid specifying total degradation of 3-chlorobenzoate by a modified ortho pathway. J. Bact. 146:639-646.
32. Don, R.H., and J.M. Pemberton (1981). Properties of six pesticide degradation plasmids isolated from Alcaligenes paradoxus and Alcaligenes eutrophus. J. Bact. 145:681-686.
33. Gaunt, J.K., and W.C. Evans (1971) Metabolism of 4-chloro-2-methylphenoxyacetate by a soil pseudomonad. Ring fission, lactonizing and delactonizing enzymes. Biochem. J. 122:533-542.
34. Evans, W.C., B.S.W. Smith, N.H. Fernley and J.I. Davies (1971) Bacterial metabolism of 2,4-dichlorophenoxyacetate. Biochem. J. 122:543-551.
35. Don, R.H., A.J. Weightman, H.-J. Knackmuss and K.N. Timmis (1984). Transposon mutagenesis and cloning analysis of the

pathways for the degradation of 2,4-dichlorophenoxyacetic acid and 3-chlorobenzoate by Alcaligenes eutrophus JMP134 (pJP4). J. Bact. (in press).

36. Bartels, I., H.-J. Knackmuss and W. Reineke (1984) Suicide inactivation of catechol 2,3-dioxygenase from Pseudomonas putida mt-2 by 3-halocatechols. App. Env. Microbiol. 47:500-505.

37. Reineke, W., D.J. Jeenes, P.A. Williams, and H.-J. Knackmuss (1982) TOL plasmid pWWO in constructed halobenzoate-degrading Pseudomonas strains: prevention of meta-pathway. J. Bact. 150:195-201.

38. Bagdasarian, M., R. Lurz, B. Rueckert, F.C.H. Franklin, M.M. Bagdasarian, J. Frey and K.N. Timmis (1981) Specific purpose cloning vectors. II. Broad host range, high copy number RSF1010-derived vectors, and a host:vector system for gene cloning in Pseudomonas. Gene 16:237-247.

39. Yen, K.-M., and I.C. Gunsalus (1982) Plasmid gene organization: naphthalene/salicylate oxidation. Proc. Natl. Acad. Sci., USA 79:874-878.

40. Timmis, K.N. (1981) Gene manipulation in vitro. Soc. Gen. Microbiol. Symp. 31:49-109.

THE PLASMID-SPECIFIED AEROBACTIN IRON UPTAKE SYSTEM OF *ESCHERICHIA COLI*

Peter H. Williams and Nicholas H. Carbonetti

Department of Genetics
University of Leicester
University Road
Leicester LE1 7RH, England

INTRODUCTION

Plasmids and Pathogenicity

Escherichia coli is a major aerobic commensal bacterial species in the intestinal flora of a healthy individual. Some strains, however, are pathogenic for man, causing disease by a variety of mechanisms, and with a severity that ranges from the temporary, if acute, discomfort of "turista" to the chronic human tragedy of widespread infant mortality in the third world. In domestic animals, too, *E. coli* infections are of significant economic importance in terms of both lost production and cost of control and treatment.

The virulence of *E. coli* strains can in many cases be attributed to the presence of plasmids. For instance, the plasmid location of the genetic determinants of the enterotoxins, of which one, the heat labile toxin, is very closely related to the chromosome-specified cholera toxin of *Vibrio* (19,36), has been known since the pioneering studies of H. Williams Smith in the 1960s (32). Moreover, it is clear from his early work with animal pathogens (34), and has since been confirmed with enterotoxigenic isolates from human disease (11,12), that plasmid-mediated surface antigens promote bacterial colonization of the intestinal mucosa, and thereby effectively concentrate toxin action at target cells. By contrast, it has been shown relatively recently that plasmids are implicated in the virulence of enteroinvasive strains of *E. coli*, which penetrate and proliferate in the mucosal epithelium and elicit dysentery-like diarrheal disease (18).

In a search for novel plasmid-specified virulence determinants among isolates of *E. coli* from cases of bacteremia of man and a range of domestic animals, H. Williams Smith reported that a significant proportion of strains carried plasmids specifying the antibacterial protein, colicin V (31). Furthermore, he showed that loss of such ColV plasmids by curing markedly reduced the virulence of clinical isolates in experimental infections of a variety of laboratory and domestic animals, and, conversely, that gain of a ColV plasmid by conjugation significantly enhanced the virulence of laboratory strains of *E. coli* (31,33). Several phenotypic properties of ColV plasmids have been proposed to contribute to this phenomenon, including increased ability to adhere to epithelial cells (9) and resistance to the bactericidal effects of serum (4). Moreover, it has been suggested tha colicin V itself might aid bacterial invasion by increasing vascular permeability or by depressing macrophage activity (25,26). Of crucial importance in the progress of extraintestinal infections, however, is the operation of a plasmid-determined high affinity mechanism for the specific active uptake of iron which allows extensive bacterial proliferation in the tissues and body fluids of an infected animal. It is this particular aspect of ColV plasmids that we discuss in detail here.

Iron and Infection

Iron is essential for microbial growth. In nature, however, iron is virtually unavailable to microorganisms because it forms insoluble aggregates at physiological pH or because it exists in complexed forms, such as hemoglobin, in animal bodies. Indeed, the level of free ferric ions in the blood and other body fluids of a healthy individual is well below that required for bacterial growth because of the presence of the iron-binding glycoproteins "transferrin" in serum and "lactoferrin" in secretions (38). Furthermore, systemic infections may provoke a host hypoferremic response in which the level of ferric ions in the serum transferrin iron pool is reduced, thereby limiting the availability of extracellular iron to invading microorganisms (38). It is to be expected, therefore, that a mechanism which permits more efficient sequestration of iron from the complexed state would be an important component in the virulence of strains of *E. coli* associated with extraintestinal infections. Evidence that ColV plasmids mediate such a mechanism is provided by the observation that they confer on the host bacterium a significant selective advantage for growth in iron-limited conditions, both in vitro, in media containing transferrin, and in vivo, in the bodies of experimentally infected animals (40).

ColV Plasmid-Mediated Iron Uptake

E. coli, and other enteric bacteria, secrete the catechol siderophore "enterochelin" (enterobactin) to solubilize and sequester iron in the medium and to actively transport it into the cell (29). Mutants defective in the biosynthesis of enterochelin rely on low affinity, nonspecific iron uptake from media in which iron is present at relatively high levels (13). Alternatively, some bacterial strains will take up externally supplied iron-chelating compounds such as citrate (13) or ferrichrome (22). However, the presence of a ColV plasmid allows growth of an enterochelin-deficient mutant of *E. coli* K-12 in iron-limited conditions in the absence of added citrate, because of an alternative mechanism for the efficient active uptake of ferric ions from the medium. Phenotypic analysis of plasmid mutants defective in iron uptake (41) suggested the secretion of an extracellular chelating compound, defined by *iuc* mutations, and a specific ferric-siderophore transport function, defined by the *iut* mutation, which, like other bacterial active transport mechanisms, requires functional *tonB* protein (40). The *iut* gene product is an outer membrane protein (M_r = 74,000) whose synthesis is induced in conditions of iron starvation (15). This protein is, fortuitously, also the receptor for cloacin DF13, a bacteriocin of *Enterobacter cloacae* (2,35). Thus, strains of *E. coli*, which are naturally insensitive to cloacin DF13, are rendered susceptible by the presence of ColV plasmids and, furthermore, mutants of the *iut* type are cloacin-resistant. The siderophore whose synthesis is defined by the *iuc* mutations was identified as aerobactin (37), a hydroxamate compound previously purified from and named for a strain of *Aerobacter aerogenes* (14). In addition, strains of *En. cloacae* (35), *Shigella flexneri* (27), *Salmonella austin*, *S. memphis*, and *Arizona hinshawii* (24) have also been shown to secrete aerobactin.

PHYSIOLOGICAL ASPECTS OF AEROBACTIN-MEDIATED IRON UPTAKE

Acquisition of Iron by Aerobactin

The association constant of ferric-aerobactin (approximately 10^{23}, Ref. 23) is many orders of magnitude lower than those of ferric-enterochelin (10^{52}, Ref. 23) and ferric-transferrin ($\geq 10^{30}$, Ref. 38). It is interesting to speculate, therefore, how such a thermodynamically inferior siderophore is able to confer significant selective advantage for growth in vivo upon bacterial cells which can also synthesize the more powerful chelator enterochelin. Exchange of iron between transferrin and aerobactin probably proceeds through the initial formation of a ternary complex in which iron is bound to both compounds (20), and in fact the rate of removal of iron from transferrin is significantly higher to aerobactin than to other hydroxamate chelators with superior affinities for ferric iron, implying that the chemical structure of the competing ligand, rather than merely its affinity for iron, plays an important role in the acquisition of iron. In this chapter, we present evidence that, in addition to particular structural features as yet undefined, aerobactin possesses physiological and regulatory properties which enhance its efficiency as a siderophore in environmental conditions of extreme iron stringency.

Intracellular Compartmentalization of Aerobactin Iron

Streptonigrin is an aminoquinone antitumor agent. It can kill bacteria by degrading the DNA, a process which depends upon the presence of ferric iron (39,42) and we have exploited its bactericidal effect as a probe of the level of free iron within bacterial cells. Figure 1 illustrates the growth curves of strains of *E. coli* in appropriately supplemented M9 minimal medium; cultures were inoculated with nutrient broth-grown cells, and subsequently subcultured into fresh minimal medium to maintain logarithmic growth; samples taken at 0, 3, and 6 hr were tested for sensitivity to streptonigrin.

Strain W3110 secretes enterochelin. Strain LG1315 is enterochelin-deficient (*entA*), but it carries plasmid ColV-K30 (37) which specifies aerobactin biosynthesis. Both strains grew at the same constant rate (doubling time, 40 min) for the duration of the experiment (Fig. 1a). Strain LG1418, which is an *iuc* derivative of LG1315 and so can synthesize neither siderophore, also grew well for several hours; prior growth in iron-rich medium such as nutrient broth allows accumulation of intracellular pools of iron which permit growth for several generations following subculture to iron-limited medium. Eventually, growth of LG1418 ceased, presumably when intracellular supplies of iron were depleted, although addition of sodium citrate promoted rapid restoration of active cell division due to the uptake of iron as ferric-citrate (Fig. 1a). Consistent with these proposed physiological changes is the observation that strain LG1418 became increasingly resistant to the bactericidal action of streptonigrin during progressive depletion of the intracellular iron pool (Fig. 1b-d). Sensitivity to streptonigrin was regained upon addition of citrate to the medium.

It is probable that depletion of iron pools occurs to some extent after a shift from nutrient to minimal medium even in cells that are actively taking up iron, since strain W3110, although remaining very sensitive to streptonigrin during growth due to secretion of enterochelin, did show a slight increase in resistance over the course of the experiment (Fig. 1b-d). By contrast, aerobactin-promoted growth of strain LG1315 at maximal rate was accompanied by rapid loss of streptonigrin-sensitivity, suggesting that aerobactin is in some way able to direct assimilated iron for bacterial growth, rather than, as with enterochelin and citrate, contribute it to a pool from which the rate of withdrawal for growth may be concentration-dependent.

Siderophore Activity at Low Concentrations

Although strain LG1418 synthesizes no siderophores, it can take up both aerobactin and enterochelin when supplied exogenously. Using a culture of LG1418 incubated in minimal medium to deplete intracellular iron pools (as described in Fig. 1a), we have demon-

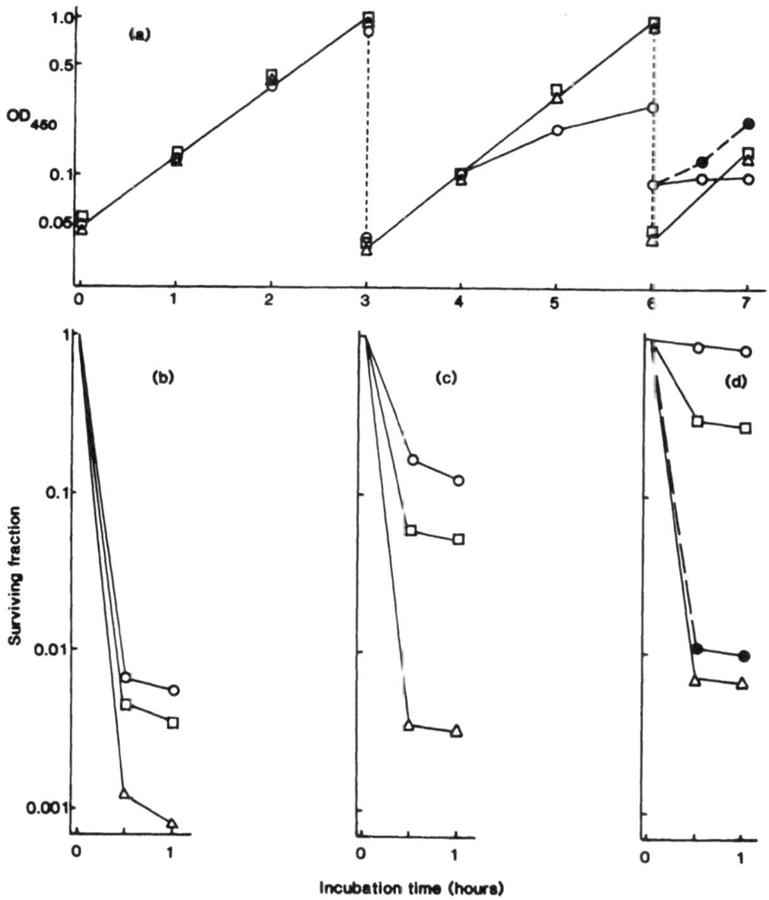

Fig. 1. Correlation of streptonigrin sensitivity with depletion of intracellular iron pools. (a) Nutrient broth-grown cells of E. coli strains W3110 (△), LG1315 (□), and LG1418 (o ●) were washed, resuspended in M9 medium, and incubated with aeration at 37°C. Bacterial growth was followed by measurement of optical density at 450 nm, and cultures were diluted with fresh medium at 3 and 6 hr. Samples were removed for determination of streptonigrin sensitivity at time 0 hr (b), 3 hr (c), and 6 hr (d). Cells were harvested by centrifugation, washed, and resuspended in phosphate-buffered saline to an optical density at 450 nm of 0.5. Streptonigrin was added at 40 µM, and samples were taken for estimation of viable cell counts at 0, 30, and 60 min during incubation at 37°C. Sodium citrate (10 mM) was added to a portion of the LG1418 culture (●) after the second subculture, and incubated for a further 1 hr before determination of streptonigrin sensitivity.

strated that levels of aerobactin in excess of 5.0 nM were adequate to stimulate growth at maximum rate, and that even a concentration as low as 0.5 nM promoted slow but significant growth (Fig. 2a). By contrast, the optimum enterochelin concentration was over 500 nM, and virtually no bacterial growth was observed at levels below 50 nM (Fig. 2b). This substantial difference may be due in part to the fact that enterochelin is enzymically degraded during intracellular release of ferric ions and therefore wasted (29); on the other hand, iron is more readily released from aerobactin and the siderophore is subsequently resecreted (5). However, there is also a major intrinsic difference in the efficiencies with which the 2 compounds transport iron into bacterial cells, particuarly at low siderophore concentrations. Figure 3 shows the maximum uptake of radioactively labeled iron into nongrowing cells of strain LG1418 in the presence of the same range of aerobactin and enterochelin concentrations used in Fig. 2. Note that dilution of aerobactin over a 10,000-fold range resulted in only a 5-fold reduction in assimilation of ^{55}Fe; enterochelin was more effective than aerobactin at high concentrations, but dilution to levels that were suboptimal for growth (see Fig. 2) resulted in a corresponding reduction in the efficiency of iron uptake into bacterial cells. It is interesting that any potential growth advantage resulting from the apparently greater effi-

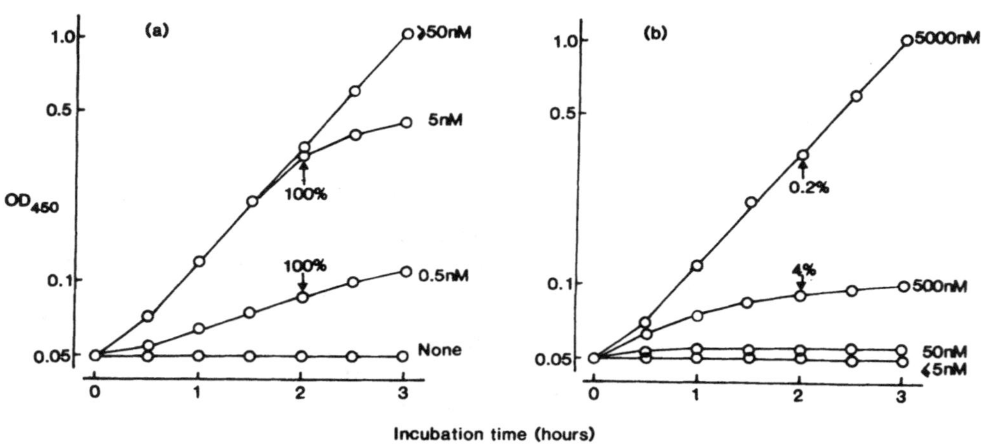

Fig. 2. Concentration optima of aerobactin and enterochelin. Strain LG1418 was preincubated in M9 medium until cessation of cell division. Aerobactin (a) or enterochelin (b) were added at the concentrations indicated, and bacterial growth was followed by measurement of optical density at 450 nm during incubation with aeration at 37°C. Percentage figures show the proportion of cells surviving streptonigrin treatment for 60 min (as described in Fig. 1) in culture samples designated by the arrows.

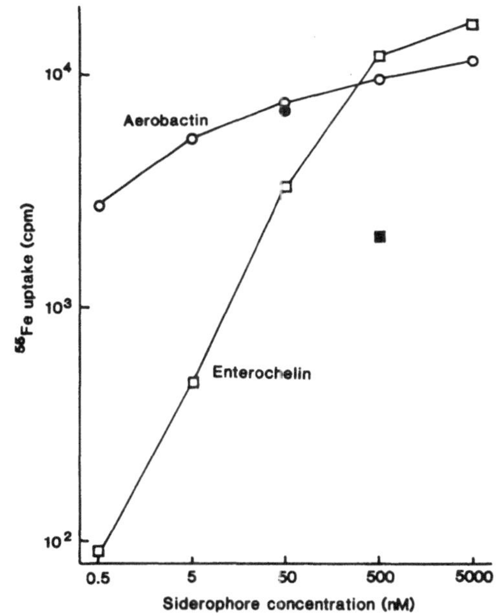

Fig. 3. Correlation of siderophore concentration with bacterial uptake of iron. Strain LG1418 was preincubated in M9 medium until cessation of growth. Cells were harvested by centrifugation, washed, and resuspended in similar medium lacking glucose and essential amino acids, but containing 100 μM sodium nitrilotriacetate and aerobactin (o) or enterochelin (□) at the concentrations indicated. Filled symbols (● ■) represent samples containing 10% (vol/vol) human serum. Carrier-free $^{55}FeCl_3$ (1 μCi/ml) was added, and suspensions were incubated with aeration for 120 min at 37°C. Cells from 5 ml of suspension were washed on membrane filters with medium containing 100 mM sodium citrate. Filters were air-dried and the level of ^{55}Fe label retained was counted; points are the average of duplicate samples.

ciency of iron uptake by high enterochelin concentrations is dissipated because assimilated iron is released to an intracellular pool. This is shown by the marked streptonigrin sensitivity of LF1418 supplied exogenously with enterochelin (Fig. 2b); even at a ferric-enterochelin concentration that is rate-limiting for growth, only 4% of the population survived streptonigrin treatment. By contrast, cells supplied with aerobactin remained streptonigrin-resistant during growth (Fig. 2a), again suggesting compartmentalization of the iron sequestered by aerobactin, rather than its contribution to a pool. Konopka et al. (20) have demonstrated that aerobactin is more effective than enterochelin, at equal molarity, in delivering iron from transferrin to bacterial cells in human serum. This may be due to the fact that enterochelin binds to serum proteins and its siderophore activity is thereby reduced. Consistent with this is our observation that addition of 10% serum to the buffer used to determine ^{55}Fe uptake into nongrowing LG1418 cells markedly reduced the effectiveness of enterochelin, although it had no significant influence on aerobactin-mediated iron uptake (Fig. 3).

Regulatory Advantages of Aerobactin

In order to study regulatory responses of the aerobactin iron uptake system, we have isolated ColV plasmid derivatives carrying the transposon-like phage Mu dl(Ap,*lac*) (8) inserted so that aerobactin synthesis is inhibited and *lacZ* gene expression is regulated by the availability of iron in the medium. With such mutants, regulation of the aerobactin system was followed simply by measuring β-galactosidase synthesis (giving the advantage of greater sensitivity and safety compared with the standard Csáky test for hydroxamates, Ref. 10), while enterochelin (and excreted degradation products) was estimated directly by the nitrite-molybdate reagent of Arnow (1). We have compared the levels of expression of both iron-uptake systems in response to increasing iron stress exerted by adding iron or an iron chelator (α,α'-dipyridyl) to M9 medium (which itself contains approximately 2 μM Fe^{3+}). Our results (Fig. 4a) indicate that enterochelin synthesis is less readily induced by conditions of iron stress than the aerobactin system, since significant β-galactosidase activity was present in cultures (up to 10 μM dipyridyl) which showed basal levels of enterochelin. Only at dipyridyl concentrations above 50 μM was there also significant synthesis of enterochelin. Note that the bacterial strain used in this experiment (LG1689) could not synthesize aerobactin due to insertional inactivation by the Mu phage, so that all iron actively entering the cells was transported by enterochelin. Nevertheless it is clear that, except under conditions of extreme iron stress, expression of the enterochelin system at only a fraction of its maximal level was adequate to support bacterial growth in defined medium.

Thus, in addition to its greater effectiveness at low concentrations, and to the more efficient, directed use of the iron it sequesters, aerobactin has the advantage that its synthesis is induced more readily in response to conditions of iron-limitation. It has been suggested that iron-uptake systems in enteric bacteria have common regulatory components. For example, mutations in the *fur* (ferric uptake regulation) locus cause constitutive expression of both the enterochelin and aerobactin systems (6,7,17). Using indicator plates, we have isolated spontaneous mutants of strain LG1689 which show high levels of β-galactosidase synthesis even in the presence of high concentrations of iron (Fig. 4b). Mutants such

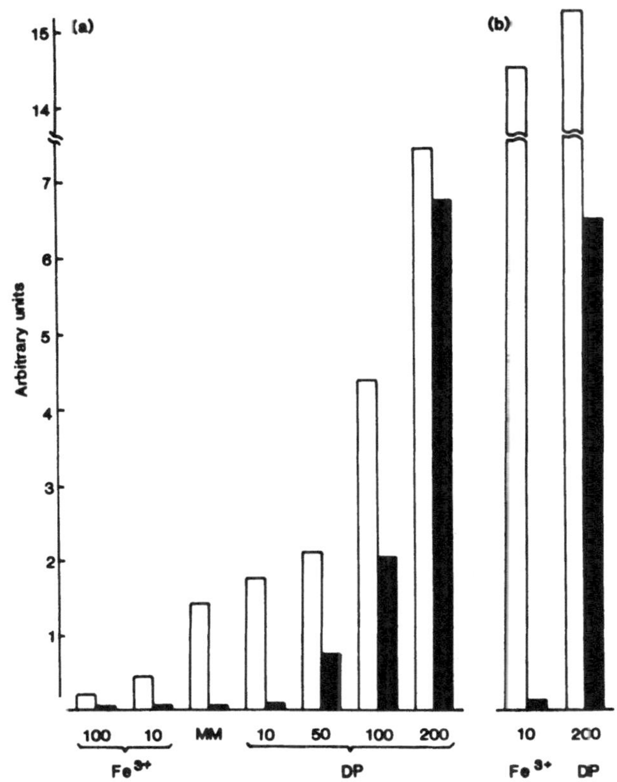

Fig. 4. Effect of iron stress on expression of the aerobactin and enterochelin iron-uptake systems. Strains LG1689 (a) and LG1712 (b) were grown overnight in M9 medium (MM) containing Fe^{3+} (as $FeCl_3$) or α,α'-dipyridyl (DP) at the concentrations indicated (μM). Cultures were diluted in the same media to an optical density at 450 nm of 0.1, incubated with aeration at 37°C for 2 hr and assayed for enterochelin and β-galactosidase. Results are displayed in arbitrary units in order to compare directly the range of detectable levels of enterochelin (approximately 1 to 50 μM) with β-galactosidase activities up to approximately 200 Units.

as LG1712 carry chromosomal mutations causing constitutive expression of the system which modulates aerobactin biosynthesis, but unlike *fur* mutants they showed apparently normal regulation of the enterochelin genes. These observations, and our data showing that the 2 systems have markedly different sensitivities to iron stress, strongly suggest the action of rather more complex regulatory mechanisms than have been proposed (6,7,17).

GENETIC ANALYSIS OF THE AEROBACTIN IRON-UPTAKE SYSTEM

Physical Mapping of the Aerobactin System

The aerobactin iron-uptake system of plasmid ColV-K30 has been cloned as the multicopy recombinant plasmid pABN1, and an initial restriction map of this plasmid has been obtained (Fig. 5, Ref. 3). In order to locate the genetic determinants for the aerobactin system on pABN1, we have used the transposon Tn1000 (γδ) to generate mutants by insertional inactivation, followed by restriction mapping of the sites of transposon insertion.

The sex factor F, which carries transposon Tn1000, can mediate low frequency transfer of nonmobilizable plasmids such as pABN1 by a process that involves plasmid cointegrate formation promoted by Tn1000 (16). Subsequent resolution of cointegrates following conjugal transfer to a recipient cell results in the duplication of Tn1000 so that each regenerated component plasmid now carries a copy of the transposon sequence. Tn1000-insertion mutants of plasmid pABN1 were isolated by transforming an F^+ strain with pABN1 DNA (which encodes ampicillin-resistance) and mating the resultant strain with an F^- recipient strain, followed by selection for ampicillin-resistance; all transconjugants should, in theory, carry a derivative of plasmid pABN1 with Tn1000 inserted at random.

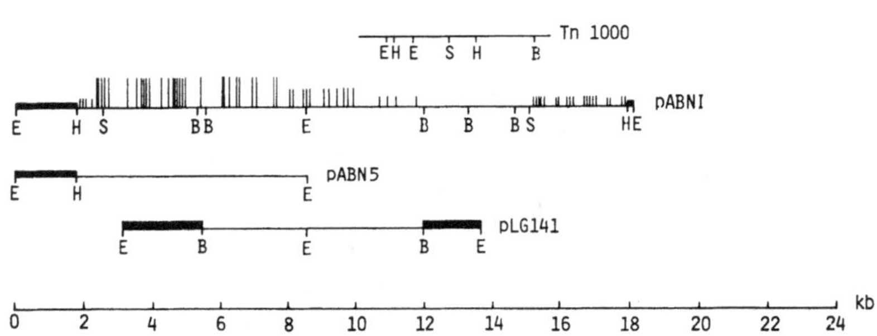

Fig. 5. Physical and genetic maps of pABN1 and of derived subclones. Restriction sites for the enzymes EcoR1 (E), HindIII (H), BamH1 (B), and SalI (S) are indicated for pABN1 and Tn1000. Long stalks on pABN1 indicate insertions of Tn1000 inactivating aerobactin production, while the intermediate-length stalks represent insertions which result in cloacin-resistance (i.e., which inactivate receptor activity). Short stalks indicate insertions which leave the whole iron-uptake system intact. Plasmids pABN5 and pLG141 are subclones of pABN1; the vector (represented by thick lines) in plasmids pABN1 and pABN5 is pPlac (3), and that in plasmid pLG141 is pACYC184.

Transconjugants were screened for aerobactin production using an indicator strain designated LG1522; this has the _iuc_ mutant plasmid from LG1418 (specifying normal aerobactin uptake) in a _fepA_ genetic background (in which ferric-enterochelin transport is defective), and will therefore grow on iron-deficient medium only when supplied with exogenous aerobactin. Of 450 strains assayed for their ability to cross-feed LG1522, 220 (49%) were found to be unable to secrete siderophore, and the transposon insertion site in 30 of these was determined by restriction analysis of plasmid DNA. The sites of insertion in these aerobactin-deficient mutant plasmids (indicated by the long stalks in Fig. 5) define a region of approximately 5.5 kb of the ColV-K30-derived _HindIII_ fragment of pABN1. This localization based on transposon mutagenesis is consistent with the observation that strains carrying plasmid pABN5, a derivative comprising only the 8.6 kb _EcoRI_ fragment of pABN1 (Fig. 5), produce aerobactin (as determined in the biological assay with strain LG1522) and so clearly contain all genes required for its biosynthesis.

Since the aerobactin receptor is also the receptor for the _En. cloacae_ bacteriocin cloacin DF13 (2,35), sensitivity to cloacin was used to assay receptor synthesis among strains carrying pABN1::-Tn1000 plasmids. Of 70 aerobactin-producing strains assayed in this way, 11 (16%) were cloacin DF13-resistant, and the sites of Tn1000 insertion in the plasmids of these and of 35 cloacin-sensitive strains were determined by restriction analysis. In cloacin-resistant strains, insertion sites (indicated by the intermediate-length stalks in Fig. 5) spanned a region of approximately 2 kb which includes the central _EcoRI_ site of pABN1. This is consistent with the observation that neither _EcoRI_ fragment of pABN1, when cloned alone, conferred cloacin DF13-sensitivity on the host cell (3). Furthermore, the defined region is within the 6.5 kb _BamHI_ fragment of pABN1, which, when cloned into a vector plasmid, was shown to specify receptor activity as indicated by sensitivity to cloacin DF13 (21).

Expression of pABN1 and Derivative Plasmids in Maxicells and Minicells

In order to determine the polypeptide products of the aerobactin system, pABN1 and derivative plasmids were expressed in maxicells (30) and minicells (28). The cloned fragment in pABN1 specified the synthesis of 5 polypeptides with apparent molecular weights of 74,000; 62,000; 50,000; 45,000; and 35,000 in both the minicell (Fig. 6) and maxicell (data not shown) expression systems. All except the 74,000 d polypeptide were also produced by pABN5; however, pABN5 encodes an additional polypeptide of 34,000 d which may be a truncated form of the 74,000 d product. The 6.5 kb _BamHI_ fragment of pABN1, which specifies receptor activity when cloned in vector

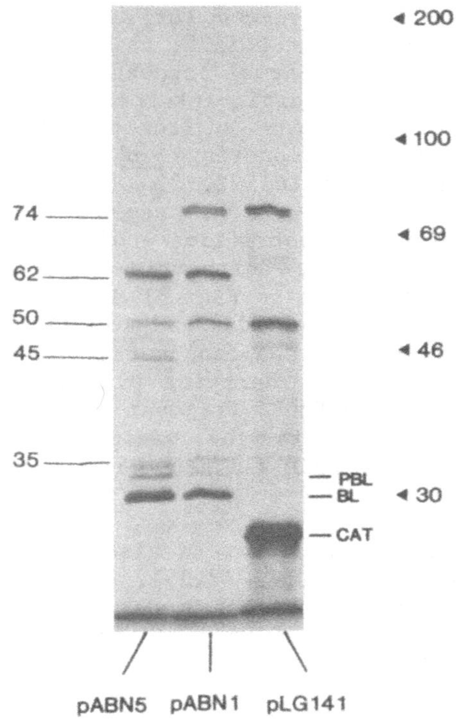

Fig. 6. Sodium dodecyl sulfate-polyacrylamide gel electrophoresis of polypeptides expressed in minicells harboring plasmid pABN1 and derivatives. Mobilities of standard molecular weight markers are indicated (in kilodaltons) on the right-hand side. Translation products of pABN5, pABN1, and pLG141 are shown with an indication of their molecular weights; the 74,000 dalton polypeptide is visible in its mature and precursor forms. BL and PBL indicate pPlac-encoded β-lactamase and pre-β-lactamase, respectively, and CAT is pACYC184-encoded chloramphenicol acetyl transferase.

plasmid pACYC184 (as in pLG141), encodes 2 polypeptides (M_r = 74,000 and 50,000).

Two approaches were used to determine the linear order of genes specifying the polypeptides required for aerobactin biosynthesis. The first involved expression of selected aerobactin-deficient pABN1::Tn1000 mutant plasmids in maxicells. The second approach was to construct derivatives of selected mutants to give a series of plasmids containing varying lengths of the siderophore biosynthesis region of pABN1. These derivative plasmids were expressed in minicells to determine their polypeptide products. Taken together, results from the 2 methods indicated a linear map of the translation products of plasmid pABN1 as shown in Fig. 7. The presence in some gels of putative truncated forms of the 74,000; 62,000; and 50,000 dalton polypeptides suggests that the direction of transcription, at least for 3 of the genes, is from left to right of the map, as drawn (Fig. 7).

Interestingly, insertions of the 5.7 kb transposon Tn1000 did not necessarily marked any polar effects. For example, mutants deficient in siderophore biosynthesis were all cloacin-sensitive, indicating some receptor expression, and moreover, expression of these mutants in maxicells and minicells generally showed the loss of only a single polypeptide. These data may indicate that the system is not expressed as a single operon, but instead has several promoter sites, possibly 1 for each gene. However, insertions of Tn1000 into a small region at the proximal end of the gene encoding the 62,000 d polypeptide did show polar effects, markedly reducing synthesis of all 5 polypeptides in the system. It is possible that these insertions identify an essential control region for the aerobactin system; the nature and possible role of this region are currently under investigation in our laboratory.

The Biosynthesis of Aerobactin

We have shown that the aerobactin system of plasmid ColV-K30 specifies 5 polypeptides, 4 of which are involved in aerobactin biosynthesis, the fifth being the 74,000 d outer-membrane protein that serves as the receptor for ferric-aerobactin. The aerobactin molecule comprises 2 N-acetylated, N-hydroxylated lysine derivatives linked to the α and γ carboxyl groups of a single residue of citric acid (Fig. 8, Ref. 14). Although the specific pathway of synthesis from existing lysine and citric acid has not yet been characterized, it is not unreasonable to speculate, in the light of our evidence, that: 4 genes are required for this synthesis, 2 specific enzymes mediate acetylation and hydroxylation of lysine, and 2 more effect condensation of these derivatives to the citric acid moiety. It is hoped that a greater understanding of the precise biochemical roles of the 4 polypeptides we have identified will allow the development of inhibitors or analogs which may be of therapeutic value in the control of invasive bacterial infections.

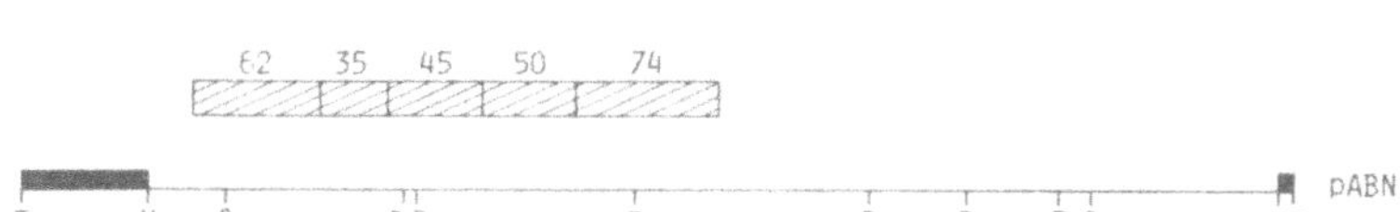

Fig. 7. Translational map of the iron-uptake region of plasmid pABN1. Numbers above the shaded boxes denote polypeptide sizes in kilodaltons.

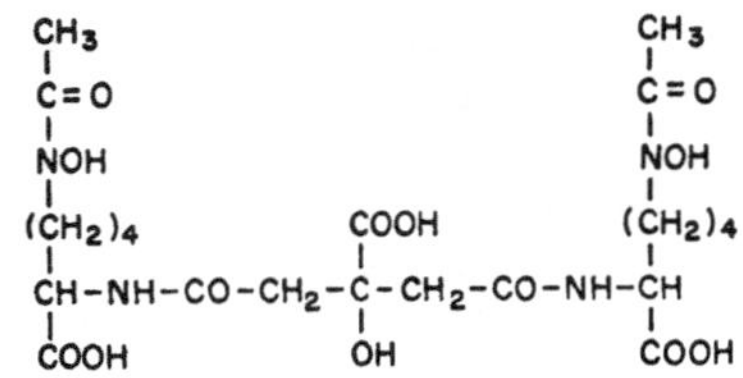

Fig. 8. The structure of aerobactin.

ACKNOWLEDGMENTS

We are very grateful to Maureen Burns for excellent technical assistance and to Carolyn Evans who isolated the ColV-Mu derivatives as part of an undergraduate research project. We thank Albrecht Bindereif and Joe Neilands for providing plasmid pABN1, and Rhone-Poulenc for the gift of a sample of streptonigrin. N.H.C. was supported by a Medical Research Council Studentship.

REFERENCES

1. Arnow, L.E. (1937) Colorimetric determination of the components of 3,4-dihydroxyphenylalanine-tyrosine mixtures. J. Biol. Chem. 118:531-537.
2. Bindereif, A., V. Braun, and K. Hantke (1982) The cloacin receptor of ColV-bearing *Escherichia coli* is part of the Fe^{3+}-aerobactin transport system. J. Bact. 150:1472-1475.
3. Bindereif, A., and J.B. Neilands (1983) Cloning of the aerobactin-mediated iron assimilation system of plasmid ColV. J. Bact. 153:1111-1113.
4. Binns, M.M., D.L. Davies, and K.G. Hardy (1979) Cloned fragments of the plasmid ColV,I-K94 specifying virulence and serum resistance. Nature (London) 279:778-781.
5. Braun, V., C. Brazel-Faisst, and R. Schneider (1984) Growth stimulation of *Escherichia coli* in serum by iron(III)aerobactin. Recycling of aerobactin. FEMS Microbiol. Lett. 21:99-103.
6. Braun, V., and R. Burkhardt (1982) Regulation of the ColV plasmid-determined iron(III)aerobactin transport system in *Escherichia coli*. J. Bact. 152:223-231.
7. Braun, V., R. Gross, W. Köster, and L. Zimmermann (1983) Plasmid and chromosomal mutants in the iron(III)-aerobactin transport system of *Escherichia coli*. Use of streptonigrin for selection. Mol. Gen. Genet. 192:131-139.
8. Casadaban, M.J., and S.N. Cohen (1979) Lactose genes fused to exogenous promoters in one step using a Mu-*lac* bacteriophage: In vivo probe of transcriptional control sequences. Proc. Natl. Acad. Sci., USA 76:4530-4533.

9. Clancy, J., and D.C. Savage (1981) Another colicin V phenotype: Adhesion in vitro of *Escherichia coli* to mouse intestinal epithelium. *Infect. Immun.* 32:343-352.
10. Csáky, T.Z. (1948) On the estimation of bound hydroxylamine in biological materials. *Acta Chem. Scand.* 2:450-454.
11. Evans, D.G., and D.J. Evans, Jr. (1978) New surface-associated heat-labile colonization factor antigen (CFA/II) produced by enterotoxigenic *Escherichia coli* of serogroups 06 and 08. *Infect. Immun.* 21:638-647.
12. Evans, D.G., R.P. Silver, D.J. Evans, Jr., D.G. Chase, and S.L. Gorbach (1975) Plasmid-controlled colonization factor associated with virulence in *Escherichia coli* enterotoxigenic for humans. *Infect. Immun.* 12:656-667.
13. Frost, G.E., and H. Rosenberg (1973) The inducible citrate-dependent iron transport system in *Escherichia coli* K-12. *Biochim. Biophys. Acta* 330:90-101.
14. Gibson, F., and D.I. Magrath (1969) The isolation and characterization of a hydroxamic acid (aerobactin) formed by *Aerobacter aerogenes* 62-1. *Biochim. Biophys. Acta* 192:175-184.
15. Grewal, K.K., P.J. Warner, and P.H. Williams (1982) An inducible outer membrane protein involved in aerobactin-mediated iron transport by ColV strains of *Escherichia coli*. *FEBS Lett.* 140:27-30.
16. Guyer, M. (1978) The γδ sequence of F is an insertion sequence. *J. Mol. Biol.* 126:347-365.
17. Hantke, K. (1981) Regulation of ferric iron transport in *Escherichia coli* K-12: Isolation of a constitutive mutant. *Mol. Gen. Genet.* 182:288-292.
18. Harris, J.R., I.K. Wachsmuth, B.R. Davis, and M.L. Cohen (1982) High molecular weight plasmid correlates with *Escherichia coli* enteroinvasiveness. *Infect. Immun.* 37:1295-1298.
19. Klipstein, F.A., and R.F. Engert (1977) Immunological interrelationships between cholera toxin and heat-labile and heat-stable enterotoxins of coliform bacteria. *Infect. Immun.* 18:110-117.
20. Konopka, K., A. Bindereif, and J.B. Neilands (1982) Aerobactin-mediated utilization of transferrin iron. *Biochem.* 21:6503-6508.
21. Krone, W.J.A., B. Oudega, F. Stegehuis, and F.K. de Graaf (1983) Cloning and expression of the cloacin DF13/aerobactin receptor of *Escherichia coli* (ColV-K30). *J. Bact.* 153:716-721.
22. Luckey, M., J.R. Pollack, R. Wayne, B.N. Ames, and J.B. Neilands (1972) Iron uptake in *Salmonella typhimurium*: Utilization of exogenous siderochromes as iron carriers. *J. Bact.* 111:731-738.
23. Neilands, J.B. (1981) Microbial iron compounds. *Ann. Rev. Biochem.* 50:715-731.
24. Neilands, J.B. (1982) Microbial envelope proteins related to iron. *Ann. Rev. Microbiol.* 36:285-309.

25. Ozanne, G., L.G. Mathieu, and J.P. Baril (1977) Production of colicin V in vitro and in vivo and observations on its effects in experimental animals. Infect. Immun. 12:240-244.
26. Ozanne, G., L.G. Mathieu, and J.P. Baril (1977) Production de colicines V et V2 in vitro et in vivo; Étude de leur action inhibitrice sur la phagocytose par des macrophages péritonéaux. Rev. Can. Biol. 36:307-316.
27. Payne, S.M. (1980) Synthesis and utilization of siderophores by Shigella flexneri. J. Bact. 143:1420-1424.
28. Reeve, J. (1979) Use of minicells for bacteriophage-directed polypeptide synthesis. In Methods in Enzymology, R. Wu, ed. Academic Press, Inc., London and New York, 68:493-503.
29. Rosenberg, H., and I.G. Young (1974) Iron transport in the enteric bacteria. In Microbial Iron Metabolism, J.B. Neilands, ed. Academic Press, Inc., New York. pp. 67-82.
30. Sancar, A., R.P. Wharton, S. Seltzer, B.M. Kacinski, N.D. Clarke, and W.D. Rupp (1981) Identification of the uvrA gene product. J. Mol. Biol. 148:45-76.
31. Smith, H.W. (1974) A search for transmissible pathogenic characters in invasive strains of Escherichia coli: The discovery of a plasmid-controlled toxin and a plasmid-controlled lethal character closely associated, or identical, with colicin V. J. Gen. Microbiol. 83:95-111.
32. Smith, H.W., and S. Halls (1968) The transmissible nature of the genetic factor in Escherichia coli that controls enterotoxin production. J. Gen. Microbiol. 52:319-334.
33. Smith, H.W., and M.B. Huggins (1976) Further observations on the association of the ColV plasmid of Escherichia coli with pathogenicity and with survival in the alimentary tract. J. Gen. Microbiol. 92:335-350.
34. Smith, H.W., and M.A. Linggood (1971) Observations on the pathogenic properties of the K88, Hly and Ent plasmids of Escherichia coli with particular reference to porcine diarrhoea. J. Med. Microbiol. 4:467-485.
35. van Tiel-Menkveld, G.J., J.M. Mentjox-Vervuurt, B. Oudega, and F.K. de Graaf (1982) Siderophore production by Enterobacter cloacae and a common receptor protein for the uptake of aerobactin and cloacin DF13. J. Bact. 150:490-497.
36. Vasil, M., R.K. Holmes, and R.A. Finkelstein (1975) Conjugal transfer of a chromosomal gene determining production of enterotoxin in Vibrio cholerae. Science 187:849-850.
37. Warner, P.J., P.H. Williams, A. Bindereif, and J.B. Neilands (1981) ColV plasmid specified aerobactin synthesis by invasive strains of Escherichia coli. Infect. Immun. 33:540-545.
38. Weinberg, E.D. (1978) Iron and infection. Microbiol. Rev. 42:45-66.
39. White, J.R., and H.N. Yeowell (1982) Iron enhances the bactericidal action of streptonigrin. Biochem. Biophys. Res. Commun. 106:407-411.

40. Williams, P.H. (1979) Novel iron uptake system specified by ColV plasmids: An important component in the virulence of invasive strains of _Escherichia coli_. _Infect. Immun._ 26:925-932.
41. Williams, P.H., and P.J. Warner (1980) ColV plasmid-mediated, colicin V-independent iron uptake system of invasive strains of _Escherichia coli_. _Infect. Immun._ 29:411-416.
42. Yeowell, H.N., and J.R. White (1982) Iron requirement in the bactericidal mechanism of streptonigrin. _Antimicrob. Agents and Chemother._ 22:961-968.

PLASMID-MEDIATED IRON SEQUESTERING SYSTEMS IN PATHOGENIC STRAINS OF *VIBRIO ANGUILLARUM* AND *ESCHERICHIA COLI*

J.H. Crosa, L.A. Actis, Y. Mitoma, J. Perez-Casal, M.E. Tolmasky, and M.A. Valvano

Department of Microbiology and Immunology
School of Medicine
Oregon Health Sciences University
3181 S.W. Sam Jackson Park Road
Portland, Oregon 97201

INTRODUCTION

Properties of the host vertebrate and invading bacteria define the conditions that precipitate the onset of an infectious process. In disseminated infections, an important bacterial attribute is the ability to grow in the host vertebrate's body fluids and tissues.

An essential element for bacterial growth is iron, which is found as a component of many bacterial cellular processes (29). However, in vertebrate fluids, iron is bound by high affinity iron-binding proteins and thus the concentration of available iron is far too low for bacterial utilization (6). Certain pathogenic isolates of *Vibrio anguillarum* and *Escherichia coli* have developed very efficient plasmid-mediated iron-sequestering mechanisms (7,8,31). These systems encoded by pJM1 in *V. anguillarum* and pColV-K30 in *E. coli*, responsible for the high-virulence phenotype of these bacteria, have been genetically characterized (9,11,12,28,32). The analysis of mutants in these 2 systems led to the demonstration of the existence of 2 components: a siderophore and a receptor for the iron-siderophore complexes. In the case of pColV-K30 the siderophore was identified as aerobactin (30), a hydroxamate compound originally found in culture supernatants of *Aerobacter aerogenes* (14). The receptor was determined to be a 74 kilodalton (kd) outer membrane protein (4,15). Recently, the genetic determinants for aerobactin and its receptor have been cloned (3).

In the case of *V. anguillarum*, concomitant with efficient plasmid-mediated iron uptake, there was an induction of the synthesis of 2 novel outer membrane proteins. One of them, the 86 kd OM2 protein was associated with the presence of pJM1 (10). Analysis of transposition-generated mutants suggested that OM2 is a receptor for complexes of iron with a siderophore produced by _V. anguillarum_ cells harboring an intact pJM1 plasmid (28).

The present work describes the molecular cloning and expression of genetic determinants of the pJM1 plasmid-mediated iron-uptake system of _V. anguillarum_. A component of this system, the iron-regulated outer-membrane protein OM2, is encoded by the pJM1 plasmid and may play a role as a receptor for complexes of iron with the _V. anguillarum_ siderophore anguibactin.

It is also reported here that although the pJM1 and pColV-K30 plasmid-mediated iron uptake systems are phylogenetically unrelated (27), they share a common feature: the iron uptake regions in both plasmids are flanked by repeated sequences. These characterisitcs may have played a role in the epidemiological spread of these virulence factors.

RESULTS AND DISCUSSION

Cloning of pJM1 DNA Genes Involved in the Iron Uptake Process.

The pJM1 DNA fragments obtained by partial digestion with the restriction endonuclease _XhoI_ were ligated with the vector pVK102 (17) cleaved with the same enzyme. The ligated DNA was in vitro packaged and the phage particles were used to transduce _E. coli_ HB101. Based upon previous knowledge of the iron-uptake region in pJM1 (28), clones carrying DNA fragments associated with the iron-uptake sequences were selected for further studies. The pJM1 DNA fragments contained in the clones pJHC-T2, pJHC-T7, pJHC-T11, and pJHC-T2612 are shown in Fig. 1. Each of these clones was conjugated into different iron uptake-deficient low-virulence _V. anguillarum_ strains as previously described (24) by using the helper plasmid pRK2013 (13). The _V. anguillarum_ recipients were the plasmidless H775-3, the 775::Tn_1_-5 strain harboring the Tn_1_ insertion derivative pJHC-91 and 775::Tn_1_-6 carrying the deletion plasmid pJHC9-8 (Fig. 1). The plasmids carried by each _V. anguillarum_ exconjugant are listed in Tab. 1. Analysis of the production of siderophore activity was carried out by testing culture supernatants in bioassays using as indicator strain _V. anguillarum_ 775:Tn_1_-5 (28). The results shown in Tab. 2 indicated that exconjugants harboring either the cosmid pJHC-T7 or pJHC-T2612, in addition to an indigenous pJM1 derivative such as pJHC-91 or the deleted plasmid pJHC9-8 exhibited a level of siderophore activity comparable to that produced by the wild-type _V. anguillarum_ 775. Exconjugants carrying only the clone pJHC-T7 or pJHC-T2612 produced a very low level of siderophore activity. No siderophore activity was detected in _V. anguillarum_ har-

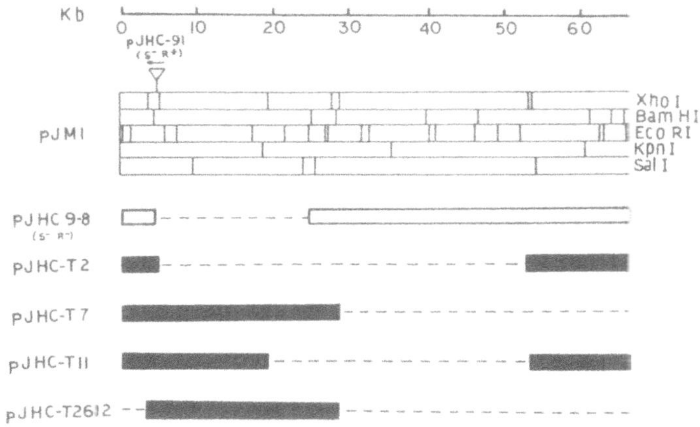

Fig. 1. Physical map of pJMl derivatives and recombinant clones. The symbol ∇ indicates the site of the Tn1 insertion in the pJMl derivative pJHC-91 while the arrow over this symbol gives the orientation of the Tn1 transposition sequence with respect to β-lactamase gene transcription. Also noted in this figure is the phenotype of the *V. anguillarum* strain carrying the pJHC-91 plasmid. Iron uptake-deficient derivatives are classified depending on whether the Tn1 insertion or deletion affected the genes associated with the production or activity of either siderophore (S^-R^+) receptor (S^-R^-). Dashed lines correspond to pJMl DNA regions that were either deleted (i.e., in pJHC9-8) or that were not included in the cloned DNA. Thicker lines represent the pJMl DNA cloned in the cosmid vector pVK102 (17). The clone pJHC-T2, pJHC-T7, pJHC-T11, and pJHC-T2612 were obtained by cloning *XhoI* partial digests of pJMl DNA.

boring either the control pVK102, the clones pJHC-T2, pJHC-T11, or the indigenous plasmids pJHC-91 or pJHC9-8. The present data suggest that an additional factor present in the indigenous plasmids and not in the recombinant clones is necessary for a production of a normal level of siderophore activity (24). The possibility of this factor to be involved in the export of siderophore outside the cell was ruled out by performing bioassays with total homogenates of the cells (24). Investigation of the production of the pJMl-encoded receptor, using each of the exconjugants strains as indicator in bioassays showed that 3 cosmid clones pJHC-T7, pJHC-T11, and pJHC-T2612 carry information for production of receptor activity while neither the clone pJHC-T2 nor the control plasmid pVK102 did (Tab. 2). Concomitantly with the presence of receptor activity, induction of the outer membrane protein OM2 was found under iron-limiting conditions. Since the ability to induce siderophore and receptor acti-

Tab. 1. Bacterial strains and properties.

STRAIN	PLASMID(S)	RELEVANT PHENOTYPE	REFERENCE
V. anguillarum			
775	pJM1	Iu^+	Crosa, 1977
H775-3	-	Iu^-	Crosa, 1980
775::Tn1-5	pJHC-91	Iu^-Ap^r	Walter et al, 1983
775::Tn1-6	pJHC9-8	Iu^-Ap^r	"
H775-3A	pJHC-T7	See table 2	Tolmasky and Crosa, 1984
H775-3D	pJHC-T2	"	"
775::Tn1-5A	pJHC-91, pJHC-T7	"	"
775::Tn1-6A	pJHC9-8, pJHC-T7	"	"
775::Tn1-5B	pJHC-91, pJHC-T2612	"	"
775::Tn1-6B	pJHC9-8, pJHC-T2612	"	"
H775-3B	pJHC-T2612	"	"
H775-3C	pJHC-T2	"	"
5822	-	Iu^+	This laboratory
E. coli			
HB101	-	*pro*, *leu*, *thi*, *lacY*, *hsr*$^-$, *hsm*$^-$, *endoI*$^-$, *recA*	Boyer and Roulland-Dussoix, 1969
C600	-	*thi*, *thr*, *leu*, *lacY*$^-$ *supE44*, *tonA*,	Appleyard, 1954
CHBA187	Five criptic plasmids	aerobactin producing	Children's Hospital Buenos Aires, Argentina.

Iu: iron uptake; Ap: ampicillin.

vity should be correlated with the iron-uptake proficiency, the iron-uptake properties of the exconjugants were tested by measuring the MIC of the iron-chelating agent ethylenediamine-di(o-hydroxyphenyl acetic acid) (EDDA). Tab. 2 shows that *V. anguillarum* harboring the indigenous plasmid pJHC-91 or pJHC9-8 increase the MIC for EDDA from 2 to 40 μM upon the introduction of either pJHC-T7 or pJHC-T2612, becoming iron-uptake proficient. Introduction of either of these clones in the plasmidless *V. anguillarum* H775-3 resulted in

Tab. 2. Properties of _V. anguillarum_ strains.

STRAIN	UPTAKE OF ^{55}Fe[a] -ang	+ang	SIDEROPHORE ACTIVITY[b]	RECEPTOR ACTIVITY[c]	PRESENCE OF OM2[d]	MIC for EDDA[e]
775	+	+	+	NA	+	40
H775-3	-	-	-	-	-	2
775::Tn1-5	-	-	-	+	+	2
775::Tn1-6	-	-	-	-	-	2
775::Tn1-5A	ND	ND	+	NA	+	40
775::Tn1-6A	ND	ND	+	NA	+	40
H775-3A	-	+	low	+	+	2
775::Tn1-5B	ND	ND	+	NA	+	40
775::Tn1-6B	ND	ND	+	NA	+	40
H775-3B	ND	ND	low	+	+	40
H775-3C	ND	ND	-	+	+	2
H775-3D	-	-	-	-	-	2
H775-3(pVK102)	ND	ND	-	-	-	2

[a] ^{55}Fe uptake was determined as described in the legend to Fig. 3.
ang: cell-free supernatant of a _V. anguillarum_ 775 culture which contains the pJM1 plasmid mediated siderophore anguibactin.

[b] Anguibactin production was determined by testing the ability of cell-free supernatant from cultures of each strain to cross-feed the indicator strain 775::Tn1-5 in bioassays as described previously (Walter et al., 1983).

[c] Receptor activity was scored by using each strain as indicator in bioassays with crude preparations of anguibactin.

[d] The presence of OM2 was tested in SDS-polyacrylamide gel elctrophoresis (SDS-PAGE) of outer membrane proteins obtained from cells of each strain cultured under iron-limiting conditions as previously described (Crosa and Hodges, 1981).

[e] Minimal inhibitory concentration (MIC) for ethylenediamine-di(o-hydroxyphenyl acetic acid) (EDDA) was determined on M9 minimal medium agarose plates containing increasing concentrations of EDDA. ND: not determined; NA: not applicable.

no modification of their MIC for EDDA, correlating with their low siderophore activity. It can also be seen in Tab. 2 that neither pJHC-T11 nor pJHC-T2 nor the control pVK102 change the iron-uptake deficient phenotype of the V. anguillarum recipients. Since these exconjugants have an active receptor, the low amount of siderophore activity must be responsible for the iron uptake-deficient phenotype. Thus, the additional factor mediated by the indigenous plasmids pJHC-91 and pJHC9-8, which is involved in the production of an increased siderophore activity, may be necessary for the ability to grow under iron-limiting conditions. This transacting factor could play a role on the regulation of the expression of the siderophore genes. Since the enhancing effect also occurs in strains harboring pJHC9-8, in which the iron-uptake region is deleted, the transacting regulatory genes must be encoded on pJM1 sequences other than the iron-uptake regions as defined by Tn1 transposition mutagenesis.

The exconjugant V. anguillarum strain 775:Tn1-5A which harbors the plasmids pJHC-91 and pJHC-T7 was also tested in experimental infections. The results showed that this strain possesses a high-virulence phenotype as compared to the original strain V. anguillarum 775::Tn1-5 which only carries pJHC-91 and was no virulent. This result demonstrated the intimate relation existing between the iron-uptake phenotype and the virulence characteristics of V. anguillarum.

Role and Origin of the OM2 Protein

The outer-membrane protein OM2 has been tentatively postulated as the receptor for the complex iron-siderophore in V. anguillarum (24,28). The OM2 protein was induced under iron-limiting conditions only if the cells harbored the pJM1 plasmid, suggesting that it was plasmid-encoded or that it had a chromosomal origin but was induced in response to a plasmid-specific substance. To determine whether the OM2 structural genes were present in pJM1, recombinant clones containing restriction fragments with the iron uptake-region sequences were introduced in E. coli HB101 cells to rule out the contribution of the V. anguillarum chromosome (1). Immunoblot analysis of either total cell lysates or outer membranes from those cells was used to detect the presence of OM2. The results, shown in Fig. 2, indicated that OM2 was present in both the total cell lysates as well as in the outer membranes of E. coli harboring clones containing the pJM1 BamHI fragment 1 (pJHC-W1), the EcoRI fragment 2 (pJHC-P100), or the SalI fragment 3 (pJHC-P102) (see physical map, Fig. 1). E. coli cells harboring clones containing other pJM1 fragments (pJHC-P101) as well as plasmidless derivatives did not synthesize the OM2 protein. Thus, these results showed that the structural genes for the OM2 protein must be present in the plasmid pJM1.

To determine the contribution of OM2 as a receptor in the iron-transport process, the kinetics of radioactive iron uptake was

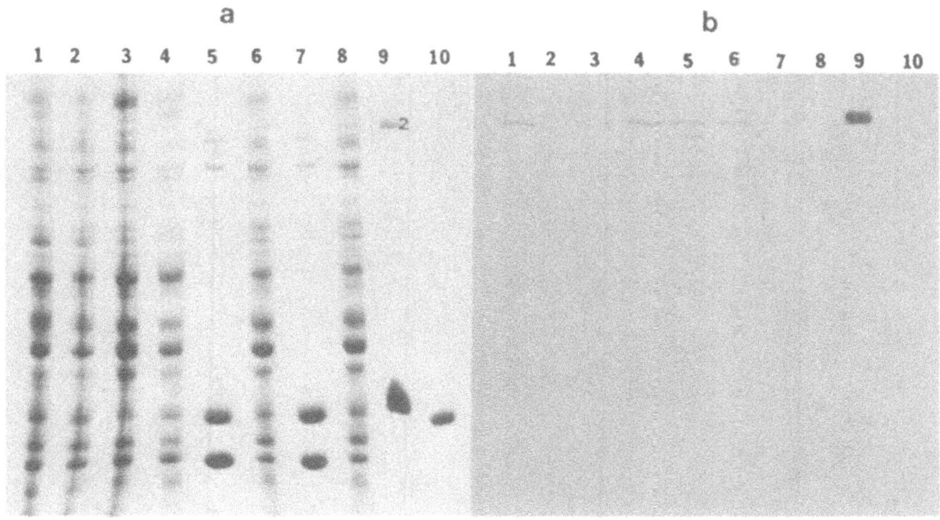

Fig. 2. SDS-PAGE and immunoblot analysis of total and outer membranes. Panel a: SDS-PAGE stained with Coomasie blue of total cell lysates (lanes 1-4, 6, and 8) and outer membranes proteins (lanes 5,7,9, and 10) of *V. anguillarum* cells (lanes 9 and 10) and *E. coli* HB101 cells (lanes 1-6) harboring recombinant plasmids containing different pJM1 fragments. Plasmidless *E. coli* HB101 cells were used as negative control (lanes 7 and 8). The pJM1 fragments cloned in HB101 cells were lane 1, *SalI* fragment 3 (pJHC-P102); lane 2, *SalI* fragment 2 (pJHC-P101); lane 3, *EcoR1* fragment 7 (pJHC-Y52); lane 4, *EcoR1* fragment 2 (pJHC-P101); and lanes 5 and 6, *BamHI* fragment 1 (pJHC-W1). Panel b: immunoblot of the gel in panel a stained with horse-radish peroxidase conjugated Protein A H_2O_2 and HRP color developing reagent (Bio-Rad, Richmond, California). The number 2 in panel a, lane 9 identifies the outer membrane protein OM2.

examined for *V. anguillarum* strains harboring indigenous and recombinant plasmids. *V. anguillarum* 775::Tn1-5 which harbors pJHC-91 and produces OM2 but not the siderophore anguibactin, was unable to take up iron under conditions in which the wild-type 775 did. However, when cell-free supernatant from *V. anguillarum* 775 which contains anguibactin was added in conjunction with ^{55}Fe, the iron-uptake kinetics was similar to that of *V. anguillarum* 775 (Fig. 3b). *V. anguillarum* cells carrying the recombinant cosmid pJHC-T7, which encodes the OM2 protein and receptor activity, were able to take up iron only in the presence of cell-free supernatant from a culture of *V. anguillarum* 775 (Fig. 3c). Conversely, *V. anguillarum* carrying

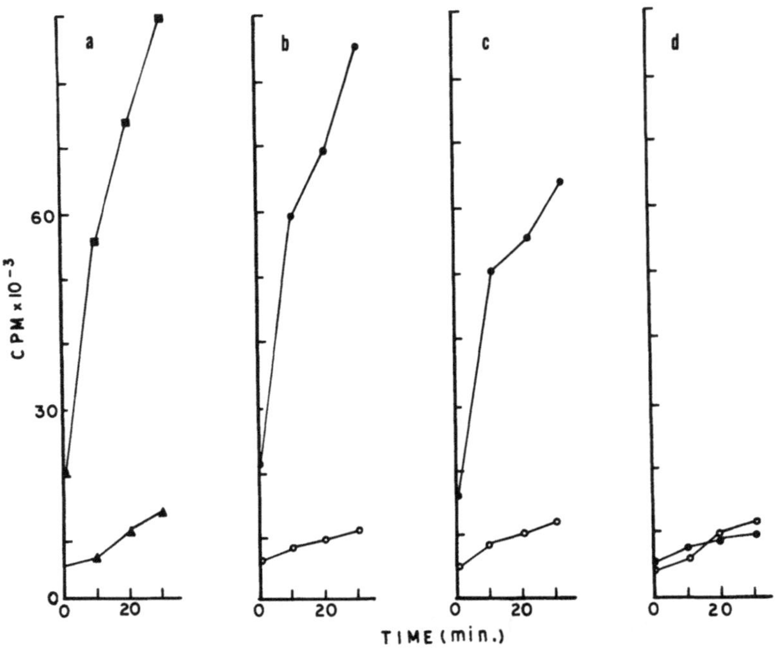

Fig. 3. Kinetics of radioactive ^{55}Fe iron uptake by nongrowing cells of various V. anguillarum strains under iron-limiting conditions. V. anguillarum strains were grown for several generations at room temperature (24°C) in M9 minimal medium containing the iron chelator EDDA at 10 μM (for iron uptake-proficient bacteria) or 2 μM EDDA (for iron uptake-deficient bacteria). Exponentially growing cultures were centrifuged and cells were washed and re-suspended in the same volume of similar medium containing the required aminoacids aspartic acid and histidine at 20 μg/ml. After further incubation for 2 hours at room temperature to deplete iron and aminoacid intracellular pools, cultures were cetrifuged and cells were washed and resuspended to an absorbance value of 1.00 at 590 nm, either in a similar medium lacking essential aminoacids but containing 100 μM sodium nitrilotriacetate or cell-free supernatant containing siderophore activity from V. anguillarum 775(pJM1) grown in the same medium. The iron-uptake kinetics was performed using 1 μCi/ml of carrier-free $^{55}FeCl_3$, as described previously (10). Panel a, strains 775(pJM1) (■) and H775-3 (▲); panel b, strain 775::Tn1-5(pJHC-91); panel c, strain H775-3A(pJHC-T7); panel d, strain H775-3D(pJHC-T2). In panels b, c, and d the experiments were carried out either in the presence (●) or absence (o) of cell-free supernatants from V. anguillarum 775(pJM1).

pJHC-T2, which did not encode receptor activity nor the OM2 protein, was unable to take up iron even in the presence of anguibactin (Fig. 2d). In all cases the iron-uptake process was inhibited by KCN (data not shown), suggesting that it must be uptake rather than simple binding to the bacterial membranes. Thus, the energy-dependent plasmid-mediated iron transport in V. anguillarum is a consequence of the interaction between anguibactin and a non-diffusible component present in the cell envelope, acting as receptor for the iron-siderophore complexes, and which may be the OM2 protein.

Repeated Sequences Flank the Iron-Uptake Region in the Plasmid pJM1

The ability of V. anguillarum to acquire iron under conditions of iron limitation was correlated with the presence of a pJM1-like plasmid in most instances, although in a few cases it was not possible to demonstrate the presence of such a plasmid (23). Nonetheless, these strains were of a high virulence phenotype and had a very active iron-sequestering system. It was speculated that the iron-uptake system present in those strains may be of a chromosomal nature and possibly related to the pJM1 plasmid-mediated iron-uptake system. Recently, Southern blot hybridization experiments indicated that pJM1 sequences were indeed present in the chromosomal DNA of these strains (20). The results suggested that the homology was, in part, due to the presence of repeated sequences in the pJM1 plasmid, that were also present in the chromosome of these strains. A detailed analysis of the pJM1 genome revealed the presence of 2 types of repeated sequences, designated RSv1 and RSv2. Two of the RSv1 sequences were found to flank the iron-uptake region of pJM1 (Fig. 4, panel a). Hybridization experiments, using a cloned pJM1 EcoR1 fragment containing the RSv1 sequences as a radiolabeled probe (plasmid pJHC-Y52), indicate that the RSv1 repeated sequence is also present in the chromosome of the plasmidless iron uptake-proficient strain 5822 (Fig. 4, panel b lanes C and D) and in the chromosome of the plasmidless iron uptake-deficient H775-3 strain (lanes A and B). The RSv2 sequences were only detected in the chromosomal DNA of the plasmidless iron uptake-deficient H775-3 strain (data not shown). The RSv1 sequences flanked the pJM1 iron-uptake region and were also present in the 5822 chromosome. Therefore, it was of interest to determine whether the pJM1 iron-uptake region was also included in the chromosome of this strain. Consequently, hybridization experiments were carried out between 5822 chromosomal DNA and the radiolabeled plasmid pJHC-P100 which contained a 9.6-kb EcoRI fragment internal to the iron-uptake region (Fig. 4, panel a). The results indicate that chromosomal DNA from strain 5822 shows homology with this probe (Fig. 4, panel B, lanes I and J), while the plasmidless iron uptake-deficient strain H775-3 did not (lanes G and H), although this strain hybridized with the RSv1 probe.

Thus, not only pJM1 iron-uptake sequences but also RSv1 sequences which flank the iron-uptake region in pJM1, occur in the

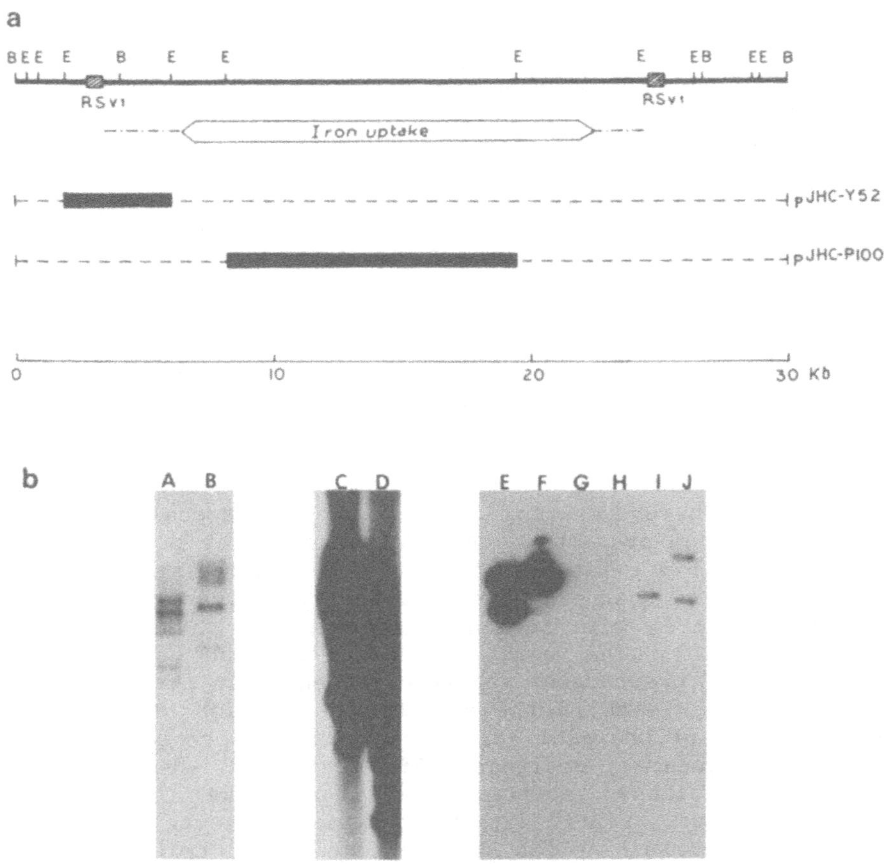

Fig. 4. Panel a: Physical map of the iron-uptake region in pJml. The position of 2 copies of the repeated sequences RSv1 is indicated by hatched boxes. Thicker lines represent cloned *EcoRI* cleaved pJM1 DNA fragments. *EcoRI* fragment 7 cloned in pKY2700 (pJHC-Y52) contains one copy of the repeated sequence RSv1. *EcoRI* fragment 2 cloned in pBR325 (pJHC-P100) is internal to the iron-uptake region. Dashed lines correspond to pJM1 DNA regions that were not included in the cloned DNA. Panel b: Southern blot hybridization using the clones pJHC-Y52 and pJHC-P100 as radioactive probes. Hybridization experiments using either pJHC-Y52 (Channels A-D), or pJHC-P100 (channels E-J), as a radioactive probe against *V. anguillarum* chromosomal DNA (A-D and G-J). (A): *EcoRI* cleaved 775-3; (B): *BamHI* cleaved 775-3; (C): *EcoRI* cleaved 5822; (D): *BamHI* cleaved 5822 DNA1; (G): *EcoRI* cleaved 775-3; (H): *BamHI* cleaved 775-3; (I): *EcoRI* cleaved 5822 and (J): *BamHI* cleaved 5822. Lanes E and F show *EcoRI* cleaved DNA of plasmids pJHC-P100 (lane E) and pJM1 (lane F).

chromosome of an iron uptake-proficient plasmidless strain of *V. anguillarum*. Can these results be construed as evidence of a recombinational capability of the pJM1 iron-uptake sequences, possibly through an interaction facilitated by the presence of the repeated sequences at the ends of the iron-uptake region? Although at this time we cannot answer this question, we would like to comment on results obtained with the pColV-K30 system that are germane to this question and to the general issue of the genetic relationships between these plasmid-mediated iron-uptake systems.

Are the pJM1 and pColV-K30 Plasmid-Mediated Systems Related?

We have recently reported that these 2 plasmid-mediated iron-uptake systems are unrelated at both the molecular as well as the functional level (27). However, we have recently found features of the pColV-K30 plasmid-mediated iron-uptake system that resemble those of the pJM1 plasmid: the pColV-K30 aerobactin sequences are flanked by repeated sequences and can also occur in the chromosome of certain enteric clinical isolates, such as, the *E. coli* K1 strain CHBA187, a cerebrospinal fluid isolate from a case of menigitis. This strain exhibited a functional aerobactin iron-transport system, as assessed by a crossfeeding bioassay and its sensitivity to cloacin a bacteriocin that recognizes the outer membrane receptor for iron-aerobactin complexes (26). Plasmid analysis of the CHBA187 strain revealed the presence of 5 plasmid classes (25). By using a variety of genetic and molecular techniques, we could not find an additional, still undetected plasmid, nor find any plasmid harboring the aerobactin genes. Hybridization of restriction endonuclease-cleaved chromosomal DNA from the CHBA187 strain with various clones containing subsets of the pColV-K30 aerobactin region, such as, pJHC-V4 and pJHC-V12 (See Fig. 5, panel a), demonstrated that the genetic determinants for the aerobactin iron-uptake system were located on a 10.5 kb *HindIII* fragment of the CHBA187 chromosome (Fig. 5, lanes B and H). Therefore, the aerobactin region in the chromosome is part of a *HindIII* restriction endounclease fragment about 6 kb smaller than the *HindIII* fragment in which they are found in the pColV-K30 plasmid (3; Fig. 5 panel a). However, the chromosomal aerobactin and receptor regions appeared to be highly conserved, as compared to the homologous regions in the plasmid pColV-K30. Thus, Fig. 5, panel b, lanes C and D, show that the distances *HindIII*-*EcoRI* and *HindIII*-*BstEII* mapped along the left end of the 16.3 kb pColV-K30 *HindIII* fragment (Fig. 5, panel a) were identical to those found in the CHBA187 aerobactin region as determined by successive digestions with various restriction endonucleases followed by hybridization with the aforementioned probes. The fact that the *BstEII* fragment, which in pColV-K30 contains the end of the receptor genes and part of the flanking sequences (Fig. 5, panel a), possessed a larger size in the case of the chromosomal *BstEII* digest (4 kb as compared to 3 kb in pColV-K30; see Fig. 5, panel b, lanes J and L) indicated that considerable rearrangement occurred in the

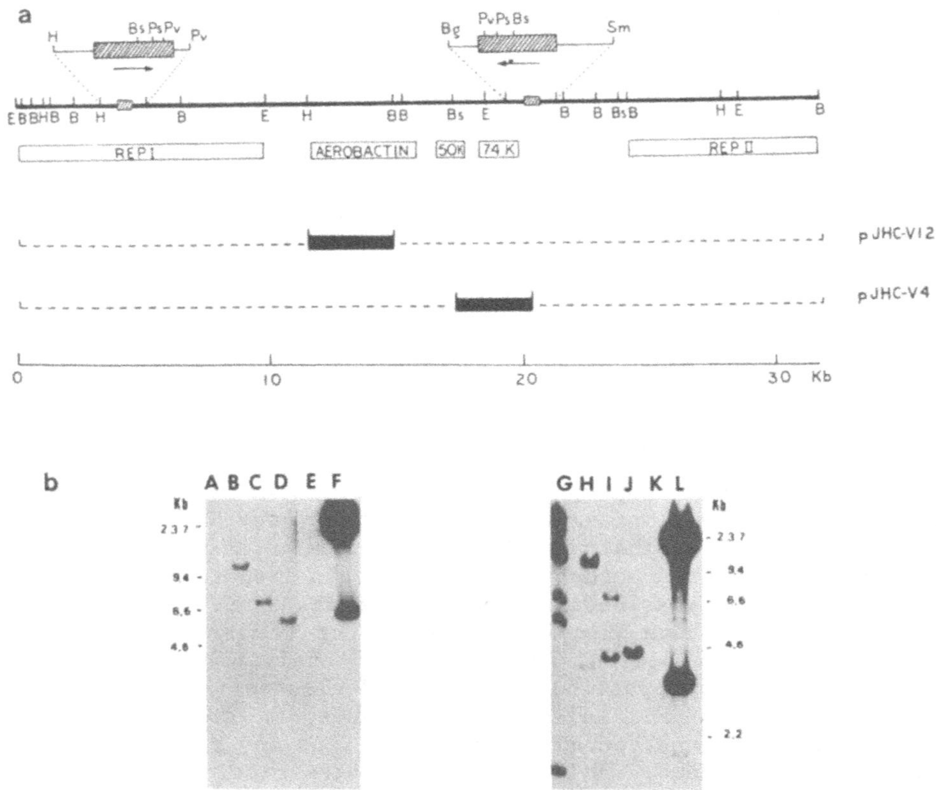

right-flanking area of the aerobactin chromosomal region. Furthermore, the recombinant plasmid pJHC-V4 containing the right-flanking sequence hybridized not only with the 10.5-kb HindIII chromosomal fragment but also with 2 other fragments, suggesting that sequences contained in this probe were repeated in the CHBA187 chromosome (Fig. 5, panel b, lanes H and I). These sequences were also found in the chromosome of the E. coli C600 strain, although its DNA did not show any hybridization with the pJHC-V12 plasmid, a specific probe for the aerobactin region (Fig. 5, panels a and b, lanes A and G), nor did the strain produce aerobactin or possess a cloacin-sensitive phenotype.

These results prompted us to study the sequences flanking the iron-uptake region in the pColV-K30 plasmid since their knowledge could shed light in understanding the high degree of conservation and ubiquity of these genes. Therefore, hybridization and restric-

Fig. 5. Panel a: Diagram showing the location of replications regions region. The top solid line represents a restriction endonuclease map of a pColV-K30 DNA fragment containing both the aerobactin and IS1-like sequences which are both flanking the pColV-K30 aerobactin region. The restriction endonuclease sites are: E: *EcoRI*, B: *BamHl*, H: *HindIII*, Bs: *BstE II*, Ps: *PstI*, Pv: *PvuII*, Bg: *BglII*, Sm: *SmaI*. *BglII*, *PvuII* and *SmaI* restriction sites are not completely mapped. Hatched boxes indicate the position of the IS1-like sequences. In an amplified scale on top are shown the restriction sites within the IS1-like elements. Arrows indicate the orientation of IS1-like elements with respect to each other. Open boxes indicate DNA regions corresponding to the aerobactin siderophore, 50 K and 74 K proteins, and also the replication regions REP I and REP II adjacent to the aerobactin region. K indicates kilodaltons. REP I region is located on a 10-kb *EcoRI* fragment whereas the REP II region is located on a 7.6-kb *BamHI* fragment. Thicker lines represent 2 cloned pColV-K30 DNA fragments. The pJHC-V12 is a recombinant plasmid containing a 3.4-kb *HindIII-BamHi* fragment which includes part of the aerobactin genes cloned in pVK101 (17). The pJHC-V4 contains a 3-kb *BstEII* fragment cloned in pJBK45, (16). Panel b: Southern blot hybridization of restriction endonuclease cleaved-chromosomal DNA from strain CHBA187 with probes containing different portions of the aerobactin region. Lanes A and G: C600 DNA; lanes B-D and H-J: CHBA187 DNA; lanes E and K: lambda DNA, lanes F and L: pColV-K30 16.3-kb *HindIII* DNA fragment containing the aerobactin region. The restriction endonucleases used were: Lanes A, B, E, G, H, and K: *HindIII*; lanes C and I: *EcoRI* and *HindIII* successive digestions; lanes D, F, J, and L: *HindIII* and *BstEII* successive digestions. Lanes A-L were hybridized with ^{32}P-labeled probes: lanes A-F, pJHC-12; lanes G-L, pJHC-V4.

tion mapping experiments were carried out to determine and characterize the nature of these flanking sequences. The results indicated that repeated sequences of about 800 bp flanked the aerobactin iron-uptake region (21; Fig. 5, panel a). Restriction endonuclease analysis and hybridization with probes containing known insertion elements demonstrated that the repeated sequences were homologous to IS1 elements (21,29). Similar findings were recently reported (18,19).

The IS1-like sequence occurring at the left end of the aerobactin region, was in reverse orientation as compared to the IS1-like sequence found at the right end of this region (21; Fig. 5, panel a).

Inverted IS1 sequences were also recently reported to flank the gene for the ST toxin, another important virulence factor that has spread to various medically important microorganisms (22). In this case the whole unit, ST-toxin genes together with the flanking IS1 elements, were shown to have transposition ability. Thus, the finding of the IS1-like elements flanking the aerobactin region in the pColV-K30 plasmid is suggestive of a potential recombinational ability because although transposition of these genes has not been demonstrated as yet, there is a potential capability for such an activity since the genes are bracketed by IS1-like elements.

Another possibility is that the presence of IS1 elements, per se, could enhance the ability of the aerobactin region to recombine by homologous recombination, via the IS1 elements, with genomes that also possess IS1 sequences. The finding of aerobactin sequences and IS1-like sequences in the same 10-kb *Hind*III fragment of chromosomal DNA from *E. coli* CHBA187 supports these hypotheses and suggests that the spread of the aerobactin sequences may be due to a high recombinational mobility (25).

As part of the genetic analysis of the pColV-K30 plasmid, its replication region was cloned by ligation of pColV-K30 DNA partially digested with *Eco*RI restriction endonuclease into the *Eco*RI site of pBR325. Since, unlike the pColV-K30 plasmid, this cloning vector requires polymerase I for its replication, we used the ability of recombinant clones to replicate in polymerase I-deficient mutants of *E. coli* to select, after transformation, for clones carrying the pColV-K30 replication regions. The results indicated that there were 2 replication regions in the pColV-K30 plasmid (21). Restriction endonuclease mapping and hybridization experiments located these replication regions, designated REP I and REP II, at the left and right end, respectively, of the aerobactin region (Fig. 5, panel a). The fact that the REP I and REP II replication regions flank the aerobactin system sequences may contribute to their conservation since recombinational events that conserve one of these replication regions, may also leave intact the aerobactin system, especially in deletion or insertion events in which new plasmids are generated. Thus, the combination of essential replication regions and insertion sequences adjacent to the aerobactin sequences may have played an important role in their preservation during evolution and may have contributed to their epidemiological spread in nature.

ACKNOWLEDGMENTS

This work was supported by U.S. Public Health Service Grant AI 19018 to J.H.C. from the National Institutes of Health. M.E.T. was supported by a postdoctoral fellowship from the Fogarty International Center, National Institute of Health of U.S.A. 1 F05 TW03345-01. Y.M. was a holder of a visiting graduate student scholarship from the Ministry of Education of Japan. L.A.A., J.P., and M.A.V. were supported by fellowships from the Consejo Nacional de Investigaciones Cientificas y Tecnicas of Argentina.

REFERENCES

1. Actis, L.A., and J.H. Crosa (1984) submitted for publication.
2. Appleyard, R.K. (1954) Genetics 39:440.
3. Bindereif, A., and J.B. Neilands (1983) J. Bact. 153:1111-1113.
4. Bindereif, A., V. Braun, and K. Hantke (1982) J. Bact. 150:1472-1475.
5. Boyer, H.W., and D. Roulland-Dussoix (1969) J. Mol. Biol. 41:459-472.
6. Bullen, J.J., H.J. Rogers, and E. Griffith (1978) Curr. Top. Microbiol. Immunol. 80:1-35.
7. Crosa, J.H. (1979) Fish Health News. 8:7-9.
8. Crosa, J.H. (1980) Nature (London) 284:566-568.
9. Crosa, J.H. (1984) Annu. Rev. Microbiol. 38:69-89.
10. Crosa, J.H., and L.L. Hodges (1981) Infect. Immun. 31:223-227.
11. Crosa, J.H., M.H. Schiewe, and S. Falkow (1977) Infect. Immun. 18:509-513.
12. Crosa, J.H., L.L. Hodges, and M.H. Schiewe (1980) Infect. Immun. 27:897-902.
13. Figurski, D.H., and D.R. Helinski (1979) Proc. Natl. Acad. Sci., USA 76:1648-1652.
14. Gibson, F., and D.I. McGrath (1969) Biochem. Biophys. Acta. 6:104-111.
15. Grewal, K.K., P.J. Warner, and P.H. Williams (1982) FEBS Lett. 140:27-30.
16. Kaper, J.B., H. Lockman, M.M. Baldini, and M.M. Levine (1984) Biotechnology 2:345-349.
17. Knauff, V., and E.W. Nester (1980) Plasmid 8:45-54.
18. Lawlor, K.M. and S.M. Payne (1984) Abstr. Ann. Meet. Am. Soc. Microbiol. St. Louis, p. 53.
19. McDougal, S., and J.B. Neilands (1984) J. Bact. In press.
20. Mitoma, Y., J. Perez-Casal, M.A. Walter, and J.H. Crosa (1984) Abstr. Ann. Meet. Am. Soc. Microbiol. St. Louis, p. 46.
21. Perez-Casal, J., and J.H. Crosa (1984) J. Bact. In press.
22. So, M., F. Heffron, and B.J. McCarthy (1979) Nature London 277:453-456.
23. Toranzo, A.E., J.L. Barja, S.A. Potter, R.R. Colwell, F.M. Hetrick, and J.H. Crosa (1983) Infect. Immun. 39:1220-1227.
24. Tolmasky, M.E., and J.H. Crosa (1984) Submitted for publication.
25. Valvano, M.A., and J.H. Crosa (1984) Submitted for publication.
26. Van Tiel-Menkveld, G.J., B. Oudega, O. Kempers, and F.K. de Graaf (1981) FEMS Microbiol. Lett. 12:373-380.
27. Walter, M.A., A. Bindereif, J.B. Neilands and J.H. Crosa (1984) Infect Immun. 43:765-767.

28. Walter, M.A., S.A. Potter, and J.H. Crosa (1983) J. Bact. 156:880-887.
29. Waring, W.S., and C.H. Werkman (1942) Arch. Biochem. 4:75-87.
30. Warner, P.J., P.H. Williams, A. Bindereif, and J.B. Neilands (1981) Infect. Immun. 33:540-545.
31. Williams, P.H. (1979) Infect. Immun. 26:925-932.
32. Williams, P.H., and P.J. Warner (1980) Infect. Immun. 29:411-416.

CHAIRMAN'S INTRODUCTION: SPECIALIZED FUNCTIONS

R.C. Clowes

Programs in Biology
The University of Texas at Dallas
P.O. Box 688
Richardson, Texas 75080

The final session on specialized functions was concerned with studies of plasmids that determine virulence in plants and humans and others that specify antibiotic resistance in human bacterial pathogens.

Dr. R. Amasino (University of Washington, Seattle) reported the analysis of insertion mutants in the Ti plasmid of *Agrobacterium tumefaciens* that is responsible for the formation of tumors in a variety of plants. Many virulence (*vir*) genes have in this way been mapped on the plasmid in a linked cluster, and have been shown to be expressed only when the bacteria are growing in the presence of plant cells. Two of the Ti-plasmid oncogenes have been shown to code for enzymes involved in the synthesis of the plant hormone, indoleacetic acid (IAA), and have been demonstrated to have extensive functional and DNA homology with similar genes in another plant pathogenic species, the *savastanoi* variety of *Pseudomonas syringae*. Plasmids in this species, which are also capable of producing tumors on plants, were described in more detail by Dr. T. Kosuge (University of California, Davis), who also reported homology between the IAA genes of *Ps. syringae* and the oncogenes of the *Agrobacterium* Ti plasmid. Homologous genes determining IAA were also located on the chromosomal DNA in other isolates of *Ps. savastanoi*.

Dr. R. Clowes (University of Texas at Dallas) reported the cloning of a gene from another *Pseudomonas* species, *aeruginosa*, which determines a toxin contributing to its pathogenicity for humans. This toxin is identical to diphtheria toxin in its mode of action on mammalian cells, which are killed by the inhibition of protein synthesis when the toxin is released in the cytoplasm, but the 2 toxins have different receptor sites. Dr. W. Goebel (Genetics and Microbiology Institute, Würzburg, West Germany) described his work on the characterization of plasmids that cause virulence in some *Escherichia coli* strains by the release of products that lyse red blood cells. The genes and products involved in the synthesis of such hemolysins have been characterized, and again, similar genes with identical functions have been mapped to the chromosome of other varieties of *E. coli*, specifically those causing acute pyelonephritis.

Finally, Dr. B. Weisblum (University of Wisconsin Medical School, Madison) reported the analysis at the molecular level of the regulation of genes conferring inducible resistance in *Staphylococcus aureus* to the antibiotics chloromycetin and erythromycin. Resistance to either antibiotic, which is due to the synthesis of methylating enzymes, has been shown to be regulated at the post-transcriptional level by the formation of double-stranded regions within each transcript, incorporating that part of the transcript that is required for ribosome binding, thereby interfering with the synthesis of the enzymes concerned.

CLONING OF AN ENZYMATICALLY ACTIVE SEGMENT OF THE EXOTOXIN-A GENE OF *PSEUDOMONAS AERUGINOSA*

Royston C. Clowes, Mark A. Mozola, Russell B. Wilson, Shin R. Hwang, and Rockford K. Draper

Programs in Biology
The University of Texas at Dallas
P. O. Box 688
Richardson, Texas 75080

Pseudomonas aeruginosa is a common, opportunistic, nosocomial pathogen. It excretes a variety of proteins, one of which, exotoxin A, appears to be a major virulence factor (9,19,36). Exotoxin A (PE) is a 68,000 molecular weight protein (2,28) with a mechanism of action identical to that of diphtheria toxin (DT) (26). That is, it inhibits protein synthesis in mammalian cells by the transfer of the ADP-ribosyl group of NAD^+ to a specific site on the peptide, elongation factor #2 (EF2), identified in mammalian-cell protein synthesis (14,35). As in diphtheria toxin, this enzymatic activity can be expressed by only part of the toxin molecule, the A (active) fragment of 26,000 to 27,000 molecular weight that is found in stored preparations (7,35), or after chymotrypsin or elastase cleavage of the intact toxin molecule (20,30). The remaining segment of the molecule of some 41,000 molecular weight, the B (binding) fragment, like the corresponding B fragment of DT, is believed to be involved in the recognition of cell receptors and the subsequent steps of endocytosis (26). The sequence of events in the pathway of receptor-mediated endocytosis remains obscure, and its resolution might be simplified if DT or PE mutants blocked at different steps were available. Such mutants might be derived and characterized more readily if the gene were cloned, and this was the main incentive for the project described, some details of which have already been published (23). A cloned toxin gene would also be an important intermediate in the construction of hybrid toxins (25).

We chose, as the source of the PE structural gene, the high-level PE-producing strain, *Ps. aeruginosa* PA103 (18). The primary cloning was carried out in HB101, a restrictionless *Escherichia coli*

K-12 mutant (3), using a pBR322 derivative as a vector. This vector, pRO1614 (Fig. 1), was derived by Olsen et al. (24) by the insertion of a 1.85-kb *PstI* fragment, that carries the replication genes from the broad host-range vector, RP1, into the *PstI* site of pBR322 (33). Since RP1 also carries a β-lactamase gene on a *Tn1* transposon, cleaved at the identical *PstI* site, this results in the retention of β-lactamase activity in pRO1614 as shown in Fig. 1, and permits the use of carbenicillin (Cb) as an alternative to tetracycline (Tc) in the selection of host cells carrying the plasmid.

The chromosomal DNA isolated from PA103 was partially digested with *Sau3A*, and following its equilibration on a 10 to 40% sucrose gradient, fractions containing fragments from 10 to 15 kb, as monitored on an agarose gel, were separated and ligated to the *BamHI*-cleaved pRO1614 vector. The ligated DNA was added to HB101 cells made competent by heat treatment with Ca^{++} (8,13) and samples producing 3,000 to 10,000 Cb^r colonies (approximately 3% to 9% of which were shown to be Tc-s) were spread over nitrocellulose filters that had been laid on the agar surface of the selection plates (15). Following incubation and the imprinting of replicas (37), the colonies were lysed and the released proteins fixed to the filter in situ (22). The filters were then probed using antibodies to PE (35), which were labeled by the addition of ^{125}I-protein A from *Staphylococcus aureus* (21). After autoradiography, clones expressing immunologically related exotoxin A molecules were readily visible (Fig. 2). Antigen-positive clones were found with a frequency of about 10^{-3} Cb^r Tc^s transformants, and one of these clones, designated 56C1, was selected after retesting (see Fig. 2) for further studies.

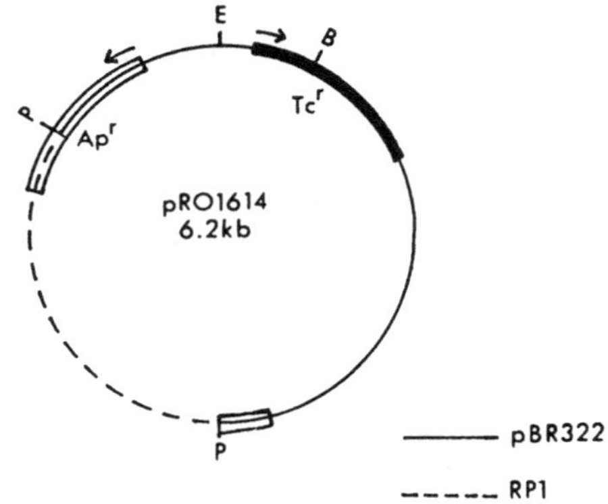

Fig. 1. Plasmid pRO1614, constructed from the insertion of a 1.85-kb segment of RP1 carrying replication genes, into the *PstI* site of pBR322 (24). This segment also carries the carboxy-terminal region of the β-lactamase gene of *Tn1*, and is oriented so that pRO1614 retains ampicillin resistance.

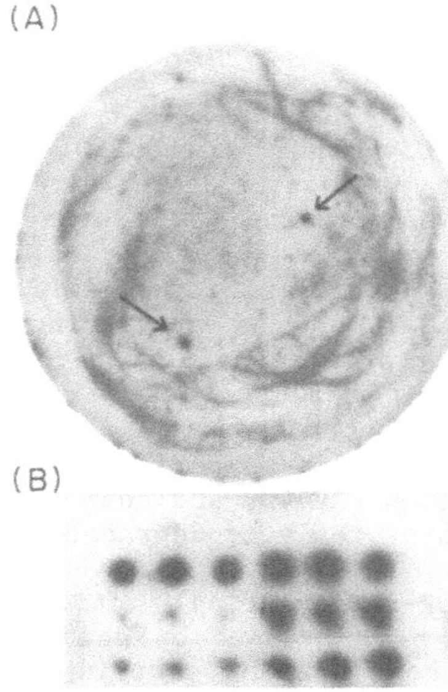

Fig. 2. Identification of antigen-positive clones by the filter-immune assay. (A) Initial identification of antigen-positive clones. Two such clones are indicated by arrows. (B) Verification of antigen-positive phenotype of clone 56C1. Top row (left to right): first 3 spots, <u>Ps. aeruginosa</u> PA0286: second 3, <u>Ps. aeruginosa</u> PA103. Middle row: first 3, <u>E. coli</u> HB101(pR01614); second 3, HB101(pRC345). Bottom row: first 3, MAM2(pR01614), second 3, MAM2(pRC345).

56C1 was found to carry a plasmid of about 23.9 kb, and, thus, an inferred 17.7 kb insert. Plasmid DNA from 56C1, designated pRC345, was then used to transform a number of nontoxigenic (Tox) mutants of <u>Ps. aeruginosa</u> PA0286. PA0286 is an auxotrophic mutant of another toxin-producing strain of <u>Ps. aeruginosa</u>, PA01 (32), the <u>Pseudomonas</u> strain on which most of the genetic studies of this species have been carried out. Nontoxigenic mutants of PA0286 were derived by nitrosoguanidine mutagenesis (1) and preliminary screening by the Elek immunodiffusion test (2). They were then tested for the lack of additional auxotrophic mutations, and finally for the CRM^- phenotype, that is, for colonies that, after lysis, failed to show the presence of products that would cross-react with antibodies to PE. All of 4 such CRM^- mutants were shown by the dot-blot test (Fig. 2) to be complemented to produce a CRM^+ phenotype when transformed with pRC345 plasmid DNA. The mutation in one of these mutants, designated MAM2, was then mapped after the MAM2 strain had been converted to a genetic donor by transfer of the <u>Psedudomonas</u> chromosomal-mobilizing factor, R68.45 (10,11). These mapping studies showed the MAM2 mutation to be located on the PAO chromosome in the 85 min region (29), where the structural toxAl gene has been mapped (12) (Tab. 1).

The activity of pRC345 in *E. coli* HB101 and in MAM2 was investigated using 2 other procedures. The first procedure involved Western-blot analysis of proteins (5,34) that were immunologically reactive with anti-PE toxin on the basis of their size (17) (Fig. 3). Neither bacterial host carrying the pRO1614 vector produced detectable bands, but either host, when carrying pRC345, gave positive results. In both cases, a series of cross-reacting bands was found with a combined molecular weight grossly in excess of that of the toxin, the largest band corresponding to about 50,000 molecular weight. We concluded from these experiments that we had cloned only a part of the toxA gene on pRC345, and that the product of this gene fragment was subject to proteolytic degradation in both hosts.

The enzymatic (ADP-ribosylating) activity associated with PE (7) was also found to be expressed by both hosts carrying pRC345, and again the control host strains carrying pRO1614 were negative. The PE specificity of this activity was verified by its removal in the presence of antitoxin (Tab. 2). Moreover, the wheat-germ fraction containing EF2 to which the radioactivity was transferred was shown by polyacrylamide-gel analysis to have a molecular weight of

Tab. 1. Mapping the CRM$^-$ mutation in *Ps. aeruginosa* MAM2.

Recipient[1)]	Selected Marker	% Coinheritance of tox allele: CRM$^-$ MAM2	% Coinheritance of tox allele: toxAl[2)]
PAO949	pur-67$^+$	25	25
PAO949	cys-59$^+$	NT	10
GMA253	hisI 5075$^+$	5	NT
PAO944	cys-54$^+$	14	28

1) The chromosome-mobilizing plasmid R68.45 was transferred to MAM2 (met-28 trp-6 CRM$^-$) and this donor was mated with one of the recipients (31,32) as shown. 100 colonies prototrophic for the selected marker were selected and tested for transfer of the CRM$^-$ phenotype.

2) Linkage data of Hanne et al. (12).

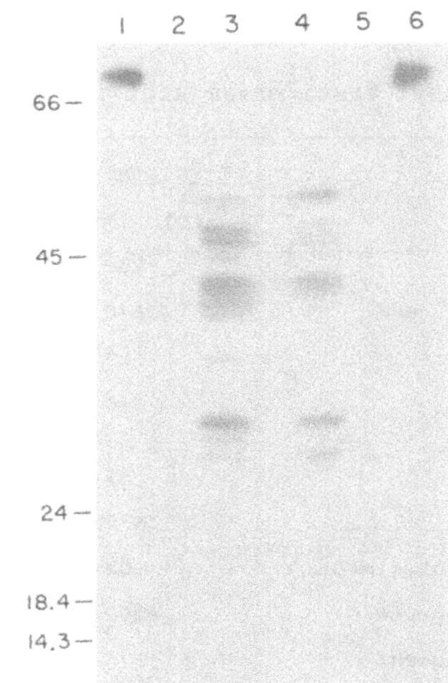

Fig. 3. Western-blot hybridization analysis of exotoxin A-related polypeptides expressed by clone 56C1. Lane 1, purified exotoxin A (0.2 μg); 2, E. coli HB101(pRC1614); 3, HB101 (pRC345); 4, P. aeruginosa MAM2(pRC345); 5, MAM2 (pRO1614); 6, exotoxin A (0.2 μg). Positions of protein molecular weight standards (BSA, 66,000; egg albumin, 45,000; trypsinogen, 24,000; β-lactoglobulin, 18,400; and lysozyme; 14,300) after electrophoresis are indicated (molecular weight x 10^{-3}).

about 90,000, in good agreement with the size of EF2 (16). These results appear to justify our conclusion that pRC345 carries a part of the structural toxA gene.

pRC345 DNA was found by restriction-enzyme cleavage to carry approximately 17.7 kb of PA103 chromosomal DNA, comprised of 8 BamHI fragments (Fig. 4). Several other antigen-positive clones like 56C1, each derived from a separate ligation and transformation procedure, were isolated during the preliminary screening and all were found to carry a similar spectrum of fragments. We, therefore, concluded that there appeared to be some barrier, either to the cloning, or to the expression of the intact toxin gene in E. coli using pRO1614. Similar cloning experiments were then undertaken using the Ps. aeruginosa MAM2 strain as a recipient. We were unable to identify any antigen-positive clones by our screening procedures, which we believe is due partly to the lower efficiency of transforming

Tab. 2. ADP-ribosyl transferase activity in cell extracts.[a]

Sample	[^{14}C]-ADP-ribose incorporated into EF-2[1] (CPM ± S.D.)
HB101(pRO1614)	256 ± 33
HB101(pRC345)	2640 ± 500
MAM2(pRO1614)	168 ± 26
MAM2(pRC345)	2060 ± 520
MAM2(pRC345) + anti-exotoxin A	61 ± 11
Exotoxin A	3470 ± 145
Exotoxin A + anti-Exotoxin A	55 ± 8
Diphtheria toxin fragment A	6500 ± 380
Diphtheria toxin fragment A + anti-exotoxin A	7500 ± 410

[a] All assays were done in triplicate. Samples treated with anti-exotoxin A were incubated with the IgG fraction (114 µg) for 10 min at 24°C prior to addition of urea (see Materials and Methods). The specificity of the inhibitory effect of anti-exotoxin A on the activity of exotoxin A was tested in controls by measuring the effect of the anti-exotoxin A on the ADP-ribosyl transferase activity of fragment A of diphtheria toxin, which is not antigenically similar to exotoxin A, but which has the same enzymatic activity as exotoxin A. Since anti-exotoxin A did not interfere with the activity of fragment A, the effect of the antitoxin on the activity of exotoxin A was specific.

[1] The component in the Ef-=2 fraction of wheat germ to which radioactivity was transferred was shown, by polyacrylamide-gel electrophoresis, to have a molecular weight of about 90,000, in good agreement with published values for EF-2 (16) (data not shown).

MAM2, but is also due to the higher background expression of the nontransformed clones.

pRC345 DNA was then subcloned by BamHI digestion, religation, and transformation to HB101. Only one antigen-positive clone was isolated. Following analysis of the plasmid DNA fragments derived by BamHI and/or EcoRI cleavage, it was found to carry only 3 of the BamHI fragments of pRC345 (Fig. 4). The proteins from this clone, designated 8E4, when analyzed by Western-blot transfer produced only a single positive band of a size 30-35,000 d, and failed to show evidence of enzymatic transfer of ADP ribose to EF2.

The plasmid of clone 8E4, pRC346, carries one BamHI-fragment of 3.15 kb that was located on pRC345 near the left-hand vector insert junction (Fig. 4), in addition to 2 smaller 0.47 and 0.16 kb fragments that were located near the center of the insert of pRC345. We conclude that the PE structural-gene segment of pRC345 must, therefore, be identified with the 0.52-kb and 3.15-kb BamHI fragments. The 3.15 fragment was cloned in pBR322 in both orientations, neither of which, when tested, showed either enzymatic or antigenic activity. EcoRI and HinfI cleavage of the 3.1 BamHI fragment led to the map as shown in Fig. 5, and end labeling using DNA polymerase I Klenow fragment with α-^{32}P-dGTP, followed by polyacrylamide-gel electrophoresis and autoradiography, led to the identification and separation of the terminal 0.22-kb and 0.6-kb fragments. The 0.22-kb fragment was isolated and the sequence of 120 nucleotides was determined from the labeled strand in the 3' to 5' direction, leading to the identification of an XhoI site at nucleotide 39.

From preliminary studies of Vasil et al. (38), it has been concluded that the enzymatic (A) fragment of PE, in contrast to the corresponding A fragment of DT, may be located at the carboxy-terminal end of the protein. If this be true, then we can rationalize our data as shown in Fig. 5 as follows. For unknown reasons, there appears to be a barrier to the expression of the intact toxA gene on this plasmid vector, whether in E. coli or Ps. aeruginosa. In pRC345, we conclude that we have cloned the carboxy terminus of PE, which is located on the 520 and 450 nucleotide fragments. Expression appears to be due to the P2 tetracycline promoter of pBR322, located at approximately 45 nucleotides from the EcoRI site (4), and leading to the initiation of a hybrid protein from the ATG initiation site of the tetracycline-resistance gene, located 86 nucleotides from EcoRI (27). Such a hybrid protein would comprise 75 amino acids of the N-terminus of the tet protein, corresponding to a molecular weight of approximately 11,000 d, which would be fortuitously in frame with approximately 39,000 d of the carboxy terminus of the PE toxin, and would be consistent with the C-terminus of the gene being close to the EcoRI site at the right hand terminus of the 450 nucleotide fragment. The inability to isolate antigen-positive

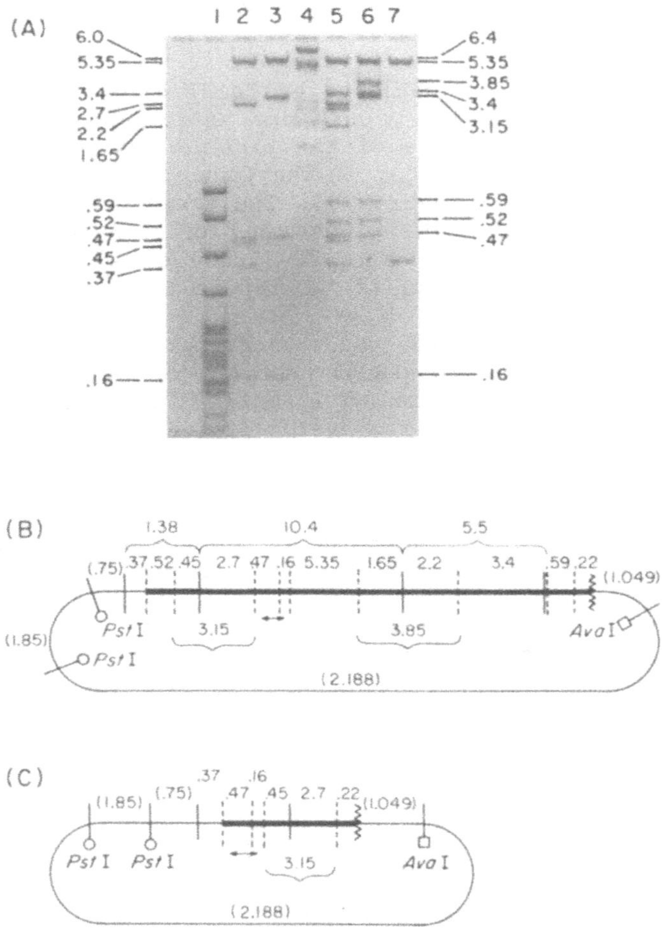

clones carrying larger segments of the gene could be due to other upstream Sau3A/BamHI junctions being out of frame with the N-terminal fragment of the tet protein. We conclude that antigenic activity by pRC346 may also be due to expression promoted from P2, but with translation initiated within the 470 nucleotide fragment. The 30 to 35,000 d peptide thus produced would be translated from less than half of the toxA coding sequence than in pRC345, consistent with its retention of antigenic activity and loss of enzymatic activity.

The pRC345 plasmid is now being used as a probe to identify the amino terminus of the PE gene. Figure 6 shows a Southern blot of

Fig. 4. BamHI and EcoRI cleavage analysis of plasmid DNA from clones 56C1 and 8E4. (A) 5% polyacrylamide gel in Tris borate buffer of cleaved DNA fragments following electrophoresis at 25 V overnight. Lanes 1 and 4; HpaII cleaved pBR322 and EcoRI plus HindIII-cleaved λ DNA size standards, respectively. Lanes 2 and 3; plasmid DNA from clone 8E4 cleaved with EcoRI plus BamHI, or BamHI, respectively. Lanes 5 and 6; pRC345 DNA cleaved with EcoRI plus BamHI, or BamHI, respectively. Lane 7: pRO1614 DNA cleaved with EcoRI plus BamHI. The sizes of the smaller (<0.6 kb) fragments shown in kb are derived from the pBR322 standard on this gel. Sizes of fragments in kb from double digestion of pRC345 with BamHI and EcoRI (lane 5) are shown at left of gel, and those from pRC345 digestion with BamHI (lane 6) at right of gel. [The sizes of the larger (>1.6 kb) fragments are derived from the λ standard following agarose-gel electrophoresis.] (B) Order of the fragments on pRC345. The thin line represents the vector pRO1614, and the thick line represents PA103 chromosomal DNA. The symbols (|), (¦) and (⌇) represent EcoRI, BamHI, and Sau3A recognition sites, respectively (only one terminal Sau3A site on the insert and vector DNA being shown). The figures represent sizes in kb derived from cleavage analysis (6) as shown above, together with BamHI-cleavage data of the isolated EcoRI fragments (data not shown). The arrow between the 0.47 and 0.16 kb fragments indicates that their relative order is not established. The location of the AvaI and PstI sites and the sizes of fragments shown in parentheses are derived from published data of pBR322 (33) and pRO1614 (24). (C) Order of fragments on pRC346 derived from data shown in (A). For details see (B).

BamHI-cleaved PA103 chromosomal DNA, hybridized with ^{32}P nick-translated, BamHI-cleaved pRC345 DNA. It should be noted that the 520 bp fragment is absent in the PA103 digest, but an approximate 1.5-kb BamHI fragment shows homology. This finding would be consistent with the creation of the upstream cleavage site of the 520-bp BamHI fragment within the 1.5-kb BamHI fragment by ligation of a Sau3A terminus of PA103 DNA with the BamHI terminus of pRO1614.

We are currently using similar radioactive probes constructed from the 520 or 450-bp fragments to identify overlapping PA103 chromosomal fragments produced by other restriction enzymes, to ligate such fragments into pRO1614 and to screen for their clones by hybridization with the radioactive 520 or 450 bp fragments. In this way we hope to clone extensive upstream fragments of the toxA gene, irrespective of their ability to express antigenicity, in E. coli or Ps. aeruginosa, and then, if necessary, to explore expression of these fragments when coupled to expression vectors of various kinds.

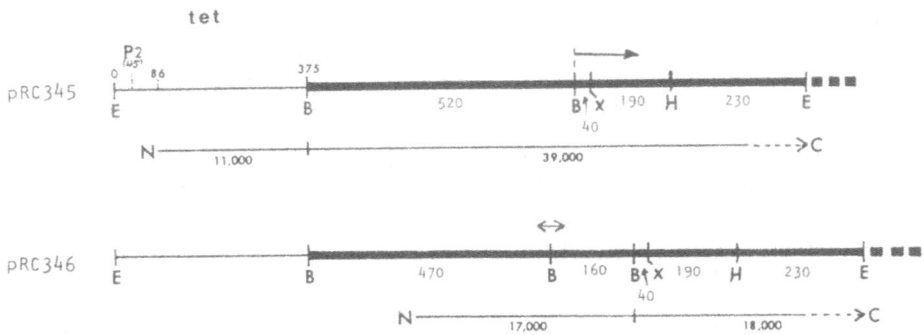

Fig. 5. Part of the toxA structural gene as cloned in pRC345 and pRC346. The thin line shows the EcoRI/BamHI segment of pRO1614 and the thick line shows PA103 chromosomal DNA. B, E, X, and H represent cleavage sites of BamHI, EcoRI, XhoI, and HinfI, respectively. The numbers above these lines represent base pairs measured rightwards from the EcoRI site of pRO1614. The numbers below these lines represent distances between cleavage sites in base pairs. P2 and tet represent the P2 (tet) promoter and part of the tetracycline-resistance gene of pRO1614. The single arrow indicates the region over which the base sequence has been determined. The double arrow indicates the alternative arrangements of the 470 and 160 nucleotide fragments. The lower line in each case terminated by N and C represents the region presumed to be translated, N representing the amino terminus of the hybrid protein and C the carboxy terminus including the toxA-encoded sequences. The numbers under these lines represent the calculated size of the part of the hybrid protein product in daltons.

ACKNOWLEDGMENTS

We thank Bruce Holloway, George Jacoby, and Ronald Olsen for supplying bacterial strains and plasmids; Stephen Leppla for the gift of purified exotoxin A; Elzora Jordan for expert technical assistance; and Beecham Laboratories for antibiotics. This work was supported by Public Health Service grants from the National Institute of Allergy and Infectious Diseases to R.C.C. (AI-10638) and to R.K.D. (AI-17400), and part of it was submitted by R.B.W. in partial fulfillment of the requirements of the M.S. degree of the University of Texas at Dallas.

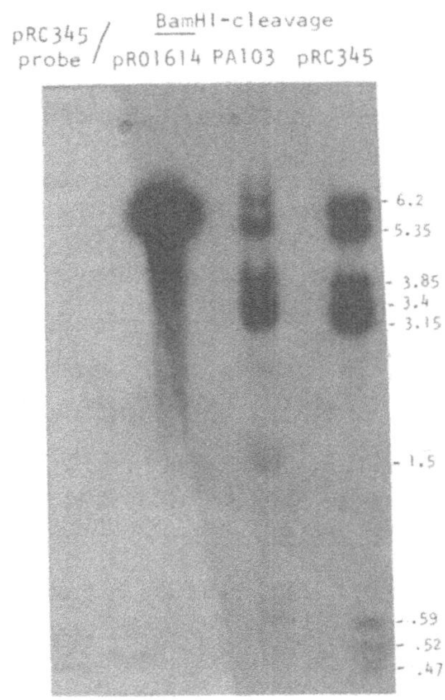

Fig. 6. Southern blot of hybridization with radioactive pRC345. BamHI-cleaved DNAs were separated on an 0.8% agarose gel for 13 hr at 20 volts. They were then transferred to a nitrocellulose sheet and hybridized with 1 μg, α-^{32}P nick-translated EcoRI-cleaved pRC345 DNA (3×10^6 c.p.m.). Slot 1, pRO1614 (10 ng); Slot 2, PA103 chromosomal DNA (20 μg); Slot 3, pRC345 (5 ng).

REFERENCES

1. Adelberg, E.A., M. Mandel, and G.C.C. Chen (1965) Optimal conditions for mutagenesis by N-methyl-N'-nitro-N-nitrosoguanidine in Escherichia coli K12. Biochem. Biophys. Res. Commun. 18:788-795.
2. Bjorn, M.J., M.L. Vasil, J.C. Sadoff, and B.H. Iglewski (1977) Incidence of exotoxin production by Pseudomonas species. Infec. Immun. 16:362-366.
3. Boyer, H.W., and D. Roulland-Dussoix (1969) A complementation analysis of the restriction and modification of DNA in Escherichia coli. J. Mol. Biol. 41:459-472.
4. Brosius, J., R.L. Cate, and A.P. Perlmutter (1982) Precise location of two promoters for the β-lactamase gene of pBR322. J. Biol. Chem. 257:9205-9210.

5. Burnette, W.N. (1981) "Western blotting": Electrophoretic transfer of proteins from sodium dodecyl sulfate-polyacrylamide gels to unmodified nitrocellulose and radiographic detection with antibody and radioiodinated protein A. Analyt. Biochem. 112:195-203.
6. Chiang, S-J., and R.C. Clowes (1980) Intramolecular transposition and inversion in plasmid R6K. J. Bact. 142:668-682.
7. Chung, D.W., and R.J. Collier (1977) Enzymatically active peptide from the adenosine diphosphate-ribosylating toxin of *Pseudomonas aeruginosa*. Infec. Immun. 16:832-841.
8. Cohen, S.N., A.C.Y. Chang, and L. Hsu (1972) Nonchromosomal antibiotic resistance in bacteria: Genetic transformation of *Escherichia coli* by R-factor DNA. Proc. Natl. Acad. Sci., USA 69:2110-2114.
9. Cross, A.S., J.C. Sadoff, B.H. Iglewski, and P.A. Sokol (1980) Evidence for the role of toxin A in the pathogensis of infection with *Pseudomonas aeruginosa* in humans. J. Infect. Dis. 142:538-546.
10. Haas, D., and B.W. Holloway (1976) R factor variants with enhanced sex factor activity in *Pseudomonas aeruginosa*. Mol. Gen. Genet. 144:243-251.
11. Haas, D., and B.W. Holloway (1978) Chromosome mobilization by the R plasmid R68:45 A tool in *Pseudomonas* genetics. Mol. Gen. Genet. 158:229-237.
12. Hanne, L.F., T.R. Howe, and B.H. Iglewski (1983) Locus of the *Pseudomonas aeruginosa* toxin A gene. J. Bact. 154:383-386.
13. Holmans, P.L., and R.C. Clowes (1979) Transposition of a duplicate antibiotic resistance gene and generation of deletions in plasmid R6K. J. Bact. 137:977-989.
14. Iglewski, B.H., and D. Kabat (1975) NAD-dependent inhibition of protein synthesis by *Pseudomonas aeruginosa* toxin. Proc. Natl. Acad. Sci., USA 72:2284-2288.
15. Iglewski, B.H., and J.C. Sadoff (1979) Toxin inhibitors of protein synthesis: Production, purification, and assay of *Pseudomonas aeruginosa* toxin A. Methods Enzymol. 68:780-793.
16. Iglewski, W.J., and H. Lee (1983) Purification and properties of an altered form of elongation factor 2 from mutant cells resistant to intoxication by diphtheria toxin. Eur. J. Biochem. 134:237-240.
17. Laemmli, U.K. (1970) Cleavage of structural proteins during the assembly of the head of the bacteriophage T4. Nature 227:680-685.
18. Liu, P.V. (1966) The role of various fractions of *Pseudomonas aeruginosa* in its pathogenesis. III. Identity of the lethal toxins produced *in vitro* and *in vivo*. J. Infec. Dis. 116:481-489.
19. Liu, P.V. (1974) Extracellular toxins of *Pseudomonas aeruguinosa*. J. Infect. Dis. 130:594-599.
20. Lory, S., and R.J. Collier (1980) Expression of enzymic activity by exotoxin A from *Pseudomonas aeruginosa*. Infec. Immun. 28:494-501.

21. Markwell, M.-A.K. (1982) A new solid-state reagent to iodinate proteins. Analyt. Biochem. 125:427-432.
22. Meyer, T.F., N. Mlawer, and M. So (1982) Pilus expression in Neisseria gonnorrhoeae involves chromosomal rearrangement. Cell 30:45-52.
23. Mozola, M.A., R.B. Wilson, E.M. Jordan, R.K. Draper, and R. C. Clowes (1984) Cloning and expression of a gene segment encoding the enzymatic moiety of Pseudomonas aeruginosa exotoxin A. J. Bact. (in press).
24. Olsen, R.H., G. DeBusscher, and W.R. McCombie (1982) Development of broad-host-range vectors and gene banks: Self-cloning of the Pseudomonas aeruginosa PAO chromosome. J. Bact. 150:60-69.
25. Olsnes, S., and A. Pihl (1982) Chimeric toxins. Pharmac. Ther. 15:355-381.
26. Pappenheimer, Jr., A.M. (1977) Diphtheria toxin. Annu. Rev. Biochem. 256:69-94.
27. Peden, K.W.C. (1983) Revised sequence of the tetracycline-resistance gene of pBR322, Gene 22:277-280.
28. Pollack, M., N.S. Taylor, and L.T. Callahan III (1977) Exotoxin production by clinical isolates of Pseudomonas aeruginosa. Infec. Immun. 15:776-780.
29. Royle, P.L., H. Matsumoto, and B.W. Holloway (1981) Genetic circularity of the Pseudomonas aeruginosa PAO chromosome. J. Bact. 145:145-155.
30. Sanai, Y., K. Morihara, H. Tsuzuki, J.Y. Homma, and I. Kato (1980) Proteolytic cleavage of exotoxin A from Pseudomonas aeruginosa: Formation of an ADP-ribosyltransferase active fragment by the action of Pseudomonas elastase. FEBS Let. 120:131-134.
31. Soldati, L., R. Crockett, J.M. Carrigan, T. Leisinger, B.W. Holloway, and D. Haas (1984) Revised locations of the hisI and pru (proline utilization) genes on the Pseudomonas aeruginosa chromosome map. Mol. Gen. Genet. 193:431-436.
32. Stanisich, V.A., and B.W. Holloway (1969) Conjugation in Pseudomonas aeruginosa. Genetics 61:327-339.
33. Sutcliffe, J.G. (1979) Complete nucleotide sequence of the Escherichia coli plasmid pBR322. Cold Spring Harb. Symp. Quant. Biol. 43:77-90.
34. Towbin, H., T. Staehelin, and J. Gordon (1979) Electrophoretic transfer of proteins from polyacrylamide gels to nitrocellulose sheets: procedure and some applications. Proc. Natl. Acad. Sci., USA 76:4350-4354.
35. Vasil, M.L., D. Kabat, and B.H. Iglewski (1977) Structure-activity relationships of an exotoxin of Pseudomonas aeruginosa. Infec. Immun. 16:353-361.
36. Young, L.S. (1980) The role of exotoxins in the pathogenesis of Pseudomonas infections. J. Infect. Dis. 142:626-630.
37. Young, R.A., and R.W. Davis (1983) Efficient isolation of genes using antibody probes. Proc. Natl. Acad. Sci., USA 80:1194-1198.

38. Vasil, M.L. (1984) Abstract, Ann. Mtg. A.S.M., Washington, D.C., D70, p. 62.

STRUCTURE, FUNCTION, AND REGULATION OF THE PLASMID-ENCODED HEMOLYSIN DETERMINANT OF *ESCHERICHIA COLI*

W. Goebel, J. Hacker, S. Knapp, I. Then, W. Wagner, C. Hughes, and A. Juarez

Institut für Genetik und Mikrobiologie
Universität Würzburg
8700 Würzburg, Federal Republic of Germany

INTRODUCTION

In 1967, H. William S. Smith first reported on the extrachromosomal nature of hemolysin formation in *Escherichia coli* (1). Shortly thereafter it was demonstrated that transmissible plasmids carrying the information for the hemolytic phenotype were present in *E. coli* strains which were isolated from among the fecal flora of various animals. Since then a large number of hemolytic (Hly) plasmids have been characterized (2-5), clearly establishing the early assumption of the extrachromosomal nature of this property. These Hly plasmids belong to various incompatibility groups although the *incJ2* and FVI groups are predominant. Their molecular sizes range from 60 kilobases (kb) to over 160 kb (2-5). As we have shown by hybridization with cloned *hly* genes, even those Hly plasmids which are structurally rather unrelated share extensive homology in their hemolysin determinants, suggesting that the hemolysin determinant may represent a discrete genetic entity that can undergo recombination without extended sequence homology. Association of the hemolytic phenotype with other known extrachromosomal phenotypes of *E. coli* such as antibiotic resistance, heavy metal resistance, enterotoxin production, or colonization factor antigens appears to be very rare and has never been observed in our laboratory. Recently it became apparent, however, that most hemolytic *E. coli* strains isolated from humans, especially from patients suffering from urinary tract infections, bear the hemolysin determinant on the chromosome (6,7). It was shown in several animal model systems (8,9,20) that chromosomal *hly*-determinants contribute to virulence whereas little effect could be demonstrated for plasmid-encoded *hly*-determinants.

In this chapter we will primarily describe the structural and functional aspects of a plasmid-encoded hemolysin determinant derived from the Hly plasmid pHly152 (3,10). In the last section we will compare our knowledge obtained on this determinant with recent data from chromosomal hly-determinants.

Structure of the Plasmid-Encoded Hly Determinant

The region of the low copy Hly plasmid pHly152 essential for complete hemolysis, e.g., synthesis of active hemolysin and its excretion, was cloned as a 14-kb SalI-partially HindIII digested DNA fragment into pACYC184. The resulting high-copy plasmid, pANN202-312, expressed in E. coli K12 similar amounts of external and internal hemolysin as the original pHly152. The same fragment was also cloned into the mobilizable broad host-range vector pKT231, and the resulting recombinant plasmid, pANN202-419, was introduced with RP4 into several gram-negative bacteria including Salmonella typhimurium, Aeromonas hydrophila, Proteus mirabilis, Pr. morganii, and Serratia marcescens. Hemolysin was expressed in all these hosts and properly excreted, indicating that the genetic information for the hemolytic phenotype was plasmid-inherited and located on the cloned HindIII-SalI fragment.

The region essential for hemolysin synthesis and excretion was further narrowed down by transposition of Tn5 or Tn3 into pANN202-312 and selecting for mutants unable to form hemolytic zones on blood agar plates (Hly^-). A region of about 7 kb, starting a few hundred base pairs (bp) to the left of the single Sma site (at coordinate 1.3 kb) and ending shortly before the PstI site (at coordinate 8.8 kb) as shown in Fig. 1, was found to be indispensable for the full hemolytic phenotype. The Hly^- mutants obtained could be divided into 2 groups (10), those which still produced internal hemolysin but were unable to secrete it (Tn3 insertions between coordinates 5.5 kb to about 8.6 kb) and those which were unable to synthesize hemolysin (Tn3 insertions from coordinate 1.0 to 5.5 kb). By complementing of both types of Tn3 mutants with cloned fragments of the hly region, we could identify 4 complementation groups. Two cistrons were shown to be essential for synthesis of active hemolysin (hlyC and hlyA) and 2 other cistrons were found to be required for excretion of hemolysin ($hlyB_a$ and $hlyB_b$) (10,15).

In addition to the coding region of the 4 cistrons (hlyC, A, B_a, and B_b) most Hly plasmids additionally seem to share homologous flanking regions. On 4 out of 5 Hly plasmids tested we found a region at the left end (proximal to hlyC) of more than 1 kb extending beyond the N-terminal end of the hlyC gene (12). Similarly all 5 Hly plasmids shared a common region at the right end of the hly-determinant (proximal to $hlyB_b$). Probes from both flanking regions were prepared and hybridized with the chromosomal DNA of E. coli K12 and hemolytic uropathogenic E. coli strains. Both probes hybridized

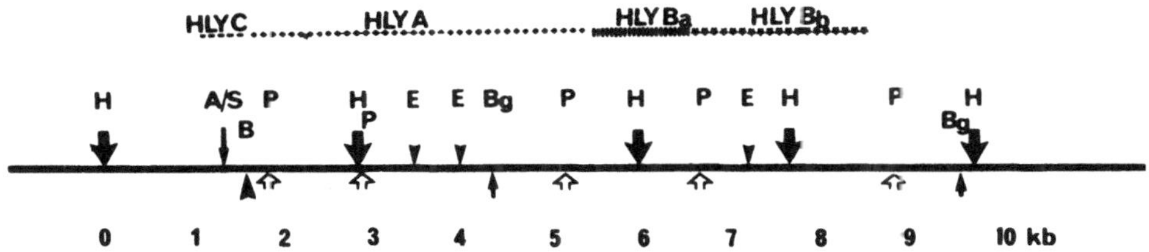

Fig. 1. Restriction map and localization of the 4 hly genes of the hemolysin determinant of plasmid pHly152. The map starts at the left (coordinate 0.06 kb) end with the HindIII site located in IS2 and ends with a HindIII site (coordinate 9.7 kb) outside of the other flanking sequence, ISx (see text for details). B (BamHI); Bg (BgIII; E (EcoRI); H (HindIII); P (PstI); A/S (AvaI/Sma).

with several chromosomal restriction fragments indicating that they may represent short, frequently occurring sequences (11,12). The hybridization patterns obtained with the 2 probes were different, however, suggesting that the 2 flanking sequences are not identical. This was further demonstrated by the complete lack of hybridization when 1 probe was tested against the other (12). DNA from the left flanking region was sequenced starting at the HindIII site at 0 kb (Fig. 1), a restriction site which is conserved in 4 out of 5 Hly plasmids. Comparison of a 216-bp sequence from this HindIII site to the right with the known sequences of the IS elements of E. coli revealed great similarity to sequence 876 (HindIII) to 660 of IS2 (Fig. 2). There are a few deviations with the reported IS2 sequence (21) which we cannot explain at the moment. The restriction map of this region fits that of IS2. These data clearly indicate that the hlyC proximal sequence is nearly identical to IS2. The other flanking sequence is localized between the PstI site at 8.8 kb and the BglII site at 9.6 kb (Fig. 1). Sequencing was performed from both restriction sites. The obtained sequences could not be correlated to any of the known IS elements. No clear-cut short repeats at the ends of this flanking sequence have yet been identified, suggesting that this sequence may be either partially deleted or may not represent a bona fide IS element.

The coding frame of the hlyC gene was revealed by fusion of the lacZ gene into the single BamHI site (13). The resulting fusion protein possesses a molecular weight of 146 kilodaltons (kd) and contains β-galactosidase activity, indicating that the N-terminal end of hlyC must be located about 360 bp upstream of the BamHI site. The DNA sequence of this region (Fig. 3) shows a single open reading frame starting with an ATG codon at the expected position. The size of this coding region (170 amino acids) matches the size of the HlyC protein (18,000 d) which was detected in mini- and maxicells of E. coli (14).

```
              .         .         .         .         .
660 CGCAGGCATACCATCAAGTTCTGCCTGTCTGCGAAGCAGCGCCCATACCC 709
    ööööööööööööööööööööööö  ööööööööööööööööööööööööö
 85 CGCAGGCATACCATCAAGTTCTGTTTGTCTGCGAAGCAGCGCCCATACCC 134

              .         .         .         .         .
710 GACGATAACCATACGTTGGCAGCTCTCCGATAACATGGTGTATACGGAGA 759
    ööööööööööööö öö öööööööööööööööööööö ööööööööö öö
135 GACGATAACCATATGTGGGCAGCTCTCCGATAACATGATGTATACGGCGA 184

              .         .         .         .         .
760 AGCACATCCGTATCATCAGTGTGACGACTGCGGCGGCCATCCATCCAGTC 809
    ööööööööööööööööö öööööööö ööööööööööö ööö ööööööö
185 AGCACATCCGTATCATCCGTGTGACGGCTGCGGCGGCCGTCCTTCCAGTC 234

              .         .         .         .         .
810 ATCGGTTCGTCTGAGAATGACGTGCAACTGCGCACGCGACACCCGGAGAC 859
    öööööööööööööööööööööööööööööööööööööööööööööööööö
235 ATCGGTTCGTCTGAGAATGACGTGCAACTGCGCACGCGACACCCGGAGAC 284

              .
860 AACGGCTGACTAAGCTT 876
    öööööööööö öööööö
285 AACGGCTGACGAAGCTT 301
```

Fig. 2. DNA sequence starting from HindIII site (coordinate 0 kb in Fig. 1) 216 bp downstream compared to positions 876 to 660 of IS2 (21).

Upstream from the start codon for HlyC, a putative promoter sequence was identified with sequences fitting almost perfectly the -10 (Pribnow box) and -35 consensus sequences for E. coli promoters (Fig. 4). There is, however, only a rather short (15 bp) distance in between, which could explain the weak transcriptional activity of this promoter (see below). Further upstream of this putative promoter sequence is an AT-rich sequence which contains 2 12-bp direct repeats and several inverted repeats which overlap the directly repeated sequences. We are investigating the possibility that these sequences may be involved in the regulation of the hly determinant (see also below).

The region between the C-terminal end of hlyC and the N-terminal end of hlyA is 11 bp long. This region likely functions as a ribosome binding site since 7 bp are homologous to the 3' end of 16 S rRNA ("Shine-Dalgarno" sequence). The N-terminal end of hlyA was identified by microsequencing of a hlyA-lacZ fusion protein which was expressed under the control of λp_L and λp_R (Ref. 17; Vogel and Goebel, unpub. results). The coding region of hlyA was deduced from the size of the primary HlyA protein (106,000 d), of a truncated HlyA protein (104,000 daltons) (deleted at the PstI site at coordinate 5.1 kb) and of a HlyA-βgal fusion protein (220,000 d). The latter protein was obtained by fusion of lacZ into the BglII site at 4.3 kb. This fusion leads to a protein which has β-galactosidase activity and hence allows the determination of the reading frame of hlyA. The putative interspace between hlyA and $hlyB_a$ gene (Fig. 4) is 60 bp long and contains several stop codons in all 3 reading frames, and a new Shine-Dalgarno sequence beginning 10-bp in front of the putative start codon of $hlyB_a$. No sequence data are as yet available for the $hlyB_a$ and $hlyB_b$ genes.

Expression and Regulation of the Plasmid Hly Determinant

Hemolysin is synthesized and excreted most actively in the late logarithmic growth phase (13). Its synthesis stops when the cells enter the stationary phase. Aerobic conditions seem to be required for its expression and rich media stimulate both synthesis and excretion of hemolysin. These observations suggest a rather complex regulation of this protein. The native active hemolysin is highly unstable and this has hindered both its purification and efforts to investigate the regulation of its synthesis and transport. In bypassing this problem we have used insertions of Mud-1 phage (16) into pHly152 which allowed fusion of the lacZ gene to the hly-specific genes and subsequent assessment of gene expression along the hly-determinant. Furthermore we have constructed recombinant plasmids which carry genes where lacZ was fused either to hlyC or to hlyA.

It was previously shown (15) that Tn5 insertions in hlyC are polar, removing hlyA and $hlyB_a$ but not $hlyB_b$ activity, suggesting that the hly-determinant is transcribed from 2 promoters, one located in front of hlyC and the other near $hlyB_b$. The data obtained with Mud-1 insertions into pHly152 agree with this notion and allow determination of the strength of the transcription of the 2 promoters by measurement of the β-galactosidase activity. As shown in Fig. 5, the operon hlyC, A, B is transcribed rather weakly whereas the transcription of $hlyB_b$ is considerably stronger. The relative weakness of the hlyC proximal promoter was confirmed by cloning the HindIII-BamHI fragment (see Fig. 1) into pKO-11, a vector allowing identification of promoter strength by insertion upstream of the galK gene. The transcriptional activity observed (Tab. 1) was less than half that of the gal promoter, regarded as average in E. coli and slightly lower than $p\lambda_{C17}$, a phage λ promoter classed as "average to low." The hly promoter was, however, considerably stronger than $p\lambda_{482}$, a synthetic promoter determining the lowest transcriptional activity yet observed in this system.

The direction of transcription from the other hly promoter (in Fig. 4, transcribing $hlyB_b$) was determined by selection of Mu-sensitive derivatives from a Hly^- mutant containing the Mud-1 insertion in $hlyB_b$. Out of 6 Mu^s revertants, 5 were found to be Ap^r and exhibited the same level of β-galactosidase activity as the original clone but failed to produce internal hemolysin. This indicates that the deletion which affected the gene C-proximal end of Mu, and thereby caused the ts^+, Mu^s phenotype, extended into the adjacent $hlyB_a$ and hlyA genes. The direction of transcription from the $hlyB_b$ promoter is therefore from right to left, i.e., opposite to transcription of the hlyC, A, B_a operon. Restriction fragments con-

```
        TAGTCACGCAATAAAACGTTCTTTAATATTAATGCAGTTATGACATTAAAGGCAAGAAAC
     1  ---------+---------+---------+---------+---------+---------+ 60
        ATCAGTGCGTTATTTTGCAAGAAATTATAATTACGTCAATACTGTAATTTCCGTTCTTTG

a:      EndSerArgAsnLysThrPhePheAsnIleAsnAlaValMetThrLeuLysAlaArgAsn
b:       SerHisAlaIleLysArgSerLeuIleLeuMetGlnLeuEndHisEndArgGlnGluThr
c:        ValThrGlnEndAsnValLeuEndTyrEndCysSerTyrAspIleLysGlyLysLysHis

                                 T
        ATAAAGGCATATTTTTGCCACAATATTTAATCATATAATTTAAGTTGTAGTGAGTTTATT
    61  ---------+---------+---------+---------+---------+---------+ 120
        TATTTCCGTATAAAAACGGTGTTATAAATTAGTATATTAAATTCAACATCACTCAAATAA
                                 A

a:      IleLysAlaTyrPheCysHisAsnIleEndSerTyrAsnLeuSerCysSerGluPheIle
b:       EndArgHisIlePheAlaThrIlePheAsnHisIleIleEndValValValSerLeuLeu
c:        LysGlyIlePheLeuProGlnTyrLeuIleIleEndPheLysLeuEndEndValTyrTyr

        ATGAATATAAACAAACCATTAGAGATTCTTGGGCATGTATCCTGGCTATGGGCCAGTTCT
   121  ---------+---------+---------+---------+---------+---------+ 180
        TACTTATATTTGTTTGGTAATCTCTAAGAACCCGTACATAGGACCGATACCCGGTCAAGA

a:      MetAsnIleAsnLysProLeuGluIleLeuGlyHisValSerTrpLeuTrpAlaSerSer
b:       EndIleEndThrAsnHisEndArgPheLeuGlyMetTyrProGlyTyrGlyProValLeu
c:        GluTyrLysGlnThrIleArgAspSerTrpAlaCysIleLeuAlaMetGlyGlnPheSer

        CCACTACACAGAAACTGGCCAGTATCTTTGTTTGCAATAAATGTATTACCCGCAATACAG
   181  ---------+---------+---------+---------+---------+---------+ 240
        GGTGATGTGTCTTTGACCGGTCATAGAAACAAACGTTATTTACATAATGGGCGTTATGTC

a:      ProLeuHisArgAsnTrpProValSerLeuPheAlaIleAsnValLeuProAlaIleGln
b:       HisTyrThrGluThrGlyGlnTyrLeuCysLeuGlnEndMetTyrTyrProGlnTyrArg
c:        ThrThrGlnLysLeuAlaSerIlePheValCysAsnLysCysIleThrArgAsnThrGly

        GCTAACCAATATGTTTTATTAACCCGGGATGATTACCCTGTCGCGTATTGTAGTTGGGCT
   241  ---------+---------+---------+---------+---------+---------+ 300
        CGATTGGTTATACAAAATAATTGGGCCCTACTAATGGGACAGCGCATAACATCAACCCGA

a:      AlaAsnGlnTyrValLeuLeuThrArgAspAspTyrProValAlaTyrCysSerTrpAla
b:       LeuThrAsnMetPheTyrEndProGlyMetIleThrLeuSerArgIleValValGlyLeu
c:        EndProIleCysPheIleAsnProGlyEndLeuProCysArgValLeuEndLeuGlyEnd

        AATTTAAGTTTAGAAAATGAAATTAAATATCTTAATGATGTTACCTCATTAGTTGCAGAA
   301  ---------+---------+---------+---------+---------+---------+ 360
        TTAAATTCAAATCTTTTACTTTAATTTATAGAATTACTACAATGGAGTAATCAACGTCTT

a:      AsnLeuSerLeuGluAsnGluIleLysTyrLeuAsnAspValThrSerLeuValAlaGlu
b:       IleEndValEndLysMetLysLeuAsnIleLeuMetMetLeuProHisEndLeuGlnLys
c:        PheLysPheArgLysEndAsnEndIleSerEndEndCysTyrLeuIleSerCysArgArg

        GACTGGACTTCAGGTGATCGTAAATGGTTCATTGACTGGATTGCTCCTTTCGGGGATAAC
   361  ---------+---------+---------+---------+---------+---------+ 420
        CTGACCTGAAGTCCACTAGCATTTACCAAGTAACTGACCTAACGAGGAAAGCCCCTATTG

a:      AspTrpThrSerGlyAspArgLysTrpPheIleAspTrpIleAlaProPheGlyAspAsn
b:       ThrGlyLeuGlnValIleValAsnGlySerLeuThrGlyLeuLeuLeuSerGlyIleThr
c:        LeuAspPheArgEndSerEndMetValHisEndLeuAspCysSerPheArgGlyEndArg

        GGTGCCCTGTACAAATATATGCGAAAAAAATTCCCTGATGAACTATTCAGAGCCATCAGG
   421  ---------+---------+---------+---------+---------+---------+ 480
        CCACGGGACATGTTTATATACGCTTTTTTTAAGGGACTACTTGATAAGTCTCGGTAGTCC

a:      GlyAlaLeuTyrLysTyrMetArgLysLysPheProAspGluLeuPheArgAlaIleArg
b:       ValProCysThrAsnIleCysGluLysAsnSerLeuMetAsnTyrSerGluProSerGly
c:        CysProValGlnIleTyrAlaLysLysIleProEndEndThrIleGlnSerHisGlnGly

        GTGGATCCCAAAACTCATGTTGGTAAAGTATCAGAATTTCATGGAGGTAAAATTGATAAA
   481  ---------+---------+---------+---------+---------+---------+ 540
        CACCTAGGGTTTTGAGTACAACCATTTCATAGTCTTAAAGTACCTCCATTTTAACTATTT

a:      ValAspProLysThrHisValGlyLysValSerGluPheHisGlyGlyLysIleAspLys
b:       TrpIleProLysLeuMetLeuValLysTyrGlnAsnPheMetGluValLysLeuIleAsn
c:        GlySerGlnAsnSerCysTrpEndSerIleArgIleSerTrpArgEndAsnEndEndThr

        CAGTTAGCGAATAAAATTTTTAAACAATATCACCACGAGTTAATAACTGAAGTAAAAAGA
   541  ---------+---------+---------+---------+---------+---------+ 600
        GTCAATCGCTTATTTTAAAAATTTGTTATAGTGGTGCTCAATTATTGACTTCATTTTTCT

a:      GlnLeuAlaAsnLysIlePheLysGlnTyrHisHisGluLeuIleThrGluValLysArg
b:       SerEndArgIleLysPheLeuAsnAsnIleThrThrSerEndEndLeuLysEndLysGlu
c:        ValSerGluEndAsnPheEndThrIleSerProArgValAsnAsnEndSerLysLysLys

        AAGTCAGATTTTAATTTTTCATTAACTGGTTAA
   601  ---------+---------+---------+--- 633
        TTCAGTCTAAAATTAAAAAGTAATTGACCAATT

a:      LysSerAspPheAsnPheSerLeuThrGlyEnd
b:       SerGlnIleLeuIlePheHisEndLeuValAsn
c:        ValArgPheEndPhePheIleAsnTrpLeuIle
```

Fig. 3. DNA sequence of the hlyC region. The open reading frame fitting the size of HlyC protein starts at position 121 and ends at position 630.

taining this promoter have not yet been identified in the pKO-11 system. A polar effect from hlyC to hlyA is also observed when lacZ is fused to one of these genes. Whereas β-galactosidase activity of the HlyC-βgal fusion protein reaches 700 units, that of HlyA-βgal fusion protein is considerably less active (200 units); the copy numbers of the plasmids carrying the fusions are the same in both cases.

Hemolysin activity is not increased with increasing gene dosage, suggesting a negative control of transcription of the hly genes. However, when the dosage of the fused genes hlyC-lacZ or hlyA-lacZ is modulated by cloning these genes together with the control region (16) into the low copy plasmid pSC101, β-galactosidase activity drops in correlation to the gene dosage. This strict gene dosage effect of β-galactosidase occurs whether the other hly genes (hlyA, B_a and B_b for hlyC-lacZ and hlyB$_a$ and B_b for hlyC, hlyA-lacZ) were present or not, ruling out possible negative regulatory functions of these hly gene products. When lacZ is under the control of the hlyC proximal promoter, either as an intact gene (by Mud-1 insertion) or as a fusion gene with hlyC or hlyA, the level of β-galactosidase in contrast to hemolysin formation increases with the cell mass and reaches a maximum in the early stationary phase. This rules out the possibility that the observed decrease in internal and excreted hemolysin activity in the late exponential phase is caused by transcriptional (or translational) effects.

The cyclic nucleotides, cAMP and cGMP, had no effect on β-galactosidase activity in strains which carried lacZ placed under the control of the phlyC promoter indicating that hemolysin synthesis is not influenced by these compounds. There was, however, a remarkable difference in β-galactosidase activity in Mud-1 insertion mutants, with Mud-1 inserted either in hlyC or hlyA, when they were grown either aerobically or anaerobically. This effect correlates with the observed reduction in hemolysin formation under anaerobic conditions. No such effect was seen in expression of β-galactosidase activity when Mud-1 was inserted into hlyB$_b$. This suggests that promoter phlyC but not phlyB$_b$ is controlled on the transcriptional level directly or indirectly by the oxygen level in the cell.

Function of the Hly Gene Products

When pANN202-312 Δ H, a recombinant plasmid carrying exclusively the entire hly-determinant from HindIII (0.0 kb) to HindIII (9.6 kb) cloned into pACYC184, is transformed into minicells or maxicells of *E. coli* and the plasmid-specific gene products are

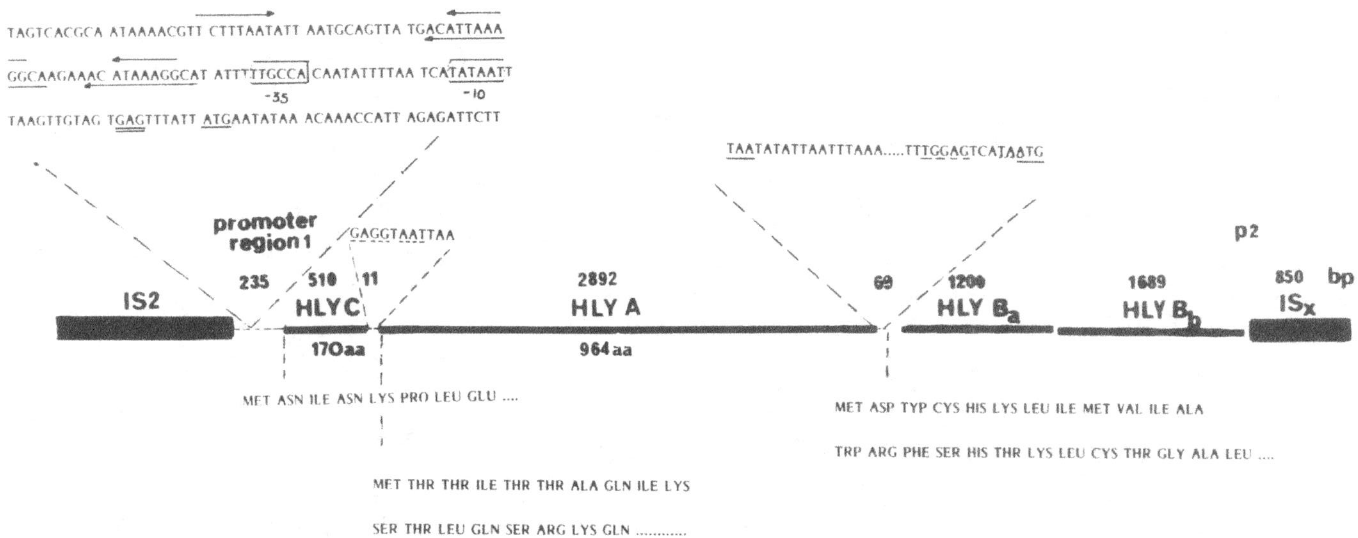

Fig. 4. Functional map of the hly determinant of pHly152 showing the interspace regions between hlyC and hlyA, and hlyA and $hlyB_a$, respectively. Characteristic sequences are specifically marked. N-terminal amino acid sequences deduced from the DNA sequences are shown for HlyC, HlyA, and $HlyB_a$.

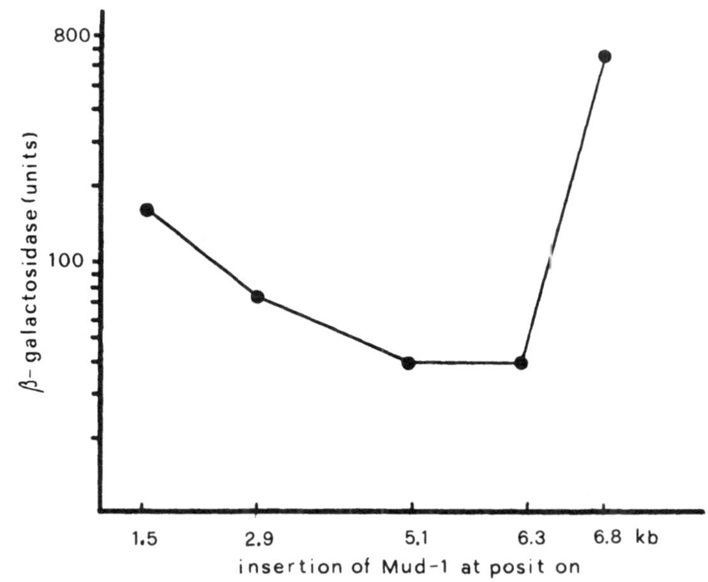

Fig. 5. β-galactosidase activity after insertion of Mud-1 (Mu-lacZ, Y, Ap^R) into various rgions of the hly determinant. Positions of mud-1 insertions follow the coordinates in Fig. 1.

Tab. 1. Transcriptional activity observed by cloning the HindIII-BamHI fragment into pKO-11, a vector allowing identification of promoter strength by insertion upstream of the galK gene.

Promoter	Galactokinase activity*
pgal	1.000
$p\lambda_{C17}$	0.54
p hlyC	0.41
$p\lambda_{482}$	0.25
no insert	< 0.05

* Galactokinase activity (assessed as galactose phosphorylated per cell number and time) was neutralized to the E. coli promoter pgal takes as 1.0

labeled, 3 major proteins (18-kd, 25-kd, and 58-kd) are identified. In addition, there are several faint bands above (up to 106-kd) and below the 58-kd protein band. Thé 25-kd protein is chloramphenicol acetyltransferase (CAT) which is determined by the vector. In order to identify the primary gene products of the 4 hly genes, recombinant plasmids were constructed which carried the separate hly genes under the control of various promoters (4). This allowed the identification of the hlyC gene product as the 18-kd protein; the $hlyB_a$ gene under the control of plac yielded a protein of 48-kd which is barely visible when expressed from its own promoter (phlyC). Cistron $hlyB_b$ expressed a protein of 60-62-kd. The primary product of hlyA was identified as a rather large protein of 106-kd by putting this gene under the control of λ_{PL} or λ_{PR} and λ_{cts} (17). Temperature induction leads to the synthesis of this protein and to a rapid loss of cell viability, indicating that the primary HlyA product may be lethal to the cell when it is accumulated inside the cell. Similar observations were made with recombinant Hly plasmids in which hlyC was inactivated by deletion or by a ts mutation.

Active hemolysin is only formed when the 2 hly genes, hlyA and hlyC, are present in the cell. As previously shown (14), this hemolysin is predominantly found in the periplasm. When mini- or maxicells containing a recombinant plasmid with hlyA and hlyC are labeled with ^{35}S-methionine, the HlyC (18-kd) protein and a protein of 58-kd are the major hly-specific proteins (the vector-specific CAT protein bands at 25-kd). There are, in addition, small amounts of the 106-kd protein and several discrete protein bands with sizes larger and smaller than the 58-kd protein (14). This suggests that the 106-kd protein is eventually proteolytically degraded to yield several intermediates and a major product of 58-kd. This could be more directly demonstrated by pulse chase experiments. Short pulse labeling of minicells containing hlyC and hlyA with ^{35}S-methionine leads mainly to the appearance of the 106-kd protein which upon chase with cold methionine is converted to the major 58-kd protein (14). This proteolytic process does not depend on the HlyC protein since a mutant which is temperature-sensitive in HlyC still shows this proteolysis of HlyA under the restrictive conditions (14) without yielding active hemolysin under these conditions. It is difficult to decide at the moment whether this proteolytic degradation of the 106-kd HlyA is an artifact of the systems used (mini- and maxicells) or whether proteolysis of HlyA may be necessary for transport of hemolytically-active protein. This latter possibility is based on the following experimental evidence: (a) Pulse labeled 106-kd HlyA protein is found in the cytoplasm. (b) Deletion of 60-bp at the C-terminal end of hlyA leads to a truncated protein of 104-kd which yields hemolytically active protein in the presence of HlyC protein, but its transport is drastically inhibited (14). (c) The above described HlyA-βgal fusion protein, which is expressed by a fusion gene consisting of lacZ fused to the BglII site (4.2 kb) which is close to the C-terminal end of the hlyA gene, remains as a large protein of 220-kd in the cytoplasm (Tab. 2). (d) The N-terminal end of HlyA (Fig. 4) does not appear to be a signal sequence, which would be necessary if the primary HlyA protein itself were transported.

The arguments in (b) and (c) further indicate that the proteolysis of HlyA depends on the intact C-terminal end of HlyA and suggest that this process occurs either autoproteolytically or that the C-terminal end is required for leading the HlyA protein to its cellular processing site.

The function of HlyC is not yet understood. This protein remains in the cytoplasm as evidenced by the absence of an obvious signal sequence at the N-terminal end of HlyC protein (Fig. 4) and the cytoplasmic localization of a HlyC-βgal protein (Tab. 2). It is required for rendering the HlyA protein and/or its proteolytic products hemolytically active, suggesting an enzymatic conversion of the HlyA product(s) before transport across the cytoplasmic membrane occurs. It is not yet clear whether activation by HlyC occurs before or after proteolysis of HlyA, but the observation that the truncated 104-kd HlyA protein becomes hemolytically active in the presence of HlyC indicates that proteolysis is not required for the HlyC action nor does activation by HlyC seem to trigger proteolysis of HlyA.

Hemolysin is only excreted through the outer membrane of the *E. coli* cell when *hlyC* and *hlyA* are complemented by *hlyB*$_a$ and *hlyB*$_b$. The 2 gene products are predominantly found in the outer membrane

Tab. 2. Protein formation and β-galactosidase activity in plasmid recombination.

Plasmid	Protein formed	β-galactosidase activity (units) External	Periplasmic	Membrane bound	Cytoplasmic
none	none	22	25	30	28
pANN651 (*)	native β-galactosidase	20	30	82	3.300
pANN653 (*)	HlyC-βgal	30	25	122	560
pANN654 (*)	HlyA-βgal	18	22	75	189

(*) These recombinant plasmids were previously described (13). Plamid pANN651 carries Mud-1 inserted into *hlyC*, pANN653 contains a fused *hlyC*::*lacZ* gene with *lacZ* being fused at the *Bam*HI site and pANN654 carries a fused *hlyA*::*lacZ* gene and an intact *hlyC*; fusion of *lacZ* occurred in pANN654 at the *Bgl*II site at coordinate 4.3 kb

fraction (15), but a substantial amount of both proteins remains in the M-band and even in the cytoplasmic membrane fraction (L-band). It was recently shown (15) that mutants defective in $hlyB_a$ tend to accumulate more hemolysin in the periplasm, whereas $hlyB_b$ mutants concentrate most hemolysin in the outer membrane without releasing it into the surroundings. Thus, $hlyB_b$ mutants form small hemolysis zones around growing colonies on blood agar plates. This has led to a model which assumes that $HlyB_a$ protein fixes hemolysin to the outer membrane and turns it inside out, whereas $HlyB_b$ protein releases hemolysin from the outer membrane.

The recent isolation of a chromosomal mutation (18) that specifically inhibits excretion of hemolysin without affecting transport of periplasmic proteins or excretion of colicin E1, E2, or E3 suggests that specific cellular protein(s) may be necessary in addition to the *hly*-specific gene products, $HlyB_a$ and $HlyB_b$, to properly insert the specific hemolysin transport machinery into the cellular envelope.

Comparison of Plasmid-encoded and Chromosomal hly Determinants

Chromosomal *hly*-determinants were cloned from uropathogenic *E. coli* strains of 4 O-serogroups (O4, O6, O18, and O75). Hybridization with plasmid-encoded *hly* genes indicated extensive homology in the 4 structural *hly* genes which were shown to be also functionally exchangeable (7,9). Differences were seen by more precise restriction mapping of the various determinants, particularly in *hlyA* which was verified by partial DNA sequences of *hlyA* from *hly*O18 and *hly*O75 (Wagner, Juarez, and Goebel, unpub. results). These differences seem to be somewhat concentrated at the C-terminal half of *hlyA*. Obvious differences between the plasmid and the chromosomal *hly* determinants exist in the flanking regions. None of the chromosomal *hly* determinants is flanked by IS2 at the *hlyC* proximal ends although several IS2 elements were identified in the chromosomes of these strains (12). The other end sequence of the *hly* determinant ($hlyB_b$ proximal) was found to be located close to both ends of *hly* determinants from several hemolytic O6 strains without flanking them directly. Evidence was obtained that this sequence is involved in the frequent loss of *hly* determinants in such strains (11).

The most dramatic difference was, however, observed in the in vivo action in various animal model systems (8,9,20, Hacker et al., unpub. results). Whereas the *hly* determinant of plasmid pHly152 proved to be nontoxic, highly toxic effects were exerted by the chromosomal *hly* determinants, especially by those of two O18 and O75 strains (9). This difference cannot be explained by different expression of hemolysin of these determinants, although there is circumstantial evidence (12,20) that the activity of the *hlyC*-proximal promoters may be different between different *hly* determinants. But even when the same promoter (phlyC, from *hly* determinant of plasmid pHly152) was used, the chromosomal hly determinants still exerted a much higher toxicity in the mouse model (9). Recent experiments in which mixed hly determinants between a chromosomal (018) and a plasmid (pHly152) determinant were constructed seem to indicate that the major cause for the difference in pathogenicity of the hly determinants is the composition of hlyA and its proper combination with hlyC (Wagner and Goebel, unpub. results).

CONCLUSION

While our knowledge on the structure of the plasmid-encoded hemolysin determinant and its relationship to the chromosomal hly determinants of uropathogenic E. coli strains has been considerably widened, the understanding of how the 4 gene products of the hly determinant interact to bring about synthesis and excretion of hemolysin is still rather incomplete and several questions remain unanswered: What is the function of HlyC protein which seems to be absolutely necessary for the production of active hemolysin? Is the secreted hemolysin identical in size to the primary product of hlyA gene or is the proteolytic processing of HlyA observed in mini cells essential for excretion? What is the precise function of the 2 gene products, $HlyB_a$ and $HlyB_b$, which are required for the excretion of hemolysin? The latter question is of general interest since little is known of how proteins are transported across the outer membrane of gram-negative bacteria.

Finally it remains to be shown whether plasmid hly determinants can undergo recombination with the chromosome and vice versa, and how this might occur.

ACKNOWLEDGMENTS

This work was supported by a grant from the Deutsche Forschungsgemeinschaft (SFB 105-A12) and by a personal grant DAAD to one of us (A.J.). The excellent technical assistance of M. Vogel and H. Düvel is acknowledged. E. Appel is thanked for typing of this manuscript.

REFERENCES

1. Smith, H.W., S. Halls (1967) The transmissible nature of the genetic factor in Escherichia coli that controls hemolysin. J. Gen. Microbiol. 47:153-167.
2. Goebel, W., and H. Schrempf (1971) Isolation and characterization of supercoiled circular DNA from β-hemolytic strains of Escherichia coli. J. Bact. 106:311-318.

3. Goebel, W., B. Royer-Pokora, W. Lindenmaier, and H. Bujard (1973) Plasmids controlling synthesis of hemolysin in Escherichia coli. I Molecular properties. J. Bact. 118:964-973.
4. Le Minor, S., and E. Le Coueffic (1975) Studies on hemolysins of enterobacteriaceae. Ann. Microbiol. (Paris) 126:313-332.
5. De la Cruz, F., J.C. Zabala, and J.M. Ortiz (1979) Incompatibility among α-hemolysin genes by transposition of Tn802. Plasmid 2:507-519.
6. Minshew, B.H., J. Jorgenson, G.W. Counts, and S. Falkow (1978) Association of hemolysin production, hemaggulutinantion of human erythrocytes and virulence for chicken embryos of extraintestinal Escherichia coli isolates. Infect. Immun. 20:50-54.
7. Berger, H., J. Hacker, A. Juarez, C. Hughes, and W. Goebel (1982) Cloning of the chromosomal determinants encoding hemolysin production and mannose-resistant hemagglutination in Escherichia coli. J. Bact. 152:1241-1247.
8. Welch, R.A., E.P. Dellinger, B. Minshew, and S. Falkow (1981) Hemolysin contributes to virulence of extraintestinal E. coli infections. Nature (London) 294:665-667.
9. Hacker, J., C. Hughes, H. Hof, and W. Goebel (1983) Cloned hemolysin genes from Escherichia coli that cause urinary tract infection determine different levels of toxicity in mice. Infect. Immun. 42:57-63.
10. Noegel, A., U. Rdest, W. Springer, and W. Goebel (1979) Plasmid cistrons controlling synthesis and excretion of the exotoxin α-hemolysin of Escherichia coli. Mol. Gen. Genet. 171:277-285.
11. Hacker, J., S. Knapp, and W. Goebel (1983) Spontaneous deletions and flanking regions of the chromosomally inherited hemolysin determinant of an Escherichia coli 06 strain. J. Bact. 154:1145-1154.
12. Knapp, S., J. Hacker, I. Then, D. Müller, and W. Goebel (1984) Multiple copies of hemolysin genes and associated sequences in the chromosome of uropathogenic Escherichia coli strains. J. Bact. (submitted for publication).
13. Juarez, A., M. Härtlein, and W. Goebel (1984) Fusion of the β-galactosidase gene (lacZ) to hemolysin genes to study regulation and transport of hemolysin. J. Bact. (submitted for publication).
14. Härtlein, M., M. Schileßl, W. Wagner, U. Rdest, J. Kreft, and W. Goebel (1984) Transport of hemolysin by Escherichia coli. J. Cell Biol. (in press).
15. Wagner, W., M. Vogel, and W. Goebel (1983) Transport of hemolysin across the outer membrane of Escherichia coli requires two functions. J. Bact. 154:200-210.
16. Casadaban, M.J., J. Chon, and S.N. Cohen (1980) In vitro gene fusions that join an enzymatically active β-galactosidase segment to amino terminal fragments of exogenous proteins: Escherichia coli vectors for the detection of translational initiation signals. J. Bact. 143:972-980.

17. Goebel, W., and J. Hedgpeth (1982) Cloning and characterization of the entire plasmid-encoded hemolysin determinant of *Escherichia coli*. *J. Bact.* 151:190-198.
18. Juarez, A., and W. Goebel (1984) Chromosomal mutations affecting excretion of hemolysin in *Escherichia coli*. *J. Bact.* (submitted for publication).
19. Müller, D., C. Hughes, and W. Goebel (1983) Relationship between plasmid and chromosomal hemolysin determinants of *Escherichia coli*. *J. Bact.* 153:846-851.
20. Welch, R.A., and S. Falkow (1984) Characterization of *Escherichia coli* hemolysins conferring quantitative differences in virulence. *Infect. Immun.* 43:156-160.
21. Ghosal, D., H. Sommer, and H. Saedler (1979) Nucleotide sequence of the transposable dna element IS2. *Nucl. Acids Res.* 6:1111-1122.

pIAA, A VIRULENCE PLASMID IN PSEUDOMONAS SAVASTANOI

Tsune Kosuge, Curtis J. Palm, Steven V. Hutcheson,
N. Louise, E. Glass, and Tetsuji Yamada

Department of Plant Pathology
University of California
Davis, California 95616

INTRODUCTION

Bacteria are responsible for many diseases of plants; however, only a few of the many species encountered are capable of establishing a pathogenic relationship with plants. Certain specialized functions therefore, must allow such bacteria to become pathogenic in plants. One component of pathogenicity of bacteria is expression of virulence which can be assessed on plants by the severity of the primary systems produced. Virulence, therefore, is used here as a quantitative term to describe the factor(s) responsible for production of primary symptoms in diseased plants. In the case to be discussed, the primary symptom of the disease is formation of tumorous outgrowths by the infected plant. Virulence, therefore is assessed by the size of the plant tumors incited by different strains of the pathogen. The aim of these studies is the development of concepts on virulence and its expression by the bacterium, *Pseudomonas syringae* p.v. *savastanoi*, (p. savastanoi), which causes tumors to form on oleander and olive plants. Expression of virulence is associated with genes borne on a plasmid in this bacterium.

Formation of tumors is the host response to production and secretion by the bacterium of high concentrations of the phytohormones, indoleacetic acid and cytokinins. These compounds are natural plant hormones required for normal development. Under normal circumstances only small amounts of the compounds are needed and regulatory mechanisms in the plant assure that their production and delivery to the sites of action occur in keeping with the developmental needs of the plant. In the diseased state, production and secretion of phytohormones by the pathogen by-passes the normal regulatory systems of the host. The resulting phytohormone imbalances cause the host tissues to respond by undifferentiated proliferation.

The work to be discussed concerns the role of a plasmid pIAA1 in indoleacetic acid (IAA) production and virulence of *P. savastanoi* on oleander plants. The plasmid bears *iaaM* and *iaaH*, genetic determinants for tryptophan 2-monooxygenase and indoleacetamide hydrolase, respectively, the enzymes involved in the conversion of tryptophan to IAA by the pathway: L-tryptophan → indoleacetamide → IAA.

Genes for IAA Production Are Plasmid Borne

The genes, *iaaM* and *iaaH* are borne on a plasmid, pIAA, in oleander strains. Loss of pIAA1 by curing attenuates virulence and results in loss of IAA production (1,2). Virulence is fully restored together with IAA production if IAA^- mutants are transformed with pIAA1.

A 2.7-kb EcoR1 restriction fragment (fragment M) bearing *iaaM* was cloned from pIAA1 into the vector plasmid RSF 1010 to construct a recombinant plasmid, pLUC1. Transformation of Iaa-mutants of *P. savastanoi* with pLUC1 restores high levels of tryptophan 2-monooxygenase activity and partial virulence (3). The locus coding indoleacetamide hydrolase is located promoter distal from *iaaM*; insertional inactivation of *iaaM* by IS-elements causes polar mutations and prevents expression of *iaaH* (4). Thus, *iaaM* and *iaaH* are part of an operon in oleander strains.

Distribution of iaaM and iaaH

By selection for resistance to α-methyl tryptophan, IAA^- mutants are readily isolated. These mutants either lack pIAA, have deletions in pIAA with loss of *iaaM* and *iaaH*, or have an insertion element, IS-51, in the *iaaM* locus. (3,4,5).

In all oleander strains examined thus far, *iaaM* and *iaaH* occur on a plasmid. However, at least 3 different forms of pIAA have been found: the 52 kb pIAA1 in California oleander strains 2009 and 2015; the 72-kb pIAA2 in Italian oleander strains (2,5); the 70-kb pIAA3, in California strain 1050. In contrast, the IAA genes in olive strains appear to be borne on the chromosome. Southern blot experiments with pIAA1 restriction fragments as probes reveal that plasmids pIAA1, pIAA2, and pIAA3 share considerable amount of homology (C. Palm and T. Kosuge, unpub. results). However, homology between pIAA1 and chromosome fragments of olive strains appears to be limited to 5 to 6 kb spanning *iaaM* and *iaaH*.

Properties of Tryptophan Monooxygenase and Its Role in Regulation of IAA Production.

The foregoing evidence stressed the importance of tryptophan monooxygenase in virulence expression and IAA production. To determine its properties, the enzyme was purified to apparent homogeneity (S. Hutcheson, T. Kosuge, unpub. results). It is a FAD-linked enzyme with a monomer weight of 62 kd. It is strikingly product-inhibited by indoleacetamide (apparent Ki 7μM) and inhibition is competitive with respect to the substrate L-tryptophan; it is less strongly inhibited by IAA (apparent Ki, 220 μM). Tryptophan production is also regulated by tryptophan inhibition of anthranilate synthetase (M. Smidt, unpub. results); but the inhibition is less strong (apparent Ki for tryptophan, 30 μM). Tryptophan pool size therefore can be limiting in the production of IAA. When grown in culture media containing L-tryptophan, the wild types of *P. savastanoi* accumulate 10-fold or higher levels of IAA than in the absence of tryptophan. Certain mutants designated "super producers" have larger tryptophan pool sizes and accumulate 2 to 3 times more IAA than parental wild types. However, both "super producers" and parental types, when grown in media supplemented with L-tryptophan, produce increased but nevertheless similar amounts of IAA. Thus, the "super producer" phenotype is expressed only under conditions of tryptophan limitation (1). The tryptophan pool size therefore is an important factor in regulating IAA production by the bacterium and regulation of tryptophan monooxygenase by competitive feedback inhibition of indoleacetamide helps assure conversion of tryptophan to IAA does not occur at the expense of conversion to protein. In this way the primary requirements for tryptophan metabolism are met before conversion to indoleacetic acid occurs.

Specificity of Tryptophan Monooxygenase

Tryptophan monooxygenase is not specific for L-tryptophan, however. The enzyme catalyzes the oxidation of several tryptophan derivatives such as 5-hydroxy-, 5-fluoro-, and 5-methyltryptophan, in addition to azatryptophan (S. Hutcheson and T. Kosuge, unpub. results). Although DL-mixtures were used as substrates in these cases, it is presumed that the L-isomer is oxidized. Oxidation of these tryptophan derivatives by tryptophan monooxygenase represents an effective detoxification mechanism since the compounds otherwise would be toxic to *P. savastanoi*. The tryptophan derivatives are potent false feedback inhibitors of anthranilate synthetase, and strains lacking tryptophan monooxygenase are inhibited by the compounds.

Loss of pIAA1 and Effects on Physiology

Loss of pIAA1 and IAA production does not affect the ability of *P. savastanoi* IAA^- mutants such as 2009-3 to grow in host tissue at least up to 16 days and does not affect the ability of the mutant to grow in culture (1). Indeed, the mutant grows as well as its parent under both conditions. Ability to grow in the host and expression of virulence therefore are separate phenomena and genes on pIAA1 are not concerned with the ability of the bacterium to grow in its host. However, it is not certain how well IAA$^-$ mutants survive in nature. The availability of such mutants allows tests for survivability to be carried out under field conditions.

The loss of IAA production and virulence without loss of viability illustrates a common phenomenon displayed by many plant pathogenic bacteria and fungi, i.e., spontaneous loss of virulence without loss of viability in culture.

The rationale for relating expression of virulence with secondary metabolism is illustrated by these studies. Loss of plasmids concerned with specialized functions likewise illustrates the concept of the "genetic plasticity" associated with plasmid borne determinants. The mobility associated with plasmid borne determinants has not been investigated in this bacterium; thus the possibility that pIAA1 is self transmissible deserves to be investigated.

It is interesting that IAA genes are on pIAA in oleander strains but on the chromosome of olive strains (5). Since oleander strains generally infect and produce tumors on both oleander and olive plants while olive strains (at least in California) generally show preference for olive plants, location of genes on the chromosome may be associated with reduced host preference. Indeed, most plant pathogenic bacteria show high host selectivity, a major exception being *Agrobacterium tumefaciens* (6). Genes determining host range in this bacterium appear to be borne on the large plasmid (Ti) that is associated with virulence and oncogenicity (7). An important experiment that remains to be done is the ability of Ti-less strains of *A. tumefaciens* to grow or survive in host tissue.

We had earlier mentioned that strains with tryptophan monooxygenase activity are not inhibited by 5-methyl tryptophan, and loss of pIAA1 confers sensitivity to the tryptophan analogue. Thus pIAA1 functionally resembles plasmids bearing genes for antibiotic resistance. Moreover, antibiotic resistance genes are commonly borne on transposons. It has not been determined if *iaaM* and *iaaH* occur on a transposon.

Other Functions of pIAA1

Wild type oleander strains convert IAA to IAA-lysine (8). The enzyme requires ATP and a divalent metal cation for activity. *P. savastanoi* mutants such as 2009-3 when lacking pIAA1 are unable to produce IAA-lysine when administered IAA; IAA$^-$ mutants, in which *iaaM* is inactivated by IS-elements such as IS-51 or IS-52, nevertheless produce IAA-lysine when grown in the presence of IAA (L. Glass and T. Kosuge, unpub. results). Olive strains, however, typically do not produce the conjugate and accumulate high concentration of IAA in culture (Glass and Kosuge, unpub. results). The possibility exists that conversion to IAA-lysine may help control the free pool of IAA in the cell and the amount of IAA excreted into host tissue. Other functions for IAA-lysine synthesis may become apparent as this system is studied in more detail.

Is There A Common Basis for Tumorigenicity?

It is clear that cytokinins and IAA are involved in tumorigenicity of plant pathogens. The earlier work by Skoog and Miller (9) had demonstrated the role of cytokinins and IAA in regulating morphogenesis of tobacco tissue. Both elevated levels and the relative proportions of the two compounds are important in modulating tobacco callus morphogenesis. The relationship between cytokinin and IAA and crowngall tumor morphology has been demonstrated (10, 11, 12, 13, 14). Tumors on tobacco plants inoculated with *A. tumefaciens* mutants constructed by Garfinkel and Nester (1) were analyzed for cytokinins and IAA (10). Mutants inactivated in the tms 1 or tms 2 locus by Tn-5 mutagenesis caused production of tobacco tumors with shoots; those with Tn 5-inactivated tm 2 locus caused formation of rooty tumors with low cytokinin and high IAA contents. Recently, Schröder et al., (15) provided evidence that the Tms 2 gene encodes an enzyme catalyzing the conversion of indoleacetamide to IAA. The enzyme activity was found in crowngall tumor tissue and in *E. coli* cells transformed with a recombinant plasmid containing the tms 2 locus. The indoleacetamide hydrolase activity was absent in untransformed tobacco callus cells and in *E. coli*. Although the in vitro conversion of L-tryptophan to indoleacetamide has not been demonstrated, wild-type crown gall tumor tissue shows a phenotype exhibited by wild-type *P. savastanoi* cells, being resistant to 5-methyltryptophan and sensitive to α-methyltryphotphan. Further, the amino acid sequences of the coding region of the *P. savastanoi* tryptophan monooxygenase gene shares 67% homology with the amino acid sequence of the tms 1 locus of T-DNA (T. Yamada and T. Kosuge, unpub. results). Further, Corbin, Yanofsky, and Nester (unpub. results) have found hybridization between *iaaM* and tms 1 occurs under low stringency conditions.

The picture is emerging that there is common basis for tumorigenicity and the similarity between *iaaM* and the tms 1 locus raises interesting questions concerning the origin of the genes concerned with auxin synthesis.

Genetics and Evolution of iaaM.

The aforementioned similarities between tryptophan monooxygenase and lysine monooxygenase raises interesting questions concerning the evolution of tryptophan monooxygenase. Lysine monooxygenase and δ-aminovaleramide hydrolase function to initiate the catabolism of lysine as a source of carbon (16). They catalyze reactions functionally similar to those catalyzed by tryptophan monooxygenase and indoleacetamide hydrolase. It is tempting to suggest that the tryptophan monooxygenase-indoleacetamide hydrolase system originated from a catabolic system for tryptophan. The loss of genes coding enzymes beyond indoleacetamide hydrolase could have occurred; _iaaM_ and _iaaH_ were conserved under selective pressure for factors conducive to parasitism (17).

Plasmids encoding catabolic functions are discussed elsewhere in this volume. The similarities between pIAA1 and plasmids conferring catabolic functions suggest that studies on plasmids with specialized functions will permit broad concepts to be developed concerning the basic biology and metabolic functions of these interesting replicons.

Recently, Akiyoshi and Nester (unpub. results) have conclusively demonstrated that the tmr locus of T-DNA encodes the transferase which catalyzes the formation of isopentenyl adenine ribotide from 5'-AMP and dimethylallyl pyrophosphate. T-DNA fragments bearing the tmr of locus show homology with a restriction endonuclease fragment from plasmids of _P. savastanoi_. (Corbin, Yanofsky, and Nester, unpub. results). The results would suggest further similarities exist between _P. savastanoi_ and _A. tumefaciens_ T-DNA genes in phytohormone synthesis.

REFERENCES

1. Smidt, M., and T. Kosuge (1978) The role of indole-3-acetic acid accumulation by alpha-methyl tryptophan resistant mutants of _Pseudomonas savastanoi_ in gall formation on oleanders. _Physiol. Plant Pathol._ 13:203-214.
2. Comai, L., and T. Kosuge (1980) Involvement of plasmid deoxyribonucleic acid in indoleacetic acid synthesis in _Pseudomonas savastanoi_. _J. Bact._ 143:950-957.
3. Comai, L., and T. Kosuge (1982) Cloning and characterization of _iaaM_, a virulence determinant of _Pseudomonas savastanoi_. _J. Bact_. 149:40-46.
4. Comai, L., and T. Kosuge (1983) Transposable element that causes mutations in a plant pathogenic _Pseudomonas_. _J. Bact._ 154:1162-1667.
5. Comai, L., G. Surico, and T. Kosuge (1982) Relation of plasmid DNA to indoleacetic acid production in different strains of _Pseudomonas syringae_ p.v. _savastanoi_. _J. Gen. Microbiol._ 128:2157-2163.
6. Nester, E.W., and T. Kosuge (1981) Plasmids in plant hyperplasias. _Ann. Rev. Microbiol._ 35:531-565.
7. Nester, E.W., M.P. Gordon, R.M. Amasino, and M.F. Yanosky (1984) Crown gall: A molecular and physiological analysis. _Ann. Rev. Plant Physiol_. 35:387-413.

8. Hutzinger, O., and T. Kosuge (1968) Microbial synthesis and degradation of indole-3-acetic acid. III. The isolation and characterization of indole-3-acetyl-Σ-L-lysine from culture filtrates of Pseudomonas savastanoi. Biochemistry 7:601-605.
9. Skoog, F., and C.O. Miller (1957) Chemical regulation of growth and organ formation in plant tissues cultured in vitro. Symp. Soc. Exp. Biol. XI, pp. 118-131.
10. Akiyoshi, D., R. Morris, R. Hinz, B.S. Mischke, T. Kosuge, D.J. Garfinkel, M.P. Gordon, and E.W. Nester (1983) Cytokinin-auxin balance in crown gall tumors is regulated by specific loci in the T-DNA. Proc. Natl. Acad. Sci., USA, 80:407-411.
11. Garfinkel, D.J., and E.W. Nester (1980) Agrobacterium tumefaciens mutants affected in crown gall tumorigenesis and octopine catabolism. J. Bact. 144:732-743.
12. Garfinkel, D.J., R.B. Simpson, L.W. Ream, F.F. White, M.P. Gordon, and E.W. Nester (1981) Genetic analysis of Crown Gall: structure map of the T-DNA by site directed mutagenesis. Cell 27:143-153.
13. Ooms, G., P.J.J. Hooykas, G. Moolenaar, and R.A. Schilperoot (1981) Crown gall plant tumors of abnormal morphology, induced by Agrobacterium tumefaciens carrying mutated octopine Ti-plasmids; analysis of T-DNA functions. Gene 14:33-50.
14. Amasino, R.M., and C.O. Miller (1982) Hormonal control of tobacco Crown Gall Tumor Morphology. Plant Physiol. 69:389-392.
15. Schröder, G., S. Waffenschmidt, E.W. Weiler, and J. Schröder (1983) The T-region of Ti plasmids codes for an enzyme synthesizing indole-3-acetic acid. European J. Biochem. 138:387-391.
16. Takeda, H., and O. Hayaishi (1966) Crystalline L-lysine oxygenase. J. Biol. Chem. 241:2733-2736.
17. Comai, L., and T. Kosuge (1983 The genetics of indoleacetic acid production and virulence in Pseudomonas savastanoi. In Molecular Genetics of the Bacteria-Plant Interaction, A. Pichler, ed. Springer-Verlag Berlin, pp. 362-366.

THE MOLECULAR BASIS OF PLANT CELL TRANSFORMATION BY

AGROBACTERIUM TUMEFACIENS

E. W. Nester,[*] R. Amasino,[*] D. Akiyoshi,[+] H. Klee,[*] A. Montoya[*] and M. P. Gordon[+]

[*] Department of Microbiology and Immunology
[+] Department of Biochemistry
University of Washington
Seattle, Washington 98195

Crown gall is an uncontrolled proliferation of tumor cells incited by Agrobacterium tumefaciens on most dicotyledonous plants. To accomplish this, the bacterium transfers and integrates a portion (the T-region) of its tumor inducing (Ti) plasmid into plant nuclear DNA (1,2,3,4). The transferred DNA (T-DNA) is expressed in the transformed cell and encodes enzymes involved in the synthesis of amino acid condensation products called opines (5,6,7,8,9,10). This paper will report data which suggest that T-DNA also encodes genes which specify the synthesis . . . also encode the synthesis of the phytohormones, Indole acetic acid, Isopentenyl adenosine, Isopentenyl adenosine 5' monophosphate, and Isopentenyl adenine.

Garfinkel et al. (11,12) and Ooms et al. (13,14) isolated Tn5 insertion mutations in the T-DNA region of the Ti-plasmid. They described three tumor morphology loci tms, tmr, tml (Fig. 1): 1.) tms mutants (tumor morphology shoot) incite tumors with shoots emerging from the center of the unorganized tissue (Fig. 1b); 2.) tmr mutants (tumor morphology root) incite tumors with roots emerging from the tumorous growth (Fig. 1c); 3.) tml mutants (tumor morphology large) incite tumors on Kalachöe leaves which are larger, greener, and more amorphous than those incited by the wild-type strain (Fig. 1d). Ooms et al. (14,15) suggested that the morphologies exhibited are associated with the level auxin and cytokinin found in the tumors, as addition of auxin to tms-induced tumors suppressed shoot formation. Subsequently, Akiyoshi et al. (16) measured the levels of auxin and cytokinin in tumors induced by tumor morphology mutants. They found that tms-induced tumors contained elevated levels of cy-

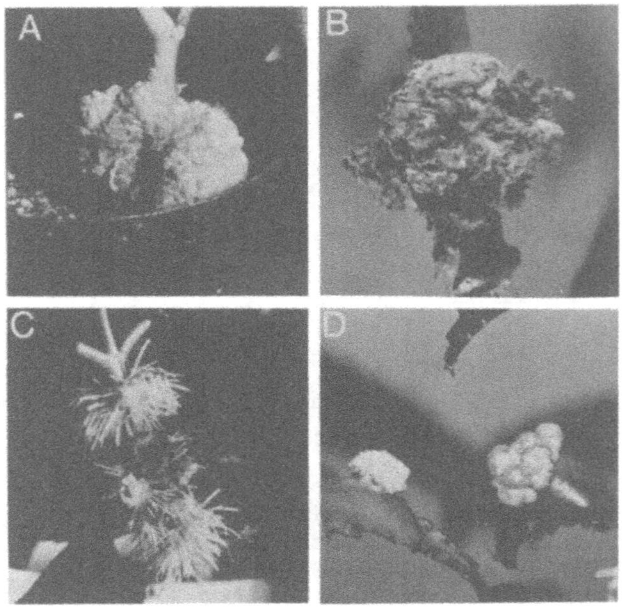

Fig. 1. Crown gall tumors: (A) A wild-type tumor incited by A. tumefaciens strain A6 on Datura stramonium stems. (B) A tumor on Nicotiana tabacum stem incited by a tms mutant of strain A6. (C) Tumors incited by a tmr mutant of strain A6 on the stem of Kalanchöe diagramontium. (D) Tumors incited by strain A6 (right) and a tml mutant of strain A6 (left) on leaves of Kalanchöe diagramontium.

tokinins and decreased levels of auxin while tmr-induced tumors had elevated levels of auxin and decreased levels of cytokinin when compared to wild-type tumors.

Willmitzer et al. (5) demonstrated that 2 polyadenylated transcripts are encoded by the tms loci. Schröder et al. (17,18) showed that the tms2 locus could be translated and transcribed in an Escherichia coli mini cell system to give a protein product of 49,000 daltons. This agrees with the protein product predicted by sequence data of the locus (19,20). The sequence data predict a protein product of 83,769 d for the tms1 locus. The sequence data of tms1 indicate a significant level of homology to the adenine binding region of p-hydroxybenzoate hydroxylase from Pseudomonas fluorescens suggesting that this protein binds adenine either as a substrate or as cofactor.

Schröder et al. (18) have demonstrated that the tms2 gene is expressed in both eukaryotic and prokaryotic systems. Extracts of

E. coli and Agrobacterium expressing the tms2 gene hydrolyzed indole-3-acetamide into indole-3-acetic acid while the same activity was not found in strains lacking this gene (18). They also demonstrated that extracts from crown gall cells exhibited an enzymatic activity which converted indole-3-acetamide to indole-3-acetic acid while no activity was detected in extracts of nontransformed cells. Inze and coworkers (21) showed that coinfection of tms1 or tms2 mutants with strains which contained only the tms2 or tms1 gene, respectively, produced wild-type tumors. These data show that diffusable gene products of the tms1 and tms2 genes can interact within the tumor cells to produce the same effect that external addition of α-naphthalene acetic acid produces, namely a wild-type tumor. The addition of α-napthalene acetamide to tumors incited by tms1 mutants restores the wild-type phenotype, suggesting that gene 1 encodes an enzyme involved in the synthesis of an intermediate in the production of the auxin indole-3 acetic acid. Thus, it can be concluded that the tms1 and tms2 loci encode enzymes involved in the synthesis of the plant phytohormone indole-3-acetic acid.

The correlation of shoot production with an elevated cytokinin to auxin ratio and the involvement of the tms loci with auxin production suggests that the tmr locus is involved with the production of cytokinin. This, indeed, is the case. Recently Akiyoshi et al. (22,23) have demonstrated that the tmr gene encodes an enzyme of cytokinin biosynthesis, dimethylallylpyrophosphate: AMP dimethylallyl transferase (DMAPP transferase). Initially they demonstrated DMAPP transferase activity in tobacco crown gall tumors incited by the strain A6 while no activity was detected in normal tobacco tissue (Tab. 1). This table also shows that DMAPP transferase activity is present in elevated levels in a tumor incited by a tms mutant of A6 (tms163) and is absent in a tumor incited by the A6 tmr mutant tmr149.

Recently, work by D. Akiyoshi in our laboratory has demonstrated DMAPP transferase activity in cell-free extracts of E. coli which contain a cloned tmr gene (23). The enzymatic activity is detected both when the gene is controlled by its own promoter and when it is fused to the lac promoter of pUC18 (24). Table 2 details the activity detected with the following T-DNA clones. pTiA6 EcoRl fragment 7 was cloned into PUC 18 in both orientations. This fragment contains the tms1, tmr, and tml loci. To determine if the activity detected was due to the tmr locus or one of the other loci, clones were constructed which removed the other loci. EcoRl fragment 7 cloned into pUC18 in orientation I, the direction of transcription of lac, tms1, tmr, and transcript 6a all being the same, was cut with the restriction endonuclease KpnI and then recircularized by ligation removing transcripts 6a and 6b of the tml locus to give clone pDQ14. The clone pDQ64 was constructed by taking EcoRI fragment 7 cloned into pUC18 in orientation II, cutting with the restriction endonuclease SalI and then recircularizing the molecule

Tab. 1. DMAPP transferase activity in tobacco crown gall tumors.

Plasmid	Enzyme Activity	
	Radioactivity in iP/iPA region (cpm)	Units*
Nicotiana tabacum var Xanthi Normal Callus	N.D.[++]	-
A6S/2[+] tumor	8,950	7.6
tmr-149 tumor	N.D.	-
tms-393 tumor	31,830	27.2

* Units are defined as the synthesis of 1 fmol (^{3}H) isopentenyladenie/isopentenyladenosine per min. at 22°C (1 fmol = 13 cpm).

\+ *N. tabacum* cv White Burley. Other tumor lines were incited on *N. tabacum* cv Xanthi nc.

++ N.D. = not detected.

thereby removing the 5' end of the *tms1* gene (25,26). The final construct, pRA120, was a fusion of the *tmr* gene to the α peptide of β-galactosidase. It was constructed by inserting a restriction fragment with a blunt end at the first nucleotide of the second condon of the *tmr* gene extending to the BamHI site between fragments 29 and 19 of pTiA6, into SmaI/BamHI cut pUC18. This plasmid was then cut with BamHI and SalI and ligated to a BamHI/SalI restriction fragment from a mutant A6 Ti-plasmid with a Tn5 insertion in transcript 6a within the BamHI fragment 19. The resulting plasmid contained the entire *tmr* gene linked to the first ten codons of the α peptide of β-galactosidase. The resulting protein has 10 additional amino acids at the amino terminus of the gene and should be under the regulation of the *lac* promoter on pUC18.

Enzyme activity was then measured in strains containing EcoRI fragment 7 cloned in the wide host range vector pRK290 (27) (Tab. 3). No activity was detected in extracts of a Tn5 insertion mutant within the *tmr* gene coding region (*tmr*149, Tab. 3). However, a low level of activity was detected in another mutant which gave a *tmr*

Tab. 2. Enzymic activity with T-DNA clones.

Plasmid	Enzyme Activity	
	Radioactivity in iP/iPA region (cpm)	Units*
pUC18	N.D.$^{+}$	-
pUC18 EcoRI 7 (orientation I)	7,730	6.6
pDQ14	14,900	12.7
pUC18 EcoRI 7 (orientation II)	4,190	3.6
pDQ64	2,750	2.4
pRA120	10,100	8.6

* Units are defined as the synthesis of 1 fmol (^{3}H) iP/iPA per min at 22°C. (1 fmol = 13 cpm).

$^{+}$ N.D. = not detected.

phenotype on plants. When this mutant 169 was mapped, it was located 20 nucleotides past the termination codon for the protein but before the polyadenylation sites of the gene. Thus, these studies establish that the _tmr_ gene produced is an enzyme involved in the biosynthesis of phytohormones with cytokinin activity.

From these studies, it has been established that the _tms1_ and _tms2_ loci encode enyzmes involved in the production of the plant phytohormone Indole-3-acetic acid and that the _tmr_ locus encodes an enzyme of cytokinin biosynthesis. Thus, the transformation to a tumorous phenotype is the result of phytohormone production induced by the T-DNA which _A. tumefaciens_ has inserted into the plant nuclear genome. The T-DNA also encodes the synthesis of compounds called opines, which can be utilized by the bacterium as sources of carbon and nitrogen creating a unique ecological niche for the bacterium. Therefore, the transformation of plant cells by _Agrobacterium tumefaciens_ is a natural case of genetic engineering.

Tab. 3. Enzyme activity measured in strains containing fragments cloned in the wide host range vector.

Plasmid	Enzyme Activity	
	Radioactivity in iP/iPA region (cpm)	Units*
pRK290	20	<<0.1
pRK290-*EcoRI7*::Tn5(349)	10,820	9.3
pRK290-*tmr*-149:Tn5	80	<0.1
pRK290-*tmr*-169:Tn5	2,400	2.0

* Units are defined as the synthesis of 1 fmol (^{3}H) iP/iPA per min at 22°C. fmol - 13 cpm).

ACKNOWLEDGMENTS

The above research was supported by a grant from Standard Oil (Indiana), National Institutes of Health grant #R01-GM32618-12, and American Cancer Society grant #NP-336A.

REFERENCES

1. Bevan, M.W., and M.-D. Chilton (1982a) T-DNA of the *Agrobacterium* Ti- and Ri-plasmids. *Annu. Rev. of Genet.* 16:357-384.
2. Depicker, A., M. Van Montagu, and J. Schell (1983) Plant cell transformation by *Agrobacterium* plasmids. In *Genetic Engineering of Plants: An Agricultural Perspective*, T. Kosuge, C.P. Meredith, A. Hollaender, eds. Plenum Publishing Corp., New York, pp. 143-176.
3. Hooykaas, P.J.J., and R.A. Schilperoort (1984) The molecular genetics of crown gall tumorigenesis. *Adv. Genet.* Vol. 22 (in press).
4. Nester, E.W., and T. Kosuge (1981) Plasmids specifying plant hyperplasias. *Ann. Rev. Microbiol.* 35:531-565.
5. Willmitzer, L., G. Simons, and J. Schell (1982) The T-DNA in octopine crown gall: The T-DNA in octopine crown gall tumors codes for seven well-defined polyadenylated transcripts. *EMBO J.* 1:139-146.

6. Gelvin, S.B., M.F. Thomashow, J.C. McPherson, M.P. Gordon, and E.W. Nester (1982) Sizes and map positions of several plasmid-DNA-encoded transcripts in octopine-type crown gall tumors. Proc. Natl. Acad. Sci., USA 79:76-80.
7. Gurley, W.B., J.D. Kemp, M.J. Albert, D.W. Sutton, and J. Callis (1979) Transcription of Ti-plasmid-derived sequences in three octopine-type crown gall tumor lines. Proc. Natl. Acad. Sci., USA 76:2828-2832.
8. Bevan, M.W., and M.-D. Chilton (1982b) Multiple transcripts of T-DNA detected in nopaline crown gall tumors. J. Mol. Appl. Genet. 1:539-546.
9. Schröder, J., G. Schröder, H. Huisman, R.A. Schilperoort, and J. Schell (1981) The mRNA for lysopine dehydrogenase in plant tumor cells is complementary to a Ti-plasmid fragment. FEBS Lett. 129:166-170.
10. Murai, N., and J.D. Kemp (1982) Octopine synthase mRNA isolated from sunflower crown gall callus is homologous to the Ti-plasmid of Agrobacterium tumefaciens. Proc. Natl. Acad. Sci., USA 79:86-90.
11. Garfinkel, D., and E.W. Nester (1980) Agrobacterium tumefaciens mutants affected in crown gall tumorigenesis and octopine catabolism. J. Bact. 144:732-743.
12. Garfinkel, D.J., R.B. Simpson, L.W. Ream, F.F. White, M.P. Gordon, and E.W. Nester (1981) Genetic analysis of crown gall: Fine structure map of the T-DNA by site-directed mutagenesis. Cell 27:143-153.
13. Ooms, G., P. Klapwijk, J. Poulse, and R.A. Schilperoort (1980) Characterization of Tn904 insertions in octopine Ti-plasmid mutants of Agrobacterium tumefaciens. J. Bact. 144:82-91.
14. Ooms, G., G. Hooykaas, G. Moolenaar, and R.A. Schilperoort (1981) Crown gall plant tumours of abnormal morphology induced by Agrobacterium tumefaciens carrying mutated octopine Ti-plasmids. Gene 14:33-50.
15. Ooms, G., L. Molendijk, and R.A. Schilperoort (1982) Double infection of tobacco plants by two complementing octopine T-region mutants of Agrobacterium tumefaciens. Plant. Mol. Biol. 1:217-221.
16. Akiyoski D., R. Morris, R. Hinz, B. Mischke, T. Kosuge, D. Garfinkel, M. Gordon, and E. Nester (1983) Cytokinin auxin balance in crown gall tumors is regulated by specific loci in the T-DNA. Proc. Natl. Acad. Sci., USA 80:407-411.
17. Schröder, G., W. Klipp, A. Hillebrand, R. Ehring, C. Koncz, and J. Schröder (1983a) The conserved part of the T-region in Ti-plasmids expresses four proteins in bacteria. EMBO J. 2:403-409.
18. Schröder, G., S. Waffenschmidt, E.W. Weiler, and J. Schröder (1983) The T-region of Ti-plasmids codes for an enzyme synthesizing indole-3-acetic acid. Eur. J. Biochem. 138:387-391.
19. Barker, R.F., K.B. Idler, D.V. Thompson, and J.D. Kemp (1983) Nucleotide sequence of the T-DNA region from Agrobacterium tumefaciens octopine Ti-plasmid pTi15955. Plant Mol. Biol. 2:335-350.

20. Klee, H., A. Montoya, F. Horodyski, C. Lichtenstein, D. Garfinkel, S. Fuller, C. Flores, J. Peschon, E. Nester, and M. Gordon (1984) Nucleotide sequence of the tms genes of the pTiA6NC octopine Ti-plasmid: Two gene products involved in plant tumorigenesis. Proc. Natl. Acad. Sci., USA 81:1728-1732.

21. Inze, D., A. Follin, M. Van Lijsebettens, C. Simoens, C. Genetello, M. Van Montague, and J. Schell (1983) Genetic analysis of the individual T-DNA genes of Agrobacterium tumefaciens; Further evidence that two gene are inolved in indole-3-acetic acid synthesis. Mol. Gen. Genet. (in press).

22. Akiyoshi, D.E., and R.O. Morris (1984a) Effect of T-DNA copy number on cytokinin content in crown gall. Planta (submitted for publication).

23. Akiyoshi, D.E., H. Klee, R.M. Amasino, E.W. Nester, and M.P. Gordon (1984) T-DNA of Agrobacterium tumefaciens codes an enzyme of cytokinin biosynthesis. Proc. Natl. Acad. Sci., USA (in press).

24. Norrander, J., T. Kempe, and J. Messing (1983) Construction of improved M13 vectors using oligodeoxynucleotide-directed mutagenesis. Gene 26:101-106.

25. Heidekamp, F., W.G. Kirkse, J. Hille, and H. van Ormondt (1983) Nucleotide sequence of the Agrobacterium tumefaciens octopine Ti-plasmid-encoded tmr gene Nucleic Acids Res. 11:6211-6223.

26. Lichtenstein, C., H. Klee, A. Montoya, D. Garfinkel, S. Fuller, C. Flores, E. Nester, and M. Gordon (1984) Nucleotide sequence and transcript mapping of the tmr gene of the pTiA6NC octopine Ti-plasmid: A bacterial gene involved in plant tumorigenesis. J. Mol. Appl. Genet. 2:354-362.

27. Ditta, G., S. Stanfield, D. Corbin, and D.R. Helinski (1981) Broad host range DNA cloning system for gram-negative bacteria: Construction of a gene bank of Rhizobium meliloti. Proc. Natl. Acad. Sci., USA 77:7347-7351.

POST-TRANSCRIPTIONAL REGULATION OF CHLORAMPHENICOL ACETYL TRANSFERASE

Woo-Hyeon Byeon and Bernard Weisblum

Pharmacology Department
University of Wisconsin Medical School
Madison, Wisconsin 53706

QUESTIONS

This report describes some of our recent work which extends previous studies of post-transcriptional control of antibiotic resistance (8,9,10). Our previous efforts were directed towards elucidating the post-transcriptional mechanism by which erythromycin induces the synthesis of an N-methyl transferase that in turn, catalyzes the post-transcriptional methylation of adenine in 23S rRNA, a process that leads to reduced affinity between erythromycin and its target, the ribosome.

The central questions that we ask in these studies are: 1) How does the cell perceive the presence of the inducing antibiotic? i.e., What molecular species serves as the sensor of the antibiotic?, and 2) Upon sensing the presence of the inducer, how does the cell link the information to the process that ultimately increases gene expression? Our present studies examine these questions in relation to the chloramphenicol-induced synthesis of chloramphenicol acetyl transferase, an enzyme that confers resistance by inactivating chloramphenicol.

SUMMARY

The +1 site for initiation of inducible chloramphenicol acetyl transferase (CAT) messenger ribonucleic acid (mRNA) encoded by plasmid pC194 was determined experimentally using γ-^{32}P-ATP-labeled run-off transcripts partially digested with T1 ribonuclease. By partial digestion of the in vitro transcripts with S1-, T1-, and cobra venom nucleases as probes of mRNA conformation, single- and double-stranded regions, respectively, were also identified. Thus, a prominent inverted complementary repeat sequence was demonstrated spanning the +14 to +50 positions which contain the complementary sequences CCUCC, and GGAGG (the Shine and Dalgarno sequence for synthesis of CAT) symmetrically apposed and paired as part of a perfect 12 bp inverted complementary repeat sequence (-19.5 kcal/mol).

The CAT mRNA was stable to digestion by T1 ribonuclease at the 4 guanosine residues in the Shine and Dalgarno sequence GGAGG, even at 60°C, suggesting that nascent CAT mRNA allows ribosomes to initiate protein synthesis inefficiently and that induction involves post-transcriptional unmasking of the Shine and Dalgarno sequence. Consistent with this model of regulation, we found that cells carrying pC194, induced with chloramphenicol (CAM), contain about the same concentration of pulse labeled CAT-specific RNA as do uninduced cells.

Induction of CAT synthesis by the nonacetylatable CAM analog fluorothiamphenicol was tested using minicells of _Bacillus subtilis_ carrying pC194 as well as minicells containing the cloned pC194 derivatives in which parts of the CAT structural gene were deleted in vitro using Bal31 exonuclease. Optimal induction of both full-length (active) and deleted (inactive) CAT required similar concentrations of fluorothiamphenicol, whereas induction by CAM required a higher concentration for the wild-type full-length (active) CAT than for the (inactive) deleted CAT. Because synthesis of deleted CAT was inducible, we infer that CAT plays no direct role in regulating its own synthesis.

EXPERIMENTAL

DNA Sequence of the Structural Gene for CAT Including the Promoter and Putative Transcription Terminator

The -35 and -10 sequences have been identified on the basis of similarity to the respective consensus sequences in _E. coli_ (17,20) and the +1 site for transcription initiation was determined experimentally in the present study. The structural gene sequence is flanked by an inverted complementary repeat sequence at the 5' end overlapping the CAT Shine and Dalgarno sequence and postulated to act as a post-transcriptional regulatory element, and by an inverted complementary repeat sequence at the 3' end postulated to act as a rho factor-independent transcription terminator. An additional inverted complementary repeat region overlapping the -35 region of the CAT promoter and of unknown function is also shown. Sites within the structural gene for digestion by _DdeI_ and _NcoI_ restriction endonucleases are indicated. Nucleotides are numbered above the text line and amino acids are numbered using unsigned bracketed numbers below the text line. Dashed lines indicate inverted complementary repeat regions. SD, denotes Shine and Dalgarno sequence.

Autoradiographic Analysis of Digestion Products Obtained Using In Vitro CAT Transcripts Partially Digested with Nucleases as Probes of Conformation

DNA template for transcription was prepared by digesting plasmid pC194 DNA sequentially with *MspI* and *DdeI*, yielding 5 fragments (1651, 609, 452, 160, and 40 bp in length) which were separated by electrophoresis on a 5% polyacrylamide gel. The third largest fragment, 452 bp in length, was located after ethidium bromide staining of the gel, electroeluted, and used as DNA template for in vitro synthesis of CAT mRNA transcripts. Synthesis of CAT mRNA transcript was performed in a 100 μl reaction mixture containing: 20 mM Tris HCl, pH 7.9; 5 mM $MgCl_2$; 100 mM KCl; 0.1 mM DTT; 0.1 mM Na_2EDTA; GTP, CTP, UTP, 0.3 mM each; 0.12 mM unlabeled ATP; γ-^{32}P- labeled ATP, 2.5 mCi, specific activity greater than 7,500 Ci/mM; and 6 μg *E. coli* RNA polymerase. Incubation was for 60 min. at 37°C. At the end of the incubation period 4 μl of 10% Na dodecylsulfonate (SDS) and an equal volume of stop buffer (9 M Urea; 25 mM Na_2EDTA; 0.1% bromophenol blue; and 0.1% xylene cyanol) was added. The sample was

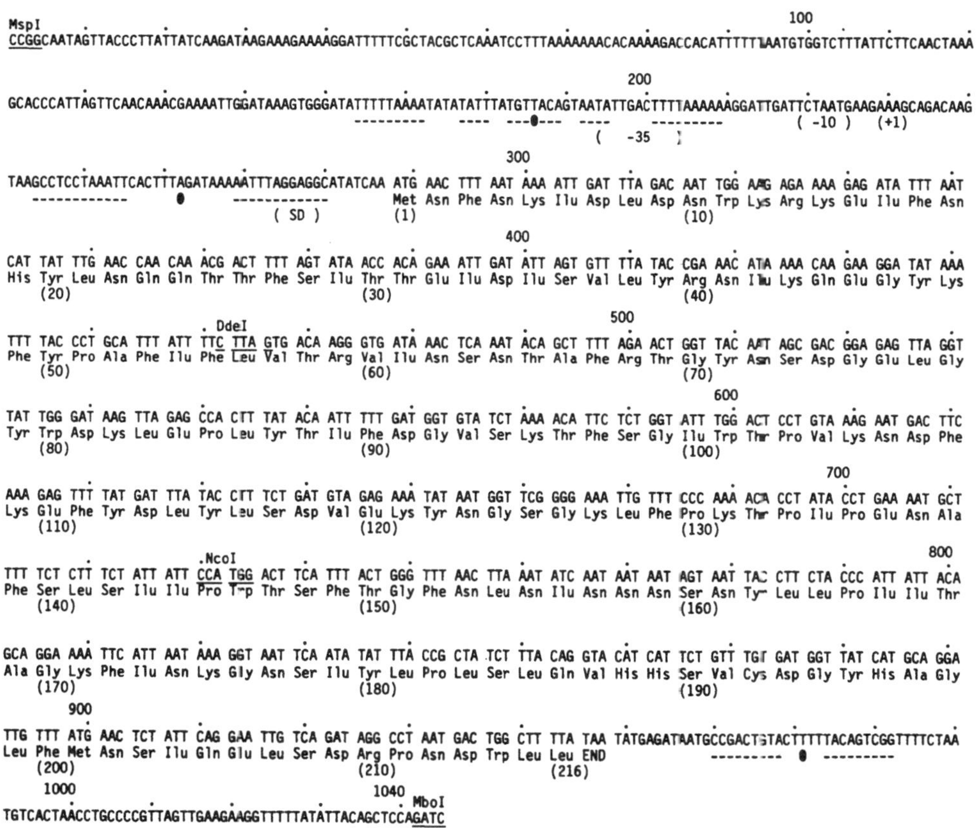

heated at 75°C for 5 min and was applied to a preparative polyacrylamide gel (8% acrylamide, 0.4% bisacrylamide) containing tris-EDTA-borate (TEB), (0.089 M tris base, 0.089 M boric acid, and 2 mM Na_2EDTA, pH 8.3).

The procedures described by Vournakis et al. 1981 (23) for analysis of end-labeled RNA transcripts were used. Lanes A-F were run on 10% polyacrylamide gel slabs to show structural features close to the CAT structural gene, whereas lanes G-K were run on 20% polyacrylamide gel slabs to show smaller fragments corresponding to structural features beginning at +1. (A) undigested control, (B,F,G, and K) T1 digests, (C and H) alkali digest standards, (D and I) S1 digests, (E and J) Cobra venom digests. Lane K contains the same sample shown in lane G, but overexposed to demonstrate the presence of 4 weak bands corresponding to the 4 inaccessible guanosine residues in the Shine and Dalgarno sequence.

The figure below summarizes, autoradiographic data presented above showing the location of transcription initiation based on T1 mapping of in vitro transcripts, and location of regions with conformation in accessible single-stranded- (T1- and S1-sensitive), or double-stranded form (CV- i.e., cobra-venom-sensitive), regions.

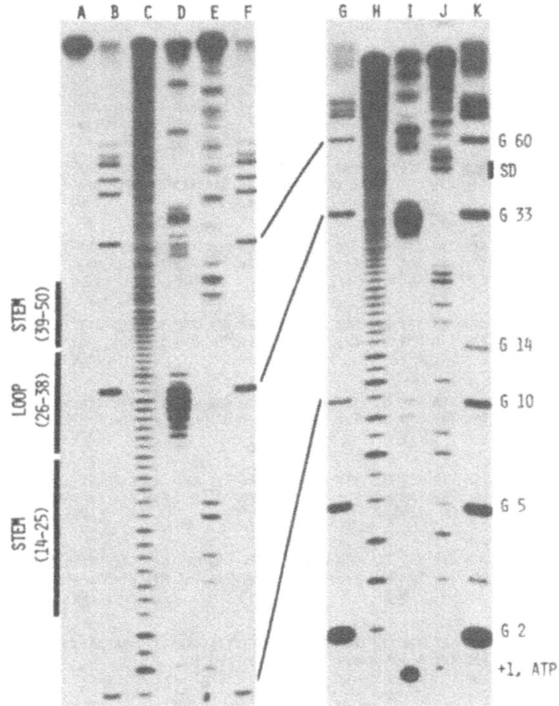

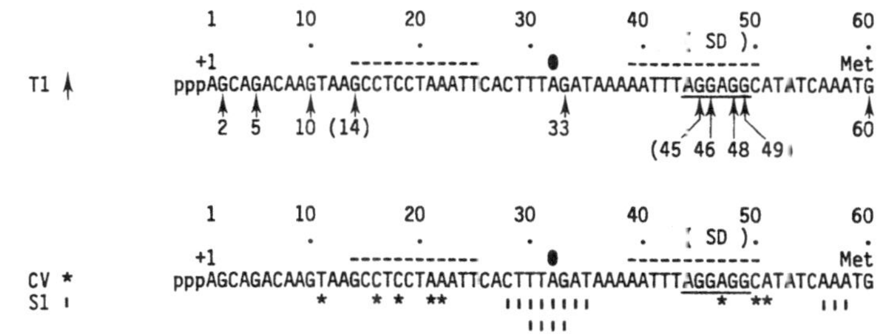

Autoradiographic Analysis of Proteins Synthesized by B. subtilis Minicells Carrying Plasmid pC194 DNA and Deletion Mutations in the CAT Structural Gene

Proteins were synthesized by *B. subtilis* minicells under the following conditions. Minicells of *B. subtilis* strain BGSC 1A197 carrying plasmids as indicated were prepared by centrifugation in sucrose density gradients as described by Mertens and Reeve (13). For analysis of the products of plasmid-directed protein synthesis, minicells were incubated for 30 min at 37°C in 0.5 ml minimal medium supplemented with 50 μCi ^{35}S-methionine, as described by Shivakumar, Hahn, and Dubnau (19). After 30 min incubation, labeled minicells were collected by centrifugation at 10,000 X g for 10 min, washed with minimal medium, and resuspended in 0.1 ml solution containing: 50 mM tris HCl, pH 8.0; 1 mM Na_2EDTA; and 25% w/v sucrose. Lysozyme, 0.1 mg was added and the mixture was incubated at 37°C for 15 min following which the minicells were lysed by addition of cracking buffer, and the resultant proteins fractionated by polyacrylamide gel electrophoresis. (A,B) pE194 *cop-6*, uninduced, and induced with 0.1 μg/ml erythromycin; (C,D) pC194, uninduced, and, induced with 0.3 μg/ml chloramphenicol; (E,F) pBW18, uninduced, and induced with 0.3 μg/ml chloramphenicol; (G,H) pBW18 del 9, uninduced, and induced with 0.3 μg/ml chloramphenicol; (I,J) pBW18 del 10, uninduced, and induced with 0.3 μg/ml chloramphenicol.

Autoradiographic Analysis of CAT and its Mutant Forms Synthesized Under Induction by Chloramphenicol and Fluorothiamphenicol

Synthesis of intact and deleted forms of chloramphenicol acetyltransferase by *B. subtilis* minicells as a function of inducer and of inducer concentration was next studied. Minicell preparations containing (A) pBW18; (B) pBW18 del 10; and (C) pBW18 del 9 were induced with a range of chloramphenicol (CAM) or fluorothiamphenicol (FTAM) concentrations at the concentrations shown. General conditions of incubation and sample workup were as in the previous section.

To the extent that protein bands other than CAT have comparable intensities in the autoradiogram, synthesis of the deleted CAT ap-

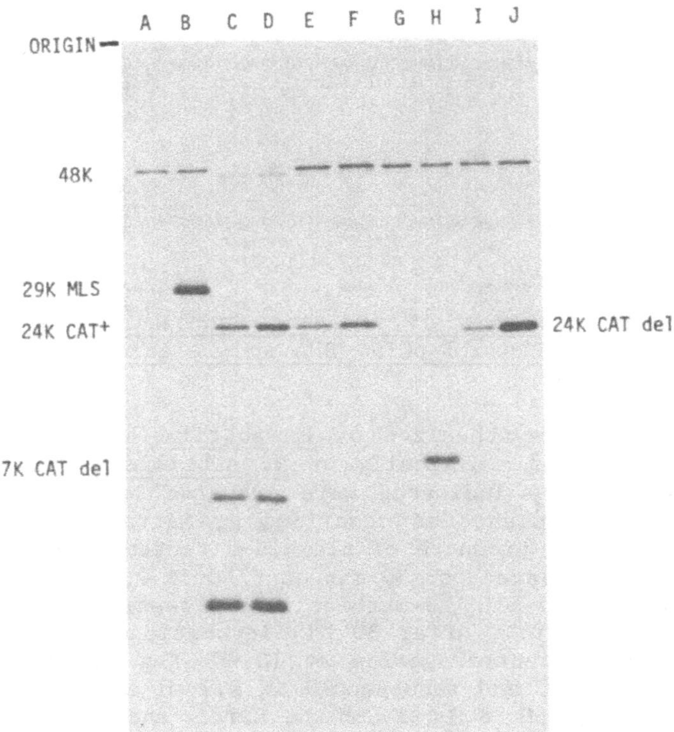

peared to be hyperinducible, first with respect to the lower background seen in the uninduced sample, and second with respect to the higher absolute intensity of the CAT band seen in the autoradiograms. A possible interpretation of this observation would be that since low (subinhibitory) concentrations of CAM were used for induction, the basal level of CAT expression in pBW18 inactivated part of the already low CAM concentration, whereas deleted forms of CAT were unable to do so for lack of enzymatic activity.

To test this hypothesis, we measured the ability of FTAM to induce CAT synthesis. FTAM can serve as a useful probe of induction because it contains $-CH_2F$ instead of $-CH_2OH$ at the 3-position and is therefore resistant to inactivation by the acetylation reaction which initially derivatizes the 3-OH position (21). In these determinations a series of inducer concentrations bracketing the minimum inhibitory concentration was used. As shown in the figure, induction by CAM, minimum inhibitory concentration (m.i.c.) = 10 μg/ml, progressively reached a maximum induction level in the 1-3 μg/ml range, whereas cells induced with FTAM (m.i.c.) = 1 μg/ml were maximally induced to a comparable level over the 0.01-0.3 μg/ml range.

If the apparent difference in inducing potency could be ascribed to acetylation of CAM, we would expect a shift in the concentration of CAM required for induction of (enzymatically inactive) deleted forms of CAT and no shift in the concentration of CAM required for induction. As shown in Fig. 5B and C for both the 24K deleted CAT (encoded by pBW18 del 10) and 17K deleted CAT (encoded by pBW18 del 9), this was found to be the case. We therefore infer that acetylation of CAM inactivates both antibiotic and inducing activities, and that both inducing activity and antibiotic activity are inseparable.

DISCUSSION

The 5' end of the CAT mRNA contains a 37-base inverted complementary repeat sequence, which in the associated form is capable of sequestering the translation initiation sequence of CAT. In this conformation the leader region would associate to form 12 perfectly matched sequential base pairs and experimental data obtained in this study indicate that the region possesses a high degree of thermal stability insofar as the 4 guanosine residues comprising the Shine and Dalgarno sequence are relatively unavailable for interaction

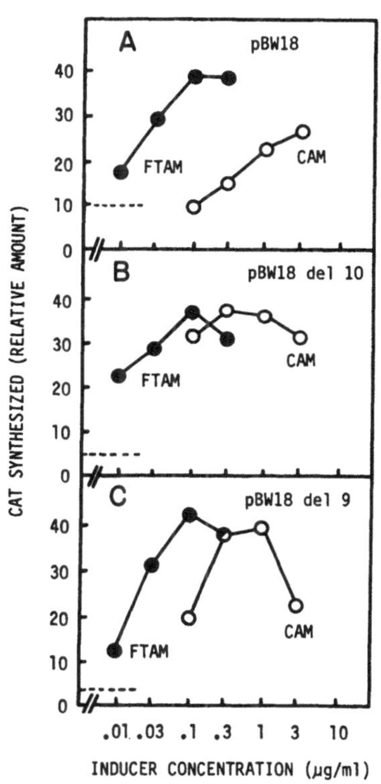

with T1 ribonuclease, even at 60°C. These findings accord with the free energy of association calculated by empirical formalisms (3,22) which yield an estimate of -19.5 kcal/mol, suggesting a high degree of thermal stability for this region of CAT mRNA.

Inverted complementary repeat sequences near the site of transcription initiation could serve as regulatory protein binding sites which, acting at the DNA level might function as a regulator of transcription. In fact, a prominent inverted complementary repeat region which spans residues 144 to 209, centering on residue 186 is shown in Fig. 1. This region overlaps the putative -35 sequence of the CAT promoter but does not extend into the messenger portion. Such a sequence could, in principle, serve as an operator for binding of a regulatory protein. However, at present we are unable to assign a primary regulatory function to this sequence. We feel that transcriptional control at the level of induction by CAM as the principal regulatory mechanism is unlikely because nascent CAT message synthesized after transcriptional induction would have a high probability of folding and assuming the stable inactive conformation shown in Fig. 8. Such a model would require 2 induction steps for regulation of CAT expression, one for transcriptional activation and one for translational activation. The simplest interpretation would be that regulation of CAT by CAM takes place at the level of translation.

A mechanism for induction must therefore describe how a comparatively stable association in the CAT message leader sequence is either disrupted or prevented from forming under inducing conditions. In the case of inducible erythromycin resistance (7,8,9), a translation initiation sequence is likewise masked and mRNA realignment is enabled by an intramolecular rearrangement of the erythromycin resistance rRNA methylase mRNA leader sequence. Unlike the control region for MLS resistance, the CAT regulatory region contains only a single inverted complementary repeat sequence with no apparent preemptor or associated control peptide characteristic of transcriptional or translational attenuators.

A second class of control mechanisms to consider, e.g.. mechanisms which autogenously regulate synthesis of phage T4 gene 32 product (5) or of ribosomal proteins (15), acts at the level of translational repression. According to this type of mechanism, CAT would bind directly to and thereby alter stability of the associated inactive conformation of CAT mRNA; CAM and CAT protein would thus function as inducer and repressor or activator, respectively. The various subcloning, resection, and insertion experiments suggest that the active CAT enzyme plays no direct role as a regulator of its own expression, which raises the question whether CAT protein plays any direct regulatory role at all. Studies of Lovett and coworkers (4,24,25) using a cognate inducible CAT from *Bacillus pumilus* subcloned into *B. subtilis* favor this idea. In their studies it was reported that trpC and DHFR inserted into the *Bacillus pumilus* CAT structural gene downstream from the CAT ribosomal binding site were inducible by CAM; on the other hand, DHFR inserted upstream from the ribosome binding site was expressed constitutively.

A further parallel in the comparison between induction of CAT and of MLS resistance is in the association between the activity of the antibiotic as inducer of resistance and activity as an inhibitor of protein synthesis. 3-deoxy CAM has been reported by Winshell and Shaw (26) to possess activity as a gratuitous inducer of CAT. Although the m.i.c. of 3-Deoxy CAM is low, (10 μg/ml), it nevertheless possesses inhibitory activity indicative of ability to interact with the ribosome. In the MLS system gratuitous inducers have not been found. Pestka et al. (16) in studying a series of erythromycin derivatives, observed that ability to induce and ability to inhibit protein synthesis were inseparable. A key feature in the translational attenuation model for regulation of MLS resistance is the binding of the inducing antibiotic to the ribosome during induction, causing the ribosome to stall at a critical site in the leader sequence of the message.

The observation that the pC194 CAT gene with its promoter retains inducible expression after introduction into *E. coli* cells suggests further that any other participants in induction might be ordinary cell constituents. The ribosome remains as the only other known cellular constituent capable of binding chloramphenicol specifically. We have examined the 23S rRNA primary structure to see whether a subsequence capable of interacting with the CAT mRNA inverted complementary repeat sequences might be present. The 23S rRNA contains 7 occurrences of the subsequence GGAGG which corresponds to the canonical Shine and Dalgarno sequence (18). One of these occurs in association with the complementary sequence CCUCC, as part of inverted complementary repeat sequence, number 78, described by Noller et al. (14). Examination of this region in 23S rRNA further reveals an exact match between both the CAT mRNA and inverted complementary repeat sequence number 78 extending over the 9 nucleotide sequence 5'CCUCCUAAA, and in both cases pairing between CCUCC and GGAGG is physically possible. We therefore speculate that 23S rRNA contains a part of the information necessary for activation of CAT mRNA.

According to the model that we propose for CAT induction: (i) The rate of CAT mRNA synthesis is unaffected by the presence of the inducer. (ii) The 50S ribosome subunit serves as the macromolecular sensor of CAM for regulation of CAT synthesis. (iii) CAM forms a specific complex with the 50S subunit which alters the conformation of the latter. (iv) As a result of CAM-ribosome complex formation, the paired sequence CCUCCUAAA:UUUAGGAGG at position 2260 of *E. coli* 23S rRNA becomes functionally accessible to interact with the nascent CCUCCUAAA during CAT mRNA synthesis. (v) The sequence GGAGG in the nascent CAT mRNA unpaired with the sequence CCUCC in the CAT mRNA would then be free to interact with CCUCC in 16S rRNA with consequent formation of initiation complex with 30S subunits. (vi) Finally, for synthesis of CAT, a 50S subunit not complexed with CAM links up with the initiation complex to form a 70S ribosome active in translation. The 50S subunit thus serves the role of mRNA selection when complexed with CAM and for synthesis of CAT if uncomplexed with CAM.

The initial formation of a complex between the 30S subunit and mRNA followed by a link-up with the 50S subunit is generally accepted as the way in which protein synthesis is initiated (6). In several models of translational regulation, the 30S ribosomal subunit, together with soluble factors and information inherent at the 5' end of mRNA select which message will be preferentially translated [see Kozak (12) for a recent review of these types of mechanisms].

Our current model contrasts with other control mechanisms insofar as it postulates that the 50S subunit makes the initial contact with and selects a particular mRNA for translation. The interaction between 23S rRNA and CAT mRNA need not involve dissociation of the 23S rRNA secondary structure. Experimental evidence for triplex and tetraplex polynucleotide conformations has been reported by Arnott and Bond (1), Arnott and Selsing (2), and Johnson and Morgan (11) make this model feasible.

REFERENCES

1. Arnott, S., and P.J. Bond (1973) Structures for Poly(U).poly(A).poly(U) triple stranded polynucleotides. Nature New Biology 244:99-101.
2. Arnott, S., and E. Selsing (1974) Structures for Poly(dA).poly(dT) and Poly(dT).poly(dA).poly(dT). J. Mol. Biol. 88:509-521.
3. Borer, P.N., B. Dengler, I. Tinoco, and O.C. Uhlenbeck (1974) Stability of ribonucleic acids double-stranded helices. J. Mol. Biol. 86:843-853.
4. Duval, E.J., D.M. Williams, P.S. Lovett, C. Rudolf, N. Vasantha, and M. Guyer (1983) Chloramphenicol inducible gene expression in Bacillus subtilis. Gene 24:170-177.
5. Gold, L., D. Pribnow, T. Schneider, S. Shinedling, B.S. Singer, and G. Stormo (1981) Translation initiation in procaryotes. Ann. Rev. Microbiol. 35:365-403.
6. Guthrie, C., and M. Nomura (1968) Initiation of protein synthesis: A critical test of the 30s subunit model. Nature 219:232-235.
7. Gryczan, T.J., G. Grandi, J. Hahn, and D. Dubnau (1980) Conformational alteration of mRNA structure and the posttranscriptional regulation of erythromycin-induced drug resistance. Nucleic Acids Res. 8:6081-6096.

8. Horinouchi, S., and B. Weisblum (1980) Posttranscriptional modification of messenger RNA conformation: Mechanism of erythromycin inducible resistance. Proc. Natl. Acad. Sci., USA 77:7079-7083.
9. Horinouchi, S., and B. Weisblum (1981) The control region for erythromycin resistance: Free energy changes related to induction and mutation to constitutive expression. Mol. Gen. Genet. 182:341-348.
10. Horinouchi, S., and B. Weisblum (1982) Nucleotide sequence and map of pC194, a plasmid that specifies inducible resistance to chloramphenicol. J. Bact. 150:815-825.
11. Johnson, D., and A.R. Morgan (1978) Unique structures formed by pyrimidine-purine DNAs which may be four-stranded. Proc. Natl. Acad. Sci., USA 77:7079-7083.
12. Kozak, M. (1983) Comparison of initiation of protein synthesis in procaryotes, eucaryotes, and organelles. Microbiol. Revs. 47:1-45.
13. Mertens, G., and J.N. Reeve (1977) Synthesis of cell envelope components by anucleate cells (minicells) of Bacillus subtilis. J. Bact. 129:1198-1207.
14. Noller, H.F., J. Kop, V. Wheaton, J. Brosius, R.R. Gutell, A.M. Kopylov, F. Dohme, W. Herr, D.A. Stahl, R. Gupta, and C.R. Woese (1981) Secondary structure model for 23S ribosomal RNA. Nucleic Acids Res. 9:6167-6189.
15. Nomura, M., S. Jinks-Robertson, and A. Miura (1982) Regulation of ribosome biosynthesis in Escherichia coli. In Interactions of Translational and Transcriptional Controls in the Regulation of Gene Expression, M. Grunbert-Manago and B. Saifer, eds. Elsevier Biomedical, New York. pp. 91-104.
16. Pestka, S., R. Vince, R. LeMahieu, F. Weiss, and J. Unowsky (1976) Induction of erythromycin resistance by erythromycin derivatives. Antimicrob. Agents Chemother. 9:128-130.
17. Rosenberg, M., and D. Court (1980) Regulatory sequences involved in the promotion and termination of RNA transcription. Ann. Rev. Genetics 13:319-353.
18. Shine, J., and L. Dalgarno (1974) The 3'-terminal sequence of Escherichia coli ribosomal RNA: Complementary to nonsense triplets and ribosome binding sites. Proc. Natl. Acad. Sci., USA 71:1342-1346.
19. Shivakumar, A.G., J. Hahn, and D. Dubnau (1979) Studies on the synthesis of plasmid-coded proteins and their control in Bacillus subtilis minicells. Plasmid 2:279-289.
20. Siebenlist, U., R.B. Simpson, and W. Gilbert (1980) E. coli RNA polymerase interacts homologously with two different promoters. Cell 20:269-281.
21. Syriopoulou, V.P., A.L. Harding, D.A. Goldman, and A.L. Smith (1981) In vitro antibacterial activity of fluorinated analogs of chloramphenicol and thiamphenicol. Antimicrob. Agents Chemother. 19:294-297.

22. Tinoco, I., P.N. Borer, B. Dengler, M. Levine, O.C. Uhlenbeck D.M. Crothers, and J. Gralla (1973) Improved estimation of secondary structure in ribonucleic acids. Nature New Biology 246:40-41.
23. Vournakis, J., J. Celantano, M. Finn, R.E. Lockard, T. Mitra, G. Pavlakis, A. Troutt, M. Van Den Berg, and R.M. Wurst (1981) Sequence and structure analysis of end-labeled RNA with nucleases. In Gene Amplification and Analysis. Volume II: Analysis of Nucleic Acids by Enzymatic Methods, J. Chirikjian and T. Papas, eds. Elsevier, New York, pp. 269-298.
24. Williams, D.M., E.J. Duvall, and P.S. Lovett (1981) Cloning restriction fragments that promote expression of a gene in Bacillus subtilis. J. Bact. 146:1162-1165.
25. Williams, D.M., R.G. Schoner, E.J. Duvall, L.H. Preis, and P.S. Lovett (1981) Expression of Escherichia coli trp genes and the mouse dihydrofolate reductase gene cloned in Bacillus subtilis. Gene 16:199-206.
26. Winshell, E., and W.V. Shaw (1969) Kinetics of induction and purification of chloramphenicol acetyltransferease from chloramphenicol-resistant Staphylococcus aureus. J. Bact. 98:1248-1257.

PLASMID AND LARVAL TOXIN OF *BACILLUS LATEROSPORUS*

Yun-liu Fan and Wan-xue Xu*

Laboratory of Molecular Biology
Chinese Academy of Agricultural Sciences
Beijing, China

Mosquitos are vectors of various diseases which constitute serious health hazards to humans in a number of tropical countries. Microbial control of the mosquito has received attention because of many problems, particularly environmental problems, that have been associated with the extensive use of chemical pesticides.

So far, only 2 species of Bacillus (*B. thuringiensis* Israeliensis and *B. sphaericus*) have been reported (1,2) as agents used against mosquito larvae. We recently discovered that *B. laterosporus* Asl.864 isolated from Heilungkiang province in China (Fig. 1-3) exhibited entomocidal activity against different species of mosquito larvae including *Culex tritaeniorhynchus* (vector of encephalitis) *C. pipiens*, *C. neovishnui*, *Aedes albopictus*, and *Anopheles sinensis*.

It has been reported that the entomocidal toxin of *B. thuringiensis* Israeliensis against mosquito larvae is located entirely in the parasporal crystalline inclusion (3), and there is a correlation between the loss of crystal production and the curing of plasmids of 4.0 - 4.4 megadalton (4). The larvicidal toxin activity of *B. sphaericus* 1593M is present in the cell wall/membrane fraction of presporulating cells (5) and the toxin has not yet been well-characterized. The toxin gene(s) is located on the *B. sphaericus* chromosome, and a DNA fragment from *B. sphaericus* coding for biocidal activity against mosquito larvae has been cloned and expressed in *Escherichia coli* (6). There was no larvicidal activity in vegetative cells of either *B. thuringiensis* Israeliensis or of *B. sphaericus*. In contrast, we have found that the larvicidal activity of *B.*

* The Institute of Microbiology, Academia Sinica, Beijing, China.

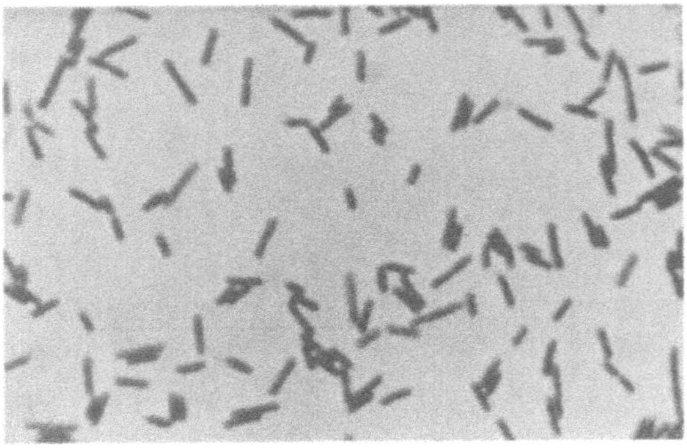

Fig. 1. Vegetative cells of *Bacillus laterosporus* As1.864.

laterosporus As1.864 was reached in exponential phase culture while the bacterial cells were vegetative, and that both bacterial suspension and the filtrates in either exponential- or stationary-phase cultures were effective against the larvae of mosquitos. Therefore, the mosquito larval toxin(s) of *B. laterosporus* As1.864 seems different from that of *B. thuringiensis* and *B. sphaericus* 1593M.

We have examined *B. laterosporus* As1.864 for presence of plasmid DNA and have found the presence of a large plasmid of 40 megadaltons in size (Fig. 4). Cells of *B. laterosporus* As1.864 were

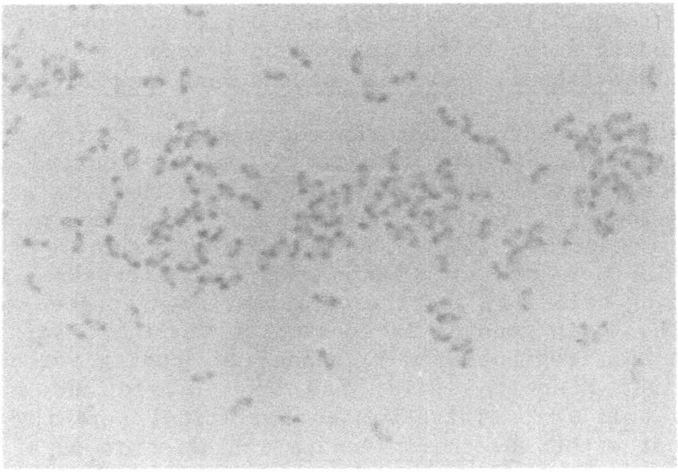

Fig. 2. Photographs of sporangium from *Bacillus laterosporus* As1.864.

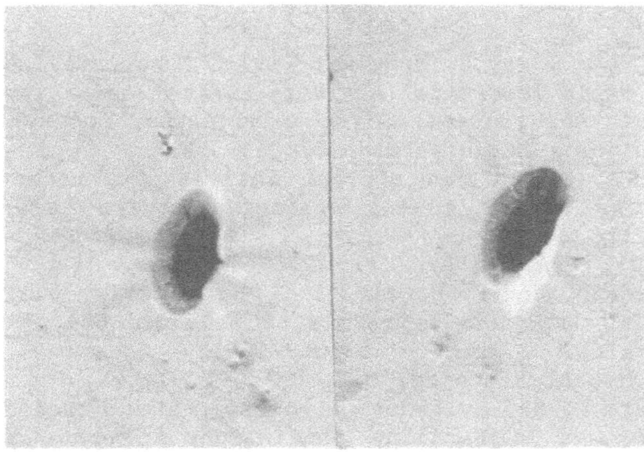

Fig. 3. Electron micrographs of sporangium from Bacillus laterosporus As1.864 (10,000X).

sensitive to antibiotics tested such as gentamicin, erythromycin, ampicillin, tetracycline, chloramphenicol, penicillin, lincomycin, kanamycin, streptomycin, and rifampicin. Establishing the location of the larvicidal toxin gene(s) either on plasmids or on chromosomes is in progress.

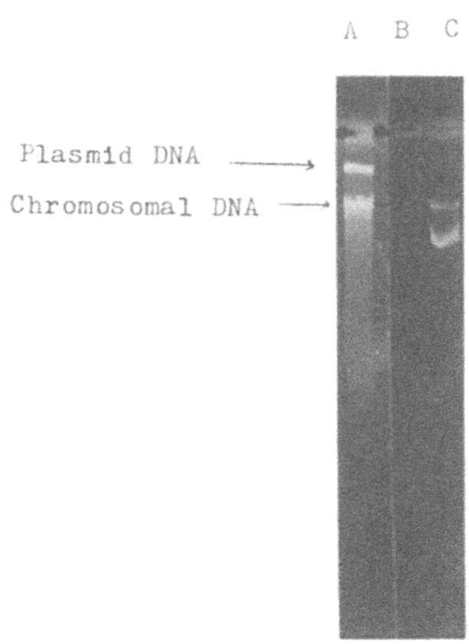

Fig. 4. Agarose gel electrophoresis of Bacillus laterosporus As1.864 plasmid (well A), RPA (well B), and PCM1 (well C).

REFERENCES

1. Goldberg, L.J., and S. Margalit (1971) A bacterial spore demonstrating rapid larvicidal activity against *Anopheles sergentii*, *Uranotaenia unguiculata*, *Culex univitattus*, *Aedes aegypti*, and *Culex pipiens*. *Mosquito News* 37:355-358.
2. Singer, S. (1973) Insecticidal activity of recent bacterial isolates and their toxins against mosquito larvae. *Nature* 224:110-111.
3. Tyrell, D., L.A. Bulla, Jr., R.E. Andrews, Jr., K.J. Kramer, L.I. Davidson, and P. Nordin (1981) Comparative biochemistry of entomocidal parasporal crystals of selected *Bacillus thuringiensis* strains. *J. Bact.* 145:1052-1062.
4. Faust, R.M., A. Kazunore, G.A. Held, T. Izuka, L.A. Bulla, and C.L. Meyers (1983) Evidence for plasmid-associated crystal toxin production in *Bacillus thuringiensis* subsp. Israelensis. *Plasmid* 9:98.
5. Myers, P., and A.A. Yousten (1980) Localization of a mosquito larval toxin of *Bacillus sphaericus* 1593. *Appl. Env. Microb.* 39:1205-1211.
6. Ganesan, S., H. Kamdar, K. Jayaraman, and J. Szulmajster (1983) Cloning and expression in *Escherichia coli* of a DNA fragment from *Bacillus sphaericus* coding for biocidal activity against mosquito larvae. *Mol. Gen. Genet.* 189:181-183.

POSTERS

REPLICATION, INCOMPATIBILITY, PARTITION

THE COMPLETE SEQUENCE OF THE P1 PLASMID MAINTENANCE REGION

*Abeles, Ann, Chattoraj, Dhruba, and Austin, Stuart

LBI-Basic Research Program, Frederick Cancer Research Facility, P.O. Box B, Frederick, Maryland, USA

The DNA sequence of a 5-kb region of P1 which encodes sufficient information for stable plasmid maintenance has been determined. This region can be divided into two functionally independent units - one for plasmid replication and the other for partitioning the daughter plasmids to daughter cells.

The unit containing the replication function is 1.5 kb long and includes an origin within a 245-bp segment, the coding sequence for a 32-kd essential protein, RepA, and a 0.3-kb nonessential incompatibility locus, incA. This locus is responsible for the low copy number of the P1 miniplasmid and consists of nine 19-bp repeated sequences. Five similar 19-bp repeats also occur in the origin region (incC). The structural similarities of these two loci suggest that they accomplish their respective functions by related mechanisms.

The partitioning functions are determined by an adjacent 3-kb segment. This region includes the open reading frames for two partitioning proteins, ParA and ParB, and a region, incB, which serves as a "centromere" for partitioning. There are similarities in DNA sequence between the incB locus and the promoter region for parA. These similarities may represent Par protein binding sites that are important for the partitoning process and for autoregulation of the the par operon.

Research sponsored by the National Cancer Institute, DHHS, under contract No. NO1-CO-23909 with Litton Bionetics, Inc.

A MUTANT OF *B. SUBTILIS* PLASMID pC194 AFFECTING SEGREGATION

Juan Carlos Alonso

Max-Planck-Institut für molekulare Genetik
Ihnestrasse 63/73, D-1000 Berlin 33, Germany

The *S. aureus*/*B. subtilis* plasmid pC194 (*seg*$^+$) is stably maintained in *B. subtilis* in the absence of selective pressure. We have characterized a spontaneous plasmid mutant, pC194-1, which segregates rapidly during cell division. Such segregation can be greatly enhanced when cells containing the mutant plasmid are briefly exposed to a pulse of MMS before growth under nonselective conditions.

The mutation causing the *seg*-phenotype, has been identified as a two base pair addition within the *Taq*I D+C fragments, extending between bases 2280 and 2790 of the Horinouchi/Weisblum sequence of pC194. The mutation falls within an open reading frame of about 80 bp. Transacting complementation of the defect of pC194-1 could be provided either by restriction fragments from pC194 containing the *seg*$^+$ allele or by some unrelated plasmids like pE194.

MMS = Methyl Methanesulfonate

MASSIVE TRANSCRIPTION OUTSIDE OF THE REPLICATION REGION CAN INTERFERE WITH REPLICATION, MAINTENANCE AND STABILITY OF COL E1 DERIVED PLASMIDS

Bujard, H., Stueber, D., Deuschle, U. and Peschke, U. Molekulare Genetik der Universitaet Heidelberg, INF 230, 69 Heidelberg, Federal Republic of Germany

Studying structure-function relationships of E.COL 1 expression signals, promoters differing in strength up to seventyfold were investigated in vitro and in vivo. Depending upon their individual strength, the cloning of promoters into plasmids containing the COL E1 or P15A replicon can have a profound effect on copy number, propagation and structural stability of the vector. We therefore, have constructed several plasmid families (1,2), of which the PDS1 system proved to be the most suitable for our purpose. The results of these studies can be summarized as follows:

- overproduction of the ROP-protein reduces the copy number of the plasmid and can contribute to its loss (2).
- readthrough transcription into the replication region from either direction interferes with the plasmids replication and again - depending upon the intensity of transcription - may just reduce the copy number or even prevent propagation of the plasmid.
- all these effects can be overcome by correct placement of strong transcriptional terminators which 'silence' the replication region from outside interference.
- tandemly repeated sequences in transcriptional units outside the replication region can undergo recombination with high frequency if efficiently transcribed.

Based on these findings, the analysis of expression signals in a 'balanced' vector system will be discussed (3).

1) Gentz, R., Langner, A., Chang, A. C. Y., Cohen, S. N., and Bujard, H. (1981), Proc. Natl. Acad. Sci. US, 78, 4936-4940.
2) Stueber, D. and Bujard, H. (1982), Embo J. 1, 1399-1404.
3) Bujard, H., Baldari, C., Brunner, M., Deuschle, U., Gentz, R., Hughes, J., Kammerer, W. and Stueber, D. in 'Gene Amplification and Analysis' Eds. T.S. Papas, M. Rosenberg and J.G. Chirikjian, Vol. 3, PP 65, Elsevier, N. Y. 1983.

ABORTED INSTALLATION OR MAINTENANCE OF MINI-F INDUCES SOS FUNCTIONS

A. Bailone, A. Brandenburger, A. Levine, M. Dutreix, M. Pierre and R. Devoret

Lab. d'Enzymologie, CNRS, 91190 Gif-sur-Yvette, France

The SOS inducing potency of UV-damaged F plasmid resides in miniF, a 9.3 kb fragment, coding for the replication and segregation functions of plasmid F. λminiF, a λ carrying miniF, damaged by about 0.5 pyrimidine dimers per kb, induced prophages Ø80 or λ in more than 50% of the infected lysogens. Expression of sfiA, a chromosomal SOS gene under the control of LexA repressor, was also efficiently induced. UV-damaged λminiF, but not the phage λ vector, produced in the host an SOS signal almost as potent as UV-damaged bacterial chromosome, indicating that only miniF is responsible for generating the SOS signal.

We located on miniF a locus, called lynA, 800 bp long, involved in the production of the SOS signal by UV-damaged miniF. We characterized six series of deletion mutants (Rep^-, SOS^-, Fid^-, $Par\text{-}2^-$, $Par\text{-}1^-$, and Par^+) related to four distinct phenotypes displayed by miniF such as partition, copy number, replication, and SOS induction.

We describe a novel mode of SOS induction, called **gratuitous** indirect induction, which is elicited when the maintenance of an intact λminiF introduced into a lysogen was inhibited by a resident plasmid or by mutations in miniF that impaired either its partition or replication. Gratuitous induction required an active RecA protein and a functional miniF $lynA^+$ gene as does UV-indirect induction.

We postulate that impairment of either replication or partition disturbs lynA expression, altering co-segregation of miniF with the chromosome. This entails RecA protein activation through a change in DNA helicity of the DNAs associated with the co-segregation apparatus.

REPLICON FUSION AND THE ORIGIN OF THE INCF GROUP OF PLASMIDS

P.L. Bergquist and H.E.D. Lane, Department of Cell Biology, University of Auckland, Auckland, New Zealand
S. Saadi and W.K. Maas, Department of Microbiology, New York University Medical School, New York, U.S.A.

IncF plasmids have been divided into subgroups on the basis of their incompatibility behaviour. The cloning of the replication regions of the F plasmid of *E. coli* has demonstrated that more than one region can express incompatibility. Four inc determinants have been identified.

Three basic replicons have been recognized in the IncFl plasmid group from cloning experiments resulting in the construction of autonomously-replicating miniplasmids. RepFlA is the primary replicon of the F plasmid. RepFlB is the secondary replicon of F and is found in nearly all plasmids of the IncFl group. RepFlC is a basic replicon showing homology with the RepA region of the incFll plasmid group (repFllA). Our results show that of the IncFl plasmids, only R453 carries all three replicons in a functional form. Plasmids of other IncF groups show homology to various extents with these three basic replicons. For example, plasmids R100-1 (IncFll) and R124 (IncFIV) possess an area of homology to repFlB. Representative plasmids of other incompatibility groups have been examined but no homology with the three basic replicons can be detected except for the IncH plasmids which show homology with gene E of repFlA of F.

These results suggest that the subgroups of IncF plasmids may have arisen from an ancestral F-like plasmid which carried only the repFlC replicon and the tra genes. Evolutionary divergence has resulted from various genetic re-arrangements such as replicon fusions, followed by deletion or inactivation of some of the replicons, only remnants of which now remain.

AUTOREGULATION OF THE pSC101 *repA* GENE

G. Churchward, P. Linder, Xia Guixan & L. Caro Department of Molecular Biology, University of Geneva, Geneva, Switzerland

We have isolated *Tn1000* insertions into the replication genes of the pSC101 plasmid, cloned on a colE1 vector. Two of the insertion plasmids can neither transform *polA*$^-$ bacteria nor replicate after a temperature shift in a *polAts* host. Both transformation and replication activities are restored if the *polA*$^-$ host strains harbor a wild-type pSC101 plasmid. The *Tn1000* insertions thus inactivate a transacting replication function *repA*. The insertions inactivate the synthesis of a 35,000 Mr protein. We have identified the transcription and translation initiation signals of the *repA* gene. Fusion of these signals to a β-galactosidase gene segment results in strong expression of β-galactosidase activity which is repressed, in trans, by plasmids carrying a cloned *repA* gene. The *repA* gene is, therefore, autoregulated. Measurements of transcription from the *repA* promotor show that this autoregulation operates primarily at the level of transcription. The analysis of *repA* promotor mutants that are no longer sensitive to autoregulation shows that transcription of the *repA* gene is repressed by the binding of the *repA* protein to a segment of DNA, which overlaps the *repA* promoter.

ELIMINATION OF RESISTANCE PLASMIDS FROM *ESCHERICHIA COLI* BY 2,2'-DIPYRIDYLAMINE

Dendinger, S. M., M. Norcia and R. Borovoy

Pfizer Central Research, Pfizer, Inc., Groton, CT 06340

Various chemical agents are able to eliminate certain plasmids from their bacterial hosts. Many of these work well against the F plasmid but are not reproducibly effective against resistance plasmids. Using a Tn951 (Lac^+) derivative of the R plasmid R100 (IncFll) we found that 2,2'-dipyridylamine at 500 µg/ml caused reproducible plasmid loss (approximately 8%) from *E. coli* K12 strain CSH26 after overnight growth in BHl broth. The resistance markers of R100 were lost simultaneously with the Lac^+ phenotype. Gel electrophoresis of lysates of cured cell lines confirmed loss of the plasmid. Untreated control cultures had <0.5% plasmid loss.

The curing effect of dipyridylamine was not confined to this particular strain, as other K12 strains (C600, JG3272) carrying R100 were also affected. Curing was dose-dependent and occurred at concentrations well below the minimum inhibitory concentration. Curing took place in minimal, as well as rich, medium and was more effective at 30C than at either 37 or 42C. Studies on the kinetics of plasmid elimination indicated that during the exponential phase the number of plasmid-free cells in the population increased at a constant rate. This rate was too rapid to be due solely to the replication of plasmid-free cells, indicating that dipyridylamine acts by interfering with plasmid maintenance, rather than by selectively enriching for pre-existing plasmid-free cells. To determine the plasmid spectrum for curing, one or two representative plasmids of nine incompatibility groups were examined. Dipyridylamine affected the stability of the IncFll plasmids Rl and R6-5. Stability of the representative of each of the other groups was not affected. Dipyridylamine should be useful in efficiently curing *E. coli* of certain R plasmids.

PLASMID R1 REPA PROTEIN PROMOTES INITIATION OF R1 REPLICATION THAT IS INDEPENDENT OF TRANSCRIPTION BY HOST RNA POLYMERASE

R. Diaz and S. Ortega

Instituto de Biologia Celular (C.I.B.)

Velazquez 144, Madrid-6, Spain

RepA is a plasmid R1 coded protein which synthesis is required for replication of this plasmid. A *repA* recombinant, pSO1, was constructed that codes for a fused RepA protein which expression is under the control of the bacteriophage lambda P_L promoter and of translation initiation signals of gene *c*II of lambda. In cells containing pSO1, derepression of the lambda P_L promoted resulted in efficient synthesis of the fused RepA protein and in enhanced replication of co-resident R1 miniplasmids. Furthermore, cell free extracts of *E. coli* prepared and assayed as described (Diaz *et al*., Nature, *289*, 326-328, 1981) promoted efficient synthesis of the fused RepA protein when supplemented with pSO1 DNA. Following synthesis and accumulation of RepA in these extracts, semiconservative replication of the R1 miniplasmid pKN182 could be assayed in the absence of further protein synthesis. RepA dependent initiation of R1 replication in this system was found to be independent of transcription by host RNA Polymerase. This was indicated by the insensitivity of pKN182 replication to Rifampicin, as well as to antibodies prepared against whole RNA Polymerase. M13mp8 based R1 recombinants assayed in this replication system, defined the region with which RepA interacts, to the 1 kb *Sau*3A fragment that contains the sequences in which R1 *in vitro* replication starts (Diaz and Staudenbauer, J. Bacteriol. *150*, 1077-1084, 1982).

NEW STABILITY FUNCTIONS OF PLASMID R1 THAT INTERFERE WITH CELL DIVISION ARE CLOSE TO THE ORIGIN OF REPLICATION

R. Diaz, A. Bravo, and G. de Torrontegui

Instituto de Biologia Celular (C.I.B.)

Velazquez 144, Madrid-6, Spain

Mutations in the 1.5 kb *Pst*I-*Eco*RI fragment adjacent to the origin of replication of plasmid R1 have been isolated that stabilize the *par*$^-$ miniplasmid pKN1562 without altering its copy number. These mutants slow down the growth of the cell by way of a reversible inhibition of cell division. The stability and interference with cell growth are simultaneously complemented *in trans* by pBR322 recombinants carrying the 1.5 kb *Pst*I-*Eco*RI fragment. A weak incompatibility was found to be specified by this fragment. Selection for fast growth in cells containing the stability mutants leads to the isolation of plasmid derivatives that are slightly more unstable than pKN1562; some of these carry insertions or deletions inside the 3.0 kb *Pst*I fragment that complements the mutations. *In vitro* mutagenesis with hydroxylamine of derivatives with insertions or deletions, give rise to conditional replication mutants of R1 with a frequence of 0.8%. However, the same mutagenesis treatment of the parental miniplasmid pKN1562 gives no replication mutants (frequence < 0.02%). It is proposed that the stability system described here requires the interplay of two diffusible factors and a site and that interactions with the cell membrane, probably at the division plane, are involved. It is also proposed that the role of this system is to prevent the appearance of plasmid free cells that result from failures in replication of the plasmid.

THE REPA2 PROTEIN OF NR1 REPRESSES TRANSCRIPTION OF ONE OF THE MESSENGERS FOR THE REPA1 REPLICATION PROTEIN

*X. Dong, D.D. Womble, and R.H. Rownd

Department of Molecular Biology
Northwestern University, The Medical School
Chicago, IL 60611 USA

The DNA encoding the *repA1* replication protein of the plasmid NR1 is transcribed into two overlapping mRNAs, RNA-A and RNA-CX. Transcription of RNA-CX begins about 350 bp upstream from the 5'-end of RNA-A. The non-overlapping 5'-end of RNA-CX also codes for the protein *repA2*, which is a repressor of RNA-A transcription. Transcription of RNA-CX is constitutive at a low level and the synthesis of *repA2* protein is gene dosage-dependent. This causes RNA-A transcription to vary inversely with plasmid copy number, which is an important component of the NR1 replication control system. The total amount of *repA1* mRNA is the sum of the constitutive RNA-CX and the regulated RNA-A.

The *lacZ* gene was placed under transcriptional control of the RNA-A promoter on a λ prophage. Transcription rates were also measured directly by quantitative hybridization of pulse-labeled RNA. We determined the *in vivo* relationship between RNA-A promoter activity and *repA2* protein concentration by providing a source of *trans* *repA2* repressor which could be regulated by *lac* control elements. We found that: 1) The repA2-RNA-A promoter interaction is cooperative, probably as a result of active tetramer formation; 2) At the normal copy number of NR1, RNA-A transcription is 96% repressed, and RNA-CX serves as the primary mRNA for the *repA1* initiator; 3) Below the normal copy number, RNA-A transcription becomes substantially derepressed, which significantly increases the amount of *repA1* mRNA.

SELECTION AND CHARACTERIZATION OF INCOMPATIBILITY MUTANTS OF ColE1 RNAI

*Dooley, Thomas P. and Barry Polisky

Dept. of Biology, Indiana University, Bloomington, IN

The bacterial plasmid ColE1 has been extensively studies as a model system for the control of DNA replication and plasmid incompatibility. RNAI has been shown previously to be the major control element of ColE1 replication, and is solely responsible for plasmid incompatibility. We have used a rapid selection to generate ColE1 RNAI impaired function mutants. Using the constructed plasmid pTF421Δ,(containing a trp-promoter-ColE1 RNAI gene fusion and Ap^r), mutants in the RNAI gene are unable to segregate away another plasmid containing a ColE1 origin of replication. Mutagenesis of pTF421Δ followed by transformation into cells containing a Cm^r plasmid containing a ColE1 ori, has allowed us to sèlect many mutants in RNAI. These mutants have been analyzed by DNA sequencing, in vitro transcriptions, segregation assays, and other methods. These point mutants are of various classes affecting 1) trp promoter, 2) transcription termination, RNAI loops I and II, 3) RNAI stem I secondary structure, 4) the 5' single stranded tail. These novel mutants allow us to determine the critical structural and functional domains of the ColE1 RNAI molecule.

A NOVEL TYPE OF COPY NUMBER MUTANT OF COL E1

*Randolph Elble, Mark Muesing, Joe Tamm, Jane Schneider and Barry Polisky. Dept. of Biology, Indiana U., Bloomington 47405

Copy number control of the plasmid Col E1 has been demonstated by mutational studies to be a function of a 108 nucleotide transcript (RNA1) which, *in vitro*, inhibits formation of an active primer of replication. RNA1 and primer transcripts converge, primer starting a -555 from the origin and RNA1 at -446. RNA1 transcription has been investigated *in vivo* for wild-type (cop+) Col E1 and a number of high copy-number mutants by gel analysis of pulse-labeled RNA and by hybridizing RNA blots with RNA1-specific probes. This analysis reveals the 108 nucleotide RNA1 seen *in vitro* anda shorter RNA (RNA1*) as well. 5' end analysis shows that RNA1* is a processing product of RNA1 from which the first 5-6 nucleotides have been removed. Both RNA1 and RNA1* are very unstable (half-life 1-2 minutes) *in vivo* despite their extensive secondary structure. Results are similar in cop- mutants. Three of these mutants (*cop*7, *cop*3, and *cop*4) however show an aberrantly stable RNA 1*. RNA1* has a half-life of about 30 minutes in these mutants and accumulates to 5-10% of total RNA. RNA1 on the other hand is no more stable than in *cop*+-containing cells. The *cop*7 phenotype is caused by a point mutation between the -10 and -35 regions of the RNA1 promoter. Through studies of plasmid segregation from cells containing both *cop*+ and *cop*7 plasmids, I have shown that the *cop*7 mutant is *trans*-recessive despite its high copy number and high level of RNA1 transcription.

A SECOND INCOMPATIBILITY FUNCTION ON RP4 INVOLVES PLASMID-HOST INTERACTIONS

K. Ellis, Biochemistry Department, St. Mary's Hospital Medical School, London W2, and P.T. Barth, ICI Corporate Bioscience Group, Runcorn, Cheshire, England.

Two plasmid maintenance mechanisms have been shown to be involved in the phenomenon of incompatibility : replication control and partition. The oriV region of plasmid RK2/RP4 determines strong incompatibility (IncP) and appears to contain the replication control system of this plasmid. We have cloned DNA fragments from the trfB/korA/korB region of RP4 into various vectors and noted that some of these also confer an IncP phenotype, which we designate IncPb. The fragment from 54.4 (SstII) to 56.4kb (EcoRI) of RP4 confers $IncPb^+$ whereas the 54.6 to 56.4kb region is inactive. Both these fragments express the $korA^+$ gene product but the latter was obtained from an RP4::Tn76 derivative which is unstable, reduces its host's growth rate and has suffered a reduction in host-range.

IncPb differs from incompatibility mediated by a replication control region in that dual replicons do not suffer loss when challenged by either replicon, whereas our $IncPb^+$ clones do. There is no evidence that the $IncPb^+$ fragment can act as a replicon, neither does it stabilize pACYC184 ie it is not a par region. We have noted, however, significant effects of $IncPb^+$ clones on the growth rates of their hosts especially when in a dual plasmid incompatibility situation. We suggest that the IncPb phenomenon is the result of plasmid-coded products interacting with host growth or replication mechanisms. The intact IncP plasmid normally provides further interacting products which relieve the host and provides a plasmid maintenance mechanism active across its broad host-range.

A RECOMBINANT PLASMID WITH INCREASED COPY NUMBER IN BACILLUS SUBTILIS Ki-2-132

Shu-Tian Fan, Yong-Hua Sun, Ming-Feng Li and Mao-Hong Tang. Institute of Genetics, Academia Sinica, Beijing China

Since cloning system of B.subtilis has special advantages, it becomes an important system for cloning and expression of foreign genes. Ehrlich, Dubnau and Lovett et al have introduced Staphylococcus aureus plasmid pUB110, pCM194, pSA2100 pE194 into B.subtilis. Among them pUB110 has been used most extensively, because its copy number and transformation efficiency is higher than others, but it is still lower than some plasmids of E.coli. It is important for cloning system of B.subtilis to reconstruct plasmid vecter having high copy and transformation efficiency. This paper reported recombinant pK209-8 constructed with plasmid pUB110 digested with restriction endonuclease BamHI and EcoRI and BamHI-EcoRI fragment of chromosomal DNA of B.subtilis 209. The copy number and transformation efficiency of the reconstructed plasmid both are higher than pUB110 and the molecular weight is 4.5×10^6 daltons. Its copy number is 101 per cell, 1.9 times that of pUB110, with transformation efficiency 10 times higher than that of pUB110.

CHARACTERIZATION OF THE REPLICATION AND STABILITY REGIONS OF THE *AGROBACTERIUM* PLASMID pTAR

D. R. Gallie, D. Zaitlin, M. Hagiya, and C. I. Kado, Davis Crown Gall Group, Department of Plant Pathology, University of California, Davis, California

The 44 kb *Agrobacterium* plasmid pTAR, from the non-oncogenic grapevine strain *1D1422*, confers on its host the ability to catabolize tartrate and is compatible with the Ti plasmids pTiC58 and pTiAch5. We have mapped the origin of replication and the partition locus in a 5.4 kb region of the plasmid. Within this region is a 1.3 kb segment that is capable of directing autonomous replication. The remaining segment contains the partition functions necessary for stable plasmid inheritance of pTAR during nonselective growth. Using a cell-free coupled transcription-translation *Agrobacterium* system, we determined that at least two proteins, 7.8 kd and 20 kd and possibly a third protein, 26 kd, are encoded by genes in a tightly clustered arrangement which defines the region involved in partitioning. Deletions within this region render the plasmid unstable with the concurrent loss of these proteins, suggesting that one or more of these proteins may be involved in efficient partitioning of pTAR.

PARTITIONING OF PLASMID R1: TWO DIFFERENT LOCI REQUIRED FOR COMPLETE STABILITY

Kenn Gerdes, Poul B. Rasmussen, and Søren Molin, The Technical University of Denmark

The low copy number plasmid R1 has been characterized very well with respect to replication control functions, whereas the analysis of the *par* genes has only just started. We present data showing, that R1 codes for two independent stability regions.

It was previously indicated, that the largest *EcoRI* fragment (*EcoRI* A) of R1 is involved in partitioning. This was verified by cloning the *EcoRI* A-fragment on a par^- miniR1 replicon: The recombinant plasmid was completely stably inherited. By a series of subclonings it was shown, that the *EcoRI* A-fragment carries two stability regions, which we designated $parA^+$ and $parB^+$. Each region stabilize R1 partially; $parA^+$ and $parB^+$ replicons are typically lost with a frequency of 10^{-4}/cell/generation, whereas miniR1 replicons containing no *par* region are lost with a frequency of 10^{-2}/cell/generation. MiniR1 replicons containing both *par* regions have loss frequencies in the range of 10^{-6}/cell/generation.

Both *par* regions exert allel-specific incompatibility (e.i. $parA^+$ plasmids dislodge $parA^+$ stabilized plasmids, and $parB^+$ stabilized plasmids dislodge $parB^+$ stabilized plasmids). No cross-incompatibility between $parA^+$ plasmids and $parB^+$ plasmids has been detected, whereas $parB^+$ plasmids show weak incompatibility towards $parA^+$ $parB^+$ stabilized plasmids.

The *par* regions were cloned on different p15 plasmid vectors, and it was shown, that $parB^+$ will confer a high degree of stability on these replicons. In contrast, the $parA^+$ region has no effect on the stability of such replicons. Whether the $parA^+$ region is able to stabilize other low copy number plasmid awaits further experimentation.

The two *par* regions were mapped more precisely using exonuclease Bal31. The $parA^+$ region is confined to a DNA fragment of 1500 bp near one end of the *EcoRI* A-fragment. The nucleotide sequence of the $parA^+$ region confirmed the presence of a 36 K protein detected by the maxicell method. By constructing transcriptional- and translational gene fusions between this gene and the *lacZ* gene, it was shown, that the *parA* gene is autogenously regulated.

The $parB^+$ region, located at the other end of the *EcoRI* A-fragment, resides within 575 bp and also in this case, the entire nucleotide sequence was determined. The region contains two open reading frames, corresponding to two very small polypeptides. In addition, regions with symmetric sequences are present.

The mechanisms by which the two *par* regions act are unknown, but we have indications, that the $parB^+$ region couples plasmid replication to the host cell cycle.

PLASMID STABILITY: INVOLVEMENT OF A PLASMID-CODED PRIMASE AND SITE-SPECIFIC RECOMBINATION/RESOLUTION SYSTEMS

N. J. Grinter, Department of Biochemistry & Soil Science, University College of North Wales, Bangor, Gwynedd, U.K. and P. T. Barth, I.C.I. Corporate Laboratory, The Heath, Runcorn, Cheshire, U.K.

Plasmid RP4 in E.coli is found predominately as monomeric DNA molecules with a copy number of 4-5 per chromosome equivalent. In this state it is stably inherited. Deletion derivatives were isolated which show high levels of instability and which form multimers by recA-dependent recombination. Complementation studies demonstrate that RP4 multimerisation is normally prevented by at least two distinct mechanisms: one involves the plasmid-coded primase. A cloned 12kb primase-encoding fragment from the Tral region of RP4 also suppresses a temperature-sensitive mutation affecting the E.coli single-stranded DNA-binding protein; this activity may limit recombination between RP4 monomers. The second mechanism acts by efficiently resolving any plasmid multimers to monomers via a site in the 28-34kb region. Inactivation of primase or deletion of the resolution site leads to low levels of plasmid instability (0.2 or 0.9% loss per generation, respectively); elimination of both primase and the resolution site causes a high rate of plasmid loss (5-7% per generation) and an accumulation of plasmid multimers. Spontaneous revertants with partially restored stability show reduced multimerisation and result from deletions or insertions close to the RP4 origin of replication. We propose that this region contains a hot spot for homologous recombination. Multimerisation probably disrupts normal partitioning and the results are consistent with only limited numbers of RP4 molecules being actively segregated at cell division.

PLASMID RSF1010 DNA REPLICATION *IN VITRO* PROMOTED BY PURIFIED RSF1010-ENCODED *repA*, *repB* and *repC* PROTEINS

Haring, V., Scholz, P., Lurz, R.,
Bagdasarian, M. and Scherzinger, E.

Max-Planck-Institut für Molekulare Genetik
Abt. Schuster, D-1000 Berlin 33, FRG

A soluble enzyme system for the study of plasmid RSF1010 DNA replication has been developed. The system consists of a partially purified mixture of *Escherichia coli* replication proteins (fraction AS[0-73]) and three highly purified plasmid-encoded proteins, the products of the RSF1010 genes *repA*, *repB* and *repC*. DNA synthesis in this system closely resembles the known characteristics of RSF1010 replication that occurs *in vivo* or *in vitro* with crude extracts of RSF1010-harboring cells. The system (i) depends completely on externally added DNA, (ii) replicates specifically supercoiled plasmid DNA or phage replicative form DNA containing the RSF1010 replication origin (*oriV*), (iii) depends on each of the three RSF1010-encoded rep proteins, (iv) depends on host replication proteins such as DNA gyrase and dnaZ protein but not on others (e.g. RNA polymerase, primase, dnaB protein and dnaC protein), as indicated by specific antibiotic or antibody inhibitions or by the activity of enzyme fractions derived from *dna*(ts) mutants of *E.coli*, and (v) creates replication 'bubbles' at or near *oriV* which are extended either bi- or unidirectionally to generate two complete supercoiled product molecules. Provided that a unique RSF1010 origin DNA sequence is present, circular single-stranded templates are also active in this *in vitro* system. In contrast to the observations with duplex DNA, the ssDNA-dependent initiation reaction requires only one of the three RSF1010-encoded rep proteins: the repB protein.

REPLICATION KINETICS OF THE *STAPHYLOCOCCUS AUREUS* PLASMID pT181

Sarah K. Highlander and Richard P. Novick, Department of Plasmid Biology, The Public Health Research Institute of the City of New York, Inc., 455 First Avenue, NY, NY, 10016

pT181 is a 4.4 kb., 20-25 copy, *S. aureus* Tc^r plasmid whose replication rate is determined by the concentration of a *trans*-active initiator protein, RepC. RepC synthesis is negatively regulated by two small RNA countertranscripts that are read from the opposite strand of the 5' end of the RepC mRNA leader. A clone that produces these small RNAs inhibits pT181 replication *in vivo*. Complementary GC to AT base changes in the interacting regions of the inhibitor and mRNA leader lower the efficiency of the inhibitory interaction and result in increased plasmid copy number. Deletion mutations within this region abolish the negative interaction and produce copy mutants with even higher copy numbers.

The kinetics of pT181 repopulation were studied using thermosensitive replication mutants (*tsr*) of the wild-type plasmid and a copy mutant having a GC to TA point mutation in the inhibitor-target region. Strains were grown in a turbidostat at 42° to lower plasmid copy number and were then shifted to 32°. Plasmid replication and copy number were followed after the downshift. Both the wild-type and the copy number mutant plasmid exhibited an initial exponential replication phase resulting in a peak copy number that was 2-3 times higher than the steady-state value. This overshoot was followed by a prolonged period of damped oscillation about the steady-state level.

Plasmid repopulation kinetics were also examined following introduction of up to six plasmid copies per naive cell by infection with the high frequency transducing phage, ϕ11. A wild-type pT181 replicon exhibited overshoot kinetics similar to those seen for the *tsr* wild-type plasmid, while a copy number mutant of pT181, having a small deletion in the inhibitor-target region, did not overshoot its steady-state copy number, but instead repopulated by hyperbolic kinetics.

The overshoot results are interpreted as being due to asynchronous cycles of RepC and inhibitor synthesis; the protein accumulates until its synthesis is inhibited by the accumulation of repressor. The overshoot and damped oscillation shown by a pT181 point mutant are consistent with its retention of inhibitor function. The hyperbolic kinetics shown by the deletion mutant indicate that this plasmid is no longer subject to negative replication control by the RNA species.

RELATIONSHIP BETWEEN PLASMID COPY NUMBER AND MOLECULAR SIZE OF THE PLASMIDS COLE1 AND NR1

Jay P. Kasner, David D. Womble, and Robert H. Rownd

Department of Molecular Biology
Northwestern University Medical School, Chicago, IL

Derivatives of the unrelated plasmids ColE1 and NR1 had a similar relationship between plasmid copy number and molecular size. ColE1 derivatives 2 to 12 kb in size had a constant copy number of 16.5 as determined by dot-blot hybridization of cesium chloride-ethidium bromide density gradient fractions. The copy numbers of ColE1 derivatives larger than 12 kb decreased as their molecular size increased. The larger ColE1 derivatives were maintained at a plasmid mass equal to 6 percent of the total cellular DNA. The location of the DNA insertions and deletions did not effect this relationship. The copy number of NR1 was constant for molecular sizes ranging from 5 to 95 kb. A point mutation in the small repressor RNA of NR1 resulted in a dependence of plasmid copy number on molecular size. Derivatives of the mutant, pRR12, were maintained at 8 percent of the total cellular DNA for the entire size range studied (5 to 95 kb). The difference between the NR1 and pRR12 relationship is identical to that found for ColE1 and its copy mutant, pOP1 (Muesing, M. and B.Polisky, p. 412 In D. Schlessinger (ed.), Microbiology-1981, ASM). pOP1 also has a single base pair alteration in its repressor RNA molecule.

These data suggest intracellular limitations of the host cell become involved in copy number control when the total plasmid mass reaches a critical level. The wild-type plasmids of NR1 and ColE1 are maintained at mass levels below this critical point. The molecular size of ColE1 had to be doubled before its copy number was size dependent. Since runaway replication mutants exist, plasmid mass is not the only factor controlling the copy number of pRR12, pOP1, and the larger ColE1 derivatives. More likely there is a balanced interaction between a limiting intracellular factor(s) and the activity of the repressor of plasmid replication.

STABILITY OF THE PLASMID COLE1 IN *ESCHERICHIA COLI*

Jay P. Kasner and Robert H. Rownd

Department of Molecular Biology
Northwestern University Medical School, Chicago, IL

ColE1 plasmids encoding a functional bacteriocin gene were stable in growing populations of *E. coli*. These cells excreted colicin into the medium which inhibited the growth of colicin sensitive cells. ColE1 plasmids encoding non-functional colicin genes were unstable and did not release the inhibitory activity. The stable ColE1 plasmids became unstable in *recA*⁻ and *tolC*⁻ bacterial strains. Colicin expression is repressed in *recA*⁻ strains. The *tolC* gene specifies a membrane protein which renders the strain resistant to the killing effects of colicin when mutated. The results suggest the low level expression of uninduced colicin killed the cured cells in the growing populations and caused the plasmid ColE1 to have a stable phenotype.

ColE1 was a comparatively stable plasmid even with a non-functional colicin gene. The number of plasmid segregating units was responsible for the relative stability of ColE1. Two factors determined the number of plasmid segregating units present in the cell. They were the average copy number of ColE1 (16.5) and the ability of ColE1 to resolve into monomeric forms after replication. The high number of plasmid segregating units caused a very low frequency of cured cells. Also important for the stability of the ColE1 plasmid was the difference in growth rate of cells with and without ColE1 plasmids. The differential growth rate caused by ColE1 was orders of magnitude greater than the rate of cured cell formation. Consequently differential growth rate dictated the downward concavity of the stability curve. The number of plasmid segregating units influenced the number of generations before the loss of the plasmid was initially observed.

Considering the number of plasmid segregating units in the cell and the differential growth rate in the cell population, ColE1 was calculated to be segregated to daughter cells by a mode of random binomial distribution. This would mean ColE1 lacks a *par* locus. Therefore, a high number of plasmid segregating units and the low level expression of its bacteriocin are adequate to stabilize the plasmid ColE1 in an exponential population of *E. coli* cells.

A PROPOSED MODEL FOR CONTROL OF MINI-F REPLICATION

B. C. Kline* and J. D. Trawick, Section of Microbiology
Department of Cell Biology, Mayo Clinic and Foundation
Rochester, MN 55905 USA

Mini-F requires an origin gene and the _repE_ plasmid gene, which encodes a 29 kilodalton (kd) protein, to form a replicon. Two copy number control genes have been identified; they are _copA_ and _copB_. Here we report that _copA_ mutations, all of which map in _repE_, inactivate autogenous regulation of the _repE_ promoter. This indicates that the 29 kd protein is an autorepressor as well as essential for replication. In order for the 29 kd protein to fulfill its indicated roles, we propose: (1) that in a simple form this protein acts as an autorepressor thereby establishing its steady state concentration; and (2) that at a rate dependent on this concentration, a small fraction of autorepressor is modified to function as an initiator. Since the _copB_ DNA sequence mimics the origin sequence, we also propose that _copB_ competitively binds initiator thereby further reducing its availability.

A MINIMAL REGION OF AN IncN-GROUP PLASMID REQUIRED FOR ITS MAINTENANCE IN *ESCHERICHIA COLI*

M. Konarska-Kozlowska and V.N. Iyer, Department of Biology and Institute of Biochemistry, Carleton University, Ottawa, Ontario, Canada K1S 5B6

The 39 kb plasmid pCU1 is a broad host range N group plasmid that is being studied in some detail. In addition to phenotypes characteristic of the group, the plasmid also confers resistance to ampicillin, streptomycin and spectinomycin. Electron microscopy of denatured molecules of pCU1 revealed four different stem-loop structures. Their dimensions, positions with respect to one another and a restriction and functional map of the plasmid have been described previously. Of particular interest was the assignment of the maintenance region of the plasmid to a location within one of the stem-loop structures.

The deletions used to locate various phenotypes of pCU1 collectively encompass all but a continuous 6 kb region of the plasmid. This and the ability of the smallest deletion derivative, pCU32, to be stably replicated and maintained in *E. coli* indicated that ampicillin-resistance (the marker on pCU32) and all of the plasmid-specified replication functions are within this 6 kb of DNA. To locate the maintenance region within pCU32, Tn5 insertions of this plasmid were isolated. Their distribution defined: (*a*) an Ampicillin-resistance region, (*b*) a region for stable maintenance, (*c*) a region insertions into which had no effect on pCU32, and (*d*) a region insertions into which were associated with deletions extending into and removing *c*. *c* could not be cloned in the absence of *d* suggesting that they may contain genes specifying functions analogous to the *kil* and *kor* (kill override) functions that have been found to be specified by plasmid RK2.

Regions *a*, *c* and *d* occupy about half of pCU32. The minimal maintenance region inferred to be on the other half was defined by Tn5 insertions delimiting a 2 kb region. Most of this region was present on one of five PvuII fragments of pCU32. Replication of pCU32 was *polA*-independent. Each PvuII fragment of pCU32 was cloned into the single PvuII site of the *polA*-dependent plasmid pBR329. Subsequent removal of the replication region of pBR329 and selection of plasmids in an *E. coli polA* mutant resulted in the isolation of plasmids that always contained PvuII fragment 2 of pCU32. Most of the maintenance region that was defined by the distribution of Tn5 insertions is within this PvuII-2 fragment. To further demonstrate that the PvuII-2 fragment contains all that is necessary for stable plasmid maintenance, PvuII-2 was purified and linked *in vitro* to the non-replicating tetracycline-resistance fragment of plasmid pMK16. Such constructed plasmids were stably maintained without antibiotic selection in wild type and in *polA* strains. We conclude that all plasmid-specified functions and sites needed for the stable maintenance of this prototype IncN group plasmid are determined by this single region. We will further dissect and sequence this region.

[Supported by the Medical Research Council of Canada and the Atkinson Foundation]

REGIONS AROUND THE *E. COLI* ORIGIN OF REPLICATION WHICH AFFECT THE STABILITY OF PLASMIDS

Helmut Kunze and Walter Messer

Max-Planck-Institut für molekulare Genetik
Ihnestrasse 63/73, D-1000 Berlin 33, Germany

Joint replicons were constructed which contain a minichromosome - a plasmid replicating from the *E. coli* chromosomal origin of replication - and miniF or pSC101, respectively. *oriC*::miniF cointegrates were completely stable, in agreement with published results (Ogura, T. et al., PNAS *77*, 3993-3997, 1980). Hybrids between pSC101 and minichromosomes pOC3 (9.5 kb *Eco*RI fragment, coordinates -5.8 kb to +3.7 kb of the *oriC* sequence) and pOC15 (6.0 kb *Hind*III-*Eco*RI fragment, coordinates -2.3 kb to 3.7 kb, Messer et al., Cold Spring Harbor Symp. Quant. Biol. *43*, 139-145, 1978) were unstable: 40-60 % of the cells were free of plasmid when grown for about 15 generations without selection. This demonstrates that not only is the partition function of pSC101 unable to stabilize these minichromosomes (Hinchcliffe, E. et al., Plasmid *9*, 286-296, 1983), but in addition, the presence of sequences around *oriC* cause unstable partition of pSC101. This effect was only observed in *cis*.

Surprisingly the region close to and including *oriC* and suggested regulatory elements like the 16 kd and 17 kd proteins with putative *oriC* specific incompatibility regions did not interfere with pSC101 stability. Cointegrates containing the *oriC*-*Hinc*II fragment (coordinates -0.18 kb to +1.5 kb) were stable.

Thus, *oriC* replication properties as well as the sequence homology in the 17 kd protein with the *par* function of pSC101 (Miller, C. A. et al., Gene *24*, 309-315, 1983) are not responsible for the instability imposed on pSC101. The destabilizing sequence is distant from *oriC*.

INFLUENCE OF REGIONS UPSTREAM THE PROMOTER FOR THE PRIMER RNA ON THE COPY NUMBER AND STABILITY OF pBR327 - DERIVED PLASMIDS

Lomelí, H., M. Zurita, F. Bolívar and X. Soberón

Centro de Investigación sobre Ingeniería Genética y Biotecnología, UNAM Apartado Posta 70-479, Cd., Universitaria D. F. 04510, Mexico

Replication of plasmid pBR327 and its derivatives proceeds by a ColE1 - type mechanism whose key elements are two RNA molecules (RNA II, and which serves as primer; RNA I, which acts as inhibitor) and the <u>rop</u> gene product (Som et al. PNAS 80: 3232 (1983).

Using recombinant DNA methodologies we have constructed a series of plasmids, derived from pBR327, which altered copy number and stability phenotypes.

One derivative resulted from a deletion that removes from position 2037 to position 2956 of pBR327 (Soberón et al., 1980) (from -43 and up the primer promoter) and eliminates two transcriptional terminators.

This plasmid has an increased copy number, probably due to transcription originating upstream the origin region. Insertion of a small (12 p. b.) DNA segment at position -43 of the primer promoter in this plasmid caused a decrease in copy number to below the levels of the parental plasmid pBR327. Another deletion that removes the most proximal of the upstream promoters also causes a decrease in the copy number. These and other results suggest that the region immediately upstream the primer promoter is important for the replication mechanism.

Our results allow us to draw the following conclusions. They confirm that transcription coming from upstream the RNA II promoter can generate primers and, this increases the copy number. Since the region comprised between -40 and -55 cannot be affected without and effect in copy number, the region may be involved in some kind of interaction with regulatory molecules.

The derivatives of pBR322 studied lack a mechanism of stability independent of copy number; its segregation appears to depend solely on the latter.

FUNCTIONAL LIMITATIONS OF THE *par* LOCUS OF pSC101 AND FURTHER CHARACTERIZATION OF ITS PHENOTYPES

C. Miller, [1]W. T. Tucker and S.N. Cohen

Department of Genetics
Stanford University
Stanford, CA 94305 USA
[1]Present address:
Research School of Biological Sciences
Australian National University
Canberra, ACT 2601, Australia

The *par* region of pSC101 has been sequenced (Miller et al., 1983, Gene 24: 309-315) and deletion analysis has been done to determine the functional *par* locus (Tucker, Miller and Cohen, Cell, 1984). Within a 100 bp sequence there are three discrete PR (partition-related) segments that contribute incrementally to expression of partition.

Deletion of any two segments within the *par* region has been correlated with the Par$^-$ phenotype. Deletion of the entire region leads to more rapid loss of the plasmid than would be predicted from a model involving random distribution of a plasmid pool of the size known to exist for pSC101. Deletions that extend into either of the lateral PR segments, but leave the internal segment and the other lateral segment intact, are stable (i.e. are Par$^+$); however, these show a phenotype described as Cmp$^-$. A Cmp$^-$ plasmid is always lost preferentially when competing with a co-existing incompatible wild-type replicon (hence the name Cmp).

AUTOREGULATION OF THE F FACTOR PIF OPERON

Miller, J.F. and M.H. Malamy

Department of Molecular Biology
Tufts University School of Medicine
Boston, Ma. 02111

The F factor *pif* operon comprises at least three structural genes, *pif*C, *pif*A and *pif*B, located near the primary origin of mini-F replication (oriV1, F coordinate 42.6kb). The *pif*A and *pif*B gene products are responsible for abortive infection of bacteriophage T7. We have studied the regulation of *pif* expression using protein fusions of *pif*C, *A* and *B* to *lac*Z. The product of *pif*C negatively controls its own expression as well as expression of *pif*A and *pif*B. A *lac*Z operon fusion to *pif*A is repressed by the *pif*C product, suggesting that *pif* regulation is at the level of transcription. A Tn5 insertion near oriV1 is polar on *pif*C, *pif*A and *pif*B, showing that these genes are co-transcribed.

We have identified an additional element, *pif*O, which is the site required *in cis* for autoregulation of *pif* expression by the *pif*C product. Spontaneous mutations in *pif*O (*pif*O36, *pif*O55, *pif*O70) result in increased expression of *pif*C-*lac*Z product in the presence of *pif*C *in trans*. Multiple copies of *pif*O (mapping within 42.9kb-42.43kb) *in trans* to F'lac result in a 140-fold increase in T7 inhibition. This is due to titration of *pif*C product by *pif*O sequences *in trans* and concomitant expression of *pif*A and *pif*B. *pif*O36, *pif*O55 and *pif*O70 sequences have little effect on F'lac *pif* expression *in trans*.

Tanimoto and Iino (Mol. Gen. Genet. (1983)192:104-109) have shown that the F factor inhibits transfer of the broad host range plasmid RP4. The gene responsible was localized to the region we assign to *pif*C. Our recent results show that RP4 titrates *pif*C product *in trans* to F'lac and that *pif*C is responsible for F factor inhibition of RP4 conjugal transfer.

IDENTIFICATION AND ANALYSIS OF REPLICATIVE INTERMEDIATES OF *E. COLI ORIC* PLASMIDS DURING *IN VIVO* AND *IN VITRO* REPLICATION

B. Munson and C. Helmstetter

Department of Experimental Biology
Roswell Park Memorial Institute
Buffalo, New York, 14263

Replication of *E. coli* minichromosomes was investigated *in vitro* and *in vivo* to identify and analyze replication intermediates. *In vitro* replication was as described by Fuller *et al.* (Proc. Natl. Acad. Sci. *78*, 7370, 1981). *In vivo* replication was analyzed in temperature-sensitive *dnaA* mutants incubated at various temperatures. Replicative intermediates in both systems were labeled with tritium in pulse-labeled and pulse-chase experiments, and examined by agarose gel electrophoresis and fluorography. The major intermediates detected were strikingly similar *in vivo* and *in vitro*. Minichromosomes replicated rapidly to an OC-OC catenated dimer form, followed by slower progression to CC-OC and CC-CC dimers and finally to CC supercoiled monomers, with ladders of forms which migrated to intermediate positions in the gels. This processing through the dimer forms required several minutes at 25-30°C in both systems. The slow manner in which one form progressed to another compared to the much shorter time necessary to polymerize the nucleotides into DNA, adds further support to the notion that this processing is rate-limiting in minichromosome replication. Further, the similarity of the *in vivo* and *in vitro* events demonstrates the fidelity of the *in vitro* replication system. However, there were differences in the complexity and quantity of some intermediates when different minichromosomes were replicated. Among the most obvious differences was the very early formation of large, open circular multimers in some minichromosomes and not in others, or at very reduced levels. This may provide evidence for initiation of replication on the supercoiled CC-CC dimer of some minichromosome species.

COLE1 PLASMID REPLICATION IN AMBER RNH (RNASE H) MUTANTS OF ESCHERICHIA COLI

S. Naito, N. Inoue and Hisao Uchida

Institute of Medical Science, University of Tokyo
4-6-1, Shirokanedai, Minato-ku, Tokyo 108, JAPAN

We have previously reported a system for isolating ColE1 mutants defective in the initiation of DNA replication in *E. coli* (Naito & Uchida, PNAS *77*, 6744 (1980)). These mutations were named *cer*, signifying a defect in ColE1 replication. One of these mutations, *cer6*, was a single base-pair alteration 160 base-pairs upstream of the ColE1 replication origin. We then proceeded to isolate and characterize *E. coli* suppressor mutants of *cer* (Naito, *et al.*, PNAS *81*, 550 (1984)). These suppressors were named *her* for host factor affecting ColE1 replication. One of the *cer6* suppressors (*herA*) was genetically located within the *rnh* (RNase H) gene. This alteration correlated with the loss of RNase H activity, leading to the suggestion that replication of *cer6* DNA *in vivo* is sensitive to RNase H activity. If this inference is correct, it should be possible to isolate *amber* mutations in the *rnh* gene by selecting for bacterial mutants capable of suppressing replication of *cer6* DNA under *supO* conditions only. Two such bacterial mutants were obtained. By DNA sequencing, both were found to bear *amber* mutation in the *rnh* gene. In these mutants, the wild type ColE1 replicon replicated normally. Therefore, we conclude that RNase H activity is dispensable for replication of ColE1 DNA as well as for the *E. coli* chromosome. This is contrary to the *in vitro* results which show that RNase H activity is essential for replication of ColE1 DNA (Itoh & Tomizawa, 1980).

Another suppressor of the *cer6* mutation (*herB*) was isolated, and was found to be closely linked to the *polA* gene on the genetic map of *E. coli*.

PT181 REPLICATION IS REGULATED INDIRECTLY BY TWO COUNTERTRANSCRIPTS

R.P. Novick, S.J. Projan, C.C.Kumar, S. Carleton, G. Adler, A. Gruss, S. Highlander, J. Kornblum, and S. Iordanescu

Department of Plasmid Biology, Public Health Research Institute of the City of New York, New York, NY USA

The replication rate of pT181 is controlled by an inhibitor-target mechanism similar to that of R1. The target consists of two tandemly initiated leader sequences for the mRNA of a plasmid replication protein, RepC. The inhibitor consists of two countertranscripts complementary to the RepC mRNA leader. These have a common start within the leader region and tandem termination sites about 85 (RNA I) and 140 (RNA II) bp upstream. Most mutations causing increased copy numbers are located in the leader-countertranscript overlap region and involve a GC to AT change in a 12 nucleotide GC-rich segment in that region. Mutations affecting termination of either RNA I or RNA II or their common promoter also cause increases in copy number. In computer-predicted secondary structures of the leader region, the 12-nt target segment forms a single-stranded loop at the end of a hairpin early during transcription of the leader; later, the leader is predicted to re-fold into a new configuration in which the 12-nt segment is largely base-paired. It is likely that complementary base pairing between inhibitor and target RNAs, initiated early during leader transcription, facilitates a secondary folding of the downstream part of the leader that occludes the _repC_ mRNA ribosome binding site within a stem-loop. A deletion that eliminates the countertranscript promoter result in a 15-fold increase in copy number; if the segment that can pair with the ribosome binding site is also deleted, the copy number increases by 50-fold, suggesting that (in the absence of the countertranscripts), the mRNA transcripts are utilized 3x more efficiently if the ribosome binding site is unable to participate in a stable secondary structure. A mutation that decreases the stability of the secondary structure involving the ribosome binding site causes a 20-fold increase in copy number even in the presence of excess RNA I and RNA II.

Because RepC is fully _trans_-active, pT181 and its copy mutants behave differently in _cis_-_trans_ incompatibility tests than do those of the directly regulated ColE1 or IncFII plasmids. Although the cloned pT181 inhibitor gene blocks replication of the WT plasmid and inhibitor-sensitive mutants, dominant copy mutants do not exclude either in the heteroplasmid state, nor do mutations that change the specificity of the inhibitor-target interaction give rise to new incompatibility types. Because all mutants share a common RepC pool, all combinations show symmetric segregational incompatibility with segregation rates that are predictable from copy numbers.

THE P1-PAR REGION CAN PARTIALLY STABILIZE OTHER REPLICONS LACKING A PARTITION SYSTEM

J.H. Pringle and M. Masters

Department of Molecular Biology, University of Edinburgh, King's Buildings, Mayfield Road, Edinburgh EH9 3JR, Scotland

A 3Kb region of P1 [the *par* region] essential for the stable inheritance of the P1 plasmid can also ensure the stable inheritance of F when joined to a *par*$^-$ F replicon. [Austin and Abeles (1983) J. Mol. Biol. *169*, 353.] In order to test whether P1 *par* is able to promote the stable inheritance of other unstable replicons we cloned the *par* fragment of P1 [as defined by Austin and Abeles] into other unstable plasmids. These include the *oriC* plasmids pEH21 and pEH22 [Hinchliffe *et al.*, (1983) Plasmid *9*, 286.], pPM22, a *par*$^-$ plasmid derived from pSC101 [Meacock and Cohen, (1980) Cell, *20*, 529.] and pBR325. The *oriC* and pPM22 derived plasmids were transformed into a *recA* host. The plasmids derived from pBR325 were transformed into WT286, a strain presumed to be partially deficient in *polA* activity because *polA* dependent replicons have a reduced copy no. [4-6]. The stability of the plasmids in these strains was then examined.

pPM22 was completely stabilized by the presence of P1-*par*. The pBR325 and *oriC* P1-*par* hybrids were more stable than their *par*$^-$ parents but were not fully stabilized. The stabilization observed could not be attributed to improved replication as the copy number of the par$^+$ derivatives was not increased. We attribute the inability of the P1-*par* region to fully stabilize the *oriC* plasmids and pBR325/WT286 to the fact that these plasmids probably fail to replicate with complete regularity.

FUNCTIONING OF A TRANSCRIPTION TERMINATOR STRUCTURE IS DEPENDENT ON METHYLATION: ROLE OF METHYLATION IN CLO DF13 PLASMID REPLICATION

A.J. van Putten*[1], R. de lang[1], J.A. van den Berg[2], P. van Solingen[2], E. Veltkamp[1], and H.J.J. Nijkamp[1]

[1] Dept. of Genetics, Biological Laboratory, Vrije Universiteit, Amsterdam, The Netherlands
[2] Gist-Brocades, P.O. Box 1, Delft, The Netherlands

Plasmid Clo DF13 (M.W. 6 Md) contains a relative small DNA region, located between 1.8 and 11.5%, essential for autonomous replication. Several replication mutants of Clo DF13 have been isolated and characterized. One of these mutants, Clo DF13 cop-1 (TS), has a temperature-dependent plasmid copy number and shows a "runaway" behaviour. We have determined the nature of this mutation, located within a region involved in the initiation of RNA II and the termination of the cloacin operon transcript. Our data show that the mutation results in a pleiiotropic effect, namely a non-conditional increase in promotor activity and a temperature dependent decrease in termination efficiency. Recently, we obtained evidence that the absence of dam-methylation in the cloacin terminator structure affects plasmid replication and leads to a loss of the cop-1 (TS) phenotype. In order to study the molecular basis of this phenomenon, we determined the efficiency of termination of the region harbouring the cop-1 (TS) terminator structure in the presence and absence of dam methylation. The data obtained indicate that the absence of methylation results in an increased termination efficiency.

HOST RANGE REPLICATION AND MAINTENANCE DETERMINANTS OF IncP GROUP PLASMID RK2

Schmidhauser, T.J. and Helinski, D.R., Department of Biology, B-022, University of California, San Diego, La Jolla, California 92093 USA

Extensive deletion analysis of the broad host range *IncP* plasmid RK2 has identified two regions of the 56 kb RK2 replicon, a 700 bp origin of replication (*oriV*) and a 1.8 kb region designated *trfA**, which together encode stable replication in *Escherichia coli* and in the distantly related species *Pseudomonas putida* (1, 2, 3). Utilizing the RK2 origin of transfer (*oriT*), the *trfA** and *oriV* segments and larger RK2 derivatives were introduced into and maintenance properties assessed in a variety of Gram-negative species. A *trfA**, *oriV* replicon encodes RK2 replication functions in the nine hosts studied yet maintenance is stable (less than 0.2% loss per generation) in only four of the nine; *Escherichia coli*, *Pseudomonas putida*, *Pseudomonas aeruginosa*, and *Azotobacter vinelandii*. Stable RK2 maintenance in the other five species studied is dependent on two regions of RK2 designated host range maintenance A (*hrmA*) and B (*hrmB*). The *hrmA* and *hrmB* regions map at RK2 coordinates 16.0 to 20.0 and 52.6 to 55.7, respectively. *HrmA* stabilizes mini-RK2 maintenance in *Acinetobacter calcoaceticus* while the *hrmB* locus stabilizes mini-RK2 replicons in *Rhizobium meliloti*, *Argobacterium tumefaciens*, *Rhodopseudomonas sphaeroides*, and *Caulobacter crecentus*.

References

1. Figurski, D.H. and Helinski, D.R. Proc. Nat. Acad. Sci. USA *76*, 1648-1652, 1979.
2. Thomas, C.M. Plasmid *5*, 277-291, 1981.
3. Schmidhauser, T.J., Filutowicz, M. and Helinski, D.R. Plasmid *9*, 325-330, 1983.

REPLICATION DETERMINANTS OF THE BROAD HOST RANGE PLASMID RSF1010

Scholz, P., Haring, V., Scherzinger, E.,
Lurz, R., Bagdasarian, M.M., Schuster, H.
and Bagdasarian, M.

Max-Planck-Institut für Molekulare Genetik,
Abt. Schuster, D-1000 Berlin 33, Ihnestrasse 63,
Federal Republic of Germany

RSF1010 is a small (8.7kb) multicopy plasmid of the IncQ group that has an exceptionally broad host range among Gram-negative bacteria. This latter property has made it attractive for the construction of DNA cloning vehicles that may be used outside enterics and has motivated studies on its mode of replication.

Cloning of specific regions of RSF1010 has shown that a 2.1 kb segment of the plasmid that contains the replication origin (*oriV*) can replicate in *E.coli* if another DNA region, that is located remote from *oriV*, is present in the same cell on an independent plasmid. Deletion analysis of the replication region has revealed three genes, *rep*A, *rep*B and *rep*C, each being essential for replication of the plasmid in *E.coli*. The *rep* genes have been cloned into plasmid ptac12 such that their expression was controlled by the *tac* promoter. Inactivation of the *lac* repressor by IPTG resulted in an overproduction of the *rep*A, *rep*B and *rep*C proteins until they constituted 4 to 20% of the total cellular protein. Direction of transcription of each *rep* gene, as deduced from the cloning experiments, is clockwise on the standard map of RSF1010. This was confirmed by nucleotide sequence determination of the *rep* region.

Our finding that RSF1010, in spite of its small size, has maintained in the evolution a set of at least three replication genes suggests that these genes make the plasmid independent of certain host determined replication functions and thus allow the plasmid to be replicated in a wide range of different bacterial species.

INTERACTION *IN VITRO* BETWEEN RNA MOLECULES INVOLVED IN COLE1 DNA REPLICATION CONTROL

*Tamm,J.,Dooley,T.,Elble,R.,Fitzwater,T.,and Polisky,B.
Indiana University, Dept. of Biology, Bloomington,In. 47405,USA

The plasmid ColE1 encodes a small RNA molecule (RNA1) which is a repressor of plasmid DNA replication. This repression involves an interaction between RNA1 and the replication primer RNA. Previous studies have shown that RNA1 consists of three stem/loop structures and a single-stranded 5'end. Thus far,only the loops have been implicated in the functioning of the molecule.

We have developed an *in vitro* system which allows detection of the complex formed when RNA1 and primer interact. Using this system,we have studied the ability of wild-type and mutant RNA1 molecules to interact with primer RNA. By testing RNA1 molecules with point mutations or small deletions at the 5' end,we have shown that,in addition to the loops,the 5' single-stranded region of RNA1 is also critical to its function. This system has also been used to study the physical parameters (Tm and nuclease susceptibility) of the complex formed by wild-type RNA1 and primer. The data suggest that the end product of the interaction is an RNA:RNA hybrid involving all of RNA1 and the 5' end of primer.

REGULATION OF REPLICATION AND MAINTENANCE FUNCTIONS OF BROAD HOST RANGE PLASMID RK2

*C.M. Thomas, C.A. Smith, V. Shingler, *M.A. Cross, A.A.K. Hussain
Genetics Department, University of Birmingham, Birmingham B15 2TT, England

Plasmid RK2 is capable of conjugal transfer between and stable maintenance in most Gram negative bacterial species. The locations of plasmid genes are given by reference to kb coordinates clockwise from the unique EcoRI site on the recently recalibrated 60 kb circular genome. For replication RK2 encodes $oriV_{RK2}$ (12.3 to 13.0 kb), the origin from which vegetative replication proceeds unidirectionally anticlockwise, and trfA (16.0 to 17.3 kb) which encodes a polypeptide(s) essential for $oriV_{RK2}$ function. In addition to these functions various kil loci which have host or plasmid deleterious effects and whose expression is controlled by kor loci have previously been mapped. By analogy with various genes recently described on plasmid F it seems likely that these kil loci are involved in plasmid stable maintenance.

We have shown that the trfA gene consists of an open reading frame which produces, from two different translational starts, polypeptides of 382 and 285 amino acids, of which at least the smaller has trfA activity. The trfA gene is the second cistron of a polycistromic operon transcribed anticlockwise from a promoter at about 17.6 to 17.7 kb. The kilD locus seems to be part of this operon and may in fact be the first cistron which produces an abundant 116 amino acid polypeptide. Transcription of this 'trfA operon' is negatively regulated by the products of both the trfB/korA/korD locus (58.7 to 59.3 kb) and the korB locus (56.8 to 57.9 kb) both of which are expressed as part of a polycistronic message, transcribed from a promoter at approximately 59.3 kb, which may represent an operon autoregulated by the trfB and korB products. A third locus, incC (57.9 to 59.3 kb), is expressed as part of the trfB, korB operon and expresses incompatibility towards mini plasmids containing korB. Near the trfA operon we have identified a transcriptional unit which we believe is responsible for kilB expression because it is negatively regulated by korB but not by trfB alone.

Deletions in the $oriV_{RK2}$ region have indicated the presence of copy number control functions adjacent to and possibly overlapping with the minimal $oriV_{RK2}$ segment. The nature of these elements is being analysed at present. However, the overriding control of copy number seems to reside in the trfB and korB loci which in conjunction appear to reduce expression of the trfA gene to levels limiting for replication. The possible role of the incC locus in this copy number control circuit remains to be elucidated.

MULTIPLE INITIATION SITES FOR DNA REPLICATION FLANKING THE ORIGIN REGION OF λdv GENOME

Tsurimoto, T., Kouhara, H., Matsubara, K.

Institute for Molecular and Cellular Biology

Osaka University, Kita-ku, Osaka 530, Japan

λdv plasmid carries four 19bp *ori* repeats in the region that is believed to be the initiation site for replication. These repeats are the target for binding of O protein, a λ-specific initiator. We have established an *in vitro* system that allows replication of any circular DNA molecule carrying the λ *ori* region. This paper describes our attempts to locate the site of primer RNA→DNA switch in the initiation process occuring at/around the *ori* region in the *in vitro* λdv replication system.

Early λdv replicative intermediates were prepared in the presence of ddCTP, an inhibitor of DNA chain elongation. Short chain DNAs produced from regions near the replication origin were purified from the intermediates, and those DNAs that are covalently linked to primer RNA were collected for mapping the transition sites from primer RNA to DNA synthesis.

Major transition sites were observed to locate on both sides of the *ori* repeats. In either side of the *ori* region, the transition sites are multiple, and are clustered in one of the two strands in such a way that DNA synthesis from either side proceeds convergently. The frequency of the "leftward" DNA synthesis is several times higher than that of "rightward" synthesis. These results reflect asymmetrical, bidirectional replication of λ.

INCOMPATIBILITY AND THE REPLICATION OF THE INCFI PLASMID COLV2-K94

P. C. Weber, C. Doty, G. Mitra, and S. Palchaudhuri

Department of Immunology and Microbiology
Wayne State University School of Medicine
Detroit, Michigan 48201

The colicin-producing plasmid ColV2-K94 is an IncFI plasmid which we have found to be capable of expressing incompatibility by both partitioning (*incD*-mediated) and replication control (*repA*-mediated) mechanisms. Two overlapping mini-ColV derivatives, pWS15 and pWS16, were obtained by digestion of the ColV plasmid pWS12 with *Eco*RI and *Bam*HI, respectively. Although these plasmids contain an IncFII replicon, the intact ColV plasmid does not express IncFII incompatibility because another replicon (cloned in plasmid pWS14) functions during the plasmid's coexistence with IncFII plasmids. This second replicon shows homology with the secondary replicon of F (f7) and is dependent on the *polA* product of the host cell. The plasmid pWS12 behaves as a member of the IncFII group in a *polA*$^-$ strain, BK202. The 75 kb *Bam*HI-generated pWS16 consisted of an IncFII replicon of ColV, an entire transfer operon, and an *incD* partitionong locus, but lacked the second replicon. It exhibited incompatibility towards both IncFI and IncFII plasmids. The 32 kb *Eco*RI-generated pWS15 lacked the second replicon, the *incD* locus and part of the transfer operon. It was therefore non-transmissible and expressed strong IncFII incompatibility, but lacked the IncFI incompatibility properties of its parent, pWS12.

MUTANT COLE1 REPLICATION PRIMERS HAVE ALTERED SECONDARY STRUCTURES

Wong, E., Ford, M., Tamm, J. and Polisky, B.

Department of Biology, Indiana University, Bloomington, IN 47405

A ColE1 thermosensitive overproducer mutant and several revertants have been isolated. The mutant loci map within the primer RNA coding region, spanning a distance of over 300 nucleotides. The idea that the mutations alter primer secondary structure--the consequences being aberrant primer processing or function--is suggested by:
(a) altered primer RNA mobilities in non-denaturing polyacrylamide gels;
(b) nuclease T1 probing studies which indicate that new cleavage sites are exposed as a result of the mutation;
(c) computer-generated models that show a gross disruption of the most-stable wild type primer structure (at the site where RNAI presumably interacts with primer) when mutant base changes are introduced into the primer sequence.

SPECIALIZED FUNCTIONS

TRANSFERABLE RESISTANCE TO CEFOTAXIME AND CEFOTAXIME IN STRAINS OF KLEBSIELLA PNEUMONIAE IS GOVERNED BY A SINGLE GENE

Martin Antal[1], Vladimír Krčméry[1], Hans Knothe[2]

[1] Research Institute for Preventive Medicine, 83301 Bratislava, Czechoslovakia, and

[2] Hygiene-Institut der Universität, Paul-Ehrlich Str.40, Frankfurt/Main, F.R.G.

We extended the genetic analysis of exconjugants that originated from our transfer experiment in which Cefotaxime, Cefuroxime and Cefamandole resistance determinants could be transferred from nosocomial strains of Klebsiella to recipient E.coli K-12 strains (Knothe et al. Infection 11 : 315-318, 1983). Exconjugant colonies from the transfer plates containing either Cefuroxime or Cefamandole, were picked into minitubes with broth, and, by means of a multiloop applicator the clones were applied on to plates containing only a single drug from the multiple drug resistance spectrum of the original Klebsiella strain. Results obtained from the indirect selection analysis of more than 600 colonies obtained in five cycles of transfer to, and between, E.coli K-12 variants showed an obligatory presence of Cefotaxime, Cefuroxime, Cefamandole and Cephalotin resistance determinants in all exconjugants selected primarily with Cefuroxime, but the presence of Cefamandole and Cefalotin resistance determinants selected with Cefamandole or Cephalotin. As resistance determinant to Cefotaxime could not be separated in as many as five cycles of transfers from that to Cefuroxime, it might be concluded that, in strains studied, the transferable resistance to both these antibiotics is regulated by a single gene.

COMPARATIVE STUDIES OF COLK AND COLE1

* Archer, J.A.C & Sherratt, D.J. Institute of Genetics, Glasgow University, Glasgow G11 5JS, Scotland

The natural multicopy plasmid ColK appears to be a "cousin" of ColE1. Though its genome is larger (7.2kb), its general organization of functions appears to be similar to that of ColE1. In an attempt to understand more about the structure and function of ColE1 and ColK plasmids, we have initiated a DNA sequencing study of ColK. Additionally we have been analysing, at the comparative level, the functions of ColK involved in conjugal mobility and plasmid stability.

At a functional level, the ColK and ColE1 transfer origin regions are interchangeable. A 155bp _Hpa_II fragment of ColK was shown to contain a fully functional transfer origin, and shows a conserved DNA sequence. Similarly the mobility regions that specify proteins required for conjugal transfer are interchangeable for conjugation mediated by the _Inc_Ix plasmid R64 _drd_II (but not for transfer by F'lac: Tacon, W. _et al_, 1981). The regions equivalent to _rop_ (Cesareni _et al_, 1981) and _cer_ (Summers, D.K., Sherratt, D.J., 1984) are present in ColK.

Tacon, W., Bhamra, S., Sunar, B. and Sherratt, D.J. (1981). Plasmid _6_, 358-359.

Cesareni, G., Muesing, M.A. and Polisky, B. (1981). Proc. Nat. Acad. Sci. USA _79_, 6313-6317.

Summers, D.K. and Sherratt, D.J. (1984). Cell _36_, 597-603.

CONTROL OF A CHROMOSOMALLY ENCODED p-CRESOL PATHWAY BY A TOL-LIKE PLASMID IN *ALCALIGENES EUTROPHUS*

R.C. Bayly, E.J.L. Hughes and R.A. Skurray
Microbiology Department, Monash University, Clayton 3168, Australia

A. eutrophus strain 345 metabolizes *m*- and *p*-toluate *via* a catechol *meta*- cleavage pathway. DNA analysis, curing studies and transfer of this phenotype showed that the degradative genes are encoded on an 85 kb plasmid, pRA1000. Restriction endonuclease analysis of pRA1000 showed it to be similar to the archetypal TOL plasmid, pWWO. In strain 345, pRA1000 prevented the expression of chromosomally encoded enzymes required for the degradation of *p*-cresol whereas these enzymes were expressed in strains cured of pRA1000. Studies with an R68.45-pRA1000 cointegrate plasmid, pRA1001, show that the gene(s) responsible for the effect on *p*-cresol degradation resides within or near the *m*- and *p*-toluate degradative region on pRA1000.

Analysis of the enzymes synthesized by wild-type and mutant strains and by strains cured of pRA1000, indicated that two or more isofunctional enzymes mediated several steps in meta-cleavage pathways for the degradation of *m*- and *p*-toluate, phenol and *p*-cresol. The formation of three 2,3-oxygenases and two semialdehyde hydrolases was indicated from an examination of the ratio of the specific activities of these enzymes against various substrates. Evidence for two semialdehyde dehydrogenases, two isomerases and decarboxylases and three hydratases was derived from the induction of these enzymes under different growth conditions. Each activity was detected when strain 345 was grown in the presence of *m*-toluate but not when grown with phenol or *p*-cresol. In strains cured of pRA1000, growth with phenol or *p*-cresol, but not with *m*-toluate, induced these enzymes.

MAPPING OF THE GENOME OF *RHODOPSEUDOMONAS SPHAEROIDES*

A.R. St.G. Bowen and J.M. Pemberton

Department of Microbiology
University of Queensland
St. Lucia, Queensland, 4067
Australia

Photosynthetic bacteria have proven popular subjects for studies on photosynthesis, photochemical nitrogen fixation and membrane biogenesis. The Gram negative, facultatively phototrophic bacterium *R. sphaeroides* has proven an ideal subject for combined biochemical and biophysical studies of photosynthesis. Recent advances in the genetics of this organism indicate that it is amenable to gene mapping and cloning techniques applicable to other Gram negative bacteria. Using the broad host range plasmid pRP1 carrying the mercury transposon Tn*501*, we have been able to obtain Hfr promoted chromosome mobilization; linkage results using this system of chromosome transfer enabled us to produce a map of the *R. sphaeroides* genome which consisted of two linkage groups. The ability of Tn*501* to convert pRP1 to an Hfr was thought to be due in part to its low resolvase activity, which trapped RP1 in a co-integrate with the main chromosome during the transposition of Tn*501*.

The idea that the activity of the resolvase function was critical to Hfr formation was examined using a mutant of Tn*21* deleted for its resolvase function (Tn*21* Δ *trp*R); Tn*21* is closely related to Tn*501* and forms co-integrates as part of the transposition process. The broad host range plasmid pR751 carrying Tn*21*Δ was shown to give Hfr mobilization of the main chromosome in a manner similar to RP1:Tn*501*; the major difference being that now markers on the two linkage groups (obtained with RP1:Tn*501*) show linkage to one another. It is concluded that *R. sphaeroides* has a single linkage group, presumably the main chromosome.

An additional feature of R751:Tn*21*Δ is its ability to form extremely stable co-integrates with the popular cloning vectors pBR322, pACYC184, and pUC8. Indeed the co-integrates are so stable that these vectors can be transferred and maintained (as co-integrates with pR751::Tn*21*Δ) in a wide range of Gram negative organisms including *R. sphaeroides, Cellvibrio mixtus, Aeromonas punctata,* and *Alcaligenes eutrophus. Rhodopseudomonas sphaeroides* genes cloned in *E. coli* using pUC8 can be transferred to and maintained in *R. sphaeroides* and the cloned segment detected by complementation of mutants of *R. sphaeroides* opening the way for more detailed analyses of the genome of this bacterium.

FUNCTION AND CHARACTERISTICS OF THICK AND THIN PILI OF TRANSFER-DEREPRESSED I COMPLEX PLASMIDS IN *E. COLI*

D.E. Bradley, Faculty of Medicine, Memorial University of Newfoundland, St. John's, Newfoundland, Canada, A1B 3V6

IncI_1 and IncI_2 plasmids determine both thin (6 nm) flexible, and thick (10.5 nm) rigid pili. Thick I_1 and I_2 pili are structurally different. Eleven transfer-derepressed plasmids from Inc groups I_1, I_5, B, K, and Z (made in a strain with the *dnaG3* mutation) determined thin and thick I_1-type pili constitutively. Plasmids made with only thick pili transferred well on plates but not in a liquid (surface mating systems), those with both kinds transferring equally in both environments (universal mating systems). IncI_2 plasmid R721*pil*^{c}.4 (made by conventional mutagenesis) determined thin pili constitutively, but very few thick pili; its low transfer frequency (universal system) corresponded with the small numbers of thick pili. The presence of thin R721*pil*^{c}.4 pili on recipient cells converted the mating system of Sa (IncW) from surface to universal. Thus, the thin pili were only "grappling hooks" thick, rigid pili alone mediating fertile cell contact. Likewise, the thick flexible pili of R402 (IncT, universal mating system) had the same effect. All thick flexible pili probably combine the stabilizing function of thin flexible I-type pili with the fertility of rigid pili.

APPARENT AMINO ACID AUXOTROPHY MEDIATED BY E4 COLICIN PLASMIDS

*Brunner, D. P., M. A. Bott, L. R. Allard and E. A. Galella

Microbiology Group, Department of Biological Sciences, Illinois State University, Normal, IL, USA

Transformation with E4 colicin plasmids, but not other E colicin plasmids, resulted in conversion of prototrophic E. coli K12 strain DB364 to apparent amino acid auxotrophy. Supplementation of minimal media with pools of amino acids allowed incomplete and variable amounts of growth; a pool containing Ile, Leu, Val, Met, Arg, Gly and Ser had the greatest growth-promoting effect. Prototrophic revertants were selected from auxotrophs (frequency ca. 3×10^{-4} per cell plated) without detection of plasmid loss. Analyses of the plasmids from such revertants revealed ca. 30% modified to increased sizes (> 4.7 Md), while ca. 70% remained unmodified (4.2 Md). Plasmids from prototrophic and auxotrophic clones were characterized by restriction endonuclease mapping and Southern transfers to localize the relevant plasmid sequence and ascertain the nature of the plasmid modification. Plasmids, modified via tandem duplication, were unable to mediate auxotrophy, while unmodified plasmids retained the ability to mediate auxotrophy. A chromosomal alteration was inferred to have arisen in prototrophic revertants with unmodified plasmids. Examination of detergent susceptibility and leakage of periplasmic enzymes of strain DB364 suggested a permeability barrier defect. Additional data indicated that E4 colicin plasmids perturbed their host's envelope by a mechanism independent of a putative lysis gene which, by analogy with other E colicin plasmids, may exist within the colicin operon. Furthermore, this additional envelope alteration by E4 colicin plasmids resulted in the apparent amino acid auxotrophy of strain DB364.

AEROBACTIN AND THE PATHOGENESIS OF EXTRAINTESTINAL INFECTIONS OF MAN

N.H. Carbonetti and P.H. Williams

Department of Genetics
University of Leicester
Leicester, LE1 7RH, England

Iron is essential for microbial growth. In nature, however, it exists predominantly in insoluble or complexed forms, and therefore various mechanisms have evolved in microorganisms for the specific active uptake of iron. Enteric bacteria secrete enterochelin, a high affinity catechol iron chelator (siderophore) whose synthesis and activity are controlled by chromosomal genes. In addition, some strains of *Escherichia coli* harbour plasmids mediating synthesis of the hydroxamate siderophore aerobactin. These plasmids significantly enhance the virulence of host bacteria in conditions of iron deficiency in experimental infections of mice.

To investigate the ubiquity of the aerobactin system among *E. coli* strains associated with human extraintestinal infections we have screened 451 isolates using two complementary approaches. A biological assay employs a mutant strain (LG1522) defective in both uptake of ferric-enterochelin and biosynthesis of aerobactin; isolates which secrete aerobactin (solely, or in addition to enterochelin) promote growth of this indicator strain by cross-feeding when inoculated onto a lawn on iron-deficient medium. In addition we have used colony hybridisation to detect the presence of specific nucleotide sequences directly rather than by their expression. Probes used in this screening were restriction fragments carrying aerobactin biosynthesis or receptor genes. Results show a significantly higher incidence of the aerobactin system among strains isolated from cases of septicaemia (68.8%), pyelonephritis (75.4%), lower urinary tract infections (58.4%), asymptomatic bacteriuria (61.0%), and cystitis (50.9%) than among human faecal isolates (34.3%). These data indicate that aerobactin-mediated bacterial iron uptake may be an important factor in the pathogenesis of extraintestinal infections of man.

CHARACTERIZATION OF PROMOTERS IN

β-LACTAMASE MUTANTS

*Chen, Shin-Tai and R. C. Clowes

The University of Texas at Dallas
Biology Programs, P.O. Box 688
Richardson, TX 75080

A mutation in an ampicillin-resistance (Ap-r) plasmid, R6K, resulting in a tenfold increase in resistance has been analysed. Previous studies showed the original R6K plasmid had undergone intramolecular transposition to result in the duplication of the transposon, Tn2660. The two transposons were separately cloned and transposed to a small, base-sequenced plasmid, pA03, to derive the constructs pA03::Tn2660, which specifies the normal (R6K) level of Ap-r, and pA03::Tn2661, which specifies a tenfold increase in Ap resistance.

Restriction fragments which carry DNA sequences immediately 'upstream' of the initiating codon for β-lactamase were excised from each construct and used to measure the size of *in vivo* RNA transcripts from each construct by the Berk and Sharp (1977) method. Two new promoter sites were identified on Tn2661. Base-sequencing of Tn2660 and Tn2661 DNA showed the same sequence as Tn3, except that Tn2661 had a single base-pair substitution at coordinate 3773.

This single base change can account for the synthesis of the two new messenger RNAs identified, in one instance by the creation of an improved '-35' region, and in the other by the creation of an improved -10 'Pribnow' box. Similar experiments have verified the sites of the two β-lactamase promoters on pBR322, and a plasmid has been constructed in which the new promoter sites of Tn2661 have been substituted in pBR322 to result in transcription from four β-lactamase promoters.

ISOLATION OF A LARGE PLASMID FROM *CLOSTRIDIUM THERMOACETICUM*

*Chou, Chih-Fong, Laura H. Carreira, and Lars G. Ljungdahl

Biochemistry Department and Center for Biological Resource Recovery, University of Georgia, Athens, GA., USA

Clostridium thermoaceticum, a thermophilic anaerobe, ferments one mol of glucose to almost three mol of acetate. It also synthesizes acetate from CO_2/H_2 or CO. The pathway of acetate synthesis is as follows: CO_2 → formate → formyltetrahydrofolate → → methyl-H_4folate → methyl-corrinoid. The methyl group of the methylcorrinoid with either carbon monoxide or the carboxyl group of pyruvate yields acetate. Two strains of this bacterium (one maintained in H. G. Wood's laboratory and one carried by us) have been found to contain a single large plasmid. It was isolated by the high pH technique of Anderson and McKay (Appl. Environ. Microbiol. 46, 549, 1983). It has 87 kb as determined by agarose gel electrophoresis. The number of copies of the plasmid is amplified by the addition of chloramphenicol (70 to 80 μg/ml, acridine orange 6 μg/ml, rifamycin 30 μg/ml and chloramphenicol, 30 μg/ml. The relationship, if any, of the plasmid to the acetate biosynthesis pathway or to the resistance of drugs is not clear. Efforts are being made to characterize the plasmid and to cure the bacterial strains of it.

MOLECULAR AND PHYSIOLOGICAL STUDIES OF TETRACYCLINE RESISTANCE MEDIATED BY *STREPTOCOCCUS FAECALIS* PLASMID pCF10

P. J. Christie and G. M. Dunny, Department of Veterinary Microbiology, Cornell University, Ithaca, New York

The sex pheromone-dependent tetracycline resistance plasmid pCF10 (36.3 megadaltons) of *Streptococcus faecalis* has been physically mapped by restriction analysis. Southern hybridization experiments have revealed that pCF10 carries two distinct tetracycline resistance determinants. One region of pCF10 shows homology to the determinant of the non-conjugative plasmid pAMα1, whereas a second region shows homology to the tetracycline resistance element, Tn916. Our evidence indicates that pCF10 may carry the entire transposon. When wild type donor cells carrying pCF10, or mutant plasmid derivatives selected for on the basis of increased donor ability or decreased levels of tetracycline resistance, were repeatedly transferred in broth containing 10 ug/ml of tetracycline, five-to-ten fold increases in levels of resistance were observed. Wild type donor cells and some mutant strains passaged in tetracycline showed an increase in cell aggregation and plasmid transfer ability, reminiscient of sex pheromone response. In the case of the wild type pCF10, addition of 10 ug/ml tetracycline to a donor culture maintained at a density of 10^7 cfu/ml (by periodic dilution) resulted in a transient decrease in donor potential, followed by an increase of 1-2 orders of magnitude after six generations in the presence of tetracycline.

GENETIC ANALYSES OF THE *VIR* GENES ON THE TIC58 PLASMID OF *AGROBACTERIUM TUMEFACIENS* - PROMOTERS AND POLYPEPTIDES

*T.J. Close, M. Hagiya, B. Tait, R.C. Lundquist,
D. Zaitlin and C.I. Kado
Plant Pathology Department
University of California
Davis, USA

We have constructed a series of new plasmids that are suitable as general cloning vectors in *Eschericia coli* and *Agrobacterium tumefaciens*. Plasmids pUCD2 and pUCD206 are amplifiable in *E. coli* and contain a broad host range *Inc*W origin of replication. Using these vectors we have shown that some genes in the virulence (*vir*) region of the tumor inducing plasmid pTiC58 are expressed constitutively in *A. tumefaciens*. RNA polymerase binding studies on the *vir* region have identified several sites of tight binding, further suggesting that there are multiple promoters within this area. We have mapped several polypeptides encoded by *vir* region genes using *E. coli* mini-cells and maxi-cells, and through a coupled in vitro transcription-translation system prepared from *A. tumefaciens*. The six kilobase restriction fragment *Kpn*I-11 encodes at least six polypeptides, four of which are abolished by a Tn5 insertion. DNA sequences analysis of this cluster has shown that this polar Tn5 insertion lies within the coding region of the first polypeptide, suggesting the existance of a *vir* operon. Fusion of a *cat* gene to various positions within this gene cluster allowed us to demonstrate its constitutive expression. In contrast, *cat* fused to an adjacent gene was expressed only in association with plant cells.

GENES FOR ANTHRANILATE SYNTHASE IN PSEUDOMONAS AERUGINOSA ARE UNUSUALLY CONDENSED

*Crawford, Irving P. and Eberly, Lee

Microbiology Department, University of Iowa, Iowa City, Iowa 52242

The adjacent genes for the two subunits of anthranilate synthase have been cloned from an R68.44 R prime plasmid originally isolated by Robert Hedges into several *E. coli* plasmid vectors. Initially weak expression was improved somewhat by growth rate selection in an *E. coli trpE* deleted host. A contiguous DNA sequence was obtained extending from 92 bp upstream to 490 bp downstream from the *trpG* coding sequence. The *trpG* gene product of *P. aeruginosa* itself is 200 amino acids in length and shows the five areas of sequence homology Brian Nichols finds to be conserved among glutamine amido-transferase subunits of anthranilate and p-aminobenzoate synthases. Surprisingly, the sequence does not resemble that of *P. putida* anthranilate synthase glutamine amidotransferase subunit any more than those of *E. coli* or other microbes. The sequence downstream from *trpG* contains no obvious candidate for an additional gene product. The sequence upstream has in one reading frame a coding sequence with 18 of 37 amino acids homologous to the C-terminus of *E. coli trpE*. This overlaps the *trpG* reading frame by 23 bp, including the termination codon (TAA). There is a typical ribosome binding sequence also included in the *trpE* gene about 10 bp ahead of the *trpG* start codon. In all, it appears that approximately 35 bp of *P. aeruginosa* DNA have a dual function, encoding the end of *trpE* and the beginning of *trpG*. This arrangement is sufficiently unlike the small, 1-4 bp overlaps reported in other prokaryotic operons to suggest that it may not favor translational coupling between the two gene products.

CONTROL REGION FROM *PSEUDOMANAS AERUGINOSA* PLASMID pMG7 WHICH REGULATES THE EXPRESSION OF THE *Pae*R7 RESTRICTION-MODIFICATION SYSTEM IN *E. COLI*

*Gingeras, T. R.[1], Theriault, G.[2], Brooks, J.E.[3]

[1]Salk Institute of Biotechnology and Industrial Associates, San Diego, CA; [2]Department of Biochemistry, Laval Université, Ste, Foy, Quebec; [3]New England Biolabs, Beverly, MA

A segment of the *Pseudomanas aeruginosa* plasmid pMG7 containing the restriction modification system *Pae*R7 has been studied. A 3.8kb DNA fragment containing both the endonuclease and methylase genes was cloned into pBR322 and expressed in *E. coli*. A series of Bal31 mutants have been generated from this original clone allowing for the identification of the control region of this system. The ability of these clones to restrict phage was shown to be variable depending on the segment of the control region deleted. In one clone, the expression of both the endonuclease and methylase genes is observed without restriction of invading phage.

DNA sequence analysis and subcloning experiments of this control region indicates a potentially novel regulation system for these genes.

GENETICS OF AMPLIFICATION OF TRANSPOSON TN9 IN ESCHERICHIA COLI

S.K. Mahajan, N.N. Pandit, and J.S. Sarkari

Biology and Agriculture Division
Bhabha Atomic Research Centre
Bombay 400085, India

We have isolated ten Tn9 insertion mutations in *E. coli* K12 in each of which the inserted transposon inactivates a gene required for its own amplification. These mutations, which have been designated *aml*, map in five well separated chromosomal loci and identify at least six different genes. They have been transferred to different genetic backgrounds and the resulting strains characterized in respect of the residual Tn9 amplification proficiency general recombination, UV sensitivity and transposition. These studies, along with investigations of the effects of different Rec mutations on Tn9 amplification, have led to the following conclusions: (i) Though wild-type alleles of all the *aml* genes are needed for amplification of a Tn9 located on the bacterial chromosome, several of these genes are not essential for amplification of the same transposon when it is located on the plasmid prophage of P1. It appears that a plasmid-borne transposon requires a simpler genetic apparatus for its amplification than a chromosomal transposon. (ii) Amplification of even a plasmid-borne transposon requires at least one host pathway of general recombination in active state. (iii) The *aml* mutations do not significantly affect UV-sensitivity, but two of them completely block the transpositional activation of the *bgl* operon suggesting a overlap between the transposition and amplification processes. Two of the *aml* mutations identify new *rec* genes. (iv) The F-plasmid carries genes which can amplify a chromosomal Tn9 even in the absence of functional host *aml* genes.

The data in support of these conclusions will be presented. Their bearing on the evolution of plasmids and a new model of gene amplification suggested by our results will be discussed.

LOCALIZATION OF A FUNCTIONAL DOMAIN IN THE COLICIN IA AND COLICIN IB MOLECULES

John A. Mankovich and J. Konisky

Department of Microbiology
University of Illinois, Urbana, USA

The colicin Ib structural and immunity genes reside on the 60 MD low copy number plasmid Col Ib-P9. The colicin Ib structural gene has been cloned into the plasmid vector pBR322. Observations of growth characteristics of strains carrying the cloned gene and analysis of insertion and deletion derivatives of the recombinant plasmid have led to the conclusion that colicin Ib production can occur without the colicin Ib immunity system. The direction of transcription for the colicin Ib gene was determined by use of an expression vector and subsequently confirmed by DNA sequence analysis of the colicin Ib promoter. The promoter was found to contain a sequence homologous to the base sequence of the binding site for Lex A protein which is involved in "SOS" regulation. Hybrid genes were constructed by ligating cloned fragments of the colicin Ib gene with fragments of the cloned colicin Ia gene and transforming them into *E. coli*. From an analysis of the immunity specificities of the chimeric colicins and the locations of the fusions, we concluded that the information necessary for immunity recognition by colicin Ia and colicin Ib resides in the C-terminal half of the colicin molecules.

FREQUENCY OF THE TetA, TetB, AND TetC CLASSES IN *SALMONELLA* AND *SHIGELLA* STRAINS ISOLATED IN MEXICO CITY

J.M. Martínez-Salazar and M.C. Gómez-Eichelmann, Instituto de Investigaciones Biomédicas, UNAM, P.O. Box 70228, 04510 México D.F. MEXICO.

In this work, we determined the level of induced tetracycline (Tc) resistance, the colicin production, the plasmid content, and the frequency of classes TetA, TetB and TetC in 20 *Salmonella* and 34 *Shigella* clinical strains isolated in Mexico City in 1978-1979. All the strains were resistant to streptomycin (Sm), Sulfonamide (Su) and Tc.

Our results showed that there were differences in the level of induced Tc resistance, colicin production and plasmid band pattern not only in *Salmonella* and *Shigella* strains, but also between the *Sh. sonnei* and *Sh. flexneri* strains. The class most frequently present in the *Salmonella* and *Sh. flexneri* analyzed was TetB. A high percentage of *Sh. sonnei* (73%) did not hybridize with the Tet classes tested, the majority of these strains were resistant to low Tc concentrations (25 μg/ml). The TetD class, represented by the plasmid RA1, conferred resistance to high Tc concentrations to *E. coli* and *Sh. sonnei*. This fact suggests that the class or classes of *Tet* genes present in *Sh. sonnei* were different from the four classes reported in Gram-negative bacteria.

CONSTRUCTION OF A SINGLE STRANDED DNA PLASMID CONTAINING A PROMOTER FOR SP6 RNA POLYMERASE

*Mead, D. A., Skorupa, E., and B. Kemper, Department of Physiology and Biophysics, University of Illinois at Urbana-Champaign, Urbana, IL 61801

A major obstacle in studying biochemical processes involving RNA transcripts has been the difficulty in obtaining sufficient quantities of purified RNA species. Two recently developed promoter plasmids, pSP64 and pSP65, are capable of directing *in vitro* transcription of large amounts of unique RNA. The value of these plasmids lies in the high specificity of purified bacteriophage SP6 RNA polymerase for the SP6 promoter DNA, and the multiple cloning site, oriented in both directions, immediately following the promoter. To further extend the usefulness of these plasmids, it would be advantageous to obtain them as single stranded DNA (ssDNA) for mutagenesis and rapid sequencing purposes. We have constructed four ssDNA promoter plasmids by inserting a 514 bp restriction fragment containing the replication origin and morphogenetic signal of the ssDNA bacteriophage fl into pSP64 and 65, in both orientations. Because the orientation of the fl fragment determines which of the two DNA strands of the plasmid will be packaged as the viral particle, the resulting plasmids, pUI+4 and +5, and pUI-4 and -5, are secreted as either sense or anti-sense strands, respectively, relative to the RNA transcript. Secretion of the pUI vectors as ssDNA upon superinfection with the interference resistant phage IR1 has been demonstrated by agarose gel electrophoresis, the ability to transduce the ampicillin marker, and dideoxy sequencing. We are presently cloning cDNA to parathyroid hormone (PTH) mRNA in pUI+4 and -4, for subsequent *in vitro* transcription and translation. In addition to the more traditional plasmid DNA manipulations possible with the pUI vectors, we will apply ssDNA mutagenesis techniques on either strand of the PTH insert and rapidly confirm the change by dideoxy sequencing. Thus, microgram quantities of mutant PTH RNA transcripts will be rapidly synthesized for subsequent functional analysis.

GENETIC PROPERTIES OF LARGE PLASMIDS CONTROLLING RESISTANCES TO HEAVY METALS IN *ALCALIGENES EUTROPHUS CH34*

M.Mergeay(1), M.Faelen(2), and H.G.Schlegel(3)

(1) Département de Radiobiologie, SCK/CEN, B-2400 Mol, Belgium (2) Service de Génétique, ULB, Rhode-St.-Genèse, Belgium (3) Institut für Mikrobiologie der Universität, Göttingen, B.R.D.

A. eutrophus CH34 is a facultative chemolithotroph carrying two large plasmids : pMOL28 (165 kb ; resistances to Cobalt and Nickel) and pMOL30 (220 kb ; resistances to Cadmium, Cobalt and Zinc). These plasmids poorly transfer from the wild type strain to the plasmid free derivate AE104, the frequencies being around 10^{-7} transconjugant per donor cell. The transfer of both plasmids was strongly stimulated by IncP1 plasmids as RP4 and pUZ8 (52 kb ; Km, Tc, Hg). RP4 mobilized either pMOL28 or pMOL30 at frequencies up to 10^{-3} transconjugant per donor cell.

DNA from nickel-resistant and zinc-resistant transconjugants obtained with RP4 or pUZ8 helper plasmids was analyzed in agarose gel electrophoresis according to Kado & Liu.

The RP4 mediated Ni^r and Zn^r transconjugants contained the pMOL28 or pMOL30 plasmid and all of them also carried an enlarged RP4 plasmid. Some of these became also sensitive to tetracycline. Rearrangements including cointegrates, deletions or insertions in pUZ8 were also observed in pUZ8 mediated Zn^r or Ni^r transconjugants.

We analyzed by digestion with endonucleases three insertions in RP4 and one in pUZ8, all coming from Zn^r transconjugants : all of them contain a similar HindIII fragment (6 kb) and a similar EcoRI fragment (∿ 2.5 kb). It suggests the existence of a common transposable element. One of these insertions in RP4 was mapped inside the 2 kb KpnI-BglII fragment of RP4 thus near oriV. Insertions in RP4 or pUZ8 isolated from Ni^r transconjugants have a distinct restriction pattern.

pUZ8 plasmids which have received an insertion in Ni^r transconjugants are also able to mobilize at frequencies of 5.10^{-7} R1-Cm, a 5 kb Tra^- Mob^- plasmid derived from IncFII plasmid R1 (L.Roueach, pers.comm.)pUZ8 is unable to mobilize this plasmid (< 10^{-8}).

These results suggest the presence of different transposable elements in pMOL28 and pMOL30.

DNA SEQUENCES OF THE MERCURIC RESISTANCE GENES OF PLASMID R100 AND TRANSPOSON TN501

Tapan K. Misra, Nigel L. Brown*, Annette Schmidt, Linda Haberstroh, George Jander, and Simon Silver. Washington University, St. Louis, MO 63130, U.S.A., and *Department of Biochemistry, University of Bristol, Bristol, U.K.

Five genes in the mercuric resistance operon of plasmid R100 have been identified by Tn5 insertion mutation studies (ref. 1). A comparison of DNA sequence data from R100 with that from Tn501 show close homologies in the plausible reading frames for the genes predicted from the genetic studies (ref.2). The sequence in the operator promoter region is most highly conserved. There are many wobble changes in the coding regions. The *merR* gene contains 180 base pairs both in R100 and Tn501. There are seven amino acid changes in these arginine-rich serine-rich *merR* polypeptides. The *merT* gene codes for a very hydrophobic polypeptide of 116 amino acids and the *merC* for a polypeptide 91 amino acids long with a hydrophobic amino-terminal end. Whereas, the *merR*, *merT* and *merC* genes are of the same length in Tn501 and R100, the *merA* genes for the mercuric reductase enzyme subunit (1.7 kb) differ slightly in length. Comparisons within the reading frames indicate an approximately 85% base pair identity between the R100 and the Tn501 systems with approximately 90% amino acid identities. However, within the genes some regions are more conserved than others.

1) N. NíBhriain, S. Silver and T.J. Foster (1983). Tn5 insertion mutations in the mercuric ion resistance genes derived from plasmid R100. J. Bacteriol. 155: 690-703.

2) T.K. Misra, N.L. Brown, D.C. Fritzinger, R.D. Pridmore, W.M. Barnes, L. Haberstroh and S. Silver (1984). Mercuric resistance operons of plasmid R100 and transposon Tn501: the beginning of the operon including the regulatory region and the first two structural genes. Proc. Natl. Acad. Sci. USA, in press.

GENETIC ENGINEERING WITH ACETOGENS AND METHANOGENS USING PLASMIDS AND PHAGES

Arun K. Misra, Department of Biology, University of Mississippi
Oxford, MS 38677

Both methanogens and acetogens constitute of several industrially important bacteria, which are source of useful chemicals and provide with means of studying microbial physiology and genetics in fascinating ways. Little attempt has been made to genetically engineer these organisms for desired and higher yields of chemicals. Physiologogical limitations of these organisms may also be genetically manipulated to enable them to produce more and desired chemicals, and make them to grow autotrophically or on agricultural wastes.

Molecular techniques have been applied to species of Methanosarcina and Methanobacterium(Reeve 1981). Cloning of genes from methanogens using phage lambda and cosmids are being attemptec. Plasmids occur in Halobacterium(Thom 1983). Clostridium butyricum and C. aceobutylicum(Minton and Morris 1981, Truffaut and Sebald 1983) contain plasmids. Transfer of genetic information between species of clsotridia have been detected by Magot(1983). Genes from species of Clostridium have been cloned in phages and made to get expressed in E. coli. Phages of clostridia have been discussed by Ogata and Hongo(1979), and Reid et al.(1983). Attempts are being made elsewhere also to isolate phages and plasmids from acetogens.

The interesting and useful enzyme complexes of these organisms make them very promising systems for applying recombinart DNA technology, plasmid biology and genetic engineering to obtain basic scientific informationon on physiology and genetics of acetogens and methanogens, especially thermophiles; as well as their potential use in industrial processes.

The enormous scope in this area has not been fully anticipated, and there are definite limitations due to the lack of proper experimental techniques related to these unusual organisms, that hinders breakthroughs in genetic engineering with these important group of anaerobes and thermophiles.

Hollaender,A.(ed) 1982 Genetic Engineering of Microorganisms for Chemicals. Plenum Press,NY.

Krumphanzl,V.,B.Sikyta and Z.Vanek(eds) 1982 Overproduction of Microbial Products. Academic Press, NY.

Lambe,D. and K.J.Mayberry-Carson(eds) Anaerobic Bacteria. Plenum,NY.

SELECTIVE EXPRESSION OF A PLASMID GENE DURING BACILLUS SUBTILIS SPORULATION

*S. Mongkolsuk and P.S. Lovett

Department of Biological Sciences
University of Maryland Baltimore County
Catonsville, USA

The transition of *Bacillus subtilis* from vegetative growth to sporulation is accompanied by changes in the predominent forms of RNA polymerase within the cells. The various enzyme forms differ in the sigma-like subunits that associate with a common core enzyme, and at least four forms of RNA polymerase have been identified in vegetative and early stationary phase cells. As cells enter sporulation a sporulation-specific RNA polymerase form, sigma-29, appears. We have observed that the *cat-86* gene in plasmid pPL603, which specifies chloramphenicol acetyltransferase, is selectively expressed in *Bacillus subtilis* at a stage in sporulation in which internal spores are first observed (approximately T8). The gene is unexpressed in vegetatively growing cells. *cat-86* expression and spore formation are both blocked when cells are grown in excess glucose. *cat-86* expression at T8 is due to selective transcription of the gene, since cat-86 mRNA is undetectable in vegetatively growing cells, but is readily demonstrated in sporulating cells. The transcription start site for *cat-86* mRNA from sporulating cells is within a 203 bp restriction fragment designated P1, which is located upstream from the *cat* coding region on pPL603. Deletion of P1 from pPL603 eliminates the sporulation-associated expression of *cat-86*. Host sporulation genes whose function is absolutely required for *cat-86* expression at T8 include six early sporulation, *spo0*, genes and *spoIIE*. pPL603 therefore provides the first system in which the *in vivo* expression of a known, plasmid-linked gene is dependent on sporulation-specific changes in *B. subtilis*.

REGULATION OF TRANSCRIPTION INITIATION OF *xyl* OPERONS ON TOL PLASMID

T. Nakazawa*, S. Inouye, and A. Nakazawa

School of Allied Health Sciences, Yamaguchi University* and Department of Biochemistry, Yamaguchi University School of Medicine, Ube 755, Japan

The TOL plasmid from *Pseudomonas putida* mt-2 specifies enzymes for complete degradation of toluene and xylenes. The genes for the enzymes are organized in two regulatory blocks. The *xylABC* operon directs the synthesis of enzymes for conversion of toluene to benzoate, and the *xylDEGF* operon for complete degradation of benzoate or *m*-toluate. Analyses using the molecular cloning technique revealed the following regulatory mechanism of gene expression: The gene product of the regulatory gene *xylR* activates the *xylABC* operon in the presence of *m*-xylene or *m*-methylbenzyl alcohol. The activation of the *xylDEGF* operon is achieved by the product of another regulatory gene *xylS* in the presence of *m*-toluate. Both *xylR* and *xylS* products are required for the *m*-methylbenzyl alcohol-dependent activation of the *xylDEGF* operon.

The nucleotide sequence of the regulatory region of each operon was determined. The transcription initiation sites *in vivo* were determined by S1 nuclease and reverse transcriptase mapping. The transcription of the two operons in *Escherichia coli* started exactly at the same site as in *P. putida*, and the amounts of mRNA synthesized in *E. coli* were substantially lower than those in *P. putida*.

The nucleotide sequence of *xylABC* promoter showed some homology with *xylDEGF* promoter in -10 and -35 regions. These sequences, however, were not homologous with the consensus sequences of -10 and -35 regions found in *E. coli* promoters. A sequence complementary to the 3' ends of 16s rRNA of *Pseudomonas aeruginosa* and *E. coli* existed in front of the predicted start codon of the first gene of each operon. These results indicated that the inefficient expression of TOL genes in *E. coli* is due to the poor start of transcription.

PLASMID AND CHROMOSOMAL INDOLEACETIC ACID GENES OF *PSEUDOMONAS SAVASTANOI*

Palm, C. J. and T. Kosuge

Department of Plant Pathology
University of California
Davis, California 95616

Pseudomonas syringae pv. *savastanoi* is a plant pathogenic bacteria which produces indoleacetic acid (IAA). The production of IAA by the bacteria is a requirement for the induction of galls in the host plants, olive and oleander. Tryptophan is converted to IAA by *P. savastanoi* in two steps with indoleacetamide as the intermediate. In strains of *P. savastanoi* isolated from oleander galls the genetic determinants for IAA production are located on plasmid DNA. Yet in strains of the bacteria isolated from olive galls, these genes are located on chromosomal DNA. From different oleander isolates three unique plasmids of 52, 73, and 70 kb (pIAA1, pIAA2 and pIAA3 respectively), have been found which carry the IAA genes. The region of DNA encoding the IAA genes has been cloned in a wide host range plasmid. Evidence obtained from restriction enzyme disgests and southern blot analysis indicate that these plasmids contain large regions of homologous DNA. These experiments show that the region of DNA which contains the IAA structural genes (and possibly other functions) is conserved in the three plasmids. There appears to be little or no difference between the IAA synthesis genes from the different plasmid and the chromosomal isolates of the bacterium. (Aided by grants from the USDA and NSF).

PLASMIDS OF A NOVEL TYPE FROM PENICILLINASE PRODUCING HAEMOPHILUS INFLUENZAE

J.C. Piffaretti, B.Borer-Bigliardi, A.Arini and R.Peduzzi. Istituto Cantonale, 6904, Lugano, Switzerland

H.influenzae, as a causitive agent of meningitis, septicemia, and epiglottitis in infants, is an important pathogen. Since 1975, penicillinase producing strains have been reported throughout the world; these strains are characterized by two types of plasmids : those belonging to the first type are larger than 45 kb and share a common core; those belonging to the second type are 7.4 kb large and are similar to the resistant genetic elements found among N.gonorrhoeae; both types of plasmids carry transposons of the Tn3 family or deletion derivatives of them.

During an epidemiological survey of the resistant H.influenzae in Switzerland, two penicillinase producing clinical isolates characterized by a novel type of resistant plasmid wer found, one in Bern and the other in Geneva. These genetic elements, pPJ301 and pPJ302, are 10.0 kb large and carry Tn3-like transposons as demonstrated by transposition assays on λ DNA, heteroduplex analysis, and Southern experiments using TN2301 (a Tn3-like transposon) as a probe. A restriction map with different endonucleases has been constructed. Transformation experiments showed that pPJ301 and pPJ302 can replicate in E.coli; however, when intrcduced into E.coli C600, these plasmids frequently originate larger derivatives characterized by DNA insertions up to 1.0 kb. Work is in progress to analyse the nature of these insertions so as the origin of these plasmids of a novel type in H.influenzae.

TRANSPOSON MUTAGENESIS AND CLONING STUDIES OF A CITRATE-UTILIZATION DETERMINANT FROM A PLASMID IN *ESCHERICHIA COLI*

Masanori Sasatsu, Ralph H. Hartke, Jr., and Simon Silver
Washington University, St. Louis, MO 63130, U.S.A.

Atypical *E. coli* strains that can utilize citrate as sole carbon source frequently harbor plasmids with the gene(s) governing citrate utilization. These plasmids code for a citrate transport system (C.H. Reynolds and S. Silver, 1983, J. Bacteriol. 156: 1019-24). The citrate utilization determinant from a 200 kilobase (kb) natural plasmid was cloned as a 9 kb *Pst* I fragment (designated pWR61). Citrate-non-utilizing mutants of pWR61 were isolated with transposon Tn*5* and the physical location of the insertions determined by restriction endonuclease mapping. Then, the citrate utilization determinant was subcloned as 4.7 kb *Pst*I-*Eco*RI fragment (pMS185). To locate the citrate utilization determinant more precisely, derivatives of pMS185 were constructed. The citrate determinant is located in a 1.7 kb *Bam*HI-*Bgl*II fragment. DNA sequencing analysis of the citrate determinant is proceeding now by methods using M13 phage. ^{14}C-citrate transport properties of these mutants were determined to correlate genetic and physiological characteristics.

CONSTRUCTION OF CHIMERIC OPERONS FOR SPECIALTY CHEMICAL PRODUCTION

T. Osslund, *M.J. Simon, M. Joyce and B. Ensley.

Amgen, 1900 Oak Terrace Lane, Thousand Oaks, CA 91320

Combining genes from different microorganisms and controlling their expression with a common regulatory element is a promising route to the biosynthesis of chemicals not ordinarily produced by any single microorganism. We have used this approach to combine the naphthalene dioxygenase genes from _Pseudomonas putida_ with the tryptophanase gene from _Escherichia coli_ to produce the dye indigo. The action of tryptophanase produces indole, which is oxidized to an indigo precursor by the naphthalene dioxygenase enzyme complex.

Construction of the Indigo Operon involved the cloning of 5 genes: the 4 structural genes for the components of naphthalene dioxygenase, namely the flavoprotein, the ferredoxin, and the two subunits of the iron-sulfur protein and the structural gene for tryptophanase. Constructions were studied with the Indigo Operon under the control of the _lac_ promoter. Higher enzyme levels were observed from promoter-proximal genes than from promoter-distal genes. The highest level of indigo production was observed when the tryptophanase gene was placed immediately downstream from the _lac_ promoter followed by the naphthalene dioxygenase genes. Our results suggest that designing chimeric operons for specialty chemical production may be a feasible alternative to current chemical syntheses.

TOXEMIA DE LA PREÑEZ EN OVINOS-TRATAMIENTO MEDIANTE HISTEROTOMIA EXCLUSIVAMENTE

J.R. Spinelli
San Martin s/n
7007 San Manuel
Pcia. de B. Aires, Argentina

La experiencia se realizó en una majada de 26 ovejas Corriedale y un carnero, en los que se detectó la epizootia. El objetivo fué comprobar los resultados obtenidos unicamente mediante operación cesarea, sin complementos terapeuticos.

Se utilizó instrumental común de cirugía, practicandose anestesia epidural con 10 cc de Novocaína al 2%. Se realizó incisión en región ventro-lateral izquierda, siendo el tiempo promedio que insumió cada operación de 12-14 min. La incidencia fué del 30, 7%. De los 8 animales operados, 3 vivieron (37, 5%) y 5 murieron (62, 5%) entre las 16 y 36 hs. posteriores a la cesarea.

Todas las enfermas eran melliceras en el último mes de gestación. En todos los casos se intervino quirurgicamente en el último período de la enfermedad cuando ya había decúbito permanente, pedaleo, opistótonos, rechinar de dientes y sialorrea.

Los resultados se consideran satisfactorios y es de esperar que la letalidad sea menor actuando ante los primeros signos.

STABILITY FUNCTIONS OF COLE1

Summers, D.K. & Sherratt, D.J. Institute of Genetics, Glasgow University, Glasgow G11 5JS, Scotland

The high copy number plasmid ColE1 which appears to be partitioned randomly at cell division, requires two independent functions for its stable maintenance. The first (*cer*) is a region of less than 280 bp required in *cis* for the resolution of plasmid multimers to monomers. Multimerisation of *cer*⁻ plasmids by homologous recombination reduces the number of segregating units per cell and hence causes instability (Summers & Sherratt, 1984). *cer* contains the site for *trans*-acting functions encoded by the chromosome of many *E. coli* strains and by ColE1 itself.

The *rop* protein (Cesareni *et al*, 1981), in addition to its activity as a repressor of plasmid replication, is also required for the stable maintenance of ColE1. *rop*$^+$ plasmids are more stable than their *rop*$^-$ counterparts, even though their copy number is lower. We propose that, in addition to reducing mean copy number, *rop* reduces the copy number variance in a cell population thus removing a sub-population of low copy number cells which would otherwise generate plasmid-free segregants at relatively high frequency.

Cesareni, G., Muesing, M.A. & Polisky, B. (1981). Proc. Nat. Acad. Sci. USA **79**, 6313-6317.

Summers D.K. & Sherratt, D.J. (1984). Cell **36**, 597-603.

THE ROLE OF PLASMIDS IN GROUP N LACTIC ACID STREPTOCOCCI FROM DAIRY STARTER CULTURES

Teuber, M., Geis, A., Andresen, A., and Neve, H., Bundesanstalt für Milchforschung, D-2300 Kiel, Fed. Rep. Germany

Lactic acid streptococci are needed for the production of fermented milk and cheese. The following functions are linked to plasmids (3): transport and metabolism of lactose, transport of citrate, proteolysis of caseins, protection against bacteriophages by restriction/modification and synthesis of bacteriocins. 2 to 15 different plasmids in sizes from 1 to 80 Mdal have been detected in 400 strains (1). The plasmid profile of a strain is stable and strain-specific in milk, allowing determination of strain composition of single, multiple and mixed strain cultures. European mixed cultures do contain up to 50 strains distinguishable by their plasmid complement. Plasmids can be exchanged by conjugation, transformation of protoplast and for smaller plasmids, by transduction (2). Many of the plasmids are cryptic but hybridize with a bacteriocin plasmid indicating homologous DNA and common genetic elements (4). The mechanism by which the large number of plasmids has evolved in these bacteria remains to be established.

The work was supported by grant Te 46/7-1 from DFG (Bonn).
(1) A.Andresen et al., Milchwissenschaft 39, 140 (1984); (2) M.J. Gasson, A. v. Leeuwenhoek 49, 275 (1983); (3) L.L. McKay, ibid. 49, 259 (1983); (4) H. Neve, et al. J. Bacteriol. 157, 833 (1984)

FUNCTIONS OF EXTRACHROMOSOMAL DNA IN THE EPIPHYTIC BACTERIUM *ERWINIA UREDOVORA*

G.Thiry(1), M.Thiry(1), J.Thiry-Braipson(1), M.Faelen(2), M.Mergeay(3,1) and L.Ledoux(3,1)

(1) Botany Department, University of Liège, Liège (2) Laboratory of Genetics, University of Brussels, Rhode-St.-Genèse (3) Department of Radiobiology, SCK/CEN, B-2400 Mol, Belgium

Erwinia uredovora is an epiphytic yellow pigmented bacterium, related to *E. herbicola*. The strains 20D3 and 20D31 contain 4 and 6 extrachromosomal DNA respectively, with molecular sizes ranging from 7.8 to 260 kb. Functions ascribed to some of these plasmids have been analysed.

1. The appearance of unpigmented (Flv^-) and Thi^- phenotypes upon curing treatments is correlated with the loss of the 260 kb plasmid (pULG1 in 20D3, pULG11 in 20D31) in both strains. The pigment lost after curing has spectral characteristics of carotenoid. pULG1 and pULG11 carry genes related to the biosynthesis of the thiazole moiety of thiamine.

2. A phage tail like bacteriocin active against *E. amylovora* and *E. carotovora* strains can be induced in both strains by mitomycin C. This induction is correlated with the amplification of an extrachromosomal DNA of 17 kb coding for bacteriocin determinants. The whole phenomenon suggests a prophage induction.

3. A 88 kb plasmid (pULG3) from 20D3 was marked by the phage D108.1 (a miniderivative of D108 carrying the marker Cm^r) and found to be self-transferable. pULG3 belongs to the Inc M group.

4. Through curing and mobilization, a 28 kb DNA from 20D31 has been identified as a conjugative plasmid conferring susceptibility to the bacteriophage GU5. This plasmid (pULG14) belongs to the Inc N group. It is one of the smallest natural conjugative plasmids yet described in gram-negative bacteria.

MOLECULAR CLONING AND EXPRESSION OF A PLASMID DNA REGION INVOLVED IN *VIBRIO ANGUILLARUM* IRON UPTAKE

Marcelo E. Tolmasky and Jorge H. Crosa Department of Microbiology and Immunology, Oregon Health Sciences University, Portland, Oregon, USA

Plasmid pJM1 harbored by *Vibrio anguillarum* mediates a very efficient iron uptake system associated with the high-virulence phenotype of this bacterium. Transposition mutagenesis analysis allowed us to localize the iron uptake region on pJM1. Iron-uptake deficient mutants generated by transposition mutagenesis were characterized as having lesions in the biosynthesis of either a siderophore or a membrane receptor for the complex siderophore-iron. Further characterization of the pJM1 iron uptake region required the molecular cloning of essential components of this system. Subsets of the iron uptake region were cloned in the cosmid pVK102, a derivative from the wide host range cloning vehicle pRK290, *in vitro* packaged and transduced into *Escherichia coli*. The recombinant DNA molecules were then mobilized to various iron uptake deficient strains of *V. anguillarum* and scored for their ability to grow under conditions of iron limitation and to express siderophore and/or receptor activity. The results showed that concomitant with the receptor activity expressed by exconjugants harboring the recombinant cosmids, there was induction of the biosynthesis of the protein OM2 under iron limiting conditions. OM2 is an 86 kd outer membrane protein which, thus, may be the presumed receptor. Some of the exconjugants harboring the cloned derivatives were able to produce siderophore activity. Iron uptake proficient derivatives obtained after introduction of the cloned pJM1 iron-uptake region regained full virulence, demostrating the intimate relatin between these two properties.

CHROMOSOMAL LOCALIZATION OF AEROBACTIN-MEDIATED IRON UPTAKE GENES COMMONLY ENCODED BY CERTAIN COL V PLASMIDS

M. A. Valvano, M. K. Wolf, L. M. Crosa, and J. H. Crosa

Department of Microbiology and Immunology, The Oregon Health Sciences University, Portland, U.S.A.

The aerobactin-mediated iron uptake system encoded by the pColV-K30 and other Col V plasmids has been associated with the ability of invasive *Escherichia coli* strains to cause disease. To study the occurrence of these genes in microorganisms responsible for nosocomial human neonatal infections, we have examined a number of enteric strains isolated from cerebrospinal fluid of human neonates. The clinical isolates *E. coli* K 1 CHBA187, *Enterobacter cloacae* CHBA31 and *Klebsiella pneumoniae* CHBA33 exhibited a functional aerobactin-mediated iron uptake system, as assessed by a cross-feeding bioassay and by their sensitivity to cloacin, a bacteriocin that recognizes the outer membrane receptor for iron-aerobactin complexes. By using a variety of techniques we could not find any plasmid harboring the aerobactin genes in these strains. Hybridization of restriction endonuclease-cleaved chromosomal DNA from these isolates with various clones containing subsets of the pColV-K30 aerobactin region showed a chromosomal location for these genes. Analysis of internal sequences demonstrated that the chromosomal aerobactin genes were conserved as compared with those present in the pCol V-K30 plasmid, although they were located on a different restriction site environment. The flanking sequences of the aerobactin genes in pColV-K30 were recently shown to be IS1-like elements (Perez-Casal and Crosa, 1984, submitted for publication). Probes containing IS1 sequences hybridized with the same chromosomal restriction endonuclease fragments that harbored the aerobactin genes. The finding of flanking and internal sequences of aerobactin genes on plasmids and chromosomes suggest a recombinational mobility of these genes which may have played a role in their epidemiological spread.

ISOLATION OF THE *A. TUMEFACIENS* TUMORIGENIC

SUPPRESSION GENES ON THE *Inc*W R-FACTOR pSa

David Zaitlin and C.I. Kado

Department of Plant Pathology
University of California
Davis, California 95616

We have previously reported that the intracellular presence of the *Inc*W plasmid pSa prevents *Agrobacterium tumefaciens* from inciting crown gall tumors on permissive host plants Loper and Kado (1979) J. Bacteriol. 139: 591-6., Farrand *et al*. (1981) Mol. gen. Genet. 181: 44-51., Tait *et al*. (1982) Mol. gen. Genet. 186: 10-15 . This phenomenon is absolute for a range of plant species tested, and operates in both octopine and nopaline strains of *A. tumefaciens*. To further examine the role of pSa in virulence suppression, it was necessary to first clone the R-factor genes involved. To this end, a 7.6 kb segment of pSa was isolated first in pUC4 as a partial Sau3A fragment and then transferred as an EcoRI fragment to pUCD105, a stable *Agrobacterium*-*E. coli* shuttle vector containing the replication and stability regions of the plasmid pTAR (see abstract by Gallie et al.). This DNA fragment retains all information required for complete virulence inhibition in vivo; no accessory pSa sequences are necessary. The suppressive DNA fragment is presently being characterized with respect to gene number and location, as well as the encoded polypeptide products. We have also found that the IncW plasmid R388 encodes biological virulence suppression, and that the related plasmids R7K and R404 contain sequences homologous to the 7.6 kb DNA fragment of pSa. This evidence indicates that virulence suppression may be a common feature of many *Inc*W group plasmids.

STRUCTURE AND EVOLUTION

SINGLE-STRANDED DNA BINDING PROTEIN ACTIVITY CODED BY PROMISCUOUS PLASMIDS

P. T. Barth, ICI Corporate Bioscience Group, Runcorn, Cheshire, England

The plasmid F has been shown to confer single-stranded DNA binding (ssb) activity on its host (Kolodkin *et al* 1983). The biological role of this ssb protein is not known: it is not coded by the *tra* operon of F, although it would be expected to be involved in protecting the separated DNA strands formed during conjugation. We have shown that the promiscuous plasmid RP4 encodes a DNA primase in its Tral region that is involved in the priming of the DNA strand transferred to the recipient during conjugation (Lanka and Barth, 1981). A 12kb fragment carrying the primase gene has been cloned (by A. Merryweather in our collaboration with B. Wilkins) from the RP4::Tn7 derivative pRP26 which overproduces primase. The vector was the multicopy promiscuous plasmid R300B. This clone (pAM4) completely suppresses the temperature-sensitive phenotype of an *E.coli ssb* mutant (KL450) at 44°C while the plasmid-free strain loses viability by over four orders of magnitude. RP4 itself suppresses *ssb*^{ts} but not as completely as pAM4. The vector R300 also gave suppression, by over two orders of magnitude.

As DNA primases presumably bind to single-stranded DNA, the ssb activity of RP4 may be due to its primase. R300B (like F) has no measurable primase activity however. The regions responsible for this suppression on the two promiscuous plasmids will be characterized and mapped.

A. L. Kolodkin, M. A. Capage, E. I. Golub and K. B. Low, 1983, Proc. Natl. Sci. USA 80: 4422-4426.

E. Lanka and P. T. Barth, 1981, J. Bacteriol. 148:769-781.

CHROMOSOMAL INTEGRATION AND EXCISION OF A MULTICOPY PLASMID JOINED TO PHAGE DETERMINANTS FOR SITE SPECIFIC RECOMBINATION

Bertani, L.E. (Biology Division) and Bertani, G. (Jet Propulsion Laboratory), California Institute of Technology, Pasadena, California 91125

We are studying the life style (cytoplasmic *vs*. chromosomally integrated) of derivatives of the small, multicopy plasmid pBR322 into which a 4.6 kb fragment of temperate bacteriophage P2 DNA has been inserted (Ljungquist, E. and Bertani, L.E., 1983, Molec. Gen. Genet. 192:87-94). This fragment contains P2 genes *int* (integrase), D and U (tail proteins) and site *att* (attachment). Gene *cox* (control of excision) is absent from this fragment. The degree of amplification of the plasmid has been followed indirectly by the level of resistance to tetracycline and the degree of complementation upon infection with a gene D (or U) defective bacteriophage.

Previous observations (Bertani, G., Ljungquist, E. and Bertani, L.E., 1981, Abstract, IUPAC Congress, Vancouver, B.C.) suggested that such plasmids could integrate chromosomally and that strains of normal viability, apparently carrying a single integrated rather than multiple cytoplasmic copies of the plasmid, could be obtained by a rather laborious procedure.

We have now confirmed these results using a *ts-polA* host strain of *Escherichia coli* C: growth at the non-permissive temperature eliminates the cytoplasmic plasmid population, while a relatively high proportion of the bacteria retain low levels of tetracycline resistance and of gene D complementation. These strains are generally stable even when cultured at low temperature (host cell phenotypically *polA*$^+$). If infected with a temperate P2 phage, however, expression of genes *int* and *cox* appears to lead to mobilization of the integrated plasmid and to the reestablishment of a cytoplasmic plasmid population. If the fragment carries a defective *int* gene, chromosomal integration does not take place spontaneously, but does occur when the P2 *int* gene product is supplied by a helping phage.

This work aims at eventually developing a method for transfer of foreign genes directly to the bacterial chromosome and their amplification *in situ*. It is sponsored by the Energy Conversion and Utilization Technologies Division of the U.S. Department of Energy, through an agreement with the National Aeronautics and Space Administration.

THE PLASMID ColE1. GENETIC ORGANIZATION AS DERIVED FROM ITS COMPLETE NUCLEOTIDE SEQUENCE AND RNA POLYMERASE BINDING STUDIES.

Chan, P.T. and Lebowitz, J. Department of Microbiology, University of Alabama in Birmingham, Birmingham, AL 35294

The primary structure of the 6646 basepairs plasmid ColE1 has been determined. The plasmid genome can be divided into eight functional regions: *cea-kil*(structural gene for colicin E1 and its possible release factor), *imm*(colicin E1 immunity), *bom*(basis of mobility), *mob*(plasmid mobilization), *exc*(plasmid entry exclusion), *rep*(replication functions including the origin of DNA replication and the RNA primer coding region), *inc*(plasmid incompetibility and copy number control), and *rom*(modulation of RNA I activity). From the nucleotide sequence, over twenty open reading frames can be identified. Overlapping of these coding sequences occured at three areas including the *rep*, *mob* and *cea* regions. Overlapping transcription units in opposite orientations also existed in the *cea-kil*/*imm*, *mob*/*rom*, and *rep*/*inc* regions. Regulation of gene expression in ColE1 is also extremely complex. These included regulation by the SOS regulon, cAMP receptor protein, possible attenuations and transcription antitermination in the *cea-kil*, *exc*, and *mob* operons. The locations of known and potential promoters, ranging from having very strong to very weak affinity to bind *Escherichia coli* RNA polymerase holoenzyme, have been mapped by a combination of nitrocellulose filter binding assays, DNase I "footprinting", and exonuclease III mapping. The overlapping of transcription and coding sequences in ColE1 showed unusual economy in genetic organization and conservation of DNA sequence.

P1 MINI PLASMIDS PRODUCED *IN VIVO*

B.J. Froehlich, J.R. Scott, K. Tatti, D.K. McLain

Department of Microbiology and Immunology, Emory University, Atlanta, Georgia, 30322, USA

E. coli recA, dnaA(Ts) strains integratively suppressed by P1 cop mutants (copy number 5-8 times P1 wild type) contain only a single integrated copy of the P1 prophage when the cells are grown at the nonpermissive temperature for *dnaA* (Froehlich et al, 1983). When the P1cop-suppressed strains are shifted to 30°C, the survival frequency is very low (10^{-3} to 10^{-5}). 90% of the 30° survivors contain supercoiled plasmid DNA as well as the integrated copy. Southern Blot analysis indicates that the integrated copy is whole P1. Restriction analysis and Southern Blot analysis show the plasmid DNA is present at high copy number. There is no common terminus in the P1 miniplasmids analyzed, indicating that they are not formed from the chromosomal P1 by recombination at a unique site. Probably, the miniplasmids are produced by premature termination of replication rounds initiated from the origin of the integrated P1.

All of the P1 miniplasmids contain the DNA region believed to serve as the P1 plasmid origin. Some of the miniplasmids are independent replicons and others are not. Those that are not independent replicons persist in cells carrying appropriate cloned fragments of P1. This allows us to define a minimal replicon and to define a DNA region encoding a trans-acting product required for replication.

THE INTERACTION OF MINI-F WITH THE PLASMIDS R124 and R124/3

Harper, S.H., Firman, K., Glover, S.W.

Department of Genetics, Ridley Building,
University of Newcastle Upon Tyne NE1 7RU,
U.K.

The Inc FIV plasmid R124 encodes a unique restriction and modification (R-M) system. R124/3 is a derivative plasmid which encodes a R-M system of different specificity. When R124 or R124/3 are present in the same cell as the sex Factor F a restriction deficient (r^-) phenotype results. This r^- phenotype reverts back to r^+ with a frequency dependent upon the particular isolate. When the F plasmids are isolated from these reverted stains they are found to encode the R-M system of the donor R124 or R124/3 plasmid. The R-M system is translocated into the region of EcoRl fragments 5 and 7 of F, which encodes the Phi and Ori VI functions of F. During this transfer of DNA some F DNA is lost.

Recent work with mini-F (cloned EcoRl fragments 5 and 7 of F) have shown that they alone do not result in the translocation of R-M genes from R124 or R124/3. EcoRl fragments 5 and 7 of F when present in the same cell as R124 or R124/3 result in a fraction of the cells expressing a r^- phenotype which reverts back to a r^+ phenotype. This reversion process is not accompanied by any DNA transfer or rearrangement between F, R124, or R124/3 plasmids, This strongly suggests that some other region of F is involved in the DNA rearrangement.

Studies with mini-F have also shown that under certain circumstances the rearrangement results in R124 expressing incFl incompatability instead of the normal Inc FIV.

THE STRUCTURE AND FUNCTION OF MINI-Rts1

Kamio, Y., Itoh,Y., Tabuchi,A., Katagiri,H., and Terawaki, Y.

Department of Bacteriology, Shinshu University School of Med., Asahi 3-1-1, Matsumoto 390 Japan

Rts1 is a large(ca. 200 Kb) stringent type plasmid of the IncT group. The basic replicon of Rts1, named mini-Rts1, is capable of autonomous replication in a polA *Escherichia coli* host and is maintained stably at 37°C in a low copy state(about 3 copies per host chromosome).

We determined the nucleotide sequence of mini-Rts1 consisting of 1,855 base pairs. Two clusters of direct repeated sequences were found on the mini-Rts1 genome, each of which contained five 24-bp repeats(left cluster) and three 21-bp repeats(right cluster). A large open reading frame that directed from right to left and could encode a 32,980 dalton polypeptide(RepA protein) consisting of 288 amino acids was located between the two clusters. A 33-Kd protein corresponding to the RepA protein was detected in the *E. coli* minicell system. Outside the right cluster, an oriC homologous region consisting of four GATC sequences accompanying TTATCCACA sequence was located.

All the eight repeats contained a consensus sequence "TTCCCC-PyPyPuPuCACACACC", and both cluster-containing fragments expressed strong incompatibility toward mini-Rts1, respectively. A fragment containing the five repeats of the left cluster exhibited a much stronger incompatibility than the four repeats-containing fragment, suggesting that the strength of incompatibility is dependent on the number of the repeating unit. In addition, the copy number of the mini-Rts1 plasmid increased three fold upon removal of the four repeats-containing region from the plasmid. Thus, the direct repeats of mini-Rts1 appear to express incompatibility and copy control functions by titrating the RepA protein or other component(s) being essential for the plasmid replication. The overall structure and function of mini-Rts1 are quite similar to those of mini-F and mini-P1.

A considerably high copy mutant(about 20 copies per host chromosome) was constructed by removing the four repeats from the "cop1" mutant of mini-Rts1, on which two base pair substitutions occurred, i.e. at the RepA coding region and near the oriC homologous region of mini-Rts1.

STREPTOMYCES PROMOTER-PROBE PLASMIDS AND CARTRIDGES

T. Kieser, M. J. Bibb, M. J. Bibb, J. M. Ward
and *D. A. Hopwood

John Innes Institute, Norwich, U.K.

The ribosome binding site and coding sequence of the kanamycin phosphotransferase gene of the *E. coli* transposon Tn5 can be separated from the promoter by cleavage with the restriction endonuclease *Bgl*II or *Bcl*I which cut 35 and 30 base pairs, respectively, ahead of the ATG start codon. *Streptomyces* plasmids carrying the promoter-less resistance determinant conferred kanamycin resistance on their host. The level of resistance varied between c. 2 and 100 µg kanamycin/ml, depending on the position and orientation of the insert. Addition of a 333 base pair *Sau*3A fragment containing a transcription terminator of phage fd in front of the kanamycin resistance determinant effectively abolished kanamycin resistance in *Streptomyces*, thus supporting the hypothesis that expression of resistance depends on upstream promoter activity. Promoter-probe plasmids (e.g. pIJ425) with a unique *Bgl*II cloning site between the kanamycin resistance gene and the fd terminator are useful for the isolation of DNA fragments with in vivo promoter activity from *Streptomyces* strains.

Transcriptional activity can also be detected by inserting the promoter-less kanamycin resistance fragment into DNA sequences of interest. Experience has shown that direct selection for kanamycin resistance is not practical for low levels of expression and can lead to enrichment for mutated (rearranged) plasmids. To avoid these problems, DNA fragments consisting of the promoter-less kanamycin resistance linked to a second resistance determinant (thiostrepton or viomycin resistance) have been constructed. These fragments, called "Km^R-cartridges", can be conveniently excised from *E. coli* vectors (which cannot replicate in *Streptomyces*) and inserted into non-essential regions of *Streptomyces* plasmids. Thiostrepton or viomycin resistance can be used to select for hybrid plasmids and the level of kanamycin resistance subsequently determined. The Km^R-cartridges have been used to probe for transcriptional activity in the *Streptomyces* plasmid pIJ101.

STRUCTURAL EVOLUTION OF SELECTED GENE SEGMENTS OF Col E2 AND E3 PLASMIDS

P.C.K. Lau and L.P. Visentin. Molecular Genetics Group, National Research Council of Canada, 100 Sussex Drive, Ottawa, Ontario, Canada, K1A 0R6

The DNA sequences encompassing the middle portions and the carboxyl terminal ends of the colicin E2 and E3 polypeptides encoded in plasmids ColE2-P9 and ColE3-CA38 respectively have been determined. Despite the differing catalytic function, the colicin E2-DNAase and the colicin E3-RNAase domains were found to contain several conserved amino acid residues. Moreover, for colicin E3, the amino acid residues which may be involved in catalysis or binding are similar to those of other microbial RNAases. Accordingly, we propose an evolutionary link between the colicin nuclease domains and the various microbial RNAases.

A portion of the receptor-binding domain of both colicins was also identified based on the occurrence of identical sequences as these two colicins share a common btu B receptor on sensitive *E. coli* cells.

We have also determined the sequences of the structural genes of the immunity proteins for colicins E2 and E3. A comparison of the nucleotide and amino acid sequences of these proteins reveals little homology. However, the sequence for the immunity protein produced by plasmid ColE2$_{imm}$-K317 differs from that of ColE2-P9 by only one amino acid change. The single amino acid change may account for the 'partial' immunity protection to colicin E2-P9.

Two regions of deletion in ColE2$_{imm}$-K317 have also been localized, one of which corresponds to a region in ColE2-P9 or ColE3-CA38 where a potential hairpin structure may be found. The structures and evolutionary perspectives of another set of genes, designated hic, whose products are required for the release of colicin from the various producing strains will also be presented.

Watson, R., Rowsome, W., Tsao, J. and Visentin, L.P. J. Bacteriol. (1981) 147, 569-577.

Watson, R. and Visentin, L.P. Gene (1982) 19, 191-200.

Watson, R., Lau, P., Vernet, T. and Visentin, L.P. Gene (1984) in press.

Lau, P., Rowsome, W., Zuker, M. and Visentin, L.P. (manuscript submitted).

PLASMIDS FROM *BORDETELLA BRONCHISEPTICA*

Lax, Alistair, J., Walker, Carol, A.

AFRC Institute for Research on Animal Diseases, Compton, Newbury, Berks. RG16 0NN. U.K.

Bordetella bronchiseptica has been implicated as a causative agent in the pig disease, atrophic rhinitis. Several possible virulence determinants have been described, though their importance in pathogenesis is not known. Plasmids have been found in some isolates, but their role has not been investigated.

We have examined 14 stains of *B. bronchiseptica* isolated from pig herds in the U.K. and found plasmids in 7 of these. The plasmids found show considerable homology both by restriction analysis and hybridisation, and we have mapped these regions. Some of the strains had previously been examined for virulence, and we found that virulence did not relate to the presence or absence of plasmids. Six non-pig strains were studied, but none contained plasmids. Strains without plasmids were examined for the presence of plasmid sequences in the genomic DNA.

Phase variation, in which all the virulence- associated factors are not made, has been described in *Bordetella*. Phase variation was observed in strains both with and without plasmids and no detectable alteration of the plasmids was found.

STRUCTURAL AND FUNCTIONAL STUDIES ON SMALL CRYPTIC PLASMIDS FROM *BACILLUS THURINGIENSIS* H-1

J. Mahillon, D. Sartenaer, B. Nizet and J. Delcour

Université Catholique de Louvain
Laboratoire de Génétique Moléculaire
Place Croix du Sud 5 (Bte 6)
B-1348 Louvain-la-Neuve (Belgium)

Except for some of the large plasmids, which in several strains have been shown to contain the δ-endotoxin (crystal) gene, little is known about the structure or the function of *Bacillus thuringiensis* (Bt) plasmids. As a first step towards tackling that problem, we focused our attention on three small plasmids (in the 10 kb range) commonly found in several Bt strains. We used the H1.1 strain as a model system (*Bacillus thuringiensis* var. *thuringiensis* serotype H1, kindly provided by H. de Barjac), harboring the following cryptic plasmids : pBT1 (8.5 kb), pBT2 (9.5 kb) and pBT3 (10.5 kb). We performed experiments dealing with the structure of these plasmids (restriction maps), their function (phenotypic analysis of cured strains, expression into *E. coli* minicells of cloned plasmids) and their distribution among several Bt strains (Southern blots). The following information has been gained from our work :

1. pBT2 and pBT3 have been cloned in *E. coli via* pBR322, and some of their genes have been expressed in this host.

2. pBT1 seems to be "unclonable" in *E. coli* using *Sal*I. This could be due to a lethal effect resulting from the expression of some of its genes.

3. The three small plasmids found in the H1.1 serotype are distinct DNA molecules bearing no strong sequence homology. These sequences are widely distributed among various strains belonging to different serotypes.

4. Whereas pBT1 and pBT3 seem to be well conserved both in size and sequence among the various strains studied, pBT2 appears as a more "labile" plasmid whose sequences can be found on plasmids widely diverging in size. The fact that pBT2 is easy to cure without any significant phenotypic change is another interesting observation with respect to the cryptic nature of that plasmid.

DNA SEQUENCE ANALYSIS OF THE FINO REGION FROM PLASMID R100

McIntire, S.A. and Dempsey, W.B.

VA Medical Center, 4500 S. Lancaster Rd., Dallas, TX 75216
and Univ. of Texas Health Science Center, Dallas, TX 75235

It was observed more than 20 years ago that the transfer of F-like plasmids is inhibited to about 0.1-1% that of the naturally derepressed plasmid F. The mechanism of this fertility inhibition (or fin) is not known. It has been proposed that Fin involves two structural genes *finO* and *finP*, whose products interfere with the transcription of the transfer operon's positive control gene, *traJ*.

Towards our goal of understanding the Fin mechanism at the molecular level, we have cloned and sequenced the *finO* gene of the F-like antibiotic resistance plasmid R100. This gene is located near R100 kb coordinate 81.3, approximately 30 kb away from *finP* and at the opposite end of the transfer operon.

The DNA sequence shows one long open reading frame of 558 nucleotides, containing several other possible start codons within the frame. We have isolated and analyzed several frame shift-generating deletions around an *XmaI* site located 120 nucleotides inside the start of the long open frame. Since some of these deletion mutants are still $FinO^+$, we conclude that, if *finO* does encode a protein, the structural gene is smaller than the 558 nucleotide maximum size mentioned above. However, other of these mutants were $FinO^-$ and we assume that some aspect controlling *finO* expression is involved.

An interesting observation was of two regions of strong homology with the λ*att* sequence (10 and 11 bases, respectively, homologous to the 15 base pairs of λ*att*). This finding explains the high incidence of λ integration into R100 *finO*.

HOMOLOGY OF PLASMID DNA TO THE *STREPTOMYCES SCABIES* CHROMOSOME

D.A. Ross McQueen[1] and Janet L. Schottel[2]

Department of Plant Pathology[1] and Department of Biochemistry[2]
University of Minnesota
St. Paul, MN 55108

Nine separate strains of *Streptomyces scabies* have been isolated from potato scab lesions on tubers from Florida, North Dakota and Northern Minnesota. Agarose gel electrophoresis of *Eco*R1 digested chromosomal DNA from these strains revealed the presence of DNA fragments which appear amplified. The lengths of the amplified bands differ between strains.

One of the *S. scabies* isolates contains a 64 kilobase plasmid, designated pRB1, which we have purified and partially characterized with restriction endonucleases. The pRB1 plasmid has been used as a probe in Southern hybridizations against the chromosomal DNA from its host strain. We found DNA fragments in the chromosomal sample that hybridize to and co-migrate with pRB1 which had been cleaved with the same endonuclease. DNA fragments from the chromosomal sample that hybridize to pRB1 but do not co-migrate with plasmid bands were also observed; the lengths and numbers of these chromosomal bands that are homologous to pRB1 differ depending upon the endonuclease used for cleavage. These results suggest that all or part of pRB1 is integrated into the chromosome. The varying intensities of the bands in the Southern blots indicate that there may be amplification of integrated plasmid sequences.

The pRB1 plasmid was also used as a probe in a Southern hybridization against chromosomal DNA from the other *S. scabies* isolates. Homology was observed between pRB1 and these other chromosomal samples; the lengths and numbers of homologous bands were different between the chromosomal DNAs tested.

ANALYSIS OF HETERODUPLEX MOLECULES BETWEEN PLASMID p15B AND PHAGES P1 and P7 DNA

*J. Meyer, M. Stålhammar-Carlemalm, M. Streiff, S. Iida and W. Arber

Microbiology Department
Biozentrum, Basel University
CH-4056 Basel, Switzerland

Plasmid p15B isolated from E. coli 15T⁻ and the prophage P1 are closely related: both code for a type III DNA restriction and modification system (1,2), they can recombine with each other (1,3) and they share DNA homology (4). We have determined by electron microscopy the size and location of homologous and non-homologous segments in heteroduplex molecules between p15B and P1 as well as between p15B and the P1-related phage P7 (5,6). The results, summarized in the drawing below, indicate that the genomes share about 80% homology. p15B has substitutions in addition to the regions A-G which are also different between P1 and P7 (5), notably p15B contains two stem-loop structures.

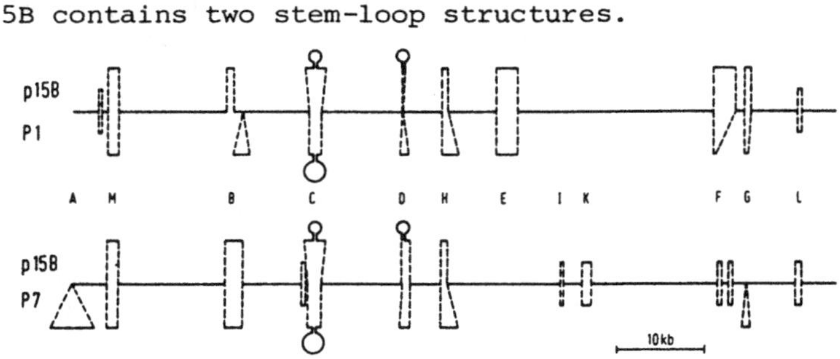

(1) Arber W & Wauters-Willems D (1970) Mol Gen Genet 108, 203. (2) Iida S et al. (1983) J Mol Biol 165, 1. (3) Meyer J et al. (1983) J Mol Biol 165, 191. (4) Ikeda H et al. (1970) J Mol Biol 50, 457. (5) Yun T & Vapnek D (1977) Virology 77, 376 (6) Iida S & Arber W (1979) Mol Gen Genet 173, 249.

QUANTITATION OF REGULATORY RNAs - THE pT181 COUNTERTRANSCRIPTS

S.J. PROJAN, J.S. KORNBLUM, AND R.P. NOVICK

Department of Plasmid Biology, The Public Health Research Institute of the City of New York, 455 First Ave. N.Y. 10016

The *Staphylococcus aureus* tetracycline resistance plasmid, pT181, is a plasmid whose rate of replication in indirectly regulated by the intracellular concentration of two small RNA species. It has previously been determined that RepC is a plasmid encoded protein required for pT181 replication and that two small RNA species complementary to the RepC mRNA leader (RNA I of 85 bases and RNA II of 140 bases) are negative effectors of pT181 replication. RNA I and RNA II have a common start within the RepC mRNA leader. When the coding sequence for these small RNAs is cloned onto a carrier plasmid the resultant plasmid inhibits pT181 replication *in vivo*. This is termed Inc3A incompatibility. Copy mutants of pT181 have been isolated, sequenced, and classed as either dominant (resistant to Inc3A incompatibility) or recessive (sensitive to Inc3A). Utilizing a method of Northern Blot analysis, RNA from whole cell lysates was separated on formaldehyde-agarose gels and the relative amounts for RNA I and RNA II were determined. For the point mutants tested, dominant copy mutants make the small RNAs in a manner proportional to their increase in copy number (gene dosage). One recessive mutant maps in the -10 promoter sequence for RNA I and RNA II, and another in the -35 region. Strains carrying these high copy mutant plasmids have the small RNAs at the same intracellular concentration as the wild-type and therefore make less of RNA I and RNA II per plasmid copy. Other recessive mutants appear to be defective in the termination of either RNA I or RNA II, thus indicating that both molecules are involved in the negative control of pT181 replication.

SEQUENCE ANALYSIS OF Tn917

J. H. Shaw and D. B. Clewell

Departments of Microbiology/Immunology and Oral Biology
The University of Michigan, Ann Arbor, MI 48109-2007

Tn917 (Em) is a 5.0 kb transposon identified in *Streptococcus faecalis* that has been shown to undergo enhanced transposition when exposed to inducing concentrations of erythromycin. DNA sequencing has revealed terminal inverted repeats which are 38 base pairs in size, but are not completely identical. They are highly homologous to the inverted repeats of Tn551 (Khan, S., and R. Novick. Plasmid 4:148-154 1980) and Em-resistance transposon isolated from *Staphylococcus aureus*; both show homolgy with the inverted repeats of Tn3.

Sequences closely resembling the MLS determinant reported by Weisblum, et al., (J. Bacteriol. 154:1252-1262 1983) for pAM77 (a 6.75 kb plasmid originally identified in *Streptococcus sanguis*) have been found near one end of the transposon. In addition, the resistance determinant is flanked by short open reading frames similar to sequences observed in pAM77. The sequence 5' to the MLS resistance determinant is believed to represent a "leader sequence" involved in attenuative translational control of the synthesis of a ribosome methylating enzyme. Although a function for the sequence on the 3' end is unknown, a possible attenuational role in Em-induced transposition is conceivable. The latter sequence contains a nearly complete copy of a 38 base pair terminal repeat in the same orientation as that located on the opposite side of the resistance determinant. Closely adjacent to this repeat is a 20 base pair sequence resembling the res site of Tn3. A similar sequence is not present on pAM77; however, the equivalent region contains two copies resembling the "repeat sequence". They are present in opposite orientation and overlap each other by 16 base pairs. It is conceivable that the evolution of pAM77 involved a Tn917 which acquired a deletion.

Evidence that the Clindamycin-Erythromycin Resistance Gene of the *Bacteroides* Plasmid pBF4 is on a Transposable Element

Nadja B. Shoemaker, Ellen P. Guthrie, Abigail A. Salyers and Jeffrey F. Gardner

Department of Microbiology
University of Illinois
Urbana, Illinois 61801

The conjugative *Bacteroides* plasmid, pBF4 (pIP410), carries a gene for clindamycin-erythromycin resistance which is expressed in *Bacteroides* (Privitera *et al.*, J. Inf. Dis. 139:97; Welch *et al.*, Plasmid 2:261). pBF4 also carries a tetracycline resistance gene which is expressed in *E. coli* but not in *Bacteroides* (Guiney *et al.*, Plasmid, in press). We have constructed a chimeric plasmid, pE5-2, which contains the *E. coli* plasmid RSF1010, plus a cryptic *Bacteroides* plasmid (pB8-51) and a 3.8 Kb *Eco*RI fragment from pBF4. pE5-2 confers tetracycline resistance on *E. coli*. It is mobilized by the broad host range plasmid R751 from *E. coli* to *Bacteroides thetaiotaomicron* or *Bacteroides uniformis* with a frequency of 10^{-6}. pE5-2 was the only plasmid which was recovered from *Bacteroides* transconjugants whereas 98% of the *E. coli* transconjugants, contained both R751 and pE5-2. The *Bacteroides* transconjugants were resistant to clindamycin-erythromycin. This is the first direct evidence that these resistances are located within the *Eco*RI-D fragment of pBF4. We have also constructed a chimeric plasmid, pSS2, which contains a much larger portion of pBF4 (33 Kb). pSS2, unlike pE5-2, was not stable in *E. coli*. Mobilization of pSS2 by R751 from *E. coli* to *E. coli*, was associated with a variety of rearrangments. In some cases a segment of pBF4 was inserted into R751. This segment included the *Eco*RI-D segment which confers tetracycline resistance in *E. coli* and clindamycin-erythromycin resistance in *Bacteroides*. In two cases, the same segment appeared to have inserted in the *E. coli* chromosome. Thus it appears that there is a transposable element which contains the *Eco*RI-D fragment of pBF4. We did not detect any transfer of pSS2 from *E. coli* to *Bacteroides*.

CONJUGATIVE MAPPING OF *PSEUDOMONAS PUTIDA* USING XYL-K

Sykes, Pamela J., Menard, Joan and Sokatch, John R.
Department of Microbiology and Immunology. University of Oklahoma Health Sciences Center, P.O. Box 26901, Oklahoma City, Oklahoma, 73190 USA

Pseudomonas putida has been used extensively for biochemical studies because of its metabolic versatility, however, genetic studies have been hampered by the lack of a conjugative system. Mylroie *et al*. (1977) reported the use of the recombinant plasmid-sex factor, XYL-K for mapping the *P. putida* chromosome and Dean and Morgan (1983) constructed the high frequency recombination plasmid R91-5::Tn501 to map and to prove the circularity of the *P. putida* chromosome. We are interested in the biochemistry and genetics of the three keto acid dehydrogenases of *P. putida*, branched chain keto acid dehydrogenase (bckad), pyruvate (pdh) and 2-ketoglutarate (kgdh) dehydrogenases. In addition, *P. putida* produces two distinct lipoamide dehydrogenases, LPD-val and LPD-glc. LPD-val is specifically required as the E3 component of bckad; LPD-glc is required as the E3 component of kgdh, probably pdh and for glycine oxidation. We have isolated mutants of *P. putida* in bckad, pdh, kgdh, LPD-val and LPD-glc as well as auxotrophic and catabolic mutants by mutagenesis with nitrosoguanidine or with plasmid pRKTV14. Conjugation was effected with XYL-K using strain AC143 as the donor. Filter matings were performed as described by Mylroie *et al*. (1977) except that streptomycin was not always used in counterselection. Times of entry for pdh and kgdh were 22 and 10 min respectively. Mutants of *P. putida* deficient in all three subunits of bckad have been obtained (E1, E2 and LPD-val) and all map at about 35 minutes. There is good agreement with map locations of Mylroie *et al*. for *put* and *hut* and the data are consistent with a single point of origin.

GENETIC ORGANIZATION AND FUNCTION OF THE PLASMID pKM101

Graham C. Walker and Stephen C. Winans

Biology Department
Massachusetts Institute of Technology
Cambridge, Massachusetts 02139

The *incN* plasmid pKM101 was derived from its clinically isolated parent R46 by an in vivo deletion of 14 kb which removed genes coding for drug and metal resistances. pKM101 has been extensively studied because it increases the susceptibility of cells to UV and chemical mutagenesis by coding for the *mucA* and *mucB* genes which analogs of the chromosomally encoded *umuD* and *umuC* genes. It has played a major role in the success of the Ames *Salmonella* strains used for detecting mutagens.

The conjugal transfer system of pKM101 consists of three regions, TRA I, TRA II, and TRA III, each separated from each other by a region not absolutely required for transfer. All the genes in the first two regions are required for sensitivity to donor-specific phage. The plasmid's origin of transfer has been localized to a region at the far end of TRA III. Using two methods we have identified eleven complementation groups needed for transfer. The TRA I and TRA II regions are separated from each other by *eex*, a gene mediating entry exclusion. The TRA II and TRA III regions are separated from each other by *nuc*, a gene coding for a periplasmically located endonuclease, and by *fip*, a gene whose product inhibits the fertility of P group plasmids.

Two pKM101-coded loci, *kilA* and *kilB*, have been identified which elaborate gene products that are potentially lethal to the bacterial cell. The lethal effect of each of these gene products is inhibited by two other plasmid-coded loci, *korA* and *korB*, both of which are required to control the lethality of either *kil* gene. In the presence of *korA* and *korB* both *kil* genes have other phenotypes: *kilB* is necessary for conjugal transfer, while *kilA* is responsible for the small colony morphology on defined media that is characteristic of pKM101-containing strains.

PLASMIDS OF BACTEROIDES FRAGILIS

Weinblum, D., Beul, H. A. and Schreiner, M., Physiologisch-Chemisches Inst., Universität of Mainz, Ger.

A plasmid screenig procedure was developed for lysozyme resistent *Bacteroides*. 24 strains of B. *fragilis, subsp. fragilis*, isolated from suppurative infections, were analysed for their plasmid content, 9 strains contained 1 to 3 plasmids. Plasmids could be divided into 5 classes of 2.8, 3.5, 3.6, 4.2 and 19 MDa size. From multiple occurence we conclude, that the small plasmids belong to different compatibility groups. The small plasmids were also observed in strains, not belonging to subsp. *fragilis*. By Southern blotting it was shown, that homology exists only between plasmids of the same size. Plasmids were digested with 17 restriction endonucleases, plasmids of the same size gave identical fragmentation patterns. Restriction maps were constructed from partial and double digestion. The 19 MDa plasmid showed no transposon like structures by homoduplex analysis. Presently no function can be assigned to the plasmids. No dependence was observed between plasmid content and bacteriocin and hemolysin production. Resistance against 9 antibiotics (penicillin, clindamycin, tetracyclin, chloramphenicol, erythromycin, ampicillin, kanamycin, streptomycin and rifampicin) showed no correlation to the occurence of particular plasmids.

SECONDARY STRUCTURE PREDICTIONS FOR RNA TRANSCRIPTS FROM THE NR1 REPLICATION CONTROL REGION

*D.D. Womble and R.H. Rownd

Department of Molecular Biology
Northwestern University, The Medical School
Chicago, IL 60611 USA

The repA1 protein of the IncFII plasmid NR1 is required for initiation of DNA replication at the plasmid origin. Synthesis of repA1 is negatively regulated at both the transcriptional and translational levels. Translation of the repA1 mRNA is inhibited by a small, complementary RNA transcribed from the opposite strand. The inhibitor RNA is synthesized in a 20-fold excess, and a majority of the repA1 messengers pair with the inhibitor RNA. The predicted secondary structure of repA1 mRNA which has paired with the inhibitor RNA has the start codon and ribosome recognition sequence for repA1 translation paired in an intramolecular double-stranded structure. This should prevent ribosome binding and inhibit repA1 synthesis. For the rare repA1 mRNA which escapes pairing with the inhibitor RNA, the predicted secondary structure has both the start codon and ribosome recognition sequence in single-stranded regions. This should allow ribosome binding and repA1 synthesis, which will then cause initiation of DNA replication at the plasmid origin. The choice between these alternate secondary structures is the "switch" which regulates initiation NR1 DNA replication.

AUTOGENOUS REGULATION OF THE π STRUCTURAL GENE OF PLASMID R6K

Filutowicz, M., Davis, G., Greener, A. and Helinski, D.R.,

Department of Biology, B-022, University of California, San Diego, La Jolla, California 92093 USA

It has been proposed that binding of the R6K π initiation protein to the seven 22 bp repeats in the γ-origin of replication and to the eighth repeat near the _pir_ (π structural gene) promoter are required for initiation of plasmid R6K replication and autoregulation of _pir_ gene expression, respectively (1). Evidence has been obtained for the autogeneous regulation of the _pir_ gene (2) and the specific binding of a β-galactosidase-π fusion protein to the tandem seven direct repeats and the 8th repeat (3). In the present study a plasmid was constructed that placed the β-galactosidase gene under control of the _pir_ promoter. Greater than 95% repression of β-galactosidase expression from this plasmid was observed in _E. coli_ cells when the complete _pir_ gene is present _in trans_. Alterations in the N-terminus sequence for the π protein substantially reduces its autoregulatory activity. Purified π protein also was shown _in vitro_ to specifically repress transcription from the _pir_ promoter. Binding sites on the _pir_ promoter for the π protein and _E. coli_ RNA polymerase were delineated by DNaseI protection experiments. Purified π protein and RNA polymerase bind to the operator-promoter region of the _pir_ gene which includes the 8th repeat. Since only a portion of π binding site is shared with the RNA polymerase binding site, the _pir_ operator-promoter region appears to be unusual. In addition, purified π protein not only prevents binding of RNA polymerase but it displaces bound RNA polymerase from the operator-promoter region of the _pir_ gene.

References

1. Stalker, D.M., Kolter, R. and Helinski, D.R. Proc. Natl. Acad. Sci. USA _76_, 1150, 1979.
2. Shafferman, A., Kolter, R., Stalker, D. and Helinski, D.R. J. Mol. Biol. _161_, 57, 1982.
3. Germino, J. and Bastia, D. Cell _34_, 125, 1983.

TRANSFER

traT GENE AND ITS SURFACE EXCLUSION CHARACTERISTICS

Paul J. Chuba*, Myron A. Leon, and Sunil Palchaudhuri

Department of Immunology and Microbiology
Wayne State University School of Medicine
Detroit, Michigan 48201

The ColV plasmid was given a genetic marker by insertion of the kanamycin resistance transposon Tn903 to yield plasmid pWS12; this derivative was found to possess all normal colV properties. Since Tn903 has a tendency to insert near an origin of replication, we used this fact in cloning restriction fragments carrying the replicons of ColV. Restriction and ligation of the 130 kb pWS12 yielded the self-replicating fragments pWS15 (using *Eco*RI) and pWS16 (using *Bam*HI), which were 32 kb and 75 kb in size, respectively. The plasmid pWS12 and its deletion derivatives pWS15 and pWS16 carry the surface exclusion (*SEx*) genes *tra*T and *tra*S of the ColV2-K94 transfer region. The parent plasmid pWS12 and its self-transmissible derivative pWS16 showed *SEx* towards the plasmids of specificity group II. Interestingly, pWS15, Tra$^-$ derivative of pWS12, surface excluded plasmids of *SEx* groups I and II. However, pWS15 and pWS12 produced a similar amount of *tra*T protein in the outer membranes of host cells. The molecular weights of the *tra*T proteins produced by pWS12 and pWS15 were identical as determined by SDS-PAGE. These proteins were also immunologically cross-reactive. Thus it appears that plasmid gene(s) within coordinates 32V to 75V outside the *tra*T and *tra*S loci regulate expression of specificity group I and II in this system.

MOLECULAR DETAILS OF THE GENES FINO AND FINP WHICH CONTROL THE INFECTIOUS TRANSFER OF THE ANTIBIOTIC RESISTANCE PLASMID R100

Walter B. Dempsey

VA Medical Center, Dallas, TX 75216 and Biochemistry Dept., Univ. of Texas Health Science Center, Dallas, TX 75235

The finO and finP genes of R100 are two of three plasmid genes that regulate fertility of R100. They both act in trans but their gene products are not yet unequivocally identified. To get more information about finO, finP and tra, all of the EcoRI fragments that comprise the R100 tra operon have been cloned into pBR322 or pBR325 and used to form a detailed restriction map of tra. From this information, finO, finP, traJ, oriT, and traOYA of R100 have been isolated on small fragments and subcloned individually into pHP34. The finO gene resided entirely on a 735bp fragment whose ends map at R100 coordinates 81.11 and 81.84. The finP gene resided entirely on a 700bp fragment whose ends mapped at R100 coordinates 46.52 and 47.22. S1 nuclease mapping with these isolated fragments was unsuccessful. Runoff transcripts from the finO fragment showed one or two RNA's of 60-80bp encoded in the far right end. No transcripts were found from the finP fragment. The isolated finO and finP genes were used in hybridization experiments to probe the tra genes for stringent and weak homologies. None were found between these genes or with any other part of the R100 molecule. Chromosomal probing showed no hybridization. Specific hybrids were formed at low stringency between the finP probe and most of the fragments of a commercially available DNA molecule weight standard. The finO probe on the other hand hybridized specifically with only two of the bands of that standard, one uniquely. The structure of the standard being unavailable, digests of pBR322 were probed. The finO fragment was found to hybridize at moderate stringency to pBR322 in the amp gene and to the region between 1800 and 2000bp on the pBR322 map.

DISSECTION OF THE TRANSFER REGION Tra1 OF PLASMID RP4 — EXPRESSION AND MOLECULAR CHARACTERIZATION OF GENE PRODUCTS

Fürste, J.P., Lurz, R., [+]Kröger, M. and Lanka, E.

Max-Planck-Institut für Molekulare Genetik, Abt.Schuster, D-1000 Berlin 33; [+]Institut für Biologie III, Universität Freiburg, D-7800 Freiburg, FRG

The conjugative broad-host-range Inc Pα plasmid RP4 contains three unlinked transfer regions (Tra) which are essential for self-transmissibility of the plasmid from donor to recipient cells. One of the regions, Tra1, is located between Km^r and trfB; it includes the origin of transfer (oriT) where initial events of transfer DNA replication requiring plasmid-specified proteins must occur. Five putative promoter sites have been localized within Tra1 by RNA polymerase binding studies. Gene products of this plasmid portion are being studied. Transcription of a major part of Tra1 appears to proceed in opposite direction to that of Km^r. Efficient expression of five polypeptides requires only one of the promoters suggesting the synthesis of a polycistronic messenger. The molecular masses of the proteins determined on denaturing gels are 16.5, 68, 77, 80 and 118 kDa. The primase gene (pri, 3.2kb) specifies two proteins of 118 and 80 kDa. As shown by deletion analysis of recombinant Pri^+-plasmids and immunological studies, the primase polypeptides represent two separate translation products which arise from an in-phase overlapping gene arrangement. Both these polypeptides mediate initiation of complementary strand synthesis on single-stranded phage DNA in vitro by synthesis of RNA primers. Two genes specifying polypeptides of 26 and 52 kDa map in close vicinity to oriT as demonstrated with a minicell system. The gene for the 26 kDa protein has been localized adjacent to the oriT sequence. Molecular cloning of the structural gene into an expression vector facilitated overproduction and purification. Subcloning of the oriT region has revealed that the fully functional origin comprises ∿320bp. The sequence consists of two regions. One includes the putative nic site on a stretch of 112 bp (Yakobson and Guiney,1983, Proc.Nat.Acad. Sci.USA 80,3595). The second region, which contains a promoter sequence and probably additional recognition sites for transfer proteins, is necessary to retain a high mobilization frequency of oriT recombinant plasmids. The location of the promoter coincides with one of the sites found by RNA polymerase binding studies.

FORMATION OF ΔtraF' PLASMIDS: SITE-SPECIFIC RECOMBINATION NEAR oriT

*Horowitz, Burton and Deonier, Richard C.

Molecular Biology, Dept. of Biological Sciences, University of Southern California, University Park, Los Angeles, CA 90089-1481, USA

Mating between Hfr strains and $recA^-$ recipients with selection for proximal Hfr markers produces ΔtraF' plasmids, which contain the proximal region of F up to one junction with chromosomal DNA. These plasmids are nonconjugative, and the entire tra operon is deleted. Restriction analysis indicates that the process can be site-specific, with the novel joints mapping near oriT.

BglII fragments containing novel joints from $\Delta traFproA^+$ or $\Delta traFpurE^+$ plasmids were cloned into a mob^- vector pED961, and the resulting plasmids were tested for mobilization from $recA^-$ strains by the F derivative pOX38. Plasmids containing oriT of F were mobilized at frequencies of 10^{-1}-10^{-2} per input donor, but plasmids containing novel joints from the ΔtraF' plasmids were not (frequencies < 10^{-7}-10^{-8}), indicating a cis-dominant defect.

The regions containing the novel joints, the corresponding region of F, and the relevant chromosomal segment near purE were sequenced. The nucleotide sequence of the novel joint region was identical for 4 independent $\Delta traFpurE^+$ isolates. This also was true for the $\Delta traFproA^+$ plasmids. The F DNA joined bacterial DNA 135-140 bp clockwise of the BglII site at the 66.6 kb F map position in both plasmid types. The chromosomal sequence involved in $\Delta traFpurE^+$ formation (F homologies are underlined) is

$$\begin{matrix} 5' & A & C & A & C & C & A & C & A & C & G & 3' \\ 3' & \underline{T} & \underline{G} & \underline{T} & \underline{G} & \underline{G} & T & \underline{G} & \underline{T} & \underline{G} & \underline{C} & 5' \end{matrix}$$

One $\Delta tra\ FpurE^+$ terminated at the same position in F but at a different chromosomal site.

During ΔtraF' formation, recombination events have occurred in the oriT region at one defined sequence on F. This sequence may include the site of nicking in oriT.

BACK TRANSFER : A PROPERTY OF SOME BROAD HOST RANGE PLASMIDS

M.Mergeay(1), P.Lejeune(1), G.Thiry(2), and M.Faelen(3)

(1) Département de Radiobiologie, SCK/CEN, B-2400 Mol
(2) Département de Botanique, Université de Liège
(3) Service de Génétique, ULB, Rhode-St.-Genèse, Belgium

Back transfer would be defined as a plasmid-mediated transfer of genes mediated in the plasmid bearing strain itself. Three lines of evidence supporting such a definition are here reported :

1) Homologous matings mediated by IncP1 plasmid pULB113(RP4::miniMu)
In such matings, an excess of recombinants carrying all the markers of the plasmid bearing strain but the counterselected one, was regularly observed : in a mating between *S. typhimurium* strains, the pULB113 bearing strain carrying seven markers (including Δ *trp* BA196) dispersed on the chromosome was mated with a single marked Δ *hisBHA* recipient. His^+ Trp^+ recombinants were selected and up to 95% of recombinants carried the markers of the plasmid bearing strain (*metA22*, *metE331*, *ilv452*, *xyl 404*, *rpsL120*, *galE493*) and were thus believed to be Trp^+ backtransconjugants instead of the expected His^+ direct transconjugants. In *P. fluorescens* 6.2, the Tra^- Mob^- IncQ plasmid pKT261 was introduced by transformation to mark the recipient strain and allowed to discriminate direct (normal) and back transconjugants.

2) Heterospecific matings mediated by RP4::miniMu
In matings between pULB113 bearing *P. fluorescens* and the unrelated strain *Alcaligenes eutrophus* CH34, R-primes were found at similar frequencies in both parents. In *P. fluorescens* the back R-primes carried up to 70 kb of genetic material coming from *A. eutrophus*.

3) Mobilization of plasmids poorly or non self transferable
Back mobilization of the Tra^- IncQ plasmid R300B was mediated in *E. coli* by plasmids RP4 and pUZ8 (IncP1), pULG14 (IncN) and R69.2 (IncM) at a frequency of 10^{-2} to 10^{-5}. Back mobilization of R300B was not induced by plasmids pSa and R339 (IncW), pIP113 and N3 (IncN). Plasmid pULG14 (*Erwinia uredovora*) attracted pMOL28 (165 kb; resistance to nickel and cobalt) from *A. eutrophus* CH34. Similarly, IncP1 plasmid pJP1 (catabolism of 2-4D) attracted in its resident strain *A. eutrophus* JMP134, pMOL28 or pMOL30 (238 kb ; resistances to Zn^{++}, Cd^{++} and Co^{++}) from *A. eutrophus* CH34. The back transcongugants resulted mainly in stable cointegrates in JMP134.

GENERALIZED TRANSDUCTION OF PLASMID REPLICONS BY BATERIOPHAGE P22 IN SALMONELLA TYPHIMURIUM

Jill Morgenstein and *Ronald Bauerle

Department of Biology
University of Virginia
Charlottesvlle, VA 22901

Plasmid pBR322 and various derivatives with inserts of the *trp* region of the *S. typhimurium* chromosome can be transferred between strains of *S. typhimurium* by generalized transduction with bacteriophage P22. The chimeric plasmids are transduced about 20-fold more efficiently than pBR322. $RecA^-$ donors as well as $RecA^-$ recipients are capable of plasmid transduction; however, the frequency is 5- to 10-fold higher with $RecA^+$ strains.

Gene 3 mutants of P22 which exhibit high frequency transduction of chromosomal markers show a corresponding increase in plasmid transducibility over that of wild type phage. Lysates of Erf^- P22 grown on $RecA^-$ donors retain the ability to transduce plasmids even though the titer of plaque forming units is severely reduced.

Plasmid DNA is packaged into transducing particles as a head to tail linear concatemer. The possibility that this DNA arises by the initiation of rolling circle replication of the plasmid upon infection of the donor strain by P22 is currently being examined. In this regard, Nic^- derivatives of pBR322 have been found to be transduced as efficiently as pBR322.

Both monomeric and a variety of multimeric plasmid replicons have been found to be established in individual transductant clones, apparently by a process of recircularization that is not necessarily RecA-dependent.

Besides serving as a convenient method for the transfer of plasmids between strains of *S. typhimurium*, plasmid transduction promises to be a model system for the analysis of the mechanism of formation of generalized transducing particles in P22 infections.

ROLE OF TRANSCRIPTION IN CLO DF13 PLASMID MOBILIZATION

*Arnold J. van Putten, Eduard Veltkamp, and H. John Nijkamp

Department of Genetics, Biological Laboratory, Free University, Amsterdam, The Netherlands

The non-conjugative, bacteriocinogenic plasmid Clo DF13 (m.w. 6 Md), can be mobilized efficiently in the presence of a conjugative plasmid, such as F or R. The mobilization process requires both gene products of the conjugative plasmid and Clo DF13 encoded proteins as well as a cis-acting Clo DF13 region, termed bom (basis of mobility). This bom region is located withi a 264 bp fragment around the unique HpaI site. Comparison with the corresponding Col E1 and pBR322 sequences showed similarities with respect to the secondary structure. With respect to a possible relationship between the origin of vegetative replication (oriV) and the origin of transfer (oriT) we found that neither distance nor orientation of the Clo DF13 bom region with respect to oriV had any significant influence or the mobilization frequency. Surprisingly, after cloning of the 264 HpaII fragment in a Bom$^-$ vector, restoration of the Bom$^+$ phenotype was only observed in one orientation. From these observations and from our sequence data we suggest that transcription into this cloned bom fragment is essential for effective mobilization of the plasmid.

TRANSFORMATION OF *CLOSTRIDIUM PERFRINGENS* WITH CHIMERIC PLASMID DNAS

C.H. Squires, D.L. Heefner, R.J. Evans, B.J. Kopp, and M. Yarus

Synergen, Inc.
1885 33rd Street
Boulder, CO 80301

L-phase variants, or autoplasts, of *C. perfringens* may be transformed with plasmid DNA in the presence of polyethylene glycol. L-phase variants are stable replicating protoplasts generated by growth of normal cells with penicillin G in osmotically protected medium. Approximately 1,000 to 10,000 transformants can be obtained per ug of *C. perfringens* DNA used; this is a frequency of 1/10,000 to 1/100,000 transformants/cell/ug DNA. *C. perfringens* will form autoplasts (self-generated protoplasts) if cultured in osmotically protective medium without penicillin G. These cells can be transformed at frequencies similar to those for L-phase variants. The cell wall of these transformants may be regenerated by growth in liquid medium containing 25% gelatin. The plasmid DNAs used to transform *C. perfringens* are chimerae of *E. coli* plasmid pBR322 and whole, cryptic, small *C. perfringens* plasmids cloned to it for replicon function. In addition, these plasmids contain 4 kb of *C. perfringens* plasmid DNA coding for resistance to tetracycline. This tetracycline resistance function is competent in both *E. coli* and *C. perfringens*.

IDENTIFICATION OF MULTIPLE SURFACE ANTIGENS ASSOCIATED WITH SEX PHEROMONE RESPONSE OF *STREPTOCOCCUS FAECALIS*.

M. L. Tortorello and G. M. Dunny, Department of Veterinary Microbiology, Cornell University, Ithaca, New York

High molecular weight surface antigens of *S. faecalis* obtained by ammonium sulfate precipitation of culture supernatants and identified in Western blots of SDS-polyacrylamide gels, have been correlated with the aggregation response of donor cells to sex pheromone ("clumping inducing agent," CIA).

CIA induced cells carrying the conjugative plasmid pCF10 produced a greater amount of a surface antigenic doublet having a subunit molecular weight in the 130,000 range than did uninduced cells or isogenic plasmid free cells. Additional unique bands having molecular weights of approximately 73,000, 134,000 and 138,000 were observed in CIA induced cells. A unique 160 kilodalton antigen was also produced by an isogenic constitutively clumpy mutant.

The 130 kilodalton antigenic complex increased in concentration with time after exposure of donor cells to CIA. Gel filtration studies indicated that this antigen exists in the native state at least as a dimer and perhaps as large as a hexameric complex. This component was susceptible to digestion by proteinase K and was not reactive with either concanavalin A or wheat germ agglutinin. The antigenicity of this component was not destroyed by treatment of blots with trypsin, chymotrypsin or papain prior to development with antibody.

VECTORS

USE OF *E. COLI* GALACTOKINASE TO PROBE FOR *STREPTOMYCES* PROMOTERS

*Brawner, M., Auerbach, J., Rosenberg, M. and Taylor, D.P.

Dept. of Molecular Genetics, Smith Kline and French Laboratories, Philadelphia, PA 19101

Plasmids designed to identify and characterize *Streptomyces* transcriptional regulatory elements have been constructed using the *Escherichia coli* galactokinase gene. The plasmids are shuttle vectors containing the pBR322 *E. coli* replicon and the pIJ350 *Streptomyces* plasmid, permitting rapid assembly and characterization of plasmid constructs in *E. coli* followed by transformation into *Streptomyces* for functional analysis. The *galK* gene is preceded by a series of unique cloning sites and a fragment, containing the tR1 transcription terminator of phage lambda, which prevents readthrough transcription into *galK*. A DNA fragment preceding the *Streptomyces* gene which encodes a β-galactosidase-like enzyme and the promoter fragment from the *S. fradiae aph* gene were cloned into a promoter probe plasmid and these fragments stimulated the synthesis of galactokinase in *Streptomyces.* The 5' end of the β-gal-galactokinase transcript was identified by S1 mapping and found to be near an AT rich region, also identified by S1 mapping of the chromosomally-encoded β-gal transcript. The level of galactokinase produced when controlled by the β-gal promoter was 10% of the total cellular protein. This promoter should prove useful in the construction of *Streptomyces* gene expression vectors.

PLASMID CLONING VEHICLES - NEW EXPRESSION VECTORS BASED ON THE P_L PROMOTER OF COLIPHAGE λ; EFFICIENT SYNTHESIS OF HUMAN INTERFERONS AND INTERLEUKIN 2

G. Buell, J. Delamarter (Biogen, S.A., Geneva, Switzerland), G. Simons, *E. Remaut and W. Fiers (Laboratory of Molecular Biology, State University of Ghent, Ghent, Belgium)

We have earlier described a plasmid expression system based on the inducible leftward promoter of coliphage λ (1). New derivatives have been constructed which incorporate a prokaryotic ribosome binding site and an easily accessible ATG initiator codon. The ATG codon is part of a NcoI site (CCATGG). Following NcoI cleavage and filling-in of the sticky ends, the ATG codon can be readily joined to any blunted coding region. Because of its much greater fidelity, the filling-in reaction is more advantageous than the previously used S1 nuclease in exposing the ATG codon (2). Ribosome binding sites were obtained from the replicase gene of phage MS2, the *ner* gene of phage Mu, or a modified attenuator region of the *E. coli trp* operon. In each case, a unique NcoI site was engineered at the initiation codon. Several unique restriction sites are situated further downstream. The vectors are small (2.8 Kb) and their complete nucleotide sequence is known.

We have used these vectors to obtain inducible, high-level synthesis in *E. coli* of human interferons and interleukin 2. β-interferon could be boosted to make up 2% of the total cellular protein. For interleukin 2 and γ-interferon, levels of 5% and 25%, respectively, were obtained.

1. Remaut et al. (1981) Gene 15, 81-93.

2. Remaut et al. (1983) Nucl. Acids Res. 11, 4677-4688.

INCLUSION OF PLASMIDS INTO LARGE LIPOSOMES: POTENTIAL FOR ELECTRIC FIELD INDUCED FUSION WITH CELLS

R.Büschl[2], *C.v.Eichel-Streiber[1], H.Ringsdorf[2], E.J.Zöllner[1]

Inst.Physiol.Chem.[1], Inst Org.Chem.[2] J.Gutenberg University
D-6500 Mainz, Federal Republic of Germany

We are interested in transfer of genes from cell-sized, DNA-carrying liposomes into target cells by electric field induced fusion described by Zimmermann[1].

I. Preparation of large liposomes: Liposomes with a diameter up to 100 µm were prepared by swelling lipid multilayer films in a medium of low ionic strength. An equimolar mixture of cholesterol and different natural and synthetic lipids [2] was used.

II. Inclusion of plasmid DNA: Large liposomes were prepared by initial swelling of lipid films with a small volume of aqueous pBR322 solution. For control of the amount of included DNA the liposomes were spun down by centrifugation, resuspended and the solution was than digested with DNase I to eliminate DNA in the outer space. After destruction of the vesicles by adding detergent solution the DNA was directly shown on agarose gel electrophoresis. The inclusion rates were up to 14% of the DNA input.

III. Fusion of empty liposomes with each other or with cells: Intimate contact between liposomes was established by the application of an alternating nonuniform field, resulting in dielectrophoresis and pearl chain formation. The fusion process was initiated by a field pulse of 3-9 $kV.cm^{-1}$ and 20-50 µs duration. The first experiments to realize a fusion of vesicles with Friend cells have been carried out successfully[3].

Fusion experiments with bacteria have thus far been unsuccessful because it is still unclear how to prepare the bacteria before the experiments. Therefore we plan to first study the electric field induced fusion between DNA-carrying liposomes and Saccharomyces cerevisiae cells.

[1] U.Zimmermann, Biochim.Biophys.Acta 694,227 (1982)
[2]. L.Gros, H.Schupp, H.Ringsdorf, Angew.Chem.Int.Ed.Engl. (1981) 20, 305
[3]. R.Büschl, Inauguraldissertation, Mainz (1984)

A NOVEL BACTERIAL-YEAST SHUTTLE VECTOR SYSTEM FOR CLONING AND INVESTIGATING STRUCTURAL OR REGULATORY DNA REGIONS

Mark C. Fagan* and John F. Scott

Department of Microbiology
University of Illinois
Urbana, IL USA.

We report the construction of several unique cloning vectors composed of bacterial and yeast DNA which allow replication and selection in both organisms. Subsequent facile manipulations transmute these shuttle vectors into Yeast Acentric Ring plasmids, (YARp's). This system has distinct advantages over other shuttle vectors. One important advantage is that the YARp's constructed will be derivatives of the TRP1 RI Circle (YARp1) which contains only the TRP1 coding region and an autonomously replicating sequence, ARS1. YARp1 has been shown to transform yeast at high frequency, to be established at a copy number of 100-200 per cell, and to have high stability as compared to most shuttle plasmids containing non-yeast DNA. Another important property of this system is that YARp1 can be isolated as a protein-associated chromatin complex relatively free from other cellular constituents. Also, various DNA sequences have been inserted into YARp1 resulting in chimeric plasmids retaining properties similar to the parental YARp1, thus establishing its validity as a cloning vector. For example, taking advantage of selection and screening techniques in bacteria, we cloned a galactose-inducible yeast promoter into a unique EcoRI site proximal to the TRP1 coding region in one of the shuttle vectors. Cleavage by one restriction enzyme released the combined yeast fragment which was then recircularized with DNA ligase. The resulting YARp may prove to be a superior plasmid for some structural and functional studies of galactose regulation in yeast.

EXPRESSION OF BOVINE GROWTH HORMONE IN *E. COLI*: ENHANCED EXPRESSION BY ALTERING SEQUENCES OF THE CODING REGION WITH SYNTHETIC DNA FRAGMENTS

*H. Hsiung, B. Schoner, D. Smith and R. Schoner
Lilly Research Laboratories, Eli Lilly and Company, 307 E. McCarty St., Indianapolis, IN 46285

The DNA sequences required for high level expression of bovine growth hormone genes in *E. coli* was investigated using temperature-sensitive copy control plasmids as cloning vectors. By changing the 5'-terminal coding region of the bovine growth hormone (BGH) gene with various synthetic DNA fragments, we were able to dramatically alter the level of expression. Some BGH gene constructions gave greater than 35 percent of the total cell protein as methionyl bovine growth hormone (met-BGH), while the original cDNA met-BGH gene gave very low espression (1 percent).

The results showed that minor changes (one or few base changes) in the 5'-terminal coding region of cloned BGH gene can have a profound effect on BGH expression. A possible correlation between the increase in the level of the BGH expression and the destabilization of potential 5'-end mRNA secondary structure in the region of the initiation codon, Shine-Dalgarno sequence, and the NH_2-terminal sequences of the BGH genes is discussed.

EXPRESSION OF THE *Escherichia coli* XYLOSE ISOMERASE GENE IN *Bacillus subtilis* USING A *Bacillus* PROMOTER

Huang, J.J. and H, N.W.Y., Laboratory of Renewable Resources Engineering, Purdue University, West Lafayette, IN 47907

A DNA fragment containing both the *E. coli* D-xylose isomerase gene (*xylA*) and D-xylulokinase gene (*xylB*) has been isolated from an *E. coli* genomic bank constructed by Clarke and Carbon. The D-xylose isomerase gene was characterized and then separated from the D-xylulokinase gene by a series of subcloning processes. D-Xylose isomerase, the enzyme product of the *xylA* gene, is responsible for the conversion of D-xylose to D-xylulose, as well as D-glucose to D-fructose. We have subcloned the intact *xylA* gene into an *E. coli* and *Bacillus* shuttle plasmid. The resulting plasmid was used to transform *Bacillus*. However, enzymatic assay indicated that the intact *E. coli xylA* gene was not functioning in *Bacillus* transformants. We believed this was due to the ineffectiveness of the *E. coli xylA* promoter in *Bacillus*.

In order to express the *xylA* gene in *Bacillus*, we have isolated the gene encoded for the synthesis of β-lactamase (penicillinase), *penP*, from *B. lichenformis*. It has been shown that the promoter of *penP* functions properly in both *E. coli* and *Bacillus*. Basically, we are trying to replace the *xylA* promoter with *penP* promoter. The promoter region of the *E. coli xylA* gene was deleted and the resulting *xylA* structural gene was inserted downstream to the promoter of the *penP* gene. The resulting hybrid gene (*penP-xylA*) complements *E. coli xylA* mutation. *Bacillus* transformants containing the hybrid gene produced elevated D-xylose isomerase activity. Improvement of the expression of D-xylose isomerase gene in *Bacillus* is in progress. (Supported by NSF PCM-83C5043, and SERI Sub-contract XX-3-03116.)

LOW-, INTERMEDIATE-, AND HIGH-COPY-NUMBER CLONING VECTORS DERIVED FROM THE *PSEUDOMONAS* PLASMID pVS1

Itoh, Y., Soldati, L., Leisinger, Th., and *Haas, D.,

Department of Microbiology, Swiss Federal Institute of Technology, CH-8092 Zürich, Switzerland

The *Pseudomonas* plasmid pVS1 was reduced to a minimal replicon of ca. 3 kb and used to construct gene cloning vehicles. Several non-mobilizable or RP1-mobilizable plasmids were obtained. Each carried two resistance markers for selection and insertional inactivation, respectively. The Mob$^-$ vectors pME290 (Cb Km) and pME291 (Hg Km) could be introduced by transformation and maintained in *P. aeruginosa*, *P. putida*, *P. fluorescens*, *P. stutzeri*, *P. acidovorans*, *P. mendocina*, *P. cepacia*, and *P. syringae*.

Since the vector copy number appears to influence critically the stability of recombinant plasmids in *P. aeruginosa* PAO, we isolated mutants of pME290 (ca. 7 copies) with different copy numbers: pME292 (1-2 copies) and pME294 (ca. 15 copies). The *argF* gene of *P. aeruginosa* cloned into these vectors was expressed according to their copy numbers and the recombinant plasmids were stable in *P. aeruginosa* during *argF* repression (in arginine medium) and derepression (in glutamate medium). However, under repressive conditions 15 *argF* copies had a strong inhibitory effect on growth and an *argF*$^+$ recombinant plasmid derived from the RSF1010 vector pKT240 ($\geq$20 copies) led to severe physiological disturbances and was entirely unstable in *P. aeruginosa*.

CONSTRUCTION OF AN IMPROVED MAXICELL STRAIN FOR THE IDENTIFICATION OF RECOMBINANT PLASMID ENCODED PROTEINS

M.C. Lorence, J.L. Alcorn and C.S. Rupert

The University of Texas at Dallas
Biology Programs, P.O. Box 830688
Richardson, TX 75083-0688, USA

An improved "maxicell" host strain of *Escherichia coli* [MCL22: *trkA405 trkD1 thi rpsL31 Δ(kdp-phr)214 Δ(gal-uvrB) Δ(srl-recA)306 rha*], has been constructed to avoid the most common problems blocking identification of plasmid-encoded proteins by the method of Sancar *et al*. (J. Bacteriol. **137**, 692, 1979). (1) Use of deletions, rather than point mutations, to inactivate genes of the DNA repair pathways [photoreactivation (*phr*), excision-repair (*uvrB*), and recombination-repair (*recA*)] prevents inadvertent reversions, and the attendant, unwanted bacterial chromosomal protein synthesis. (2) Absence of nutritional deficiencies allows a more vigorous growth than that shown by previously used strains (45 minute doubling time in Luria broth, and growth in minimal media without amino acid supplement) facilitating establishment of cultures free of slow-growing cells. The strain also shows high efficiency of transformation by plasmid DNA ($>10^6$ transformants per μg pBR322 DNA). Efficacy of these features is shown in tests to distinguish plasmid-encoded protein products of pBR322.

CLONING VECTORS DERIVED FROM THE LOW COPY *STREPTOMYCES* PLASMID SCP2*: USE OF SUCH A VECTOR IN ISOLATING THE ENTIRE SET OF BIOSYNTHETIC GENES FOR AN ANTIBIOTIC

D.J. Lydiate, F. Malpartida, and D.A. Hopwood*

John Innes Institute
Norwich, U.K.

SCP2* (31 kb), a variant of the natural plasmid SCP2 of *Streptomyces coelicolor* A3(2), exists in 1-2 copies per host chromosome. By *in vitro* deletion of non-essential DNA, and addition of one or more marker genes, we have derived a series of useful cloning vectors. As well as sequences essential for replication, they all retain a "partition" function needed for stable plasmid inheritance. They include pIJ903 with unique cloning sites for several enzymes: *Bam*HI, *Eco*RI, *Hin*dIII, etc; pIJ916, with sites for *Bgl*II, *Sst*I, etc; pIJ922, with sites for *Bam*HI, *Pst*I, etc; and pIJ940, with sites for *Bam*HI, *Bgl*II, *Pst*I, etc. All four vectors carry the *tsr* gene for thiostrepton resistance and pIJ940 carries hygromycin resistance in addition. pIJ903, pIJ922, and pIJ940 are transmissible by conjugation, while pIJ916 is non-transmissible.

These vectors are very suitable for the stable cloning at low copy number of large segments of *Streptomyces* DNA. For example, pIJ922 (24.5 kb) has been used to clone DNA from the wild-type *S. coelicolor* into mutants blocked in actinorhodin biosynthesis (*act*). Size-fractionated donor chromosomal DNA (15-30 kb), produced by partial digestion with MboI, was inserted into the *Bam*HI site of pIJ922 and introduced by transformation into protoplasts of an *act* mutant. Two Act$^+$ clones were obtained, with overlapping inserts of 16 kb and 34 kb. By *in vitro* reconstruction, a single clone (pIJ2303 with an insert of 32.5 kb) was obtained which complemented all known classes of *act* mutants. It also promoted actinorhodin synthesis in an unrelated species, *Streptomyces parvulus*. It therefore appears to carry the entire set of genetic information needed to make this polyketide antibiotic from acetate. Even plasmids as large as PIJ2303 can transform *Streptomyces* protoplasts at good molar efficiency.

λPL-COLE1 ROP GENE FUSION:

POSITIVE SELECTION EXPRESSION VECTOR-ANTIGEN OVERPRODUCTION

*Muesing, M., Carpenter, C., Klein, W. and B. Polisky
Program in MCD Biology   Department of Biology
Indiana University   Bloomington, Indiana 47405

A plasmid vector has been constructed which has several features that facilitate high-level expression of foreign DNA in *E. coli*. These include: 1) a powerful, inducible promoter (Remaut *et al.*, 1981), efficient translation initiation signals and 2) a positive genetic selection for recombinant plasmids carrying an insert. The vector, designated pMAM17, carries the ColE1 *rop* gene (Cesareni et al., 1982) under control of the thermally inducible λPL promoter. The *rop* gene product is a negative regulator of ColE1 DNA replication, and its high-level expression is lethal to cells. However, cells harboring a plasmid with an insert in the *rop* gene grow normally under these conditions. This provides a convenient method for the genetic selection and translational amplification of previously cloned DNA sequences. Our results demonstrate that induced extracts from cells harboring pMAM17 recombinants can produce 20-25% of their total soluble protein as *rop* hybrid protein. Moreover, unfractionated *E. coli* extracts containing *rop*-fusion protein can be used to obtain antibodies capable of specific recognition of authentic polypeptide present in cells from which the insert DNA originated (Muesing et al., 1984).

Remaut, E., Stanssens, P. and W. Fiers (1981) Gene *15*: 81-93.

Cesareni, G., Muesing, M. and B. Polisky (1982) Proc. Natl. Acad. Sci. USA *79*: 6313-6317.

Muesing, M., Carpenter, C., Klein, W. and B. Polisky (1984) submitted for publication.

A NEW HIGH-EFFICIENCY VECTOR FOR CLONING DNA IN

HAEMOPHILUS INFLUENZAE

N.K. Notani*, V.P. Joshi, and R. Kanade

Biology and Agriculture Division
Bhabha Atomic Research Centre
Bombay - 400 085, India

RSF0885 is a multicopy plasmid of *Haemophilus* which carries an *amp*^{r} marker. Transformation with RSF0885 DNA of a highly competent *H. influenzae* culture for *amp*^{r} marker is surprisingly low. Absence in RSF0885 DNA of a specific 11-bp sequence thought to be required for uptake of any DNA into *H. influenzae* was considered to be an explanation for the low transformation efficiency. Splicing to RSF0885 of chromosomal DNA segments from *H. influenzae* genome which normally contains 600 copies of uptake sequences, improves the transformation 100-1000 fold. However, the chimeric plasmid DNA still requires *rec* gene expression for peak transformation.

RSF0885 is cut once with *Pvu*II and chromosomal DNA can be spliced to it and the inserts detected efficiently. A chimeric plasmid pD7 was isolated which carries a large segment of *H. influenzae* chromosomal DNA. pD7 has 4 *Eco*RI sites, all in the chromosomal segment. In two steps, a plasmid pJ1-8 was derived from pD7 which by itself gives little or no transformation and is cut only once with *Eco*RI. Its molecular weight is 2.5×10^6 versus 3.7×10^6 of RSF0885. Thus, pJ1-8 offers some advantages over RSF0885 such as (1) cutting can be done with an enzyme (*Eco*RI) that produces 'sticky' ends versus blunt-ends produced by *Pvu*II, (2) detection of colonies carrying inserts is more efficient since there are virtually no 'background' colonies and (3) a bigger insert can be put because of the plasmid being smaller now. Using this vector, two genes from *H. influenzae* have been cloned. Transformation with these chimeric plasmids suggests the involvement of recombination in DNA entry.

MOBILIZABLE VECTOR PLASMIDS FOR GENETIC MANIPULATIONS IN GRAM-NEGATIVE BACTERIA

Ursula Priefer, Simon, R. and Pühler, A.
Lehrstuhl für Genetik, Fakultät für Biologie, Universität Bielefeld, Postfach 8640, D-4800 Bielefeld 1, FRG

Genetic engineering vectors commonly used in *E.coli* (e.g. pBR325, pACYC184) have been modified to be applicable in a broad variety of gram-negative bacterial species (1,2).

There are two groups of plamids. Both are efficiently mobilizable in the presence of RP4 or other IncP plasmids, but they differ with respect to their replication host range.

One group, constructed by cloning the *ori*T region of RP4 into conventional *E.coli* vectors, can be introduced into all gram-negative species tested so far, but fails to replicate in bacteria outside the enterobacteria. This type of vector has for example been succesfully applied for general transposon mutagenesis and homogenotisation experiments in a variety of gram-negative strains.

The other group of mobilizable vector plasmids combines usual *E.coli* vector plasmids with the broad host range replication and mobilization functions of plasmid RSF1010. These vectors can therefore be introduced and also maintained in all gram-negative bacteria tested. They have been used for complementation experiments in *Rhizobium meliloti*.

We also constructed cosmid libraries of different Rhizobium strains with vector derivatives containing the λ *cos*-site. These experiments revealed that the replicative type of vector becomes unstable in *E.coli* and Rhizobium when carrying large inserts of foreign DNA. The cosmids are either completely lost (under non-selective conditions) or they retain their resistance marker (with selective pressure) but are degradated to plasmids of different sizes. This instability is markedly reduced in *E.coli* *rec*A hosts.

(1) R. Simon, U. Priefer and A. Pühler (1983): Proceedings of the 1st Internaitonal Symposium on "Molecular Genetics of the Bacteria-Plant Interaction (ed. A. Pühler), Springer-Verlag, Heidelberg
(2) R. Simon, U. Priefer and A. Pühler (1983): BIO/TECHNOLOGY 1, pp 784-791

PROTEIN EXCRETION IN *ESCHERICHIA COLI*. APPLICATION OF THE PLASMID-ENCODED CLOACIN DF13 EXPORT SYSTEM

* A.J. van Putten[1], C.P.P.M. Süter[1], B. Oudega[2], F.K. de Graaf[2], and E. Veltkamp[1]
[1] Department of Genetics
[2] Department of Microbiology; Biological Laboratory, Vrije Universiteit, Amsterdam The Netherlands

The bacteriocin, cloacin DF13, excreted by the gram-negative bacterium *Escherichia coli*, has been studied extensively in our laboratory. Cloacin DF13 is encoded by the bacteriocinogenic plasmid Clo DF13. It consists of two subunits designated cloacin (m.w. 59,000) and immunity protein (m.w. 10,000). Both subunits are synthesized without a signal sequence and form an equimolar complex in the cytoplasm. Another pClo DF13 encoded polypeptide, designated protein H (m.w. 2,800), is responsible for the translocation of the bacteriocin molecules across the cytoplasmic and outer membrane.

The genes encoding the cloacin protein, the immunity protein and protein H are organized in one, mitomycin C inducible, operon with a common promotor in front of the cloacin gene. A model for the excretion mechanism of claocin DF13 will be presented. Furthermore the strategy for engineering protein excretion in *E. coli* by using this excretion system as well as preliminary results will be discussed.

NOVEL PLASMID-MEDIATED CATABOLIC PATHWAYS FOR HETEROCYCLIC COMPOUNDS

O.P. Shukla

Division of Biochemistry
Central Drug Research Institute
Lucknow-226 001, INDIA

A strain of Nocardia, capable of growing on pyridine-N-oxide (PNO), 2-hydroxypyridine and pyridine, catabolised PNO via 2-hydroxypyridine while pyridine was degraded via succinic semialdehyde; PNO and 2-hydroxypyridine did not serve as intermediates in this pathway. PNO-degrading ability was lost on cultivation in nonselective media, but such cells retained the ability to degrade pyridine and 2-hydroxypyridine. Pyridine-degrading ability of another Nocardia sp. growing on pyridine was also unstable, and lost spontaneously on storage or cultivation in non-selective media, suggesting the involvement of plasmids.

Two distinct pathways of quinoline catabolism have been deciphered employing four different organisms. A strain of Pseudomonas testosteroni and a Nocardia sp that arose as a contaminant in this culture, metabolised quinoline via 2-hydroxyquinoline and 2,6-dihydroxyquinoline. However, 8-hydroxycoumarin has been characterised as the key intermediate of quinoline catabolism in Pseudomonas stutzeri and another Pseudomonas sp.. These organisms metabolise quinoline via 2-hydroxyquinoline, 2,8-dihydroxyquinoline, 8-hydroxycoumarin and 2,3-dihydroxyphenylpropionic acid as intermediates. Coumarin was not utilised as a growth substrate by these organisms, but quinoline-grown resting cells converted coumarin to melilotic acid which was not further metabolised.

TRANSFER OF SHUTTLE PLASMID pTZ22 BY PROTOPLAST FUSION BETWEEN BACILLUS SUBTILIS Ki-2-132 AND ESCHERICHIA COLI C600

Mao-Hong Tang, Ming-Feng Li, Yue-Qin Yang,
Shu-Tian Fan, Yong-Hua Sun, and Yin-Mei Zhang

Institute of Genetics, Academia Sinica
Beijing, China

Bacillus subtilis Ki-2-132 (Thr^-, Ile^-, Val^-) is a highly transformable mutant of Ki-2 which was screened from chinese soil. Its competence developed faster and stability of plasmids was higher than that of the commonly used strain 168. Plasmid pTZ22 is a shuttle plasmid between *B. subtilis* and *E. coli* and could be transferred by protoplast fusion from *E. coli* into *B. subtilis* and from *B. subtilis* into *E. coli*. The fusion frequency was 10^{-4}-10^{-6}. The secondary colonies, resembled *B. subtilis*, of fusion product colonies resembled *E. coli* were observed. The proteinase of *B. subtilis* Ki-2-132 did not seem to obstruct protoplast fusion, plasmid transfer, and gene expression in both directions of plasmid transfer.

RECOMBINANT PLASMIDS FOR THE ANALYSIS OF TRANSPOSITION

*Thorpe, P.A. and R.C. Clowes

The University of Texas at Dallas
Biology Programs, P.O. Box 688
Richardson, TX 75080, U.S.A.

Recombinant plasmids have been constructed carrying inverse and direct repeats of sites believed to be involved in recombination during transposition. These plasmids carry the RES site (TTATAA), identified as the site of action of the *tnpR* gene product, or a second site, termed "IRS", which contains a sequence of 23 bp, 19 of which are homologous with the 38 bp terminal inverse repeat of Tn*2660*. These two sites have been cloned separately, and jointly. (RES and IRS are respectively found at coordinates 2998 to 3003, and 3021 to 3043 of the Tn*3* sequence, and Tn*2660* appears to be identical in all respects to Tn*3*.

Rearrangements within the transposable element Tn*2660* have been constructed to produce mutants in either the *tnpA* or the *tnpR* gene. Host cells carrying one of the mutant Tn*2660* plasmids and a plasmid with an inverse or direct repeat of RES or IRS sites have been constructed in *recA* and *recA*$^+$ hosts and are being analyzed to determine if recombination events related to transposition occur. Deletions or inversions in the recombinant plasmids are expected as a consequence of the predicted activities of the *tnpR* and *tnpA* gene products. Identification of these events would further refine the current model of transposition.

HUMAN PROINSULIN EXPRESSION USING A COLICIN E3 DERIVATIVE VECTOR FOR THE DIRECT SELECTION OF RECOMBINANT DNA

T. Vernet and L.P. Visentin. National Research Council of Canada, Div. of Biological Sciences, Ottawa, Ont., Canada, K1A 0R6

The screening of *E. coli* cells in cloning experiments can be done in a fast and efficient way by using a vector designed to allow a direct selection for recombinant plasmids. pVT25, a 3.94 kb plasmid, is a hybrid between the ampicillin resistance gene, the origin of replication of pBR327 and the colicin E3 gene (fig. 1). Since it does not carry the colicin E3 immunity gene such a plasmid becomes lethal for the host cells unless the bacteriocin is inactivated by insertion of a foreign DNA fragment. The number of recombinant clones obtained using pVT25 as a vector is proportional to the concentration of the DNA to be inserted. DNA fragments generated by ClaI, TagI, HpaII or AccI enzymes can be cloned into the single ClaI site of pVT25. This vector can be maintained within *E. coli* if complemented with another plasmid which expresses the colicin E3 immunity gene as well as the tetracycline phenotype.

The synthetic human proinsulin gene was fused in phase with the colicin gene of pVT25. Antigenic determinants against the C-peptide of proinsulin were shown to be expressed only after treatment of the cell culture with mitomycin C, a classic inducer of colicin synthesis. The maximum synthesis of the colicin-proinsulin hybrid protein was reached after three hours (fig.2).

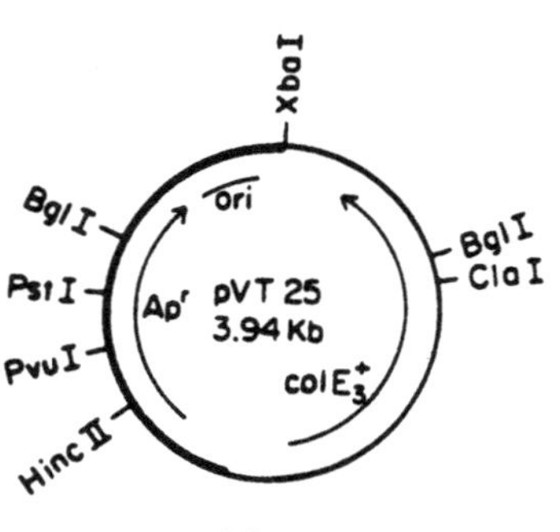

Fig.I

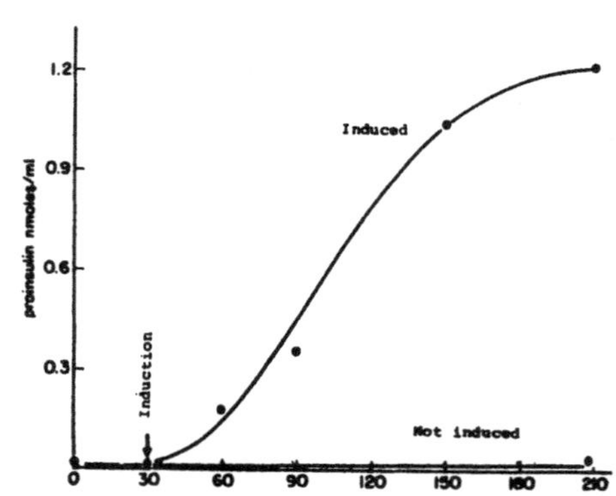

Fig.2

DEVELOPMENT OF SHUTTLE VECTORS BETWEEN STREPTOMYCES SPECIES AND *ESCHERICHIA COLI*

Wohlleben, W. and Pühler, A.

Lehrstuhl für Genetik, Fakultät für Biologie, Universität Bielefeld, Postfach 8640, D-4800 Bielefeld 1, FRG

Endogenous plasmids of *Streptomyces ghanaensis* and *Streptomyces venezuelae* were used to develop various shuttle vectors which replicate in Streptomyces as well as in *Escherichia coli*. The three Streptomyces plasmids which were characterized are suitable basic replicons for cloning experiments. They have a size between 12 and 14 kb, a copy number of about 20, and useful cloning sites for various restriction endonucleases. The shuttle vectors were obtained by fusing *E.coli* cloning vehicles like pBR325 and pACYC184 to the above described Streptomyces plasmids. In addition, antibiotic resistance genes for thiostrepton and neomycin (1) were cloned into these shuttle vectors. These resistance genes are expressed in a variety of Streptomyces strains. Further investigations in *S.lividans* 1326 revealed that these shuttle vectors can be transferred by conjugation between *S.lividans* strains. Transfer in *S.lividans* is accompanied by the typical pock structure described for *S.lividans* and *S.coelicolor* plasmids (2). The hybrids based on pACYC184 are stably maintained in *E.coli* and in the Streptomyces strains, so far investigated.

The shuttles are now used to study and to manipulate Streptomyces-DNA in *E.coli* and Streptomyces.

(1) C. J. Thompson, T. Kieser, J. M. Ward, D. A. Hopwood: Physical analysis of antibiotic-resistance genes from Streptomyces and their use in vector construction. Gene *20*, 51-62 (1982)

(2) M. Bibb, J. L. Schottel, S. N. Cohen: A DNA cloning system for interspecies gene transfer in antibiotic-producing Streptomyces. Nature *284*, 526-531 (1980)

THE ESCHERICHIA COLI INITIATOR PROTEIN, dnaA, IS INDISPENSABLE FOR THE MAINTENANCE OF λ PLASMIDS

J.Kur, I.Górska and [x]K.Taylor
Departments of Microbiology and [x]Molecular Biology,
University of Gdańsk, 24 Kładki, 80-822 Gdańsk, Poland

The plasmid pIG7 obtained by cloning the λ replication region (derived from the λdvi plasmid) into the HindIII site of pBR322 transforms the polAts dnaAts Hfr cells with normal efficiency at 30°C. Due to polAts mutation, this hybrid replicon should be constrained to the λ-type of replication at 42°C although one cannot eliminate a possible cooperation of the origin of pBR322. At this temperature plasmid-less cells start to appear exponentially after a lag of about 2 hours, and the half-life in the loss of pIG7 is very close to one generation time. Hybridization of ^{3}H-thymidine pulse-labeled pIG7 DNA to the DNA of pBR322 points to the inhibition of replication of the hybrid plasmid at 42°C. The lack of dnaA activity may inhibit the replication of pIG7 directly, or indirectly - affecting the process of partitioning of the plasmid copies.

We have not been able to transform dnaAts Hfr cells with standard λ plasmids, even at 30°C. We obtained extremely rare transformants only for pRLM4 (1), using selection for kanamycin resistance. Plasmids isolated from these transformants revealed normal efficiency of transformation of dnaAts or dnaA$^+$ cells at 30°C, and were shown to bear the πA type of mutation in the λ replication gene P (2, 3). When shifted to 42°C, the dnaAts Hfr bacteria harboring λπ plasmids start to segregate plasmid-less cells, as in the previously-described system. The results presented demonstrate the participation of functional dnaA protein in the maintenance of λ plasmids.

Since dnaA-independence of λ phage is well established, we postulate that λ phage may bypass the dnaA-dependent step (replication and/or partitioning) specific for λ plasmids.

(1) Wold, Mallory, Roberts, LeBowitz and McMacken, 1982. P.N.A.S. 79 : 6176.
(2) Georgopoulos and Herskowitz, 1971. Lambda Book I, p.553.
(3) Kellenberger-Gujer and Podhajska, 1978. M.G.G. 162 : 17.

Speaker Roster

AMASINO, RICHARD, Dept. of Microbiology, University of Washington, Seattle, WA
ARBER, WARNER, Dept. of Microbiology, University of Basel, Klingelbergstrasse, Switzerland
AUSTIN, STUART, Frederick Cancer Research Facility, Frederick, MD
BASTIA, DEEPAK, Dept. of Microbiology, Duke University, Durham, NC
BERG, DOUGLAS E., Dept. of Microbiology, Washington University, School of Medicine, St. Louis, MO
BUJARD, HERMAN, Molecular Genetics, University of Heidelberg, Federal Republic of Germany
CARLTON, BRUCE, Dept. of Molecular and Population, Genetics University of George, Athens, GA
CHAKRABARTY, A., Dept. of Microbiology, University of Illinois Medical School, Chicago, IL
CHANDLER, M., Centre de Recherche de Biochimie et de Genetique Cellulaires, C.N.R.S., Toulouse, France
CLEWELL, DON, Dept. of Oral Biology, University of Michigan, School of Dentistry, Ann Arbor, MI
CLOWES, ROYSTON, Biology Programs, The University of Texas, Dallas, Richardson, TX
COHEN, AMIKAN, Dept. of Microbiological Chemistry, Hebrew University, Hadassah Medical School, Jerusalem, Israel
COHEN, STANLEY N., Dept. of Genetics, Stanford University Medical Center, Stanford, CA
COLLINS, JOHN, Gessellschaft für Biotechnologische Forschung mblt., Mascheroder Weg, West Germany
CROSA, JORGE, Dept. of Microbiology L220, Oregon Health Sciences University, Portland, OR
DATTA, NAOMI, Dept. of Bacteriology, Royal Postgraduate Medical School, Hammersmith Hospital, London, England
EHRLICH, STANISLAV D., Institut de Biologie Moléculaire, Faculty de Science, Paris, France
FAN, YUN-LIU, Institute of Microbiology, Academia Sinica, Beijung, China
FIGURSKI, DAVID H., Dept. of Microbiology, Columbia University, New York, NY
FLYNN, JAMES E., Biochemistry Dept., Stanford University Medical School, Stanford, CA (deceased)
GAWRON-BURKE, CYNTHIA, 300 N. Ingalls, Ann Arbor, MI

GOEBEL, WERNER, Institute für Genetik und Microbiologie, Universitat Wurzburg, Wurzburg, Germany
GUINEY, DONALD, Dept. of Infectious Disease, University Hospital--Univ. of California Medical Center, San Diego, CA
GUNSALUS, I.C., Dept. of Biochemistry, University of Illinois, RAL, Campus, Champaign/Urbana, IL
HELINSKI, DON, Dept. of Biology, University of California at San Diego, La Jolla, CA
HERSHBERGER, CHARLES L., Biochemical Development Division, Eli Lilly & Company, Indianapolis, IN
HIRAGA, SOTA, Institute for Virus Research, Kyoto University, Sakyo-Ku, Kyoto, Japan
HOPWOOD, DAVID, Dept. of Genetics, John Innes Institute, Norwich, NR England
IPPEN IHLER, KAREN, Dept. of Microbiology, Texas A&M University, Medical School, College Station, TX
IKE, YASUYOSHI, Dept. of Microbiology, Gunma Medical School, Maebashi City, Japan
JACKSON, DAVID A., Genex Corp., Gaithersburg, MD
KOSUGE, TSUNE, Dept. Plant Pathology, University of California, Davis, CA
MACRINA, FRANCIS, Dept. of Microbiology & Immunology, Virgina Commonwealth University, Medical College of Virginia, Richmond, VA
MATSUBARA, KENICHI, Laboratory of Molecular Genetics, Osaka University Medical School, Kitaoku, Osaka, Japan
MEYER, RICHARD, Dept. of Microbiology, Univ. of Texas at Austin, Austin, TX
NAKAZAWA, TERUKO, School of Allied Health Sciences, Yamaguchi University, Ube, Japan
NIJKAMP, H.J.J., Dept. of Molecular Genetics, Free University, Biological Lab, Amsterdam, The Netherlands
NORDSTRÖM, KURT, Dept. of Microbiology, University of Uppsala, Biomedical Center, Uppsala, Sweden
NOVICK, RICHARD, The Public Health Research Institute of the City of New York, Inc., New York, NY
PEMBERTON, JOHN, Dept. of Microbiology, University of Queensland, St. Lucia, Brisbane
POLISKY, BARRY, Department of Microbiology, Indiana University, Bloomington, IN
PRITCHARD, R.H., Dept. of Genetics, University of Leicester, Leicester, England
ROWND, R.H., Dept. of Molecular Biology, Northwestern University, Chicago, IL
SAKAGUCHI, KENJI, Mitsubishi-Kasei Institute of Life Sciences, Tokyo, Japan
SCHMITT, RUDIGER, Dept. of Biochemistry and Genetics, University of Regensburg, Federal Republic of Germany
SETLOW, JANE, Dept. of Biology, Brookhaven National Lab, Upton Long Island, NY

SHAPIRO, JAMES, Dept. of Microbiology, University of Chicago, Chicago, IL
SILVER, SIMON, Washington University, St. Louis, MO
SUMMERS, ANNE, Dept. of Microbiology, University of Georgia, Athens, GA
TIMMIS, KENNETH, Dept. Biochimie Medcale, Geneva, Switzerland
WEISBLUM, BERNARD, Pharmacology Department, Univ. of Wisconsin Med. School, Madison, WI
WILKINS, BRIAN M., Dept. of Genetics, University of Leicester, Leicester, England
WILLETTS, NEIL, Dept. of Molecular Biology, University of Edinburgh, Edinburgh, Scotland
WILLIAMS, PETER, Dept. of Genetics/School of Medicine, University of Leicester, Leicester, England
YARMOLINSKY, MICHAEL, Frederick Cancer Research Center, Frederick, MD

Participants Roster

ABELES, ANN, Frederick Cancer Research Facility, Frederick, MD
ADSIT, JEAN, NYS College Vet. Med., Veterinary Microbiology, Ithaca, NY
ALLARD, LLOYD R., Illinois State University, Dept. of Biological Sciences, Normal, IL
ALONSO, JUAN CARLOS, Max-Planck-Institut für Molekulare, Genetile, Berlin, Germany
ALVAREZ, ANNE, University of Hawaii, Dept. Plant Pathology, Maile Way, HI
ANDERSON, DOUGLAS G., KELCO, San Diego, CA
ANNICK, MERCENIER, c/o Transgene S.A., Strasbourg, France
ARCHER, JOHN, University of Glasgow, Institute of Genetics, Scotland, UK
ARCHER, GORDON L., Medical College of Virginia, Richmond, VA
ARLITSCH, HELLA S., Purdue University, W. Lafayette, IN
ATTRIDGE, MICHAEL, GIBCO Labs, Madison, WI
BAGDASARIAN, MICHAEL, Max-Planck-Institut für Molekulare, Genetik, Berlin, Germany
BANG, SOOKIE S., University of Notre Dame, Dept. of Biology, Notre Dame, IN
BARRY, GERARD F., Monsanto Company, St. Louis, MO
BAUERLE, RONALD, University of Virginia, Dept. of Biology, Charlottesville, VA
BEACH, MICHAEL J., Purdue University, Dept. of Biochemistry, West Lafayette, IN
BERGQUIST, P.L., University of Auckland, Dept. of Cell Biology, Auckland, New Zealand
BERTANI, GIUSEPPE, Cal Tech., Pasadena, CA
BEST, ELAINE, University of Michigan, Ann Arbor, MI
BHUTA, PRAKASH, Eastern Washington University, Dept. of Biology, Cheney, WA
BITTNER, MICHAEL, Monsanto Company-U4D, St. Louis, MO
BOTT, MARTHA A., Illinois State University, Dept. of Biological Sciences, Normal, IL
BRADLEY, D.E., Memorial Univ. of Newfoundland, Faculty of Medicine, St. John's Newfoundland, Canada
BRAWNER, MARY E., Smith Kline & French Laboratories, Philadelphia, PA

BRUNNER, DAVID P., Illinois State University, Normal, IL
BUCHHOLZ, STEVEN E., Rutgers University, Dept. Biochemistry & Microbiology, Cook College, New Brunswick, NJ
BUCKLAND, R. MARTIN, Celltech, Ltd., Slough, Berkshire, UK
BUZAN, JENNY, Dept. of Pathology, Washington University Medical School, St. Louis, MO
CAPAGE, MIKE, Synergen, Inc., Boulder, CO
CARREIRA, LAURA, University of Georgia, Dept. of Biochemistry, Athens, GA
CARSIOTIS, MICHAEL, University of Cincinnati, Dept. of Microbiology & Molecular Genetics, Cincinnati, OH
CHAN, PETER, T., Dow Chemical Co., Midland, MI
CHANG, MING, University of Iowa, Dept. of Microbiology, Iowa City, IA
CHAPMAN, JOHN, University of Georgia, Dept. of Molecular & Population Genetics, Athens, GA
CHASSY, BRUCE M., NIH, Bethesda, MD
CHEN, SHIN-TAI, University of Texas at Dallas, Richardson, TX
CHIANG, SHU-JEN, Abbott Laboratories, North Chicago, IL
CHRISTIE, PETER, NYS College Vet Medicine, Veterinary Microbiology, Ithaca, NY
CHURCHWARD, GORDON, Dept. de Biologie Moléculaire, Geneva, Switzerland
CLEARY, JOSEPH M., Kelco, San Diego, CA
CLOSE, TIMOTHY J., University of California, Davis, CA
CONDINO, JANET, Beckman Instruments, Inc. Spino, Palo Alto, CA
COPLIN, DAVID, Ohio State University, Dept. Plant Pathology, Wooster, OH
CRAWFORD, IRVING P., University of Iowa, Dept. of Microbiology, Iowa City, IA
CROSS, MICHAEL, University of Birmingham, Dept. of Genetics, Birmingham, England
CROUZET, JOEL, GENETICA, Paris, France
DELAPPE, IRVING P., NIH, Bethesda, MD
DEMPSEY, WALT, VA Hospital, Dallas, TX
DEMPSEY, LAURIE A., The Public Health Research Inst., Dept. of Microbiology, New York, NY
DENDINGER, SUSAN M., Groton, CT
DeSIMONE, SUSAN M., Smith College, Dept. of Biological Sciences, Northampton, MA
DEVORET, R., Radiobiologie Cellulaire, Laboratoire d'Enzymologie, GIF-sur-YVETTE, France
DeWITT, JANET, Abbott Labs, Northern Chicago, IL
DOHERTY, DANIEL H., 1885 33rd Street, Boulder, CO
DONG, XINNIAN, NUMS, Dept. of Molec. Biology, Chicago, IL
DOOLEY, MARGARET, Rutgers University, Biochemistry & Microbiol., Cook College, New Brunswick, NJ
DOOLEY, THOMAS P., Indiana University, Dept. of Biology, Bloomington, IN
DORSHORST, JUNE, PROMEGA BIOTEC, Madison, WI

DUNNY, G., NYS College Vet. Medicine, Veterinary Microbiology, Ithaca, NY
DURLAND, ROSS, UCSD, Dept. Biology, La Jolla, CA
EHRENFELD, ELIZABETH, University of Michigan, Ann Arbor, MI
ELBLE, RANDOLPH, 433 E. Southern Drive, Bloomington, IN
ENNS, ROBERT, University of California-San Diego, Dept. of Biology, La Jolla, CA
ENSLEY, BURT D., Amgen, Thousand Oaks, CA
ESSAR, DAVID, University of Iowa, Dept. of Microbiology, Iowa City, IA
EUELEIGH, D., Rocky Hill, NJ
FAYERMAN, JEFFREY T., Eli Lilly and Company, Indianapolis, IN
FILUTOWICZ, MARCIN, University of California, Dept. Biology, La Jolla, CA
FINK, CYNTHIA, Monsanto Company U4D, St. Louis, MO
FIRSHEIN, WILLIAM, Wesleyan University, Department of Biology, Middletown, CT
FISCHER, MEIR, Biotechnology General (ISRAEL) Ltd., Rehovot, Israel
FITZGERALD, GERALD F., University of Michigan, Ann Arbor, MI
FRANKE, ARTHUR, Pfizer, Inc., Groton, CT
FROEHLICH, BARBARA, Emory University, Dept. of Microbiology & Immunology, School of Medicine, Atlanta, GA
GAFFNEY, THOMAS D., University of Massachusetts, Amherst, MA
GALELLA, ELIZABETH A., Illinois State University, Dept. of Biological Sciences, Normal, IL
GEORGES, THIRY, University of Liege, Botany Dept., Liege, Belgium
GERDES, KENN, The Technical University of Denmark, Laboratory of Microbiology, Lyngby, Denmark
GHOSAL, D., University of Illinois at Chicago, Dept. of Microbiology/Immunology, Chicago, IL
GILL, JR., JOHN F., Purdue University, Dept. of Biochemistry, West Lafayette, IN
GINGERAS, THOMAS, c/o SIBIA, San Diego, CA
GINSBURGH, CHARLES, Corn Products, Moffett Technical Center, Summit-Argo, IL
GRAHAM, TERRY, Monsanto, St. Louis, MO
GREGORY, EILEEN, Rollins College, Dept. of Biology, Winter Park, FL
GRINDLEY, JUNE N., Biogen, Inc., Cambridge, MA
GUTEKUNST, KAREN A., Dunwoody, GA
GUTTERSON, NEAL, Advanced Genetic Sciences, Inc., Oakland, CA
HAAS, D., Mikrobiologisches Institut, Zurich, Switzerland
HARPER, SHEENA, University of New Castle-upon-Tyre, Dept. of Genetics, Newcastle-upon-Tyre, England
HELMER, GEORGIA, CIBA GEIGI, Research Triangle Park, NC
HESPELL, ROBERT B., University of Illinois, Urbana, IL
HICKMAN, ROBERT K., Tufts University Medical School, Dept. of Microbiology, Boston, MA
HIEBSCH, RONALD R., The Upjohn Company, Kalamazoo, MI
HIGHLANDER, SARAH, Public Health Res. Institute, New York, NY
HOROWITZ, AMI, B.F. Goodrich, Brecksville, OH

HOROWITZ, BURTON, University of Southern California, Dept. of Molecular Biology, Los Angeles, CA
HSIUNG, HANSEN, Indianapolis, IN
IKE, YASUYOSHI, Gunma Medical School, Dept. of Microbiology, Maebashi City, Japan
IYER, V.N., Carleton University, Dept. of Biology, Ottawa, Ontario, Canada
JONCZYK, PIOTR, University of California-San Diego, La Jolla, CA
JORGENSEN, RICHARD, Advanced Genetic Sciences, Inc., Oakland, CA
KADO, CLARENCE, Davis Crown Gall Group, University of California, Davis, CA
KANG, SUK-HEE, University of Notre Dame, Dept. of Biology, Notre Dame, IN
KASNER, JAY P., NUMS, Dept. of Molec. Biol., Chicago, IL
KELLOGG, SCOTT T., SIBIA, San Diego, CA
KHAN, SALEEM, University of Pittsburgh School of Medicine, Dept. of Microbiology, Pittsburgh, PA
KHANNA, SUNIL, University of Cincinnati, Dept. of Microbiology & Molecular Genetics, Cincinnati, OH
KIBANE, JOHN, University of Illinois Medical School, Dept. of Microbiology, Chicago, IL
KIRSCHNER, RICHARD J., The Upjohn Company, Portage, MI
KLINE, BRUCE, Mayo Clinic, Dept. Cell Biology Sect., Microbiology, Rochester, MN
KNIGHT, ROYCE G., VA Medical Center, Cleveland, OH
KOKJOHN, TYLER, Loyola University of Chicago, Maywood, IL
KORNACKI, JON, Wesleyan University, Dept. Biology, Middletown, CT
KORÑBLUM, JOHN, Public Health Res. Institute, New York, NY
KOSLOV, DR., Institute of Genetics & Selection of Industrial Micro-organisms, Gordzhmaya, Moscow, USSR
KOZLOWSKI, MAYA, Allelix, Inc., Mississauga, Ontario, Canada
KREMER, FRAN, Argonne National Laboratory, Argonne, IL
KREMERY, V., Research Institute of Preventive Medicine, Bratislava, Czechoslovakia
KUNZE, HELMUT, Max-Planck-Institut für Molekulare, Genetile, Berlin, Germany
KWANG-MU, YEN, AMGEN, Thousand Oaks, CA
LAM, STEPHEN, CIBA GEIGI, Research Triangle Park, NC
LANKA, ERICH, Max-Planck-Institut für Molekulare, Genetik, Berlin, Federal Republic of Germany
LAU, PETER C.K., National Res. Council of Canada, Ottawa, Ontario, Canada
LAX, ALISTAIR J., AFRC IRAD, Compton, Newbury, Berkshire, England
LEDERBERG, ESTHER M., Stanford Univ. School of Medicine, Stanford, CA
LEJEUNE, ANDRE, Rutgers University, Dept. Biochem. & Microbiol., Cook College, New Brunswick, NJ
LEJEUNE, PHILIPPE, Université Catholique de Louvain, Louvain-La-Neuve, Belgium
LEVANON, AVIGDOR, Biotechnology General (ISRAEL) Ltd., Rehovot, Israel

LOCKER-GILADI, HILLA, Hadassah Medical School, Jerusalem, Israel
LORENCE, MATTHEW C., University of Texas at Dallas, Richardson, TX
MAHILLON, JACQUES, U.C.L., Laboratoire de Genetique Molécularie, Louvain-La-Neuve, Belgium
MARTARANO, LUIS, Charlottesville, VA
MARTINE, COUTURIER, Université Libre de Bruxelles, Laboratoire de Genetique, Rhode St. Genese, Belgium
MARTINEZ S., JAIME M., Inst. Investig. Biomedicas, Mexico, Mexico
MASTERS, MILLICENT, Edinburgh University, Dept. of Molecular Biology, Edinburgh, UK
McCARTHY, MARY ELLEN, McGraw-Hill Biotechnology Newswatch, New York, NY
McCARTY, CYNDEE S., Indiana University, Bloomington, IN
McCOY, JOHN, Genetics Institute, Boston, MA
McCULLOUGH, JOHN E., E.R. Squibb & Sons, Princeton, NJ
McEACHERN, MICHAEL, University of California-San Diego, Dept. of Biology, La Jolla, CA
McINTIRE, SARAH, VA Medical Center, Dallas, TX
McQUEEN, ROSS, 304 Stakman Hall, St. Paul, MN
MERGEAY, MAX, SCK/CEN, Dept. of Radiology, B2400 NOL, Belgium
MEYER, JURG, Dept. Microbiology, Basel, Switzerland
MILLER, JEFF F., Tufts University School of Medicine, Dept. of Molecular Biology, Boston, MA
MILLER, ROBERT V., Loyola University of Chicago, Maywood, IL
MILLER, CHRISTINE, Stanford University, Dept. of Genetics, Stanford, CA
MISRA, ARUN, P.O. Box 4723, Oxford, MS
MJOLI, NOZIBELE P., University of Notre Dame, Dept. Microbiology, Notre Dame, IN
MONAHAN, THOMAS R., Schering Corporation, Bloomfield, NJ
MONGKOLSUK, S., U.M.B.C., Dept. of Bio Science, Catonsville, MD
MOORE, JR., ROBERT E., Monsanto Co.-U4G, St. Louis, MO
MORRIS, CHARLES F., Amgen, Thousand Oaks, CA
MORTENSON, LEONARD E., Exxon Research & Engineering Co., Annandale, NJ
MUESING, MARK, Indiana University, Prog. In MCD Biology, Bloomington, IN
MUKHOPADHYAY, PRADIP, University of California, Dept. of Biology, San Diego, CA
MUNSON, BENJAMIN, Roswell Park Memorial Institute, Buffalo, NY
OBUKOWICZ, MARK, Monsanto, St. Louis, MO
ORIEL, PATRICK, Michigan Biotech Institute, E. Lansing, MI
PALM, CURTIS J., University of California, Dept. Plant Pathology, Davis, CA
PIERCE, GEORGE, Battelle Institute, Columbus, OH
PIFFARETTI, JEAN-CLAUDE, Institute Cant. Batteriologico, Lugano, Switzerland
POHL, MANFRED, University of Cincinnati, Dept. of Microbiology and Molecular Genetics, Cincinnati, OH
POLAK, JUNE, Public Health Research Institute, Dept. Plasmid Bacteriology, New York, NY

PRINGLE, JAMES HOWARD, Dept. of Molecular Biology, Edinburgh, Scotland
PROJAN, STEVEN, Public Health Res. Institute, New York, NY
RABERT, DOUGLAS K., University of Michigan, Dept. of Microbiology, Ann Arbor, MI
ROEHL, RANDALL, Celanese Research Company, Summit, NJ
ROSENFELD, IRA S., General Foods Corporation, Tarrytown, NY
ROSKEY, MARK T., University of Notre Dame, Notre Dame, IN
ROSS, HOPE, Public Health Research Institute, Dept. Plasmid Bacteriology, New York, NY
ROY, PAUL H., Université Laval, Dept. Biochimie, Fac. Sci., Quebec, Canada
RUBY, CAROLYN, Merck & Co., Inc., Rahway, NJ
RUSSELL, PAUL F., Allied Corporation, Crop Science Laboratory, Solvay, NY
RYAN, M.J., CELLTECH Limited, Berkshire, England
SAYLER, GARY, University of Tennessee, Dept. of Microbiology, Knoxville, TN
SCHERZINGER, EBERHARD, Max-Planck-Institut für Molekulare, Genetik, Berlin, Federal Republic of Germany
SCHMIDHAUSER, TOM, University of California-San Diego, La Jolla, CA
SCORDILIS, GAIL E., University of Massachusetts, Dept. of Microbiology, Amherst, MA
SHAW, JAY, University of Michigan, Ann Arbor, MI
SHIMEI, BARBARA A., University Patents, Inc., Norwalk, CT
SIMON, MARY J., AMGEN, Thousand Oaks, CA
SIROTKIN, KARL, University of Tennessee, Dept. of Microbiology, Knoxville, TN
SMITH, DOUGLAS W., University of California-San Diego, La Jolla, CA
SOKATCH, JOHN R., Univ. of Oklahoma, Dept. of Microbiology & Immunology, Oklahoma City, OK
SQUIRES, CHUCK, 1885 33rd Street, Boulder, CO
STROMAN, PER, Karlsburg Laboratory, Copenhagen, Denmark
SUMMERS, DAVID K., Institute of Genetics, Glasgow UK, Scotland
SYKES, JOHN, University of Sheffield, Dept. of Biochemistry, Sheffield UK, England
TAMM, JOSEPH, Indiana University, Dept. Biology, Bloomington, IN
TERAWAKI, YOSHIRO, Shinshu University, Dept. of Biology, School of Medicine, Asahi, Matsumoto, Japan
TEUBER, MICHAEL, Kiel, W. Germany
THIERRY, VERNET, National Res. Council Canada, Ottawa, Canada
THOMAS, CHRISTOPHER M., University of Birmingham, Birmingham, England
THOMASL, BILL, University of Notre Dame, Dept. of Biology, Notre Dame, IN
THORNE, GRACE M., New England Medical Center, Boston, MA
THORPE, PATRICK A., University of Texas at Dallas, Richardson, TX
TOLMASKY, MARCELO, Oregon Health Sciences University, Portland, OR
TORTORELLO, MARY LOU, NYS College Vet Medicine, Veterinary Microbiology, Ithaca, NY
TROTTER, KAREN, NY College Vet. Medicine, Veterinary Microbiology, Ithaca, NY

UCHIDA, HISAO, University of Tokyo, Inst. of Medical Science, Minato-ku, Tokyo, Japan
VALVANO, MICHAEL, Oregon Health Sciences University, Portland, OR
VAN de POL, H., 46 Route des Acasias, Carouge, Geneva, Switzerland
VAN PUTTEN, ARNOLD G., University of Amsterdam, Amsterdam, De Boelelaan, Holland
VANDENBOSCH, JAMES, University of Michigan, Dept. of Microbiology, Ann Arbor, MI
VARY, PATRICIA S., 315 W. Prairie, Wheaton, IL
VON EICHEL, CHRISTOPH, Naheweinstrabe 5, Schweppenhausen BRD, West Germany
WAKARCHUK, WARREN W., University of British Columbia, Vancouver B.C., Canada
WALEH, NAHID, S.R.I. International, Menlo Park, CA
WALKER, GRAHAM C., Massachusetts Institute of Technology, Cambridge, MA
WALON, CORINE, Université Catholique de Louvain, Laboratoire de Genetique, Microbienne, Louvain-la-Neuve, Belgium
WATRUD, LIDIA S., Monsanto (1750), St. Louis, MO
WAUGH, DAVID S., Indiana University, Dept. MCD Biology, Bloomington, IN
WELCH, RODNEY A., University of Wisconsin, Dept. of Medical Microbiology, Madison, WI
WHITE (McGUIRE), PATRICIA, 830 Lake Hollow Boulevard, Marietta, GA
WILLIAMS, PETER A., University College of North Wales, Dept. Biochemistry, Angor Gwyneed, UK
WILLIAMS, STEVEN A., Smith College, Northampton, MA
WILLSON, WILLIAM, University of Illinois, Dept. of Animal Science, Urbana, IL
WILSON, RON, Dow Chemical Company, Bioproducts Lab, Midland, MI
WIRTH, REINHARD, Dental Research Institute, Ann Arbor, MI
WOMBLE, DAVID D., NUMS, Dept. of Molec. Biol., Chicago, IL
WONG, EDWIN, Indiana University, Bloomington, IN
WOOD, DAVID O., University of South Alabama, Mobile, AL
YABLONSKY, Michael, 4-2A Petunia Drive, North Brunswick, NJ
YAGI, YOSHI, The Upjohn Company, Kalamazoo, MI
YAMAMOTO, MITSUYO, University of Michigan, Ann Arbor, MI
YARRANTON, G.T., Celltech Limited, Slough, Berkshire, UK
YOU, IN-SOON, University of Illinois at Chicago, Chicago, IL
ZIEGENKOFF, DR., Institute of Genetics and Selection of Industrial Micro-organisms, Gordzhmaya, Moscow, USSR
ZYLSTRA, GERBEN, 430 Kellogg, Apt. 100, Ann Arbor, MI
ZYSKIND, JUDITH, San Diego State University, San Diego, CA

INDEX OF POSTER AUTHORS

INDEX

www.ingramcontent.com/pod-product-compliance
Lightning Source LLC
LaVergne TN
LVHW081957200926
848481LV00010B/2755

* 9 7 8 1 4 6 1 2 9 4 8 7 0 *